José R. Dorronsoro (Ed.)

Artificial Neural Networks – ICANN 2002

International Conference
Madrid, Spain, August 28-30, 2002
Proceedings

Springer

Series Editors

Gerhard Goos, Karlsruhe University, Germany
Juris Hartmanis, Cornell University, NY, USA
Jan van Leeuwen, Utrecht University, The Netherlands

Volume Editor

José R. Dorronsoro
ETS Informática
Universidad Autónoma de Madrid
28049 Madrid, Spain
E-mail: jose.dorronsoro@uam.es

Cataloging-in-Publication Data applied for

Die Deutsche Bibliothek - CIP-Einheitsaufnahme

Artificial neural networks : international conference ; proceedings /
ICANN 2002, Madrid, Spain, August 28 - 30, 2002. Dorronsoro (ed.). -
Berlin ; Heidelberg ; New York ; Barcelona ; Hong Kong ; London ;
Milan ; Paris ; Tokyo : Springer, 2002
 (Lecture notes in computer science ; Vol. 2415)
 ISBN 3-540-44074-7

CR Subject Classification (1998): F.1, I.2, I.5, I.4, G.3, J.3, C.2.1, C.1.3

ISSN 0302-9743
ISBN 3-540-44074-7 Springer-Verlag Berlin Heidelberg New York

Springer-Verlag Berlin Heidelberg New York
a member of BertelsmannSpringer Science+Business Media GmbH

http://www.springer.de

© Springer-Verlag Berlin Heidelberg 2002
Printed in Germany

Typesetting: Camera-ready by author, data conversion by PTP-Berlin, Stefan Sossna e.K.
Printed on acid-free paper   SPIN: 10873756   06/3142   5 4 3 2 1 0

Preface

The International Conferences on Artificial Neural Networks, ICANN, have been held annually since 1991 and over the years have become the major European meeting in neural networks. This proceedings volume contains all the papers presented at ICANN 2002, the 12th ICANN conference, held in August 28–30, 2002 at the Escuela Técnica Superior de Informática of the Universidad Autónoma de Madrid and organized by its Neural Networks group.

ICANN 2002 received a very high number of contributions, more than 450. Almost all papers were revised by three independent reviewers, selected among the more than 240 serving at this year's ICANN, and 221 papers were finally selected for publication in these proceedings (due to space considerations, quite a few good contributions had to be left out). I would like to thank the Program Committee and all the reviewers for the great collective effort and for helping us to have a high quality conference.

I would also like to thank the European Neural Networks Society (ENNS, on whose behalf the ICANN conferences are organized) for its support and also ICANN 2002's other main sponsors: the Universidad Autónoma de Madrid, Spain's Ministerio de Ciencia y Tecnología, the Instituto de Ingeniería del Conocimiento, and the European Commission under the FET Program. I also thank our cooperating institutions, the IEEE Neural Network Council, the Japanese Neural Network Society, the Asia Pacific Neural Network Assembly, and the Spanish RIG of the IEEE NNC for the very good job they did in promoting ICANN among their membership. Particular thanks go to the Neural Networks group of the E.T.S. de Informática, i.e., Ramón Huerta, Juan A. Sigüenza, Alberto Suárez, Alejandro Sierra, Francisco Rodríguez, Pablo Varona, Carlos Aguirre, Ana González, and Eduardo Serrano. I would also like to thank Werner Horn for his suggestions about the submission and review software. And finally, I must mention and thank ICANN 2002 technical staff, Juana Calle and Angel González: without their tremendous effort and dedication the conference would not have been possible.

June 2002 José R. Dorronsoro

Organization

ICANN 2002 was organized by the Neural Network group of the Escuela Técnica Superior de Ingeniería Informática of the Universidad Autónoma de Madrid in cooperation with the European Neural Network Society. Other sponsoring institutions were the Universidad Autónoma de Madrid, Spain's Ministerio de Ciencia y Tecnología, the Instituto de Ingeniería del Conocimiento, and the European Commission under the FET Program. Other Cooperating Institutions were the IEEE Neural Network Council, the Japanese Neural Network Society, the Asia Pacific Neural Network Assembly, and the Spanish RIG of the IEEE NNC. We also acknowledge the collaboration of the EUNITE project for the Special Session on Adaptivity in Neural Computation

Executive Committee

Conference Chair: José R. Dorronsoro
Co–chairs: Carme Torras (Barcelona)
Senén Barro (Santiago)
Javier de Felipe (Madrid)
Juan A. Sigüenza (Madrid)
Local Organizing Committee: Alejandro Sierra (Workshop Chairman)
Alberto Suárez (Tutorial Chairman)
Ramón Huerta
Francisco Rodríguez
Pablo Varona

Program Committee

Luis Almeida (Portugal)
Daniel Amit (Israel)
Pierre Baldi (U.S.A.)
Peter Bartlett (Australia)
Joan Cabestany (Spain)
Trevor Clarkson (United Kingdom)
José María Delgado (Spain)
Kostas Diamantaras (Greece)
Georg Dorffner (Austria)
Gerard Dreyfus (France)
Wlodek Duch (Poland)
Péter Érdi (Hungary)
Aníbal Figueiras (Spain)
Patrick Gallinari (France)

Wulfram Gerstner (Germany)
Stan Gielen (The Netherlands)
Marco Gori (Italy)
Kurt Hornik (Austria)
Bert Kappen (The Netherlands)
Stefanos Kollias (Greece)
Vera Kurkova (Czech Republic)
Anders Lansner (Sweden)
Hans–Peter Mallot (Germany)
Thomas Martinetz (Germany)
José Mira (Spain)
Lars Niklasson (Sweden)
Erkki Oja (Finland)
Masato Okada (Japan)

Günther Palm (Germany)
Néstor Parga (Spain)
Helge Ritter (Germany)
Olli Simula (Finland)
John Taylor (United Kingdom)
Werner von Seelen (Germany)

Johan Suykens (Belgium)
Michel Verleysen (Belgium)
Paul Verschure (Switzerland)
David Willshaw (United Kingdom)
Rodolfo Zunino (Italy)

Additional Referees

Agostinho Rosa
Alberto Pascual
Alessandro Sperduti
Alessandro Villa
Alexander Ypma
Alfonso Renart
Alfonso Valencia
Alfred Ultsch
A. Alonso Betanzos
Ana González
Andrea Boni
Andreas Stafylopatis
Andreas Weingessel
Andreas Ziehe
Andrés Pérez–Uribe
Angel Navia–Vázquez
Angel P. del Pobil
Angelo Cangelosi
Anna Maria Colla
Antonio Turiel
Ari Hämäläinen
Arthur Flexer
Bart Bakker
Bartlomiej Beliczynski
Ben Kröse
Benjamin Blankertz
Carl van Vreeswijk
Carlos Aguirre
Carlos G. Puntonet
C. Ortiz de Solórzano
Charles–Albert Lehalle
Christian Goerick
Christian Jutten
Christian W. Eurich
Claudio Mattiussi

Conrad Pérez
Constantine Kotropoulos
Cristiano Cervellera
Daniel Polani
David De Juan
David Meyer
David Musicant
Davide Anguita
Davide Marocco
Dietmar Heinke
Dirk Husmeier
Don Hush
Edmondo Trentin
Eduardo Serrano
Elena Bellei
Elka Korutcheva
Esa Alhoniemi
Evgenia Dimitriadou
F. Xabier Albizuri
Federico Abascal
Felipe M. G. França
Felix Gers
Fernando Corbacho
Fernando Díaz de María
Francesco Camastra
Francisco Sandoval
Franco Scarselli
Frank Meinecke
Fred Hamker
Frederic Piat
Friedrich Leisch
Gabriela Andrejkova
Giorgos Stamou
Gladstone Arantes Jr
Gonzalo Joya

Gunnar Rätsch
Gustavo Deco
Heiko Wersing
Hélène Paugam–Moisy
Hiroyuki Nakahara
Horst Bischof
Hubert Dinse
Iead Rezek
Igor Mokris
Jaakko Hollmen
Javier Torrealdea
J. Hellgren Kotaleski
Jean–Pierre Nadal
Jens Kohlmorgen
Jesús Cid–Sueiro
Jimmy Shadbolt
J.J. Merelo
Joaquín Dopazo
Joaquín J. Torres
John Rinzel
Jordi Madrenas
Jorma Laaksonen
José del R. Millán
José L. Bernier
José M. Ferrández
José María Carazo
José R. Álvarez Sánchez
Juan Cires
Juan M. Corchado
Jürgen Schmidhuber
Juha Vesanto
Julian Eggert
Julio Ortega
Karl Goser
Kimmo Raivio

Klaus Obermayer

Klaus–Robert Müller

Koji Tsuda

Krista Lagus

L.M. Reyneri

Laurenz Wiskott

Leon Bobrowski

Lorenzo Sarti

Maksim Bazhenov

Marc Toussaint

Marcello Sanguineti

Marco Balsi

Markus Peura

Martijn Leisink

Masashi Sugiyama

Massimo Conti

Maurizio Mattia

Michael Schmitt

Michele Giugliano

Mikael Bodén

Mikel L. Forcada

Mira Trebar

Misha Rabinovich

Motoaki Kawanabe

Naonori Ueda

Neep Hazarika

Neill Taylor

Nicolas Brunel

Nicolas Tsapatsoulis

Nikolaos Vassilas

Nikos Nikolaidis

Norbert Krueger

Oscar Herreras

Osvaldo Graña

Panagiotis Tzionas

Paolo Coletta

Paolo del Giudice

Paul Kainen

Pedro J. Zufiría

Peter Auer

Peter Geczy

Peter König

Peter Protzel

Peter Sincak

Peter Tino

Petra Philips

Philippe Gaussier

Piotr Suffczynski

Ralf Der

Ralph Neuneier

Ramón Alonso

Reinoud Maex

René Schüffny

Richard Everson

Richard J. Duro

Richard Kempter

Rolf Würtz

Roman Neruda

Samuel Kaski

Sander Bohte

Sepp Hochreiter

Sergio Bermejo

Sethu Vijayakumar

Shin Ishii

Shotaro Akaho

Stavros J. Perantonis

Stefan Harmeling

Stefano Fusi

Stephan ten Hagen

Stephen Roberts

Sylvie Thiria

Taichi Hayasaka

Terezie Sidlofova

Thierry Artières

T. Martini Jørgensen

Thomas Natschläger

Thomas Villmann

Tom Heskes

Tom Ziemke

Ulrich Ramacher

Ulrich Rueckert

Vicente Ruiz de Angulo

Vladimir Nekorkin

Vladimir Olej

Vlado Kvasnicka

Volker Roth

Volker Steuber

Volker Tresp

Volkmar Sterzing

Walter Senn

Wee Sun Lee

Werner Hemmert

Werner M. Kistler

Wim Wiegerinck

Yannis Avrithis

Ying Guo

Younès Bennani

Zhijun Yang

Table of Contents

Computational Neuroscience

Connectionist Cognitive Science

Data Analysis and Pattern Recognition

Kernel Methods

Robotics and Control

Selforganization

Signal and Time Series Analysis

Vision and Image Processing

Special Session: Adaptivity in Neural Computation

Special Session: Recurrent Neural Systems

Part I

Computational Neuroscience

A Neurodynamical Theory of Visual Attention: Comparisons with fMRI- and Single-Neuron Data

Gustavo Deco[1] and Edmund Rolls[2]

[1] Siemens AG, Computational Neuroscience, Munich, Germany.
[2] University of Oxford, Experimental Psychology, Oxford, England.

Abstract. We describe a model of invariant visual object recognition in the brain that incorporates different brain areas of the dorsal or 'where' and ventral or 'what' paths of the visual cortex. The dorsal 'where' path is implemented in the model by feedforward and feedback connections between brain areas V1, V2 and a PP module. The ventral 'what' path is implemented in a physiologically plausible four-layer network, corresponding to brain areas V1, V2, V4 and IT, with convergence to each part of a layer from a small region of the preceding layer, with feature-based attentional feedback connections, and with local competition between the neurons within a layer implemented by local lateral inhibition. In particular, the model explains the gradually increasing magnitude of the attentional modulation that is found in fMRI experiments from earlier visual areas (V1, V2) to higher ventral visual areas (V4, IT). The model also shows how the effective size of the receptive fields of IT neurons becomes smaller in natural cluttered scenes.

1 Introduction

In the research described here, we follow a computational neuroscience approach in order to investigate the processes that underlie high-level vision, and in particular visual object recognition and attention. We describe a computational neuroscience model for invariant visual object recognition that combines a hierarchical architecture with convergence from stage to stage and competition within each stage, with feedback biasing effects of top-down attentional mechanisms. In particular, we focus in this paper on the locally implemented but gradually increasing global character of the competition that is produced in a hierarchical network with convergent forward connectivity from area to area, and on the interaction between space-based and object-based attentional top-down feedback processes.

2 A Large-Scale Visual Cortex Neurodynamical Model

The neurophysiological findings (see Rolls and Deco, 2002 for a review), wider considerations on the possible computational theory underlying hierarchical feedforward processing in the visual cortical areas with layers of competitive networks

J.R. Dorronsoro (Ed.): ICANN 2002, LNCS 2415, pp. 3–8, 2002.

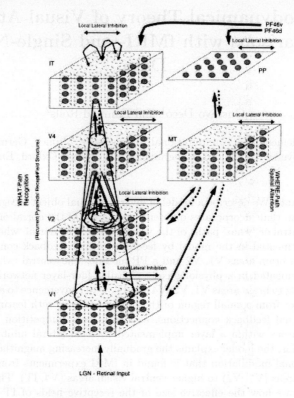

Fig. 1. Neurodynamical Architecture.

trained with a trace learning rule, and the analysis of the role of attentional feedback connections and interactions between an object and a spatial processing stream, lead to the neurodynamical model that we present in this section for invariant hierarchical object recognition and selective visual attention. Figure 1 shows the overall systems-level diagram of the multi area neurodynamical architecture used for modeling the primate visual cortical areas. The system is essentially composed of five modules or networks structured such that they resemble the two known main visual paths of the mammalian visual cortex. Information from the retino-geniculo-striate pathway enters the visual cortex through area V1 in the occipital lobe and proceeds into two processing streams. The occipital-temporal stream leads ventrally through V2, V4 to IT (the inferior temporal cortex), and is mainly concerned with object recognition, independently of position and scaling. The occipito-parietal stream leads dorsally into PP (posterior parietal complex) and is responsible for maintaining a spatial map of a object's location and/or the spatial relationship of an object's parts as well as moving the spatial allocation of attention.

The ventral stream consists of the four modules V1, V2, V4 and IT. These different modules allow combinations of features or inputs that occur in a given spatial arrangement to be learned by neurons, ensuring that higher-order spatial properties of the input stimuli are represented in the network (Elliffe et al, 2002). This is implemented via convergent connections to each part of a layer from a small region of the preceding layer, thus allowing the receptive field size of cells to increase through the ventral visual processing areas, as is observed in the primate ventral visual stream. An external top-down bias, coming it is postulated from a short-term memory for shape features or objects in the more ventral part of the prefrontal cortex area v46, generates an object-based attentional component that is fed back down through the recurrent connections from IT through V4 and V2 to V1. The V1 module contains hypercolumns, each covering a pixel in a topologically organized model of the scene. Each hypercolumn contains orientation columns of orientation-tuned (complex) cells with Gabor filter tuning at octave intervals to different spatial frequencies. V1 sends visual inputs to both the ventral and dorsal streams, and in turn receives backprojections from each stream, providing a high-resolution representation for the two streams to interact. This interaction between the two streams made possible by the backprojections to V1 is important in the model for implementing attentional effects. All the feedforward connections are trained by an associative (Hebb-like) learning rule with a short term memory (the trace learning rule) in a learning phase in order to produce invariant neuronal responses. The backprojections between modules, a feature of cortical connectivity, are symmetric and reciprocal in their connectivity with the forward connections. The average strength of the backprojections is set to be a fraction of the strength of the forward connections so that the backprojections can influence but not dominate activity in the input layers of the hierarchy (Rolls and Deco, 2002). Intramodular local competition is implemented in all modules by lateral local inhibitory connections between a neuron and its neighboring neurons via a Gaussian-like weighting factor as a function of distance. The cortical magnification factor is explicitly modelled by introducing larger numbers of high spatial resolution neurons in a hypercolumn the nearer the hypercolumn is to the fovea. The density of these fine spatial resolution neurons across the visual field decreases in the model by a Gaussian function centered on the fovea. The dorsal stream includes a PP module which receives connections from V1 and V2, and which has reciprocal backprojections. An external top-down bias to the PP module, coming from a spatial short-term memory and denoted as prefrontal cortex area d46 in the model, generates a spatial attentional component. The backprojections from PP influence the activity in the V2 and V1 modules, and thus can indirectly influence activity in the ventral stream modules. A lattice of nodes provides topological organization in module PP. The main neurodynamical equations are given in the Appendix.

3   fMRI Data

Functional magnetic resonance imaging (fMRI) studies show that when multiple stimuli are present simultaneously in the visual field, their cortical representations within the object recognition pathway interact in a competitive, suppres-

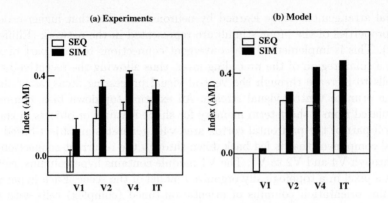

Fig. 2. Experimental and theoretical simulation of fMRI data.

sive fashion (Kastner et al., 1999). Directing attention to one of the stimuli can counteract the suppressive influence of nearby stimuli. The model we describe here was able to simulate and account for these results. In the first experimental condition Kastner et al. (1999) showed the presence of suppressive interactions among stimuli presented simultaneously (SIM) within the visual field in the absence of directed attention (UNATT).

The comparison condition was sequential (SEQ) presentation. An Attentional Modulation Index (AMI) was defined as $AMI = \frac{[ATT-UNATT]}{ATT}$ where ATT = the fMRI response in the attended condition. In a second experimental condition they showed that spatially directed attention increased the fMRI signal more strongly for simultaneously presented stimuli than for sequentially presented stimuli. Thus, the suppressive interactions were partially cancelled out by attention. This effect was indicated by a larger increase of the AMI_{SIM} in comparison to AMI_{SEQ} caused by attention. The results further showed that attention had a greater effect (the AMI was higher) for higher (IT, V4 and V2) than for earlier (V1) visual areas, as shown in Figure 2a. In a third experimental condition the effects of attention were investigated in the absence of the visual stimuli. Figure 2b shows the results of our simulations. The simulations show a gradually increasing magnitude of attentional modulation from earlier visual areas (V1, V2) to higher ventral stream visual areas (V4, IT), which is similar to that found in the experiments. This attentional modulation is location-specific, and its effects are mediated by the PP attentional biasing input having an effect via the backprojections in V2 and V1, from which the effect is fed up the ventral stream in the forward direction. The gradually increasing influence of attentional modulation from early visual cortical areas to higher ventral stream areas is a consequence of the gradually increasing global character of the competition between objects and/or parts of objects as one ascends through the ventral visual system, and the locally implemented lateral inhibition becomes effectively more global with respect due to the convergence in the forward direction in the hierarchical pyramidal architecture of the ventral stream.

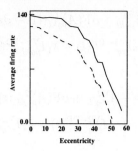

Fig. 3. Average firing activity of an IT neuron in the model vs eccentricity. Solid line – plain background. Dashed line – complex background

4 Single-Cell Data: IT-Receptive Field Size

Rolls and colleagues (2001) (see Rolls and Deco 2002) showed that the receptive fields of IT neurons were large (65 deg) with a single stimulus in a blank background, and were greatly reduced in size (to 36.6 deg) when the stimulus was presented in a complex natural scene. We simulate this effect. In a first experiment we placed only one object on the retina at different distances from the fovea. This corresponds to the blank background condition. In a second experiment, we also placed the object at different eccentricities relative to the fovea, but on a cluttered natural background. Figure 3 shows the average firing activity of an inferior temporal cortex neuron in the model specific for the test object as a function of the position of the object on the retina relative to the fovea (eccentricity). In both cases relatively large receptive fields are observed, because of the translation invariance obtained with the trace learning rule and the competition mechanisms implemented within each layer of the ventral stream. However, when the object was in a blank background, larger receptive fields were observed (upper curve). The decrease in neuronal response as a function of distance from the fovea is mainly due to the effect of the magnification factor implemented in V1. On the other hand, when the object was in a complex cluttered background (lower curve), the effective size of the receptive field of the same inferior temporal cortex neuron shrinks because of competitive effects between the object features and the background features in each layer of the ventral stream. In particular, the global character of the competition present in the inferior temporal cortex module (due to the large receptive field of the neurons there) is the main cause of the reduction of the receptive fields in the complex scene.

5 Appendix: Neurodynamical Equations

Let us denote by A^{V1}_{kpql}, $A^{V2}_{p'q'l'}$, $A^{V4}_{p''q''l''}$, $A^{IT}_{p'''q'''l'''}$, and A^{PP}_{ij} the activity of neural pools in the module V1, V2, V4, and IT, and PP module, respectively. The main neurodynamical equations that regulate the temporal evolution of the whole system are:

$$\tau\frac{\partial A_{pqkl}^{V1}(t)}{\partial t} = -A_{pqkl}^{V1} + \alpha F(A_{pqkl}^{V1}(t)) - \beta I_{pq}^{\text{inh},V1}(t) + I_{pqkl}^{V1}(t)$$
$$+\gamma_b I_{pq}^{V1-PP}(t) + \lambda_1 I_{pqkl}^{V1-V2}(t) + I_0 + \nu \tag{1}$$

$$\tau\frac{\partial A_{p'q'l'}^{V2}(t)}{\partial t} = -A_{p'q'l'}^{V2} + \alpha F(A_{p'q'l'}^{V2}(t)) - \beta I_{pq}^{\text{inh},V2}(t)$$
$$+\gamma_b I_{p'q'}^{V2-PP}(t) + \lambda_2 I_{p'q'l'}^{V2-V4}(t) + I_0 + \nu \tag{2}$$

$$\tau\frac{\partial A_{p''q''l''}^{V4}(t)}{\partial t} = -A_{p''q''l''}^{V4} + \alpha F(A_{p''q''l''}^{V4}(t)) - \beta I_{pq}^{\text{inh},V4}(t)$$
$$+\lambda_3 I_{p''q''l''}^{V4-IT}(t) + I_0 + \nu \tag{3}$$

$$\tau\frac{\partial A_{p'''q'''l'''}^{IT}(t)}{\partial t} = -A_{p'''q'''l'''}^{IT} + \alpha F(A_{p'''q'''l'''}^{IT}(t)) - \beta I_{pq}^{\text{inh},IT}(t)$$
$$+I_{l'''}^{IT,A} + I_0 + \nu \tag{4}$$

$$\tau\frac{\partial A_{ij}^{PP}(t)}{\partial t} = -A_{ij}^{PP} + \alpha F(A_{ij}^{PP}(t)) - \beta I_{ij}^{\text{inh},PP}(t) + \gamma_f I_{ij}^{PP-V1}(t)$$
$$+\gamma_f I_{ij}^{PP-V2}(t) + I_{ij}^{PP,A} + I_0 + \nu \tag{5}$$

where I_{pqkl}^{V1} is the sensory input activity to a pool in V1, I_{pq}^{V1-PP}, I_{pq}^{V2-PP}, I_{pq}^{PP-V1} and I_{ij}^{PP-V2} are the spatial attentional biasing couplings due to the intermodular 'where' connections with the pools in the parietal module PP, I_{pqkl}^{V1-V2}, I_{pqkl}^{V2-V1}, I_{pqkl}^{V2-V4}, I_{pqkl}^{V4-V2}, I_{pqkl}^{V4-IT}, I_{pqkl}^{IT-V4} are the feature based attentional top-down biasing terms due to the intermodular 'what' connections of pools between two immediate modules in the ventral stream, $I_{pq}^{\text{inh},VE}$ and $I_{ij}^{\text{inh},PP}$ are local lateral inhibitory interactions in modules in the ventral stream, and PP. The external attentional spatially-specific top-down bias $I_{ij}^{PP,A}$ is assumed to come from prefrontal area 46d, whereas the external attentional object-specific top-down bias $I_{l'''}^{IT,A}$ is assumed to come from prefrontal area 46v. Both of them are associated with working memory.

6   References

Elliffe, M. C. M., Rolls, E. T. and Stringer, S. M. (2002) Biological Cybernetics, 86, 59–71.

Kastner, S., Pinsk, M., De Weerd, P., Desimone, R. and Ungerleider, L. (1999). Neuron, 22, 751–761.

Rolls, E. and Deco, G. (2002). Computational Neuroscience of Vision. Oxford: Oxford University Press.

Rolls, E. T, Zheng, F., and Aggelopoulos, N. C. (2001). Society for Neuroscience Abstracts, 27.

A Neural Model of Spatio Temporal Coordination in Prehension

Javier Molina-Vilaplana, Jorge Feliu Batlle, and Juan López Coronado

Departamento de Ingeniería de Sistemas y Automática. Universidad Politécnica de Cartagena. Campus Muralla del Mar. C/ Dr Fleming S/N. 30202. Cartagena. Murcia. Spain.
{Javi.Molina, Jorge.Feliu, Jl.Coronado}@upct.es

Abstract. The question of how the transport and grasp components in prehension are spatio–temporally coordinated is addressed in this paper. Based upon previous works by Castiello [1] we hypothesize that this coordination is carried out by neural networks in basal ganglia that exert a sophisticated gating / modulatory function over the two visuomotor channels that according to Jeannerod [2] and Arbib [3] are involved in prehension movement. Spatial dimension and temporal phasing of the movement are understood in terms of basic motor programs that are re–scaled both temporally and spatially by neural activity in basal ganglia thalamocortical loops. A computational model has been developed to accommodate all these assumptions. The model proposes an interaction between the two channels, that allows a distribution of cortical information related with arm transport channel, to the grasp channel. Computer simulations of the model reproduce basic kinematic features of prehension movement.

1 Introduction

Prehension is usually divided into two distinct components: hand transport and grip aperture control. The transport component is related with the movement of the wrist from an initial position to a final position that is close to the object to be grasped. The grip aperture component is related to the opening of the hand to a maximum peak aperture and then, with fingers closing until they contact the object. There is a parallel evolution of reaching and hand preshaping processes, both of them initiating and finishing nearly simultaneously.

Temporal invariances exist in prehension, in terms of a constant relative timing between some parameters of the transport and grasp components. For instance, the time to maximum grip aperture occurs between 60–70% of movement duration despite large variations in movement amplitude, speed, and different initial postures of the fingers [2], [4], [5]. Time of maximum grip aperture is also well correlated with time of maximum deceleration of the transport component [6]. Jeannerod [2] and Arbib [3] suggest that the transport and hand preshaping components evolve independently trough two segregated visuomotor channels, coordinated by a central timing mechanism. This timing mechanism ensures the temporal alignment of key moments in the evolution of the two components. In this way, Jeannerod [2] suggests that the central timing mechanism operates such that peak hand aperture is reached at

J.R. Dorronsoro (Ed.): ICANN 2002, LNCS 2415, pp. 9–14, 2002.
© Springer-Verlag Berlin Heidelberg 2002

the moment of peak deceleration of the reaching component. Castiello et al [1] in reach to grasp perturbation experiments with Parkinson's disease subjects, conclude that there are indications that the basal ganglia might be seen in the context of the neural networks that actively gate the information to and between the channels, needed for the coordination in time of the two components in prehension movement.

Haggard and Wing [7] proposed a model of coordination during prehension movement. This model is based on coupled position feedback between the transport and grasp components. Hoff and Arbib [8] proposed a model for temporal coordination during prehension movement based on constant enclosed time. In this paper, a neural model of prehension movement coordination is proposed. Vector Integration To Endpoint (VITE) [9] dynamics is used to model the grasp and transport channels. Movement execution is modulated by a basal ganglia neural network model that exerts a sophisticated gating function [10] over these channels.

2 Neural Model of Reach to Grasp Coordination

The visuomotor channels related with transport and grasp components have been simulated using the VITE model of Bullock and Grossberg [9]. VITE gradually integrates the difference between the desired target finger aperture (T) and the actual finger aperture (P) to obtain a difference vector (V). For transport component, (T) models the wrist desired final position near the object to be grasped and (P) the actual wrist position. The difference vectors code information about the amplitude and direction of the desired movement. These vectors are modulated by time-varying G(t) gating signals from basal ganglia neural networks [10], producing desired grip aperture velocity and wrist velocity commands (V*G(t)). Temporal dynamics of (V) and (P) vectors are described by equations (1) – (2).

$$dV/dt = 30.(-V + T - P) \tag{1}$$

$$dP/dt = G(t).\ V \tag{2}$$

According to Jeannerod [2], we propose a biphasic programming of the grasp component; therefore the motor program for the movement of fingers has been modeled as consisting of two sequential subprograms (G1, G2), where G1 is related to the maximum grip aperture and G2 equals the object size. VITE model has been modified in the grasp channel to account for the apparent gradual specification of target amplitude [11]. The model shown in Figure 1 (this figure only shows the structure of one of the two visuomotor channels) assumes that the target aperture is not fully programmed before movement initiation; rather, it is postulated that target acquisition neuron (T) in grasp channel, sequentially and gradually specify, in a first phase the desired maximum grip aperture (G1), and in a second phase the aperture corresponding to the object size (G2). Dynamics of the target acquisition neuron in grasp channel is described by equation (3):

$$dT/dt = \alpha.\ (G - T) \tag{3}$$

In this model, time of peak deceleration (Tpdec, the instant when the acceleration in the transport channel is minimum) triggers the read-in of G2 by the target acquisition neuron in grasp channel. In principle, proprioceptive information could be used by CNS to derive tpdec or a related measurement [12]. Tpdec detection generates a signal in a SMA (Supplementary Motor Area) cortical neuron (Fig1). This neuron projects its activity to the putamen neurons in the grasp channel triggering the modulation of the final enclosing phase. No other interactions between transport and grasp components were assumed.

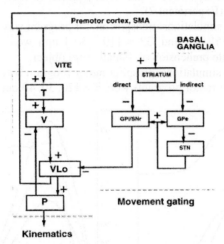

Fig. 1. Schematic diagram of the computational structure of the model. Movement execution is modeled by VITE dynamics (left part of Figure 1). Modulatory gating functions over movement execution is carried out by basal ganglia neural networks (Right part of Figure 1). Inhibitory projection activity from striatum to GPi and GPe is affected by dopamine levels (see text for details). Arrows finishing with + / - symbol, mean excitatory / inhibitory connections respectively. Connections from VLo to P in both channels are multiplicative gating connections.

Neural circuits of basal ganglia depicted in Figure 1 have been modeled (like in previous models [10], [13]), with, non linear, ordinary differential equations. Dynamics of basal ganglia neurons are described by equations (4) - (5):

$$dS_k/dt = \beta_k[- A_k S_k + (B_k - S_k)E_k - (D_k + S_k)I_k] \tag{4}$$

$$dN_J/dt = b(BN_J(D) - N_J) - cS^J_I N_J \tag{5}$$

where S_k represents neural population activity of Putamen (k=1), internal globus pallidus (GPi, k=2), external globus pallidus (GPe, k=3), subthalamic nucleus (STN, k=4) and VLo (k=5). Appendix A shows the numerical values of the parameters used in our simulations and an extended mathematical description of the model.

Thalamus (VLo) neuronal activity is identified with G(t) gating signals of the VITE model. Dopamine appears as an explicit parameter (D), modulating the functioning of basal ganglia thalamocortical loops. In equation (5), N_J models the amount of available neurotransmitter in direct (J=1) and indirect pathway (J=2) (GABA/ Substance P in direct pathway, GABA/Enkephalin in indirect pathway). Maximum level of available neurotransmitter (BN_J) is a function of dopamine level ($BN_J(D) = 1$ for D = 1 and J=1,2).

3 Simulation Results

Two different simulated experiments were performed. In first of them, 'subjects' carried out single reach to grasp movements of different width objects placed at he same distance (Experiment 1). In the second simulated experience, 'subjects' had to move to different amplitudes from initial wrist position to final wrist placement in order to grasp an object (Experiment 2).

Experiment 1: In these simulations, transport motor program was named as T and set to 30 cm. G1 = [35, 75] mm and G2 = [10 , 50] mm were the biphasic motor programs used to simulate prehension of [small , big] object.

Experiment 2: In these simulations, grasping motor program was set to G1 = 75 mm / G2 = 50 mm. Transport motor programs were T = [20, 30, 40]cm.

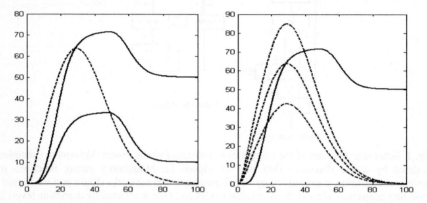

Fig. 2. Wrist velocity and aperture profiles for grasping as a function of distances and object size. Experiment 1 (Left) simulates a 'subject' that reaches out to grasp two different width objects at the same distance away. No difference is seen in wrist velocity profiles but differences appear in grip aperture profiles. Experiment 2 (Right) simulates a 'subject' that reaches out to grasp an object of constant size at three different distances away. The velocity profile shows a systematic increase with distance but no difference is seen in the grip aperure. Abcisas represents normalized movement time. Wrist velocity (cm/s) in dashed lines. Finger aperture (mm) in solid lines.

In all simulations initial finger aperture and initial wrist position was set to zero. The biphasic motor program in the grasp channel (G1,G2) was designed taking into account the fact that the amplitude of maximum grip aperture covaries linearly with object size.

Transport velocity exhibits a bell shaped but asymmetrical velocity profile typical of point to point arm movements [9]. The plot of hand aperture shows the opening of the hand until it gets to maximum peak aperture, then it shows a decreasing of grip aperture until it reaches object size. As seen in Figure 2, the distance of the object away from the subject affects the transport component (peak velocity increases with the distance to be moved) but not the grasping component (Experiment 2). Figure 2 also shows how the object size affects the grasping component (maximum aperture covaries linearly with object size), while the transport component remains unaltered (Experiment 1).

4 Discussion

The question of how the transport and grasp components in prehension are spatio–temporally coordinated is addressed in this paper. In the proposed model this coordination is carried out by neural networks in basal ganglia that exert a sophisticated gating/modulatory function over the two visuomotor channels that are involved in the reach to grasp movement. Simulation results show how the model is able to address temporal equifinality of both channels (i.e. wrist transport and grip closure finishing simultaneously) and how it reproduces basic kinematic features of single prehension movements.

One advantage of the model is the explicit inclusion of a parameter such like dopamine level (D) which affects whole model performance, allowing us to simulate prehension tasks under simulated Parkinson Disease (PD) symptoms. These symptoms are characterized by dopamine level depletions (D<1 in the model). In those simulation conditions, we could match our results with observed impairments in PD patients during reach to grasps movements. Future research will use PD symptoms as a window to test the hypothesis in which the model is based, the hypothesis that points to the basal ganglia as the neural networks that actively gate the information to and between the channels, needed for the coordination in time of the two components in reach to grasp movement [1].

The proposed computational model does not explicitly account for the modelling of the inherent visuomotor transformations from object intrinsic properties (size and shape) into motor programs for grasp. Further developments of the model proposed in this paper should take into account this transformations in order to offer a more complete explanation of prehension movement.

References

1. Castiello U, Bennett K, Bonfligioli C, Lim S, Peppard RF. (1999) The reach to grasp movement in Parkinson's Disease: response to simultaneous perturbation of object position and object size. Exp Brain Res 125: 453 - 462
2. Jeannerod M (1984) The timing of natural prehension movements. J. Motor Behav, 16: 235-254.
3. Arbib MA (1981) Schemas for the temporal organization of behaviour. Hum Neurobiol 4:63– 72.
4. Saling M, Mescheriakov S, Molokanova E, Stelmach GE, Berger M (1995) Grip reorganization during wrist transport: the influence of an altered aperture. Exp Brain Res 108:493-500
5. Wallace SA, Weeks DL, Kelso JAS (1990) Temporal constraints in reaching and grasping behavior. Hum Mov Science, 9:69-93.
6. Gentilucci M, Castiello U, Corradini ML, Scarpa M, Umiltà C, Rizzolatti G (1991) Influence of different types of grasping on the transport component of prehension movements. Neuropsychologia 29:361 – 378
7. Haggard P, Wing AM (1995) Coordinated responses following mechanical perturbations of the arm during prehension. Exp Brain Res, 102: 483-494.
8. Hoff B, Arbib MA (1993) Models of trajectory formation and temporal interaction of reach and grasp. J. Motor Behav, 25: 175-192

9. Bullock D, Grossberg S (1988) Neural dynamics of planned arm movements: emergent invariants and speed-accuracy properties during trajectory formation. Psychol Rev, 95: 49-90
10. Contreras-Vidal JL, Stelmach GE (1995) A neural model of basal ganglia- thalamocortical relations in normal and Parkinsonian movement. Biological Cybernetics, 73:467-476.
11. Fu QG, Suarez JI, Ebner TJ (1993) Neuronal specification of direction and distance during reaching movements in the superior precentral area and primary motor cortex of monkeys. J. Neurophysiol., 70: 2096-2126.
12. Cordo P, Schieppati M, Bevan L. Carlton LG, Carlton MJ (1993) Central and peripheral coordination in movement sequences. Psychol Res, 55:124-130.
13. Fukai T. (1999) Sequence generation in arbitrary temporal patterns from theta-nested gamma oscillations: a model of the basal ganglia – thalamocortical loops. Neural Networks 12: 975 – 987.

Appendix A: Mathematical Description of the Model

In equations (4)–(5) E_k represents net excitatory input to population S_k and I_k net inhibitory input to the same population. Equations are integrated using a Runge Kutta numerical method of fourth order with a time step of 1 ms. The simulation time was 1.2 seconds. Numerical values used in simulations were (k=1,2,3,4,5): $\beta_k=(1,2,1,1,2)$, $A_k=(10,1,1,10,2)$, $B_k=(1,3,2,2,2)$, $D_k =(0,0.8,0.8,0.8,0.8)$. Net excitatory (E_k) and inhibitory (I_k) inputs to neural populations are described by: $E_1 = I(t) + I_{ACh} + f(S_1)$, $E_2 = 3 S_4 + f(S_2) + 0.5$, $E_3 = 5 S_4 + f(S_3) + 2.0$, $E_4 = I_s + f(S_4)$, $E_5 = 0.9$, $I_1 = 0.0$, $I_2 = 50 S^1_1 N_1 + 0.5 S_3 I_3 = 50 S^2_1 N_1 + 0.3 S_2 I_4 = 1.5 S_3 I_5 = 2.5 S_2$ where $f(x) = x^2 / (0.3 + x^2)$ and S^1_1 and S^2_1 represent two putamen neurons, first of one (S^1_1) projects to the GPi and the second one (S^2_1) projects to the GPe. I_{Ach} is a tonic constant input from cholinergic cells in striatum with 0.5 value. I_s is a tonic activity from cortex. $I_s = 25$ in the grasp channel and $I_s = 75$ in the transport related channel. I(t) is a signal from SMA . In the grasp channel I(t) starts with a burst of 5 during 200 ms and then goes to zero. Tpdec detection in transport channel burst activity of I(t) in grasp channel with a value of 5 during another 200 ms. I(t) in transport channel has a constant value of 5 during the simulation. Neurotransmitter dynamics parameters are set to b = 2.0, c = 8.0. The integration rate α for the target acquisition neuron in grasp channel is set to 10 in all simulations.

Stabilized Dynamics in Physiological and Neural Systems Despite Strongly Delayed Feedback

Andreas Thiel, Christian W. Eurich, and Helmut Schwegler

Institut für Theoretische Physik, Universität Bremen
Postfach 330 440, D-28334 Bremen, Germany
{athiel,eurich,schwegler}@physik.uni-bremen.de

Abstract. Interaction delays are ubiquitous in feedback systems due to finite signal conduction times. An example is the hippocampal feedback loop comprising excitatory pyramidal cells and inhibitory basket cells, where delays are introduced through synaptic, dendritic and axonal signal propagation. It is well known that in delayed recurrent systems complex periodic orbits and even chaos may occur. Here we study the case of distributed delays arising from diversity in transmission speed. Through stability considerations and numerical computations we show that feedback with distributed delays yields simpler behavior as compared to the singular delay case: oscillations may have a lower period or even be replaced by steady state behavior. The introduction of diversity in delay times may thus be a strategy to avoid complex and irregular behavior in systems where delayed regulation is unavoidable.

1 Introduction

Recurrent inhibition, in which one population of neurons excites a second one that in turn inhibits the first, is often encountered in the nervous system. However, this feedback sets in after a considerable delay, caused by the time needed for synaptic transmission and conduction of postsynaptic potentials and spikes. Because of the diversity of transmission velocities within a neuronal population, it is reasonable to assume that there will be a temporal dispersion of signals depending on which individual cells are activated along the feedback loop. Therefore, recurrent inhibition is not determined by the state of the population at a single instant in the past, but rather by its states during a past interval, as signals arriving synchronously may have been created during a whole time span. In a model of such a system, the amount of feedback can be calculated by convolving the population's activity with a delay distribution function.

Here, we investigate the stability of two feedback systems after substituting a singular delay with a distribution of delays, namely the hippocampal mossy fibre–pyramidal cell–basket cell complex, a prominent example of recurrent inhibition in neural systems [5], and the Mackey–Glass equation [6]. By numerical and analytical computations we find that increasing the width of the distribution function, i. e. the amount of temporal dispersion in the feedback process, has a profound simplifying influence on both systems' dynamics.

J.R. Dorronsoro (Ed.): ICANN 2002, LNCS 2415, pp. 15–20, 2002.

2   Distributed Delayed Regulation in the Hippocampus

The system. As an example of a neural feedback loop, the hippocampal mossy fibre–CA3 pyramidal cell–basket cell complex is chosen. In this system, the mossy fibres provide excitatory presynaptic input to the CA3 pyramidal cells. The interneuronal population of basket cells is activated by the pyramidal axon collaterals and in turn generates inhibitory input to the CA3 neurons via GABAergic synapses. A vast amount of anatomical, physiological and pharmacological data is available for this system, and there have been numerous modeling efforts as well (see references in [2]). In the present context, the model of Mackey and an der Heiden [5] is preferable because of its simplicity. It essentially consists of a delay–differential equation describing the normalized membrane potential v of an average pyramidal neuron. Here, we extend this model by considering the fact that the feedback signal at a particular instant depends on an interval of past states:

$$\frac{dv(t)}{dt} = -\Gamma v(t) + \Gamma e - \beta \frac{\mathcal{F}_\xi(v(t))}{1 + \mathcal{F}_\xi(v(t))^n} , \tag{1}$$

$$\mathcal{F}_\xi(v(t)) = f_0 \int_0^\infty [v(t-\tau) - \theta] H[v(t-\tau) - \theta] \xi(\tau) d\tau . \tag{2}$$

In Eq. (1), Γ denotes the inverse membrane time constant of the CA3 neurons. The inhibitory signal $\mathcal{F}_\xi$ mediated by the basket cells is a functional of v, denoting the system's past states convolved with a delay distribution $\xi(\tau)$. Its computation further includes a threshold operation and rectification to consider the fact that neurons communicate via the generation of action potentials and firing rates (thus the Heaviside function $H[\cdot]$ in Eq. (2)). The feedback signal acts inhibitorily, which makes it necessary to add the excitatory drive e from the mossy fibres that is assumed to be constant throughout the following. The delay distribution $\xi(\tau)$ is normalized and restricted to $\tau \geq 0$, since the system's behavior can only be influenced by its past or present states. For reasons of mathematical tractability, we choose a uniform function:

$$\xi(\tau) = \begin{cases} \frac{1}{2\sigma} & \text{if } \tau_m - \sigma \leq \tau \leq \tau_m + \sigma \\ 0 & \text{otherwise} \end{cases} , \tag{3}$$

where τ_m is the mean delay and $\sigma \leq \tau_m$ is the half width of the distribution. This corresponds to the fact that the system's states during a past interval ranging from $\tau_m - \sigma$ to $\tau_m + \sigma$ all contribute equally to the feedback signal. The length of this interval σ is used as a bifurcation parameter, whereas the average delay τ_m is kept constant.

Regular membrane potential fluctuations despite strong inhibition. The time course of the pyramidal cell's membrane potential is obtained by numerically solving Eq. (1) applying the Runge–Kutta method with step size $\Delta t = 0.01$. We start by choosing parameters causing irregular behavior in the

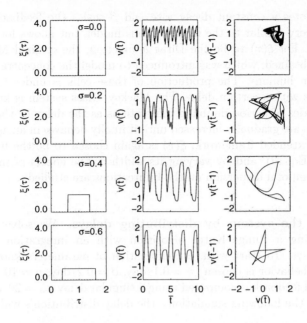

Fig. 1. Delay distribution function (left column) used to compute the excitatory membrane potential v as a function of rescaled time $\bar{t} = t/\tau_m$ (middle column) during simulations of the hippocampus model with distributed delays (Eq. (1)). The interval plotted corresponds to $1.0\,\mathrm{s}$. Also shown are phase plane portraits $v(\bar{t} - \tau_m)$ vs. $v(\bar{t})$ (right column) generated by two–dimensional embedding using delay coordinates. $\sigma = 0.0$ denotes the case of all connections having the same delay $\tau_m = 1.0$. Transients are omitted.

singular delay case ($\Gamma = 10.0$, $e = 1.6$, $\beta = 54.0$, $n = 3$, $f_0 = 9.0$, $\theta = 1.0$, $\tau_m = 1.0$, initial value $v_0 = 0.0$) and then gradually increase σ. Results are shown in Fig. 1. With singular delay ($\sigma = 0$), the potential exhibits irregular fluctuations, which become more and more regular the wider the delay distribution is. Thus, in presence of a fixed feedback strength β, regular oscillations may be obtained by considering a broader interval of past states in the generation of the inhibitory signal.

3   The Mackey–Glass Equation with Distributed Delays

The system. To demonstrate the generality of the effect described above, we next consider an extension of the Mackey–Glass equation [6,3,4]:

$$\frac{dv(t)}{dt} = -\gamma v(t) + \beta \frac{\mathcal{V}_\xi(v(t))}{1 + \mathcal{V}_\xi(v(t))^n} , \qquad (4)$$

$$\mathcal{V}_\xi(v(t)) \equiv \int_0^\infty v(t - \tau)\xi(\tau)\,d\tau , \qquad (5)$$

where γ denotes a constant decay rate and β scales the feedback gain. This equation is very similar to the hippocampus model, but allows for a analytical investigation. For $\xi(\tau)$ to be the Dirac δ–function, the original Mackey–Glass equation is obtained, which was introduced to model the concentration of white blood cells in humans. The production of these cells is under feedback control that acts with a certain delay. The Mackey–Glass system is known to pass through a series of period doubling bifurcations as the delay or the gain in the feedback loop are gradually increased until it finally behaves in an aperiodic way. Within our extended framework, $\xi(\tau)$ is again chosen to be the uniform delay distribution Eq. (3), and by varying its width σ, the effects of increasing the feedback's temporal dispersion on the bifurcations are studied.

Stabilizing the system by distributing delays. We solve Eq. (4) numerically using a Runge–Kutta algorithm with an integration step size of $\Delta t = 0.01$ days. To start with, a combination of parameters known to result in aperiodic behavior is chosen ($\gamma = 0.1$/day, $\beta = 0.2$/day, $n = 10$, initial value $v_0 = 0.1$) and feedback is computed using a singular delay $\tau_m = 20$ days (top row of Fig. 2). In the following simulations, the delay distribution's width σ is grad-

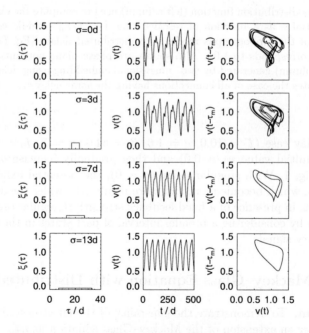

Fig. 2. Delay distribution function (left column) used to compute the blood cell concentrations $v(t)$ (middle column) during simulations of the Mackey–Glass equation with distributed delays (Eq. (4)). Also shown are phase plane portraits $v(t - \tau_m)$ vs. $v(t)$ (right column) generated by two–dimensional embedding using delay coordinates. $\sigma = 0$d denotes the singular delay case $\tau_m = 20$ days. Transients are omitted.

ually increased up to the maximum value $\sigma = 19$ days while all other parameters are left unchanged. Fig. 2 shows three examples of the resulting time courses and the corresponding phase plane plots, together with the distributions used to obtain them. Evidently, the larger the diversity of delay times is, the more regular the blood cell concentration $v(t)$ gets and the aperiodic behavior of the singular delay case is eventually turned into regular oscillations, the transitions resembling an inverse bifurcation cascade.

Feedback strength versus distribution width. To systematically investigate the phenomenon described above, the feedback strength β and the delay distribution's width σ are varied simultaneously. For each combination of the two parameters, the oscillation period is determined and plotted as a grey scale value in Fig. 3. Increasing β destabilizes the system's behavior which manifests itself in an increase in the oscillation period number, but only up to a certain value of $\beta \approx 4.0$/day, where oscillations become regular again even in the singular delay case. The destabilization caused by strong feedback can be counterbalanced by increasing σ. Aperiodic behavior is abolished completely in the regime $\sigma > 8.0$ days.

A local stability analysis yields the minimum values of σ needed for dynamics to be stable for given β. By linearizing Eq. (4) around its second fixed point, a single complex eigenvalue equation is obtained which in turn yields the following conditions after separating real and imaginary parts:

$$\gamma \sin \omega \tau_m + \omega \cos \omega \tau_m = 0 \tag{6}$$

$$\frac{n\gamma}{\frac{\sigma \omega}{\cos(\omega \tau_m) \sin(\omega \sigma)} + n - 1} = \beta. \tag{7}$$

Solving Eq. (6) numerically yields values $\omega(\gamma, \tau_m)$ that mark the system's transition from stability to oscillations of period 2. Next, $\omega(\gamma, \tau_m)$ is inserted into Eq. (7) and thereby $\sigma(\beta)$ can be computed. The resulting function is also plotted in Fig. 3 and perfectly fits with the numerical findings, thereby confirming

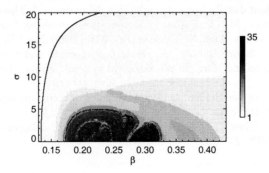

Fig. 3. Oscillation period of the Mackey–Glass system depending on feedback strength β and delay distribution width σ. The solid black curve shows the transition from a single fixed point to oscillating behavior of period 2 obtained from Eqs. (6) and (7).

that the onset of instability is shifted towards higher feedback strengths β the broader the delay distribution is.

4 Conclusion

We cross–checked our results by choosing Gaussian and Gamma distribution functions for both the Mackey–Glass and the hippocampus simulations, but the qualitative behavior did not change (data not shown). Furthermore, stabilization due to distributed delays is also observed in a linear feedback differential equation [1]. Thus, the effect observed does neither depend on the exact shape of the delay distribution function nor on the particular system. Rather, we suggest that increasing the temporal dispersion smoothes large fluctuations in the feedback signal that would otherwise deteriorate the system's current state.

The mechanism proposed here may explain why aperiodic behavior does not occur in neural systems as often as expected given the ubiquity of delayed regulation. Due to the diversity in signal transduction times, regular behavior is still possible even if strong and delayed feedback is unavoidable. In general, we conclude that ignoring the variability in inherent in neural populations and only focusing on average properties may be inappropriate to account for the system's dynamics.

Acknowledgments. The authors thank Prof. Michael C. Mackey for fruitful discussions on the issue of delayed feedback.

References

1. Bernard, S., Bélair, J., Mackey, M. C.: Sufficient conditions for stability of linear differential equations with distributed delays. Discrete and Continuous Dynamical Systems B **1** (2001) 233–256
2. Eurich, C. W., Mackey, M. C., Schwegler: Recurrent inhibitory dynamics: The role of state dependent distributions of conduction delay times. J. Theor. Biol. (in press)
3. Glass, L., Mackey, M. C.: Pathological conditions resulting from instabilities in physiological systems. Ann. N. Y. Acad. Sci. **316** (1979) 214–235
4. Mackey, M. C., an der Heiden, U.: Dynamical diseases and bifurcations: Understanding functional disorders in physiological systems. Funkt. Biol. Med. **1** (1982) 156–164
5. Mackey, M. C., an der Heiden, U.: The dynamics of recurrent inhibition. J. Math. Biol. **19** (1984) 211–225
6. Mackey, M. C., Glass, L.: Oscillation and chaos in physiological control systems. Science **197** (1977) 287–289

Learning Multiple Feature Representations from Natural Image Sequences

Wolfgang Einhäuser, Christoph Kayser, Konrad P. Körding, and Peter König

Institute of Neuroinformatics, University / ETH Zürich
Winterthurerstr. 190, 8057 Zürich, Switzerland
{weinhaeu,kayser,koerding,peterk}@ini.phys.ethz.ch

Abstract. Hierarchical neural networks require the parallel extraction of multiple features. This raises the question how a subpopulation of cells can become specific to one feature and invariant to another, while a different subpopulation becomes invariant to the first but specific to the second feature. Using a colour image sequence recorded by a camera mounted to a cat's head, we train a population of neurons to achieve optimally stable responses. We find that colour sensitive cells emerge. Adding the additional objective of decorrelating the neurons' outputs leads a subpopulation to develop achromatic receptive fields. The colour sensitive cells tend to be non-oriented, while the achromatic cells are orientation-tuned, in accordance with physiological findings. The proposed objective thus successfully separates cells which are specific for orientation and invariant to colour from orientation invariant colour cells.

Keywords. Learning, visual cortex, natural stimuli, temporal coherence, colour UTN: I0109

1 Introduction

In recent years there has been increasing interest in the question on how the properties of the early visual system are linked to the statistics of its natural input. Regarding primary visual cortex (V1), the spatial properties of simple [[1]] as well as complex [[2]] cells could be explained on the basis of sparse coding. A different coding principle based on the trace rule originally proposed by Földiak [[3]], namely temporal coherence or stability, has also been shown to lead to the emergence of simple [[4]] and complex [[5]] type receptive fields when applied to natural image sequences. These studies use greyscale images and thus address spatial properties only. However, a considerable fraction of primate V1 neurons is sensitive to colour and their spatial properties deviate from those of achromatic cells [[6,7]]. We here address whether temporal coherence leads to the emergence of colour-sensitive cells, how their spatial receptive fields are related to their chromatic properties, and how their fraction relative to the whole population is determined.

J.R. Dorronsoro (Ed.): ICANN 2002, LNCS 2415, pp. 21–26, 2002.
© Springer-Verlag Berlin Heidelberg 2002

2   Methods

Natural Stimuli. Sequences of natural stimuli are recorded using a removable lightweight CCD camera (Conrad electronics, Hirschau, Germany) mounted to the head of a freely behaving cat, while the animal is taken for walks in various local environments. The output of the camera is recorded via a cable attached to the leash onto a VCR carried by the experimenter and digitized offline at 25 Hz and 320x240 pixels. A colour-depth of 24-bit is used and encoded in standard RGB format. In each of the 4900 image frames thus obtained, at 20 randomly chosen locations a 30x30 pixel wide patch is extracted. This patch together with the patch from the same location of the following frame constitutes a stimulus pair. Each colour channel of each patch is smoothed with a Gaussian kernel of width 12 pixel to reduce boundary effects. For computational efficiency a principal component analysis (PCA) is performed on the stimuli to reduce the input dimensionality of 2700 (30x30 pixels times 3 colours). Unless otherwise stated, the first 200 principal components are used for further analysis, which explain 97 % of the variance. In order to process the stimuli independently of the global illumination level, the mean intensity is discarded by excluding the first principal component.

    Objective function. We analyse a population of N=200 neurons. The activity of each neuron is computed as

$$A_i(t) = \left| \sum_j W_{ij} * I_j(t) \right| \tag{1}$$

where I is the input vector and W the weight matrix. We define the temporal coherence of each neuron as

$$\psi_i^{stable} := -\frac{\left\langle \left(\frac{\mathrm{d}}{\mathrm{d}t} A_i(t)\right)^2 \right\rangle_t}{var_t\left(A_i(t)\right)} \tag{2}$$

where $\langle \rangle$ denotes the mean and var_t the variance over time. The temporal derivative is calculated as the difference in activity for a pair of stimuli from consecutive frames. We refer to this objective as 'stability', since it favours neurons, whose responses vary little over time. The total stability is defined as sum over the individual stabilities:

$$\Psi^{stable} := \sum_i \psi_i^{stable} \tag{3}$$

Furthermore, we define the decorrelation objective as

$$\Psi^{decorr} = -\frac{1}{(N-1)^2} \left\langle \sum_i \sum_{j \neq i} \left(\sigma_{ij}^2(t)\right) \right\rangle_t \tag{4}$$

where $\sigma_{ij}(t)$ denotes the correlation coefficient between $A_i(t)$ and $A_j(t)$. Combining these objectives we define

$$\Psi^{total} = \Psi^{stable} + \beta \Psi^{decorr} \tag{5}$$

where β is a constant.

The network is trained from random initial conditions by gradient ascent: For each iteration Ψ^{total} is computed over the complete natural stimulus set and the weightmatrix W is changed in the direction of the analytically given gradient, $\frac{d\Psi^{total}}{dW}$ to maximise Ψ^{total}. All presented results are taken after 60 iterations, when the network has converged under all analysed conditions. Simulations are performed using MATLAB (Mathworks, Natick, MA).

Analysis of neuron properties. By inverting the PCA the receptive field representation in input space is obtained from the rows of W. For further analysis we convert this colour channel (RGB) representation into a representation separating the hue, the saturation and the value (brightness) of each pixel (HSV representation) by the standard MATLAB function rgb2hsv using default mapping. To illustrate the colour properties of a receptive field abstracted from its topography, we plot the projection of each of its pixels onto the isoluminant plane. We quantify the colour content of a receptive field by the mean saturation over all pixels and define a cell to be colour sensitive if its mean saturation falls above 0.2. The isotropy of a receptive field is assessed using a standard method [[8]]: The tensor of inertia is computed on the pixel values and anisotropy is defined as the ratio of the difference between the tensor's long and short principal axis divided by the sum of these axes. This measure is 0 for an isotropic structure and approaches 1 for a perfectly oriented structure.

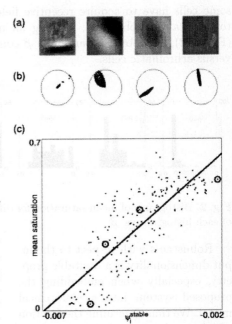

Fig. 1. (a) Four examples of receptive fields of a simulation using $\beta = 5$ after convergence (60 iterations) sorted (from left to right) by increasing ψ_i^{stable}.
(b) All pixels of each receptive field from (a) projected onto the isoluminant plane. Points close to the center indicate low saturation, i.e. little colour content, points to the periphery indicate saturated (coloured) pixels. The scaling is identical for all plots.
(c) Dependence of mean saturation on ψ_i^{stable}. Encircled points represent examples from (a) and (b).

3   Results

We train the network to optimize Ψ^{total} on natural stimuli. After convergence one observes an about equal fraction of chromatic and achromatic cells. The achromatic cells have lower individual values of ψ_i^{stable} than the colour sensitive cells, indicating that the stability objective favours chromatic cells (Figure 1).

The relative contribution of the decorrelation and the stability objective can be regulated by changing the parameter β. In order to investigate the influence of the stability objective alone, we set $\beta = 0$. In this case nearly all cells become colour-sensitive (Figure 2 left). This shows that colour-sensitive cells have more stable responses to natural stimuli than achromatic cells.

Strengthening of the decorrelation objective – by increasing β – on the other hand forces the cells to acquire more dissimilar receptive fields. In this case some cells have to acquire receptive fields which are suboptimal with respect to the stability objective, yielding an increasing fraction of achromatic cells (Figure 2). Thereby the parameter β controls the relative fraction of chromatic versus achromatic cells.

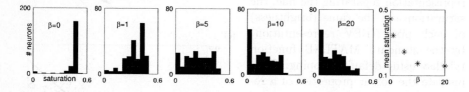

Fig. 2. Histograms of mean saturation for different values of β. Rightmost panel: mean of each histogram vs. β

Robustness with respect to the input dimensionality is a desirable property, especially when embedding the proposed system into a hierarchical model. We thus determine the fraction of colour selective cells in dependence on the PCA dimension used. We find the results to be over a wide range independent of the input dimension (Figure 3). This is in contrast to studies using independent component analysis (ICA), where the fraction of colour sensitive cells strongly depends on the PCA dimension [[9]].

Unlike achromatic cells, most colour sensitive neurons in primate V1 are non-oriented, i.e. they do not exhibit a pronounced orientation preference [[6, 7]]. We thus analyse the dependence of the model cells' anisotropy (see methods) to their mean saturation. We indeed find a strong tendency for chromatic cells to be non-oriented. On the other

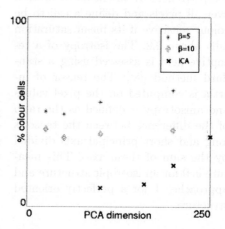

Fig. 3. Percentage of colour cells (mean saturation > 0.2) vs. PCA dimension for two values of β compared to ICA results of reference [[9]].

hand most achromatic cells show the orientation tuning typical of V1 simple cells (Figure 4), in compliance with the application of the stability objective on greyscale images.

Concluding, unoriented chromatic cells exhibit optimally stable responses on natural stimuli. When cells are forced by decorrelating to achieve suboptimal stability, a second subpopulation emerges, oriented achromatic cells. This implies that colour is the most stable feature in natural scenes, but that on the other hand achromatic edges are more stable than chromatic edges.

4   Discussion

A common problem in hierarchical networks is the separation of different variables. Here we obtain two distinct populations, one specific for colour and invariant to orientation and the other vice versa. The individual stability ψ_i^{stable} of each cell provides a system inherent measure to distinguish between both populations. Furthermore, the relative size of both populations is regulated by a single parameter. These properties make the proposed stability objective promising for the use at different stages of hierarchical systems.

We showed in a previous study how the stability objective can be implemented in a physiologically realistic framework [[10]]. A possible implementation of the decorrelation objective in neural circuits is mediated by strong lateral inhibition. Due to response la-

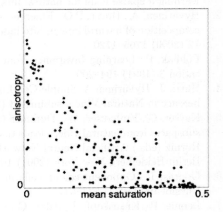

Fig. 4. Value anisotropy vs. mean saturation for a simulation of $\beta = 5$.

tencies this mechanism would have larger impact with increasing distance from the input layer. The fact that stronger decorrelation leads to less chromatic cells thus is in compliance with the physiological finding that in primate V1 most chromatic cells are found in layer 4.

A number of studies recently addressed the application of ICA on standard colour natural stimuli [[9]] and on hyperspectral images [[11,12]]. These studies find colour-sensitive cells, similar to the ones described here. However, none of the studies quantifies the relation of the cells' spatial to their chromatic properties, leaving an important issue for the comparison to physiology unaddressed. Here we find a strong correlation between spatial and chromatic properties comparable to physiological findings. Another remarkable difference between stability and ICA is the dependence of the latter on the input dimension. This implies that the relative size of the emerging chromatic and achromatic subpopulations might be influenced by changes in feedforward connectivity. In the case of stability on the other hand, the subpopulations' size is regulated by the relative strength of decorrelation, which might be mediated by lateral connections.

Acknowledgments. This work was financially supported by Honda R&D Europe (Germany), the Center for Neuroscience Zurich (ZNZ), the Collegium Helveticum and the Swiss national fund (SNF – grantNo: 31-65415.01). All procedures are in accordance with national and institutional guidelines for animal care.

References

1. Olshausen, B.A., Field, D.J.: Emergence of simple-cell receptive field properties by learning a sparse code for natural images. Nature 381 (1996) 607–609
2. Hyvärinen, A., Hoyer, P.O.: Emergence of phase and shift invaraint features by decomposition of natural images into independent feature subspaces. Neural Comput. 12 (2000) 1705–1720
3. Földiak, P.: Learning Invariance from Transformation Sequences. Neural Computation 3 (1991) 194–200
4. Hurri, J., Hyvärinen A. Simple-Cell-Like Receptive Fields Maximize Temporal Coherence in Natural Video. submitted (2002)
5. Kayser, C., Einhäuser, W., Dümmer O., König P., Körding K.P.: Extracting slow subspaces from natural videos leads to complex cells. In G. Dorffner, H. Bischoff, K. Hornik (eds.) Artificial Neural Networks – (ICANN) LNCS 2130, Springer-Verlag, Berlin Heidelberg New York (2001) 1075–1080
6. Gouras, P.: Opponent-colour cells in different layers of foveal striate cortex. J. Physiol 199 (1974) 533–547
7. Lennie, P., Krauskopf, J., Sclar, G.: Chromatic Mechanisms in Striate Cortex of Macaque. J. Neurosci. 10 (1990) 649–669
8. Jähne, B.: Digital Image Processing - Concepts, Algortihms and Scientific Applications, 4th compl. rev. edn. Springer-Verlag, Berlin Heidelberg New York (1997)
9. Hoyer, P.O., Hyvärinen, A.: Independent Component Analysis Applied to Feature Extraction From Colour and Stereo Images. Network 11 (2000) 191–210
10. Einhäuser, W., Kayser, C., König, P., Körding, K.P.: Learning of complex cells properties from their responses to natural stimuli. Eur. J. Neurosci. 15 (2002) in press.
11. Lee, T.W., Wachtler, T., Sejnowski, T.: Color Opponency Constitutes A Sparse Representation For the Chromatic Structure of Natural Scenes. NIPS. 13 (2001) 866–872.
12. Wachtler, T., Lee, T.W., Sejnowski, T.: Chromatic Structure of natural scenes. J. Opt. Sco. Am. A 18 (2001) 65–77

Analysis of Biologically Inspired Small-World Networks

Carlos Aguirre[1], Ramón Huerta[1,2], Fernando Corbacho[1], and Pedro Pascual[1]

[1] Escuela Técnica Superior de Informática, Universidad Autonoma de Madrid,
28049 Madrid, Spain
{Carlos.Aguirre, Ramon.Huerta, Fernando.Corbacho,
Pedro.Pascual}@ii.uam.es
[2] Institute for Nonlinear Science, UCSD,
San Diego 92093 CA, USA

Abstract. Small-World networks are highly clusterized networks with small distances between their nodes. There are some well known biological networks that present this kind of connectivity. On the other hand, the usual models of Small-World networks make use of undirected and unweighted graphs in order to represent the connectivity between the nodes of the network. These kind of graphs cannot model some essential characteristics of neural networks as, for example, the direction or the weight of the synaptic connections. In this paper we analyze different kinds of directed graphs and show that they can also present a Small-World topology when they are shifted from regular to random. Also analytical expressions are given for the cluster coefficient and the characteristic path of these graphs.

1 Introduction

Graph theory [1] provides the most adequate theoretical framework in order to characterize the anatomical connectivity of a biological neural network. To represent neural networks as graphs allows a complete structural description of the network and the comparison with different known connection patterns. The use of graph theory for modeling of neural networks has been used in theoretical neuroanatomy for the analysis of the functional connectivity in the cerebral cortex. In [2] it is shown that the connection matrices based on neuroanatomical data that describes the macaque visual cortex and the cat cortex, present structural characteristics that coincide best with graphs whose units are organized in densely linked groups that were sparsely but reciprocally interconnected. These kind of networks also provide the best support for the dynamics and high complexity measures that characterize functional connectivity.

There are some well known biological neural networks [3,4] that present a clear clustering in their neurons but have small distances between each pair of neurons. These kind of highly clusterized, highly interconnected sparse networks are known as Small-World (SW) networks. SW topologies appear in many real life networks [6,7], as a result of natural evolution [4] or a learning process [8]. In

J.R. Dorronsoro (Ed.): ICANN 2002, LNCS 2415, pp. 27–32, 2002.

[5] it is shown that on SW networks coherent oscillations and temporal coding can coexist in synergy in a fast time scale on a set of coupled neurons.

In [9] a method to study the dynamic behavior of networks when the network is shifted from a regular, ordered network to a random one is proposed. The method is based on the random rewiring with a fixed probability p for every edge in the graph. We obtain the original regular graph for $p = 0$, and a random graph for $p = 1$. This method shows that the characteristic path length (the average distance between nodes measured as the minimal path length between them) decreases with the increasing value of p much more rapidly than the clustering coefficient (that average number of neighbors of each node that are neighbors between them) does. It was found that there is a range of values of p where paths are short but the graph is highly clustered.

As initial substrate for the generation of SW, the use of undirected and un-weighted ring-lattices or grids is usually proposed [9]. They are used because these graphs are connected, present a good transition from regular to random, there are not specific nodes on them and model a high number of real networks. Biological or artificial neural networks are not accurately represented by these models as neural networks present a clear directionality and a different coupling in their connections. For these reasons is necessary to develop new models of regular networks that take in account the direction and the weight of the neuronal synapses and to explore if these models can also present a SW area when they are shifted from regular to random.

2   Models of Networks

In the existing literature about SW, only graphs that are connected, sparse, simple, undirected and unweighted are considered. Even though the graphs that conform the previous conditions can be accurate models for many networks, there are many other networks where relations are often directed or can have a value associated with the connection. In biological neural networks the information flows mainly from the presynaptic neuron to the postsynaptic neuron and the connections between different neurons have different coupling efficacies. As pointed in [9] and [2] it is necessary to explore models of networks where the undirected and unweighted conditions are relaxed. It is not clear, a priory, that relaxing those two conditions, the resulting networks present a similar SW area when they are shifted from regular to random.

As initial substrate we are going to consider directed rings-lattices and grids both weighted and unweighted. These substrates are selected since they are connected, regular and do not have special nodes. The directed unweighted ring-lattices will be explored for two different distributions of the neighbors of each node. The forward-backward distribution makes each node to connect to neighbors that are both at its left and right sides. In the forward networks each node only connects with nodes that are at the right side. Ring-lattices are depicted in fig. 1.

For grids we explore two different configurations. In the case of forward-backward distribution, each node connects with neighbors that are in each of

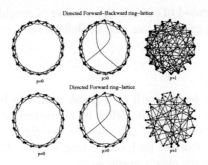

Fig. 1. Directed ring lattices

the four possible directions in the grid. In the Forward networks each node only connect with nodes that are at its right and top sides.

In the case of weighted graphs we are going to consider a forward connection pattern. In our model the weight of each connection will follow an uniform random distribution w, with values $0 < w \leq 1$. Models of weighted graphs are depictured in fig. 2.

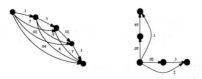

Fig. 2. Weighted graphs, left: Forward Ring lattice $k = 4$, right: Forward Grid $k = 4$

3   Length and Cluster Scaling

In this section we give analytical expressions for both the cluster coefficient and the characteristic path for the regular graphs presented in the previous section.

We denote by n as the number of nodes in the graph and k as the number of neighbors of each node. For a directed graph we say that node b is a neighbor of node a if the edge (a, b) exists. Note that b neighbor of a does not imply a neighbor of b. The neighborhood of a node is the set of nodes that a given node is connected to.

Intuitively the cluster coefficient is the average number of neighbors of each node that are neighbors between them. More precisely, for a vertex v let us define $\Gamma(v)$ as the subgraph composed by the neighbors of v (not including v itself). Let us define the cluster coefficient for a given node v as $C_v = |E(\Gamma(v))|/(k_v(k_v - 1))$ where $|E(\Gamma(v))|$ is the (weighted in the case of weighted graphs) number of edges in the neighborhood of v, k_v is the number of neighbors of v. The cluster coefficient for a given graph G can now be defined as

$$C = \frac{\sum_{i=1}^{n} C_i}{n}.$$

(1)

The characteristic path length of a graph indicates how far the nodes are among them. For a vertex v let us define its characteristic path length as $L_v = \sum_{i=1}^{n} d(v,i)/(n-1)$ where $d(v,i)$ indicates the (weighted in the case of weighted graphs) length of the shortest path connecting v and i. Using L_v we define the characteristic path length over a graph as

$$L = \frac{\sum_{i=1}^{n} L_i}{n}.$$

(2)

Simple counting for each type of graph provides with the following expressions for L and C:

– Unweighted Forward-Backward ring-lattices

$$L = \frac{n(n+k-2)}{2k(n-1)} = O(n) \quad C = \frac{3(k-2)}{2(k-1)} = O(1).$$

(3)

– Unweighted Forward ring-lattices

$$L = \frac{n(n+k-2)}{2k(n-1)} = O(n) \quad C = \frac{1}{2} = O(1).$$

(4)

– Unweighted Forward-Backward grids

$$L = \frac{2\sqrt{n}+k-4}{k} = O(n^{\frac{1}{2}}) \quad C = \frac{3(k-2)}{2(k-1)} = O(1).$$

(5)

– Unweighted Forward grids

$$L = \frac{2\sqrt{n}+k-4}{k} = O(n^{\frac{1}{2}}) \quad C = \frac{1}{2} = O(1).$$

(6)

– Weighted Forward ring-lattices

$$L = O(n) \quad C = O(1)$$

(7)

– Weighted Forward grids

$$L = O(n^{\frac{1}{2}}) \quad C = O(1)$$

(8)

The values of L for unweighted grids are obtained noting that $d(v_{ij}, v_{i'j'}) = d(v_{ij}, v_{ij'}) + d(v_{ij'}, v_{i'j'})$ and using the results for unweighted ring-lattices. The results for weighted graphs are due to the relation $\delta d_u(u,v) < d_w(u,v) < d_u(u,v)$ and $\delta > 0$ where d_w is the distance between two nodes in weighted graphs and

d_u is the distance between two nodes in unweighted graphs. We also use the fact that the weight for each connection has a value greater than 0.

In a random graph it can be shown that L scales as $log(n)$ and C scales to 0 as n tends to infinity [10]. This means that our models have a different scaling regime that the random graph model. This makes us expect that at some point, when we shift from these regular models to a random graph, there must be a phase change both in the value of L and C. If this change of phase is produced at different values of p for L and C we can build SW graphs using this regular substrates, and therefore, there exist SW graphs for these models.

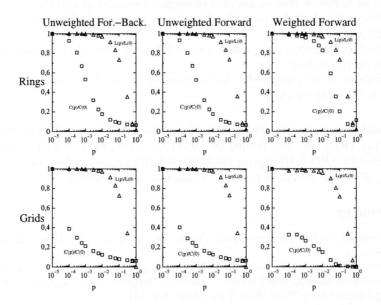

Fig. 3. Normalized values of L (squares) and C (triangles). Ring-lattice $n = 1000$, $k = 10$. Grids $n = 250000$, $k = 12$. Plots are the average of 100 experiments each

4 Metrics Behavior

In this section we explore the values of L and C when the models described in the previous sections are shifted from regular to random.

If we apply the procedure described in [11] to the models described in the previous sections we can observe that all the unweighted models present a clear SW area. For all of the unweighted models in consideration, there is a clear range of values of p where L is low but C maintains high. In figure 3 the normalized values of L and C are depictured for unweighted rings and grids.

In the case of weighted graphs, grids also present a clear SW area, but rings present a significative smaller range of values where L is low but C is high. This is due to the low number of edges rewired in the ring-lattice for low values of

p and to the fact that in the case of weighted graphs a rewired edge can make paths longer. When a small amount of edges are rewired, path length keeps almost constant, with a very small decrement. Rings with a higher number of nodes produce a wider SW area. The normalized values of L and C for weighted directed ring-lattices and grids can be seen in fig. 3.

5 Conclusions

The previous results allow us to establish the following conclusions.

- There exist regular, directed and weighted substrates that present a similar scaling behavior that undirected unweighted equivalent substrates.
- These new substrates have the directionality present in biological and artificial neural networks.
- The directed substrates present a clear SW area when they are shifted from regular to random.
- Weighted substrates need a higher number of nodes in order to present a clear SW area for rings.

Acknowledgments. We thank the Ministerio de Ciencia y Tecnología (BFI 2000-015). (RH) was also funded by DE-FG03-96ER14092, (CA) was partially supported by ARO-MURI grant DAA655-98-1-0249 during a four month stay in UCSD. (PP) and (CA) are partially supported by PB98-0850.

References

1. Chartrand, G.: Introductory Graph Theory. Dover Publications, Mineola New York (1985).
2. Sporns, O., Tononi, G., Edelman, G.M.: Relating Anatomical and Functional Connectivity in Graphs and Cortical Connection Matrices. Cerebral Cortex, Vol. 10. Oxford University Press, New York (2000) 127–141
3. White, J.G., Southgate, E., Thompson, J.N., Brenner,S.: The structure of the nervous system of the nematode Caenorhabditis elegans. Philosphical Transactions of the Royal Society of London. Series B **314** (1986) 1–340
4. Achacoso, T.B., Yamamoto, W.S.: AY's Neuroanatomy of C.elegans for Computation. CRC Press, Boca Raton Florida (1992).
5. Lago L.F. Huerta R. Corbacho F. and Siguenza J.A. Fast Response and Temporal Coding on Coherent Oscillations in Small-world Networks, Physical Review Letters, **84** (12) (2000), 2758–2761.
6. A. G. Phadke A.G. and Thorp J.S.: Computer Relaying for Power systems Wiley, New York, (1988).
7. Milgram S.: The Small World Problem, Psychology today, **2** (1967), 60–67.
8. Araújo T. and Vilela Mendes R.: Function and Form in Networks of Interacting Agents, Complex Systems **12** (2000) 357–378.
9. Watts, D.J.: Small Worlds: The dynamic of Networks between Order and Randomness, Princeton University Press, Princeton, New Jersey (1999).
10. Bollobas, B.: Random Graphs. Harcourt Brace Jovanovich, Orlando Florida (1985).
11. Watts, D.J., Strogatz, S. H.: Collective dynamics of small-world networks, Nature. **393** (1998) 440.

Receptive Fields Similar to Simple Cells Maximize Temporal Coherence in Natural Video

Jarmo Hurri and Aapo Hyvärinen

Neural Networks Research Centre
Helsinki University of Technology, P.O.Box 9800, 02015 HUT, Finland
{jarmo.hurri,aapo.hyvarinen}@hut.fi

Abstract. Recently, statistical models of natural images have shown emergence of several properties of the visual cortex. Most models have considered the non-Gaussian properties of static image patches, leading to sparse coding or independent component analysis. Here we consider the basic statistical time dependencies of image sequences. We show that simple cell type receptive fields emerge when temporal response strength correlation is maximized for natural image sequences. Thus, temporal response strength correlation, which is a nonlinear measure of temporal coherence, provides an alternative to sparseness in modeling simple cell receptive field properties. Our results also suggest an interpretation of simple cells in terms of invariant coding principles that have previously been used to explain complex cell receptive fields.

1   Introduction

The functional role of simple cells has puzzled scientists since the structure of their receptive fields was first mapped in the 1950s (see, e.g., [1]). The current view of sensory neural networks emphasizes learning and the relationship between the structure of the cells and the statistical properties of the information they process. In 1996 a major advance was achieved when Olshausen and Field showed that simple-cell-like receptive fields emerge when sparse coding is applied to natural image data [2]. Similar results were obtained with independent component analysis (ICA) shortly thereafter [3,4]. In the case of image data, ICA is closely related to sparse coding [5].

In this paper we show that an alternative principle called *temporal coherence* [6,7,8] leads to the emergence of simple cell type receptive fields from natural image sequences. This finding is significant because it means that temporal coherence provides a complementary theory to sparse coding as a computational principle behind simple cell receptive field properties. The results also link the theory of achieving invariance by temporal coherence [6,8] to real world visual data and measured properties of the visual system. In addition, whereas previous research has focused on establishing this link for complex cells, we show that such a connection exists even on the simple cell level.

Temporal coherence is based on the idea that when processing temporal input the representation changes as little as possible over time. Several authors have

J.R. Dorronsoro (Ed.): ICANN 2002, LNCS 2415, pp. 33–38, 2002.
© Springer-Verlag Berlin Heidelberg 2002

demonstrated the usefulness of this principle using simulated data – examples include the emergence of translation invariance [6] and the discovery of surface depth from random dot stereograms [9]. The contribution of this paper is to show that when the input consists of *natural* image sequences, the linear filters that maximize *temporal response strength correlation* are similar to *simple cell receptive fields*.

2   Temporal Response Strength Correlation

In this paper we restrict ourselves to linear spatial models of simple cells. Our model uses a set of filters (vectors) $w_1, ..., w_K$ to relate input to output. Let vector $x(t)$ denote the input at time t. (A vectorization of image patches can be done by scanning images column-wise into vectors.) The output of the kth filter at time t is given by $y_k(t) = w_k^T x(t)$. Let $W = [w_1 \cdots w_K]^T$ denote a matrix with all the filters as rows. Then $y(t) = Wx(t)$, where $y(t) = [y_1(t) \cdots y_K(t)]^T$. Temporal response strength correlation, the objective function, is defined by

$$f(W) = \sum_{k=1}^{K} E_t\{g(y_k(t))g(y_k(t - \Delta t))\} , \qquad (1)$$

where g is a differentiable even (rectifying) convex function, such as $g(x) = \ln \cosh x$ or $g(x) = x^2$, and Δt is a delay. Thus g measures the strength of the response of the filter, emphasizing large responses over small ones. In a set of filters with a large temporal response strength correlation, the same filters often *respond strongly at consecutive time points*, outputting large (either positive or negative) values, thereby expressing temporal coherence of a population code.

To keep the outputs bounded we enforce unit variance constraint on $y_k(t)$'s. We also force the outputs to be uncorrelated to keep the filters from converging to the same solution. These constraints can be expressed by $W E_t\{x(t)x(t)^T\}W^T = I$. The optimization algorithm used for this constrained optimization problem is a variant of the classic gradient projection method. The algorithm employs whitening – a temporary change of coordinates – to transform the constraint into an orthonormality constraint. After whitening a gradient projection algorithm employing optimal symmetric orthogonalization can be used [10].

Note that if $g(x) = x^2$ and $\Delta t = 0$, the objective becomes $f(W) = \sum_{k=1}^{K} E_t y_k^4(t)$. Then optimization under the unit variance constraint is equivalent to optimizing the sum of kurtoses of the outputs. Kurtosis is a commonly used measure in sparse coding. Similarly, in the case $g(x) = \ln \cosh x$ and $\Delta t = 0$, the objective function can be interpreted as a measure of the non-Gaussianity of outputs. We will return to this issue in the next section.

3   Experiments on Natural Image Sequences

The natural image sequences used were a subset of those used in [11]. For our experiments some video clips were discarded to reduce the effect of human-made

objects and artifacts (black bars resulting from wide-screen format). The preprocessed data set consisted of 200,000 pairs of consecutive 11 × 11 image windows
at the same spatial position, but 40 ms apart from each other. Preprocessing
consisted of temporal decorrelation, subtraction of local mean, and normalization. The use of temporal decorrelation can be motivated in two different ways:
as a model of temporal processing at the lateral geniculate nucleus [12], and as
a means for enhancing temporal changes. As discussed above, for $\Delta t = 0$ the
objective can be interpreted as a measure of sparseness. Therefore, if there was
hardly any change in short intervals in video data, our results could be explained
in terms of sparse coding. This can be ruled out by using temporal decorrelation, as illustrated in Fig.1. After temporal decorrelation the local mean was
subtracted from each window. Finally, the vectorized windows were normalized
to have unit Euclidean norm, which is a form of contrast gain control. No spatial
low-pass filtering or dimensionality reduction was performed during preprocessing. A second data set was needed to compute the corresponding (static) ICA
solution for comparison. This set consisted of 200,000 11 × 11 images sampled
from the video data. Preprocessing consisted of removal of local mean followed
by normalization.

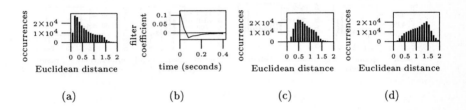

(a)   (b)   (c)   (d)

Fig. 1. Temporal decorrelation enhances temporal changes. (a) Distribution of Euclidean distances between normalized consecutive samples at the same spatial position
but 40 ms apart from each other. (b) The temporally decorrelating filter. (c) Distribution of distances between samples 40 ms apart *after* temporal decorrelation. (d) Distribution of distances (in a control experiment) between samples *120 ms apart* after
temporal decorrelation. See [10] for details.

The main experiment consisted of running the gradient projection algorithm
50 times using different random initial values, followed by a quantitative analysis
against results obtained with an ICA algorithm (see [10] for details). Figure
2 shows the results. The filters in Fig. 2(a) resemble Gabor filters. They are
localized, oriented, and have different scales. These are the main features of
simple-cell receptive fields [1]. The distributions of descriptive parameters of
receptive fields obtained by either temporal response correlation or ICA, shown
in Fig. 2(c), are very similar. This supports the idea that ICA / sparse coding
and temporal coherence are complementary theories in that they both result
in the emergence of simple-cell-like receptive fields. As for the differences, the
results obtained using temporal response strength correlation have a slightly

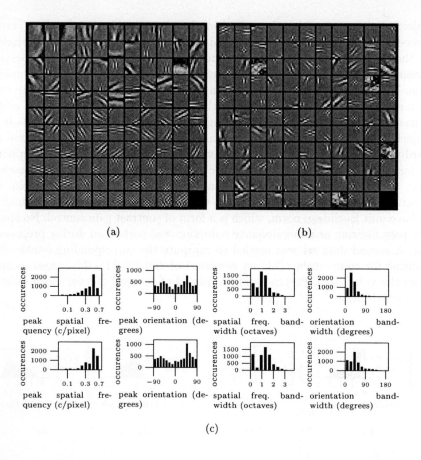

(a)   (b)

(c)

Fig. 2. Main results. (a) Filters estimated during the first run of the main experiment. Filters were estimated from natural image sequences by optimizing temporal response strength correlation (here $g(x) = \ln \cosh x$, $\Delta t = 40$ ms, and the number of extracted filters $K = 120$). The filters have been ordered according to $E_t\{g(y_k(t))g(y_k(t - \Delta t))\}$, that is, according to their "contribution" into the final objective value (filters with largest values top left). (b) For comparison, (static) ICA filters estimated from the same data. Note that these filters have a much less smooth appearance than in most published ICA results; this is because for the sake of comparison, we show here ICA filters and not ICA basis vectors, and further, no low-pass filtering or dimension reduction was applied in the preprocessing. (c) Distributions of descriptive parameters of 6000 receptive fields obtained from 50 runs of the algorithm optimizing response strength correlation (top row), and 6000 receptive fields obtained from 50 runs of an ICA algorithm (bottom row). See [4] for definitions of these properties.

smaller number of high-frequency receptive fields, and they are somewhat better localized both with respect to spatial frequency and orientation.

In addition to the main experiment we also made five control experiments: (i) no temporal decorrelation, (ii) compensation of camera movement by correlation-

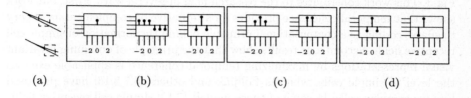

Fig. 3. A simplified model of local temporal coherence in natural image sequences shows why filters such as those presented in Figure 2(a) maximize temporal response strength correlation. Edges and lines are universal local visual properties of objects, so we examine their dynamics. A 3-D transformation of an object typically induces a different transformation in a 2-D image sequence. For example, a translation towards the camera induces a change in object size. Our hypothesis is that for local edges and lines, and short time intervals, most 3-D object transformations result in approximate local translations in image sequences. This is true for 3-D translations of objects, and is illustrated in (a) for two other transformations: a planar rotation (top) and bending of an object (bottom). Local area is marked with a dashed square. Edges, lines, and most filters in Figure 2(a) are symmetric along their longer axis, so we will consider the interaction of features and filters in one dimension: (b) illustrates the prototypes of a line (left), an edge (middle), and a Gabor-like filter (right). The response strength of the filter is strongly positive for lines and edges in spite of small translations, as shown in (c), in which the response strengths for the line (left) and the edge (right) have been calculated (as functions of spatial displacement). Therefore the response strength at time t (spatial displacement zero) is positively correlated with response at time $t + \Delta t$ even if the line or edge is displaced slightly. In contrast, (d) illustrates the fact that the raw output values, resulting from a line (left) or edge (right), might correlate either positively or negatively, depending on the displacement. Thus we see why ordinary linear correlation is not maximized for Gabor-like filters, whereas the rectified (nonlinear) correlation is.

based tracking, (iii) longer time delay $\Delta t = 120$ ms (see Fig. 1(d)), (iv) "consecutive" window pairs chosen randomly, (v) maximization of ordinary linear correlation instead of correlation of rectified outputs. The resulting filters in cases (i)–(iii) were similar to the main results, confirming that the main results are not a consequence of (i) temporal decorrelation, (ii) camera (observer) movement, or (iii) a particularly chosen Δt. The filters in (iv) showed no structure, confirming that the main results reflect dynamics of natural image sequences, and the filters in (v) resembled Fourier basis vectors, confirming that higher-order correlation is needed for the emergence of simple-cell-like receptive fields. See [10] for details.

4 Discussion

We have shown that simple cell type receptive fields maximize temporal response strength correlation at cell outputs when the input consists of natural image sequences. A simplified, intuitive illustration of why this happens is shown in

Fig.3. This work contributes to the research field in several ways. First, our work shows that temporal coherence provides an alternative or complementary theory to sparse coding as a computational principle behind the formation of simple cell receptive fields. Second, our results show that the principle of obtaining invariant visual representations by maximizing temporal coherence is applicable even on the level of simple cells, whereas Földiák and others [6,7,8,13] have proposed this for complex cells. In some of those models [7,13] simple cell receptive fields are obtained as by-products, but learning is strongly modulated by complex cells, and the receptive fields seem to lack the important properties of spatial localization and multiresolution. Furthermore, the results published in this paper have been computed from natural image sequence data. To our knowledge this is the first time that localized and oriented receptive fields with different scales have been shown to emerge from natural data using the principle of temporal coherence.

References

1. Palmer, S.E.: Vision Science – Photons to Phenomenology. The MIT Press (1999)
2. Olshausen, B.A., Field, D.: Emergence of simple-Cell receptive field properties by learning a sparse code for natural images. Nature 381 (1996) 607–609
3. Bell, A., Sejnowski, T.J.: The independent components of natural scenes are edge filters. Vision Research 37 (1997) 3327–3338
4. van Hateren, J.H., van der Schaaf, A.: Independent component filters of natural images compared with simple cells in primary visual cortex. Proceedings of the Royal Society of London B 265 (1998) 359–366
5. Hyvärinen, A., Karhunen, J., Oje, E.: Independent Component Analysis. John Wiley & Sons (2001)
6. Földiak, P.: Learning invariance from transformation sequences. Neural Computation 3 (1991) 194–200
7. Kayser, C., Einhäuser, W., Dümmer, O., König, P., Körding, K.: Extracting slow subspaces from natural Videos leads to complex cells. In Dorffner, G., Bischof, H., Hornik, K., eds.: Artificial Neural Networks – ICANN 2001, Springer (2001) 1075–1080
8. Wiskott, L., Sejnowski, T.J.: Slow feature analysis: Unsupervised learning of invariances. Neural Computation 14 (2002) 715–770
9. Stone, J.: Learning visual Parameters using spatiotemporal smoothness constraints. Neural Computation 8 (1996) 1463–1492
10. Hurri, J., Hyvärinen, A.: Simple-cell-like receptive fields maximize temporal coherence in natural Video. Submitted manuscript, electronie version available at http://www.cis.hut.fi/jarmo/publications/ (2001)
11. van Hateren, J.H., Ruderman, D.L.: Independent component analysis of natural image sequences yields spatio-temporal filters similar to simple cells in primary visual cortex. Proceedings of the Royal Society of London B 265 (1998) 2315–2320
12. Dong, D.W., Atick, J.: Temporal decorrelation: a theory of lagged and nonlagged responses in the lateral geniculate nucleus. Network: Computation in Neural Systems 6 (1995) 159–178
13. Kohonen, T., Kaski, S., Lappalainen, H.: Self-organized formation of various invariant-feature filters in the adaptive-subspace SOM. Neural Computation 9 (1997) 1321–1344

Noise Induces Spontaneous Synchronous Aperiodic Activity in EI Neural Networks

Maria Marinaro and Silvia Scarpetta

Dipartimento di Fisica "E. R. Caianiello", Salerno University, Baronissi (SA), Italy
INFM, Sezione di Salerno, IT, and IIASS, Vietri sul mare (SA), Italy

Abstract. We analyze the effect of noise on spontaneous activity of a excitatory-inhibitory neural network model. Analytically different regimes can be distinguished depending on the network parameters. In one of the regimes noise induces synchronous aperiodic oscillatory activity in the isolated network (regime B). The Coherent Stochastic Resonance phenomena occur. Activity is highly spatially correlated (synchrony), it's oscillatory on short time scales and it decorrelates in time on long time scale (aperiodic). At zero noise the oscillatory activity vanishes in this regime. Changing parameters (for example increasing the excitatory-to-excitatory connection strength) we get spontaneous synchronous and periodic activity, even without noise (regime C). The model is in agreement with the measurements of spontaneous activity of two-dimensional cortical cell neural networks placed on multi-electrode arrays performed by Segev et.al [2].

1 Introduction and Motivations

Recently much effort has been devoted to study in-vitro grown neural networks in isolation. In particular, Segev et al. [2] have done long term measurements of spontaneous activity of cortical cells neural network placed on multi-electrodes arrays. They observed both aperiodic synchronized bursting activity and periodic synchronized bursting activity, depending on values of Ca concentration. Some of their observations, and in particular some Power Spectrum Density (PSD) features, differ from the prediction of current computational neural network models and as far as we know, they have not been explained before. The aim of this paper is to show that a variant of the EI neural network model, introduced in a previous paper by one of us and collaborators [1] to model driven-activity and plasticity in cortical areas, is able to reproduce experimental results of spontaneous activity when we consider the model in isolation with intrinsic noise. Recently effects of noise on dynamics of various neural network models have been investigated in literature, using different thecniques like mean field theory, numerical simulations etc (see e.g. [3,4] and references therein). Here analytical solution of the stochastic equations in the linear approximation are studied; then noise and nonlinearity effects are taken into account, both analytically and with numerical simulations. In sec 2 the model is described and in sec 3 the analytical and numerical results are reported together with comments and conclusions.

J.R. Dorronsoro (Ed.): ICANN 2002, LNCS 2415, pp. 39–44, 2002.

2   The Model

The state variables u_i, v_i of the many interacting excitatory and inhibitory neurons evolve throw time, following the stochastic equations of motion:

$$\dot{u}_i = -\alpha u_i - \sum_j H_{ij} g_v(v_j) + \sum_j J_{ij} g_u(u_j) + \bar{F}_i(t), \tag{1}$$

$$\dot{v}_i = -\alpha v_i + \sum_j W_{ij} g_u(u_j) + F_i(t). \tag{2}$$

where $g(u_i(t))$ is the instantaneous rate of spiking of the ith neuron at time t, $i = 1 \ldots N$, α^{-1} is a time constant (about few milliseconds, for simplicity it is assumed equal for excitatory and inhibitory units), J_{ij}, W_{ij} and H_{ij} are the synaptic connection strengths. All these parameters are non-negative; the inhibitory character of the second term on the right-hand side of Eqn. (1) is indicated by the minus sign preceding it. $\bar{F}_i(t)$ and $F_i(t)$ model the intrinsic noise, respectively on the excitatory and inhibitory units, not included in the definition of u, v. In a (cultured) interacting neurons systems noise can be due to several reasons, like thermal fluctuation, ion channel stochastic activities, and many others. We take the noise $\bar{F}_i(t), F_i(t)$ to be uncorrelated white noise, such that $< F_i(t) >=< \bar{F}_i(t) >= 0$ and $< F_i(t)F_j(t') >=< \bar{F}_i(t)\bar{F}_j(t') >= \Gamma\delta_{ij}\delta_{t-t'}$. Note that even if the connection-matrices J, H, W, are symmetric (plus random fluctuations), the connections between excitatory and inhibitory neurons are still highly asymmetric in this scenario, and the total connection matrix $\left(\begin{smallmatrix} J & -H \\ W & 0 \end{smallmatrix}\right)$ is still asymmetric. An oscillatory state vector $\mathbf{u}(t) = \boldsymbol{\xi}\cos(\omega_0 t) = \boldsymbol{\xi}e^{i\omega_0 t} + c.c.$, where $\boldsymbol{\xi}$ have real positive-or-null elements correspond to a periodic synchronous bursting activity, and ξ_i describe the amplitudes of bursting on the different excitatory units. Then the local power spectrum density, PSD_i, and the global power spectrum, $PSD = 1/N \sum_i PSD_i$, will have a peak at $\omega = \omega_0$. We can describe both synchronous and phase-locked periodic activities cases by writing $u_i(t) = \xi_i e^{-i\omega_0 t} + c.c.$, taking the ξ_i real in the first case and complex ($\xi_i = |\xi_i|e^{i\phi_i}$) in the second.

3   Model Analysis

Since the connectivity formation of the cultured system we want to model has grown randomly and spontaneously, we can reasonably assume that strength of each connection is a function only of the type of pre-synaptic and post-synaptic neurons (Excitatory or Inhibitory) and eventually of the distance between them (plus some random quenched fluctuations). In particular, since we are modelling cortical area, we will assume the W and J connections to be long range connections, $J_{ij} = \frac{j_0}{N}(1 + \epsilon\eta_{ij})$, $W_{ij} = \frac{W_0}{N}(1 + \epsilon\eta_{ij})$ and the inhibitory-to-inhibitory connections H to be local, $H_{ij} = h_0\delta(i-j)(1 + \epsilon\eta_{ij})$, where $\epsilon << 1$ and η_{ij} is a random quenched value between -1 and 1. Therefore the three connection matrices have the common principal eigenvector $\boldsymbol{\xi}_0 = (1/\sqrt{N}, \ldots, 1/\sqrt{N})$ (apart from a correction of order $O(\epsilon/\sqrt{N})$), with eigenvalues j_0, W_0, h_0 respectively (with corrections $O(\epsilon/N)$). In the subspace orthogonal to the principal eigenvector the eigenvalues of J and W are negligible (of order ϵ/N). Here we take $\epsilon = 0$. Let's call $\{\bar{\mathbf{u}}, \bar{\mathbf{v}}\}$ the fixed point determined by $\dot{\mathbf{u}} = 0, \dot{\mathbf{v}} = 0$ with $\bar{\mathbf{F}}(t) = 0, \mathbf{F}(t) = 0$. Linearizing the equations (1,2) around the fixed point $\{\bar{\mathbf{u}}, \bar{\mathbf{v}}\}$, eliminating $\mathbf{v}$ from the equations, and assuming noise to be only on the $\mathbf{v}$ units, we obtain

$$\ddot{\mathbf{u}} + (2\alpha - \mathsf{J})\dot{\mathbf{u}} + [\alpha^2 - \alpha\mathsf{J} + \mathsf{H}\mathsf{W}]\mathbf{u} = -\mathsf{H}\mathbf{F}(t) \tag{3}$$

where $\mathbf{u}$ is now measured from the fixed point value $\bar{\mathbf{u}}$, and nonlinearity enters only thou the redefinition of the elements of J H and W: $J_{ij}g'_u(\bar{u}_j) \to J_{ij}$, $H_{ij}g'_u(\bar{v}_j) \to H_{ij}$ and $W_{ij}g'_u(\bar{u}_j) \to W_{ij}$. We use bold and sans serif notation (e.g. $\mathbf{u}$, J) for vectors and matrices respectively. The fixed point $(\bar{\mathbf{u}}, \bar{\mathbf{v}})$ is stable if the homogeneous associate equation of the Eq. (3) has only decaying solutions $\mathbf{u}$. In particular, when J, H and W share the same set of eigenvectors $\boldsymbol{\xi}_n$, denoting with j_n, W_n and h_n their eigenvalues, then eigensolutions are $\mathbf{u}_n = e^{\lambda_n t}\boldsymbol{\xi}_n + \text{c.c.}$, where

$$\lambda_n = -\frac{2\alpha - j_n}{2} \pm \frac{\sqrt{j_n^2 - 4h_n W_n}}{2} \tag{4}$$

and stability condition is $\text{Re}[\lambda_n] < 0$ i.e., the real parts $\text{Re}[\lambda_n]$ of the eigenvalues λ_n of the homogeneous system must be negative. Since the state vector $\mathbf{u}$ is a linear combination of all the eigenmodes, given by $\mathbf{u} = \sum_n c_n e^{\lambda_n t}\boldsymbol{\xi}_n + \text{c.c.}$ in the absence of noise $\Gamma = 0$, if all the n modes have $\lambda_n = \text{Re}[\lambda_n] < 0, (\text{Im}[\lambda_n] = 0)$, then the activity $\mathbf{u}$ simply decays toward the fixed point stationary state (regime A). However, if $\text{Re}[\lambda_n] < 0$ for each n, but $\exists n$ (call it $n = 0$) such that the imaginary part $\text{Im}[\lambda_0] \neq 0$, then, spontaneous collective aperiodic oscillations are induced by noise (regime B) as shown in the following section. Finally if $\text{Im}[\lambda_0] \neq 0$ and $\text{Re}[\lambda_0] > 0$ we get the regime C that will be analyzed in sec 3.2.

3.1   Regime B: Coherent Stochastic Resonance Phenomena

Let's analyze the dynamics of the stochastic system (3), when $\text{Re}[\lambda_n] < 0$ for each n and $\text{Im}[\lambda_n] = 0$ except for n=0. The state vector $\mathbf{u}$ is a linear combination of all the eigenmodes $\mathbf{u} = \sum_n(m_n(t) + \text{c.c.})\boldsymbol{\xi}_n$, in particular, when $\lambda_0 = -1/\tau_0 \pm i\omega_0$, it is

$$m_0(t) = [a_0 \cos(\omega_0 t) + b_0 \sin(\omega_0 t)]e^{-t/\tau_0} + \frac{1}{\omega_0}\int_0^t ds\, e^{-(t-s)/\tau_0}\sin(\omega_0(t-s))F_0(s) \tag{5}$$

where $F_0(s)$ is the projection of the noisy term $-\mathsf{H}\mathbf{F}$ on the eigenvector $\boldsymbol{\xi}_0$, and $< F_0(s)F_0(s') > = h_0^2\Gamma\delta(s-s')$. The correlation function of excitatory unit i, $\langle u_i(t)u_i(t')\rangle - \langle u_i(t)\rangle\langle u_i(t')\rangle = C_i(t-t')$ and the global correlation function $C(t-t')$ are the sum of the contributions of each mode, in particular $C(t-t') = \sum_i C_i(t-t') = \sum_n\langle(m_n + \text{c.c.})(m_n + \text{c.c.})\rangle - \sum_n\langle(m_n + \text{c.c.})\rangle\langle(m_n + \text{c.c.})\rangle \equiv \sum_n C^n(t-t')$. The PSD, i.e. the Fourier Transform $\tilde{C}(\omega)$, is just the sum of the

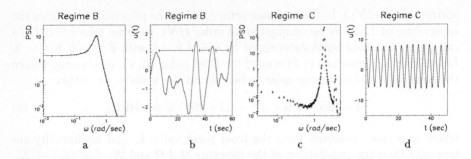

Fig. 1. a. Power Spectrum Density of excitatory units, of the linear system in the regime B, with $\Gamma = 0.0004$. Dots are numerical simulations results while solid line is our theoretical prediction. We use the following parameters: $N = 10$, $\alpha = 50sec^{-1}$, $J_{ij} = \frac{j_0}{N} = 9.98\ sec^{-1}$, $W_{ij} = W_0/N$, $H_{ij} = h_0\delta(i - j)$, and $W_0 = h_0 = \sqrt{(j_0/2)^2 + 0.5^2} = 49.9sec^{-1}$, so that $\omega_0 = 0.5 rad/sec$. b. The time behavior of the state variable $u_i(t)$, $i = 1, \ldots, 10$ of linear system in the regime B, with $\Gamma = 0.004$. All units $u_i(t)$ shows synchronous oscillatory activity. Differences in the activity between the state variables $u_i(t)$ are so small that all the lines $u_i(t)$ overlaps each other. c. Power Spectrum Density of excitatory units in regime C with $\Gamma = 0.0004$, using the saturating activation function shown in Fig. 2.A in dashed red line. In particular, $N = 10$, $\alpha = 50sec^{-1}$, $J_{ij} = \frac{j_0}{N} = 10.014$, $W_0 = h_0 = 50.07$, so that $\omega_0 = 0.5 rad/sec$ and oscillation frequency is $\omega = 1.9 rad/sec$ following eq. (11). d. State variable $u_i(t)$, $i = 1, \ldots, 10$ versus time in the regime C, with $\Gamma = 0.004$. All units $u_i(t)$ shows synchronous oscillatory periodic activity.

n contributions: if λ_n is real, $\lambda_n = -1/\tau_n < 0$, then its contribution to the PSD is just

$$\tilde{C}^n(\omega) \propto \frac{h_0^2 \Gamma \tau_n^4}{(1 + \omega^2 \tau_n^2)^2} \qquad (6)$$

peaked at $\omega = 0$, however if $Im[\lambda_0] = \omega_0 \neq 0$ and $Re[\lambda_0] = -1/\tau_0 < 0$, its contribution to the time correlation is

$$C^0(t - t') = h_0^2 \Gamma \frac{\tau_0^3}{4(1 + \omega_0^2 \tau_0^2)} e^{-|t-t'|/\tau_0} [cos(\omega_0|t - t'|) + \frac{1}{\omega_0 \tau_0} sin(\omega_0|t - t'|)] \qquad (7)$$

and the contribution to the PSD is

$$\tilde{C}^0(\omega) = h_0^2 \Gamma \frac{\tau_0^4}{4(1 + \omega_0^2 \tau_0^2)} [\frac{2 + \omega/\omega_0}{1 + \tau_0^2(\omega + \omega_0)^2} + \frac{2 - \omega/\omega_0}{1 + \tau_0^2(\omega - \omega_0)^2}] \qquad (8)$$

peaked at ω close to ω_0 (if $\frac{1}{\tau_0} > \omega_0$). Therefore the linear analysis predict that noise induces a collective oscillatory behavior in the neurons activity, in the parameter region when $\lambda_0 = -1/\tau_0 \pm i\omega_0$ (and $\frac{1}{\tau_0} > \omega_0$). This collective oscillatory behavior correspond to a broad peak in the power-spectrum of the neurons activity close to the characteristic frequency ω_0. Numerical simulation results are

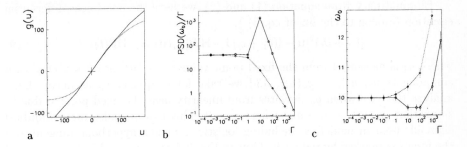

Fig. 2. a. Two activation functions for excitatory units u_i. Dashed line shows a saturating activation function (used in most of the following simulations), while solid line shows a piece-wise-linear function, whose slope increases before decreasing when u is raised from its stationary value (cross). b. Ratio $R = PSD(\omega_0)/\Gamma$ as a function of the noise level Γ. Dashed line correspond to the saturating activation function shown in A, while the solid line to the solid line activation function in A. We used the following parameters in the simulations: $N = 10$, $\alpha = 50sec^{-1}$ $J_{ij} = \frac{j_0}{N} = 9.98sec^{-1}$, $W_0 = h_0 = 50.89sec^{-1}$, so that $\omega_0 = 10rad/sec$. c. The frequency of the PSD peak versus the noise level (same simulations as in B). Effect of nonlinearity is evident in both fig. b and c, indeed for linear system R and ω_0 do not change with noise level.

shown in Fig.1.ab Oscillations are synchronous because $\boldsymbol{\xi_0}$ has real positive elements. The signal looks aperiodic on long time scale, since the noise decorrelates over long time scale the signal. This spontaneous aperiodic synchronous activity mimic the spontaneous aperiodic synchronous activity observed under 0.5 and 2.0 mM Ca concentration in [2]. Effects of nonlinearity and noise in this regime B have being investigated numerically. Fig. 2.b shows the ratio $R = \tilde{C}(\omega_0)/\Gamma$ between the output power at th peak close to the characteristic frequency ω_0 and the power of the noise Γ, versus the noise level, using two different activation functions. The first activation function deviates from linearity near the fixed point due to saturation (red dashed line), while in the second (black solid line) the slope increase before decreasing when the u is raised from its stationary value. Solid line in Fig. 2.b shows the typical maximum that has become the fingerprint of *Stochastic Resonance* phenomena. Fig. 2.c shows that oscillation frequency change slowly with the noise level Γ. If the excitatory-to-excitatory connections are more strong so that $Re[\lambda_{n=0}] = \alpha - j_0/2 \geq 0$, but not too much strong so that it is still $Im[\lambda_0] \neq 0$, then interesting spontaneous activity arise also without noise. In particular, spontaneous periodic oscillations arise in the linear approximation if $Re[\lambda_0] = 0$ and $Im[\lambda_0] \neq 0$, this is a critical point separating the regimes $Re[\lambda_0] < 0$, $Im[\lambda_0] \neq 0$ (regime B) and $Re[\lambda_0] > 0$, $Im[\lambda_0] \neq 0$ (regime C).

3.2   Regime C: Spontaneous Synchronous Periodic Activity

In the regime C linear analysis predict synchronous periodic activity with diverging amplitude, that become stable when nonlinearity is taken into account.

Eliminating **v** from equation (1) and (2), we have the non-linear differential equation (analog to the linear eq.(3)),

$$[(\alpha + \partial_t)^2]\mathbf{u} - [(\partial_t + \alpha)\mathsf{J} - \mathsf{HW}]g_u(\mathbf{u}) = -\mathsf{H}\mathbf{F}(t) \tag{9}$$

where u_i is measured from the fixed point value $\bar{u}_i$, $g_u(\mathbf{u})$ means a vector with components $[g_u(\mathbf{u})]_i = g_u(u_i)$, and we take for simplicity g_v as linear. Let's analyze the case when g_u deviates from linearity near the fixed point $\bar{u}$ due to saturation: $g_u(u_i) \sim u_i - au_i^3$ where $a > 0$. This case includes most standard sigmoids used in modeling, including logistic sigmoids, hyperbolic tangent and the function marked by red dashed line in Fig. 2.A that we used in our numerical simulations. We look self-consistently for solutions like $\mathbf{u} = q\boldsymbol{\xi}_0 e^{-i\omega t} + $ c.c.+ higher order harmonics, where $\boldsymbol{\xi}_0$ is the spatial pattern corresponding to the principal eigenvector of the connection matrices, and q is a macroscopic variable (the order parameter) measuring how much the system activity is similar to the periodic oscillating activity of interest. Our calculations predict that periodic synchronous oscillations ($q \neq 0$) arise only when the excitatory-to-excitatory strength is in the proper range

$$2\alpha < j_0 < 2(h_0 W_0)/\alpha \tag{10}$$

with both an upper and lower limit, and that the oscillation frequency is given (self-consistently) by

$$\omega^2 = 2h_0 W_0 \alpha / j_0 - \alpha^2. \tag{11}$$

Since the extracellular Ca is known to affect the synapse probability of neuro-transmitter release, the analytical prediction (10) can explain why only for a critical interval of concentrations of Ca2+ the periodic synchronous oscillations has been observed in cultured neural networks (it arise at $\sim 1mM$, not at *lower* or *higher* concentrations [2]). We have performed numerical simulations using the nonlinear saturating function shown in Fig. 2.a (dashed line). The noiseless nonlinear numerical simulations shows stable synchronous periodic spontaneous activity **u**, with frequency in agreement with the nonlinear analysis predictions of eq. (11) (results not shown). Fig.1.cd shows simulation results with noise $\Gamma = 0.0004$, the two high peaks at first and second harmonic in the PSD and the broad band at low frequency (indicating positive long-term time correlations) mimic the experimental results of [2].

References

1. S. Scarpetta, Z. Li, J. Hertz. NIPS 2000. Vol 13. T. Leen, T. Dietterich, V. Tresp (eds), MIT Press (2001).
2. R. Segev, Y. Shapira, M. Benveniste, E. Ben-Jacob. Phys. Rev. E **64** 011920, 2001. R. Segev, M. Benveniste, E. Hulata, N. Cihen, A. Palevski, E. Kapon, Y. Shapira. Phys. Rev. Lett. 2002 (in press).
3. N. Brunel Jour. of Comp. Neurosc. 8, 183-208, 2000. N. Brunel FS. Chance, N. Fourcaud, LF. Abbot , Phys. Rev. E, 86 (10), p.2186-9, 2001
4. P.C. Bressloff , Phys. Rev. E, 60,2, 1999. Y. Shim, H. Hong, M.Y. Choi, Phys. Rev. E, 65, p.36114, 2002.

Multiple Forms of Activity-Dependent Plasticity Enhance Information Transfer at a Dynamic Synapse

Bruce Graham

Department of Computing Science and Mathematics
University of Stirling, Stirling FK9 4LA, U.K.
b.graham@cs.stir.ac.uk
http://www.cs.stir.ac.uk/~bpg/

Abstract. The information contained in the amplitude of the postsynaptic response about the relative timing of presynaptic spikes is considered using a model dynamic synapse. We show that the combination of particular forms of facilitation and depression greatly enhances information transfer at the synapse for high frequency stimuli. These dynamic mechanisms do not enhance the information if present individually. The synaptic model closely matches the behaviour of the auditory system synapse, the calyx of Held, for which accurate transmission of the timing of high frequency presynaptic spikes is essential.

1   Introduction

A chemical synapse is not simply the point of connection between two neurons. It also acts as a filter for the signals passing from the presynaptic to the postsynaptic neuron. The magnitude of the postsynaptic response produced by the arrival of an action potential in the presynaptic terminal is subject to both short and long-term changes. Long-term changes, on the order of minutes or longer, can be likened to the setting of connection weights in artificial neural networks. Synapses also undergo short-term changes, on the order of milliseconds to seconds, in response to the arrival pattern of presynaptic action potentials [11]. The nature of these changes is synapse specific, and may even vary between synapses of the same presynaptic neuron, depending on the postsynaptic target [5].

Information theory provides tools for quantifying the ability of synapses to transfer information about presynaptic stimuli to the postsynaptic neuron [2]. Recent work has used information theory to examine how particular features of synaptic transmission affect information transfer. Zador [10] considered how the probabilistic nature of transmitter release and the number of release sites at a synapse affected the spiking response of a neuron being driven by a large number of synapses. Fuhrmann et al [4] considered the information contained in the amplitude of the postsynaptic response about the timing of presynaptic action potentials when the synapse was either depressed or facilitated by presynaptic activity. They showed that for Poisson distributed presynaptic spike trains, there

J.R. Dorronsoro (Ed.): ICANN 2002, LNCS 2415, pp. 45–50, 2002.
© Springer-Verlag Berlin Heidelberg 2002

is an optimal mean firing rate at which information transfer through the synapse is maximal. Forms of short-term depression and facilitation may act to filter presynaptic signals in different ways. Bertram [1] has demonstrated that different forms of depression result in either low- or high-pass filtering of spike trains.

In this paper we consider how multiple forms of facilitation and depression may combine to enhance the information contained in the postsynaptic response about the relative timing of presynaptic spikes. The synapse model is based upon data from the calyx of Held, a giant excitatory synapse in the mammalian auditory system [3]. The calyx forms a component of circuitry underlying interaural time and amplitude difference calculations. Consequently, the transfer of spike timing information through this synapse is crucial.

2   The Model Synapse

The calyx of Held exhibits significant depression in the amplitude of the post-synaptic current (EPSC) produced in response to trains of presynaptic action potentials due to depletion of the readily-releasable vesicle pool (RRVP) [7]. However, the steady-state EPSC amplitude at high frequencies is larger than expected given the estimated time constant of recovery of the RRVP of around 5 seconds. This is putatively due to an activity-dependent enhancement of the rate of replenishment of the RRVP, as has also been proposed for the neuromuscular junction [9]. Background and activity-dependent replenishment should combine to produce a steady-state EPSC amplitude that is frequency-independent. However, there is a further depression in the amplitude at stimulation frequencies above 10Hz. Recent new experimental techniques demonstrate that postsynaptic receptor desensitisation plays a major role in this extra depression [8].

The calyx of Held consists of hundreds of release sites operating in parallel and largely independently. A simple model that describes the average postsynaptic response at this synapse is as follows. The presynaptic response to an action potential is determined by the availability of vesicles at each release site and their probability of release. The availability of vesicles is given by:

$$\frac{dn}{dt} = \frac{(1-n)}{\tau_n} + (n_s - P_v n)\delta(t - t_s) \tag{1}$$

where n is the average number of releasable vesicles available per release site ($n = 1$ corresponds to every site having a single releasable vesicle). The background replenishment rate is given by the time constant τ_n. On the arrival of a presynaptic spike at time t_s, a vesicle may release with probability P_v and an average n_s new vesicles are mobilised to replenish each site. The probability of release at each site at the time of arrival of a spike is thus: $P_r(t_s) = P_v n(t_s)$.

The amplitude of the postsynaptic response depends both on the probability of vesicle release and the state of the postsynaptic receptors. Following the binding of neurotransmitter, the receptors may desensitise and are not again available for contributing to the postsynaptic response until they recover. The average receptor desensitisation, r_d is given by:

$$\frac{dr_d}{dt} = -\frac{r_d}{\tau_d} + Dr_d\delta(t - t_s) \qquad (2)$$

where D is the fraction of receptors that enter the desensitised state on the binding of neurotransmitter and τ_d specifies their rate of recovery. The average amplitude of the postsynaptic response is the product of the number of released vesicles and the available receptors: $PSR(t_s) = P_r(t_s)(1 - r_d(t_s))$.

This model adds activity-dependent vesicle replenishment and a significant desensitisation term to the basic model of Tsodyks and Markram [6]. It is sufficient to capture the major characteristics of the amplitude and time course of depression at the calyx of Held [8]. Depression of the postsynaptic response when subject to regular presynaptic stimulation is shown in Fig. 1. Note that unlike the postsynaptic response, the steady-state release probability is essentially independent of the stimulation frequency (Fig. 1(a)).

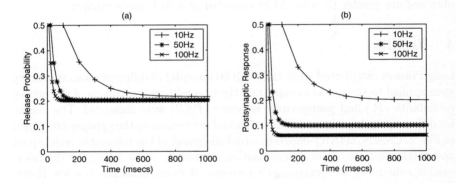

Fig. 1. Depression of (a) the probability of vesicle release, and (b) amplitude of the postsynaptic response for regular presynaptic stimulation at 10, 50 and 100Hz. Model parameters: $P_v = 0.5$, $\tau_n = 5secs$, $n_s = 0.2$, $D = 0.5$, $\tau_d = 50msecs$

3   Measuring Information Transfer

To determine the amount of information about the relative timing of the presynaptic action potentials contained in the postsynaptic response we follow the approach of Fuhrmann [4]. The synapse is stimulated with Poisson distributed spike trains of particular mean frequency. We then measure the mutual information between the interspike interval distribution (ISI) and the probability of release (P_r) or the amplitude of the postsynaptic response (PSR). The mutual information between two random variables, X and Y is given by the difference between the unconditional and conditional entropies of X: $I(X;Y) = H(X) - H(X|Y)$.

In conditions where X is uniquely determined by Y, the conditional entropy, $H(X|Y) = 0$, and thus $I(X;Y) = H(X)$, the unconditional entropy of X. For a discrete random variable, $H(X) = -\sum_x p(x)\log_2 p(x)$, where $p(x)$ is the probability that X will take the value x.

The synapse model is fully deterministic, thus the probability of release and the postsynaptic response are uniquely determined by the sequence of presynaptic spike times. Thus the information contained in P_r and PSR about the preceding $ISIs$ of long spike trains is given solely by the unconditional entropy of the P_r and PSR distributions. However, they are continuous variables and as a consequence their entropies are infinite and the synapse transmits infinite information about the $ISIs$. The information in this deterministic model becomes finite if we restrict the accuracy with which all values can be measured. Spike times are measured to an accuracy of 1ms and P_r and PSR are discretised using a bin size of 1% of their maximum values [4]. Fuhrmann et al [4] have demonstrated that with this approach, a deterministic model yields qualitatively similar results to more realistic probabilistic models of synaptic transmission, when considering relative values of information. However, the absolute values of information obtained are greater than would be expected at a real, noisy synapse.

4   Results

Long Poisson distributed spike trains (1000 seconds) of different mean frequency were applied to the model synapse and the unconditional entropies of the release probability (P_r) and postsynaptic response (PSR) were measured. The results for different combinations of facilitation and depression at the synapse are shown in Fig. 2. Clearly, activity-dependent replenishment of the releasable vesicle pool and postsynaptic receptor desensitisation combine to greatly enhance the information content of the postsynaptic response at frequencies above a few Hertz. These mechanisms by themselves do not enhance the information transfer.

To examine the sensitivity of information transfer to the model parameters, the desensitisation recovery rate, τ_d, and the amount of desensitisation, D, were varied. Fig. 3(a) indicates that there is an optimum recovery rate for a particular frequency of stimulation, with the optimum time constant decreasing as frequency increases. However, information is close to maximal at all frequencies considered for τ_d ranging from around 10ms to 50ms. Information transfer is relatively insensitive to the amount of desensitisation for D greater than about 0.5 (Fig. 3(b)). With smaller amounts of desensitisation the optimal recovery time constant increases and information transfer for fast τ_d is decreased. Also, smaller D results in lower information transfer below 100Hz, but slightly greater information transfer at higher stimulation frequencies (not shown).

5   Discussion

Recent work [4] has demonstrated that short-term depression at a synapse can convey information about the relative timing of presynaptic spikes in the amplitude of the postsynaptic response, in a frequency-dependent manner. In that work, and in previous studies (e.g. [5,6]), only a simple model of vesicle depletion

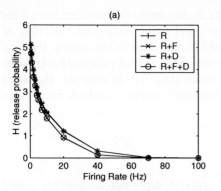
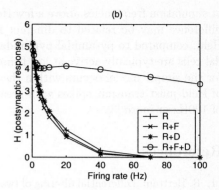

Fig. 2. Entropy of (a) the probability of release and (b) the postsynaptic response when stimulated with Poisson distributed spike trains of different mean frequencies. Legend: R - only background replenishment of vesicles ($n_s = 0$, $D = 0$); $R + F$ - with activity-dependent replenishment ($n_s = 0.2$, $D = 0$); $R + D$ - with desensitisation ($n_s = 0$, $D = 0.5$); $R + F + D$ - full synapse model ($n_s = 0.2$, $D = 0.5$)

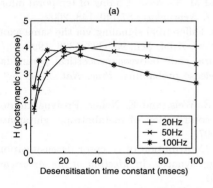
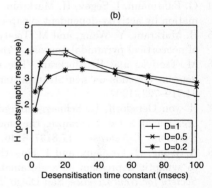

Fig. 3. Entropy of the postsynaptic response as a function of desensitisation recovery time constant, τ_d, with (a) Poisson distributed presynaptic stimulation at 20, 50 and 100Hz, and (b) different amounts of desensitisation, D, with Poisson distributed presynaptic stimulation at 100Hz

and recovery was used for depression. Here we extend this model to include a facilitatory mechanism in the form of activity-dependent replenishment of the releasable vesicle pool, and a postsynaptic depression mechanism in the form of receptor desensitisation. This model provides a good match for the amplitude and time course of depression at the calyx of Held, a giant synapse in the mammalian auditory system. The effect of these extra facilitatory and depressive mechanisms is to greatly enhance the information contained in the postsynaptic response about the timing of presynaptic spikes for stimulation frequencies up to 100Hz and above. This contrasts with the rapid decline in information transfer at simulation frequencies above a few Hertz in the simple depression model. This difference may be related to different functional requirements for the calyx of Held, compared to pyramidal-pyramidal cell connections in the cortex. Pyramidal cells are typically active at low firing rates, and optimal information transfer for the simple model occurs within their active range [4]. In contrast, the calyx of Held must transmit spikes with great temporal accuracy at high frequencies of 100Hz or more [3].

References

1. R. Bertram. Differential filtering of two presynaptic depression mechanisms. *Neural Camp.*, 13:69–85, 2000.
2. A. Borst and F. Theunissen. Information theory and neural coding. *Neuroscience*, 2:947–957, 1999.
3. I. Forsythe, M. Barnes-Davies, and H. Brew. The calyx of Held: a model for transmission at mammalian glutamatergic synapses. In H. Wheal and A. Thomson, editors, *Excitatory Amine Acids and Synaptic Transmission*, chapter 11, pages 133–144. Academic Press, London, 2 edition, 1995.
4. G. Fuhrmann, I. Segev, H. Markram, and M. Tsodyks. Coding of temporal information by activity-dependent synapses. *J. Neurophys.*, 87:140–148, 2002.
5. H. Markram, Y. Wang, and M. Tsodyks. Differential signaling via the same axon of neocortical pyramidal neurons. *Proc. Nat. Acad. Sci.*, 95:5323–5328, 1998.
6. M. Tsodyks and H. Markram. The neural code between neocortical pyramidal neurons depends on neurotransmitter release probability. *Proc. Nat. Acad. Sci.*, 94:719–723, 1997.
7. H. von Gersdorff, R. Schneggenburger, S. Weis, and E. Neher. Presynaptic depression at a calyx synapse: the small contribution of metabotropic glutamate receptors. *J. Neurosci.*, 17:8137–8146, 1997.
8. A. Wong, B. Graham, and I. Forsythe. Contribution of receptor desensitization to synaptic depression at a glutamatergic synapse. In *Meeting of the American Society for Neuroscience*, San Diego, 2001.
9. M. Worden, M. Bykhovskaia, and J. Hackett. Facilitation at the lobster neuromuscular junction: a Stimulus-dependent mobilization model. *J. Neurophys.*, 78:417–428, 1997.
10. A. Zador. Impact of synaptic unreliability on the information transmitted by spiking neurons. *J. Neurophys.*, 79:1219–1229, 1998.
11. R. Zucker. Short-term synaptic plasticity. *Ann. Rev. Neurosci.*, 12:13–31, 1989.

Storage Capacity of Kernel Associative Memories

B. Caputo and H. Niemann

Department of Computer Science, Chair for Pattern Recognition
University of Erlangen-Nuremberg,
Martenstrasse 3, 91058 Erlangen, Germany

Abstract. This contribution discusses the thermodynamic phases and storage capacity of an extension of the Hopfield-Little model of associative memory via kernel functions. The analysis is presented for the case of polynomial and Gaussian kernels in a replica symmetry ansatz. As a general result we found for both kernels that the storage capacity increases considerably compared to the Hopfield-Little model.

1   Introduction

Learning and recognition in the context of neural networks is an intensively studied field. A lot of work has been done on networks in which learning is Hebbian, and recognition is represented by attractor dynamics of the network. Particularly, the Hopfield-Little model (H-L, [6],[7]) is a network exhibiting associative memory based on the Hamiltonian

$$H = -\frac{1}{2N} \sum_{\mu=1}^{M} \sum_{i \neq j}^{N} s_i s_j \xi_i^{(\mu)} \xi_j^{(\mu)}, \tag{1}$$

where the s_i are N dynamical variables taking on the values ± 1 and the $\xi_i^{(\mu)}$ (with $\xi_i^{(\mu)} \pm 1$) are M fixed patterns which are the memories being stored. The storage capacity and thermodynamic properties of this model have been studied in detail within the context of spin glass theory [3]. Many authors studied generalizations of the H-L model which include interactions between p (> 2) Ising spins ([1],[5]), excluding all terms with at least two indices equal (symmetric terms). As a general result, higher order Hamiltonians present an increase in the storage capacity compared to the H-L model.

It was pointed out in [4] that if the symmetric term is included in the Hamiltonian (1), it can be written as a function of the scalar product between $\boldsymbol{\xi}^{(\mu)}$ and $\mathbf{s}$. The Euclidean scalar product can thus be substituted by a Mercer kernel [9], providing a new higher order generalization of the H-L energy. We call this new model Kernel Associative Memory (KAM). This new energy was used in [4] within a Markov Random Field framework for statistical modeling purposes. There are several reasons for considering this generalization. First, we will show

J.R. Dorronsoro (Ed.): ICANN 2002, LNCS 2415, pp. 51–56, 2002.
© Springer-Verlag Berlin Heidelberg 2002

in this paper that this model presents a higher storage capacity with respect to the H-L model. Second, the higher order generalization of the H-L model via Mercer kernel gives the possibility to study a much richer class of models, due to the variety of possible Mercer kernels [9]. Here we study the storage capacity and thermodynamic properties of KAM for polynomial and Gaussian kernels, in a replica symmetry ansatz. To our knowledge, no previous works has considered the storage capacity of such a generalization of the H-L model.

The paper is organized as follows: Section 2 presents KAM; in Section 3 we compute the free energy and the order parameters within a replica symmetry ansatz, and in Section 4 we study in detail the zero temperature limit. The paper concludes with a summary discussion.

2   Kernel Associative Memories

The H-L energy (1) can be rewritten in the equivalent form

$$H = -\frac{1}{2} \sum_{\mu=1}^{M} \left[\left[\frac{1}{\sqrt{N}} \sum_{i=1}^{N} s_i \xi_i^{(\mu)} \right]^2 - 1 \right]. \tag{2}$$

This energy can be generalized to higher order correlations via Mercer kernels [9]:

$$H_{KAM} = -\frac{1}{2} \sum_{\mu=1}^{M} K \left(\frac{1}{\sqrt{N}} s, \xi^{(\mu)} \right). \tag{3}$$

The possibility to kernelize the H-L energy function was recognized first in [4]. We call this model Kernel Associative Memory (KAM). It is fully specified once the functional form of the kernel is given. In this paper we will consider *polynomial* and *Gaussian kernels*:

$$K_{poly}(x, y) = (x \cdot y)^p, \qquad K_{Gauss}(x, y) = \exp(-\rho||x - y||^2). \tag{4}$$

Our goal is to study the storage capacity of energy (3) for kernels (4), using tools of statistical mechanics of spin glasses. To this purpose, we note that the study of energy (3) can be done for both kernels (4) considering the general case

$$H = -\frac{1}{N^{1-p/2}} \sum_{\mu=1}^{M} F \left(\frac{1}{\sqrt{N}} \sum_{i=1}^{N} s_i \xi_i^{(\mu)} \right), \qquad F(x) = A_p x^p. \tag{5}$$

For $A_p = 1$, p finite, equations (5) represent the KAM energy (3) for polynomial kernels (4). For $A_p = \rho^p/p!, p \to \infty$, equations (5) describe the behavior of the KAM energy (3) for Gaussian kernels (4). This can be shown as follows: note first that

$$\exp(-\rho||s-\xi^{(\mu)}||^2) = \exp(-\rho[s \cdot s + \xi^{(\mu)} \cdot \xi^{(\mu)} - 2s \cdot \xi^{(\mu)}]) = \exp(-2\rho[N - s \cdot \xi^{(\mu)}]).$$

The multiplicating factor can be inglobated in the ρ, and the constant factor can be neglected. The Gaussian kernel function thus becomes

$$\exp(-\rho||\boldsymbol{s}-\boldsymbol{\xi}^{(\mu)}||^2) \longrightarrow \exp(\rho[\boldsymbol{s}\cdot\boldsymbol{\xi}^{(\mu)}]) \simeq 1+\rho\boldsymbol{s}\boldsymbol{\xi}^{(\mu)}+\frac{\rho^2}{2!}(\boldsymbol{s}\boldsymbol{\xi}^{(\mu)})^2+\frac{\rho^3}{3!}(\boldsymbol{s}\boldsymbol{\xi}^{(\mu)})^3+\ldots$$

A generalization of the H-L energy function in the form (5) was proposed first by Abbott in [1]. In that paper, storage capacity and thermodynamic properties were derived for a simplified version of (5), where all terms with at least two equal indices were excluded; a particular choice of A_p was done. Other authors considered this kind of simplifications for higher order extension of the H-L models (see for instance [5]). The analysis we present here include all the terms in the energy and is not limited to a particular choice of the coefficient A_p; thus is more general. To the best of our knowledge, this is the first analysis on the storage capacity and thermodynamic properties of a generalization of the H-L model via kernel functions.

3   Free Energy and Order Parameters

We study the overlap of a configuration s_i with one of the stored patterns, arbitrarily taken to be $\xi_i^{(1)}$,

$$m = \left\langle\!\!\left\langle \frac{1}{N}\sum_i^N \langle s_i\rangle\xi_i^{(1)} \right\rangle\!\!\right\rangle, \tag{6}$$

where the angle bracket $\langle ...\rangle$ represents a thermodynamic average while the double brackets $\langle\!\langle\ldots\rangle\!\rangle$ represents a quenched average over the stored patterns $\xi_i^{(\mu)}$. The quenched average over patterns is done using the replica methods [8]; in the mean-field approximation, the free energy depends on m and on the order parameters

$$q_{ab} = \left\langle\!\!\left\langle \frac{1}{N}\sum_i^N \langle s_i^a\rangle\langle s_i^b\rangle \right\rangle\!\!\right\rangle, r_{ab} = \frac{1}{MN}\sum_{\mu=2}^M \left\langle\!\!\left\langle \sum_i^N \langle s_i^a\rangle\xi_i^{(1)} \sum_j^N \langle s_j^b\rangle\xi_j^{(1)} \right\rangle\!\!\right\rangle, \tag{7}$$

with a,b replica indices. The calculation is analog to that done in [1],[2]. Considering a *replica symmetry ansatz* [8], $q_{ab} = q, r_{ab} = r$ and the free energy at temperature $T = 1/\beta$ is given by

$$f = (p-1)A_p m^p + \frac{\alpha\beta}{2}[r(1-q)-G(q)] - \frac{1}{\beta}\int Dz\ln[2\cosh\beta(\sqrt{\alpha r}z) + pA_p m^{p-1}], \tag{8}$$

with $M = \alpha N^{p-1}$ and $Dz = \frac{dz}{\sqrt{2\pi}}e^{-z^2/2}$. The function $G(q)$ is given by

$$G(q) = A_p^2 \sum_{c=1}^p \frac{1}{2}[1+(-1)^{p+c}]\frac{1}{c!}\left[\frac{p!}{(p-c)!!}\right]^2(1-q^c) \tag{9}$$

The free energy leads to the following *order parameters*:

$$m = \int Dz \tanh \beta(\sqrt{\alpha r}z + pA_p m^{p-1}), r = -\frac{\partial G(q)}{\partial q}, q = \int Dz \tanh^2 \beta(\sqrt{\alpha r}z + pA_p m^{p-1})$$

(10)

At all temperatures, these equations have a paramagnetic solution $r = q = m = 0$.

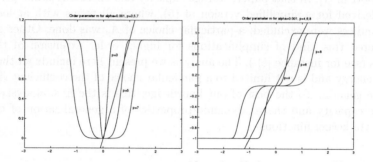

Fig. 1. Graphical representation of the solutions of the retrieval equation for $p = 3, 5, 7$ (left) and $p = 4, 6, 8$ (right), $\alpha = 0.001$.

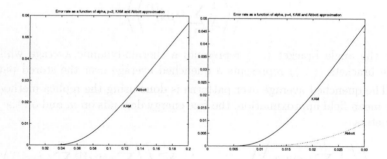

Fig. 2. The error rate $(1 - m)/2$ as a function of α, for KAM and Abbott model, for $p = 3$ (left) and $p = 5$ (right). For $p = 3$ the two curves are indistinguishable, for $p = 5$ the Abbott approximation gives the lower error rate.

4   The Zero–Temperature Limit: $\beta \to \infty$

In the zero-temperature limit $\beta \to \infty$, equations (10) simplifies to $q \to 1$,

$$m \to \mathrm{erf}\left(\frac{pA_p m^{p-1}}{\sqrt{\alpha r}}\right), \qquad r \to \frac{1}{2}A_p^2 \sum_{c=1}^{p} R_c(p)$$

(11)

with

$$R_c(p) = [1 + (-1)^{p+c}] \frac{c}{c!} \left[\frac{p!}{(p-c)!!} \right]^2 \tag{12}$$

Note that, in this approximation, m does not depend on A_p:

$$m = \mathrm{erf}\left(\frac{pA_p m^{p-1}}{\sqrt{\alpha r}} \right) = \mathrm{erf}\left(\frac{pm^{p-1}}{\sqrt{\alpha}} \frac{A_p}{A_p \sqrt{\sum_{c=1}^{p} R_c(p)/2}} \right). \tag{13}$$

Graphical solutions of equation (13) are shown in Figure 1, for several values of p (odd and even). From these graphics we can make the following considerations: first, higher order Hamiltonians (5) with p odd eliminate the symmetry between the stored memory and their complementary states, in which the state of each neuron is reversed (see Figure 1). Second, as p increases, the value of α for which there is more than one intersection ($m = 0$, the spin-glass solution), as to say the storage capacity, goes to zero (see Figure 1). Figure 2 shows the percentage of errors $(1 - m)/2$ made in recovering a particular memory configuration as a function of α. The percentage of errors for the KAM model, for $p = 3, 5$, is compared with the same quantity obtained by Abbott [1], considering a higher order Hamiltonian. In that case m is given by:

$$m^{Abbott} = \mathrm{erf}\left[\left(\frac{p}{2\alpha p!} \right)^2 m^{p-1} \right] \tag{14}$$

in the zero temperature limit. In both cases (p=3,5) and for both models (KAM and Abbott), the overlap with the input pattern m remains quite close to 1 even for $\alpha \to \alpha_c$. This can be seen because the fraction of errors is always small. Thus the quality of recall is good. Nevertheless, the fraction of errors is smaller for the Abbott model as p increase.

Even if $\alpha_c \to 0$ for large p, the total number of memory states allowed is given by $M = \alpha N^{p-1}$, thus it is expected to be large. In the limit $p \to \infty$, α_c and M can be calculated explicitly [2]: for large values of the argument of the erf function, it holds

$$m \approx 1 - \frac{\sqrt{\alpha/2 \sum_{c=1}^{p} R_c(p)}}{\sqrt{\pi} p m^{p-1}} \exp\left[-\frac{p^2 m^{p-1}}{\alpha/2 \sum_{c=1}^{p} R_c(p)} \right]. \tag{15}$$

For stability considerations, $m \approx 1$, thus the second term in equation (15) must go to zero. This leads to the critical value for α

$$\alpha_c = \frac{2p^2}{[\sum_{c=1}^{p} R_c(p)] \ln p} \approx \frac{2p^2}{p! 2^p \ln p \left[\sum_{c=1}^{p} c(1 + (-1)^{c+p}) \right]}. \tag{16}$$

The total number of memory states that is possible to store is (using Stirling's approximation):

$$M = \alpha N^{p-1} \to \left(\frac{eN}{p} \right)^{p-1} \sqrt{\frac{2p}{\pi}} \frac{e}{2^p \ln p \sum_{c=1}^{p} (1 + (-1)^{c+p})}. \tag{17}$$

This result must be compared with the one obtained by Abbott [1]: $\alpha_c^{Abbott} = \frac{p}{2p!lnp}$; it is easy to see that $\alpha_c = \alpha_c^{Abbott} \cdot \frac{p}{2^p \sum_{c=1}^{p} c[1+(-1)^{c+p}]}$; thus, the introduction of terms in which more than one indices is equal in the generalized Hamiltonian leads to a decrease in the storage capacity; the decrease will be higher the higher is the number of these terms introduced in the energy formulations, according with the retrieval behavior (see Figure 2).

5 Summary

In this paper we presented kernel associative memories as a higher order generalization of the H-L model. The storage capacity of the new model is studied in a replica symmetry ansatz. The main result is that the storage capacity is higher than the H-L's one, but lower than the storage capacity obtained by other authors for different higher order generalizations. This work can be developed in many ways: first of all, the mean field calculation presented here is only sensitive to states that are stable in the thermodynamic limit. We expect to find in simulations correlations between spurious states and the memory patterns even for $\alpha > \alpha_c$. Simulations should also provide informations about the size of the basin of attraction. Second, replica symmetry breaking effects should be taken in account. Third, it should be explored how kernel properties can be used in order to reduce the memory required for storing the interaction matrix [5]. Finally, this study should be extended to other classes of kernels and to networks with continuous neurons. Future work will be concentrated in these directions.

References

1. L. F. Abbot, Y. Arian, "Storage capacity of generalized networks", *Physical Review A*, 36 (10), pp 5091-5094, November 1987.
2. D. J. Amit, *"Modeling Brain Function"*, Cambridge University Press, Cambridge, USA, 1989.
3. D. J. Amit, H. Gutfreund, H. Sompolinsky, "Spin glass models of neural networks", *Physical Review A*, 32, 1007, 1985.
4. B. Caputo, H. Niemann, "From Markov Random Fields to Associative Memories and Back: Spin Glass Markov Random Fields", *Proc. of IEEE workshop on Statistical and Computational Theories of Vision*, Vancouver, CA, July 2001.
5. E. Gardner, "Multiconnected neural network models", *Journal of Physics A*, 20, pp 3453-3464, 1987.
6. J. J. Hopfield, "Neural networks and physical systems with emergent collective computational abilities", *Proc. Natl. Acad. Sci. USA*, Vol. 79, pp 2554-2558, April 1982.
7. W. A. Little, G. L. Shaw, "Analytical study of the memory storage capacity of a neural network", *Math. Biosci.*, 39, 281, 1981.
8. M. Mezard, G. Parisi, M. Virasoro, *Spin Glass Theory and Beyond*, World Scientific, Singapore, 1987.
9. V. Vapnik, *Statistical learning theory*, J. Wiley, New York, 1998.

Macrocolumns as Decision Units

Jörg Lücke[1], Christoph von der Malsburg[1,2], and Rolf P. Würtz[1]

[1] Institut für Neuroinformatik, Ruhr-Universität Bochum, Germany
[2] Computer Science Dept., University of Southern California, Los Angeles, USA

Abstract. We consider a cortical macrocolumn as a collection of inhibitorily coupled minicolumns of excitatory neurons and show that its
dynamics is determined by a number of stationary points, which grows
exponentially with the number of minicolumns. The stability of the stationary points is governed by a single parameter of the network, which
determines the number of possibly active minicolumns. The dynamics
symmetrizes the activity distributed among the active columns but if
the parameter is increased, it forces this symmetry to break by switching off a minicolumn. If, for a state of maximal activity, the parameter
is slowly increased the symmetry is successively broken until just one
minicolumn remains active. During such a process minor differences between the inputs result in the activation of the minicolumn with highest
input, a feature which shows that a macrocolumn can serve as decision
and amplification unit for its inputs. We present a complete analysis of
the dynamics along with computer simulations, which support the theoretical results.

1 Introduction

The cerebral cortex can be subdivided into *neural modules*, which are associated
to different magnitudes of spatial scale ranging from *areas* of size of approximately $20cm^2$ (in humans), *maps* ($\approx 5cm^2$) to *macrocolumns* ($\approx 0.5mm^2$) and
minicolumns of about $0.003mm^2$. The minicolumns are considered the smallest
neural modules containing several tens up to a few hundred neurons, which are
stacked orthogonal to the cortical surface. The grouping into minicolumns can
be revealed by Nissl stains, as was first done by Cajal, by stimulus-response
experiments or, more recently, by direct measurements of neural connectivity,
e.g. [1]. The minicolumns themselves can be grouped into macrocolumns, neural
modules, which are best studied in primary sensory areas and which are considered to process stimuli from the same source such as an area of the visual
field or a patch of the body surface [2]. Although different features of columns
of regions concerned with different levels of information processing can vary significantly it is widely believed that (1) the neural circuits, at least for neural
modules of small scales, are of a common design which makes them universal for
various computational tasks and (2) that the understanding of the interplay of
the circuitries of the different modules presents the key to the understanding of

J.R. Dorronsoro (Ed.): ICANN 2002, LNCS 2415, pp. 57–62, 2002.
© Springer-Verlag Berlin Heidelberg 2002

information processing of the brain of vertebrates.

Several models of neural networks reflecting the modular organization of the brain have been suggested. They range from models based on random inter-connections [3] and Hopfield-like models [4] to models based on self-organizing interconnections, e.g. [5]. In this paper we study a model of a single macrocol-umn that consists of a collection of coupled minicolumns and we show that it can serve as a *decision unit*, which changes its activity state by a process of symmetry breakings delicately depending on the relative inputs to the minicolumns. The model shows a dynamic behavior that differs from so far suggested ones and that makes possible the construction of networks, in which macrocolumns as principal units communicate via symmetry differences of their inputs. In such networks a macrocolumn can individually change its activity state if these differences are sufficiently non-ambiguous. Here we will be concerned with the dynamics of a single macrocolumn: In Sec. 2 we study the dynamics of a minicolumn, in Sec. 3 the macrocolumn dynamics as a coupled system of minicolumn dynamics is investigated and its properties are discussed, and in Sec. 4 we summarize the results and give a short outlook to future work.

2   Dynamics of Minicolumns

We consider a minicolumn as a network of N neurons, which are excitatorily interconnected. The neurons are modeled as threshold devices with refraction time. Thresholds and refraction times are equal for all neurons. The state of a neuron at time t, $n_i(t)$, is *one* if the neuron is active and *zero* if it is not. A non-refractory neuron is active at time $(t+1)$ if the input it receives ¿from other neurons at time t exceeds Θ. The refraction time is chosen as one time step. The input from an active neuron j to a neuron i is given by the synaptic strength T_{ij}. The resulting dynamics is given by ($i = 1, \ldots, N$):

$$n_i(t+1) = \mathcal{S}(\sum_{j=1}^{N} T_{ij}\, n_j(t) - \Theta) \cdot \underbrace{\mathcal{S}(1 - n_i(t))}_{refraction}, \quad \mathcal{S}(x) := \begin{cases} 0 \text{ if } x \leq 0 \\ 1 \text{ if } x > 0 \end{cases}. \quad (1)$$

The connectivity is random. Each neuron has s synapses on its axon, and each synapse connects to any post-synaptic neuron with probability $\frac{1}{N}$. All synaptic weights are equal to a constant $c > 0$. Note that the resulting interconnection can include multiple connections between neurons, i.e. $T_{ij} > c$. Considering the dynamics (1) it is always possible to compensate any value of $c > 0$ by an ap-propriate choice of Θ. Without loss of generality, we choose c to be equal to $\frac{1}{s}$ in oder to normalize the sum over all T_{ij}, $\frac{1}{N}\sum_{i,j=1}^{N} T_{ij} = 1$. Together with the constant probability for a synapse to connect to any post-synaptic neuron we can describe (1) by a simplified dynamics for a single global observable of the network — the probability $p(t)$ of a neuron to be active at time t. The probability, $p_i(t+1)$, of a neuron i to be active at time $t+1$ depends on the probability, $P_i^A(t)$, to receive enough input and on the probability, $P_i^B(t)$, of the neuron to be non-refractory, and we assume these probabilities to be approxi-mately independent, $p_i(t+1) = P_i^A(t)\, P_i^B(t)$. The probability $P_i^B(t)$ is simply given by the complement of the probability of the neuron i to be active at time t, $P_i^B(t) = (1 - p_i(t))$. For the computation of $P_i^A(t)$ we have to estimate the number of excitatory post-synaptic potentials (EPSPs) received by neuron i at time t. Due to the assumptions made above the probability $P_e(x)$ of the neuron i to receive exactly x EPSPs is given by the binomial distribution,

$$P_e(x) = \binom{N s\, p(t)}{x} (\frac{1}{N})^x (1 - \frac{1}{N})^{N s\, p(t) - x}. \tag{2}$$

In order to facilitate later calculations we approximate (2) by a Gaussian distribution ($\bar{x} = s\, p(t)$, $\sigma^2 = s\, p(t)$ for $N \gg 1$). Integrating all probabilities for numbers $x > s\Theta$ we finally receive a compact and easy to handle description of the dynamics (1) in terms of the activation probability $p(t)$,

$$p(t+1) = \Phi_s(\frac{p(t) - \Theta}{\sqrt{p(t)}}) (1 - p(t)) \tag{3}$$

where $\Phi_s(x) = \frac{1}{\sqrt{2\pi}} \int_{-\infty}^{\sqrt{s}\, x} e^{-\frac{1}{2} y^2}\, dy$. Equation (3) can be used to reproduce calculation results of [3] where (2) was approximated by a Poisson distribution. Here we will exploit (3) to introduce a special kind of inhibitory feedback.

Inhibitory neurons differ more substantially in their properties than just generating negative post synaptic potentials. On average, inhibitory neurons have thicker axons than excitatory ones and their post-synaptic targets are concentrated on the cell bodies and the proximal dendrites. This suggests to model inhibition as being generated faster than excitation. We will therefore model the inhibitory influence on the excitatory neurons to be present already in the next time-step. The dependency of the inhibition $I(t)$ on the activity of the excitatory neurons we choose to be proportional to the global activity $B(t) = \sum_{i=1}^{N} n_i(t)$ and further demand that it is equally sensed by all neurons. Such a dependency turns out to stabilize the activity in the most efficient way. Replacing Θ in (3) by $I(t) + \Theta_o$ with $I(t) = \mu \frac{B(t)}{N} = \mu\, p(t)$ yields:

$$p(t+1) = \Phi_s(\frac{(1-\mu)\, p(t) - \Theta_o}{\sqrt{p(t)}}) (1 - p(t)). \tag{4}$$

It is now possible to determine the maximal values of stationary activity $\mathcal{P}$ for parameters s, Θ_o, and μ by numerically computing the function

$$\mathcal{P}_{s,\Theta_o}(\mu) := \max \{ p \mid p = \Phi_s(\frac{(1-\mu) p - \Theta_o}{\sqrt{p}}) (1 - p) \}. \tag{5}$$

Function (5) can be compared to the values of stable stationary activity obtained by directly simulating equation (1) with $\Theta = \mu \frac{B(t)}{N} + \Theta_o$. For the simulations one has to consider coherence effects which are due to possibly cycling neuron activities but which can be suppressed by noise or by varying the number of possibly active synapses after each time-step (see [3]). For $s = 20$, $\Theta_o = \frac{1}{20}$, and $\mu \in [0, 2]$, e.g., it turns out that the predicted activity rates match the measured ones with absolute errors around 0.01 for values of p between 0.5 and 0.05. For lower activity rates the approximation of (2) by a Gaussian distribution gets too coarse. If the coherence effects are not suppressed, the dynamic behavior can differ significantly from the computed one.

3   Inhibitorily Coupled Minicolumns

We now consider a system of k inhibitorily coupled minicolumns. The calculations are independent of the number of minicolumns such that the results are applicable to relatively small macrocolumns ($k \approx 2, ..., 10$) as suggested, e.g., by short-ranging lateral inhibiting cells [6] [7] or to macrocolumns of about $0.5mm^2$ for k of size of several hundreds. Each minicolumn consists of M excitatory neurons with interconnection as above. In analogy to (1) the dynamics is described by $N = kM$ difference equations ($\alpha = 1, ..., k;\ i = 1, ..., M$):

$$n_i^\alpha(t+1) = \mathcal{S}(\sum_{j=1}^{M} T_{ij}^\alpha n_j^\alpha(t) - I(t) - \Theta_o) \cdot \mathcal{S}(1 - n_i^\alpha(t)), \qquad (6)$$

where the inhibitory feedback $I(t)$ is equal for all neurons. We want to have stable stationary mean activity in the macrocolumn and therefore again choose the inhibition to be proportional to the over-all activity $B(t) = \sum_{i,\alpha} n_i^\alpha(t)$. We get in this case $I(t) = \mu \frac{B(t)}{N} = \frac{\mu}{k} \sum_{\alpha=1}^{k} p_\alpha(t)$, where $p_\alpha(t) = \frac{1}{M} \sum_{i=1}^{M} n_i^\alpha(t)$ is the probability of a neuron in column α to be active. The direct simulation of dynamics (6) shows a complex behavior and for a wide range of parameters s and Θ_o we get stable ongoing activity. The dynamics favors to activate only a subset of minicolumns whereas the others are switched off. Hereby, the number of minicolumns which can be activated strongly depends on the proportionality factor of the inhibition μ. The points of stable activity and their dependency on μ can be studied again by the reformulation of (6) in terms of the activation probabilities $p_\alpha(t)$ of the different minicolumns. Calculations in analogy to above yield a system of $\alpha = 1, \dots, k$ difference equations:

$$p_\alpha(t+1) = \Phi_s(\frac{p_\alpha(t) - \frac{\mu}{k}\sum_{\beta=1}^{k} p_\beta(t) - \Theta_o}{\sqrt{p_\alpha(t)}})\ (1 - p_\alpha(t)) =: G_\alpha(\boldsymbol{p}(t)) \quad (7)$$

Equations (7) can be studied by a stability analysis and we just give the relevant results: For a macrocolumn with k minicolumns we get a family of 2^k potentially stable stationary points of the form,

$$\boldsymbol{q}_\gamma = (\underbrace{q_o, q_o, \dots, q_o}_{\text{l-times}}, \underbrace{0, 0, \dots, 0}_{\text{(k-l)-times}}), \quad q_o = \mathcal{P}(\frac{l}{k}\mu), \qquad (8)$$

and all permutations. Their stability is determined by the eigenvalues of the Jacobian of $\boldsymbol{G}(\boldsymbol{p})$ (see (7)) at these points which can be computed to be

$$\lambda_{1,2} = \frac{1 - \mathcal{P}(\frac{l}{k}\mu)}{2\sqrt{\mathcal{P}(\frac{l}{k}\mu)}}\ (1 \pm \frac{l}{k}\mu + \frac{\Theta_o}{\mathcal{P}(\frac{l}{k}\mu)})\ \Phi_s'(h(\frac{l}{k}\mu)) - \Phi_s(h(\frac{l}{k}\mu)), \quad \lambda_3 = 0, \quad (9)$$

where $h(\mu) = \frac{(1-\mu)\mathcal{P}(\mu) - \Theta_o}{\sqrt{\mathcal{P}(\mu)}}$. λ_1 is of multiplicity $(l - 1)$, λ_2 of multiplicity 1, and λ_3 of multiplicity $(k - l)$. We get eigenvalues of magnitude greater than one

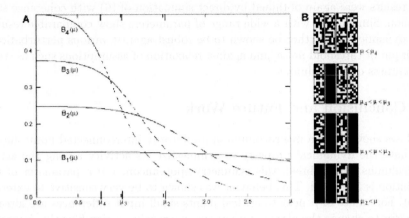

Fig. 1. A Stationary over-all activity in a macrocolumn ($k=4$) plotted against μ for $s = 20$ and $\Theta_o = \frac{1}{20}$. The four plots correspond to states of four to one active minicolumns (dotted parts mark unstable stationary activity). **B** Screenshots of a macrocolumn with four minicolumns. Each picture shows one activity-configuration of the respective activity probability (white pixels mark active neurons).

if and only if $\frac{l}{k}\mu$ gets greater than a critical value μ_c. Hence, the stability of a stationary point (8) with $l \geq 2$ is determined by its critical value $\mu_l := \frac{k}{l}\mu_c$. For a macrocolumn consisting, e.g., of $k = 4$ minicolumns we get a collection of 15 non-zero stationary points of type (8), whose stability is determined by the three critical points $\mu_4 = \mu_c$, $\mu_3 = \frac{4}{3}\mu_c$, and $\mu_2 = \frac{4}{2}\mu_c$. For $s = 20$ and $\Theta_o = \frac{1}{20}$ their values are $\mu_4 \approx 0.76$, $\mu_3 \approx 1.01$, and $\mu_2 \approx 1.52$. In Fig. 1A the stationary over-all activities $B_l(\mu) = lM\,P(\frac{l}{k}\mu)$ are plotted for $l = 4,\ldots,1$ active minicolumns together with the points μ_4, μ_3, and μ_2, which mark their intervals of stability. The macrocolumn's dependency on μ can be used to force the network to perform successive symmetry breakings: if we start for $\mu < \mu_4$ with the totally symmetric stable stationary point $(P(\mu),\ldots,P(\mu))$ and slowly increase the parameter, the macrocolumn is forced to break the symmetry by switching off one of the minicolumns as soon as $\mu > \mu_4$. The activity is then symmetrized between the minicolumns which remain active. But as soon as $\mu > \mu_3$ this symmetry is broken again. The process of symmetrizing the activity among the active columns and breaking the symmetry again continues until just one column remains active (see Fig. 1B). If the macrocolumn is exposed to input in form of externally induced EPSPs, it will keep the minicolumn with highest relative input active while successively switching off the others. After each symmetry breaking the network symmetrizes the minicolumn activities again and the next decision can be made relative to the inputs to the columns with non-zero activity. For a macrocolumn with four minicolumns of $N = 100$ neurons, with $s = 20$, $\Theta_o = \frac{1}{20}$, and μ increased from zero by 0.01 per time-step already an average difference of three EPSPs per neuron every 10 time-steps is sufficient to select the corresponding minicolumn with a probability of more than 99%.

The results were again obtained by direct simulation of (6) with coherence suppression. Simulations with a wide range of parameters show comparable results. The dynamics can further be shown to be robust against various perturbations, e.g. input or threshold noise, and against relaxation of assumptions such as strict disjointness of minicolumns.

4   Conclusion and Future Work

We have shown that a macrocolumn of inhibitorily interconnected minicolumns can have the dynamical property to symmetrize its activity among the active minicolumns and to break this symmetry spontaneously if a parameter of the inhibition is increased. This behavior was shown to be very sensitive to external input. For values of μ near to critical points small input differences are already sufficient to change the global state of the macrocolumn significantly. A macrocolumn can therefore serve to select and amplify inputs to its minicolumns. It can do so either directly with the parameter μ set near to critical points or indirectly by a succession of symmetry breakings with increasing μ. The repeated activity-suppression of the column with weakest input together with the symmetrization of the remaining activities presents a property not observed in usual winner-take-all mechanisms. Networks of interconnected macrocolumns can be expected to converge from a state of maximal activity to a state of minimal one by a process in which each macrocolumn makes a decision only if its input is sufficiently non-ambiguous. The networks can be applied to problems such as signal integration or classification and are subject of our current investigations.

Acknowledgments. Partial funding by the EU in the RTN MUHCI (HPRN-CT-2000-00111), the German BMBF in the project LOKI (01 IN 504 E9), and the Körber Prize awarded to C. von der Malsburg in 2000 is gratefully acknowledged.

References

1. A. Peters and E. Yilmaze. Neuronal organization in area-17 of cat visual-cortex. *Cereb. Cortex*, 3(1):49–68, 1993.
2. O.V. Favorov and D.G. Kelly. Minicolumnar organization within somatosensory cortical Segregates 1. *Cereb. Cortez*, 4:408–427, 1994.
3. P.A. Anninos, B. Beek, T.J. Csermely, E.M. Harth, and G. Pertile. Dynamics of neural structures. *J. Theo. Biol.*, 26:121–148, 1970.
4. E. Fransen and A. Lansneer. A model of cortical associative memory based on a horiz. netw. of connected columns. *Netw.-Comp. Neur. Sys.*, 9(2):235–264, 1998.
5. O.V. Favorov and D.G. Kelly. Stimulus-response diversity in local neuronal populations of the cerebral cortex. *Neuroreport*, 7(14):2293–2301, 1996.
6. J. DeFelipe, M.C. Hendry, and E.G. Jones. Synapses of double bouquet cells in monkey cerebral cortex.*Bruin Res.*, 503:49–54, 1989.
7. J.M.L. Budd and Z.F. Kisvarday. Local lateral connectivity of inhibitory clutch cells in layer 4 of tat visual cortex. *Exp. Bruin Res.*, 140(2):245–250, 2001.

Nonlinear Analysis of Simple Cell Tuning in Visual Cortex

Thomas Wennekers

Max-Planck-Institute for Mathematics in the Sciences,
Inselstr. 22–26, D-04103 Leipzig, Germany,
Thomas.Wennekers@mis.mpg.de
http://www.mis.mpg.de

Abstract. We apply a recently developed approximation method to two standard models for orientation tuning: a one-layer model with difference-of-Gaussians connectivity and a two-layer excitatory-inhibitory network. Both models reveal identical steady states and instabilities to high firing rates. The two-field model can also loose stability through a Hopf-bifurcation that results in rhythmically modulated tuning widths around 0 to 50Hz. The network behavior is almost independent of the relative weights and widths of the kernels from excitatory to inhibitory cells and back. Formulas for tuning properties, instabilities, and oscillation frequencies are given.

1 Introduction

Neural field models are used frequently in studies approaching spatio-temporal activation patterns in extended cortical tissue [1,2,3,4,5]. They describe a piece of cortex by one or more spatial layers of locally interconnected neurons. Spatio-temporal impulse response functions in such networks provide approximations for cortical dynamic receptive fields. Unfortunately, nonlinear neural field models are difficult to study analytically. Therefore, we recently introduced a new approximation scheme for localized solutions in such networks [4]. In the present work we apply this method to two standard models for orientation tuning: A one-field model with "difference-of-Gaussians" coupling kernel (DOG) and a two-field excitatory-inhibitory model with Gaussian shaped mutual kernels (cf. [1,2, 3,4]). We determine and compare tuning and stability properties of these models. An extended version of this paper can be found in [5].

2 Field Models and Reduced Approximate Equations

We first consider a 2-layer model of excitatory and inhibitory cells (cf., [4,5])

$$\tau_1 \dot{\phi}_1 = -\phi_1 + I_1 + k_{11} \otimes f_1(\phi_1) - k_{12} \otimes f_2(\phi_2) \tag{1}$$
$$\tau_2 \dot{\phi}_2 = -\phi_2 + k_{21} \otimes f_1(\phi_1) . \tag{2}$$

J.R. Dorronsoro (Ed.): ICANN 2002, LNCS 2415, pp. 63–68, 2002.
© Springer-Verlag Berlin Heidelberg 2002

We assume $x \in R$, $f_i(\phi) = [\phi]_+ := \max(0, \phi)$, and vanishing initial conditions. Coupling kernels, $k_{ij}(x)$, are Gaussians with variance σ_{ij}^2 and amplitude $K_{ij}/\sqrt{2\pi}$. The input, $I_1 = I_1(x, t)$, is a spatial Gaussian with variance σ_{10}^2 and amplitude $I_{01}T_1(t)$, where $T_1(t)$ is the time-course of the stimulus. '$\otimes$' denotes spatial convolution and τ_1, τ_2 are membrane time constants. Note, that in (2) we neglect input into as well as feedback between inhibitory cells. Moreover, the firing threshold of all cells is zero as in [1,3,4]. Steady state solutions of (1) and (2) are obviously given by $\phi_2^*(x) = k_{21} \otimes f_1(\phi_1^*(x))$ and

$$\phi_1^*(x) = I_1(x) + (k_{11} - k_{12} \otimes k_{21}) \otimes f_1(\phi_1^*(x)) . \tag{3}$$

The effective coupling kernel between the excitatory cells in (3) is a difference of Gaussians (DOG), and, clearly, the same steady state equation also results for a model that lumps excitatory and inhibitory cells into a single layer [3]:

$$\tau_1 \dot{\phi}_1 = -\phi_1 + I_1 + k_{DOG} \otimes f_1(\phi_1) , \tag{4}$$

where $k_{DOG} = k_{11} - k_{12} \otimes k_{21}$. Thus, we can identify the effective DOG-parameters $K_+ = K_{11}$, $\sigma_+ = \sigma_{11}$, $K_- = K_{21}K_{12}\sigma_{12}\sigma_{21}/\sigma_-$, $\sigma_-^2 = \sigma_{12}^2 + \sigma_{21}^2$.

The most interesting firing rate profiles in the context of cortical tuning are spatially localized peaks or "bumps". The approximation method developed in [4,5] replaces those rate profiles by spatial Gaussians with amplitude $f_i(\phi_i(0, t))$ and variance σ_{pi}^2, and derives a set of $2n$ ordinary differential equations for the response amplitudes and tuning widths alone. The resulting ODEs per network layer $i = 1, \cdots, n$ are (cf. [4,5] for details)

$$\tau_i \dot{\phi}_{0i} = -\phi_{0i} + I_{0i}T_i(t) + \sum_{j=1}^{n} c_{ij}f_j(\phi_{0j}) \tag{5}$$

$$\tau_i \dot{\phi}_{2i} = -\phi_{2i} - \frac{I_{0i}T_i(t)}{\sigma_{i0}^2} - \sum_{j=1}^{n} \frac{c_{ij}f_j(\phi_{0j})}{\Sigma_{ij}^2} . \tag{6}$$

Here, the ϕ_{0i} are the amplitudes $\phi_i(0, t)$, whereas the ϕ_{2i} are second order derivatives of the potential profiles in layer i at zero. Thus, peak firing rates are $h_i := f_i(\phi_i(0, t)) = [\phi_{0i}]_+$. Moreover, the c_{ij} are "effective couplings", $c_{ij} := K_{ij}\sigma_{ij}\sigma_{pj}/\Sigma_{ij}$, where $\Sigma_{ij}^2 := \sigma_{ij}^2 + \sigma_{pj}^2$. Tuning widths can further be estimated as $\sigma_{pi}^2 = -f_i(\phi_{0i})/(f_i'(\phi_{0i})\phi_{2i})$ or $\sigma_{pi}^2 = -[\phi_{0i}]_+/\phi_{2i}$ for $f_i = [\cdot]_+$ (cf., [4,5]). Notice, that the c_{ij} in general depend on time since they are functions of the $\sigma_{pj}(t)$. Near steady states one may assume $\sigma_{pj} \approx \sigma_{pj}^* = const$. Then, (5) decouples from (6) and determines the amplitudes in the network alone. We call this lowest order approximation "meanfield", because equations as (5) also describe the average behaviour in n pools of homogenously coupled neurons.

3   Steady State Properties

In [4] we studied equation (3) for a whole class of nonlinear rate functions. This revealed that only the semilinear function, $f = [\cdot]_+$, generically leads to contrast-independent tuning in arbitrary feedback as well as feedforward networks (cf. also

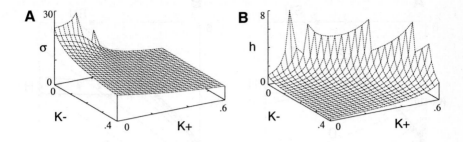

Fig. 1. Numerical solution of the fixed point equations (7).

[1,2,3]). In the sequel we restrict attention to such functions and define $\sigma :=$ σ_{p1}, $c_+ := c_{11}$, $c_- := c_{12}c_{21}$, $\gamma := c_+ - c_-$, and $\Sigma_{\pm}^2 := \sigma_{\pm}^2 + \sigma_{p1}^2$. From (5) and (6) one then gets in steady staes $h_2 = c_{21}h_1$, $\sigma_{p2} = \Sigma_{21}$, and

$$h := h_1 = \frac{I_0}{1 - \gamma} \quad , \quad \frac{1}{\sigma^2} := \frac{1}{\sigma_{p1}^2} = \frac{1 - \gamma}{\sigma_0^2} + \frac{c_+}{\Sigma_+^2} - \frac{c_-}{\Sigma_-^2} \quad , \qquad (7)$$

where the $h_i := [\phi_{0i}^*]_+$ are steady state peak rates and the σ_{pi} the tuning widths. Observe that the equation for σ_{p1} neither contains I_0 nor h_1. Therefore, the tuning must be independent of the absolute input intensity (e.g. contrast). Similarly, only K_- and σ_- enter in (7) and can influence the tuning, but not the precise $K_{12}, K_{21}, \sigma_{12}$, and σ_{21}.

Equation (7) is easy to solve numerically. Figure 1 displays solutions for the parameters $\sigma_+ = 7.5$, $\sigma_- = 60$, $\sigma_0 = 23$, $I_0 = 1$. Couplings used in [2, 3,4] correspond with $K_+ = .23$ and $K_- = .092$. Figure 1A reveals a point singularity at $K_- = 0$, $K_+ \approx .14$, but in large parameter regions the excitatory tuning is relatively sharp and comparable to $\sigma_+ = \sigma_{11}$. Figure 1B plots the "cortical amplification factor", $m = h_1/I_0$ (because $I_0 = 1$). The region of cortical amplification $m > 1$ is only rather small. In large parts m is smaller than one, although tuning is typically sharp in these regions. Moreover, m becomes infinity along a line-singularity.

4   Stability of Steady States

Stability of steady states as indicated in Fig. 1 can be investigated using linear stability theory applied to the reduced dynamic equations (5) and (6) restricted to the 2-field model under consideration. The resulting expressions for the characteristic polynomials etc. are somwhat messy and not given here. They can, however, be easily solved numerically. Results are presented in Fig. 2 where we used the additional parameters $\tau_1 = \tau_2 = 10ms$, $\sigma_{21} = \sigma_{21} = \sigma_-/\sqrt{2}$ and fixed K_{21} arbitrarily at 1. (As long as K_- is kept constant changes of K_{21} only scale $\phi_2(x,t)$ proportionally but have no other impact.) The labeled thin lines

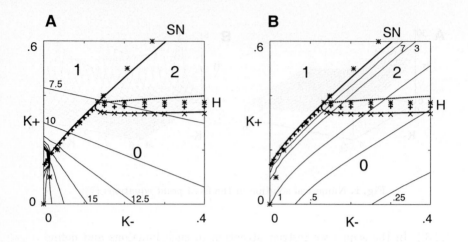

Fig. 2. Bifurcation diagram of the excitatory-inhibitory two-field model.

in Fig. 2A&B are contour lines of the surfaces in Fig. 1. Other lines and symbols are identical in A and B and indicate two branches of instabilities: First, a line of saddle-node bifurcations, 'SN', which coincides with the line of infinite amplification (cf. Figs. 1B and 2B). Second, a Hopf-bifurcation line, 'H', which separates a region of stable fixed points (indicated as '0') from a region where stable limit cycles appear ('2'). Thick lines in Fig. 2A&B are derived numerically by means of a linear stability analysis of the reduced equations (5) and (6). Crosses, '×', are derived from direct simulations of (5) and (6). Plus signs furthermore show simulation results for the full neural field equations (1) and (2). As can be seen, the approximation scheme predicts the stability boundaries of the full system quite well. The stars, '∗', further result from the equivalent meanfield model, i.e., (5) with steady state effective couplings, $c_{ij} = c_{ij}^* = c_{ij}(\sigma_{pj}^*)$, for which one easily obtains the characteristic polynomial $C_{MF}(\lambda) = \tau_1\tau_2\lambda_2^2 + \left(\tau_2(1 - c_+^*) + \tau_1\right)\lambda + (1 - c_+^* + c_-^*)$ as well as the stability conditions $c_+^* = c_-^* + 1$ for the saddle-node line and $c_+^* = 1 + \tau_1/\tau_2$, $c_-^* \geq \tau_1/\tau_2$ for the Hopf-bifurcation. Furthermore, the theory provides an estimate of the frequency of oscillations appearing along the Hopf-branch: $\omega^2 = c_-^*/(\tau_1\tau_2) - 1/\tau_2^2$. Now, away from the point singularity $\sigma \approx \sigma_+ \ll \sigma_-$ and $\sigma_+^2 \ll \sigma_0^2$ (cf., Fig. 2A). So in (7) one can skip the terms containing σ_0 and Σ_- and assume that $\Sigma_+^2 = \sigma_+^2 + \sigma^2 \approx 2\sigma_+^2$. This results in $\sigma \approx \sqrt{2}\sigma_+/\left(K_+\sigma_+\right)^{1/3}$ which, inserted into the meanfield stability conditions, further yields $K_-^{SN} \approx \frac{1}{\sqrt{2}}\left(K_+ - \frac{(K_+\sigma_+)^{1/3}}{\sigma_+}\right)$, $K_+^H \approx \frac{\sqrt{2}}{\sigma_+}\left(1 + \frac{\tau_1}{\tau_2}\right)$, $K_-^H \geq K_-^c := \frac{\tau_1}{\sigma_+\tau_2}$, and $\omega^2 \approx \left(K_- - K_-^c\right)\frac{\sigma_+}{\tau_1\tau_2}$. These simple explicit conditions are plotted with stars, '∗', in Fig. 2A&B and are well in accord with the other results. Finally, the stability analysis of the one-field model (4) yields a saddle-node line as in Fig. 2, but no Hopf-bifurcation. That Hopf-bifurcation in any case would require coupling strengths roughly two times

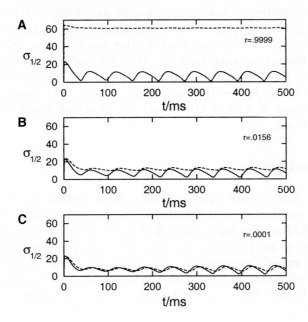

Fig. 3. Oscillatory tuning for differently tuned kernels k_{12} and k_{21}.

larger than the biologically motivated values given in [2,3,4]: $K_+ = .23$ and $K_- = .092$ (cf., Fig. 2). Oscillation frequencies near the Hopf-bifurcation line are in the alpha- and beta-range for $\tau_1 = \tau_2 = 10ms$, $K_- < .4$, and may reach the gamma-range for shorter time-constants or even stronger inhibition.

5   Influence of the Mutual Kernels k_{12} and k_{21}

It can be shown rigorously by simple scaling arguments that steady states and their stability properties are independent of K_{12} and K_{21} in the two-field model as long as $K_- = K_{12}K_{21}\sigma_{21}\sigma_{21}/\sigma_-$ and $\sigma_-^2 = \sigma_{12}^2 + \sigma_{21}^2$ are kept fix. Interestingly, simulations in the parameter ranges studied here indicate that also the precise values of σ_{12} and σ_{21} have almost no influence (simulations not shown). Thus, somewhat surprisingly, Figures 1 and 2 virtually don't change, if the relative weights or widths of k_{12} and k_{21} are modified. Figures 3A to C demonstrate that even asymptotic limit cycle properties of the excitatory cells are universal in that sense. Displayed are $\sigma_{p1/2}(t)$ as derived from the reduced set of equations in the oscillatory regime for $K_- = .2$, $K_+ = .35$ and different $r = (\sigma_{21}/\sigma_-)^2$: Figure 3C, $r \to 0$, corresponds with strictly local connections from excitatory to inhibitory cells, $k_{21} \sim \delta(x)$; B, $r = .0156$, with $\sigma_{21} = \sigma_+$; and, A, $r \to 1$, with $k_{12} \sim \delta(x)$. Since σ_- is fixed, $\sigma_{12}^2 = (1-r)\sigma_-^2$. Observe that the tuning of the excitatory cells (solid lines in Fig. 3A to C) is oscillatory but asymptotically in time virtually independent of r; only transients last longer when r gets smaller.

In contrast, the inhibitory tuning, σ_{p2}, changes drastically with r: for $r > .0156$ it is quite large (because k_{21} is broad) and almost independent of time, $\sigma_{p2} \approx \Sigma_{21} \approx \sigma_{21} = \sqrt{r}\sigma_- = constant$. On the other hand, for $r < .0156$ the kernel k_{21} becomes sharper localized than k_{11}. Then the tuning of the inhibitory cells gets sharp and time-dependent similar to that of the excitatory cells. In fact, for $r \to 0$ in (2) or $k_{21} \to \delta(x)$, $[\phi_2(x,t)]_+ = \phi_2(x,t)$ becomes a temporally low-passed version of $f_1(\phi_1(x,t)) = [\phi_1(x,t)]_+ \geq 0$ and, hence, should have a comparable tuning.

6   Conclusions

In summary, using a new approximation scheme derived in [5] I have compared tuning properties of a standard one-layer model with DOG-kernel and an excitatory-inhibitory two-layer neural field model. Steady states of both models are equivalent, as are branches of instabilities leading from tuned steady solutions to exponentially growing ones. The two-layer model, in addition, reveals a Hopf-bifurcation line. To lowest order already a meanfield approximation predicts all steady state tuning and stability properties reasonably.

References

1. R. Ben-Yishai, R.L. Bar-Or and H. Sompolinsky, Theory of orientation tuning in visual cortex, Proc. Natl. Acad. Sci. USA 92 (1995) 3844–3848.
2. D.C. Somers, S.B. Nelson and M. Sur, An emergent model of orientation selectivity in cat visual cortex simple cells, J. Neurosci. 15 (1995) 5448–5465.
3. M. Carandini and D.L. Ringach, Predictions of a Recurrent Model of Orientation Selectivity, Vision Research 37 (1997) 3061-3071.
4. T. Wennekers (2001) Orientation tuning properties of simple cells in area V1 derived from an approximate analysis of nonlinear neural field models. Neural Computation 13, 1721-1747.
5. T. Wennekers (2002) Dynamic approximation of spatio-temporal receptive fields in nonlinear neural field models. Neural Computation, in press.

Clustering within Integrate-and-Fire Neurons for Image Segmentation

Phill Rowcliffe, Jianfeng Feng, and Hilary Buxton

School of Cognitive and Computing Sciences, University of Sussex, Brighton, BN1 9QH, England

Abstract. An algorithm is developed to produce self-organisation of a purely excitatory network of Integrate-and-Fire (IF) neurons, receiving input from a visual scene. The work expands on a clustering algorithm, previously developed for Biological Oscillators, which self-organises similar oscillators into groups and then clusters these groups together. Pixels from an image are used as scalar inputs for the network, and segmented as the oscillating neurons are clustered into synchronised groups.

1 Introduction

Integrate-and-Fire neurons have been studied within the field of computer vision for some time now. One aspect of IF neurons is that they can act as natural oscillators and can be thought of as pulsed-coupled oscillators (oscillators which fire a pulse when a threshold is reached, then reset to zero).

One application of such a network of oscillators is that of clustering [1]. In [6], Rhouma and Frigui introduced a clustering algorithm which was both insensitive to the initialisation of the network and did not require the optimising of a sophisticated iterative objective function. They took pulse-coupled Integrate & Fire (IF) oscillators and partitioned them into synchronised groups, with a simple objective function based on a dissimilarity measure between the oscillators and the input data. Networks of oscillators were self-organised into clusters where the number was a product of the coupling function.

Here the algorithm developed in [6] is extended for use on a fully connected network of IF neurons, and used to achieve image segmentation on face image inputs. The neurons are based on the Gerstner model [3] which is a simplified model of the single state equation [5] derived from the Hodkin-Huxely model [4].

The rest of the paper is outlined as follows. Section 2 will consist of a brief review of related literature, in particular [6], which forms the basis on which this research has been developed. Section 3 will look at the Integrate-and-Fire model used and the clustering algorithm. In Section 4 the results from the application of this model, using images of faces as inputs, are presented. Finally, in section 5 conclusions on findings and discussions of how this model will be taken further, are explained.

J.R. Dorronsoro (Ed.): ICANN 2002, LNCS 2415, pp. 69–74, 2002.

2   Related Work

Clustering, as an unsupervised method of classification, is a means by which data is partitioned into naturally present subsets, or groups. Wang et als [7] LEGION model segments data using a locally connected network, but was found to be sensitive to noisey data and its initialisation stage. Rhuoma and Frigui [6] produced an algorithm which was not computationally complex and which also addresses some of these inherent problems within clustering. In [6] a model was developed which was both insensitive to the initialisation of the network and which used a simple dissimilarity measure to update the weight values. It took pulse-coupled oscillators, with excitatory and inhibitory connections, and partitions them into synchronised groups, with an objective function based around a simple dissimilarity measure. Networks of oscillators were self-organised into clusters (the number a product of the coupling function).

The oscillators used here though, are IF neurons, with spike response. They are based on the Gerstner model [3] which is a simpler description of the Hodgkin-Huxley model [4] previously simplified as a single state equation [5]. He has shown how the IF model is a special case of the *Spike Response Model* developed in [3] from the results Gerstner et al showed in [5]. However Feng in [2] recently showed that the IF and HH models do behave differently, infact in opposite ways when correlated inputs are considered.

3   Model

3.1   Neurons

The neurons in this network are IF Neurons, with purely excitatory connections. The basic model of an IF neuron has each neuron consisting of a current $I(t)$, a capacitance C and resistance R, similar in fact to a circuit. Just like in a circuit the current input into a unit $I(t)$, is composed of two components, the current going through the capacitor I_c and that going through the resistor I_r,i.e. $I(t) = I_c + I_r$ Using Ohm's law $V = I \times R$, and the definition of *capacitance* (i.e. $C = q/v$, where q is the unit charge and V the voltage, or membrane potential), we can rewrite the current equation as:

$$I(t) = C\frac{dv}{dt} + \frac{V(t)}{R} \qquad (1)$$

Allowing C and R to have a value of 1, and integrating over δt this becomes:

$$V(t + \delta t) = V(t) - \delta t V(t) + \delta t I(t) \qquad (2)$$

The neurons receive inputs from two different sources: the *Input image* (x_i being the pixel input to neuron i) and *other neurons* in the network, which are in the form of *spikes*. A process has been used to simulate a spike train with l_j being the output spike from neuron j.

$$l_j(t) = (t - t_{thres})exp(\frac{-(t - t_{thres})}{T}) \qquad (3)$$

Here t_{thres} is the time that the neuron firing the spike reached threshold, and t, the present time. T is the time constant with which the membrane potential decays. Each neuron has a weighted coupling w connecting them to each other. At any point in time a neuron receives inputs from both other neurons in the network, and the pixels in the input scene. The voltage change is therefore:

$$V_i(t + \delta t) = V_i(t) - \delta t V_i(t) + \delta t \left(\frac{x_i}{S} + \sum w_{ji} l_j(t) \right) \tag{4}$$

where w_{ji} is the weighted connection between neurons j and i, and l_j the output spike from neuron j. S is a scaling factor used to normalise the pixel input.

3.2   Algorithm

The algorithm operates in two main parts. The first part works on the network of oscillating neurons and adjusts the lateral weight connections between the neurons. The clustering aspect of synchronised groups of neurons is the second part of the algorithm.

Oscillatory Adjustment. The adjustment to the connections between the neurons is designed to force similar groups to threshold at the same time. The method used to cluster neurons is that described in [6] as *Clustering of Object Data*. The purpose of the algorithm is to partition the neurons into groups. Each pixel in the input image is represented by a neuron and similar inputs are grouped together into the same cluster as follows:

Let X be the set of N neurons, and P be the set of M prototypes. The *prototypes* represent the clusters within the network, [6] recommends this to be around N/4 for best results. The next neuron to reach threshold is identified, say X_j. The closest prototype to this neuron is also identified, say P_k. The distances are then calculated between all the other neurons and this prototype. These distances d will be used to update the weight values on neurons connected to neuron X_j using the new updated (normalised) distance $\tilde{d}$ as follows:

$$d_{jk} = \begin{cases} lld_{jk} & \text{if neuron does not belongs to any group,} \\ \frac{\sum_{l \in G} d_{lk}}{|G_m|} & \text{if neuron belongs to group } G_m. \end{cases} \tag{5}$$

where G is the group of synchronised neurons G= G_1,..., G_m (grouping is explained in the next section), m the number of synchronised groups so far. The current state or *phase* of neuron i, at a particular point in time, is represented by ϕ_i. In synchronising neurons the phases will be equal and the firing of one neuron will bring the others to their membrane threshold. Therefore, the coupling of neruon i is updated by $\epsilon_i(\phi_j)$, where:

$$\epsilon_i(\phi_j) = \begin{cases} llC_E[1 - (\frac{d_{jk}}{\delta_0})^2] & \text{if } d_{jk} \le \delta_0, \\ C_I[(\frac{d_{jk}-1}{\delta_0-1})^2 - 1] & \text{if } d_{jk} > \delta_0. \end{cases} \tag{6}$$

Here, C_E and C_I are the excitatory and inhibitory coupling strengths, respectively and δ_0 is the dissimilarity measure between the input pixels.

Grouping. At the end of each iteration the groups of synchronised neurons are updated. Firstly all the other neurons which had been brought to threshold are identified and then one of the following cases is carried out:

- Create a new group if none of the identified neurons, which have reached threshold, have been assigned to a group and update the number of groups by 1.
- If several of the neurons which have reached threshold belong to an existing group, say G_m, then assign all these neurons to G_m.
- If several of the neurons which have reached threshold belong to groups $G_{m_1}, \ldots G_{m_q}$ then delete their association with the q-1 groups and assign these neurons to G_{m_1}.
- if no neurons reach threshold, then do nothing.

The iterative cycles are stopped when the number of groups stabilise.

4   Results

The algorithm discussed in section 3 has been implemented using a network of Integrate-and-Fire neurons. Here results are presented using a variety of faces as input. The images are greyscale and fed in as a bitmap. All the experiments took between 5-8 iterations before the grouping stabilised. The weight updating values C_E and C_I used were 0.1 and 0.01 respectively and the dissimilarity measure $\delta_0 = 0.001$.

The initial images sampled were approximately 280 x 380 pixels square. These were reduced to input images 44 x 58 by simply removing rows and columns of pixels, no sampling or other technique was used.

Figure (1)(a) shows the original photograph sampled for input. Image (b) shows the initial state of the network with all the neurons black, indicating they have not been assigned to any group. Images (c) to (f) show the progress of the algorithm through iterations 2 to 5. It can be seen that more and more of the black, unassigned neurons are grouped together. Neurons which are clustered into the same group are assigned the same colour until at the end of the iterative cycle all neurons are assigned to a group.

Figures (2) and (3) shows 2 faces which took 8 iterations until stabilisation.

In Figure (2) the algorithm finally merged small, similar groups together which had in the previous iterations been kept separate. This effect is seen to a greater degree in Figure (3) where the presence of noise is treated in the same way. The input image used in Figure (3) contained lots of noise and the algorithm appeared to be particularly robust when such noise was present, clustering these stray pixels together with similar ones. However, this does have the effect, as in Figure (2), of absorbing segments of the image which would ideally have been kept separate. This effect is a result of the scalar input. Adjusting scale parameter S so that small variations in pixel value are preserved, allows the algorithm to segment the image to achieve a finer granularity. The algorithm, as in Rhouma and Frigui's experiments [6], has shown a robustness to the effects of noise.

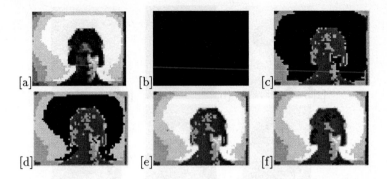

Fig. 1. Clustered neurons tested on an image of a face over 5 iterations. (c) to (f) shows the iterative states of the network on input image (a) from starting state (b).

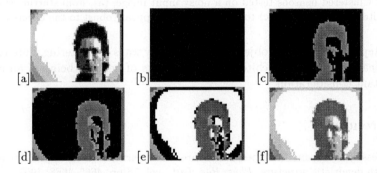

Fig. 2. From input image (a) and starting state (b), figures (c) to (f) show the segmentation of the image over 8 iterative steps. Some groups are merged together over the iterations resulting in the final segmented image (f).

5   Conclusion

The implementation of Rhouma and Frigui's algorithm on more realistic biological neurons has produced positive results when applied to image segmentation. It works effectively with single images and produces positive results even when noise is present in the data.

The IF neurons, though more representative of biological neurons than those used in [6], are still only approximates of more realistic biological models currently being used in today's research. A natural progression is to develop the model of the neuron further, and produce a closer representation of those seen in biology. This might be achieved by developing the algorithm on the FHN model of neurons as shown in [2] to more closely approximate the HH model than the IF model does.

The algorithm here has only been tested on a single layer of neurons and has yet to be developed to deal with multiple layers. This would seem an inevitable

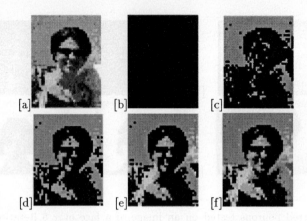

Fig. 3. Clustered neurons tested on a noisy input image, (a), from starting state (b), over 8 iterations shown in (c) to (f). The results appear robust to the effects of noise.

next step especially if our objective is to identify ways of clustering more realistic biological neurons, the architecture of networks of neurons being of an equally important consideration.

References

1. Bressloff, B.C., and Coombes, S. Synchrony in an array of integrate and fire neurons with dendrictic structure. *Phys. Rev. Lett.,* vol. 78 pp. 4665–4668, 1997.
2. Feng, J. Is the integrate-and-fire model good enough? – a review. *Neural Networks,* vol. 14, pp.955–975, 2001.
3. Gerstner, W. The spike response model. *In The Handbook of Biological Physics* vol. 4 (Ch.12), pp. 469–516, Moss, F., and Gielen, S. Eds. Elsevier Science, 2001.
4. Hodgkin, A.L. and Huxley, A.F. A quantitative description of ion currents and its applications to conductance and excitation in nerve membranes. *J. Physiol. (London),* no. 117, pp. 500–544, 1952.
5. Kistler, W.M., Gerstner, W., and van Hemmen, J.L. Reduction of Hodgkin-Huxely equations to a single-variable threshold model. *Neural Comput.,* no. 9, pp. 1015–1045, 1997.
6. Rhouma, M.B.H., and Frigui, H. Self-organisation of biological oscillators with application to clustering. *IEEE Trans. Patt. Analysis Mach. Intell.,* vol. 23, no. 2, Feb. 2001.
7. Chen, K., Wang, D.L., and Liu, X. Weight adaptation and oscillatory correlation for image segmentation. *IEEE Trans. Neural Network,* vol. 11, no. 5, pp. 1106–1123, Sept. 2000.

Symmetry Detection Using Global-Locally Coupled Maps

Rogério de Oliveira and Luiz Henrique Alves Monteiro

Universidade Presbiteriana Mackenzie, Pós-Grad. Eletrical Eng.
Rua da Consolação , 895 – 6⁰ andar
01302 – 907, São Paulo, SP - Brasil
{roger_io,luizm}@mackenzie.com.br

Abstract. Symmetry detection through a net of coupled maps is proposed. Logistic maps are associated with each element of a pixel image where the symmetry is intended to be verified. The maps are locally and globally coupled and the reflection-symmetry structure can be embedded in local couplings. Computer simulations are performed by using random gray level images with different image sizes, asymmetry levels and noise intensity. The symmetry detection is also done under dynamic scene changing. Finally the extensions and the adherence of the present model to biological systems are discussed.

1   Introduction

Symmetry is a basic feature in shapes, images and objects and plays an important role in computer vision, specially in recognition, matching, inspection and reasoning. In this paper we are interested in detecting symmetries with respect to a central axis in $2D$ images. As usual the symmetry is considered as a binary feature where an object is either symmetric or not symmetric. A brief review of the recent research and methods in reflection-symmetry detection can be found in [10].

Although many methods are available for symmetry detection, few of them are based on neural models [10]. Two exceptions, supervised and an unsupervised method, are presented in [9] and [7]. The reason for this lack is that, in general, symmetry is a combinatorial problem that astronomically grows with the number of elements of the image. Such a complexity suggests that models based on training sets are not appropriate to deal with symmetry classification [7]. A common strategy is to have the structure of the symmetry problem embedded in the neural net architecture. This approach was successfully used in [7] to classify reflection-symmetry patterns using a Dynamic Link Architecture net. Here our choice lies otherwise over coupled systems as a general architecture because of their ability to represent spatiotemporal patterns.

Coupled systems have been used to solve a large class of problems like scene segmentation [12] [1], pattern recognition [6] [3], pattern formation [8] and associative memories [4]. Coupled systems, different from models based on threshold

J.R. Dorronsoro (Ed.): ICANN 2002, LNCS 2415, pp. 75–80, 2002.

neurons, preserve spatial information [6] [12]. It means that states of the net depend on the spatial relationship among their elements. This property can be used to have the spatial symmetry relationship embedded in the net. The learning can be done through the synchronization which exempt the learning process from training sets. So coupled systems offer a class of networks that can be adapted in a simple and general architecture to symmetry detection with a minimum of scene statistics.

Coupled maps are coupled systems in discrete time. In the present model global-locally coupled maps are used for reflection-symmetry detection in an unsupervised manner. Each pixel from a $2D$ image is associated with a single map of the system. The local couplings embed the mirror-symmetry structure between pixels and enforce the map synchronization. The system detects whether the image is symmetric or not symmetric by revealing synchronized elements.

2   Model Description

Globally coupled maps (GCM) can be described by the following system of equations [6]:

$$x_i(t+1) = (1 - \varepsilon)f_\mu(x_i(t)) + \frac{\varepsilon}{N}\sum_{i=1}^{N} f_\mu(x_i(t)) \qquad (1)$$

Here $x_i(t)$ are the state variables over the discrete time t. The global coupling is given by $\sum_{i=1}^{N} f_\mu(x_i(n))$ which interconnects all maps in the system. In the present study the map $f_\mu(u)$ is the logistic map, $f_\mu(x) = \mu x(1 - x)$, with the global bifurcation parameter μ. The constant ε is the constant of coupling of the system.

In some applications of GCMs to neural networks, a unit of the GCM corresponds to a single neuron [5]. Here we adopt such a representation. Each map i, with state variable $x_i(t)$, corresponds to a single network unit that represents a pixel in the original scene.

Two main modifications were made in this model for symmetry detection. We introduced the external inputs I_i for representing the values of each pixel of the initial image, and the local coupling, $\lambda(x_i(t))$, that allows to synchronize only the maps associated to symmetrical pixel values. Thus:

$$x_i(t+1) = (1 - \varepsilon)f_\mu(x_i(t)) + \frac{\varepsilon}{N}\lambda(x_i(t))\sum_{i=1}^{N} f_\mu(x_i(t)) + I_i \qquad (2)$$

In such models, where the long-range and local coupling coexist, the global coupling gives an easier synchronization of the elements while, on the other hand, local coupling can be used to preserve spatial information [12]. Then local coupling was used to embed the symmetry structure.

We are only interested in reflectional symmetries. A $2D$ image is said to be reflection-symmetric if it is invariant with respect to one or more straight lines, denoted as reflection-symmetric axes [10]. $2D$ images are done in pixel matrices. Thus the equation (2) is re-written in a more suitable way:

$$x_{ij}(t+1) = (1-\varepsilon)f_\mu(x_{ij}(t)) + \frac{\varepsilon}{N}\lambda(x_{ij}(t))\sum_{i=1}^{m}\sum_{j=1}^{m}f_\mu(x_{ij}(t)) + I_{ij} \qquad (3)$$

Now each pixel in the position ij corresponds to a single map with the state variable $x_{ij}(t)$. The local coupling function $\lambda(x_{ij}(t))$ can be introduced to reflection-symmetric detection with respect to vertical center axis of a $m \times m$ pixel matrix:

$$\lambda(x_{ij}(t)) = \lambda(x_{i,j}(t)) = x_{i,j}(t)x_{i,m-j+1}(t) + \Delta x_{ij}(t) \qquad (4)$$

where $\Delta x_{i,j}(t)$ is the Laplace operator:

$$\Delta x_{ij}(t) = x_{i+1,j}(t) + x_{i-1,j}(t) + x_{i+1,m-j+1}(t) + x_{i-1,m-j+1}(t) - 4x_{ij}(t) \quad (5)$$

$x_{ij}(t) = 0$ if either $i > m$ or $j > m$ in the equations above. For simplicity we use square matrices and m is an even dimension.

This local coupling uses two common choices for local coupling. The product on the equation (4) simply exhibits the oscillatory correlation [5] between the symmetric maps $x_{i,j}(t)$ and $x_{i,m-j+1}(t)$, and the product assumes the maximum value if they are synchronized, i.e. $|x_{i,j}(t) - x_{i,m-j+1}(t)| \to 0$ when time grows. The equation (5) refers to the neighborhood of the symmetric maps.

The model drives to synchronize only those maps that represent symmetric pixels and have the same external input values. Symmetric pixels with different values I_{ij} are led to desynchronize or synchronize in opposite phase. From the definition of synchronization in maps we summarized the synchrony of the system as follow:

$$y_i(t) = \sum_{j=1}^{m} |x_{ij}(t) - x_{i,m-j+1}(t)| \qquad (6)$$

which exhibits the vertical reflection-symmetry over the lines of the image and $z(t) = \sum_{i=1}^{m} |y_i(t)|$ which resumes the vertical reflection-symmetry of whole scene. So the image is symmetric only if each line is symmetric, it means $y_i(t) \to 0$, and therefore $z(t) \to 0$ for $t \to \infty$. The figure 1 shows how the system works.

3   Simulation Results

Three classes of simulations were done. One, random images, with pixels assuming gray level feature values, were classified as vertical reflection-symmetric or not. Two, we verified the robustness of the model in the presence of noise. Three, the capacity of the model to deal with dynamical scene changing was checked by using images varying with the time.

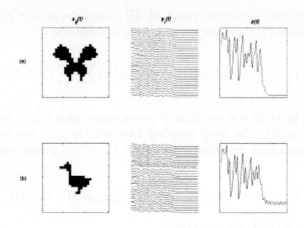

Fig. 1. Single images in black-and-white 30×30 pixel matrices. (a) the butterfly exhibits vertical reflection-symmetry, all maps x_{ij} and $x_{i,m-j+1}$ synchronize and both $y_i(t)$ and $z(t)$ vanish. (b) the duck exhibits some non-vertical reflection-symmetry lines, $y_i(t)$ and $z(t)$ oscillate.

In all experiments the global bifurcation parameter is $\mu = 3.5$, the constant of coupling is $\varepsilon = 0.6$ and the initial values of the system are random inside of the interval $[0, 0.1[$.

Most experiments were done with random images. Initially vertical reflection-symmetric images are randomly created in a square matrices of pixels. Black-white and gray level images are used. Pixels in black-white scenes (see figure 1) assume just two different values. In gray level images they can assume 256 different values (figure 2). Over the random symmetric images different levels of asymmetry are obtained by changing randomly the pixel values. Matrices with dimensions of $m = 6, 8, 10, 20, 30$ were used to form the images. For each dimension more than 50 experiments were done with gray level and black-white scenes. In all simulations we successfully verified the presence of the vertical reflection-symmetry or not, in the individual lines and the whole image. Some of these results are shown in the figure 2 where 8×8 of gray level pixel matrices exhibit different levels of vertical symmetry that was successfully detected by the present model.

Practical use of symmetry detection depends on its ability to deal with noise. Here also this ability of the model was observed. We correctly detected vertical symmetries over symmetrical images where gaussian noise with deviations $\sigma^2 < 10^{-5}$ was applied. For higher deviations it is clear the absence of symmetry (see figure 2(b)).

Finally the present model is suitable to deal with temporal information. Other fully-connected neural network models, such as Hopfield models, require the re-initialization of the network to new inputs. Our model allows dynamic scene inputs without the re-initialization. It means we can write the external inputs

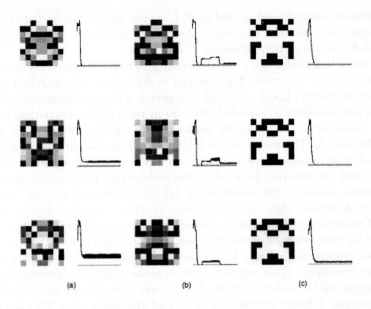

(a)   (b)   (c)

Fig. 2. 8×8 pixel matrices **(a)** Gray level images with different levels of vertical asymmetry successfully classified. **(b)** A dynamical scene changing to higher and lower symmetry levels. Here just the last scene is showed. **(c)** Different scenes with crescent gaussian noise were checked.

in the equation (3) as a dynamic feature $I_{ij}(t)$. The figure 2(b) shows how the model reacts to scene changing to higher and lower symmetry levels.

4   Discussion and Perceptual Modeling

The hypothesis that synchronization may be a basic component of many perceptual and cognitive processes is quite accepted. Neuronal oscillations and synchronization have been observed in perceptual brain functions involving audition [1], vision and olfaction (see [2] to a review). Spatiotemporal codes, such as synchronization of neuronal activity, offer significant computational advantages over traditional rates codes, specially in biological modeling. A natural implementation of temporal correlation models based on synchronization is the use of neural oscillators, whereby each oscillator represents some feature (here a pixel) of an object [1].

Usually, oscillatory networks are implemented as continuous time systems. Some examples are given by [12] [8] and [3]. In the present model coupled maps are employed to implement neural oscillators in discrete time. This approach is more suitable for digital processing and computer simulations can be made very efficient.

Particularly the excitatory-inhibitory (E-I) neuron models are widely spread in networks for biological modeling. The correspondence between the maps used in the present network and a kind of *E-I* neuron model can be found in [11]. That allows us to see each element in the present model, i.e. a single map, as a pair of *E-I* neurons or populations of *E-I* neurons.

The described model still preserves many properties of symmetry perception that can be easily observed. The learning is done in an unsupervised manner and is not stochastic. Local couplings also permit a fast synchronization. In fact it just a few iterations are needed (20 − 30 in most of the simulations), which is more plausible in the biological models than long time ranges from general stochastic models. The model also exhibits high-dimensionality, maintains their abilities despite of presence of noise and can deal with dynamical scene changing without reset the network. In addition the architecture is inherently parallel.

Moreover, although the equations (3) (4) (5) concern just vertical reflection-symmetry detection, they can be adapted to other symmetry detections. The simplest example may be horizontal reflection-symmetry detection which can be achieved by just reversing index matrices. The diagonal symmetry and others rotational symmetries can also be reached with appropriate index changing and boundary conditions. This suggests that to make a dynamic structure of couplings, changing the index matrices dynamically through the time, to identify simultaneously different symmetries or to find symmetry axes. This possibility will be treated elsewhere.

References

1. Campbell, S., Wang, D.: Synchronization and desynchronization in a network of locally coupled Wilson-Cowan oscillators. IEEE Trans. Neural Networks **7** (1996) 541–554
2. Freeman, W.J.: Tutorial on neurobiology: From single neurons to brain chaos. Int. J. Bif. Chaos **2** (1992) 451–482
3. Hayashi, Y.: Oscillatory neural network and learning of continuously transformed patterns. Neural Networks **7** (1994) 219–231
4. He, Z., Zhang, Y., Yang, L.: The study of chaotic neural network and its applications in associative memory. Neural Process. Lett. **9** (1999) 163–175
5. Ito, J., Kaneko, K.: Self-organized hierarchical structure in a plastic network of chaotic units. Neural Networks **13** (2000) 275–281
6. Kaneko, K.: Overview of coupled map lattices. Int. J. Bif. Chaos **2** (1992) 279–282
7. Konen, W., Malsburg, C.v.d.: Learning to generalize from single examples in the dynamic link architecture. Neural Computation **5** (1993) 719–735
8. Nishii, J.: The learning model goes oscillatory networks. Neural Networks **11** (1998) 249–257
9. Sejnowski, T.J., Keinker, P.K., Hinton, G.E.: Learning symmetry groups with hidden units: Beyhond the perceptron. Physica **22D** (1986) 260–275
10. Shen, D., Ip, H.H.S., Cheung, K.K.T., Teoh, E.K.: Symmetry detection by gener-alied complex (GC) moments: A close-form solution. IEEE Trans. Pattern Analysis and Machine Intelligence **21** (1999) 466–476
11. Wang, X.: Period-doublings to grounds in the simple neural network: an analytical proof. Complex Syst. **5** (1991) 425–441
12. Zhao, L., Macau, E.E.N., Nizam, O.: Scene segmentation of the chaotic oscillator network. Int. J. Bif. Chaos **10** (2000) 1697–1708

Applying Slow Feature Analysis to Image Sequences Yields a Rich Repertoire of Complex Cell Properties

Pietro Berkes and Laurenz Wiskott

Institute for Theoretical Biology,
Humboldt University Berlin,
Invalidenstraße 43, D - 10 115 Berlin, Germany,
{p.berkes,l.wiskott}@biologie.hu-berlin.de,
http://itb.biologie.hu-berlin.de/

Abstract. We apply Slow Feature Analysis (SFA) to image sequences generated from natural images using a range of spatial transformations. An analysis of the resulting receptive fields shows that they have a rich spectrum of invariances and share many properties with complex and hypercomplex cells of the primary visual cortex. Furthermore, the dependence of the solutions on the statistics of the transformations is investigated.

1   Introduction

In the past years there has been an increasing interest in understanding the computational principles of sensory coding in the cortex. One of the proposed principles is known as *temporal coherence* or *temporal smoothness* [1,2,3,4]. It is based on the assumption that the *sources* of the sensory input (e.g. objects in vision) vary on a slower time scale than the sensory signals themselves, which are highly sensitive even to small transformations of the sources (e.g. rotation or translation). By extracting slow features from the raw input one can recover information about the sources independently of these irrelevant transformations. We focus here on vision and apply Slow Feature Analysis (SFA) [4,5] to image sequences for a comparison with receptive fields of cells in the primary visual cortex V1.

2   Methods

The problem of extracting slow signals from time sequences can be formally stated as follows: given an input signal $\mathbf{x}(t) = (x_1(t) \ldots x_N(t))$, $t \in [t_0, t_1]$ and a set of real-valued functions $\mathcal{F}$, find a function $\mathbf{g}(\mathbf{x}) = (g_1(\mathbf{x}), \ldots, g_M(\mathbf{x}))$, $g_j \in \mathcal{F}$ so that for the output signals $y_j(t) := g_j(\mathbf{x}(t))$

$$\Delta(y_j) := \langle \dot{y_j}^2 \rangle \quad \text{is minimal} \tag{1}$$

J.R. Dorronsoro (Ed.): ICANN 2002, LNCS 2415, pp. 81–86, 2002.
© Springer-Verlag Berlin Heidelberg 2002

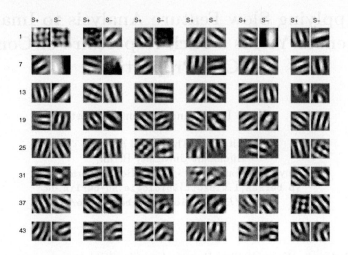

Fig. 1. Optimal excitatory and inhibitory stimuli for Run #1.

$$\text{under the constraints:} \quad \langle y_j \rangle = 0 \quad \text{(zero mean)}, \tag{2}$$
$$\langle y_j^2 \rangle = 1 \quad \text{(unit variance)}, \tag{3}$$
$$\forall i < j, \quad \langle y_i y_j \rangle = 0 \quad \text{(decorrelation)}, \tag{4}$$

with $\langle . \rangle$ indicating time averaging. Here we choose $\mathcal{F}$ to be the set of all polynomials of degree 2 and use Slow Feature Analysis (SFA) [4,5] to find the optimal input-output functions $g_j(\mathbf{x})$.

Like in electrophysiological studies of neurons in V1, we are interested in characterizing the receptive fields (RFs) of the single components g_j being extracted. We do that by determining the input vector with norm r that maximizes and the one that minimizes the output signal y_j, yielding the optimal excitatory stimulus (S⁺) and the optimal inhibitory stimulus (S⁻) (cf. Fig. 1). We choose r to be the mean norm of the training input vectors, since we want the optimal stimuli to be representative of the typical input.

Of interest are also the invariances learned by the system, which correspond tothe directions in which a variation of S⁺ has the least effect on the output. We extract them by computing the Hesse matrix of the function $g_j(\mathbf{x})$ restricted to the sphere of radius r in $\mathbf{x} = $S⁺and then choosing the directions corresponding to the smallest eigenvalues. For visualization we move the S⁺ vector on the sphere of points with norm r in the direction of the invariance, thereby producing image sequences as those shown in Figure 2[1].

Training data were taken from 36 natural images from van Hateren's natural stimuli collection and preprocessed as suggested in the original paper [6]. The end resolution was 2 minutes of arc per pixel. We produced input sequences by

---

[1] Animations corresponding to the image sequences shown in Figure 2 can be found at http://itb.biologie.hu-berlin.de/~berkes/ICANN02/results.html

choosing an initial position in a randomly chosen image, cutting a 16×16 pixels patch (ca. 0.5×0.5 degrees of arc) and moving it around according to different transformations: translation, rotation and zoom. The transformations were performed simultaneously, so that each frame differed from the previous one by position, orientation, and magnification. Patches were computed by bilinear interpolation. In the default settings, the translation speed v was chosen uniformly between 1 and 5 pixel/frame, the rotation speed ω between 0 and 0.1 rad/frame and the magnification factor z between 0.98 and 1.02 per frame. The parameters were varied every 30 frames, for a total of 150,000 frames per simulation.

A run with these settings requires the computation of a covariance matrix having $\mathcal{O}(N^4)$ elements, where N is the input dimension. We thus performed a standard preprocessing step using PCA to reduce the dimensionality of the input patches to 50 components.

3 Results

3.1 Receptive Field Analysis

Figure 1 shows the S^+/S^- pairs of the first 48 components for a typical run in decreasing order of temporal slowness. Notice that these optimal stimuli cannot be interpreted as linear filters and only give a first hint at the response properties of the RFs. Additional information can be extracted from their invariances; examples of the main types of invariances found are shown in Figure 2. The analysis of the units led to the following observations:

Gabor-like optimal stimulus. The optimal excitatory stimulus for most of the components looked like an oriented Gabor wavelet and tended to be localized, i.e. it did not fill the entire patch. This property is linked to rotation and zoom, as localized receptive fields are less sensitive to them (since they are less localized in the Fourier space). If rotation and zoom are absent, localization disappears (cf. Sec. 3.2).

Phase invariance. The first invariance for almost all analyzed units was *phase shift* invariance (Fig. 2a). In fact, the response of the units never dropped by more than 20% changing the phase of S^+. As a consequence the units responded well to an oriented edge but were not selective for its exact position, and thus matched the properties of *complex cells* in V1 [7]. Some units showed in addition to the phase insensitive response a phase sensitive response at a lower frequency and different orientation. These cells thus showed complex cell as well as *simple cell* behavior, which might be difficult to detect in an experimental situation, since the optimal simple and complex cell responses have different frequency *and* orientation. The simple cell component, being linear, was relatively stronger for stimuli of low contrast. The clear dichotomy between simple and complex cells in V1 has already been questioned in the experimental literature (see [8] for a discussion).

Complex cells with phase invariant optimal responses to Gabor stimuli have been modeled earlier [3,9]. In addition we found properties that have to our knowledge not yet been reproduced through unsupervised learning in a computational model:

End-inhibition. Some S^+ only filled one half of the patch, while the missing half was filled by S^- (e.g. Units 28, 41, 44 in Fig. 1). These units responded to edges with a specific orientation in the S^+ half of the patch but failed to respond if the stimulus was extended into the S^- half. Complex cells with this behavior are called *end-inhibited* or *hypercomplex cells* [7].

Orientation tuning. The optimal inhibitory stimulus was typically also a wavelet. Its orientation was often non-orthogonal to the preferred one (e.g. Units 14, 15, 24 in Fig. 1) resulting in sharpened or bimodal orientation tuning. This is often the case for cells in V1 and thought to contribute to the orientation selectivity of complex cells [10]. On the other side, some units showed invariance to orientation changes (Fig. 2e) and had a broad orientation tuning, a feature also found in V1 [10].

Frequency tuning. Similar mechanisms can lead to sharp frequency tuning. In some units S^- had the same orientation and shape as S^+, but had a different frequency (e.g. Units 20, 36, 40 in Figure 1). Such units reacted to a change in frequency by an abrupt drop in their response, in contrast to other units which showed an invariance to frequency changes (Fig. 2d). This suggests that our results can account for the wide range of spatial frequency bandwidths found in complex cells [11].

Non-oriented receptive fields. The first two units in all 6 runs with these settings responded to the mean and the squared mean pixel intensity, as was inferred directly from the learned functions. These units are comparable to the *tonic cells* described in [12]. A few other units responded to edges and had phase invariance but showed identical responses for all orientations. A small percentage of the neurons in V1 consists of *non-oriented complex cells* with these characteristics [10].

Additional invariances. For each unit we found 4 to 7 highly significant invariances. These included the invariances mentioned above (phase shift, orientation change and frequency change) and also change in position, size, and curvature of the S^+ wavelet (Fig. 2b,c,f). Units corresponding to faster signals showed also more complex invariances, which were sometimes difficult to interpret. Some of the units were found to be responsive to corners or T-shaped stimuli (Fig. 2g-h).

Although the precise shape and order of the components can vary in different simulations, we observed a systematic relationship between the slowness of the output of a unit and its behavior. For example, slower functions have usually a non-structured or orthogonal S^-, while inhibition at non-orthogonal orientations was typical for units with a faster output. It is possible that this kind of dependence also holds for neurons in V1.

3.2   The Role of Transformations

The characteristics of optimal stimuli and invariances apparently depend more on the statistics of the transformations than on the statistics of the images.

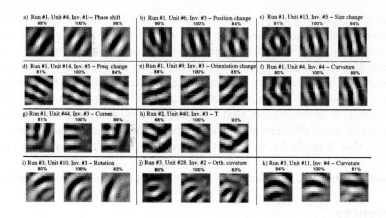

Fig. 2. Invariances. The central patch is S^+, while the other are produced by applying an invariance as described in Section 2. Each patch elicits in the considered unit the percent of the maximal output indicated over it.

Results similar to the ones described above were found by replacing the natural images by a colored noise image with a natural power spectrum of $1/f^2$. However, when changing the transformation settings during training we found significant differences:

- If we increased the influence of rotation by choosing ω to lie between 0.15 and 0.25 rad/frame, many optimal stimuli assumed a curved shape and often included in their invariance spectra (in addition to the previously discussed transformations) rotation and curvature changes both parallel and orthogonal to the orientation of the wavelet (Fig. 2i-k). Wavelets of this shape have been proposed in context of object and face recognition as *banana wavelets*. [13].
- Units in simulations involving only translation did not show any localization and had fewer invariances.
- Optimal stimuli in simulations involving only rotations or only zooms did not look like Gabor wavelets anymore, but assumed a circular and a star-like shape, respectively.

4   Conclusion

We have shown that SFA applied to image sequences generated by translation, rotation, and zoom yields a rich repertoire of complex cell properties. We found receptive fields with optimal stimuli in the shape of Gabor wavelets and invariance to wavelet phase. These properties were found also in earlier modeling studies (e.g. [3,9]). However, since SFA provides a more general functional architecture than these earlier models, we were able to reproduce additional complex cell properties, such as end-inhibition, inhibition at non-orthogonal orientations, inhibition at different frequencies, and non-oriented receptive fields. The units also showed additional invariances, such as invariances with respect to position, size, frequency, orientation, and/or curvature. Our experiments suggest that there could be a relation between the slowness of the output of complex cells and their behavior. They also suggest that some complex cells could exhibit a simple cell behavior at non-optimal frequency and orientation, particularly at low contrast. It is remarkable that the temporalsmoothness principle is able to reproduce so many different properties found in the primary visual cortex, which indicates that it might be an important learning principle in cortex in general.

Acknowledgments. Supported by a grant from the Volkswagen Foundation.

References

1. Földiák, P.: Learning invariance from transformation sequences. Neural Computation 3 (1991) 194–200
2. Stone, J.V.: Learning perceptually salient visual parameters using spatiotemporal smoothness constraints. Neural Computation 8 (1996) 1463–1492
3. Kayser, C., Einhäuser, W., Dümmer, O., König, P., Körding, K.: Extracting slow subspaces from natural Videos leads to complex cells. In: Artificial Neural Networks – ICANN 2001 Proceedings, Springer (2001) 1075–1080
4. Wiskott, L., Sejnowski, T.: Slow feature analysis: Unsupervised learning of invariarnces. Neural Computation 14 (2002) 715–770
5. Wiskott, L.: Learning invariance manifolds. In: Proc. Computational Neuro–science Meeting, CNS'98, Santa Barbara. (1999) Special issue of Neurocomputing, 26/27:925–932.
6. van Hateren J.H., van der Schaaf A.: Independent component filters of natural images compared with simple cells in primary visual cortex. Proc. R. Sec. Lond. B (1998) 359–366
7. Hubel, D., Wiesel, T.: Receptive fields, binocular interaction and functional architecture in the cat's visual cortex. Journal of Physiology 160 (1962) 106–154
8. Mechler, F., Ringach, D.L.: On the classification of simple and complex cells. Vision Research (2002) 1017–1033
9. Hyvärinnen, A., Hoyer, P.: Emergence of Phase and shift invariant features by decomposition of natural images into independent features subspaces. Neural Computation 12 (2000) 1705–1720
10. De Valois, R., Yund, E., Hepler, N.: The orientation and direction selectivity of cells in macaque visual cortex. Vision Res. 22 (1982) 531–44
11. De Valois, R., Albrecht, D., Thorell, L.: Spatial frequency selectivity of cells in macaque visual cortex. Vision Res. 22 (1982) 545–559
12. Schiller, P., Finlay, B., Volman, S.: Quantitative studies of single-cell properties in monkey striate cortex. 1. Spatiotemporal organization of receptive fields. J. Neurophysiol. 39 (1976) 1288–1319
13. Krüger, N., Peters, G.: Object recognition with banana wavelets. In: Proceedings of the ESANN97. (1997)

Combining Multimodal Sensory Input for Spatial Learning

Thomas Strösslin[1], Christophe Krebser, Angelo Arleo[2], and Wulfram Gerstner[1]

[1] Laboratory of Computational Neuroscience, EPFL, Lausanne, Switzerland
[2] Laboratoire de Physiologie de la Perception et de l'Action, Collège de France-CNRS, Paris, France

Abstract. For robust self-localisation in real environments autonomous agents must rely upon multimodal sensory information. The relative importance of a sensory modality is not constant during the agent-environment interaction. We study the interrelation between visual and tactile information in a spatial learning task. We adopt a biologically inspired approach to detect multimodal correlations based on the properties of neurons in the superior colliculus. Reward-based Hebbian learning is applied to train an active gating network to weigh individual senses depending on the current environmental conditions. The model is implemented and tested on a mobile robot platform.

1 Introduction

Multimodal information is important for spatial localisation and navigation of both animals and robots. Combining multisensory information is a difficult task. The relative importance of multiple sensory modalities is not constant during the agent-environment interaction, which makes it hard to use predefined sensor models.

The hippocampal formation of rats seems to contain a spatial representation which is important for complex navigation tasks [1]. We propose a spatial learning system in which external (visual and tactile) and internal (proprioceptive) processed signals converge onto a spatial representation. Here we focus on the dynamics of the interrelation between visual and tactile sensors and we put forth a learning mechanism to weigh these two modalities according to environmental conditions. Our system is inspired by neural properties of the superior colliculus, a brain structure that seems to be involved in multimodal perception [2,3,4].

2 Related Work

The rat's hippocampal formation receives highly processed multimodal sensory information and is a likely neural basis for spatial coding [1,5,6,7]. Hippocampal place cells discharge selectively as a function of the position of the rat in the environment.

J.R. Dorronsoro (Ed.): ICANN 2002, LNCS 2415, pp. 87–92, 2002.

The superior colliculus (SC) is involved in oculomotor responses and in the processing of multimodal information (visual, tactile and auditory) [2]. There is also evidence that SC contains neurons with spatial firing properties [3,4].

Robotic models of multimodal integration [8,9,10] are mostly based on probabilistic sensor fusion techniques in the framework of occupancy grids which cannot easily be transposed into biological models. Most current biological models of localisation and navigation [11,12] focus on a single external modality and neglect the problem of combining multimodal information.

3   Proposed Model

We adopt a hippocampal place code similar to [12] as a spatial map. Hebbian learning is used to correlate idiothetic (path integration) and allothetic (visual and tactile) stimuli with place cell activity.

Here we model the integration of visual and tactile signals into a common allothetic representation. The weight of each sense is modulated by a gating network which learns to adapt the importance of each sense to the current environmental condition. Intermodal correlations are established using uni- and multimodal units inspired by neurons in the superior colliculus. The model is implemented on a Khepera mobile robot platform. Figure 1 shows the architecture of the system.

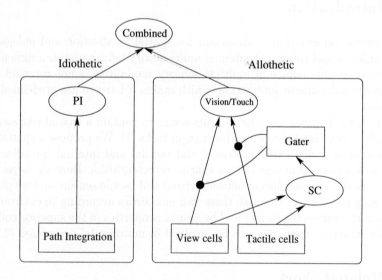

Fig. 1. Architecture of our spatial localisation system

3.1   Neural Coding of Sensory Input

During exploration, the hippocampal place code is established. At each new location, *view cells* (VCs) and *tactile cells* (TCs) convert the agent's sensory input to neural activity.

Sensory cell (VC and TC) activity r_i depends on the mean distance between the current sensor values x_i and the stored values w_{ij} at creation of cell i.

$$r_i = \exp\left(-\frac{\left(\frac{1}{N}\sum_{j=1}^{N}|w_{ij} - x_j|\right)^2}{2\sigma^2}\right) \tag{1}$$

Visual input from the greyscale camera is processed using a set of Gabor filters. The magnitudes of the complex filter responses are stored in the weights w_{ij} of the created VC (see [13] for more details).

Tactile input from each of the eight infrared proximity sensors is scaled to $[0, 1]$ and stored in the weights w_{ij} of the created TC.

3.2   Unimodal and Multimodal Cells in Superior Colliculus

Intermodal correlations between visual and tactile input are established in the exploration phase using uni- and multimodal neurons inspired by the superior colliculus (SC) [2,3]. Sensory cells project to the input layer of SC which consists of unimodal visual (UVs) and tactile (UTs) cells. Those unimodal cells project to multimodal cells (MMs) in the output layer of SC. The architecture of our SC model is shown in Figure 2 (a).

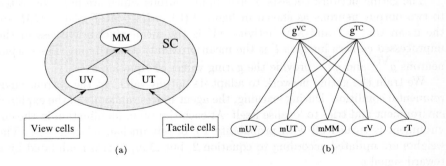

(a)   (b)

Fig. 2. Architecture of our superior colliculus model (a) and the gating network (b)

Whenever the agent receives strong visual and tactile input simultaneously, it creates a tactile and a visual unimodal cell. Synapses between sensory cells (TCs and UTs) are established and adapted using a Hebbian learning rule

$$\Delta w_{ij} = \eta\, r_i(r_j - w_{ij}) \tag{2}$$

where r_i is the postsynaptic unimodal cell, r_j is the sensory cell and η the learning rate. The same happens for VCs connecting to UVs. The firing rate r_i of an unimodal cell is given by the weighted mean activity of its presynaptic neurons j.

$$r_i = \frac{\sum_j w_{ij} r_j}{\sum_j w_{ij}} \tag{3}$$

The most active unimodal cells connect to a new multimodal output cell and synapses are learnt according to equation 2. The firing rate r_i of a multimodal cell i differs from equation 3 in that both UTs and UVs need to be active to trigger the firing of a multimodal cell

$$r_i = \tanh \left(k \; (\frac{\sum_{j \in UV} w_{ij} r_j}{\sum_{j \in UV} w_{ij}})(\frac{\sum_{j \in UT} w_{ij} r_j}{\sum_{j \in UT} w_{ij}}) \right) \tag{4}$$

where k is a constant.

3.3   Learning the Gating Network

During exploration, the most active sensory cells establish connections to a newly created allothetic place cell. The synaptic strengths evolve according to equation 2.

The firing rate r_i of an allothetic place cell i is the weighted mean activity of its presynaptic neurons j where all inputs from the same modality are collectively modulated by the gating value g^{VC} or g^{TC} respectively.

$$r_i = g^{VC} \left(\frac{\sum_{j \in VC} w_{ij} r_j}{\sum_{j \in VC} w_{ij}} \right) + g^{TC} \left(\frac{\sum_{j \in TC} w_{ij} r_j}{\sum_{j \in TC} w_{ij}} \right) \tag{5}$$

The gating network consists of five input neurons which are fully connected to two output neurons as shown in figure 2 (b). mUV, mUT and mMM are the mean UV, UT and MM activity. rV is the mean pixel brightness of the unprocessed camera image. rT is the mean proximity sensor input. The output neurons g^{VC} and g^{TC} provide the gating values of equation 5

We train the gating network to adapt its output values to the current environmental condition. During learning, the agent moves randomly in the explored environment and tries to localise itself. At each timestep, illumination in the environment is turned off with probability P_L or left unchanged otherwise. The weights are updated according to equation 2, but Δw_{ij} is also modulated by a reward signal q.

The reward q depends on two properties of the allothetic place code: (a) variance around centre of mass σ_{pc} and (b) population activity act_{pc}. Positive reward is given for compact place cell activity (ie. small variance σ_{pc}) and reasonable mean population activity. Negative reward corresponds to very disperse place coding or low activity. The equation for the reward q is as follows

$$q = \left[\exp(-\frac{(\sigma_{pc} - \sigma_{opt})^2}{2\sigma_{size}^2}) - d_{size} \right] \left[\exp(-\frac{(act_{pc} - act_{opt})^2}{2\sigma_{act}^2}) - d_{act} \right] \tag{6}$$

where $\sigma_{opt}, \sigma_{size}, d_{size}, act_{opt}, \sigma_{act}, d_{act}$ are constants.

4   Results and Conclusions

Experiments are conducted on a Khepera mobile robot. An $80 \times 80cm$ boarded arena placed on a table in a normal office serves as environment. A rectangular-shaped object is placed in the arena to increase the amount of tactile input to the system.

Figure 3 (a) shows the gating values for the visual and tactile senses after learning the gating network. Most of the time, visual input is the only activated modality. Everytime the robot is near an obstacle however, the tactile sense is assigned a slightly higher importance than vision. The abrupt changes are due to the binary nature of the tactile sensors.

Figure 3 (b) shows the gate values when the illumination is reduced by 80%. Most of the time, vision and tactile senses receive equal importance. Whenever an obstacle is near, however, the agent relies mostly on its tactile input.

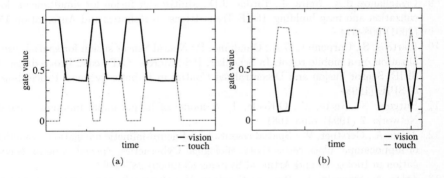

Fig. 3. Gate values in openfield and border positions. (a) good illumination. (b) almost no light

The main difficulty in learning the importance of sensory input lies in determining the reliability and uncertainty of a percept. We use the mean place cell activity and the activity variance around the centre of mass as a quality measure to change the weights of the gating network. Accessing the variance in spatial representations might be difficult to motivate biologically. Plausible neural mechanisms that measure place code accuracy should be found.

The brain certainly uses various methods to evaluate the relevance of sensorial input. We are working on more biologically plausible methods to assess place code quality.

References

1. O'Keefe, J., Nadel, L.: The Hippocampus as a Cognitive Map. Clarendon Press, Oxford (1978)

2. Stein, B.E., Meredith, M.A.: The Merging of the Senses. MIT Press, Cambridge, Massachusetts (1993)
3. Cooper, B.G., Miya, D.Y., Mizumori, S.J.Y.: Superior colliculus and active navigation: Role of visual and non-visual cues in controlling cellular representations of space. Hippocampus 8 (1998) 340–372
4. Natsume, K., Hallworth, N.E., Szgatti, T.L., Bland, B.H.: Hippocampal thetarelated cellular activity in the superior colliculus of the urethane-anesthetized rat. Hippocampus **9** (1999) 500–509
5. Quirk, G.J., Muller, R.U., Kubie, J.L.: The firing of hippocampal place cells in the dark depends on the rat's recent experience. Journal of Neuroscience **10** (1990) 2008–2017
6. Save, E., Cressant, A., Thinus-Blanc, C., Poucet, B.: Spatial firing of hippocampal place cells in blind rats. Journal of Neuroscience **18** (1998) 1818–1826
7. Save, E., Nerad, L., Poucet, B.: Contribution of multiple sensory information to place field stability in hippocampal place cells. Hippocampus **10** (2000) 64–76
8. Elfes, A.: Using occupancy grids for mobile robot perception and navigation. IEEE Computer **22** (1998) 46–57
9. Castellanos, J.A., Neira, J., Tardos, J.D.: Multisensor fusion for simultaneous localization and map building. IEEE Transattions on Robotics and Automation **17** (2001) 908–914
10. Martens, S., Carpenter, G.A., Gaudiano, P.: Neural Sensor susion for spatial visualization on a mobile robot. In Schenker, P.S., McKee, G.T., eds.: Proceedings of SPIE, Sensor Fusion and Decentralized Control in Robotic Systems, Proceedings of SPIE (1998)
11. Burgess, N., Recte, M., O'Keefe, J.: A model of hippocampal function. Neural Networks **7** (1994) 1065–1081
12. Arleo, A., Gerstner, W.: Spatial cognition and neuro-mimetic navigation: A model of hippocampal place cell activity. Biological Cybernetics, Special Issue on Navigation in Biological and Artificial Systems 83 (2000) 287–299
13. Arleo, A., Smeraldi, F., Hug, S., Gerstner, W.: Place cells and spatial navigation based on 2d visual feature extraction, path integration, and reinforcement learning. In Leen, T.K., Dietterich, T.G., Tresp, V., eds.: Advances in Neural Information Processing Systems 13, MIT Press (2001) 89–95

A Neural Network Model Generating Invariance for Visual Distance

Rüdiger Kupper and Reinhard Eckhorn

Philipps-University Marburg, Neurophysics Group, D-35037 Marburg, Germany

Abstract. We present a neural network mechanism allowing for distance-invariant recognition of visual objects. The term *distance-invariance* refers to the toleration of changes in retinal image size that are due to varying view distances, as opposed to varying real-world object size. We propose a biologically plausible network model, based on the recently demonstrated spike-rate modulations of large numbers of neurons in striate and extra-striate visual cortex by viewing distance. In this context, we introduce the concept of *distance complex cells*. Our model demonstrates the capability of distance-invariant object recognition, and of resolving conflicts that other approaches to size-invariant recognition do not address.

1 Introduction

Invariant object recognition means the ability of the human visual system to recognize familiar objects appearing in varying poses in the visual field, such as varying position, size, or three-dimensional view. It can be argued that positional invariance may mostly be achieved by fixational eye movements. Nonetheless, some sort of neural computation must be performed along the ventral pathway, to achieve invariance to size, view, or other transformations involving a change in retinal projection. Among these, transformation of size plays a special role, as it includes changes solely in extent, but not in shape of the visual input.

1.1 Size-Invariance vs. Distance-Invariance

Size-invariant object recognition demands closer investigation, regarding the possible causes that make a familiar shape appear in different sizes on the retina:

Viewing Distance. One reason for varying retinal image extent is, that the same or identical objects appear at different viewing distances. A possible source for this type of size variation is a change in object-observer distance. The resulting images are perceived as being instances of the very same object even if there are huge differences in the extent of their retinal projections. We will refer to this type of invariant recognition as distance-invariance. It is unconsciously perceived.

J.R. Dorronsoro (Ed.): ICANN 2002, LNCS 2415, pp. 93–98, 2002.

Real-world object size. Another reason for varying retinal image extent can be, that the observer is facing different objects of the same shape, but of different physical size, measured in real-world coordinates (e.g. a car and its toy model). These are perceived as being different objects, though possibly belonging to a common class.

Under normal viewing conditions, the two types of size variation can perceptually be well distinguished, although the images at the retina may be identical. The retinal image of a nearby toy car can match that of a far away real car. Nonetheless, perceptually the two objects are not confused. That is, *invariant recognition is achieved in the case of varying viewing distance, but not for varying real-world size*. We regard this as being of major importance. To our knowledge, this difference is not accounted for by other models of size invariant object recognition, making use of neurally implemented two-dimensional image transformations [7,11], or of cascaded local feature pooling [9,6]. There is, however, first evidence, that the distinction between those two viewing modes is also based on neural properties recently found in V1, V2 and V4 of monkeys [4,10].

2   Distance Estimation by the Visual System

2.1   Psychophysical Evidence

Psychophysical investigations give rise to the notion of separate neural modules engaged in the judgment of physical size, shape, and distance [2,3]. While these modules seem to operate largely independent from each other, consistency between the judgments is normally maintained by the consistency in the input from the environment and by the common use of certain measures [2]. One of these common measures apparently is a neural representation of the viewing distance [3], as can be provided from a variety of sources, including ocular vergence, lens accommodation, angle below horizon [8], or pictorial cues such as contrast, texture gradient and motion parallax. It is found that information on real-world size might be present as early in the visual pathway as at the level of primary visual filters, i.e., the striate visual cortex (V1) [1]. Furthermore, learning that is specific to visual objects transfers effortlessly across changes in image size [5].

2.2   Physiological Evidence

Distance dependent modulation of single cell spike rates has been found to high abundance (64–85% of neurons) in visual cortical areas V1, V2, and V4, making it a very common property of cells in the ventral pathway [4,10]. While tuning profiles to two-dimensional stimulus properties (edge orientation, contrast, speed, etc.) stay unchanged, the cells exhibit a modulation of firing rate with fixation distance [4]. The results can be interpreted as viewing distance being a further property coded by the respective neurons, in addition to the preferred classical receptive field properties.

What functional purpose could the modulation of such large portions of neurons along the ventral pathway serve? We suggest, that viewing distance information is used to select subsets of local feature detectors, which represent visual elements at a preferred viewing distance. The actual object representation is then made up from the neural signals of detectors coding for the fixation distance currently in question.

3   Model

3.1   Extending the Concept of Visual Features

A widely accepted class of network models, known as hierarchical models for object recognition, adopt the view that increasingly complex features constitute the representation of objects in the visual system. Our present model, belonging to this class, extends this concept by introducing the experimental finding of spike rate modulation by viewing distance into the hierarchy. Our model consists of the following parts (or modules, Fig. 1), A) a linear neural chain representing the current fixation distance by a single activated blob, B) distance modulated feature detectors, C) distance complex cells, and D) an object knowledge base.

A: Neural Chain Representing Distance. The exact origin and type of the distance signal is unknown (see Sect. 2.1 for suggestions). We model its action by a linear chain of coupled neurons, like a discretized one-dimensional neural field, in which the position of a single activation blob represents the current distance estimate of the ocularly fixated object (Fig. 1, A).

B: Distance Modulated Feature Detectors. The retinal image is represented by the activation of low- and higher-level visual filters, each coding for their preferred features. Coding for distance is introduced here by modulating their activities by a distance signal [4,10] (see Fig. 1, A and B). Owing to this modulation, feature detectors of *compatible* distance preference will predominantly represent the visual scene, while activity of *incompatible* detectors is diminished. Their distance tuning corresponds to the activation blob in (A:).

C: Distance Complex Cells. Feature detector signals converge systematically onto next stage neurons, yielding what we term *distance complex cells*. Input to a single distance complex cell is provided by all those feature detectors that are activated by the same object element (e.g. a part of a contour or a surface) when it approaches or moves away from the observer fixating this object. This means that the receptive-field-properties of distance complex cells reflect the distance-variant transformation that a distinct visual feature undergoes, when the distance between observer and object changes (see Fig. 1, B and C, connections shown for one cell only). Throughout such a movement, the same distance complex cell would be kept activated. Maximum pooling [9] is used to disambiguate the input to the distance complex cells, if distance tuning of the input feature detectors to these complex cells is broad.

5   Discussion

The presented model belongs to the class of hierarchical models for object recognition. These are known to produce invariance when constructed accordingly [9, 6], but do so implicitly, losing information, e.g. on object size and position, during recognition. Other models use control signals to set up image transformations [7,11], but act in the two-dimensional domain, unable to exploit distance information to resolve viewing conflicts. Our model can be seen as an extension to both strategies, using pooling operations guided by an explicit distance signal.

A possible drawback of our approach is the large number of required feature detectors. Detectors need to be present, which share the same preferred two-dimensional feature, but are modulated differently with viewing distance. The quality of invariance generation depends on the width and overlap of tuning profiles. We will investigate, to what multiplicity detectors are required to allow for stable operation, and what constraints are imposed thereon by biological cell numbers.

Note, finally, that this is only a partial approach to invariant object recognition, addressing distance- (or size-) invariance, but it is likely to blend well with other mechanisms for invariant recognition. Many more setups of our model can be investigated to test the proposed mechanism, including attention to distance and real-world size, attention to known objects, and operation in reduced cue environments (i.e., no distance signal available).

References

1. Bennett, P. J., Cortese, F.: Masking of spatial frequency in visual memory depends on distal, not retinal frequency. Vis. Res. **36(2)** (1996) 233–238
2. Brenner, E., van Damme, W. J. M.: Perceived distance, shape and size. Vis. Res. **39** (1999) 975–986
3. van Damme, W. J. M., Brenner, E.: The distance used for scaling disparities is the same as the one used for scaling retinal size. Vis. Res. **37(6)** (1997) 757–764
4. Dobbins, A. C., Jeo, R. M., Fiser, J., Allman, J. M.: Distance modulation of neural activity in the visual cortex. Science **281** (1998) 552–555
5. Furmanski, C. S., Engel, S. A.: Perceptual learning in object recognition: object specificity and size invariance. Vis. Res. **40** (2000) 473–484
6. Mel, B. W., Fiser, J.: Minimizing binding errors using learned conjunctive features. Neur. Comp. **12** (2000) 247–278
7. Olshausen, B. A., Anderson, C. H., van Essen, D. C.: A neurobiological model of visual attention and invariant pattern recognition based on dynamic routing of information. J. Neurosci. **13(11)** (1993) 4700–4719
8. Ooi, T. L., Wu, B., He, Z. J.: Distance determined by the angular declination below the horizon. Nature **414** (2001) 197–200
9. Riesenhuber, M., Poggio, T.: Hierarchical models of object recognition in cortex. Nature Neurosci. **2(11)** (1999) 1019–1025
10. Rosenbluth, D., Allman, J. M.: The effect of gaze angle and fixation distance on the responses of neurons in V1, V2, and V4. Neuron **33** (2002) 143–149
11. Salinas, E., Abbott, L.: Invariant visual responses from attentional gain fields. J. Neurophysiol. **77** (1997) 3267–3272

Modeling Neural Control of Locomotion: Integration of Reflex Circuits with CPG

Ilya A. Rybak[1,3], Dmitry G. Ivashko[1], Boris I. Prilutsky[2], M. Anthony Lewis[3], and John K. Chapin[4]

[1] School of Biomedical Engineering, Science and Health Systems, Drexel University,
Philadelphia, PA 19104, USA
{rybak, ivashko}@cbis.ece.drexel.edu
[2] Center for Human Movement Studies, Georgia Institute of Technology,
281 Ferst Drive, Atlanta, GA 30332, USA
boris.prilutsky@hps.gatech.edu
[3] Iguana Robotics, Inc., P.O. Box 628, Mahomet, IL 61853, USA
tlewis@iguana-robotics.com
[4] Department of Physiology and Pharmacology, Health Science Center at Brooklyn,
State University of New York, 450 Clarkson Avenue, Brooklyn, NY 11203, USA
john_chapin@netmail.hscbklyn.edu

Abstract. A model of the spinal cord neural circuitry for control of cat hindlimb movements during locomotion was developed. The neural circuitry in the spinal cord was modeled as a network of interacting neuronal modules (NMs). All neurons were modeled in Hodgkin-Huxley style. Each NM included an α-motoneuron, Renshaw, Ia and Ib interneurons, and two interneurons associated with the central pattern generator (CPG). The CPG was integrated with reflex circuits. Each three-joint hindlimb was actuated by nine one- and two-joint muscles. Our simulation allowed us to find (and hence to suggest) an architecture of network connections within and between the NMs and a schematic of feedback connections to the spinal cord neural circuitry from muscles (Ia and Ib types) and touch sensors that provided a stable locomotion with different gaits, realistic patterns of muscle activation, and kinematics of limb movements.

1 Introduction

The central nervous system controls locomotion and other automatic movements in a hierarchical fashion. The lower-level controller in the spinal cord generates the motor program for the neuromuscular apparatus. This low-level controller interacts with proprioceptive feedback and receives descending signals from the higher-level (supra-spinal) centers. The higher centers, in turn, select and initiate the appropriate motor program from the repertoire of the low-level controller (spinal cord). The descending commands from supra-spinal centers to spinal interneurons are automatically integrated into the current state of proprioceptive and exteroceptive information [6].

The neuronal circuits in the mammalian spinal cord can generate rhythmic motor patterns that drive locomotor movements even in the absence of descending inputs from higher brain centers and sensory feedback [2], [5]. This supports the concept of

J.R. Dorronsoro (Ed.): ICANN 2002, LNCS 2415, pp. 99–104, 2002.
© Springer-Verlag Berlin Heidelberg 2002

the *central pattern generator* (CPG), which presumably is located in the spinal cord and generates a basic locomotor rhythm (for review see [4]). According to the contemporary biological view, the CPG is a complex, distributed network of interneurons in the spinal cord integrated into the system of multiple reflex circuits [6]. The basic locomotor pattern, generated by the CPG, provides a coordinated activation of functionally different muscles, which in turn control and coordinate joint movement within and between the limbs. Therefore, locomotion results from a complex interplay between the CPG, reflex circuits and multiple feedback and feedforward modulatory signals. The proprioceptive signals strongly influence the locomotor rhythm by providing necessary correction of the locomotor rhythm and pattern to maintain the walking animal in a proper relationship to the environment [12]. They regulate the timing of phase transitions and reinforce the generation of motoneuronal activity during ongoing phases of locomotion [9]. Previous modeling studies have demonstrated that a stable and adaptive locomotion involves a global entrainment between the CPGs and musculoskeletal system [7], [14]. The objective of this work was to develop and analyze a comprehensive model of neural control of locomotion at the spinal cord level using realistic models of the network of neurons (in the Hodgkin-Huxley style), muscles, and limb biomechanics.

2   Model

The neural model of the locomotory CPG was constructed using the hypothesis that each limb is controlled by one complex CPG, which in turn is connected with the other CPGs via a coordinating neural network [8]. The CPG was incorporated into the spinal cord neural circuitry and integrated with the circuits of spinal reflexes via direct synaptic interconnections and through multiple proprioceptive feedbacks. The schematic of reflex circuits was modified from the previous models, [1] and [3], and applied to each antagonistic group of muscles. Each hindlimb was modeled as a system of three rigid segments interconnected by three joints: hip, knee and ankle. Two hindlimbs were connected to the common segment (pelvis) (Fig. 1A). A trunk segment was connected with the pelvis. The distal end of the trunk was held at the necessary distance from the ground to compensate for the lack of forelimbs. Each hindlimb was controlled by nine one- and two-joint muscles. The dynamics of muscle contraction was described by a Hill-type model that incorporates the muscle force-length-velocity properties, muscle geometry, and the properties of the tendon.

The exact network of interneurons in the mammalian spinal cord responsible for the generation of the basic locomotor rhythm has not been identified yet. Therefore, in addition to the existing data on the spinal cord neural architecture, we used the suggestion that the mechanism for the locomotor pattern generation in the spinal cord is functionally similar to the brainstem mechanisms providing generation and control of the respiratory motor pattern [8]. Specifically, we assumed that some general architectural principles and particular neural schematics discovered in studies of the respiratory CPG (e.g. those for phase transitions) might be useful and applicable for the construction of the locomotory CPG (see also [8], [12]). In respect to the respiratory CPG, both experimental [11] and modeling [13] studies have demonstrated that, in addition to the "principal" CPG elements (whose activity explicitly defines each phase of the cycle), the CPG may contain special "switching" neural elements that fire during phase transitions and, in fact, produce these transitions via inhibition of the corresponding principal CPG elements. Moreover, it appears that the switching interneurons operate (fire) under control of various proprioceptive and descending control signals and hence significantly contribute to the shaping of the locomotor pattern (timing of phase transitions, shaping motoneuronal firing busts, etc.).

The developed model of the spinal cord neural circuitry has a modular structure. The schematic of a single Neuronal Module (NM) is shown in Fig. 2A. This schematic is considered as a minimal network structure necessary for integration of basic reflexes with the CPG. The NM contains a part of the reflex circuitry and two CPG elements. Each NM controls one muscle. Specifically, the NM includes the output α-motoneuron (α-MN), that actuates the controlled muscle, and several interneurons, including the Renshaw cell (R-In), Ia interneuron (Ia-In) receiving Ia proprioceptive feedback, Ib interneuron (Ib-In) receiving force-dependent Ib proprioceptive feedback, and two interneurons associated with the locomotory CPG. The CPG elements within the NM include the principal CPG neuron (CPG-N) providing activation to the α-MN and the switching interneuron (CPG-In) controlling the principal CPG neuron. The entire neural circuitry for control of locomotion comprises a network of NMs (interconnected directly and via mutual proprioceptive afferents). The CPG, in turn, is formed as a network of all CPG elements located in all participating NMs. Fig. 2B shows an example of two interconnected NMs, controlling a pair of antagonistic flexor and extensor muscles actuating the same joint. The synaptic connections within and between the NMs and the structure of Ia and Ib proprioceptive afferents provide for the classical flexor and extensor stretch reflexes.

The locomotor movement (see Fig. 1B) could be initiated by applying the "descending" drive to all principal CPG neurons (CPG-Ns). Switching the locomotor phases was performed by the firing of the corresponding "switching" CPG interneuron (CPG-In). The timing of phase transitions was controlled by multiple control signals (Ia, Ib, touch sensors) to the "switching" CPG-Ins. These signals provided a necessary adjustment of the duration of each locomotor phase. Interestingly, during the extension phase of locomotion ("stance"), the active extensor CPG-N neuron inhibited the extensor Ib neuron (Ib-In) and hence broke the "classical" negative feedback loop of Ib fibers to the extensor α-motoneurons (α-Mn) (see left side of Fig. 2B). At the same time, the same extensor CPG-N neuron received input from Ib fibers and provided excitation of the extensor α-Mn. Therefore during locomotion the Ib feedback loop to the extensor α-Mn changed from negative to the positive, which is consistent with the experimental data [9], [10].

The developed model was able to provide control of stable locomotor movements (see Fig. 1B). The model demonstrated the flexibility necessary for the adaptive adjustment of locomotor movements to characteristics of the environment.

The above modeling studies allowed us to hypothesize the possible structure of the locomotory CPG, the architecture of network connections within the spinal cord circuitry, and the schematic of feedback connections, which may be tested in further experimental studies.

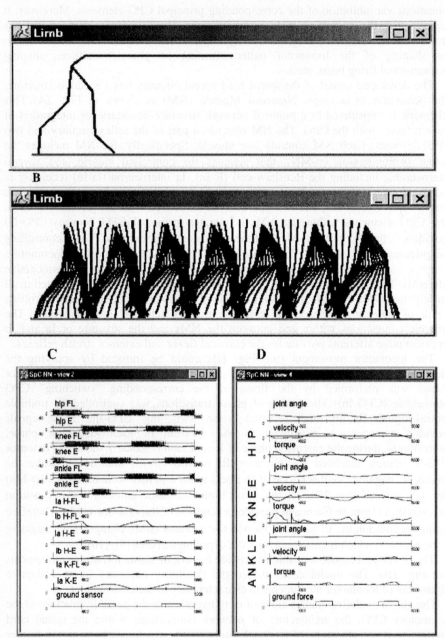

Fig. 1. A. The model of two hindlimbs with the trunk. **B.** Stick diagram of movement of one hindlimb. **C.** Activity of selected motoneurons and proprioceptive feedbacks. **D.** Dynamics of some key biomechanical variables.

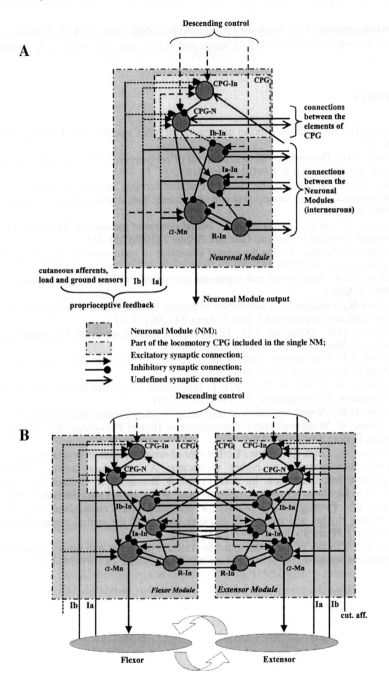

Fig. 2. A. Neuronal Module (NM) of CPG-based neural sub-system. **B**. Two NMs controlling a pair of antagonistic muscles.

clear that, whatever its exact relationship with the underlying neuronal activity, the BOLD fMRI signal, due to its vascular origin, can only reflect a spatial average of the neuronal signal with a resolution that at best is of the order of fractions of mm. Information is transmitted in the brain by the spiking electrical activity of populations of individual neurons. Characterizing neuronal information processing from a spatially averaged signal may lead to distortions if the region of the cortex over which the average is performed contains neurons with dissimilar stimulus-response profiles. In this paper we try to shed some light on this issue by investigating how spatially averaging the neuronal signal may influence the reconstruction of sensory representations in the cortex from 'computational' fMRI experiments. We make use of a computational model to address quantitatively the relationships between the information carried by neuronal population firing, the changes in the average of the population activity over a voxel, and the information carried by the voxel averaged population activity. We then use these relationships to investigate how the stimulus encoding properties of the underlying neuronal population can be inferred from the measurement of the BOLD signal.

In this work we do not consider explicitly the nature of the relationship between the imaging signal and the average activity of the voxel neuronal population. We focus on the relationship between the average activity and the encoded information, which is already complex enough. If the relationship between the imaging signal and the average activity is linear then the results of this work apply directly to the relationship between the imaging signal and the encoded information. Under linearity the signal change is proportional to the average activity change and the signal information equals the average activity information. Otherwise one has to take into account non-linearities between signal and average activity. Furthermore, a more general approach should include additional sources of noise in the signal apart from the noise in the average activity.

In a previous work [6] we investigated the suitability of estimating the information that a certain voxel conveys about two different stimuli using the fMRI signal. We used Fisher information as measure of the information which is a stimulus dependent quantity. In this occasion we explore the possibility of using the imaging signal to compare the information that two different voxels convey about a certain stimulus set. In the present case we calculate Shannon mutual information which is a measure of the information conveyed about the whole set of stimuli.

2   The Model

We defined the neuronal response as the number of spikes fired by the neuron in a given time window. We hence ignored non-stationarities in the neuronal response which may have some role on the dynamics of blood oxygenation changes [3]. The neuronal response was assumed to follow a Poisson distribution whose mean followed a Gaussian tuning function $f(\theta)$:

$$f(\theta) = m \exp\left(-\frac{(\theta - \theta_p)^2}{2\sigma_f^2}\right) + b \tag{1}$$

The tuning function (1) was characterized by the following parameters: preferred stimulus (θ_p), stimulus modulation (m), spontaneous firing rate (b), and width of the tuning function (σ_f). This Gaussian tuning curve model is a good description of the tuning properties of e.g. MT neurons to angular variables, such as motion direction. In addition, we assumed that the distribution of preferred stimuli of the neurons within a voxel was a Gaussian of width σ_p and centered at a certain preferred orientation $\hat{\theta}_p$:

$$p(\theta_p) = \frac{1}{\sqrt{2\pi\sigma_p^2}} \exp\left(-\frac{(\theta_p - \hat{\theta}_p)^2}{2\sigma_p^2}\right) \tag{2}$$

This models how neurons are organized, for example, in some sensory areas, where neurons with similar response properties are clustered together into columns.

3   Quantities

In order to quantify the information encoded in the neuronal responses about the set of stimuli we use Shannon mutual information which is defined as follows [2]:

$$I = \int ds dr \rho(\theta, r) log \frac{\rho(\theta, r)}{\rho(\theta)\rho(r)} \tag{3}$$

$\rho(\theta, r)$ is the joint probability distribution function of the stimulus and the response, while $\rho(\theta)$ and $\rho(r)$ are the marginal probability distributions of stimulus and response respectively.

In our model neurons fire independently of each other given the stimulus and each voxel has a large number of neurons. In this case the information can be expressed as [1]:

$$I = \int d\theta \rho(\theta) \log\left(\frac{1}{\rho(\theta)}\right) - \int d\theta \rho(\theta) \log\left(\sqrt{\frac{2\pi e}{J_{voxel}(\theta)}}\right) \tag{4}$$

where (see [6])

$$J_{voxel}(|\theta - \hat{\theta}_p|) = \int d\theta_p N p(|\hat{\theta}_p - \theta_p|) J_{neuron}(|\theta - \theta_p|) \tag{5}$$

$$J_{neuron}(\theta) = \frac{\left(\frac{\partial}{\partial\theta} f(|\theta - \theta_p|)\right)^2}{f(|\theta - \theta_p|)} \tau \tag{6}$$

N is the number of neurons in the population and τ the time window in which spikes are counted. $\rho(\theta)$ is taken to be constant between $-\theta_{lim}$ and θ_{lim}. J_{voxel} is the Fisher information jointly encoded in the activity of all the single neurons in the voxel. Since, for our case, neurons fire independently given the stimulus it can be expressed as the sum of the Fisher information conveyed by each neuron J_{neuron} which takes the above form for a Poisson process.

To understand the effect of the spatial average of the neuronal signal intrinsic in fMRI measures, it is useful to compare the information carried by the ensemble of neurons in the voxel, Eq. (4), to quantities constructed from the measures of the fMRI signal. Two quantities are of obvious interest. The first quantity is the change of the averaged activity with respect to the baseline firing rate. We evaluate it for the preferred stimulus of the voxel, $\hat{\theta}_p$

$$\Delta r = (f_{voxel}(\hat{\theta}_p) - Nb) \tag{7}$$

where f_{voxel} is the averaged activity in the voxel:

$$f_{voxel}(\hat{\theta}_p) = \int d\theta_p p(|\hat{\theta}_p - \theta_p|) f(|\hat{\theta}_p - \theta_p|) \tag{8}$$

In the present work we assume that Δr is proportional to the percentage changes in the BOLD signal, and hence corresponds to the traditional way to quantify the fMRI response to different stimuli.

To calculate the mutual information carried by the average activity we apply [1]:

$$I_{mr} = \int d\nu \rho(\nu) log \frac{1}{\rho(\nu)} - \int d\nu \rho(\nu) \log \left(\sqrt{2\pi e \frac{\nu}{\tau}} \right) \tag{9}$$

where ν is the averaged firing rate of the voxel. This expression applies when the response is continuous and the noise is low. Since the number of neurons in the voxel is large and a large number of spikes are fired in a finite time window the average activity can be taken as continuous and we are in the limit of low noise. $\rho(\nu)$ can be calculated from the probability of stimulus presentation $\rho(\theta)$ and the tuning function $f_{voxel}(\theta)$

$$\rho(\nu) = \int d\theta \rho(\theta) \delta(\nu - f_{voxel}(\theta)) \tag{10}$$

Instead of assessing the goodness of the estimation of I with Δr and I_{mr} we will compare $\exp(I)$ with Δr and $\exp(I_{mr})$. The reason for this is that in our model I grows logarithmically with the number of neurons N whereas $\Delta\nu$ grows linearly with N.

Obviously the comparison of the spatially averaged fMRI quantities with the information transmitted by the neuronal populations is going to tell us how the quantities derived from fMRI experiments may be used to draw conclusions about the information transmitted by the neurons generating the BOLD signal. Since the relationship between mean firing rates and tuning parameters is complex [8], and the relationship between tuning parameters and information is complex [10], we expected that the relationship between mean firing rates and information processing could be very complex as well, and worth a systematic investigation.

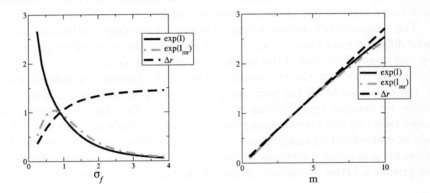

Fig. 1. Exponential of mutual information (*solid line*), exponential of signal information (*dotted-dashed line*) and signal change (*dashed line*) for the population of neurons in a voxel. The units in the y-axis are arbitrary and different for the 3 curves to allow for a better assessment of their proportionality. *Left plot:* The information and the signal change behave very differently as a function of the tuning width σ_f. The signal information is approximately proportional to the information only for high σ_f. *Right plot:* The three curves behave similarly vs the stimulus modulation m. The values of the parameters were $b = 1$, $\sigma_p = 1$, $\theta_{lim} = 2$, $\sigma_f = 1$ (*right plot*), m=4 (*left plot*).

4   Results and Conclusions

We used the formulae above to investigate whether the voxel that undergoes a higher activity change also corresponds to the voxel that conveys more information about the stimuli. In Figure 1 we plot some examples of our results in this investigation. I, Δr and I_{mr} are plotted as a function of σ_f, the width of the neuronal tuning function (*left plot*), and m the modulation (*right plot*). I and I_{mr} are not expressed in the same units so that they have similar values and it is easier to asses their proportionality. Except for the normalization I_{mr} is always equal or smaller than I because the average can always be obtained from the individual firing rates.

The information, I and the signal change Δr show a different dependence on the neuronal tuning width. This implies that the signal change is not a good predictor of the information when comparing the information encoded by voxels with different neuronal tuning widths. While the signal change is larger for voxels with high σ_f the voxels which encode more information are those with small σ_f. The signal information, on the other hand, is approximately proportional to the information for large enough σ_f.

The situation is different if we want to compare the information transmitted by voxels with different neuronal modulation (*right plot*). In this case both the signal information and the signal change are proportional to a good approximation to the information and can thus be used as estimators to compare the information carried by two different voxels with different values of the stimulus modulation and the rest of the parameters equal.

The relationship between information and its candidate estimators is even more different from linear if we consider the dependence on other parameters (eg. σ_p) or combinations of the parameters. (data not shown).

We conclude that the relationship between the information carried jointly by the responses of all the neurons in the population and the imaging signal change or imaging signal information is complex. Therefore a computational model based on the known tuning properties of the single neurons can greatly help in understanding imaging results in terms of information processing. Furthermore, additional quantities to the signal change, as the signal information can provide a better estimation of the information in some cases.

Acknowledgments. We thank K. Friston for many useful discussions. This research was supported by the Wellcome Trust, by the Comunidad Autonoma de Madrid and by an MRC Research Fellowship to SP.

References

1. N. Brunel and J.-P. Nadal, Mutual information, Fisher information and population coding. Neural Computation 10 (1998) 1731-1757.
2. T. Cover and J. Thomas, Elements of information theory (1991). John Wiley.
3. K.J. Friston, A. Mechelli, R. Turner and C.J. Price, Nonlinear responses in fMRI: Balloon models, Volterra kernels and other hemodynamics. Neuroimage 12 (2000) 466-477.
4. D.J. Heeger, A.C. Huck, W.S. Geisler and D.G. Albrecht, Spikes versus BOLD: what does neuroimaging tell us about neuronal activity? Nature Neuroscience 3 (2000) 631–633
5. N.K. Logothetis, J. Pauls, M. Augath, T. Trinath and A. Oeltermann, Neurophysiological investigation of the basis of the fMRI signal, Nature 412 (2001) 150–157
6. A. Nevado, M. P. Young and S. Panzeri, Functional imaging and information processing. Neurocomputing, in press.
7. G. Rees, K. Friston and C. Koch, A direct quantitative relationship between the functional properties of human and macaque V5. Nature Neuroscience 3 (2000) 716–723
8. J.W. Scannell and M.P. Young, Neuronal population activity and functional imaging, Proc. R. Soc. B 266 (1999) 875–881
9. B.A. Wandell, Computational neuroimaging of human visual cortex. Ann. Rev. Neurosci. 22 (1999) 145–173.
10. K. Zhang and T.J. Sejnowski, Neuronal tuning: to sharpen or to broaden? Neural Computation 11 (1999) 75–84.

Mean-Field Population Dynamics of Spiking Neurons with Random Synaptic Delays

Maurizio Mattia and Paolo Del Giudice

Physics Laboratory, Istituto Superiore di Sanità,
v.le Regina Elena 299, 00161 Roma, Italy,
{mattia,paolo.delgiudice}@iss.infn.it
http://neural.iss.infn.it

Abstract. We derive a dynamical equation for the spike emission rate $\nu(t)$ of a homogeneous population of Integrate-and-Fire (IF) neurons, in an "extended" mean-field approximation (*i.e.*, taking into account both the mean and the variance of the afferent current). Conditions for stability and characteristic times of the population transient response are investigated, and both are shown to be naturally expressed in terms of single neuron current-to-rate transfer function. Finite-size effects are incorporated by a stochastic extension of the mean-field equations and the associated Fokker-Planck formalism, and their implications for the frequency response of the population activity is illustrated through the power spectral density of $\nu(t)$. The role of synaptic delays in spike transmission is studied for an arbitrary distribution of delays.

1 Dynamic Mean-Field Equations for the Population Activity

In [7] we introduced a method to derive (*via* a population density approach, [6, 8]) the equations which govern the time evolution of the population activity of an interacting ensemble of IF neurons in the *extended* mean field approximation (*i.e.*, taking into account the instantaneous fluctuation of the afferent current). The general approach is also amenable to an approximate treatment, by which we could characterize the transient response and the power spectrum of the activity $\nu(t)$ of the neural population in a specific context, elucidating a close and interesting relationship between the time course of $\nu(t)$ and the "transfer function" characterizing the static mean field properties of the system. The synaptic transmission delays turn out to play a major role in the above analysis. In the present work we remove a somewhat unnatural constraint assumed in [7], that the synaptic delays are the same for all neurons, and we show how a *distribution* of synaptic delays can be easily embedded in the analysis, and briefly discuss some phenomenological implications.

In the diffusion approximation, the sub-threshold dynamics of the membrane depolarization V of a general class of IF neurons is given by [11] $\dot{V} = f(V) +$

J.R. Dorronsoro (Ed.): ICANN 2002, LNCS 2415, pp. 111–116, 2002.

$\mu(V,t) + \sigma(V,t)\Gamma(t)$, where the afferent current is described as a Gaussian white noise with mean $\mu(V,t)$ and variance $\sigma^2(V,t)$ and $f(V)$ is a leakage term.

In the *extended* mean field approach [3,2] all neurons in a homogeneous population are driven by stochastic currents with the same mean and variance, both depending on the recurrent ν and external ν_{ext} emission rates, equal for all neurons: $\mu = \mu(V, \nu, \nu_{ext})$ and $\sigma^2 = \sigma^2(V, \nu, \nu_{ext})$. In this approximation, the set of evolving Vs is seen as a sample of independent realizations drawn from the p.d.f $p(v,t)$, governed by the Fokker-Planck equation:

$$\partial_t p(v,t) = L\, p(v,t) = \left[-\partial_v \left[f(v) + \mu(v,t) \right] + \frac{1}{2} \partial_v^2 \sigma^2(v,t) \right] p(v,t), \qquad (1)$$

complemented by boundary conditions accounting for the realizations disappearing on the threshold and re-appearing at the reset potential H [10,1,6,4,5]. Since L depends on μ and σ^2, it is an implicit function of the emission rate ν. The latter in turn expresses the flux of realizations crossing the threshold for spike emission, or the fraction of neurons emitting spikes per unit time:

$$\nu(t) = -\frac{1}{2}\sigma^2(v,t)\, \partial_v\, p(v,t)\big|_{v=\theta} . \qquad (2)$$

It is convenient to expand $p(v,t)$ in Eqs. (1) and (2) onto the complete set of eigenfunctions ϕ_n of L: $p(v,t) = \sum_n a_n(t)\,\phi_n(v,t)$. In stationary conditions ν is the inverse of the mean inter-spike interval, and it also equals the single neuron transfer function $\Phi(\mu, \sigma^2)$, given by (2) with $p(v,t) = \phi_0(v)$, the eigenfunction of L with zero eigenvalue, stationary solution of Eq. (1). The time evolution of $p(v,t)$ is then described in terms of dynamical equations for the $a_n(t)$ (see e.g. [9,6]); taking into account Eq. (2), allows us to write the equations governing the time evolution of $\nu(t)$ (the "emission rate equation") as:

$$\begin{cases} \dot{a}(t) = [\Lambda(t-\delta) + C(t-\delta)\,\dot{\nu}(t-\delta)]\,a(t) + c(t-\delta)\,\dot{\nu}(t-\delta) \\ \nu(t) = \Phi(t-\delta) + f(t-\delta)\cdot a(t) \end{cases}, \qquad (3)$$

where f_n is the contribution to the flux due to the mode ϕ_n ($n \neq 0$); $c_n = \langle \partial_\nu \psi_n | \phi_0 \rangle$, $C_{nm} = \langle \partial_\nu \psi_n | \phi_m \rangle$; ψ_n are the eigenfunctions of the adjoint operator L^+ and $\langle . | . \rangle$ is a suitable inner product. C_{nm} and c_n are coupling terms, in that for uncoupled neurons μ and σ do not depend on the recurrent frequency ν, and $\partial_\nu \psi_n$ vanishes. Λ is the diagonal matrix of the common eigenvalues of L and L^+. For simplicity in Eq. (3) a single allowed synaptic delay δ appears. However, taking into account a *distribution* of delays is relevant both in order to relax a somewhat implausible assumption and because it might provide an effective treatment of non-instantaneous synaptic currents provoked by each spike (see [4]).

Going from a single δ to a distribution $\rho(\delta)$ amounts to substitute the convolution $\int \nu(t-\delta)\rho(\delta)d\delta$ for $\nu(t-\delta)$. We will show later the implications of a non-trivial $\rho(\delta)$.

2 Stability, Spectral Analysis, and Transients

The system (3) has fixed points ν_0 given by the self-consistency equation [3,2] $a = 0$ and $\nu_0 = \Phi(\nu_0)$; we assess their stability using a local analysis Eq. (3).

To this end we study the poles of the Laplace transform $\nu(s)$ of the linearized form of $\nu(t)$ in Eq. (3). The real and imaginary parts of these poles describe the characteristic times and the oscillatory properties of the collective activity $\nu(t)$.

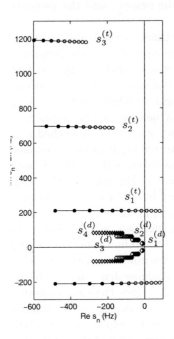

Fig. 1. Poles distribution for a recurrent inhibitory populations with different coupling strengths in a drift-dominated regime. Diamonds: first 4 diffusion poles $(s_n^{(d)})$; Circles: first 3 transmission poles $(s_n^{(t)})$ (poles are complex conjugate pairs). The darker the marker, the smaller (in module) the slope of the transfer function, and than the coupling strength. $s_1^{(t)}$ (with its complex conjugate) is the first pole crossing the imaginary axes, determining the instability of the population dynamics. For different coupling strengths the external currents are adjusted in order to have the same fixed point at $\nu_0 = 20Hz$. It is clearly seen in the Figure that the diffusion and the transmission poles move in opposite directions with respect to the imaginary axis, when Φ' is varied. For positive real part of the $s_n^{(t)}$ the local analysis is no longer suited to describe the dynamics [4].

Fig. 1 shows a subset of poles of $\nu(s)$ for an inhibitory population of linear (constant leakage) IF neurons ('LIF') [5]. Such poles can be grouped in two classes. The first is related to the transmission delays ("transmission poles", circles in Fig. 1), appearing only in coupled networks, approximately given by:

$$s_n^{(t)} \simeq \frac{1}{\delta} \ln |\Phi'| + i \frac{n\pi}{\delta}, \tag{4}$$

where n is any odd (even) integer for inhibitory (excitatory) populations and $\Phi' = \partial_\nu \Phi|_{\nu=\nu_0}$. The fixed point becomes unstable when $\mathrm{Re}(s_n^{(t)}) > 0$, which happens exactly when

$$\Phi'(\nu_0) > 1, \tag{5}$$

for an excitatory population. A sufficient condition for a weakly coupled inhibitory population to be stable is $\Phi'(\nu_0) > -1$. The "diffusion poles" $\{s_n^{(d)}\}$, (diamonds in Fig. 1) have negative real parts and do not contribute to the stability conditions. Far from instability $-1/\mathrm{Re}(s_n^{(d)})$ set the time scale of the transient of the network relaxation to the fixed point. For weak coupling, in a drift-dominated (supra-threshold) regime, $\{s_n^{(d)}\}$ are a perturbation of the eigenvalues λ_n of L; for the longest time scale:

$$s_1^{(d)} \simeq \lambda_1 \left(1 + \frac{f_1 c_1}{1 - \Phi' + \lambda_1 \delta}\right). \tag{6}$$

It is worth noting how, despite the fact that there is no obvious *a priori* relation between the single neuron properties and characteristic times, and the dynamics of the collective activity, the single neuron transfer function Φ emerges in a leading role in determining both the stability of the system, and the response times.

Fig. 2, left, shows how both the characteristic times of the transient population response to a stepwise change in the afferent current, and the frequency of the transient oscillations match well the predictions generated by the first diffusion poles. The (stationary and transient) spectral content of $\nu(t)$ is embodied in the power spectrum $P(\omega)$, a proper treatment of which requires taking into account the effects of the finite number N of neurons in the population under consideration. Besides incoherent fluctuations, *e.g.* due to quenched randomness in the neurons' connectivity and/or to external input (which are taken into account in the variance σ^2 of the afferent current entering the Fokker-Planck equation) two additional finite-size effects contribute: 1) For finite N, the number of spikes emitted per unit time is well described by a Poisson process with mean and variance $N\nu(t)$, such that the fraction ν_N of firing neurons is approximated by $\nu_N(t) = \nu(t) + \sqrt{\nu(t)/N}\Gamma(t)$, with Γ a memoryless, white noise [4]. In the mean field treatment, the fluctuating ν_N enters the infinitesimal mean and variance of the afferent current, so that μ_N and σ_N^2 are now stochastic variables, thereby making the Fokker-Planck operator itself N-dependent and stochastic (L_N). 2) While the above transition $L \to L_N$ still describes the time evolution of an *infinite* number of realizations, though driven by N-dependent fluctuating currents, the finite size of the neurons sample has to be explicitly taken into account in the boundary condition expressing the conservation of the total number of realizations crossing the threshold and re-appearing at the reset potential. The latter effect was not considered in previous treatments of finite-size effects. The combined finite-N effects result in the following modified FP equation

$$\partial_t\, p(v,t) = L_N\, p(v,t) + \delta(v - H) \sqrt{\frac{\nu(t - \tau_0)}{N}}\, \Gamma(t - \tau_0) \tag{7}$$

In the framework of the local analysis, the resulting expression for the power spectral density of the population activity is:

$$P(\omega) = \frac{\left| 1 + \boldsymbol{f} \cdot (i\,\omega\,\mathbf{I} - \boldsymbol{\Lambda})^{-1}\, \boldsymbol{\psi}\, e^{-i\,\omega\,\tau_0} \right|^2}{\left| e^{i\,\omega\,\bar{\delta}} - \Phi'\, \rho(i\omega) - i\,\boldsymbol{f} \cdot (i\,\omega\,\mathbf{I} - \boldsymbol{\Lambda})^{-1}\, \boldsymbol{c}\,\omega\rho(i\omega) \right|^2}\, \frac{\nu_0}{N}. \tag{8}$$

$\rho(i\omega)$ is the Fourier transform of the delays distribution centered around the mean $\bar{\delta}$; the elements of $\boldsymbol{\psi}$ are the eigenfunctions of L^+, evaluated at H, and the ψ-dependent term accounts for the finite-N effects on the boundary. We remark that: 1) the asymptotic $P(\omega)$ oscillates around the white-noise flat spectrum ν_0/N; 2) $P(\omega)$ has resonant peaks centered in the imaginary parts of the poles s_n (the zeros of the denominator of Eq. 8). Those peaks disappear for uncoupled neurons, since in that case $\Phi' = 0 = \boldsymbol{c}$; 3) the numerator of Eq. (8) modulates the spectrum and introduces peaks corresponding to the imaginary part of the eigenvalues of L. We conjectured, and checked by explicit calculation for the LIF neuron [7], that the eigenvalues are real for noise-dominated regimes, and complex conjugate pairs for drift-dominated regimes, in which case the imaginary part approximates the stationary mean emission rate ν_0. The finite-N effects on the boundary H reflects itself in the low-ω peaks in $P(\omega)$, approximately centered in multiples of ν_0. We stress that the latter peaks are present even for uncoupled neurons, and they are inherently related to the finite number of neurons in the population.

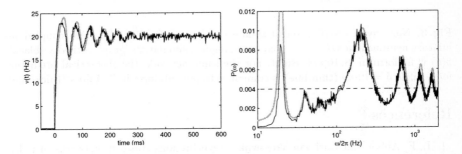

Fig. 2. Left: Transient response to a step change in the afferent current of a population of inhibitory neurons in a drift-dominated regime: Simulations *vs* theory. Starting from an asynchronous stationary state with $\nu = 0.2Hz$, an instantaneous increase of the external current, thereafter kept constant, drives the activity towards a new stable state with $\nu = 20Hz$. The solid black line is the mean of the activity from 10 simulations of a coupled network (5000 inhibitory LIF neurons). The thick gray line is the theoretical prediction, obtained from the first 4 pairs of diffusion poles. Right: Power spectrum of the activity of a population of inhibitory neurons in a stationary, drift-dominated regime: Simulations *vs* theory. The solid black line is the power spectrum from a 60 seconds simulation; the thick gray line is the theoretical prediction; the dashed line is the power spectrum of the white noise with variance ν_0/N, being $\nu_0 = 20Hz$ and $N = 5000$.

In Fig. 2, right, the predicted power spectrum is compared to the estimate from simulation for the stationary population activity of the same network as in Fig. 2, left; it is seen that the agreement is remarkably good. For the network of Fig. 2 all the spikes are propagated with the same delay; for a distribution of delays, it can be seen from Eq. 8 that 1) the absolute value of the real part of the transmission poles is an increasing function of the width Δ of $\rho(\delta)$; in other words, for given network parameters the transmission poles are driven far away from the imaginary axis as Δ increases, so that a distribution of delays improves the network stability. 2) as Δ grows, the power spectrum $P(\omega)$ gets more and more smoothed, and only the lowest frequency peaks survive for high enough Δ. As an illustration, Fig. 3 reports the theoretical predictions *vs* simulations for a uniform $\rho(\delta)$ (left) and the smoothing effect predicted by the theory depending on the width Δ of $\rho(\delta)$ (right).

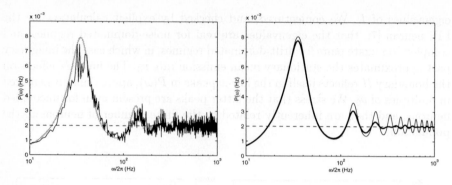

Fig. 3. Non-trivial distribution of synaptic delays. Left: $P(\omega)$ for 2000 coupled inhibitory neurons in a $4Hz$ noise-dominated regime: simulation (gray) *vs* theory (black). $\rho(\delta)$ is uniform in $[8,10]ms$. Right: For the same network, the theoretical prediction for a single $\delta = 10ms$ (thin line) is compared to $\rho(\delta)$ uniform in $[7,13]ms$ (thick line).

References

1. L. F. Abbot and Carl van Vreeswijk. Asynchronous states in networks of pulse-coupled oscillators. *Phys. Rev. E*, 48(2):1483–1490, 1993.
2. Daniel J. Amit and Nicolas Brunel. Model of global spontaneous activity and local structured (learned) delay activity during delay periods in cerebral cortex. *Cerebral Cortex*, 7:237–252, 1997a.
3. Daniel J. Amit and Misha Tsodyks. Quantitative study of attractor neural network retrieving at low spike rates: I. substrate–spikes, and neuronal gain. *Network*, 2:259, 1991.
4. Nicolas Brunel and Vincent Hakim. Fast Global Oscillations in Netwoks of Integrate-and-Fire Neurons with Low Firing Rates. *Neural Comput.*, 11:1621–1671, 1999.
5. Stefano Fusi and Maurizio Mattia. Collective Behavior of Networks with Linear (VLSI) Integrate-and-Fire Neurons. *Neural Comput.*, 11(3):633–653, apr 1999.
6. Bruce W. Knight, Dimitri Manin, and Lawrence Sirovich. Dynamical models of interacting neuron propulations in visual cortex. In E. C. Gerf, editor, *Proceedings of Symposium on Robotics and Cybernetics, Lille-France, July 9-12*, 1996.
7. Maurizio Mattia and Paolo Del Giudice. On the population dynamics of interacting spiking neurons. *submitted (draft available in* http://neural.iss.infn.it*)*, 2001.
8. Duane Q. Nykamp and Daniel Tranchina. A population density approach that facilitates large-scale modeling of neural networks: analysis and an application to orientation tuning. *J. Comput. Neurosci.*, 8:19–50, 2000.
9. Hannes Risken. *The Fokker-Planck Equation: Methods of Solution and Applications*. Springer-Verlag, Berlin, 1984.
10. Alessandro Treves. Mean-field analysis of neuronal spike dynamics. *Network*, 4:259–284, 1993.
11. Henry C. Tuckwell. *Introduction to Theoretical Neurobiology*, volume 2. Cambridge University Press, 1988.

Stochastic Resonance and Finite Resolution in a Network of Leaky Integrate-and-Fire Neurons

Nhamo Mtetwa, Leslie S. Smith, and Amir Hussain

Department of Computing Science and Mathematics, University of Stirling, Stirling,
FK9 4LA, Scotland

Abstract. This paper discusses the effect of stochastic resonance in a
network of leaky integrate-and-fire (LIF) neurons and investigates its
realisation on a Field Programmable Gate Array (FPGA). We report
in this study that stochastic resonance which is mainly associated with
floating point implementations is possible in both a single LIF neuron
and a network of LIF neurons implemented on lower resolution integer
based digital hardware. We also report that such a network can improve
the signal-to-noise ratio (SNR) of the output over a single LIF neuron.

1 Introduction

Stochastic Resonance (SR) occurs when a bistable nonlinear system is driven
by a weak periodic signal, and provision of additional noise improves the system's detection of the periodic signal (see [7] for a review). Here "weak" means
that the input signal is so small that when applied alone it is undetected [11].
Simulation of SR in "continuous" systems using floating point based models is
well established [3], [13] and [12]. Such models are fairly accurate but do not
always result in real-time performance. This work aims to show that SR still
occurs with lower resolution integer based representations which allow real-time
performance on digital hardware such as FPGAs. The FPGA allows us to vary
word length for activation and investigate the effect of this on SR. FPGAs are
reconfigurable programmable logic devices [4]. The FPGA used here is a Xilinx
Virtex XCV1000 [1]. A 7 neuron network implementation on this FPGA occupied 147 625 gates and a single neuron implementation occupied 10 718 gates
[10]. We investigate the effect of varying the resolution of the numbers inside the
simulation on SR: we do this by examining SR in a 64 bit floating point model
in Java, and in an integer model where the integer length is varied downwards
from 16 bits on the FPGA. The SR effect is assessed by applying a subthreshold
periodic signal plus noise to the system, and examining its output SNR. For both
models we calculate the SNR at different input noise amplitudes. For the integer
model we calculate the SNR values at different resolutions. This work extends
[10] to a network (see figure 1). Our results show that integer lengths of at least
10 bits produce usable results, and that the network produces a performance
improvement for six input neurons at integer length of at least 11 bits.

J.R. Dorronsoro (Ed.): ICANN 2002, LNCS 2415, pp. 117–122, 2002.
© Springer-Verlag Berlin Heidelberg 2002

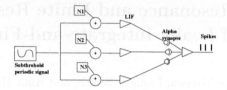

Fig. 1. Network model. Layer 1: identical, parallel input LIF neurons receive the same 20Hz subthreshold sinusoidal input plus independent uncorrelated Gaussian white noise (N1, N2, N3). Output: Same LIF neuron model receives spiking input via an Alpha synapse.

2   The Model

Each neuron is modelled by the LIF neuron model (equation (1)) which idealises the neuron as a capacitor C, in parallel with a resistor of resistance R. The effective current $I(t)$ may hyperpolarise or depolarise the membrane.

$$C\frac{dV}{dt} = I(t) - \frac{V(t)}{R} \tag{1}$$

For the input neurons, $I(t)$ is composed of a periodic subthreshold sinusoidal part and a stochastic component made of Gaussian white noise [2]. For the output neuron, $I(t)$ is the input layer's spikes passed through an Alpha synapse [8]. A network model similar to ours using a depressing synapse with 4 input layer neurons was studied in [14]. They report that a depressing synapse improves the SNR at the output. Similar results were also reported in [6] using a rate coding based FitzHugh-Nagumo model network with a similar topology. Both these networks used floating point numbers.

The floating point model uses the Euler approximation of equation (1) given by equation (2) .

$$V(t + \Delta t) = V(t) \times (1 - \frac{\Delta t}{\tau}) + \Delta t \times I(t) \tag{2}$$

where Δt is the time step, and $\tau = CR$ is the membrane time constant.

Equation (2) is implemented on an FPGA platform by (i) making all values integers and (ii) making division only by powers of 2. $\frac{\Delta t}{\tau}$ is expressed as 2^{-k} (we call k the leakage factor) and Δt is expressed as 2^{-m}. This is because full integer division is expensive to implement (both in number of gates and speed) whereas bit-shifting is very fast. The synaptic response function was implemented using a look-up table. In integer form equation (2) becomes

$$\bar{V}(t + 1) \approx \bar{V}(t) - \bar{V}(t) >> k + \bar{I}(t) >> m \tag{3}$$

where $>>$ denotes bit-shifting to the right.

3   Methodology

The floating point model can take inputs directly. However the integer model requires integer inputs which means the input signals must be quantised. This is achieved by multiplying by 2^n (where n is the resolution i.e. integer length). The floating point model was implemented in Java and the integer model in Handel-C. Handel-C code is compiled into a netlist file which is converted to a bitmap file by the placing and routing software under the Xilinx Foundation Series 3.1i package.

The main outputs of interest are the spike trains from both the "input" and "output" neurons (see figure 1) from which the power spectra were computed in Matlab using the method of Gabbiani and Koch in chapter 9 of [9]. The SNR is computed using the method of Chapeau-Blondeau in [5] (see equation (4)).

$$SNR = \frac{\log_{10} S(\omega)}{\log_{10} N(\omega)} \tag{4}$$

where $S(\omega)$ is the power spectrum of the spike train at the signal frequency and $N(\omega)$ is the power spectrum of the background noise around the signal frequency. Each SNR value is an average from 10 different spike trains.

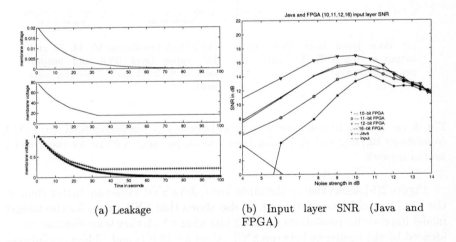

(a) Leakage

(b) Input layer SNR (Java and FPGA)

Fig. 2. (a) Decay due to leakage of membrane from a high start potential without input. Top: Java model decay. Middle: 12-bit FPGA model decay. Bottom: normalised decay for Java (*) and FPGA (+). (b) Input layer SNR for Java and FPGA (resolutions 10,11,12 and 16)

Quantisation affects the evolution of activation and its effect is that the changes in $V(t)$ over Δt implemented by equation (3) are forced into integer values and that the integer activation value ceases to decrease when $V(t) >> 4 + I(t) >> 10) < 1$. This leakage problem in the integer model is also coupled to the size of the time step Δt. Decreasing the time step makes the problem worse as the decrement $\frac{V(t) \times \Delta t}{\tau}$, is proportional to Δt. Yet increasing Δt is not normally an option, as we need $\Delta t \ll T$. For the chosen values of $\tau = 0.02s$ and $\Delta t = 0.001s$, the resolution needs to be at least 10 bits.

4   Results

The results are summarised in figures 2 to 4. They show that there is stochastic resonance at the input layer and at the output neuron in both models.

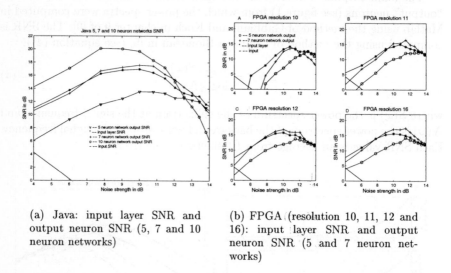

(a) Java: input layer SNR and output neuron SNR (5, 7 and 10 neuron networks)

(b) FPGA (resolution 10, 11, 12 and 16): input layer SNR and output neuron SNR (5 and 7 neuron networks)

Fig. 3. (a) 5, 7 and 10 neuron network input layer and output neuron SNR. (b) FPGA (resolutions 10,11,12 and 16) input layer and output neuron SNRs for the 5 and 7 neuron network

Figure 2(b) shows that at the input layer, Java SNR values are better than all the FPGA resolutions considered. It also shows that SNR values for the integer model improve for resolutions up to 12 bits after which they stay constant as evidenced by the similarity between SNR values for 12 bits and 16 bits resolutions. This result applies to resolutions up to 32 bits [10].

The SNR at the output also increases with increase in resolution (see figure 4) just like it does at the input layer. In both figures the SNR values for the integer model resolutions 12 and 16 bits slightly overtake the Java SNR between 11dB and 13dB noise values. This is interesting because at the input layer SNR values for the Java model are always better than at all the FPGA resolutions considered (see figure 2(b)).

Figure 3(b) shows that the SNR at the output neuron is influenced by both the resolution and number of input neurons. The SNR value at the output increases with an increase in the number of input neurons. The 10-bit resolution

SNR result shows that the input layer SNR is always better than the output neuron result for both 5 and 7 neurons networks. For 11 bits, 12 bits and 16 bits, only the output for the 5 neurons network is less than the input layer neuron SNR, the output neuron SNR for the 7 neurons network overtakes that of the input layer between 8dB and 13dB noise values. This suggests there is complex interaction between the output neuron SNR, the number of input neurons and the resolution of the integer model.

Figure 3(a) shows that as the network increases in size so does the SNR of the output neuron. This result was only proven on a limited scale on the FPGA (see 3(b)) because we were unable to produce a network with more than 7 neurons due to computing equipment limitations.

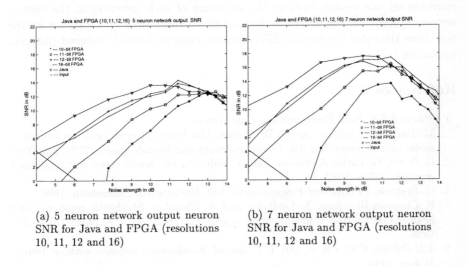

(a) 5 neuron network output neuron SNR for Java and FPGA (resolutions 10, 11, 12 and 16)

(b) 7 neuron network output neuron SNR for Java and FPGA (resolutions 10, 11, 12 and 16)

Fig. 4. (a): 5 neuron network output neuron SNR (both Java and FPGA). (b) 7 neuron network output neuron SNR (both Java and FPGA)

5   Discussion

Figures 2(b), 3 and 4 show that the differences in SNR values for the resolutions considered in the low noise regime are much higher than the differences at the high noise end. This applies to both input and output neurons. The difference in SNR in the lower noise levels at the input layer (see figure 2(b)) are as much as 3dB and the same differences for the output layer are as much 6dB (see figure 4(b)). The decrease in the differences in SNR (as noise strength is increased) may suggest that quantisation effects are more pronounced at lower signal plus noise levels. This could be attributed to loss of changes in the activation. The maximum input layer improvement in SNR due to SR on the FPGA platform never reaches the same level as that of the floating point model for lower noise levels.

6   Conclusion

The results presented here show that SR in an LIF network can occur in low resolution integer based implementations for both single neuron and the topology used in figure 1. It does not require a continuous system. This is important because such systems can permit digital electronic implementation with real-time performance, for example through FPGA implementations. The amount of SR may not be as high as in high resolution implementations. It has also been shown that in integer systems SNR saturates as we increase the bit length at which activation is computed. For the parameters chosen in this simulation, SNR was not found to improve with an increase in the resolution beyond 12 bits. The improvement in output SNR when the output of a parallel array of neurons converge on one neuron justifies the existence of such networks in the central nervous system confirming the findings of [6]. In addition, we have shown for the first time that this increase in SR continues even with much reduced (integer) resolution.

References

1. *Virtex 2.5 V Field Programmable Gate Arrays.*
2. M. Barbi, S. Chillemi, and A.D. Garbo. The leaky integrate-and-fire neuron: a useful tool to investigate SR. *Chaos, solutions and fractals*, pages 1273–1275, 2000.
3. R. Benzi, G. Parisi, A. Sutera, and A. Vulpiani. Stochastic resonance in climatic changes. Tellus, 1982.
4. G. Bestock. *FPGAs and Programmable LSI.* Butterworth Heinemann, 1996.
5. F. Chapeau-Blondeau, X. Godivier, and N. Chambet. Stochastic resonance in a neuronal model that transmits spike trains. *Physical Review E*, pages 1273–1275, 1996.
6. J.J. Collins, C.C. Chow, and T.T. Imhoff. Stochastic resonance without tuning. *Nature*, 1995.
7. L. Gammaitoni, P. Hanggi, P. Jung, and P. Marchesoni. Stochastic resonance. *Review Modern Physics*, 70:223–287, 1998.
8. C. Koch. *Biophysics of Computation.* Oxford University Press, NY, 1999.
9. C. Koch and I. Segev, editors. *Methods in Neural modeling: from ions to networks.* The MIT Press, 1999.
10. N. Mtetwa, L.S. Smith, and A. Hussain. Stochastic resonance and finite resolutions in a leaky integrate-and-fire neuron. In *ESANN2002: Proceedings of the European Symposium on Artificial Neural Network, Bruges, Belgium*, pages 343–348, April 24-26, 2002.
11. D. Petracchi, I.C. Gebeshuber, L.J. DeFelice, and A.V. Holden. Stochastic resonance in biological Systems. *Chaos, Solutions and Fractals*, pages 1819–1822, 2000.
12. M. Stemmler. A single spike suffices: the simplest form of stochastic resonance in model neurons. *Network: Computation in Neural Systems*, 7(4):687–716, November 1996.
13. K. Weisenfeld and F. Moss. Stochastic resonance and the benefits of noise: from ice ages to the crayfish and squids. *Nature*, 373:33–36, 1995.
14. L. Zalanyi, F. Bazso, and P. Erdi. The effect of synaptic depression on stochastic resonance. *Neurocomputing*, pages 459–465, 2001.

Reducing Communication for Distributed Learning in Neural Networks

Peter Auer, Harald Burgsteiner, and Wolfgang Maass

Institute for Theoretical Computer Science
Technische Universität Graz
A-8010 Graz, Austria
{pauer,harry,maass}@igi.tu-graz.ac.at

Abstract. A learning algorithm is presented for circuits consisting of a single layer of perceptrons. We refer to such circuits as *parallel perceptrons*. In spite of their simplicity, these circuits are universal approximators for arbitrary boolean and continuous functions. In contrast to backprop for multi-layer perceptrons, our new learning algorithm – the *parallel delta rule* (*p*-delta rule) – only has to tune a single layer of weights, and it does not require the computation and communication of analog values with high precision. Reduced communication also distinguishes our new learning rule from other learning rules for such circuits such as those traditionally used for MADALINE. A theoretical analysis shows that the *p*-delta rule does in fact implement gradient descent – with regard to a suitable error measure – although it does not require to compute derivatives. Furthermore it is shown through experiments on common real-world benchmark datasets that its performance is competitive with that of other learning approaches from neural networks and machine learning. Thus our algorithm also provides an interesting new hypothesis for the organization of learning in biological neural systems.

1 Introduction

Backprop requires the computation and communication of analog numbers (derivatives) with high bit precision, which is difficult to achieve with noisy analog computing elements and noisy communication channels, such as those that are available in biological wetware. We show that there exists an alternative solution which has clear advantages for physical realization: a simple distributed learning algorithm for a class of neural networks with universal computational power that requires less than 2 bits of global communication. In order to get universal approximators for arbitrary continuous functions it suffices to take a single layer of perceptrons in parallel, each with just binary output. One can view such single layer of perceptrons as a group of voters, where each vote (with value -1 or 1) carries the same weight. The collective vote of these perceptrons can be rounded to -1 or 1 to yield a binary decision. Alternatively one can apply a squashing function to the collective vote and thereby get universal approximators for arbitrary continuous functions.

J.R. Dorronsoro (Ed.): ICANN 2002, LNCS 2415, pp. 123–128, 2002.

Parallel perceptrons and the p-delta rule are closely related to computational models and learning algorithms that had already been considered 40 years ago (under the name of committee machine or MADALINE; see chapter 6 of [7] for an excellent survey). At that time no mathematical tools were available to show the the universal approximation capability of these models. A major advantage of the p-delta rule over algorithms like MADALINE (and backprop for multiplayer perceptrons) is the fact that the p-delta rule requires only the transmission of less than 2 bits of communication (one of the three possible signals "up", "down", "neutral") from the central control to the local agents that control the weights of the individual perceptrons. Hence it provides a promising new hypothesis regarding the organization of learning in biological networks of neurons that overcomes deficiencies of previous approaches that were based on backprop. The p-delta rule consists of a simple extension of the familiar delta rule for a single perception. It is shown in this article that parallel perceptrons can be trained in an efficient manner to approximate basically any practically relevant target function. The p-delta rule has already been applied very successfully to learning for a pool of spiking neurons [6]. The empirical results from [6] show also that the p-delta rule can be used efficiently not just for classification but also for regression problems.

2   The Parallel Perceptron

A perceptron (also referred to as threshold gate or McCulloch-Pitts neuron) with d inputs computes the following function f from $\mathbb{R}^d$ into $\{-1, 1\}$:

$$f(\mathbf{z}) = \begin{cases} 1, & \text{if } \boldsymbol{\alpha} \cdot \mathbf{z} \geq 0 \\ -1, & \text{otherwise} \end{cases},$$

where $\boldsymbol{\alpha} \in \mathbb{R}^d$ is the weight vector of the perceptron, and $\boldsymbol{\alpha} \cdot \mathbf{z}$ denotes the usual vector product. (We assume that one of the inputs is a constant bias input.)

A *parallel perceptron* is a single layer consisting of a finite number n of perceptrons. Let $f_1, \dots, f_n$ be the functions from $\mathbb{R}^d$ into $\{-1, 1\}$ that are computed by these perceptrons. For input $\mathbf{z}$ the output of the parallel perceptron is the value $\sum_{i=1}^{n} f_i(\mathbf{z}) \in \{-n, \dots, n\}$, more precisely the value $s(\sum_{i=1}^{n} f_i(\mathbf{z}))$, where $s : \mathbb{Z} \to \mathbb{R}$ is a squashing function that scales the output into the desired range.

In this article we will restrict our attention to binary classification problems. We thus use as squashing function a simple threshold function

$$s(p) = \begin{cases} -1 & \text{if } p < 0 \\ +1 & \text{if } p \geq 0. \end{cases}$$

It is not difficult to prove that *every* Boolean function from $\{-1, 1\}^d$ into $\{-1, 1\}$ can be computed by a such parallel perceptron.

For regression problems one can use a piecewise linear squashing function. Parallel perceptrons are in fact universal approximators: *every* continuous function $g : \mathbb{R}^d \to [-1, 1]$ can be approximated by a parallel perceptron within any given error bound ε on any closed and bounded subset of $\mathbb{R}^d$. A prove is shown in the full paper [1].

3   The p-delta Learning Rule

3.1   Getting the Outputs Right

We discuss the p-delta rule for incremental updates where weights are updated after each presentation of a training example. The modifications for batch updates are straightforward. Let $(\mathbf{z}, o) \in \mathbb{R}^d \times \{-1, +1\}$ be the current training example and let $\boldsymbol{\alpha}_1, \ldots, \boldsymbol{\alpha}_n \in \mathbb{R}^d$ be the current weight vectors of the n perceptrons in the parallel perceptron. Thus the current output of the parallel perceptron is calculated as

$$\hat{o} = \begin{cases} -1 \text{ if } \#\{i : \boldsymbol{\alpha}_i \cdot \mathbf{z} \geq 0\} < \#\{i : \boldsymbol{\alpha}_i \cdot \mathbf{z} < 0\} \\ +1 \text{ if } \#\{i : \boldsymbol{\alpha}_i \cdot \mathbf{z} \geq 0\} \geq \#\{i : \boldsymbol{\alpha}_i \cdot \mathbf{z} < 0\}. \end{cases}$$

If $\hat{o} = o$ then the output of the parallel perceptron is correct and its weights need not be modified. If $\hat{o} = +1$ and $o = -1$ (we proceed analogously for $\hat{o} = -1$ and $o = +1$) the number of weight vectors with $\boldsymbol{\alpha}_i \cdot \mathbf{z} \geq 0$ needs to be reduced. Applying the classical delta rule to such a weight vector yields the update $\boldsymbol{\alpha}_i \leftarrow \boldsymbol{\alpha}_i - \eta\mathbf{z}$, where $\eta > 0$ is the learning rate. However it is not obvious which weight vectors with $\boldsymbol{\alpha}_i \cdot \mathbf{z} \geq 0$ should be modified by this update rule. There are several plausible options:

1. Update only one of the weight vectors with $\boldsymbol{\alpha}_i \cdot \mathbf{z} \geq 0$. For example choose the weight vector with minimal $|\boldsymbol{\alpha}_i \cdot \mathbf{z}|$.
2. Update N of the weight vectors with $\boldsymbol{\alpha}_i \cdot \mathbf{z} \geq 0$, where N is the minimal number of sign changes of individual perceptrons that are necessary to get the output $\hat{o}$ of the parallel perceptron right. This is the MADALINE learning rule discussed in [7, Section 6.3].
3. Update all weight vectors with $\boldsymbol{\alpha}_i \cdot \mathbf{z} \geq 0$.

For our p-delta rule we choose the third option. Although in this case too many weight vectors might be modified, this negative effect can be counteracted by the "clear margin" approach, which is discussed in the next section.

Note that the third option is the one which requires the least communications between a central control and agents that control the individual weight vectors $\boldsymbol{\alpha}_i$: each agent can determine on its own whether $\boldsymbol{\alpha}_i \cdot \mathbf{z} \geq 0$, and hence no further communication is needed to determine which agents have to update their weight vector once they are told whether $\hat{o} = o$, or $\hat{o} = +1$ and $o = -1$, or $\hat{o} = -1$ and $o = +1$.

3.2   Stabilizing the Outputs

For any of the 3 options discussed in the previous section, weight vectors are updated only if the output of the parallel perceptron is incorrect. Hence weight vectors remain unmodified as soon as the output $\hat{o}$ of the parallel perceptron agrees with the target output o. Thus at the end of training there are usually quite a few weight vectors for which $\boldsymbol{\alpha}_i \cdot \mathbf{z}$ is very close to zero (for some training input $\mathbf{z}$). Hence a small perturbation of the input $\mathbf{z}$ might change the sign of $\boldsymbol{\alpha}_i \cdot \mathbf{z}$. This reduces the generalization capabilities and the stability of the parallel perceptron. Therefore we modify the update rule of the previous section to keep $\boldsymbol{\alpha}_i \cdot \mathbf{z}$ away from zero. In fact, we try to keep a *margin*[1] γ around zero clear from any dot products $\boldsymbol{\alpha}_i \cdot \mathbf{z}$.

The idea of having a clear margin around the origin is not new and is heavily used by support vector machines [5,3]. In our setting we use the clear margin to stabilize the output of the parallel perceptron. As is known from the analysis of support vector machines such a stable predictor also gives good generalization performance on new examples. Since our parallel perceptron is an aggregation of simple perceptrons with large margins (see also [4]), one expects that parallel perceptrons also exhibit good generalization. This is indeed confirmed by our empirical results reported in Section 4.

Assume that $\boldsymbol{\alpha}_i \cdot \mathbf{z} \geq 0$ has the correct sign, but that $\boldsymbol{\alpha}_i \cdot \mathbf{z} < \gamma$. In this case we increase $\boldsymbol{\alpha}_i \cdot \mathbf{z}$ by updating $\boldsymbol{\alpha}_i$ by $\boldsymbol{\alpha}_i \leftarrow \boldsymbol{\alpha}_i + \eta\mu\mathbf{z}$ for an appropriate parameter $\mu > 0$. The parameter μ measures the importance of a clear margin: if $\mu \approx 0$ then this update has little influence, if μ is large then a clear margin is strongly enforced. Observe that a larger margin γ is effective only if the weights $\boldsymbol{\alpha}_i$ remain bounded: one could trivially satisfy condition $|\boldsymbol{\alpha}_i \cdot \mathbf{z}| \geq \gamma$ by scaling up $\boldsymbol{\alpha}_i$ by a factor $C > 1$, but this would have no impact on the sign of $\boldsymbol{\alpha}_i \cdot \mathbf{z}$ for new test examples. Thus we keep the weights $\boldsymbol{\alpha}_i$ bounded by the additional update $\boldsymbol{\alpha}_i \leftarrow \boldsymbol{\alpha}_i - \eta\left(||\boldsymbol{\alpha}_i||^2 - 1\right)\boldsymbol{\alpha}_i$, which moves $||\boldsymbol{\alpha}_i||$ towards 1. Concluding, we can summarize the p-delta rule for binary outcomes $o \in \{-1, +1\}$ as follows:

---

p-delta rule

For all $i = 1, \ldots, n$:

$$\boldsymbol{\alpha}_i \leftarrow \boldsymbol{\alpha}_i - \eta\left(||\boldsymbol{\alpha}_i||^2 - 1\right)\boldsymbol{\alpha}_i + \eta \begin{cases} o \cdot \mathbf{z} & \text{if } \hat{o} \neq o \text{ and } o \cdot \boldsymbol{\alpha}_i \cdot \mathbf{z} < 0 \\ +\mu \cdot \mathbf{z} & \text{if } \hat{o} = o \text{ and } 0 \leq \boldsymbol{\alpha}_i \cdot \mathbf{z} < \gamma \\ -\mu \cdot \mathbf{z} & \text{if } \hat{o} = o \text{ and } -\gamma < \boldsymbol{\alpha}_i \cdot \mathbf{z} < 0 \\ 0 & \text{otherwise .} \end{cases}$$

The rather informal arguments in this and the previous section can be made more precise by showing that the p-delta rule performs gradient descent on an appropriate error function. This error function is zero iff $\hat{o} = o$ and all weight vectors $\boldsymbol{\alpha}_i$ satisfy $|\boldsymbol{\alpha}_i \cdot \mathbf{z}| \geq \gamma$ and $||\boldsymbol{\alpha}_i|| = 1$. The details of the error function are given in the full paper [1].

3.3   Application to Networks of Spiking Neurons

One can model the decision whether a biological neuron will fire (and emit an action potential or "spike") within a given time interval (e.g., of length 5 ms)

---

[1] The margin needs to be set appropriately for each learning problem. In the full paper [1] we define a rule for setting this parameter automatically.

quite well with the help of a single perceptron. Hence a parallel perceptron may be viewed as a model for a population P of biological neurons (without lateral connections inside the population), where the current output value of the parallel perceptron corresponds to the current firing activity of this pool of neurons. The p-delta learning rule for parallel perceptrons has already been tested in this biological context through extensive computer simulations of biophysical models for populations of neurons [6]. The results of these simulations show that the p-delta learning rule is very successful in training such populations of spiking neurons to adopt a given population response (i.e. regression problem), even for transformations on time series such as spike trains. We are not aware of any other learning algorithm that could be used for that purpose.

Table 1. Empirical comparison. Accuracy on test set (10 times 10-fold CV).

Dataset[a]	#exam.	#attr.	p-delta ($n=3$)	MADA-LINE	WEKA MLP+BP	WEKA C4.5	WEKA SVM[b]
BC	683	9	96.94 %	96.28 %	96.50 %	95.46 %	96.87 %
CH	3196	36	97.25 %	97.96 %	99.27 %	99.40 %	99.43 %
CR	1000	24	71.73 %	70.51 %	73.12 %	72.72 %	75.45 %
DI	768	8	73.66 %	73.37 %	76.77 %	73.74 %	77.32 %
HD	296	13	80.02 %	78.82 %	82.10 %	76.25 %	80.78 %
IO	351	34	84.78 %	86.52 %	89.37 %	89.74 %	91.20 %
SI	2643	31	95.72 %	95.73 %	96.23 %	98.67 %	93.92 %
SN	208	60	74.04 %	78.85 %	81.63 %	73.32 %	84.52 %

[a] BC = Wisconsin breast-cancer, CH = King-Rook vs. King-Pawn Chess Endgames, DI = Pima Indian Diabetes, GE = German Numerical Credit Data, HD = Cleveland heart disease, IO = Ionosphere, SI = Thyroid disease records (Sick), SN = Sonar.
[b] MADALINE: n=3, MLP: 3 hidden units, SVM: 2nd degree polynomial kernel

4   Empirical Evaluation

For another empirical evaluation of the p-delta rule we have chosen eight datasets with binary classification tasks from the UCI machine learning repository [2]. We compared our results with the implementations in WEKA[2] of multilayer perceptrons with backpropagation (MLP+BP), the decision tree algorithm C4.5, and support vector machines (with SMO). We also compared our results with MADALINE. We added a constant bias to the data and initialized the weights of the parallel perceptron randomly[3]. The results are shown in Table 1. Results are averaged over 10 independent runs of 10-fold crossvalidation. The p-delta rule was applied until its error function did not improve by at least 1% during the

[2] A complete set of Java Programs for Machine Learning, including datasets from UCI, available at http://www.cs.waikato.ac.nz/~ml/weka/.
[3] For a more detailed description of our experiments see [1].

second half of the trials. Typically training stopped after a few hundred epochs, sometimes it took a few thousand epochs.

The results show that the performance of the p-delta rule is comparable with that of other classification algorithms. We also found that for the tested datasets small parallel perceptrons ($n = 3$) suffice for good classification accuracy.

5 Discussion

We have presented a learning algorithm — the p-delta rule — for parallel perceptrons, i.e., for neural networks consisting of a single layer of perceptrons. It presents an alternative solution to the credit assignment problem that neither requires smooth activation functions nor the computation and communication of derivatives. Because of the small amount of necessary communication it is argued that this learning algorithm provides a more compelling model for learning in biological neural circuits than the familiar backprop algorithm for multi-layer perceptrons. In fact it has already been successfully used for computer simulations in that context.

We show in the full paper [1] that the parallel perceptron model is closely related to previously studied Winner-Take-All circuits. With nearly no modification the p-delta rule also provides a new learning algorithm for WTA circuits.

References

1. Auer, P., Burgsteiner, H. & Maass, W. *The p-delta Learning Rule*, submitted for publication, http://www.igi.TUGraz.at/maass/p_delta_learning.pdf.
2. Blake, C.L. & Merz, C.J. (1998). *UCI Repository of machine learning databases*, http://www.ics.uci.edu/~mlearn/MLRepository.html. Irvine, CA: University of California, Department of Information and Computer Science.
3. N. Cristianini, N., Shawe-Taylor J. (2000). *An Introduction to Support Vector Machines*, Cambridge University Press.
4. Freund, Y., and Schapire, R. E. (1999). *Large margin classification using the Perceptron algorithm*, Machine Learning 37(3):277–296.
5. Guyon I., Boser B., and Vapnik V. (1993). *Automatic capacity tuning of very large VC-dimension classifiers*, Advances in Neural Information Processing Systems, volume 5, Morgan Kaufmann (San Mateo) 147–155.
6. Maass, W., Natschlaeger, T., and Markram, H. (2001). *Real-time computing without stable status: a new framework for neural computation based on perturbations*, http://www.igi.TUGraz.at/maass/liquid_state_machines.pdf.
7. Nilsson, N. J. (1990). *The Mathematical Foundations of Learning Machines*, Morgan Kauffmann Publishers, San Mateo (USA).

Flow Diagrams of the Quadratic Neural Network

David R.C. Dominguez[1], E. Korutcheva[2*], W.K. Theumann[3], and
R. Erichsen Jr.[3]

[1] E.T.S. Informática, Universidad Autónoma de Madrid,
Cantoblanco 28049 Madrid, Spain, david.dominguez@ii.uam.es
[2] Departamento de Física Fundamental, UNED,
c/Senda del Rey No 9, 28080 Madrid, Spain
[3] Instituto de Física, Universidade Federal do Rio Grande do Sul,
C.Postal 15051, 91501-970 Porto Alegre, Brazil

Abstract. The macroscopic dynamics of an extremely diluted three-state neural network based on mutual information and mean-field theory arguments is studied in order to establish the stability of the stationary states. Results are presented in terms of the pattern-recognition overlap, the neural activity, and the activity-overlap. It is shown that the presence of synaptic noise is essential for the stability of states that recognize only the active patterns when the full structure of the patterns is not recognizable. Basins of attraction of considerable size are obtained in all cases for a not too large storage ratio of patterns.

1 Introduction

In a recent paper [1], information theory was used to obtain an effective energy function that maximizes the Mutual Information of a three-state neural network. Neurons which have other structure than the binary state neurons are relevant from both the biological and technological point of view. The performance of the stationary states of an extremely diluted version of the network in mean-field theory revealed the presence of either retrieval, quadrupolar or zero phases.

A three-state neural network is defined by a set of $\mu = 1, ..., p$ embedded *ternary* patterns, $\{\xi_i^\mu \in [0, \pm 1]\}$ on sites $i = 1, ..., N$, which are assumed here to be independent random variables that follow the probability distribution

$$p(\xi_i^\mu) = a\delta(|\xi_i^\mu|^2 - 1) + (1 - a)\delta(\xi_i^\mu), \tag{1}$$

where a is the activity of the patterns ($\xi_i^\mu = 0$ are the inactive ones). Accordingly, the neuron states are three-state dynamical variables, defined as $\sigma_i \in \{0, \pm 1\}$, $i = 1, ..., N$ and coupled to other neurons through synaptic connections, for our purpose, of a Hebbian-like form [1]. The active states, $\sigma_i = \pm 1$, become accessible by means of an effective threshold.

---

* Permanent Address: G.Nadjakov Inst. Solid State Physics, Bulgarian Academy of Sciences, 1784 Sofia, Bulgaria

J.R. Dorronsoro (Ed.): ICANN 2002, LNCS 2415, pp. 129–134, 2002.
© Springer-Verlag Berlin Heidelberg 2002

The pattern retrieval task becomes successful if the state of the neuron $\{\sigma_i\}$ matches a given pattern $\{\xi_i^\mu\}$. The measure of the quality of retrieval that we use here is the mutual information, which is a function of the conditional distribution of the neuron states given the patterns, $p(\sigma|\xi)$. The order parameters needed to describe this information are the large-N (thermodynamic) limits of the standard overlap of the μth pattern with the neuron state,

$$m_N^\mu \equiv \frac{1}{aN} \sum_i \xi_i^\mu \sigma_i \to m = \langle\langle\sigma\rangle_{\sigma|\xi}\frac{\xi}{a}\rangle_\xi, \tag{2}$$

the neural activity,

$$q_{Nt} \equiv \frac{1}{N} \sum_i |\sigma_{it}|^2 \to q = \langle\langle\sigma^2\rangle_{\sigma|\xi}\rangle_\xi, \tag{3}$$

and the so called *activity-overlap*[4,5],

$$n_{Nt}^\mu \equiv \frac{1}{aN} \sum_i^N |\sigma_{it}|^2 |\xi_i^\mu|^2 \to n = \langle\langle\sigma^2\rangle_{\sigma|\xi}\frac{\xi^2}{a}\rangle_\xi. \tag{4}$$

The brackets denote averages over the probability distributions.

A study of the stability of the stationary states, as well as the dynamic evolution of the network, with particular emphasis on the size of the basins of attraction, has not been carried out so far, and the purpose of the present work is to report new results on these issues.

2 Mutual Information

The Mutual Information between patterns and neurons is defined as $I[\sigma;\xi] = \langle\langle\ln[p(\sigma|\xi)/p(\sigma)]\rangle_{\sigma|\xi}\rangle_\xi$, regarding the patterns as the input and the neuron states as the output of the channel at each time step [2,3]. $I[\sigma;\xi]$ can also be written as the entropy of the output substracted from the conditional entropy (so-called equivocation term). For the conditional probability we assume the form [1]

$$p(\sigma|\xi) = (s_\xi + m\xi\sigma)\delta(\sigma^2 - 1) + (1 - s_\xi)\delta(\sigma) \tag{5}$$

with $s_\xi = s + (n - q)\xi^2/(1 - a)$ and $s = (q - na)/(1 - a)$.

We search for an energy function which is symmetric in any permutation of the patterns ξ^μ, since they are not known initially to the retrieval process, and we assume that the initial retrieval of any pattern is weak, i.e. the overlap $m^\mu \sim 0$ [1]. For general a and q, also the variable σ^2 is initially almost independent of $(\xi^\mu)^2$, so that $n^\mu \sim q$. Hence, the parameter $l^\mu \equiv (n^\mu - q)/(1 - a) = <\sigma^2\eta^\mu>$, $\eta^\mu \equiv ((\xi^\mu)^2 - a)/(a(1 - a))$, also vanishes initially in this limit. Note that this parameter is a recognition of a fluctuation in $(\xi^\mu)^2$ by the binary state variable σ^2.

An expansion of the mutual information around $m^\mu = 0$ and $l^\mu = 0$ thus gives, initially, when $a \sim q$, $I^\mu \approx \frac{1}{2}(m^\mu)^2 + \frac{1}{2}(l^\mu)^2$, and the total information of the network will be given by summing over all the patterns $I_{pN} = N \sum_{\mu} I^{\mu}$. An energy function $H = -I$ that rules the network learning and retrieval dynamics is obtained from this expansion[1]. The dynamics reads $\sigma_i^{t+1} = sign(h_i^t)\Theta(|h_i^t|+\theta_i^t)$ with the neuron potential and the effective threshold given respectively by

$$h_i^t = \sum_j J_{ij}\sigma_{jt}, \quad J_{ij} \equiv \frac{1}{a^2 N} \sum_{\mu=1}^{p} \xi_i^{\mu}\xi_j^{\mu}, \tag{6}$$

$$\theta_i^t = \sum_j K_{ij}\sigma_{jt}^2, \quad K_{ij} \equiv \frac{1}{N} \sum_{\mu=1}^{p} \eta_i^{\mu}\eta_j^{\mu}. \tag{7}$$

3   Macrodynamics for the Diluted Network

The asymptotic macrodynamics for the parameters in Eqs.(2),(3) and (4) in the extremely diluted version of the model, follows the single-step evolution equations, exact in the large-N limit, for each time step t [1],

$$m_{t+1} = \int D\Phi(y) \int D\Phi(z) F_{\beta}(\frac{m_t}{a} + y\Delta_t; \frac{l_t}{a} + z\frac{\Delta_t}{1-a}), \tag{8}$$

$$q_t = an_t + (1-a)s_t, \tag{9}$$

$$n_{t+1} = \int D\Phi(y) \int D\Phi(z) G_{\beta}(\frac{m_t}{a} + y\Delta_t; \frac{l_t}{a} + z\frac{\Delta_t}{1-a}), \tag{10}$$

for the overlap, the neural activity, and the activity overlap, respectively, where

$$s_{t+1} \equiv \int D\Phi(y) \int D\Phi(z) G_{\beta}(y\Delta_t; -\frac{l_t}{1-a} + z\frac{\Delta_t}{1-a}), \tag{11}$$

defined above describes the wrong matches between the active neurons and patterns. Here, $D\Phi(y)$ and $D\Phi(z)$ are Gaussian probability distributions for random variables y and z with zero mean and unit variance, whereas $\Delta_t^2 = \alpha q_t/a^2$, in which $\alpha = p/N$ is the storage ratio of patterns. The functions

$$F_{\beta}(h,\theta) = \frac{1}{Z}2e^{\beta\theta}\sinh(\beta h), \quad G_{\beta}(h,\theta) = \frac{2}{Z}e^{\beta\theta}\cosh(\beta h) \tag{12}$$

with $Z = 1 + 2e^{\beta\theta}\cosh(\beta h)$, are the mean $\overline{\sigma_t}$ and the square mean $\overline{\sigma_t^2}$ of the neuron states over the synaptic noise with parameter $\beta = a/T$, respectively.

4   Results

There are two main aspects in which we are interested in the present work. One, is the nature of the possible phases in the order-parameter space (m, l, q), for given pattern activity a, storage ratio α and synaptic noise parameter (temperature) T. These are results to be extracted from the stationary states of the network and some have been presented before [1]. They concern a retrieval phase $R(m \neq$

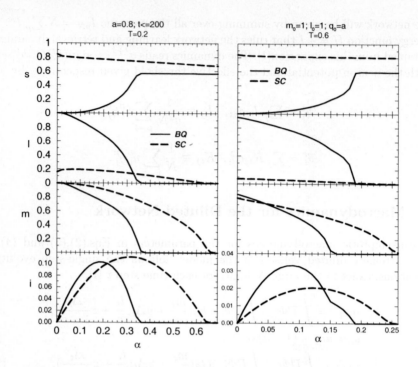

Fig. 1. The information i and order parameters m, l, q against α for small (left) and high (right) noise levels. The BQN and SCN models are compared.

$0, l \neq 0)$, a quadrupolar phase $Q(m = 0, l \neq 0)$ and a zero phase $Z(m = 0, l = 0)$, all for $q \sim a$. The second aspect concerns the dynamic evolution of the network and the stability of the possible phases.

As far as the phases is concerned, one may argue that whenever $m \neq 0$ also $l \neq 0$ but the inverse is not necessarily true. When $T = 0$ but even with some stochastic noise due to the macroscopic number of stored patterns, if the states do not recognize the patterns, that is, if $m = 0$, σ^2 may not recognize $(\xi^\mu)^2$ either and as a result also l may be zero, excluding thereby the phase Q. However, one cannot rule out, *a priori*, that with increased stochastic noise, l remains finite while m becomes zero making place for a Q phase.

Indeed, as shown in Fig. 1 (solid lines), when $a = 0.8$, the overlap m is always larger than l for the small $T = 0.2$, but for higher synaptic noise, say $T = 0.6$, m vanishes for $\alpha \sim 0.15$ while $l \neq 0$ up to $\alpha \sim 0.19$. In Fig. 1 we also compare the present biquadratic neural (BQN) model with the self-control network (SCN). We have extended the SCN model for the synaptic noisy case, using the linear correction $\tilde{\theta}_t = \theta_t - T$, where θ_t is the self-control neural threshold at $T = 0$[4,5]. One can see that for $T = 0.2$ (left) the SCN, even with a small l for low storage ratio α, has a peak in the information as large as that of the BQN model, but for the higher noise $T = 0.6$ (right), the information for the BQN is much larger

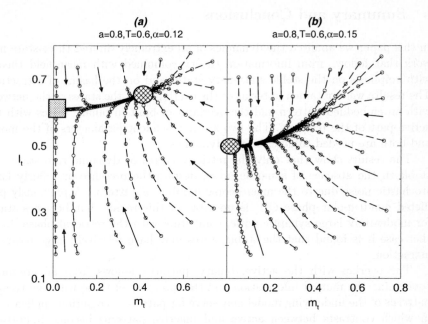

Fig. 2. Flow diagrams in the order-parameter space l,m, for a noisy network. The stable attractor is the R-phase for small α (a) and the Q-phase for larger α (b).

than that for the SCN. Hence, for high T, the activity-overlap plays a crucial role in the network dynamics, meaning that the BQN is robust with respect to high noise levels. The basis for this argument requires a detailed discussion that will be taken up elsewhere. We present next our results for the dynamics of the network based on eqs.(8)-(11).

In Fig. 2 we show the flux lines, for various initial values of l and m for two points in Fig. 1(b) with large synaptic noise, $T = 0.6$. In Fig. 2(a), for $\alpha = 0.12$, we find a stable (the ellipse) R state and a metastable (the square) Q state. The Q state attracts points on the line $m = 0$ and then repeals them in direction to the R state. There is a large basin of attraction. However, the time the network spends on the line between the saddle-point Q and the stable R states, is usually as large as $t = O(10^4)$, meaning that a fraction of the network could be trapped in the Q-phase, as one would suppose for finite systems. On the other hand, there is a changeover to a stable Q phase at slightly higher stochastic noise, as shown in Fig. 2(b) for $\alpha = 0.15$. For higher activity, $a = 0.9$, a stable Q-phase already appears for shorter convergence times at $T = 0.4$ for $0.23 \leq \alpha \leq 0.28$, and there is a stable retrieval phase for lower α. As expected, there is a decrease in the size of the basins of attraction with increasing α, leading finally to the convergence of the flows towards the zero (Z) phase, for which $m, l = 0$ and there is no information processed[1].

5   Summary and Conclusions

In this paper we analised the dynamics of an extremely diluted three-state network that follows from information theory combined with mean-field theory, with emphasis on the stable stationary states and on the basins of attraction. The results were obtained in terms of the overlap m of the states of the network with an embedded pattern and the overlap l of built in binary states with the active part of the patterns. The two are independent parameters of the model and both are needed to describe the dynamics.

The results show that only the retrieval states or dynamic zero states are stable in the absence of synaptic noise, despite the presence of a fairly large stochastic noise due to the macroscopic number of patterns. A previously predicted quadrupolar phase $Q[1]$ is unstable in this regime but becomes stable for moderately large synaptic noise, competing with the retrieval phase. In either case it is found that the stable states may have relatively large basins of attraction.

The overlap with the active, binary, patterns follows from an argument concerning the mutual information and the interplay of these with the ternary patterns of the underlying model may serve for pattern-recognition applications, in which contrasts between active and inactive patterns become important. Further results for the performance of other three-state networks as well as the stability of states will be reported elsewhere.

Acknowledgments. We thank Desiré Bollé for stimulating discussions. D.D. is supported by the program Ramón y Cajal from the MCyT, Spain. E.K. gratefully acknowledges financial support from the MCyT, grant BFM2001-291-C02-01. The research of W.K.T. is supported, in part, by CNPq, Brazil. Both, W.K.T. and R.E.Jr., acknowlege financial support from FAPERGS, RS, Brazil.

References

1. D.Dominguez and E.Korutcheva, Three-state neural network: From mutual Information to the Hamiltonian. Phys.Rev.E **62(2)**, (2000) 2620-2628.
2. C.E.Shannon, *A Mathematical Theory of Communication*, The Bell System Technical Journal, **27**, (1948).
3. R.E.Blahut, *Principles and Practice of Information Theory*, Addison-Wesley, Reading, MA, (1990), Chapter 5.
4. D.Dominguez and D.Bollé, Self-control in Sparsely Coded Networks. Phys.Rev.Lett. **80(13)**, (1998) 2961-2964.
5. D.Bollé, D.Dominguez and S.Amari, Mutual Information of Sparsely-Coded Associative Memory with Self-Control and Ternary Neurons. Neural Networks **13(4-5)**, (2000) 455-462.

Dynamics of a Plastic Cortical Network

Gianluigi Mongillo[1] and Daniel J. Amit[2]

[1] Dept. of Human Physiology and Pharmacology, University "La Sapienza", Rome
[2] Dept. of Physics, University "La Sapienza", Rome
and
Racah Institute of Physics, Hebrew University, Jerusalem

Abstract. The collective behavior of a network, modeling a cortical module, of spiking neurons connected by plastic synapses is studied. A detailed spike-driven synaptic dynamics is simulated in a large network of spiking neurons, implementing the full double dynamics of neurons and synapses. The repeated presentation of a set of external stimuli is shown to structure the network to the point of sustaining selective delay activity. When the synaptic dynamics is analyzed as a function of pre- and post-synaptic spike rates in functionally defined populations, it reveals a novel variation of the Hebbian plasticity paradigm: In any functional set of synapses between pairs of neurons - (stimulated-stimulated; stimulated-delay; stimulated-spontaneous etc...) there is a finite probability of potentiation as well as of depression. This leads to a saturation of potentiation or depression at the level of the ratio of the two probabilities, preventing the uncontrolled growth of the number of potentiated synapses. When one of the two probabilities is very high relative to the other, the familiar Hebbian mechanism is recovered.

1 Introduction

Long-term modifications of the synaptic efficacies are believed to affect information processing in the brain. The occurrence of such modifications would be manifest as a change in neural activity for particular input patterns. Both in IT and PF cortex of monkeys, trained to perform a delayed response task, small neural assemblies have been found to exhibit selective, persistent enhanced spike rates during the delay interval between successive stimulations (see e.g. [1], [2]). This kind of activity is related to the ability of the monkey to actively hold an item in memory in absence of the eliciting stimulus [3]. It will be referred to as working memory (WM) activity.

Only stimuli repeatedly presented are able to evoke WM activity. Thus, WM appears only after a substantial training stage, during which, presumably, the local synaptic structure is modified by the incoming stimuli. Spiking neural network models indicate that WM activity could be produced by coherent modifications of the synaptic efficacies, see e.g. [4]. Because each selective delay population overlaps significantly with the population of visually responsive neurons to the same stimulus (i.e. neurons with enhanced activity during stimulus presentation), one expects that the synaptic dynamics underlying long-term modifications strengthens synapses connecting pairs of visually responsive neurons and

J.R. Dorronsoro (Ed.): ICANN 2002, LNCS 2415, pp. 135–140, 2002.

(eventually) weakens those from visually responsive to visually non-responsive neurons.

A simulation like the one presented here provides a more complete benchmark (relative to simulations with fixed synapses), in the bridge between experiment and theory [5]. The synaptic dynamics, originally proposed by Fusi et al. ([6]), is propelled purely by the actual spikes emitted by the neurons, as a consequence of a preassigned protocol of stimulus presentation. It is shown that WM is actually formed in the process, and that its slow formation can be qualitatively understood.

2   A Model of Plastic Cortical Network

The model system we study is a recurrent network of excitatory and inhibitory linear integrate-and-fire (LIF) neurons. The synapses among the excitatory cells are plastic: their efficacy can get two different values, potentiated J_p and depressed $J_d (< J_p)$, according to a synaptic internal state, described by an analog variable $X(t)$. If X is above a given threshold, the synapse is in its potentiated state. Otherwise, synaptic efficacy is J_d. Crossing the threshold results in a synaptic transition: from below to above $(J_d \rightarrow J_p)$ corresponds to LTP, while from above to below $(J_p \rightarrow J_d)$ corresponds to LTD.

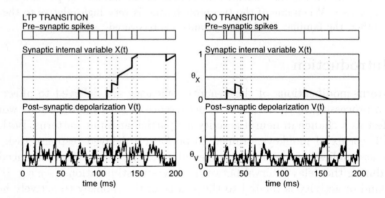

Fig. 1. Synaptic dynamics. The time evolution of $X(t)$ (center); following pre-synaptic emission (top), $X(t)$ is regulated up(down) if the post-synaptic depolarization $V(t)$ is greater(smaller) than a critical level θ_V (bottom). In the time intervals between spikes, $X(t)$ drifts linearly up or down (refresh term). Left: LTP; Right: No transition; (reproduced from [6])

The dynamics of X is driven by pre- and post-synaptic neural activity. Each pre-synaptic spike induces the up-/down-regulation of X according to the instantaneous post-synaptic depolarization level. High depolarization levels (indicating high emission rate for the post-synaptic cell) cause up-regulation, while low de-polarization levels (indicating low post-synaptic activity) cause down regulation. The effect of a single regulation is progressively damped out by a refresh term, responsible for long-term state preservation. The spike-driven synaptic dynamics described exhibits, for a suitable choice of the parameters, both long-term poten-tiation and homosynaptic long-term depression under experimental stimulation protocols (for a discussion on biological plausibility see [7]). Figure 1 sketches a specific example of such a synaptic dynamics.

If several regulation of X are required to cross synaptic threshold, the plastic synapse behaves in a Hebbian-like way. A synapse between two high-rate neurons tends to be potentiated. On the other hand, a synapse connecting a high-rate pre-synaptic neuron to a low-rate post-synaptic one tends to be depressed. However, as a consequence of stochastic neural activity, synaptic transitions are themselves stochastic. In other words, a synapse where one would expect LTP, may undergo LTD and vice versa.

3   Effects of Stimulation: The Structuring Process

To the network are repeatedly presented a set of stimuli as in a typical Delay-Match-to-Sample (DMS) protocol, i.e a stimulus appears for T_{stim} milliseconds and, following a delay period, during which none of the stimuli is presented, another stimulus is presented. The effects of the stimulus presentation on neu-ral activity are obtained by injecting an external depolarizing current into a pool of 'visually responsive' excitatory neurons. Visually responsive pools are selected non-overlapping, i.e. a neuron responds to a single stimulus. There are also neurons which do not respond to any stimulus. Note that, even in this basic case, the structuring process is not trivial. To track the structuring, we monitor both the fraction $C_p^{(HH)}$ of potentiated synapses among the neurons belonging to the same selective population and the fraction $C_p^{(HL)}$ of potentiated synapses among neurons responsive to different stimuli. $C_p^{(HH)}$ increases monotonically with stimulus presentation, while $C_p^{(HL)}$ decreases with stimulus presentation, as expected. Indeed, neurons belonging to the same selective population are si-multaneously active at high rates, when stimulus is presented. On the other hand, the synapses connecting neurons responsive to different stimuli see high pre-synaptic and low post-synaptic activity, during stimulus presentation.

Figure 2 reports the evolution of $C_p^{(HH)}$ and $C_p^{(HL)}$ as a function of the number of presentations per stimulus. When the synaptic structuring reaches a critical level, selective delay activity appears, see Figure 3.

As Figure 4 shows, there are depressions among the synapses connecting neu-rons responsive to the same stimulus as well as potentiations among the synapses connecting neurons responsive to different stimuli. This is a consequence of the previously mentioned stochasticity. The evolution of the fraction C_p of poten-tiated synapses in a given synaptic population as a function of the number of

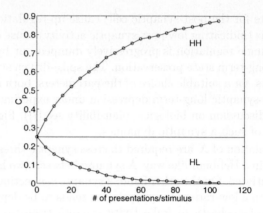

Fig. 2. Synaptic structuring: C_p as a function of the number of presentations (non-overlapping populations). Horizontal line: the unstructured, initial, state. $C_p^{(HH)}$ (between visually responsive neurons) increases monotonically with stimulus presentation; C_p^{HL} (between visually responsive to non-responsive neurons) decreases monotonically. Synaptic structuring allowing for WM is reached after 105 presentations of each stimulus.

stimulus presentations can be described by

$$C_p(n+1) = C_p(n)[1 - q_-] + q_+[1 - C_p(n)] \tag{1}$$

where n represents the number of presentations and $q_+(q_-)$ the probability that a depressed(potentiated) synapse undergoes LTP(LTD). The probabilities q_+ and q_- can be estimated from synaptic and network parameters, see [7].

In the asymptotic state, i.e. after a large number of presentations, we obtain

$$C_p(\infty) = \frac{q_+}{q_+ + q_-} \tag{2}$$

The fact that both q_+ and q_- are different from zero prevents the saturation of structuring, i.e. all synapses either potentiated or depressed. The asymptotic state can be seen as a detailed balance between potentiating and depressing synapses.

4   Discussion and Perspectives

The results we report represent a proof that in the rich and complex space of the system (neurons and synapses) parameters there exists a zone in which the fundamental structuring for working memory can take place. Moreover, the dynamics of the system exhibits, in a completely natural way, a mechanism preventing the uncontrolled growth (or decrease) of the number of potentiated synapses. This helps to keep firing rates within functional boundaries and to

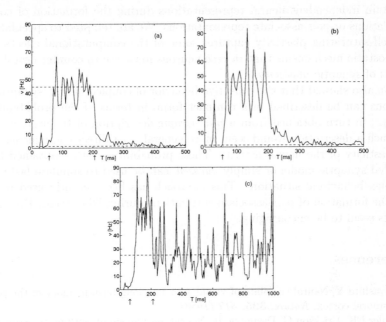

Fig. 3. Selective population activity before, during and following stimulation at 3 structuring stage. Arrows indicate 150ms stimulation interval. (a) 55 presentations per stimulus; (b) 70 presentations per stimulus. Horizontal dashed lines represent average emission rate during stimulus presentation and spontaneous activity (SA). In both (a) and (b) no WM: population returns to SA after removal of stimulus. (c) Appearance of WM state: 105 presentations per stimulus. Horizontal dashed line represents average emission rate during WM activity.

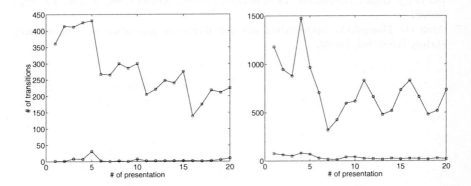

Fig. 4. Number of synaptic transitions per presentation vs presentation number. Squares: number of potentiations. Circles: number of depressions. Left: synapses among neurons responsive to the same stimulus - number of potentiations greater than the number of depressions. Right: synapses among neurons responsive to different stimuli - number of depressions greater than the number of potentiations.

maintain independent neural representations during the formation of context-correlations or pair-associate representations. We are tempted to speculate that the self-saturating plasticity captures some of the computational effects of the homeostatic mechanisms that *in vivo* neurons must use to counter destabilizing effects of synaptic plasticity.

We also showed that the collective behavior of coupled neural/synaptic populations can be described in a compact form, in terms of the probabilities q_+ and q_-, in turn obtained from a *microscopic* description of the system. A priori, such a description would seem more general, as a direct consequence of the stochasticity of the neural activity. These probabilities, either obtained from a detailed synaptic model or simply *guessed*, can be used to simulate faster more complex behavioral situations. This approach has been recently used to simulate the formation of pair-associate correlations during DMS trials. Preliminary results seem to be encouraging.

References

1. Miyashita Y, Neural correlate of visual associate long-term memory in the primate temporal cortex, *Nature*, **335**, *817 (1988)*
2. Miller EK, Erickson C, Desimone R, Neural mechanism of working memory in prefrontal cortex of macaque, *Journal of Neuroscience*, **16**, *5154 (1996)*
3. Amit DJ, The Hebbian paradigm reintegrated: Local reverberations as internal representations, *Behavioral and Brain Sciences*, **18**, *617 (1995)*
4. Amit DJ, Brunel N, Model of global spontaneous activity and local structured activity during delay periods in the cerebral cortex, *Cerebral Cortex*, **7**, *237 (1997)*
5. Amit DJ, Simulation in Neurobiology - Theory or Experiment?, *TINS*, **21**, *231 (1998)*
6. Fusi S, Annunziato M, Badoni D, Salamon A, Amit DJ, Spike-driven synaptic plasticity: theory, simulation, VLSI implementation, *Neural Computation*, **12**, *2227 (2000)*
7. Amit DJ, Mongillo G, Spike-driven synaptic dynamics generating working memory states, *Submitted, (2002)*

Non-monotonic Current-to-Rate Response Function in a Novel Integrate-and-Fire Model Neuron

Michele Giugliano*, Giancarlo La Camera, Alexander Rauch,
Hans-Rudolf Lüscher, and Stefano Fusi

Institute of Physiology, University of Bern, CH-3012 Bühlplatz 5, Switzerland.
{giugliano,lacamera,rauch,luescher,fusi}@pyl.unibe.ch

Abstract. A novel integrate-and-fire model neuron is proposed to account for a non-monotonic *f-I* response function, as experimentally observed. As opposed to classical forms of adaptation, the present integrate-and-fire model the spike-emission process incorporates a state - dependent inactivation that makes the probability of emitting a spike decreasing as a function of the mean depolarization level instead of the mean firing rate.

1 Introduction

Recent experimental evidence indicates that, at relatively high firing rates (*e.g.* 50 Hz), the *in vitro* response of neocortical neurons to the injection of a noisy current, emulating the heavy barrage of *in vivo* presynaptic bombardment (see e.g. [1,2]), shows a non-stationary behavior. In particular, such neurons cannot sustain high rates for a prolonged time (*e.g.* 30 s), ultimately reducing their spiking frequency or stopping firing [1]. Furthermore, in *in vitro* disinhibited cultured networks of dissociated spinal cord neurons a similar phenomenon seems to characterize the intrinsic firing properties of a specific neuronal subclass (see [3]), whose activity decreases or completely shuts down as the excitatory afferent synaptic current from the network becomes large. Interestingly, the last phenomenon has been hypothesized to be involved in determining and shaping the slow spontaneous rhythmic collective activity, characterizing such an *in vitro* neurobiological system at the network level. Moreover, previous theoretical studies also pointed out that similar single-neuron response behaviors might substantially improve performances in attractor neural networks [4].

Although the underlying detailed sub-cellular mechanisms are not yet fully understood, and it is still not clear if these are *in vitro* artifacts or if they play any *in vivo* physiological role at the network level, the described phenomena result in a steady-state non-monotonic *f-I* response function (RF) for individual neurons.

* We are grateful to Anne Tscherter, Dr. Pascal Darbon and Dr. Jürg Streit for fruitful discussions and comments. M.G. is supported by the Human Frontier Science Program Organization (grant LT00561/2001-B).

J.R. Dorronsoro (Ed.): ICANN 2002, LNCS 2415, pp. 141–146, 2002.
© Springer-Verlag Berlin Heidelberg 2002

Such an excitability modulation should not be regarded as a form of classical frequency-dependent adaptation, by which neurons are known to accommodate their responses to transient as well as sustained stimulations. Actually, it is widely accepted that such class of adapting behaviors relies on the accumulation of intracellular ion-species (*e.g.* calcium ions), whose concentration reflects on a first approximation the mean firing rate, and can only account for a saturating RF and not for a non-monotonic relationship [5].

A feasible alternative interpretation may trace such an activity-dependent reduction of excitability back to the biophysical bases of action potentials generation. Specifically, the progressive voltage-dependent reduction of recovery from inactivation in voltage-dependent membrane inward currents (*i.e.* such as the fast-inactivating TTX-sensitive sodium currents) might play a substantial role in decreasing the output spike-rate and ultimately affecting the generation of further action potentials, as the total number of *non-inactive* channels becomes small. Such a working hypothesis can be tested in the framework set by a conductance-based mathematical description of neuronal excitability, incorporating for instance only Hodgkin-Huxley-like fast-inactivating sodium currents and delayed-rectifier potassium currents. Although under noisy current injection such a model qualitatively reproduces a non-monotonic dependence (see Fig 1), an extensive analysis performed at the network level is not possible, given the complexity of the model and the consequent heavy computational loads of large-scale simulations.

In order to investigate at the network level the relevance and the impact of a non-monotonic response function, in the present contribution we propose a novel integrate-and-fire (IF) model neuron, reproducing the described phenomena and well-suited to undergo a full statistical analysis.

2   The Model

2.1   *Integrate-and-May-Fire* (IMF) Model Neuron: Uncertain Spike Emission

Below the excitability threshold θ, the behavior of the novel IMF neuron schematically resembles a linear integrator of the afferent current $I(t)$, as for the linear integrate-and-fire model (LIF) [6,1]. As a consequence, the subthreshold membrane depolarization $V(t)$ evolves in time according to the following differential equation:

$$\frac{dV(t)}{dt} = -\beta + I(t) \tag{1}$$

where β is a constant decay ($\beta > 0$) that, in the absence of afferent currents, drives the depolarization to the resting potential $V_{rest} = 0$. The resting potential is a reflecting barrier, i.e. the depolarization cannot be driven below zero [6]. Compared to conventional IF dynamics described in the literature, when V crosses the threshold θ, the IMF neuron does not always emit a spike. In other words, for the IMF neuron the spike emission process has been assumed

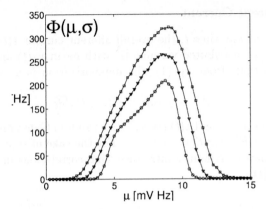

Fig. 1. Conductance-based model neuron: the mean firing rates, assessed across 60 s of simulation time after discarding a transient of 1 s, have been plotted versus the mean injected current μ, for increasing values of the amplitude of its fluctuations σ (circles, triangles and squares) in a Hodgkin-Huxley-like model. Spikes were detected by a positive threshold-crossing criterium either in the amplitude vs time or in the slope vs time domains: for large input currents, such a criterium let us discriminate between large fluctuations of the membrane potential and spikes.

to be not fully determined by the dynamics of the membrane depolarization alone but to depend on additional intrinsic biophysical mechanisms, not explicitly modeled. More precisely, if $V(t_0) = \theta$ the emission of a spike at t_0 is an event occurring with an activity-dependent probability $q \leq 1$. After the spike emission, V is clamped to the value H_1 ($0 < H_1 < \theta$), for an absolute refractory time τ_{arp}, after which the current integration starts again. However, each time the excitability threshold θ is crossed and no spike has been generated (*i.e.* an event with probability $1 - q$), V is reset to H_2 ($0 < H_1 < H_2 < \theta$) and no refractoriness entered.

In order for the IMF neuron to have a non-monotonic RF, we made the additional assumption that q is a decreasing function of a slow voltage-dependent variable w, reminiscent of the sigmoidal voltage-dependence of the fast inactivation state variables that characterize conductance-based model neurons:

$$q = \left(1 + e^{\frac{(w - w_o)}{\sigma_w}}\right)^{-1} \tag{2}$$

where w evolves by eq. (3) below, corresponding on a first approximation to the average transmembrane electric field experienced by individual ion channels and affecting their population-level activation and inactivation.

$$\tau_w \frac{dw(t)}{dt} = \frac{V(t)}{\theta} - w \tag{3}$$

We note that $0 < w < 1$ and that, in the limit $\tau_w \to +\infty$, it approximates the expected normalized depolarization $\langle V \rangle / \theta$, providing a negative-feedback on the spike-emission mechanisms that depends on the statistics of the membrane voltage.

2.2   The Afferent Current

We assume that at any time t, the overall afferent current $I(t)$ can be approximated by a Gauss distributed variable [7] with mean $\mu_I(t)$ and variance $\sigma_I^2(t)$ in unit time, so that, from eq. (1), the depolarization is a stochastic process obeying:

$$dV = \mu(t)dt + \sigma_I(t)z(t)\sqrt{dt} \tag{4}$$

where $\mu(t) \equiv -\beta + \mu_I(t)$, and $z(t)$ represents a random Gauss distributed process with $E[z] = 0$ and $E[z(t)z(t')] = \delta(t - t')$. For the sake of simplicity, we neglect that realistic synaptic dynamics introduce time correlations in the resulting afferent current $I(t)$.

3   Results

3.1   IMF Steady-State Response Function

Taking advantage of the simplified mathematical expression of Eq. 1, we analytically characterized the statistical properties of the IMF model neuron under a noisy current injection, first assuming q to be a constant. For comparison with the numerical simulations, we included the dependence of q on w, by solving self-consistent equations at the steady-state, accounting for the behavior of the full model. We note that an equation of motion can be derived for the probability density that at time t the neuron has a depolarization $V = v$, under the diffusion approximation (see e.g. [8]):

$$\frac{\partial p}{\partial t} = -\frac{\partial}{\partial v}\left(\mu p - \frac{1}{2}\sigma^2\frac{\partial p}{\partial v}\right) \tag{5}$$

This is the Fokker-Planck equation for $V(t)$ and it must be complemented by appropriate boundary conditions at $v = 0$, $v = \theta$, $v = H_1$ and $v = H_2$ in the case of the IMF neuron (see [6,8]). At the steady-state, eq. 5 turns into a second-order ordinary linear differential equation and it gives rise to closed-form expressions for the stationary RF $\Phi_{\mu,\sigma,q}$ and the steady-state expected value of the membrane voltage $\langle V \rangle_{\mu,\sigma,q}$.

$$\Phi_q(\mu, \sigma) = q\,\nu \tag{6}$$

$$\langle V \rangle_{\mu,\sigma,q} = \frac{1}{2\mu}\left[\nu\left(\theta^2 - \overline{H^2}\right) + \sigma^2\left(q\,\nu\,\tau_{arp} - 1\right)\right] + q\,\nu\tau_{arp}\,H_1 \tag{7}$$

with ν given by:

$$\nu = \left[q\tau_{arp} + \frac{1}{\mu\lambda}\left(e^{-\lambda\theta} - \overline{e^{-\lambda H}}\right) + \frac{(\theta - \overline{H})}{\mu}\right]^{-1} \tag{8}$$

and $\lambda = \frac{2\mu}{\sigma^2}$, $\overline{e^{-\lambda H}} = qe^{-\lambda H_1} + (1 - q)e^{-\lambda H_2}$, $\overline{H} = qH_1 + (1 - q)H_2$ and $\overline{H^2} = qH_1^2 + (1 - q)H_2^2$.

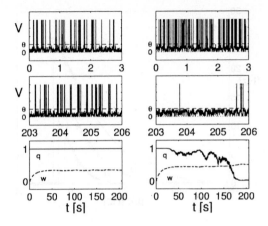

Fig. 2. Transient temporal evolution of the depolarization V of the IMF neuron: although at low firing rates (left panels, $\mu = -10\theta mVHz$, $\sigma^2 = 15\theta^2 mV^2Hz$) a stationary behavior is maintained indefinitely, higher rates are only transiently sustained (right panels, $\mu = 10\theta mVHz$, $\sigma^2 = 15\theta^2 mV^2Hz$) as the IMF ultimately relaxes to a much lower firing regime; bottom panels show the temporal evolution of q (continuous line) and w (dashed line) in the two situations. The characteristic time scales depend on the choice of τ_w (i.e. $10s$ in the reported simulations).

3.2   Numerical Results vs. Simulations

We have simulated the IMF model neuron (Eqs. 1, 2 and 3) to evaluate its RF for different values of μ and σ. Firing rates above a certain value (set by w_o) cannot be sustained for periods larger than a few τ_w, after which the rate decreases or the firing completely shuts down (see Fig. 2). Fig. 3 shows the agreement between theory (self-consistent solutions of Eqs. 2, 6, 7 and 8, replacing w with $\langle V \rangle/\theta$) and simulations. The decrease in the agreement for large μ is due to the decrease in accuracy while approximating w (i.e. the temporal average over a time window $\sim \tau_w$ of V/θ) with the expected value $\langle V \rangle/\theta$. Furthermore, for large σ and large firing rates, the agreement between theory and simulations slightly decreases, resulting in an underestimation of the simulated firing rates. Such a loss of accuracy is mainly due to the finite numerical integration time step employed in eq. 4, corresponding to a worse numerical approximation of a delta-correlated Gaussian stochastic process.

4   Conclusions

In the present contribution, we have introduced a novel IF model characterized by a non-monotonic f-I response function. The spike-emission process is not regarded as relying on the instantaneous temporal evolution of the membrane depolarization alone, but it shows a state-dependent *inactivation*, accounting for the recent evidence that *in vitro* cortical and spinal cord neurons cannot sustain high-frequency regimes for a long time [1,3]. The statistical steady-state behavior

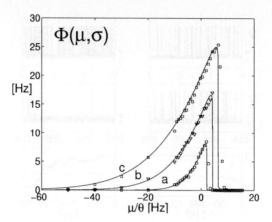

Fig. 3. Current-to-rate response function as predicted by the theory (continuous lines) and numerically simulated, assessing mean-rates across 50 s, after discarding a transient of $7\tau_w$: simulations refers to three increasing values of σ^2 (a: $5\theta^2$ - circles, b: $10\theta^2$ - triangles and c: $15\theta^2$ - squares).

of the depolarization was computed analytically and the agreement between the theory and the simulation of the IMF neuron is excellent, under white noise afferent current injection. Together with its simpler analytical tractability, this makes the IMF neuron an ideal candidate as a realistic reduced point-neuron model to investigate the impact of a non-monotonic transfer function and more complex collective phenomena at the network level.

References

1. Rauch A., La Camera G., Lüscher H., Senn W., Fusi S., Neocortical pyramidal cells respond as integrate-and-fire neurons to *in vivo*-like input currents, *submitted* (2002)
2. Silberberg G., Bethge M., Markram H., Tsodyks M., Pawelzik K., Rapid signalling by variance in ensembles of neocortical neurons, *submitted* (2002)
3. Darbon P., Scicluna L., Tscherter A., Streit J., Mechanisms controlling bursting activity induced by disinhibition in spinal cord networks, *Europ. J. Neurosci.* **15**:1–14 (2002)
4. Brunel N. and Zecchina R., Response functions improving performance in attractor neural networks, *Physical Review E*, **49**: R1823-1826 (1994)
5. La Camera G., Rauch A., Senn W., Lüescher H., Fusi S., Firing rate adaptation without losing sensitivity to input fluctuations, *Proceedings of ICANN 2002, Int. Conf. on Artificial Neural Networks, LNCS series, Springer* (2002)
6. Fusi S. and Mattia M., Collective behavior of networks with linear (VLSI) Integrate and Fire Neurons, *Neural Computation* **11**: 643–662 (1999)
7. Amit D.J. and Tsodyks M.V., Effective neurons and attractor neural networks in cortical environment, *NETWORK* **3**: 121–137 (1992)
8. Cox D.R and Miller H.D., The theory of stochastic processes, *London: METHUEN & CO LTD* (1965)

Small-World Effects in Lattice Stochastic Diffusion Search

Kris De Meyer, J. Mark Bishop, and Slawomir J. Nasuto

Department of Cybernetics, University of Reading, Whiteknights,
PO Box 225, Reading, RG6 6AY, United Kingdom
k.demeyer@rdg.ac.uk

Abstract. Stochastic Diffusion Search is an efficient probabilistic best-fit search technique, capable of transformation invariant pattern matching. Although inherently parallel in operation it is difficult to implement efficiently in hardware as it requires full inter-agent connectivity. This paper describes a lattice implementation, which, while qualitatively retaining the properties of the original algorithm, restricts connectivity, enabling simpler implementation on parallel hardware. Diffusion times are examined for different network topologies, ranging from ordered lattices, over small-world networks to random graphs.

1 Introduction

Stochastic Diffusion Search (SDS), first introduced in [1], is a population-based best-fit pattern matching algorithm. It shares many similarities with e.g. Evolutionary Algorithms, Memetic Algorithms and Ant Algorithms [2]. During operation, simple computational units or *agents* collectively construct a solution by performing independent searches followed by diffusion through the population of potentially relevant information. Positive feedback promotes better solutions by allocating more agents for their investigation. Limited resources induce strong competition from which a large population of agents corresponding to the best-fit solution rapidly emerges.

SDS has been successfully applied to a variety of real-world problems: locating eyes in images of human faces [3]; lip tracking in video films [4]; self-localisation of an autonomous wheelchair [5]. Furthermore, a neural network model of SDS using Spiking Neurons has been proposed [6]. Emergent synchronisation across a large population of neurons in this network can be interpreted as a mechanism of *attentional amplification* [7]; the formation of dynamic clusters can be interpreted as a mode of *dynamic knowledge representation* [8].

The analysis of SDS includes the proven convergence to the globally optimal solution [9] and linear time complexity [10]. Recently it has been extended to the characterisation of its steady state resource allocation [11].

As search is applied to ever more complex problems with larger search spaces, even the most efficient algorithms begin to require some form of dedicated hardware to meet real-world performance demands. The standard formulation of

J.R. Dorronsoro (Ed.): ICANN 2002, LNCS 2415, pp. 147–152, 2002.

SDS is parallel in nature and thus implementing it on parallel hardware seems straightforward. However, the requirement for efficient communication links between all agents means that it is difficult to implement efficiently, both on dedicated hardware (e.g. FPGA's) or general purpose parallel computers.

This paper describes the effects of restricting communication between agents. In particular, the effects of the number of connections and the topology of the underlying connection graph on search performance are investigated empirically. It will be shown that, even for a modest number of connections, the performance of randomly connected networks of agents is close to the performance of standard SDS, and much better than performance of ordered lattices with the same average number of connections. However, small-world networks [12], based on regular lattices with a few long-range connections, perform almost as good as random networks. Two important conclusions can be drawn from the results:

1. Inter-agent communication in SDS can be significantly restricted without decreasing the performance of the algorithm too much, given that either a random or small-world network topology is used. However, the limited number of long-range connections in small-world networks facilitates the layout of the connections, making them the preferred network topology for hardware implementation.
2. Independent from the actual search process of SDS, the paper seems to confirm results in several epidemiological models using the small-world network topology, e.g. [13,14]: namely that information or disease spreads much easier on small-world networks and random graphs than on ordered lattices.

2   Stochastic Diffusion Search

SDS utilises a population of *agents* to process information from the *search space* in order to find the best fit to a specified target pattern, the *model*. Both the search space and model are composed of *micro-features* from a pre-defined set. For instance, in a string matching problem, both the search space and model are composed of a one-dimensional list of characters.

In operation each agent maintains a hypothesis about the location and possible transformations (the *mapping*) of the model in the search space. It evaluates this hypothesis by testing how a randomly selected micro-feature of the model, when mapped into the search space, compares to the corresponding micro-feature of the search space. This part of the algorithm is called the *testing phase*. Based on the outcome of this test, agents are divided into two modes of operation: *active* and *inactive*. An active agent has successfully located a micro-feature from the model in the search space; an inactive agent has not.

During the *diffusion phase* the information about potential solutions may spread through the entire population. This is because each inactive agent chooses at random another agent for communication. If the selected agent is active, the selecting agent copies its hypothesis: *diffusion* of information. Conversely, if the selected agent is also inactive, then there is no information flow between agents; instead, the selecting agent adopts a new random hypothesis.

By iterating through test and diffusion phases agents will stochastically explore the whole search space. However, since tests will succeed more often in regions having a large overlap with the model than in regions with irrelevant information, an individual agent will spend more time examining 'good' regions, at the same time attracting other agents, which in turn attract even more agents. Potential matches to the model are thus identified by concentrations of a substantial population of agents.

Two important performance criteria for SDS are *convergence time* and *steady-state resource allocation*. Convergence time can in general be defined as the number of iterations until a stable population of active agents is formed and is very clearly defined when a single, perfect match of the model is present in the search space: it is then simply the number of iterations until all agents become active. Resource allocation is a measure for robustness in the case of imperfect matches and presence of noise: it is defined as the average number of active agents during steady-state behaviour, and is dependent on the quality of the match.

Examples of search behaviour, resource allocation and a more detailed description of the algorithm can be found in [11,15].

3   Lattice Stochastic Diffusion Search

SDS gains its power from the emergent behaviour of a population of communicating agents and as such is inherently a parallel algorithm - notionally each agent is independent and its behaviour can be computed by an independent processor. However, a fundamental difficulty in implementing standard SDS efficiently on either a parallel computer or dedicated hardware is its requirement that each agent is able to directly communicate with all others. An obvious alteration to the algorithm is thus to restrict agent communication to a smaller number of agents. In the resulting algorithm, Lattice Stochastic Diffusion Search (LSDS), agents are assigned to spatial locations (e.g. on a 2D square grid) and connections between agents are specified. During the diffusion phase, agents will only communicate with agents they are connected to. Regular, local connections lead to an ordered lattice; or connections can be specified at random, thus effectively constituting a random graph.

An important question is how the performance and robustness of LSDS compares to standard SDS. The performance of standard SDS has previously been extensively analysed using Markov chain theory [9,10,11]. However, in LSDS the probability distribution determining communication between agents defines a neighbourhood structure over the entire set of agents. Analysis of this kind of process as a Markov chain is extremely complex: the process is not characterised by a simple integer denoting the number of active agents, but by the exact topological location of both active and inactive agents. These types of Markov processes are also known as Markov random fields. Work on a mathematical model incorporating the effects of restricted connectivity is ongoing, but at present performance measures for LSDS are investigated through simulations.

Table 1. T_d in iterations for 4 different populations sizes N. Results are reported for random graphs with a mean number of connections per agent k; and for regular 2-dimensional square lattices with k-nearest neighbours connections and periodic boundary conditions. The case where $k = N$ corresponds to standard SDS. All results are averaged over 1000 runs, and for random graphs over 10 different graphs each.

	$k = 4$		$k = 8$		$k = 12$		$k = 24$		$k = N$
	random	lattice	random	lattice	random	lattice	random	lattice	
$N = 64$	15.5	15.7	11.5	13.4	10.9	11.8	10.3	10.4	10.2
$N = 256$	21.3	29.5	15.0	23.8	13.9	20.1	13.1	16.3	12.5
$N = 1024$	25.8	55.5	18.1	44.1	16.7	36.0	15.6	27.3	14.8
$N = 4096$	32.3	106.9	21.1	83.4	19.5	66.9	18.1	49.3	17.1

4   Experimental Results

4.1   Convergence Time

[16] introduced the terms 'time to hit' (T_h) and 'diffusion time' (T_d) in the analysis of convergence time (T_c) of standard SDS. T_h is the number of iterations before at least one agent of the entire population 'guesses' the correct mapping and becomes active. T_d is the time it takes for this mapping to spread across the population of agents. It is clear that T_h is independent of the connectivity of the population and only depends on search space size M and number of agents N. T_d, on the other hand, is very much dependent on the connectivity within the population and on population size N, but independent of M. To focus attention on the effect of connectivity, experimental results for T_d are reported. It could be argued that for complex, high-dimensional problems $T_h \gg T_d$, and thus that the effect of T_d on T_c can be neglected with respect to T_h. However, T_d should not just be regarded as a measure for rate of convergence, but more as a measure for 'ease of information spread'. As such, it is also indirectly a measure for robustness: experiments indicate that the more freely information spreads through the network, the more robust the algorithm is in the case of imperfect matches or noise [15].

T_d is studied by initialising one randomly chosen agent with the correct mapping and recording the number of iterations until this mapping has spread to all other agents. Results are reported in Table 1. T_d for regular lattices does not scale very well with population size for a fixed number of connections k. For random graphs, T_d scales much better with population size and performance remains close to performance of standard SDS, even for a small number of connections and large population sizes.

4.2   Small-Worlds: Between Order and Randomness

Regular lattices have poorer T_d than random graphs, but are easier implemented in hardware, since connections are local and thus shorter, and regular. However, diffusion of information in a population of searching agents shows an obvious

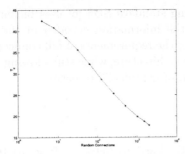

Fig. 1. T_d in iterations for $N = 1024$ and variable number of random extra connections x. Small-world networks are constructed starting from an ordered lattice with $k = 8$ and with x extra connections added at random. Note that for $x = 2048$, the last measurement, the mean connectivity in the network is $k = 12$. All results are averaged over 1000 runs, and over 10 different networks each.

correspondence with epidemiological models. Such models of disease or information spread have recently received much attention, due to the interest in the so called 'small-world effect'. It was shown in e.g. [12,17] that only a limited amount of long-range connections is necessary to turn a lattice with k-nearest neighbour connections into a small-world network, in which spreading of disease or information behaves much more like spreading on random graphs. To test whether the same is true for LSDS, small-world networks were generated as described in [18]: a number of random links is added to an ordered lattice, and no connections are removed. T_d is recorded for various numbers of random connections; the results can be seen in Fig. 1. Randomly adding connections decreases T_d more or less exponential for a wide interval of parameter x, leading to an almost linear curve in the semilog plot. The benefits of adding relatively few long range connections seem obvious: a small-world network with only 256 extra connections (mean connectivity $k = 8.5$) outperforms a regular lattice with $k = 24$; a small-world network with 512 extra connections (thus $k = 9$) diffuses information twice as fast as the underlying regular lattice with $k = 8$, and is only 1.5 times slower in diffusing than fully connected SDS. Note that, even when adding much more connections, T_d will never become less than the value for standard SDS, in this case 14.8 (see Table 1).

5   Conclusions

The effect of mean number of connections and connection topology on diffusion time T_d was investigated empirically. T_d is an important performance parameter, not just because of its effect on T_c, but more importantly because it is also an indicator for resource allocation stability [15].

The good performance of 'small-world' LSDS has wider implications than just implementation in hardware. It has been suggested (e.g. in [12]) that biological neural structures can show small-world connectivity. The neural network architecture implementing standard SDS [6] uses biologically inspired neurons operating as filters on the information encoded in the temporal structure of the spike trains. Relaxing the requirements of full connectivity in these networks leads to a more plausible architecture, while still allowing for self-synchronisation across a large population of neurons [7] to occur.

References

1. Bishop, J.M.: Stochastic Searching Networks. Proc. 1st IEE Conf. ANNs, London (1989) 329–331
2. Corne, D., Dorigo, M., Glover, F.: New Ideas in Optimisation. McGraw-Hill (1999)
3. Bishop, J.M., Torr, P.: The Stochastic Search Network. In Lingard, R., Myers, D.J., Nightingale, C.: Neural Networks for Images, Speech and Natural Language. Chapman & Hall, New York (1992) 370–387
4. Grech-Cini, E.: Locating Facial Features. PhD Thesis, University of Reading (1995)
5. Beattie, P.D., Bishop, J.M.: Self-Localisation in the SENARIO Autonomous Wheelchair. Journal of Intellingent and Robotic Systems **22** (1998) 255–267
6. Nasuto, S.J., Dautenhahn, K., Bishop, J.M.: Communication as an Emergent Methaphor for Neuronal Operation. Lect. Notes Art. Int. **1562** (1999) 365–380
7. De Meyer, K., Bishop, J.M., Nasuto S.J.: Attention through Self-Synchronisation in the Spiking Neuron Stochastic Diffusion Network. Consc. and Cogn. **9(2)** (2000)
8. Bishop, J.M., Nasuto, S.J., De Meyer, K.: Dynamic Knowledge Representation in Connectionist Systems. ICANN2002, Madrid, Spain (2002)
9. Nasuto, S.J., Bishop, J.M.: Convergence Analysis of Stochastic Diffusion Search. Parallel Algorithms and Applications **14:2** (1999) 89–107
10. Nasuto, S.J., Bishop, J.M., Lauria, S.: Time Complexity of Stochastic Diffusion Search. Neural Computation (NC'98), Vienna, Austria (1998)
11. Nasuto, S.J., Bishop, J.M.: Steady State Resource Allocation Analysis of the Stochastic Diffusion Search. Submitted (2002) cs.AI/0202007
12. Watts, D.J., Strogatz, S.H.: Collective Dynamics of 'Small-World' Networks. Nature **393** (1998) 440–442
13. Zanette, D. H.: Critical Behavior of Propagation on Small-World Networks. Physical Review E **64:5** (2001) 901–905
14. Kuperman, M., Abramson, G.: Small-World Effect in an Epidemiological Model. Physical Review Letters **86:13** (2001) 2909–2912
15. De Meyer, K.: Explorations in Stochastic Diffusion Search. Technical Report KDM/JMB/2000-1, University of Reading (2000)
16. Bishop, J.M.: Anarchic Techniques for Pattern Classification, Chapter 5. PhD Thesis, University of Reading (1989)
17. Moukarzel, C. F.: Spreading and Shortest Paths in Systems with Sparse Long-Range Connections. Physical Review E **60:6** (1999) R6263–R6266
18. Newman, M.E.J., Watts, D.J.: Scaling and Percolation in the Small-World Network Model. Physical Review E **60:6** (1999) 7332–7342

A Direction Sensitive Network Based on a Biophysical Neurone Model

Burkhard Iske[1], Axel Löffler[2], and Ulrich Rückert[1]

[1]Heinz Nixdorf Institute, System and Circuit Technology,
Paderborn University, Germany
{iske,rueckert}@hni.upb.de, http://wwwhni.upb.de/sct/
[2]Earth Observation & Science, Dep. of Systemdynamics and Simulation, Astrium GmbH
Axel.Loeffler@astrium-space.com
http://www.astrium-space.com/

Abstract. To our understanding, modelling the dynamics of brain functions on cell level is essential to develop both a deeper understanding and classification of the experimental data as well as a guideline for further research. This paper now presents the implementation and training of a direction sensitive network on the basis of a biophisical neurone model including synaptic excitation, dendritic propagation and action-potential generation. The underlying model not only describes the functional aspects of neural signal processing, but also provides insight into their underlying energy consumption. Moreover, the training data set has been recorded by means of a real robotics system, thus bridging the gap to technical applications.

1 Introduction

In this work, the focus is on the iplementation and training of a direction sensitive neural network based on a biophysical neurone model. The first part includes a brief description of the synaptic excitation, the propagation of this excitation along a dendritic branch and finally the axon hillock with the generation of an action potential, which results from a phase correct superposition of several synaptic excitations propagated along different dendritic branches. Based on this model described in [3] a direction sensitive network is implemented and trained with a set of sensor data recorded with a real robot. The fundamental difference to conventional training algorithms for neural networks is that the lengths of the dendritic branches are adapted here instead of the synaptic strengths leading to a spatio-temporal synchronisation of membrane potentials. The coding and training of neural networks based on phase correlations and timing conditions respectively is explained in [2]. The one-layered direction sensitive network provides a means for a very fast signal processing and therefore appears to play a major role in both animal orientation and in the area of autonomous navigation of mobile robots.

J.R. Dorronsoro (Ed.): ICANN 2002, LNCS 2415, pp. 153–159, 2002.

2 The Neurone Model

The neurone model introduced in [3] and in a more detailed form in the PhD thesis [4] describes the concentration changes of sodium (Na^+) and potassium (K^+) ions across the membrane of a biological neurone occurring during signal propagation. Within the membrane of a biological neurone there are separate channels for Na^+ and K^+ ions. These channels are opened for signal propagation but also in quiscet state some channels are open. In order to maintain the quiescent concentration gradient between the intra-cellular and extra-cellular space a transport protein called NaK-ATPase exists, which can be regarded as an ion-pump pumping 3 Na^+ ions from inside the membrane to outside and in the same cycle 2 K^+ ions from outside to inside. To energetically drive this process approximately of energy is required. The mentioned physiological processes can be described through differential equations. In order to solve them with reasonable effort the extra-cellular concentrations are assumed to remain constant, what is approximately given through the biological environment and a linear regime in the vicinity of the operating point is considered.

2.1 The Base Model

For modelling the so called channel currents, the channels are considered as ohmic resitors and the potentials are described through linearised Nernst equilibrium equations. For the total potential U the potentials U_n (n = Na, K) have to be superposed, but additionally the NaK-ATPase (pump rate $f = 3/2$) and the relative permeability P have to be considered. This can be described through the so called steady state equation. Since the potentials depend non-linearly on the concentrations c_n (n = Na, K) they are linearised around the operating point using a first order taylor series.

2.2 The Synaptic Excitation

The non-linear processes of the synaptic excitation are based on the time-dependent conductances and controlled by the release of neurotransmitters. If a neurotransmitter is bonded to a channel the channel opens and its conductivity increases. It should be noted that the ratio of the two conductivities is considerably changed in this process resulting in a transiently increasing potential, a higher pump rate and energy consumption respectively.

2.3 Action Potential Generation

An action potential is generated if an incoming potential profile e.g. passively propagated by the dendritic tree of the neurone from the synapses to the axon hillock, causes the local potential to exceed the threshold voltage. In contrast to the synaptic excitaion, additional time-dependent conductivities are modelled as being ion-specific here and are determined by the so called empiric functions m, n, and h, which are voltage dependent themselves. For each of the empiric functions a potential dependent transition probability equation exists and requires solving.

2.4 The Dendritic Propagation

The fundamental difference to the synapse and action potential generation consists in the propagation of the signal, which is based upon a change in the concentration gradient across the dendritic branch, whereas the conductances remain constant. Therefore, the model has to be extend with a term describing the location dependence. The concentration change leads to a diffusion of the ions along the dendritic branch additional to the membrane currents. Apart from that, an electric field evolves through the local concentration gradient along the dendritic branch. Due to the complexity of the resulting partial nonlinear differential equation, the effects of the electric field are linearised and considered as an additional diffusion.

3 The Direction Sensitive Network

One application among technical neural networks are direction sensitive networks, with which symbolic angular information can be extracted from sub symbolic sensor data [1]. With the mini-robot Khepera for example a direction sensitive network can be used to map the position of light-sources. This functionality is now to be reproduced with the new biophysical neurone model introduced above. The recognition of different angles is thereby based on different run-times of signals, which result from different lengths of dendritic branches. Therefore, the measured real sensor data is transformed to a corresponding time delay, which determines when a signal is transferred from the synapse to the dendritic branch. Through the dendritic run-time differences the maxima of the signals are superposed phase correctly for a trained neurone and the threshold value for the generation of an action potential is exceeded. With an non-matched neurone the maxima will not synchronise and the threshold value will not be exceeded and thus the neurone not activated i.e. generating an action potential.

3.1 Recording the Sensor Data

To record the training data the robot is positioned at a distance of approx. 1cm to a light-source and turned by 180° (Fig. 1). While turning, the 6 sensor values are recorded with the corresponding angles. The resulting sensor values are shown in Fig. 2.

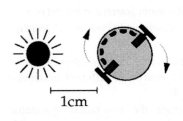

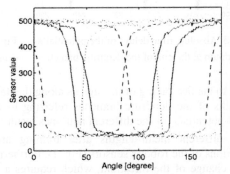

Fig. 1. Schematic of the robot recording the sensor data with ist six light sensors.

Fig. 2. Sensor values developing at the six Khepera sensors. A small value corresponds to a high light intensity.

3.2 The Training Algorithm

The fundamental difference of this training method to others is an adaptation of the length of the dendritic branch instead of a change of the strength of the synaptic excitation. To ensure a proper functioning of the neurones, the lengths of the dendritic branches of the different neurones have to be adapted to the time delays resulting from the different sensor values. The sensor values S_i (i=0..5) range from 0 to 512 and the lengths of the dendritic branches is to be taken out of the interval [200µm; 250µm]. The maximal delay between the potential maxima with dendritic branches of length 200µm and 250µm is about 6.5ms. To maintain the possibility of slight variations the maximum time delay is set to 5ms. The formula to transfer the sensor values into time delays is given in Eq. (1). A large sensor value results with this coding in a small time delay, thus a longer branch length and finally a smaller contribution to the total potential at the axon hillock.

In our case 12 neurones are employed to recognise angles from 0° to 180°. The 12 neurones are equally distributed along the 180°, which means that the lengths of the dendritic branches is exactly set to input vectors of the angles from 7.5° to 172.5° in steps of 15°. This is achieved by firstly calculating the time delays from the sensor data vectors and followingly choosing the length of the dendritic branch such that an overall time delay of 5ms results (Fig. 3) with respect to the time delay coming from the first 200µm of a dendritic branch. After execution of this process all 12 neurones are trained each to exactly one specific angle, with the minimal length of a dendritic branch of 200µm and the maximal length resulting from the total time delay of 5ms.

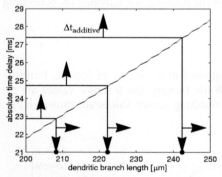
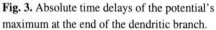

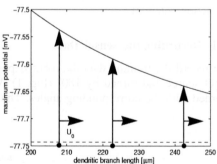

Fig. 3. Absolute time delays of the potential's maximum at the end of the dendritic branch.

Fig. 4. Maximum potential at the end of a dendritic branch for different branch lengths.

Due to the fast fading of the signals along the dendritic branch and the different mean-lengths of the dendritic branches referring to the 12 neurones, different maximum total-potentials for the different neurones result (Fig. 4).

The different mean-lengths after training arise from the non-radial symmetric positions of the robot's sensors (Fig. 1). Furthermore, the neurones react very sensible to a change of the threshold, which requires a different threshold for each neurone. This requirement can be avoided using a modified training algorithm. In the modification a suitable upper bound for the maximum total-potential for all neurones is chosen (here 0.8mV). Then the neurones are trained with the procedure described above. If a neurone's maximum total-potential exceeds the chosen upper bound, the minimum length of the dendritic branches is incremented by 1μm and the training is repeated. This leads to an increase of the mean branch length and lowers the maximum potential thus. If the upper bound is still exceeded the minimum length of the dendritic branches is again incremented by 1μm until the bound condition is met. Thereby, maximum potentials out of the interval [0.7709mV; 0.7981mV] can be achieved, were the difference to 0.8mV can be explained by the insufficient numerical resolution. Due to the variation of the maximum total-potentials a separate threshold is applied to each neurone to achieve better results.

4 Simulation Results

After training, the sensor data recorded with the Khepera mini-robot is presented to the direction sensitive network. Only with the neurones being trained for a specific angle, the potentials of the six dendritic branches arriving at the axon hillock will superpose phase correctly and the threshold for generating an action potential is exceeded. Fig. 5 and Fig. 6 show the time-dynamics at the axon hillock of two neurones after presenting the sensor data with a light-source at 54°.

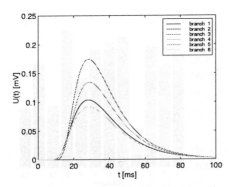
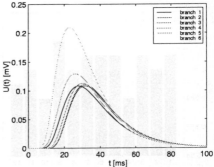

Fig. 5. Time-dynamics of the potential at the axon hillock of neurone 4 (52.5°). Here the different single potentials superpose almost phase correctly and the treshold is exceeded.

Fig. 6. Potential at the axon hillock of neurone 12 (172.5°). The maxima of the different single potentials do not superpose phase correctly and the threshold is not exceeded.

The response of the direction sensitive network to the complete sensor data set recorded with the Khepera mini-robot is shown in Fig. 7 and Fig. 8. It can clearly be seen that the angular extension of the receptive fields of single neurones decreases non-linearly with an increase of the threshold. This becomes particularly obvious when one for example concentrates on the first three neurones that are marked with the dashed lines.

Additional to the functionality of the direction sensitive network, the biophysical model also allows statements about the energy consumption, by evaluating the number of pump cycles of the NaK-ATPase. Fig. 9 shows the total additionally dissipated energy of all 12 neurones. Moreover, a correlation between the threshold potential of a neurone (Fig. 10) and its mean additional energy consumption is visible. The higher the threshold is set the higher the corresponding energy consumption becomes.

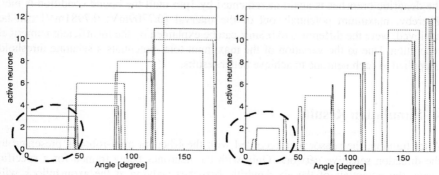

Fig. 7. Receptive fields of the 12 neurones with a threshold of 99% of the maximum total-pontential.

Fig. 8. Potential at the axon hillock of neurones with a threshold of 99.99% of the maximum total-potential.

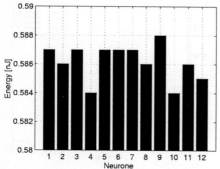

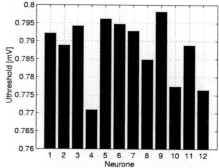

Fig. 9. Total additionally (i.e. activation dependent) dissipated energy of the 12 neurones.

Fig. 10. Threshold potential of the 12 neurones. A correlation to the additional energy consumption can clearly be seen.

5   Discussion and Conclusion

In this paper, it is shown that a network of biophysical neurones can be used to implement a direction sensitive network, working with real sensor data. In contrast to conventional neural networks the sensor data here were time-coded and accordingly the training algorithm is based on a phase synchronisation of dendritic end-potentials. The analysis of the direction sensitive network showed that the natural phenomenon of receptive fields could be reproduced with a sensibility control through threshold adaptation. The synaptic strengths are not used to further influence the result, which could also be envisaged if a higher accuracy was required. A side effect of the biophysical model is the possibility to also calculate the energy consumption of single neurones as well as networks of neurones.

Acknowledgement. This work has been partly supported by the Deutsche Forschungsgemeinschaft (German Research Council) DFG Graduate Center "Parallele Rechnernetzwerke in der Produktionstechnik".

References

1. A. Löffler, J. Klahold, U. Rückert, *The Mini-Robot Khepera as a foraging Animate: Synthesis and Analysis of Behaviour*, Autonomous Minirobots for Research and Edutainment AMiRE 2001, Proceedings of the 5th International Heinz Nixdorf Symposium, pp. 93–130, 2001

2. J. J. Hopfield, *Pattern recognition computation using action potential timing for stimulus representation*, Nature, **376**, pp. 33 to 36, 6 July 1995

3. A. Löffler, B. Iske, U. Rückert, *A New Neurone Model Describing Biophysical Signal Processing and Energy Consumption,* submitted to ICANN 2002, Madrid, Spain

4. A. Löffler, *Energetische Modellierung Neuronaler Signalverarbeitung*, PhD thesis (in German), HNI-Verlagsschriftenreihe 72, Paderborn, 2000

Characterization of Triphasic Rhythms in Central Pattern Generators (I): Interspike Interval Analysis

Roberto Latorre, Francisco B. Rodríguez, and Pablo Varona

GNB. E.T.S. Informática, Universidad Autonóma de Madrid,
Ctra. Colmenar Viejo, Km. 15, 28049 Madrid, Spain.
{Roberto.Latorre, Francisco.Rodriguez,Pablo.Varona}@ii.uam.es

Abstract. Central Pattern generators (CPGs) neurons produce patterned signals to drive rhythmic behaviors in a robust and flexible manner. In this paper we use a well known CPG circuit and two different models of spiking-bursting neurons to analyze the presence of individual signatures in the behavior of the network. These signatures consist of characteristic interspike interval profiles in the activity of each cell. The signatures arise within the particular triphasic rhythm generated by the CPG network. We discuss the origin and role of this type of individuality observed in these circuits.

1 Introduction

Central Pattern Generators (CPGs) are assemblies of neurons that act cooperatively to produce regular signals to motor systems. CPGs are responsible for activities like chewing, walking and swimming [1]. The inner properties of every neuron in the CPG, together with the connection topology of the network and the modulatory inputs, determine the shape and phase relationship of the electrical activity. A single CPG (acting alone or together with other CPGs) can generate many different patterns of activity that control a variety of motor movements [2]. An essential property for these neural assemblies is the presence of robust and regular rhythms in the membrane potentials of their member neurons. Paradoxically, some of these cells can display highly irregular spiking-bursting activity when they are isolated from the other members of the CPG [3, 4] manifesting a rich individual dynamics.

Recently, the presence of characteristic interspike interval profiles in the activity of individual CPG neurons has been revealed using *in vitro* preparations of the pyloric CPG of the lobster [5,6]. These individual *signatures* of the CPG cells can have important implications for the understanding of the origin of the rhythms, their fast response to modulation and the signaling mechanisms to the muscles that the CPGs control. In this paper we want to study the origin of the individual neuron signatures. In particular, we want to determine their dependence on the network architecture and on the properties of the individual neural dynamics. For this goal, we will use two different types of spiking-bursting

J.R. Dorronsoro (Ed.): ICANN 2002, LNCS 2415, pp. 160–166, 2002.
© Springer-Verlag Berlin Heidelberg 2002

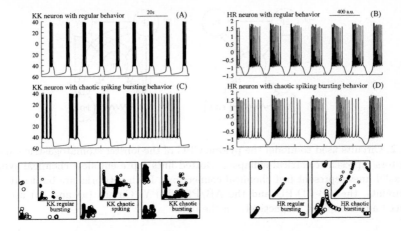

Fig. 1. Top panel: time series of the isolated model neurons (chaotic spiking in KK neurons not shown). Units are s and mV for KK neurons. Units are dimensionless for HR neurons. Bottom panel: ISI return maps of isolated KK and HR neurons. Axis length is 14s for KK neurons and 160 a.u. for HR neurons. Insets show a blowup of the region close to the origin (axis length is 4s for KK and 50 a.u. for HR) that will be compared with the ISIs obtained from the network activity.

neuron models: Komendantov-Kononenko [7] neurons (a Hodgkin-Huxley type model with eight dynamical variables), and Hindmarsh-Rose [8] neurons with three dynamical variables. These two models have in common a rich dynamical behavior with the ability to generate the characteristic chaotic spiking-bursting activity observed in isolated CPG cells.

2   Single Neuron Models

Komendantov-Kononenko (KK) type neurons [7] can have several patterns of activity as a function of the parameters used in the model. In this paper, we use three different behaviors: regular spiking-bursting, chaotic spiking and chaotic spiking-bursting activity (see Fig. 1). Each behavior is the result of a particular choice for the values of the maximum conductances of the ionic channels used in the model. The equations that describe the dynamics of the model and the parameters used to obtain each type of behavior can be found in [7]. The inter-spike interval (ISI) return maps (ISI_i vs. ISI_{i+1}) of the isolated KK neurons can be seen in Fig. 1 (left bottom panel) for the three different behaviors used in this paper. These return maps, which can be considered as signatures of the individual activity, constitute a useful tool to analyze the regularity of the spiking activity for each mode of behavior.

The Hindmarsh-Rose (HR) neurons [8] can also display a wide variety of behaviors. Right panels in Fig. 1 show two time series corresponding to the regular and chaotic spiking-bursting activity in this model, respectively. The equations of the HR model and the values of the parameters used in our simulations can

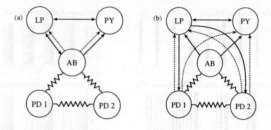

Fig. 2. Circuits used to model the pyloric CPG of the lobster stomatogastric ganglion. Resistors represent electrical synapses. Dotted lines represent slow chemical synapses and solid lines represent fast chemical connections. Both networks are built out of two subcircuits: the AB-PD-PD and the AB/PD-LP-PY. Circuit (a) and (b) are called reduced and complete, respectively, in the text.

be found in [9]. The right bottom panels in Fig. 1 show the ISI return maps of the isolated HR neurons for regular and chaotic spiking-bursting activity.

3   CPG Network

The well-known pyloric CPG of crustacean [10] is formed by 14 different neurons: one anterior buster (AB), two pyloric dilators (PDs), one lateral pyloric (LP), one inferior cardiac (IC), one ventricular dilator (VD) and eight pyloric neurons (PYs). In our models we have simplified this circuit by building a network of five neurons: the AB, the two PDs, the LP and one PY (a single PY neuron represents the eight electrically coupled PYs). We exclude the action of the IC and VD neurons since this two neurons do not innervate the pyloric muscles but the cardio-pyloric valve. For the analysis we have used two different network topologies depicted in Fig. 2. Both networks are built out of two subcircuits: the AB-PD-PD, with electrical coupling, and the AB/PD-LP-PY, with chemical connectivity. The difference between the two topologies is the way the two subnetworks are connected. The circuit in Fig. 2a has a single union point between the two subcircuits, the AB neuron. This neuron sends and receives signals of both subcircuits. This is a simplification often used as a consequence of the synchronization of AB and PD neurons (see below). Fig. 2b is a more realistic circuit. The PD neurons in this case are presynaptic cells for all the slow chemical synapses between AB-PD-PD and AB/PD-LP-PY subcircuits; while the LP neuron is connected to both PD neurons with a fast chemical connection. We will use both KK and HR type neurons to build these two different topologies of the CPG. We will also consider *damaged* circuits without the presence of slow chemical connections. In this way, we will be able to analyze the dependence of the spiking behavior on the individual neuron model and/or the architecture of connections.

In the AB-PD-PD subnetwork (often called the pacemaker group), the neurons are electrically coupled with symmetric connections [3,4] (see table 1). As a consequence, the membrane potentials of these neurons are synchronized. Physiologists have always recorded a regular spiking-bursting activity in the AB neuron even when isolated from the rest of the circuit. For this reason, in all our models the AB model neuron has a regular behavior. Conversely, the isolated PD model neurons are set into a non-regular regime since this is the behavior observed in the experiments on isolated cells.

The AB/PD-LP-PY network comprises the pacemaker group and the LP and PY neurons (Fig. 2). The later have connections to and from the pacemaker group by means of fast and slow chemical synapses [10,11]. With this architecture of connections, the living pyloric CPG generates a robust triphasic rhythm: there is an alternating regular bursting activity between the pacemaker group, the LP and the PY neuron.

In our simulations we model the chemical synaptic currents as follows:

$$I_{fastX} = \sum_Y \frac{g_{fastYX}(V_X - E_{syn})}{1.0 + exp(s_{fast}(V_{fast} - V_Y))}; I_{slowX} = \sum_Y g_{slowYX} \, m_{slowX} \, (V_X - E_{syn})$$

(1)

The total synaptic current $I_{synX} = I_{slowX} + I_{fastX}$. Here m_{slowX} is given by:

$$\frac{dm_{slowX}}{dt} = \frac{k_1 X(1.0 - m_{slowX})}{1.0 + exp(s_{slow}(V_{slow} - V_{AB}))} - k_2 X \, m_{slowX}$$

(2)

where X and Y are the postsynaptic and presynaptic neuron, respectively. The values of the synapse parameters are shown in table 2. The values of maximal conductances used in our simulations are given in tables 3 and 4.

4   Results

We have built several networks using the two topologies shown in Fig. 2 (both for KK and HR neuron models) to analyze the interspike interval return plots of the individual neurons when the network generates triphasic rhythms.

Before establishing the connections in all circuits we have placed each neuron (except AB) in a chaotic regime. The AB is tuned into a regular spiking-bursting behavior. For KK networks we have chosen chaotic spiking for the PY and one of the PD neurons, while the second PD and the LP are set into a chaotic spiking-bursting behavior. For the HR networks, all neurons but the AB are set in the chaotic spiking-bursting regime. When we connect all circuits with the parameters specified above, a characteristic triphasic rhythm arises independently of the mode of activity in isolation, the particular connection topology and the neuron model used in the network. The triphasic rhythm consists of an alternation of regular bursts in the activity of the AB, LP and PY neurons in this sequence (PDs follow the AB phase as mentioned before). This can be seen in Fig. 1 of the second part of this paper [12].

Fig. 3 shows the interspike interval return plots for the different connection configurations and the two neural dynamics used to model the CPG neurons. In this paper we concentrate our attention in the analysis of the regularity of the timing of the spikes within the bursts, rather than the timing between the last

Table 1. Values of maximal conductances of electrical synapses in the AB-PD-PD subnetwork. g_{XY} represents strength of electrical connection between X and Y neuron. Units are μS for KK and dimensionless for HR. (r) and (c) denote values for the reduced and complete circuits displayed in Fig. 2, respectively. **Table 2.** Values of AB/PD-LP-PY chemical synapse parameters for all topologies. **Table 3.** Values of AB/PD-LP-PY network maximal conductances of fast chemical connections. **Table 4.** Values of AB/PD-LP-PY network maximal conductances for slow chemical connections.

Table 1

		g_{ABPD1}	g_{PD1AB}	g_{ABPD2}	g_{PD2AB}	g_{PD1PD2}	g_{PD2PD1}
KK	(r)	0.0096	0.0096	0.0223	0.0223	0.0151	0.0151
	(c)	0.0096	0.0096	0.0223	0.0223	0.0151	0.0151
HR	(r)	0.325	0.325	0.548	0.548	0.332	0.332
	(c)	0.325	0.325	0.548	0.548	0.332	0.332

Table 2

	E_{syn}	V_{fast}	s_{fast}	s_{slow}	V_{slow}	$k_1 LP$	$k_2 LP$	$k_1 PY$	$k_2 PY$
KK	-65.0	-44.7	0.31	1.0	-49.0	1.0	0.01	1.0	0.0275
HR	-1.92	-1.66	0.44	1.0	-1.74	0.74	0.007	0.74	0.015

Table 3

		g_{ABLP}	g_{ABPY}	g_{LPAB}	g_{LPPD1}	g_{LPPD2}	g_{LPPY}	g_{PYLP}
KK	(r)	0.0446	0.0556	0.0578	—	—	0.0398	0.0311
	(c)	0.0446	0.0556	—	0.0211	0.0269	0.0398	0.0311
HR	(r)	0.112	0.120	0.585	—	—	0.241	0.186
	(c)	0.112	0.120	—	0.208	0.432	0.241	0.186

Table 4

		g_{ABLP}	g_{ABPY}	g_{PD1LP}	g_{PD1PY}	g_{PD2LP}	g_{PD2PY}
KK	(r)	0.0043	0.0056	—	—	—	—
	(c)	—	—	0.0015	0.0023	0.0033	0.0028
HR	(r)	0.032	0.029	—	—	—	—
	(c)	—	—	0.046	0.065	0.038	0.035

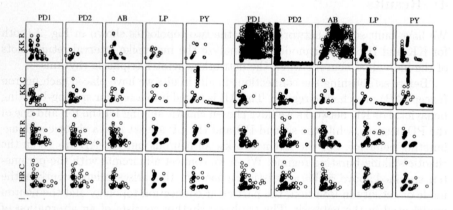

Fig. 3. ISI return plots or *signatures* of the neurons for all network models considered in this study. Left panel shows the ISIs in the circuits with the slow chemical connections. Right panel correspond to the ISIs in circuits without the slow synapses (damaged circuits). Axis length is 0.8s and 30 a.u. for KK and HR neurons, respectively. *C* and *R* row labels stand for networks with complete or reduced topology, respectively (see figure 2).

spike in one burst and the first in the following burst. For this reason the ISIs depicted in Fig. 3 omit these events. Left panel of Fig. 3 displays the ISIs for the circuits with slow chemical synapses. Right panel shows the ISIs for damaged circuits without slow connections. The ISIs of network elements dramatically change from those corresponding to isolated neurons (cf. Fig. 1, note the different time scales in the axis). When the connections are established, the neuron signatures change even for the AB that also had a regular dynamics in isolation. In all simulations the networks display a regular triphasic rhythms (cf. Fig. 1 in [12]). Regular in this context means that the burst width remains constant from burst to burst. Each neuron in the network has a characteristic ISI signature. As can be seen in Fig 3, the signatures depend both on the particular neuron model used to implement the circuit and on the topology of connections. In general, the presence of slow chemical connections produced a more precise ISI return plot than those corresponding to damaged circuits (for both neuron types and for both neuron topologies: reduced R and complete C). KK pacemaker group neurons lost ISI regularity in the reduced circuits without the slow connections. Looking at the neuron types we can see that HR pacemaker neurons had less precise ISIs than KK pacemaker groups (C circuits). This difference points out the dependence of the signatures on the particular individual dynamics. This can also be seen in the PD neuron profiles. In KK networks, PD1 and PD2 neurons were placed in two different modes in activation, and had major differences in the ISI plots in the damaged circuits. Simulations also show that in spite of its intrinsic regular bursting behavior, the AB can have complex signatures. Interestingly, LP and PY neurons had, in general, very similar and precise signatures in our simulations. These neurons are the only units that receive at least a double inhibition in all circuits (up to 4 inhibitory connections in the more realistic circuit).

5   Discussion

In this paper we have shown that CPG networks of intrinsically irregular elements can generate robust and regular triphasic rhythms in several network configurations using two different neuron models. Even though the triphasic rhythms are stable and regular in all cases (see the analysis in the second part of this paper [12]), there exists a presence of structure in the interspike interval of the individual cells in each case. These signatures depend on the individual dynamics and also on the particular connections of the network. CPGs are thought to be multifunctional circuits. Different modulatory input can modify the internal dynamics of the cells and the performance of the connections inducing different rhythms in the CPG[1]. In addition, several CPGs can cooperate together in different tasks depending on the particular circumstances under which these circuits are working. The presence of individual neuron signatures within the overall rhythmic activity of the network can be of particular importance both for the functioning of the circuit and for the behavior of the muscles that operate with these commands.

Acknowledgments. This work has been supported by MCT BFI2000-0157, CICyT TIC98-0247-C02-02 and TIC2002-572-C02-02 grants.

References

1. Selverston A.: What invertebrate circuits have taught us about the brain. *Brain Research Bulletin*, 50(5-6) (1999) 439-40.
2. Marder E., Calabrese R.L.: Principles of rhythmic motor pattern production. *Physiol. Rev.*, 76 (1996): 687-717.
3. Elson R.C., Selverston A.I., Huerta R., Rulkov N.F., Rabinovich M.I., Abarbanel H.D.I.: Synchronous Behavior of Two Coupled biological Neurons, *Physical Review Letters*, 81 (1988) 5692.
4. Varona P., Torres J.J., Huerta R., Abarbanel H.D.I., Rabinovich M.I.: Regularization mechanims of spiking-bursting neurons. *Neural Networks*, 14 (2001) 865-875.
5. Szücs A., Pinto R.D., Rabinovich M.I., Abarbanel H.D.I., Selverston A.I.: Aminergic and synaptic modulation of the interspike interval signatures in bursting pyloric neurons. Submitted (2002).
6. Pinto R.D., Szucs A., Huerta R., Rabinovich M.I., Selverston A.I., Abarbanel, H.D.I.: Neural information processing: Analog simulations and experiments. *Soc. for Neurosc. Abs.*, 27(2) (2001).
7. Komendantov A.O., Kononenko N.I.: Deterministic Chaos in Mathematical Model of Pacemaker Activity in Bursting Neurons of Snail, Helix Pomatia. *J. theor. Biol.*, 183 (1996) 219-230.
8. Hindmarsh, J.L., Rose, R.M.: A Model of Neuronal Bursting Using Tree Coupled First Order Differential Equations. *Philos. Trans. Royal Soc. London*, B221 (1984) 87-102.
9. Rodríguez F.B., Varona P., Huerta R., Rabinovich M.I., Abarbanel H.D.I.: Richer network dynamics of intrinsically non-regular neurons measured through mutual information. *Lect. Notes Comput. Sc.*, 2084 (2001): 490-497.
10. Selverston A.I., Moulins M.: The Crustaceam Stomatogastric System: a Model for the Study of Central Nervous Systems. Berlin; New York: Springer-Verlag (1987).
11. Golowasch J., Casey M., Abbott L.F., Marder E.: Network stability from activity-dependent regulation of neuronal conductances. *Neural Computation*, 11(N5) (1999) 1079-96.
12. Rodríguez F.B., Latorre R., Varona P.: Characterization of triphasic rhythms in central pattern generators (II): burst information analysis. ICANN'02 proceedings. LNCS (2002).

Characterization of Triphasic Rhythms in Central Pattern Generators (II): Burst Information Analysis

Francisco B. Rodríguez, Roberto Latorre, and Pablo Varona

GNB. E.T.S. Ingeniería Informática, Universidad Autonóma de Madrid,
Ctra. Colmenar Viejo, Km. 15, 28049 Madrid, Spain.
{Francisco.Rodriguez, Roberto.Latorre, Pablo.Varona}@ii.uam.es

Abstract. Central Pattern generators (CPGs) are neural circuits that produce patterned signals to drive rhythmic behaviors in a robust and flexible manner. In this paper we analyze the triphasic rhythm of a well known CPG circuit using two different models of spiking-bursting neurons and several network topologies. By means of a measure of mutual information we calculate the degree of information exchange in the bursting activity between neurons. We discuss the precision and robustness of different network configurations.

1   Introduction

Central Pattern Generator (CPG) neurons act cooperatively to produce regular signals to motor systems. The CPGs of the stomatogastric ganglion of crustacean is one of the most studied and best known neural circuits. The pyloric CPG is responsible for the triphasic sequence of activation of the muscles of the pylorus. Each cycle begins with simultaneous contractions of the extrinsic dilator muscles followed by contractions of the intrinsic constrictors and finally the posterior intrinsic constrictors [1]. An essential property of the neural assemblies that control this motor activity is the presence of robust and regular rhythms in the membrane potentials of their member neurons. Paradoxically, and as mentioned in the first part of our study [2], some of these cells can display highly irregular spiking-bursting activity when they are isolated from the other members of the CPG showing a rich individual dynamics. After studying the presence of structure in the interspike intervals within the triphasic signals generated by the CPG models in [2], we now address the analysis of the precision and robustness of the bursts. For this task we use a measure of mutual information that characterizes the slow dynamics of these rhythms.

2   The Neural Models

To study the information exchange between neurons in CPG circuits we will use the same single neuron models – Hindmarsh-Rose (HR) and Komendantov-Kononenko (KK) type neurons – and the same complete and reduced connection

J.R. Dorronsoro (Ed.): ICANN 2002, LNCS 2415, pp. 167–173, 2002.

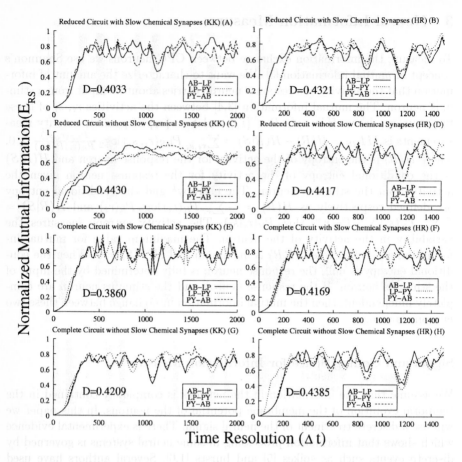

Fig. 2. Transmission of information between neurons $AB-LP$, $LP-PY$ and $PY-AB$ corresponding to the simulations shown in Fig. 1. The quantity E_{RS} is displayed as a function of the time resolution, Δt. The size of word chosen is $L = 10$. Each time series analyzed had 625 and 750 bursts for KK and HR neurons, respectively.

$N - L + 1$ words of L bits in each activity time series. We map these activity time series to activity words for fixed values of the parameters Δt and L.

Normalized Information Measure

In order to calculate the mutual information between two neurons we consider one neuron in each connection as the output neuron (response neuron) and the other neuron as the input neuron (stimulus neuron). Then, for a fixed value of Δt and L, we calculate $P_{RS}(w_{r_i}, w_{s_i})$ to estimate the average of mutual information, where the activity words w_{r_i} and w_{s_i} are drawn from sets $WR_{\Delta t}^{L} = \{w_{r_i}\}$ and $WS_{\Delta t}^{L} = \{w_{s_i}\}$ of all possible words for the response and the stimulus neurons. We calculate the activity words for all time series shown in Fig. 1 following the criterion explained above. The normalized average mutual information is given by the expression $E_{RS} = \frac{MI_{RS}}{H(S)}$. This quantity measures the efficiency of the information transmission from the stimulus neuron to the response neuron. The normalized average mutual information is dimensionless and its value range is $0 \leq E_{RS} \leq 1$. Due to the fact that $H(S)$ is the maximal information amount that can be transferred to the response neuron from the stimulus neuron, $E_{RS} = 0$ means that all information is lost (response and stimulus are independent quantities). On the other hand, $E_{RS} = 1$ means that there is a perfect matching between the neurons (complete synchronization).

4   Results

We have measured the information exchange between the three neuron pairs $AB - LP$, $LP - PY$ and $PY - AB$ as a function of the time resolution, Δt for each topology and neuron model whose time series are shown in Fig. 1. For all simulations we have chosen an optimal word size of $L = 10$ (see detailed analysis in [9]). We are interested in comparing the activity of three different pairs of neurons at the same time. Thus, we need to define a normalized distance to give a global measure of information transfer in the sub-circuit formed by the neurons AB, LP and PY. Let us define $D_{R-S-R'}$ as the normalized distance between connection from neuron R to S and connection from neuron R' to S. When complete synchronization occurs, $E_{RS} = E_{R'S} = 1$ for the two neuron pairs of the sub-circuit. For this case we want the distance $D_{R-S-R'}$ to have a value 0. When there is no matching at all among the bursting membrane potentials of the neurons $E_{RS} = E_{R'S} = 0$. Then, our normalized distance has to be $D_{R-S-R'} = 1$. Lastly, we must penalize those cases in which $E_{RS} \neq E_{R'S}$ for a particular time resolution even if the values are close to one. A measure of the distance that satisfies the above properties for two pairs of connections is $D_{R-S-R'} = 0.5 \times \sum_{\Delta t}\{|E_{RS} - E_{R'S}| + (1 - E_{RS}) + (1 - E_{R'S})\}$. This distance is normalized to the number of different Δt's in our calculations. We name the normalized average mutual information values for the three pairs of neurons in this sub-circuit as E_{ABLP}, E_{LPPY}, E_{PYAB}, and the partial distances between two connections $D_{AB-LP-PY}$, $D_{LP-PY-AB}$, $D_{PY-AB-LP}$. A function, D that satisfies the above properties for our global measure of the distance in the sub-circuit formed by the neurons AB, LP and PY is given by

$$D = \frac{1}{3}[D_{AB-LP-PY} + D_{LP-PY-AB} + D_{PY-AB-LP}]. \quad (1)$$

In the above plot we can see graphically the first term of equation 1: $D_{AB-LP-PY} = 0.5 \times \sum_{\Delta t}\{|E_{ABLP} - E_{LPPY}| + (1 - E_{ABLP}) + (1 - E_{LPPY})\}$. The normalized distance, D, can serve as a measure of the precision of the triphasic rhythms generated by the CPG since it keeps the history (words of discrete activity, see section 3) of temporal relations (information between words) among the bursting activities of AB, LP, and PY neurons. The normalized information for each connection ($AB - LP$, $LP - PY$ and $PY - AB$) and the values of the distance D defined by equation 1 are shown in Fig. 2. We have selected a set of representative examples out of a large number of simulations where the triphasic rhythm was generated (see Fig. 1). We found that the value of normalized entropies and, in consequence, the normalized distance D strongly depends on the values of the synaptic conductances chosen in the different models. The ratio between the different conductances has been kept the same for both KK and HR models (see tables 1,3,4 in [2]). In Fig. 2 we can see that the KK networks (panels A, C, E, G) and the HR networks (panels B, D, F, H) generate triphasic rhythms with an overall similar precision. However in general, triphasic patterns generated by HR networks are less precise, which is due to the fact that the bursting activities of the three neurons overlap in these networks (see panels B, D, F, H in Fig. 1).

In general, for reduced topologies the network is capable to generate rhythms, but they are less precise than those generated by the complete circuit (see D values and also Fig. 1). When we suppress the slow chemical synapses (damaged circuits) in complete and reduced circuits, the triphasic rhythm is worse (see panels C, D, G, H in Fig. 2) than when the circuits have these connections. In Fig.s 1C and 2C a phase inversion in the triphasic rhythm takes place, and the shape of the AB bursts changes completely from second 940 to the end of the simulation. Here, the precision of the rhythm is lower than that displayed by the network with the slow chemical synapses (see Fig.s 1C, 2C).

5   Discussion

In this paper we have analyzed the ability of different CPG networks to generate triphasic rhythms. We have used a measure of mutual information to characterize the precision of the rhythms in realistic networks and to analyze the robustness of damaged circuits. We have employed two different neuron models to validate the generality of our results. CPGs are assemblies of neurons that produce regular signals that control motor systems. The rhythms must be robust but at the same time flexible to perform different tasks or to adapt to different situations. In many simulations, we have seen major instant changes of the triphasic rhythms produced by slight changes in the conductance of the connections or the parameters that shape the dynamics of the neuron model (e.g. the maximum conductances of the ionic channels in the KK model). This is a desirable feature in a CPG with multiple mechanisms for robustness as it acts as a dynamic controller.

Acknowledgments. This work has been supported partially by CICyT TIC98-0247-C02-02, TIC2002-572-C02-02 grants and MCT BFI2000-0157 grant.

References

1. Selverston A.I., Moulins M.: The Crustaceam Stomatogastric System: a Model for the Study of Central Nervous Systems. Berlin; New York: Springer-Verlag (1987).

2. Latorre R., Rodriguez F.B., Varona P.: Characterization of triphasic rhythms in central pattern generators (I): interspike interval analysis. ICANN'02 proceedings. LNCS (2002).
3. Shannon, C.E.: A Mathematical Theory of Communication. Bell Sys. Tech. J. **27** (1948) 379–423 623–656.
4. Cover, T.M., Thomas, J.A.: Elements of Information Theory. Wiley and Sons (1991).
5. Rieke, F., Warland, D., de Ruyter van Steveninck, R., Bialek, W.: Spikes: Exploring the Neuronal Code. A Bradford Book. MIT Press Cambridge. Massachusetts, London, England (1997).
6. Harris–Warrick, R.M.: Dynamic Biological Network: The Stomatogastric Nervous System. Canbridge, Mass.: MIT Press. (1992).
7. Strong, S.P., Koberle, R., de Ruyter van Steveninck, R., Bialek, W.: Entropy and Information in Neural Spike Train. Physical Review Letters **80** 1 (1998) 197–200.
8. Eguia, M.C., Rabinovich, M.I., Abarbanel, H.D.I.: Information Transmission and Recovery in Neural Communications Channels. Phys. Rev. E **62** (2000) 7111-7122.
9. Rodríguez, F.B., Varona, P., Huerta, R., Rabinovich, M.I., Abarbanel, H.D.I.: Richer Network Dynamics of Intrinsically Non-Regular Neurons Measured through Mutual Information. In: Lecture Notes in Computer Science (Connectionist Models of Neurons, Learning Processes, and Artificial Intelligence), Mira, J., Prieto, A (Eds.), **2084** (2001), 490–497.

Neural Coding Analysis in Retinal Ganglion Cells Using Information Theory

J.M. Ferrández[1,2], M. Bongard[1,3], F. García de Quirós[1], J.A. Bolea[1], and E. Fernández[1]

[1]Instituto de Bioingeniería, U. Miguel Hernández, Alicante,
[2]Dept. Electrónica y Tecnología de Computadores, Univ. Politécnica de Cartagena,
[3]Dept. Neurobiologie, Univ. Oldenburg, Germany
Corresponding Author: jm.ferrandez@upct.es

Abstract. Information Theory is used for analyzing the neural code of retinal ganglion cells. This approximation may quantify the amount of information transmitted by the whole population, versus single cells. The redundancy inherent in the code may be determined by obtaining the information bits of increasing cells datasets and by analyzing the relation between the joint information compared with the addition the information achieved by aisle cells. The results support the view that redundancy may play a crucial feature in visual information processing.

1. Introduction

Information transmission and information coding in neural systems is one of the most interesting topics in neuroscience, nowadays. As technology grows new neural acquisition tools become available. They permit recording simultaneously tens, even hundreds of cells stimulated under different conditions. This process produces enormous data files, which need new tools for extracting the underlying organization of the neural principles, captured in the recordings.

First of all, it is necessary to determine if there exists an aisle cell coding, or a population coding. An aisle cell fires with limited action potentials, in a restricted interval, so it is difficult to code, numerically speaking, a broad spectrum of stimuli on its parameters. On the other hand a population of cells have more coding capabilities, and it can provide robustness to the representation of the stimuli by using redundancy on the population firing patterns. Adding more cells for coding the same stimuli will produce a fault tolerance system.

There have been published studies about auditory [1] and olfactory coding [2], however understanding visual coding is a more complex task due to and number and characteristics of the visual stimulation parameters, and the difficulty of achieving a filtered population ganglion cells firing pattern database with a considerable number of cells recorded simultaneously. New tools are also required for understanding this vast database. Initially it has been applied statistical analysis, [3][4][5] for obtaining insights about the parameters used in the visual system for coding and transmitting the

J.R. Dorronsoro (Ed.): ICANN 2002, LNCS 2415, pp. 174–179, 2002.

information to higher centers in the visual hierarchy. Artificial neural networks are also another tool, which can provide, supervised or autoorganizative, new insights about the way visual parameters are encoded and the inherent organization principles. [5][6]

Recent approaches use Information Theory for quantifying the code transmission. It can be used for comparing, also, the aisle cells coding capability versus the population code. It looks at the disorder, entropy, of a system, assuming that a system with more variability (disorder) will be able to transmit more symbols that a system with zero variance. Some studies [8] replace symbols with a list of action potential firing times for analyzing the neural data.

This paper uses information theory for quantifying the information transmitted by single retinal ganglion cells compared with the information conveyed by the whole population. The number of neurons in the datasets is also changed for determining if information grows linearly with number of neurons involved in the coding or it saturates, producing a redundancy phenomenon. Finally the redundancy effect will be observed for assuming if it is consequence of the saturation limit on the information, or if it exists for lower number of cells, aspect that will produce the desired robustness in the code. The results show that information is transmitted by the population code mainly, there exist some saturation on the information provided, determined by the stimuli dataset, and redundancy appears for all numbers of cells involved in the coding.

2. Methods

Registers were obtained on isolated turtle (Trachemy scripta elegans) retinas. The turtle was dark adapted for a few hours, before it was sacrificed and decapitated. Then the head was stored half an hour under 4° Celsius in order to ease the removing of the vitreous. The eye was enucleated and hemisected under dim illumination, and the retina was removed using bubbled Ringer solution taking care to keep intact the outer segment when removing the pigment epithelium. The retina was flatted in an agar plate, and fixed using a Millipore filter with a squared window where the electrodes will be placed. The agar plated with the retina was placed in a beam splitter with the ganglion cell on the upper side and bubbling Ringer solution flowed through the filter.

Light stimulation was applied using a halogen light lamp, selecting the wavelength by means of narrow band pass interference filters. Intensity was fixed by using neutral density filters, and a shutter provides flashes of the stimuli to the preparation. For each stimuli (the wavelength, the intensity, and the spot was varied) seven consecutive flashes, with 250 msec. length, were applied, using a lens to focus the stimulus on the photoreceptor layer of the whole retina.

The Utah microelectrode array was used for obtaining the extracellular recordings (Figure 1). It consists in an array of 100 (10x10); 1.5 mm long niddles with a platinized tip 50 microns long. The distance between each niddle is 400 mm, and the rest of the array is insulated with polyamide for providing biocompatibility. It was mounted on a micromanipulator, and each electrode was connected to a 25000 gain band pass (filtered from 250 to 7500 Hz) differential amplifier. The analog signal was digitized using a multiplexer and an A/D converter and stored in a computer.

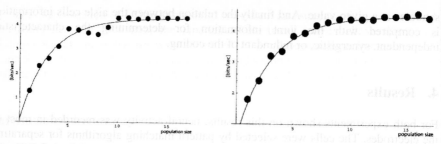

Fig. 2. Information transmitted for an increasing population of retinal ganglion cells in discriminating intensities.

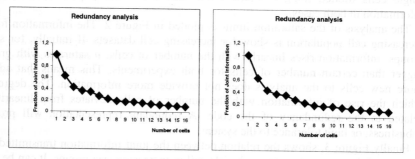

Fig. 3. Redundancy analysis for an increasing population of retinal ganglion cells.

5. Conclusions

Information Theory has been applied for analyzing the neural code of retinal ganglion cells. The information transmitted by the population of cells is always higher than the information contained in the responses of aisle cell, suggesting a population code instead of a single cell code. The overall information registered may vary, it depends on the experiment, number of cells recorded, discriminative character of this cells, location of the multielectrode array, size of the bins, etc., however the overall behavior does not change significantly.

The information on the population grows linearly with number of cells until certain limit, where information saturates. Adding more cells to the set does not provide more information. The redundant characteristic of this scenario has been proved by relating the joint transmitted information versus the addition of the singles cell transmitting rate. Redundancy could be useful to achieve the robustness required for the retina to maintain its performance under adverse environments.

This analysis may be also applied to color coding as well as coding of natural scenarios, where synergistic behaviors may appear. The cooperation between new multichannel acquisition system and innovative analysis tools will provide new insights in neural coding.

Acknowledgements. This research is being funded by E.C. QLK-CT-2001-00279, Fundación Séneca PI-26/00852/FS/01, a grant from the National Organization of the Spanish Blind (ONCE), and Fellowship from the Spanish Ministerio de Educacion to M.B.

References

1. Secker H. and Searle C.: "Time Domain Analysis of Auditory-Nerve Fibers Firing Rates". J. Acoust. Soc. Am. 88 (3) pp. 1427–1436, 1990.
2. Buck L.B.: "Information Coding in the vertebrate olfactory system". Ann. Rev. Neurosci. 19, pp. 517–544, 1996.
3. Fitzhugh, R. A.: "A Statistical Analyzer for Optic Nerve Messages". J. Gen. Physiology 41, pp. 675–692, 1958.
4. Warland D., Reinagel P., Meister M.: " Decoding Visual Information from a Population of Retinal Ganglion Cells". J. Neurophysiology 78, pp. 2336–2350, 1997.
5. Fernández E., Ferrández J.M., Ammermüller J., Normann R.A.: "Population Coding in spike trains of simultaneously recorded retinal ganglion cells Information". Brain Research, 887, pp. 222–229, 2000.
6. Ferrández J.M., Bolea J.A., Ammermüller J. , Normann R.A. , Fernández E.: "A Neural Network Approach for the Analysis of Multineural Recordings in Retinal Ganglion Cells: Towards Population Encoding". Lecture Notes on Computer Science 1607, pp. 289–298, 1999.
7. Shoham S., Osan R., Ammermuller J. Branner A., Fernández E., Normann R.: "The Classification of the Chromatic, and Intensity Features of Simply Visual Stimuli by a Network of Retinal Ganglion Cells". Lecture Notes on Computer Science 1607, pp. 44–53, 1999.
8. Rieke F., Warland D., van Steveninck R., Bialek W.: "Spikes: Exploring the Neural Code". MIT Press. Cambridge, MA, 1997.
9. Jones, K.E., Campbell, P.K. and Normann, R.A.:"A glass/silicon composite intracortical electrode array". Annals of Biomedical Engineering 20, 423-437 (1992).
10. Shannon C.E..: "A Mathematical Theory of Communication". Bell Sys Tech J. 27. pp. 379–423 , 1948.

Firing Rate Adaptation without Losing Sensitivity to Input Fluctuations

Giancarlo La Camera, Alexander Rauch, Walter Senn, Hans-R. Lüscher, and Stefano Fusi

Institute of Physiology, University of Bern, CH-3012 Bühlplatz 5, Switzerland
{lacamera,rauch,senn,luescher,fusi}@pyl.unibe.ch

Abstract. Spike frequency adaptation is an important cellular mechanism by which neocortical neurons accommodate their responses to transient, as well as sustained, stimulations. This can be quantified by the slope reduction in the *f-I* curves due to adaptation. When the neuron is driven by a noisy, *in vivo*-like current, adaptation might also affect the sensitivity to the fluctuations of the input. We investigate how adaptation, due to calcium-dependent potassium current, affects the dynamics of the depolarization, as well as the stationary *f-I* curves of a white noise driven, integrate-and-fire model neuron. In addition to decreasing the slope of the *f-I* curves, adaptation of this type preserves the sensitivity of the neuron to the fluctuations of the input.

1 Introduction

Many *in vivo* phenomena, like spontaneous activity, or the selective, delay activity observed in many cortical areas of behaving animals, are characterized by sustained spike activity throughout long intervals. During these intervals, the statistics of the input current is likely to be stationary or quasi-stationary, although originated by very irregular synaptic activity. When the activity of a neural network can be characterized in terms of mean output spike frequencies of subpopulations of neurons, the *f-I* response function (RF) has proved an invaluable tool to describe the network's stationary activity (see e.g. [1,2]). At the same time, spike frequency adaptation (see e.g. [3,6]) is an important cellular mechanism by which neocortical neurons accommodate their responses to transient, as well as sustained, stimulations. Adaptation should then be considered in the above framework, especially after that it proved to be essential for two models of integrate-and-fire (IF) neurons to fit the *in vitro* response of rat pyramidal neurons [5]. In addition to reducing the stationary spike frequency, adaptation could also affect the neuron's sensitivity to the fluctuations of the input. In this work we analyze this problem for the *linear* IF (LIF) model neuron with adaptation [5], driven by white noise, emulating the intense, irregular synaptic activity driving a cortical neuron *in vivo*.

J.R. Dorronsoro (Ed.): ICANN 2002, LNCS 2415, pp. 180–185, 2002.

2   The Model

The sub-threshold behavior of the LIF neuron is fully described by the depolarization V, which obeys

$$CdV = -\lambda dt + I_\xi dt + I_\alpha dt \tag{1}$$

Here C is the capacitance of the membrane, λ a constant leak term, I_ξ the synaptic input and I_α a calcium dependent potassium current, responsible for spike frequency adaptation. I_α is proportional to intra-cellular calcium concentration $[Ca]$, $I_\alpha = \bar{g}[Ca]$. Upon emission of an action potential, an amount A_{Ca} of calcium immediately enters the cell and decays exponentially to zero with a slow time constant τ_{Ca} [5]:

$$\frac{d[Ca]}{dt} = -\frac{[Ca]}{\tau_{Ca}} + A_{Ca} \sum_k \delta(t - t_k) \tag{2}$$

where the sum goes over all spikes emitted by the neuron up to time t.

The synaptic current I_ξ is the result of the Poisson activation of (many) independent excitatory and inhibitory inputs, with average m, variance s^2 and time correlation length τ', corresponding to the decay time constant of a single post-synaptic potential. If τ' is very short, the correlation length of the current becomes negligible, and the current can be replaced by white noise [7,5]:

$$I_\xi dt \to mdt + s\sqrt{2\tau'}\xi_t\sqrt{dt} \tag{3}$$

where ξ_t is a Gauss distributed variable with $E[\xi_t] = 0$ and $E[\xi_t\xi_t'] = \delta(t - t')$. Hereafter we set $\tau' = 1$ ms.

Equation (1) must be completed by boundary conditions: a spike is emitted when V is driven above a threshold θ, after which is reset to a value $0 < V_r < \theta$ and clamped there for a refractory time τ_r. V is confined in the range $[0, \theta]$ by a reflecting barrier at 0 (see [4]). Without adaptation ($I_\alpha \equiv 0$), the f-I curve of (1) is known and reads [4] (see also Fig. 1)

$$f = \Phi(m, s) = \left[\tau_r + \frac{\tau's^2}{(m-\lambda)^2}\left(e^{-\frac{C\theta(m-\lambda)}{\tau's^2}} - e^{-\frac{CV_r(m-\lambda)}{\tau's^2}}\right) + \frac{C(\theta - V_r)}{m - \lambda}\right]^{-1} \tag{4}$$

The f-I curve gives the mean output spike frequency f as a function of m, s in stationary conditions. When adaptation is present, its effect can be taken into account in the RF using the fact that its dynamics is much slower compared to the other dynamic variables [3]. In particular, if $\tau_{Ca} \gg 1/f$, the fluctuations of $[Ca]$ can be neglected, $\bar{g}[Ca] \sim \bar{g}A_{Ca}\tau_{Ca}f \equiv \alpha f$. As a consequence, an additional negative current $\langle I_\alpha \rangle = -\alpha f$, proportional to the neuron's own frequency f, affects the mean current m which, in equation (4), has to be replaced by [3,5]

$$m \to m - \alpha f \equiv m(f) \tag{5}$$

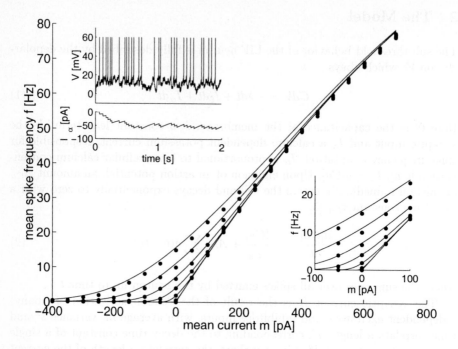

Fig. 1. *f-I* curves of the adapted LIF neuron, theory vs simulations. Mean output frequency plotted as a function of the mean current m at constant s (stepwise increasing from 0 to 600 pA). Lines: self-consistent solutions of equation (4-5) (f_{th}). Dots: simulations (f_{sim}). Adaptation parameters: $\tau_{ca} = 500$ ms, $\bar{g}A_{Ca} = 8$ pA (so that $\alpha = 4$ pA s). Neuron parameters: $\tau_r = 5$ ms, $C = 300$ pF, $\theta = 20$ mV, $V_r = 10$ mV, $\lambda = 0$. The right inset shows an enlargement around the rheobase $\tilde{m} = 0$. Left inset: sample of depolarization and I_α for the point $(m, s) = (100, 200)$ pA.

The new spike frequency is found by iterating the equation $f = \Phi(m(f), s)$ until a fixed point is reached.[1] We have simulated the LIF neuron with calcium dynamics, equations (1-2-3), for different values of m, s and τ_{Ca}. Fig. 1 shows the agreement between theory (self-consistent solutions of equation (4-5)) and simulations. The good agreement for high s justifies the assumption that I_α affects only the mean current felt by the neuron or, equivalently, that fluctuations of I_α are negligible. In addition, calcium dynamics slower than ~ 100 ms leads to an error below 3% for four sample points of the (m, s)-plane (Fig. 2).

3   Effect of Adaptation on the *f-I* Curves

We next show that, contrary to the effect of the other parameters, adaptation allows the neuron to reduce its spike frequency, retaining most of its sensitivity

---

[1] The condition for any fixed point f^* to be stable is $\partial\Phi(m(f), s)/\partial f|_{f^*} = -\alpha\partial\Phi/\partial m|_{f^*} < 1$, which holds since Φ is an increasing function of m.

to fluctuations. Given a change in C, there is always a RF-equivalent change in the couple $\{\theta, V_r\}$, because the RF (4) is invariant under the scaling $C \to Ch$, $\{\theta, V_r\} \to \{\theta, V_r\}/h$, $h > 0$ constant. As a consequence, from now on we consider only changes in C, keeping θ and V_r fixed and such that their difference is finite.

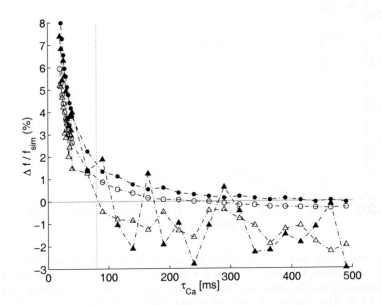

Fig. 2. Adapted LIF neuron, dependence of $(f_{sim} - f_{th})/f_{sim}$ (Fig. 1) on τ_{Ca}, theory vs simulations. As τ_{Ca} is varied, A_{Ca} is rescaled so that the total amount of adaptation ($\alpha = 4$ pA s) is kept constant. Parameters of the current: $m = 100$ pA (full symbols) and $m = 300$ pA (empty symbols); $s = 0$ (circles) and $s = 300$ pA (triangles). All other parameters as in Fig. 1. Mean spike frequencies assessed across 50 s, after discarding a transient of $10\tau_{ca}$. For $s > 0$ deviations from a monotonic behavior have to be expected, but the error is always below 3%. For $\tau_{Ca} < 80$ ms (vertical dotted line) the error is positive, meaning that $f_{sim} > f_{th}$: the neuron only slightly adapts because calcium decay is too fast.

We focus on two observables, the slope ρ of the RF at the rheobase[2] ($\tilde{m} = m - \lambda = 0$) for $s = 0$ (which quantifies the amount of frequency reduction), and the *distance* d between two RFs at different s, or equivalently the dependence of the RF on s at $\tilde{m} = 0$, taken as a measure of the sensitivity to the fluctuations of the input current.[3] In the following it will be useful to use the notation

---

[2] The rheobase current I_{th} is the threshold current, i.e. $f = 0$ if $m < I_{th}$ when $s = 0$. For the LIF neuron $I_{th} = \lambda$.

[3] Since we are interested in the behavior at $\tilde{m} \approx 0$, the refractory period does not have an effect, and we set $\tau_r = 0$. The following formulae are also valid for $\tau_r > 0$ provided that the frequencies at the rheobase are not too large, say $f < 30$ Hz.

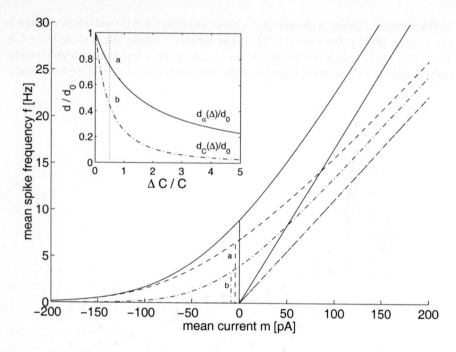

Fig. 3. Adaptation only slightly affects the sensitivity to fluctuations. Full lines: RFs without adaptation, $C = 300$ pF, $\theta_1 = \theta = 20$ mV, $\tau_r = 0$, $s = 0$ and 400 pA. Vertical line: distance at the rheobase d_0. Dashed lines: same as full lines but with adaptation $\alpha = 3$ pA s. The distance is reduced to only 80% of d_0 (see inset, a). Dot-dashed lines: $C = 450$ pF, no adaptation, a slope-equivalent change to the previous case (a). The distance is approximately halved (inset, b). The inset shows $d_\alpha(\Delta)/d_0$ (full line) and $d_C(\Delta)/d_0$ (dot-dashed line) as a function of slope-equivalent changes expressed in $\Delta C/C$ for the parameters of the main figure. Inset's vertical line: $\Delta C/C = 0.5$. Note that $d_C(\Delta) \leq d_\alpha(\Delta)$ approximately by a factor $(1 + \rho_0 \alpha \eta)^{-1} = (1 + \Delta C/C)^{-1}$ (see text).

$\theta_n \equiv \theta^n - V_r^n$. For $s = 0$ the RF is linear in $\tilde{m}$, $f = \rho_\alpha \tilde{m}$ with

$$\rho_\alpha = \frac{1}{C\theta_1 + \alpha} = \frac{\rho_0}{(1 + \rho_0 \alpha)}$$

where $\rho_0 \equiv 1/C\theta_1$ is the slope with no adaptation ($\alpha = 0$). Whereas both C and α have a 'first order' effect on the slope, their effect on the distance d is very different. By expanding the f-I curve (4) around the rheobase, introducing adaptation and solving for f at $\tilde{m} = 0$, one gets the distance as a function of C and α:

$$d_\alpha = \frac{\sigma^2}{\theta_2} \frac{1}{1 + \rho_\infty \alpha} \equiv \frac{d_0}{1 + \rho_\infty \alpha}$$

where $\sigma^2 \equiv 2\tau' s^2/C^2$, $\rho_\infty \equiv 2\theta_3/3\theta_2^2 C$.

This exposes the difference between the 'quadratic' $(1/C^2)$ behavior in C (or θ because of the scaling properties of the RF), and the 'first order' behavior in α: the price to be payed, to reduce the slope without adaptation, is a decreased sensitivity to fluctuations, as implied by a reduced distance between the curves. This is apparent by comparing the change in d caused by slope equivalent changes in α $(\alpha \rightarrow \alpha(1+\eta))$ and C $(\Delta C/C = \rho_0 \alpha \eta)$ respectively (see inset of Fig. 3):

$$d_\alpha(\Delta) = \frac{d_0}{1 + \rho_\infty \alpha + \rho_\infty \alpha \eta} \quad , \quad d_C(\Delta) = \frac{d_0}{(1 + \rho_0 \alpha \eta)(1 + \rho_\infty \alpha + \rho_0 \alpha \eta)}$$

4   Conclusions

We have analyzed a simple model of spike frequency adaptation, due to a *slow* calcium dependent potassium current I_α, for the IF neuron with a linear decay. I_α enters the RF only as a negative, feedback current which depends on the neuron's own spike frequency. We showed that there is an excellent agreement between the theoretical frequencies, as predicted through the RF, and the ones obtained by simulations of the neuron dynamics, up to large values of the amplitude of the input fluctuations, and for slow enough calcium dynamics. In addition, we have shown that adaptation reduces the slope of the RF retaining most of neuron's sensitivity to the fluctuations of the input current. We have also found that the same results hold for the classical IF neuron with leakage proportional to V (not shown here). These two model neurons with adaptation have been recently proved able to fit the spike frequencies of rat neocortical pyramidal neurons *in vitro* injected with Gauss distributed current resembling the white noise (3) [5].

References

1. Amit D.J. and Tsodyks M.V., Quantitative study of attractor neural network retrieving at low rates I: Substrate – spikes, rates and neuronal gain, *NETWORK* **2**: 259 (1991)
2. Brunel, N., Persistent activity and the single-cell frequency-current curve in a cortical network model, *NETWORK* **11**: 261–280 (2000)
3. Ermentrout, B., Linearization of *f-I* curves by adaptation, *Neural Comp.* **10**: 1721–1729 (1998)
4. Fusi S. and Mattia M., Collective behavior of networks with linear (VLSI) Integrate and Fire Neurons, *Neural Comp.* **11**: 643–662 (1999)
5. Rauch A., La Camera G., Lüscher, H-R., Senn W. and Fusi S., Neocortical pyramidal cells respond as integrate-and-fire neurons to *in vivo*-like input current, *submitted* (2002)
6. Wang, X.J., Calcium coding and adaptive temporal computation in cortical pyramidal neurons, *J. Neurophysiol.* **79**: 1549–1566 (1998)
7. Tuckwell, H.C., Introduction to theoretical neurobiology, vol II, *Cambridge: Cambridge University Press* (1988)

Does Morphology Influence Temporal Plasticity?

David C. Sterratt[1] and Arjen van Ooyen[2]

[1] Institute for Adaptive and Neural Computation, Division of Informatics,
University of Edinburgh, 5 Forrest Hill, Edinburgh EH1 2QL, Scotland, UK
dcs@anc.ed.ac.uk
[2] Netherlands Institute for Brain Research, Meibergdreef 33, 1105 AZ Amsterdam,
The Netherlands A.van.Ooyen@nih.knaw.nl

Abstract. Applying bounded weight-independent temporal plasticity rule to synapses from independent Poisson firing presynaptic neurons onto a conductance-based integrate-and-fire neuron leads to a bimodal distribution of synaptic strength (Song et al., 2000). We extend this model to investigate the effects of spreading the synapses over the dendritic tree. The results suggest that distal synapses tend to lose out to proximal ones in the competition for synaptic strength. Against expectations, versions of the plasticity rule with a smoother transition between potentiation and depression make little difference to the distribution or lead to all synapses losing.

1 Introduction

This paper deals with two phenomena. The first, spike-timing dependent plasticity (STDP), is observed in various preparations (Markram et al., 1997; Bi and Poo, 1998; Abbot and Nelson, 2000). The relative timing of individual pre- and postsynaptic spikes can alter the strength of a synapse ("weight"). The size of the change also depends on the weight, with stronger synapses being depressed more and potentiated less. The second phenomenon is dendritic filtering of synaptic inputs and action potentials. Excitatory postsynaptic potentials (EPSPs) evoked at a synapse are attenuated and spread out temporally on their way to the soma. The dendritic tree also filters backpropagating action potentials (Spruston et al., 1995). The effects of filtering are more pronounced at more distal locations on the dendritic tree.

Theoretical work has suggested that when STDP along with upper and lower bounds on each weight operates at synapses onto a single postsynaptic neuron, the synapses compete with each other to grow strong, leading to a bimodal weight distribution (Song et al., 2000). This model used a conductance-based integrate-and-fire neuron. The aim of this paper is to discover whether the synapse's location on the dendritic tree affects its ability to change its strength. To do this, we incorporate dendritic filtering and the delay of backpropagating action potentials into the Song et al. (2000) model. A key assumption is that changes to a synapse depend only on signals local to it.

J.R. Dorronsoro (Ed.): ICANN 2002, LNCS 2415, pp. 186–191, 2002.

In Sect. 2 we present the neuronal model. In Sect. 3 describe the weight-independent synaptic plasticity rule we use. We look at the behaviour of the rule in Sect. 4. Finally, in Sect. 5, we discuss the implications of this work.

2   The Neuron Model

We study a single conductance-based integrate-and-fire neuron with N_E modifiable excitatory synapses and N_I non-modifiable inhibitory synapses onto it. Presynaptic spikes are generated according to a Poisson process at a rate f_E for the excitatory inputs and f_I for the inhibitory inputs. In this paper $f_E = 40$ Hz and $f_I = 10$ Hz.

In order to model the effects of forward and backward filtering, each excitatory input i is allocated a distance x_i away from the soma. The distance of each synapse is chosen randomly from a uniform distribution between $100\mu m$ and $300\mu m$. The distance of synapse away from the soma affects the attenuation, delay and time constants of conductance changes evoked at the synapse, as well as the time taken for a backpropagating action potential to reach the synapse from the soma. Inhibitory inputs are assumed to be directly onto the soma.

When a spike arrives at an excitatory synapse there is a delay of $\Delta^{\text{orth}}(x_i)$ before it increases the conductance $g_{\text{ex},i}(t)$ in the soma by an amount $a(x_i)\bar{g}_i$ where $\bar{g}_i$ is the synapse's maximum conductance (or "weight") and $a(x_i)$ is a distance-dependent attenuation factor. Inbetween spikes the conductances obey the first-order differential equation

$$\tau_{\text{ex}}(x_i)\frac{dg_{\text{ex},i}}{dt} = -g_{\text{ex},i} \tag{1}$$

where $\tau_{\text{ex}}(x_i)$ is the time constant of the conductance, which depends on location on the dendritic tree.

Similar equations are obtained for the inhibitory conductances $g_{\text{in},i}$ except that there are no delays or attenuation factor, the time constant is uniformly τ_{in} and the non-modifiable weights are set to $\bar{g}_{\text{in}} = 0.05$.

The membrane potential V is governed by the update rule

$$\tau_{\text{m}}\frac{dV}{dt} = V_{\text{rest}} - V + \sum_i g_{\text{ex},i}(E_{\text{ex}} - V) + \sum_i g_{\text{in},i}(E_{\text{in}} - V) \tag{2}$$

where $\tau_{\text{m}} = 20$ ms is the membrane time constant, $V_{\text{rest}} = -70$ mV is the resting potential, $E_{\text{ex}} = 0$ mV is the reversal potential of the excitatory synapses and $E_{\text{in}} = -70$ mV is the reversal potential of the inhibitory synapses. When the membrane reaches a threshold $\theta = -54$ mV, the neuron fires an action potential and is reset to $V_{\text{reset}} = -60$ mV.

In order to find how the attenuation factor $a(x_i)$, the dendritic delay $\Delta^{\text{orth}}(x_i)$ and the conductance time constant $\tau_{\text{ex}}(x_i)$ depend on distance we used the data of Magee and Cook (2000). From their measurement of the relative size of EPSPs at the site of initiation and at the soma, we described the attenuation factor by

$$a(x) = 1 - \frac{x}{375} \; . \tag{3}$$

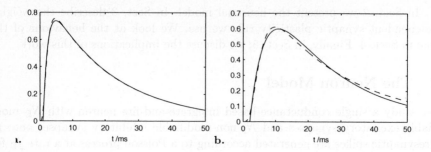

Fig. 1. a. Fit (solid line) of proximal ($100\mu m$) EPSP to experimental EPSP (dashed line). Parameters of experimental EPSP were $\tau_r = 0.5$ ms and $\tau_d = 20$ ms. Parameters of fit curve were $\tau = 1.33$ ms, $\Delta = 0.97$ ms, $k = 0.96, \tau_m = 20$ ms. **b**. Fit of distal ($300\mu m$) EPSP to experimental EPSP. Parameters of experimental EPSP were $\tau_r = 2.5$ ms, $\tau_d = 25$ ms. Parameters of fit curve were $\tau = 4.62$ ms, $\Delta = 2.07$ ms, $k = 1.21, \tau_m = 20$ ms.

Magee and Cook (2000) fitted the EPSPs recorded at the soma by the expression $(1 - \exp(-t/\tau_r))^5 \exp(-t/\tau_d)$ where the rise time τ_r and decay time τ_d at the soma depend on the distance of the synaptic input along the dendritic tree. In order to incorporate their data into a point neuron model, we used the Levenberg-Marquardt nonlinear regression algorithm (MATHWORKS ftp site) to fit these curves to double-exponential alpha functions of the form

$$
k \left(e^{-\frac{t-\Delta^{\mathrm{orth}}}{\tau_m}} - e^{-\frac{t-\Delta^{\mathrm{orth}}}{\tau_{\mathrm{ex}}}} \right) \quad t > \Delta^{\mathrm{orth}}
$$
$$
0 \qquad\qquad\qquad\qquad \text{otherwise}
$$
(4)

where k is a constant. The decay time constant is identical to the membrane time constant $\tau_m = 20$ ms of the integrate-and-fire model; this is close to the experimental values. Figure 1 shows the fits for proximal and distal dendritic inputs. We assume that $\tau_{\mathrm{ex}}(x)$ varies linearly between its endpoint values of $\tau_{\mathrm{ex}}(100) = 1.33$ ms and $\tau_{\mathrm{ex}}(300) = 4.62$ ms and that $\Delta^{\mathrm{orth}}(x)$ varies linearly between its endpoint values of $\Delta^{\mathrm{orth}}(100) = 0.97$ ms and $\Delta^{\mathrm{orth}}(300) = 2.07$ ms.

3   Synaptic Plasticity

We modified the phenomenological mechanism for the observed plasticity used by Song et al. (2000) so that there was a continuous transition between depression and potentiation at the crossover point. The amount of potentiation or depression depends on signals at the synapse. When a presynaptic spike arrives at the synapse it increases the amount of a substance P_i^* by A_+/τ_+^*. In the absence of spikes, the substance P^* decays with a time constant τ_+^* and catalyses synthesis of a substance P_i, which in turn decays at a rate $\tau_+^* = 20$ ms:

$$
\tau_+^* \frac{dP_i^*}{dt} = -P_i^* \quad \text{and} \quad \tau_+ \frac{dP_i}{dt} = -P_i + P_i^* \ .
$$
(5)

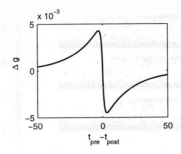

Fig. 2. The effective temporal plasticity rule. t_{pre} is the time of arrival of the presynaptic spike at the synapse, and t_{post} is the time of the backpropagating spike at the synapse. Parameters: $\tau_+^* = \tau_-^* = 1$ ms, $A_+ = 0.005$, $A_- = 1.05A_+$.

When the postsynaptic neuron fires, an action potential starts to backpropagate through the dendritic tree. This backpropagating action potential arrives $\Delta^{\mathrm{anti}}(x) = x/300$ ms later at the synapse and releases an amount A_-/τ_-^* of a substance M_i^* which decays with a time constant τ_-^*. This substance synthesises substance M_i, which decays at a rate $\tau_- = 20$ ms. The expression for $\Delta^{\mathrm{anti}}(x)$ is based on the observation that the propagation speed of the peak of the backpropagating action potential is constant up to about $300\mu m$ and that the delay at this point is about 1 ms (Spruston et al., 1995).

The presynaptic spike also decreases the weight by an amount of $M_i\bar{g}_{\mathrm{max}}$ where $\bar{g}_{\mathrm{max}}$ is the maximum weight. Similarly, when a backpropagating action potential arrives at the synapse, the weight increases by $P_i\bar{g}_{\mathrm{max}}$. The weight is prevented from growing larger than $\bar{g}_{\mathrm{max}}$ or smaller than zero. The effective temporal plasticity rule is shown in Fig. 2.

In the simulations we set $A_+ = 0.1$, $A_- = 1.05A_+$ and $\bar{g}_{\mathrm{max}} = 0.06$.

4　Results

We first ran the model with very small values of τ_+^* and τ_-^* (0.001 ms) which gave an approximation to the discontinuous case. All simulations lasted 5000 s. After this time the weight distribution was as shown in the top panel of Fig. 3a. This distribution is bimodal; synapses tend to be winners or losers. We can also see that proximal synapses tend to "win" whereas distal ones tend to "lose".

This result fits with the competitive nature of weight-independent temporal plasticity with bounded weights (Song et al., 2000; van Rossum et al., 2000). Because distal inputs are attenuated, they are less likely to help make the postsynaptic neuron fire and therefore less likely to be potentiated. Temporal factors may also disadvantage distal inputs. Firstly, distal inputs that do help to fire the postsynaptic neuron have to fire earlier than contributing proximal inputs. Secondly, the backpropagating action potential takes longer to reach the distal

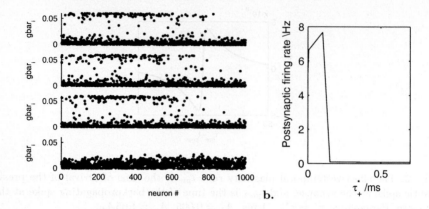

Fig. 3. a. The weight distribution for various values of τ_+^* and τ_-^*. The abscissa of each panel shows the presynaptic neuron number; lower numbers are more proximal and higher numbers more distal. The ordinate shows the strength of the corresponding synapse. From top to bottom the values of τ_+^* and τ_-^* are 0.001 ms, 0.006 ms, 0.1 ms and 0.15 ms. **b.** The postsynaptic firing rate as a function of τ_+^* and τ_-^*.

synapses. Thus there is less of the substance P_i when the backpropagating action potential arrives at distal synapses, and therefore less potentiation.

Could a smoother transition from depression to potentiation alter this behaviour? Bi and Poo's (1998) data shows maximum potentiation and depression peaks separated by ±5 ms. If the time of the potentiation peak matches the interval between a contributing presynaptic spike and a backpropagating spike arriving at the synapse, the synapse should be potentiated more.

Figure 3a shows the weight scatter plots for various times of the depression and potentiation peaks. For small values of τ_+^* the dependence of the weight distribution on distance does not change much, though a shorter transition phase seems to favour distal synapses. For $\tau_+^* > 0.15$ ms, all the weights lose, contrary to our hypothesis that smoother curves should favour distal synapses.

The behaviour with higher values of τ_+^* is presumably because the peaks of the plasticity curve overlap with presynaptic spikes that are uncorrelated to the postsynaptic spikes, and depression is stronger than potentiation. This idea is confirmed by the dependence of firing rate on the width of the central section (Fig. 3b) which shows that the firing rate of the neuron is around 6 Hz and 8 Hz for smaller values of τ_+^* but almost zero for larger ones.

5   Discussion and Conclusions

In short, morphology does influence temporal plasticity — at least with the neuron model, temporal plasticity rule and input firing statistics we used. In this section we discuss these caveats.

Our point neuron model with morphology modelled by location-dependent delays and EPSP rise time constants could not include factors that could be modelled using compartmental models, such as dendritic conductances, branching and dendritic width. Although this does limit the scope of the conclusions, we believe that the simpler model gives a first-order understanding of the problem.

Different firing statistics, for example temporally-correlated inputs, affect the weight distribution in pure point neuron models (Song et al., 2000; van Rossum et al., 2000). It seems unlikely that they could affect our results, unless the correlated inputs occured preferentially over the distal dendritic tree, and were therefore more effective than proximal correlated inputs. Also, successful synapses hit the same upper bound regardless of their location.

The weight-dependent rule studied by van Rossum et al. (2000) leads to a unimodal weight distribution in which synapses cluster around the weight at which depression becomes more potent than potentiation. Temporal properties are much less important, so in our model we would not expect the synaptic weights to depend on location. Preliminary simulations bear out this hypothesis.

A uniform weight distribution leads to distal inputs being less effective. Magee and Cook (2000) have shown that distal synapses in CA1 pyramidal neurons compensate for dendritic filtering by being stronger. Neither weight-dependent nor weight-independent rules appear to lead to this weight distribution. A location-dependent plasticity rule might account for location-dependent weights. Further work will investigate location-dependent plasticity rules derived from synapse-local signals in compartmental models (Rao and Sejnowski, 2001).

References

1. Abbott, L. F., Nelson, S. B.: Synaptic plasticity: taming the beast. Nature Neuroscience **3** (2000) 1178–1183
2. Bi, G.-q., Poo, M.-m.: Synaptic modifications in cultured hippocampal neurons: Dependence on spike timing, synaptic strength, and postsynaptic cell type. Journal of Neuroscience **18** (1998) 10464–10472
3. Magee, J. C., Cook, E. P.: Somatic EPSP amplitude is independent of synapse location in hippocampal pyramidal neurons. Nature Neuroscience **3** (2000) 895–903
4. Markram, H., Lübke, J., Frotscher, M., Sakmann, B.: Regulation of synaptic efficiency by coincidence of postsynaptic APs and EPSPs. Science **275** (1997) 213–215
5. Rao, R. P. N., Sejnowski, T. J.: Spike-timing-dependent Hebbian plasticity as temporal difference learning. Neural Computation **13** (2001) 2221–2237
6. Song, S., Miller, K. D., Abbott, L. F.: Competitive Hebbian learning through spike-timing-dependent synaptic plasticity. Nature Neuroscience **3** (2000) 919–926
7. Spruston, N., Schiller, Y., Stuart, G., Sakmann, B.: Activity-dependent action potential invasion and Calcium influx into hippocampal CA1 dendrites. Science **268** (1995) 297–300
8. van Rossum, M. C. W., Bi, G. Q., Turrigiano, G. G.: Stable Hebbian learning from spike timing-dependent plasticity. Journal of Neuroscience **20** (2000) 8812–8821

Attractor Neural Networks with Hypercolumns

Christopher Johansson, Anders Sandberg, and Anders Lansner

Department of Numerical Analysis and Computing Science, Royal Institute of
Technology, 100 44 Stockholm, Sweden
{cjo, asa, ala}@nada.kth.se

Abstract. We investigate attractor neural networks with a modular structure,
where a local winner-takes-all rule acts within the modules (called hypercolumns). We make a signal-to-noise analysis of storage capacity and noise
tolerance, and compare the results with those from simulations. Introducing
local winner-takes-all dynamics improves storage capacity and noise tolerance,
while the optimal size of the hypercolumns depends on network size and noise
level.

1 Introduction

A problem that arises in many kinds of recurrent neural networks (NNs) is that of
properly regulating network activity, i.e. keeping an appropriate level of activity in
the network. This is one motivation for investigating a local winner-takes-all
dynamics that regulates the total activity within network modules. This form of
activity control, while not as sophisticated as other control strategies is easy to
implement and also somewhat biologically plausible. Here, these local winner-take-all
modules are called *hypercolumns* in analogy with the hypercolumns described in
cortex [1]. A unit in such a hypercolumn is assumed to correspond to a cortical
minicolumn rather than to an individual neuron.

Another motivation for exploring networks composed of modules with internal
activity control comes from the study of Bayesian Confidence Propagation Neural
Networks (BCPNN), a class of networks derived from statistical considerations [2, 3].
The hypercolumnar structure is a natural consequence of their derivation [4]. The
units of a hypercolumn typically represent the different values of a discrete feature,
e.g. color. Computational experiments have shown that such a hypercolumnar
structure improves noise tolerance, convergence speed and capacity compared to a
non-modular structure [5]. It thus becomes relevant to examine whether this effect is
universal among different types of attractor NNs, which is what we contribute to here.

The hypercolumn architecture used here is somewhat related to the Potts neuron
models [6], although the latter focus on combinatorial optimization without learning
rather than associative memory. There are also some similarities to other kinds of
modular networks [7, 8]; the difference being that we do not consider autoassociation
within the hypercolumns.

J.R. Dorronsoro (Ed.): ICANN 2002, LNCS 2415, pp. 192–197, 2002.

Neural Networks with Hypercolumns

The hypercolumn NNs are constructed with N units divided equally into H hypercolumns each containing U units. For simplicity we let all hypercolumns consist of the same number of units; $N=HU$. Within each hypercolumn the activity always sums to one, which is ensured by the use of a winner-take-all function.

We use patterns, $\xi \in \{0,1\}^N$, constructed with a fraction $a=1/U$ of the positions set to 1. The patterns are – as the NNs – divided into H hypercolumns and in each hypercolumn only one of the positions is set to 1 and the rest are set to 0.

Assuming that each hypercolumn in the patterns are independent to all others and each unit is equally likely to be active, the information I_h contained in one hypercolumn becomes

$$I_h = -\sum_i P(x_i)\log_2 P(x_i) = -\sum^U \frac{1}{U}\log_2\frac{1}{U} = \log_2 U \tag{1}$$

and if we multiply eq. (1) with the number of hypercolumns in the NN we get the amount of information I_p in one pattern:

$$I_p = I_h H = H\log_2 U \tag{2}$$

Learning and Update Rule

The learning rule used is that of the sparse Hopfield network (eq. (3)). After learning P patterns the weights become

$$w_{ij} = \frac{1}{N}\sum_\mu (\xi_i^\mu - a)(\xi_j^\mu - a) \tag{3}$$

The introduction of the hypercolumns does not affect the learning rule. All weights w_{ij} where $i=j$ are set to 0.

The update-rule consists of summing the contributions of other units

$$h_i = \sum_j w_{ij}o_j \tag{4}$$

and applying a winner-takes-all function to the units within each hypercolumn: the unit with the highest h_i has its output set to 1, the others to 0.

Analysis of NN with Hypercolumns

We define a stochastic variable $X \equiv (\xi_i-a)(\xi_j-a)$. Evaluation of the product $(\xi_i-a)(\xi_j-a)$ yields three different results $x_1=(1-a)(1-a)$, $x_2=(1-a)a$ and $x_3=a^2$. These three values are attained with the probabilities $p_X(x_1)=1/U^2$, $p_X(x_2)=2(U-1)/U^2$ and $p_X(x_3)=(U-1)^2/U^2$. The mean and variance of X can now be identified:

$$E(X) = \sum_{i=1}^3 x_i p_X(x_i) = 0 \tag{5}$$

$$V(X) = \sum_{i=1}^3 (x_i - E(X))^2 p_X(x_i) = \frac{(U-1)^2}{U^4} \tag{6}$$

If we consider the case where the activity is set to that of a pattern (retrieval cue) v: $o=\xi^v$, the support of each unit can then be written as

$$h_i = \sum_j w_{ij} o_j = \frac{1}{N} \sum_j^N \sum_\mu^P (\xi_i^\mu - a)(\xi_j^\mu - a)\xi_j^v = \frac{1}{N} \sum_j \sum_\mu X_{ij}^\mu \xi_j^v \qquad (7)$$

where we make the important assumption that each instance of X is independent. Now we want to estimate the mean and variance of the support values h^- of the units that do not participate in the active pattern:

$$E(h^-) = E(\frac{1}{N}\sum_j^N\sum_\mu^P X_j^\mu \xi_j^v) = \frac{1}{N}\sum_j^H\sum_\mu^P E(X_j^\mu) = \frac{HP}{N}E(X) = \frac{P}{U}E(X) = 0 \qquad (8)$$

$$V(h^-) = V(\frac{1}{N}\sum_j^N\sum_\mu^P X_j^\mu \xi_j^v) = \frac{1}{N^2}\sum_j^H\sum_\mu^P V(X_j^\mu) = \frac{HP}{N^2}V(X) = \frac{P}{NU}V(X) = \frac{P(U-1)^2}{NU^5} \qquad (9)$$

Next we want to estimate the mean and variance of all the support values h^+ that participate in the active pattern v. The parameter r corresponds to the number of hypercolumns that are not participating in building up the support of the active pattern. The default value of r is 1 since the units have no recurrent connections to themselves. Increasing the value of r can simulate noise in the input patterns.

$$E(h^+) = E(\frac{1}{N}\sum_j^N\sum_\mu^P X_j^\mu \xi_j^v) = \frac{1}{N}\left(\sum_j^{H-r} E(X_j^v) + \sum_j^H\sum_{\mu\neq v}^{P-1} E(X_j^\mu) \right) = \qquad (10)$$

$$= \frac{1}{N}\left((H-r)E(x_1) + H(P-1)E(X)\right) = \frac{(H-r)x_1}{N} = -\frac{(rU-N)(U-1)^2}{NU^3}$$

$$V(h^+) = V(\frac{1}{N}\sum_j^N\sum_\mu^P X_j^\mu \xi_j^v) = \frac{1}{N^2}\left(\sum_j^{H-r} V(X_j^v) + \sum_j^H\sum_{\mu\neq v}^{P-1} V(X_j^\mu) \right) = \qquad (11)$$

$$= \frac{1}{N^2}\left((H-r)V(x_1) + H(P-1)V(X)\right) = \frac{H(P-1)}{N^2}V(X) = \frac{(P-1)(U-1)^2}{NU^5}$$

Now we are in a position were we can formulate the mean and standard deviation of a normal distribution that expresses the probability that the support value of a unit participating in the active pattern will be larger than that of a non-participating unit. We form the stochastic variable $Y=h^+-h^-$ which has the mean and standard deviation;

$$m_Y = E(h^+) - E(h^-) = \frac{(rU-N)(U-1)^2}{NU^3} \qquad (12)$$

$$\sigma_Y = \sqrt{V(h^+)+V(h^-)} = \sqrt{\frac{(2P-1)(U-1)^2}{NU^5}} \qquad (13)$$

which has the normal distribution $Y \in N(m_Y, \sigma_Y)$.

We can now calculate the probability of occurrence of an error, i.e. the probability that the support of a unit not participating in a pattern will be larger than that of a participating unit:

$$P_{uerr} = P(Y \leq 0) = \Phi\left(\frac{-m_Y}{\sigma_Y}\right) = \frac{1}{\sqrt{2\pi}} \int_{-\infty}^{-m_Y/\sigma_Y} e^{-t^2/2} dt \qquad (14)$$

We now turn to consider a single hypercolumn. We know that $P(Y\leq 0)$ must not occur if we want the support of the unit participating in a pattern to be the largest one in the hypercolumn. We can thus write the probability of a hypercolumn producing a correct output as $p_{Hcorr}=(1-p_{uerr})^{U-1}$.

The probability that the network will produce a correct output i.e. that the active pattern will remain stable can now be stated. Let Z represent the number of hypercolumns with a correct output, $Z \in Bin(H, p_{Hcorr})$. If we consider the case where no errors are allowed in the retrieved pattern, the probability of a correct output from the NN is; $p_{netcorr} = p_Z(0) = p_{Hcorr}^H$.

In order to simplify the calculations we rewrite $H = N^b$ and $U = N^{1-b}$ where $b \in (0,1)$. This reduces the number of parameters in the expression for $p_{netcorr}$ to three; N, P and b. The storage capacity of a particular network (N and b) can be estimated by solving the equation $p_{netcorr} = 0.9$ for P.[1] The amount of information stored in a network is attained by multiplying the information content in one pattern with the number of patterns stored in the NN. Numerical solutions of $p_{netcorr} = 0.9$ for P when $a > 0$ and $N \rightarrow \infty$ indicates that the number of stored patterns roughly increases as N^{2-b}.

2 Simulation Results

The storage capacity of four different types of NNs is presented in Fig. 1a. A retrieved pattern without any errors was classified as a correct. The number of stored patterns was measured as the point were 90% of the patterns were correctly retrieved. The sparse Hopfield NN stored more patterns than the standard Hopfield NN and the Hopfield NN with hypercolumns stored more patterns than the sparse Hopfield NN. The information content was smallest in a hypercolumn pattern, it was slightly larger in a pattern used with the sparse Hopfield NN and it was largest in a pattern used with the Hopfield NN. This meant that the amount of stored information in the three

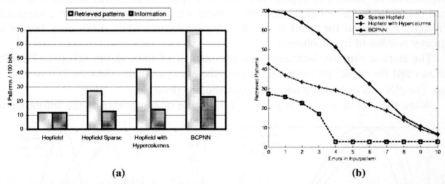

(a)   (b)

Fig. 1. The storage capacity of four different NNs each consisting of 100 units. The sparse Hopfield NN was used with sparse random patterns with a fraction a of the units set to 1. The sparse Hopfield NN is very sensitive to how the threshold is set (threshold was set to 0.04). The hypercolumn NNs were partitioned into 10 hypercolumns (b=0.5). **(a)** The left bar represents the number of patterns stored in the NN and the right bar represents the amount of information stored in the NN. **(b)** Successfully retrieved patterns as a function of noise. The number of errors in the retrieval cues was varied from zero to ten.

---

[1] The factor 0.9 represents the probability of retrieving a pattern and this probability roughly corresponds to the fraction of patterns that can be retrieved out of the P patterns stored in the NN.

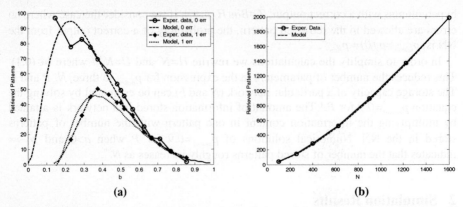

Fig. 2. Empirical fit of network capacity to the analytic model. **(a)** Capacity as a function of b (number of hypercolumns = N^b). **(b)** Capacity as a function of network size N for $b=0.5$.

different types of Hopfield NN was almost equal. Fig. 1b shows how the number of retrieved patterns depends on the number of errors (n:o flipped active units) in the retrieval cue. The sparse Hopfield NN has a sharp drop in the number of retrieved patterns when the number of errors is increased beyond 3; this can be avoided by using an active threshold regulation.

Empirical data was fitted against the analytic model of the sparse Hopfield NN with hypercolumns (Fig. 2). Fig 2a shows how the number of retrieved patterns depends on the number of hypercolumns. A NN with 120 units was studied with zero and one error in the retrieval cues. When there are no errors in the cue the maximal capacity is reached for small b, i.e. few hypercolumns, while noisy cues favour a greater number of hypercolumns.

The storage capacity increases with the number of units in the network. Fig. 2b shows that the model provides a tight fit to the empirical data when the NN is scaled up. The NN was partitioned into $\sqrt{N}$ hypercolumns ($b=0.5$).

When the size of the NN is increased the value of b maximizing the information

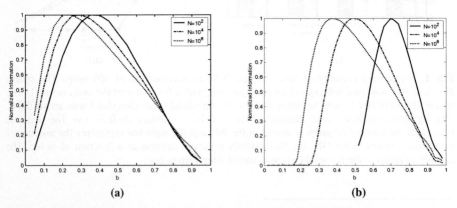

Fig. 3. The normalized amount of information stored plotted against the size and number of hypercolumns. **(a)** The optimal value of b is shifted towards zero as the size is increased. **(b)** When there is noise in the retrieval cue the optimal value is shifted towards one.

stored is shifted towards zero if there is no noise in the NN (Fig. 3a). If noise is introduced the optimal value of b is shifted towards one (Fig. 3b). In this simulation the input patterns had a constant level of 10 errors.

3 Discussion

The introduction of hypercolumns into the sparse Hopfield NN improved its performance significantly, beyond the performance increment due to the more structured hypercolumn patterns. The hypercolumn structure increased the noise tolerance and capacity of the network. The theoretical model provides a good fit to the empirical data even though it does not consider the iterative relaxation process.

A conclusion from the results is that the optimal partitioning of the NN, in terms of hypercolumns, depends on the noise level of the cues expected. If cues are noiseless, large networks achieve optimal capacity through very large hypercolumns, while more numerous and smaller hypercolumns are more efficient in dealing with noisy cues.

The three different types of the Hopfield NN roughly have an equal storage capacity measured in bits. There are learning rules that achieve a higher storage capacity, e.g. the BCPNN.

Finally, such a division into smaller modules is also relevant for parallel implementation of large networks. Partitioning of the NN into modules makes it possible to create a computational grain size that suits the hardware and also it helps to reduce the communication between the units.

References

1. Hubel, D.H. and T.N. Wiesel: Uniformity of monkey striate cortex: A parallel relationship between field size, scatter and magnification factor. J. Comp. Neurol., 1974(158): p. 295–306.
2. Holst, A. and A. Lansner: A Higher Order Bayesian Neural Network for Classification and Diagnosis, in Applied Decision Technologies: Computational Learning and Probabilistic Reasoning, A. Gammerman, Editor. 1996, John Wiley & Sons Ltd.: New York. p. 251–260.
3. Lansner, A. and Ö. Ekeberg: A one-layer feedback artificial neural network with a Bayesian learning rule. Int. J. Neural Systems, 1989. Vol. 1: p. 77–87.
4. Sandberg, A., et al.: A Bayesian attractor network with incremental learning. to appear in Network: Computation in Neural Systems, 2002. Vol. 13(2).
5. Ström, B.: A Model of Neocortical Memory, Using Hypercolumns and a Bayesian Learning Rule. 2000, Nada, KTH: Stockholm.
6. Peterson, C. and B. Söderberg: A New Method for Mapping Optimization Promlems onto Neural Networks. International Journal of Neural Systems, 1989. Vol. 1(3).
7. Levy, N., D. Horn, and E. Ruppin: Associative Memory in a Multi-modular Network. Neural Computation, 1999. Vol. 11: p. 1717–1737.
8. O'Kane, D. and A. Treves: Short- and long-range connections in autoassociative memory. J. Phys. A, 1992. Vol. 25: p. 5055–5069.

Edge Detection and Motion Discrimination in the Cuneate Nucleus

Eduardo Sánchez[1], S. Barro[1], and A. Canedo[2]

[1] Grupo de Sistemas Intelixentes (GSI)
Departamento de Electrónica e Computación, Facultade de Físicas,
Universidade de Santiago de Compostela,
15706 Santiago de Compostela, Spain
{eduardos, elsenen}@usc.es
[2] Departamento de Fisioloxía, Facultade de Medicina,
Universidade de Santiago de Compostela,
15706 Santiago de Compostela, Spain
{fsancala}@usc.es

Abstract. In this paper we investigate how the cuneate nucleus could perform edge detection as well as motion discrimination by means of a single layer of multi-threshold cuneothalamic neurons. A well-known center-surround receptive field organization is in charge of edge detection, whereas single neuronal processing integrates inhibitory and excitatory inputs over time to discriminate dynamic stimuli. The simulations show how lateral inhibition determines a sensitized state in neighbouring neurons which respond to dynamic patterns with a burst of spikes.

1 Introduction

The cuneate nucleus is a structure of the somato-sensory system that is located within the brain stem, in the most rostral area of the spinal chord. The middle zone of the cuneate, where the experimental work reported in this study has been carried out, can in turn be divided into two areas [1]: a core (central region), and a shell (peripheral region). The core is basically made up of cuneothalamic neurons, which are also referred to as projection or relay neurons. The shell is basically made up of local neurons, also known as interneurons or non-thalamic projection neurons.

The neuronal behaviour in the middle cuneate has been recently studied by means of intracellular and whole-cell techniques *in vivo* [2,3]. Under both cutaneous stimulation and intracellular injection of positive current pulses [3] the cuneothalamic neurons show a tonic activity characterized by a single-spike firing pattern (figure 1 (left)). A sodium ionic current, I_{Na}, and a potassium current, I_K, have been suggested to explain such dynamics. On the other hand, under hyperpolarizing current pulses (figure 1 (right)): (1) a depolarizing sag is uncovered, and (2) if the hyperpolarization is deep enough, a slow rebound potential appears inducing a burst of spikes. The former could be explained by

J.R. Dorronsoro (Ed.): ICANN 2002, LNCS 2415, pp. 198–203, 2002.
© Springer-Verlag Berlin Heidelberg 2002

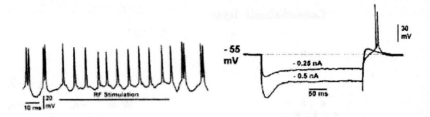

Fig. 1. Intracellular recordings in cuneothalamic neurons. Response to cutaneous stimulation of the excitatory receptive field (RF) (left). Response to hyperpolarizing current pulses (right).

means of a hyperpolarization-activated cationic current, I_h, while the later could be presumably induced by a low-threshold calcium current, I_T [3,4].

The main cuneate inputs originate from primary afferent (PAF) and corticocuneate fibers. The PAFs establish synaptic contact with projection cells and interneurons [1]. Stimulating the excitatory receptive field center of cuneothalamic neurons originated excitatory action potentials [5]. This excitatory response was delayed, and finally changed to an inhibitory type, when the stimulation was applied at increasing distances from the center of the excitatory receptive field. This behaviour can be explained by means of a receptive field showing an excitatory center as well as an inhibitory surround mediated by appropriate interneurons [4,5].

It is well-known that such center-surround connectivity can perform contrast detection [6]. Furthermore, if the synaptic weights are chosen to approximate a mexican-hat gaussian function, a network of McCulloch-Pitts neurons with a threshold output function can perform edge detection, as it was already shown in a previous model of the cuneate circuitry [4,7]. However, the dynamics of a cuneothalamic cell is much more complex than a simple McCulloch-Pitts neuron. The later is a single-threshold unit, whereas the former shows multiple thresholds associated with each ionic current. In this paper we investigate the processing capabilities that the cuneate could perform by combining the center-surround circuitry with these complex cuneothalamic neurons.

2   The Model

The model consists on a single bidimensional layer of cuneothalamic neurons with center-surround receptive fields (figure 2). Each neuron n_{ij} has been modeled as a single compartment with the following firing condition:

$$y_{ij}(t) = \begin{cases} 1 \text{ if } v_{ij}(t) > \Theta_{spike} \\ 0 \text{ otherwise} \end{cases} \qquad (1)$$

where $v_{ij}(t)$ is the membrane potential at time t and Θ_{spike} a positive threshold value. The membrane potential is updated by the following equation:

$$v_{ij}(t) = v_{ij}(t-1) + I_{ij}^{total}(t) - I_{ij}^{ionic}(t) \qquad (2)$$

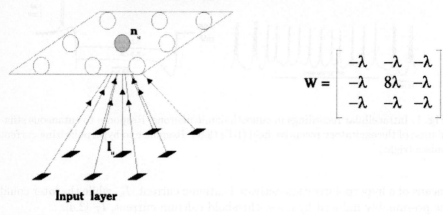

Fig. 2. Center-surround organization and synaptic weights. Each cuneothalamic neuron receives one excitatory input from the center and eight inhibitory inputs from the surround (left). The connectivity matrix W describes the weight of each synapse, with λ meaning any positive integer value. The excitatory synapse is set to 8λ, thus W approximates a Mexican-hat gaussian function, in order to perform an edge detection operation over the input.

The total afferent input $I_{ij}^{total}(t)$ of the cuneothalamic neuron n_{ij} is computed by multiplying every input I_{ij} inside its receptive field by its corresponding synaptic weight w_{ij}, thus $I^{total} = [W][I]$. The synaptic weight matrix W, shown in figure 2, is an approximation of a Mexican-hat gaussian function with an excitatory center and an inhibitory surround.

The ionic current term I_{ij}^{ionic} is a linear combination of the contributions of a sodium current, I_{Na}, which is activated under depolarization and generates sodium spikes; a potassium current, I_K, which repolarizes the membrane potential; a hyperpolarization-activated cationic current I_h, which impedes a greater hyperpolarization; and a low-threshold calcium current I_T, which is deinactivated after hiperpolarization and generates after further depolarization an slow rebound potential. Each current was computed accordingly the generalized Ohm law $I = g * (v - v_{ion})$, where g denotes the current conductance and v_{ion} the equilibrium potential for such ion. In a realistic model g is usually described by means of Hodgkin-Huxley formalism but here we have used simplified descriptions of the conductance dynamics:

$$g_{ij}^{Na}(t) = \begin{cases} 0 \text{ if } y_{ij}(t - k \triangle t) = 1 \ \forall k; 0 < k \le a \\ 1 \text{ otherwise} \end{cases} \tag{3}$$

$$g_{ij}^{K}(t) = \begin{cases} \frac{g_{ij}^{K}(t-1)}{\alpha} \text{ if } y_{ij}(t - k \triangle t) = 1 \text{ and } v_{ij}(t) > 0 \ \forall k; 0 < k \le a \\ 0 \qquad \text{otherwise} \end{cases} \tag{4}$$

$$g_{ij}^h(t) = \begin{cases} 1 \text{ if } v_{ij}(t) < v_{ij}(t-1) \text{ and } v_{ij}(t) < \Theta_h \\ 0 \text{ otherwise} \end{cases} \tag{5}$$

$$g_{ij}^T(t) = \begin{cases} 1 \text{ if previous hyperpolarization and } v_{ij}(t) > v_{ij}(t-1) \\ 0 \text{ otherwise} \end{cases} \tag{6}$$

3   Results

3.1   Static Patterns

Figure 3 shows the results of processing static patterns. At time t_0 the network performs an edge detection operation over the input. In the following two iterations no output activity is observed because of the activation of I_K as well as the inactivation of sodium conductances. If the input persists a single-spike firing behaviour is observed as it is illustrated in the lower part of figure 3.

3.2   Dynamic Patterns

In addition to participate in the edge detection operation, lateral inhibition also hyperpolarizes all the neighbouring neurons that do not receive excitatory input from matched receptive fields. This hyperpolarization can (1) activate I_h currents, thus stabilizing the membrane potential to some hyperpolarized value; and (2) de-inactivate I_T currents, thus preparing I_T for activation. This state characterizes what we can call a sensitized neuron. At this state, if further

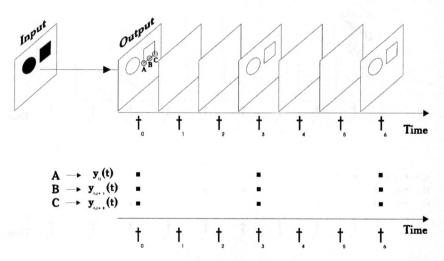

Fig. 3. Static pattern processing. The network performs an edge detection operation over the input (top). Neurons at positions A, B, and C, show single-spike activity (black squares) after detecting an edge in their receptive fields (bottom).

stimulation is applied at the excitatory center of the receptive field, a bursting discharge will appear as a result of both I_{Na} and I_T activation.

Sensitized neurons are the key to explore a new possible role of the cuneate nucleus: motion discrimination. Figure 4 shows the results when a dynamic pattern is presented to the cuneate network. When the pattern is at its initial position, single-spike firing is observed in the neuron n_{ij} located at position A. At the same time, the neuron $n_{i,j+1}$ at position B is sensitized through lateral inhibition. When motion is initiated, such neuron responds with a burst of spikes at times t_3 and t_4. It must be noticed that at time t_4, only those neurons that received inhibitory-excitatory input sequences are currently firing. The bursting activity is again observed in neuron $n_{i,j+2}$, at position C, and all neurons in subsequent positions along the direction of motion.

There are some conditions that need to be fulfilled for motion discrimination: (1) the hyperpolarization has to be deep enough to de-inactivate I_T currents, (2) the dynamics of the ionic currents have to be faster than those of the synaptic currents in order to observe bursting activity, and (3) the input pattern needs to follow a continuous trajectory in order to sensitize neurons in neighbouring positions.

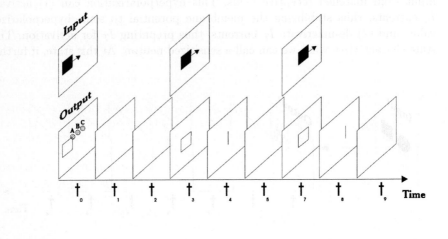

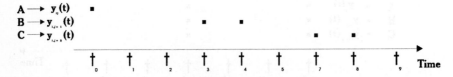

Fig. 4. Dynamic pattern processing. The network detects the edges as well as thelocations in which changes take place (top). Neurons at positions B and C, generate a burst of two spikes (black squares) indicating the presence of the dynamic pattern (bottom).

As a result of the modeling work, some predictions can be stated: (1) if I_h is blocked, no bursting response can be observed after sustained lateral inhibition; (2) if I_T is blocked, no bursting response can be observed after lateral inhibition; and (3) directional selectivity requires asymmetric center-surround receptive fields.

4 Conclusions

Edge detection and motion discrimination are two computational capabilities that the cuneate nucleus can perform. Edge detection can be understood as a coding reduction mechanism as well as the first step for further texture processing in the ascending somatosensory pathway. Motion discrimination can be the basis for directional selectivity and a driving mechanism for reactive motor actions. The key for both static and dynamic processing is the concept of a sensitized neuron or sensitized stated, which can be reached by a suitable hyperpolarization induced by means of an appropriate lateral inhibition. This sensitized mechanism could also be used by the somatosensory cortex in order to recruit cuneothalamic neurons facilitating the processing of ascending signals from selected regions or particular cutaneous receptive fields.

References

1. Fyffe, R. E., Cheema, S. S., Rustioni, A.: Intracelular Staining Study of the Feline Cuneate Nucleus. I. Terminal Patterns of Primary Afferent Fibers. Journal of Neurophysiology. Vol. 56. **5** (1986) 1268–1283
2. Canedo, A.: Primary motor cortex influences on the descending and ascending systems. Progress in Neurobiology. Vol. 51 (1997) 287–335
3. Canedo, A., Martinez, L., Mariño, J.: Tonic and bursting activity in the cuneate nucleus of the chloralose anesthetized cat. Neuroscience. Vol. 84 **2** (1998) 603–617
4. Sánchez, E.: Modelos del cuneatus: de la biología a la computación. Ph. D. Thesis. University of Santiago de Compostela (2000)
5. Canedo A., Aguilar J.: Spatial and cortical influences exerted on cuneothalamic and thalamocortical neurons of the cat. European Journal of Neuroscience. Vol. 12 **2** (2000) 2513–2533
6. Ratliff, F.: Mach bands. Holden-day (1965)
7. Sánchez E., Mucientes M., Barro, S.: A cuneate-based network and its application as a spatio-temporal filter in mobile robotics. Lecture Notes in Computer Science. (2001)

Encoding the Temporal Statistics of Markovian Sequences of Stimuli in Recurrent Neuronal Networks

Alessandro Usseglio Viretta[1], Stefano Fusi[2], and Shih-Chii Liu[1]

[1] Institute of Neuroinformatics, University and ETH Zurich
Winterthurerstrasse 190, CH-8057 Zurich, Switzerland
[ale, shih]@ini.phys.ethz.ch
[2] Institute of Physiology, University of Bern
Bühlplatz 5, CH-3012 Bern, Switzerland
fusi@cns.unibe.ch

Abstract. Encoding, storing, and recalling a temporal sequence of stimuli in a neuronal network can be achieved by creating associations between pairs of stimuli that are contiguous in time. This idea is illustrated by studying the behavior of a neural network model with binary neurons and binary stochastic synapses. The network extracts in an unsupervised manner the temporal statistics of the sequence of input stimuli. When a stimulus triggers the recalling process, the statistics of the output patterns reflects those of the input. If the sequence of stimuli is generated through a Markov process, then the network dynamics faithfully reproduces all the transition probabilities.

1 Introduction

Our current understanding of the neuronal mechanisms that permit biological systems to encode and recall temporal sequences of stimuli is still marginal. In the last decade many studies attempted to relate the problem of encoding temporal sequences to the generation of associations between visual stimuli [1, 2,3]. Interestingly, in one simple case [4], in which the stimuli were presented in a fixed temporal order, it was possible to study the neural correlate of this type of associative memory. Cortical recordings displayed significant correlations between the patterns of activity elicited by neighboring stimuli in the temporal sequence. Hence these internal representations of the visual stimuli encode the temporal context in which stimuli were repeatedly presented during training. These patterns of activities were stable throughout long time intervals and have been interpreted as global attractors of the network dynamics [3]. Despite the fact that the learning rule used by Griniasty [3,5] makes use of only the information about the contiguity of two successive stimuli, these attractors are correlated up to a distance of 5 in the temporal sequence, similar to that observed in the experiment of Miyashita [4]. Here we extend these mechanisms to a more general situation. First, we show that the network in the presence of noise

J.R. Dorronsoro (Ed.): ICANN 2002, LNCS 2415, pp. 204–209, 2002.
© Springer-Verlag Berlin Heidelberg 2002

of sufficient amplitude can spontaneously jump to a pattern of activity representing a different stimulus. The pattern of connectivity between the internal representations of the stimuli encodes the transition probability, and the presentation of a single stimulus can trigger the recalling of a sequence of patterns of activity corresponding to temporally correlated stimuli. Noise is exploited as in [2], and time is essentially encoded in the escape rates from the attractors. Second, the pattern of connectivity encoding the transition probabilities can be learned when the network is repeatedly exposed to the temporal statistics of the stimuli. The transition probabilities are automatically extracted during this "training phase" and encoded in the synaptic matrix. The learning rule was inspired by the one introduced in [3] and it makes use of the information carried by the current stimulus and by the pattern of activity elicited by the previous stimulus. A possible mechanism for making this rule local in time has been suggested in [6,5] and relies on the stable activity that is sustained by the network in the interval between two successive stimuli.

2   The Model

We implemented a recurrent neural network with N excitatory neurons, labeled by index i, $i = 1 \ldots N$. The state of neuron i is described by the variable S_i: $S_i = 1$ ($S_i = 0$) corresponds to a firing (quiescent) neuron. The network is fully connected with binary excitatory synapses J_{ij} from neuron j to neuron i [6]. The neuron's state is updated using the *Glauber dynamics* (see, for example, [7]), in which $S_i=1$ with probability $g_\beta(h_i) \equiv \frac{1}{1+\exp(-2\beta h_i)}$, where $1/\beta$ is the *pseudo temperature* of the network and $h_i = \sum_j J_{ij} S_j - I$ the synaptic input, or *field*, to the neuron. The global inhibition I dynamically adjusts the activity of the network. I depends on the fraction $F = \frac{1}{N} \sum_k S_k$ of neurons that are active as expressed by the following equation:

$$I(t+1) = \begin{cases} s_0(F(t) - s_1), & \text{If } F > f_m \\ I_m & \text{If } F \le f_m \end{cases} \tag{1}$$

where f_m is the threshold for the global activity, and I_m the minimum inhibition. The parameters s_0 and s_1 are chosen so that $s_0(f_0 - s_1) = I_0$, where I_0 is usually chosen between the maximum input to quiescent neurons and the minimum input to active neurons when the network state corresponds to a learned pattern of activity, and f_0 is the average activity of the learned patterns.

2.1   Learning Rule

Following [8], we implemented learning as a stochastic process. During the presentation of a pattern η_i^μ, the μ-th pattern of a temporal sequence, the neuron states are set to $S_i = \eta_i^\mu$ and the binary synapses are updated according to the following rules:

- If both the pre and the post-synaptic neurons are active, then a transition to the potentiated state occurs with a probability q_+.

- If only one of the two neurons connected by the synapse is active, then a transition to the depressed state of the synaptic weight occurs with a probability q_-.
- If both the neurons are inactive, the synapse is left unchanged.
- If the pre-synaptic activity imposed by the previous stimulus in the sequence is high and the post-synaptic activity induced by the current stimulus is also high, then a transition to the potentiated state of the synaptic weight $J_{ij} : 0 \rightarrow 1$ occurs with a probability $q_\times = \lambda_f q_+$, where $\lambda_f < 1$ (see [6, 5]). This part of the rule allows one to connect events that are separated in time and to encode the information about the temporal context in which the stimuli are presented.

In the case of random patterns, if the average activity is f, the probability for two randomly chosen neurons to be both active is f^2, whereas the probability for the two neurons to have different activities is $2f(1-f)$. In order that the probability for long term depression is approximately equal to the probability of long term potentiation, we choose $q_- = \frac{fq_+}{2(1-f)}$.

The learning of a transition results in making neurons that belong to the same pattern have a finite probability of providing a nonzero synaptic input to neurons belonging to a different pattern. In the case where several patterns and transitions have been learned, the transition probability is shown, for a suitable choice of the network parameters, to be a monotonically increasing function of the relative frequency of presentation of the patterns during the learning phase.

2.2   Learning Markov Processes

We assume that the sequence of patterns to be learned is generated according to a Markov process, that is to a random process in which the probability for which a pattern is chosen depends only on the immediate preceding chosen pattern. Given a Markov matrix M, the element $M_{\mu\nu}$ is the probability of transition from state s_μ to state s_ν. To teach the network to encode a Markov process, a pattern η_i^μ is randomly chosen from a pool of p patterns. The next pattern η_i^ν is chosen according to the transition probability of the Markov process from state s_μ to state s_ν. The two patterns are presented to the network, which thus learns them and the transition from the first to the second. This process is repeated until the synaptic matrix has reached its asymptotic configuration. Alternatively, the probability for each synapse to be potentiated can be analytically calculated in the limit for vanishing q_+ and a pattern sequence of infinite length [9]. Calculating the weight matrix in this way is computationally less expensive than on-line learning. Given the auxiliary variables P_{ij} and Q_{ij}, which are respectively proportional to the number of events leading to synapse potentiation and depression,

$$\begin{cases} P_{ij} = \sum_{\mu=1}^p (\eta_i^\mu \eta_j^\mu q_+ + \sum_{\nu=1}^p (\eta_i^\nu \eta_j^\mu M_{\mu\nu} q_\times)) \\ Q_{ij} = \sum_{\mu=1}^p ((1 - \eta_i^\mu)\eta_j^\mu q_- + q_- \eta_i^\mu (1 - \eta_j^\mu)) \end{cases} \quad (2)$$

the probability for the synapse J_{ij} to be 1 is given by $p_{ij} = \frac{P_{ij}}{P_{ij}+Q_{ij}}$.

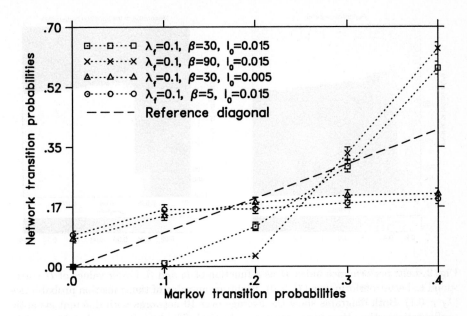

Fig. 1. Transition probabilities as a function of the Markov process transition probabilities for different combinations of β and I_0. Upper and lower error bars have been calculated using the formula in Meyer [10] , with $k = 1$ and $n = 700$ (490 neurons, 7 patterns). An increase in β decreases the transition probabilities corresponding to low Markov transition probabilities and increases those corresponding to high Markov probabilities. A low I_0 or β makes the dynamics less dependent on the connectivity pattern. The corresponding plot is shallower, indicating a poor capability in reproducing the statistics of the input sequences.

3   Results

The performance of the network was evaluated on a sequence generated by a Markov chain. The network was able to reproduce the statistics of the input sequences without an external input. Given an initial pattern of activity, the network made spontaneous transitions to other patterns with a probability close to the corresponding Markov chain probability. The performance of the network strongly depended on the temperature $1/\beta$ and inhibition I_0. The on-line learning procedure and the analytical derivation of the synaptic matrix led to qualitatively equivalent results (data not shown).

To evaluate the performance we computed the transition probability matrix $T_{\mu\nu}$. This matrix describes the probability that the network makes at some time t a transition from an initial state highly resembling pattern μ (the overlap $m^\mu = \frac{1}{N} \sum_i S_i \eta_i^\mu$ [7] between the network state and pattern η^μ is the maximum), into a state that resembles maximally pattern η^ν.

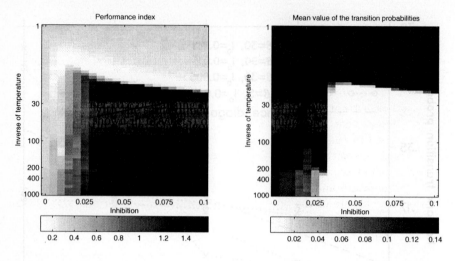

Fig. 2. Left: performance index Π as a function of I_0 and β. Lower values of Π correspond to better performance ($\Pi \geq 0$). Right: mean value of the transition probabilities ($\lambda_f = 0.1$). Both diagrams show the average over 10 networks with different synaptic configurations: the other parameters are identical. The transition probability matrix was measured after 100 transitions per pattern from the starting state and allowing a maximum of 100 updates of all neurons to make a transition. The performance of the network is maximal in a region around $I_0 = 0.01$ and $\beta = 14$. For I_0 up to 0.04 the optimal β is roughly proportional to I_0 and for higher inhibition the best performance lies in the area just above the sharp boundary in the plot. In the right-hand plot, the sharp boundary goes together with a sudden decrease in the mean value of the transition probability for an increasing β. Very low temperatures trap the network in the initial basin of attraction.

3.1   Performance Evaluation

The Markov matrix M used to generate the input patterns has been compared with the transition probability matrix T. The closer the elements of the transition probability matrix $T_{\mu\nu}$ are to the corresponding values of the Markov matrix $M_{\mu\nu}$, the better the network is at reproducing the statistics of the Markov process. We have analyzed the performance of the network as a function of I_0 and β using a Markov matrix whose rows are randomly reshuffled versions of $\{0, 0, 0, 0.1, 0.2, 0.3, 0.4\}$. Figure 1 shows the performance of the network for four combinations of λ_f, β and I_0. Depending on the combination of the parameters, the network reproduces more or less faithfully the statistics of the temporal sequence used for training. Π has been chosen as the performance index of a network that has learned P patterns.

$$\Pi = \frac{1}{P^2} \sum_{\substack{\mu,\nu = 1 \\ M_{\mu\nu} \neq 0}}^{P} \frac{|M_{\mu\nu} - T_{\mu\nu}|}{\frac{M_{\mu\nu} + T_{\mu\nu}}{2}} \qquad (3)$$

Figure 2 shows the performance index Π and the mean value of the transition probabilities as a function of I_0 and β for $\lambda_f = 0.1$.

4   Conclusions

We described a recurrent neuronal network consisting of binary neurons and binary synapses that is able to learn and reproduce the statistics of the input sequences used in the learning phase. The performance has been evaluated by comparing the transition probability matrix with the Markov matrix used in generating the input sequences. The performance of the network strongly depends on the temperature and global inhibition. The binary coding of the variables, the stochastic nature of the network and its scalability makes this architecture attractive for implementation in hardware.

References

1. D. Kleinfeld and H. Sompolinsky. Associative neural network model for the generation of temporal Patterns. *Biophysical Journal*, 54:1039–1051, 1988.
2. J. Buhmann and K. Schulten. Noise-driven temporal association in neural networks. *Europhysics Letters*, 4(10):1205–1209, 1987.
3. M. Griniasty, M. V. Tsodyks, and D. J. Amit. Conversion of temporal correlations between stimuli to spatial correlations between attractors. *Neural Computation*, 5:1–17, 1993.
4. Y. Miyashita. Neuronal correlate of visual associative long-term memory in the primate temporal cortex. *Nature*, 335(6193):817–820, October 1988.
5. V. Yakovlev, S. Pusi, E. Berman, and E. Zohary. Inter-trial neuronal activity in inferior temporal cortex: a putative vehicle to generate long-term visual associations. *Nature neuroscience*, 1(4):310–317, August 1998.
6. N. Brunel. Hebbian learning of context in recurrent neural network. *Neural Compututation*, 8:1677–1710, 1996.
7. D. J. Amit. *Modeling Bruin Function*. Cambridge University Press, New York, 1989.
8. D. J. Amit and S. Fusi. Learning in neural networks with material synapses. *Computation*, 6:957–982, 1994.
9. N. Brunel, F. Carusi, and S. Fusi. Slow stochastic hebbian learning of classes of Stimuli in a recurrent neural network. *Network*, 9:123–152, 1998.
10. P. L. Meyer. Introductory probability and statistical applications. Addison-Wesley, Reading, MA, 1965.

Multi-stream Exploratory Projection Pursuit for the Formation of Complex Cells Similar to Visual Cortical Neurons

Darryl Charles[1], Jos Koetsier[2], Donald MacDonald[2], and Colin Fyfe[2]

[1] Faculty of Informatics, University of Ulster, UK.
dk.charles@ulster.ac.uk
[2] University of Paisley, UK.
{koet-ci0, macd-ci0}@paisley.ac.uk

Abstract. A multi-stream extension of Exploratory Projection Pursuit is proposed as a method for the formation of local, spatiotemporal, oriented filters similar to complex cells found in the visual cortex. This algorithm, which we call the Exploratory Correlation Analysis (ECA) network, is derived to maximise dependencies between separate, but related, data streams. By altering the functions on the outputs of the ECA network we can explore different forms of shared, higher order structure in multiple data streams.

1 Introduction

Recently, an information-maximisation learning rule for the generation of cells similar to those in visual cortex [1] has been proposed. The model architecture on which this rule is based is well-known to researchers in this area: the output value from a complex cell is the weighted sum of the activations from simple cells. The activation of a simple cell is normally modelled as a function of a weighted linear sum of input values. An appropriate function to use on the output of a simple cells is thought be a half-wave positive (or negative) rectification, or a squared half-wave positive rectification. The function that is used on the ECA network in this paper is similar to the squared half-wave positive rectification function.

A multi-cell extension of this model [1] is illustrated in Figure 1a, beside the Exploratory Correlation Analysis (ECA) network (Figure 1b). The ECA network may thought of as being composed of multiple, co-operating Exploratory Projection Pursuit [2] networks, and the learning rules have been derived to maximise the dependence between pairs of outputs. The network outputs co-operate in the search for shared, higher order structure in data. ECA is proposed as an alternative algorithm for the formation of spatio-temporal, localised and oriented complex cells similar to those found in the visual cortex. The results prove to be similar to those found in [1].

J.R. Dorronsoro (Ed.): ICANN 2002, LNCS 2415, pp. 210–215, 2002.
© Springer-Verlag Berlin Heidelberg 2002

2   Exploratory Projection Pursuit

Exploratory Projection Pursuit (EPP) is a statistical technique that is used to visualise structure in high dimensional data. Data is projected to a lower dimensional space in order to identify interesting structure. The projection should capture all of the aspects that we wish to visualise, which means it should maximise an index that defines a degree of interest of the output distribution [2].

One measure of interestingness is based on an argument that random projections tend to result in Gaussian distributions [3]. Therefore, we can define an interesting projection as one that maximises the non gaussianity of the output distributions. In this paper we concentrate on measures that are based on kurtosis and skewness.

2.1   Neural EPP

The ECA network is related to the single stream, neural EPP algorithm [2]. The feed-forward step of this network is described in (1), in which the inputs are multiplied by the weights and summed to activate the outputs. This is followed by a feedback phase (2) in which the outputs are fed back and subtracted from the input to form a residual. This residual is then used in the weight update rule (3).

$$\mathbf{y} = W\mathbf{x} \tag{1}$$
$$\mathbf{r} = \mathbf{x} - W^T\mathbf{y} \tag{2}$$
$$\Delta W = \mathbf{r}^T \mathbf{f}(\mathbf{y}) \tag{3}$$

The function $\mathbf{f}(\mathbf{y})$ in (3) causes the weight vectors to converge to directions that maximise a function whose derivative is $\mathbf{f}(\mathbf{y})$. Thus, if $\mathbf{f}(\mathbf{y})$ is linear, i.e. $\mathbf{f}(\mathbf{y}) = \mathbf{y}$, the EPP algorithm performs identically to Oja's subspace algorithm [4]. If the function is $\mathbf{f}(\mathbf{y}) \propto \mathbf{y}^2$, the third moment in the data is maximised and if $\mathbf{f}(\mathbf{y}) \propto \mathbf{y}^3$ is used, the fourth moment in the data is maximised.

3   Exploratory Correlation Analysis

The Neural EPP algorithm is extended to allow for multiple input streams (Figure 1b). Both streams are assumed to be have a set of common underlying factors. Mathematically we can write this as

$$\mathbf{y}_1 = W\mathbf{x}_1$$
$$\mathbf{y}_2 = V\mathbf{x}_2$$

The input streams are denoted as $\mathbf{x}_1$ and $\mathbf{x}_2$, the projected data as $\mathbf{y}_1$ and $\mathbf{y}_2$ and the basis vectors are rows of the matrices W and V. Each input stream can be analysed separately by performing EPP and finding common statistical features that have maximum non-gaussianity. However, as we know that the features we

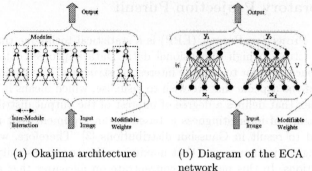

(a) Okajima architecture   (b) Diagram of the ECA
                                      network

Fig. 1. A comparison of Okajima's multiple complex cell architecture with that of the
ECA network

are looking for have the same statistical structure, we can add another constraint
which maximises the dependence between the outputs. The simplest way to
express this formally is by maximising $E(\mathbf{g}(\mathbf{y}_1)^T \mathbf{g}(\mathbf{y}_2))$. We also need to ensure
the weights do not grow indefinitely, which we can achieve by adding weight
constraints $W^T W = A$ and $V^T V = B$. Writing this as an energy function with
Lagrange parameters $\lambda_{i,j}$ and $\mu_{i,j}$ [5] we obtain (4).

$$J(W,V) = E(\mathbf{g}(W\mathbf{x}_1)^T \mathbf{g}(V\mathbf{x}_2)) +$$
$$\frac{1}{2} \sum_{i=1}^{N} \sum_{j=1}^{N} \lambda_{i,j}(\mathbf{w}_i^T \mathbf{w}_j - a_{i,j}) +$$
$$\frac{1}{2} \sum_{i=1}^{N} \sum_{j=1}^{N} \mu_{i,j}(\mathbf{v}_i^T \mathbf{v}_j - b_{i,j}) \qquad (4)$$

The energy function (4) may be differentiated with respect to the weights $v_{i,j}$
and $w_{i,j}$ [6] to obtain the following stochastic gradient descent rules:

$$\Delta W = \eta[(\mathbf{g}(\mathbf{y}_2) \otimes \mathbf{g}'(\mathbf{y}_1))(\mathbf{x}_1^T - \mathbf{y}_1^T W)] \qquad (5)$$
$$\Delta V = \eta[(\mathbf{g}(\mathbf{y}_1) \otimes \mathbf{g}'(\mathbf{y}_2))(\mathbf{x}_2^T - \mathbf{y}_2^T V)] \qquad (6)$$

We have set the A and B matrices to the identity matrix, which causes the
weights W and V to converge to orthonormal weight vectors. As with the neural
EPP algorithm, we need to replace the output functions with stable versions
for the ECA algorithm. In contrast to the neural EPP algorithm, we not only
require the derivative of the function to be maximised, but also the function
itself. We therefore need an additional stable function, whose truncated Taylor
expansion is $\mathbf{g}(\mathbf{y}) = \mathbf{y}^4$. The function we chose for the experiments in this paper
is $\mathbf{g}(\mathbf{y}) = 1 - \exp(-\mathbf{y}^4)$.

Table 1. Artificial data set. S_1 and S_2 are more kurtotic than the common source S_3. S_4 is a normal data source.

$$x_{1,1} = S_1 + N(0, 0.2) \; , \; x_{2,1} = S_2 + N(0, 0.2)$$
$$x_{1,2} = S_3 + N(0, 0.2) \; , \; x_{2,2} = S_3 + N(0, 0.2)$$
$$x_{1,3} = S_4 + N(0, 0.2) \; , \; x_{2,3} = S_4 + N(0, 0.2)$$

3.1 Artificial Data Set

A simple artificial data-set is used to test the network, generated from a combination of kurtotic and a normal data sources. The inputs are two 3-dimensional input vectors, $\mathbf{x_1}$ and $\mathbf{x_2}$ (Table 1), each of which have different kurtotic values. Input S_1 and S_2 were generated by taking a value from a normal distribution and raising it to the power of 5. Input S_3 was generated from a normal distribution raised to the power of 3. The common data source S_3 is therefore less kurtotic than input S_1 or S_2. The last data source we used is S_4, which was simply generated from a normal distribution. In order to show the robustness of the network zero mean Gaussian noise was added with variance 0.2 to each of the inputs independently. After training the network for 50000 iterations with a learning rate of 0.003, the weights converged to the values shown in Table 2. The network has clearly identified the common kurtotic data source and has ignored the common normal input and the independent input sources S_1 and S_2, although they are more kurtotic than S_3.

Table 2. Weight vectors after training the ECA network on artificial data.

w	0.0029	1.0000	0.0028
v	0.0043	1.0000	-0.0182

4 Contextual Guidance in Early Vision

Natural images contain a great deal of structure, for example lines, surfaces edges and a variety of textures are present in most images. The human brain appears to have learned to code these structures efficiently. In an attempt to understand how our brains can achieve this we may adopt a statistical approach. From this perspective we can try to describe the relationship between neighbouring pixels. It has been argued that although there is a strong second order relationship between pixels, we cannot adequately capture the interesting structure that is necessary to analyse images efficiently by only considering second order statistics. Therefore a compact coding such as the well-known principal components analysis does not suffice.

4.1   The EPP Algorithm and Sparse Coding

As the EPP algorithm searches for codes with high kurtosis, it is suitable for coding natural images. Due to the large scale nature of the experiments, the performance may be improved by using a different form of weight constraint. In the case of EPP the weight update rule is simplified to a Hebbian update rule.

$$\Delta W = \mathbf{x}^T \mathbf{f}(\mathbf{y})$$

and for ECA the weight update rules become:

$$\Delta W = \eta[(\mathbf{g}(\mathbf{y}_2) \otimes \mathbf{g}'(\mathbf{y}_1))\mathbf{x}_1^T]$$
$$\Delta V = \eta[(\mathbf{g}(\mathbf{y}_1) \otimes \mathbf{g}'(\mathbf{y}_2))\mathbf{x}_2^T]$$

As we have eliminated the weight constraints, the weights can grow indefinitely and two weights can learn the same feature. We ensure a bounded solution by using symmetric decorrelation given by:

$$W(t+1) = (W(t)W(t)^T)^{-\frac{1}{2}}W(t)$$

Because the EPP and ECA networks require data to be whitened we need to pre-process the images. First the data is mean-centered and then it is filtered by a filter with frequency response $R(f) = f exp(-(f/f_0)^4)$. This is a widely used whitening/low-pass filter that ensures the Fourier amplitude spectrum of the images is flattened. It also decreases the effect of noise by eliminating the highest frequencies. As a reference point we first carry out an experiment with the standard EPP network in which 9 natural images were chosen and preprocessed as described above. We randomly sampled the pre-processed images by taking 12 by 12 pixel patches, which were used as inputs to the network. Figure 2 shows a sample of 10 weight vectors, after the network was fully trained.

Fig. 2. 10 converged weight vectors when training then EPP network with natural images

4.2   Stereo Images

Stereo images consist of two images: one part as seen through the left eye and another as seen through the right. As both images are different views of the same scene, they share a number of features, which can be extracted with the ECA network.

For this experiment we chose 9 natural pre-processed stereo images. The images were sampled by randomly taking 12 by 12 patches, which were used

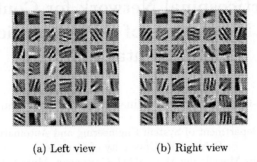

(a) Left view   (b) Right view

Fig. 3. Converged weight vectors when training the ECA network with stereo images

pair-wise as input to the ECA network. The resulting weight vectors of the trained network are displayed in Figure 3. We find interesting differences between the filters obtained from standard images and those obtained from stereo images. The first difference is that there are significantly less shared codes for stereo images. This can be explained by the fact that stereo images are not only views of a scene at slightly different angles, but the two images are also shifted. This makes the 'overlap' between two patches smaller and the amount of shared information less. The ECA filters are also more complex.

5   Conclusion

A multi-stream version of Exploratory Projection Pursuit has been presented as a method for generating local, spatiotemporal, oriented filters similar to complex cells found in the visual cortex. The ECA network proves successful in this regard and provides an alternative, intuitive and flexible approach for exploring multi-data streams such as stereo image pairs.

References

1. K. Okajima, "An information-based learning rule that generates cells similar to visual cortical neurons," *Neural Networks*, no. 14, pp. 1173–1180, 2001.
2. C. Fyfe, "A general exploratory projection pursuit network," *Neural Processing Letters*, vol. 2, no. 3, pp. 17–19, May 1995.
3. Persi Diaconis and David Freedman, "Asymptotics of graphical projections," *The Annals of Statistics*, vol. 12, no. 3, pp. 793–815, 1984.
4. E. Oja, "Neural networks, principal components and subspaces," *International Journal of Neural Systems*, vol. 1, pp. 61–68, 1989.
5. Juha Karhunen and Jyrki Joutsensalo, "Representation and separation of signals using nonlinear pca type learning," *Neural Networks*, vol. 7, no. 1, pp. 113–127, 1994.
6. J. Koetsier, D. Charles, D. MacDonald, and C. Fyfe, "Exploratory correlation analysis," *ESANN 2002*, to appear 2002.

A Corticospinal Network for Control of Voluntary Movements of a Physiologically Based Experimental Platform

Francisco García-Córdova, Javier Molina-Vilaplana, and Juan López-Coronado

Department of System Engineering and Automatic
Polytechnic University of Cartagena
Campus Muralla del Mar, 30202, Cartagena, Murcia, SPAIN
{francisco.garcia, javi.molina, jl.coronado}@upct.es

Abstract. In this paper, we present a corticospinal network for control voluntary movements within constraints from neurophysiology. Neural controller is proposed to follow desired joint trajectories of a single link controlled by an agonist-antagonist pair of actuator with muscle-like properties. This research work involves the design and implementation of an efficient biomechanical model of the animal muscular actuation system. In this biomechanical system the implementation of a mathematical model for whole skeletal muscle force generation on DC motors is carried out. Through experimental results, we showed that neural controller exhibits key kinematic properties of human movements, dynamics compensation and including asymmetric bell-shaped velocity profiles. Neural controller suggests how the brain may set automatic and volitional gating mechanisms to vary the balance of static and dynamic feedback information to guide the movement command and to compensate for external forces.

1 Introduction

In recent years, the interface between biology and robotics is carried out by biorobotic researches. Biorobotics tries to emulate the very properties that allow humans to be successful. Each component of a biorobotic system must incorporate the knowledge of diverse areas as neuromuscular physiology, biomechanics and neuroscience to name a few, into the design of sensors, actuators, circuits, processors and control algorithms [1].

Neurophysiological experiments have made progress in characterizing the role played by various neural cell types in the control of the voluntary movements. These experiments have addressed issues such as the coordinate frames used in motor cortex and post-central areas [2], the relation of cell activity to movement variables [3], preparatory activity, load sensitivity, the latencies of various responses and equilibrium point control [4]. As part of an attempt to unify these diverse experimental data, Bullock et al. [5] proposed a computational model that incorporates model neurons corresponding to identified cortical cell types in a circuit that reflects known anatomical connectivity. Computer simulations

J.R. Dorronsoro (Ed.): ICANN 2002, LNCS 2415, pp. 216–222, 2002.
© Springer-Verlag Berlin Heidelberg 2002

in Bullock et al. [5] showed that properties of model elements correspond to the dynamic properties of many known cell types in area 4 and 5 of the cerebral cortex. Among these properties are delay period activation, response profiles during movement, kinematic and kinetic sensitivities and latency of activity onset [6].

In this work, a corticospinal network controller for control of voluntary reaching movements is proposed. The controller is developed of biologically inspired neural model [5], [7]. A new aspect of this research work is to apply knowledge of human neuro-musculo-skeletal motion control to a biomechanically designed, neural controlled, robotic system and to demonstrate that such system is able to respond voluntary similar human movements on an experimental platform. The neural controller applies a strategy of trajectory control using an extension and revision [8],[9] of the Vector Integration to Endpoint (VITE) model [10], which exhibits key kinematic properties of human movements, including asymmetric bell-shaped velocity profiles.

In addition, the research presented in this paper describes the emulation of the animal musculoskeletal system to a biomechanically designed robotic system. We use appropriate planetary gearboxes and multi-radial flexible couplings (in order to pull and be pushed), force and position sensors, and tendons, in order to approach the behavior of the DC motors like the animal muscles [1].

The following section describes the emulation of the whole musculoskeletal system in an experimental platform. The next section presents the corticospinal network controller for motion control of a biomechanically-designed robotic system. Section IV describes the experiments on the proposed scheme for control of voluntary reaching movements on an anthropomorphic single link. Finally, in Section V, conclusions based on experimental results are given, and future works are indicated.

2   Biomechanical System

Experimental platform has been designed in form of an anthropomorphic system and applies the 2N configuration, in which a single joint is driven for two auto-reversible DC motors (Micro-motors of Maxon) and by means of cables, which emulates the system of tendons. The application of DC motors presents in "mimic" form one of the properties of the system musculo-skeletal [11]. This is the Force-velocity relationship of the muscle. With an appropriate planetary gearboxes and multi-radial flexible couplings (in order to pull and be pushed), and a series of cables (tendons) with lineal properties, the actuators approach their behaviour to of a muscle. The muscle spindle is simulated with a resistor potentiometer located on joint, and force sensors on the tendons of the system represent the tendon organs. The transmission of the force to the link is carried out by special tendons make of polystyrene highly molecular (see Table 1). In addition, this system integrate a force FSR (Force sensing resistance) sensor at the end of the link, and encoder sensors in the DC motors. A specific electronic hardware to drive the platform has been developed.

Table 1. Elements of the anthropomorphic single link

Biological system	An anthropomorphic single link
Bones	Light plastic link
Joints	Revolution joints
Tendons	Polystyrene highly molecular
Muscles	DC Motors like artificial muscles

We first implemented the model of the muscle and the forces generated by this model are passed to a force controller as is shown in Fig.1. A PID controller as force control implemented in a microcontroller was used here. The model of the muscle allows to carry out the force-length relationship with which we covered the property that lacked in the DC Motors.

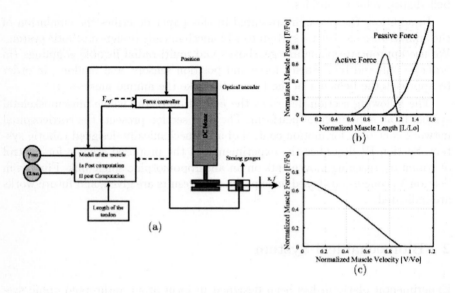

Fig. 1. Muscle model applied in the biomechanical system. (a) Schematic diagram of artificial muscle model using DC motors. (b) and (c) are the dimensionless relationships between force-length and force-velocity under isometric conditions, respectively.

3   Architecture of the Neural Control System

The proposed neural control system is a muti-channel central pattern generator capable of generating desired joint movement trajectories by smoothly interpolating between initial and final muscle length commands for the antagonist muscles (dc micro-motors) involved in the movement. The rate of interpolation is controlled by the product of a difference vector, which continuously computes the difference between the desired and present position of the joint. Basic properties of this circuit, notably its allowance for movement priming, and for performing the same movement at various speeds while maintaining synchrony among synergists. Figure 2 depicts the neural controller, which uses several representations of neural variables postulated to be coded by activity levels distributed across corticospinal populations.

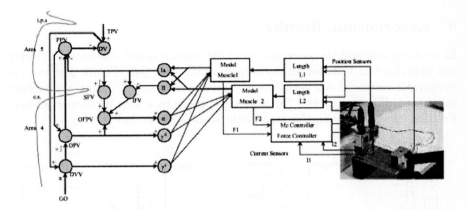

Fig. 2. Architecture of the corticospinal control system. This system presents GO-scalable gating signal; DVV- desired velocity vector; OPV-outflow position vector; OFPV-outflow force+position vector; SFV- static force vector; IFV-inertial force vector; PPV-perceived position vector; DV-difference vector; TPV target position vector; γ^d - dynamic gamma motor neuron; γ^s- static gamma motor neuron; α -alpha motor neuron; Ia- type -Ia afferent fiber; II- type II afferente fiber; c.s.-central sulcus; i.p.s.-intraparietal sulcus. The symbol + represents excitation, - represents inhibition, x represents multiplicative gating, and $+\int$ represents integration.

In the corticospinal network, the OPV is the average firing rate of a population of area 4 tonic cells, which represents the position that the link would be without presence of obstacles.The GO is scalable gating signal and exhibits a sigmoidal growth during the movement generation interval. DVV is the activity of area 4 phasic movement-time (MT) cells and their activity profiles resemble a bell-shaped velocity profile, they are tuned to direction of movement and show little load-sensitivity. The DVV in area 4 is interpreted to be a gated and scaled version of a movements command that is continuously computed in posterior area 5 like the vector difference (DV) between the target and the perceived limb position vectors. Perceived Position Vector (PPV) are anterior area 5 tonic cells that are assumed to receive an efference copy input from area 4 and position error feedback from muscle spindles. In the neural controller, an Inertial Force Vector (IFV), identified with activity of area 4 phasic reaction-time (RT) cells, extracts velocity errors from the primary and secondary spindle feedback. To compensate for static loads such gravity, the neural controller integrates positional errors reported by the spindles and adds them to the alpha motor neuron command. Spindle error integration is performed by a Static Force Vector (SFV). The activity of the phasic-tonic cells constitutes an Outflow Force+Position Vector (OFPV). OFPV produces the desired kinematic result under variable external forces. The system can be used to generate voluntary reaching movements at variable speeds while compensating for external perturbations including inertial and static loads, to maintain posture against perturbations and exert forces against objects that obstruct a reaching movement.

4   Experimental Results

In order to evaluate the performance of the cortico-spinal controller, we carry out experimental tests on an anthropomorphic sigle link. Figure 3 shows experimental results of cell activities in cortical area 4 and 5 during a voluntary movement. Figure 4(a) and (b) depict responses to a step position reference. Position control with impulse load disturbance is shown in Fig. 4(c).

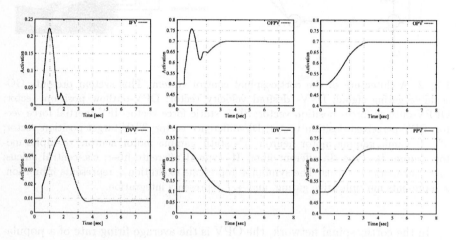

Fig. 3. Activity of the cortical cells during a simple voluntary reaching task.

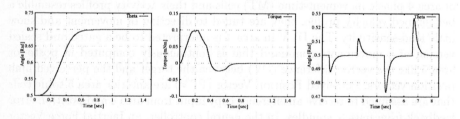

Fig. 4. Experimental results: Response of the single link for a step position reference (a) and (b). In (c) is shown the response to static loads. Two loads were applied the first of 25 gr and the second was the double of the first. Note that the excursions of the single link are small and are quickly compensated

5   Conclusions

In this paper, we have presented a corticospinal network controller for control of voluntary movements. This proposed neural controller suggests how the brain may set automatic and volitional gating mechanism to the balance of static and dynamic feedback information to guide the movement command and to compensate for external forces. For example, with increasing movement speed, the system shifts from a feedback position controller to a feedforward trajectory generator with superimposed dynamics compensation. Modulation of a corticospinal network controller for trajectory generation and dynamics compensation with automatic and volitional mechanism allows the system to achieve a high degree of task sensitivity. The proposed control strategy can also compensate for unexpected disturbances, which is as essential capability for compliance control. The neural controller exhibits key kinematic properties of human movements, including asymmetric bell-shaped velocity profiles. Future work is aimed in extending the spinal cord inside the corticospinal network.

References

1. García-Córdova, F., Guerrero, A., Pedreño, J.L, López-Coronado, J. (2001). "Emulation of the animal muscular actuation system in an experimental platform". Proceedings of the IEEE Conference on Systems, Man, and Cybernetics, October 7–10, Tucson, AZ.
2. Kettner, R.E., Schwartz, A.B., and Georgopoulos, A.P., "Primate motor cortex and free arm move-ments to visual targets in three-dimensional space. III. Positional gradients and population coding of movement direction from various movements origin," The journal of Neuroscience, 8(8):2938–2947, 1988.
3. Riehle, A., Mackay, W.A., and Renqui, J., "Are extent and force independent movement parameter? Preparation and movement-related neuronal activity in the monkey cortex", Experimental Brain research, 99(1), 56–74, 1994.
4. Bizzi, E., Accornero, N., Chapple, W., and Hogan, N., "Posture control and trajectory formation during arm movement," The Journal of Neuro-science, 4(11):2738–2744, 1984.
5. Bullock, D., Cisek, P., and Grossberg, S., "Cortical networks for control of voluntary arm movements under variable force conditions," Cere-bral Cortex, 8, 48–62, 1998.
6. Kalaska, J.F., and Crammond, D.J., "Cerebral cortical mechanics of reaching movements," Science, 255:1517–1527, 1992.
7. García-Córdova, F., Guerrero, A., Pedreño, J.L, López-Coronado, J., (2001). "A cortical network for vcontrol of voluntary movements in robotics sytems". Proceedings of the IEEE Conference on Systems, Man, and Cybernetics, October 7–10, Tucson, AZ.
8. García-Cordova, F., Coronado, J.L., Guerrero, A., "A Neural Controller for an artificial finger built with SMA actuators", Symposium on Intelligent Robotic Systems (SIRS99). July 1999. Coimbra (Portugal).
9. García-Cordova, F., Coronado, J.L., Guerrero, A., "Design of an anthropomorphic finger using shape memory alloy springs", Proceedings of the IEEE on System, Man and Cybernetic (SMC99). September 1999. Tokyo (Japan).

10. Bullock, D. and Grossberg, S., "The VITE model: A neural command circuit for generating arm and articulatory trajectories," In J.A.S. Kelso, A.J. Mandell, and M.F. Shlesinger (Eds), Dynamic patters in complex systems. Singapore: World Scientific Publishers, 1988.

11. Hannaford, Blake and Winters, Jack , "Actuator properties and movement control: Biological and technological models", Multiple Muscle systems: Biomechanics and Movement Organization, J.M. Winters and S.L-Y. Woo (eds.), pp 101–120, 1990 Springer-Verlag.

Firing Rate for a Generic Integrate-and-Fire Neuron with Exponentially Correlated Input

Rubén Moreno and Néstor Parga

Universidad Autónoma de Madrid, Cantoblanco, Madrid 28049, SPAIN,
rmoreno@delta.ft.uam.es,
http://ket.ft.uam.es/~neurociencia/

Abstract. The effect of time correlations in the afferent current on the firing rate of a generalized integrate-and-fire neuron model is studied. When the correlation time τ_c is small enough the firing rate can be calculated analytically for small values of the correlation amplitude α^2. It is shown that the rate decreases as $\sqrt{\tau_c}$ from its value at $\tau_c = 0$. This limit behavior is universal for integrate-and-fire neurons driven by exponential correlated Gaussian input. The details of the model only determine the pre-factor multiplying $\sqrt{\tau_c}$. Two model examples are discussed.

1 Introduction

Synchrony is a form of temporal correlation in the neuronal activity that occurs when a set of neurons tend to fire together within a time window of small size. It has been experimentally recorded throughout the cortex [1,2]. Both the size of the population of neurons firing synchronously and the temporal precision of the synchronization may have important effects on the activity of the post-synaptic neurons. As a consequence of synchronization, the output rate of a post-synaptic neuron typically increases. This has been shown analitucally for example in [3, 4] and in simulations of leaky integrate-and-fire neurons in [5]. However, most of these studies obtain analytical results only for perfectly synchronized inputs where neurons tend to fire at the same time. The response of more complicated integrate-and-fire neuron models to a perfect synchronized input or to an input synchronized with finite temporal precision has, to our knowledge, only been treated in [4].

In this work we show that the rate of a generalized integrate-and-fire neuron (see eq. (1) below) increases with the precision of the synchronization in a universal way. Whithin this class of neuron models this increase in the response behaves as the squared root of the time window τ_c where the inputs are synchronized. The $\sqrt{\tau_c}$ behavior indicates that corrections to the perfectly synchronized case ($\tau_c = 0$) are important even for small τ_c. The universal $\sqrt{\tau_c}$ behavior of integrate-and-fire neurons in obtained using the diffusion approximation and the Fokker-Planck formalism. The constant pre-factor multiplying $\sqrt{\tau_c}$ depends on the details of the particular neuron model selected and can also be computed with our formalism. An example of this, corresponding to a leak

J.R. Dorronsoro (Ed.): ICANN 2002, LNCS 2415, pp. 223–228, 2002.
© Springer-Verlag Berlin Heidelberg 2002

proportional to the membrane potential, can be found in some previous work [4]. Here we will also present a direct solution of another model corresponding to a simple non-leaky neuron which will be used to test the previous analytical results derived for small τ_c. We close this paper with a comparison between the two models.

2   The Integrate-and-Fire Neuron and Input Statistics

Here we describe the generalized integrate-and-fire neuron model and present the statistic properties of the input current.

The integrate-and-fire neuron model. The depolarization membrane potential $V(t)$ evolves from the reset voltage H according to the stochastic equation [7,8, 9]

$$\dot{V}(t) = -f(V) + I(t) \tag{1}$$

where $f(V)$ is a voltage dependent term and $I(t)$ is the afferent current . When the potential reaches a threshold value Θ, a spike is emitted and the neuron is reset to H, from where it restarts to integrate the input signal after a refractory time τ_{ref}. Within the refractory time the voltage is fixed at H.

The input statistics. The neuron is driven by a stationary, correlated random current $I(t)$ coming from upstream neurons that can be written as a sum of spike events

$$I(t) = \sum_{i=1}^{N} J_i \sum_k \delta(t - t_i^k) \tag{2}$$

Here i denotes the i-th pre-synaptic neuron, the index k labels the arrival times of the individual pre-synaptic spikes, N is the number of connections, and J_i is the strength of the connection with neuron i. We assume that the cross-correlations between pairs of neurons have exponential decay with range τ_c, so that the two-point correlation of the current can be written as (for further details see [4])

$$C(t - t') \equiv\; < I^2(t) > - < I(t) >^2 = \sigma^2 \delta(t - t') + \sigma^2 \frac{\alpha^2}{2\tau_c} e^{-\frac{|t-t'|}{\tau_c}} \tag{3}$$

where σ^2 is a white noise variance and α^2 is the amplitude of the correlations (in the useful units of $\sigma^2/2\tau_c$) and measures the relative importance of the temporal correlations compared to white noise fluctuations. The exponential form in eq. (3) has been chosen first because it fully characterizes correlations by a timescale range τ_c (intuitively, it can be interpreted as the time window around which the synchronized spikes occur) and by the intensity factor α^2 and second because it makes the computations feasible. Note that the correlation function $C(t - t')$ describes a stationary property of the input statistics, a fact reflected in that it only depends on the difference between the two times.

The input current is completely described by its mean $\mu =< I(t) >$, its variance σ^2, and its second order correlations. This is justified because a typical neuron in the cortex receives a large barrage of spikes per second [6, 8]. Each spike induces a very small membrane depolarization J compared to the distance between the threshold and reset potentials. Moreover if the condition $\frac{J(1+\alpha^2)}{\Theta-H} < \frac{1}{10}$ holds, the correlations are weak enough for the diffusion approximation to be valid and thus the input current can be considered as Gaussian.

3   The Analytical Solution

A random Gaussian current $I(t)$ as described above, with correlations as in eq.(3), can be obtained as [10,4]

$$I(t) = \mu + \sigma\eta(t) + \sigma\frac{\alpha^2}{\sqrt{2\tau_c}}z(t) \qquad (4)$$

$$\dot{z}(t) = -\frac{z}{\tau_c} + \sqrt{\frac{2}{\tau_c}}\zeta(t) \qquad (5)$$

where $\eta(t)$ and $\zeta(t)$ are two independent white noise random variables with unit variances ($< \eta(t) >=< \zeta(t) >= 0$, $< \eta(t)\eta(t') >=< \zeta(t)\zeta(t') >= \delta(t - t')$, $< \eta(t)\zeta(t') >= 0$) and $z(t)$ is an auxiliary random variable.

The diffusion process defined by eqs. (1, 4, 5) can be studied by means of the stationary Fokker-Planck equation [10,4]

$$[\frac{\partial}{\partial V}(f(V)-\mu+\frac{\sigma^2}{2}\frac{\partial}{\partial V})+\frac{1}{\tau_c}\frac{\partial}{\partial z}(z+\frac{\partial}{\partial z})-\sqrt{\frac{2\sigma^2\alpha^2}{\tau_c}}\frac{\partial}{\partial V}]P = -\delta(V-H)J(z) \quad (6)$$

where $P(V, z)$ is the steady state probability density of having the neuron in the state (V, z) and $J(z)$ is the escape probability current, which integrated over z gives the firing rate of the neuron. It appears as a source term representing the reset effect: whenever the potential V reaches the threshold Θ, it is reset to the value H keeping the same distribution in the auxiliary variable z. If we suppose that the correlation time τ_c is very small compared to the refractory time τ_{ref} ($\tau_c << \tau_{ref}$) the escape current can be rewritten as $J(z) = \nu_{out}e^{-z^2/2}/\sqrt{2\pi}$ [11], where ν_{out} is the output firing rate of the neuron. This means that after one spike the variable z has enough time to relax to its stationary distribution (a normal distribution). The Fokker-Planck equation has to be solved with the normalization condition $\nu_{out}\tau_{ref} + \int_{-\infty}^{\Theta} dV \int_{-\infty}^{\infty} dzP(V, z) = 1$ (meaning that with probability $\nu_{out}\tau_{ref}$ the neuron is at the reset value, and with the complementary probability it is at some state (V, z)) and the condition that P vanishes at the threshold $P(\Theta, z) = 0$ [10,4].

Eq.(6) is solved using an extension of the tecnique described in [4] for the leaky integrate-and-fire model. The probability density $P(V, z)$ is expressed in powers of $k = \sqrt{\tau_c}$ keeping the orders $O(k^0)$ and $O(k)$. At the end of a rather lengthy calculation one obtains that the output rate expands as $\nu_{out} = \nu_0 + \nu_1\sqrt{\tau_c}$ where

$$\nu_0^{-1} = \tau_{ref} + \int_H^{\Theta} due^{\frac{2}{\sigma_{eff}^2}\int_{\Theta}^u dr(\mu - f(r))} \int_{-\infty}^u dve^{-\frac{2}{\sigma_{eff}^2}\int_{\Theta}^v dr(\mu - f(r))}$$

$$\nu_1 = -\frac{\sqrt{2}\alpha^2\nu_0^2}{\sigma}\int_{-\infty}^{\Theta} dve^{-\frac{2}{\sigma^2}\int_{\Theta}^v dr(\mu - f(r))} \tag{7}$$

Here appears the effective variance which is given by $\sigma_{eff}^2 = \sigma^2(1+\alpha^2)$. Note that the analytical expression for the rate ν_0 is the same as in the white noise input case with mean input μ and variance σ_{eff}^2. At order $O(k)$, ν_1 has been calculated up to order $O(\alpha^2)$ in the amplitude of the correlations, so the expression is valid only for small values of α^2. However, the order $O(k^0)$ is exact for all α^2. The rate has been obtained in the Fokker-Planck formalism only for positive correlations ($\alpha^2 > 0$), which is the relevant case in this work. The mathematical tools for studying negative correlations are harder but with them it can be seen [4] that the expression for ν_0 is exact even for negative correlations.

The first term, ν_0, gives the response to a perfectly synchronized input. As was announced before, the correction to perfect synchronization behaves as $\sqrt{\tau_c}$ for any function $f(.)$, only the pre-factor ν_1 depends of the particular neuron model chosen. To exemplify this universal behavior we now discuss and compare two different models, the leaky and the non-leaky integrate-and-fire neurons.

The non-leaky neuron. In this model there is no voltage dependent decay, thus $f(V) = 0$, and we find

$$\nu_0^{-1} = \tau_{ref} + \frac{\Theta - H}{\mu}$$

$$\nu_1 = -\frac{\alpha^2\nu_0^2\sigma}{\sqrt{2}\mu} \tag{8}$$

The rate of the non-leaky neuron is known to be insensitive to input fluctuations [9], and here it is shown that the rate is also insensitive to perfectly synchronized inputs ($\tau_c = 0$), in fact ν_0 does not dependent on α. However, when synchronization is not perfect ($\tau_c \neq 0$), the first correction to the rate is negative. Since the rate at order $O(k^0)$ is the same as if the neuron received a white noise input, the effect of correlations is to decrease the rate as τ_c increases, what is a peculiarity of the non-leaky neuron with refractory time.

The interest in such a simple model comes from the fact that eq. (6) can be solved exactly for all τ_c in powers of α^2 when $J(z) = \nu_{out}e^{-z^2/2}/\sqrt{2\pi}$ (valid for $\tau_{ref} \gg \tau_c$). This allows us to test the solution in eqs.(8) derived from the general expressions given in eqs. (7). The rate obtained in this way is

$$\nu_{out} = \nu_0 - \frac{\alpha^2\nu_0^2[1 - e^{(\gamma-\lambda)(\Theta-H)}]}{\mu(\gamma + \lambda)} + O(\alpha^4) \tag{9}$$

where $\gamma = \frac{\mu}{\sigma^2}$, $\lambda = \sqrt{\gamma^2 + \frac{2}{\sigma^2 \tau_c}}$ and ν_0 is the same as in eq. (8). Expanding eq. (9) for small τ_c one obtains the same universal $\sqrt{\tau_c}$ behavior as predicted, and the same prefactor as in eq. (8).

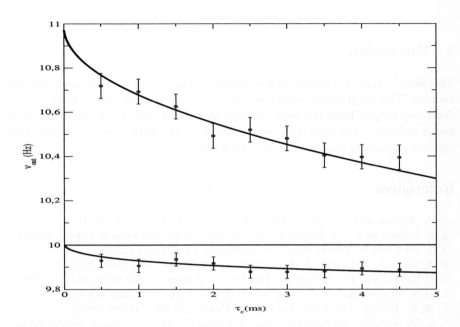

Fig. 1. Firing rates for both a non-leaky and leaky integrate-and-fire as a function of τ_c. Upper curve and data set: theoretical curve using eqs. (10) for a leaky neuron, and comparison with simulation data using eqs. (1, 4, 5). After each spike, the auxiliary variable z is reinitialized with a normal distribution (valid for $\tau_c \ll \tau_{ref}$). The parameters are: $\Theta = 1$ (in arbitrary units), $H = 0$, $\tau = 10ms$, $\tau_{ref} = 5ms$, $\mu = 81.7$, $\sigma^2 = 2.2$, $\alpha^2 = 1.10$. Bottom curve and data set: the same as above for a non-leaky neuron. The different parameters are: $\mu = 10.53$, $\sigma^2 = 1.2$, $\alpha^2 = 1.44$; the rest of parameters are the same as above. Intermediate straight line: firing rates in both models when there is no correlations ($\alpha^2 = 0$).

The leaky neuron. This neuron is characterized by a voltage dependent decay $f(V) = V/\tau$, where τ is the membrane time constant ($\tau \sim 10ms$). From eq.(8) we obtain

$$\nu_0^{-1} = \tau_{ref} + \sqrt{\pi}\tau \int_{\hat{H}_{eff}}^{\hat{\Theta}_{eff}} du\, e^{u^2}(1 + erf(u))$$

$$\nu_1 = -\alpha^2 \nu_0^2 \sqrt{\frac{\pi\tau}{2}} e^{\hat{\Theta}^2}(1 + erf(\hat{\Theta})) \qquad (10)$$

Here $erf(t)$ is the error function. Besides, $\hat{\Theta}_{eff} = \frac{\Theta - \mu\tau}{\sqrt{\sigma^2_{eff}\tau}}$, $\hat{H}_{eff} = \frac{H - \mu\tau}{\sqrt{\sigma^2_{eff}\tau}}$ and $\hat{\Theta} = \frac{\Theta - \mu\tau}{\sqrt{\sigma^2\tau}}$. This result agrees with what was found in [4]. For positive correlations the firing rate grows, meaning that correlated spikes are more effective in driving the leaky neuron than uncorrelated ones. Some numerical results and comparison between leaky and non-leaky rates are shown in Fig. 1.

4   Discussion

This work points to a better understanding of the role of correlations in driving neurons. The main result is that the firing rate of an integrate-and-fire neuron decreases as $\sqrt{\tau_c}$ from the rate obtained at $\tau_c = 0$. This behavior is universal and it indicates that even small departures from the perfect synchronized input case are important and need to be considered.

References

1. A. Aertsen and M. Arndt. Curr. Opin. Neurobiol. **3**, 586-594 (1993).
2. E. Salinas and T. J. Sejnowski. Nature Reviews Neuroscience **2**: 539-550 (2001).
3. J. Feng and D. Brown, Neural Computation **12**, 671-692 (2000).
4. R. Moreno, A. Renart, J. de la Rocha and N. Parga, submitted, 2002.
5. E. Salinas and T. J. Sejnowski, The Journal of Neuroscience **20**, 6193-6209 (2000).
6. M.N. Shadlen and W.T. Newsome, J. Neurosci. **18**, 3870-3896 (1998).
7. B. W. Knight, The Journal of General Physiology **59**, 734-778 (1972)
8. L. M. Ricciardi, *Diffusion processes and related topics in biology* (Springer-Verlag, Berlin, 1977).
9. H. C. Tuckwell, *Introduction to Theoretical Neurobiology II* Cambridge University Press, Cambridge, UK, 1988
10. H. Risken, *The Fokker-Planck equation* (Springer-Verlag, Berlin. Second ed., 1989)
11. N. Brunel and S. Sergi, J. Theor. Biol. **195**, 87-95 (1998).

Iterative Population Decoding Based on Prior Beliefs

Jens R. Otterpohl and K. Pawelzik

Institute of Theoretical Physics, University of Bremen
28334 Bremen, Germany

Abstract. We propose a framework for investigation of the modulation of neural coding/decoding by the availability of prior information on the stimulus statistics. In particular, we describe a novel iterative decoding scheme for a population code that is based on prior information. It can be viewed as a generalization of the Richardson-Lucy algorithm to include degrees of belief that the encoding population encodes specific features. The method is applied to a signal detection task and it is verified that – in comparison to standard maximum-likelihood decoding – the procedure significantly enhances performance of an ideal observer if appropriate prior information is available. Moreover, the model predicts that high prior probabilities should lead to a selective sharpening of the tuning profiles of the corresponding recurrent weights similar to the shrinking of receptive fields under attentional demands that has been observed experimentally.

1 Introduction

The view that the process of perception involves carrying out statistical inferences – dating back as early as the works of Helmholtz [4] – has been inspiring to psychophysicists and neurobiologists alike ([5], [1]). However, the question as to how and where in the brain probabilistic quantities are acquired and represented and by which mechanisms they are put into computational action has remained an elusive issue to neuroscientists.

Neural Coding / Decoding

This question is closely related to the quest for the appropriate understanding of neural coding. Neural tuning functions (receptive fields) are typically broad and strongly overlapping. Moreover, in addition to the randomness of the *external* world, neural activity is generally subject to some *intrinsic* noise. Thus, closely related stimuli are *encoded* into closely similar patterns of population activity. However, the brain is able to perform discrimination of very similar stimuli. This requires a read out from their common representation that is able to disentangle information about the stimuli to be discriminated. We refer to this process as *decoding*. As pointed out in [1], this is also a requirement for learning (prior) probabilities of stimuli that could in turn be used for carrying out Bayesian

J.R. Dorronsoro (Ed.): ICANN 2002, LNCS 2415, pp. 229–234, 2002.

inference. Following this line of argument, it seems necessary for the brain to explicitly represent the decoded information. In the following, we refer to this representation as *decoding population.*

However, since neural decoding is often seen purely as a method for quantifying the information contained in the activity of a population from the experimenter's point of view rather than an essential function of neuronal processing itself (e.g. [9]), little attention has been devoted to the question how an encoding and a decoding population should interact in order to purport statistical inference. A direct step in this direction has been made by showing that – for single stimulus elements – a particular type of recurrent networks is suit to perform *maximum likelihood* (ML) estimation when driven by the activity of the encoding population [2]. Nevertheless, ML–estimators are efficient (in the sense of attaining the Cramer-Rao bound) only in case of infinite sampling data. In a finite time, however, the decoding population receives a finite amount of data from the encoding population and the ML–estimator becomes very sensitive to noise. If prior information is available, it could be used to make the estimate more robust, e.g. by performing *maximum a posteriori* (MAP) estimation.

It is obvious that correct prior information should benefit the disentanglement of the ambiguities arising from noisy representations with large overlap. However, the main question is what kind of prior information can be implemented and how it is to affect the interactions between encoding and decoding representations on a neural level.

2   Neural Representations of Prior Information

In order to investigate this question, one needs an experimental paradigm which allows to induce prior beliefs in the animal in a controlled manner. In psychophysics, one typically considers signal detection and discrimination tasks. As a paradigm that combines both demands we propose a task that requires detection of a target signal among a set of distractor signals.

Signal Detection Task with Prior Information

We allow the stimulus to contain elements or features i. In each trial, the presence of each feature i is chosen independently with probabilities p_i. The contrasts $\widehat{H}_i$ of the present features are randomly and independently chosen from the uniform distribution over $\left[0, \widehat{H}_M\right]$. For each possible feature i, a series of training trials is performed in which $T = i$ is cued as the target (the remaining features $k \neq T$ are distractors) such that the animal might be able to learn the correct prior probabilities p_i for all i .

Encoding Model

As common in models of population codes [9], we assume a linear model with Poisson noise for encoding the stimulus into activities $V_j = \lambda_j + noise$ with

$\lambda_j = \sum_i^M W_{ji}\widehat{H}_i + V_B$; $(j = 1, ..., N)$. The model is fully specified by the receptive fields $f_i(j) = W_{ji}$ of each neuron V_j and the background rate V_B that accounts for the spontaneous firing observed in cellular recordings.

Iterative Decoding Using Prior Beliefs

Now, interpreting V_j in terms of specific features i amounts to estimation of $\widehat{H}_i$ based on a decoding model and a priori probabilities π_i regarding these features. We specify the decoding model as $V_j = \sum_i^M W_{ji}\widehat{H}_i + MH_B/N + noise$. Note that this is the same as the encoding model only if $NV_B = MH_B$ and the implicit prior beliefs π_i are the same as the true probabilities p_i.

In principle, prior knowledge could be incorporated via MAP-estimation using the prior distribution over $\widehat{H}_i$.

Instead, we directly modify the well-known Richardson-Lucy algorithm for ML-estimation [6] in order to incorporate the prior beliefs.

We propose the following iterative algorithm for determining the estimate H_i of $\widehat{H}_i$:

$$\widetilde{H}_i^{t+1} = \widetilde{H}_i^t \sum_j \frac{V_j\widetilde{W}_{ji}}{\sum_k \widetilde{W}_{jk}\widetilde{H}_k^t} \qquad (1)$$

$$\widetilde{H}_i^t = H_i^t + H_B \; ; \; \widetilde{W}_{ji} = \frac{[H_B/N + W_{ji}\pi_i H_M/2]}{H_B + \pi_i H_M/2} \qquad (2)$$

This procedure can be shown to converge, and the resulting estimate is given by $H_i = \lim_{t \to \infty} H_i^t$. All the prior information is contained in the $\widetilde{W}_{ji}$. The latter can be interpreted as effective recurrent weights, since they project the activities $\widetilde{H}_i$ down to the encoding layer V_j .

Note that, if $H_B = 0$, the update rule (1) reduces to the original Richardson-Lucy algorithm [6]:

$$H_i^{t+1} = H_i^t \sum_j \frac{V_j W_{ji}}{\sum_k W_{jk}H_k^t} \qquad (3)$$

The advantage of the decoding procedure (1) can be easily seen when setting $\forall i \notin S : \pi_i = 0$, where S denotes the set of potential causes of the input. Then, the recurrent weights become $\widetilde{W}_{ji} = 1/N \; \forall i \notin S$, i.e. the corresponding $\widetilde{H}_i$ are only used to fit the background noise and cannot take part in the reconstruction of the input signal $\sum_i^M W_{ji}\widehat{H}_i$.

Hence, setting prior beliefs in specific feature values to zero forces the activity to be distributed among the remaining neurons $H_{i \in S}$.

As a consequence, the prior information enables the decoder to resolve ambiguities that are due to starkly overlapping receptive fields W_{ji} and high background noise H_B.

3   Modeling a Signal Detection Task

In order to evaluate the performance of the decoding scheme, we modeled a signal detection task of the type described above.

We used a population of $M = 5$ decoding neurons with homogeneous, bell-shaped tuning functions (receptive fields) evenly spaced among $N = 25$ encoding neurons V_j (with periodic boundary conditions). Neighbouring receptive fields were strongly overlapping (variance of tuning functions as measured on the V layer was 20). In all the trials, all $\widehat{H}_i$ were set to zero except of three neighbouring units $\widehat{H}_{1,2,3}$ that contributed to the signal. In each trial, the presence of a signal at i ($\widehat{H}_{i \in \{1,2,3\}} > 0$) was chosen independently with probabilities $p_1 = p_3 = 0.7$, $p_2 = 0.3$, and the contrasts of the present signals were independently chosen from the uniform distribution over $[0, H_M = 1000]$. In each trial, a Poisson realization was drawn using the rates $\lambda_j = \sum_i^M W_{ji} \widehat{H}_i + V_B$. The background rate ($V_B = 80$) was chosen such that the activity encoding a single stimulus element (e.g. $\widehat{H}_k = C_k > 0$) was very similar to the activity encoding presentation of multiple stimuli (if $\widehat{H}_1 + \widehat{H}_2 + \widehat{H}_3 \approx C_k$). The resulting patterns were presented to the iterative decoder (1) and the responses were recorded. In a series of trials, one of the three possible signals was defined as target ($i = T$), the remaining were distractors ($i \neq T$, $i \in \{1,2,3\}$).

Model Observer

How should an outer observer (resp. a higher brain area) read out this decoding population? Since – as we have argued in the introduction – the purpose of the decoding is to represent the correlated activities of the encoding population as largely independent features, the decoding procedure should be judged by the performance of an observer who decides upon presence of a target solely based on the activity $H_{i=T}$, that is, who considers all ambiguities as being resolved by the decoding procedure. Therefore, as a model observer we use the ideal Bayesian observer of H_T who knows the correct prior probability p_T of the target and chooses the alternative of *maximum a posteriori* probability [3].

Results

In order to evaluate the decoding scheme, 15 conditions with different prior beliefs were tested (see caption of Fig. 1). After 300 iterations, the decoding has nearly converged in any of these conditions. The performance of the model observer for detecting the target $i = T = 1$ in the output of the decoding procedure after 300 iterations is shown in the figure. While the performance of the standard maximum likelihood (ML) decoding (Eqn. 3) was not significantly better than the 'blind observer' who achieves 70 % correct by simply choosing the a priori most likely alternative, performance was slightly above this level when using our procedure with correct background rate but incorrect priors $\pi_i = 1$, for all i. As to be expected, setting prior beliefs to zero for nodes

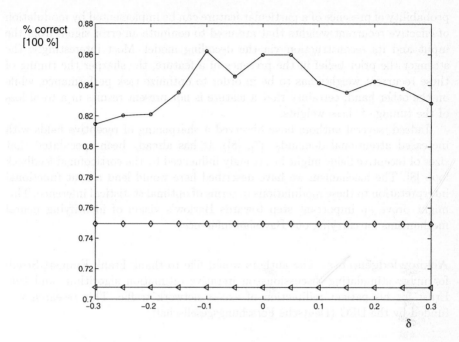

Fig. 1. Performance as computed from the simulation trials. The top curve (circles) gives the percentage of correct responses as a function of the divergence δ from the correct priors p_i [$\pi_{i\in\{1,3\}} = p_i + \delta$; $\pi_2 = p_2 - \delta$; $\pi_{i>3} = 0$]. $\delta = 0$ corresponds to the correct prior information $\pi_1 = \pi_3 = 0.7$, $\pi_2 = 0.3$. For comparison, the straight lines show the performance level in the conditions with no prior information [$\pi_i = 1$, for all i] (triangles); and with prior beliefs $\pi_{i\in\{1,2,3\}} = 1$, $\pi_{i>3} = 0$ (diamonds). [40000 trials were used in each condition.]

that were never present in the input ($\pi_{i>3} = 0$) improved performance above this level, but only to 75 % correct. However, using approximately correct prior information significantly improved detection performance up to 86 % in case of exactly correct beliefs.

4   Conclusions

We have proposed a heuristically derived algorithm for decoding of population activities using a linear encoding model with high background noise and inhomogeneous distribution over signals. As we have shown, incorporation of the correct prior information regarding presence of signals into the decoding scheme significantly improves signal detection performance based on the most relevant channel of the decoding population.

While we have not given a detailed neural implementation of the algorithm, the form of the update rule strongly suggests that prior information regarding the probability of presence of a particular feature can be implemented by modulation of effective recurrent weights that are used to compute an error signal from the input and its reconstruction via the decoding model. Most interestingly, the stronger the prior belief in the presence of a feature, the sharper the tuning of these recurrent weights has to be in order to optimize task performance, while on the other hand, certainty that a feature is not present results in a total loss of the tuning of these weights.

Indeed, several authors have observed a sharpening of receptive fields with increased attentional demands ([7], [8]). It has already been speculated that sizes of receptive fields might be strongly influenced by the corticofugal feedback loop [8]. The mechanism we have described here would lend a neat functional interpretation to these modulations in terms of optimal statistical inference. This might prove an important step towards Barlow's vision of identifying neural mechanisms for carrying out Bayesian inference.

Acknowledgements. The authors would like to thank Frank Emmert-Streib for many stimulating discussions on iterative estimation algorithms and Udo Ernst for his patient redirection of excess network traffic. This research was funded by the DFG (Deutsche Forschungsgesellschaft).

References

1. Barlow, H. Redundancy reduction revisited. (2001) Network 12, 241–253.
2. Deneve, S., Latham, P.E. and Pouget, A. Reading population codes: a neural implementation of ideal observers. (1999). Nature Neuroscience. 2(8):740–745.
3. Green, D., Swets, J. Signal Detection Theory and Psychophysics (Wiley, New York, 1966)
4. Helmholtz, H. von (1867/1925). Physiological optics. Vol. 3. Optical Society of America.
5. Marr, D. (1982): Vision. Freeman.
6. Richardson, W. H. Bayesian-Based Iterative Method of Image Restoration. (1972) Jour Opt Soc Am 62, 1, 55
7. Spitzer, H. et al. Increased attention enhances both behavioral and neuronal performance (1988). Science 240, 338–340.
8. Wörgötter F., Eysel, U.T. Context, state and the receptive fields of striatal cortex cells (2000). Trends in Neurosciences 23 (10): 497–503.
9. Zemel, R. S., Dayan, P., Pouget, A. Probabilistic Interpretation of Population Codes. (1998) Neural Computation, 10(2), 403–430.

When NMDA Receptor Conductances Increase Inter-spike Interval Variability

Giancarlo La Camera, Stefano Fusi, Walter Senn, Alexander Rauch, and
Hans-R. Lüscher

Institute of Physiology, University of Bern, Bühlplatz 5, 3012 Bern, Switzerland
{lacamera,fusi,senn,rauch,luescher}@pyl.unibe.ch

Abstract. We analyze extensively the temporal properties of the train
of spikes emitted by a simple model neuron as a function of the statistics of the synaptic input. In particular we focus on the asynchronous
case, in which the synaptic inputs are random and uncorrelated. We
show that the NMDA component acts as a non-stationary input that
varies on longer time scales than the inter-spike intervals. In the sub-
threshold regime, this can increase dramatically the coefficient of variability (bringing it beyond one). The analysis provides also simple guidelines for searching parameters that maximize irregularity.

1   Introduction

Spike trains recorded from single cells in vivo appear rather irregular. The origin
of this high degree of variability has been debated in the last decade (see e.g.
[3,4]) and many authors suggested that input synchronization is necessary to
achieve such a high variability (see e.g [2,5]). While it is easy to show that input
synchronization leads to high irregularity, it is difficult to rule out the possibility
that the same variability can be achieved with asynchronous inputs. Proving this
kind of negative results requires clear guidelines that allow to explore systematically the parameters space. Here we start by simulating the *in vitro* experiment
described in [2] in which the authors were not able to obtain high variability
with asynchronous inputs. We show that it is possible to achieve high irregularity without synchronicity and that synaptic NMDA receptor conductance may
play an important role.

2   Methods

We simulate a single compartment leaky integrate-and-fire neuron whose dynamics is fully described by the membrane depolarization V:

$$\frac{dV}{dt} = -\frac{V(t) - V_{rest}}{\tau_m} + \frac{I_{syn}}{C_m}$$

where I_{syn} is the total synaptic current flowing into the cell, V_{rest} is the resting potential, τ_m is the membrane time constant and C_m is the membrane

J.R. Dorronsoro (Ed.): ICANN 2002, LNCS 2415, pp. 235–240, 2002.

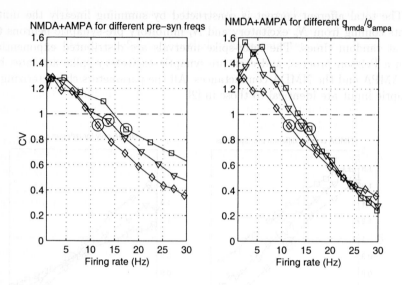

Fig. 3. NMDA and AMPA components. CV vs output rate for different pre-synaptic frequencies (left) and for different ratios $\bar{g}_{ampa}/\bar{g}_{nmda}$ (right). The curves are generated as in Fig. 2. The arrow indicates the experimental point of [2]. Left: The pre-synaptic frequencies are: diamonds: $\nu_e N_e = 800Hz$, triangles: $\nu_e N_e/2$, squares: $\nu_e N_e/4$. The ratio $g_{ampa}/g_{nmda} = 10$. CVs are smaller than in Fig. 2 but the curves are qualitatively the same. The differences between the three curves are more evident in the supra-threshold regime, where higher σ/μ ratios due to low pre-synaptic frequencies allow for a wider range of high CVs. Since the relative strength of AMPA and NMDA components is kept fixed, so is the correlation length and the maximal CV (achieved in the sub-threshold regime) is not affected much by a change in the pre-synaptic frequencies. Right: the three curves correspond to different ratios $r = \bar{g}_{ampa}/\bar{g}_{nmda}$: diamonds: $r = 10$, triangles: $r = 5$, squares: $r = 2.5$. As the NMDA component increases, the maximal CV also increases. The supra-threshold regime is not affected by r: the NMDA component does not change the variability at high rates where the temporal aspects of the input current are less important than the relative strength of the fluctuations. The ranges of $\bar{g}_{nmda}/C_m$ (nS) are (left panel) $(1.7, 2.3)$ (diamonds), $(3.2, 4.7)$ (triangles) and $(5.8, 9.4)$ (squares); (right panel) $(1.7, 2.3)$, $(2.1, 2.7)$, and $(2.3, 3.1)$.

the neurons should operate in a sub-threshold, fluctuations dominated regime, i.e. μ should stay below the threshold current μ_θ and σ should be large enough to drive the neuron across the threshold at a reasonable rate. The maximal ratio between the amplitude of the fluctuations and the average current is achieved when μ is just below μ_θ. To further increase this ratio one has to jump onto another line, i.e. to change one of the parameters that determine the slope. Smaller numbers N of afferents, shorter time constants τ and lower frequencies permit larger fluctuations.

NMDA conductance only. Complete simulations of the model confirm this picture (Fig. 2). As $\nu_e N_e$ is progressively reduced (diamonds: $\nu_e N_e = 800Hz$, triangles: $\nu_e N_e/2$, squares: $\nu_e N_e/4$, left part of the figure) the curves of the coefficient of variability (CV, ratio of standard deviation to the average of the

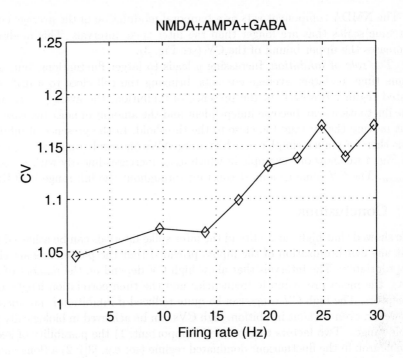

Fig. 4. CV vs output rate in the presence of inhibition. $\nu_e N_e = 800Hz$, $\nu_i N_i = 600Hz$ as in [2]. $\bar{g}_{nmda}/Cm$ range: $(0.09, 9)$ nS, with $\bar{g}_{ampa} = 10\bar{g}_{nmda}$, $\bar{g}_{gaba} = 4.7\bar{g}_{nmda}$ throughout.

inter-spike intervals) shift up, more prominently in the high rates region. Halving ν corresponds to jumping on a steeper line in Fig. 1 which allows to extend the fluctuations dominated regime to higher $\bar{g}_{nmda}$. Note that a reduction of $\nu_e N_e$ must be compensated by an increase of $\bar{g}_{nmda}$ to preserve the same output rate. Hence the intervals of variation for $\bar{g}_{nmda}$ are different for the three different curves. Note that this is equivalent to an increase of spatial correlation in the input. The dependence of the CV on τ_{nmda} is illustrated in the right part of the figure: diamonds, triangles and squares correspond respectively to $\tau_{nmda} = 150, 75, 37.5ms$. Smaller τ_{nmda} lead to steeper lines, which is reflected by an upwards shift of the CV in the supra-threshold region, as expected. In the sub-threshold regime instead, the temporal structure of the input current is dominant: long τ_{nmda} give rise to a slowly modulating component which increases the maximal CV [6,7], although the ratio σ/μ is decreased (see also below). For $\tau_{nmda} \to 0$ these curves tend to the AMPA case for which the upper bound is $CV = 1$.

NMDA and AMPA conductances. The relative contribution of the AMPA component to the fluctuations grows quadratically with g and hence it usually dominates over the NMDA part. The AMPA component permits higher σ/μ ratios and extends to higher frequencies the fluctuations dominated regime. However the relatively short time correlation length limits the maximal CV to

1. The NMDA component acts as a temporal modulation of the average current on time scales that are longer than the inter spike intervals. This modulation increases the upper bound of the CV (see Fig. 3).

The role of inhibition. Increasing g leads to larger fluctuations, but, at the same time, to larger average currents, bringing the cell close to a drift dominated regime. However, in the presence of inhibition, the average current and the fluctuations can become independent and the amount of noise increase without moving the average too close to the threshold. In the presence of inhibition it is therefore much simpler to achieve large CVs on a wide range of frequencies. In Fig. 4 we show one example in which $\bar{g}_{gaba}$ increases linearly with $\bar{g}_{nmda}$ and $\bar{g}_{ampa}$. The CV remains almost constant throughout the full range 1-30 Hz.

4   Conclusions

We showed that high variability in the inter-spike intervals can be achieved without any synchronization of the inputs provided that the parameters are chosen appropriately. The intervals that allow high CV depend on the number of afferents, the mean pre-synaptic frequencies and the time correlation length of the receptors. The high CV range can be quite reduced if inhibition is not included. However, even without inhibition, high CVs can be achieved in biologically plausible ranges. Two factors turn out to be important: 1) the possibility of keeping the neuron in the fluctuations dominated regime (see e.g. [3]); 2) a (long enough) time correlation length, as that of NMDA receptors. For large fluctuations, in a sub-threshold regime, the CV can always reach 1, although the corresponding frequencies might be too low to be measurable. The NMDA component is equivalent to a non-stationary input current that varies on longer time scales than inter-spike intervals. This non-stationarity can bring the CV above 1 [6,7], surpassing the upper bound for uncorrelated processes.

References

1. Amit D.J. and Tsodyks M.V., Effective neurons and attractor neural networks in cortical environment, *NETWORK* **3**:121–137 (1992)
2. Harsch A. and Robinson H.P.C., Postsynaptic variability of firing in rat cortical neurons: the roles of input synchronization and synaptic NMDA receptor conductance, *J. Neurosci.* **16**:6181–6192 (2000)
3. Shadlen M.N. and Newsome W.T., The variable discharge of cortical neurons: implications for connectivity, computation and information coding, *J. Neurosci.* **18**(10):3870–3896 (1998)
4. Softky W.R. and Koch C., The highly irregular firing of cortical cells is inconsistent with temporal integration of random EPSPs, *J. Neurosci.* **13**(1):334 (1993)
5. Stevens C.F. and Zador A.M., Input synchrony and the irregular firing of cortical neurons, *Nature Neuroscience* **1**:210–217 (1998)
6. Svirskis G. and Rinzel J., Influence of temporal correlation of synaptic input on the rate and variability of firing in neurons, *Biophys. J.* **5**:629–637 (2000)
7. Troyer T.W. and Miller K.D., Physiological gain leads to high ISI variability in a simple model of a cortical regular spiking cell, *Neural Comput.* **9**:971–983 (1997)

Spike-Driven Synaptic Plasticity for Learning Correlated Patterns of Asynchronous Activity

Stefano Fusi

Institute of Physiology, University of Bern, CH-3012 Bühlplatz 5, Switzerland
`fusi@cns.unibe.ch`

Abstract. Long term synaptic changes induced by neural spike activity are believed to underlie learning and memory. Spike-driven long term synaptic plasticity has been investigated in simplified situations in which the patterns of asynchronous activity to be encoded were statistically independent. An extra regulatory mechanism is required to extend the learning capability to more complex and natural stimuli. This mechanism is provided by the effects of the action potentials that are believed to be responsible for spike-timing dependent plasticity. These effects, when combined with the dependence of synaptic plasticity on the post-synaptic depolarization, produce the learning rule needed for storing correlated patterns of asynchronous neuronal activity.

1 Introduction

It is widely believed that synaptic plasticity is the phenomenon underlying learning and memory. In the last years many models of spike-driven plastic synapses have been proposed (see e.g. [1,2]). Most of these theoretical works were devoted to modeling experimental results on single pairs of synaptically connected neurons [2], to studying regulatory mechanisms based on spike-timing dependent plasticity [3], or to relating the long term modifications to the neuronal activity to be encoded [1]. Only very recently a few investigators started to study the formation of memory and the collective dynamical behavior of networks of integrate-and-fire neurons connected by spike-driven plastic synapses [4,5]. Here we make a further step in the direction of understanding the computational relevance of spike-based synaptic dynamics by extending the model introduced in [1]. The Hebbian paradigm there implemented was good enough to embed in the synaptic matrix an extensive number of random, uncorrelated patterns of asynchronous activity. However going beyond these simplistic patterns requires a regulatory mechanism that ensures a balanced redistribution of the memory resources among the different stimuli to be stored. This is usually achieved by introducing a mechanism that guarantees a global normalization of the synaptic weights (see e.g. [6]). One possible local mechanism is studied in [7,8] where the internal state of the synapse if changed according to a Hebbian paradigm only if it is necessary, i.e. if the pattern of activity to be encoded is not yet correctly memorized. For instance, when a stimulus to be learnt imposes a pattern of elevated activity to a subset of neurons, the synapses connecting these

J.R. Dorronsoro (Ed.): ICANN 2002, LNCS 2415, pp. 241–247, 2002.
© Springer-Verlag Berlin Heidelberg 2002

neurons are potentiated only if the activity of the post-synaptic (output) neuron is below some threshold value. Otherwise the synapse is left unchanged because the pattern of activity imposed by the external stimulus is already stable and the memory resources can be exploited in a more efficient way by other stimuli. Learning prescriptions based on this principle permit to memorize complex correlated patterns [7], and to classify real data, like handwritten digits [8], provided that learning is slow, i.e. that every pattern has to be presented several times. These non-monotonic learning rules are achieved here by introducing in the synaptic dynamics of [1], which depends solely on the sub-threshold depolarization of the post-synaptic neuron, an extra dependence on the effects of the post-synaptic action potential. In particular an extra depression occurs when the pre-synaptic spike follows the post-synaptic action potential within some time interval. This combination of depolarization and spike-timing dependence makes the model more realistic and captures the observations of recent experiments on long term synaptic modifications [9].

2   The Synaptic Dynamics

The synaptic dynamics can be fully described in terms of a single internal variable X which represents the state of the synapse and determines the synaptic efficacy. In the absence of any pre or post-synaptic spike the synapse is made bistable by a recall force that drives the internal state to one of the two stable states: to $X = 1$, that corresponds to the upper bound, if X is above some threshold θ_X, or to the lower bound $X = 0$ if $X < \theta_X$. Above θ_X the synapse is potentiated and synaptic efficacy on the post-synaptic neuron is elevated, whilst below θ_X the synapse is depressed (low efficacy). These dynamics can be expressed as follows:

$$\frac{dX}{dt} = -\alpha\Theta(-X + \theta_X) + \beta\Theta(X - \theta_X)$$

where α and β represent the two refresh currents. Pre- and post-synaptic spikes trigger temporary modifications of the internal variable. They depend on the depolarization V of the post-synaptic neuron [1] and on the time that passed since the last occurrence of a pre or a post-synaptic spike [2]. When the depolarization is high ($V > V_H$) every pre-synaptic spike pushes X upwards, weakly if the last post-synaptic spike has been emitted within the previous time interval T_-, strongly otherwise. If the depolarization is low ($V < V_L$), the pre-synaptic spikes induce a depression which is either weak, if there is no recent post-synaptic spike (within a time interval T_-), or strong otherwise. Post-synaptic spikes preceding pre-synaptic action potentials provoke small upwards jumps. The table below summarizes the effects of the spikes and the corresponding probability of occurrence when the pre- and post-synaptic neurons fire asynchronously, with a Poisson distribution, at a mean frequency ν_{pre} and ν_{post} respectively. Q_a (Q_b) is the probability that $V > V_H$ ($V < V_L$) upon the arrival of a pre-synaptic spike. The modifications induced by the spikes leave a long term memory trace only if

X crosses the threshold θ_X during the stimulation. Otherwise the state prior to the stimulation is restored by the refresh currents.

Effects of a pre-synaptic spike that arrives at time t		
Condition	Effect	Prob. of occurrence
$V > V_H$ and no spike in $[t - T_-, t]$	$X \to X + a$	$\nu_{pre} Q_a e^{-\nu_{post} T_-}$
$V > V_H$ and k spikes in $[t - T_-, t]$	$X \to X + a - ka'$	$\nu_{pre} Q_a e^{-\nu_{post} T_-} \frac{(\nu_{post} T_-)^k}{k!}$
$V < V_L$ and no spike in $[t - T_-, t]$	$X \to X - b$	$\nu_{pre} Q_b e^{-\nu_{post} T_-}$
$V < V_L$ and k spikes in $[t - T_-, t]$	$X \to X - b - kb'$	$\nu_{pre} Q_b e^{-\nu_{post} T_-} \frac{(\nu_{post} T_-)^k}{k!}$
Effects of a post-synaptic spike that arrives at time t		
k spikes in $[t - T_+, t]$	$X \to X + ka'$	$\nu_{post} e^{-\nu_{pre} T_+} \frac{(\nu_{pre} T_+)^k}{k!}$

2.1   Mean Firing Rates vs. Distribution of the Depolarization

The probabilities of occurrence of the temporary modifications a, b, a', b' control the direction in which the synapse is modified by the activity of the pre- and post-synaptic neurons. Q_a and Q_b depend on the statistics of the depolarization of the post-synaptic neuron under stimulation and can be calculated analytically when the neuron is embedded in a large network of interacting neurons and is subject to the heavy bombardment of synaptic inputs. Such a in-vivo like situation is emulated by injecting into an integrate-and-fire neuron a gaussian current characterized by its mean μ and its variance σ^2. The stationary distribution $p(V)$ can be computed as in [11,1] and is given by:

$$p(V) = \frac{\nu}{\mu} \left[\Theta(V - H) \left(1 - e^{-\frac{2\mu}{\sigma^2}(\theta - V)} \right) + \Theta(H - V) \left(e^{-\frac{2\mu}{\sigma^2} H} - e^{-\frac{2\mu}{\sigma^2} \theta} \right) e^{\frac{2\mu}{\sigma^2} V} \right]$$

where H is the reset potential, θ is the threshold for emitting a spike and ν is the mean firing frequency:

$$\nu = \left[\tau_r + \frac{\sigma^2}{2\mu^2} \left(e^{-\frac{2\mu\theta}{\sigma^2}} - e^{-\frac{2\mu H}{\sigma^2}} \right) + \frac{\theta - H}{\mu} \right]^{-1}$$

τ_r is the absolute refractory period. Q_a and Q_b are given by the integral of $p(V)$ in the interval $[V_H, \theta]$ and $[0, V_L]$ respectively. It is straightforward to compute these integrals analytically. μ and σ characterize the synaptic input and depend on the network interactions. We assume that ν_{post} is changed by increasing or decreasing the average spike frequency of a subpopulation of pre-synaptic neurons [1]. If the recurrent feedback of the post-synaptic neurons does not affect much the network activity, then the parameters of the input current move along a linear trajectory in the (μ, σ^2) space. We chose μ as an independent parameter, and $\sigma^2 = J\mu + K$. In a network of excitatory and inhibitory neurons, in which in a spontaneous activity state the recurrent input is as large as the external input, we have that $J = J_E$ (the average coupling between excitatory neurons) and $K = \nu_0^I N_I J_I (J_I + J_E)$, where ν_0 is the spontaneous activity of the N_I inhibitory neurons that are projecting to the post-synaptic cell (mean coupling J_I). Q_a and

Q_b are plotted in Fig. 1. As the external stimulus increases ν_{post}, the distribution of V changes in such a way that Q_b decreases and Q_a increases. This makes it rather likely that when the two neurons are simultaneously activated most of the numerous temporary modifications of the synapse are upwards and the synapse tends to potentiate. On the contrary, when the pre-synaptic neuron is active and the post-synaptic neuron has low spontaneous activity, most of the jumps are downwards and the synapse is eventually depressed.

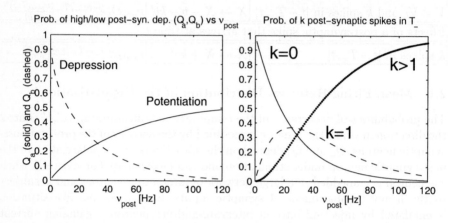

Fig. 1. Left: Probability of occurrence of upwards (Q_a) and downwards (Q_b) jumps upon the arrival of a pre-synaptic spike for different post-synaptic activities. Right: Probability that k post-synaptic spikes are within a time window T_-. It determines the depressing and regulatory effects of the action potential on the synaptic dynamics.

When ν_{pre} is low, the few jumps are usually not enough to drive the internal state across the threshold. It can be proven that this scheme of modifications (a Hebbian paradigm) maximizes the signal of the stimuli that can be extracted by the network dynamics from the synaptic matrix [12]. Hence Q_a and Q_b contain most of the relevant information to encode the patterns of activities imposed by the stimulus. The extra jumps (a', b') due to spike-timing depend essentially on the product of the mean frequencies of the pre and the post-synaptic neuron. The ratio between the number of potentiating events and the number of depressing events is roughly constant when the pre- and the post-synaptic frequencies are multiplied by the same factor. This activity dependence does not contribute much to the signal, but has an important regulatory role. Indeed, if $T_- \gg T_+$, as the frequency of the post-synaptic neuron increases the fraction of cases in which the direction of the temporary modification is controlled by the depolarization V decreases and depression becomes predominant, especially if one considers the possibility of multiple spikes within the time interval T_-.

3   The Non-monotonic Learning Rule

The next step is to compute the probability that a stimulus induces a stable modification in the synaptic couplings, i.e. that during the stimulation the synapse makes a transition from one stable state to another stable state. The transition probabilities control the learning process. Low probabilities correspond to a slow learning scenario in which the stimuli need to be presented a large number of times to be learnt. This is the price to be paid if one wants an optimal redistribution of the memory synaptic resources among the stimuli to be learnt [12, 7,8]. Indeed, high transition probabilities allow fast acquisition of information about the stimuli, but limit dramatically the storage capacity of the network because past stimuli are forgotten very quickly [12]. For uncorrelated patterns of activities the optimal storage capacity is achieved when the coding level is low (a small fraction of neurons is activated by each stimulus) and the LTD probability is small enough, compared to the LTP probability, to preserve the balance between potentiations and depressions [12]. For correlated patterns a control mechanism is needed to stop learning in case the stimulus is already correctly encoded in the synaptic matrix [7,8].

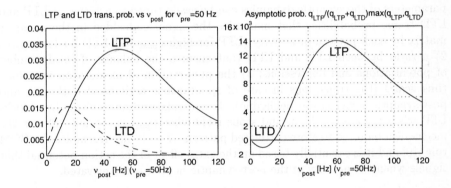

Fig. 2. Left: LTP and LTD transition probabilities (q_{LTP}, q_{LTD}) as a function of postsynaptic frequency for $\nu_{pre} = 50Hz$. Right: The normalized asymptotic probability which combines q_{LTP} and q_{LTD} to summarize the behavior of the synapse for a wide range of post-synaptic activities.

3.1   Solving the Long Term Synaptic Dynamics

The transition probabilities can be approximately calculated by using the density approach introduced in [1]. Since the pre and the post-synaptic neuron fire asynchronously, the Takács equations that govern the density function $p(X,t)$ of the synaptic internal variable can be easily generalized to incorporate the extra jumps a' and b' of the new model. For example, for $X < \theta_X$ we have:

$$\frac{\partial p(X,t)}{\partial t} = \alpha \frac{\partial p(X,t)}{\partial^{(+)} X} + \nu_{pre}[A(X,t) - p(X,t)] + \nu_{post}Q'_a[p(X-a',t) - p(X,t)]$$

where the term with α accounts for the refresh current and A contains the temporary changes induced by the spikes:

$$A(X,t) = Q_a P_0(t)\left[\sum_{k=0}^{\infty}\delta(X - a + kb')Q'_b(k)\right] + Q_a\left[\sum_{k=0}^{\infty}p(X - a + kb')Q'_b(k)\right]$$

$$+Q_b\left[\sum_{k=0}^{\infty}p(X + b + kb',t)Q'_b(k)\right]$$

where $Q'_b(k) = e^{-\nu_{post}T_-}(\nu_{post}T_-)^k/k!$. We assume that $p(x,t) = 0$ for $x \notin [0,1]$. The equation for the discrete probability $P_0 = prob(X = 0)$ is modified in a similar way. The equations are approximated because the temporal auto-correlations of the post-synaptic V are not considered (see [1] for more details).

4   Conclusions

The numerical solution of the density equations provides the LTP and LTD transition probabilities q_{LTP} and q_{LTD} for any pair of ν_{pre}, ν_{post} (Fig. 2). The parameters of the synaptic dynamics can be easily tuned to achieve: 1) LTP and LTD trans. probabilities that maximize the signal embedded in the synaptic matrix by each pattern of activity (LTP for high ν_{post} and LTD for low ν_{post}); 2) a ratio between LTP and LTD that ensures the balance of total number of potentiations and depressions; 3) the preservation of memory on very long time scales (the transitions in case of spontaneous activity of both pre- and post-synaptic neurons are negligible); 4) a regulatory mechanism that blocks LTP when ν_{post} is too high, and stops LTD when ν_{post} is too low. This is the mechanism required to store correlated patterns [7,8]. It also resembles the BCM rule [10] and it is essentially based on the extra effect of the action potential that signals when the activity of the post-synaptic neuron is too elevated.

References

1. S. Fusi, M. Annunziato, D. Badoni, A. Salamon, D.J.Amit (2000) Spike-driven synaptic plasticity: theory, simulation, VLSI implementation, *Neural Comp.*, **12**, 2227–2258
2. W. Senn, H. Markram and M. Tsodyks (2001) An algorithm for modifying neurotransmitter release probability based on pre- and post-synaptic spike timing, *Neural Comp.* **13**, 35–68
3. L. Abbott and S. Nelson, Synaptic plasticity: taming the beast, *Nature Neuroscience* **3** Supp. 1178–1183
4. P. Del Giudice and M. Mattia, (2001) Long and short-term synaptic plasticity and the formation of working memory: A case study, *Neurocomputing*, **38**, 1175–1180
5. D.J. Amit and G. Mongillo (2002) Spike-driven Synaptic Dynamics Generating Working Memory States, *Neural Comp.*, in press

6. W. Gerstner and W. Kistler (2002) Spiking neuron models: single Neurons, populations, plasticity, *Cambridge University Press*

7. S. Diederich and M. Opper (1987) Learning correlated patterns in spin-glass networks by local learning rules. *Phys. Rev. Lett.*, **58**, 949–952

8. Y. Amit and M. Mascaro (2001) Attractor Networks for Shape Recognition, *Neural Comp.* **13** 1415–1442

9. P.J. Sjöström, G.G. Turrigiano and S.B. Nelson, (2001) Rate,Timing and Cooperativity Jointly Determine Cortical Synaptic Plasticity,*Neuron* **32**, 1149–1164

10. E.L. Bienenstock, L.N. Cooper, and P.W. Munro (1982) Theory for the development of neuron selectivity: Orientation specificity and binocular interaction in visual cortex. *J. Neurosci.*, **2**, 32

11. S. Fusi, and M. Mattia (1999) Collective behavior of networks with linear (VLSI) integrate-and-fire neurons *Neural Comp.*, **11**, 633–652

12. D.J. Amit, and S. Fusi (1994). Dynamic learning in neural networks with material synapses *Neural Comp.*, **6**, 957–982

A Model of Human Cortical Microcircuits for the Study of the Development of Epilepsy

Manuel Sánchez-Montañés[1], Luis F. Lago-Fernández[1],
Nazareth P. Castellanos[1], Ángel Merchán-Pérez[2,3], Jon I. Arellano[3], and
Javier DeFelipe[3]

[1] E.T.S. de Informática, Universidad Autónoma de Madrid, 28049 Madrid, Spain
[2] Universidad Europea de Madrid, Villaviciosa de Odón, Madrid, Spain
[3] Instituto Cajal, Consejo Superior de Investigaciones Científicas, 28002 Madrid, Spain

Abstract. Brain lesions and structural abnormalities may lead to the development of epilepsy. However, it is not well known which specific alterations in cortical circuitry are necessary to create an epileptogenic region. In the present work we use computer simulations to test the hypothesis that the loss of chandelier cells, which are powerful inhibitory interneurons, might be a key element in the development of seizures in epileptic patients. We used circuit diagrams based on real data to model a $0.5\,mm^3$ region of human neocortical tissue. We found that a slight decrease in the number of chandelier cells may cause epileptiform activity in the network. However, when this decrease affects other cell types, the global behaviour of the model is not qualitatively altered. Thus, our work supports the hypothesis that chandelier cells are fundamental in the development of epilepsy.

1 Introduction

Epilepsy is one of the most frequent human neurological diseases and appears as a result of a synchronous, excessive neuronal discharge. A variety of brain lesions, tumors or other structural abnormalities are commonly found in the cerebral cortex of epileptic patients [10]. These structural abnormalities are associated with neuronal loss and the subsequent synaptic reorganization, which is thought to contribute to the establishment of recurrent seizures. However, there does not seem to exist lesions which are intrinsically epileptogenic, since some patients are epileptic whereas others with similar abnormalities are not, or they develop epilepsy after a variable delay. In other words, cortical circuits in an affected brain may undergo a series of changes that in the end result in epilepsy. Thus, the principal question in epilepsy pathology is: What alterations are necessary in the neuronal circuitry to create an epileptogenic region?

There is a large number of studies and hypotheses about the mechanisms that generate epilepsy. One of the hypotheses emphasizes the importance of the axo-axonic inhibitory neurons ("chandelier cells") [2]. These cells (among other types of neurons) have been found to be lost in the neocortex and/or

J.R. Dorronsoro (Ed.): ICANN 2002, LNCS 2415, pp. 248–253, 2002.
© Springer-Verlag Berlin Heidelberg 2002

hippocampus of epileptic patients independently of the primary pathology [3, 8]. The source of GABAergic input on dendrites and somata of cortical pyramidal cells (projection neurons) originates from numerous and various types of interneurons (axo-dendritic cells and basket cells, or neurons that form synapses preferentially with dendrites or somata, respectively). However, all synapses at the axon initial segment of the pyramidal cells come from just a few chandelier cells (five or less) [2]. Therefore, a loss of a few axo-dendritic and/or basket cells might have little impact on the inhibitory control of the pyramidal cell, whereas if chandelier cells are affected, this inhibitory control could be strongly altered.

In this work we try to verify this hypothesis using computer simulations. We model a 3-dimensional region of neocortical tissue of $0.5\,mm^3$, and study the effect of eliminating neurons in a non-specific way. Our results show that the loss of a few cells does not alter the global behaviour of the model, as long as the chandelier cells are not significantly affected. However, the loss of a few chandelier cells strongly alters the balance between excitation and inhibition, leading to uncontroled, epileptic-like, activity.

2 Methods

2.1 Model Description

We model a region of layers II/III human neocortex of dimensions $1 \times 1 \times 0.5\,mm$. To simplify computations, we assume only three groups of neurons: *pyramidal* neurons (p); axo-axonic inhibitory (*chandelier*) neurons (c); and the rest of inhibitory (*SD-inhibitory*) neurons (i), which project to the soma and/or the dendrites. We use a three-dimensional model in which the different neuron populations are randomly distributed according to uniform distributions with the following average densities: $\rho_p = 13158\,mm^{-3}$, $\rho_c = 131.58\,mm^{-3}$, $\rho_i = 7830\,mm^{-3}$. These values have been calculated taking into account the average density of neurons in the human cerebral cortex ($\rho = 21120\,mm^{-3}$, measured[1]), the percentage of gabaergic neurons in humans (37.7%, [5]), and some additional considerations on chandelier to pyramidal innervation [11]. According to the above criteria, each neuron is assigned a position (which we assume to coincide with the center of its soma) that will be used to determine the connection weight between any pair of potentially interconnected neurons (all connections but $c \to i$ and $c \to c$ are possible).

Neuron model: We use an integrate and fire neuron model, which captures the essence of the generation of action potentials while allowing fast simulation of large neural networks [7]. The parameter values for each neuron group are chosen within biologically plausible ranges, and are described in detail in [11].

---

[1] Neuronal countings were performed on 4 individuals (3 males and 1 female) with ages ranging from 23 to 49 (mean 33.5), using physical disectors in 1 micron-thick sections of layers II/III of the temporal cortex. For each individual 4 disectors of $307200\,\mu m^2$ were taken.

Synaptic interactions: Synaptic conductances are modeled as α-functions [7], with the maximum conductance given by $g_{max} = nw$, where n is the number of synaptic contacts, and w is the maximum conductance for a single contact. For any pair of potentially interconnected neurons, the connection probability is modeled as a gaussian on the distance between them, whose scaling factor and width are calculated so that the average connectivity k (i.e. the number of pre-synaptic cells that project onto a single post-synaptic cell) for each synapsis type responds to biological constraints. Once the existence of the connection is decided, the number n of synaptic contacts is drawn from a poisson distribution whose average $(\bar{n})$ is taken as 10 for all kind of synapses [4], with the exception of $\bar{n}(c \rightarrow p)$, which will be explicitly calculated. The rest of synaptic parameters have been tuned so that, on average, the post-synaptic response (EPSP/IPSP) resembles that found in physiological experiments (explicit values can be found in [11]). Connectivities among the different neural populations have been calculated according to the following discussion (anatomical data taken from [5,9,2], unless otherwise specified).

Chandelier to pyramidal: The number of synaptic contacts on the axon initial segment of pyramidal cells is around 20 [4], and all of them come from chandelier cells. Furthermore, the number of chandelier cells that innervate a single pyramidal neuron has been estimated to be between 1 and 5. Therefore, we will consider $k(c \rightarrow p) = 3$. The identity $k(c \rightarrow p)\,\bar{n}(c \rightarrow p) = 20$ allows us to calculate $\bar{n}(c \rightarrow p) = 20/3 = 6.67$.

SD-inhibitory to pyramidal: The percentage of inhibitory connections in layers II/III of human cortex is approximately 15%. This leaves an 85% of excitatory connections. In pyramidal neurons, most excitatory connections are established on basal dendrites, where around 13000 spines have been estimated. Therefore we have $k(i \rightarrow p) \cdot \bar{n}(i \rightarrow p) = 15 \cdot 13000/85 - k(c \rightarrow p) \cdot \bar{n}(c \rightarrow p)$, which leads to $k(i \rightarrow p) = 227$.

SD-inhibitory to SD-inhibitory/chandelier: There are not enough experimental data to explicitly calculate the parameters for these connections. Hence we assume that all the neurons receive a similar level of inhibition, that is, $k(i \rightarrow i) = k(i \rightarrow c) = k(i \rightarrow p) = 227$.

Pyramidal to chandelier: Again, we do not have enough experimental data for this kind of connection. However, we assume a reciprocity principle, that is, pyramidal neurons tend to connect to the chandelier cells that innervate them. With this assumption, we can formulate the following equation:

$$\rho_c k(p \rightarrow c) = \rho_p k(c \rightarrow p) \tag{1}$$

We know $k(c \rightarrow p)$ and the densities, so we obtain $k(p \rightarrow c) = 300$.

Piramidal to SD-inhibitory: We know that $k(i \rightarrow i) \cdot \bar{n}(i \rightarrow i)$ represents only the 15% of connections to a SD-inhibitory cell. So the average number of excitatory connections a SD-inhibitory neuron receives is $85 \cdot k(i \rightarrow i) \cdot \bar{n}(i \rightarrow i)/15 = 12886$. Only the 23.3% of these connections comes from local circuits [11], so we can conclude that $k(p \rightarrow i) = 300$.

Piramidal to pyramidal: The number $k(p \to p)$ is calculated as $k(p \to i)$, but taking into account the 20 inhibitory synapses that come from chandelier cells. We obtain $k(p \to c) = 300$.

Finally, we describe the synaptic interactions between chandelier and pyramidal cells. It is known that chandelier cells do not affect the input resistance of pyramidal neurons, but instead modulate their spiking threshold [1]. Thus, this kind of synapses must be modeled in a different way. We assume that the spiking threshold of any pyramidal neuron evolves according to the following equation:

$$V_{th}(t) = V_{th0} + \sum_{i \in \Gamma} W_i(t), \tag{2}$$

where V_{th0} is the base threshold; and the sum extends to the subset Γ of chandelier cells that contact the pyramidal neuron. $W_i(t)$ is the threshold increment produced by a presynaptic spike, and is modeled as an α-function [7] (parameters in [11]).

2.2   Simulations Performed

We performed simulations with a duration of $300\,ms$ under different anatomical conditions. In all of them neuron membrane potentials are initialized to values at rest. However, after a few milliseconds the network has evolved to a random basal condition which will be used as control. At $t = 50\,ms$ a stimulus of duration $100\,ms$ is injected to the network. The stimulus is characterized by its center of application (neurons in the model that are maximally affected) and its width, σ. Its effect on the network is simulated by increasing the spontaneous activity of any network neuron by an amount $f_{est} = \kappa \exp\left(-d^2/2\sigma^2\right)$, where $\kappa = 150\,Hz$, $\sigma = 0.1\,mm$, and d is the distance of the neuron to the center of application of the stimulus. The following anatomical regimes are studied: a) Normal tissue, under the anatomical and physiological assumptions of the previous section, will be used as control situation; b) The same tissue, eliminating all chandelier cells, will be simulated to study the importance of these neurons in the control of recurrent excitation in the cortex; c) Idem for the SD-inhibitory neurons; and d) Non-selective lesions that will be produced in a normal tissue so that all the neurons in the network are subject to a potential death with a probability p that measures the severity of the injury.

3   Results

First we simulate the response of a normal tissue to the external stimulation. In Fig. 1a we see the response of the pyramidal cells. It can be observed that the mean firing rate increases, going back to the basal level once the stimulus has finished. As a control, we perform the same simulation both without chandelier cells (Fig. 1b) and without SD-inhibitory neurons (Fig. 1c). In both cases the cortical tissue becomes unstable, that is, the activity of the tissue does not go back to the basal level once the stimulus has finished. Therefore, neither the chandelier cells alone nor the SD-inhibitory cells keep the system stable. Hence the cooperation of these populations is the key in the stability control, which is in accordance with [6].

Next we investigate whether an unspecific lesion could be responsible of the loss of stability in the circuit. In Fig. 1d we show the response of the same network to the stimulus, but now assuming that a lesion has caused the death of 20 % of the population. We see that the network gets unstable. In this case there are 1347 of 6539 pyramidal, 14 of 73 chandelier and 766 of 3907 SD-inhibitory cells lost. However, the model is stable when we repeat exactly the same lesion but keeping all the chandelier cells intact (data not shown). Therefore, the loss of a few chandelier cells is the responsible of instability generation in our model.

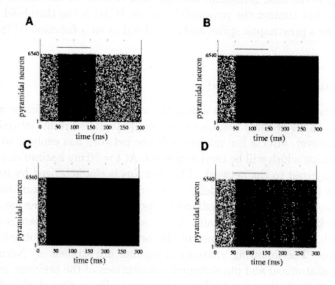

Fig. 1. Rasters of activity of the pyramidal population under different conditions. The external stimulus is applied from $t = 50\,ms$ to $t = 150\,ms$ (horizontal line). A: Normal tissue. B: Tissue without chandelier cells. C: Tissue without SD-inhibitory cells. D: Tissue with a 20 % of degree of non specific lesion. See text for further details.

4   Discussion

The cerebral cortex is composed of millions of neurons which are interconnected through synapses in an ordered fashion, forming highly intricate neuronal networks. The disruption of these networks may induce epilepsy or other neurological or psychiatric diseases.

In this paper we have used a realistic model of the human cortex to study under which conditions the system becomes unstable (epileptic). First we have shown that the cooperation of the chandelier cells and the rest of inhibitory cells is the key factor in the control of stability, which is in agreement with previous work [6]. Second, we have shown that the loss of a few chandelier cells caused by an unspecific lesion, can lead to the appeareance of instability in the circuit (epilepsy), as was suggested in anatomical studies [2].

The more detailed circuit diagrams become available, the more we will learn with computer simulations about the role of each element of the circuit and, thus, better undestand the kind of alterations that cause the disease. On the other hand, the parameters measured and derived in this paper can be useful for the development of more accurate models of the cortex. Future work will address the study of the probability of becoming epileptic as a function of the degree and form of the lesion, as well as the effect of plasticity in the development of epilepsy.

References

1. Colbert, C.: Personal communication.
2. DeFelipe, J.: Chandelier cells and epilepsy. Brain **122** (1999) 1807–1822
3. DeFelipe, J., Sola, R.G., Marco, P., del Río, M.R., Pulido, P., Ramón y Cajal, S.: Selective changes in the microorganization of the human epileptogenic neocortex revealed by parvalbumin immunoreactivity. Cerebral Cortex **3** (1993) 39–48
4. DeFelipe, J.: Microcircuits in the brain. In Mira, J., Moreno-Díaz, R., Cabestany, L., eds. Lecture Notes in Computer Science **1240** (Springer, Berlin, 1997) 1–14
5. del Río, M.R., DeFelipe, J.: Colocalization of calbindin D-28k, calretinin and GABA immunoreactivities in neurons of the human temporal cortex. J. Comp. Neurol. **364** (1996) 472–482
6. Douglas, R.J., Martin, K.A.: Control of neural output by inhibition at the axon initial segment. Neural Comp. **2** (1990) 283–92
7. Koch, C., Segev, I.: Methods in neuronal modeling, from ions to networks. (MIT Press, Cambridge, Massachusetts, 1998).
8. Marco, P., Sola, R.G., Pulido, P., Alijarde, M.T., Sánchez, A., Ramón y Cajal, S., DeFelipe, J.: Inhibitory neurons in the human epileptogenic temporal neocortex: an immunocytochemical study. Brain **119** (1996) 1327–1347
9. Marco P., DeFelipe, J.: Altered synaptic circuitry in the human temporal epileptogenic neocortex. Exp. Brain Res. **114** (1997) 1–10
10. Meldrum, B.S., Bruton, C.J.: Epilepsy. In Adams, J.H., Duchen, L.W., eds. Greenfield's Neuropathology (Arnold, London, 1992) 1246–1283
11. Sánchez-Montañés M.A., Lago-Fernández L.F., Castellanos N.P., Merchán-Pérez A., Arellano J., DeFelipe J.: A Model of Human Cortical Microcircuits for the Study of the Development of Epilepsy. Internal Report, (2002), ETS de Informática, Universidad Autónoma de Madrid, Spain. Available at http://www.ii.uam.es/~msanchez/papers.html

On the Computational Power of Neural Microcircuit Models: Pointers to the Literature

Wolfgang Maass

Institute for Theoretical Computer Science
Technische Universität Graz
A-8010 Graz, Austria
maass@igi.tu-graz.ac.at
http://www.igi.tugraz.at/maass

Abstract. This paper provides references for my invited talk on the computational power of neural microcircuit models.

Biological neural microcircuits are highly recurrent and consist of heterogeneous types of neurons and synapses, which are each endowed with an individual complex dynamics [4], [6], [3], [7], [8], [23], [20]. Hence neural microcircuits are as different as one can imagine from the familiar boolean circuits in our current generation of computers, but also very different from common artificial neural network models. This has given rise to the question how neural microcircuits can be used for purposeful computations.

There are two quite different ways of approaching this question. One way is to construct circuits consisting of biologically realistic components that can simulate other models for general-purpose computers such as Turing machines [13] or general-purpose artificial neural network models [14], [16], or to construct circuits that carry out specific computations such as for example simpified speech recognition [10]. Another way, which is discussed in my talk, is to recruit – with the help of suitable adaptive mechanisms – biologically realistic "found" or emerging models for neural microcircuits for purposeful computations. An inspiring first example for this approach is given in [2]. It became the basis of our new approach towards real-time computing in neural systems in [17]. The underlying computational theory is presented in [15], and discussed from a biological point of view in [19]. Particular computational consequences of the high dimensionality of neural microcircuits are discussed in [9]. It turns out that this approach yields superior performance in terms of noise robustness, computing speed, and size compared with special-purpose neural circuits that have been constructed by hand for a specific computational task [18]. Publicly available software for generating and simulating generic neural microcircuit models, and for evaluating their computational power, is discussed in [21].

Herbert Jaeger discovered independently quite similar phenomena of temporal integration in recurrent circuits in the context of artificial neural network models [11].

Current work addresses the application of the resulting new principles for neural computation to online processing of real-world time-varying inputs, such

J.R. Dorronsoro (Ed.): ICANN 2002, LNCS 2415, pp. 254–256, 2002.
© Springer-Verlag Berlin Heidelberg 2002

as movement prediction for visual inputs [12], speech recognition in real-time, and real-time processing of sensory inputs on a robot. Another line of current research explores the computational role of specific details of biological neural microcircuits, and the role of learning principles in this context.

References

1. Auer, P., Burgsteiner, H., and Maass, W. (2002). Reducing communication for distributed learning in neural networks. In *Proc. ICANN'2002*. Online available as # 127 on http://www.igi.tugraz.at/maass/publications.html.
2. Buonomano, D.V., and Merzenich, M.M. (1995). Temporal information transformed into spatial code by a neural network with realistic properties. *Science* 267, 1028-1030.
3. Braitenberg, V., and Schuez, A. (1998). *Cortex: Statistics and Geometry of Neuronal Connectivity*, 2nd ed., Springer Verlag, Berlin.
4. Cajal, S. Ramón y (1911). Histologie du système nerveux de l'homme et des vertébrés, translated by L. Azoulay. Consejo superior de investigaciones científicas. Instituto Ramon y Cajal, Madrid, edn. 1972.
5. Douglas, R., and Martin, K. (1998). Neocortex. In *The Synaptic Organization of the Brain*, G.M. Shepherd, Ed. (Oxford University Press), 459–509.
6. De Felipe, J. (1993). Neocortical neuronal diversity: chemical heterogeneity revealed by colocalization studies of classic neurotransmitters, neuropeptides, calcium-binding proteins, and cell surface molecules. *Cerebral Cortex* 7, 476–486.
7. Gupta, A., Wang, Y., and Markram, H. (2000). Organizing principles for a diversity of GABAergic interneurons and synapses in the neocortex. *Science* 287, 273–278.
8. Gupta, A., Silberberg, G., Toledo-Rodriguez, M., Wu, C.Z., Wang, Y., and Markram, H. (2002). Organizing principles of neocortical microcircuits. *Cellular and Molecular Life Sciences*, in press.
9. Häusler, S., Markram, H., and Maass, W. (2002). *Low dimensional readout from high dimensional neural circuits*, submitted for publication. Online available as # 137 on http://www.igi.tugraz.at/maass/publications.html.
10. Hopfield, J.J., and Brody, C.D. (2001). What is a moment? Transient synchrony as a collective mechanism for spatio-temporal integration. *Proc. Natl. Acad. Sci.*, USA, 89(3), 1282.
11. Jaeger, H. (2001). *The "echo state" approach to analyzing and training recurrent neural networks*, submitted for publication.
12. Legenstein, R.A., Markram, H., and Maass, W. (2002). *Input prediction and autonomous movement analysis in recurrent circuits of spiking neurons*, submitted for publication. Online available as # 140 on http://www.igi.tugraz.at/maass/publications.html.
13. Maass, W. (1996). Lower bounds for the computational power of networks of spiking neurons. *Neural Computation* 8(1):1-40. Online available as # 75 on http://www.igi.tugraz.at/maass/publications.html.
14. Maass, W. (1997). Fast sigmoidal networks via spiking neurons. *Neural Computation* 9:279-304. Online available as # 82 on http://www.igi.tugraz.at/maass/publications.html.
15. Maass, W., and Markram, H. (2002). *On the computational power of recurrent circuits of spiking neurons*, submitted for publication. Online available as # 135 on http://www.igi.tugraz.at/maass/publications.html.

16. Maass, W., and Natschläger, T. (1997). Networks of spiking neurons can emulate arbitrary Hopfield nets in temporal coding. *Network: Computation in Neural Systems* 8(4):355-372. Online available as # 93 on http://www.igi.tugraz.at/maass/publications.html.

17. Maass, W., Natschläger, T., and Markram, H. (2002). Real-time computing without stable states: A new framework for neural computation based on perturbations. *Neural Computation*, in press. Online available as # 130 on http://www.igi.tugraz.at/maass/publications.html.

18. Maass, W., Natschläger, T., and Markram, H. (2002a). Real-time computing with emergent neural microcircuit models, submitted for publication.

19. Markram, H., Ofer, M., Natschläger, T., and Maass, W. (2002) Temporal integration in neocortical microcircuits. *Cerebral Cortex*, in press. Online available as # 142 on http://www.igi.tugraz.at/maass/publications.html

20. Mountcastle, V.B. (1998). *Perceptual Neuroscience: The Cerebral Cortex*, Harvard University Press (Cambridge).

21. Natschläger, T., Markram, H., and Maass, W. (2002). Computer models and analysis tools for neural microcircuit models. In *A Practical Guide to Neuroscience Databases and Associated Tools*, R. Kötter, Ed., Kluver Academic Publishers (Boston), in press.

22. Shepherd, G.M. (1988). A basic circuit for cortical organization. In *Perspectives in Memory Research*, M. Gazzaniga, Ed., MIT-Press, 93-134.

23. Thomson, A., West, D.C., Wang, Y., and Bannister, A.P. (2002). Synaptic connections and small circuits involving excitatory and inhibitory neurons in layers 2 to 5 of adult rat and cat neocortex: triple intracellular recordings and biocytin-labelling in vitro. *Cerebral Cortex*, in press.

Part II

Connectionist Cognitive Science

Networking with Cognitive Packets*

Erol Gelenbe, Ricardo Lent, and Zhiguang Xu

School of Electrical Engineering and Computer Science
University of Central Florida, Orlando FL 32816
{erol,rlent,zgxu}@cs.ucf.edu

Abstract. This paper discusses a novel packet computer network architecture, a *"Cognitive Packet Network (CPN)"*, in which intelligent capabilities for routing and flow control are moved towards the packets, rather than being concentrated in the nodes. The routing algorithm in CPN uses reinforcement learning based on the Random Neural Network. We outline the design of CPN and show how it incorporates packet loss and delay directly into user Quality of Service (QoS) criteria, and use these criteria to conduct routing. We then present our experimental test-bed and report on extensive measurement experiments. These experiments include measurements of the network under link and node failures. They illustrate the manner in which neural network based CPN can be used to support a reliable adaptive network environment for peer-to-peer communications over an unreliable infrastructure.

1   Introduction

In recent papers [2,3,4] we have proposed a new network architecture called "Cognitive Packet Networks (CPN)". These are store-and-forward packet networks in which intelligence is constructed into the packets, rather than at the routers or in the high-level protocols.

CPN is a reliable packet network infrastructure which incorporates packet loss and delays directly into user QoS criteria and use these criteria to conduct routing. The practical issue of measuring packet loss and delay is discussed and we present the QoS based routing algorithm that we have designed and implemented. We then present our test-bed and report on extensive measurement experiments. These experiments illustrate the manner in which CPN can be used to support a reliable adaptive network infrastructure.

2   Cognitive Packets and CPNs

CPNs carry three major types of packets: smart packets (SP), dumb packets (DP) and acknowledgments (AP). Smart or cognitive packets route themselves, they learn to avoid link and node failures and congestion, and to avoid being

---

* The research reported in this paper was supported by U.S. Army Simulation and Training Command and by Giganet Technologies, Inc.

lost. They learn from their own observations about the network and from the experience of other packets. They rely minimally on routers. When a smart packet arrives to a destination, an AP packet is generated by the destination and the AP heads back to the source of the smart packet along the inverse route stored in a Cognitive Map (CM). As it traverses successive routers, it is used to update mailboxes in the CPN routers, and when it reaches the source node it provides source routing information for dumb packets. Dumb CPN packets of a specific QoS class use successful routes which have been selected in this manner by the smart packets of that class.

A node in the CPN acts as a storage area for CPs and for mailboxes which are used to exchange data between CPs, and between CPs and the node. It has an input buffer for CPs arriving from the input links, a set of mailboxes, and a set of output buffers which are associated with output links.

3   Routing Algorithm Using Reinforcement Learning

We use Random Neural Networks (RNN) with reinforcement learning (RNNRL) in order to implement the Smart packet routing algorithm. A recurrent RNN is used both for storing the Cognitive Map, and making decisions. The weights of the RNN are updated so that decisions are reinforced or weakened depending on how they have been observed to contribute to the success of the QoS goal.

A RNN [1,7] is an analytically tractable spiked random neural network model whose mathematical structure is akin to that of queuing networks. It has "product form" just like many useful queuing network models, although it is based on non-linear mathematics. The state q_i of the $i - th$ neuron in the network is the probability that it is excited. The q_i, with $1 \leq i \leq n$ satisfy the following system of non-linear equations:

$$q_i = \lambda^+(i)/[r(i) + \lambda^-(i)], \tag{1}$$

where

$$\lambda^+(i) = \sum_j q_j w_{ji}^+ + \Lambda_i, \qquad \lambda^-(i) = \sum_j q_j w_{ji}^- + \lambda_i. \tag{2}$$

Here w_{ji}^+ is the rate at which neuron j sends "excitation spikes" to neuron i when j is excited, w_{ji}^- is the rate at which neuron j sends "inhibition spikes" to neuron i when j is excited, and $r(i)$ is the total firing rate from the neuron i. For an n neuron network, the network parameters are these n by n "weight matrices" $\mathbf{W}^+ = \{w^+(i,j)\}$ and $\mathbf{W}^- = \{w^-(i,j)\}$ which need to be "learned" from input data. Various techniques for learning may be applied to the RNN. These include Hebbian learning (which will not be discussed here since it is too slow and relatively ineffective with small networks), and Reinforcement Learning which we have implemented for CPN.

Given the Goal G that the CP has to achieve as a function to be minimized (i.e. Transit Delay or Probability of Loss, or a weighted combination of the two), we formulate a reward R which is simply $R = G^{-1}$. Successive measured values of the R are denoted by R_l, $l = 1, 2, ..$ These are first used to compute a decision threshold:

$$T_l = aT_{l-1} + (1 - a)R_l, \tag{3}$$

where a is some constant $0 < a < 1$, typically close to 1.

An RNN with as many nodes as the decision outcomes is constructed. Let the neurons be numbered 1, ... , n. Thus each decision i corresponds to some neuron i. Decisions in this RL algorithm with the RNN are taken by selecting the decision j for which the corresponding neuron is the most excited, i.e. the one with the largest value of q_j. Note that the $l - th$ decision may not contribute directly to the $l - th$ observed reward because of time delays between cause and effect.

Suppose we have now taken the $l - th$ decision which corresponds to neuron j, and that we have measured the $l - th$ reward R_l. Let us denote by r_i the firing rates of the neurons before the update takes place. We first determine whether the most recent value of the reward is larger than the previous "smoothed" value of the reward which we call the threshold T_{l-1}. If that is the case, then we increase very significantly the excitatory weights going into the neuron that was the previous winner (in order to reward it for its new success), and make a small increase of the inhibitory weights leading to other neurons. If the new reward is not better than the previously observed smoothed reward (the threshold), then we simply increase moderately all excitatory weights leading to all neurons, except for the previous winner, and increase significantly the inhibitory weights leading to the previous winning neuron (in order to punish it for not being very successful this time). This is detailed in the algorithm given below. We compute T_{l-1} and then update the network weights as follows for all neurons $i \neq j$:

- If $T_{l-1} \leq R_l$
    - $w^+(i, j) \leftarrow w^+(i, j) + R_l$,
    - $w^-(i, k) \leftarrow w^-(i, k) + \frac{R_l}{n-2}$, $if\ k \neq j$.
- Else
    - $w^+(i, k) \leftarrow w^+(i, k) + \frac{R_l}{n-2}, k \neq j$,
    - $w^-(i, j) \leftarrow w^-(i, j) + R_l$.

After the values for the weights are re-normalized, the probabilities q_i are computed using the non-linear iterations (1), (2), leading to a new decision to send the packet forward to the output link which corresponds to the neuron which has the largest excitation probability.

4   An Experimental CPN Test-Bed: Protocol Design and Software Implementation

The CPN algorithm has been integrated into the Linux kernel 2.2.x., providing a connectionless service to the application layer through the BSD4.4 socket layer. The CPN test-bed consists of a set of hosts interconnected by links, where each host can operate both as an end node of communication and as a router. The addressing scheme utilizes a single number of 32 bits to represent the CPN address of each node.

We are experimenting with various topologies where each CPN router can be connected up to four other routers. For the results discussed later we have used two nodes exchanging data packets through a full-mesh of four nodes [8].

5   Network Measurements

The purpose of the measurements we describe on the test-bed is to evaluate the CPN architecture with respect to its dynamic behavior and ability to adapt to changing network conditions.

We have conducted measurements for a main flow of traffic from one end (source) to another (destination). For the first set of experiments, the input rate was fixed to 5 packets second. In several of the experiments we have conducted there is a flow of "obstructing" traffic over and above the main traffic flow.

Each time we have added this obstructing traffic it has been on one or more specific links, and at a relatively high traffic rate of 5700 packets per second in each direction (i.e. bi-directional on each selected link).

In Figure 1 we plot the delays and routes experienced by individual packets on the testing traffic flow. We have obstructing traffic which perturbs this traffic flow, with 20% of SPs in all traffic flows. The x-axis in the plots refers to successive packets and is scaled in packet counts. In each plot, the y-axis presents delay in milliseconds (left) or route number (right). An "X" under the x-axis indicates an instance of packet loss of either a smart or dumb packet. The additional traffic is introduced when the packet count reaches 30 on the traffic flow. The obstructing traffic flows on one of the links in the full mesh We see that when the obstructing traffic is initialized, the main traffic flow encounters significant delays as well as losses. Then, thanks to the CPN algorithm, the network determines a new route and delays go back to a low level. Further spurious increases in delay occur but are short-lived each time the SPs probe the network for better routes and some losses do occur again.

In the next set of results, we show the effectiveness of the CPN to adapt to different quality of service requirements, for this, we have introduced jitter control as a new component for the routing decision of the smart packets in addition to the average delay. Along with the main flow of packets we introduce a second flow to produce obstructing traffic to the former. The resulting average delay after the transmission of 200 thousand packets is shown in Figure 2. The x-axis represent the inter-packet time for the main flow of packets. There is a significant improvement in the overall standard deviation of the packets when the jitter control is active as shown in Figure 3.

6   Conclusions

CPN simplifies router architecture by transfering the control of QoS based best-effort routing to the packets, away from the routers. Routing tables are replaced

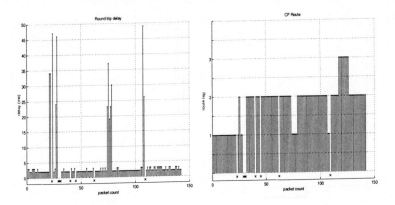

Fig. 1. Network with obstructing traffic. Route tags indicate different paths from the source to the destination

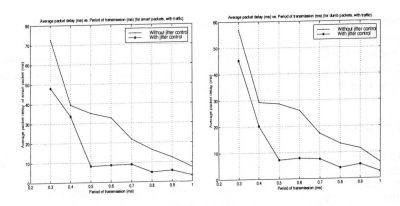

Fig. 2. Average round-trip delay for smart (left) and dumb (right) packets under regular routing conditions and routing with jitter control

by reinforcement algorithm based routing functions. A CPN carries three distinct types of packets: Smart or cognitive packets which search for routes based on a QoS driven reinforcement learning algorithm, ACK packets which bring back route information and measurement data from successful smart packets, and Dumb packets which do source routing.

In this paper we have summarized the basic principles of CPN. Then we have described the Reinforcement Learning (RL) algorithm which taylors the specific routing algorithm to the QoS needs of a class of packets. We have then described in some detail the design and implementation of our current test-bed network which uses ordinary Linux PC-based workstations as routers.

We have provided extensive measurement data on the test-bed to illustrate the capacity of the network to adapt to changes in traffic load. Some of the measurements we report present traces of short sequences (under 200 packets)

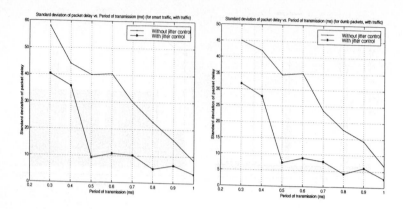

Fig. 3. Standard deviation of delay for smart (left) and dumb (right) packets under regular routing conditions and routing with jitter control

of packet transmissions showing individual packet delays, and paths taken by the packets.

Ongoing work in this project includes the deployment of a larger test-bed, the inclusion of wireless links, and the design of single-card routers leading to very low cost routing technologies which use CPN.

References

1. E. Gelenbe (1993) "Learning in the recurrent random neural network", *Neural Computation*, Vol. 5, No. 1, pp. 154–164, 1993.
2. E. Gelenbe, E. Seref, Z. Xu "Towards networks with intelligent packets", *Proc. IEEE-ICTAI Conference on Tools for Artificial Intelligence*, Chicago, November 9–11, 1999.
3. E. Gelenbe, R. Lent, Z. Xu "Towards networks with cognitive packets," Opening Key-Note Paper, *Proc. IEEE MASCOTS Conference*, ISBN 0-7695-0728-X, pp. 3–12, San Francisco, CA, Aug. 29–Sep. 1, 2000.
4. E. Gelenbe, R. Lent, Z. Xu, "Towards networks with cognitive packets", Opening Invited Paper, *International Conference on Performance and QoS of Next Generation Networking*, Nagoya, Japan, November 2000, in K. Goto, T. Hasegawa, H. Takagi and Y. Takahashi (eds), "Performance and QoS of next Generation Networking", Springer Verlag, London, 2001.
5. R. Viswanathan and K.S. Narendra "Comparison of expedient and optimal reinforcement schemes for learning systems", *J. Cybernetics*, Vol. 2, pp 21–37, 1972.
6. P. Mars, J.R. Chen, and R. Nambiar, "Learning Algorithms: Theory and Applications in Signal Processing, Control and Communications", CRC Press, Boca Raton, 1996.
7. E. Gelenbe, Zhi-Hong Mao, Y. Da-Li (1999) "Function approximation with spiked random networks" *IEEE Trans. on Neural Networks*, Vol. 10, No. 1, pp. 3–9, 1999.
8. E. Gelenbe, R. Lent, Z. Xu, "Design and performance of Cognitive Packet Networks", to appear in *Performance Evaluation*.

Episodic Memory: A Connectionist Interpretation

J.G. Wallace and K. Bluff

Swinburne University of Technology, P.O. Box 218, Hawthorn, Victoria, Australia 3122.

Abstract. The strengths and weaknesses of episodic memory and its analysis when represented in symbolic terms are indicated. Some aspects may be more easily implemented in a connectionist form. Those which pose a challenge to a PDP system are presented in the context of recent neurobiological work. A theoretical, connectionist account of episodic memory is outlined. This assigns a critical role to hebbosomes in individual neurons as a means of establishing episodic memory records and relies on interaction between episodic memories in temporal and entorhinal cortices and the hippocampus as a determinant of modification of semantic memory.

1 Introduction

There is general agreement that cognition is functionally a hybrid system involving both symbolic and sub-symbolic processes. Controversy surrounds the parts played by the two types of processes and their interactions. It is particularly lively regarding the capability of associationistic/connectionist/sub-symbolic mechanisms to produce fundamental, natural language characteristics such as systematicity. Like Reilly [1] our approach involves investigating the view that "while we may have a functionally hybrid system, appearing to combine associationistic and symbolic capabilities, its computational infrastructure may not be that hybrid at all, merely an extension of existing associationistic mechanism". In this paper we will consider a particular aspect of symbolic memory in the light of recent neurobiological work and explore the extent to which it may be attributable to connectionist mechanisms. The nature of long term memory for autobiographical events has been a research topic since Penfield's [2] demonstration that electrical stimulation of the exposed temporal lobes of the cortices of epilepsy patients produced highly detailed verbal accounts of sequences of personal experience from the distant past. The relationship of this episodic memory [3] and our structured knowledge of the world or semantic memory is of particular significance.

2 The BAIRN System

How does a complex sequence of individual experience give rise to a heterarchical structure of declarative and procedural knowledge? We have attempted to contribute to answering this question in earlier work [4]. BAIRN, a self-modifying information processing system, is an integrated model of performance and learning. Performance

J.R. Dorronsoro (Ed.): ICANN 2002, LNCS 2415, pp. 265–270, 2002.

is generated by a world model that represents BAIRN's knowledge in symbolic terms. It takes the form of a network of interconnected nodes in long term memory (LTM). Each node represents a basic unit of knowledge about a feature of the world and comprises symbolic productions in addition to information about its connectivity within the network.

BAIRN does not represent declarative and procedural knowledge in a separate propositional network and production repertoire [5]. There is a single network of nodes. Each node contains representations of procedural knowledge and related declarative knowledge. When a node is activated, details of the information that triggered its activation and of the end state produced by its operation become an entry in an episodic memory or time line that provides a totally comprehensive, sequential record of processing activity. Additions and modifications to the network or semantic memory are made by learning processes capable of detecting significant and/or consistently recurring segments of the time line record. A brief indication of the fundamental processes involved in this symbolic system provides a framework for assessing the capabilities of a connectionist system in representing episodic memory and its relationship with semantic memory. A detailed description of BAIRN is presented elsewhere [4, 6].

BAIRN's episodic memory provides a single sequential record of symbolic processing occurring in several parallel channels. For reasons indicated later, a limited amount of trace data is, also, stored on an experience list included in the structure of each node in semantic memory. Separate time lines for each processing channel were rejected since detection of the interaction between streams of processing is critical for the learning of fundamental forms of human performance such as oral language, reading and writing.

Establishing the record of each processing episode is relatively straightforward since it includes the input and output symbols directly derived from an activation of an individual node in semantic memory. In a similar fashion, reliance on symbols facilitates determination of the existence of a match between the outputs in two episodic records. This is the fundamental operation in analysing stretches of the time line in the search for repetition and consistency. Unfortunately, total reliance on exact matching of symbols is unlikely to discover consistency in a system's environmental interaction or further its knowledge of functional relationships and class equivalences that exist in its environment. The nodes which have produced the episodic records being compared embody the current class structure of the network in the expression of their activation conditions. BAIRN adopts these class levels as those to be used in seeking a match by substituting them for the specific class members featuring in the episode.

The interweaving of sequences of episodes derived from separate processing channels facilitates the discovery of interaction but complicates the discovery of consistent sequences in a single channel since episodes reflecting processing in other channels become 'noise'. Partial matching is adopted in tackling this difficulty. The level of match required is determined by the value of a parameter, the Matching Index Criterion (MIC), varying between 0 and 1 with a constant value in any version of the system. Partial matching, also, enables the detection of new classes for addition to semantic memory identified as functionally equivalent mismatches in sequences satisfying the MIC criterion.

As in the case of the MIC, the decisions to add to or modify semantic memory as a result of episodic memory analysis depend on parameters crossing threshold levels of successful matching. Three parameters are involved. The Critical Consistency Level (CCL) is the criterion for deciding whether a detected consistent sequence warrants semantic memory modification. It is a value bandwidth constant for each version of the system. The CCL varies within the bandwidth as a function of the consistency level detected in environmental interaction. The CCL must be exceeded by the Current Consistency Level (ccl) before semantic memory modification occurs. The ccl of a consistent sequence basically reflects its length and frequency of occurrence. If ccl > CCL, the consistent sequence gives rise to a new node added to semantic memory. The third parameter (ECCL), a less demanding criterion than CCL, sets a trigger value for the addition of consistent sequences to the experience list of already existing nodes if ccl > ECCL.

3 A Connectionist Approach

Many of the fundamental operational features of BAIRN may be more easily implemented in a connectionist form than in a symbolic representation. The generalisation capability and noise tolerance exhibited by parallel distributed processing (PDP) systems lend themselves to detection of consistent activation patterns in a much more straightforward fashion than can be achieved by symbol matching and use of the MIC, CCL and other BAIRN parameters. Consequently, the remainder of our paper will address aspects which are relatively straightforward in symbolic systems but pose a challenge to a PDP system. Implementing an episodic memory providing a comprensive sequential record of a system's processing is such an aspect. More specifically, achieving the functional equivalents of an episodic record and establishing its location in a temporal sequence lies in a somewhat under-emphasized zone of PDP system development.

Some recent neurobiological work provides interesting lines of attack on these problems. A significant, current focus is investigation of the basis of long term declarative memory. This is typically attributed to structural and strengthening modifications of synaptic connections between neurons. The synaptic plasticity hypothesis of memory has been questioned on the grounds that newly formed patterns of neural activity can only provide permanent memory traces if they remain stable throughout life. As Ortiz and Arshavsky [7] point out, there are several sources of instability including repetitive synapse use during recall, spontaneous activity of neurons and participation of the same neurons and synapses in several neural ensembles. They offer the intriguing alternative possibility that "elements of lifetime memory are stored within particular neurons as permanent changes in their DNA and that de novo-generated proteins are carriers of memory traces".

4 Hebbosomes and Episodic Memory

For our present purpose the most directly relevant alternative is provided by the work of Seth Grant and his colleagues [8, 9]. They have identified, in mouse forebrain extract, protein complexes involving multiple inputs, internal pathways and outputs and containing proteins associated with synaptic and non-synaptic effects such as cytoskeletal changes, synaptic and dendritic structural changes, signals to the cell nucleus for transcription, translational activation and synaptic tagging. Since many of the proteins are definitely not part of the synapses, these complexes offer a new, intraneuronal mechanism. Grant suggests that they may be miniature information processing centres making decisions about memory storage and retrieval. The ability to represent varied memories arises from the addition of new proteins to the complexes as a result of interaction with nuclear DNA which creates different structures from a standard set of components. Since they include, among other characteristics, all of the functionality required for Hebbian learning the complexes have been named 'hebbosomes'. From our perspective, hebbosomes offer a possible means of establishing episodic memory records and a basis for processing leading to semantic memory modification. In contrast to BAIRN, a hebbosome based episodic memory would be distributed across a range of neurons in areas associated with long term declarative memory such as the temporal cortex. A possible sequence of events begins with registration of a dendritic spine profile of activation by one or more hebbosomes during neural activation in a single neuron. Successive activations result in Hebbian learning, if appropriate, by pre-existing hebbosomes and/or creation of new hebbosomes to reflect the registration of new dendritic spine profiles. Over time the hebbosome population includes both hebbosomes reflecting consistently recurring dendritic profiles and others representing alternative members of functional categories. Settlement of the process giving rise to semantic memory modification results from the interaction between the hebbosomes and nuclear DNA. This broadly resembles the mode of operation of the immune system or a dynamic, modular ANN.

5 The Locus of Episodic Memory

The process described could occur entirely in the temporal cortex. Experience from operating BAIRN and neurobiological evidence suggest otherwise. BAIRN has highlighted the problems posed by the sheer volume of declarative memory data. This is tackled in two ways. The episodic memory sequence being analysed is initially subjected to a process for detecting novel segments. This involves comparison with the previous sequences stored on the experience lists of the relevant nodes. Familiar stretches are substituted by a single identifying token in the remainder of the analysis. The second strategy hinges on the inclusion of motivational and emotional processes in BAIRN. After the search for novel stretches interaction between repertoires of emotions and motives identifies episodic records that are regarded as sufficiently significant to be selected as starting points for processing the episodic memory sequence.

The neurobiological evidence centres on the role of the hippocampus in memory consolidation and its relationship to the entorhinal cortex and neocortex. Having reviewed a wide range of evidence Eichenbaum and his colleagues [10] conclude that non-spatial as well as spatial memory is dependent on the hippocampus. They provide the following account of the memory processes involved. „First, the combination of diverse and distinctive input gradients plus Hebbian mechanisms for encoding coactive inputs mediates the development of a range of specificities for an exceedingly broad range of attended information encoded by hippocampal cells. Second, the organizing principle for these codings is the temporal sequence of events that make up behavioral episodes. Those cells whose activity reflects the most highly specific conjunctions of cues and actions encode rare events that are elements of unique behavioral episodes. Other hippocampal cells whose activity reflects sequences of events serve to link large sets of successive events into representations of episodes that are unique in behavioral significance ... The bidirectional connections between the hippocampal network, the parahippocampal region, and the cortex could mediate the gradual development of sequence and nodal representations in the cerebral cortex".

More recently, Eichenbaum [11] has added to this theoretical account. The hippocampus may contain a network of episodic memories but it is unlikely that they contain many details of episodes. They are connected to areas of the entorhinal cortex encoding more detailed information. Repetitive interactions between the hippocampus and entorhinal cortex could fix episodic representations and their links in the entorhinal area. It is expected that the greatest level of detail about episodes is contained in the neocortex and that networks in the entorhinal area are connected to these neocortical representations. Repetitive interactions via these connections between the entorhinal area and directly connected widespread neocortical areas lead to a fixation of links between neocortical representations.

Combining this account with use of the hebbosome mechanism produces the outline of a theory which copes with the data volume problem and incorporates a number of other positive characteristics. It assumes that hebbosome based episodic memory capabilities are available in neurons in the temporal cortex, entorhinal cortex and hippocampus. Neural activity initially gives rise to a completely comprehensive episodic record in the episodic memory network in the temporal cortex. Here a connectionist process similar to the novelty assessment in BAIRN takes place before an attenuated version of the episodic sequence is created in the entorhinal cortex. Entorhinal cortex produces a further reduction by sparse coding and this is represented in the hippocampal episodic network. It is here that the influence of motivational and emotional states impacts on the process of analysing episodic content. The nature of the connectivity involved and an outline of a possible mechanism have been provided in an earlier paper [12]. The ascending sequence begins with the constraining of processing in entorhinal cortex to the zones highlighted by the hippocampal activity. This constraint, in turn, passes upwards to facilitate the consistency detection activity in temporal cortex.

In addition to tackling the data volume problem, this account deals with another fundamental difficulty in the approach. This originates in the nature of Hebbian learning. Although some variant appears to be the dominant form of learning in natural nervous systems it cannot be used to learn directly even a trivial higher-order function such as XOR. Experimental evidence suggests that the solution is to present the learning mechanism initially with a simplified version of the material then, when learning has occurred, to expose it to successively more detailed and complex versions until the desired performance level is achieved. [1] We propose that Hebbian learning occurs in each of the episodic memory areas and concludes with the comprehensive version in temporal cortex. Connections between temporal cortex and other cortical zones translate results of episodic memory processing into modifications of the world knowledge in semantic memory.

References

1. Reilly, R. G.: Evolution of Symbolisation: Signposts to a Bridge between Connectionist and Symbolic Systems. In: Wermter, S, and Sun, R. (eds.): Hybrid Neural Systems LNAI 1778, (2000) 363–371
2. Penfield, W.: The Permanent Record of the Stream of Consciousness: In Technical Report 486. Montreal: Montreal Neurological Institute, 14th International Congress of Psychology, (1954)
3. Tulving, E.: Episodic and Semantic Memory. In Tulving, E., Donaldson, W. (eds), Organization of memory, New York, Academic Press (1972)
4. Wallace, I., Klahr, D., Bluff, K.: A Self-Modifying Production System Model of Cognitive Development. In: Klahr, D., Langley, P., Neches, R. (eds.): Production System Models of Learning and Development (1987) 359–435
5. Anderson, J.R. and Lebiere, C.: The Atomic Components of Thought. Lawrence Erlbaum Associates, New Jersey (1998)
6. Bluff , K. A Preliminary Investigation of a General Learning System. Unpublished Ph.D thesis, Deakin University, Australia. (1988)
7. de Ortiz, S.P., Arshavsky, Y.I.: DNA Recombination as a possible mechanism in declarative memory: A hypothesis. Journal of Neuroscience Research, Vol. 63, Issue 1, 2001, 72–81
8. Husi, H., Ward, M.A., Choudary, J.S., Blackstock, W.P., Grant, S.G.N.: Proteomic analysis of NMDA receptor-adhesion protein signaling complexes. Nature Neuroscience, Vol. 3, no. 7, July 2000
9. Grant, S.G.N., O'Dell: Multiprotein Complex signaling and the plasticity problem. Current Opinion in Neurobiology 11, July 2001, 363–368
10. Eichenbaum, H., Dudchenko, P., Wood, E., Shapiro, M., Tanila, H.: The Hippocampus, Memory, and Place Cells: Is It Spatial Memory or a Memory Space? Neuron, Vol. 23 (June 1999) 209–226
11. Eichenbaum, H.: The Long and Winding Road to Memory Consolidation. Nature Neuroscience, Vol. 4, no. 11, November 2001, 1057–1058
12. Wallace, J.G. and Bluff, K.: The Borderline Between Subsymbolic and Symbolic Processing: A Developmental Cognitive Neuroscience Approach. Proceedings of the 20th Annual Conference of the Cognitive Science Society 1998, Lawrence Erlbaum Associates, New Jersey (1998) 1108–1112

Action Scheme Scheduling with a Neural Architecture: A Prefrontal Cortex Approach

Hervé Frezza-Buet

Supélec, 2 rue Edouard Belin,
F-57070 Metz Cedex, France
http://www.ese-metz.fr/~ersidp

Abstract. This paper presents a computational model addressing behavioral learning and planning with a fully neural approach. The prefrontal functionality that is modeled is the ability to schedule elementary action schemes to reach behavioral goals. The use of robust context detection is discussed, as well as relations to biological views of the prefrontal cortex.

1 Addressing Neural Planning

Planning is a crucial process in beings or systems that actually behave. Behaving requires perceptive and motor skills, to provide complex servo-control of actions, but also the ability to schedule these actions to reach a goal, that can be the satisfaction of a need. The first point, related to servo-control of actions, is closer to the problems that are usually addressed by current neural techniques, because it may be based in some kind of matching function to be learned [1]. This function deals with current perception, the "desired" one, current actuator state, and returns the motor command that allow to reach the "desired" perceptive state [1]. Providing such skills in artificial systems from neural techniques is still an open problem, because difficulties such as robustness to temporal distortions in servo controls still stands. Nevertheless, the learning of this matching function seems grounded on a regression paradigm, which is actually the one underlying many neural network researches.

The second point, scheduling of actions, has been successfully addressed by symbolic Artificial Intelligence techniques, allowing to solve problems by scheduling the use of predicates and inference rules. These techniques use mainly stacks to split goals into subgoals and algorithmic methods to find a way from available facts to desired ones, exploring the huge graph of predicates that can be derived through inference rules. Even if these techniques can endow an artificial system with powerful reasoning skills, symbolic data are difficult to ground on analogical data from sensor or actuators of a system like a robot. Attempts have been made to build hybrid neural and symbolic systems [11], but algorithms providing robust reasoning from analogical data are still to be found.

The purpose of this paper is to present one way to deal with action scheduling, viewed as a motor reasoning, in a fully neural architecture. Biological inspiration from prefrontal cortex studies that underlies this approach will be first presented. Then, a computational model instantiating this paradigm will be discussed.

J.R. Dorronsoro (Ed.): ICANN 2002, LNCS 2415, pp. 271–276, 2002.

2   Hints from Prefrontal Cortex

The prefrontal cortex is referred as a neural structure of the brain involved in hierarchical scheduling of actions toward reward [6]. This structure is associated to the concept of working memory required to perform complex behavior. The role of working memory is to store information temporarily during the time period it is relevant for achieving current behavior. This storage can concern perceptive cues that may help to perform some further choice, this is the sense of *working memory* defined by Goldman-Rakic (see her article in [7]). It can also be used to link past events to further ones. This is the Fuster's view of working memory, the *provisional memory*, allowing the *biding of cross-temporal contingencies* [6].

These memory skills are said to be supported by sustained neural activities, provided by loops including prefrontal cortex, basal ganglia and thalamus (see article by Cummings in [7], and [2]). Moreover, the loops, at the level of basal ganglia, play a crucial role in competition between possible actions [9,12]. Last, the crucial reward information is actually provided to the prefrontal cortex by specific sub-cortical structures [10].

Norman and Shallice have described the previous processes as the schedule of hierarchical action schemes (see article by Stuss *et al.* in [7]). Their model is grounded on non neural but symbolic processing. It deals with action schemes, competing each with the others for execution. The model avoids the execution of exclusive schemes, implementing the *contention scheduling* function of the prefrontal cortex. The triggering of a scheme consists of performing an action, or, if not feasible, activating some lower level action schemes before, allowing the former to be performed then.

The model we present in this paper has to be considered as first steps toward a fully unsupervised neural version of the Norman and Shallice scheme scheduling paradigm. The units of the model are inspired from biological cortical column model by Burnod [8], and *the emerging properties of the model are the result of strict local computation at the level of the units*. It is rather a computational model than a biological one, and it is neither dedicated to biological data fitting nor to data prediction. It is actually designed to show that biological paradigms can be reused in a computer science purpose. Therefore, the description of the cortex in this paper is a functional one, and it is very rough compared to biological works it is inspired from. *The purpose of the model is to show that robust action scheme processing can be provided by a neural architecture, and can be used outside biological framework,* as opposed to often experiment-dependent biological models. For that purpose, we have tested the model on a simulated robotic task. *This paper only sums up our architecture, rather insisting on the behavior of our central context detection mechanism.*

3   A Model of Action Scheme Scheduling

The simulated robot lives in a flat environment where some areas are colored. The robot can move continuously (rotation and translation), and it gets an artificial video image of the scene. The colored areas can be red or blue, respectively corresponding to food and water areas. When it is over an area, corresponding action (eat or drink) satisfies the internal need for food or water. For example, when the robot feels "hungry", it has to trigger "eat" action scheme. As eating only succeed on blue areas, the robot has to *stack* this action scheme and invoke "reaching" action scheme, until it detects that action scheme of eating becomes feasible (the robot is over the right area). This detection reactivates the previously stacked scheme, eat is then performed and leads to satisfaction of the need. This small story illustrates the approach. The architecture has to deal with the ability to call the good scheme, according to the feeling of current need. It also has to be able to stack unfeasible schemes, and invoke sub-schemes to make the stacked ones feasible. Sub-schemes may also be stacked if not directly feasible, recursively. Feasibility has to be detected from perceptive state. In the following, the perceptive state allowing to conclude that an action scheme is feasible will be called the *context* of this scheme. Context use is a crucial point in the model. During sub-scheme invocation, the stacked scheme must stay in some kind of "ready-to-apply" state, waiting for its context to be right. This is the mentioned stacked state, and the related sustained activity is an interpretation of Fuster's provisional memory [6] mentioned above. It also has been modeled in a biological framework as a bistable activity [8]. The actual architecture used in this simulated robotic context is described in [3] and only general principles are presented in the following.

3.1   An Architecture Inspired from Prefrontal Cortex

The neural architecture consists of two mirrored sets of units: the frontal and the posterior map. Each unit of one set is connected to the corresponding unit in the other set with one-to-one reciprocal connectivity. The posterior map contains P_i units, that can be in three states : inactive, called and excited, as suggests the cortical column model by Burnod [1]. Each P_i unit corresponds to an elementary action, that can be complex, as arm reaching for example, but that doesn't require splitting into sub-tasks. These units provide the servo-control skills mentioned at the beginning of this paper. The way this control is performed is out of the scope of this paper, but it has been addressed by previous work [5]. What must be retained here is that excitation activity in P_i means that some specific perceptive event is occurring, and call activity in P_i triggers servo controlled actions allowing to make this perceptive event occur. The role of a prefrontal unit F_i is to call the corresponding P_i when P_i is likely to succeed in getting the perception. The F_is also have to schedule calls in P_i in the right order, so that current needs get satisfied. Few required mechanisms are described in the following, but it has to be remembered that they are all running at the same time, without any supervision of their execution.

First mechanism is random spontaneous call of the corresponding P_i unit. It allows to perform random actions when nothing is known about the world. Let π_i the probability that F_i sets call activity in P_i. Spontaneous activity means that default π_i is not null. In the model, it is proportional to current need intensity.

Second mechanism is the learn of rewarded actions. A reward signal is send to all F_is when a need is satisfied. Reward signal only tells the system that *some* need has just been satisfied, without telling which specific one it is, as suggested by Taylor [12]. When reward occurs, and only in that case, the F_i units increase their sensitivity to need if they have called their P_i successfully, i.e. if servo control performed by P_i has allowed to get P_i detection of the perceptive event it is tuned on. We use temporal competitive learning adapted from Rescorla-Wagner rule to select such F_is, considering that reward occurrence can be predicted by their successful calls. Then, when *any* need occur, the F_i units that have been detected to be significantly able to provide some reward increase π_i. At the beginning, the call from such F_i may not systematically succeed, because requirements for the call to be rewarded are not fulfilled in the world. Nevertheless, this allows the F_i to learn these requirements, i.e. the configuration of excitations (current detected perceptive events) in P_i that actually predicts rewarded satisfaction of the call. This is what is called *context* of this F_i. Context matching is then used to increase π_i, so that when need occurs, the best matching rewarding F_is have higher probability to trigger call. To take benefit from this, it is necessary to have need detectors (e.g. for hunger an thirst) within P_is so that current need can favor context of units related to its satisfaction. This allows the units to be indirectly need dependent whereas learning and rewards deal with non specific need and reward signals. The P_is that are excited consecutively to one specific need are the somatic markers (see article by Damasio in [7]) of the present model.

Third, as each F_i knows from its context which perceptions P_j are required to ensure efficiency of call in elementary action P_i, it can ask the corresponding F_j units to make the P_j that are not yet detecting perceptive events trigger appropriate action. In that case, the F_i has a null probability π_i to call its P_i unit, but it contributes to sustain raise of all the π_j, until each F_j succeed in getting its perception. This non-calling state of F_i, that sustain the π_j in its context units, is the actual stacking effect of the model. During stacking of F_i, the favored F_j can recursively stack to obtain their own context. The stacking state may stop by itself spontaneously, the F_i giving up, but any improvement of the F_i context tends to reduce this effect, in order to persevere in successful searches. This can be done by *non selective context detection*, providing a context matching value sensitive to minor improvements.

Fourth, stacking of a F_i has to stop when context is detected, so that call in P_i, that is now supposed to succeed, is triggered. This context detection has to be much more *selective*, ensuring reliably the success of the call.

Last, as each F_i can learn which excitations in P_i are needed to ensure success of call in P_i, it can learn the same way another context, in order to detect which *calls* in the P_js are responsible for the failure of calling P_i. This allows to detect that some calls in P_j, i.e. some elementary actions, are exclusive to the P_i action. This endows F_i with the ability to inhibit the spontaneous activity of disturbing P_js by decreasing their π_j. This mechanism in our model implements the contention scheduling function reported as a major prefrontal cortex role (see article by Stuss *et al.* in [7]).

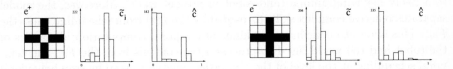

Fig. 1. Test of the contexts $\tilde{c}$ and $\hat{c}$ learning and recognition rule described in [4]. 25 possible perceptive inputs P_i are used. They are gathered in a 5×5 matrix for illustration convenience. Slow learning rate and large number of trials (3000) have been used to get the mean effect. A trial consists in choosing with equal probability to present event e^+ and e^-. Before computation of contexts, the perceptive patterns (0/white or 1/black) are added noise as follows. First, each perception is changed with probability 0.15. Then each value at level 1 is replaced by a value between 0.8 and 1, simulating the temporal trace effect due to asynchronousness of signals. After this learning stage, learning is artificially frozen, and 1000 supplementary trials are performed to compute histograms of $\tilde{c}$ and $\hat{c}$ when e^+ and e^- occur. Simulation parameters are $\theta = 1$, $\rho = 0$, $\tau = 0.01$ (see [4]).

3.2   Context Detection for Scheme Triggering

Context learning and detection have to be robust to temporal and spatial noise to be reusable in robotic framework. Two kinds of context matching values are used, a non selective $\tilde{c}$ and a selective one $\hat{c}$, as mentioned previously. The learning and detection rules related to contexts are described elsewhere [4], only computational behavior is reported here. Context computation occurs when some "positive" events e^+ occur, as success of a call, but also in the case of "negative" ones e^-, as call failure. At both these events, a F_i takes snapshots of P_j activities, and use them for learning. Figure 1 shows histogram of $\tilde{c}$ and $\hat{c}$ context values in both noisy e^+ and e^-. It can be seen that, even if e^+ and e^- share similar perceptions, both $\tilde{c}$ and $\hat{c}$ detect e^+ but do not strongly respond to e^-. Moreover, selectivity of $\hat{c}$ regards to $\tilde{c}$ when e^+ occurs is confirmed by histograms that shows distribution corresponding to noise for $\tilde{c}$ and sharp splitting into quite pure e^+ and too much noisy ones for $\hat{c}$.

4   Discussion

The above described model have been applied to a simulated toy robotic task, and its ability to be efficient in a real robot control framework has still to be improved. This is actually the purpose of current work, addressing the planning of ocular saccades for visual scene recognition by a real Koala robot (distributed by K-Team). Nevertheless, the model allows to understand how planning can emerge without any supervision within a neural architecture *by the use of contexts presented in this paper*. Context allow to get rid of wiring relations between an action scheme and possible sub-schemes. Then, the choice of the right action scheme to perform is the result of competition and robust numerical context detection. The learning paradigm associated with such a frontal model is incremental learning [8], which consist of presenting more and more complex tasks to a freely behaving animal (and robot in present work). Moreover, the model separates reactive complex skills (provided by the P_is) from scheduling ones (the F_is). This allows to conciliate two kinds of temporal computation: dynamics of the robot and the world, used to process servo controls from the P_is, and temporal scheduling at the level of the F_is, rather related to step by step behavioral reasoning. The later point is actually what is specific to frontal cortex computation, according to Fuster [6]. The F_i units of the model are still complex, because they have to manage competition, sustained activity during stacking, and context detection. This complexity becomes critical if the task requires many F_is and high level sub-schemes hierarchy. Inspiration from loops including cortex, thalamus and basal ganglia for sustained activities [2,12] and competition [9] may help to distribute some computations in the F_is among dedicated non-cortical neural architectures, thus simplifying local computation and making the whole architecture more reliable for complex behaviors.

References

1. Burnod, Y. : An adaptive neural network: the cerebral cortex. Masson (1989)
2. Dominey, P. F. : Complex sensory-motor sequence learning based on recurrent state representation and reinforcement learning. Biol Cybernetics. **73** (1995) 265–274
3. Frezza-Buet, H., Alexandre, F. : Modeling prefrontal functions for robot navigation. International Joint Conference on Neural Networks (1999)
4. Frezza-Buet, H., Rougier, N. P., Alexandre, F. : Integration of Biologically Inspired Temporal Mechanisms into a Cortical Framework for Sequence Processing. In Neural, symbolic and Reinforcement methods for sequence learning. Springer (2000)
5. Frezza-Buet, H., Alexandre, F. : From a biological to a computational model for the autonomous behavior of an animat. Information Sciences. In press
6. Fuster, J. M. : The Prefrontal Cortex: Anatomy, Physiology, and Neurophysiology of the Frontal Lobe. Lippincott Williams and Wilkins Publishers (1997)
7. Grafman, J., Holyoak, K. J., Boller, F., (Eds.) : Structure and Functions of the Human Prefrontal Cortex. Annals of the New York Academy of Sciences. **769** (1995)
8. Guigon, E., Dorizzi, B., Burnod, Y., Schultz, W. : Neural correlates of learning in the prefrontal cortex of the monkey: A predictive model. the monkey. Cerebral Cortex. **5** (1995) 2:135–147
9. Mink, J. W. : The basal ganglia: Focused selection and inhibition of competing motor programs. Progress in Neurobiology. **50** (1996) 381–425
10. Schultz, W. : Dopamine neurons and their role in reward mechanisms. Current Opinion in Neurobiology. **7** (1997) 191–197
11. Sun, R., Alexandre, F. : Connectionist-Symbolic Integration : From Unified to Hybrid Approaches. Lawrence ErlBaum Associates (1997)
12. Taylor, J. G. : Learning to Generate Temporal Sequences by Models of Frontal Lobes. International Joint Conference on Neural Networks (1999)

Associative Arithmetic with Boltzmann Machines: The Role of Number Representations

Ivilin Stoianov[1], Marco Zorzi[1], Suzanna Becker[2], and Carlo Umilta[1]

[1] University of Padova (Italy)
[2] McMasters University (Canada)

Abstract. This paper presents a study on associative mental arithmetic with *mean-field Boltzmann Machines*. We examined the role of number representations, showing theoretically and experimentally that cardinal number representations (e.g., numerosity) are superior to symbolic and ordinal representations w.r.t. learnability and cognitive plausibility. Only the network trained on numerosities exhibited the problem-size effect, the core phenomenon in human behavioral studies. These results urge a reevaluation of current cognitive models of mental arithmetic.

1 Simple Mental Arithmetic

Research on mental number processing has revealed a specific substrate for *number representations* in the inferior parietal cortex of the human brain, where numbers are thought to be encoded in a *number-line* (NL) format (Fig.2a,b) [1]. Simple mental arithmetic, however, is thought to be based on an associative network storing arithmetic facts in a verbal (symbolic) format [2]. Psychometric studies show that in production or verification of single-digit arithmetic problems (e.g., addition or multiplication), reaction times (RTs) and errors increase as a function of the size of the problem (*problem-size effect*) [3]. Problem size can be indexed by some function of the two operands, such as their sum or the square of their sum. The latter is the best predictor of the reaction time data for simple addition problems [4].

The first connectionist attempt to model simple multiplication was based on the autoassociative Brain-State-in-a-Box (BSB) network [5]. Numbers were represented jointly as NL and symbolic codes. Learning performance was far from optimal, in spite of the fact that the problem was simplified to computing "approximate" multiplication (to reduce the computational load). A later model of simple multiplication was MATHNET [6]. This model used NL representations and was implemented with a Boltzmann Machine (BM) [7]. The network was exposed to the arithmetic problems according to a schedule that roughly followed the experience of children when learning arithmetic, i.e., facts with small operands came before larger facts. However, fact frequency was manipulated in a way that did not reflect the real distribution: small facts were presented up to seven times as often as large problems. Therefore, MATHNET exhibited a (weak) problem-size effect, which was entirely produced by the specific training schedule and by the implausible frequency manipulation.

In the present study, we used mean-field BMs trained with the contrastive divergence learning algorithm [8] to model the simple addition task and to contrast three different hypotheses about the representation of numbers. We found that numerosity-based representations facilitate learning and provide the best match to hu

J.R. Dorronsoro (Ed.): ICANN 2002, LNCS 2415, pp. 277–283, 2002.

man reaction times. We conclude that the traditional view of symbolic mental arithmetic should be reevaluated and that number representations for arithmetic should incorporate the basic property of cardinal meaning.

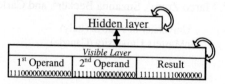

Hidden layer		
Visible Layer		
1st Operand	2nd Operand	Result
111000000000000	111110000000000	1111111110000000

Fig. 1. BMs for Mental Arithmetic. Patterns are encoded at the visible layer. To recall a fact, its two arguments are fixed at the visible layer and the network iterates until convergence.

2   Mean-Field BMs with Contrastive Divergence Learning

In line with previous connectionist attempts to model mental arithmetic, we assume that arithmetic facts are learned and stored in an associative NN. One typical associative network is the Boltzmann Machine, consisting of binary neurons with stochastic dynamics, fully connected with symmetric weights that store correlations of activations between connected units [7]. Data are encoded by visible neurons, whereas hidden neurons capture high-order statistics (Fig.1). BMs recall stored patterns by synchronous or asynchronous iterations, starting from initial activations representing partial patterns. Learning derives from the probabilistic law governing the net free-running state and targets a weight set W^* resulting in free-running distribution $P_{BM}{}^\sim(s)$ similar to the <u>data distribution</u> $P_{BM}{}^0(s) = Q(s)$. The learning procedure minimizes the Kullback-Liebler divergence between the distributions at time zero and equilibrium by computing derivatives w.r.t. the weights and applying gradient descent:

$$\Delta w_{ij} = \eta \ (\ <s_i \ s_j>^0 - <s_i \ s_j>^\sim) \tag{1}$$

The update of a weight connecting two units is proportional to the difference between the average of the correlations between these two units, computed at time 0 (positive, or fixed phase) and after reconstructing the pattern (negative, or free-running phase). Since the stochastic BM is computationally intractable, [9] replaced the correlations $<s_i \ s_j>$ with mean field approximation: $<s_i \ s_j> = m_i \ m_j$, where m_i is the mean field activity of neuron i and is given by the solution of a set of n coupled mean-field equations (2). Such an approximation turns the stochastic BM into a discrete NN since we can operate entirely with mean-field values, which also allows graded values.

$$m_i = \sigma (\ \Sigma_j \ w_{ij} \ m_j + \theta_j) \tag{2}$$

Hinton [10] replaced the correlations computed in the free-running phase with correlations computed after one-step data reconstruction (contrastive divergence learning), which was shown also to drive the weights toward a state in which the data will be reproduced according to their distribution. This was followed by the fast Contrastive Divergence Mean-Field learning (3) [8], that we use for our simulations.

$$\Delta w_{ij} = \eta \ (\ m_i{}^0 \ m_j{}^0 - m_i{}^1 \ m_j{}^1) \tag{3}$$

To learn a set of patterns, the learning algorithm presents them to the network in batches. For each pattern, the network performs a positive (wake) phase, when only the hidden layer settles, and a negative (sleep) phase, in which the network further reconstructs the visible pattern and then once again settles the hidden layer. After each phase, statistics for the correlations between the activations of each pair of connected neurons is collected. The weights can either be updated after each pattern or at the end of a batch. Here we used batch learning. The network recalls patterns by initializing the visible layer with a part of a pattern and iterating by consequent updating of the hidden and the visible layer until convergence. The number of steps to converge corresponds to RTs. We used the unsupervised learning mode, which in the sleep phase allows the network to reconstruct the entire input pattern. To improve the stability of learning, the weights were decayed toward zero after each learning iteration with a decay of 0.001. Every 20 epochs the network performance was tested.

3   Properties of Cardinal, Ordinal, and Symbolic Representations

This section studies cardinal and ordinal number codes, showing that from a statistical point of view cardinal representations inherently bias associative architectures toward longer processing time for larger numbers. The analysis is based on the fact that BMs learn to match the probability distribution of the data.

We examine two factors that can affect network RTs. One is the <u>empirical distribution</u> P_i^C of activations of individual neurons i, for a number encoding C. This is a "static" factor causing a neuronal bias b_i toward the mean $E(P_i^C)$ of that distribution. The time to activate a neuron i to a target state t_i, starting from rest state r_i (zero), is proportional to $(t_i - r_i)$ and is inversely proportional to its bias b_i. Since activating the representation $R^{C,n}$ of a number n needs convergence of all bits $R_i^{C,n}$ in this code, the time to activate $R^{C,n}$ depends on the time to activate its "slowest" bit. In addition, for the result-part of addition facts, the neuron activation distribution will also reflect the frequency of a given sum (e.g., for table $[1...9]+[1...9]$, the most frequent sum is 10).

The second factor is <u>pattern overlap</u>. Addition arithmetic facts consist of the combination of two numbers $(n_1,n_2)_{n1,n2=1...9}$, appended with their sum (n_1+n_2). There is a great a deal of pattern overlap among these facts, hence, some interference (i.e., cross-talk) will arise at the time of pattern recall. To account for this, we will measure the mean pattern overlap $D_{n1,n2}$ for all addition facts $<n_1,n_2,n_1+n_2>$ with a normalized vector dot-product and then project it to the sum (n_1+n_2): $D_{(n1+n2)}$. The network RT for a given sum is expected to be inversely proportional to the overlap $D_{(n1+n2)}$.

In **Cardinal** number codes N (e.g, *Numerosity* [11], Fig.2a), the representation of a number n embodies the pattern of numbers $m<n$: $R_n^N \supset R_m^N$ (in terms of active neurons). Therefore, the mean of the empirical distribution P_i^N of the activation of neurons i active in most of the numbers is close to 1, in contrast to the mean of the distribution P_j^N of neurons j active in larger numbers only, which is close to 0. Hence, networks trained to learn numbers with an encoding N and uniform frequency would develop a monotonically decreasing bias, larger for neurons active in most of the representations and smaller for neurons active in bigger numbers only. As a consequence, the time to activate representations $R^{N,n}$ should monotonically increase with n. However, the sum-frequency in addition-facts additionally modifies the distribution, resulting in the probabilities shown in Fig.3a (numerosity). When we also accounted

for the pattern-overlap factor, we obtained estimates for the time to produce results 2-18 encoded with N code (see Fig.3b, numerosity).

In **ordinal** number representations O, (e.g., the popular *Number Line (NL)*, Fig.2b,c), the representation of a number n partially overlaps with representations of neighboring numbers $[n-k_1...n-1$ & $n+1...n+k_2]$, so the distribution P_i^O of activations for each bit i of this code depends on the type of overlap. If there is symmetric overlap (as in a *shifting-bar* code; used in [6]), the distributions will be almost the same for each bit, so there is no bias towards longer processing time for larger numbers (Fig.3a,b). Two different versions of O coding have been proposed. One is the *Linear Number Line with Scalar Variability* [12], where the representation $R_i^{O_1,n}$ of a number n is a gaussian centered at n with variance proportional to n (here, $\sigma=n/8$) (Fig.2b). The other is the *Compressed Number Line* [1], whereby the representation $R_i^{O_2,n}$ of a number n is a gaussian centered according to a logarithmic scale (here, $\mu=18\cdot\ln(n/3+1)$) and with fixed variance (here, $\sigma=1.7$) (Fig. 2c). We can follow the same reasoning as before and compute the empirical density functions of neuron activation (Fig.3a), and next, estimate the reaction time for every number (Fig.3b). The predicted RTs for the NL codes initially increase with n until 10, but then decrease – in spite of the bigger overlap among gaussians for larger numbers (both for linear and compressed number lines), RTs here are primarily affected by the degree of pattern interference D_{sum} among addition facts – which is greatest for sums around 10.

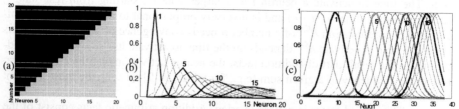

Fig. 2. Meaningful encoding for numbers: (a) Cardinal: *Numerosity* (thermometer code). Large numbers include small numbers (b) Ordinal: *Linear number-line with scalar variability*. (c) Ordinal: *Compressed number line*. In (b) and (c) numbers partially overlap with neighbors

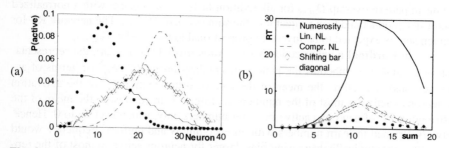

Fig. 3. (a) P(bit$_i$=1) for the result-part of the pattern, for *Numerosity*, *Linear NL* with scalar variability, *Compressed NL*, *Shifting-bar*, and *Symbolic* number codes (b) Estimated RTs.

With regard to **symbolic** encodings, we assume that they do not overlap. No bias for the activation time can be expected from the nature of the pattern encoding (Fig. 3a), therefore, only sum frequency and pattern interference should affect RTs (Fig.3b).

To conclude, our analysis of the three different types of number representations predicts that only cardinal number codes should result in the desired problem size effect, when employed in associative neural networks to learn simple arithmetic.

4   The Role of Number Representations: Simulations

Support to our analysis of different number representations for simple arithmetic is provided by simulations with BMs. As before, we contrasted symbolic codes (implementing the idea of verbally stored facts[2]), ordinal codes (used in MATH-NET and BSB), and cardinal codes (i.e., numerosity, as recently proposed in [11]).

Experiment 1: Mental Arithmetic with Symbolic Number Representations

Cognitive models of mental arithmetic suggest that simple arithmetic facts are stored in a verbal code [1,2]. To examine learning of symbolically encoded arithmetic facts, we abstracted from verbal and/or Arabic numbers by using simple symbolic representations – a localistic diagonal code. A BM with 36 visible and 40 hidden neurons was trained on simple addition facts (1+1 to 9+9). Frequency of the addition facts was not manipulated, to assess independently the computational properties of the representations. Learning these 'simple' problems was slow – about 120000 cycles. The network's ability to 'calculate' was then tested with facts presented with their two arguments; the result-part and the hidden neurons were initialized with zeros. The network was allowed to iterate until converge (threshold of 0.0001) by synchronous update of, first, all hidden neurons, and then all visible neurons. For each fact, the response time was recorded. The RT profile against the problem size was uniform; the net did not exhibit the problem-size effect. The correlation of the network RTs against the classical structural predictors for human RTs (the sum of the arguments and the square of the sum [4]) was very low: 0.017 and 0.014. Therefore, the hypothesis of purely symbolic representations for mental arithmetic is not supported.

Experiment 2: Ordinal Semantic Number Representations

Psychometric studies [1] have suggested a representation of numbers that is spatially organized from left to right in a continuous fashion, i.e. a mental *number line*. A NL code was also used in previous network models of mental arithmetic (MATHNET and BSB). In this experiment we assessed the property of NL representations, in the absence of the implausible frequency manipulation employed in [6]. The same BM was trained on addition facts encoded with the NL. As before, we tested the simple shifting-bar code (MATHNET and BSB), the linear NL with scalar variability, and the compressed NL. Learning in these experiments was facilitated by the more informative code. In general, however, RTs behaved as predicted: they initially increased with the sum, but then drastically decreased for larger problems (Fig.4a,b). The structural variables - the sum and the square of the sum - did not predict the RTs and accounted for less than 5% of the variance. NL representations failed to produce a problem-size effect, the hallmark phenomenon of human performance in mental calculation.

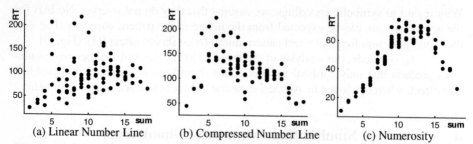

Fig. 4. *RT* of the BM (in cycles) for each addition fact against the *sum*. Only the network trained with the numerosity code exhibited the problem-size effect.

Experiment 3: Numerosity Number Representations

As far as addition is concerned, computing with numerosities is as easy as adjoining sets. Indeed, learning was unexpectedly fast for BMs: just 4000 cycles. Importantly, the network for the first time exhibited the problem-size effect (Fig. 4c): the sum and the square of the sum explained a significant proportion of the variance in the network's RTs (29% and 14%, respectively; p<0.001). This result (replicated in multiple simulations) supports our earlier prediction that cardinal codes should produce a problem-size effect. Note that the only difference with the earlier experiments is the encoding of numbers; therefore, we conclude that the faster learning and the similarity of the RTs to the human data is entirely produced by the numerosity representation.

5   Conclusions

A number of cognitive models of mental arithmetic [e.g., 2, 13] can be taken to support the popular idea that arithmetic facts can be, and perhaps should be, learned by rote: after all, it is simply another example of common or garden-variety associative learning. Our theoretical and computational results, however, point to the opposite conclusion – mental arithmetic is grounded in a representation of numbers that has the basic property of cardinal meaning (numerosity [11]). This kind of representation not only makes learning arithmetic facts easier, but it is also the only representational format that forces BM networks – all other things being equal – to retrieve arithmetic fact with a processing time course that matches the pattern of human reaction times (where the benchmark phenomenon is the effect of problem-size). Our analysis shows that the latter is the result of the distribution of active bits/neurons in cardinal representations: neurons active in small and large numbers are more frequently active and will be biased towards one, whereas neurons active in larger numbers only are less frequently active and will be biased towards zero. Therefore, the processing time that is required to activate a given number is correlated to its numerosity. Although still far from a comprehensive account of the cognitive processes in mental arithmetic, our findings represent a fundamental step towards understanding the computational bases of numerical processing in the human brain.

Acknowledgements. This research was funded through grants from the European Commission (RTN), the Royal Society of London, and San Raffaele University (Milan). We are grateful to Geoff Hinton and Max Welling for helpful suggestions on BMs.

References

1. Dehaene, S., L. Dehaene-Lambertz, & L. Cohen (1998). Abstract representations of numbers in the animal and human brain. *TINS*, **21**, 355–361.
2. Dehaene, S., & L. Cohen (1995). Toward an Anatomical and Functional Model of Number Processing. *Mathematical Cognition*, **1**, 83–120.
3. Groen, G. J., & J.M. Parkman (1972). A chronometric analysis of simple addition. *Psychological Review*, **79**, 329–343.
4. Butterworth, B., M. Zorzi, L. Girelli, & A.R. Jonckheere (2001). Storage and retrieval of addition facts: The role of number comparison. *Quart.J.Exp.Psych.*, **54A**,1005–1029.
5. Viscuso,S., J.Anderson & K.Spoehr(1989). Representing simple arithmetic in neural networks. In (G. Tiberghien, ed.), *Adv. in Cog.Sci. Vol.2: Theory and Applications*, 144-164,.
6. McCloskey & Lindemann (1992). MATHNET: preliminary results from a distributed model of arithmetic fact retrieval. In (Cambpell, ed.)*The Nature and Origin of Math.Skills,*365–409
7. D. Ackley, G. Hinton & T. Sejnowski (1985). A Learning Algorithm for Boltzmann Machines. *Cog. Sci.*, **9**, 147–169
8. Welling, M. & G. Hinton (2001). A New Learning Algorithm for Mean Field Boltzmann Machines. *GCNUTR2001-002*
9. Peterson,C.&Anderson,J. (1987). A mean field theory learning algorithm for Neural Networks. *Comp. Syst.*,**1**,995–1019
10. Hinton, G. (2000). Training Products of Experts by Minimizing Contrastive Divergence. *GCNU TR 2000-004.*
11. Zorzi, M. & B. Butterworth (1999). A computational model of number comparison. In (Hahn & Stoness, eds.), *Proc. 21st Ann. Meeting of the Cog. Sc. Soc.*, 778-783.
12. Gallistel & Gelman (2000). Non-verbal numerical cognition:from reals to int. *TICS,***4**,59–65
13. Ashcraft,M.(1992).Cognitive arithmetic: a review of data and theory. *Cognition,***44**,75–106.

Learning the Long-Term Structure of the Blues*

Douglas Eck and Jürgen Schmidhuber

Istituto Dalle Molle di Studi sull'Intelligenza Artificiale (IDSIA)
Galleria 2, CH 6928 Manno, Switzerland
{doug,juergen}@idsia.ch, http://www.idsia.ch/

Abstract. In general music composed by recurrent neural networks (RNNs) suffers from a lack of global structure. Though networks can learn note-by-note transition probabilities and even reproduce phrases, they have been unable to learn an entire musical form and use that knowledge to guide composition. In this study, we describe model details and present experimental results showing that LSTM successfully learns a form of blues music and is able to compose novel (and some listeners believe pleasing) melodies in that style. Remarkably, once the network has found the relevant structure it does not drift from it: LSTM is able to play the blues with good timing and proper structure as long as one is willing to listen.

1 Introduction

The most straight-forward way to compose music with an RNN is to use the network as single-step predictor. The network learns to predict notes at time $t+1$ using notes at time t as inputs. After learning has been stopped the network can be seeded with initial input values—perhaps from training data–and can then generate novel compositions by using its own outputs to generate subsequent inputs. This note-by-note approach was first examined by [12,1] and later used by others [11,8].

A feed-forward network would have no chance of composing music in this fashion. With no ability to store past information, such a network would be unable to keep track of where it is in a song. In principle an RNN does not suffer from this limitation. In practice, however, RNNs do not perform very well at this task. As Mozer [8] aptly wrote about his attempts to compose music with RNNs, "While the local contours made sense, the pieces were not musically coherent, lacking thematic structure and having minimal phrase structure and rhythmic organization." In short, RNNs do not excel at finding long-term dependencies in data — the so-called *vanishing gradient* problem [5] — and so are relatively insensitive to the global structure that defines a particular musical form.

Long Short-Term Memory (LSTM) [6] solves the problem of vanishing gradients by enforcing constant error flow. In doing so, LSTM is able to find long-term dependencies in data. Recent research [3] has already shown that LSTM can solve certain rhythmical timing and counting tasks. In the current study we

---

* Work supported by SNF project 21-49144.96

ask the question, "Can LSTM compose music?" The answer is a guarded "Yes." Though the experiments presented here are very preliminary, LSTM was able to find global musical structure and use it to compose new pieces. Section 2 describes the music composition model, Section 3 presents two experiments in music composition and Section 4 discusses the results.

2   An LSTM Music Composer

LSTM Architecture. Due to space constraints it is impossible to describe LSTM in depth. See [4] for details. In summary, LSTM is designed to obtain constant error flow through time and to protect this error flow from undesired perturbations. LSTM uses linear units to overcome the problem of error decay or error explosion. Each unit has a fixed self-connection and is surrounded by a cloud of nonlinear gating units responsible for controlling information flow. Learning is done by a gradient descent method that is a combination of modified BPTT and modified RTRL [10].

   Data Representation. Inputs and targets are represented in a simple local form (similar to [12]). We use one input/target unit per note, with 1.0 representing ON and 0.0 representing OFF. Inputs are then adjusted to have mean=0.0 and a standard deviation=1.0. The representation is multi-voice and makes no distinction between chords and melodies. Time is encoded implicitly, with one input vector representing a slice of real time. This is a major difference from Mozer's CONCERT system [8] which used two different distributed representations and processed a single *note* per timestep regardless of duration.

   Note that the start (or end) of a note is not marked. This means that (with eighth-note quantization) a series of eight identical eighth notes are encoded the same way as four identical quarter notes. This can be remedied by decreasing the stepsize of quantization and marking note offsets with a zero. Alternatively, special unit(s) can indicate the start of note(s) [12]. 12 notes (starting an octave below middle C and ascending by half steps) were allocated for chords and 13 notes for melodies (continuing up by half steps from middle C). For these simulations, melody notes never occurred in the range for chords or vice-versa.

   Training Data. For the experiments in this study, a form of 12-bar blues popular among bebop jazz musicians is used 1. With 8 time steps per bar, each song was 96 time steps long. For Experiment 1, only these chords were presented.

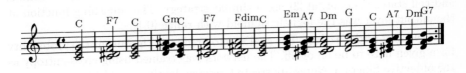

Fig. 1. Bebop-style blues chords used for training data (transposed up one octave).

For Experiment 2, a single melody line was also presented. This melody was built using the pentatonic scale 2 commonly used in this style of music.

Fig. 2. Pentatonic scale used for training data melodies.

Training melodies were constructed by concatenating bar-long segments of music written by the first author to fit musically with each chord. Datasets were constructed by choosing randomly from the space of unique complete pieces ($n = 2^{12} = 4096$). Only quarter notes were used. No rests were used. Space constraints make it impossible to include examples. However several of these training pieces are provided as sheet music and audio at www.idsia.ch/~doug/blues/index.html.

3   Experiments

EXPERIMENT 1 – Learning Chords. We test the ability for LSTM to learn and reproduce the global structure of bebop-jazz style blues. Musical structure is presented to the network in the form of chords only. This ensures that the model is truly sensitive to long time lag structure and is not exploiting local regularities that may be found in the melody line. With a quantization of eight time steps per bar, this dataset could not be learned using RTRL or BPTT [7].

Architecture and Parameters. The chords used are the ones described in 2. No melodies are presented. The quantization timestep is eight events per whole note. In the network four cell blocks containing 2 cells each are fully connected to each other and to the input layer. The output layer is fully connected to all cells and to the input layer. Forget gate, input gate and output gate biases for the four blocks are set at -0.1, -0.3, -0.5 and -0.7. This allows the blocks to come online one by one. Output biases were set at 0.5. Learning rate was set at .00001. Momentum rate was set at .9. Weights are burned after every timestep. Experiments showed that learning is faster if the network is reset after making one (or a small number) of gross errors. Resetting went as follows: on error, burn existing weights, reset the input pattern and clear partial derivatives, activations and cell states. Gers et. al [3] use a similar strategy. The squashing function at the output layer was the logistic sigmoid with range [0,1].

Training and Testing. The goal was to predict at the output the probability for a given note to be on or off. The network was trained using cross-entropy as the objective function. Notes are treated independently, as multiple notes are on at once. The network was trained until it could successfully predict the entire chord sequence multiple times. This performance was tested by starting the network with the inputs from the first timestep of the dataset and then using network predictions to generate the next input. Notes were predicted using a decision threshold of 0.5.

Results. LSTM easily handled this task under a wide range of learning rates and momentum rates. As it is already well documented that LSTM excels at timing and counting tasks [3], success at this task is not surprising. Fast convergence was not a goal of this study, and learning times were not carefully collected.

EXPERIMENT 2 – Learning Melody and Chords. In this experiment both melody and chords are learned. Learning continues until the chord structure is learned and cross-entropy error is relatively low. Note that there are far too many melodies for the network to learn them all. Once learning has been stopped, the network is started with a seed note or series of notes and then allowed to compose freely. The goal of the study was to see if LSTM could learn chord structure and melody structure and then use that structure to advantage when composing new examples.

Architecture and Parameters. The network topology for this experiment differs from the previous task in that some cell blocks processed chord information while other cell blocks processed melody information. Eight cell blocks containing 2 cells each are used. Four of the cell blocks are fully connected to the input units for chords. The other four cell blocks are fully connected to the input units for melody. The chord cell blocks have recurrent connections to themselves and to the melody cell blocks. However, melody cell blocks are only recurrently connected to other melody cell blocks. That is, melody information does not reach the cell blocks responsible for processing chords. At the output layer, output units for chords are fully connected to cell blocks for chords and to input units for chords. Output units for melody are fully connected to cell blocks for melody and to input units for melody. Forget gate, input gate and output gate biases for the four blocks dedicated to processing chords are set at -0.1, -0.3, -0.5 and -0.7. Gates for processing melodies are biased in exactly the same way. All other parameters are identical to those in Experiment 1.

Training and Testing. The goal was to predict at the output the probability for a given note to be on or off. For chords, the same method as Experiment 1 is used: the network applies a decision threshold of 0.5 for all chord notes. For melodies we restrict the network to choosing a single note at any given timestep. This is achieved by adjusting melody output activations so that they sum to 1.0 and then using a uniform random number in the range [0,1] to choose the appropriate next note. The network was trained until it had learned the chord structure and until objective error had reached a plateau. Then the network was allowed to freely compose music. Music was composed by seeding the network with a single note or series of notes (up to 24) as was done in Experiment 1.

Results. LSTM composed music in the form of blues. It learned the long-term chord structure in training and used that structure to constrain its melody output in composition mode. Because it is difficult to evaluate the performance objectively—this point is commonly made in AI art research, e.g.,

[8]—we urge the reader to visit www.idsia.ch/~doug/blues/index.html. On that page are examples of network blues composition in sheet music and audio. The network compositions are remarkably better sounding than a random walk across the pentatonic scale (this is also available for listening at the website). Unlike a random walk, the network compositions follow the structure of

the musical form. They do diverge from the training set, sometimes significantly. But due to the influence of the chord structure, they never "drift" too far away from the form: the long-term structure reflected by chord changes always bring them back. Also, an informal survey in our lab indicates that the compositions are at times quite pleasant. It is striking, to the authors at least, how much the compositions sound like real (albeit not very good) bebop jazz improvisation. In particular, the network's tendency to intermix snippets of known melodies with less-constrained passages is in keeping with this style.

4 Discussion

These experiments were successful: LSTM induced both global structure and local structure from a corpus of musical training data, and used that information to compose in the same form. This answers Mozer's [8] key criticism of RNN music composition, namely that an RNN is unable to compose music having global coherence. To our knowledge the model presented in this paper is the first to accomplish this. That said, several parts of the experimental setup made the task easier for the model. More research is required to know whether the LSTM model can deal with more challenging composition tasks.

Training Data. There was no variety in the underlying chord structure. This made it easier for LSTM to generate appropriately-timed chord changes. Furthermore, quantization stepsize for these experiments was rather high, at 8 time steps per whole note. As LSTM is known to excel at datasets with long time lags, this should not pose a serious problem. However it remains to be seen how much more difficult the task will be at, say, 32 time steps per whole note, a stepsize which would allow two sixteenth notes to be discriminated from a single eighth note.

Architecture. There network connections were divided between chords and melody, with chords influencing melody but not vice-versa. We believe this choice makes sense: in real music improvisation the person playing melody (the soloist) is for the most part following the chord structure supplied by the rhythm section. However this architectural choice presumes that we know ahead of time how to segment chords from melodies. When working with jazz sheet music, chord changes are almost always provided separately from melodies and so this does not pose a great problem. Classical music compositions on the other hand make no such explicit division. Furthermore in an audio signal (as opposed to sheet music) chords and melodies are mixed together.

These are preliminary experiments, and much more research is warranted. A more interesting training set would allow for more interesting compositions. Finally, recent evidence suggests [2,9] that LSTM works better in similar situations using a Kalman filter to control weight updates. This should be explored.

5 Conclusion

A music composition model based on LSTM successfully learned the global structure of a musical form, and used that information to compose new pieces in the form. Two experiments were performed. The first verified that LSTM was able to learn and reproduce long-term global music structure in the form of chords. The second experiment explored the ability for LSTM to generate new instances of a musical form, in this case a bebop-jazz variation of standard 12-bar blues. These experiments are preliminary and much more work is warranted. However by demonstrating that an RNN can capture both the local structure of melody and the long-term structure of a of a musical style, these experiments represent an advance in neural network music composition.

References

1. J. J. Bharucha and P. M. Todd. Modeling the perception of tonal structure with neural nets. *Computer Music Journal*, 13(4):44–53, 1989.
2. F. A. Gers, J.A. Perez-Ortiz, D. Eck, and J. Schmidhuber. DEKF-LSTM. In *Proc. 10th European Symposium on Artifical Neural Networks, ESANN 2002*, 2002.
3. F. A. Gers and J. Schmidhuber. Recurrent nets that time and count. In *Proc. IJCNN'2000, Int. Joint Conf. on Neural Networks*, Como, Italy, 2000.
4. F. A. Gers and J. Schmidhuber. LSTM recurrent networks learn simple context free and context sensitive languages. *IEEE Transactions on Neural Networks*, 12(6):1333–1340, 2001.
5. S. Hochreiter, Y. Bengio, P. Frasconi, and J. Schmidhuber. Gradient flow in recurrent nets: the difficulty of learning long-term dependencies. In S. C. Kremer and J. F. Kolen, editors, *A Field Guide to Dynamical Recurrent Neural Networks*. IEEE Press, 2001.
6. Sepp Hochreiter and Juergen Schmidhuber. Long Short-Term Memory. *Neural Computation*, 9(8):1735–1780, 1997.
7. Michael. C. Mozer. Induction of multiscale temporal structure. In D. S. Lippman, J. E. Moody, and D. S. Touretzky, editors, *Advances in Neural Information Processing Systems 4*, pages 275–282. San Mateo, CA: Morgan Kaufmann, 1992.
8. Michael C. Mozer. Neural network composition by prediction: Exploring the benefits of psychophysical constraints and multiscale processing. *Cognitive Science*, 6:247–280, 1994.
9. Juan Antonio Pérez-Ortiz, Juergen Schmidhuber, Felix A. Gers, and Douglas Eck. Improving long-term online prediction with decoupled extended kalman filters. In *Artificial Neural Networks – ICANN 2002 (Proceedings)*, 2002.
10. A. J. Robinson and F. Fallside. The utility driven dynamic error propagation network. Technical Report CUED/F-INFENG/TR.1, Cambridge University Engineering Department, 1987.
11. C. Stevens and J. Wiles. Representations of tonal music: A case study in the development of temporal relationship. In M.C. Mozer, P. Smolensky, D.S. Touretsky, J.L Elman, and A. S. Weigend, editors, *Proceedings of the 1993 Connectionist Models Summer School*, pages 228–235. Erlbaum, Hillsdale, NJ, 1994.
12. Peter M. Todd. A connectionist approach to algorithmic composition. *Computer Music Journal*, 13(4):27–43, 1989.

Recursive Neural Networks Applied to Discourse Representation Theory

Antonella Bua[1], Marco Gori[2], and Fabrizio Santini[2]

[1] Facoltà di Scienze della Comunicazione, Università di Siena
bua@media.unisi.it
[2] Dipartimento di Ingegneria dell'Informazione, Università di Siena
{marco,santini}@dii.unisi.it

Abstract. Connectionist semantic modeling in natural language processing (a typical *symbolic* domain) is still a challenging problem. This paper introduces a novel technique, combining Discourse Representation Theory (DRT) with Recursive Neural Networks (RNN) in order to yield a neural model capable to discover properties and relationships among constituents of a knowledge-base expressed by natural language sentences. DRT transforms sequences of sentences into directed ordered acyclic graphs, while RNNs are trained to deal with such structured data. The acquired information allows the network to reply on questions, the answers of which are not directly expressed into the knowledge-base. A simple experimental demonstration, drawn from the context of a fairy tales is presented. Finally, on-going research direction are pointed-out.

1 Introduction

Discourse Representation Theory [6] (DRT) is a semantic theory for the computational treatment of natural language which solves problems that can not be handled by strict compositional semantic theories (see [7], [2]). Given a syntactic structure, the DRT transforms sequences of phrases in conditions rules [6] over a set of reference markers, called *discourse representation structures* (DRSs) [6]. DRSs rules are appropriate expressions of a meta-language and can be easily represented by directed ordered acyclic graphs (DOAGs).

Motivated by this reason, in this paper we present a research in progress, in which we apply recursive neural networks ([8], [1]) to DOAGs expressing a knowledge-base analyzed with DRT. Our novel technique guarantees the network to learn semantic relations within the discourse. Moreover, the neural network is able to answer questions (again expressed with DOAGs) concerning the knowledge-base, that are not explicitly codified.

Our approach differs from previous works (see [4], [3]) on context sensitivity, because we train the network to perform semantic generalization instead of predictions over syntactic constituents.

J.R. Dorronsoro (Ed.): ICANN 2002, LNCS 2415, pp. 290–295, 2002.
© Springer-Verlag Berlin Heidelberg 2002

2   Discourse Representation Theory

Starting from a whole discourse act, DRT analyzes and decomposes the text in a progressive and incremental manner. The major accomplishment is that each sentence enriches a given DRS with its information content and, as a result, each new sentence S belonging to the discourse is interpreted within the context provided by all the previous sentences. A simple example will help us to better explain the construction of DRSs. Let us consider the sentence:

Fabrizio sings.

The first step is to place the sentence in a box (fig. 1-A). In the second step a variable x is introduced: it is a reference marker of DRT. The subject of the sentence, *Fabrizio*, is replaced by this variable and an identity assertion is added to the DRS (fig. 1-B). Now, let us suppose to extend the context with a second sentence, so that the discourse comes to be:

Fabrizio sings. [He] is a grasshopper

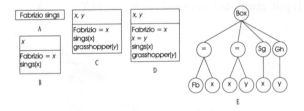

Fig. 1. Development of a context using DRT.

The DRS in fig. 1-B serves as context for the latter sentence. We only need to add another reference marker y (fig. 1-C). But, since x was previously attached to the pronoun *he*, the interpretation of *he* relates the reference marker y to the marker x. The relation is expressed by the condition $y = x$. Therefore the DRS in fig. 1-C becomes as shown in fig. 1-D, and the corresponding graphical representation is shown in fig. 1-E.

3   Adaptive Computation on Discourse Contexts

The DRS representation can be easily expressed with a specific class of DOAGs, namely trees. A DOAG can be formally defined as a pair (V, E), where V is the set of vertices, and $E = \{(v_i, v_j) \mid v_i, v_j \in V\}$ is the set of ordered pairs representing the directed edges. A *label* denoted with $L_v \in \mathbb{R}^m$ is attached to each node v and in our case it is represented by a m-dimensional array of binary values. Given a node v, we can define the set of its outgoing edges as $\mathcal{O}(v) = \{(v, z) \in$

$E \mid z \in V\}$. The cardinality of this set, $o(v) = |\mathcal{O}(v)|$, is called *node out-degree*. We can refer to the k-th child of node v as $ch_k[v]$ or as $q_k^{-1}v$. The set of incoming edges to a node v, $\mathcal{I}(v) = \{(w, v) \in E \mid w \in V\}$, is the set of the edges entering a node, and its cardinality $i(v)$ is the *node in-degree*. The set of nodes v having no descendants is the *graph frontier* $\mathcal{F} = \{v \in V \mid o(v) = 0\}$. The set of nodes having no ancestors is the set of the *graph supersources*, $\mathcal{S} = \{v \in V \mid i(v) = 0\}$. Trees are an example of DOAGs where the single supersource is also called *root node*. The forward recursive algorithm which we use to process DOAGs realizes a *Frontier to Root* computation. For each node v in the input graph we compute a state vector $X_v \in \mathbb{R}^n$ and an output $Y_v \in \mathbb{R}^k$ as follows:

$$X_v = f\left(X_{ch[v]}, L_v, v\right)$$
$$Y_v = g\left(X_v, L_v, v\right) \tag{1}$$

where n and k are the dimensionalities of the state space and the output space, respectively. The function $f : \mathbb{R}^{o \cdot n} \times \mathbb{R}^m \to \mathbb{R}^n$ is the *state transition function* and $g : \mathbb{R}^n \times \mathbb{R}^m \to \mathbb{R}^k$ is the *output function*. This definition requires each node to have the same out-degree o. The function f computes the state of node v given the states of its children $X_{ch[v]}$ and the node label L_v. When this does not apply and some of the children of the node are not specified, a given initial state X_0 is used for the missing links (*nil pointers*). In particular, the state for the leaves in the input graph is computed as $X_{v_l} = f(X_0, \ldots, X_0, L_{v_l}, v_l)$.

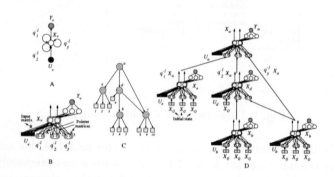

Fig. 2. The encoding network is obtained by the application of the recursive network to the vertices of a DOAG. (A) The recursive model stating the dependencies among the variables. (B) The implementation of the model with a neural network. (C) The input DOAG. (D) The encoding network unfolded on the input graph.

The nodes must be processed following a topological sorting, in which each node is processed before its parents. Thus, the computation starts from the frontier nodes since the state of their children is the initial state X_0. Then the recursive equations (1) are applied to the nodes following a bottom-up strategy. Our assumption that the graph is acyclic guarantees that the value of the state in each node does not depend neither on itself nor on other nodes not yet updated. The simplest *first-order* neural network realization of the state transition function in equation (1) can be written as:

$$X_v = f \left(\sum_{i=1}^{o} W'_i \cdot X_{q_i^{-1}v} + W'' \cdot L_v \right) \tag{2}$$

where f is the activation function, e.g., a sigmoid, the parameters $W'_i \in \mathbb{R}^{n,n}$, $i = 1, \ldots o$, express the dependence of the state for node v on the state of its children $ch_i[v]$, and the weights $W'' \in \mathbb{R}^{n,m}$ determine the contribution of the information attached to the node. Such a model, called *encoding network* [8], is shown in fig. 2. Similarly, another neural network can be trained to realize the output function of equation (1).

The learning of the matrices W' and W'' is performed by Back-propagation Through Structures (see [5], [1], [8]).

4   Generation of Context and Learning Environment

There are many kinds of domains in which a recursive neural network can be used to handle a context learning problem. We chose to describe a *context* adapted from a fairly tale. The reason is that it is often easy to find a tale with few characters, elementary relationships and simple verbs, reducing in this way the complexity of the learning space. Let us consider a grammar G which is able to generate sentences coherent with the chosen context, and the corresponding language $\mathcal{L}(G)$. We introduce a function h (called *context model*):

$$h : \mathcal{L}(G) \to \{\textbf{true, false}\} \tag{3}$$

such that, for each $s \in \mathcal{L}(G)$, the value of $h(s)$ assumes *true* if and only if the corresponding fact is true in the context, *false* otherwise.

However, in order to successfully produce a set, which is able to represent more complex sentences of the tale, we need to apply a composition process on all $s \in \mathcal{L}(G)$ using a finite set of boolean operators, which we suppose to belong to two classes: unary and binary operators.

Definition 1: The set $\mathcal{F}$ of the well formed sentences on $\mathcal{L}(G)$ is built as follows:

- If $s \in \mathcal{L}(G)$, then $s \in \mathcal{F}$;
- If $s \in \mathcal{F}$ and $\diamond$ is a unary operator, then $\diamond s \in \mathcal{F}$ and $h(\diamond s) = \diamond h(s)$;
- If $s_1 \in \mathcal{F}$, $s_2 \in \mathcal{F}$ and $\circ$ is a binary operator, then $(s_1 \circ s_2) \in \mathcal{F}$ and $h(s_1 \circ s_2) = h(s_1) \circ h(s_2)$;

It can be proven that knowing the DRS representation of a $s \in \mathcal{L}(G)$, the corresponding representation of $s \in \mathcal{F}$ is the result of the same composition process.

Definition 2: Given a set $\mathcal{F}$, the *learning environment* $\mathcal{E}$, from which we extract the training and the test set, is defined as:

$$\mathcal{E} = \{(s \in \mathcal{F}, h(s)), \forall\, s \in \mathcal{F}\}. \tag{4}$$

5   The Fairy Tale Task

In the present, illustrative task, we defined a grammar which is able to generate very simple sentences like: *Antonella is an ant, Marco doesn't help who isn't an ant* and so on. Therefore, a good grammar over some selected tale constituents is $G = \langle V_T, V_N, P_R, S \rangle$ where:

$$V_T = \{"Antonella", "Fabrizio", "Marco", "ant", "grasshopper",$$
$$"helps", "mokes", "reproaches", "sings", "saves", "works"$$
$$"is", "a", "an", "whoisn't", "doesn't"\}$$
$$V_N = \{N, T, I, P, \neg P, \neg T\}$$
$$P_R = \{R_0 = R_1 \mid R_2 \mid R_3 \mid R_4$$
$$\quad N = Antonella \mid Fabrizio \mid Marco$$
$$\quad T = helps \mid mokes \mid reproaches$$
$$\quad I = sings \mid saves \mid works$$
$$\quad P = ("a" \; grasshopper \mid "an" \; ant)$$
$$\quad \neg P = "whoisn't" \; P$$
$$\quad \neg T = "doesn't" \; T$$
$$\quad R_1 = NP \mid N \neg P \mid NTP \mid NT \neg P \mid NTN \mid PTP \mid PT \neg P \mid PTN$$
$$\quad R_2 = N \neg TP \mid P \neg T \neg P \mid N \neg TN \mid P \neg TP \mid P \neg T \neg P \mid P \neg TN$$
$$\quad R_3 = \neg PTP \mid \neg PT \neg P \mid \neg PTN \mid \neg P \neg TP \mid \neg P \neg T \neg P \mid \neg P \neg TN$$
$$\quad R_4 = PI \mid NI \mid P \neg I \mid N \neg I \mid \neg PI \mid \neg P \neg I\}$$
$$S = R_0$$

Then, we fixed the context model h assigning either *true* or *false* to each $s \in \mathcal{L}(G)$ strictly on the base of the chosen story. Afterwards, we used the definition 1 with the application of the unary boolean operator NOT($\neg$), and the two boolean binary operators AND($\wedge$) and IMPLICATION ($\rightarrow$). We first used the operator NOT to the $\mathcal{L}(G)$, producing a new set $\mathcal{L}'(G) \supset \mathcal{L}(G)$; then we applied just once the binary operators to $\mathcal{L}'(G)$ yielding another new set $\mathcal{L}''(G) \supset \mathcal{L}'(G)$. Finally, we built the learning environment $\mathcal{E}$, simply dividing all pairs $\{s \in \mathcal{L}''(G), h(s)\}$ into training and test sets.

6   Results

In order to prove that a trained neural network is able to make inference on the knowledge base supplied, we explicitly moved from the training set to the test set some elementary facts (e.g. *Fabrizio is a grasshopper, Fabrizio is not an ant*), that can be anyhow extrapolated by others.

On the training set we made a simulation using a simple BPTS algorithm (see [5]) with an adaptive learning rate, a momentum factor fixed to 0.1, 2000 epochs and a variable number of trees as training set. We used an encoding neural network with two layers, three neurons representing the internal state, and 5 neurons representing all grammar elements and operators codified with binary labels of 5 bit. Moreover, we defined a one-layer output neural network with one linear output neuron corresponding to the question answer (TRUE or FALSE).

Results shown in Table 1 are the relevant average results on 30 experiments with a test set of 1500 questions, and are obtained counting correct answers yielded by the neural network. It can be observed that the percentage of success increases with the dimension of the trees in the knowledge base. Moreover, the answers to the elementary facts removed from the training set are always correct.

Table 1. Average results on increasing-size learning sets.

Trees in the learning set	Percentage of success
1	49.13 %
500	90.78 %
1000	91.65 %
1500	91.80 %

7 Conclusions

We discussed a novel application of recursive neural networks to the computational treatment of natural language, relying on the DRT semantic theory. Unlike other connectionist models, which process either flat representations or syntactic constituents forecast, recursive neural networks fully exploit the data coming from the DRT. Preliminary results on a simple illustrative task, rooted in fairy tales domain, were shown. On-going research is focused on (i) the extension of the technique, e.g., the modeling of temporal relations within a given discourse, and (ii) its application to a broader class of natural language processing problems.

References

1. Frasconi, P., Gori, M., Sperduti, A., A general framework for adaptive processing of data structures, IEEE Transaction on Neural Networks, vol. 9, pp. 768–786, Sept. 1998.
2. Fodor, J. A., The Language of Thought. Language and Thought. New York, Crowell(NY), 1975.
3. Cleeremans, A., Mechanisms of implicit Learning. MIT press, 1993.
4. Elman, J.L., Finding structure in time. Cognitive Science, vol. 14, pp. 179-211, 1990
5. Goller, C., Küchler, A., Learning task-dependent distributed structure-representations by Back-propagation Through Structure. IEEE International Conference on Neural Networks, pp. 347-352, 1996.
6. Kamp, H., Reyle, U., From Discourse to Logic. Vol. I. Kluwer, Dordrecht, 1993.
7. Partee, B., Montague Grammar and Transformational Grammar. Linguistic Inquiry vol. 6:2, pp. 203-300, 1975.
8. Sperduti, A., Starita, A., Supervised neural networks for the classification of structures, IEEE Transaction on Neural Networks, vol. 8, pp. 714–735, May 1997.

Recurrent Neural Learning for Helpdesk Call Routing*

Sheila Garfield and Stefan Wermter

University of Sunderland, Sunderland, SR6 0DD, United Kingdom
{stefan.wermter,sheila.garfield}@sunderland.ac.uk
http://www.his.sunderland.ac.uk/

Abstract. In the past, recurrent networks have been used mainly in neurocognitive or psycholinguistically oriented approaches of language processing. Here we examine recurrent neural networks for their potential in a difficult spoken language classification task. This paper describes an approach to learning classification of recorded operator assistance telephone utterances. We explore simple recurrent networks using a large, unique telecommunication corpus of spontaneous spoken language. Performance of the network indicates that a semantic SRN network is quite useful for learning classification of spontaneous spoken language in a robust manner, which may lead to their use in helpdesk call routing.

1   Introduction

Language is not only extremely complex and powerful but also ambiguous and potentially ill-formed [1]. Problems associated with recognition of this type of speech input can result in errors due to acoustics, speaking style, disfluencies, out-of-vocabulary words, parsing coverage or understanding gaps [5]. Spontaneous speech also includes artifacts such as filled pauses and partial words. Spoken dialogue systems must be able to deal with these as well as other discourse phenomena such as anaphora and ellipsis, and ungrammatical queries [5,9].

In this paper we describe an approach to the classification of recorded operator assistance telephone utterances. In particular, we explore simple recurrent networks. We describe experiments in a real-world scenario utilising a large, unique corpus of spontaneous spoken language.

2   Description of the Helpdesk Corpus

Our task is to learn to classify real incoming telephone utterances into a set of service level classes. For this task a corpus from transcriptions of 4 000 recorded operator assistance telephone calls was used [2]. The utterances range from simple direct requests for services to more descriptive narrative requests for help as shown by the following examples:

---

* The authors thank Mark Farrell and David Attwater of BTexact Technologies for their helpful discussions.

J.R. Dorronsoro (Ed.): ICANN 2002, LNCS 2415, pp. 296–301, 2002.
© Springer-Verlag Berlin Heidelberg 2002

1. *"could I um the er international directory enquiries please"*
2. *"can I have an early morning call please"*
3. *"could you possibly give me er a ring back please I just moved my phone I want to know if the bell's working alright"*

Examination of the utterances reveals that the callers use a wide range of language to express their problem, enquiry or to request assistance [3].

2.1   Call Transcription

The focus of the investigation was a corpus from transcriptions of the first utterances of callers to the operator service. Analysis of the utterances identified a number of service levels or call class categories, primary move types and request types [2]. The primary move is a subset of the first utterance and is like a dialogue act and gives an indication of which dialogue act is likely to follow the current utterance.

Four separate call sets of about 1 000 utterances each were used in this study. The call sets are split so that about 80% of utterances are used for training and approximately 20% of utterances used for testing. The average length of an utterance in the training set is 16.05 words and in the test set the average length of an utterance is 15.52 words. Each call class is represented in the training and test set. An illustrative example is given in Table 1, however not all call classes are shown. The part of the utterance identified as the primary move was used for both the training and test sets. At this stage some utterances were excluded from the training and test sets because they did not contain a primary move utterance.

Table 1. Breakdown of utterances in training and test sets from call set 1. Note: For illustration purposes not all classes are shown

917 utterances									
Total of 712 utterances in Training set									
Total of 205 utterances in Test set									

Categories:	class 1	class 2	class 3	class 4	class 5	class 6	class 7	class 8	class 9	class n
in train set:	261	11	41	3	85	32	6	16	28	...
in test set:	59	3	21	1	29	11	2	4	7	...

The number of call class categories in the task is 17 and call set 1 has an entropy of 3.2.

$$entropy = \sum_{i=1}^{17} P(c_i) log_2(P(c_i)) \tag{1}$$

2.2   Vectors for Semantic SRN Network

The experiments use a semantic vector representation of the words in a lexicon [8]. These vectors represent the frequency of a particular word occurring in a call class category and are independent of the number of examples observed in each category. The number of calls in a class can vary substantially. Therefore we *call normalize* the frequency of a word w in class c_i according to the number of calls in c_i (2). A *value* $v(w, c_i)$ is computed for each element of the semantic vector as the *normalized* frequency of occurrences of word w in semantic class c_i, divided by the *normalized* frequency of occurrences of word w in all classes. That is:

$$Norm.\ freq.\ of\ w\ in\ c_i = \frac{Freq.\ of\ w\ in\ c_i}{Number\ of\ calls\ in\ c_i} \quad (2)$$

where:

$$v(w, c_i) = \frac{Norm.\ freq.\ of\ w\ in\ c_i}{\sum_j Norm.\ freq.\ for\ w\ in\ c_j}, \ j \in \{1, \cdots n\} \quad (3)$$

Each call class is represented in the semantic vector. An illustrative example is given in Table 2, however not all call classes are shown. As can be seen in the illustrative example, domain-independent words like 'can' and 'to' have fairly even distributions while domain-dependent words like 'check' and 'order' have more specific preferences.

Table 2. Example of semantic vectors. Note: For illustration purposes not all classes are shown. There are 17 classes used in this study

Word	Call Class									
	class 1	class 2	class 3	class 4	class 5	class 6	class 7	class 8	class 9	class n
CAN	0.03	0.06	0.09	0.07	0.05	0.08	0.04	0.04	0.04	...
YOU	0.07	0.01	0.01	0.05	0.09	0.10	0.05	0.06	0.05	...
JUST	0.18	0.00	0.00	0.00	0.03	0.03	0.00	0.07	0.00	...
CHECK	0.53	0.00	0.00	0.00	0.00	0.00	0.11	0.00	0.06	...
TO	0.08	0.06	0.01	0.05	0.02	0.01	0.05	0.08	0.13	...
SEE	0.36	0.00	0.00	0.00	0.00	0.00	0.00	0.00	0.00	...
IF	0.12	0.08	0.02	0.00	0.06	0.10	0.05	0.08	0.03	...
IT	0.19	0.00	0.00	0.06	0.04	0.06	0.01	0.04	0.09	...
OUT	0.10	0.00	0.00	0.00	0.02	0.12	0.00	0.12	0.07	...
OF	0.09	0.00	0.00	0.11	0.14	0.01	0.03	0.00	0.06	...
ORDER	0.24	0.00	0.54	0.00	0.00	0.00	0.00	0.00	0.00	...

3   Learning and Experiments

A semantic SRN network with input, output, hidden and context layers was used for the experiments. Supervised learning techniques were used for training [4,7].

The input to a hidden layer L_n is constrained by the underlying layer L_{n-1} as well as the incremental context layer C_n. The activation of a unit $L_{ni}(t)$ at time t is computed on the basis of the weighted activation of the units in the previous layer $L_{(n-1)i}(t)$ and the units in the current context of this layer $C_{ni}(t)$ limited by the logistic function f.

$$L_{ni}(t) = f(\sum_k w_{ki} L_{(n-1)i}(t) + \sum_l w_{li} C_{ni}(t)) \qquad (4)$$

This provides a simple form of recurrence that can be used to train networks to perform sequential tasks over time. Consequently, the output of the network not only depends on the input but also on the state of the network at the previous time step; events from the past can be retained and used in current computations. This allows the network to produce complex time-varying outputs in response to simple static input which is important when generating complex behaviour. As a result the addition of recurrent connections can improve the performance of a network and provide the facility for temporal processing.

3.1   Training Environment

In one epoch, or cycle of training through all training samples, the network is presented with all utterances from the training set and the weights are adjusted at the end of each utterance. The input layer has one input for each call class category. During training and test utterances are presented sequentially to the network one word at a time. Each input receives the value of $v(w, c_i)$, where c_i denotes the particular class which the input is associated with. Utterances are presented to the network as a sequence of word input and category output representations, one pair for each word. Each unit in the output layer corresponds to a particular call class category. At the beginning of each new sequence the context layers are cleared and initialised with 0 values. The output unit that represents the desired call class category is set to 1 and all other output units are set to 0. An utterance is defined as being *classified* to a particular call class category if at the end of the sequence the value of the output unit is higher than 0.5 for the required category. This output classification is used to compute the recall and precision values for each utterance. These values are also used to compute the recall and precision rates for each call class category as well as the overall rates for the training and test sets.

The network was trained for 1 000 epochs on the training transcribed utterances using a fixed momentum term and a changing learning rate. The initial learning rate was 0.01, this changed at 600 epochs to 0.006 and then again at 800 epochs to 0.001. The results for this series of experiments are shown in Table 3.

3.2   Recall, Precision, and F-Score

The performance of the trained network in terms of recall, precision and F-score on the four call sets is shown in Table 3. Recall and precision are common evaluation metrics [6]. The F-score is a combination of the precision and recall rates and is a method for calculating a value without bias, that is, without favouring either recall or precision. There is a difference of 3.54% and 5.11% between the highest and the lowest test recall and precision rates respectively.

Table 3. Overall results for the semantic SRN network using semantic vectors

	Training Set			Test Set		
	Recall	Precision	F-Score	Recall	Precision	F-Score
Call Set 1:	85.53%	93.84%	89.49	79.02%	90.50%	84.37
Call Set 2:	84.26%	92.76%	88.31	75.48%	87.22%	80.93
Call Set 3:	87.06%	93.72%	90.27	76.00%	85.39%	80.42
Call Set 4:	85.47%	93.15%	89.14	76.38%	85.39%	80.63

4   Analysis of Neural Network Performance

The focus of this work is the classification of utterances to service levels using a semantic SRN network. In general, the recall and precision rates for the semantic SRN network are quite high given the number of service levels available against which each utterance can be classified and the ill-formed input. The semantic SRN network achieved an average test recall performance of over 76% of all utterances. This result is calculated based on the overall performance figures for the semantic SRN network shown in Table 3.

In other related work on text classification [8] news titles were used to classify a news story as one of 8 categories. A news title contains on average about 8 words. As a comparison, the average length of the first caller utterance is 16.44 words and is subject to more ambiguity and noise. On the other hand, the size of the vocabulary used in the text classification task was larger than that used for our classification of call class categories. The performance of the simple recurrent network is significant when this factor is taken into consideration because a larger vocabulary provides more opportunity for the network to learn and therefore generalise on unseen examples. While on an 8 category *text* classification task we reached about 90%, in this study presented in this paper here for a much more ill-formed *spoken language* classification task and 17 categories we reached above 75% (recall) and 85% (precision) for unseen examples.

We have compared our results also with a feedforward network without recurrent connections. Recurrent networks performed significantly better. The better performance of the SRN network shows that the network does make use of the memory introduced by the context layer to improve both its recall and precision rates. This shows that the information stored in the context layer, which is passed back to the first hidden layer, does assist the network in assigning the correct category to an utterance.

5 Conclusions

In conclusion the main aim of this research is to identify indicators about useful semantic SRN architectures that can be developed in the context of a larger hybrid symbolic/neural system for helpdesk automation. A description has been given of a recurrent neural architecture, the underlying principles and an initial evaluation of the approach for classifying the service level of operator assistance telephone utterances. The main result from this work is that the performance of the SRN network is quite good when factors such as noise in the utterance and the number of classes are taken into consideration. This work makes a novel contribution to the field of robust learning classification using a large, unique corpus of spontaneous spoken language. From the perspective of connectionist networks it has been demonstrated that a connectionist network, in particular a semantic SRN network, can be used under *real-world* constraints for spoken language analysis.

Acknowledgments. This research has been partially supported by the University of Sunderland and BTexact Technologies under agreement ML846657.

References

[1] Abney, S.: Statistical Methods and Linguistics. In: Klavans, J., and Resnik, P. (eds): The Balancing Act. MIT Press, Cambridge, MA (1996)

[2] Durston, P.J., Kuo, J.J.K., et al.: OASIS Natural Language Call Steering Trial. Proceedings of Eurospeech, Vol 2. (2001) 1323–1326

[3] Edgington, M., Attwater, D., Durston, P.J.: OASIS - A Framework for Spoken Language Call Steering. Proceedings of Eurospeech '99. (1999)

[4] Elman, J.L., Bates, E.A., Johnson, M.H., Karmiloff-Smith, A., Parisi, D., Plunkett, K.: Rethinking Innateness. MIT Press, Cambridge, MA (1996)

[5] Glass, J.R.: Challenges for Spoken Dialogue Systems. Proceedings of IEEE ASRU Workshop. Keystone, CO (1999)

[6] Salton, G., McGill, M.: Introduction to Modern Information Retrieval. McGraw Hill, New York (1983)

[7] Wermter, S.: Hybrid Connectionist Natural Language Processing. Chapman and Hall, Thomson International, London, UK (1995)

[8] Wermter, S., Panchev, C., Arevian, G.: Hybrid Neural Plausibility Networks for News Agents. Proceedings of the National Conference on Artificial Intelligence. Orlando, USA (1999) 93–98

[9] Clark, H.: Speaking in Time. Speech Communication 36 (2002) 5–13

An Approach to Encode Multilayer Perceptrons

Jerzy Korczak and Emmanuel Blindauer

Laboratoire des Sciences de l'Image, de l'Informatique et de la Télédétection, CNRS, Université Louis Pasteur, 7, Bld S. Brant, 67400 Illkirch, France.
{jjk,blindauer}@lsiit.u-strasbg.fr

Abstract. Genetic connectionism is based on the integration of evolution and neural network learning within one system. An overview of the Multilayer Perceptron encoding schemes is presented. A new approach is shown and tested on various case studies. The proposed genetic search not only optimizes the network topology but shortens the training time. There is no doubt that genetic algorithms can be used to solve efficiently the problem of network optimization considering not only static aspects of network architecture but also dynamic ones.

1   Introduction

One of the major domains of application on genetic algorithms (GA) is searching in a large space for good solutions and the optimization of hard problems [1], [4], [6], [14]. Compared to other methods, GA is perfectly capable to explore discontinued spaces of solutions. The space exploration is guided by a fitness function, and enables to evolve the population in parallel, rather than focusing on a single best solution. Among many types of network models, Multilayer Perceptron networks are the most common, not only because of their universality but also because of their good performance [12], [17]. It is known that the efficiency of learning and the quality of generalization are strongly related to the neural network topology. The number of neurons, their organization in layers, as well as their connection scheme have a considerable influence on network learning, and on the capacity for generalization [5], [18].

2   Network Representation and Encoding Schemes

The most common practice in neural network representation is to store only a network topology, but one may also encode the initialization parameters, connection weights, transfer functions, learning coefficients, etc. The more parameters are encoded, the bigger is the search space. In consequence, taking too many parameters will at least slow down the convergence of the genetic process. This excess cost will decrease the quality of solutions, especially if the inserted parameters are irrelevant.

The main goal of an encoding scheme is to represent neural networks in a population as a collection of chromosomes. There are many approaches to

J.R. Dorronsoro (Ed.): ICANN 2002, LNCS 2415, pp. 302–307, 2002.

genetic representation of neural networks [3], [8], [9], [10], [13], [16]. A perfect network coding scheme does not exist, because of a number of requirements that should be fulfilled by a good encoding scheme. A detailed discussion of all such requirements may be found in [2], [16].

To introduce the problem of encoding, let us consider one of most common methods used to encode networks. The direct encoding scheme consists of mapping the network structure onto a binary connection matrix which determines whether a connection between two neurons exists or not. The complete network structure is represented by the list of neurons with their incoming connections

Besides the direct encoding scheme many other interesting methods exist. Several authors proposed an encoding scheme based on grammar rules. Kitano [10] developed a method founded on rewriting rules. A more topology-oriented method is proposed by Korczak and Dizdarevic [3], [11].

Classical methods for encoding neural network use to encode the network topology into a single string [3], [8], [9], [10], [13], [16]. But frequently, for large-size problems, these methods do not generate satisfactory results, in terms of learning performance and simplicity. We have observed in the experiments that computing new weights to get satisfactory networks is very costly. Even using fast computing methods did not help a lot because thousands of back propagations had to be done on large datasets. After evaluating all networks, crossover and mutations, the population is composed of new individuals, but none of them have inherited from the weights previously computed. So a lot of computing time can be lost here.

Our idea is to keep the networks in the learned state and to transmit the knowledge acquired to their children. In several cases, choosing bad initial weights causes the learning procedure converges to a local minimum and not the global. In consequence, the network is usually considered not efficient. Therefore in the genetic process, we should reduce the fact that some networks may disappear from the potential search space, due to bad starting points.

Encoding a phenotype in a string is also an essential point in the evolution. Operators, such as mutation or crossover, are very dependent on the genotype. It should be ensuring that splitting a string implies an interesting cut in the network. Taking into consideration the above mentionned problems, a new encoding method based on the matrix encoding is proposed.

Let n be a maximal number of neurons who can be used in all networks. The network can be represented by a matrix $M \in M_{p,p}(\mathbb{R})$ where p is the number of neurons in the current network. $M_{i,j}$ equals to zero if there is no connection between the neuron i and the neuron j, else $M_{i,j}$ represents the weight of the connection.

The matrix is symmetrical, thus it is only needed to store the right superiors. Layers in the neural network are visible in the matrix as blocks of zeros over the diagonal. Using this encoding scheme, any network can be represented in a unique way assumed that neurons are sorted by weight within a layer. The matrix representation can be used as the genotype of the network. This encoding scheme respects all properties proposed by [8], [16] for evaluating the quality of an encoding: completeness, closure, proximity, short schemata, compactness, non isomorphism and modularity. These properties do not claim to be complete, and more criteria may be proposed. But these properties can be considered as the minimal set of requirements for a genetic encoding to be efficient.

Two operators for a genotype have been proposed: a crossover operator, and a mutation operator. For the crossover operation, a new matrix $M^{1,0}$ is created from two other matrices $M^{0,0}$ and $M^{0,1}$. $M^{1,0}$ is created by extracting l_1 columns from $M^{0,0}$ and $p - l_1$ columns from $M^{0,1}$. In result, the network is splitted vertically into two parts, and the offspring get two different parts, one from each parent. A more generalized crossover could be an exchange of a sub-matrix between the parents. For the mutation, a few random values in the matrix are changed. This operation can create or remove several connections, depending on the mutation coefficient. So are weights modified during the mutation. Thus, all solutions in the search space are available by the genetic algorithm.

3   Weight Initialization and Updating

With this algorithm, the weights are included within the genetic encoding. Thus, after the learning stage, trained networks are available, and the weights are reused in the next population. In the evolution process, after crossover and mutation, resulting networks have weights inherited from their parents. It can be expected that the knowlegde issued from the parents would help to speed up the convergence of the gradient based algorithms.

Gradient-based methods look for the best gradient in the surface of error. It should be pointed out that the weight adapting only occurs when weight is changed. When no connection exists between two neurons, current methods based on gradient do not create the connection, because the search space from the weights $\mathbb{R}^p$ is included in the space of all weight available $\mathbb{R}^n$. Thus, we have the inclusion $\mathbb{R}^p \subset \mathbb{R}^n$. The computed gradient is done in the $\mathbb{R}^p$ space, and thus it indicates the best slope. But nothing can guarantee that in the $\mathbb{R}^n$ space does not exist another vector with a better slope: when the slope of the gradiant in $\mathbb{R}^p$ is close to zero, it is considered to be close to a minimum. But, computing the gradiant in $\mathbb{R}^n$, we could get another vector indicating the best slope allowing to converge faster to the minimum. Below, an algorithm to find the best slope and to determine qualitativly the importance of connections between neurons is proposed.

The connection weights are computed in the following way:

Let $\bigcup_{i=1}^{q} \bigcup_{k=1}^{s} \{(X_i, S_k(X_i))\}$ be the set of data, where q is the number of elements, S_k is the projector onto the k-th element, and s the number of outputs. $R_k(\omega, X_i)$ is the k-th output of the network. So the total quadratic error, on all of the dataset, is computed according to the formula:

$$E = \sum_{i=1}^{q} \sum_{k=1}^{s} [R_k(\omega, X_i) - S_k(X_i)]^2$$

thus

$$\frac{\partial E}{\partial \omega} = 2. \sum_{i=1}^{q} \sum_{k=1}^{s} [R_k(\omega, X_i) - S_k(X_i)] \frac{\partial R}{\partial \omega}(\omega, X_i)$$

To compute $\dfrac{\partial R}{\partial \omega}$, the Jacobian matrix from E is computed. The term $\dfrac{\partial R}{\partial \omega}(\omega, X_i)$ can be evaluated as:

- if only one path drives from ω to the output, then $\dfrac{\partial R}{\partial \omega}(\omega, X_i) = z. \displaystyle\prod_{j}(\omega_j.f'(z_j))$, where z is the incoming value in the connection, $\prod \omega_j.f'(z_j)$ the product of all weight and the derived number from the the activation function on the path to R.
- if several paths drive from ω to the output, we just have to add each computed value for each path, as previously.

Example: A four neurons network is presented in Fig.1, where f is the activation function, x and y the input values, and A, B, C, D are values after the activation of respective neurons, $a, b, \dots m$ are the weights. R is the output value.
The values of $\dfrac{\partial R}{\partial a}$, and $\dfrac{\partial R}{\partial c}$ are computed as follows:

$$
\begin{aligned}
- \frac{\partial R}{\partial a} &= xf'(ax+by).e.f'(e.A+m.D).g.f'(g.(f(e.A+m.D)+h.D)) \\
&= x.A'.e.B'.g.C' \\
- \frac{\partial R}{\partial c} &= x.D'(m.B'.g.C'+h.C')
\end{aligned}
$$

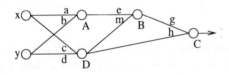

Fig. 1. Computing weights

To compute weights, all dataset is presented to the network, and all partial derivatives are added to construct the Jacobian. Each element in the matrix indicates if a connection is important or not, a large number indicates that the connection has a large impact on the output. Contrary to the normal gradient method, in our algorithm, only one step has to be applied for changing weights. Note that the method changes the topology of the network, because new connections may be created or deleted. But constraints may be defined in order to keep the network topology unchanged.
Using the Jacobian, relevance of the inputs can be take into account. Small values indicates an irrelevant entry. In the same way, we use the Jacobian to

eliminate connection between neurons: with a small value in the Jacobian and a small weight, we can take out the connection and perhaps the neuron, when it has no more input or output.

4   Experimentation

To evaluate the performance of the new encoding scheme, we used several classical problems: xor, parity 3,4,5, Heart, and Sonar. These case studies have been chosen based on the growing complexity of the problem to solve. Each population had 200 individuals. For each individual, 100 epochs were carried out for training. An maximal number of hidden neurons is fixed for each problem. 20 runs were done for each problem, only 4 runs were carried out for heart, because too much time was needed. In the genetic part, the crossover percent is set to 80%, and 5% for the mutation (the mutation rate was decreasing with the number of generation).

The learning was stopped when the best individual reached a test error under 5%, or the number of generation exceeded 500 iterations (in this case, the last test error is written in brackets). In the experiments, the fitness function was a non-linear combination of the learning error, the test error, the topology of the network, and of the number of generation. Compared with other results

Table 1. Results of experimentations

	XOR	Parity 3	Parity 4	Parity 5	Heart	Sonar
Number of hidden neurons	2	3	5	8	12	30
Number of connections	6	11	23	38	354	1182
Number of epochs (error)	13	23	80	244	209 (9%)	120 (13%)

from [7], [15], this new method has shown the best, not only in term of network complexity, but also in quality of learning (for the Heart problem, the error is only 9%, compared to other methods with at least 15%, and for the Sonar problem, 13% against 30%).

5   Conclusion

In this paper, application of genetic algorithms for the topological optimization of Multilayer Perceptron is presented. This approach is more flexible to the design of different topologies than heuristic approaches.

The new encoding scheme improves the efficiency of evolutionary process, because the weights of the neural network have been included in the genetic encoding scheme. The experiments have confirmed that firstly by encoding the network topology and weights the search space is affined; secondly, by the inheritence of connection weights, the learning stage is speeded up considerably. The presented method generates efficient networks in a shorter time compared to actual methods. The evolution process improved the quality of network populations gradually and discovered near-optimal solutions. The genetic connnectionism approach has been shown as a robust optimization method with a large spectrum of applications. Of course, the scope of GA is restricted to those problems only where it is possible to encode the set of networks as chromosomes, and where a fitness function may be defined.

References

1. R.F. Albrecht, C. R. Reeves and N. C. Steele (eds), *Artificial Neural Nets and Genetic Algorithms*, Springer Verlag, 1993.
2. E. Blindauer, *Méthodes d'encodage génétique de réseaux neuronaux*, MSc thesis in Computer Science, Louis Pasteur University, Strasbourg, 1998.
3. E. Dizdarevic, *Optimisation génétique de réseaux*, MSc thesis in Computer Science, Louis Pasteur University, Strasbourg, 1995.
4. L. Eshelman, Proc. of the Sixth Intern. Conf. on *Genetic Algorithms and Their Applications*, Morgan Kaufmann, 1995.
5. D.B. Fogel, L.J. Fogel and V.M. Porto, Evolving Neural Networks, *Biological Cybernetics* 63: 487–493, 1990.
6. D.E. Goldberg, *Genetic Algorithms in Search, Optimization and Machine Learning*, Addison-Wesley, 1989.
7. M.A. Grönroos, *Evolutionary Design Neural Networks*, PhD thesis, Department of Mathematical Sciences, University of Turku, 1998.
8. F. Gruau, *Neural Networks Synthesis using Cellular Encoding and the Genetic Algorithm*, PhD thesis, LIP, Ecole Normale Superieure, Lyon, 1992.
9. F. Gruau, Genetic Synthesis of Modular Neural Networks, [in] S. Forrest, (eds), *Genetic Algorithms: Proc. of the 5th International Conference*, M. Kaufman, 1993.
10. H. Kitano, Designing Neural Networks using Genetic Algorithms with Graph Generation System, *Complex Systems*, 4: 461–476, 1990.
11. J. Korczak and E. Dizdarevic, *Genetic Search for Optimal Neural Networks*, [in] Summer School on Neural Network Applications to Signal Processing, Kule, 1997.
12. Y. Le Cun, *Modeles Connexionnistes de l'Apprentissage*. PhD thesis, Paris, 1987.
13. M. Mandischer, Representation and Evolution of Neural Networks, [in] R.F. Albrecht, C.R. Reeves and U.C. Steele (eds), *Artificial Neural Nets and Genetic Algorithms*, pp. 643–649, 1993.
14. Z. Michalewicz, *Genetic Algorithms+Data Structures=Evolution Programs*, 3rd ed. Springer, Berlin, 1996.
15. J.C.F. Pujol, R. Poli, *Evolving the Architecture and Weights of Neural Networks Using a Weight Mapping Approach*, School of Computer Science, Birgmingham, 1999.
16. R. Salustowicz, *A Genetic Algorithm for the Topological Optimization of Neural Networks*, Diplomarbeit TU Berlin, 1995.
17. D.M. Skapura, *Building Neural Networks*, Addison-Wesley, 1996.
18. X. Yao, *Evolving Artificial Neural Networks*, Proc. of the IEEE, 87(9):1423–1447, 1999.

Dynamic Knowledge Representation in Connectionist Systems

J. Mark Bishop, Slawomir J. Nasuto, and Kris De Meyer

Department of Cybernetics, University of Reading, Whiteknights,
PO Box 225, Reading, RG6 6AY, United Kingdom
J.M.Bishop@Reading.ac.uk

Abstract. One of the most pervading concepts underlying computational models of information processing in the brain is linear input integration of rate coded uni-variate information by neurons. After a suitable learning process this results in neuronal structures that statically represent knowledge as a vector of real valued synaptic weights. Although this general framework has contributed to the many successes of connectionism, in this paper we argue that for all but the most basic of cognitive processes, a more complex, multi-variate dynamic neural coding mechanism is required - knowledge should not be spacially bound to a particular neuron or group of neurons. We conclude the paper with discussion of a simple experiment that illustrates dynamic knowledge representation in a spiking neuron connectionist system.

1 Introduction

"Nothing seems more certain to me than that people someday will come to the definite opinion that there is no copy in the ... nervous system which corresponds to a particular thought, or a particular idea, or memory", (Wittgenstein, 1948).

Over the hundred years since the publication of James' Psychology [8], neuroscientists have attempted to define the fundamental features of the brain and its information processing capabilities in terms of mean firing rates at points in the brain cortex' (neurons) and computations. After Hubel and Wiesel [7], the function of the neuron as a specialized feature detector was treated as established doctrine. From this followed the functional specialization paradigm, mapping different areas of the brain to specific cognitive function, reincarnating an era of modern phrenology.

Connectionism mapped well onto the above assumptions. Its emergence is based on the belief that neurons can be treated as simple computational devices. The initial boolean McCulloch-Pitts model neuron [10] was quickly extended to allow for analogue computations. Further, the assumption that information is encoded in the mean firing rate of neurons was a central premise of all the sciences related to brain modelling.

Over the last half century such 'classical' connectionist networks have attracted significant interest. They are routinely applied to engineering problems

J.R. Dorronsoro (Ed.): ICANN 2002, LNCS 2415, pp. 308–313, 2002.

[19], and as metaphors of concepts drawn from neuroscience, have also been offered as models of both high [16] and low level [2] [21] cognition. However the classical connectionist models of high level cognition have also been strongly criticized [5] and the situation at the domain of low level neural modelling is little better [1].

More recently spiking neuron, pulsed neural networks have begun to attract attention [9]. Such models no longer aggregate individual action potentials as a mean firing rate but act on temporal sequences of spikes. Like the classical connectionist models spiking neuron neural networks have been studied both for their computational/engineering properties and as models of neurological processes [20].

Although temporal coding of spike trains lends itself more readily to multivariate information encoding than rate encoding, both are typically discussed in a uni-variate framework - an observation that also applies to classicial connectionist frameworks. However uni-variate knowledge representation is limited to the representation of arity zero predicates and, following Dinsmore [5] and Fodor [6], we suggest that this is too strong a restriction for representing the complexity of the real world.

2   Types, Tokens, and Arity Zero Predicates

An arity zero predicate is one without an argument, eg. representation of the class of Morgan cars by the arity zero predicate, MORGAN (). Such predicates easily express different 'types' of entities. eg. A car producer may produce half a dozen 'types' of cars in a year, (here 'types' equates to the different models marketed such as the Morgan 4/4 car), but manufacture many thousand individual cars for sale ('tokens'). Knowledge of individual 'tokens', in this case individual cars, is more clumsily expressed in a predicate of arity zero. eg. To represent a particular Morgan car (eg. registration YUY405W), the arity zero predicate MORGANYUY405W () is necessary.

Although a conventional connectionist network can represent 'type' knowledge, of the form 'Morgan 4/4', by the activation of a single processing node or group of nodes, because it processes uni-variate information it can only easily instantiate tokens in a similar manner (eg. by an activation on a particular node or group of nodes). A more elegant method of representing the specific member of a class (a token) is by the use of the arity one predicate CLASS (INDIVIDUAL). However, in general this requires the use of bi-variate information to identify both CLASS and INDIVIDUAL.

However, we do not consider representations of arity zero predicates as sufficient for representation of many complex relationships. Such limitations make it difficult to interpret and analyze the network in terms of causal relationships. In particular, (cf. classical symbolic/connectionist divide), it is difficult to imagine how such a system could develop symbolic representations and quantified logical inference [17]. Such deficiencies in the representation of complex knowledge by classical neural networks have long been recognized [18] [6] [3] [15].

3   Spiking Neurons

Taking into account the above considerations we propose to investigate a spiking neuron connectionist architecture whose constituent neurons inherently operate on rich (bi-variate) information encoded in spike trains, rather than as a simple mean firing rate. NESTER, a network of such neurons, was first proposed in [13] and is further investigated herein. The task of NESTER is to locate an object (memory) projected onto an artificial retina.

The NEural STochastic diffusion search nEtwoRk (NESTER) consists of an artificial retina, a layer of fully connected matching neurons and retino-topically organized memory neurons. The bi-variate information output from retina/memory cells is encoded as a spike train consisting of two qualitatively different parts: a tag determined by its relative position on the retina/memory and a tag encoding the feature signalled by the cell. This information is processed by the matching neurons which act as spatiotemporal coincidence detectors.

It is important to note that matching neurons obtain input from both retina and memory and thus their operation is influenced by both bottom-up and top-down information. As Mumford notices [11], systems which depend on interaction between feedforward and feedback loops are quite distinct from models based on Marr's feedforward theory of vision.

Thus matching neurons are fully connected to both retina and memory neurons and accept for processing new information, contingent on their internal state (defined by the previously accepted spike train).

Each matching neuron maintains an internal representation (a hypothesis) defining a potential location of the memory on the retina and in operation simply conjoins the positional tags of the incoming spike trains from the retina/memory, (corresponding to their retinotropic positions), with its own hypothesis and, dependent upon the result, distributes its successful or unsuccessful hypothesis to other matching neurons.

Effectively NESTER is a connectionist implementation of Stochastic Diffusion Search, (SDS) [4], a simple matching algorithm whose operation depends on co-operation and competition in a population of agents which are realised in NESTER as the matching neurons. Therefore, in the next section we will describe the network operation in terms of the simpler underlying generic mechanism of SDS.

4   Stochastic Diffusion Search

In SDS a group of independent agents processes information from the search space in order to find the best-fit to a specified target pattern. Each agent searches for a micro-feature of the target and once found competes to attract other agents to evaluate this position. In this way the SDS explores the whole search space. Due to the emergent co-operation of agents pointing to the same solution, interesting areas in the search space (those that share many micro-features with the target) are more thoroughly exploited than background areas.

Agents are divided into two classes: active and inactive. An active agent has successfully found a micro-feature from the target in the search space; an inactive agent has not. Thus, the activity label identifies agents more likely to point to an instantiation of the target than to the background. Inactive agents utilise this activity information when deciding whether to communicate with a randomly selected agent in a subsequent phase of processing. Communication only occurs if the selected agent is active and results in the flow of 'where' information from the active agent to the inactive one. Conversely, if the selected agent is also inactive, then there is no information flow between agents; instead, a new random 'where' position is adopted. In this way active agents attract more resources to examining promising regions of the search space.

5   Experiment Using NESTER

NESTER was configured with 100 matching neurons, 6 memory neurons and 90 retina neurons. The content of the target memory is defined by 6 symbols from the ASCII character set and the retina by 90 symbols.

The following experiment illustrates NESTER utilizing a dynamic assembly encoding as it locates the best fit of the target memory on the retina. Finding this on the retina causes the onset of time locked activity in an assembly of matching neurons, resulting in a characteristic frequency spectrum of their spike trains.

In this experiment two patterns with 5/6 correct symbols (ie. 16.6% noise) were projected at different locations onto the retina. Matching neurons first located one pattern and formed its representation - a dynamic assembly of neurons with time locked activity. Figure 1 shows matching neuron activity against time while locked to the first pattern. The activity is periodic, as indicated by the dominant frequency in the power spectrum shown alongside. After some time a

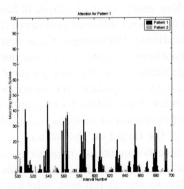
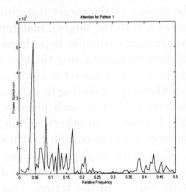

Fig. 1. Network activity and its frequency spectrum while locked to the first pattern

second assembly of matching neurons emerges reflecting the presence of the second pattern at a different location on the retina. The activity of a newly formed assembly corresponding to the second pattern and the resulting power spectrum are shown in Figure 2. It is clear that this spectrum is very different from that shown in Figure 1 although the patterns constitute equal instantiations of the memory.

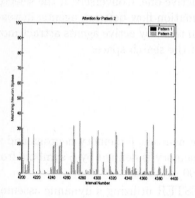 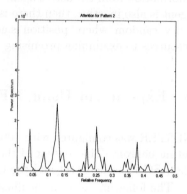

Fig. 2. Network activity and its frequency spectrum while locked to the second pattern

6   Conclusions

NESTER uses dynamic assembly knowledge encoding. The experiment illustrates that locating the target memory on the retina results in the onset of time locked activity in an assembly of matching neurons, as indicated by the characteristic frequency spectrum of matching neuron spiking. Dependent on the quality of the target instantiation (i.e. how many micro-features it has in common with the target), matching neurons will spend different periods of time maintaining particular hypotheses (retinal locations). On average, those matching neurons maintaining the best location hypothesis will spend a longer period examining the same retinal location than matching neurons with a poor location hypothesis (i.e. pointing to areas of the retina with few symbols in common with the target). Hence, such neurons will have more possibilities to communicate their hypothesis to others, and in this way a population of neurons will rapidly converge onto the current best instantiation of the target on the retina.

Continuing exploration of the retina by inactive matching neurons ensures that this process will eventually discover, and converge to, the best-possible fit of the target on the retina. This convergence to the global-best solution occurs because NESTER implements Stochastic Diffusion Search. This is formally demonstrated in [14] and the time complexity discussed in Nasuto et al. [12].

References

1. Abbot, L.: Learning in Neural Network Memories, Network: Computation in Neural Systems **1**, (1990) 105–122
2. Ahmad, S.: VISIT: An efficient computational model of human visual attention, PhD Thesis, University of Illinois, USA (1991)
3. Barnden, J., Pollack, J. (eds.): High-Level Connectionist Models, Ablex, Norwood NJ, USA (1990)
4. Bishop, J.M.: Stochastic Searching Networks, Proc. 1st IEE Int. Conf. on Artificial Neural Networks, London, UK (1989) 329–331
5. Dinsmore, J.: Symbolism and Connectionism: a difficult marriage, Technical Report **TR 90-8**, Southern University of Carbondale, USA (1990)
6. Fodor, J.A., Pylyshyn, Z.W.: Connectionism and Cognitive Architecture: a critical analysis, Cognition **28**, (1988) 3–72
7. Hubel, D.H., Wiesel, T.N.: Receptive fields, binocular interaction and functional architecture in the cat's visual cortex, Journal of Physiology **160** (1962) 106–154
8. James, W.: Psychology (briefer course), Holt, New York, USA (1891)
9. Maass, Bishop, C. (eds.): Pulsed Neural Networks, The MIT Press, Cambridge MA, USA (1999)
10. McCulloch, W.S., Pitts, W.: A logical calculus immanent in nervous activity, Bulletin of Mathematical Biophysics **5** (1943) 115–133
11. Mumford, D.: Neural Architectures for Pattern-theoretic Problems. In: Koch, Ch., Davies, J.L. (eds.): Large Scale Neuronal Theories of the Brain. The MIT Press, London, England (1994)
12. Nasuto, S.J., Bishop, J.M., Lauria, S.: Time Complexity Analysis of the Stochastic Diffusion Search, Neural Computation '98, Vienna, Austria (1998)
13. Nasuto, S.J., Bishop, J.M.: Neural Stochastic Diffusion Search Network - a theoretical solution to the binding problem, Proc. ASSC2, Bremen, Germany (1988) 19–20
14. Nasuto, S.J., Bishop, J.M.: Convergence of the Stochastic Diffusion Search, Parallel Algorithms **14:2**, (1999) 89–107.
15. Pinker, S., Prince, A.: On Language and Connectionism: Analysis of a Parallel Distributed Processing Model of Language Acquisition. In: Pinker, S., Mahler, J. (eds.): Connections and Symbols, The MIT Press, Cambridge MA, (1988)
16. Rumelhart, D.E., McClelland, J.L.: Parallel Distributed Processing: Explorations in the Microstructure of Cognition, The MIT Press, Cambridge MA, (1986)
17. Russel, B.: On Denoting, Mind, (1905); repr. in Marsh, R.C. (ed.): Bertrand Russel: Logic and Knowledge. Essays 1901–1950. London, UK (1965)
18. Sejnowski, T.J., Rosenberg, C.R.: Parallel networks that learn to pronounce English text. Complex Systems **1** (1987) 145–168
19. Tarassenko, L.: A Guide to Neural Computing Applications, Arnold, London (1998)
20. Van Gelder, T., Port, R.F.: It's About Time: an overview of the dynamical approach to cognition. In Port, R.F., Van Gelder, T. (eds.): Mind as Motion. The MIT Press, Cambridge MA (1995) 1–45
21. Van de Velde, F.: On the use of computation in modelling behaviour, Network: Computation in Neural Systems **8** (1997) R1-R32
22. Wittgenstein, L.: Last Writings on the Philosophy of Pyschology (vol. 1). Blackwell Oxford, UK (1948)

Generative Capacities of Cellular Automata Codification for Evolution of NN Codification

Germán Gutiérrez, Inés M. Galván, José M. Molina, and Araceli Sanchis

SCALAB. Departamento de Informática. Universidad Carlos III de Madrid.
Avda. Universidad 30, 28911-Leganés (Madrid)-SPAIN.
{ggutierr,igalvan, masm}@inf.uc3m.es,molina@ia.uc3m.es

Abstract. Automatic methods for designing artificial neural nets are desired to avoid the laborious and erratically human expert's job. Evolutionary computation has been used as a search technique to find appropriate NN architectures. Direct and indirect encoding methods are used to codify the net architecture into the chromosome. A reformulation of an indirect encoding method, based on two bi-dimensional cellular automata, and its generative capacity are presented.

1 Introduction

The architecture design is a fundamental step in the successful application of Artificial Neural Networks (ANN), and it is unfortunately still a human experts job. Most of the methods are based on evolutionary computation paradigms, Evolutionary Artificial Neural Networks (EANN). A wide review of using evolutionary techniques to evolve different aspects of neural networks can be find in (Yao, 1999).

The interest of this paper is focused on the design of Feedforward Neural Networks (FNN) architectures using genetic algorithms. There are two main representation approaches for codification of FNN in the chromosome to find the optimal FNN architecture. One, based on the complete representation of all the possible connections, direct encoding, relatively simple and straightforward to implement. But large architectures, for complex tasks, requires much larger chromosomes (Miller et al., 89; Fogel, 1990; Alba et al., 1993). Other based on an indirect representation of the architecture, indirect encoding schemes. Those schemes consists of codifying, not the complete network, but a compact representation of it, avoiding the scalability problem and reducing the length of the genotype. (Kitano, 1990; Gruau, 1992;Molina et al., 2000).

In this work, an indirect constructive encoding scheme, based on cellular automata (Wolfram, 1998), is reformulated. Two bi-dimensional cellular automata are used to generate FNN architectures proposed. It is inspired on the idea that only a few seeds for the initial configuration of cellular automata can produce a wide variety of FNN architectures, generative capacity. And this generative capacity, the search space of NN covered by cellular encoding, is shown too.

J.R. Dorronsoro (Ed.): ICANN 2002, LNCS 2415, pp. 314–319, 2002.

2 Description of Cellular System

The global system is composed of three different modules: the Genetic Algorithm Module, the Cellular Module and the Neural Network Module (Fig 1). All the modules are related to make a general procedure. The cellular module is composed of two bi-dimensional cellular systems and takes charge of generating FNN architectures. Initial configurations of cellular systems are given by several seeds and the rules of the systems are applied to generate final configurations, which correspond to a FNN architecture. The generated FNN is trained and relevant information about the goodness of FNN is used as the fitness value for the genetic module. The genetic algorithm module takes charge of generating the positions of the seeds (codified in the chromosome) in the two-dimensional grid of cellular systems, which determine initial configurations of cellular systems. In [Gutierrez et. al., 2001] a detailed description of Neural Network and Genetic Algorithm Modules can be found. In the next section, the new formulation of the Cellular Module is presented.

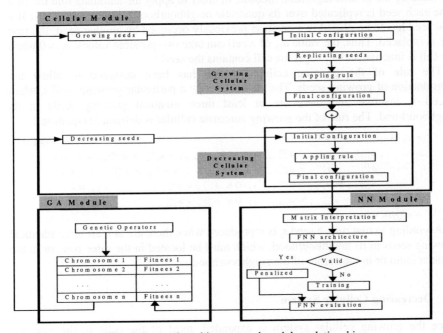

Fig. 1. System's architecture and modules relationship.

3 Cellular Module

The cellular module (Fig 1) is composed by two bi-dimensional cellular system. "Growing cellular system" and "decreasing cellular system". The bi-dimensional cellular systems consist of a regular grid of cells. Each cell takes different values, denoted by a_{ij}, and the system is updated in discrete time steps according to some rule that depends on the value of sites in some neighbourhood around it. In this work, the neighbourhood structure is defined by the square region around a cell. The size of the grids, $Dim_x x Dim_y$, for the cellular systems is previously fixed. Dim_x (rows) is equal to the number of input neurons N, plus the number of output neurons M, and Dim_y (columns) corresponds with the maximum number of hidden neurons. In the next, the initial configurations, possible values of each cell and the evolution rules of the growing and decreasing cellular systems are presented.

3.1 Growing Cellular System

The growing cellular system is designed in order to obtain FNN architectures with a large number of connections between the input and hidden layer and between hidden and output layer. The initial configuration of the growing cellular system is given by n seeds, $(s_1, s_2, ..., s_n)$, called "growing seeds" (GS). Each seed is defined by two coordinates which indicates the positions of the seed in the grid. That positions are provided by the genetic algorithm module. In order to apply the automata rule the first time each seed is replicated over its quadratic neighbourhood, in such a way that if a new seed has to be placed in a position previously occupied by another seed, the first one is replaced. Thus, the value $a_{i,j}$ of a cell can take two possible values: $a_{i,j}=0$ when the cell is inactive and $a_{i,j}=s_k$ if the cell contains the seed s_k.

The rule of the growing cellular system has been designed to allow the reproduction of growing seeds. The idea is to copy a particular growing seed s_k when a cell is inactive and there are at least three identical growing seeds in its neighbourhood. The rule of the growing automata cellular is defined in equation 1:

$$a_{i,j}^{(t+1)} = s_k \text{ if } a_{i,j}^{(t)} = 0 \text{ AND} \qquad\qquad\qquad (\text{eq. 1})$$

$$a_{i-1,j-1}^{(t)} = a_{i-1,j}^{(t)} = a_{i-1,j+1}^{(t)} = s_k \text{ OR } a_{i+1,j-1}^{(t)} = a_{i+1,j}^{(t)} = a_{i+1,j+1}^{(t)} = s_k \text{ OR}$$

$$a_{i-1,j-1}^{(t)} = a_{i,j-1}^{(t)} = a_{i+1,j-1}^{(t)} = s_k \text{ OR } a_{i-1,j+1}^{(t)} = a_{i,j+1}^{(t)} = a_{i+1,j+1}^{(t)} = s_k \text{ OR}$$

$$a_{i-1,j-1}^{(t)} = a_{i-1,j}^{(t)} = a_{i,j-1}^{(t)} = s_k \text{ OR } a_{i-1,j}^{(t)} = a_{i-1,j+1}^{(t)} = a_{i,j+1}^{(t)} = s_k \text{ OR}$$

$$a_{i,j-1}^{(t)} = a_{i+1,j-1}^{(t)} = a_{i+1,j}^{(t)} = s_k \text{ OR } a_{i+1,j}^{(t)} = a_{i+1,j+1}^{(t)} = a_{i,j+1}^{(t)} = s_k$$

$$a_{i,j}^{(t+1)} = a_{i,j}^{(t)} \text{ in other case}$$

According to that rule, a seed s_k is reproduced when there are at least three identical growing seeds in its neighbourhood, which must be located in the same row, or in the same column or in the corner of the neighbourhood.

3.2 Decreasing Cellular System

Once the growing cellular system is expanded, most of the cells in the grid are occupied by growing seeds. If the presence of a growing seed is considered as the presence of a connection in the network, could be convenient to remove seeds in the grid in order to obtain a large variety of architectures. Hence, the decreasing cellular system is incorporated to remove connections. The initial configuration of the decreasing cellular system is given by the final configuration of the growing cellular system and by m seeds $(d_1, ... d_m)$, called "decreasing seeds" (DS). Each seed is defined also by two coordinates and they are provided by the genetic algorithm module. The value $a_{i,j}$ of a cell in this automata can be: $a_{i,j}=0$ when the cell is inactive; $a_{i,j}=s_k$ when the cell contains the growing seed s_k and $a_{i,j}=d_r$ if the cell contains the decreasing seed d_r.

The rule of the decreasing cellular system is designed to remove growing seeds in the grid. A growing seed s_k is removed when two contiguous neighbouring cells contain identical growing seeds and another neighbouring cell contain a decreasing seed. The rule of the decreasing cellular system is defined as:

$$a_{i,j}^{(t+1)} = d_r \text{, if } (a_{i,j}^{(t)} = a_{i-1,j-1}^{(t)} = a_{i,j-1}^{(t)} = s_k \text{ AND } a_{i-1,j}^{(t)} = d_r) \text{ OR} \qquad (\text{eq. 2})$$
$$(a_{i,j}^{(t)} = a_{i-1,j-1}^{(t)} = a_{i-1,j}^{(t)} = s_k \text{ AND } a_{i,j-1}^{(t)} = d_r) \text{ OR}$$
$$(a_{i,j}^{(t)} = a_{i-1,j+1}^{(t)} = a_{i,j+1}^{(t)} = s_k \text{ AND } a_{i-1,j}^{(t)} = d_r) \text{ OR}$$
$$(a_{i,j}^{(t)} = a_{i-1,j}^{(t)} = a_{i-1,j+1}^{(t)} = s_k \text{ AND } a_{i,j+1}^{(t)} = d_r) \text{ OR}$$
$$(a_{i,j}^{(t)} = a_{i,j-1}^{(t)} = a_{i+1,j-1}^{(t)} = s_k \text{ AND } a_{i+1,j}^{(t)} = d_r) \text{ OR}$$
$$(a_{i,j}^{(t)} = a_{i,j+1}^{(t)} = a_{i+1,j+1}^{(t)} = s_k \text{ AND } a_{i+1,j}^{(t)} = d_r) \text{ OR}$$

$a_{i,j}^{(t+1)} = a_{i,j}^{(t)}$ in other case

Similar rules could be used, but the design must enforce that not all growing seed in the grid are removed.

3.3 Evolving Growing and Decreasing Cellular System

In order to evolve and to combine both growing and decreasing cellular systems, a special procedure that allows the convergence toward a final configuration is proposed (see Fig 1):
1. All cells in the grid are set to the inactive state and the growing seeds provided by the genetic module are located in the grid. The growing seeds are replicated over their quadratic neighbourhood.
2. The rule of the growing cellular system is applied until no more rule conditions could be fired and a final configuration is reached.
3. The decreasing seeds are placed in the grid.
4. The rule of the decreasing cellular system is applied until the final configuration is reached.
5. A binary matrix M is finally obtained, replacing the growing seeds by an 1 and the decreasing seeds or inactive cells by a 0. That matrix will be used by the neural network module to obtain a FNN architecture.

4 Experimental Results

In this paper, the cellular approach has been tested for the parity problem. The fitness function provided to the genetic algorithm module is the inverse of computational effort, equation 3 (a). Where "c" is the number of connections in the FNN architecture and "t_c" the number of training cycles carried out. If the network doesn't reach the defined error, it is trained a maximum of cycles and the fitness value associated is given by the equation 3 (b), where "$e_{reached}$" is the error reached and "e_{fixed}" the error previously fixed.

The parity problem is a mapping problem where the domain set consists of all distinct N-bit binary vectors and the results of the mapping indicates whether the sum of N components of the binary vector is odd o even. In most current studies (Sontag, 1992) shown that a sufficient number of hidden units for the network is (N/2) +1 if N is even and (N+1)/2 if N is odd. In this work parity seven has been considered as a study case. Hence, the network will have 7 input neurons and 1 output. Thus, the size of the grid would be 8x64.

The number of growing and decreasing seeds has been modified and the architectures obtained with the cellular approach for the different number of seeds after 100 generations have four hidden neurons and most of them are fully connected. Only in one case (5-5), the architecture is not fully connected, the first input neurons is only connected to one hidden neuron, without connections to the rest of hidden neurons. All of then obtain percentage of train and test errors around 90% and 80 %, respectively. When the direct encoding is used to find the optimal architecture, the length of the chromosome is 343 (7inputs x 49 hidden) and more complex architectures are obtained. After 300 generations the architecture has 48 hidden neurons and 48% of connectivity.

For the generative capacity of the method 10000 chromosomes are randomly generated, with 7 GS and 7 DS. And the nets obtained, indicating how much hidden nodes and connections has each one, are shown in Fig 2. For a FNN, with "H" hidden nodes, N inputs and M outputs there is a maximum $(H(N+M))$ and a minimum (H) number of connections, and it is displayed in Fig 2. The nets obtained cover the search space of FNN on the whole.

$$F = \frac{1}{(c \cdot t_c)} \text{ (a)} \qquad\qquad F = \frac{1}{c \cdot t_c} \cdot \left(e_{fixed} / e_{reached}\right) \text{ (b)} \qquad\qquad \text{(eq. 3)}$$

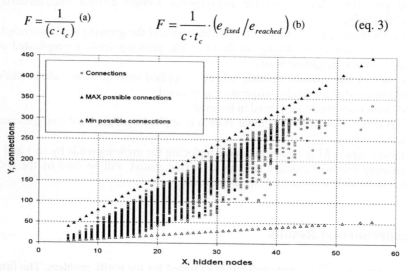

Fig. 2. Generative capacity for parity problem. A point represents a NN with X hidden nodes and Y connections

5 Conclusions and Future Work

Cellular automata are good candidates for non-direct codification's. The final representation has a reduced size and could be controlled by the number of seeds used. The results shown that the cellular scheme presented in this paper is able to find appropriate FNN architectures, and the nets obtained are independent of how many seeds (GS and DS) are placed in the CA. In addition, the number of generations over the population is less when the indirect encoding approach is used instead of direct codifications.

In future works not any individual in the population will have the same number of growing and decreasing seeds, i.e. a seed in the chromosome could be a growing seed or decreasing seed. Besides, some issues about Neural Network Module and fitness function used, i.e. how punish the nets to increase the search, will be studied in future works.

References

[Alba 1993] Alba, E., Aldana, J. F., and Troya, J. M.: Fully automatic RNA design: a genetic approach. X. Yao: Evolutionary Artificial Neural Networks 45 In Proc. of Int'l Workshop on Artificial Neural Networks (IWRNA'93), pages 399–404. Springer-Verlag, 1993. Lecture Notes in Computer Science, Vol. 686.

[Fogel et al, 1990] D.B. Fogel, Fogel L.J. and Porto V.W. Evolving Neural Network, Biological Cybernetics, 63, 487–493, 1990.

[Gruau, 1992] Gruau F. "Genetic Synthesis of Boolean Neural Networks with a Cell Rewriting Developmental Process". Proceedings of COGANN-92 International Workshop on Combinations of Genetic Algorithms and Neural Networks, pp. 55–74, IEEE Computer Society Press, 1992.

[Gutierrez et. al., 2001] G. Gutiérrez1, P. Isasi, J.M. Molina, A. Sanchís and I. M. Galván. Evolutionary Cellular Configurations for Designing Feed-Forward Neural Networks Architectures. Connectionist Models of neurons, Learning Processes and Artificial Intelligence. 6th International Work-Conference on Artificial Neural Networks, IWANN 2001. Proceedings, Part I. LNCS 2084. J. Mira, A Prieto (Eds.) Springer.

[Kitano, 1990] Kitano, H.: Designing Neural Networks using Genetic Algorithms with Graph Generation System, Complex Systems, 4, 461–476, 1990.

[Miller et al., 1989] Geoffrey F. Miller, Peter M. Todd, and Shailesh U. Hegde. Designing Neural Networks using Genetic Algorithms. In J. David Schaffer, editor, Proceedings of the Third International Conference on Genetic Algorithms, pages 379–384, San Mateo, California, 1989. Philips Laboratories, Morgan Kaufman Publishers, Inc.

[Molina et al., 2000] Molina, J. M., Torresano, A., Galván, I. M., Isasi, P., Sanchis, A.: Evolution of Context-free Grammars for Designing Optimal Neural Networks Architectures. GECCO 2000, Workshop on Evolutionary Computation in the Development of ANN. USA. Julio, 2000

[Sontag, 92] E. D. Sontag. Feedforward nets for interpolation and classification. Journal of Computer and System Sciences, 45, 1992.

[Wolfram, 1988] S. Wolfram. Theory and applications of cellular automata. World Scientific, Singapore, 1988.

[Yao, 1999] Yao, X. 1999. Evolving artificial neural networks, Proceedings of the IEEE vol. 87, no. 9, p.1423–1447.

Part III

Data Analysis and Pattern Recognition

Part III

Data Analysis and Pattern Recognition

Entropic Measures with Radial Basis Units

J. David Buldain [1]

Dept. Ingeniería Electrónica & Comunicaciones
Centro Politécnico Superior. C/ Maria de Luna 3.
50015 - Zaragoza. Spain
buldain@posta.unizar.es

Abstract. Two new entropic measures are proposed: the A-entropy and E-entropy, which are compared during competitive training processes in multiplayer networks with radial basis units. The behavior of these entropies are good indicators of the orthogonality reached in the layer representations for vector quantization tasks. The proposed E-entropy is a good candidate to be considered as a measure of the training level reached for all layers in the same training process. Both measures would serve to monitorize the competitive learning in this kind of neural model, that is usually implemented in the hidden layers of the Radial Basis Functions networks.

1 Introduction

The training of many neural models consists in the optimization of certain cost, error or energy functions, the mean square error is the most popular one. There exists an infinite variety of possibilities [4], but many of them are only applicable in supervised models. The unsupervised models present dynamics that intuitively can be observed as processes minimizing some kind of energy, till now this general description remains undiscovered. Other approach comes from the termodinamic field: the entropic measures, that are used at least with three purposes in the neural field. The first one appears in statistical descriptions of an ensemble of neural networks [9]. The weights are considered an estocastic result of the training process, and the properties of the ensemble of weights are analyzed from a termodinamic point of view, measuring the generalization error [2], or is used for selecting architectures [7]. The second entropic interpretation corresponds to the information entropy. The information content of a symbol 'q' in a codification, $\{\mathbf{Q}\}$, is given by: $I(q)=-\log(p(q))$. Where $p(q)$ is the probability of occurrence of 'q' in the transmitted or stored messages. The expected value of $I(q)$, is the informational entropy:

$$H(Q) = \langle I(q) \rangle = \langle -\log_2(p(q)) \rangle = - \sum_{q \in \{Q\}} p(q)\log_2(p(q)) \tag{1}$$

[1] Acknowledgment: this work has been partially supported by the spanish government and the european commission CDER, as part of the project n° 2000/0347, entitled: "Teleobservation and Movig Pattern Identification System".

This entropy is used in vector quantization tasks, where the data space is divided in several regions specified by reproduction vectors. The entropy of a neural binary layer doing this task measures the level of distortion provided by this neural codification [1]. Another possibility is the maximization of the informational entropy trying to use all the symbols (neurons) with the same frequency.

The third kind of entropy appears when we use the normalized responses of a group of neurons as a probabilistic distribution. The resulting entropic expression reveals the orthogonality of the neural representation [8]. Two new entropic measures with this interpretation are defined in the next section: the A-entropy and the E-entropy. Section 3 is, the simulated network architectures are described and an hypothetic training level is defined. Section 4, the results show the useful behavior of both entropies to monitorize the evolving representations during the training processes.

2 Entropies with Competitive Radial Basis Units

The neural networks dealt in this work have layers of Radial Basis Units (RBU) are trained with a competitive algorithm similar to the FSCL [1]. Their kernels present adaptive centroids and widths [5] [6] and the neurons give responses in the interval $(0,1)$ with a sigmoidal-kernel function. To optimize the number of RBUs for a given problem, the kernels of the RBUs must cover completely those areas of the input space where patterns appear, but trying to avoid excessive interference among the kernels. Entropic measures represent the interference among kernels: where entropy is low, only one neuron has an elevated response, in other zones where several neurons give high responses simultaneously, there appears a local high entropic value.

Consider a layer 'k' with $N^{(k}$ units dividing the input-space with their kernels and giving the output vector $\mathbf{y}'(\mathbf{x})$, normalized by their sum. The absolute entropy (Ha) for an input pattern $\mathbf{x}$ is defined with the expression:

$$Ha^{(k}(\mathbf{x}) = -\sum_{i=1}^{N^{(k}} (y_i') \cdot \log(y_i'(\mathbf{x})) \tag{2}$$

The mean absolute entropy is calculated by averaging the entropic measures for all patterns in the training set $\{\mathbf{X}\}$, with number of patterns P:

$$Ha^{(k} = \frac{1}{P} \sum_{\mathbf{x} \in \{\mathbf{X}\}} Ha^{(k}(\mathbf{x}) \tag{3}$$

At the beginning of the training process, the mean absolute entropy is high, because of the great extension of the initial kenerls, but it cannot exceed the maximum value:

$$Max(Ha^{(k}) = \log(N^{(k}) \tag{4}$$

We define the *A-entropy* as the mean absolute entropy divided by the maximum value of eq.4, so it becomes independent of the logarithmic base and presents the same variation interval $[0,1]$, independently of the number of units.

The euclidean entropy (He) evaluates the neural responses normalized by the euclidean norm, therefore it does not give a probabilistic distribution. Like A-entropy, the *E-entropy* corresponds to this Euclidean entropy normalized with the maximum value of the mean Euclidean entropy:

$$\text{Max}(\text{He}^{(k)}) = \frac{1}{2}\sqrt{N^{(k}} \cdot \log(N^{(k)}) \tag{5}$$

3 Vector Quantization in Parallel Layers

This section describes parallel layers that receive the same input stimulation during the training process, in two different networks to compare their behaviors and results. All simulations use a set containing 2500 patterns uniformly distributed in the 1x1 area, that are randomly presented along 50 cycles. The first network has nine parallel layers (fig.1a) with number of units from 10 to 90, varying 10 units between consecutive layers. The second one has a hidden layer with N units (N= 4, 6, 9) and 20 parallel output layers with numbers of units varying from 2 to 20 (fig. 2b).

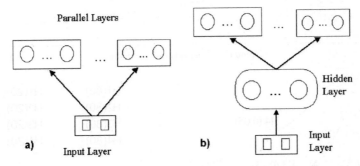

Fig. 1. The network in the figure a) will be called *the first network*, with 9 parallel layers with 10 to 90 units. The network in figure b) is *the second network*, formed by 19 parallel layers with 2 to 20 units and N hidden units (N = 4, 6, 9)

To obtain relations of the learning factors, $\eta^{(m}$, assigned to each parallel layer, we define a certain "training level" (T), for each layer 'm':

$$T = \eta^{(m} \cdot \frac{P}{N^{(m}} \tag{6}$$

Where the whole number of patterns presented during the training process is P, and the number of units in the layer is $N^{(m}$. If we assign a learning factor, $0 < \eta^{(m} < 1$, for the layer 'm' with the largest number of units, then , supposing P and T common for all parallels layers, the learning factors in the rest of layers must follow the relation:

$$\eta^{(k} = \left(N^{(k}/N^{(m}\right) \cdot \eta^{(m} \quad \text{(all k included in the network} | k \neq m) \tag{7}$$

The main effect of the learning factor is the training velocity, but it is less decisive in the generated representations.

4 Results

Figure 2 represents the evolution of A-entropy (top graph) in four parallel layers and E-entropy (bottom graph, H1) in two layers, all of them measured during the same training session of the first network with a learning factor of value 0.9. The bottom graph also shows the evolution of E-entropies (H2, H3 and H4) in other three training session with learning factors: 0.45, 0.15 and 0.05.

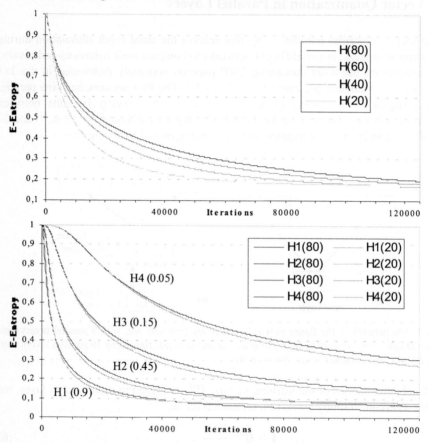

Fig. 2. The top graph presents separated curves of *A-entropies* measured in four layers of the first nework (80, 60, 40 and 20 units) during one training session (with learning factor 0.9). The bottom graph shows *E-entropies* measured in two layers (80 and 20 units) during four different training sessions (*H1, H2, H3 and H4*) with learning factors: *0.9, 0.45, 0,15 and 0.05*. The curves of the other parallel layers descend close to those two, and are not represented to maintain the picture clear. Notice that E-entropies measured with the same training factor descend following similar curves. The group of E-entropic curves for different training factors are separated; the greater factor generates a quicker descend, but finally the curves for the layers with same number of units reach similar E-entropic values. E-entropy presents a suited behavior for being considered as a common training-level measure to all the parallel layers.

Both entropies descend to steady values where the layer representations become stable. The A-entropic measures follow separated curves during the training session depending on the number of units of the layer and the learning factor. However, E-entropies of the pararallel layers tend to vary simultaneously, giving curves with less slope when the learning factor decreases, as it can be noticed in the bottom graph. It shows the curves in four training sessions of only two parallel layers (to maintain the picture clear) represented with H1, H2, H3 and H4, that were generated with the learning factors: 0.9, 0.45, 0.15 and 0.05, respectively. The concept of a training level (T in the eq.6) to monitorize the training session can be identified in the behavior of E-entropies better than using A-entropies, as E-entropic measures almost follow the same descending curve, although not exactly the same, in all the parallel layers.

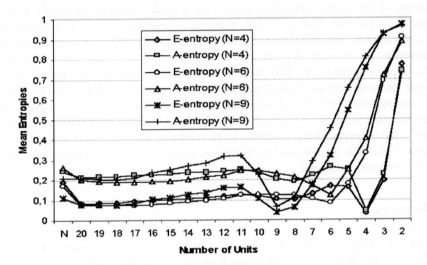

Fig. 3. Final *mean entropic values* measured in the layers of the second network with different number of units in the hidden layer (*N=4, 6, 9*). The first position *N* represents the hidden layer, while the output layers are represented by their number of units from 20 to 2

The second network was trained ten times with three different numbers of hidden units: N = 4, 5, 6. The final entropic measures averaged are represented in the fig.3. The standard deviations where low and are not shown to maintain clear the picture. Notice that the minimum entropic values of the output layers appear when the hidden number of units match the number of units of the output layer.

This configuration, where each output unit only captures high responses of one hidden unit, is called exact capture. When the number of output units is near the exact capture, some output units learn to recognize intersections of the hidden kernels. For example, in the network with N=4, another local minimum appears in the output layer with 8 units. This happens because four units produce an exact capture, but the other four output units capture high responses of two hidden units, generating kernels that recognize the intersection of two hidden kernels. A larger number of units in the output layers does not reduces further the entropic measures obtained in the exact capture. The output representations, with more units than the hidden one, mainly present groups of output units capturing the responses of a common hidden unit, but in separated intervals of its response. Entropy reveals the optimum quantization of the input space, that occurs when the number of hidden units is a square. In this case, 9 and 4, allow the representations 3x3 and 2x2 kernels, while the number of 6 hidden units can be arranged in 2x3 or 3x2, but generating more interference.

5 Conclusions

The proposed entropies show behaviors useful for monitorize the training processes of layers, or networks, with radial basis units; a model encountered mainly in the Radial Basis Functions network. Both entropies are reduced during the competitive training processes, indicating that the kernels of the units are being separated and their interference is minimized. This characteristic is interesting for the selection of the hidden representation of the RBF networks, as the minimum values obtained in the entropic measures indicate the best number of units needed to cover the training data.

E-entropy results a better choice than A-entropy when dealing with a group of layers, because it seems more easily identified as a training-level measure for the whole group of layers. Both entropies can be compared between layers with different numbers of units as the imposed normalization makes them to vary between the zero and the maximum value one.

References

1. Ahalt, S.C., Krishnamurthy, A.K., Chen, P., & Melton, D.E. (1990) Competitive learning algorithms for Vector Quantization. Neural Networks. Vol. 3, pp277-290.
2. Amari, S.&Murata, N. Statistical theory of learning curves under entropic loss criterion. Neural Comp. vol1, pp 140-153, 1993
3. Banzhaf, W & Haken, H. (1990) Learning in a Competitive Network. Neural Networks Vol. 3, pp423-435.
4. Barnard, E. & Casasent, D. A comparison between criterion functions for linear classifiers with an application to neural nets. IEEE Trans.Syst., Man and Cybern. vol.19, pp 1030-1041, Sept-Oct. 1989
5. Buldain, J.D. (2001) Divided-data Analysis in a Financial Case Classification with Multi-Dendritic Neural Networks. Proc. of 6th Int.Work-Conference on Artificial Neural Networks. Vol.1, pp 253-268.
6. Buldain, J.D. & Roy, A (1999) Association with Multi-Dendritic Radial Basis Units. Proc. of 5th Int.Work-Conference on Artificial Neural Networks. Vol.1, pp 573-582
7. Levin, E., Tishby, N. & Solla, S.A. A statistical approach to learning and generalization in layered NN. Proc.of IEEE vol78, n°10, pp1568-1574, 1990.
8. Webber, C.J. (1991) Competitive Learning, Natural Images and Cortical Cells. Network vol.2, n° 2, pp169-187.
9. White, H. Learning in artificial neural networks: a statistical perspective. Neural Comp. vol1, pp 425-464. 1989

New Methods for Splice Site Recognition[*]

Sören Sonnenburg[1], Gunnar Rätsch[2], Arun Jagota[3], and Klaus-Robert Müller[1,4]

[1] Fraunhofer FIRST, Kekuléstr. 7, 12489 Berlin, Germany
[2] Australian National University, Canberra, ACT 0200, Australia
[3] University of California at Santa Cruz, CA 95064, USA
[4] University of Potsdam, August-Bebel-Str. 89, 14482 Potsdam, Germany

Abstract. Splice sites are locations in DNA which separate protein-coding regions (exons) from noncoding regions (introns). Accurate splice site detectors thus form important components of computational gene finders. We pose splice site recognition as a classification problem with the classifier learnt from a labeled data set consisting of only local information around the potential splice site. Note that finding the correct position of splice sites without using global information is a rather hard task. We analyze the genomes of the nematode Caenorhabditis elegans and of humans using specially designed support vector kernels. One of the kernels is adapted from our previous work on detecting translation initiation sites in vertebrates and another uses an extension to the well-known Fisher-kernel. We find excellent performance on both data sets.

1 Introduction

Splice sites are locations in DNA at the boundaries of exons (which code for proteins) and introns (which do not). The more accurately a splice site can be located, the easier and more reliable it becomes to locate the genes – hence the coding regions – in a DNA sequence. For this reason, splice site "detectors" are valuable components of state-of-the-art gene finders [4,13,21,14,5]. Furthermore, since ever-larger chunks of DNA are to be analyzed by gene finders the problem of accurate splice site recognition has never been more important.

Although present-day splice site detectors are reported to perform at a fairly good level [13,12,7], several of the reported performance numbers should be interpreted with caution, for a number of reasons. First of all, these results were based on small data sets of a limited number (one or two) organisms. Now that large, complex genomes have been fully sequenced, these results will need to be re-evaluated. Second, issues in generating negative examples (decoys) were, if recognized, not adequately documented.[1] Third, the results are expected to be highly dependent on the chosen window size. (The

---

[*] We thank for valuable discussions with A. Zien, K. Karplus and T. Furey. G.R. would like to thank UC Santa Cruz for warm hospitality. This work was partially funded by DFG under contract JA 379/9-2, JA 379/7-2, MU 987/1-1, and NSF grant CCR-9821087. This work was supported by an award under the Merit Allocation Scheme on the National Facility of the Australian Partnerschip for Advanced Computing.
[1] To our knowledge, on the splice site recognition problem, only the work of [13] explicitly documented the care it exercised in the design of the experiments.

J.R. Dorronsoro (Ed.): ICANN 2002, LNCS 2415, pp. 329–336, 2002.
© Springer-Verlag Berlin Heidelberg 2002

window defines the extent of the context around a site's boundary used during training and classification.) Since the different studies [13,12,7] chose different window sizes, and we choose a fourth different one; unfortunately no pair of these studies is directly comparable. Finally, some works [12] highlighted their accuracy ($\sim$error-rate) results. These can paint an overly optimistic picture on this problem (also [2]).

Support vector machines (SVMs) (e.g. [20,11,15]) with their strong theoretical roots are known to be excellent algorithms for solving classification problems. To date, they have been applied only to a handful of Bioinformatics problems (see e.g. [10,22,8,9]). In this paper we apply SVMs to two binary classification problems, the discrimination of donor sites (those at the exon-intron boundary) from decoys for these sites, and the discrimination of acceptor sites (those at the intron-exon boundary) from decoys for these sites. We evaluate our SVMs on two data sets: (a) on the IP-data data set (a relatively old data set of human splice sites with weak decoys) where the SVM method outperforms nine other methods, including a recent one [12] and (b) on splice sites extracted from the *complete* C. Elegans genome [1]. Also here SVM methods with a good kernel are able to achieve remarkably high accuracies. Apart from pure bioinformatics and experimental insights we obtain a better understanding about the important issue of whether and when SV-kernels from probabilistic models are preferable over specially engineered kernels. Note that both kernel choices are intended to incorporate prior knowledge into SVM learning.

2   Basic Ideas of the Methods

The essence of SVMs is that a very rich feature space is used for discrimination purposes while at the same time the complexity of the overall classifier is controlled carefully. This allows to give guarantees for high performance on unseen data. The key is a good choice of the so-called support vector kernel k which implicitly defines the feature space in which one classifies. Three particularly successful options for the kernel choice in DNA analysis exist: (a) available biological prior knowledge is directly engineered into a polynomial-type kernel, for example the so-called "locality improved (LI) kernel" as proposed in [22] or a probabilistic model that encodes the prior knowledge, e.g. an HMM is extracted from the data and is used for constructing a kernel: (b) the so called fisher-kernel [19] or its recent extension (c) the TOP kernel [9]. In some cases, one can reinterpret (a) in the probabilistic context. For instance the locality improved kernel corresponds to a higher order Markov model of the DNA sequence [9,18].

Support Vector Machines. For a given data set $\mathbf{x}_i \in \mathbb{R}^n$ ($i = 1, \ldots, N$) with respective labels y_i, a SVM classifier yields $f(\mathbf{x}) = \text{sign}\left(\sum_{i=1}^{N} y_i \alpha_i \, k(\mathbf{x}, \mathbf{x}_i) + b\right)$ given an input vector $\mathbf{x}$. For learning the parameters α_i a quadratic program is to be solved by (cf. [20]): $\max_{\boldsymbol{\alpha}} \sum_{i=1}^{N} \alpha_i - \frac{1}{2} \sum_{i,j=1}^{N} \alpha_i \alpha_j y_i y_j \, k(\mathbf{x}_i, \mathbf{x}_j)$ such that $\alpha_i \in [0, C], i = 1, \ldots, N$ and $\sum_{i=1}^{N} \alpha_i y_i = 0$.

The Locality Improved Kernel [22] is obtained by comparing the two sequences locally, within a small window of length $2l + 1$ around a sequence position, where we count matching nucleotides. This number is then multiplied with weights $p_{-l}, \ldots, p_{+l}$ increasing linearly from the boundaries to the center of the window. The resulting weighted counts are taken to the d_1^{th} power, where d_1 reflects the order of local correlations (within the window) that we expect to be of importance: $\text{win}_p(\mathbf{x}, \mathbf{y}) = \left(\sum_{j=-l}^{+l} p_j \text{ match}_{p+j}(\mathbf{x}, \mathbf{y}) \right)^{d_1}$. Here, $\text{match}_{p+j}(\mathbf{x}, \mathbf{y})$ is 1 for matching nucleotides at position $p + j$ and 0 otherwise. The window scores computed with win_p are summed over the whole length of the sequence. Correlations between up to d_2 windows are taken into account by finally using the SV-kernel $k_{\text{LI}}(\mathbf{x}, \mathbf{y}) = \left(\sum_{p=1}^{l} \text{win}_p(\mathbf{x}, \mathbf{y}) \right)^{d_2}$.

Fisher and TOP Kernel. A further highly successful idea is to incorporate prior knowledge via probabilistic models $p(\mathbf{x}|\hat{\boldsymbol{\theta}})$ of the data (e.g. HMMs) into SVM-kernels [19]. This so-called *Fisher Kernel* (FK) is defined as $k_{\text{FK}}(\mathbf{x}, \mathbf{x}') = s(\mathbf{x}, \hat{\boldsymbol{\theta}})^\top Z^{-1}(\hat{\boldsymbol{\theta}}) s(\mathbf{x}', \hat{\boldsymbol{\theta}})$, where s is the Fisher score $s(\mathbf{x}, \hat{\boldsymbol{\theta}}) = \nabla_{\boldsymbol{\theta}} \log p(\mathbf{x}, \hat{\boldsymbol{\theta}})$ and where Z is the Fisher information matrix: $Z(\boldsymbol{\theta}) = \mathrm{E}_\mathbf{x} \left[s(\mathbf{x}, \boldsymbol{\theta}) s(\mathbf{x}, \boldsymbol{\theta})^\top \mid \boldsymbol{\theta} \right]$ (for further details see [19]). Our recent extension of the FK uses the Tangent vectors Of Posterior log-odds (TOP) leading to the TOP kernel $k_{\text{TOP}}(\mathbf{x}, \mathbf{x}') = \boldsymbol{f}_{\hat{\boldsymbol{\theta}}}(\mathbf{x})^\top \boldsymbol{f}_{\hat{\boldsymbol{\theta}}}(\mathbf{x}')$ [9], where $\boldsymbol{f}_{\hat{\boldsymbol{\theta}}}(\mathbf{x}) := (v(\mathbf{x}, \hat{\boldsymbol{\theta}}), \partial_{\theta_1} v(\mathbf{x}, \hat{\boldsymbol{\theta}}), \ldots, \partial_{\theta_p} v(\mathbf{x}, \hat{\boldsymbol{\theta}}))^\top$ with $v(\mathbf{x}, \boldsymbol{\theta}) = \log(P(y = +1|\mathbf{x}, \boldsymbol{\theta})) - \log(P(y = -1|\mathbf{x}, \boldsymbol{\theta}))$ [9]. (We do not use Z^{-1} but a diagonal matrix for FK & TOP such that the variance in each coordinate is 1.) The essential difference between both kernels is that the TOP kernel is explicitly designed [9] for discrimination tasks. In fact, on protein family classification experiments we have shown that it performs significantly better than the Fisher kernel. As probabilistic models we employ Hidden Markov Models (HMMs), with several biologically motivated architectures (description below). We use the implementation described in [17] of the Baum-Welch algorithm [6] for training.

3 Data Sets, Experiments, and Results

Two data sets are analysed with different purposes. For the first IP benchmark set the goal is a comparison to other machine learning methods and we will see that our approaches easily outperform existing entries. For C. elegans we cannot compare to existing algorithms for systematic reasons (see section 1) and since our results on the IP data tell us that SVMs with our kernels are the method of choice, we focus in the second part of this section on the evaluation of different SVM kernels: locality improved vs. probabilistic model based kernels. Whereas the IP data is a fixed benchmark, for C. elegans, the decoys were chosen to be windows of the same length as the true sites from -25 to +25 of the site with two additional constraints: (i) the decoy windows were limited to those near the true sites and (ii) the decoy windows were forced to contain the conserved dinucleotide (GT or AG) centered in the same location in the window as in the true sites (donor and acceptor, respectively). This made the decoys not only harder than random ones from throughout the genome but also modeled the use of a splice site detector in a gene finder somewhat more realistically since it is more likely that a gene finder invokes a splice site detector in the proximity of true sites than at an arbitrary place in the genome (vast amounts of intergenic regions are already filtered out before any splice site prediction needs to be done). Not surprisingly, the performance reported in [13], where the decoys were similarly constructed, though in the proximity -40 and +40, was significantly poorer than in [12].

IPData is a benchmark set of human splice site data from the UC Irvine machine learning repository [3]. It contains 765 acceptor sites, 767 donor sites and 1654 decoys; the latter are however of low quality as they do not have a true site's consensus dint centered except by chance. The task is a donor/acceptor classification given a position in the middle of a window of 60 DNA letters as input.

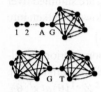

Fig. 1. Acceptor and Donor Model

In our experiments, first a careful model selection of the hyper-parameters of the HMMs and SVM is performed (cf. [11]). This is done separately on each of ten random (train, test) split of the data of size (2000,1186) (a single same-sized split was used in [12]). As HMM architecture we used (a) a combination of a linear model and a fully connected model for the acceptor sites (cf. Fig. 1, upper), (b) a combination of two fully connected model for the donor sites (cf. Fig. 1, lower) and (c) a fully connected model for modeling decoys. (These architectures can be biologically motivated.) The corresponding number of states in the components, as well as the regularization parameter of the SVM, are found by 10-fold cross validation.

For our comparison we computed the plain HMM, SVMs with locality improved kernel and with FK and TOP kernel (based on the HMMs for each class). Each classifier was then evaluated on the test set and results are averaged over the 10 runs and the standard deviation is given in Table 1 (shows errors on each of the classes).

Comparing our classifiers we observe that SVMs with TOP and FK (total error 5.4% and 5.3%) cannot improve the HMM, which performs quite well (6.0%), but has a quite large error in classifying the acceptor sites. The SVM with locality improved kernel does not suffer from this problem and achieves the best total error of of 3.7%.

System	Neither	Donor	Acceptor
HMM	2.6±0.5%	1.0±0.4%	2.4±0.7%
LI-SVM	2.0±0.3%	0.8±0.2%	0.9±0.3%
TOP-SVM	2.2±0.4%	1.5±0.4%	1.7±0.3%
FK-SVM	2.1±0.4%	1.6±0.5%	1.6±0.4%
NN-BRAIN	n.d.	2.6%	4.3%
BRAIN	4.0%	5.0%	4.0%
KBANN	4.6%	7.6%	8.5%
BackProp	5.3%	5.7%	10.7%
PEBLS	6.9%	8.2%	7.6%
Perceptron	4.0%	16.3%	17.4%
ID3	8.8%	10.6%	14.0%
COBWEB	11.8%	15.0%	9.5%
Near. Neigh.	31.1%	11.7%	9.1%

Table 1: Test-set errors on the IPData data set. All except the first 4 results are cited from [12], Table 6. (n.d.=not documented)

We observe that the SVM methods outperform all other documented methods on the IP data set (taken from [12]). These include not only the BRAIN algorithms of [12] published recently, but also established machine learning methods such as nearest-neighbor classifiers, neural networks and decision trees. The SVM achieves test-set errors that are half of the best other methods, but only if the kernel is suitable.[2]

The C. Elegans data set was derived from the Caenorhabditis Elegans genome [1], specifically from the chromosome and GFF files at http://genome.wustl.edu/gsc/C_elegans. From these files, windows of -50 to

---

[2] If one uses RBF kernels, one gets worse results than the BRAIN method [11].

+50 around the true sites were extracted.[3] This resulted in 74,455 acceptor sites and 74,505 donor sites, each of which we clipped to -25 to +25 (i.e. length 50) with the consensus dinucleotide centered. For the decoys, we extracted, from the -50/+50 windows around each site, all windows w (except the true site's window) of length 50 as the true site windows, with the consensus dint GT or AG centered in w in the same offset as in a true site's window. This resulted in 122,154 acceptor decoys and 177,061 donor decoys. The complete data is available at our web-site http://mlg.anu.edu.au/~raetsch/splice related to this paper. In this paper we will only use subsets of at most 25000 examples.

In our study, we consider the classification of C. el-egans acceptor sites only. We expect similar results on the donor sites. As probabilistic models for true accep-tor sites we use the HMM described in Fig. 1 (upper). Since we assume that the decoy splice sites are close to a true splice site (in fact, in our data they are never fur-ther away then 25 base pairs), we can exploit this in our

Fig. 2. The HMM architecture for modeling C. elegans decoys

probabilistic model for the decoys. We propose to append the previously obtained HMM for the true sites to a linear strand of length 25 (as in Fig. 2). Then we allow all new states $1', 2', \ldots, 25'$ and all states in the positive model (except the first state) to be the starting states of this model. Hence, true sites not centered are detected as decoys. (Only the emission probabilities of the new states and the start state distribution are optimized.)

For training of HMMs and SVMs we use 100, 1000 and 10000 examples. For simplicity we use additional 5000 examples for model selection (to select number

	HMM	Loc.Imp.	FK	TOP
100	10.7±2.8%	**7.6±1.0%**	9.4±4.2%	20.8±3.0%
1000	**2.8±0.1%**	5.2±0.1%	3.5±0.2%	4.6±0.4%
10000	2.6±0.2%	3.9±0.2%	2.5±0.3%	**2.3±0.1%**

Table 2: Test errors of our 4 methods on 100-10000 examples

of states, regularization constant; not possible in practice, but makes the comparison easier). This is done on each of 5 realizations and then the best classifiers are chosen and evaluated on the test set (10000 examples) leading to the results in Table 2. Our first result reveals that the test error of the SVM classifier decreased consistently as the training set size was increased. This means that although the larger data set certainly contains redundant information, the SVMs can still extract some additional insights for the classification. We conjecture that there is useful statistical signal away from the conserved portion of the sites that the SVM classifier is eventually picking up from the larger training set. Also observe that the locality improved kernel starts with a very good result on 100 examples and then cannot gain much from more examples. We conjecture that it profited from the weak prior information incoded in the kernel, but then only improves slowly. The TOP kernel, however, starts with a poor result (possibly due to numerical problems) and improves much with more examples. The HMM reaches a plateau at 1000 examples (using 100.000 examples the HMMs achieve 2.4%), whereas TOP and FK SVMs can improve when seeing more patterns (preliminary results show error rates even below 2%).

Figure 3 shows the ROC plot of the performance of our four classifiers on the C. elegans *acceptor* data set of size 10,000 on a test set of size 10,000. The predictive perfor-

---

[3] We thank David Kulp and others at University of California, Santa Cruz for preparing these datasets and David Haussler for granting us permission to use them. As an extra step, we verified their extracted sites by matching them to the chromosome DNA sequences.

mance was plotted as a function of the classification threshold swept to vary the trade-off between false positives and false negatives. From the perspective of gene finding as well as researchers wanting to locate the sites, it is important to keep the false negative rate as low as possible. But since the number of true negatives (non-sites) when scanning even the regions of the genome in the proximity of the true sites will vastly outnumber the true sites, it is also important to keep the false positive rate down. Since we cannot keep both down simultaneously, we should look at the performance of the classifier at least at two extremes – at low false positive rate and at low false negative rate.

We see that TOP- and FK-SVM classifier achieves a simultaneous 1% false-positive rate (i.e., a sensitivity of 0.99) and a 5% and 8% false-negative rate (i.e., a specificity of about 0.95 and 0.92), respectively. While conclusive comparisons are inadvisable owing to experiments having been done on different data sets, some comparisons with the results of [13] are still helpful. In [13] a similar methodology as ours was applied to similar data sets, in particular, the procedure to construct the decoys is similar (although, as already indicated above, in the proximity -40 and +40 of the site instead of -25 and +25). The

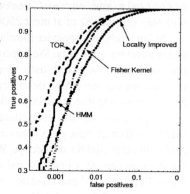

Fig. 3: ROC Curves

result in [13] could achieve a simultaneous 1% false-positive rate and 20% false-negative rate, which is worse than our result.

We would also like to highlight an interesting outcome concerning the issue of SV-kernel choice. The experiments show that the locality improved kernel, where biological knowledge has been directly engineered into the kernel geometry, works very nicely. Nevertheless this approach can be outperformed by a SV kernel derived from a probabilistic model like fisher or TOP kernel. The important point is, however, that this additional improvement holds only for very problem-specific probabilistic models, like the specially taylored negative and positive HMMs used here (cf. section 2). Already as stand-alone classifiers those HMMs perform very competitive. If less fine-tuned HMMs are used as a basis for discriminative training with FK or TOP kernel, the performance decreases considerably (cf. splice site recognition results in [10]). So the incorporation of detailed biological prior knowledge makes the difference in performance.

4   Conclusions

In this paper we successfully applied SVMs to the problem of splice site detection. The key for obtaining our excellent results was a smart inclusion of prior knowledge into SVMs, more precisely into their kernels. A general problem in assessing classification performance on bioinformatics data is that, while there is a lot of publically available molecular data, at present there are few standardized data sets to evaluate new classifiers with. (We contribute to overcoming this problem by making all data and detailed results publically available on the previously mentioned website.) Another issue is that the problems we address here (as well as many other classification problems in bioinformatics) involve separating signal (one class) from noise (the other class). The noise class is generally far denser than the signal class. Both due to this imbalance, and because the noise class is ill-defined, classifiers have to be designed and evaluated with special care.

Our first set of experiments used the well-known but small IP benchmark data set and showed that our SVMs compare favourably over existing results. For the C. Elegans study, we could not find any preprocessed benchmark data, and therefore could not compare the performance directly against existing methods (except HMMs). Therefore, we decided to study more closely the SVM-based learning itself, and in particular the quality of probabilistic vs. engineered kernels. Clearly, including biological prior knowledge like in locality improved kernels gives an excellent performance which cannot be surpassed by a *straight forward* probabilistic kernel (e.g. a first order Markov model as used in [10]). However, if we use sophisticated probabilistic modeling like in specific HMMs that are fine-tuned for splice site recognition, then an additional discriminative training on top of the probabilistic model provides a further improvement.

Future research will focus on the construction of better probabilistic models and SV kernels. We furthermore plan to train our classifiers on larger problems (we used only 10.000 out of 180.000 examples), for which some additional practical problems have to be solved.[4] And finally we would like to apply our insights to splice site detection on the complete human genome.

References

1. Genome sequence of the Nematode Caenorhabditis elegans. *Science*, 282:2012–2018, 1998.
2. P. Baldi, S. Brunak, Y. Chauvin, C.A.F. Andersen, and H. Nielsen. Assessing the accuracy of prediction algorithms for classification: an overview. *Bioinformatics*, 16(5):412–424, 2000.
3. C.L. Blake and C.J. Merz. UCI repository of machine learning databases, 1998.
4. C. Burge and S. Karlin. Prediction of complete gene structures. *J. Mol. Biol.*, 268:78–94, 1997.
5. A.L. Delcher, D. Harmon, S. Kasif, O. White, and S.L. Salzberg. Improved microbial gene identification with GLIMMER. *Nucleic Acids Research*, 27(23):4636–4641, 1999.
6. R. Durbin, S. Eddy, A. Krogh, and G. Mitchison. Biological sequence analysis probabilistic models of proteins and nucleic acids. Cambridge University Press, 1998.
7. D. Cai et al. Modeling splice sites with Bayes networks. *Bioinformatics*, 16(2):152–158, 2000.
8. M.P.S. Brown et al. Knowledge-based analysis by using SVMs. *PNAS*, 97:262–267, 2000.
9. T.S. Jaakkola, M. Diekhans, and D. Haussler. *J. Comp. Biol.*, 7:95–114, 2000.
10. T.S. Jaakkola and D. Haussler. Exploiting generative models in discriminative classifiers. In M.S. Kearns et al., editor, *Adv. in Neural Inf. Proc. Systems*, volume 11, pages 487–493, 1999.
11. K.-R. Müller, S. Mika, G. Rätsch, K. Tsuda, and B. Schölkopf. An introduction to kernel-based learning algorithms. *IEEE Transactions on Neural Networks*, 12(2):181–201, 2001.
12. S. Rampone. Recognition of splice junctions on DNA. *Bioinformatics*, 14(8):676–684, 1998.
13. M.G. Reese, F. H. Eeckman, D. Kulp, and D. Haussler. *J. Comp. Biol.*, 4:311–323, 1997.
14. S. Salzberg, A.L. Delcher, K.H. Fasman, and J. Henderson. *J. Comp. Biol.*, 5(4):667–680, 1998.

---

[4] When using the TOP or FK kernel, one has to handle vectors of length about 5000 per example. This leads to quite large matrices, which are more difficult to handle. We plan on following the approach of [16] to overcome this problem with a more efficient algorithm and also more computing power.

15. B. Schölkopf and A. J. Smola. *Learning with Kernels*. MIT Press, Cambridge, MA, 2002.
16. A.J. Smola and J. MacNicol. Scalable kernel methods. Unpublished Manuscript, 2002.
17. S. Sonnenburg. *Hidden Markov Model for Genome Analysis*. Humbold University, 2001. Proj. Rep.
18. S. Sonnenburg. New methods for splice site recognition. Master's thesis, 2002. Forthcoming.
19. K. Tsuda, M. Kawanabe, G. Rätsch, S. Sonnenburg, and K.R. Müller. A new discriminative kernel from probabilistic models. In *Adv. in Neural Inf. proc. systems*, volume 14, 2002. In press.
20. V.N. Vapnik. The nature of statistical learning theory. Springer Verlag, New York, 1995.
21. Y. Xu and E. Uberbacher. Automated gene identification. *J. Comp. Biol.*, 4:325–338, 1997.
22. A. Zien, G. Rätsch, S. Mika, B. Schölkopf, T. Lengauer, and K.-R. Müller. Engineering svm kernels that recognize translation initiation sites. *Bioinformatics*, 16(9):799–807, 2000.

A Weak Condition on Linear Independence of Unscaled Shifts of a Function and Finite Mappings by Neural Networks

Yoshifusa Ito

Aichi-Gakuin University,
Nisshin-shi, Aichi-ken 470-0195, Japan,
ito@psis.aichi-gakuin.ac.jp

Abstract. Let $1 \leq c \leq d$ and let g be a slowly increasing function defined on $\mathbf{R}^c$. Suppose that the support of the Fourier transform $\mathcal{F}_c g$ of g includes a converging sequence of distinct points y_k which sufficiently rapidly come close to a line as $k \to \infty$. Then, any mapping of any number, say n, of any points $x_1, \cdots, x_n$ in $\mathbf{R}^d$ onto $\mathbf{R}$ can be implemented by a linear sum of the form $\sum_{j=1}^{n} a_j g(W x_i + z_j)$. Here, W is a $d \times c$ matrix having orthonormal row vectors, implying that g is used without scaling, and that the sigmoid function defined on $\mathbf{R}$ and the radial basis function defined on $\mathbf{R}^d$ are treated on a common basis.

1 Introduction

Even under a strong restriction that the activation function cannot be scaled, a neural network is often sufficiently versatile for many purposes. Neural networks having unscaled activation functions can approximate continuous functions on compact sets [2] and those on the compactification of the whole space [1], [4]. Using an unscaled activation function is sometimes beneficial because the parameters are partly confined in a compact set, while at the same time it is of theoretical interest to realize a task under such a strong restriction. We have been interested in using unscaled activation functions since [1], [2], [4], [8].

Any mapping of any finite number, say n, of points in $\mathbf{R}^d$ onto $\mathbf{R}$ can be implemented by a three-layer neural network having n hidden layer units with a nonlinear activation function [3]. This mapping can also be realized with an unscaled activation function [8].

Three-layer neural networks are often classified depending on the spaces, on which the input-output function of the hidden layer units is defined [9]. Another way of treating them is not to classify them. Ito [6] has shown that the concept of activation functions can be extended to slowly increasing functions defined on $\mathbf{R}^c$, $1 \leq c \leq d$, so that the activation function defined on $\mathbf{R}$ and the radial basis function defined on $\mathbf{R}^d$ can be treated on a common basis.

The paper [6] includes a proof that a slowly increasing function g defined on $\mathbf{R}^c$ can be an activation function for the finite mapping without scaling, if the support of the Fourier transform $\mathcal{F}_c g$ includes a nonempty open set. Later this

J.R. Dorronsoro (Ed.): ICANN 2002, LNCS 2415, pp. 337–343, 2002.

condition was weakened: if the support of $\mathcal{F}_c g$ includes a converging sequence of points aligned on a line, it can be an activation function for the finite mapping without scaling [7]. The identity theorem of holomorphic functions is used in [7]. This is the reason why the converging sequence of points must be on a line.

In this paper, we weaken this condition and, moreover, prove the result without using the identity theorem. As in [6], [7], the mapping realized by our three-layer neural networks is of the form

$$f(x_i) = \sum_{j=1}^{n} a_j g(Wx_i + z_j), \qquad i = 1, \cdots, n, \tag{1}$$

where $z_j \in \mathbf{R}^c$, W is a $d \times c$ matrix having orthonormal row vectors and g is a slowly increasing function defined on $\mathbf{R}^c$, implying that g is used without scaling, and that the sigmoid function defined on $\mathbf{R}$ and the radial basis function defined on $\mathbf{R}^d$ are treated on a common basis. The key idea for the proof is to use Taylor's theorem.

2   Linear Sum of Exponential of Functions

Let $1 \leq c \leq$ d. Denote by $\mathbf{S}^{d,c}$ a set of $c \times d$ matrices, each of which has orthonormal row vectors. Note that $\mathbf{S}^{d,1}$ is the set $\mathbf{S}^{d-1}$ of unit vectors and $\mathbf{S}^{d,d}$ is the group $\mathbf{O}(d)$ of orthogonal matrices of order d. For a vector $\omega \in \mathbf{S}^{c-1}$, $\mathbf{L}_\omega = \{t\omega| - \infty < t < \infty\}$ is a line in $\mathbf{R}^c$. We regard t as the coordinate of a point $t\omega$ on $\mathbf{L}_\omega$ and set $\mathbf{L}_\omega^+ = \{t\omega| t \geq 0\}$. We denote by P_ω the projection operator onto $\mathbf{L}_\omega$.

Lemma 1. Let $\omega \in \mathbf{S}^{c-1}$, let $x_1, \cdots, x_n \in \mathbf{R}^c$ such that $P_\omega x_1, \cdots, P_\omega x_n$ are distinct, and let $\{u_k\}_{k=1}^{\infty}$ be a sequence of distinct points of $\mathbf{R}^c$ converging to the origin. If

$$\lim_{k \to \infty} \frac{\|u_k - P_\omega u_k\|}{\|u_k\|^n} = 0 \tag{2}$$

then,

$$\sum_{i=1}^{n} a_i e^{\sqrt{-1} x_i \cdot u_k} = 0, \qquad k = 1, 2, \cdots, \tag{3}$$

imply $a_1 = \cdots = a_n = 0$.

Proof. By taking a subsequence of $\{u_k\}_{k=1}^{\infty}$ and/or replacing ω by $-\omega$, we may suppose that $P_\omega u_k \in \mathbf{L}_\omega^+$. Let ζ_i be the coordinate of $P_\omega x_i$ on $\mathbf{L}_\omega$. Then, $\zeta_1, \cdots, \zeta_n$ are distinct. We prove n equalities below by induction:

$$\sum_{i=1}^{n} a_i \zeta_i^q = 0, \qquad q = 0, \cdots, n - 1. \tag{4}$$

If (3) holds, we immediately have (4) for $q = 0$ because the left hand side of (3) converges to $\sum_{i=1}^{n} a_i$ as $k \to \infty$. Suppose that (4) holds for $q < p \leq n - 1$. Since $x_i \cdot P_\omega u_k = \zeta_i \|P_\omega u_k\|$, we have that

$$\left| \frac{1}{\|u_k\|^p} \{ e^{\sqrt{-1}x_i \cdot u_k} - \sum_{m=0}^{p-1} \frac{1}{m!} (\sqrt{-1}\zeta_i \|P_\omega u_k\|)^m \} - \frac{1}{p!} (\sqrt{-1}\zeta_i)^p \right|$$

$$\leq \frac{1}{\|u_k\|^p} \left| \sum_{m=p+1}^{\infty} \frac{1}{m!} (\sqrt{-1}x_i \cdot u_k)^m \right| + \frac{1}{p!} \left| \left(\frac{\sqrt{-1}\zeta_i \|P_\omega u_k\|}{\|u_k\|} \right)^p - (\sqrt{-1}\zeta_i)^p \right|$$

$$+ \frac{1}{\|u_k\|^p} \sum_{m=0}^{p} \frac{1}{m!} \left| (\sqrt{-1}x_i \cdot u_k)^m - (\sqrt{-1}x_i \cdot P_\omega u_k)^m \right|$$

$$\leq \|x_i\|^{p+1} \|u_k\| e^{|x_i \cdot u_k|} + \frac{1}{p!} |\zeta_i|^p \left| \frac{\|P_\omega u_k\|^p}{\|u_k\|^p} - 1 \right|$$

$$+ \sum_{m=1}^{p} \frac{1}{(m-1)!} \frac{\|u_k - P_\omega u_k\|}{\|u_k\|^{p-m+1}} \|x_i\|^m \longrightarrow 0 \tag{5}$$

as $k \to \infty$, because $\|u_k\| \to 0$, $\|P_\omega u_k\|/\|u_k\| \to 1$ and $\|u_k - P_\omega u_k\|/\|u_k\|^n \to 0$. Hence,

$$\lim_{k \to \infty} \sum_{i=1}^{n} a_i \frac{1}{\|u_k\|^p} \{ e^{\sqrt{-1}x_i \cdot u_k} - \sum_{m=0}^{p-1} \frac{1}{m!} (\sqrt{-1}\zeta_i \|P_\omega u_k\|)^m \}$$

$$- \sum_{i=1}^{n} a_i \frac{1}{p!} (\sqrt{-1}\zeta_i)^p = 0. \tag{6}$$

By (3) and (4) for $q < p$, the first sum on the left hand side of (6) is equal to zero. Hence, we obtain (4) for $q = p$, which in turn implies (4) for all $q < n$. We regard (4) as simultaneous equations with respect to $a_1, \cdots, a_n$. As $\zeta_1, \cdots, \zeta_n$ are distinct, Vandermonde's determinant with these constants is nonzero. Accordingly, we obtain that $a_1 = \cdots = a_n = 0$ by Cramer's formula.

In [7], we have proved that if the points u_k are aligned on the line $\mathbf{L}_\omega$, (3) implies $a_1 = \cdots = a_n = 0$, using the identity theorem of holomorphic function. Here, we have obtained the same conclusion under a weaker condition (2) without using the identity theorem.

Lemma 2. Let $\{y_k\}_{k=1}^{\infty}$ be a converging sequence of distinct points of $\mathbf{R}^c$ and let y_0 be the limit of the sequence. Suppose that $u_k = y_k - y_0$, $k = 1, 2, \cdots$, satisfy the condition (2) for a unit vector $\omega \in \mathbf{S}^{c-1}$. Then, for any n distinct points $x_1, \cdots, x_n \in \mathbf{R}^c$, there exists a matrix $W \in \mathbf{S}^{c,c}$ such that

$$\sum_{i=1}^{n} a_i e^{\sqrt{-1}W x_i \cdot y_k} = 0, \qquad k = 1, 2, \cdots, \tag{7}$$

imply $a_1 = \cdots = a_n = 0$.

Proof. There exists a matrix $W \in \mathbf{S}^{c,c}$ such that $P_\omega W x_i$, $i = 1, \cdots, n$, are distinct. Set $a_i' = a_i e^{\sqrt{-1}W x_i \cdot y_0}$. Then,

$$\sum_{i=1}^{n} a_i e^{\sqrt{-1}W x_i \cdot y_k} = \sum_{i=1}^{n} a_i' e^{\sqrt{-1}W x_i \cdot u_k} = 0, \qquad k = 1, 2, \cdots. \tag{8}$$

By Lemma 1, (8) imply $a'_1 = \cdots = a'_n = 0$. Since $e^{\sqrt{-1}Wx_i \cdot y_0} \neq 0$, we finally obtain that $a_1 = \cdots = a_n = 0$.

3   Theorems

If a slowly increasing distribution ([10]) is a function, we call it a slowly increasing function. Roughly speaking, if $(1 + \|z\|^2)^{-m}g$ is integrable for an $m > 0$, g is a slowly increasing function. In this paper, activation functions are slowly increasing functions defined on $\mathbf{R}^c$, $1 \leq c \leq d$. For an integrable function g defined on $\mathbf{R}^c$, we define its Fourier transform by $\mathcal{F}_c g(y) = \int e^{-ix \cdot y} g(x) dx$. This transform can be extended to slowly increasing distributions as well as slowly increasing functions. The lemma below is proved in [7].

Lemma 3. Let T be a distribution on $\mathbf{R}^c$ and let h be an infinitely continuously differentiable function on $\mathbf{R}^c$. If $hT = 0$, then $h = 0$ on the support of T.

Lemma 4. Let g be a slowly increasing function defined on $\mathbf{R}^c$. Suppose that the support of $\mathcal{F}_c g$ includes a converging sequence of distinct points y_k, $k = 1, 2, \cdots$, for which $u_k = y_k - y_0$, $k = 1, 2, \cdots$, satisfy the condition (2) for a unit vector $\omega \in \mathbf{S}^{c-1}$, where y_0 is the limit of the sequence. Then, for any distinct points $x_1, \cdots, x_n$ of $\mathbf{R}^c$, there is $W \in \mathbf{S}^{c,c}$ for which $g(Wx_i + x)$, $i = 1, \cdots, n$, are linearly independent.

Proof. Suppose that

$$\sum_{i=1}^n a_i g(Wx_i + x) = 0 \qquad \text{on} \qquad \mathbf{R}^c.$$

Then,

$$\sum_{i=1}^n a_i \mathcal{F}_c g(Wx_i + \cdot) = \sum_{i=1}^n a_i e^{\sqrt{-1}Wx_i \cdot y} \mathcal{F}_c g = 0.$$

By Lemma 3, this implies that (7) holds. Hence, there is a matrix $W \in \mathbf{S}^{c,c}$ for which $a_1 = \cdots = a_n = 0$. Thus, the lemma follows.

We denote by $\det(a_{ij})_{i,j=1}^n$ a determinant of order n having a_{ij} as its (i,j)-th element. If $\det(g(Wx_i + z_j))_{i,j=1}^n \neq 0$, we can realize any finite mapping from $\{x_1, \cdots, x_n\}$ onto $\mathbf{R}$ by (1), determining the coefficients a_i by Cramer's formula. This idea was used in [1]. The proof of Lemma 5 is described in [8].

Lemma 5. Let g_{ij}, $i = 1, \cdots, n$, $j = 1, \cdots, n$, be functions defined on $\mathbf{R}^c$. Suppose that $g_{1j}, \cdots, g_{jj}$ are linearly independent for $j = 1, \cdots, n$. Then, there are points $z_1, \cdots, z_n$ for which $\det(g_{ij}(z_j))_{i,j=1}^n \neq 0$.

Theorem 6. Let $1 \leq c \leq d$ and let g be any slowly increasing function defined on $\mathbf{R}^c$. Suppose that the support of $\mathcal{F}_c g$ includes a converging sequence of distinct points y_k, $k = 1, 2, \cdots$, for which $u_k = y_k - y_0$, $k = 1, 2, \cdots$, satisfy the condition

(2) for a unit vector $w \in \mathbf{S}^{c-1}$, where y_0 is the limit of the sequence. Then, for any distinct points $x_1, \cdots, x_n$ of $\mathbf{R}^d$ and any constants $b_1, \cdots, b_n$, there are $W \in \mathbf{S}^{d,c}$, $a_1, \cdots, a_n \in \mathbf{R}$ and $z_1, \cdots, z_n \in \mathbf{R}^c$ for which

$$\sum_{j=1}^{n} a_j g(W x_i + z_j) = b_i, \qquad i = 1, \cdots, n. \tag{9}$$

Proof. There exists a matrix $W_1 \in \mathbf{S}^{d,c}$ for which $W_1 x_i \in \mathbf{R}^c$, $i = 1, \cdots, n$, are distinct. Hence, by Lemma 4, there exists a matrix $W_2 \subset \mathbf{S}^{c,c}$ for which $g(W_2 W_1 x_i + x)$, $i = 1, \cdots, n$. are linearly independent. By Lemma 5, there exist $z_1, \cdots, z_n \in \mathbf{R}^c$ for which $\det(g(W x_i + z_j))_{i,j=1}^{n} \neq 0$, where $W = W_2 W_1$. Consequently, there exist $a_1, \cdots, a_n$ for which (9) holds.

If $c = 1$, the condition (2) is always satisfied for any n. Hence:

Corollary 7. Let g be a slowly increasing function defined on $\mathbf{R}$. Suppose that the support of $\mathcal{F}_1 g$ includes a converging sequence of distinct points y_k, $k = 1, 2, \cdots$. Then, for any number n, any distinct points $x_1, \cdots, x_n$ of $\mathbf{R}^d$ and any constants $b_1, \cdots, b_n$, there are a unit vector $v \in \mathbf{S}^{d-1}$ and points $z_1, \cdots, z_n \in \mathbf{R}$ for which

$$\sum_{j=1}^{n} a_j g(v \cdot x_i + z_j) = b_i, \qquad i = 1, \cdots, n.$$

Proof. When $c = 1$, the condition (2) is satisfied for any n because $P_w u_k = u_k$. Hence, there is a matrix $W \in \mathbf{S}^{d,1}$ for which (9) hold. Regarding the W as a vector v in $\mathbf{S}^{d-1}$, we obtain the corollary.

Under a stronger condition (10) than (2), the theorem below holds.

Theorem 8. Let $1 \leq c \leq d$ and let g be any slowly increasing function defined on $\mathbf{R}^c$. Suppose that the support of $\mathcal{F}_c g$ includes a converging sequence of distinct points y_k, $k = 1, 2, \cdots$, for which $u_k = y_k - y_0$, $k = 1, 2, \cdots$, satisfy

$$\|u_k - P_w u_k\| < a e^{-\|u_k\|^{-1}} k = 1, 2, \cdots, \tag{10}$$

for a unit vector $w \in \mathbf{S}^{c-1}$, where a is a positive constant and y_0 is the limit of the sequence. Then, for any number n, any distinct points $x_1, \cdots, x_n$ of $\mathbf{R}^d$ and any constants $b_1, \cdots, b_n$, there are $W \in \mathbf{S}^{d,c}$, $a_1, \cdots, a_n \in \mathbf{R}$ and $z_1, \cdots, z_n \in \mathbf{R}^c$ for which

$$\sum_{j=1}^{n} a_j g(W x_i + z_j) = b_i, \qquad i = 1, \cdots, n.$$

Proof. The condition (10) implies (2) for any n. Hence, this theorem follows.

The corresponding theorem in [6] and the main theorem in [7] are now corollaries to Theorem 6 or 8. In the case of $c = 1$, the condition (2) or (10) can be removed.

4   Discussion

Now the key condition on the activation function g defined on $\mathbf{R}^c$, $1 \leq c \leq d$, is that the support of $\mathcal{F}_c g$ includes a sequence of converging distinct points y_1, $y_2, \cdots$ which come close to a line as $k \to \infty$. This is weaker than the condition stated in [7], where the points must be strictly on a line. It may be also a progress that the results in this paper are proved without using the identity theorem of holomorphic functions.

Seeing the condition of Theorem 8, we can construct a function which satisfies neither the condition in [6] nor that in [7], but can be an activation function for any finite mapping from $\mathbf{R}^d$ onto $\mathbf{R}$ without scaling. Of course, most of familiar functions such as sigmoid functions and ordinary radial basis functions satisfy not only the condition in this paper but also those stated in [6] and [7]. It is not easy to give an intuitive characterization of the functions which satisfy only the condition in this paper. Many periodic functions on $\mathbf{R}$, such as triangle-wave or saw-wave functions, satisfy the condition of Theorem 8 as well as that of the main theorem in [7].

If there are only a finite number of points $y_1, \cdots, y_n$ for which $\mathcal{F}_c g(y_k) \neq 0$, $k = 1, \cdots, n$, it is easy to find a set of points $x_1, \cdots, x_n$ for which $\sum_{i=1}^n a_i e^{\sqrt{-1} x_i \cdot y_k} = 0$, $k = 1, \cdots, n$, do not imply that $a_1 = \cdots = a_n = 0$. Hence, it is necessary for the support of $\mathcal{F}_c g$ to include infinitely many distinct points to realize arbitrary finite mappings. However, this is not sufficient. We can construct a counter example in which (3) for a nonconverging sequence $\{y_k\}_{k=1}^\infty$ do not imply $a_1 = \cdots = a_n = 0$. Whether or not the condition obtained in this paper can be further weakened is still a problem to be solved.

In this paper, the shifts of an activation function are used. However, rotations of an function can also be linearly independent (see [5], [11] for the case $c = 1$). Implementation of finite mappings using the rotations will be discussed somewhere else.

References

1. Ito, Y.: Representation of functions by superpositions of a step or sigmoid function and their applications to neural network theory. Neural Networks **4**, (1991a) 385–394.
2. Ito, Y.: Approximation of functions on a compact set by finite sums of a sigmoid function without scaling. Neural Networks **4**, (1991b) 817–826.
3. Ito, Y.: Finite mapping by neural networks and truth functions. Math. Scientist **17**, (1992a) 69–77.
4. Ito, Y.: Approximation of continuous functions on $\mathbf{R}^d$ by linear combinations of shifted rotations of a sigmoid function with and without scaling. Neural Networks **5**, (1992b) 105–115.
5. Ito, Y.: Nonlinearity creates linear independence. Adv. Compt. Math. **5**, (1996) 189–203.
6. Ito, Y.: What can a neural network do without scaling activation functions? Proceedings of ICANNGA2001, 1–6. eds. V. Kurková et al. Springer, (2001a)

7. Ito, Y.: Independence of unscaled basis functions and finite mappings by neural networks. Math. Scientist **26**, (2001b) 117–126.
8. Ito, Y. and K. Saito: Supperposition of linearly independent functions and finite mapping by neural networks. Math. Scientists **21**, (1996) 27–33.
9. Scarselli, F. and A.C. Tsoi: Universal approximation using feedforward neural networks: a survey of some existing methods, and some new results. Neural Networks **11**, (1998) 15–37.
10. Schwartz, L.: Théorie des distributions. Hermann, Paris. (1966)
11. Sussmann, H.J.: Uniqueness of the weights for minimal feedforward nets with a given input-output map. Neural Networks **5**, (1992) 589–593.

Identification of Wiener Model Using Radial Basis Functions Neural Networks

Ali Syed Saad Azhar and Hussain N. Al-Duwaish

King Fahd University of Petroleum and Minerals,
Department of Electrical Engineering,
Dhahran 31261, Saudi Arabia
{saadali, hduwaish}@kfupm.edu.sa
http://www.kfupm.edu.sa

Abstract. A new method is introduced for the identification of Wiener model. The Wiener model consists of a linear dynamic block followed by a static nonlinearity. The nonlinearity and the linear dynamic part in the model are identified by using radial basis functions neural network (RBFNN) and autoregressive moving average (ARMA) model, respectively. The new algorithm makes use of the well known mapping ability of RBFNN. The learning algorithm based on least mean squares (LMS) principle is derived for the training of the identification scheme. The proposed algorithm estimates the weights of the RBFNN and the coefficients of ARMA model simultaneously.

1 Introduction

System identification and modeling is a very important step in control applications since it is a prerequisite for analysis and controller design. Due to the nonlinear nature of most of the applications there has been extensive research covering the field of nonlinear system identification [1]. One of the most promising and simple nonlinear models is the Wiener model which is characterized by a linear dynamic part and a static nonlinearity connected in cascade as shown in Fig. 1 where, $u(t)$ is the input to the system, $y(t)$ is the output and $x(t)$ is

Fig. 1. Structure of Wiener Model.

the intermediate nonmeasureable quantity. The identification of Wiener systems involves estimating the parameters describing the linear and the nonlinear parts from the input-output data. Examples of applications includes pH control [2], fluid control flow [3], identification of biological systems [4], control systems [5] and identification of linear systems with nonlinear sensors [6]. These examples show and apparent need for algorithms able to recover the nonlinearities in the systems of various kinds. Many identification methods have been developed to identify the Wiener model [7], [8], [9], [10], [11], [5], [12], [13], [14], and [15]. Generally Wiener model is represented by the following set of equations:

$$y(t) = f(x(t), \rho), \tag{1}$$

the intermediate nonmeasureable variable $x(t)$ is the output of the linear part as,

$$x(t) = \frac{B(q^{-1})}{A(q^{-1})} q^{-d} u(t), \tag{2}$$

with

$$B(q^{-1}) = b_o + b_1 q^{-1} + \cdots + b_m q^{-m},$$
$$A(q^{-1}) = 1 + a_1 q^{-1} + \cdots + a_n q^{-n}$$

where q^{-1} is the unit delay operator, $u(t)$ is the input, $y(t)$ is the output, (m, n) represent the order of the linear dynamics, d is the system delay and ρ corresponds to the parameters of the nonlinearity $e.g.$ weights of the RBFNN. Thus, the problem of the Wiener model identification is to estimate the parameters $a_1, \ldots, a_n, \ b_o, \ldots, b_m$ and ρ from input-output data. The algorithm presented in [5] used multi-layered feed forward neural networks (MFNN) to identify the static nonlinearities in the Wiener model. The back-propagation algorithm is used to update the weights of the MFNN. It is well known in the literature that the convergence of the back-propagation algorithm is very slow compared to LMS when used for training radial basis functions neural networks (RBFNN) [16], [17], [18], [19]. Moreover, the RBFNN has the same universal approximation capabilities as the MFNN [16]. This motivated the use of RBFNN instead of MFNN to model the static nonlinearity. The linear part is modeled by an ARMA model.

2   ARMA/RBFNN Identification Method Structure

The proposed identification structure for the Wiener model consists of an ARMA model in series with RBFNN as shown in Fig. 2.

Fig. 2. ARMA/RBFNN identification structure for Wiener model.

The linear part of the Wiener model is modeled by an ARMA model, whose output is given by,

$$y(t) = \sum_{i=1}^{n} a_i y(t-i) + \sum_{j=0}^{m} b_j u(t-j) \tag{3}$$

The static nonlinearity in the Wiener model is identified using RBFNN. RBFNN is a type of feed forward, multi-layered network. In these networks the learning involves only one layer comprising of basis functions, normally taken as a Gaussian function. The output of the RBFNN is a weighted sum of the outputs of the basis functions, given by

$$y(t) = W\varPhi(t),$$

$$y(t) = \sum_{i=1}^{n_o} w_i \phi(\|x(t) - c_i\|),$$

$y(t)$ is the output, w_i is the weight corresponding to the ith basis function ϕ in the hidden layer.

3   Training Algorithm for Wiener Model

Considering Fig. 2, the objective is to develop a recursive algorithm that adjusts the parameters of the ARMA model and the weights of the RBFNN in such a way, that the set of inputs produces the desired set of outputs. This goal is achieved by developing a new parameter estimation algorithm based on the LMS approach. The parameters (weights of RBFNN and the coefficients of the ARMA) are updated by minimizing the performance index I given by,

$$I = \frac{1}{2} e^2(t),$$

where, $e(t) = y(t) - \hat{y}(t)$ and $\hat{y}(t)$ is the estimated output of the Wiener model. According to LMS algorithm the coefficients of the ARMA model and the weights of the RBFNN should be updated in the negative direction of the gradient. Keeping the coefficients in a parameter vector θ as $\theta = [a_1 \ \dots \ a_n \ b_o \ \dots \ b_m]$ and the regressions of the quantities $x(t)$ and $u(t)$ in regression vector $\psi(t)$ as $\psi(t) = [\hat{x}(t-1) \ \dots \ \hat{x}(t-n) \ \hat{u}(t-d) \ \dots \ \hat{u}(t-m-d)]$, finding partial derivative.

$$\frac{\partial I}{\partial \theta} = \frac{1}{2} \frac{\partial e^2(t)}{\partial \theta},$$

$$= e(t) \frac{\partial}{\partial \theta} \left(y(t) - W\varPhi(t) \right),$$

$$= -e(t) \frac{\partial}{\partial \theta} \left(w_1 \phi(\|\hat{x}(t) - c_1\|) + \dots + w_{n_o} \phi(\|\hat{x}(t) - c_{n_o}\|) \right),$$

$$= -e(t) \frac{\partial}{\partial \theta} \left(w_1 \exp\left(-\frac{\|\hat{x}(t) - c_1\|^2}{\beta^2}\right) + \dots + w_{n_o} \exp\left(-\frac{\|\hat{x}(t) - c_{n_o}\|^2}{\beta^2}\right) \right),$$

$$= -e(t) \frac{\partial}{\partial \theta} \left(w_1 \exp\left(-\frac{(\hat{x}(t) - c_1)^2}{\beta^2}\right) + \dots + w_{n_o} \exp\left(-\frac{(\hat{x}(t) - c_{n_o})^2}{\beta^2}\right) \right). \tag{4}$$

Let $\Omega_i = w_i \exp(-\frac{(\hat{x}(t) - c_i)^2}{\beta^2})$, where $i = [1, 2, \ldots n_o]$. Considering the partial derivative of Ω_j term $w.r.t.$ any a_i in the parameter vector θ,

$$\frac{\partial \Omega_j}{\partial a_i} = \frac{\partial}{\partial a_i} w_j \exp(-\frac{(\hat{x}(t) - c_j)^2}{\beta^2}),$$

$$= -2(\hat{x}(t) - c_j)\frac{w_j}{\beta^2} \exp(-\frac{(\hat{x}(t) - c_j)^2}{\beta^2})\hat{x}(t - i),$$

$$\frac{\partial \Omega_j}{\partial a_i} = -2\hat{x}(t - i)(\hat{x}(t) - c_j)\frac{w_j}{\beta^2} \exp(-\frac{(\hat{x}(t) - c_j)^2}{\beta^2}). \tag{5}$$

Putting the $\frac{\partial \Omega_j}{\partial a_i}$ term from Eq. 5 in Eq. 4 for a_i,

$$= -\frac{2e(t)}{\beta^2}\hat{x}(t - i) \sum_{j=1}^{n_o}(\hat{x}(t) - c_j)w_j\phi(\|\hat{x}(t) - c_j\|).$$

Similarly for any b_i,

$$\frac{\partial I}{\partial b_i} = -\frac{2e(t)}{\beta^2}\hat{u}(t - i - d) \sum_{j=1}^{n_o}(\hat{x}(t) - c_j)w_j\phi(\|\hat{x}(t) - c_j\|),$$

and finally stacking the derivatives in Eq. 4 again,

$$\frac{\partial I}{\partial \theta} = \frac{2e(t)}{\beta^2}\psi(t) \sum_{j=1}^{n_o}(\hat{x}(t) - c_j)w_j\phi(\|\hat{x}(t) - c_j\|).$$

Now, this gradient is used in finding the updated parameters.

$$\theta(K + 1) = \theta(K) - \frac{2\alpha e(t)}{\beta^2}\psi(t) \sum_{j=1}^{n_o}(\hat{x}(t) - c_j)w_j\phi(\|\hat{x}(t) - c_j\|). \tag{6}$$

The partial derivatives for the weights are derived as follows,

$$\frac{\partial I}{\partial W} = \frac{1}{2}\frac{\partial e^2(t)}{\partial W},$$

$$= e(t)\frac{\partial}{\partial W}(y(t) - \hat{y}(t)),$$

$$= e(t)\frac{\partial}{\partial W}(y(t) - W\Phi(t)),$$

$$\frac{\partial I}{\partial W} = -e(t)\Phi(t). \tag{7}$$

The gradient in Eq. 7 is used to find the updated weights.

$$W(K + 1) = W(K) + \alpha\, e(t)\Phi(t). \tag{8}$$

4   Simulation Results

In this example, the proposed identification algorithm is applied to a model that describes a valve for control of fluid flow described in [5] and [9], in presence of output additive white Gaussian noise with SNRs 20dB and 30dB.

The linear part is described by,

$$x(t) = 0.4x(t-1) + 0.35x_2(t-2) + 0.1x_3(t-3) + 0.8u(t) - 0.2b_1u(t-1),$$

$$(9)$$

and the nonlinear part is given by,

$$y(t) = \frac{x(t)}{\sqrt{0.10 + 0.90x^2(t)}}.$$

$$(10)$$

In this model, $u(t)$ represents the pneumatic control signal applied to the stem of the valve and $x(t)$ represents the the stem position. The linear dynamics describe the dynamic balance between the control signal, a counteractive spring force and friction. The resulting flow through the valve is given by the nonlinear function of the stem position $x(t)$ reflected by $y(t)$.

The proposed identification algorithm is applied to estimate the linear and nonlinear parts of the model. An ARMA model structure was used given by,

$$x(t) = a_1x(t-1) + a_2x_2(t-2) + a_3x_3(t-3) + b_ou(t) + b_1u(t-1).$$

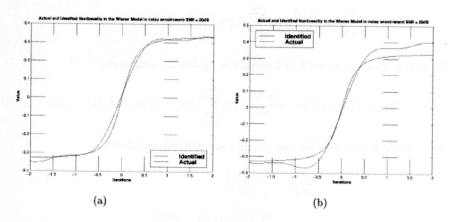

(a)   (b)

Fig. 3. Actual and identified nonlinearities for SNRs 30dB and 20dB

The nonlinearity is modeled by an RBFNN centered at $[-0.2, -0.1, 0.1, 2]$. The width of the basis functions and the learning rate are set to 1.1 and 0.04.

Using random inputs $u(t)$ uniformly distributed in the interval $[-2, 2]$, the desired outputs are generated by employing the process model given by Eq. 9 and Eq. 10.

The identification algorithm identified the static nonlinearity and the parameters of the linear part accurately. The actual and identified static nonlinearity in the Wiener model for different cases of SNRs 20dB and 30dB are shown in Fig. 3(a) and Fig. 3(b).

Conclusions

A new method for identification of Wiener systems has been developed. The new method consists of an ARMA in series with an RBFNN model. Training algorithm is developed to update the weights of the RBFNN and coefficients of ARMA model. The new method is suitable for standard linear control design because the nonlinearity can be cancelled by inserting the inverse of the nonlinearity into the appropriate location in the control loop. The inverse of the nonlinearity can be modeled by the another RBFNN. The algorithm is derived for the single-input single-output systems. However, it can be generalized to multi-input multi-output systems. Exhaustive simulation studies suggest robust and satisfactory identification behaviour of the developed method and algorithm.

Acknowledgement. The authors would like to acknowledge the support of King Fahd University of Petroleum & Minerals, Dhahran, Saudi Arabia.

References

1. S. A. Billings. Identification of non-linear systems – a survey. *IEE proceedings – part D*, 127(6):272–285, 1980.
2. J. N. Sandra, A. Palazaglu, and Romagnoli J. A. Application of wiener model predictive control (wmpc) to a ph neutralization experiment. *IEEE Transactions on Control Systems Technology*, 7:437–45, 1999.
3. M. Singh, J. Elloy, r. Mezencev, and N. Munro. *Applied Industrial Control*. Pergamon Press, Oxford, UK, 1978.
4. Kenneth J. Hunt, Marko Munih,Nick de Donaldson, and Fiona M. D. Barr. Investigation of hammerstein hypothesis in the modeling of electrically simulated muscle. *IEEE Transactions on Biomedical Engineering*, 45, 1998.
5. H. Al-Duwaish, M. Nazmul Karim, and V. Chandrasekar. Use of multilayer feedforward neural networks in identification and control of wiener model. *IEE Proceedings: Control Theory and Applications*, 143(3):225–258, 1995.
6. E. Q. Doeblin. *Measurement Systems*. McGraw Hill Int., Auckland, New Zealand, 1983.
7. A. Billings and S. Y. Fakhouri. Identification of non-linear systems using correlation analysis and pseudorandom inputs. *International Journal of Systems Science*, 11:261–279, 1980.
8. W. Greblicki. Nonparametric approach to wiener systems identification. *IEEE Transactions on Information Theory*, 44:538–545, 1997.
9. T. Wigren. Recursive prediction error identification using the nonlinear wiener model. *Automatica*, 29:1011–1025, 1993.

10. D. T. Westwick and R. E. Kearney. Identification of multiple input wiener system. April 1990.
11. D. W. Hu and Z. Z. Wang. An identification method for the wiener model of nonlinear systems. December 1991.
12. Marco Lovera and Miche Verhaegen. Recursive subspace identification of linear and nonlinear type wiener type models. September 1998.
13. Anna Hagenblad and Lennart Ljung. Maximum likelihood estimation of wiener models. *IEEE Proceedings of the 39th Conference on Decision & Control,* 3:2417–2418, 2000.
14. Yong Fang and T. W. S. Chow. Orthogonal wavelet neural network applying to identification of wiener model. *IEEE transactions Circuit and Systems,* 47(4):591–593, 2000.
15. Seth L. Lacy and R. Scott Erwin Dennis S. Bernstein. Identification of wiener systems with known nonivertible nonlinearities. June 2001.
16. S. Haykin. *Neural Networks: A Comprehensive Foundation II.* Macmillan/IEEE Press, 1994, 1999.
17. W. L. Jun-Dong, P. Jones. Comparison of neural network classifiers for nscat sea ice flag. 1998.
18. L. Fortuna, G. Muscato, and M. G. Xibilia. A comparison between hmlp and hrbf for attitude control. 122:318–328, March 2001.
19. R. A. Finan, A. T. Sapeluk, and R. I. Damper. Comparison of multilayer and radial basis functions neural networks for the text dependent speaker recognition. 1996.

A New Learning Algorithm for Mean Field Boltzmann Machines

Max Welling and Geoffrey E. Hinton

Depart. of Computer Science, Univ. of Toronto
10 King's College Road, Toronto, M5S 3G5 Canada.
{welling,hinton}@cs.toronto.edu

Abstract. We present a new learning algorithm for Mean Field Boltzmann Machines based on the *contrastive divergence* optimization criterion. In addition to *minimizing* the divergence between the data distribution and the equilibrium distribution, we *maximize* the divergence
between one-step reconstructions of the data and the equilibrium distribution. This eliminates the need to estimate equilibrium statistics, so we
do not need to approximate the multimodal probability distribution of
the free network with the unimodal mean field distribution. We test the
learning algorithm on the classification of digits.

1 Introduction

A network of symmetrically-coupled binary (0/1) threshold units has a simple
quadratic energy function that governs its dynamic behavior [4].

$$E(\mathbf{v}, \mathbf{h}) = -(\frac{1}{2}\mathbf{v}^T\mathbf{V}\mathbf{v} + \frac{1}{2}\mathbf{h}^T\mathbf{W}\mathbf{h} + \mathbf{v}^T\mathbf{J}\mathbf{h}) \qquad (1)$$

where $\mathbf{v}$ represent visible units whose states are fixed by the data $\{\mathbf{d}_{1:N}\}$, $\mathbf{h}$
represent hidden units, and where we have added one unit with value always 1,
whose weights to all other units represent the biases. The energy function can
be viewed as an indirect way of defining a probability distribution over all the
binary configurations of the network [2] and if the right stochastic updating rule
is used, the dynamics eventually produces samples from this Boltzmann distri-
bution, $P(\mathbf{v}, \mathbf{h}) = e^{-E(\mathbf{v},\mathbf{h})}/Z$ where Z denotes the normalization constant or
partition function. This "Boltzmann machine" (BM) has a simple learning rule
[2] which minimizes the Kullback-Leibler divergence between the data distribu-
tion $P_0(\mathbf{v}, \mathbf{h}) = P(\mathbf{h}|\mathbf{v})\tilde{P}_0(\mathbf{v})$ (where $\tilde{P}_0(\mathbf{v})$ is the empirical data distribution)
and the equilibrium distribution $P_{\mathrm{EQ}}(\mathbf{v}, \mathbf{h})$.

$$\delta\mathbf{W} \propto \langle\mathbf{h}\mathbf{h}^T\rangle_0 - \langle\mathbf{h}\mathbf{h}^T\rangle_{\mathrm{EQ}} \qquad \delta\mathbf{V} \propto \langle\mathbf{v}\mathbf{v}^T\rangle_0 - \langle\mathbf{v}\mathbf{v}^T\rangle_{\mathrm{EQ}} \qquad \delta\mathbf{J} \propto \langle\mathbf{v}\mathbf{h}^T\rangle_0 - \langle\mathbf{v}\mathbf{h}^T\rangle_{\mathrm{EQ}}$$
$$(2)$$

This learning rule is both simple and local, but the settling time required to get
samples from the right distribution and the high noise in the estimates of the
correlations make learning slow and unreliable.

J.R. Dorronsoro (Ed.): ICANN 2002, LNCS 2415, pp. 351–357, 2002.
© Springer-Verlag Berlin Heidelberg 2002

To improve the efficiency of the BM learning algorithm Peterson and Anderson [6] introduced the mean field (MF) approximation which replaces the averages in eqn. 2 with averages over factorized distributions.

$$\langle \mathbf{h}\mathbf{h}^T \rangle_0 \to \frac{1}{N} \sum_{n=1}^{N} \mathbf{q}_{0,n}\mathbf{q}_{0,n}^T \quad \langle \mathbf{v}\mathbf{v}^T \rangle_{\mathrm{EQ}} \to \mathbf{r}_{\mathrm{EQ}}\mathbf{r}_{\mathrm{EQ}}^T \quad \langle \mathbf{h}\mathbf{h}^T \rangle_{\mathrm{EQ}} \to \mathbf{q}_{\mathrm{EQ}}\mathbf{q}_{\mathrm{EQ}}^T \quad (3)$$

where the parameters $\mathbf{q}_{0,n}$, $\mathbf{r}_{\mathrm{EQ}}$ and $\mathbf{q}_{\mathrm{EQ}}$ are computed through the mean field equations,

$$\mathbf{q}_{0,n} = \sigma\left(\mathbf{W}\mathbf{q}_{0,n} + \mathbf{J}^T\mathbf{d}_n\right) \; \mathbf{r}_{\mathrm{EQ}} = \sigma\left(\mathbf{V}\mathbf{r}_{\mathrm{EQ}} + \mathbf{J}\mathbf{q}_{\mathrm{EQ}}\right) \; \mathbf{q}_{\mathrm{EQ}} = \sigma\left(\mathbf{W}\mathbf{q}_{\mathrm{EQ}} + \mathbf{J}^T\mathbf{r}_{\mathrm{EQ}}\right)$$
$$(4)$$

and σ denotes the sigmoid function. These learning rules perform gradient descent on the cost funtion

$$F^{\mathrm{MF}} = F_0^{\mathrm{MF}} - F_{\mathrm{EQ}}^{\mathrm{MF}} \qquad \text{with} \qquad F_Q^{\mathrm{MF}} = \langle E \rangle_Q - H(Q) \qquad (5)$$

where $Q = \prod_i q_i^{s_i}(1-q_i)^{1-s_i}$ is a factorized MF distribution, E is the energy in eqn. 1 and H denotes the entropy of Q.

The main drawback of training BMs using MF distributions is that we are approximating distributions which are potentially highly multimodal with a unimodal factorized distribution. This is especially dangerous in the negative phase where no units are clamped to data and the equilibrium distribution is expected to have many modes.

2   Contrastive Divergence Learning

Contrastive Divergence (CD) learning was introduced in [1], to train "Products of Experts" models from data. We start by recalling that the KL-divergence between the data distribution and the model distribution can be written as a difference between two free energies,

$$\mathrm{KL}[P_0(\mathbf{v})\|P_{\mathrm{EQ}}(\mathbf{v})] = \mathrm{KL}[P_0(\mathbf{v},\mathbf{h})\|P_{\mathrm{EQ}}(\mathbf{v},\mathbf{h})] = F_0 - F_{\mathrm{EQ}} \geq 0 \qquad (6)$$

To get samples from the equilibrium distribution we imagine running a Markov chain, first sampling the hidden units with the data clamped to the visible units, then fixing the hidden units and sampling the visible units and so on until we eventually reach equilibrium. It is not hard to show that at every step of Gibbs sampling the free energy decreases on average, $F_0 \geq F_i \geq F_{\mathrm{EQ}}$. It must therefore be true that if the free energy hasn't changed after i steps of Gibbs sampling (for any i), either $P_0 = P_{\mathrm{EQ}}$ or the Markov chain does not mix (which must therefore be avoided). The above suggests that we could use the following contrastive free energy (setting $i = 1$),

$$\mathrm{CD} = F_0 - F_1 = \mathrm{KL}\left[P_0(\mathbf{v},\mathbf{h})\|P_{\mathrm{EQ}}(\mathbf{v},\mathbf{h})\right] - \mathrm{KL}\left[P_1(\mathbf{v},\mathbf{h})\|P_{\mathrm{EQ}}(\mathbf{v},\mathbf{h})\right] \geq 0 \quad (7)$$

as an objective to minimize. The big advantage is that we do not have to wait for the chain to reach equilibrium. Learning proceeds by taking derivatives with

respect to the parameters and performing gradient descent on CD. The derivative
is given by,

$$\frac{\partial \text{CD}}{\partial \theta} = \left\langle \frac{\partial E}{\partial \theta} \right\rangle_0 - \left\langle \frac{\partial E}{\partial \theta} \right\rangle_1 - \frac{\partial F_1}{\partial P_1} \frac{\partial P_1}{\partial \theta} \tag{8}$$

with $\theta = \{\mathbf{V}, \mathbf{W}, \mathbf{J}\}$. The last term is hard to evaluate, but small compared with
the other two. Hinton [1] shows that this awkward term can be safely ignored.
For the BM, this results in the following learning rules,

$$\delta \mathbf{W} \propto \langle \mathbf{hh}^T \rangle_0 - \langle \mathbf{hh}^T \rangle_1 \quad \delta \mathbf{V} \propto \langle \mathbf{vv}^T \rangle_0 - \langle \mathbf{vv}^T \rangle_1 \quad \delta \mathbf{J} \propto \langle \mathbf{vh}^T \rangle_0 - \langle \mathbf{vh}^T \rangle_1 \tag{9}$$

Intuitively, these update rules decrease any systematic tendency of the one-step
reconstructions to move away from the data-vectors.

Although some progress has been made, this algorithm still needs equi-
librium samples from the conditional distribution $P(\mathbf{h}|\mathbf{v})$ [1]. Unfortunately, this
implies that in the presence of lateral connections among hidden units further
approximations remain desirable.

3   Contrastive Divergence Mean Field Learning

In this section we formulate the deterministic mean field variant of the contrastive divergence learning objective. First, let's assume that the MF equations
minimize the MF free energy $F_Q^{\text{MF}} = \langle E \rangle_Q - H(Q)$. Imagine N independent systems where data-vectors $\mathbf{d}_n$ are clamped to the visible units and MF equations
are run to solve for the means of the hidden units $\mathbf{q}_{0,n}$. The sum of the resultant
MF free energies is denoted with $F_0^{\text{MF}} = \sum_n F_{0,n}^{\text{MF}}$. Next, we fix the means of
the hidden units, initialize the means of the visible units at the data and take
a few steps downhill on the MF free energy. For convenience we will assume
that a few iterations of the MF equations achieves this[2] but alternative descent
methods are certainly allowed. Finally, we fix these *reconstructions* of the data
$\mathbf{r}_{1,n}$, initialize the means of the hidden units at $\mathbf{q}_{0,n}$ and run the MF equations
to compute $\mathbf{q}_{1,n}$. Call the sum of the resultant free energies $F_1^{\text{MF}} = \sum_n F_{1,n}^{\text{MF}}$.
Summarizing the above with equations we have

$$\mathbf{q}_{0,n} = \sigma(\mathbf{W}\mathbf{q}_{0,n} + \mathbf{J}^T \mathbf{d}_n) \rightarrow \mathbf{r}_{1,n} = \sigma(\mathbf{V}\mathbf{r}_{1,n} + \mathbf{J}\mathbf{q}_{0,n}) \rightarrow \mathbf{q}_{1,n} = \sigma(\mathbf{W}\mathbf{q}_{1,n} + \mathbf{J}^T \mathbf{r}_{1,n}) \tag{10}$$

The last argument in the sigmoid is fixed and acts as a bias term. By the assumption that the MF equations minimize the MF free energy, we may interpret
the above procedure as *coordinate descent* on the MF free energy in the variables
$\{\mathbf{q}, \mathbf{r}\}$. When this coordinate descent procedure is performed until convergence,
each chain, initialized at a particular data-vector, ends up in some local mini-
mum. The sum of the resultant free energies will be called $F_\infty^{\text{MF}} = \sum_n F_{\infty,n}^{\text{MF}}$. The

---

[1] However, when initialized at the data, brief sampling from $P(\mathbf{v}|\mathbf{h})$ is sufficient.
[2] By running the MF equations sequentially, or by damping them sufficiently this can
easily be achieved.

global minimum is denoted as $F_{\text{EQ}}^{\text{MF}}$. It is now easy to verify that the following inequalities must hold.

$$F_0^{\text{MF}} \geq F_1^{\text{MF}} \geq F_\infty^{\text{MF}} \geq N \, F_{\text{EQ}} \tag{11}$$

By analogy with the stochastic contrastive divergence objective we now propose the following 1-step MF contrastive divergence (CD^{MF}) objective,

$$\text{CD}^{\text{MF}} = F_0^{\text{MF}} - F_1^{\text{MF}} = \text{KL}[Q_0(\mathbf{v}, \mathbf{h}) \| P_{\text{EQ}}(\mathbf{v}, \mathbf{h})] - \text{KL}[Q_1(\mathbf{v}, \mathbf{h}) \| P_{\text{EQ}}(\mathbf{v}, \mathbf{h})] \geq 0 \tag{12}$$

where $Q_0(\mathbf{v}, \mathbf{h}) = \tilde{P}_0(\mathbf{v})Q_0(\mathbf{h}|\mathbf{v})$ and Q_1 is the MF distribution after one step of coordinate descent in the variables $\{\mathbf{q}, \mathbf{r}\}$. Due to the inequalities in eqn. 11 this objective is always positive. Notice that the only difference with the usual MF objective (eqn. 5) is the fact that we have replaced Q_{EQ} with Q_1. The above cost-function is minimized when the distribution of reconstructions $Q_1 \sim \{\mathbf{r}_{1,n}, \mathbf{q}_{1,n}\}$ after one step of MF coordinate descent does not show any average tendency to drift away from the data distribution $Q_0 \sim \{\mathbf{d}_n, \mathbf{q}_{0,n}\}$. One could envision balls initialized at the data which roll down towards their respective local minima in the MF free energy surface over a distance $\mathbf{F}\delta t$. When the shape of the surface is such that the outer products of all *forces* $\mathbf{F}$ (instead of *distances* to the minima) cancel, learning stops.

To compute the update rules we take the derivatives of the CD^{MF} objective with respect to the weights,

$$\frac{\partial \text{CD}^{\text{MF}}}{\partial \theta} = \left\langle \frac{\partial E}{\partial \theta} \right\rangle_{Q_0} - \left\langle \frac{\partial E}{\partial \theta} \right\rangle_{Q_1} - \frac{\partial F^{\text{MF}}}{\partial Q_1} \frac{\partial Q_1}{\partial \theta} \tag{13}$$

where $\theta = \{\mathbf{V}, \mathbf{W}, \mathbf{J}\}$. The last term represents the effect that the parameters $\{\mathbf{q}_{1,n}, \mathbf{r}_{1,n}\}$ will have different values when we change the shape of the surface on which we perform coordinate descent to compute them. This term vanishes for the usual MF objective since in that case we have $\partial F^{\text{MF}}/\partial Q_\infty = 0$. Although this term is awkward to compute, it turns out to be much smaller than the other two in eqn. 13 and can be safely ignored. In section 4 we show experimental evidence to support this claim. Thus, the following update rules can be derived to minimize the CD^{MF} objective eqn. 12,

$$\delta \mathbf{W} \propto \sum_n \left(\mathbf{q}_{0,n} \mathbf{q}_{0,n}^T - \mathbf{q}_{1,n} \mathbf{q}_{1,n}^T \right) \qquad \delta \mathbf{V} \propto \sum_n \left(\mathbf{d}_n \mathbf{d}_n^T - \mathbf{r}_{1,n} \mathbf{r}_{1,n}^T \right)$$

$$\delta \mathbf{J} \propto \sum_n \left(\mathbf{d}_n \mathbf{q}_{0,n}^T - \mathbf{r}_{1,n} \mathbf{q}_{1,n}^T \right) \tag{14}$$

The main advantage of the above learning algorithm is that it only runs MF equations (until convergence) over the hidden units *conditioned* on data-vectors or one-step reconstructions of these data-vectors[3]. Most importantly, MF equations on the highly multimodal energy surface of the free network are entirely avoided.

---

[3] In fact, for the $\mathbf{q}_{1,n}$ a few steps downhill on the MF free energy is sufficient.

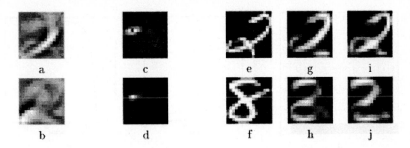

Fig. 1. (a, b) Two examples of the weights from one hidden unit to all visible units (features) which can be interpreted as thinning (a) and shifting (b) operators. (c) Visible-to-visible weights for one unit. (d)-ZCA-whitening filter for the same unit as in (c), providing evidence that the visible weights decorrelate the data. (e,f) Two data vectors. (g,h) One-step reconstructions of (e,f) by the two-model. (i,j)-Local minima of the two-model corresponding to (e,f). Note that the "8" is being reconstructed as a "2".

4 Experiments

In the experiments described below we have used 16×16 real valued digits from the "br" set on the CEDAR cdrom # 1. There are 11000 digits available equally divided into 10 classes. The first 7000 were used for training, while we cycled through the last 4000, using 3000 as a validation set and testing on the remaining 1000 digits. The final test-error was averaged over the 4 test-runs. All digit-images were separately scaled (linearly) between 0 and 1, before presentation to the algorithm.

Separate models were trained for each digit, using 700 training examples. Each model was a fully connected MF-BM with 50 hidden units. A total of 2000 updates were performed on mini-batches of 100 data-vectors using a small weight-decay term and a momentum term. When training was completed, we computed the free energy F_0^{MF} for all data on all models (including validation and test data). Since it is very hard to compute the term $F_{\mathrm{EQ}}^{\mathrm{MF}}$, we fit a multinomial logistic regression model to the training data *plus* the validation data, using the 10 free energies F_0^{MF} for each model as "features". The prediction of this logistic regression model on the test data is finally compared with ground truth, from which a confusion matrix is calculated (figure 2-a). The total averaged classification error is 2.5% on this data set, which is a significant improvement over simple classifiers such as a 1-nearest-neighbor (5.5%) and multinomial logistic regression (6.4%). By comparison, a (stochastic) RBM with 50 and 100 hidden units, trained and tested using the same procedure, score 3.1% and 2.4% misclassification respectively. Figures 1 and 2 show some further results for this experiment (see figure captions for explanation).

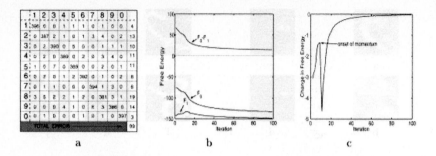

a   b   c

Fig. 2. (a) Confusion matrix for the digit classification task. (b) Contrastive MF free energy (computed every 20 iterations). (c) Change in contrastive MF free energy. Note that this change is always negative supporting our claim that the ignored term in eqn. 13 is much smaller than the other two.

5   Discussion

In this paper we have shown that efficient *contrastive divergence learning* can be used for BMs with lateral connections by replacing expensive Gibbs sampling with MF equations. During learning the negative phase is replaced with a "one-step-reconstruction" phase, for which the unimodal mean field approximation is expected to be appropriate. Recently (see [7] in this volume) this algorithm has been successfully applied to the study of associative mental arithmetic.

The approach presented in this paper is straightforwardly extended to supervised learning (see [5] for related work) but seems less successful on the digit recognition task.

CD-learning is a very general method for training undirected graphical models from data. The ideas presented in this paper are easily modified to more sophisticated deterministic approximations of the free energy like the TAP and Bethe approximations. Also, both the stochastic and deterministic versions are easily extended to discrete models with an arbitrary number of states per unit. We have recently also applied CD-learning to models with continuous states, where Hybrid Monte Carlo sampling was used to compute the one-step reconstructions of the data [3].

References

1. G.E. Hinton. Training products of experts by minimizing contrastive divergence. Technical Report GCNU TR 2000-004, Gatsby Computational Neuroscience Unit, University College London, 2000.
2. G.E. Hinton and T.J. Sejnowski. *Learning and relearning in Boltzmann machines*, volume Volume 1: Foundations. MIT Press, 1986.
3. G.E. Hinton, M. Welling, Y.W. Teh, and K. Osindero. A new view of ICA. In *Int. Conf. on Independent Component Analysis and Blind Source Separation*, 2001.

4. J.J. Hopfield. Neural networks and physical systems with emergent collective computational abilities. In *Proceedings of the National Academy of Sciences*, volume 79, pages 2554–2558, 1982.
5. J.R. Movellan. Contrastive hebbian learning in the continuous hopfield model. In *Connectionist Models, Proceedings of the 1990 Summer School*, pages 10–17, 1991.
6. C. Peterson and J. Anderson. A mean field theory learning algorithm for neural networks. *Complex Systems*, 1:995–1019, 1987.
7. I. Stoianov, M. Zorzi, S. Becker, and C. Umilta. Associative arithmetic with Boltzmann machines: the role of number representations. In *International Conference on Artificial Neural Networks*, 2002.

A Continuous Restricted Boltzmann Machine with a Hardware-Amenable Learning Algorithm

Hsin Chen and Alan Murray

Dept. of Electronics and Electrical Engineering,
University of Edinburgh, Mayfield Rd.,
Edinburgh, EH9 3JL, UK,
{Hsin.Chen,A.F.Murray}@ee.ed.ac.uk

Abstract. This paper proposes a continuous stochastic generative model that offers an improved ability to model analogue data, with a simple and reliable learning algorithm. The architecture forms a continuous restricted Boltzmann Machine, with a novel learning algorithm. The capabilities of the model are demonstrated with both artificial and real data.

1 Introduction

Probabilistic generative models offer a flexible route to improved data modelling, wherein the stochasticity represents the natural variability of real data. Our primary interest is in processing and modelling analogue data close to a sensor interface. It is therefore important that such models are amenable to analogue or mixed-mode VLSI implementation.

The Product of Experts (PoE) has been shown to be a flexible architecture and "Minimising Contrastive Divergence" (MCD) can underpin a simple learning rule [1]. The Restricted Boltzmann Machine (RBM) [2] with an MCD rule has been shown to be amenable to further simplification and use in real applications [3]. The RBM has one hidden and one visible layer with only inter-layer connections. Let s_i and s_j represent the states of the stochastic units i, j, and w_{ij} be the interconnect weights. The MCD rule for RBM replaces the computationally-expensive relaxation search of the Boltzmann Machine with:

$$\Delta w_{ij} = \eta(< s_i s_j >_0 - < \hat{s}_i \hat{s}_j >_1) \tag{1}$$

$\hat{s}_i$ and $\hat{s}_j$ correspond to one-step Gibbs sampled "reconstruction" states, and $<>$ denotes expectation value over the training data. By approximating the probabilities of visible units as analogue-valued states, the RBM can model analogue data [1][3]. However, the binary nature of the hidden unit causes the RBM to tend to reconstruct symmetric analogue data only, as will be shown in Sect. 3.

The rate-coded RBM (RBMrate) [4] removes this limitation by sampling each stochastic unit for m times. The RBMrate unit thus has discrete-valued states, while retaining the simple learning algorithm of (1). RBMrate offers an improved ability to model analogue image [4], but the repetitive sampling will cause more spiking noise in the power supplies of a VLSI implementation, placing the circuits in danger of synchronisation [5].

J.R. Dorronsoro (Ed.): ICANN 2002, LNCS 2415, pp. 358–363, 2002.

2 The Continuous Restricted Boltzmann Machine

2.1 A Continuous Stochastic Unit

Adding a zero-mean Gaussian with variance σ^2 to the input of a sampled sigmoidal unit produces a continuous stochastic unit as follows:

$$s_j = \varphi_j \left(\sum_i w_{ij} s_i + \sigma \cdot N_j(0, 1) \right), \tag{2}$$

$$\text{with} \quad \varphi_j(x_j) = \theta_L + (\theta_H - \theta_L) \cdot \frac{1}{1 + \exp(-a_j x_j)} \tag{3}$$

where $N_j(0, 1)$ represents a unit Gaussian, and $\varphi_j(x)$ is a sigmoid function with lower and upper asymptotes at θ_L and θ_H, respectively. Parameter a_j controls the steepness of the sigmoid function, and thus the nature of the unit's stochastic behaviour. A small value of a_j renders input noise negligible and leads to a near-deterministic unit, while a large value of a_j leads to a binary stochastic unit. If the value of a_j renders the sigmoid linear over the range of the added noise, the probability of s_j remains Gaussian with mean $\sum_i w_{ij} s_i$ and variance σ^2. Replacing the binary stochastic unit in RBM by this continuous form of stochastic unit leads to a continuous RBM (CRBM).

2.2 CRBM and Diffusion Network

The model and learning algorithms of the Diffusion Network (DN) [6][7] arise from its continuous stochastic behaviour, as described by a stochastic differential equation. A DN consists of n fully-connected units and an $n \times n$ real-valued matrix W, defining the connection-weights. Let $x_j(t)$ be the state of neuron j in a DN. The dynamical diffusion process is described by the Langevin equation:

$$dx_j(t) = \kappa_j \left(\sum_i w_{ij} \varphi_i(x_i(t)) - \rho_j x_j(t) \right) \cdot dt + \sigma \cdot dB_j(t) \tag{4}$$

where $1/\kappa_j > 0$ and $1/\rho_j > 0$ represent the input capacitance and resistance of neuron j. $dB_j(t)$ is the Brownian motion differential [7]. The increment, $B_j(t + dt) - B_j(t)$, is thus a zero-mean Gaussian random variable with variance dt. The discrete-time diffusion process for a finite time increment Δt is:

$$x_j(t + \Delta t) = x_j(t) + \kappa_j \sum_i w_{ij} \varphi_i(x_i(t)) \Delta t - \kappa_j \rho_j x_j(t) \Delta t + \sigma z_j(t) \sqrt{\Delta t} \tag{5}$$

where $z_j(t)$ is a Gaussian random variable with zero mean and unit variance. If $\kappa_j \rho_j \Delta t = 1$, the terms in $x_j(t)$ cancel and writing $\sigma \sqrt{\Delta t} = \sigma'$, this becomes:

$$x_j(t + \Delta t) = \kappa_j \sum_i w_{ij} \varphi_i(x_i(t)) \Delta t + \sigma' z_j(t) \tag{6}$$

If $w_{ij} = w_{ji}$ and κ_j is constant over the network, the RHS of (6) is equivalent to the total input of a CRBM as given by (2). As $s_j = \varphi_j(x_j)$, the CRBM is simply a symmetric restricted DN (RDN), and the learning algorithm of the DN is thus a useful candidate for the CRBM.

2.3 M.C.D. Learning Algorithms for the CRBM

The learning rule for the parameter λ_j of the DN is [6]:

$$\Delta\lambda_j = < S_{\lambda_j} >_0 - < S_{\lambda_j} >_\infty \tag{7}$$

where $<>_0$ refers to the expectation value over the training data with visible states clamped, and $<>_\infty$ to that in free-running equilibrium. S_{λ_j} is the system-covariate [6], the negative derivative of the DN's energy function w.r.t. parameter λ_j. The restricted DN can be shown to be a PoE [8] and we choose to simplify (7) by once again minimising contrastive divergence [1].

$$\Delta\hat{\lambda}_j = < S_{\lambda_j} >_0 - < \hat{S}_{\lambda_j} >_1 \tag{8}$$

where $<>_1$ indicates the expectation values over one-step sampled data. Let $\varphi(s)$ represent $\varphi_j(s)$ with $a_j = 1$. The energy function of CRBM can be shown to be similar to that of the continuous Hopfield model [9][6].

$$U = -\frac{1}{2}\sum_{i \neq j} w_{ij}s_i s_j + \sum_i \frac{\rho_i}{a_i}\int_0^{s_i} \varphi^{-1}(s)ds \tag{9}$$

(8) and (9) then lead to the MCD learning rule for the CRBM's parameters:

$$\Delta\hat{w}_{ij} = \eta_w(< s_i s_j >_0 - < \hat{s}_i\hat{s}_j >_1) \tag{10}$$

$$\Delta\hat{a}_j = \eta_a\left(\frac{\rho_j}{a_j^2}\left\langle\int_{\hat{s}_j}^{s_j}\varphi^{-1}(s)ds\right\rangle\right) \tag{11}$$

where $\hat{s}_j$ denotes the one-step sampled state of unit j, and $<>$ in (11) refers to the expextation value over the training data. To simplify the hardware design, we approximate the integral term in (11) as

$$\int_{\hat{s}_j}^{s_j}\varphi^{-1}(s)ds \propto (s_j + \hat{s}_j)(s_j - \hat{s}_j) \tag{12}$$

The training rules for w_{ij} and a_j thus require only adding and multiplying calculation of local units' states.

3 Demonstration: Artificial Data

Two-dimensional data were generated to probe and to compare the performance of RBM and CRBM on analogue data (Fig.1(a)). The data include two clusters

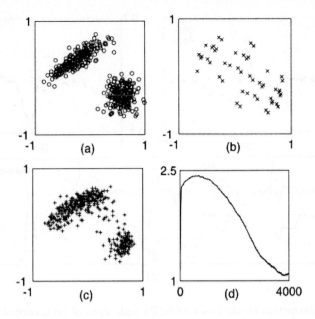

Fig. 1. (a) Artificial-generated analogue training data. (b) Reconstruction by the trained RBM (c) Reconstruction by the trained CRBM (d)Learning trace of a_j.

of 200 data. Figure 1(b) shows points reconstructed from 400 random input data after 20 steps of Gibbs' sampling by an RBM with 6 hidden units, after 4000 training epochs. The RBM's tendancy to generate data in symmetric patterns is clear. Figure 1(c) shows the same result for a CRBM with four hidden units, $\eta_w = 1.5$, $\eta_a = 1$ and $\sigma = 0.2$ for all units. The evolution of the gain factor a_j of one visible unit is shown in Fig.1(d) and displays a form of 'autonomous annealing', driven by (11), indicating that the approximation in (12) leads to sensible training behaviour in this stylised, but non-trivial example.

4   Demonstration: Real Heart-Beat (ECG) Data

To highlight the improved modelling richness of the CRBM and to give these results credence, a CRBM with four hidden units was trained to model the ECG data used in [3] and [10]. The ECG trace was divided into one training dataset of 500 heartbeats and one test dataset of 1700 heartbeats, each of 65 samples. The 500 training data contain six Ventricular Ectopic Beats (VEB), while the 1700 test data contain 27 VEBs. The CRBM was trained for 4000 epochs with $\eta_w = 1.5$, $\eta_a = 1$, $\sigma = 0.2$ for visible units and $\sigma = 0.5$ for hidden units.

Figure 2 shows the reconstruction by the trained CRBM, from initial visible states as 2(a) an observed normal QRS complex 2(b) an observed typical VEB, after 20 subsequent steps of unclamped Gibbs' sampling. The CRBM models both forms of heartbeat successfully, although VEBs represent only 1% of the

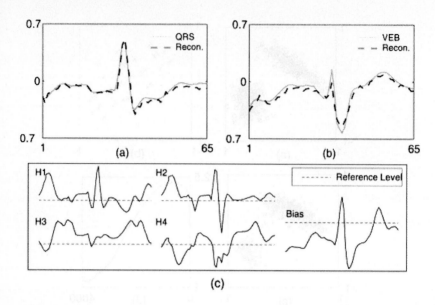

Fig. 2. Reconstruction by the trained CRBM with input of (a) a normal QRS and (b) a typical VEB. (c) The receptive fields of the hidden bias and the four hidden units

training data. Following [3], Fig.2(c) shows the receptive fields of the hidden bias unit and the four hidden units. The Bias unit codes an "average" normal QRS complex, and H3 adds to the P- and T- waves. H1 and H2 drive a small horizontal shift and a magnitude variation of the QRS complex. Finally, H4 encodes the significant dip found in a VEB. The most principled detector of VEBs in test data is the log-likelihood under the trained CRBM. However, log-likelihood requires complicated hardware. Figure 2(c) suggests that the activities of hidden units may be usable as the basis of a simple novelty detector. For example, the activities of H4 corresponding to 1700 test data are shown in Fig.3, with the noise source in equation (2) removed. The peaks indicate the VEBs clearly. VL in Fig.3 indicates the minimum H4 activity for a VEB and QH marks the datum with maximum H4 activity for a normal heartbeat. Therefore,

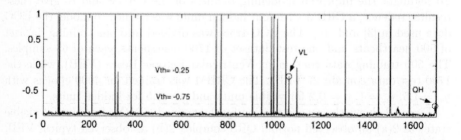

Fig. 3. The activities of H4 corresponding to 1700 test data.

even a simple linear classifier with threshold set between the two dashed line will detect the VEBs with an accuracy of 100%. The margin for threshold is more than 0.5, equivalent to 25% of the total value range. A single hidden unit activity in a CRBM is, therefore, potentially a reliable novelty detector and it is expected that layering a supervised classifier on the CRBM, to "fuse" the hidden unit activities, will lead to improved results.

5   Conclusion

The CRBM can model analogue data successfully with a simplified MCD rule. Experiments with real ECG data further show that the activities of the CRBM's hidden units may function as a simple but reliable novelty detector. Component circuits of the RBM with the MCD rule have been successfully implemented [5][11]. Therefore, the CRBM is a potential continuous stochastic model for VLSI implementation and embedded intelligent systems.

References

1. Hinton, G.E.: Training Products of Experts by minimising contrastive divergence. Technical Report: Gatsby Comp. Neuroscience Unit, no.TR2000-004. (2000)
2. Smolensky, P.: Information processing in dynamical systems: Foundations of harmony theory. Parallel Distributed Processing. Vol.1 (1986) 195–281
3. Murray, A.F.: Novelty detection using products of simple experts-A potential architecture for embedded systems. Neural Networks 14(9) (2001) 1257–1264
4. Teh, Y.W., Hinton, G.E.: Rate-coded Restricted Boltzmann Machine for face recognition. Advances in Neural Information Processing Systems, Vol.13 (2001)
5. Woodburn, R.J., Astaras, A.A., Dalzell, R.W., Murray, A.F., McNeill, D.K.: Computing with uncertainty in probabilistic neural networks on silicon. Proc. 2nd Int. ICSC Symp. on Neural Computation. (1999) 470–476
6. Movellan, J.R.: A learning theorem for networks at detailed stochastic equilibrium. Neural Computation, Vol.10(5) (1998) 1157–1178
7. Movellan, J.R., Mineiro, P., William, R.J.: A Monte-Carlo EM approach for partially observable diffusion process: Theory and applications to neural network. Neural Computation (In Press)
8. Marks, T.K., Movellan, J.R.: Diffusion networks, products of experts, and factor analysis. UCSD MPLab Technical Report, 2001.02. (2001)
9. Hopfield, J.J.: Neurons with graded response have collective computational properties like those of two-state neurons. Proc. Nat. Academy of Science of the USA, Vol.81(10). (1984) 3088–3092
10. Tarassenko, L., Clifford G.: Detection of ectopic beats in the electrocardiogram using an Auto-associative neural network. Neural Processing Letters, Vol.14(1) (2001) 15–25
11. Fleury, P., Woodburn, R.J., Murray, A.F.: Matching analogue hardware with applications using the Products of Experts algorithm. Proc. IEEE European Symp. on Artificial Neural Networks. (2001)63–67

prototypes, $\bigcup_{c^i=c} R^i_\lambda$, is as small as possible. This is achieved by a stochastic gradient descent on the cost function

$$C_{\text{GRLVQ}} := \sum_{i=1}^m \text{sgd}\left(\mu_\lambda(x^i)\right) \text{ where } \mu_\lambda(x^i) = \frac{d^+_\lambda(x^i) - d^-_\lambda(x^i)}{d^+_\lambda(x^i) + d^-_\lambda(x^i)}.$$

$\text{sgd}(x) = (1 + \exp(-x))^{-1}$ denotes the logistic function. $d^+_\lambda(x^i) = |x^i - w^{i+}|^2_\lambda$ is the squared weighted Euclidian distance of x^i to the nearest prototype w^{i+} of the same class as x^i, and $d^-_\lambda(x^i) = |x^i - w^{i-}|^2_\lambda$ is the squared weighted Euclidian distance of x^i to the nearest prototype w^{i-} of a different class than x^i. The learning rule of GRLVQ is obtained by taking the derivatives of the above cost function [2]:

$$\Delta w^{i+} = \frac{4\epsilon^+ \text{sgd}'(\mu_\lambda(x^i))d^-_\lambda(x^i)}{(d^+_\lambda(x^i) + d^-_\lambda(x^i))^2} \cdot \Lambda \cdot (x^i - w^{i+})$$

$$\Delta w^{i-} = \frac{-4\epsilon^- \text{sgd}'(\mu_\lambda(x^i))d^+_\lambda(x^i)}{(d^+_\lambda(x^i) + d^-_\lambda(x^i))^2} \cdot \Lambda \cdot (x^i - w^{i-})$$

$$\Delta\lambda_j = \frac{-2\epsilon \, \text{sgd}'(\mu_\lambda(x^i))}{(d^+_\lambda(x^i) + d^-_\lambda(x^i))^2} \left(d^-_\lambda(x^i)(x^i_j - w^{i+}_j)^2 - d^+_\lambda(x^i)(x^i_j - w^{i-}_j)^2 \right)$$

where Λ is the diagonal matrix with entries $\lambda_1, \ldots, \lambda_n$. $\epsilon, \epsilon^-, \epsilon^+ > 0$ are learning rates. GLVQ only updates the prototypes w^i, keeping the weighting factors λ_i fixed to $1/n$.

NG adapts a set of *unlabeled* prototypes $W = \{w^1, \ldots, w^K\}$ in $\mathbb{R}^n$ such that they represent a given set of unlabeled data $X = \{x^1, \ldots, x^m\}$ in $\mathbb{R}^n$. The cost function of NG,

$$C_{\text{NG}} = \sum_{i=1}^m \sum_{j=1}^K h_\sigma(k_j(x^i, W))(x^i - w^j)^2/h(K)$$

is minimized with a stochastic gradient descent where $h(K) = \sum_{i=0}^{K-1} h_\sigma(i)$. Thereby, $h_\sigma(x) = \exp(-x/\sigma)$. $\sigma > 0$ denotes the degree of neighborhood cooperation and is decreased to 0 during training. $k_j(x^i, W)$ denotes the rank of w^j in $\{0, \ldots, K-1\}$ if the prototypes are ranked according to the distances $|x^i - w^k|_\lambda$. The according learning rule becomes
$$\Delta w^j = 2\epsilon \, h_\sigma(k_j(x^i, W))(x^i - w^j)$$
where $\epsilon > 0$ is the learning rate. Note that this update constitutes an unsupervised learning rule which minimizes the quantization error. The initialization of the prototypes is not crucial because of the involved neighborhood cooperation through the ranking k_j.

3 Supervised Relevance Neural Gas Algorithm

The idea of supervised relevance neural gas (SRNG) is to incorporate neighborhood cooperation into the cost function of GRLVQ. This neighborhood cooperation helps to spread prototypes of a certain class over the possibly multimodal data distribution of the respective class. We use the same notation as for GRLVQ. Denote by $W(y^i)$ the set of prototypes labeled with y^i and by K_i its cardinality. Then we define the cost function

$$C_{\text{SRNG}} = \sum_{i=1}^m \sum_{w^j \in W(y^i)} h_\sigma(k_j(x^i, W(y^i))) \cdot \text{sgd}(\mu_\lambda(x^i, w^j))/h(K_i)$$

with $\mu_\lambda(x^i, w^j) = \left(|x^i - w^j|^2_\lambda - d^-_\lambda(x^i)\right) / \left(|x^i - w^j|^2_\lambda + d^-_\lambda(x^i)\right)$. As above, $h(K_i) = \sum_{j=0}^{K_i-1} h_\sigma(j)$. $k_j(x^i, W(y^i)) \in \{0, \ldots, K_i - 1\}$ denotes the rank of w^j if the prototypes in $W(y^i)$ are ranked according to the distances $|w^k - x^i|_\lambda$. $d_\lambda^-(x^i)$ is the distance to the closest wrong prototype w^{i-}. Note that $\lim_{\sigma \to 0} C_{\text{SRNG}} = C_{\text{GRLVQ}}$. If σ is large, typically at the beginning of training, the prototypes of one class share their responsibility for a given data point. Hence neighborhood cooperation is taken into account such that the initialization of the prototypes is no longer crucial. The learning rule is obtained by taking the derivatives. Given a training example (x^i, y^i), all $w^j \in W(y^i)$, the closest wrong prototype w^{i-}, and the factors λ_k are adapted:

$$\triangle w^j = \frac{4\epsilon^+ \text{sgd}'(\mu_\lambda(x^i, w^j)) h_\sigma(k_j(x^i, W(y^i))) d_\lambda^-(x^i)}{(|x^i - w^j|_\lambda + d_\lambda^-(x^i))^2 h(K_i)} \cdot \Lambda \cdot (x^i - w^j)$$

$$\triangle w^{i-} = -4\epsilon^- \sum_{w^j \in W(y^i)} \frac{\text{sgd}'(\mu_\lambda(x^i, w^j)) h_\sigma(k_j(x^i, W(y^i))) |x^i - w^j|_\lambda^2}{(|x^i - w^j|_\lambda^2 + d_\lambda^-(x^i))^2 h(K_i)} \cdot \Lambda \cdot (x^i - w^{i-})$$

$$\triangle \lambda_k = -2\epsilon \sum_{w^j \in W(y^i)} \frac{\text{sgd}'(\mu_\lambda(x^i, w^j)) h_\sigma(k_j(x^i, W(y^i)))}{(|x^i - w^j|_\lambda^2 + d_\lambda^-(x^i))^2 h(K_i)} \cdot$$
$$\left(d_\lambda^-(x^i)(x_k^i - w_k^j)^2 - |x^i - w^j|_\lambda^2 (x_k^i - w_k^{i-})^2 \right)$$

In comparison to the GRLVQ learning rule, all prototypes responsible for the class of x^i are taken into account according to their neighborhood ranking. For $\sigma \to 0$, GRLVQ is obtained. With respect to SOM-LVQ, this update allows an adaptive metric like GRLVQ, and shows more stable behavior due to the choice of the cost function and the fact that neighborhood cooperation is only included within the prototypes of one class. Note that adaptation of the prototypes without adapting the metric, i.e. the factors λ_i, is possible, too. We refer to this modification as supervised neural gas (SNG).

4   Experiments

We tested the algorithm on artificial multimodal data sets. Data sets 1 to 6 consist of two classes with 50 clusters with about 30 points for each cluster which are located on a two dimensional checkerboard. Data sets 1, 3, and 5 differ with respect to the overlap of the classes (see Figs. 1,2). Data sets 2, 4, and 6 are copies of 1, 3, and 5, respectively, where 6 dimensions have been added: A point (x_1, x_2) is embedded as $(x_1, x_2, x_1 + \eta_1, x_1 + \eta_2, x_1 + \eta_3, \eta_4, \eta_5, \eta_6)$ where η_i is uniform noise with $|\eta_1| \leq 0.05$, $|\eta_2| \leq 0.1$, $|\eta_3| \leq 0.2$, $|\eta_4| \leq 0.1$, $|\eta_5| \leq 0.2$, and $|\eta_6| \leq 0.5$. Hence dimensions 6 to 8 contain pure noise, dimensions 3 to 5 carry some information. Data set 7 comprised 3 classes of different cardinality in two dimensions (see Fig. 3). Data set 8 consists of an embedding of set 7 in 8 dimensions as above. The sets are randomly divided into training and test set. All prototypes are randomly initialized around the origin. Training has been done for 3000 cycles with 50 prototypes for each class for sets 1 to 6 and 5 or 6 prototypes for each class for sets 7 and 8. The algorithms converged after about 2000 cycles in all runs. Learning rates are $\epsilon^+ = 0.1$, $\epsilon^- = 0.05$, $\epsilon = 0.0001$. The initial neighborhood size $\sigma = 100$ is decreased by 0.995 after each epoch. The reported results have been validated in several runs. For comparison, a classification with prototypes set by hand in the cluster centers (opt) and a nearest neighbor classifier (NN) are provided.

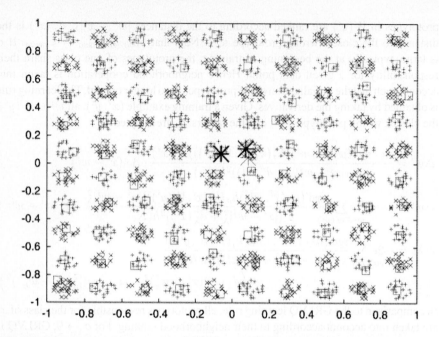

Fig. 1. Artificial multimodal data set used for training: training set 1 and prototypes found by SRNG (□, spread over the clusters) and GRLVQ (∗, only within two clusters in the middle).

Simple GLVQ and GRLVQ without neighborhood cooperation are not capable of learning data sets 1 to 6. The prototypes only represent the clusters closest to the origin (see Fig. 1) and classification is nearly random. Nevertheless, the weighting factors obtained from GRLVQ for data set 2 indicate the importance of the first two dimensions with $\lambda_1 \approx \lambda_3 = 0.23$, $\lambda_2 = 0.33$. For data sets 4, 6, and 8, some less important dimensions are emphasized by GRLVQ, as well. SNG produces good results for data sets 1, 3, and 5, and slightly worse results for their 8-dimensional counterparts. SRNG shows very stable behavior and a good classification accuracy in all runs. The relevance terms clearly emphasize the important first two dimensions with vectors $\lambda = (\geq 0.34, \geq 0.4, \leq 0.18, \approx 0, \approx 0, \approx 0, \approx 0, \approx 0)$. The prototypes spread over the clusters and

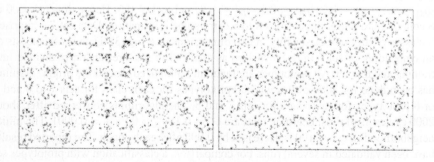

Fig. 2. Artificial multimodal data sets: training set 3 (left) and training set 5 (right).

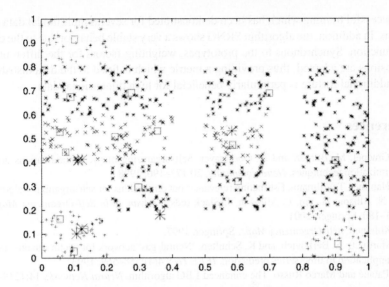

Fig. 3. Training set 7 and prototypes found by SRNG (□) and GRLVQ (∗).

mostly only 2 or 3 out of 100 clusters are missed in the runs for data sets 1 to 4. Since data sets 5 and 6 are nearly randomly distributed (see Fig. 2), no generalization can be expected in these runs. Nevertheless, SNG and SRNG show comparably small test set errors in these runs, too. A similar behavior can be observed for the smaller sets 7 and 8, where SRNG achieves close to 100% accuracy, SNG performs slightly worse if additional noisy dimensions are added, and GRLVQ and GLVQ are not capable of representing all clusters of the data (see Fig. 3). Experiments with real life data for GRLVQ in comparison to GLVQ can be found e.g. in [2]. Tehreby, runs for SRNG for these data sets provide only a further slight improvement since these data, unlike the above artificial data sets, are not highly multimodal.

5   Conclusions

We have proposed a modification of GRLVQ which includes neighborhood cooperation within prototypes of each class. Hence initialization of prototypes is no longer crucial

Table 1. Training/test set accuracy (%) of the obtained clustering for multimodal data sets.

	data1	data2	data3	data4	data5	data6	data7	data8
opt	100	100	97.3	97.3	58	58	100	100
NN	100	95.4	93.1	77.1	67.3	50.1	96.3	84.7
GLVQ	49.4/48.9	50/49	51.5/49.5	50/49.5	50/50	51/50	62/65	62/61
GRLVQ	50.2/50	50/50	55/49	50.5/50	51/50	51/49	62/65	65/65
SNG	99/98	80/72	90/90	75/64	61/50	70/50	98/97	88/85
SRNG	99/99	95/94	92/91	92/92	66/54	67/55	98/97	99/98

for successful training which has been demonstrated for several multimodal data distributions. In addition, the algorithm SRNG shows a very stable behavior due to the chosen cost function. Synchronous to the prototypes, weighting factors for the different input dimensions are adapted, thus providing a metric which is fitted to the data distribution. This additional feature is particularly beneficial for high dimensional data.

References

1. T. Graepel, M. Burger, and K. Obermayer. Self-organizing maps: generalizations and new optimization techniques. *Neurocomputing*, 20:173–190, 1998.
2. B. Hammer, T. Villmann. Estimating relevant input dimensions for self-organizing algorithms. In: N. Allison, H. Yin, L. Allinson, J. Slack (eds.), *Advances in Self-Organizing Maps*, pp. 173–180, Springer, 2001.
3. T. Kohonen. *Self-Organizing Maps*. Springer, 1997.
4. T. Martinetz, S. Berkovich, and K. Schulten. 'Neural-gas' network for vector quantization and its application to time-series prediction. *IEEE TNN* 4(4):558–569, 1993.
5. G. Patané and Marco Russo. The enhanced LBG algorithm. *Neural Networks* 14:1219–1237, 2001.
6. A. S. Sato, K. Yamada. Generalized learning vector quantization. In G. Tesauro, D. Touretzky, and T. Leen, editors, *Advances in NIPS*, volume 7, pp. 423–429. MIT Press, 1995.
7. T. Villmann, R. Der, M. Herrmann, and T. M. Martinetz, Toplogy Preservation in Self-Organizing Feature Maps: Exact Definition and Precise Measurement, *IEEE TNN* 8(2):256–266, 1997.
8. N. Vlassis and A. Likas. A greedy algorithm for Gaussian mixture learning. *Neural Processing Letters* 15(1): 77–87, 2002.

Learning the Dynamic Neural Networks with the Improvement of Generalization Capabilities

Mirosław Galicki[1,2], Lutz Leistritz[1], and Herbert Witte[1]

[1] Institute of Medical Statistics, Computer Sciences and Documentation
Friedrich Schiller University Jena
Jahnstrasse 3, D–07740 Jena, Germany
galicki@imsid.uni-jena.de
[2] Institute of Organization and Management, University of Zielona Góra
Podgórna 50, 65-246 Zielona Góra, Poland

Abstract. This work addresses the problem of improving the generalization capabilities of continuous recurrent neural networks. The learning task is transformed into an optimal control framework in which the weights and the initial network state are treated as unknown controls. A new learning algorithm based on a variational formulation of Pontryagin's maximum principle is proposed. Numerical examples are also given which demonstrate an essential improvement of generalization capabilities after the learning process of a recurrent network.

1 Introduction

Artificial neural networks are now commonly used in many practical tasks such as the classification, identification and/or control, fault diagnostic etc. of dynamic processes. A general problem for a neural network in all of the aforementioned tasks is to find an underlying mapping which correspondingly relates their input-output spaces based on a given (usually finite) set called a learning set of input-output pairs. As a rule, the learning pairs present some functions of time (trajectories). It is well known that the problem of finding such a mapping is ill posed [1] because its solution is not unique. Consequently, some criterion reflecting a network performance which in turn influences the realization quality of the above tasks, should be introduced in order to obtain a unique solution. The capability of a neural network to generalize seems to be the most important from both the theoretical and practical point of view. The generalization property is strongly related to the bias and variance – components of the generalization error [2]. In the case of neural networks, these components depend in general on a training algorithm used that finds an optimal solution with respect to the weights, a network structure (an optimal balance between bias and variance) and the power of learning set. In this context, several approaches may be distinguished [3,4,5,6,7] which influence in fact only variations of the bias. An alternative way to controlling the generalization capability is the use of regularization techniques which are based on the addition of a penalty term, called a regularizer to the error function. The penalty term influences variations

J.R. Dorronsoro (Ed.): ICANN 2002, LNCS 2415, pp. 377–382, 2002.

of the variance. To the best of our knowledge, they have been only applied so far to feed-forward neural networks [8,9,10,11]. The generalized dynamic neural networks (GDNN's), first introduced and theoretically analyzed in the work [6] seem to be suitable to extend the static regularization theory to the dynamic case. Our aim is to construct a learning algorithm producing a small global error between output trajectories and desired (target) ones, which in addition, results in good generalization capabilities of GDNN's. The approach is based on using the sensitivity of GDNN's mapping. Its role is to damp out the rapid variations in the network mapping that generally are needed to fit the noisy learning set. As opposed to static networks, the (dynamic) sensitivity is not directly given. An adjoint system of ordinary differential equations is introduced to find the network sensitivity. In order to fully specify them, initial conditions must be found during the learning of dynamic network. Thus, the learning task defined above may be treated as a problem of dynamic programming with state constraints in which the weights and initial network state are treated as unknown controls. In order to solve it, the variational formulation of Pontryagin's maximum principle is used here which makes it possible to handle state constraints efficiently. The paper is organized as follows. Section 2 presents the learning task in terms of the optimal control problem. Section 3 describes how to employ the variational formulation of Pontryagin's maximum principle to determine optimal weights. A numerical example confirming theoretical results is presented in section 4.

2   Problem Formulation

A neural network of the Hopfield type is considered whose state evolves according to the following differential and algebraic equations

$$
\begin{aligned}
\dot{\boldsymbol{y}} &= \boldsymbol{f}(\boldsymbol{y}, \boldsymbol{w}, \boldsymbol{u}) \\
\boldsymbol{z} &= \boldsymbol{g}(\boldsymbol{y})
\end{aligned}
\tag{1}
$$

where $\mathbb{R}^N \ni \boldsymbol{y} = (y_1, \dots, y_N)^T$ is a state vector of the network; N denotes the number of all neurons; $\boldsymbol{w} = (w_{(i-1)N+j} : 1 \le i, j \le N)$; $w_{(i-1)N+j}$ represents the weight connection out of neuron i to neuron j; $\boldsymbol{u} = \boldsymbol{u}(t) = (u_1, \dots, u_N)^T$ stands for the vector of external inputs to the network; $\boldsymbol{f} = (f_1, \dots, f_N)^T$; $f_i(\boldsymbol{y}, \boldsymbol{w}, \boldsymbol{u}) = -y_i + h(\sum_{j=1}^{N} w_{(i-1)n+j} y_j) + u_i$; $i = 1 : N$; $t \in [0, T]$; T is a fixed time of network evolution; $\boldsymbol{y}(0) = \boldsymbol{A} = (A_1, \dots, A_N)^T$ denotes an initial network state; $h(\cdot)$ is strictly increasing neural activation function; $\boldsymbol{g} : \mathbb{R}^N \longrightarrow \mathbb{R}^L$; $\boldsymbol{g}(\boldsymbol{y}) = (y_1, \dots, y_L)^T$ stands for a given mapping which represents outputs of the network; $N \ge L$ is a number of neurons producing the outputs. The learning task of multiple continuous trajectories is to modify the weight vector $\boldsymbol{w}$ and the initial state $\boldsymbol{y}(0)$ so that the output trajectories $\boldsymbol{z}_p(t) = \boldsymbol{g}(\boldsymbol{y}_p(t)) = (y_{p,1}(t), \dots, y_{p,L}(t))^T$ corresponding to external inputs $\boldsymbol{u}_p(t) = (u_{1,p}(t), \dots, u_{N,p}(t))^T$ follow differentiable desired (target) trajectories $\boldsymbol{d}_p(t) \in \mathbb{R}^L$, where $p = 1 : P$; $1 \le P$ denotes a given number of training pairs. On account of (1) and by analogy to the static regularization theory, the following performance index may be introduced in the dynamic case to improve the generalization capability of a dynamic network

$$J = \frac{1}{2} \int_0^T \left(\sum_{p=1}^P \|\boldsymbol{g}(\boldsymbol{y}_p) - \boldsymbol{d}_p\|^2 + \lambda \left\| \frac{\partial \boldsymbol{g}}{\partial \boldsymbol{y}_p} \frac{\partial \boldsymbol{y}_p}{\partial \boldsymbol{u}_p} \right\|^2 \right) dt \qquad (2)$$

where $\|\cdot\|$ denotes the standard Euclidean norm; λ is the regularization parameter which controls the compromise between the degree of smoothness of the network output mapping and its closeness to the data; $\frac{\partial \boldsymbol{y}_p}{\partial \boldsymbol{u}_p} = \left[\frac{\partial y_{p,i}}{\partial u_{p,j}} \right]_{1 \leq i,j \leq N}$. The computation of J from (2) requires the knowledge of the first derivative $\frac{\partial \boldsymbol{y}_p}{\partial \boldsymbol{u}_p}$ which may be found by differentiating the upper equation of (1) with respect to $\boldsymbol{u}_p$ (assuming that the corresponding partial derivatives are continuous). The result is

$$\frac{d}{dt} \left(\frac{\partial \boldsymbol{y}_p}{\partial \boldsymbol{u}_p} \right) = \frac{d\boldsymbol{f}}{d\boldsymbol{u}_p} \qquad (3)$$

where $\frac{d\boldsymbol{f}}{d\boldsymbol{u}_p} = \left[\frac{df_i}{du_{p,j}} \right]_{1 \leq i,j \leq N} = \frac{\partial \boldsymbol{f}}{\partial \boldsymbol{y}_p} \frac{\partial \boldsymbol{y}_p}{\partial \boldsymbol{u}_p} + \frac{\partial \boldsymbol{f}}{\partial \boldsymbol{u}_p}$. In order to fully specify differential equation (3), initial value for derivative $\frac{\partial \boldsymbol{y}_p}{\partial \boldsymbol{u}_p}$, that is, $\frac{\partial \boldsymbol{y}_p}{\partial \boldsymbol{u}_p}(0) = \boldsymbol{B}$, where $\boldsymbol{B} = [B_{ij}]_{1 \leq i,j \leq N}$ denotes some constant matrix, should be given. However, it is not known in practice. Therefore, an additional task arises which consists in determining $\boldsymbol{B}$ when learning dynamic network (1). Expressions (1)–(3) present a relatively complex learning problem whose solution by means of classical networks (with constant weights) seems to be difficult. Generalized dynamic neural networks (GDNN's), first introduced in the work [6] offer a possibility to tackle this problem. They are defined by introducing time-dependent weights into a Hopfield type network. The time-dependent weights, $\boldsymbol{A}$ and $\boldsymbol{B}$ may be treated as controls. Thus, dependencies (1)–(3) formulate the learning task as an optimal control problem. In order to solve it, an approach based on the application of Pontryagin's maximum principle will be proposed in the next section.

3   Learning with Smoothing the Network Mapping

Before presenting the algorithm, variables $\boldsymbol{y}_p$ and $\frac{\partial \boldsymbol{y}_p}{\partial \boldsymbol{u}_p}$ will be transformed into a vector form which is more convenient for further considerations. Therefore, the following state vectors are introduced $\boldsymbol{x}_p = (x_{p,1}, \ldots, x_{p,N+N^2})^T = \left(y_{p,1}, \ldots \right.$ $\left. , y_{p,N}, \frac{\partial y_{p,1}}{\partial u_{p,1}}, \ldots, \frac{\partial y_{p,N}}{\partial u_{p,N}} \right)^T$ and $\boldsymbol{x}_0 = (A_1, \ldots, A_N, B_{11}, \ldots, B_{NN})^T$, $p = 1 : P$. Using the quantities $\boldsymbol{x}_0$, $\boldsymbol{x}_p$, the optimal control problem given by expressions (1)–(3) is transformed into a functional form (which is equivalent to the previous one) as follows. Minimize:

$$J(\boldsymbol{w}, \boldsymbol{x}_0) = \int_0^T \sum_{p=1}^P C_p dt \qquad (4)$$

where $C_p = \frac{1}{2}\|(x_{p,1}, \ldots, x_{p,L})^T - d_p\|^2 + \lambda\frac{1}{2}\sum_{m=N+1}^{N+LN}(x_{p,m})^2$, subject to the constraints

$$\dot{x}_p = F(x_p, w, u_p) \quad x_p(0) = x_0 \tag{5}$$

where $F = \left(f_1, \ldots, f_p, \frac{df_1}{du_{p,1}}, \ldots, \frac{df_N}{du_{p,N}}\right)^T$. The application of variational formulation of Pontryagin's maximum principle to solve the corresponding optimal control problem (see e.g. the work [6] for details regarding variational formulation of Ponryagin's maximum principle) requires a specification of initial weight vector $w^0 = w^0(t)$, $t \in [0, T]$. The initial trajectories $x_p^0(t)$ are obtained by solving (5) for each u_p, $p = 1 : P$. By assumption, w^0 and x_0 do not minimize the performance index (4). The use of the variational formulation of Pontryagin's maximum principle requires incrementation of functional (4). Therefore, it is assumed that the weight vector $w^0 = w^0(t)$ and initial state x_0 are perturbed by small variations $\delta w = \delta w(t) = (\delta w_m : 1 \leq m \leq N^2)$ and $\delta x_0 = (\delta x_{1,0}, \ldots, \delta x_{N+N^2,0})^T$, where $\|\delta w\|_\infty = \max_{t\in[0,T]}\{\max_{1\leq m\leq N^2}|\delta w_m|\} \leq \rho$; $\|\delta x_0\|_\infty = \max_{1\leq i\leq N+N^2}\{|\delta x_{i,0}|\} \leq \xi$; ρ and ξ are given small numbers ensuring the correctness of the presented method. On account of the fact that $\delta x_p(0) = \delta x_0$, the value of the functional J for the perturbed weight vector $w^0 + \delta w$ and initial state $x_0 + \delta x_0$ may be expressed to the first order (by neglecting the higher order terms $o(\delta w, \delta x_0)$) by the following dependence

$$J(w^0 + \delta w, x_0 + \delta x_0) = J(w^0, x_0) + \delta J(w^0, x_0, \delta w, \delta x_0) \tag{6}$$

where $\delta J(w^0, x_0, \delta w, \delta x_0) = \sum_{p=1}^{P}\left(\int_0^T \langle(F_w(p,t))^T\Psi_p, \delta w\rangle dt + \langle\Psi_p(0), \delta x(0)\rangle\right)$ is the Frechet differential of the functional $J(\cdot, \cdot)$; $\Psi_p(\cdot)$ stand for conjugate mappings which are determined by the solution of the Cauchy problem $\dot{\Psi}_p + (F_x(p,t))^T\Psi_p = -C_x(p,t)$; $\Psi_p(T) = 0$; $F_w(p,t) = \left(\frac{\partial F}{\partial w}\right)_{\substack{x_p=x_p^0 \\ w=w^0}}$; $F_x(p,t) = \left(\frac{\partial F}{\partial x}\right)_{\substack{x_p=x_p^0 \\ w=w^0}}$; $C_x(p,t) = \left(\frac{\partial C_p}{\partial x}\right)_{x_p=x_p^0}$. For properly selected variations δw and δx_0, the Frechet differential of functional (4) approximates the increment of this functional with any desired accuracy. Consequently, the variational formulation of Pontryagin's maximum principle takes the following form

$$J(w^0 + \delta w^*, x_0 + \delta x_0^*) = J(w^0, x_0) + \min_{(\delta w, \delta x_0)}\{\delta J(w^0, x_0, \delta w, \delta x_0)\} \tag{7}$$

subject to the constraints

$$\|\delta w^*\|_\infty \leq \rho \quad \|\delta x_0^*\|_\infty \leq \xi. \tag{8}$$

The assumptions of non-optimality of weight vector w^0 and initial network state x_0 imply the existence of sufficiently small variations δw^* and δx_0^* such that $J(w^0 + \delta w^*, x_0 + \delta x_0^*) < J(w^0, x_0)$. Consequently, the task of minimizing functional (7) is well-posed and variations δw^*, δx_0^* are results of its solution. A finite-dimensional approximation of the optimal control problem defined by dependencies (7)–(8) seems to be very effective for solving it numerically. The process of approximation for weight variation δw may be accomplished e.g. by using quasiconstant or quasilinear functions or splines. The present paper looks for a sequence of weight variations in a class of functions $\delta w = V \cdot C$, where $V \in \mathbb{R}^{N^2 \times M}$ is the matrix of coefficients to be found; M denotes a given number of basic functions (accuracy of computations); $C = C(t)$ stands for the M-dimensional vector of specified basic functions (e.g. Fourier trigonometric ones). The computationally efficient procedure to solve (7)-(8) in such a case has been proposed in work [12]. After the determination of optimal variations δw^*, δx_0^* from the minimization task (7)–(8), the computations are repeated for the revised quantities $w^1 = w^0 + \delta w^*$ and $x_0^1 = x_0 + \delta x_0^*$. A sequence of pairs (w^k, x_0^k), $k = 0, 1, ..., x_0^0 = x_0$ may thus be obtained as a result of solving the iterative approximation scheme (7)–(8). Each element of this sequence corresponds to the state trajectory $x_p^k(\cdot)$. The convergence of the above sequence to an optimal solution follows from the work [6].

4   Computer Example

The purpose of this section is to show on the basis of a chosen classification task, the essential improvement of the generalization capability of GDNN's trained with the algorithm presented in Section 3. The results are compared with the same GDNN, however learned with conventional algorithm ($\lambda = 0$). The task is to learn a two-dimensional ($L = 2$), two-class problem ($P = 2$) – a circle (target trajectory $d_1 = (\cos(t), \sin(t))^T$) and a point attractor (target trajectory $d_2 = (0,0)^T$). In order to solve the classification task, we have chosen a fully connected network structure with three neurons ($N = 3$) and $h = \arctan$, as the activation function. The first and second neurons are not stimulated by input signals, that is, $u_1 = u_2 = 0$. The constant external inputs to the third neuron (learning set) took in the training the following values $u_{3,1}(t) \in \{1.25, 1, 0.75\}$ and $u_{3,2}(t) \in \{0.25, 0, -0.25\}$, respectively, where $t \in [0, 10\pi]$. The network output trajectory was required to follow the circle fivefold (one circuit corresponded to 2π) if $u_{3,1}$ was presented; in the case of $u_{3,2}$, the network was forced to take constant value d_2. To achieve a satisfactory minimum value for the learning error, $M = 11$ has been assumed. A sample test set of 900 perturbed external inputs $\{\mu + \sigma \cdot X(t) : \mu \in [0.25, 0.75], \sigma \in [0, 2]\}$, where $X(t) \sim N(0, 1)$ i.i.d., had been randomly generated by means of Gaussian distribution. Then the estimated generalized error was used as an evaluation criterion for both algorithms. The results of computer simulations are presented in Table 1, where N_{Ri}, N_{Fi} stand for the numbers of right and false classified elements, respectively of the i-th class, $i = 1 : 2$. It shows unambiguously that the network learned with regularization term generalizes better than that with $\lambda = 0$ (see the first column). Moreover, the mean squared deviations of the network outputs to their class conditional targets d_1 and d_2 were computed. It results in mean errors of 6.297 in the case of $\lambda = 0$, and 5.825 for $\lambda = 10^{-2}$. The improvement by the regularization technique is highly significant ($p < 0.0001$, Wilcoxen test for paired samples, $n = 900$).

Table 1. Results of classification both without and with regularization

The value of λ	N_{R1}	N_{R2}	N_{F1}	N_{F2}
$\lambda = 0$	380	406	70	44
$\lambda = 10^{-2}$	399	405	51	45

5   Concluding Remarks

In this work, the learning procedure was formulated as an optimal control problem with state dependent constraints. The convergence of the proposed approximation scheme to an optimal solution was also discussed. The generality of the presented approach carries with it a penalty; somewhat larger amount of computations is required to train a GDNN subject to performance index (2). However, in our opinion, the capability of good generalization seems to be the most important when applying GDNN's to many practical dynamic tasks.

Acknowledgement. This work was supported by the DFG, Ga 652/1-1.

References

1. Morozov, V.A.: Methods for solving incorrectly posed problems, Springer-Verlag, Berlin, (1984)
2. Geman, S., Bienenstock, E., Doursat, R.: Neural networks and the bias/variance dillema, Neural Comp., vol. 4, no.1, (1992) 1–58
3. Giles, C.L., Chen, D., Sun, G.Z.: Constructive learning of recurrent neural networks. Limitations of recurrent cascade correlation and a simple solution, IEEE Trans. Neural Networks, vol. 6, no. 4, (1995), 829–836
4. Pearlmutter, B.A.: Gradient calculation for dynamic recurrent neural networks: A survey, IEEE Trans. Neural Networks, vol. 6, (1995), 1212–1228
5. Cohen, B., Saad, D., Marom, E.: Efficient training of recurrent neural network with time delays, Neural Networks, vol. 10, no.1, (1997), 51–59
6. Galicki, M., Leistritz, L., Witte, H.: Learning continuous trajectories in recurrent neural networks with time-dependent weights, IEEE Trans. Neural Networks, vol. 10, (1999), 741–756
7. Holden S.B., Niranjan, M.: On the practical applicability of VC dimension bounds, Neural Comp., vol. 7, no. 6, (1995), 1265–1288
8. Czernichow, T.: A double gradient algorithm to optimize regularization, in Proc. Artificial Neural Networks–ICANN'97, (1997), 289–294
9. Bishop, C.: Improving the generalization properties of radial basis function neural networks, Neural Comp., 3, (1991), 579–588
10. Girosi, F., Jones, F., Poggio, T.: Regularization theory and neural networks architectures, Neural Comp., 7, (1995), 219–269
11. Bishop, C.: Training with noise is equivalent to Tikhonov regularization, Neural Comp., 7, (1995), 108–116
12. Galicki, M., Leistritz, L., Witte, H.: Improved learning of multiple continuous trajectories with initial network state, in Proc. IJCNN'2000, (2000), 15–20

Model Clustering for Neural Network Ensembles

Bart Bakker and Tom Heskes

SNN, University of Nijmegen
Geert Grooteplein 21, 6525 EZ Nijmegen
The Netherlands
{bartb,tom}@mbfys.kun.nl

Abstract. We show that large ensembles of (neural network) models, obtained e.g. in bootstrapping or sampling from (Bayesian) probability distributions, can be effectively summarized by a relatively small number of representative models. We present a method to find representative models through clustering based on the models' outputs on a data set. We apply the method on models obtained through bootstrapping (Boston housing) and on a multitask learning example.

1   Introduction

In neural network analysis we often find ourselves confronted with a large ensemble of models trained on one database. One example of this is resampling, which is a popular approach to try and obtain better generalization performance with non-linear models. Individual models are trained on slightly different samples of the available data set, which are generated e.g. by bootstrapping or cross-validation (see e.g. [1]).

A considerable number of network representations, or models, may be needed to catch the fine nuances of the data. For large problems, the number of models may even be too high to keep a good overview of the ensemble, and special transformations will be required to summarize it, preferably without loss of information. The clustering procedure that we will present will help to understand such problems better, and may be a valuable tool in their analysis.

In Section 2 we will review the clustering method outlined by Rose *et al* [2]. Their method, based on the principles of deterministic annealing as first described by Jaynes [3], was shown to yield good results for the clustering of two-dimensional data with a Euclidean distance function. We will generalize this method for use with other data types and distance functions, and use it to cluster models (e.g. neural networks). In Section 3 we describe the algorithms that we use to implement the method.

We apply our methods on two databases in Section 4. The first, containing the well-known Boston housing data, serves as a benchmark problem to study the effect of bootstrapping. The other database concerns the prediction of single-copy newspaper sales. This can be represented as a series of parallel tasks (outlets at different locations), which we optimize through multitask learning. We show that model clustering can give valuable new insights into the nature of this problem.

J.R. Dorronsoro (Ed.): ICANN 2002, LNCS 2415, pp. 383–388, 2002.
© Springer-Verlag Berlin Heidelberg 2002

2   Clustering by Deterministic Annealing

Suppose we have N_W models, fully characterized through their parameters $\mathbf{w}_i$, $i = 1 \ldots N_W$. We define N_M other models, which we will refer to as cluster centers, denoted $\mathbf{m}_\alpha$, $\alpha = 1 \ldots N_M$. $D(\mathbf{w}_i, \mathbf{m}_\alpha)$ is the distance of model i to cluster center α. The models considered in the present article are feedforward models, which are optimized through supervised training.

We assume distances of the form

$$D(\mathbf{w}_i, \mathbf{m}_\alpha) = \sum_\mu d(y(\mathbf{w}_i, \mathbf{x}^\mu), y(\mathbf{m}_\alpha, \mathbf{x}^\mu)) \,,$$

for some distance measure $d(y_1, y_2)$, where $y(\mathbf{w}_i, \mathbf{x}^\mu)$ is the output of a model with parameters $\mathbf{w}_i$ on an input $\mathbf{x}^\mu$. Each model is supposed to have the same input. Since, once the inputs $\mathbf{x}^\mu$ are given, the clustering procedure depends on the models $\mathbf{w}_i$ only through the outputs $y(\mathbf{w}_i, \mathbf{x}^\mu)$, we can compute these in advance.

We define variables $p_{i\alpha}$ as the probability that model i belongs to cluster α. The distances between models $\mathbf{w}_i$ and cluster centers $\mathbf{m}_\alpha$ are given by $D(\mathbf{w}_i, \mathbf{m}_\alpha)$. We assume that the models $\mathbf{w}_i$ are given, but that the probabilities $p_{i\alpha}$ and cluster centers $\mathbf{m}_\alpha$ are still free to choose. One of the goals of clustering is to put the cluster centers such as to minimize the average distance of the models to the cluster centers, i.e., to find a low average energy $E(\{m\}, \{p\}) = \sum_{i\alpha} p_{i\alpha} D(\mathbf{w}_i, \mathbf{m}_\alpha)$, where $\{m\}$ and $\{p\}$ represent the full sets $\mathbf{m}_\alpha$ and $p_{i\alpha} \ \forall_{i\alpha}$. In this framework the average energy is the average over the distances between models and cluster centers, weighted by $p_{i\alpha}$. For fixed cluster centers $\mathbf{m}_\alpha$, minimizing the average energy would correspond to assigning each model to its nearest cluster center with probability one. A proper way to regularize this is through a penalty term of the form $S(\{p\}) = -\sum_{i\alpha} p_{i\alpha} \ln p_{i\alpha}$. Maximizing $S(\{p\})$ favors a state of total chaos, i.e. $p_{i\alpha} = p_{i'\alpha'} \ \forall_{i,i',\alpha,\alpha'}$, which corresponds to the notion that we have no prior knowledge about the structure of the clusters.

We introduce a regularization parameter T, weighting the two different terms, to arrive at the 'free energy' $F(\{m\}, \{p\}) = E(\{m\}, \{p\}) - TS(\{p\})$. Minimizing $F(\{m\}, \{p\})$ can be seen as a search for the best compromise between a low average distance (minimizing $E(\{m\}, \{p\})$) and keeping a reasonable amount of chaos in the system (maximizing $S(\{p\})$). For any choice of model centers $\mathbf{m}_\alpha$, the probabilities $p_{i\alpha}$ minimizing the free energy $F(\{m\}, \{p\})$ read (under the constraint $\sum_\alpha p_{i\alpha} = 1 \ \forall_i$)

$$p_{i\alpha}(m) = \frac{e^{-\beta D(\mathbf{w}_i, \mathbf{m}_\alpha)}}{\sum_{\alpha'} e^{-\beta D(\mathbf{w}_i, \mathbf{m}_{\alpha'})}} \tag{1}$$

with $\beta = 1/T$. Substitution of this result into the free energy then yields

$$F(\{m\}) = F(\{m\}, \{p(\{m\})\}) = \sum_i \log \sum_\alpha e^{-\beta D(\mathbf{w}_i, \mathbf{m}_\alpha)} \,.$$

3   Annealing and the EM Algorithm

The annealing process finds cluster centers $\mathbf{m}_\alpha$ minimizing the free energy $F(\{m\})$ for increasing values of β. We start with β close to zero and a large number of cluster centers $\mathbf{m}_\alpha$. Such a low value of β will strongly favor the entropy part of the free energy, resulting in a solution where all $\mathbf{m}_\alpha$ are identical (so $p_{i\alpha} = p_{i'\alpha'} \ \forall_{i,i',\alpha,\alpha'}$). When the new $\mathbf{m}_\alpha$ have been found, we increase β and again minimize the free energy. These steps are repeated until the balance between average energy and entropy has shifted enough to warrant multiple clusters, and the cluster centers are divided over two separate solutions. More and more clusters will appear when β is increased further, until we have reached a satisfactory number of clusters (estimated e.g. through cross validation) and we terminate the process.

We find cluster centers $\mathbf{m}_\alpha$ (for each value of β) through minimization of the free energy $F(\{m\})$. This corresponds to solving a series of coupled equations:

$$\frac{\partial F(\{m\})}{\partial \mathbf{m}_\alpha} = \sum_i p_{i\alpha} \frac{\partial D_{i\alpha}}{\partial \mathbf{m}_\alpha} = 0 \ \ \forall_\alpha \ ,$$

where the equations for different $\mathbf{m}_\alpha$ are interdependent through the normalization of $p_{i\alpha}$, which is a function of all $\mathbf{m}_\alpha$. We solve this system of equations using an EM algorithm, a full description of which can be found in [4].

In the expectation step of the EM algorithm the probabilistic assignments $p_{i\alpha}$, as given by Equation 1, are calculated. In the maximization step we find new cluster centers $\mathbf{m}_\alpha'$ that maximize the free energy $F(\{m\})$ given these assignments. A solution for $\mathbf{m}_\alpha'$ can be obtained from any gradient descent algorithm on $\sum_i p_{i\alpha} D(\mathbf{w}_i, \mathbf{m}_\alpha)$ starting from the current $\mathbf{m}_\alpha$.

Our aim is to summarize an ensemble of neural networks through a smaller set of networks with a similar architecture. However, the above description also applies to 'model free' clustering, solely based on the outputs on examples $\mathbf{x}^\mu$. The maximization step for model free clustering (where we now maximize with respect to $\mathbf{y}_\alpha = [y(\mathbf{m}_\alpha, \mathbf{x}^1), y(\mathbf{m}_\alpha, \mathbf{x}^2), \ldots, y(\mathbf{m}_\alpha, \mathbf{x}^N)]$) can be very simple: for suitable distance functions (e.g. the sum-squared error or the cross-entropy error) it is simply $\mathbf{y}_\alpha' = \sum_i p_{i\alpha} \mathbf{y}_i$. Note that for this type of model free clustering the cluster centers $\mathbf{m}_\alpha$ have the same dimensionality as the individual network outputs $\mathbf{y}_i$.

Model based clustering can still benefit from this simplified maximization step. We simply ignore the model parameters $\mathbf{w}_i$ during (parts of) the clustering process, and use the corresponding model outputs $\mathbf{y}_i$ for model free clustering. Obviously, the 'cluster outputs' need to be translated back to model parameters in the end: we split the cluster outputs and the corresponding inputs into a training set and a validation set. Retraining then proceeds in exactly the same way that the original networks were trained: we minimize the mean squared error on the training set, and stop training when the error on the validation set starts to increase. Although this takes some extra time, we may still gain a tremendous speed-up of the annealing process.

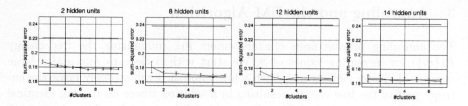

Fig. 1. Boston housing data: explained variance as a function of the number of clusters. The upper horizontal lines in each plot indicate the variance that is explained by the full bootstrap ensemble. The lower horizontal lines indicate the average variance explained by the single models in the ensemble.

4   Results

Boston housing. The clustering algorithm was applied to ensembles of 50 models, obtained by bootstrapping. Larger ensembles did not improve performance. We predicted the outputs on new inputs $\mathbf{x}^{\nu}$ through

$$\tilde{y}^{\nu} = \sum_{\alpha} p_{\alpha} y(\mathbf{m}_{\alpha}, \mathbf{x}^{\nu}), \text{ with } p_{\alpha} = \frac{1}{N} \sum_{i=1}^{N} p_{i\alpha} \,,$$

a weighted sum over the representative model outputs, where the weights are determined by the 'effective' number of models in each cluster.

We compare these predictions to prediction through bagging (taking the average over the outputs of all models in the ensemble on a new input). Figure 1 shows the explained variance (averaged over 10 independent choices of training and test set) as a function of the number of cluster centers for ensembles of models with 2, 8, 12 and 14 hidden units (one type of model in each ensemble). It can be seen that for more complicated models (more hidden units) less cluster centers are needed to match the performance of the full ensemble. This progression from many cluster centers to one can be understood as follows. For one simple model it is impossible to give an adequate representation of the hidden (complex) function underlying the data. Bootstrapping in this case serves to find an ensemble of models that, when put together, can closely approximate this function. Since in this case the bootstrapping ensemble contains multiple significantly different models, multiple cluster centers are required to represent the ensemble. Bootstrapping then serves to reduce the bias in the model.

One sufficiently complex model however, may be enough to represent the unknown function that generated the data. In this case, bootstrapping serves to reduce overfitting. Although for complicated models many bootstrap samples may be needed to obtain a low variance estimation, in the end the full ensemble can be represented by just one cluster center. More discussion on the effectiveness of bootstrapping and ensemble learning can be found in [5].

Figure 1 illustrates the bias/variance reducing properties of bootstrapping. The improvement of the one cluster solution over the average single model is

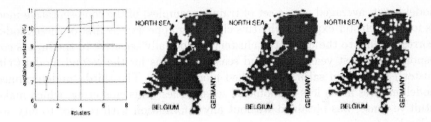

Fig. 2. Left: explained variance as a function of the number of clusters. The upper horizontal line represents the variance that is explained by the multitask learning network. Right three panels: geographical clustering of the newspaper outlets. Circles mark outlets assigned with weight larger than 0.9 to either the 'seasonal' cluster (left panel), the 'short term' cluster (middle panel) or the 'undetermined' cluster (right panel).

due to variance reduction, since in both cases the same type of network is used for prediction. The improvement obtained with increasing numbers of clusters is (mainly) due to bias reduction, since the summation over multiple networks in fact yields a more complex model.

Prediction of newspaper sales. For the newspaper data we only trained ensembles of networks with two hidden units, since we know from past experience that these yield the best results for this particular data set (see [6] for more information). We also trained one network similar to those in the ensemble, with two hidden units, but with one output for each task (outlet). This means that all tasks share the same input-to-hidden weights of the network, but had independent hidden-to-output weights. See e.g. [7] for similar work. We predicted the outputs on new inputs $\mathbf{x}^\nu$ through

$$\tilde{y}_t^\nu = \sum_\alpha p_{t\alpha} y(\mathbf{m}_\alpha, \mathbf{x}_t^\nu), \text{ with } p_{t\alpha} = \frac{1}{n_t} \sum_{i \in t} p_{i\alpha} ,$$

where n_t is the number of models trained for task t, and where the weights $p_{t\alpha}$ for task t depend only on the corresponding models trained on this task.

Explained variance for the newspaper data increases strongly until the 3-cluster solution, at which point it is significantly higher than the variance explained by the multitask learning network (one output for each outlet). In this case model clustering is used to regulate the hidden-to-output weights of the network ('shrinkage to the mean'), and to prevent overfitting. The clustering obtained on the newspaper data also reveals an interesting structure, which was hidden in the data. Figure 2 shows the outlets in Holland that are assigned to clusters 1, 2 and 3 with probability $p_{i\alpha} > 0.9$. It can be seen that the outlets in the first cluster tend to be near the beach, and in the eastern part of Holland, touristic spots without many large cities. The outlets in cluster 2 center around the 'Randstad' (Amsterdam and other relatively large Dutch cities). The outlets in cluster 3 are spread all around, and can be considered 'undetermined'. Calculation of the 'sensitivity' (the derivative of the model output with respect to a model input, averaged over a set of training samples) of the representative models for each cluster explains why this clustering appears: the first of these models (corresponding to the 'touristic' cluster) is especially sensitive to the inputs corresponding to last year's sales and season, whereas for the second model ('city cluster') short term sales are weighed more heavily. The third, 'undetermined' model features much less pronounced sensitivities. This clustering, which makes intuitive sense, was obtained without any information with respect to city size or level of tourism.

5   Discussion

In the present article we have presented a method to summarize large ensembles of models to a small number of representative models. We have shown that predictions based on a weighted average of these representatives can be as good as, and sometimes even better than, predictions based on the full ensemble. We believe that this method provides an extremely useful addition to any method featuring an ensemble of models, such as bootstrapping, sampling of Bayesian posterior distributions or multitask learning. The method is not only valuable in terms of predictive quality, but also on a more abstract level. This improvement was apparent on the newspaper data, where different clusters of models brought out different aspects of the data.

In the present article we have chosen the clustering procedure outlined by Rose *et al* [2] to perform model clustering. Alternative clustering procedures can of course be considered. We do however stress the importance of using a 'natural' distance function based on model outputs rather than a more arbitrary distance based on model parameters.

Acknowledgments. This research was supported by the Technology Foundation STW, applied science division of NWO and the technology programme of the Ministry of Economic Affairs.

References

1. B. Efron and R. Tibshirani. *An Introduction to the Bootstrap.* Chapman & Hall, London, 1993.
2. K. Rose, E. Gurewitz, and G. Fox. Statistical mechanics of phase transitions in clustering. *Physical Review Letters*, 65:945–948, 1990.
3. E. Jaynes. Information theory and statistical mechanics. *Physical Review*, 106:620–630, 1957.
4. Donald B. Rubin. EM and beyond. *Psychometrika*, 56(2):241–254, 1991.
5. P. Domingos. Why does bagging work? A Bayesian account and its implications. *Proceedings of the KDD*, pages 155–158, 1997.
6. T. Heskes. Solving a huge number of similar tasks: a combination of multi-task learning and a hierarchical Bayesian approach. In *ICML*, 1998.
7. R. Caruana. Multitask learning. *Machine Learning*, 28:41–75, 1997.

Does Crossover Probability Depend on Fitness and Hamming Differences in Genetic Algorithms?

José Luis Fernández-Villacañas Martín and Mónica Sierra Sánchez

Universidad Carlos III, Madrid, Spain

Abstract. The goal of this paper is to study if there is a dependency between the probability of crossover with the genetic similarity (in terms of hamming distance) and the fitness difference between two individuals. In order to see the relation between these parameters, we will find a neural network that simulates the behavior of the probability of crossover with these differences as inputs. An evolutionary algorithm will be used, the goodness of every network being determined by a genetic algorithm that optimizes a well-known function.

1 Introduction

Genetic algorithms are searching algorithms based on the mechanisms of natural selection and genetics (Goldberg, [1]). The basic mechanisms that alter the initial population of candidate solutions are reproduction, crossover and mutation. In most variations of genetic algorithms, the crossover probability is constant during the course of evolution. In this paper, we will attempt to verify the following hypothesis: Does the crossover probability of every pair of individuals selected for the crossing operation depend on their genotypic distance and their difference in the value of fitness?.

The relationship between crossover probability and these two parameters will be given by a neural network that receives the Hamming distance and the difference of fitness for two individuals as inputs obtaining a crossover probability for the same pair of individuals. To find this neural network, we will use a genetic algorithm in which every individual is a neural network with a structure previously defined. The chromosome of each neural network is a weight vector and the algorithm will try to optimise these weights (Montana and Davis, 1989, [2]).The fitness of each network will be evaluated using another genetic algorithm on the Saddle function [3].

The work will be presented as follows: Section 2 will describe the topology of the neural networks and representation issues for the neural network genetic algorithm. In Section 3, we will study the parameters used for the genetic algorithm for the minimization of the Saddle function. In Section 4, we will discuss the results obtained with the meta-genetic algorithm, and analyze the resulting "winning" networks. Lastly, in Section 5, we will extract the conclusions and suggest paths for further investigation.

J.R. Dorronsoro (Ed.): ICANN 2002, LNCS 2415, pp. 389–394, 2002.

2   The Neural Network

The individuals of the first (or outer) genetic algorithm are neural networks (Haykin, [4]); their structure consisting basically in an input layer with two neurons, an output layer with a neuron a hidden layer with three neurons. The network is fully connected. The genetic algorithm's job is to determine which are the values of the connections.

Each network chromosome is a vector of real numbers, and every position of the vector is a weight connection of the network. Each neural net need two inputs (Hamming distance and fitness). These inputs are normalized by dividing the number of different bits by the total number of bits in the individual. Also, the network output will be normalized. The level of activation of each neural net will be given by a standard sigmoid function (for every neuron except the neurons in the input layer).

3   The Meta-genetic Algorithm

The outer GA evolves a population of neural networks. First of all, we will need a mechanism to evaluate the fitness of each network. The solution opted for here is the performance of a genetic algorithm on a function with a known optimum. We will use the Saddle function from De Jong set [3], whose minimum is zero. In the evolution of this "inner" genetic algorithm, the neural network will give the crossover probability for every pair of solutions selected to cross. We also need to specify how to measure network fitness in the Saddle optimization algorithm. A meta-GA (see [5] for a general description of the idea) is used to find optimum parameters of a GA and the fitness of the inner algorithm is evaluated using the average final objective function value from two independent runs. We have considered three possibilities: 1) the sum of the mean fitness of every generation; 2) the best fitness achieved in the last generation; 3) the best fitness achieved in the last generation using elitism, that is, saving the best solution for the next generation. The meta-GA consists of two genetic algorithms: an outer genetic algorithm with the population of neural networks and an inner genetic algorithm that serves to evaluate each network of the outer algorithm. In the following subsections we describe the parameters that we are going to use in each algorithm.

3.1   Saddle Optimization Genetic Algorithm

This algorithm is used to evaluate the network fitness. The individuals of the population are vectors of two real numbers. Each vector is a a 64-bit chromosome. The Hamming distance between two individuals is calculated as the different bits in the same position. The network inputs are normalized, so the Hamming distance is calculated as: number of different bits / 64. Fitness and their differences must be in the [0,1] interval.

The probability of crossover for each pair of solution vectors is the output of the network being evaluated. In order to see which type of crossover and mutation (and mutation probability) is applied, a study has been made with random networks looking at the each network's fitness histogram (fifty executions of the Saddle algorithm). One would expect distributions and mean "peak" values that differ from one network to another. An estimation of the peak and the width of the histograms was made for each network, and the difference between peaks and the square root of the product of the estimation of widths represented for each pair of networks used. This analysis was conducted for three different ways of estimating network fitness, and mutation probabilities of 0.01, 0.001 and 0.0001. These results are shown in Figure 2.

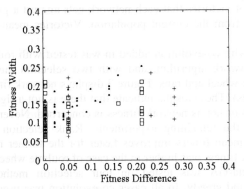

Fig. 1. Network fitness differences and histogram widths for pairs of randomly chosen networks. The legend is as follows: dots (.) sum of mean fitness, pluses (+) best fitness without elitism, squares () best fitness with elitism.

Comparing similar graphs as the one above, with different mutation probabilities, we came to the conclusion that a probability of 0.0001 gave greater differences among the networks, though the histograms were not very narrow.

Similar experiments were conducted for selecting a reproduction method for the Saddle "inner" GA. Mutation probability was fixed to 0.0001 and four selection methods were trailed: roulette wheel, tournament of size 4, rank and greedy. The results favored the choice of roulette wheel and rank methods. Both will be considered for the meta-GA.

3.2 Neural Networks Genetic Algorithm

Network fitness will be calculated as the sum of mean fitness of the generations. This choice of fitness calculation is justified as there are more differences between one network and another in fitness. Nevertheless, this is not totally fair. We can observe that network fitness distributions for the best solution fitness in the Saddle algorithm (pluses in Figure 2) show a wider spread than mean fitness. The reason behind this is that most of the networks get the optimal solution and saturate the right part of the histograms (fitness close to 1); the problem is that the fitness differences (between adjusted "mean" peaks) are not reliable. In general, we will calculate network fitness on 10 different executions of the algorithm, and will use the average to discern results from different networks.

The neural networks individuals are represented as real number vectors; we will only the use of mutation will explore new paths to solutions. Mutation is achieved

individuals. The main limitation is that, in most cases, networks return crossover rates near the extremes (0 and 1); we did not find many networks returning rates spreading over the whole range. This means we cannot study this dependence fully, but only in their limits.

Nevertheless, a number of qualitative conclusions was made: high values of fitness and hamming distances bring nearly-zero crossover rate; high crossover rates are achieved for both small fitness and hamming distances.

For the particular function (Saddle) used in this work, there was no relation between Hamming distance and the fitness of the solutions. The fitness landscape is rugged and uncorrelated with a high degree of epistasis (Davidor, [9]). Under this binary representation, crossover has the same effect of a random jump from one point of the genotypic space to another .

Future work could include the extension to other types of functions, particularly those for which we can calculate their epistasis, and see what is the relation between the performance of the networks and epistasis. Along the same lines, we could look at different ways of representation for the solutions. The question if our initial hypothesis is right or wrong seems to be tightly linked to that if crossover is useful or not, and this one, in turn, to that if we can have some predictability from the genotype to our fitness landscape.

References

1. Goldberg, D. E.: Genetic Algorithms in Search, Optimization and Machine Learning, Addison Wesley (1989)
2. Montana, D.K. and Davis, L.D.: Training feedforward networks using genetic algorithms. In Proceedings of the International Joint Conference on Artificial Intelligence, Morgan Kauffman, (1989).
3. De Jong, K.: An analysis of the behaviour of a class of genetic adaptive systems, PhD. Thesis, University of Michigan, Ann Arbor (1975).
4. Haykin, S.: Neural Networks: a comprehensive foundation, Prentice Hall (1999), 2nd ed
5. Bäck, T.: Parallel Optimization of Evolutionary Algorithms, in Y. Davidor, H.-P. Schwefel and R. Männer, editors: Parallel Problem solving from Nature – PPSN III, International Conference on on Evolutionary Computation, pp 418–427, Springer, Berlin, (1994).
6. Fernández-Villacañas, J. L., Amin S.: Simulated Jumping in Genetic Algorithms for a set of test functions, Proceedings of the IASTED International Conference on Intelligent Information Systems, Grand Bahama Island, December, 1997.
7. Rosin, C. and Belew, R.: Methods for competitive co-evolution: Finding opponents worth beating, in Proceedings of the Sixth International Conference on Genetic Algorithms (1995).
8. Kallel, L., Naudts, B., Schoenauer, M.: On Functions with a Fixed Fitness Versus Distance-to-Optimum Relation, Proceedings of CEC'99, p. 1910–1916, IEEE press, (1999).
9. Davidor, Y.: Epistasis Variance: A Viewpoint on GA-Hardness, in Foundations of Genetic Algorithms 1, ed. G. J. E. Rawlins, Morgan Kaufmann Publishers, San Mateo (1991).

Extraction of Fuzzy Rules Using Sensibility Analysis in a Neural Network

Jesús Manuel Besada-Juez and Miguel A. Sanz-Bobi

Universidad Pontificia Comillas. Escuela Técnica Superior de Ingeniería
Instituto de Investigación Tecnológica
Alberto Aguilera, 23. 28015 Madrid (Spain)
Phone: 34 91 542 28 00. Fax: 34 91 542 31 76
jesus.besada@iit.upco.es; masanz@iit.upco.es

Abstract. This paper proposes a new method for the extraction of knowledge from a trained type feed-forward neural network. The new knowledge extracted is expressed by fuzzy rules directly from a sensibility analysis between the inputs and outputs of the relationship that model the neural network. This easy method of extraction is based on the similarity of a fuzzy set with the derivative of the tangent hyperbolic function used as an activation function in the hidden layer of the neural network. The analysis performed is very useful, not only for the extraction of knowledge, but also to know the importance of every rule extracted in the whole knowledge and, furthermore, the importance of every input stimulating the neural network.

1 Introduction

This paper proposes a new method for the extraction of knowledge from an artificial neural network (NN) that is represented by fuzzy rules. The quality of the rules extracted and the easy method used to obtain them improve the current procedures used in this field until now [4].

The rules derived from the NN can be used for several purposes:

- Knowledge representation in fuzzy terms of the relationships modelled by the NN
- Suggestion of modifications for the architecture of the NN studied by adding new knowledge required or eliminating some knowledge not needed.
- Retraining of particular zones of the NN with less knowledge.
- Possible complement of the knowledge of an expert system by the addition of the knowledge extracted from the NN

This method is based on a set of procedures that analyse how and when the neurons are activated by a particular input, the importance of the inputs for the NN prediction and the monitoring of its performance.

An important parameter obtained from the new method is the reliability of the knowledge included in the rules extracted. This is key information to be taken into account in many applications such as failure detection, or adaptive control.

J.R. Dorronsoro (Ed.): ICANN 2002, LNCS 2415, pp. 395–400, 2002.

2 Type of NN Analysed

The methodology proposed for knowledge extraction has been analysed and tested using *feed forward* NN [1]. In particular, this paper pays attention to the *feed forward* NN consisting of *three layers*: input, hidden and output. The neurons located in the hidden layer have a *hyperbolic tangent* activation function and the neurons in the output layer have a *linear* activation function. The notation that will be used for the analysis of this NN is represented as follows.

Let $\vec{e}$ be a vector with n components corresponding to the inputs exciting the NN. Let m be the number of neurons at the NN hidden layer. The activation weights between the input and hidden layers are represented by the mxn matrix W_1.

The activation weights between the hidden and the output layers are represented by the pxm matrix W_2, where p is the number of outputs in the NN.

The vectors $\vec{b}_1$ and $\vec{b}_2$ will represent the bias of the hidden and output layers, with dimensions m and p respectively.

3 Method for Fuzzy Explanation of a NN

According to the notation described, the mathematical equation 1 represents the ith output of the NN.

$$\vec{S}_i = \sum_{j=1}^{m} W_{2,i,j} \cdot tanh\ (W_{1,j} \cdot \vec{e} + \vec{b}_{1,m}) + b_{2,i} \tag{1}$$

$W_{1,j}$ represents row j in the matrix W_1, and $W_{2,i,j}$ the element ixj in the matrix W_2. The partial derivative of the outputs (equation 2) in respect to every input h is:

$$\frac{\partial S_i}{\partial \vec{e}_h} = \sum_{j=1}^{m} W_{2,i,j} \cdot W_{1,j,h}\ sec\ h^2\ (W_{1,j} \cdot \vec{e} + \vec{b}_{1,m}) \tag{2}$$

Equation 2 shows that the partial derivative of output i in respect to input h is the addition of a set of weighted squared hyperbolic secants.

The hyperbolic secant (see fig. 1) recalls the shape of a membership function corresponding to a fuzzy set that could be defined as "approximately zero" or "near to zero". This similarity is the key point for the new method proposed for the extraction of rules from a NN.

Coming back to equation 2, note that the right side of the argument of every hyperbolic secant is a hyperplane and the left side can be considered a measurement of the degree of activation of the output neuron i when any input $\vec{e}$ is exciting the NN. According to this, it is possible to formulate the knowledge included in a NN by fuzzy rules.

An example of NN will be used to explain the procedure used to extract the knowledge and without loss of generality. This NN example has two inputs, one single output, two neurons in the hidden layer and one neuron in the output layer. The activation function of the neurons at the hidden layer is $f(x)=\tanh(x)$ and the linear function is used as activation function in the output neuron.

The output equation of this NN will be the following:

$$\bar{S} = w_{2,1,1} * tanh\,(w_{1,1,1} * e_1 + w_{1,1,2} * e_2 + b_{11}) + \qquad (3)$$
$$w_{2,1,2} * tanh\,(w_{1,2,1} * e_1 + w_{1,2,2} * e_2 + b_{12}) + b_{2,1}$$

Deriving equation 3 with respect to the inputs of the NN, equations 4 and 5 are obtained.

$$\frac{\partial S}{\partial e_1} = w_{1,1,1} * w_{2,1,1} * sec\; h^2\,(w_{1,1,1} * e_1 + w_{1,1,2} * e_2 + b_{11}) + \qquad (4)$$
$$w_{1,2,1} * w_{2,1,2} * sec\; h^2\,(w_{1,2,1} * e_1 + w_{1,2,2} * e_2 + b_{12})$$

$$\frac{\partial S}{\partial e_2} = w_{1,1,2} * w_{2,1,1} * sec\; h^2\,(w_{1,1,1} * e_1 + w_{1,1,2} * e_2 + b_{11}) + \qquad (5)$$
$$w_{1,2,2} * w_{2,1,2} * sec\; h^2\,(w_{1,2,1} * e_1 + w_{1,2,2} * e_2 + b_{12})$$

Let E_1, E_2, E_3 and E_4 be the arguments of the hyperbolic secants in equations 4 and 5. It is possible to define the following four fuzzy sets: "E_i near to zero" ($i=1$ to 4) of which the membership function can be defined as $\eta(E_i)= sech(E_i)^2$. The maximum membership degree is reached when $E_i = 0$, $\eta(E_i)=1$.

Once the NN has been trained and taking into account the previous concepts, it is possible to very easily define effective fuzzy rules where their hypothesis depend on the input variables of the NN.

In the case of the simple NN used as an example, the following fuzzy rules can be extracted according to equations 4 and 5:

Rule 1: If "$w_{1,1,1}*e_1+w_{1,1,2}*e_2+b_{11}$ is near to 0" then the output increases the values $w_{1,1,1}*w_{2,1,1}$ and $w_{1,1,2}*w_{2,1,1}$ in respect to the increasing of inputs e_1 and e_2,

Rule 2: If "$w_{1,2,1}*e_1+w_{1,2,2}*e_2+b_{12}$ is near to 0" then the output increases the values $w_{1,2,1}*w_{2,1,2}$ and $w_{1,2,2}*w_{2,1,2}$ in respect to the increasing of inputs e_1 and e_2,

These rules can be reformulated in a compact form defining two additional concepts: *direction of knowledge generalisation* and *activation strength*.

The direction of knowledge generalisation (d_i) is defined by the hyperplanes which are the arguments in the hyperbolic secants in equations 4 and 5. As an example, rule 1 defines the straight line "$w_{1,1,1}*e_{1+}w_{1,1,2}*e_2+b_{11}$" in a plane defined by the inputs e_1 and e_2. This is a direction of knowledge generalisation (d_1) and any combination of both inputs e_1 and e_2 close to d_1 will reach high values of membership degree to the fuzzy set "$w_{1,1,1}*e_{1+}w_{1,1,2}*e_2+b_{11}$ near to 0", which means, that the NN has knowledge about this combination of inputs. However if the membership degree reaches a low value, then the NN does not store much information about the current inputs.

The activation strengths f_{ij} correspond to the coefficients before the hyperbolic secants in equations 4 and 5. As an example, from equations 4 and 5 the activation strength $f_{1,1}$ would be $w_{1,1,1}*w_{2,1,1}$ and the activation strength $f_{1,2}$ would be $w_{1,1,2}*w_{2,1,1}$.

The two fuzzy rules can be rewritten as follows:

Rule 1: If d_1 near to $0 \to f_{1,1}/e_1$ and $f_{1,2}/e_2$, Rule 2: If d_2 near to $0 \to f_{2,1}/e_1$ and $f_{2,2}/e_2$

The directions of knowledge generalisation are very important to know where the NN has stored its knowledge and where it has not. Furthermore, the activation strengths are important in order to know the influence of the inputs on the outputs.

4   Understanding the Information Extracted from a NN

Once a NN has been trained using a training set named E, and the method of knowledge extraction applied, it is possible to obtain very helpful information about the knowledge extracted. This is presented in the following three subsections.

4.1   Amount and Importance of Knowledge in a Fuzzy Rule

The hypothesis of the fuzzy rules extracted are a fuzzy measurement of the proximity of the inputs observed to the direction of knowledge generalisation obtained after the training process. In order to consider a particular proximity to be important, a threshold can be adopted. This threshold has an immediate interpretation in our fuzzy sets as an α-cut of a particular degree of membership. A rule will be activated if the membership degree of a set of inputs overpasses a pre-defined α-cut. Below this α-cut there is no reliable knowledge and its sensibility output/input will be poor.

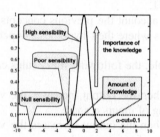

Fig. 1. Knowledge in a fuzzy rule

According to this, the amount of knowledge included in an extracted fuzzy rule will correspond to the support of the pre-defined α-cut. Figure 1 summarises the previous ideas using an α-cut= 0.1.

The set of inputs reaching the maximum membership degree in every fuzzy set will be the most representative of the knowledge included in the rule.

4.2   Similarity among Fuzzy Rules Extracted

The method proposed allows for the elimination of redundant knowledge extracted measuring the similarity among rules. In order to do this, a symmetric squared similarity matrix is constructed where each component represents the similarity between two rules i and j. Every component $i.j$ of this matrix will be calculated according to equation 6. If $S_{i,j} = 0$ then rules i and j are identical (maximum similarity). If $S_{i,j} = 1$ then rules i and j do not have common knowledge.

$$S_{i,j} = \frac{\sum_{i=1}^{n} (abs \ (\eta \ (d_i(\bar{e}_i)) - \eta \ (d_j(\bar{e}_i))))}{n} \tag{6}$$

4.3 Membership Degree of Knowledge Included in the Set of Inputs E to the Different Fuzzy Rules

This membership degree can be calculated using equation 7.

$$G_i = \frac{\sum_{j=1}^{n} \eta_i(\vec{e}_j)}{n} \tag{7}$$

5 Certainty of the Neural Network Estimation

If a new input $\vec{e}$ is presented to a trained NN, it will estimate the corresponding output but the certainty of the estimation could be a risk, if the input is unknown or similar inputs were not used in the training set.

Certainty is easy to obtain in the method proposed by the simple measurement of the proximity between inputs and directions of knowledge generalisation d_i. This certainty is the direct fuzzyfication of the input through every fuzzy set included in the hypothesis of the fuzzy rules extracted $\eta_{Ri}(\vec{e})$, or equivalent, the membership degree to every fuzzy set defined. So a membership degree is obtained for each rule or for each neuron. These n values can be combined into only one for example using the expression: H=max($\eta_{Ri}(\vec{e})$), with i=1: n number of rules.

Figure 2 shows an example using two inputs and three neurons. In this figure the three directions of knowledge generalisation are shown (regions of maximum certainty and membership degree). Also, figure 3 shows a region where the NN has poor or null knowledge for the combination of inputs e1 and e2.

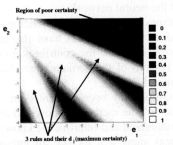

Fig. 2. Certainty of the estimation

6 Example

This example is based on a mathematical function defined as follows: $y=e_2$ $e_1 <= -1$, $y=e_1+e_2$ $-1<e_1<1$, $y=2e_2$ $e_1 > =1$ .This function has as output variable y which depends on two inputs variables e_1 and e_2 where e_1 is a variable uniformly distributed between -2 and 2, and $e_2 = 0.2 * sinh(e_1) + \varepsilon$ with ε a Gaussian noise of 5%. This mathematical function creates three differentiated classes (C1, C2 and C3).

Using an architecture such as that presented in the previous section, a NN was trained using a different

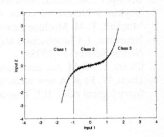

number of neurons in the hidden layer (2, 3 and 4 neurons), where a NN with 2 neurons is insufficient and 4 neurons produce an oversized network with worse qualities of generalisation.

The figures on the left show the hyperbolic secant squared of each d_i stimulated by each input used in the training of the NN and at the bottom the maximum value of the membership degrees of the previous plots. The figures on the right show the distribution of knowledge in each rule for cases using 3 and 4 neurons.

Due to the simplicity of the example, it is not necessary to calculate the measurements described in previous sections. A visual observation is sufficient to extract conclusions.

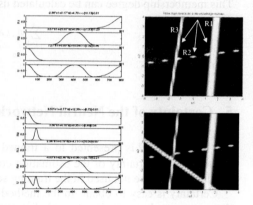

7   Conclusions

A new method for the extraction of fuzzy rules from a trained NN has been explained in this paper. It is based on the sensibility analysis of the outputs in respect to the inputs of the neural network

Guidelines for a better understanding and improvement of the knowledge included in the trained neural network have also been established. Using the method proposed, it is possible to know the contribution of each rule to the whole knowledge of the NN, the regions of the domain where there is no information and the importance of each input in NN.

An example has been presented in order to confirm the quality and the main capabilities of the method proposed to convert a NN in a set of symbolic rules, and also to understand and improve the knowledge inside a NN.

The next steps will include an intensive application in industrial processes and investigation in more sophisticated architectures of NNs

References

1. Bishop, C. M. Neural Networks for Pattern Recognition, Clarendon Press. Oxford. (1995).
2. Haykin, S. Neural Networks: A Comprehensive Foundation, IEEE PRESS. (1994).
3. Mitchell, T. M. Machine Learning, McGraw-Hill International Editions. (1997).
4. Neumann, J. Classification and Evaluation of Algorithms for Rule Extraction from Artificial Neural Networks. (1998).
5. Cloete, I. y J. M. Zurada. Knowledge-Based Neurocomputing, MIT Press. (2000).
6. Duch, W. Et al. A new methodology of extraction, optimization and application of crisp and fuzzy logical rules. IEE Transactions on Neural Networks (2000)

A Simulated Annealing and Resampling Method for Training Perceptrons to Classify Gene-Expression Data*

Andreas A. Albrecht[1], Staal A. Vinterbo[2], C.K. Wong[3], and
Lucila Ohno-Machado[2,4]

[1] Univ. of Hertfordshire, Computer Science Dept., Hatfield, Herts AL10 9AB, UK
`A.Albrecht@herts.ac.uk`
[2] Harvard Medical School, Decision Systems Group, 75 Francis Str., Boston, MA,
USA
`{staal,machado}@dsg.harvard.edu`
[3] CUHK, Dept. of Computer Science and Engineering, Shatin, N.T., Hong Kong
`wongck@cse.cuhk.edu.hk`
[4] MIT, Division of Health Sciences and Technology, Cambridge, MA, USA

Abstract. We investigate the use of perceptrons for classification of microarray data. Small round blue cell tumours of childhood are difficult to classify both clinically and via routine histology. Khan et al. [10] showed that a system of artificial neural networks can utilize gene expression measurements from microarrays and classify these tumours into four different categories. We used a simulated annealing-based method in learning a system of perceptrons, each obtained by resampling of the training set. Our results are comparable to those of Khan et al., indicating that there is a role for perceptrons in the classification of tumours based on gene expression data. We also show that it is critical to perform feature selection in this type of models.

1 Introduction

Measuring gene expression levels is important for understanding the genetic basis of diseases. The simultaneous measurement of gene expression levels for thousands of genes is now possible due to microarray technology. Data derived from microarrays are difficult to analyze without the help of computers, as keeping track of thousands of measurements and their relationships is overwhelmingly complicated. Several authors have utilized unsupervised learning algorithms to cluster gene expression data [4]. In those applications, the goal is to find genes that have correlated patterns of expression, in order to facilitate the discovery of regulatory networks. Recent publications have begun to deal with supervised classification for gene expression derived from microarrays [6]. The goal in these

---

* Research partially supported by EPSRC Grant GR/R72938/01, by CUHK Grant SRP 9505, by HK Government RGC Earmarked Grant CUHK 4010/98E, and by the Taplin award from the Harvard/MIT Health Sciences and Technology Division.

J.R. Dorronsoro (Ed.): ICANN 2002, LNCS 2415, pp. 401–407, 2002.
© Springer-Verlag Berlin Heidelberg 2002

applications is usually to classify cases into diagnostic or prognostic categories that are verifiable by a "gold-standard". Additionally, researchers try to determine which genes are most significantly related to the category of interest. Since the number of measurements is very large compared to the number of arrays, there is tremendous potential for overfitting in models that do not utilize a preprocessing step for feature selection. The feature selection process itself is of interest, as it helps to determine the relative importance of a gene in the classification. In this paper, we propose an algorithm for learning perceptrons based on simulated annealing [1,3,11] and a resampling strategy and show that it can be successfully applied to the analysis of gene expression, which implies a new method for feature selection.

2　Methods

Let $D \subseteq \mathbb{Q}^n$ be our input data table where each of the columns corresponds to expression measurements for a particular gene over the tissues investigated. Further let $c : \mathbb{Q}^n \to \{1, 2, \ldots, m\}$ be a partial function that for D returns the tumor class associated with each row.

We would like to find a realization of a function $F : \mathbb{Q}^n \to 2^{\{1,2,\ldots,m\}}$ that represents an extension of c that we can use to classify new, unseen expression measurement vectors.

We do this as follows. For each class $i \in \{1, 2, \ldots, m\}$, we construct a classifier $F_i : \mathbb{Q}^n \to [0, 1]$. These F_i are then combined to form F as:

$$F(x) = \{j | F_j(x) = \max_i F_i(x)\}.$$

The value $|F(x)|$ gives us an indication of how uniquely we were able to classify x.

We now turn to the construction of the functions F_i. A *perceptron* p is a function $p : \mathbb{R}^n \times \mathbb{R}^n \times \mathbb{R} \to \{0, 1\}$ such that

$$p(x, w, \vartheta) = p_{w,\vartheta}(x) = \tau_\vartheta(wx) = \begin{cases} 0 \text{ if } wx < \vartheta, \\ 1 \text{ otherwise.} \end{cases}$$

Given k_i perceptrons, we define F_i to be

$$F_i(x) = \frac{1}{k_i} \sum_{j=1}^{k_i} p_{w_j^i, 0}(x).$$

where the w's are restricted to be rational (for simpler notations, we assume $\vartheta = 0$ and always $x_{n+1} = 1$, i.e., w_{n+1} represents the threshold).

2.1　Perceptron Training

Let $T = T^+ \cup T^-$ be a training set composed of positive and negative examples, respectively. We want to find the parameters w of the perceptron p that maximize the separation of the positive and negative examples in T.

Höffgen, Simon, and van Horn [9] have shown that finding a linear threshold function that minimizes the number of misclassified vectors is NP-hard in general. Hence we need to apply heuristics to the problem of training a perceptron.

Simulated annealing [1,3,11] is our choice of optimization strategy. Given a search space W in which we want to find an element that minimizes an objective function $o : W \to \mathbb{N}$, an initial "current location" $w_s \in W$, a function $s : W \to W$ that in a stochastic fashion proposes a next "current location" that we hope will lead to the wanted optimum, a function $a : \mathbb{N} \times W \times W \to \{0,1\}$ that accepts or reject the proposed next current location, and finally a stopping criterion. We can illustrate the strategy by the following simple pseudo-code skeleton:

```
k ← 0; w ← w_s;
while not stop(k, w)
    w' ← s(w); k ← k + 1;
    if a_k(w, w') = 1
        w ← w'
```

The idea of the algorithm is to, while initially allowing locally suboptimal steps, become more restrictive in allowing locally sub-optimal steps to be taken over time. The hope is to avoid premature convergence to local minima. We define our objective function to be the number of misclassified elements in T. Let

$$M_T(w) = \{x \in T^+ | p_w(x) < 1\} \cup \{x \in T^- | p_w(x) > 0\},$$

then we can define our objective function as $o(w) = |M_T(w)|$. The set $M_T(w)$ can be viewed as a neighborhood of w containing all the possible next steps we can take from w. As a *first key feature* of our heuristic, we now construct a probability mass function u over $M_T(w)$ as

$$q(x) = |xw| / \sum_{y \in M_T(w)} |yw|.$$

Elements that are "further away" from being classified correctly by p_w are assigned higher values by q. We now define

$$s(w) = w - \chi(x) \cdot \text{sample}(M_T(w), q) / \sqrt{ww},$$

where sample stochastically selects one element from the set $M_T(w)$ with the probability given by q, and $\chi(x) = 1$ for $x \in T^-$, $\chi(x) = -1$ for $x \in T^+$. The acceptance function at step k (the k'th time through the while-loop) is defined as

$$a_k(w, w') = \begin{cases} 1 \text{ if } \pi_k(w, w') > \rho, \\ 0 \text{ otherwise,} \end{cases}$$

where

$$\pi_k(w, w') = \begin{cases} 1 & \text{if } o(w') - o(w) \leq 0, \\ e^{-(o(w')-o(w))/t(k)} & \text{otherwise,} \end{cases}$$

and $\rho \in [0, 1]$ is uniformly randomly sampled at each step k. The function t, motivated by Hajek's Theorem[7] on convergence of inhomogenous Markov chains for big enough constants Γ, is defined as

$$t(k) = \Gamma / \ln(k + 2), \quad k \in \{0, 1, \ldots\}$$

and represents the "annealing" temperature (*second key feature*). As t decreases, the probability of accepting a w that does not decrease the objective function decreases. We empirically chose $\Gamma = (|T^+| + |T^-|)/3$, essentially using the same method as in [2].

Finally our stopping criterion is given by a pre-determined number of iterations through the while-loop.

2.2   Perceptron Training Set Sampling

In order to generate (a large number of) different hypotheses (*Epicurean learning* [2,8]) as well as to achieve zero learning error in a short time, the following sampling scheme for training sets is applied (*third key feature*):

Let $C_i^+ = \{x \in D | c(x) = i\}$, be the positive examples for class i in D, and let $C_i^- = D - C_i^+$, be the negative examples of class i in D. Further, let for two parameters $\alpha, \beta \in (0,1]$, $s_i^+ = \lfloor \alpha | C_i^+ | \rfloor$, and let $s_i^- = \lfloor \beta | C_i^- | \rfloor$. For each j in $\{1, 2, \ldots, k_i\}$ we randomly sample $T_{i,j}^+ \subseteq C_i^+$ and $T_{i,j}^- \subseteq C_i^-$ such that $|T_{i,j}^+| = s_i^+$, and $|T_{i,j}^-| = s_i^-$. The set $T_{i,j} = T_{i,j}^+ \cup T_{i,j}^-$ is then the training set used to train perceptron $p_{w_j^i}$ in F_i.

The approach is different from the Boosting method [5] where one tries to reduce the training error by assigning higher probabilities to "difficult" samples in a recursive learning procedure.

2.3   Dimensionality Reduction

In our case where there are many more gene expression measurements than there are tissue samples, experiments have shown that reducing the dimensionality of the data is beneficial [10]. The scheme we applied is based on selecting the genes that incur the coefficients with the biggest absolute values in the perceptrons after having completed training on all dimensions in the data (*fourth key feature*). Let $g(w, q)$ be the set of q positions that produce the q biggest values $|w_l|$ in $w = (w_1, w_2, \ldots, w_n)$ (ties are ordered arbitrarily). Let $G_i = \cap_{j=1}^p g(w_j^i, q)$ be the set of dimensions selected for class i, i.e., $k_i = p$ for $i = 1, \ldots, m$. Each G_i is truncated to $\kappa = \min_{i \in \{1, \ldots, m\}} |G_i|$ positions with the largest associated values (called priority genes). Training each F_i is then repeated with the data D projected onto the dimensions in G_i for $k_i = K$ perceptrons, $i = 1, \ldots, m$.

3   Data

The data used in this paper are provided by Khan et al. in [10]. Given are gene-expression data from cDNA microarrays containing 2308 genes for four types of small, round blue cell tumours (SRBCTs) of childhood, which include neuroblastoma (NB), rhabdomyosarcoma (RMS), Burkitt lymphoma (BL) and the Ewing family of tumours (EWS), i.e., $m = 4$. The number of training samples is as follows: 23 for EWS, 8 for BL, 12 for NB, and 20 for RMS. The test set consists of 25 samples: 6 for EWS, 3 for BL, 6 for NB, 5 for RMS, and 5 "others". The split of the data into training and test sets was the same as in the paper by

Khan et al., where 3750 ANNs are calculated to obtain 96 gene for training the final ANN.

4 Results

The algorithm described in Section 2 has been implemented in C^{++} and we performed computational experiments for the dataset from Section 3 on SUN Ultra 5/333 workstation with 128 MB RAM. We present three types of results from computational experiments.

In Table 1, computations include all 2308 input variables (gene-expression data). The parameter settings are $\alpha = 0.75$ and $\beta = 0.25$ for balanced but still sufficiently large numbers of positive and negative examples. The entries in the table are the errors on classes in the order [EWS,BL,NB,RMS]. The values are typical results from three to five runs for each parameter setting; the deviation is very small and therefore average values are not calculated.

Table 1.

$K = k_i, i = 1, \ldots, 4.$	11	33	99	297	891
Error Distr.	[1,1,5,0]	[1,1,5,0]	[0,1,5,0]	[0,1,5,0]	[0,1,5,0]
Total errors	35%	35%	30%	30%	30%

In Table 2, the training procedure has been performed on priority genes only. The number of priority genes is indicated by κ. The parameters are $p = 5, 9, 11$, and $q = 250$ (q is large enough to have non-empty intersections).

Table 2.

$p \, / \, K = k_i, i = 1, \ldots, 4.$	11	33	99	297	891
5 ($\kappa = 23$)	[1,0,4,0]	[1,0,4,0]	[0,0,4,0]	[0,0,4,0]	[0,0,4,0]
Total errors	25%	25%	20%	20%	20%
9 ($\kappa = 10$)	[1,0,2,0]	[1,0,2,0]	[1,0,0,0]	[0,0,0,0]	[0,0,0,0]
Total errors	15%	15%	5%	0%	0%
11 ($\kappa = 8$)	[1,0,1,0]	[1,0,1,0]	[1,0,0,0]	[1,0,0,0]	[1,0,0,0]
Total errors	10%	10%	5%	5%	5%

The results are stable for values close to $p = 9$ and larger $K \geq 300$. The rating of the five "other" examples is below 0.25. Thus, our method is sensitive to the right choice of p. The run-time ranges from 1 min (for $p = 5$ and $K = 11$) up to 113 min (for $p = 11$ and $K = 891$).

Since we are using $|w_l|$ for the selection of priority genes, it is interesting to know whether the largest absolute values of weights correspond to the most significant average values of gene-expression data. However, this is not the case:

Table 3 provides the rank of the average value for priority genes calculated for $K = 297$ and the EWS cancer type. In this run, we have $p = 9$ and $\kappa = 10$

Table 3.

Gene number	246	469	509	545	1708	1781	1834	1841	1961	2223
Rank number	2280	2290	1580	2284	2186	2306	2166	1770	1544	2049

(see Table 2). We recall that $q = 250$; for the largest average values of gene-expression data this means the range of ranks $2059, \ldots, 2308$. In Table 3, 40% of the genes do not belong to this range; thus, there is no direct correspondence between the rank of the average value of gene-expression data and the absolute value of weights in classifying threshold functions.

5 Discussion

Initial reports on the analysis of gene expression data derived from microarrays concentrated on unsupervised learning, in which the main objective was to determine which genes tend to have correlated patterns of expression. Supervised leraning models are increasingly being used, and classification of cases into diagnostic or pronostic categories has been attempted. Besides classification performance, an important goal of these models is to determine whether a few genes can be considered good markers for disease. It is important to investigate whether simple supervised learning models such as perceptrons can be successfully used for this purpose. In this paper, we showed an algorithm based on simulated annealing that is used to train a system of perceptrons, each obtained by resampling of the training set. The model was able to successfully classify previously unseen cases of SBRCTs using a small number of genes.

Acknowledgement. The authors would like to thank the anonymous reviewers for their careful reading of the manuscript and helpful suggestions that resulted in an improved presentation.

References

1. E.H.L. Aarts. *Local Search in Combinatorial Optimization.* Wiley&Sons, 1998.
2. A. Albrecht, E. Hein, K. Steinöfel, M. Taupitz, and C.K. Wong. Bounded-Depth Threshold Circuits for Computer-Assisted CT Image Classification. *Artificial Intelligence in Medicine,* 24(2):177–190, 2002.
3. V. Černy. A Thermodynamical Approach to the Travelling Salesman Problem: An Efficient Simulation Algorithm. Preprint, Inst. of Physics and Biophysics, Comenius Univ., Bratislava, 1982 (see also: *J. Optim. Theory Appl.,* 45:41–51, 1985).
4. M.B. Eisen, P.T. Spellman, P.O. Brown, D. Botstein. Cluster Analysis and Display of Genome-wide Expression Patterns. *Proc. Natl. Acad. Sci. USA,* 95(25):14863–8, 1998.
5. Y. Freund and R.E. Schapire. A Decision-Theoretic Generalization of On-Line Learning and an Application to Boosting. *J. of Computer and System Sciences,* 55:119–139, 1997.
6. T.S. Furey, N. Cristianini, N. Duffy, D.W. Bednarski, M. Schummer, and D. Haussler. Support Vector Machine Classification and Validation of Cancer Tissue Samples Using Microarray Expression Data. *Bioinformatics,* 16:906–914, 2000.

7. B. Hajek. Cooling Schedules for Optimal Annealing. *Mathem. of Operations Research*, 13:311–329, 1988.
8. D. Helmbold and M.K. Warmuth. On Weak Learning. *J. of Computer and System Sciences*, 50:551–573,1995.
9. K.-U. Höffgen, H.-U. Simon, and K.S. van Horn. Robust Trainability of Single Neurons. *J. of Computer System Sciences*, 50:114–125, 1995.
10. J. Khan, J.S. Wei, M. Ringner, L.H. Saal, M. Ladanyi, F. Westermann, F. Berthold, M. Schwab, C.R. Antonescu, C. Peterson, and P.S. Meltzer. Classification and Diagnostic Prediction of Cancers Using Gene Expression Profiling and Artificial Neural Networks. *Nature* [Medicine], 7(6):673–679, 2001.
11. S. Kirkpatrick, C.D. Gelatt, Jr., and M.P. Vecchi. Optimization by Simulated Annealing. *Science*, 220:671–680, 1983.

Neural Minimax Classifiers[*]

Rocío Alaiz-Rodríguez[1] and Jesús Cid-Sueiro[2]

[1] Dpto. Ingeniería Eléctrica y Electrónica
Universidad de León, Campus de Vegazana, León, Spain.   dierar@unileon.es
[2] Dpto. de Tecnologías de las Comunicaciones
Universidad Carlos III de Madrid , Leganés, Madrid, Spain   jcid@tsc.uc3m.es

Abstract. Many supervised learning algorithms are based on the assumption that the training data set reflects the underlying statistical model of the real data. However, this stationarity assumption may be partially violated in practice: for instance, if the cost of collecting data is class dependent, the class priors of the training data set may be different from that of the test set. A robust solution to this problem is selecting the classifier that minimize the error probability under the worst case conditions. This is known as the *minimax* strategy. In this paper we propose a mechanism to train a neural network in order to estimate the minimax classifier that is robust to changes in the class priors. This procedure is illustrated on a softmax-based neural network, although it can be applied to other structures. Several experimental results show the advantages of the proposed methods with respect to other approaches.

1   Introduction

A common assumption on supervised classification is that class probabilities are stationary. Unfortunately, class priors are frequently unknown, vary with time or differ from one environment to others. To overcome this problem, several classification rules that dynamically update themselves to changes in future data have been proposed [5]. The approach of Saerens [7] is based on re-estimating the prior probabilities in an unsupervised way with real data (that is, data generated by the true statistical model), and adjusting the outputs of the classifier according to the new a *priori* probabilities estimated in an unsupervised way.

Under changes on the class priors, the classifier should guarantee an upper bound of the error rate. Thus, our proposal is to build a neural network classifier minimizing the maximum expected error, i.e, assure the best possible classifier under the least favorable conditions [9]. We will refer to it as a *minimax* classifier. Unlike [7], our approach does not require unlabeled samples, that may be not available.

Our work has some overlap with learning from unbalanced datasets since both deal with the minimization of the differences in performance between the individual classes. If we assure similar performance between classes, the classifier will be more robust to changes in priors. Most research on unbalanced datasets,

[*] This paper has been partially supported by project CAM-07T/0046/2000

J.R. Dorronsoro (Ed.): ICANN 2002, LNCS 2415, pp. 408–413, 2002.
© Springer-Verlag Berlin Heidelberg 2002

focuses on altering the interclass probabilities in order to get a balanced training dataset with uniform priors by means of undersampling (remove examples of the majority class) or oversampling (duplicate examples from the minority class)[10, 6,8]. The same underlying principle guides the prior scaling techniques, that use different learning rates for different class patterns[8].

Our research, unlike [6,10], is not only restricted to learning from unbalanced datasets, but any classification problem with the aim to avoid poor performance under the least favorable conditions. Therefore, a *minimax* classifier with error rate (or risk) small and invariant to class priors is our goal.

2   Minimax Neural Network Training

Consider sample feature set $S = \{(\mathbf{x}^k, \mathbf{d}^k), k = 1, \ldots, K\}$ where $\mathbf{x}^k$ is an observation vector and $\mathbf{d}^k \in \mathcal{U}_L = \{\mathbf{u}_0, \ldots, \mathbf{u}_{L-1}\}$ is an element of the set of possible target classes. Label of class i is a unit L-dimensional vector, $\mathbf{u}_i = \delta_{ij}$, with every component equal to 0, unless the i-th component which is equal to 1.

We assume a learning process that finds the non-linear mapping of the input data into the soft decision space $\mathcal{P} = [0, 1]$. *Soft* output is given by $\mathbf{y}^k = \mathbf{f_w}(\mathbf{x}^k)$, where $\mathbf{f_w}$ is a non-linear function with parameters $\mathbf{w}$ and $\mathbf{y}$ is an L-dimensional vector with components $y_i \in \mathcal{P}$. Output components y_i satisfy $\sum_{i=0}^{L-1} y_i = 1$, what can be interpreted in a probabilistic way. The hard output of the classifier $\widehat{\mathbf{d}}$ is computed as a *Winner Takes All* decision over the soft output.

We focus our study in a binary classification problem with classes $\mathbf{u}_0$, $\mathbf{u}_1$ and training prior probabilities $P\{\mathbf{d} = \mathbf{u_0}\}$, $P\{\mathbf{d} = \mathbf{u_1}\}$, respectively. Standard training of the neural network pursues the minimization of some empirical estimate of $E\{C(\mathbf{y}, \mathbf{d})\} = P_0 E_0 + P_1 E_1 = P_0 E\{C(\mathbf{y}, \mathbf{d})|\mathbf{d} = \mathbf{u}_0\} + P_1 E\{C(\mathbf{y}, \mathbf{d})|\mathbf{d} = \mathbf{u}_1\}$ where $P_i = P\{\mathbf{d} = \mathbf{u}_i\}$ and $C(\mathbf{y}, \mathbf{d})$ may be any cost function.

When the priors probabilities are exposed to variation, the error function can take values between E_0 and E_1. It will be equal to E_0 in the case $P_1 = 0$, $P_0 = 1$ and will take the value E_1 in the opposite case. Error function against class priors is just a straight line.

Consider $\widehat{E} = \dfrac{1}{K} \sum_{k=1}^{K} C(\mathbf{y}^k, \mathbf{d}^k) = \widehat{P}_0 \widehat{E}_0 + \widehat{P}_1 \widehat{E}_1$ as the empirical cost function estimation where $C(\mathbf{y}^k, \mathbf{d}^k)$ is the error function for pattern k, $\widehat{E}_i = \frac{1}{K_i} \sum_{\mathbf{d} = \mathbf{u}_i} C(\mathbf{y}, \mathbf{d})$ and $\widehat{P}_i = \frac{K_i}{K}$. The number of training examples is equal to K and K_i are the examples from class i.

The idea behind our minimax approach consists in training the neural network to minimize the worst case performance, that is, to minimize $max(\widehat{E}_0, \widehat{E}_1)$.

We approximate the maximum function by $max(a, b) \approx a\frac{a^r}{a^r + b^r} + b\frac{b^r}{a^r + b^r}$ as r tends to infinite.

Neural network training (minimax learning) is carried out in two steps:

1. A first learning process to minimize $\widehat{E}$.

2. A second training in order to minimize $\widehat{E}_{minimax} = max(\widehat{E}_0, \widehat{E}_1) \approx \frac{\widehat{E}_0^{r+1} + \widehat{E}_1^{r+1}}{\widehat{E}_0^r + \widehat{E}_1^r}$ with r increasing from $r = 0$ to infinite in successive steps. Only those networks parameters sensitive to changes in priors are updated in this training.

The stopping criterion for this second training is satisfied when the absolute difference between $\widehat{E}_0$ and $\widehat{E}_1$ is under a minimum threshold.

2.1   Softmax Neural Network

We illustrate the minimax training algorithm on a *Generalized Softmax Perceptron* (GSP)[1] with the architecture represented by Fig.1.

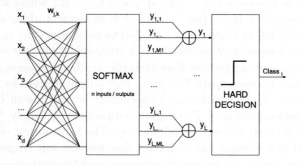

Fig. 1. GSP (Generalized Softmax Perceptron) Network

The soft-decisions are given by $y_i = \sum_{j=1}^{M_i} y_{ij}$ being $y_{ij} = \dfrac{\exp(\mathbf{w}_{ij}^T \mathbf{x} + w_{ij0})}{\sum_{k=0}^{L-1} \sum_{l=1}^{M_k} \exp(\mathbf{w}_{kl}^T \mathbf{x} + w_{kl0})}$

where L is the number of classes, M_j the number of softmax outputs used to compute y_j and $\mathbf{w}_{ij}$ are weight vectors.

The cross entropy cost function $C(\mathbf{y}, \mathbf{d}) = -\sum_{i=0}^{L-1} d_i \log y_i$ was employed in our experiments. We center on binary classification problems, being $L = 2$, $\mathbf{y} = (y_0, y_1)$ and $\mathbf{d} = (d_0, d_1)$

TRAINING PHASE

First Training, minimizing the *cross entropy* cost function with stochastic gradient learning rule given by $\mathbf{w}_{ij}^{n+1} = \mathbf{w}_{ij}^n + \mu \dfrac{y_{ij}^n}{y_i^n}(d_i^n - y_i^n)\mathbf{x}^n$ where μ is a decreasing learning step as iterations progress. Learning rate at iteration k corresponds to $\mu_k = \frac{\mu_0}{1+k/\tau}$ where $\mu_0 = 0.02$ and $\tau = 1000$. Learning carried out in the way described above will be referred as standard learning.

Second Training in order to minimize the *minimax* error function $\widehat{E}_{minimax}$. It is easy to see that only w_{ij0} parameters are affected by changes in priors and they are updated according to the standard gradient descent rule given by $w_{ij0}^{n+1} = w_{ij0}^n - \mu \frac{\partial \widehat{E}_{minimax}}{\partial w_{ij0}}$

3 Experimental Results

3.1 Performance Measures

It is widely known that the error rate is not an adequate criterion for many classification tasks, especially when the interclass prior probabilities vary largely or there is a great variance in the error rate among different classes.

As our goal is to get a classifier that performs uniformly on the individual classes, we use the g performance criterion [6] given by $g = \sqrt{A_0 \cdot A_1 \cdot \ldots A_{L-1}}$ where A_i is the accuracy of *class-i* (proportion of *class i* patterns correctly classified).

3.2 Artificial Dataset

We generated an artificial data set with two classes $(\mathbf{u_0}, \mathbf{u_1})$ and prior probabilities of class membership $(P\{\mathbf{d} = \mathbf{u_0}\} = 0.60, P\{\mathbf{d} = \mathbf{u_1}\} = 0.40)$ and the following distributions: $p(\mathbf{x}|\mathbf{d} = \mathbf{u_0}) = \frac{1}{3}N(m_{00}, \sigma_0^2) + \frac{1}{3}N(m_{01}, \sigma_0^2) + \frac{1}{3}N(m_{02}, \sigma_0^2)$ and $p(\mathbf{x}|\mathbf{d} = \mathbf{u_1}) = N(m_1, \sigma_1^2)$ where $N(\mu, \sigma^2)$ is the normal distribution with mean μ and standard deviation σ. Mean values correspond to $m_1 = (0,0)$, $m_{00} = (0,2)$, $m_{01} = (2,2)$, $m_{02} = (2,0)$ and $\sigma_0 = 0.9$, $\sigma_1 = 0.2$.

Neural network is trained using 75% of the data and tested on the remaining 25%. Initial learning rates were fixed to 0.02 for the first training phase and 0.18 for the second.

Table 1. Comparison Results between Standard learning and Minimax learning in the artificial dataset.

	Standard Learning	Minimax Learning
g	0.857	**0.934**
$\mathbf{u_0}$ class Accuracy , $\mathbf{u_1}$ class Accuracy	(0.98 , 0.75)	(0.93 , 0.94)
Training set (E_0, E_1)	(0.45 , 0.19)	(0.28 , 0.28)
Test set (E_0, E_1)	(0.50 , 0.11)	(0.35 , 0.06)

Due to space constraints, we express the experimental results of this dataset in Table 1. First row presents g measure exhibiting a performance for minimax training higher than **0.93** compared with nearly 0.86 for standard learning. While the accuracy of $\mathbf{u_1}$ class and $\mathbf{u_0}$ class are very unbalanced for standard training, minimax training achieves a nearly balanced performance $(0.93, 0.94)$.

In the same way, minimax learning succeeds in minimizing the worst case performance, that would happen with a priori probabilities: $P\{\mathbf{d} = \mathbf{u_0}\} = 0$ and $P\{\mathbf{d} = \mathbf{u_1}\} = 1$. Minimax learning reduces the maximum cost function from 0.45 to **0.28** in training and from 0.50 to **0.35** in test.

3.3   Ionosphere Dataset

We evaluate the performance of the minimax approach on the ionosphere dataset [2]. The test set has been constructed by selecting, at random, 25% of examples from both classes ("good" and "bad", denoted as *class1* and *class0*, respectively).

To explore the impact of changing priors, we artificially vary the prior distribution in the train and test set from 0 to 1 as shown in Fig.2 and Fig.3. With minimax learning, cost function in the training set is invariant to class priors (straight line). The classifier trained with minimax learning is more robust to changes in priors as can be seen in Fig 3. Evolution of performance criterion g with training cycles is shown in Fig4. Minimax and standard learning agree for the first 40 training cycles being remarkable the differences in performance once minimax learning begins the second training.

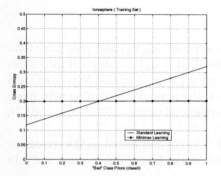

Fig. 2. Cost function in the training set for possible different a priori probabilities of class 0. Ionosphere dataset

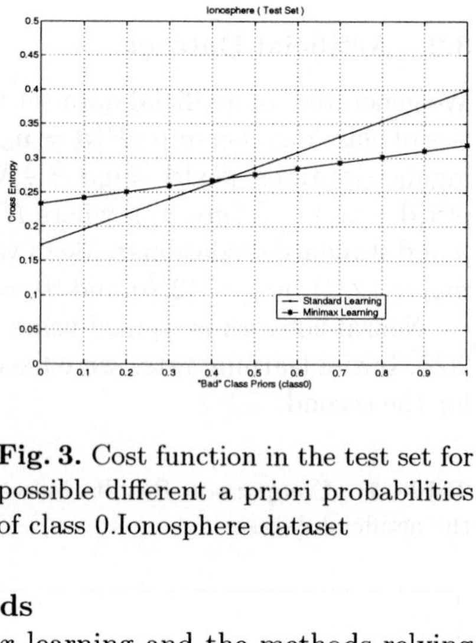

Fig. 3. Cost function in the test set for possible different a priori probabilities of class 0.Ionosphere dataset

Comparative Study with other Methods

We present a comparative study of *minimax* learning and the methods relying in using a uniform prior training dataset [10,6,8] that were briefly described in the introduction. Experimental results are compared in Fig. 5 by means of the classification error rate. Due to priors change, the worst case performance would happen with a priori probabilities of *class0* equal to 1. Using a balanced dataset, the performance with the priors mentioned above are improved but it is the *minimax* learning method the one with minimum error in these worst case conditions and with hardly scarifying the performance for the opposite priors conditions.

4   Conclusions and Further Work

In this paper, we propose a new neural network training approach in order to get a minimax classifier. This classifier is characterized by its robustness to changes

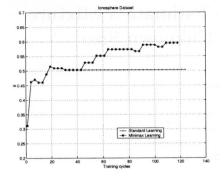

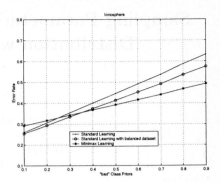

Fig. 4. Evolution of **g** performance measure with the number the number of training cycles. Ionosphere dataset

Fig. 5. Error rate vs. a priori probabilities of *class0* in the Ionosphere dataset

in priors and by achieving higher performance than a standard classifier under the worst case conditions. We showed empirically, that it does this remarkably well and we consider to probe it on large scale learning task like text categorization at the same time we extend it to risk minimizing problems.

References

1. J.I. Arribas, J.Cid-Sueiro, T. Adali, A.R. Figueiras-Vidal, "Neural Architectures for parametric estimation of a posteriori probabilities by constrained conditional density functions", *NNSP IX*, pp. 263–272, Aug. 99.
2. C.L. Blake, C.J. Merz, Repository of machine learning databases [http://www.ics.uci.edu/mlearn/MLRepository.html], Irvine, CA: University of California, Department of Information and Computer Science.
3. J.Cid-Sueiro, A.R. Figueiras-Vidal, "On the Structure of Strict Sense Bayesian Cost Functions and its Applications", *IEEE Trans. Neural Networks*, May 2001.
4. P.K. Chan, S.J. Stolfo. "Learning with Non-uniform Class and Cost Distributions: Effects and a Multi-Classifier Approach". *Machine Learning Journal*, 1999, 1–27.
5. M. G. Kelly, D. J. Hand, and N. M. Adams. "The impact of changing populations on classifier performance." *In Proceedings of Fifth International Conference on SIG Knowledge Discovery and Data Mining (SIGKDD)*, 1999.
6. M. Kubat and S. Matwin. "Addressing the curse of imbalanced training sets: One-sided selection". *Proceedings of the 14th ICML* , 179–186. Morgan Kaufmann, 1997.
7. M. Saerens, P. Latinne and Ch. Decaestecker, "Adjusting a Classifier for New A Priori Probabilities: A Simple Procedure", *Neural Computation*, Jan. 2002.
8. S. Lawrence, I. Burns, A. Back, A. Tsoi and C. Giles, "Neural Networks Classification and Prior Class Probabilities", *Tricks of the trade, Lecture Notes in Computer Science State-of-the-Art Surveys*, Springer Verlag, pp. 299–314, 1998.
9. H.L. Van Trees, *Detection, Estimation and Modulation Theory*, vol. I, John Wiley, New York, 1968.
10. G.M. Weiss and F. Provost. "The effect of class distribution on classifier learning." *Tech. Report ML-TR 43*, Dept. Computer Science, Rutgers University, 2001.

Sampling Parameters to Estimate a Mixture Distribution with Unknown Size

Martin Lauer

Lehrstuhl Informatik I
Universität Dortmund
D-44221 Dortmund, Germany
martin.lauer@udo.edu

Abstract. We present a new view on the problem of parameter estimation in the field of Gaussian Mixture Models. Our approach is based on the idea of stable estimates and is realized by an algorithm which samples appropriate parameters. It further determines the adequate complexity of the model without extra effort. We show that the sampling approach avoids overfitting and that it outperforms maximum likelihood estimates in practical tasks.

1 Introduction

Gaussian Mixture Models (GMMs) have been widely used in the field of unsupervised learning [11,5]. Most approaches to determine the model parameters from data, e. g. the EM-algorithm [2], are based on the maximization of the data likelihood or the posterior function in a Bayesian modeling.

Although these algorithms are used in many applications they have several shortcomings, namely a) the maximum likelihood approach tends to overfit the data since the likelihood function is not bounded above, b) the maximum a posteriori approach needs informative priors which are difficult to determine, and c) both approaches are based on the assumption of a fixed number of components, i.e. the model complexity is fixed. Therefore the parameters of differently sized models have to be estimated in several runs and the appropriate model has to be chosen afterwards using certain criteria like Akaike's information criterion (AIC) [1] or the Bayesian information criterion (BIC) [9], among others.

Another approach for the estimation of GMM parameters are Markov Chain Monte Carlo methods like the Gibbs sampler [3,6] or Data Augmentation [10, 3]. These algorithms are created to randomly sample a sequence of possible parameters, whereby the distribution of these parameters fulfills some optimality condition. In the literature up to now the emphasis has been put on reversible jump algorithms based on the Gibbs sampler that permanently jump between mixtures of different size and thus implicitly estimate a whole bunch of differently sized GMMs [6]. In contrast, in this paper we want to propose the Data Augmentation approach in its special form with non-informative priors. We will derive its good generalization abilities and extend the algorithm to implicitly determine the adequate size of the GMM deleting all superfluous components.

J.R. Dorronsoro (Ed.): ICANN 2002, LNCS 2415, pp. 414–419, 2002.
© Springer-Verlag Berlin Heidelberg 2002

In the subsequent sections we want to present the key idea of an estimate's stability, its application to GMM estimation and its realization by Data Augmentation. We conclude with an experimental comparison of Data Augmentation with classical estimation techniques.

2   Basic Idea: Stability

Think of an estimation task: given some data $x_1, \ldots, x_n$ and a simple class of parameterized models $\mathcal{M}(\theta)$ (e.g. $\mathcal{M}(\theta)$ is the class of Gaussian distributions parameterized by its mean and covariance matrix). The data are assumed to be generated by a certain true model $\mathcal{M}(\theta^*)$. An estimator $\hat{\theta}$ is usually a function of the data, i.e. $\hat{\theta} = \hat{\theta}(x_1, \ldots, x_n)$.

What can we conclude about the true parameter θ^* given $\hat{\theta}$? Assuming that the available training data are an independent random sample of $\mathcal{M}(\theta^*)$ we get that the estimate itself is random with distribution $P(\hat{\theta}|\theta^*, n)$. With Bayes' rule we compute: $P(\theta^*|\hat{\theta}, n) = P(\hat{\theta}|\theta^*, n) \cdot P(\theta^*|n) \cdot P(\hat{\theta}|n)^{-1}$. Since the training data are given and fixed we can omit the third factor and since we don't want to make prior assumptions on θ^*, i.e. we set $P(\theta^*|n) = const$, we get:

$$P(\theta^*|\hat{\theta}, n) \propto P(\hat{\theta}|\theta^*, n) \qquad (1)$$

Thus it does not matter whether we regard $\hat{\theta}$ as fixed and θ^* as random or vice versa.

If we want to know whether the estimate $\hat{\theta}$ is trustworthy we have to consider the distribution $P(\hat{\theta}|\theta^*, n)$. If the probability mass is concentrated in a small area around $\hat{\theta}$ the estimate is trustworthy. On the other hand, if the distribution is very widespread there is not much evidence for the estimate and thus $\hat{\theta}$ is not reliable.

Now we can derive the relationship between stability and overfitting which forms the basis for our further argumentation: if we want to check whether an estimate overfits the data we compute the likelihood on a test set (cross-validation). Assuming that θ^* is fixed an estimate that does not overfit the training data yields a high likelihood on the test sample. Another way of testing whether an estimate is overfitted would be to compute a second estimate $\tilde{\theta}$ on the test set and to compare $\hat{\theta}$ with $\tilde{\theta}$. If we use simple models for which similar distributions are related to similar parameters (e.g. Gaussians) both approaches are equivalent since a) in the case of overfitting $\hat{\theta}$ results in a low likelihood on the test set and $\hat{\theta}$ and $\tilde{\theta}$ are very different, while b) in the case of a non-overfitted estimate $\hat{\theta}$ we get a high likelihood on the test set and both estimates $\hat{\theta}$ and $\tilde{\theta}$ resemble, i.e. the distance in the parameter space between $\hat{\theta}$ and $\tilde{\theta}$ is small.

Observing that both $\hat{\theta}$ and $\tilde{\theta}$ are independent samples from the distribution $P(\hat{\theta}|\theta^*, n)$ we conclude that sampling parameters from that distribution is a kind of implicit cross-validation. Due to (1) we can equivalently sample from the distribution $P(\theta^*|\hat{\theta}, n)$ which can be computed from the training set. Thus we evade the use of a test set and we can use all available data for training.

3   Decomposition of GMM Estimates

Our reasoning thus far cannot directly be transfered to more complex models since for these model classes similar distributions may be described by very different parameters and thus our argumentation of the preceding two paragraphs cannot be applied directly. But if we can interpret a complex model as composition of simple models we can use the idea of stability for the simpler components and thus build up an estimation procedure for the complex model.

Such a class of models are GMMs [11]. They consist of a weighted sum of Gaussian distributions. If we denote with $\phi_{\mu,\Sigma}$ the density of a Gaussian distribution with mean μ and covariance matrix Σ we define the density of a GMM with $k \geq 1$ components as:

$$p_{GMM}(x) := \sum_{j=1}^{k} \left(w_j \cdot \phi_{\mu_j, \Sigma_j}(x) \right) \tag{2}$$

Each component j is defined by its mean μ_j and covariance matrix Σ_j and is weighted by the mixing weight $w_j \geq 0$, whereby the mixing weights of all components sum to 1.

We can interpret each component of a GMM as a local expert which is able to explain a subset of the training data. Thus we should be able to partition the training data into subsets, one for each component, and estimate the parameters of each component only from the corresponding subset of data. On the other hand, if we would know the true parameters of the GMM we were faced with the problem of assigning each data point to a certain component and thus to partition the training data. To denote the assignments we introduce assignment variables z_i, one for each data point, that indicate to which component the respective data point x_i is assigned. The task of determining the assignment variables is indeed an estimation task for a multinomial random process.

Thus far we stated that a GMM can be understood as a combination of simple models, namely a) a Gaussian distribution for each component and b) a multinomial distribution for the assignment variables. Now we want to show how the idea of stability from section 3 can be used for GMM parameter estimation.

4   Data Augmentation for GMMs

Data Augmentation has originally been developed to solve incomplete data problems [10]. It has been transfered to the task of GMMs in [3]. Other descriptions of this method can be found in [7] and [8]. In this paper we want to give an idea of its behavior in the special case of non-informative priors while the formulae can be found in the literature.

To use Data Augmentation for GMM parameter estimation we decompose the task into the simpler subproblems 'estimation of the parameters' and 'determining the assignments' as described above. Starting with an arbitrary parameter vector $\theta^0 = (w_1^0, \mu_1^0, \Sigma_1^0, \ldots, w_k^0, \mu_k^0, \Sigma_k^0)$ the algorithm alternately samples random assignments $z_1^{t+1}, \ldots, z_n^{t+1}$ from the distribution $P_I^{t+1} :=$

$P(z_1, \ldots, z_n | \theta^t, x_1, \ldots, x_n)$ (imputation step) and random parameters θ^{t+1} from the distribution $P_P^{t+1} := P(\theta | z_1^{t+1}, \ldots, z_n^{t+1}, x_1, \ldots, x_n)$ (posterior step). Thus we get a dependent sequence of parameters $\theta^0, \theta^1, \theta^2, \ldots$ which converges in distribution to[1] $P(\theta | x_1, \ldots, x_n)$.

Now we want to extend the Data Augmentation principle using the idea of stability from section 2. Thereto we choose non-informative priors for the parameters of the GMM so that we can identify P_P^t with the distribution $P(\theta^* | \hat{\theta}, n)$ from section 2. Then the posterior step implements the sampling approach from section 2 applied to the determination of the GMM parameters given the assignments. Likewise we can understand the imputation step as a realization of the same sampling approach to determine the assignment variables.

Since the imputation and posterior steps are performed alternately the sequence of parameters becomes stable only when both steps become stable. Then we get an estimate which avoids overfitting. But what happens otherwise?

Assuming that the posterior step is unstable for a certain component j due to a very small number of assigned data points. Then the sampled parameters of component j are almost completely random. As consequence it is very unlikely that the new parameters fit to any data point and thus the component runs out of assignments in the subsequent imputation step with high probability. Any such situation allows us to delete the respective component since a component without assignments does not contribute anything to the mixture.

If, on the other hand, the imputation step is unstable there are several components which overlap a lot in a certain area of data space, i.e. they explain the same data points. Then the partitioning of the concerned data points is almost completely random. In these situations we can observe that by random sometime one of the following situations occurs: a) the partitioning allows each component to specialize on a certain subset of the data and the overlap reduces such that the assignments become stable or b) the number of data points that are assigned to one component becomes so small that the subsequent posterior step becomes unstable. In any case either one component is deleted or the components specialize on disjoint data sets. Thus the estimation process sometime becomes stable for any data set.

The entire algorithm consists of the following steps:

1. `Start with arbitrary initial parameters`
2. `Compute` P_I^{t+1} `and sample random assignments (imputation step)`
3. `Remove components without assignments`
4. `Compute` P_P^{t+1} `and sample random parameters (posterior step)`
5. `Goto 2 until stabilization`

After the process has been stabilized we get the final estimate computing the mean of the sampled parameters. Label switching does not occur any more and the result does not overfit the data. Thus we can start with a very large GMM and let the algorithm itself delete all superfluous components.

---

[1] for the proof see [10] and [3]

Table 1. Description of the benchmarks we used from the IDA benchmark repository

Benchmark	Size of training set	Size of test set	Input dimension
banana	400	4900	2
breast-cancer	200	77	9
diabetes	468	300	8
thyroid	140	75	5

5   Experiments and Discussion

We compared the Data Augmentation algorithm with non-informative priors on several different benchmarks: a) univariate and trivariate Gaussian data, b) trivariate uniform data distributed in a cube and c) four benchmarks from the IDA benchmark repository [4]. For the experiments in a) we generated 1000 (100 in case b) random training and test sets each with 1000 test points. In c) we used the given splits of the data into 100 different training and test sets described in table 1. The task was to estimate the probability distribution of the input variables. In all experiments we performed 1000 iterations of Data Augmentation starting with a GMM of 10 components. For the benchmarks of group a) we compared the results of Data Augmentation with the maximum likelihood estimate. The results in table 2 show even in this simple task a better performance of Data Augmentation than the maximum-likelihood approach in particular for multivariate data and small training sets.

For the benchmarks in the cases b) and c) we compared the results of Data Augmentation with the result of the EM-algorithm (1000 iterations) where we used both the AIC [1] as well as the BIC [9] to determine the number of components. The results are listed in table 3 and 4, respectively. In all tests the Data Augmentation approach was at least as good as the EM-algorithm and outperformed it in six (AIC) and five (BIC) of seven experiments.

Table 2. Experimental results on the univariate (left) and trivariate (right) Gaussian data. The given averaged log-likelihood was determined on independent test sets. A P-value less than 0.05 indicates significant better results of the Data Augmentation (DA) approach versus the ML-algorithm comparing the log-likelihood on the test sets.

size of training set	ave. log-likelihood DA	ML	P-value t-test
10	-1550.49	-1596.48	0
30	-1454.00	-1457.38	0.06
50	-1440.15	-1440.99	0.28

size of training set	ave. log-likelihood DA	ML	P-value t-test
10	-4942.80	-5492.28	0
30	-4425.08	-4457.26	0
50	-4354.12	-4364.72	0

The results from the experiments show that the principle of sampling parameters can be fruitfully applied to GMM fitting and it is able to outperform the maximum-likelihood approach. It does not only avoid overfitting on the training set as was argued in the theoretical analysis but also yields a good fit to the data as can be seen from the experiments. Additionally, the proposed exten-

Table 3. Comparison of the results on the trivariate uniform data

size of	ave. log-likelihood test sets			P-value t-test		ave. number components		
training set	DA	EM AIC	EM BIC	EM AIC	EM BIC	DA	EM AIC	EM BIC
50	-605.388	-842.884	-606.377	0	0.44	1.04	1.97	1.00
100	-569.844	-689.913	-567.627	0	0.64	1.16	2.83	1.01
200	-511.072	-576.846	-547.696	0	0	2.19	4.61	1.09

Table 4. Comparison of the results on the IDA benchmark data

benchmark	ave. log-likelihood test sets			P-value t-test		ave. number components		
	DA	EM AIC	EM BIC	EM AIC	EM BIC	DA	EM AIC	EM BIC
banana	-13112.3	-13221.4	-13435.9	0	0	4.54	5.77	2.62
br.-cancer	-403.478	-600.807	-676.138	0	0	3.62	2.97	2.05
diabetes	-2259.49	-2252.64	-2472.17	0.56	0	3.98	3.84	2.75
thyroid	-216.329	-360.752	-331.774	0	0	3.09	4.15	2.83

sion of Data Augmentation with component deletion and non-informative priors allows to determine the adequate model complexity without any extra effort. Thus, we neither need a validation set nor prior knowledge on the parameters. In further investigations we currently compare Data Augmentation with other approaches for mixture estimation.

References

[1] H. Akaike. Information theory and an extension of the maximum likelihood principle. In B. N. Petrov and F. Csaki, editors, *Second International Symposium on Information Theory*, pages 267–281, 1973.

[2] A. P. Dempster, N. M. Laird, and D. B. Rubin. Maximum likelihood from incomplete data via the EM algorithm. *Journal of the Royal Statistical Society Series B*, 39:1–38, 1977.

[3] Jean Diebolt and Christian P. Robert. Estimation of finite mixtures through bayesian sampling. *Journal of the Royal Statistical Society Series B*, 56(2):363–375, 1994.

[4] IDA benchmark repository.
cf. http://ida.first.gmd.de/~raetsch/data/benchmarks.htm.

[5] Geoffrey McLachlan and David Peel. *Finite Mixture Models*. Wiley, 2000.

[6] Sylvia Richardson and Peter J. Green. On bayesian analysis of mixtures with an unknown number of components. *Journal of the Royal Statistical Society Series B*, 59:731–792, 1997.

[7] Christian P. Robert. *The Bayesian Choice*. Springer, 1994.

[8] J. L. Schafer. *Analysis of Incomplete Multivariate Data*. Chapman & Hall, 1997.

[9] G. Schwarz. Estimating the dimension of a model. *The Annals of Statistics*, 6:461–464, 1978.

[10] Martin A. Tanner and Wing Hung Wong. The calculation of posterior distributions by data augmentation. *Journal of the American Statistical Society*, 82(398):528–550, 1987.

[11] D. M. Titterington, A. F. M. Smith, and U. E. Makov. *Statistical Analysis of Finite Mixture Distributions*. Wiley, 1985.

Selecting Neural Networks for Making a Committee Decision

Antanas Verikas[1,2], Arunas Lipnickas[2], and Kerstin Malmqvist[1]

[1] Halmstad University, Box 823, S-30118 Halmstad, Sweden,
av@ide.hh.se,
[2] Kaunas University of Technology, Department of Applied Electronics,
Studentu 50, 3031, Kaunas, Lithuania

Abstract. To improve recognition results, decisions of multiple neural networks can be aggregated into a committee decision. In contrast to the ordinary approach of utilizing all neural networks available to make a committee decision, we propose creating adaptive committees, which are specific for each input data point. A prediction network is used to identify classification neural networks to be fused for making a committee decision about a given input data point. The jth output value of the prediction network expresses the expectation level that the jth classification neural network will make a correct decision about the class label of a given input data point. The effectiveness of the approach is demonstrated on two artificial and three real data sets.

1 Introduction

It is well known that a combination of many different neural networks can improve classification accuracy. A variety of schemes have been proposed for combining multiple classifiers. The approaches used most often include the majority vote, averaging, weighted averaging, the fuzzy integral, the Dempster-Shafer theory, the Borda count, aggregation through order statistics, probabilistic aggregation, the fuzzy templates, and aggregation by a neural network [7,8,9].

Prevail two main approaches to utilizing neural networks, or ordinary classifiers, for building a committee. The most predominant one is to use all the networks available for making a committee decision. The alternative approach selects a single network, which is most likely to be correct for a given sample. Thus only the output of the selected network is considered in the final decision.

In this paper, we propose building adaptive, data-dependent, committees, which are specific for each input data point, in the way that, depending on an input data point, different networks and a different number of them may be chosen to make a committee decision about the data point. A prediction network is used to identify classificaiton neural networks to be fused for making the committee decision. To obtain diverse networks comprising a committee, we use the half & half bagging approach [1] to collect data for training neural networks of the committee. We compare the technique developed with the predominant aggregation approach. Two artificial and three real world problems are used to evaluate the approach proposed.

J.R. Dorronsoro (Ed.): ICANN 2002, LNCS 2415, pp. 420–425, 2002.

2   Data Sampling Approach

Numerous previous works on neural network committees have shown that an efficient committee should consist of networks that are not only very accurate, but also diverse in the sense that the network errors occur in different regions of the input space. Bootstrapping [3], Boosting [4], and AdaBoosting [4] are the most often used approaches for data sampling when training members of neural network committees. Breiman has recently proposed a very simple algorithm, the so called half & half bagging approach [1]. When tested on decision trees the approach was competitive with the AdaBoost algorithm. We use the half & half bagging approach to collect data for training neural networks.

Half & Half Bagging. It is assumed that the training set contains N data points. Suppose that k networks have been already trained. To obtain the next training set, randomly select a data point $\mathbf{x}$. Present $\mathbf{x}$ to that subset of k networks, which did not use $\mathbf{x}$ in their training sets. Use the majority vote to predict the classification result of $\mathbf{x}$ by that subset of networks. If $\mathbf{x}$ is misclassified, put it in set MC. Otherwise, put $\mathbf{x}$ in set CC. Stop when the sizes of both MC and CC are equal to M, where $2M \leq N$. Usually, CC is filled first but the sampling continues until MC reaches the same size. In [1], $M = N/4$ has been used. The next training set is given by a union of the sets MC and CC.

Diversity of Networks. To assess the diversity of the obtained neural networks, we used the κ-error diagrams. The κ-error diagrams display the accuracy and diversity of the individual networks. For each pair of networks, the accuracy is measured as the average error rate on the test data set, while the diversity is evaluated by computing the so-called *degree-of-agreement* statistic κ. Each point in the diagrams corresponds to a pair of networks and illustrates their diversity and the average accuracy. The κ statistic is computed as

$$\kappa = (\theta_1 - \theta_2)/(1 - \theta_2) \tag{1}$$

with $\theta_1 = \sum_{i=1}^{Q} c_{ij}/N$ and $\theta_2 = \sum_{i=1}^{Q} \{\sum_{j=1}^{Q} \frac{c_{ij}}{N} \sum_{j=1}^{Q} \frac{c_{ji}}{N}\}$, where Q is the number of classes, $\mathbf{C}$ is a $Q \times Q$ square matrix with c_{ij} containing the number of test data points assigned to class i by the first network and into class j by the second network, and N stands for the total number of test data. The statistic $\kappa = 1$ when two networks agree on every data point, and $\kappa = 0$ when the agreement equals that expected by chance.

3   Neural Networks Selection Procedure

The *networks selection* procedure is encapsulated in the following six steps.

1. Divide the data available into `Training`, `Test`, and `Cross-Validation` sets.
2. Train L neural networks using the half & half bagging technique.
3. Classify the `Training` set data by all networks of the committee.
4. For each training data vector $\mathbf{x}_i$ form a L-dimensional target vector $\mathbf{t}_i = [t_{i1}, ..., t_{iL}]^T$, with $t_{ij} = 1$, if the $\mathbf{x}_i$ data vector was correctly classified by the jth network and $t_{ij} = 0$, otherwise.

5. Using the `Training` data set and the target vectors obtained in Step 4 train a network to predict whether or not the classification result obtained from the L networks for an input data point $\mathbf{x}$ will be correct. The prediction network consists of L output nodes, one for each network, and n input nodes, where n is the number of components in $\mathbf{x}$.

6. Determine the optimal threshold value β for including neural networks into a committee. The jth network is included into a committee if $z_j > \beta$, where z_j is the jth output of the prediction network. The value β is the value yielding the minimum `Cross-Validation` data set classification error obtained from a committee of the selected networks.

Having the threshold β determined, *data classification* proceeds as follows.

1. Present a data point $\mathbf{x}$ to the prediction network and calculate the output vector $\mathbf{z}$.

2. Classify the data point by the networks satisfying the condition $z_j > \beta$.

3. Aggregate the outputs of the selected networks into a committee decision according to a chosen aggregation scheme.

Note that the optimal threshold value is determined in the training phase and then fixed for the classification phase. Note also that the committee build is specific for each input data point. We investigate three schemes for aggregating the outputs of selected networks. In the context of the aggregation schemes used, we compare the proposed concept with an ordinary decision aggregation approach, when all the trained networks are utilized to make a committee decision.

4   Aggregation Schemes Used

To test the approach we used three simple aggregation schemes that do not utilize any aggregation parameters, namely the *majority vote*, *averaging*, and the *median* aggregation rule.

Majority vote. The correct class is the one chosen by the most neural networks. Ties can be broken randomly.

Averaging. The output yielding the maximum of the averaged values is chosen as the correct class c:

$$c = \arg \max_{j=1,\ldots,Q} \left(y_j(\mathbf{x}) = \frac{1}{L} \sum_{i=1}^{L} y_{ij}(\mathbf{x}) \right) \tag{2}$$

where Q is the number of classes, L is the number of neural networks, $y_{ij}(\mathbf{x})$ represents the jth output of the ith network given an input pattern $\mathbf{x}$, and $y_j(\mathbf{x})$ is the jth output of the committee given an input pattern $\mathbf{x}$.

Median rule. It is well known that a robust estimate of the mean is the median [6]. Median aggregation leads to the following rule:

$$c = \arg \max_{j=1,\ldots,Q} \left(y_j(\mathbf{x}) = \frac{1}{L} \mathbf{MED}_{i=1}^{L} y_{ij}(\mathbf{x}) \right) \tag{3}$$

5   Experimental Testing

The data sets used have been taken from the *ELENA* project – the sets *Clouds, Phoneme, Satimage* [5] and a collection called *PROBEN 1* – the sets *Ringnorm* and *Thyroid* [2].

All comparisons presented here have been performed by leaving aside 10% of the data available as a `Cross-Validation` data set and then dividing the rest of the data into `Training` and `Test` sets of equal size. In all the tests, one hidden layer MLPs with 10 sigmoidal hidden units served as committee members. This architecture was adopted after some experiments. Since we only investigate different aggregation schemes, we have not performed expensive experiments for finding the optimal network size for each data set used. We run each experiment ten times, and the *mean* errors and *standard deviations* of the errors are calculated from these ten trials. In each trial, the data set used is randomly divided into `Training`, `Cross-Validation`, and `Test` parts.

First, we investigated the ability of the half & half bagging technique to create diverse and accurate neural networks. We compared the half & half technique with the bootstrapping sampling approach [3]. Two training techniques have been employed in these tests: 1) the Bayesian inference technique to obtain regularized networks and 2) the standard backpropagation training technique without regularization, which was run for a large number of training iterations.

These tests have shown that, on average, the half & half bagging technique outperformed the bootstrapping approach by creating more accurate neural network committees. For both sampling techniques the regularized committees, on average, were more accurate than the non-regularized ones. Fig. 1 (**Left**) presents the κ-error diagrams for the *Phoneme* data set illustrating the diversity of the networks of the committees made of 20 members, where BS_R stands for the bootstrapping sampling and regularized training case, BS stands for the bootstrapping sampling and training without regularization, $H\&H_R$ means the half & half bagging approach and regularized training, and H&H stands for the half & half bagging approach and training without regularization. As can be seen from the κ-error diagrams, the networks created by bootstrapping form a much tighter cluster than they do with the half & half bagging approach. This is expected, since with the bootstrapping technique each network is trained on a sample drawn from the same distribution. This explains why half & half sampling outperforms bootstrapping. The lower accuracy of networks produced by the half & half bagging approach is well compensated for by the increased diversity. The same pattern of accuracy and diversity was observed across the other data sets. We can, therefore, conclude that the half & half bagging technique is capable of creating diverse and sufficiently accurate neural networks.

In the next set of experiments, we investigated the effectiveness of the networks selection technique in creating accurate neural network committees. The regularized training and half & half bagging techniques have been employed in these tests. Table 1 summarizes the `Test` data set classification error obtained in the tests, where *Mean* stands for the percentage of the average test set classification error, *Std* is the standard deviation of the error, and *Best* means the

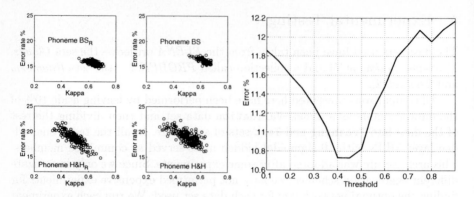

Fig. 1. Left: The κ-error diagrams for the *Phoneme* data set using bootstrapped (top) and half & half sampled (bottom) committees. **Right:** The test data set classification error rate of the committee for the *Phoneme* data set as a function of the neural network selection threshold β.

Table 1. The test data set classification error rate obtained from the neural network committees fused by the Majority Vote, Averaging, and Median aggregation rules.

Database	Best		Majority		Averaging		Median	
	Mean	Std	Mean	Std	Mean	Std	Mean	Std
Without Selection								
1. **Phoneme**	15.07	0.86	11.20	0.20	11.10	0.11	11.06	0.18
2. **Clouds**	10.77	0.18	10.42	0.20	10.38	0.26	10.45	0.22
3. **Satimage**	11.87	0.21	10.37	0.19	10.23	0.18	10.34	0.22
4. **Ringnorm**	7.01	0.46	1.65	0.22	1.48	0.21	1.47	0.22
5. **Thyroid**	1.46	0.19	0.74	0.09	0.72	0.11	0.72	0.12
With Selection								
1. **Phoneme**	15.07	0.86	10.21	0.24	10.36	0.20	10.11	0.22
2. **Clouds**	10.77	0.18	10.13	0.21	10.18	0.24	10.16	0.20
3. **Satimage**	11.87	0.21	9.86	0.15	9.76	0.23	9.85	0.26
4. **Ringnorm**	7.01	0.46	1.36	0.12	1.34	0.12	1.37	0.13
5. **Thyroid**	1.46	0.19	0.53	0.06	0.53	0.07	0.53	0.07

single neural network with the best average performance. As can be seen from the table, the proposed neural networks selection and aggregation technique is superior to the ordinary aggregation approach. In the ordinary decision aggregation approach, without the proposed neural networks selection procedure, we utilized committees consisting of 20 members. The actual average size of the committees created by the procedure proposed was considerably smaller – 12.5.

Therefore, the technique proposed allows reducing both classification error and time. Fig. 1 (**Right**) plots the Test data set classification error rate of the committee for the *Phoneme* data set as a function of the neural network selection threshold β. The graph show the strong dependence between the threshold value and the classification error rate.

6   Conclusions

In this paper, a new approach to creating neural network committees for classification was proposed. The approach banks on the idea of having a committee specific for each input data point. Different networks and a different number of them may be fused into a committee to make decisions about different input data points. The networks utilized are determined by those outputs of a prediction network, the output value at which exceeds a particular selection threshold. The jth output value expresses the expectation level that the jth classification neural network will make a correct decision about the class label of a given input data point. The effectiveness of the proposed approach in creating accurate neural network committees for classification was investigated using two artificial and three real data sets. The approach proposed was compared with the ordinary decision fusion scheme. In all the tests performed, the proposed way of generating neural network committees was superior to the ordinary decision fusion scheme utilizing all the networks available to make a committee decision.

References

1. Breiman, L.: Half & Half bagging and hard boundary points. *Technical report* **534**, Statistics Departament, University of California, Berkeley (1998)
2. Delve. <http://www.cs.toronto.edu/~delve/>
3. Efron, B., Tibshirani, R.: *An introduction to the bootstrap.* Chapman and Hall, London, (1993)
4. Freund, Y., Schapire, R. E.: A decision-theoretic generalization of on-line learning and an application to boosting. *Journal of Computer and System Sciences* **55** (1997) 119–139
5. ELENA. <ftp.dice.ucl.ac.be/pub/neural-net/ELENA /databases>
6. Kittler, J., Hatef, M., Duin, R. P. W., Matas, J.: On combining classifiers. *IEEE Trans Pattern Analysis and Machine Intelligence* **20**(3) (1998) 226–239
7. Verikas, A., Lipnickas, A., Malmqvist, K., Bacauskiene, M., Gelzinis, A.: Soft combination of neural classifiers: A comparative study. *Pattern Recognition Letters* **20** (1999) 429–444
8. Verikas, A. et al.: Fusing neural networks through fuzzy integration. In H. Bunke, A. Kandel, editors, *Hybrid Methods in Pattern Recognition*, World Scientific (2002)
9. Xu, L., Krzyzak, A., Suen, C. Y.: Methods for combining multiple classifiers and their appli-cations to handwriting recognition. *IEEE Trans. Systems, Man, and Cybernetics* **22**(3) (1992) 418–435

High-Accuracy Mixed-Signal VLSI for Weight Modification in Contrastive Divergence Learning

Patrice Fleury, Alan F. Murray, and Martin Reekie

Department of Electrical Engineering,
The University of Edinburgh, Scotland - UK
http://www.ee.ed.ac.uk
Patrice.Fleury, Alan.Murray, Martin.Reekie@ee.ed.ac.uk

Abstract. This paper presents an approach to on-chip, unsupervised learning. A circuit capable of changing a neuron's synaptic weight with great accuracy is described and experimental results from its aVLSI implementation in a $0.6\mu m$ CMOS process are shown and discussed. We consider its use in the "contrastive divergence" learning scheme of the Product of Experts (PoE) architecture.

1 Introduction

Implementing on-chip learning in aVLSI (analogue Very Large Scale Integration) presents fundamental difficulties. Storing and modifying synaptic weight is not trivial and the circuits which implement these tasks are subject to noise and offset errors inherent to aVLSI [1]. Although some of the imperfections due to aVLSI can be trained out [2], careful design is still required. In the following, we describe an approach, algorithm and circuit form that addresses these issues.

2 Analogue Weight Storage and Modification

Ideally, weight changes would be rapid and weight information retained for long periods of time. Unfortunately one must be traded against the other. Long-term storage circuits (e.g. floating gates) make the weight change complex and slow (very slow charge injection). Fast dynamic weight modification circuits (e.g. charging/discharging capacitor) do not allow long-term memory, since the weight capacitor discharges through the drain (or source) junction of the switch transistor controlling the learning [3]. Mixed signal implementation could be used to store analogue weights in digital storage [4]. This method has, however, a major drawback. It is area-hungry and networks with analogue weights require an ADC per synapse if simultaneous weight changes are to be achieved. Another method would be to use a mix of both dynamic and non-volatile analogue memories [5] to benefit from fast learning capabilities and from reliable long-term storage. Unfortunately floating gates involve the use of high-voltages [6] and rely on undocumented features of the process [3].

Regardless of the form of long-term storage that is chosen (in the recall phase), fast reliable dynamic weight change circuits are crucial elements for the learning (training phase). Therefore our focus in this paper is firmly on learning circuitry.

J.R. Dorronsoro (Ed.): ICANN 2002, LNCS 2415, pp. 426–431, 2002.

3   Products of Experts

The PoE is an unsupervised, stochastic ANN comprising a set of probabilistic generative models ("Experts"). A full explanation of the PoE is given in [7] [8].
The probability of each binary stochastic neuron being "on" is described by :

$$P(S_j = 1) = \sigma(\sum W_{ij}.S_i) \tag{1}$$

where σ is the activation function (Gaussian, Sigmoidal, etc...), W_{ij} the synaptic weights of a neuron, S_i and S_j the states of the input (visible) and hidden layers.
Learning minimises the "contrastive divergence" – a measure of model quality based upon the difference between the data and one-step reconstructions of the data by the experts. This weight update is calculated using equation 2.

$$\Delta W_{ij} = \epsilon(< S_i^+ S_j^+ > - < S_i^- S_j^- >) \tag{2}$$

where ϵ is the learning rate or step, $< S_i^+ S_j^+ >$ represents the expectation value of the products of S_i and S_j that results when a data vector $\{d\}$ is applied to the input. Similarly, $< S_i^- S_j^- >$ represents the expectation value that results from a fantasy (one-step reconstruction) vector $\{f\}$ being input.
We are however planning to use a slightly different, more hardware friendly, learning rule [9]. Instead of applying synaptic weight changes directly, the weight to be updated is incremented/decremented by fixed-size steps (ϵ) in the direction of the expectation value (Eq.3). The time needed to converge to a solution is longer [10] but since it is to be implemented in aVLSI the extra computational time is an acceptable price to pay for a simpler circuit.

$$\Delta W_{ij} = \epsilon \, sign(< S_i^+ S_j^+ > - < S_i^- S_j^- >) \tag{3}$$

4   Hardware Implementation

4.1   Learning Rule

The learning rule (Eq.2) is local, increasing hardware amenability. Simple computations underpin the weight update, leading to less latency and faster learning. For analogue hardware implementation Eq.3 is even more attractive since now only the sign of the weight change is calculated. In effect we just need to know in which direction the weight needs to be "pushed". The hardware requirements therefore become delightfully simple and only two multipliers and one comparator are needed to implement the learning rule.

4.2   Weight Update Circuit

The actual operation of weight-modification is not straightforward. Small changes must be made to a weight voltage without introducing extraneous noise or spurious increments/decrements. When minor charge injection cannot be avoided, it is essential that the weight is still pushed in the right direction [1].

Bit-accuracy is not a meaningful term with respect to analogue weights. However, we have designed circuits capable of making weight adjustments to the equivalent of at least 5 bit accuracy (resolution used in [4] by a similar network [7]).

Our design approach is to isolate the input weight of the weight-modification circuit so that noise and coupling do not alter it. To do so the synaptic weight (W) is "copied" so that a reference weight voltage (W') is obtained. This is done by forcing the currents Id_{N_1}, Id_{N_2} and Id_{P_3} to be equal (Fig.1), therefore causing Vgs_1 and Vgs_2 to have the same value and thus making W and W' also equal (only true in saturation). To insure the mismatch between the differential pair (N1 & N2) is minimal, hence the difference between W and W' is negligible, large interlaced transistors have been laid out. Then from the reference voltage W', two voltages are derived, one slightly larger (W+) and one slightly smaller (W-).

To add control over the learning step (ϵ in Eq.3), i.e. the difference between W' and W+,W-, some current mirrors have been added to the circuit shown in Fig.1 (see Fig.2). They inject an additional current, controlled by an external resistor (Rref_out), through the resistors R_{high} and R_{low}. The resistor (Rref_in) controlling the current Id_{N_1}, Id_{N_2} and Id_{P_3} is also off chip. Finally, either W+ or W- is selected and fed back to the input, with respect to the learning rule. The 'up' and 'down' switches control the path of the feedback, therefore the learning direction ($sign$ of Eq.3).

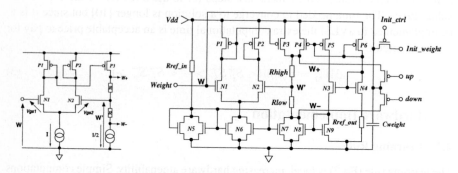

Fig. 1. Basic weight change circuit.

Fig. 2. Enhanced circuit in which the learning rate and direction (ϵ & $sign$ in Eq. 3) are controllable by Rref_out and 'up' & 'down'.

The direct feedback to the input of the circuit allows the weight changes to be very accurate, smooth and continuous over the period of the learning. The external variable resistors Rref_in and Rref_out can be used to adapt the learning rate. Eventually for the VLSI implementation of a full network, the two same external variable resistors would be used to generate two currents corresponding to the global learning rate required. These currents would then be mirrored to every single weight update circuit. This way, a uniform learning can be obtained throughout all of the network.

To summarise this section; the weight is altered by allowing a current proportional to the learning rate to charge/discharge the capacitor (Cweight) on which the weight voltage is stored. The learning rule (Eq.3) controls the 'up' and 'down' switches, i.e. the direction of the learning. Our circuit requires few control signals, no digital control circuitry and no clocking. This makes its hardware implementation easier and also takes it one step closer to biology, where computational elements directly interact with one another.

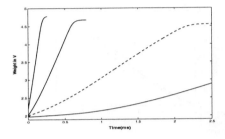
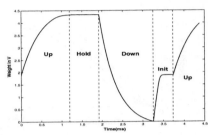

Fig. 3. Different learning rates obtainable on-chip.

Fig. 4. On-chip learning sequence – shows the range of the weight obtainable.

5   Experimental Results

The weight update circuit previously described has been implemented in the AMS $0.6\mu m$ CUP (3 metal, 2 poly) CMOS process. The circuit size is $195*95\mu m$.

The learning rule is currently off-chip and the weight update circuit is controlled by a test station (i.e. a PC). The operating region of our circuit, i.e. the region where the upper weight voltage W+ is truly bigger than the input weight W, is from 1.6V to 4.5V. This is the reason for having the 'init_ctrl' switch (Fig. 2). When the weight voltage is out of the operating region (e.g. when first powered up) and cannot recover, it can be 'kicked' in by charging up the weight to a value within the working region. An example of a weight being initialised to 1.9V can be found in Fig. 4.

Different learning rates can be obtained by varying Rref_out off-chip. For example, a weight initialised to 2V and updated in the 'up' direction until it reached its maximum ranges is shown in Fig. 3 & Fig. 4. This process was repeated with different values of Rref_out ranging from 500KΩ to 5MΩ, see Fig. 3. It is clear that different learning rates can be attained and that the weight change can vary from almost instantaneous to a much more steady pace. In our measurements the learning was shown to vary from $100\mu s$ to 6ms to reach the top of the weight range. Some specific applications may require a faster or a slower learning rate. It can be achieved by reducing or augmenting the capacitor size at the design stage.

We have shown that our weight update circuit can change the weight to any value in the range 1.6-4.5V by "pushing" it up or down. Changes as small as $100\mu V$ and less can be realised. This means that over the range 1.6-4.5V the accuracy of our circuit can go up to 15 bit. Once again some specific applications may require a better accuracy. Higher accuracy can be obtained by increasing the range over which the weight can be changed, i.e. instead of limiting the weight to 1.6-4.5V we could use the range 0-4.5V. A bit accuracy could be gained by doing so. Values from 0 to 1.6V can be reached by initialising the weight to a value in the working region (e.g. 2V) and letting it learn 'down' until it reaches the required value. Fig. 4 shows how any value from the range 0 to 4.5V can be obtained.

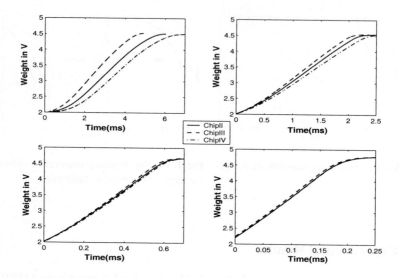

Fig. 5. Effects of process variations on the learning.

Tests included 3 different chips. It was found that for identical settings the learning rate varied slightly from one chip to another and for different speed of learning (Fig.5). These effects due to the process variations are slightly more important for slower learning rates, but the results are still acceptable. In fact the differences are so small that we would qualify our design as being process-tolerant.

6   Conclusions

Accurate and reliable synaptic weight change are difficult to realise on-chip due to many VLSI constraints. We have presented a circuit which can be used for implementing on-chip learning in artificial neural networks, in particular for contrastive divergence training in the Product of Experts. We have shown that the accuracy in the weight change of our circuit is sufficient for its use in the hardware implementation of the PoE. Experimental results show that extremely accurate synaptic weights changes can be obtained at different easily tunable learning rates. Therefore allowing the possible use of our weight change circuit for the VLSI implementation of more demanding (in terms of weight accuracy) neural networks, e.g. the MLP.

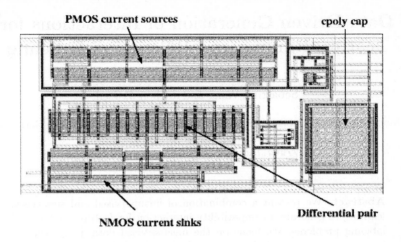

Fig. 6. Layout of the weight change circuit described in Fig. 2.

References

1. T. Lehmann, R. Woodburn, and A.F. Murray, "On-Chip Learning in Pulsed Silicon Neural Networks," in *Proceedings of the IEEE International Symposium on Circuits and Systems*, 1997.
2. B. Tam, S.M. and Gupta, H.A. Castro, and M. Holler, "Learning on an Analog VLSI Neural Network Chip," in *IEEE International Conference on Systems, Man and Cybernetics*, 1990, pp. 701–703.
3. G.M. Bo, D.D. Caviglia, and M. Valle, "An On-Chip learning Neural Network," in *International Joint Conference on Neural Networks (IJCNN)*. IEEE-INNS-ENNS, 2000, vol. 4, pp. 66–71.
4. J.Alspector and A. Jayakumar, "A cascadable Neural Network Chip Set with On-Chip Learning using noise and Gain Annealing," in *Custom Integration Circuits Conference*. IEEE, 1992, pp. 19.5.1–19.5.4.
5. A.J. Montalvo, R.S. Gyurcsik, and J.J. Paulos, "Building Blocks for a Temperature-Compensated Analog VLSI Neural Network with On-Chip Learning," in *International Symposium on Circuits and Systems (ISCAS)*. IEEE, 1994, vol. 6, pp. 363–366.
6. J. Galbraith and W.T. Holman, "An Analog Gaussian Generator Circuit for Radial Basis Function Neural Networks with Non-Volatile Center Storage," in *39th Midwest symposium on Circuits and Systems*. IEEE, 1996, vol. 1, pp. 43–46.
7. G.E. Hinton, "Products of Experts," in Proceedings of the Ninth International Conference on Artificial Neural Networks (ICANN'99), Edinburgh, Scotland, Sept. 1999, pp. 1–6.
8. G.E Hinton, "Training Products of Experts by maximizing contrastive likelihood," Tech. Rep., Gatsby Computational Neuroscience Unit, 1999.
9. A.F. Murray, "Novelty Detection using Products of Simple Experts - A Potential Architecture for Embedded Systems," *Neural Networks*, 2001.
10. P. Fleury, R.J. Woodburn, and A.F. Murray, "Matching Analogue Hardware with Applications using the Products of Experts Algorithm," *Proceedings of the IEEE European Symposium on Artificial Neural Networks (ESANN)*, pp. 63–67, 2001.

Data Driven Generation of Interactions for Feature Binding and Relaxation Labeling

Sebastian Weng and Jochen J. Steil

Neuroinformatics Department, P.O.-Box 100131, D-33501 Bielefeld, Germany,
{sweng,jsteil}@techfak.uni-bielefeld.de,
http://www.TechFak.Uni-Bielefeld.DE/ags/ni/

Abstract. We present a combination of unsupervised and supervised learning to generate a compatibility interaction for feature binding and labeling problems. We focus on the unsupervised data driven generation of prototypic basis interactions by means of clustering of proximity vectors, which are computed from pairs of data in the training set. Subsequently a supervised method recently introduced in [9] is used to determine coefficients to form a linear combination of the basis functions, which then serves as interaction. As special labeling dynamic we use the competitive layer model, a recurrent neural network with linear threshold neurons, and show an application to cell segmentation.

1 Introduction

Feature binding and relaxation labeling have widely been used to solve many segmentation tasks occurring in early vision, perceptual grouping, and pattern matching. To bind features to groups, a priori knowledge on the degree of compatibility of pairs of features is encoded in an interaction function, which in most cases is derived heuristically and fixed beforehand. Successful application are reported in the areas of point clustering, contour grouping [10], texture segmentation [3], and cell segmentation [4]. However, in more complex and more abstract feature spaces a heuristic and suitable encoding of contextual knowledge in the interaction becomes increasingly difficult and motivates the search for learning methods.

One of the main bottlenecks when deriving the interaction directly from a low number of training patterns is, that the number of feature pairs may be large, but nevertheless can span only a small subspace of all possible pairwise interactions. Several approaches to learning the compatibility function have been proposed [5],[6],[7], but lead to very high dimensional nonlinear optimization problems, which suffer from local minima and respective poor generalization. In Wersing [9], this problem is solved by reduction to the task of finding weighting coefficients for a linear combination of a number of interaction prototypes, however at the cost of heuristically choosing these prototypic basis functions. In this contribution we show, how appropriate basis functions can be generated using unsupervised learning.

J.R. Dorronsoro (Ed.): ICANN 2002, LNCS 2415, pp. 432–437, 2002.
© Springer-Verlag Berlin Heidelberg 2002

Starting point for the formal derivation of the learning problem is the concept of consistency [8] to derive a set of linear inequalities describing the learning problem. To segment a pattern consisting of feature vectors m_r[1] a label $\alpha = 1, \ldots, L$ has to be assigned to each m_r, employing pairwise interactions $f_{rr'} = f(m_r, m_{r'})$. Positive values of $f_{rr'}$ indicate that m_r and $m_{r'}$ have a tendency to get the same label, while negative give preference to separate them. Let $0 \leq x_{r\alpha} \leq 1$ represent the grade of assignment of feature m_r to the label α, then a consistent labeling implies, that the mutual support a feature receives from the assignment to the wining label $\hat{\alpha}$ is larger than in case of using a different label β:

$$\sum_{r'} f_{rr'} x_{r'\beta} < \sum_{r'} f_{rr'} x_{r'\hat{\alpha}} \quad \text{for all r, } \beta \neq \hat{\alpha}(r). \tag{1}$$

The consistency conditions (1) have been shown to imply that the corresponding labeling is an asymptotically stable attractor of the common relaxation labeling dynamics [8]. However, instead of iterating a classical relaxation dynamics, in this contribution we employ the competitive layer model (CLM), a large-scale, topographically ordered, linear threshold network, because it's stable states lead to the same consistency conditions [10]. This is especially attractive, because recently developed efficient hardware implementations [1] are available.

In Section 2, we formulate the learning problem based on the consistency approach (1) and introduce the decomposition of $f_{rr'}$ in basis prototypes. In Section 3, we employ a self organizing map for data driven generation of suitable prototypes. In Section 4, we apply this approach to a figure-background segmentation of cell images and give a discussion in the final Section 5.

2 The Learning Problem

The goal is to generate the desired interactions $f_{rr'}$ from a set of M labeled training patterns $\mathcal{P}^i$, $i = 1, \ldots, M$, where each pattern consists of a subset $\mathcal{R}^i = \{r_1^i, \ldots, r_{N_i}^i\}$ of N^i different features and their corresponding labels $\hat{\alpha}^i(r_j^i)$. For each labeled training pattern a target vector y^i is constructed by choosing

$$y_{r\hat{\alpha}^i(r)}^i = 1, \; y_{r\beta}^i = 0, \quad \text{for all r} \in \mathcal{R}^i, \; \beta \neq \hat{\alpha}^i(r), \tag{2}$$

where columns for features, which are not contained in the training pattern are filled with zeros according to $y_{s\alpha}^i = 0$ for all α, $s \notin \mathcal{R}^i$. We obtain the following set of $(L-1) \sum_i N^i$ consistency inequalities:

$$\sum_{r'} f_{rr'} y_{r'\beta}^i < \sum_{r'} f_{rr'} y_{r'\hat{\alpha}}^i \quad \text{for all i, r} \in \mathcal{R}^i, \; \beta \neq \hat{\alpha}, \tag{3}$$

Fixing a number of K basis interactions $g_{rr'}^j = g^j(m_r, m_r')$ we represent $f_{rr'}$ as linear combination $f_{rr'} = \sum_j c_j g_{rr'}^j$, $j = 1, \ldots, K$ with weight coefficients c_j.

---

[1] m_r are elements of a problem specific domain $\mathcal{F}$, e.g. the set of all possible local edge elements (x_r, y_r, φ_r) in an image

Inserting into (3) we obtain the reduced problem in K new parameters c_j as:

$$\sum_j c_j \sum_{r'} g^j_{rr'}(y^i_{r'\beta} - y^i_{r'\hat{\alpha}}) + \kappa < 0, \quad \text{for all i, } r \in \mathcal{R}^i, \ \beta \neq \hat{\alpha}, \qquad (4)$$

where an additional a positive margin variable $\kappa > 0$ was added to achieve a better robustness. Inconsistent training data or higher values of κ imply that not all inequalities can be fulfilled and thus in Wersing [9] a quadratic approach to minimize their overall violation was utilized.

In addition to the standard relaxation labeling approach we use a special label with index $\alpha = 1$ to collect all features, which have only a weak binding to the main groups and can be considered noise. To this aim the inequalities

$$\sum_j c_j \sum_{r'} g^j_{rr'} y^i_{r\beta} - m + \kappa < 0, \quad \text{for all } i, \beta, r \in \mathcal{R}^i | \hat{\alpha}(r) = 1, \qquad (5)$$

$$m - \sum_j c_j \sum_{r'} g^j_{rr'} y^i_{r\hat{\alpha}} + \kappa < 0, \quad \text{for all } i, r \in \mathcal{R}^i | \hat{\alpha}(r) \neq 1, \qquad (6)$$

introduce a threshold m. In the case a feature r is labeled noisy, the first L inequalities in (5) express that the mutual support it receives in the layer β is below the threshold. In case that feature r is assigned a label $\hat{\alpha} > 1$ the one remaining inequality in (6) enforces that the mutual support in the desired label is above the threshold m. Once the prototypes $g^j(m_r, m_{r'})$ are chosen, the overall learning problem consists of $L \sum_i N^i$ linear inequalities in (4), (5) and (6) with $K + 1$ free parameters.

3   Unsupervised Learning of Interaction Prototypes

We turn now to the problem to generate suitable symmetric interaction prototypes $g^j_{rr'} = g^j(m_r, m_{r'})$. As $f_{rr'}$ and thus as well $g_{rr'}$ express some degree of mutual compatibility or proximity of a feature pair $(m_r, m_{r'})$, we first transform it into a generalized proximity space $\mathcal{D}$ by a vector function

$$d_{rr'} = [d_1(m_r, m_{r'}), \dots, d_P(m_r, m_{r'})]^T. \qquad (7)$$

Here $d_i(\cdot), i \geq 1$ may be an arbitrary norm, but as well more general problem specific compatibility function taking also negative values. Then we specify $g_{rr} = g(d_{rr'})$ as function with arguments in $\mathcal{D}$ (and thus implicitly $f_{rr'}$ as well). The elementary example is to directly use the euclidian distance $f_{rr'} = g_{rr'} = d_{rr'} = |(x, y) - (x, y)'|$ between image coordinates $m_r = (x, y)$ for point clustering.

The most direct approach to specify a set of $g^j_{rr'}$ is to decompose the proximity vector space $\mathcal{D}$ into disjunct regions $D^i, \cup D^i = \mathcal{D}$. In Wersing [9] this was done by division of each dimension $d^j_{rr'}$ in a set of heuristically specified disjunct intervals. But in case of more complex features and corresponding proximity data there is need for data driven learning methods.

We use a variation of the Self-Organizing Map, the activity equilibration AEV [2], to reduce the proximity vectors $d_{rr'}$ to a set of K proximity prototypes $\bar{d}_j$

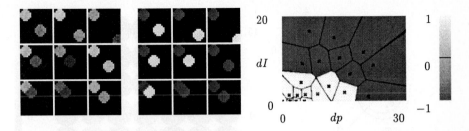

Fig. 1. Left: Training data: nine point cluster, which are grouped together only if they are close and have similar grey values. Middle: The desired labels. Right: Voronio cell prototypes in proximity space spanned by the pixel distance dp and grey value distance dI. Gray shading displays the magnitude of the respective weighting coefficients.

and choose as $j-th$ basis function g^j the corresponding $j-th$ multidimensional Voronoi cell

$$g^j_{rr'} = \begin{cases} 1 & : \quad \| \, \boldsymbol{d}_{rr'} - \tilde{\boldsymbol{d}}_j \, \| \leq \| \, \boldsymbol{d}_{rr'} - \tilde{\boldsymbol{d}}_i \, \| \text{ for all i} \neq \text{j, i,j} = 1,\dots,\text{K} \\ 0 & : \quad \text{else} \end{cases} \qquad (8)$$

Figure 1 illustrates the test case of features $m_r = (x_r, y_r, I_r)$ consisting of the two dimensional image position (x_r, y_r) and a second property I_r that is displayed in form of a grey value. A two-dimensional proximity vector is formed as $d_{rr'} = [d_p, d_I]^T$, where $d_p = ||(x_r, y_r) - (x_{r'}, y_{r'})||$ and $d_I = |I_r - I_{r'}|$.

4   Application to Cell Segmentation

In a more complex domain, we apply the learning approach to the problem of segmentation of fluorescence cell images. The training set of ten images each showing one cell is displayed at the top Fig. 3. The desired output is a figure against background segmentation of the cell body from its surrounding area. Thus the interaction function has to bind features within the cell body together and to adjust the threshold m, such that the surrounding is treated as noise.

The segmentation is based on local edge feature

$$m_r = (x_r, y_r, \varphi_r), \quad x_r, \ y_r \in [1, 30], \ \varphi_r \in [0, 2\pi], \qquad (9)$$

where x_r and y_r describe the position of an image point and φ_r describes the direction of the intensity gradient at this position. The construction of the proximity vectors $\boldsymbol{d}_{rr'} = [\|\, \boldsymbol{d}\, \|, \ \theta_1, \ \theta_2]^T$ is shown in Fig. 2 and follows [4]. It uses the euclidian distance between two features and angles between their connecting vector $\boldsymbol{d}$ and the gradients vectors of the two features.

For the unsupervised learning we use 30 SOM-nodes to restrict the subsequenet quadratic optimization problem to the corresponding 30 coefficients plus the threshold m for the background. The assignment inequalities are computed according to (4), (5) and (6) and the interaction obtained for $\kappa = 100$ is shown in Fig. 2. The result shows remarkable generalization, the segmentation of 10 out of 90 test patterns in displayed in Fig. 3.

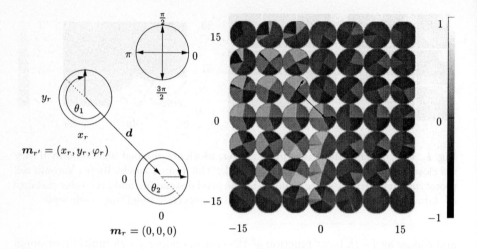

Fig. 2. Left: The feature vectors $m_r(x_r, y_r, \varphi_r)$ contain position and orientation of the local grey value gradient. The proximity vectors contain $d_{rr} = (\| \, d \, \|, \theta_1, \theta_2)$ use the point position distance and relative angles θ_1, θ_2. Right: Learned interaction: For relative feature positions on a five pixel wide lattice, the strength of the interaction with the fixed feature $m_r = (0, 0, 0)$ is visualized as grey level. The threshold m was set to 10 by the optimization algorithm.

5 Discussion

We present an approach to decompose the problem of learning a suitable compatibility interaction for feature binding and labeling problems into a new data driven step of learning prototypic basis functions, which is followed by an optimization step earlier introduced in [9]. This method removes a heuristic step in the definition of the compatibility measure and has several advantages, for instance that the number of parameters in the optimization step is restricted by the number of SOM-prototypes used. For the case of complex features it is even more important, that an arbitrary number of different compatibility measures can be used when transforming the data into the proximity space, because it may not be known beforehand, which are the relevant dimensions. This is clearly superior to the manual choice of a holistic proximity measure or manual tuning of coefficients or basis functions.

There are still a number of open issues, for instance more sophisticated methods like kernel or linear inequality optimization may obtain better coefficients from the consistency equations. Also Voronoi cell based basis functions could be replaced by radial basis functions or a more non-local approximation of the data in the proximity space. However, such approaches generate more interdependencies between the choice of the basis functions and the optimization of their combination, which renders the overall learning process less tractable and controllable. Nevertheless may future work provide results in this direction.

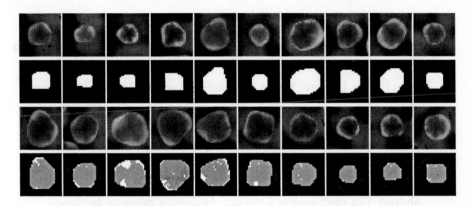

Fig. 3. First row: ten training images. Second row: the target segmentation into body and background. Third row: ten typical test images. Last row: Superimposition of target answer (light color) and CLM-segmentation with learned interaction (dark color) results in gray color for correct labels. The average proportions over 90 test images are 71% for correct labels and 9% and 20% for white and dark.

References

1. Hahnloser, R., Sarpeshkar, R., Mahowald, M.A., Douglas, R.J., Seung, H.S.: Digital selection and analogue amplification coexist in a cortex-inspired silicon circuit, Nature 405 (2000) 947–951.
2. Heidemann, G., Ritter, H.: Efficient Vector Quantization using the WTA-rule with Activity Equalization. Neural Processing Letters vol. 13, no. 1 (2001) 17–30.
3. Hofmann, T., Puzicha, J., Buhmann, J.: Unsupervised texture segmentation in a deterministic annealing framework. IEEE Trans. PAMI 20(8) (1998) 803–818.
4. Nattkemper, T.W., Wersing, H., Schubert, W., Ritter, H.: Fluorescence Micrograph Segmentation by Gestalt-Based Feature Binding. Proc. of the Int. Joint Conf. on Neur. Netw. (IJCNN), vol. 1 (2000) 248–254, Como, Italy.
5. Pelillo, M., Refice, M.: Learning compatibility coefficients for relaxation labeling processes. IEEE Trans. PAMI, 16(9) (1994) 933–945.
6. Pelillo, M., Abbattista, F., Maffione, A.: An evolutionary approach of training relaxation labeling processes. Pattern Recognition Letters, 16, (1995), 1069–1078.
7. Pelillo, M., Fanelli, A.: Autoassociative learning in relaxation labeling networks. Pattern Recognition Letters, 18, (1997), 3–12.
8. Rosenfeld, A., Hummel, R., Zucker, S.W.: Scene labeling by relaxation operations. IEEE Trans. Systems, Man Cybernetics 6(6) (1976) 420–433.
9. Wersing, H.: Learning Lateral Interactions for Feature Binding and Sensory Segmentation. NIPS (2001)
10. Wersing, H., Steil, J.J., Ritter,H.: A Competitive Layer Model for Feature Binding and Sensory Segmentation, Neural Computation vol. 13 (2001) 357–387.

A Hybrid Two-Stage Fuzzy ARTMAP and LVQ Neuro-fuzzy System for Online Handwriting Recognition

Miguel L. Bote-Lorenzo, Yannis A. Dimitriadis, and Eduardo Gómez-Sánchez

School of Telecommunications Engineering, University of Valladolid
Camino del Cementerio s/n, 47011 Valladolid, Spain
{migbot, yannis, edugom}@tel.uva.es
http://pireo.tel.uva.es

Abstract. This paper presents a two-stage handwriting recognizer for classification of isolated characters that exploits explicit knowledge on characters' shapes and execution plans. The first stage performs prototype extraction of the training data using a Fuzzy ARTMAP based method. These prototypes are able to improve the performance of the second stage consisting of LVQ codebooks by means of providing the aforementioned explicit knowledge on shapes and execution plans. The proposed recognizer has been tested on the UNIPEN international database achieving an average recognition rate of 90.15%, comparable to that reached by humans and other recognizers found in literature.

1 Introduction

Handwriting recognition is widely regarded as one of the most difficult problems in the field of pattern recognition because of the great variations present in input patterns [9]. Three main sources of variation can be identified in handwriting generation: allograph variation, execution plan variation and instance variability.

Allograph variation refers to the large amount of different shapes (i.e. *allographs*) used by individuals to represent character concepts (e.g. the letter concept {a} can be written in different ways, such as an upper case, a block printed, or a cursive variant). The shapes used by the writer depend mainly on his education at primary school and personal preferences. *Execution plan variation* is related to the different possible ways (i.e. *execution plan*) of drawing a given shape (e.g. the shape of a zero can be drawn clockwise or counterclockwise). Again execution plans used by writers depend on education as well as on the context of neighboring letters. Finally, *instance variability* holds for a given writer, and refers to the noise (e.g. different slants and sizes, movement noise) introduced by the author when writing (i.e. instantiating) the character. Handwriting recognition systems in general try to avoid instance variability while learning both allograph and execution plan variation. However, while most recognition systems found in literature do this by simply pouring large amounts of data into a single method, [10] remarks the need to provide explicit knowledge on handwriting shapes and execution plans in order to improve recognition performance.

J.R. Dorronsoro (Ed.): ICANN 2002, LNCS 2415, pp. 438–443, 2002.
© Springer-Verlag Berlin Heidelberg 2002

According to this idea, a Fuzzy ARTMAP based automatic prototype extraction method was presented and studied in [1]. This method is able to identify groups of character instances sharing the same allograph and execution plan, as well as to extract a prototype for them. Thus the extracted prototypes are intended to provide explicit knowledge about all the allographs and execution plans found in training data.

Within this framework, this paper presents a two-stage neuro-fuzzy system for handwriting recognition. In the first stage, the Fuzzy ARTMAP based method introduced in [1] is used to extract prototypes from the training data. The use of the explicit knowledge extracted by this method improves the performance of the system's second stage, consisting of a series of LVQ codebooks. This point may be shown by comparing the recognition rates yielded using prototypes extracted by the first stage of the recognizer with those achieved with other prototypes generated by two widely used LVQ initialization methods that are known to provide no explicit knowledge on shape and execution plans. The good performance of the proposed recognizer can also be realized when taking as a reference the rates achieved by other systems and human recognizers.

The organization of this paper is as follows. Section 2 briefly describes the neuro-fuzzy handwriting recognition system proposed in this paper. The UNIPEN data and the extracted prototypes used for the experiments are presented in section 3. Section 4 first shows that the explicit knowledge provided by the prototypes extracted using the first stage of our recognizer does improve performance. Next, the recognition results of our system are compared to those of some other systems from the literature. Finally, in section 5 conclusions and current research are discussed.

2   The Handwriting Recognition System

Since the prototypes extracted in [1] are computed as the mean of a cluster of vectors sharing the same allograph and execution plan, it seems reasonable to apply recognizers based on the comparison of distances between prototypes and test instances. For this purpose, Learning Vector Quantization (LVQ) codebooks [6] can be used.

LVQ is a supervised version of vector quantization that moves codevectors to define near-optimal decision borders between the classes, even in the sense of classical Bayesian decision theory [6]. Knowledge to LVQ codebooks is provided through the initial prototypes that are later refined by the LVQ algorithm.

The handwriting recognizer thus proposed consists of two stages. In the first one, knowledge on handwriting shapes and execution plans is obtained from the training data using the prototype extraction method described in [1]. This method employs Fuzzy ARTMAP neural networks to group character instances according to classification criteria. Next, a simple but effective algorithm finds clusters of instances within these groups having the same allograph and execution plan and computes a prototype for each of the clusters. One of the most outstanding properties of this extraction method is that the prototypes are extracted automatically (i.e. the number of prototypes is not fixed *a priori*).

The second stage of the handwriting recognizer comprises a series of LVQ codebooks initialized by the prototypes extracted in the previous stage. Different codebooks are employed to classify the characters according to their number of strokes.

Prior to training and test phases, raw handwriting data are first preprocessed and segmented into strokes according to the method presented in [4]. Feature vectors are built as described in [1].

3   UNIPEN Data and Prototypes Employed

UNIPEN versions 2 and 7 [5] have been used for the experiments in this paper. The use of the character database provided by UNIPEN ensures a large amount of data, author-independence and comparability to other systems.

Experiments have been carried out using three different sets of isolated characters of both versions: digits, upper-case letters and lower case letters. The number of labels found in each set is 10, 26 and 26 respectively, according to the English alphabet. Each of these sets was in turn divided in two subsets of the same size guaranteeing the presence of samples by any writer in both of them. The first subset was employed both to extract the prototypes that are used to initialize the LVQ codebooks and to further train the recognition system. The second subset was only used for recognition tests. Table 1 shows the distribution of employed data.

Table 1. Data distribution in subsets used for prototype extraction, learning and test of UNIPEN database and number of prototypes extracted from the training subsets

	Version 2			Version 7		
	Digits	Upper-c.	Lower-c.	Digits	Upper-c.	Lower-c.
Prot. Extraction / Learning	1916	2109	6100	7245	12105	23710
Test	1917	2109	6101	7242	12104	23714
Prototypes extracted	108	186	558	278	723	1577

Prototypes were extracted from the corresponding subsets using the extraction method and parameters described in [1]. The distribution of the extracted prototypes for each set is also shown in Table 1. A discussion on the performance of this prototype extraction method was held in [1] showing that a reasonable number of prototypes can be extracted from a large multi-writer amount of samples. Furthermore, the reconstructions of these prototypes are easily recognizable by humans.

4   Experiments and Discussion

Once the recognition system has been introduced, in this section, we will first show that the prototypes extracted by the Fuzzy ARTMAP based method improve recognition performance by providing explicit knowledge on shapes and execution plans. Next, the recognition rates achieved are compared with those yielded by other relevant classifiers.

4.1   Improving Performance by Providing Explicit Knowledge

Propinit and *eveninit*, are proposed in [7] as the standard initialization methods for LVQ codebooks. The *propinit* and *eveninit* initializations choose randomly the initial codebook entries (i.e. prototypes) from the training data set, making the number of entries allocated to each class be proportional or equal, respectively. Both methods try to assure that the chosen entries lay within the class edges, testing it automatically by *k-NN* classification. Thus, both *propinit* and *eveninit* can be said to provide no explicit knowledge of the allograph and execution plans found in the training data set.

Given this background, two experiments can be made to show that the explicit knowledge provided by character prototypes does improve performance. In the first experiment, the prototypes obtained with the three aforementioned methods (i.e. prototypes extracted in [1] and prototypes generated by *propinit* and *eveninit*) were used to classify the test data sets without any kind of LVQ training. In order to make comparisons as fair as possible, the number of prototypes generated by *propinit* and *eveninit* (which must be set *a priori*) was equal to the distribution of prototypes extracted by the Fuzzy ARTMAP based method. The results of the experiment is shown in the first 3 rows of Table 2.

It is noteworthy that the achieved recognition rates employing the extracted prototypes before training the system are significantly higher than using the *propinit* and *eveninit* methods in all cases. This is because the prototypes extracted by the Fuzzy ARTMAP based method are placed in the "middle" of every cluster in the training data, thus providing explicit knowledge of all the allographs and execution plans. On the contrary, the *propinit* and *eveninit* methods, given their random nature, do not assure the existence of a prototype in every cluster.

Table 2. Recognition rates achieved in experiment 1 using prototypes without LVQ training (first 3 rows) and in experiment 2 employing prototypes with LVQ training (last 3 rows)

	Version 2			Version 7		
	Digits	Upper-c.	Lower-c.	Digits	Upper-c.	Lower-c.
Extracted prot. / No training	92.80	86.96	83.87	91.12	87.28	83.53
Propinit / No training	75.85	70.65	67.30	83.97	75.78	75.50
Eveninit / No training	75.74	58.04	65.25	79.51	70.86	70.63
Extracted prot. / Training	93.84	87.81	86.76	95.04	89.68	87.76
Propinit / Training	88.47	78.38	76.71	89.23	80.92	83.49
Eveninit / Training	85.08	73.11	75.40	89.42	80.02	82.28

A second experiment can be carried out by training the three kind of prototypes with the LVQ algorithm. Actually, this training was made employing the OLVQ1 algorithm [7] using the parameter values recommended in the same paper.

The recognition rates using the extracted prototypes increase slightly after carrying out the training, as shown in Table 2. Since the prototypes are computed as the mean of the cluster vectors, the initial codebook vectors are already quite well placed from the classification point of view and the LVQ training just contributes to refine the prototypes' positions in order to minimize the classification error. The increase in recognition rates after training using *propinit* and *eveninit* initialization methods is much higher. In this case, the training phase moves the codebook entries towards more suitable positions in the feature space according to classification criteria. However, the obtained recognition rates using *propinit* and *eveninit* are still lower than using the prototype initialization. In addition, it must be also noticed that they are even lower than the achieved recognition rates using prototypes without any training. This is again because *propinit and eveninit* prototypes cannot be found in every cluster of characters sharing the same allograp and execution plan. These results support again the idea that the explicit knowledge provided by the prototypes extracted with the Fuzzy ARTMAP method improve recognition performance.

4.2 Comparison with other Handwriting Recognizers

In order to evaluate the performance of our system, recognition rates are compared in Table 3 with those achieved by some other relevant classifiers: the two neuro-fuzzy classifiers studied in [3]; a 1-NN classifier with prototypes computed using the unsupervised k-means algorithm, as described in [8] (again the number of prototypes is set according to the distribution shown in Table 1); a 1-NN classifier using all the training data as prototypes (this gives the asymptotic performance of the 1-NN rule, which was proved in [2] to be bounded by twice the Bayesian error rate); and human recognizers (results reported in [3]).

Table 3. Comparison of the proposed system's performance with other relevant classifiers

	Version 2			Version 7		
	Digits	Upper-c.	Lower-c.	Digits	Upper-c.	Lower-c.
Extracted prototypes	**93.84**	**87.81**	**86.76**	**95.04**	**89.68**	**87.76**
System 1 proposed in [3]	85.39	66.67	59.57	-	-	-
System 2 proposed in [3]	82.52	76.39	58.92	-	-	-
k-means + 1-nn	90.40	85.90	84.51	93.22	87.58	87.54
1-nn asymptotic performance	96.04	92.13	88.48	96.52	91.11	-
Human recognition	96.17	94.35	78.79	-	-	-

The proposed system exceeds the recognition rates achieved with the two systems proposed in [3] and the k-means with 1-NN classifier. It is also noticeable that the rates of our recognition system are quite near to the computed asymptotic performance. This is especially remarkable for version 7 digits and upper-case letters where the differences are under 1.5%.

The recognition rates achieved by humans give us an idea of the expected number of unrecognizable data for the different test sets. It is quite surprising to notice that the LVQ recognizer performs better than humans do in lower-case recognition. This can be due to different facts: first, humans did not spend too much time on studying the training data; second, humans get tired after some hours on the computer; and third, humans do not exploit movement information, while the recognizer does.

Finally, it can be said that the main sources of misclassification in the LVQ-based recognizer are erroneously labeled data, ambiguous data, segmentation errors and insufficient feature set. These problems affect the recognizer because of the appearance of incorrect prototypes. In addition, the presentation of erroneous patterns during the training phase may cause a deficient learning. The improvement of these aspects in the prototype extraction method should reduce the number of codebook vectors used and the increase of accuracy recognition.

5 Conclusions

In this paper, a two stage neuro-fuzzy system that exploits explicit knowledge on character's shape and execution plans was presented for on-line handwriting recognition. The first stage extracts prototypes using the Fuzzy ARTMAP based extraction method that was proposed and discussed in [1]. These prototypes provide the explicit knowledge about shapes and execution plans found in training data and are used to initialize the second stage of the recognizer consisting of a series of LVQ codebooks. It has been shown that the aforementioned explicit knowledge extracted by the first stage improves the rates of the handwriting recognizer according to the idea found in [10]. The comparison of our system's performance with other relevant recognizers showed interesting results that may foster further improvements.

References

1. Bote-Lorenzo, M. L., Dimitriadis, Y. A., Gómez-Sánchez, E.: Allograph Extraction of Isolated Handwritten Characters. Proc. of the Tenth Biennial Conference of the International Graphonomics Society, 2001. IGS'01, Nijmegen, The Netherlands (2001) 191-196
2. Devijver, P. A., Kittler, J.: Pattern Recognition: a Statistical Approach. Prentice-Hall International, London (1982)
3. Gómez-Sánchez, E., Dimitriadis, Y. A., Sánchez-Reyes Mas, M., Sánchez García, P., Cano Izquierdo, J. M., López Coronado, J.: On-Line Character Analysis and Recognition With Fuzzy Neural Networks. Intelligent Automation and Soft Computing. 7 (3) (2001)
4. Gómez-Sánchez, E., Gago González, J. Á., Dimitriadis, Y. A., Cano Izquierdo, J. M., López Coronado, J.: Experimental Study of a Novel Neuro-Fuzzy System for on-Line Handwritten UNIPEN Digit Recognition. Pattern Recognition Letters. 19 (3) (Mar. 1998) 357-364
5. Guyon, I., Schomaker, L., Plamondon, R., Liberman, M., Janet, S.: UNIPEN Project of on-Line Data Exchange and Recognizer Benchmarks. Proc. of the 12th International Conference on Pattern Recognition, Jerusalem, Israel (1994) 9-13
6. Kohonen, T.: Self-Organizing Maps. 2nd edn. Springer-Verlag, Heidelberg (1997)
7. Kohonen, T., Kangas, J., Laaksonen, J., Torkkola, K.: LVQ-PAK: The Learning Vector Quantization Program Package. Helsinki University of Technology, Finland (1995)
8. Liu, C.-L., Nakagawa, M.: Evaluation of Prototype Learning Algorithms for Nearest-Neighbor Classifier in Application to Handwritten Character Recognition. Pattern Recognition. 34 (2001) 601-615
9. Plamondon, R., Srihari, S. N.: On-Line and Off-Line Handwriting Recognition: a Comprehensive Survey. IEEE Trans. on Pattern Analysis and Machine Intelligence. 22 (1) (Jan. 2000) 63-84
10. Vuurpijl, L., Schomaker, L.: Finding Structure in Diversity: a Hierarchical Clustering Method for the Categorization of Allographs in Handwriting. Proc. of the International Conference on Document Analysis and Recognition, 1997. ICDAR'97 (1997) 387-393

A New Learning Method for Piecewise Linear Regression

Giancarlo Ferrari-Trecate[1] and Marco Muselli[2]

[1] INRIA, Domaine de Voluceau
Rocquencourt - B.P.105, 78153 Le Chesnay Cedex, France
Giancarlo.Ferrari-Trecate@inria.fr
[2] Istituto per i Circuiti Elettronici - CNR
via De Marini, 6 - 16149 Genova, Italy
muselli@ice.ge.cnr.it

Abstract. A new connectionist model for the solution of piecewise linear regression problems is introduced; it is able to reconstruct both continuous and non continuous real valued mappings starting from a finite set of possibly noisy samples. The approximating function can assume a different linear behavior in each region of an unknown polyhedral partition of the input domain.

The proposed learning technique combines local estimation, clustering in weight space, multicategory classification and linear regression in order to achieve the desired result. Through this approach piecewise affine solutions for general nonlinear regression problems can also be found.

1 Introduction

The solution of any learning problem involves the reconstruction of an unknown function $f : X \to Y$ from a finite set S of samples of f (*training set*), possibly affected by noise. Different approaches are usually adopted when the range Y contains a reduced number of elements, typically without a specific ordering among them (*classification problems*) or when Y is an interval of the real axis with the usual topology (*regression problems*).

However, applications can be found, which lie on the borderline between classification and regression; these occur when the input space X can be subdivided into disjoint regions X_i characterized by different behaviors of the function f to be reconstructed. One of the simplest situations of such kind is piecewise linear regression (PLR): in this case X is a polyhedron in the n-dimensional space $\mathbb{R}^n$ and $\{X_i\}_{i=1}^s$ is a polyhedral partition of X, i.e. $X_i \cap X_j = \emptyset$ for every $i, j = 1, \ldots, s$ and $\bigcup_{i=1}^s X_i = X$. The target of a PLR problem is to reconstruct an unknown function $f : X \to \mathbb{R}$ having a linear behavior in each region X_i

$$f(\boldsymbol{x}) = w_{i0} + \sum_{j=1}^n w_{ij}x_j \quad \text{if } \boldsymbol{x} \in X_i \tag{1}$$

when only a training set S containing m samples $(\boldsymbol{x}_k, y_k)$, $k = 1, \ldots, m$, is available. The output y_k gives a noisy evaluation of $f(\boldsymbol{x}_k)$, being $\boldsymbol{x}_k \in X$; the region

J.R. Dorronsoro (Ed.): ICANN 2002, LNCS 2415, pp. 444–449, 2002.
© Springer-Verlag Berlin Heidelberg 2002

X_i to which x_k belongs is not known in advance. The scalars $w_{i0}, w_{i1}, \ldots, w_{in}$, for $i = 1, \ldots, s$, characterize the function f and their estimate is a target of the PLR problem; for notational purposes they will be included in a vector w_i.

Since regions X_i are polyhedral, they are defined by a set of l_i linear inequalities, which can be written in the following form:

$$A_i \begin{bmatrix} 1 \\ x \end{bmatrix} \geq 0 \tag{2}$$

where A_i is a matrix with l_i rows and $n + 1$ columns, whose estimate is still an output of the reconstruction process for every $i = 1, \ldots, s$.

The target of the learning problem is consequently twofold: to generate both the collection of regions X_i and the behavior of the unknown function f in each of them, by using the information contained in the training set. In these cases, classical learning algorithms for connectionist models cannot be directly employed, since they require some knowledge about the problem, which is not available a priori.

Several authors have treated this kind of problems [2,3,4,7], providing algorithms for reaching the desired result. Unfortunately, most of them are difficult to extend beyond two dimensions [2], whereas others consider only local approximations [3,4], thus missing the actual extension of regions X_i. In this contribution a new connectionist model for solving PLR problems is proposed, together with a proper learning algorithm that combines clustering, multicategory classification, and linear regression to select a reduced subset of relevant training patterns and to derive from them suitable values for the network weights.

2   The Proposed Learning Algorithm

Following the general idea presented in [7], a connectionist model realizing a piecewise linear function f can be depicted as in Fig. 1. It is formed by three layers: the *hidden layer*, containing a linear neuron for each of the regions X_i, the *gate layer*, whose units delimit the extension of each X_i, and the *output layer*, that provides the desired output value for the pattern given as input to the network. The task of the gate layer is to verify inequalities (2) and to decide which of the terms z_i must be used as the output y of the whole network. Thus, the i-th unit in the gate layer has output equal to its input z_i if the corresponding constraint (2) is satisfied and equal to 0 in the opposite case. All the other units perform a weighted sum of their inputs; the weights of the output neuron, having no bias, are always set to 1.

As previously noted, the solution of a PLR problem requires a technique that combines classification and regression: the first has the aim of finding matrices A_i to be inserted in the gate layer of the neural network in Fig. 1, whereas the latter provides the weight vectors w_i for the input to hidden layer connections. The method we propose is summarized in Fig. 2; it is composed of four steps, each one devoted to a specific task.

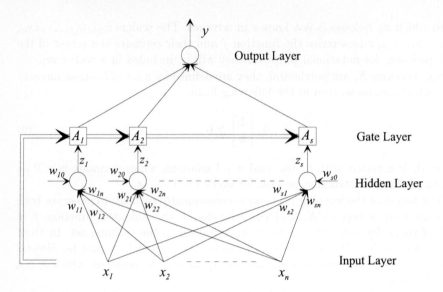

Fig. 1. Connectionist model realizing a piecewise linear function.

The first one (Step 1) aims at obtaining a first estimate of the weight vectors w_i by performing local linear regressions based on small subsets of the whole training set S. In fact, points x_k that are close to each other are likely to belong to the same region X_i. Then, for each sample (x_k, y_k), with $k = 1, \ldots, m$, we build a local dataset C_k containing (x_k, y_k) and the $c-1$ distinct pairs $(x, y) \in S$ that score the lowest values of the distance $\|x_k - x\|$. It can be easily seen that most sets C_k, called *pure*, contain samples belonging to the same region X_i, while the remaining ones, named *mixed*, include input patterns deriving from different X_i. These lead to wrong estimates for w_i and consequently their number must be kept minimum.

Denote with v_k the weight vector produced through the local linear regression on the samples (x_k^1, y_k^1), (x_k^2, y_k^2), $\ldots$, (x_k^c, y_k^c) of C_k. If Φ_k and ψ_k are defined as

$$\Phi_k = \begin{bmatrix} x_k^1 & x_k^2 & \cdots & x_k^c \\ 1 & 1 & \cdots & 1 \end{bmatrix}', \qquad \psi_k = [y_k^1 \ y_k^2 \ \cdots \ y_k^c]'$$

being $'$ the transpose operator, v_k is generated through the well known formula

$$v_k = (\Phi_k' \Phi_k)^{-1} \Phi_k' \psi_k$$

A classical result in least squares theory allows also to obtain the empirical covariance matrix V_k of the vector v_k and the scatter matrix [5] Q_k

$$V_k = \frac{S_k}{c - n + 1} (\Phi_k' \Phi_k)^{-1}, \qquad Q_k = \sum_{i=1}^{c} (x_k^i - m_k)(x_k^i - m_k)'$$

being $S_k = \psi_k' \left(I - \Phi_k (\Phi_k' \Phi_k)^{-1} \Phi_k' \right) \psi_k$ and $m_k = \sum_{i=1}^{c} x_k^i / c$

ALGORITHM FOR PIECEWISE LINEAR REGRESSION

1. *(Local regression)* For every $k = 1, \ldots, m$ do

   1a. Build the local dataset C_k containing the sample $(\boldsymbol{x}_k, y_k)$ and the pairs $(\boldsymbol{x}, y) \in S$ associated with the $c - 1$ nearest neighbors $\boldsymbol{x}$ to $\boldsymbol{x}_k$.

   1b. Perform a linear regression to obtain the weight vector $\boldsymbol{v}_k$ of a linear unit fitting the samples in C_k.

2. *(Clustering)* Perform a proper clustering process in the space $\mathbb{R}^{n+1}$ to subdivide the set of weight vectors $\boldsymbol{v}_k$ into s groups U_i.

3. *(Classification)* Build a new training set S' containing the m pairs $(\boldsymbol{x}_k, i_k)$, being U_{i_k} the cluster including $\boldsymbol{v}_k$. Train a multicategory classification method to produce the matrices A_i for the regions $\hat{X}_i$.

4. *(Regression)* For every $i = 1, \ldots, s$ perform a linear regression on the samples $(\boldsymbol{x}, y) \in S$, with $\boldsymbol{x} \in \hat{X}_i$, to obtain the weight vector $\boldsymbol{w}_i$ for the i-th unit in the hidden layer.

Fig. 2. Proposed learning method for piecewise linear regression.

Then, consider the *feature vectors* $\boldsymbol{\xi}_k = [\boldsymbol{v}'_k \ \boldsymbol{m}'_k]'$, for $k = 1, \ldots, m$; they can be approximately modeled as the realization of random vectors with covariance matrix

$$R_k = \begin{bmatrix} V_k & 0 \\ 0 & Q_k \end{bmatrix}'$$

If the generation of the samples in the training set is not affected by noise, most of the $\boldsymbol{v}_k$ coincide with the desired weight vectors $\boldsymbol{w}_i$. Only mixed sets C_k yield spurious vectors $\boldsymbol{v}_k$, which can be considered as outliers. Nevertheless, even in presence of noise, a clustering algorithm (Step 2) can be used to determine the sets U_i of feature vectors $\boldsymbol{\xi}_k$ associated with the same $\boldsymbol{w}_i$. If the number s of regions is fixed beforehand, a proper version of the K-means algorithm [6] can be adopted. It uses the following cost functional

$$J\left(\{U_i\}_{i=1}^s, \{\boldsymbol{\mu}_i\}_{i=1}^s\right) = \sum_{i=1}^s \sum_{\boldsymbol{\xi}_k \in U_i} \|\boldsymbol{\xi}_k - \boldsymbol{\mu}_i\|_{R_k^{-1}}^2$$

where $\boldsymbol{\mu}_i$ is the center of the cluster U_i. This choice allows to recover the influence of poor initialization and of outliers on the clustering process.

The sets U_i generated by the clustering process induce a classification on the input patterns $\boldsymbol{x}_k$ belonging to the training set S, due to the chain of bijections among the set of input patterns $\boldsymbol{x}_k$, the collection of local datasets C_k, and the class of feature vectors $\boldsymbol{\xi}_k$. Now, if $\boldsymbol{\xi}_k \in U_i$ for a given i, it is likely that the local dataset C_k is fitted by the linear neuron with weight vector $\boldsymbol{w}_i$ and consequently $\boldsymbol{x}_k$ is located into the region X_i. An estimate $\hat{X}_i$ for each of these regions can then be determined by solving a linear multicategory classification problem (Step 3), whose training set S' is built by adding as output to each input pattern $\boldsymbol{x}_k$ the index i_k of the set U_{i_k} to which the corresponding feature vector $\boldsymbol{\xi}_k$ belongs.

To avoid the presence of multiply classified points or of unclassified patterns in the input space, multicategory techniques [1] deriving from the Support Vector Machine approach and based on linear and quadratic programming can be employed. In this way the s matrices A_i for the gate layer are generated. Finally, weight vectors $\boldsymbol{w}_i$ for the neural network in Fig. 1 can be directly obtained by solving s linear regression problems (Step 4) having as training sets the samples $(\boldsymbol{x}, y) \in S$ with $\boldsymbol{x} \in \hat{X}_i$, being $\hat{X}_1, \ldots \hat{X}_s$ the regions found by the classification process.

3   Simulation Results

The proposed algorithm for piecewise linear regression has been tested on a two-dimensional benchmark problem, in order to analyze the quality of the resulting connectionist model. The unknown function to be reconstructed is

$$f(x_1, x_2) = \begin{cases} 3 + 4x_1 + 2x_2 & \text{if } 0.5x_1 + 0.29x_2 \geq 0 \text{ and } x_2 \geq 0 \\ -5 - 6x_1 + 6x_2 & \text{if } 0.5x_1 + 0.29x_2 < 0 \text{ and } 0.5x_1 - 0.29x_2 < 0 \\ -2 + 4x_1 - 2x_2 & \text{if } 0.5x_1 - 0.29x_2 \geq 0 \text{ and } x_2 < 0 \end{cases} \quad (3)$$

with $X = [-1, 1] \times [-1, 1]$ and $s = 3$. A training set S containing $m = 300$ samples (x_1, x_2, y) has been generated, according to the model $y = f(x_1, x_2) + \varepsilon$, being ε a normal random variable with zero mean and standard deviation $\sigma = 0.1$. The behavior of $f(x_1, x_2)$ together with the elements of S are depicted in Fig. 3a. Shaded areas in the (x_1, x_2) plane show the polyhedral regions X_i.

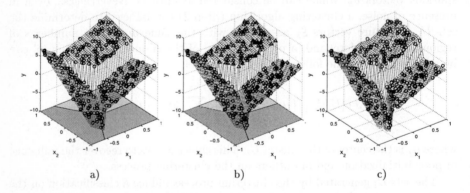

Fig. 3. Two dimensional benchmark problem: a) behavior of the unknown piecewise linear function and training set, b) simulation results obtained with the proposed algorithm and c) with a conventional two layer neural network.

The method described in Fig. 2 has been applied by choosing at Step 1 the value $c = 8$. The resulting connectionist model realizes the following function, graphically represented in Fig. 3b:

$$f(x_1, x_2) = \begin{cases} 3.02 + 3.91x_1 + 2.08x_2 & \text{if } 0.5x_1 + 0.28x_2 \geq 0 \text{ and } 0.12x_1 + x_2 \geq 0.07 \\ -5 - 5.99x_1 + 6.01x_2 & \text{if } 0.5x_1 + 0.28x_2 < 0 \text{ and } 0.5x_1 - 0.26x_2 < 0.04 \\ -2.09 + 4.04x_1 - 2.08x_2 & \text{if } 0.5x_1 - 0.26x_2 \geq 0.04 \text{ and } 0.12x_1 + x_2 < 0.07 \end{cases}$$

As one can note, this is a good approximation to the unknown function (3); the generalization Root Mean Square (RMS) error, estimated through a 10-fold cross validation, amounts to 0.296. Significant differences can only be detected at the boundaries between two adjacent regions X_i; they are mainly due to the effect of mixed sets C_k on the classification process.

As a comparison, a two layer neural network trained with a combination of Levenberg-Marquardt algorithm and Bayesian regularization yields the nonlinear approximating function shown in Fig. 3c. Discontinuities between adjacent regions are modeled at the expense of reducing the precision on flat areas. It is important to observe that neural networks are not able to reconstruct the partition $\{X_i\}_{i=1}^s$ of the input domain, thus missing relevant information for the problem at hand.

Seven hidden nodes are needed to produce this result; their number has been obtained by performing several trials with different values and by taking the best performance. A 10-fold cross validation scores an estimate of 0.876 for the generalization RMS error, significantly higher than that obtained through the PLR procedure.

References

1. E. J. BREDENSTEINER AND K. P. BENNETT, Multicategory classification by support vector machines. *Computational Optimizations and Applications*, **12** (1999) 53–79.
2. V. CHERKASSKY AND H. LARI-NAJAFI, Constrained topological mapping for nonparametric regression analysis. *Neural Networks*, **4** (1991) 27–40.
3. C.-H. CHOI AND J. Y. CHOI, Constructive neural networks with piecewise interpolation capabilities for function approximation. *IEEE Transactions on Neural Networks*, **5** (1994) 936–944.
4. J. Y. CHOI AND J. A. FARRELL, Nonlinear adaptive control using networks of piecewise linear approximators. *IEEE Transactions on Neural Networks*, **11** (2000) 390–401.
5. R. O. DUDA AND P. E. HART, *Pattern Classification and Scene Analysis.* (1973) New York: John Wiley and Sons.
6. G. FERRARI-TRECATE, M. MUSELLI, D. LIBERATI, AND M. MORARI, A Clustering Technique for the Identification of Piecewise Affine Systems. In *HSSC 2001*, vol **2304** of *Lecture Notes in Computer Science* (2001) Berlin: Springer-Verlag, 218–231.
7. K. NAKAYAMA, A. HIRANO, AND A. KANBE, A structure trainable neural network with embedded gating units and its learning algorithm. In *Proceedings of the International Joint Conference on Neural Networks* (2000) Como, Italy, III–253–258.

Stable Adaptive Momentum for Rapid Online Learning in Nonlinear Systems

Thore Graepel and Nicol N. Schraudolph

Institute of Computational Science
ETH Zürich, Switzerland
{graepel,schraudo}@inf.ethz.ch

Abstract. We consider the problem of developing rapid, stable, and scalable stochastic gradient descent algorithms for optimisation of very large nonlinear systems. Based on earlier work by Orr et al. on adaptive momentum—an efficient yet extremely unstable stochastic gradient descent algorithm—we develop a stabilised adaptive momentum algorithm that is suitable for noisy nonlinear optimisation problems. The stability is improved by introducing a forgetting factor $0 \leq \lambda \leq 1$ that smoothes the trajectory and enables adaptation in non-stationary environments. The scalability of the new algorithm follows from the fact that at each iteration the multiplication by the curvature matrix can be achieved in $O(n)$ steps using automatic differentiation tools. We illustrate the behaviour of the new algorithm on two examples: a linear neuron with squared loss and highly correlated inputs, and a multilayer perceptron applied to the four regions benchmark task.

1 Introduction

Optimisation problems arise in a wide variety of fields including science, engineering, and business. We focus on optimisation algorithms for problems with a twice-differentiable objective function that are large in scale, non-linear, non-convex, that might be non-stationary, and for which only noisy measurements of the objective function value and its first and second derivatives w.r.t. the parameters are available. This is the realm of online stochastic gradient methods.

2 Adaptive Momentum

We consider the problem of minimising a function $f : \mathbb{R}^n \to \mathbb{R}$ with respect to its parameters $\mathbf{w} \in \mathbb{R}^n$. In online optimisation the weight vector $\mathbf{w}_t$ at time step t is updated according to

$$\mathbf{w}_{t+1} = \mathbf{w}_t + \mathbf{v}_t. \tag{1}$$

The simplest form of gradient descent is recovered by choosing $\mathbf{v}_t := -\mu \mathbf{g}_t$ where μ is a scalar learning rate parameter and $\mathbf{g}_t := \nabla_{\mathbf{w}} f_t(\mathbf{w})|_{\mathbf{w}=\mathbf{w}_t}$ is the

J.R. Dorronsoro (Ed.): ICANN 2002, LNCS 2415, pp. 450–455, 2002.

instantaneous gradient measurement of the objective function f with respect to the parameters $\mathbf{w}$ evaluated at $\mathbf{w}_t$.

Various improvements over this simple scheme are conceivable. Decreasing $\mu := \mu_t$ deterministically as a function of time aims at quickly traversing the search space at the beginning of the optimisation and to smoothly converge at the end. Replacing μ by a diagonal matrix allows the use of individual learning rates for each parameter. Replacing μ by a full matrix makes it possible to include local curvature information into the optimisation process.

We follow a different line of development that suggests to introduce a momentum term,

$$\mathbf{v}_t := -\mu \left(\rho \mathbf{g}_t - \mathbf{v}_{t-1} \right) . \tag{2}$$

The intuition behind this update rule is to stabilise the trajectory through search space by choosing the new update $\mathbf{v}_t$ as a linear combination of the previous update $\mathbf{v}_{t-1}$ and the new gradient $\mathbf{g}_t$. The adaptive version of this idea, incorporating a curvature matrix $\mathbf{C}$ into the update, has been introduced in [2,4],

$$\mathbf{v}_t := -\mu_t \left(\rho_t \mathbf{g}_t - \left(\mathbf{C}_t + \frac{1}{\mu_t} \mathbf{I} \right) \mathbf{v}_{t-1} \right) , \tag{3}$$

where $\mu_t < \lambda_{\max}^{-1} (\mathbf{C}_t)$ and ρ_t is annealed from 1 down to 0. Let us assume $\mu_t := \mu$, $\rho_t = \rho$, $\mathbf{g}_t := \mathbf{g}$, and $\mathbf{C}_t := \mathbf{C}$ constant for the sake of argument. Setting $\mathbf{v}_t = \mathbf{v}_{t-1} =: \mathbf{v}_\infty$ and solving for $\mathbf{v}_\infty$ it is easily seen that the update rule (3) has a fixed-point at

$$\mathbf{v}_\infty = -\rho \mathbf{C}^{-1} \mathbf{g} . \tag{4}$$

Thus, using $\mathbf{v}_\infty$ instead of $\mathbf{v}_t$ in (1) and setting $\rho := 1$ we recover Newton's methods if $\mathbf{C}$ is the Hessian, $\mathbf{C}_t := \mathbf{H}_t := \nabla_{\mathbf{w}}^2 f_t (\mathbf{w}) \big|_{\mathbf{w}=\mathbf{w}_t}$. For nonlinear problems it is advisable to use the Gauss-Newton approximation to the Hessian which is guaranteed to be positive semidefinite while retaining as much information as possible [6].

3   Stable Adaptive Momentum

The above algorithm turns out to be very efficient, but unstable for practical non-convex problems such as multilayer perceptrons. As a stabilising mechanism we suggest to use an exponential average over past values of $\mathbf{v}_t$ governed by a forgetting factor $0 \leq \lambda \leq 1$, resulting in an update rule

$$\mathbf{v}_t := -\mu_t \left(\rho_t \mathbf{g}_t - \lambda \left(\mathbf{C}_t + \frac{1}{\mu_t} \mathbf{I} \right) \mathbf{v}_{t-1} \right) . \tag{5}$$

Clearly, for $\lambda = 1$ we recover the matrix momentum update (3) and for $\lambda = 0$ standard gradient descent. Again assuming $\mu_t := \mu$, $\rho_t = \rho$, $\mathbf{g}_t := \mathbf{g}$, and $\mathbf{C}_t := \mathbf{C}$ constant, the new update rule (5) has a fixed point at

$$\mathbf{v}_\infty = -\rho \left(\lambda \mathbf{C} + (1 - \lambda) \frac{1}{\mu} \mathbf{I} \right)^{-1} \mathbf{g} . \tag{6}$$

Thus in the limit the effective curvature matrix is a convex combination of $\mathbf{C}$ and $\lambda_{\max}(\mathbf{C})\mathbf{I}$, comparable to the Levenberg-Marquardt approach [3]. In particular, if we choose $\frac{1}{\mu} := \lambda_{\max}(\mathbf{C})$ as suggested in [1] the condition number $\mathcal{N}(\mathbf{C}) := \lambda_{\max}(\mathbf{C})/\lambda_{\min}(\mathbf{C})$ is improved to

$$\mathcal{N}\left(\lambda\mathbf{C} + (1-\lambda)\frac{1}{\mu}\mathbf{I}\right) = \frac{\lambda\lambda_{\max}(\mathbf{C}) + \frac{1-\lambda}{\mu}}{\lambda\lambda_{\min}(\mathbf{C}) + \frac{1-\lambda}{\mu}} = \frac{\lambda_{\max}(\mathbf{C})}{\lambda\lambda_{\min}(\mathbf{C}) + (1-\lambda)\lambda_{\max}(\mathbf{C})}.$$

The denominator of the condition number is proportional to a convex combination of the largest and smallest eigenvalues of $\mathbf{C}$.

For convenient and efficient implementation of this algorithm *automatic differentiation* tools[1] can be used to calculate gradient $\mathbf{g}_t$ and curvature matrix-vector product $\mathbf{C}_t\mathbf{v}$ in $\mathcal{O}(n)$ as described in [5]. The Gauss-Newton approximation of the Hessian should be used to ensure positive semidefiniteness [6].

Algorithm 1 Stable Adaptive Momentum (SAM)

Require: A twice-differentiable objective function $f : \mathbb{R}^n \to \mathbb{R}$

Require: instantaneous gradient $\mathbf{g}_t = \nabla_\mathbf{w} f_t(\mathbf{w})|_{\mathbf{w}=\mathbf{w}_t}$ and curvature matrix vector product $\mathbf{C}_t\mathbf{v}$

Require: Forgetting factor $0 \leq \lambda \leq 1$, learning rates $0 < \rho \leq 1$ and $\mu > 0$, and objective function convergence criterion $\varepsilon_f > 0$.

Ensure: Storage for $\mathbf{v}_{\mathrm{new}}$, $\mathbf{v}_{\mathrm{old}}$, $\mathbf{w}_{\mathrm{new}}$, $\mathbf{w}_{\mathrm{old}}$.

    Initialise $\mathbf{w}_{\mathrm{old}}$ with small random numbers, and $\mathbf{v}_{\mathrm{old}} \leftarrow g_0$

    repeat

        $\mathbf{v}_{\mathrm{new}} \leftarrow -\mu\left(\rho\mathbf{g}_t - \lambda\left(\mathbf{C}_t + \frac{1}{\mu}\mathbf{I}\right)\mathbf{v}_{\mathrm{old}}\right)$

        $\mathbf{w}_{\mathrm{new}} \leftarrow \mathbf{w}_{\mathrm{old}} + \mathbf{v}_{\mathrm{new}}$

    until Time-averaged error $\langle|f_t(\mathbf{w}_t) - f_t(\mathbf{w}_t)|\rangle_t < \varepsilon_f$

---

4 Numerical Experiments

4.1 The Quadratic Bowl

In order to study the influence of the parameter λ on the behaviour of the algorithm and to understand its interaction with the learning rate parameter ρ we considered the unconstrained quadratic optimisation problem of minimising the function

$$f(\mathbf{w}) = \frac{1}{2}\mathbf{w}^T\mathbf{H}\mathbf{w}, \quad \mathbf{H} = \mathbf{J}\mathbf{J}^T, \quad \mathbf{J} = \begin{pmatrix} 2 & 1.6 \\ 1.2 & 3.2 \end{pmatrix}.$$

With eigenvalues of $\mathbf{H}$ given by 16.89 and 1.19 this configuration models the situation of long diagonal valleys in the objective function that often occur in neural network learning. Obviously the optimal solution is $\mathbf{w} = \mathbf{0}$ with $f(\mathbf{0}) = 0$. Samples for online learning are drawn from a normal distribution $\mathcal{N}(\mathbf{0}, \mathbf{H})$.

---

[1] see http://www-unix.mcs.anl.gov/autodiff/

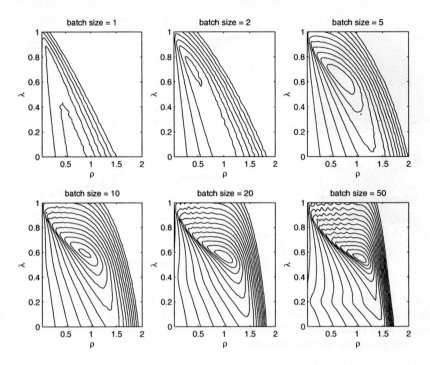

Fig. 1. Log-contour plots of objective function value $f(\mathbf{w}) = \mathbf{w}^T \mathbf{H} \mathbf{w}$ after 100 steps of SAM, averaged over starting points $\{0.0, 0.01, \ldots, 4\pi\}$ on the unit circle, as a function of the ρ and λ parameters. Each contour corresponds to a 100-fold improvement in f.

Figure 1 shows log-contour plots of the value of the objective function $f(\mathbf{w}_t)$ after 100 update steps, averaged over different starting points on the unit circle at angles $\{0.0, 0.01, \ldots, 4\pi\}$, as a function of the ρ and λ parameters, for different batch sizes. When learning fully online (batch size 1), simple gradient descent ($\lambda = 0$) works best: in this regime the curvature measurements are so noisy that they do not accelerate convergence. For batch sizes ¿ 1, however, SAM clearly outperforms both simple gradient descent and standard adaptive momentum ($\lambda = 1$). (Note that each contour on the plot corresponds to a 100-fold improvment in objective function value.) As the batch size increases, the optimal regime emerges at $\lambda \approx 0.5 \ldots 0.6$, $\rho = 1$ — the latter value illustrating that correctly stabilized and conditioned (through λ and μ parameters, respectively) SAM in fact does not require the ρ parameter. Finally, the plot for batch size 50 shows how lowering λ serves to dampen oscillations observable (as a function of ρ) for high values of λ.

4.2   The Four Regions Task

As a more realistic and difficult test example we use the four regions classification task [7] to be solved by a multi-layer perceptron, as shown in Figure 2 (left). The network has two input units, four output units and two layers of

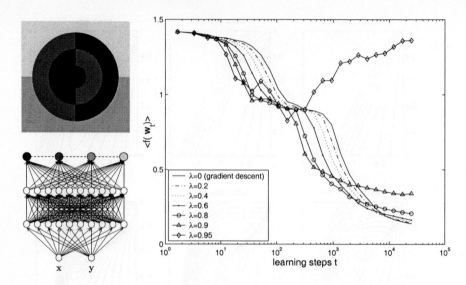

Fig. 2. Four regions task (top left) and corresponding learning curves (right) using stable adaptive momentum with different values of λ applied to a 2-10-10-4 fully connected feed-forward neural network (bottom left) with tanh activation functions, softmax output, and cross-entropy loss function.

10 hidden units (with tanh transfer function) each, with neighbouring layers fully connected, resulting in $n = 184$ weights. The classification is implemented through a softmax function applied to the outputs and a cross-entropy loss function. For each run the weights are initialised to uniformly random values in the interval $[-0.3, 0.3]$. Training patterns are generated online by drawing independent, uniformly random input samples; since each pattern is seen only once, the empirical loss provides an unbiased estimate of generalisation ability. Patterns are presented in mini-batches of ten each.

We compare SAM (with $\rho = 1$) to standard gradient descent with constant learning rate μ. The learning rate $\mu = 0.01$ was chosen such that standard gradient descent achieves the lowest value of $f(\mathbf{w})$ after 25 000 update steps. Figure 2 (right) shows learning curves for different values of λ in the range $[0.0, 0.95]$. To put these results in perspective, note that the original adaptive momentum algorithm (corresponding to $\lambda = 1$) diverges immediately everytime, as observed already in [4]. This can also be seen from the curve for $\lambda = 0.95$ in Figure 2 (right) that illustrates this problem of divergence for values greater than $\lambda = 0.9$. Compared to simple gradient descent ($\lambda = 0$), increasing λ generally results in a faster transient reduction of f while the asymptotic quality of the solution becomes worse.

We found that the ad-hoc annealing schedule $\lambda(t) = 0.9/(1 + \exp \frac{t - 5000}{1000})$, which decreases λ logistically from 0.9 in the search phase to zero in the convergence phase, is capable of combining the advantages of fast transient and good

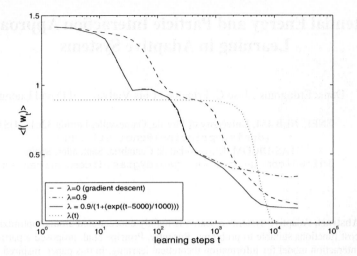

Fig. 3. Learning curves for the four regions task illustrating an ad-hoc annealing scheme for λ that accelerates convergence but retains good asymptotic performance.

asymptotic performance (Figure 3). We are now investigating how to control λ adaptively so as to automate this annealing process.

References

1. Y. LeCun, P. Y. Simard, and B. Pearlmutter. Automatic learning rate maximization in large adaptive machines. In S. J. Hanson, J. D. Cowan, and C. L. Giles, editors, *Advances in Neural Information Processing Systems*, volume 5, pages 156–163. Morgan Kaufmann, San Mateo, CA, 1993.
2. T. K. Leen and G. B. Orr. Optimal stochastic search and adaptive momentum. In J. D. Cowan, G. Tesauro, and J. Alspector, editors, *Advances in Neural Information Processing Systems*, volume 6, pages 477–484. Morgan Kaufmann, San Francisco, CA, 1994.
3. D. Marquardt. An algorithm for least-squares estimation of non-linear Parameters. *Journal of the Society of Industrial and Applied Mathematics*, 11(2):431–441, 1963.
4. G. B. Orr. *Dynamics and Algorithms for Stochastic Learning*. PhD thesis, Department of Computer Science and Engineering, Oregon Graduate Institute, Beaverton, OR 97006, 1995. ftp://neural.cse.ogi.edu/pub/neural/papers/orrPhDchi-5. ps.Z, orrPhDch6-9.ps.Z.
5. B. A. Pearlmutter. Fast exact multiplication by the Hessian. *Neural Computation*, 6(1):147–160, 1994.
6. N. N. Schraudolph. Fast curvature matrix-vector products for second-order gradient descent. *Neural Computation*, 14(7), 2002. http://www.inf.ethz.ch/~schraudo/pubs/mvp.ps.gz.
7. S. Singhal and L. Wu. Training multilayer perceptrons with the extended Kalman filter. In D. S. Touretzky, editor, *Advances in Neural Information Processing Systems. Proceedings of the 1988 Conference*, pages 133–140, San Mateo, CA, 1989. Morgan Kaufmann.

Potential Energy and Particle Interaction Approach for Learning in Adaptive Systems

Deniz Erdogmus[1], Jose C. Principe[1], Luis Vielva[2], and David Luengo[2]

[1] CNEL, NEB 454, University of Florida, Gainesville, Florida 32611, USA
{deniz, principe}@cnel.ufl.edu
[2] GTAS-DICOM, Universidad de Cantabria, Santander, Spain
luis@dicom.unican.es, david@gtas.dicom.unican.es

Abstract. Adaptive systems research is mainly concentrated around optimizing cost functions suitable to problems. Recently, Principe et al. proposed a particle interaction model for information theoretical learning. In this paper, inspired by this idea, we propose a generalization to the particle interaction model for learning and system adaptation. In addition, for the special case of supervised multi-layer perceptron (MLP) training we propose the interaction force backpropagation algorithm, which is a generalization of the standard error backpropagation algorithm for MLPs.

1 Introduction

Adaptive system training algorithms research has long been driven by pre-defined cost functions deemed suitable for the application. For instance, mean-square-error (MSE) has been extensively utilized as the criterion in supervised learning and adaptation, although alternatives have been proposed and investigated relatively less frequently [1]. Second order statistics, by definition, have also been the cost function for principal component analysis [2]. Other higher order statistics, including higher order cumulants like the kurtosis, high order polyspectra, etc., and information theoretic cost functions have mainly been studied in the context of blind signal processing with applications to independent component analysis (ICA), blind source separation (BSS), and blind deconvolution [3-5]. The commonality of all the research on these is that the analyses are mainly motivated by the corresponding selected adaptation criterion.

Working on the same problems, Principe et al. have utilized Renyi's quadratic entropy definition and introduced the term *information theoretical learning* to the adaptive systems literature [6]. Their nonparametric estimator for Renyi's quadratic entropy, which is based on Parzen windowing with Gaussian kernels, incited the idea of particle interactions in adaptation. Specifically considering the blind source separation problem, they have defined and demonstrated the *quadratic information forces* and the *quadratic information potential* at work in this context. Their insight on the adaptation process as an *interaction between information particles* deserves further investigation. Erdogmus and Principe have recently extended the entropy estimator to any entropy order and kernel function in Parzen windowing [7]. This generalization of the entropy estimator also led to the extensions of the definitions of

J.R. Dorronsoro (Ed.): ICANN 2002, LNCS 2415, pp. 456–461, 2002.
© Springer-Verlag Berlin Heidelberg 2002

information potential and *force*. Successful applications of this entropy estimator in supervised and unsupervised learning scenarios have increased confidence and interest on *information theoretic learning* [8,9].

Inspired by the above-mentioned *information-particle interaction model for learning* proposed in [6], we investigate in this communication the possibility of generalizing the concept of particle interaction learning. Our aim is to determine a unifying model to describe the learning process as an interaction between *particles*, where for some special case these may be the *information particles* or for some other special case, we may end up with the commonly utilized second order statistics of the data. The formulations to be presented in the sequel will achieve these objectives and we will call this general approach the *potential energy extremization learning* (PEEL). Also, specifically applied to supervised learning, we will obtain the *minimum energy learning* (MEL). In addition, we will propose a generalized backpropagation algorithm to train MLPs under MEL principle. For the specific choice of the *potential field* (to be defined later) that reduces the minimum energy criterion to MSE, we will observe that the generalized backpropagation algorithm reduces to the standard backpropagation algorithm.

2   Adaptation by Particle Interactions

Traditionally, the adaptation process is regarded as an optimization process, where a suitable pre-defined performance criterion is maximized or minimized. In this alternative view, we will treat each sample of the training data set as a particle and let these particles interact with each other according to the interactions laws that we define. The parameters of the adaptive system will then be modified in accordance with the interactions between the particles.

2.1   Particle Interaction Model

Suppose we have the samples $\{z_1,...,z_N\}$ generated by some adaptive system. For simplicity, assume we are dealing with single dimensional random variables; however, note that extensions to multi-dimensional situations are trivial. In the particle interaction model, we assume that each sample is a particle and a potential field is emanated from it. Suppose z_i generates a potential energy field. If the potential field that is generated by each particle is $v(\xi)$, we require this function to be continuous and differentiable, and to satisfy the even symmetry condition $v(\xi) = v(-\xi)$. Notice that due to the even symmetry and differentiability, the gradient of the potential function at the origin is zero. With these definitions, we observe that the potential energy of particle z_j due to particle z_i is $V(z_j|z_i) = v(z_j - z_i)$. The total potential energy of z_j due to all the particles in the training set is then given by

$$V(z_j) = \sum_{i=1}^{N} V(z_j \mid z_i) = \sum_{i=1}^{N} v(z_j - z_i).$$ (1)

Defining the interaction force between these particles, in analogy to physics, as

$$F(z_j \mid z_i) \overset{\Delta}{=} \partial V(z_j \mid z_i) / \partial z_j = \partial v(\xi) / \partial \xi \big|_{\xi = (z_j - z_i)} = v'(z_j - z_i) . \tag{2}$$

we obtain the total force acting on particle z_j

$$F(z_j) = \sum_{i=1}^{N} F(z_j \mid z_i) = \sum_{i=1}^{N} v'(z_j - z_i) . \tag{3}$$

Notice that the force applied to a particle by itself is $F(z_j \mid z_j) = v'(0) = 0$.

Finally, the total potential energy of the sample set is the sum (possibly weighted) of the individual potentials of each particle. Assuming that each particle is weighted by a factor $\gamma(z_j)$ that may depend on the particle's value, which may as well be independent from the value of the particle, but different for each particle, the total energy of the system of particles is found to be

$$V(z) = \sum_{j=1}^{N} \gamma(z_j) \sum_{i=1}^{N} v(z_j - z_i) . \tag{4}$$

Assuming that $\gamma(z_j) = 1$ for all samples, we can determine the sensitivity of the overall potential of the particle system with respect to the position of a specific particle z_j. This is given by

$$\frac{\partial V(z)}{\partial z_k} = \frac{\partial}{\partial z_k} \sum_{j=1}^{N} \sum_{i=1}^{N} v(z_j - z_i) = \dots = 2F(z_k) . \tag{5}$$

In the adaptation context, since the samples are generated by a parametric adaptive system, the sensitivity of the total potential with respect to the weights of the system is also of interest. This sensitivity is directly related to the interaction forces between the samples as follows

$$\frac{\partial V}{\partial w} = \frac{\partial}{\partial w} \sum_{j=1}^{N} \sum_{i=1}^{N} v(z_j - z_i) = \dots = \sum_{j=1}^{N} \sum_{i=1}^{N} F(z_j \mid z_i) \left(\frac{\partial z_j}{\partial w} - \frac{\partial z_i}{\partial w} \right) . \tag{6}$$

2.2 Some Special Cases

Consider for example the potential function choice of $v(\xi) = \xi^2 / (2N^2)$ and weighting function choice of $\gamma(z_j) = 1$ (i.e. unweighted) for all samples. Then upon direct substitution of these values in (4), we obtain $V(z)$ equals the biased sample variance, i.e. minimization of this potential energy will yield the minimum variance solution for the weights of the adaptive system. In general, if we select potential functions of the form $v(\xi) = |\xi^p|$, where $p > 1$, with no weighting of the particles we obtain cost functions of the form

$$V(z) = \sum_{j=1}^{N} \sum_{i=1}^{N} \left| (z_j - z_i)^p \right| . \tag{7}$$

which are directly related to the absolute central moments of the random variable Z,

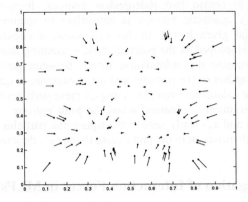

Fig. 1. A snapshot of the information particles (output vector samples) and the instantaneous information forces acting on these particles in two-dimensional BSS.

for which z_j's are samples. Each value of p corresponds to a different choice of the distance metric between the particles from the family of Minkowski norms.

The information potential estimators of [6] and [7] also fall into this same category of cost energy functions. The quadratic information potential (based on Renyi's quadratic entropy) estimator in [6], which uses Gaussian kernels $G_\sigma(.)$ with standard deviation σ (named the kernel size), is

$$V_2(z) = \frac{1}{N^2} \sum_{j=1}^{N} \sum_{i=1}^{N} G_\sigma(z_j - z_i). \tag{8}$$

The generalized information potential estimator in [7], on the other hand is

$$V_\alpha(z) = \frac{1}{N^\alpha} \sum_{j=1}^{N} \left(\sum_{i=1}^{N} \kappa_\sigma(z_j - z_i) \right)^{\alpha-1}. \tag{9}$$

In (9), α is the entropy order for Renyi's definition and $\kappa_\sigma(.)$ is the kernel function, which must be a valid pdf. Notice that for the potential function choice $v(\xi) = G_\sigma(\xi)/N^2$ and $\gamma(z_j) = 1$ in (4), we obtain the quadratic information potential of (8). Additionally, for $v(\xi) = \kappa_\sigma(\xi)/N^2$ and $\gamma(z_j) = \hat{p}(z_j)^{\alpha-2}$, we obtain (9) from (4). In the latter, $\hat{p}(z_j)$ is the Parzen window estimate, using kernel $\kappa_\sigma(.)$, of the probability density of particle z_j [10].

2.3   Illustration of Information Forces in Independent Component Analysis

As an example consider the quadratic information forces acting on the samples in a two-dimensional ICA/BSS scenario where the topology is a square matrix of weights followed by nonlinearities matched to the cumulative densities of the sources as described in [6]. Renyi's quadratic joint entropy of the outputs of the nonlinearities is to be maximized to obtain two independent sources. It is shown in [6] that maximizing Renyi's quadratic entropy is equivalent to minimizing the quadratic information potential given in (8). In this expression, a circular two-dimensional Gaussian kernel is employed as the potential field emanating from each particle and this is used to evaluate the information forces between particles. Under these circumstances, a snapshot of the particles and the instantaneous quadratic information forces, which can be calculated from (3), acting on these particles are shown in Fig. 1. Since the optimal solution is obtained when the joint entropy is maximized, these forces are repulsive and as clearly seen in the figure, the particles repel each other to arrive at a uniform distribution in the unit square in the two-dimensional output space.

3   BackPropagation of Interaction Forces in MLPs

In this section, we will derive the backpropagation algorithm for an MLP trained supervised under the MEL principle. This extended algorithm backpropagates the interaction forces between the particles through the layers instead of the error, as is the case in the standard MSE criterion case. For simplicity, consider the unweighted potential of the error as the cost function. For multi-output situations, we simply sum the potentials of the error signals from each output. Assume the MLP has l layers with m_o processing elements (PE) in the oth layer. We denote the input vector with layer index zero. Let w_{ji}^o be the weight connecting the ith input to the jth output in the oth layer. Let $v_j^o(s)$ be the synapse potential of the jth PE at o^{th} layer corresponding to the input sample $x(s)$, where s is the sample index. Let $\varphi(.)$ be the sigmoidal nonlinearity of the MLP, same for all PEs, including the output layer. Assume $v(.)$ is the potential function of choice and we have N training samples. The total energy of the error particles, where $e_k(t)$ is the error at the kth output for training sample t is then

$$V = \sum_{s=1}^{N}\sum_{t=1}^{N}\sum_{k=1}^{m_l} v(e_k(s) - e_k(t)) \sum_{s=1}^{N}\sum_{t=1}^{N} \varepsilon(s,t).$$ (10)

In order to save space, we skip the derivation and present the algorithm. It suffices to tell that the derivation of the algorithm follows the same lines as the derivation of the standard backpropagation, which can be found in numerous textbooks on neural networks [1]. In the algorithm below, η is the learning rate and $\varphi'(.)$ is the derivative of the MLP's sigmoid function.

Algorithm. Let the interaction force acting on sample s due to the potential field of sample t be $F(e_j(s) \mid e_j(t)) = v'(e_j(s) - e_j(t))$ in the jth output node of the MLP. These interactions will minimize the energy function in (10).

1.   Evaluate local gradients for the output layer for $s,t=1,\ldots,N$ and $j=1,\ldots,m_l$ using

$$\delta_j^l(s \mid t) = -F(e_j(s) \mid e_j(t)) \cdot \varphi'(v_j^l(s)), \quad \delta_j^l(t \mid s) = -F(e_j(t) \mid e_j(s)) \cdot \varphi'(v_j^l(t))$$

2.   For layer index o going down from l-1 to 1 evaluate the local gradients

$$\delta_j^o(s \mid t) = \varphi'(v_j^o(s)) \sum_{k=1}^{m_{o+1}} \delta_k^{o+1}(s \mid t) w_{kj}^{o+1}, \quad \delta_j^o(t \mid s) = \varphi'(v_j^o(t)) \sum_{k=1}^{m_{o+1}} \delta_k^{o+1}(t \mid s) w_{kj}^{o+1}$$

3.   For each layer index o from 1 to l evaluate the weight updates (to minimize V)
$$\Delta w_{ji}^{o} = -\eta \left(\delta_{j}^{o}(s \mid t) y_{i}^{o-1}(s) + \delta_{j}^{o}(t \mid s) y_{i}^{o-1}(t) \right)$$

Notice that for the squared error criterion with $v(\xi) = \xi^{2}$, the interaction force becomes $F(e_{j}(s) \mid e_{j}(t)) = 2(e_{j}(s) - e_{j}(t))$ and the algorithm reduces to the backpropagation of error values.

4   Discussion

Adaptive systems research is traditionally motivated by the optimization of suitable cost functions and is centered on the investigation of learning algorithms that achieve the desired optimal solution. In this paper, inspired by the idea of *information theoretic learning through particle interactions*, we have proposed an alternative approach to adaptation and learning. This new approach allows us to regard this process in analogy with interacting particles in a force field in physics. Besides the intellectual appeal of this viewpoint provides us for further theoretical study on learning, it may be promising in designing real systems that utilize physical forces to change its state and eventually adapt to its environment to need. It might also fascilitate self-organization in distributed systems, through pairwise interactions.

References

1.   Haykin, S.: Neural Networks: A Comprehensive Foundation. 2nd edn. Prentice Hall, New Jersey (1999)
2.   Oja, E.: Subspace Methods for Pattern Recognition. Wiley, New York (1983)
3.   Haykin, S. (ed.): Unsupervised Adaptive Filtering, Vol. 1: Blind Source Separation. Wiley, New York (2000)
4.   Haykin, S. (ed.): Unsupervised Adaptive Filtering, Vol. 2: Blind Deconvolution. Wiley, New York (2000)
5.   Hyvarinen, A., Karhunen, J., Oja, E.: Independent Component Analysis. Wiley, New York (2001)
6.   Principe, J.C., Xu, D., Fisher, J.W.: Information Theoretic Learning. In: Haykin, S. (ed.): Unsupervised Adaptive Filtering, Vol. 1: Blind Source Separation. Wiley, New York (2000)
7.   Erdogmus, D. Principe, J.C.: Generalized Information Potential Criterion for Adaptive System Training. To appear in IEEE Trans. in Neural Networks (2002)
8.   Santamaria, I., Erdogmus, D., Principe, J.C.: Entropy Minimization for Digital Communications Channel Equalization. To appear in IEEE Trans. on Signal Processing (2002)
9.   Torkkola, K., Campbell, W.M.: Mutual Information in Learning Feature Transformations. In: Proceedings of the International Conference on Machine Learning. Stanford (2000)
10. Parzen, E.: On Estimation of a Probability Density Function and Mode. In: Time Series Analysis Papers. Holden-Day, California (1967)

Piecewise-Linear Approximation of Any Smooth Output Function on the Cellular Neural Network

Víctor M. Preciado

Departamento de Electrónica e Ingeniería Electromecánica,
Universidad de Extremadura,E-06071 Badajoz, Spain
{vpdiaz@unex.es}

Abstract. The Cellular Neural Network Universal Machine (CNNUM)[7] is a novel hardware architecture which make use of complex spatio-temporal dynamics performed in the Cellular Neural Network (CNN)[1] for solving real-time image processing tasks. Actual VLSI chip prototypes [6] have the limitation of performing a fixed piecewise-linear (PWL) saturation output function. In this work, a novel algorithm for emulating a piecewise-linear (PWL) approximation of any nonlinear output function on the CNNUM VLSI chip is presented.

1   Introduction

The Cellular Neural Network (CNN) topology is essentially characterized by a *local* interaction between nonlinear dynamical cells distributed in a regular 2-D grid [1]. This fact makes the CNN an useful computation paradigm when the problem can be reformulated as a task where the signal values are placed on a regular 2-D grid, and interaction between signal values are limited within a finite local neighborhood [4]. Besides, local interaction facilitates the implementation of this kind of networks as efficient and robust VLSI chips [5] ,[6]. The Cellular Neural Network Universal Machine (CNNUM) [7] is a programmable neuroprocessor based on CNN dynamics and implemented alongside photosensors which sense and process the image in a single VLSI chip. The main drawbacks encountered when using this chip in image processing application is the limitation in filtering capabilities to 3 × 3 dimension templates and the restricted possibilities provided by the fixed piecewise-linear (PWL) saturation output function in nonlinear filtering.

In this work, the CNN dynamical model and the architecture of the Cellular Neural Network Universal Machine (CNNUM) prototyping system are introduced. Then, we describe a novel and general algorithm for achieving any type of PWL approximation of an arbitrary output function on the framework of the CNNUM chipset.

J.R. Dorronsoro (Ed.): ICANN 2002, LNCS 2415, pp. 462–467, 2002.

2   CNN Dynamical Model and CNNUM Architecture

The simplest electronic implementation of a CNN cell consists in a single capacitor which is coupled to neighbouring cells through nonlinear controlled sources. The dynamic of the array can be described by the following set of differential equations

$$\frac{d}{dt}x_{i,j}(t) = -x_{i,j}(t) + \sum_{k,l \in N_r} A_{k,l} y_{i+k,j+k}(t) + \sum_{k,l \in N_r} B_{k,l} u_{i+k,j+k}(t) + I \quad (1)$$

$$y(x) = \frac{1}{2}\left[|x-1| - |x+1|\right] \quad (2)$$

The input,state and output, represented by $u_{i,j}, x_{i,j}$ and $y_{i,j}$ are defined on $0 \le i \le N_1$ and $0 \le i \le N_2$ and N_r represents the neighborhood of the cell with a radius r as $N_r = \{(k,j) : \max\{|k-i|, |l-j|\} \le r\}$. The B template and I coefficient form a simple feedforward filtered (FIR) version of the input. On the other hand, the temporal evolution of the *dynamics network* is mathematically modeled by the A template operating in a *feedback* loop along with the fixed saturation nonlinearity previously defined.

The CNNUM architecture is an *analogic (analog+logic)* spatio-temporal array computer wherein analog spatio-temporal phenomena provided by the CNN and logic operations are combined in a programmable framework to obtain more sophisticated operation mode [7]. Every implemented neuron includes circuitry for CNN processing, binary and gray scaled images storage supplied by a *local analog memory (LAM)* and *local logic memory (LLM)*, logic operation among binary images provided by a *local logic unit (LLU)* and the necessary configuration circuitry for electrical I/O and control of the different operations by means of a *local communication and control unit (LCCU)*.

3   PWL Approximation by the Infinity Norm Criterion

When considering the VLSI CNN chip model, we deal with a rigid PWL saturation output function due to difficulties in implementing flexible non-linearities on silicon. In this Section, we present a general method to approximate any nonlinear output function on current CNNUM chips by superposition of piecewise-linear (PWL) saturation blocks as defined in (2).

3.1   Previous Definitions and Notation

The following notation is used: δ_{ij} denotes Kronecker delta, $B_{zo,r}$ denotes de open ball $B_{zo,r} := \{z \in Z : \|z - z_o\| < r\}$, $\|\cdot\|$ is the weighted Euclidean norm defined as $\|z\| = \left(\sum_{i=1}^{n} \omega_i z_i^2\right)^{1/2}$, with $\omega_i > 0$, $\|\cdot\|_\infty$ the weighted infinity norm. Increment and sum of successive function in an indexed list is denoted by $\Sigma h_i := h_{i+1} + h_i$ and $\Delta h_i := h_{i+1} - h_i$, and the symbol ' denotes differentiate on variable x.

3.2   Direct PWL Approximating Method by the Infinite Norm Criterion

In this paper, a superposition of piecewise-linear saturation functions are considered in the following structure:

$$\overline{f}(x) = \sum_{i=1}^{\sigma} \left(\frac{1}{b_i} y(a_i x - c_i) + m_i \right) \tag{3}$$

where $y(x)$ is defined in (2). Each summing term consists in a linear transformation of the same PWL original saturation function; thus, $\overline{f}(x)$ yields an adjustable PWL function where $a_i, b_i, c_i, m_i \in \mathbb{R}$, are the parameters of the structure. Basically, (3) is a nonlinear combination of linear affine lines, $\Pi_i := \frac{1}{2} \left[|(a_i x - c_i) - 1| - |(a_i x - c_i) + 1| \right], i \in [1, \sigma]$.

Now, the problem under study can be stated as follows: Given a smooth function $f : S \to \mathbb{R}$, where $S \subset \mathbb{R}$ is compact, we want to design a PWL function $\overline{f}$ that minimizes the error between f and $\overline{f}$ in some sense. Formally, given a fixed number ε^* we want to find the optimal parameter vector $\theta^* = [a_i^*, b_i^*, c_i^*, m_i^*]$ that makes the objective functional $\mathfrak{J} := \left\| f(x) - \overline{f}(x) \right\|_\infty = \varepsilon^* \quad \forall x \in S$, with the most efficient shape.

The functional based on the infinite norm $\|\cdot\|_\infty$ is supported by a physical effect observed in the implementation of CNNUM VLSI chips: *the analog signals involved in the processing task are computed with an analog accuracy of 7 bits of equivalent digital accuracy [5]*. Thus, this fact gives us a fixed value for $\varepsilon \leq 2^{-7}$, which assure that the error introduced by the approximation procedure is less or equal that the error introduced by the performance of the implemented chip.The functional proposed in this paper is an alternative to the $\left\| f(x) - \overline{f}(x) \right\|$ functional studied in several papers [3], [2]. This quadratic criterion yields a nonlinear optimization problem characterized by the existence of several local minima. One practical technique used to undertake this serious problem consist in the use of iterative algorithms which produce new random search direction when a local minimum in reached.

The point of departure used to obtain the approximating PWL function based on the infinity norm is the following

Theorem 1 (Minimax). *Let $f(x)$ be a function defined in the open subset (x_i, x_{i+1}), $x_i, x_{i+1} \in \mathbb{R}$ and $P_n(x)$ a polynomial with grade n. Then $P_n(x)$ minimizes $\|f(x) - P_n(x)\|_\infty$ if and only if $f(x) - P_n(x)$ takes the value $\varepsilon := \max(|f(x) - P_n(x)|)$ at least in $n+2$ points in the interval (x_i, x_{i+1}) with alternating sign.*

Theorem 2. *Let $f(x)$ be a function with $f'' > 0$ in the interval (x_1, x_2), $x_1, x_2 \in \mathbb{R}$ and $P_1(x) := Mx + B$. Then $P_1(x)^1$ minimizes $\|f(x) - P_1(x)\|_\infty$ if and only if $M = \Delta f / \Delta x_1$; $B = \frac{1}{2} [f(x_2) + f(x_a) - \Delta f_i / \Delta x_i (x_a + x_2)]$ where x_a is obtained by solving $f'(x_a) = \Delta f_i / \Delta x_i$*

---

[1] This straight line is called *Chebyshev line* in the literature.

Proof. It follows from minimax theorem that it must be three points x_l, x_c, x_r in (x_1, x_2) which maximize $E(x) := f(x) - P_1(x)$. This condition implies that x_c is an intermediate point in the interval (x_1, x_2) with $E'(x)|_{x_c} = 0$; this is the same that $f'(x)|_{x_c} = M$. Since $f''(x) > 0$, $f'(x)$ is a strictly growing function and can equate M only once, this means that x_c is the only one intermediate point which minimizes E in the interval; thus $x_l = x_1$ and $x_r = x_2$. Applying the minimax condition we obtain $E(x_l) = -E(x_c) = E(x_r)$ and by solving these equations we can conclude $M = \Delta f_i / \Delta x_i$; $B = \frac{1}{2} [f(x_{i+1}) + f(x_a) - \Delta f_i / \Delta x_i (x_a + x_{i+1})]$

Corollary 1. *Under the previous conditions,* $\varepsilon := \|f(x) - P_1(x)\|_\infty$ *is given by*

$$\varepsilon = f(x) - \left[\frac{\Delta f_i}{\Delta x_i} x + \frac{1}{2} \left(f(x_{i+1}) + f(x_{ai}) - \frac{\Delta f_i}{\Delta x_i} (x_{ai} - x_{i+1}) \right) \right]$$

Remark 1. From the proof of this theorem it can be advised that in the case of $f'' < 0$, $\varepsilon = -E(x_l) = E(x_c) = -E(x_r)$

Theorem 3. *Let* $f(x)$ *be a function with* $f'' > 0$ *in the interval* (x_a, x_b), $x_a, x_b \in \mathbb{R}$, ε^* *an arbitrary small real number and* $\overline{f}(x) = \sum_{i=1}^{\sigma} \Pi_i$, *where* $\Pi_i := \frac{1}{2} [|(a_i x - c_i) - 1| - |(a_i x - c_i) + 1|]$, $i \in [1, \sigma]$; $a_i, b_i, c_i, m_i \in \mathbb{R}$, $i \in [1, \sigma]$. *Then* $\overline{f}(x)$ *makes* $\|f(x) - \overline{f}(x)\|_\infty = \varepsilon^*$ *minimizing the number of summing terms* σ *if the parameters of* $\overline{f}(x)$ *fulfill the following conditions:*

$$a_i = 2/ \Delta x_i, \ b_i = 2/ \Delta f_i, \ c_i = \Sigma x_i / 2, \ i \in [1, \sigma];$$
$$m_1 = \Sigma f_i / 2 - \varepsilon^*, \ m_j = \Sigma f_j / 2 - f(x_j) - \varepsilon^*, \ j \in [2, \sigma] \qquad (4)$$

where x_i *is obtained from the following set of discrete equations:*

$$\varepsilon^* - \frac{1}{2} \left[x_i + \frac{\Delta f_i}{\Delta x_i} (x_{ai} - x_i) - f(x_{ai}) \right] = 0 \ being \ f'(x_{ai}) = \frac{\Delta f_i}{\Delta x_i}, \ i \in [1, \sigma] \quad (5)$$

Proof. In order to demonstrates this theorem we can express Π_i as

$$\Pi_i := \begin{cases} m_i - b_i^{-1}, \ \forall x \in [c_i + a_i^{-1}, \infty) \\ m_i + \frac{\Delta f_i}{\Delta x_i} (x - c_i), \ \forall x \in B_{c_i, a_i^{-1}} \\ m_i + b_i^{-1}, \ \forall x \in (-\infty, c_i - a_i^{-1}] \end{cases}$$

Replacing the values of the parameters given in the statement of the theorem

$$\Pi_i := \begin{cases} \delta_{1i} (f(x_i) - \varepsilon^*), \ \forall x \in [c_i + a_i^{-1}, \infty) \\ \delta_{1i} (f(x_i) - \varepsilon^*) + \frac{\Delta f_i}{\Delta x_i} (x - x_i), \ \forall x \in B_{c_i, a_i^{-1}} \\ \delta_{1i} (f(x_i) - \varepsilon^*) + \Delta f_i, \ \forall x \in (-\infty, c_i - a_i^{-1}] \end{cases}$$

If we consider $x_a \in (x_j, x_{j+1})$ and expand $\overline{f}(x_a)$ taking into account the value of ε^* given in Corollary 1, it is obtained

$$\begin{aligned} \overline{f}(x_a) &:= \Pi_1 + \sum_{i=2}^{j-1} \Pi_i + \Pi_j + \sum_{i=j+1}^{\sigma} \Pi_i \\ &= (f(x_1) - \varepsilon^*) + \sum_{i=2}^{j-1} \Delta f_i + \left[\frac{\Delta f_j}{\Delta x_j} (x - x_j) \right] \\ &= f(x_j) - \varepsilon^* + \frac{\Delta f_j}{\Delta x_j} (x - x_j) \\ &= \frac{\Delta f_i}{\Delta x_i} x + \frac{1}{2} \left[f(x_{i+1}) + f(x_a) - \frac{\Delta f_i}{\Delta x_i} (x_a + x_{i+1}) \right] \end{aligned}$$

this is the equation of the *Chebyshev line* that approximated $f(x)$ in the interval (x_j, x_{j+1}) with $\|f(x) - P_1(x)\|_\infty = \varepsilon^*$ as it was expressed in Theorem 2.

Corollary 2. *Since the PWL function is continuous in the intervals (x_i, x_{i+1}) and the term $\sum_{i=j+1}^{\sigma} \Pi_i$ is null in the expansion of $\overline{f}(x_a)$ performed in the previous proof, it can be affirmed that $\lim_{\Delta x \to 0} f(x_i + \Delta x) = \lim_{\Delta x \to 0} f(x_i - \Delta x)$, and $\overline{f}(x)$ is a PWL continuous function.*

Remark 2. Theorem 3 gives us the possibility of approximating any continuous function $f(x)$ with $f'' > 0$ by means of a piecewise-linear function with an arbitrarily small infinite norm ε^*. Besides, the intervals of the approximation function can be obtained in a forward way if we know the analytical expression of $f(x)$, by means of solving the uncoupled set of discrete equations stated in (5). This fact supplies a direct method to design the intervals of approximations in comparison with the annealing iterative method needed in the minimization of the quadratic norm.

3.3   Implementation of the Approximation on the CNNUM

The stages in the approximation process are: (i) modification of the PWL original saturation as defined in (2) to adopt it to every affine plane Π_i as stated in Theorem 3; (ii) superposition of these modified saturation functions in order to obtain the approximating function $\overline{f}(x)$ as defined in (3).

The original saturation function $y(x)$ can be modified by the affine transformation $\frac{1}{b_i} y(a_i x - c_i) + m_i$. This reshaping translates the corners locates at (-1,-1) and (1,1) in the original saturation (2) to $(c_i - \frac{1}{a_i}, m_i - \frac{1}{b_i})$, $(c_i + \frac{1}{a_i}, m_i + \frac{1}{b_i})$ in the modified one. This transformation is performed on the CNN by means of the following two templates runned in a sequential way

$$T_{k,1}^{PWL} = \{A_{ij}^2 = 0; B_{ij}^1 = a_k \delta_{2j} \delta_{2i}); I^k = -a_k c_k \ , \forall i,j\} \tag{6a}$$

$$T_{k,2}^{PWL} = \{A_{ij}^2 = 0; B_{ij}^1 = b_k^{-1} \delta_{2j} \delta_{2i}); I^k = m_k \ , \forall i,j\} \tag{6b}$$

where δ_{ij} denotes Kronecker delta.

Then, we can obtain the optimal parameter vector $\theta^* = [a_i^*, b_i^*, c_i^*, m_i^*]$ that makes the objective functional $\mathfrak{J} = \varepsilon^*$ by means of calculating the intervals obtained by recursively solving the uncoupled set of discrete equations (5), and applying (4). Thus, we figure the templates given by (6) taking into account the values of θ^* in order to obtain each summing term Π_i needed to define the approximating function $\overline{f}(x)$.

The most efficient procedure to add each term Π_i in the framework of CN-NUM computing is introducing the input image into the CNN and operating the image by means of the nonlinear dynamics resulting from the connectivity defined in $T_{1,1}^{PWL}$ and $T_{2,1}^{PWL}$. After this, we save the result of the analog computation into the *LAM*. The same operation must be accomplished with templates $T_{1,2}^{PWL}$ and $T_{2,2}^{PWL}$ on the original input image. The result of the second step is accumulated in the *LAM* with the previous result. Making this process through every stage we finally obtain the image processed by a point operator that performs desired approximating function $\overline{f}(x)$.

Lastly, it will be used a value $\varepsilon^* = 2^{-7}$ in the analytical deduction of the parameter vector θ^* because of the physical implementation of the CNN-UM chip allows an analog accuracy of this magnitude. In the case of $f(x) = \ln(x)$, the discrete equation in Theorem 3 yields the following implicit discrete equation $\ln\left(\frac{\Delta x_i}{\Delta \ln_i}\right) + \left(\frac{\Delta \ln_i}{\Delta x_i}\right) x_i - \ln(x_i) - 1 = 2\varepsilon^*$ and in the approximation of an exponential function it can be similarly deduced the following condition $\frac{\Delta \exp_i}{\Delta x_i}\left[\ln\left(\frac{\Delta \exp_i}{\Delta x_i}\right) + x_i + 1\right] - \exp(x_i)$ where $\Delta \ln_i = \ln(x_{i+1}) - \ln(x_i)$, $\Delta \exp_i = \exp(x_{i+1})\exp(x_i)$ and $\varepsilon^* = 2^{-7}$. Both equation can be easily solved by standard numerical methods in order to obtain the neighboring points of the intervals that construct the PWL approximating function $\overline{f}(x)$ in a recursive and forward way.

4   Conclusions

In this paper, it has been introduced the equations that govern the complex CNN spatio-temporal dynamics and the CNNUM computational infrastructure implemented on silicon. After this, we have presented a general technique that allows us to approximate any nonlinear output function on the CNNUM VLSI Chip. For this purpose, we have given a theoretical analysis of an approximation technique based on the infinity norm criterion. Also, it has been comment the advantages of this technique in comparison with the quadratic error criterion. The main motivation of this work is to release CNNUM *analogic (analog+logic)* architecture from using *digital* computers when CNN image processing computing capabilities are unable to perform any required nonlinear filtering step.

References

1. L. O. Chua, L. Yang. "Cellular neural networks: Theory". *IEEE Trans. Circuits Syst.*,vol. 35, no. 10. pp. 1257–1272, 1988.
2. L.O. Chua, and R.L.P. Ying, "Canonical piecewise-linear analysis," *IEEE Trans. Circuits Syst.*, vol. CAS-30, pp. 125–140, 1983.
3. L.O. Chua, and A. Deng, "Canonical piecewise-linear modeling," *IEEE Trans. Circuits Syst.*, vol. CAS-33, pp. 511–525, 1986.
4. K.R. Crounse and L.O. Chua, "Methods for image processing and pattern formation in cellular neural networks: A tutorial," *IEEE Trans. Circuits Syst.*, Vol. 42, no. 10, 1995.
5. S. Espejo, R. Domínguez-Castro, G. Liñan, and A. Rodríguez-Vázquez, "64 x 64 CNN universal chip with analog and digital I/O", *IEEE Int. Conf. on Electronic Circuits and Systems*, 1998.
6. G. Liñan, R. Domínguez-Castro, S. Espejo and A. Rodríguez-Vazquez, "CNNUC3 user guide," *Instituto de Microelectronica de Sevilla Technical Report*, 1999.
7. T. Roska and L.O. Chua, "The CNN universal machine: An analogic array computer," *IEEE Trans. Circuits Syst.-II*, vol. 40, pp. 163–173, 1993.

MDL Based Model Selection for Relevance Vector Regression

Davide Anguita and Matteo Gagliolo

Dept. of Biophysical and Electronic Engineering
University of Genova, Via Opera Pia 11a
I–16145 Genova, Italy
{anguita,gagliolo}@dibe.unige.it

Abstract. Relevance Vector regression is a form of Support Vector regression, recently proposed by M.E.Tipping, which allows a sparse representation of the data. The Bayesian learning algorithm proposed by the author leaves the partially open question of how to automatically choose the optimal model.

In this paper we describe a model selection criterion inspired by the Minimum Description Length (MDL) principle. We show that our proposal is effective in finding the optimal kernel parameter both on an artificial dataset and a real–world application.

1 Introduction

In a *kernel model*, regression estimate at a value x is given by:

$$\hat{y}(x) = w_0 + \sum_i w_i K(x, x_i; \rho_k) \qquad (1)$$

that is, a weighted sum of the kernel function K, with parameter ρ_k, and centered in x_i, $i = 1 \ldots n$.

Given a training set, represented by a set of points $\{x_i\}$ and target values $\{t_i\}$, we may want to know which is the best interpretation of the set in terms of the family of models (1). This problem can be posed at different levels of abstraction. Namely, we may want to choose:

a) a value of $\{w_i\}$ for a given kernel K with fixed parameter ρ_k.
b) a value for the kernel parameter ρ_k.
c) a kernel function K.

The Minimum Description Length (MDL) principle [3,2] can be applied to any of the problems above. This principle consists in picking, from a family of models for a data set, the one that gives the shortest description of the data.

In the present article we will show how MDL can be used to select the optimal kernel parameter (problem *b*) when the Relevance Vector kernel machine (RVM) [4,1] is used to optimize $\{w_i\}$ (problem *a*). The same framework presented here can be adopted to select between different kernel functions (problem *c*)

J.R. Dorronsoro (Ed.): ICANN 2002, LNCS 2415, pp. 468–473, 2002.

2   MDL Based Model Selection

2.1   The MDL Principle

Given a data vector $\mathbf{t}$, of length n, with elements discretized with precision δ_t, we want to encode it using a parameterized model $p(\mathbf{t} \mid \boldsymbol{\theta})$, p being a density function and $\boldsymbol{\theta}$ a parameter with k elements discretized with precision δ_θ. The code length needed to describe the data will be at least [3, par. 3.1]:

$$L(\mathbf{t}) = L(\mathbf{t} \mid \boldsymbol{\theta}) + L(\boldsymbol{\theta}) \tag{2}$$

where $L(\mathbf{t} \mid \boldsymbol{\theta}) = -\log p(\mathbf{t} \mid \boldsymbol{\theta}) - n \log \delta_t$ is the ideal code length paid to encode $\mathbf{t}$, and $L(\boldsymbol{\theta}) = -k \log \delta_\theta$ "nats" are needed to encode the k elements of $\boldsymbol{\theta}$.

At this point, we can introduce a prior $p(\boldsymbol{\theta})$ on the value of the parameter $\boldsymbol{\theta}$, as is usual in Bayesian methods; the difference here is that the prior does not reflect any *a priori* knowledge about $\boldsymbol{\theta}$, it is just a part of the coding scheme [3, pp. 54, 67, 84]. Thus, the second term in (2) becomes:

$$L(\boldsymbol{\theta}) = -\log p(\boldsymbol{\theta}) - k \log \delta_\theta \tag{3}$$

The "two-stage" implementation [2] of the MDL principle consists in choosing, for the data $\mathbf{t}$, the model $p(\mathbf{t} \mid \boldsymbol{\theta})$ indexed by the value of $\boldsymbol{\theta}$ which minimizes the data length (2).

2.2   Relevance Vector Machines

In Relevance Vector regression [4,1], the value of $\mathbf{w}$ in (1) is obtained using a *type-II Maximum Likelihood* method, consisting in the introduction of Bayesian prior probabilities of the form:

$$p(w_i \mid \alpha_i) = \mathcal{N}(w_i \mid 0, \alpha_i^{-1}) \tag{4}$$

and consequent optimization of a vector of hyperparameters $\boldsymbol{\alpha}$; for each α_i, a corresponding hyperprior $p(\alpha_i)$ is chosen uniform over a logarithmic scale. The errors on the training set $\mathbf{t}$ are modeled as additive Gaussian noise:

$$p(t_i \mid \hat{y}(x_i)) = \mathcal{N}(t_i \mid \hat{y}(x_i), \sigma_N^2) \tag{5}$$

where $\hat{y}(\mathbf{x}) = \boldsymbol{\Phi}\mathbf{w}$ is the output of the RVM, $\boldsymbol{\Phi}$ being the design matrix, and σ_N^2 represents an automatic estimate of noise variance, whose value is optimized as an additional hyperparameter.

The value of the parameters $(\mathbf{w}, \boldsymbol{\alpha}, \sigma_N^2)$ can be found using Tipping's algorithm [1,4]. Obviously, the obtained values change depending on the choice of the kernel parameter ρ_k, and the machine outputs different estimates $\hat{y}(x_i)$, thus giving different interpretations of training data. We suggest the use of MDL to select among them. Hence we shall evaluate the data length (2) for the RVM, and then choose the value of ρ_k for which a minimum of this quantity is achieved.

The use of a different prior for each parameter does not fit well within the MDL framework, in which the introduced hyperparameters should have a small impact on the data length (2) [2]. Nonetheless, we can use the quantity (2) to compare the compression of information achieved by the RVM using different kernels, or kernel parameters, in a relative manner, thus obtaining a quantitative criterion for model selection, even if absolute performance may be poor. We must also note that, during learning, many α_i overflow the machine precision, so the corresponding weights can be "pruned" [1] by setting them to 0, and do not need to be transmitted nor taken into account in (2).

2.3   Data Length Evaluation

We have to choose between two possible ways of describing the data in the training set using the RVM parameters: in fact we can see the output $\hat{\mathbf{y}}(\mathbf{x})$ as a function of $\mathbf{w}$, thus using $\boldsymbol{\theta} \equiv \mathbf{w}$ in (2), or we can integrate out the weights and use $\boldsymbol{\alpha}$ instead ($\boldsymbol{\theta} \equiv \boldsymbol{\alpha}$). In the first case we obtain:

$$p(\mathbf{t}\,|\,\mathbf{w}) = \mathcal{N}(\mathbf{t}\,|\,\boldsymbol{\Phi}\mathbf{w},\sigma_N^2) = (2\pi\sigma_N^2)^{-n/2}\exp\{-\|\mathbf{t}-\boldsymbol{\Phi}\mathbf{w}\|^2/2\sigma_N^2\} \tag{6}$$

$$p(\mathbf{w}) = \prod_i \int p(w_i\,|\,\alpha_i)p(\alpha_i)d\alpha_i \tag{7}$$

With these quantities, the two terms in Eq. (2) become respectively:

$$L(\mathbf{t}\,|\,\mathbf{w}) = -\log p(\mathbf{t}\,|\,\mathbf{w}) - n\log\delta_t = \frac{n}{2}\log(2\pi\sigma_N^2) + \frac{\|\mathbf{t}-\boldsymbol{\Phi}\mathbf{w}\|^2}{2\sigma^2} - n\log\delta_t \tag{8}$$

$$L(\mathbf{w}) = -\log p(\mathbf{w}) - k\log\delta_w = \sum_i \log|w_i| - k\log\delta_w + O(1) \tag{9}$$

the second term obtained evaluating the integrals in (7) over a logarithmic scale on α_i [1]. Omitting the constant terms we can write:

$$L(\mathbf{t}) = \frac{n}{2}\log(\sigma_N^2) + \frac{\|\mathbf{t}-\boldsymbol{\Phi}\mathbf{w}\|^2}{2\sigma_N^2} + \sum_i \log|w_i| - k\log\delta_w + L(\boldsymbol{\gamma}) \tag{10}$$

where k is the number of relevant weights, $1 < k < k_0$, and $L(\boldsymbol{\gamma})$ is the number of nats needed to transmit *which* of the initial k_0 weights are being used. We will label the quantity (10) *MDL-w*.

The same can be done choosing to transmit $\boldsymbol{\alpha}$ instead of $\mathbf{w}$, whose value, and hence the output $\hat{\mathbf{y}}(\mathbf{x})$, can be easily reconstructed. Also in this case we use the logarithmic scale on α_i [1]. Combination of (4) and (5) yields [1]:

$$p(\mathbf{t}\,|\,\boldsymbol{\alpha}) = \mathcal{N}(\mathbf{t}\,|\,0,\mathbf{C}) = (2\pi)^{-n/2}|\mathbf{C}|^{-1/2}\exp\{-\frac{1}{2}\mathbf{t}^T\mathbf{C}^{-1}\mathbf{t}\} \tag{11}$$

where $\mathbf{C} = \sigma_N^2\mathbf{I}_n + \boldsymbol{\Phi}\mathbf{A}^{-1}\boldsymbol{\Phi}^T$ and $\mathbf{A} = \mathrm{diag}(\boldsymbol{\alpha})$. Then:

$$L(\mathbf{t}\,|\,\boldsymbol{\alpha}) = -\log p(\mathbf{t}\,|\,\boldsymbol{\alpha}) - n\log\delta_t = \frac{1}{2}\log[(2\pi)^n|\mathbf{C}|] + \frac{1}{2}\mathbf{t}^T\mathbf{C}^{-1}\mathbf{t} - n\log\delta_t \tag{12}$$

$$L(\boldsymbol{\alpha}) = -k\log\delta_e \tag{13}$$

where δ_e is the precision used to describe each $\log \alpha_i$, and we eventually obtain the formula for *MDL-α*:

$$L(\mathbf{t}) = \frac{1}{2} \log |\mathbf{C}| + \frac{1}{2} \mathbf{t}^T \mathbf{C}^{-1} \mathbf{t} - k \log \delta_e + L(\boldsymbol{\gamma}) \tag{14}$$

In both methods only relevant parameters are transmitted, so the "relevance flags" $\boldsymbol{\gamma}$ are needed to reconstruct the output. These can be transmitted in various ways: as a bit string of length k_0, which can be compressed using an adaptive Bernoulli model, with $p(1) = k/k_0$ (this yields to $L(\boldsymbol{\gamma}) = \log k_0 - k \log \frac{k}{k_0} - (k_0 - k) \log(1 - \frac{k}{k_0})$ [2]); or as a sequence of $k + 1$ natural numbers from the interval $[1, k_0]$ ($L(\boldsymbol{\gamma}) = (k + 1) \log k_0$ [2]). The first method yields better compression, with $L(\boldsymbol{\gamma})$ always bounded by k_0, but the obtained values are symmetric around $k_0/2$, so large models are not penalized: this is why we adopted the second method, which gives similar results for small values of k but exhibits a linear growth, thus penalizing non-sparse models. A more rigorous approach would have required the use of a Bernoulli model with fixed p, whose value should also have been optimized to minimize the resulting data length [2]. The quantities σ_N^2 and ρ_k should also be transmitted, but their contribution to code length is negligible and can be considered constant.

3   Experiments

To validate the proposed method, a Relevance Vector Machine with Gaussian kernel $K(x_i, x_j, \rho_k) = \exp\{-\|x_i - x_j\|^2/2\rho_k^2\}$ has been trained on different data sets, using a logarithmic grid of values for its parameter (ρ_k), and the MDL quantities described above were evaluated on another (coarser) logarithmic grid of values for δ_w and δ_e. We present two examples here, an artificial problem and a real-world application.

The first consists of 200 equally spaced samples of the function $f(x) = \cos(x^2)$ on the interval [-3,3], with added uniform [-0.1,0.1] noise. A good choice of the parameter is crucial here, as small values would overfit the noise, while large values would cut the high frequency ends. The real data set represents the flux of vehicles on a highway, measured at a constant rate during a whole day, for a total of 720 samples.

We must remark that the MDL indicators (10) and (14) may assume negative values, which may sound strange for a length measure; this is because they are evaluated omitting all constant terms. This is not an issue for our purposes because we are interested in the *location* of the minima and not in the overall compression performance [3, p. 56].

In the first experiment (Fig. 1), both MDL criteria select a single value, $\rho_k = 10^{-0.8}$, which allows the RVM to track the narrow waves without overfitting the noise and corresponds, with good accuracy, to the minimum error with respect to the noiseless original function. The result is obtained using only 25 Relevance Vectors (RVs) out of the original 200 samples. Note that the RVs are placed at the extremes of the interval, while the bias alone is used in the flat zone around the origin (Fig. 1.c, middle).

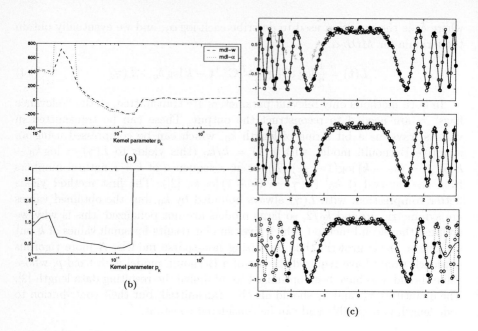

Fig. 1. $\cos(x^2)$ with added uniform noise. a) MDL criteria: MDL-w, $\delta_w = 0.1$ (*slashed line*) and MDL-α, $\delta_\alpha = 1$ (*dotted line*). b) Squared error on noiseless set. c) RVM output for $\rho_k = 10^{-0.9}$ (up), $\rho_k = 10^{-0.8}$ (middle), $\rho_k = 10^{-0.7}$ (low); *circles* are training set points, *black circles* are RVs, *dotted line* is the noiseless $\cos(x^2)$ function.

We have depicted the behavior of the RVM for the two adjacent values of the parameter ρ_k, for comparison. As expected, at a smaller value ($\rho_k = 10^{-0.9}$) the RVM overfits the noise and the number of RVs increases, while at $\rho_k = 10^{-0.7}$ or greater we obtain a sparser model (15 RVs), but the data is underfitted.

With the traffic example (Fig. 2), the MDL curves present two local minima: the lowest one (Fig. 2.b) is located at $\rho_k \approx 50$ (11 RVs), and the second one (Fig. 2.c) at a slightly higher value, in $\rho_k \approx 200$ (5 RVs). The interesting feature of this result is that both outputs provide a valuable interpretation of the physical phenomena underlying the data: the first one allows to spot temporary variations in traffic flow, so is useful for accident detection, while the second one gives an overall idea of the main traffic fluxes during the day.

In these and other experiments, not presented here due to space constraints, different choices of discretization steps δ_w and δ_α did not affect the final outcome of the model selection, resulting just in small vertical shifts of equally shaped curves. In addition, both criteria produced similar curves and selected the same value, or adjacent ones. It must be also noted that MDL-α in some occasions did attribute relevance to a smaller number of vectors, giving coarser results.

4   Conclusions

Both implementations of the MDL criterion, as presented in this work, do perform well, achieving the minimum of the test error, when available, while main-

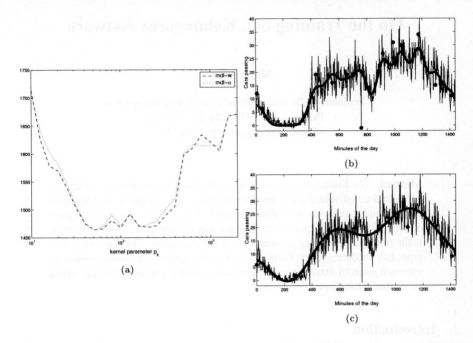

Fig. 2. Traffic data. a) MDL criteria: MDL-w, $\delta_w = 1$ (*slashed line*) and MDL-α, $\delta_\alpha = 1$ (*dotted line*). b) Best fit (*thick line*) of data (*thin line*): $\rho_k = 10^{1.7}$ (*black circles* are RVs). c) Sub-optimal fit: $\rho_k = 10^{2.3}$.

taining a strong preference for sparse models. A major drawback of the presented method is, obviously, the large computational requirement. The "brute force" technique adopted, which requires the exhaustive training of the RVM on the entire grid of different values for the parameter, could be replaced by heuristic methods. These may allow the extension of this method to multi-parameter kernels.

References

[1] Tipping, M.E.: Sparse Bayesian Learning and the Relevance Vector Machine. Journal of Machine Learning Research, 1 (2001) 211–244.

[2] Hansen, M., Yu, B.: Model Selection and the Principle of Minimum Description Length. Technical Memorandum, Bell Labs, Murray Hill, N.J. (1998),
http://cm.bell-labs.com/who/cocteau/papers/index.html

[3] Rissanen, J.: Stochastic Complexity in Statistical Inquiry. World Scientific Publishing Co., Singapore (1989).

[4] Tipping, M.E.: The relevance vector machine. Advances in Neural Information Processing Systems 12, The MIT Press (2000) 652–658.

On the Training of a Kolmogorov Network

Mario Köppen

Fraunhofer IPK Berlin, Department Pattern Recognition
Pascalstr. 8–9, 10587 Berlin, Germany
mario.koeppen@ipk.fhg.de

Abstract. The Kolmogorov theorem gives that the representation of continuous and bounded real-valued functions of n variables by the superposition of functions of one variable and addition is always possible. Based on the fact that each proof of the Kolmogorov theorem or its variants was a constructive one so far, there is the principal possibility to attain such a representation. This paper reviews a procedure for obtaining the Kolmogorov representation of a function, based on an approach given by David Sprecher. The construction is considered in more detail for an image function.

1 Introduction

This paper gives a study of the superposition of a continuous, bounded real-valued function of n variables by the superposition of functions of one variable and addition. The universal possibility of such a representation is granted by the Kolmogorov theorem [1]. It was Hecht-Nielsen [2], who rediscovered the importance of the Kolmogorov theorem for the theoretical understanding of the abilities of neural networks. The Kolmogorov theorem was also pointed out to be of importance for other designs of soft computing, as e.g. normal forms in fuzzy logic [3] [4]. Here, we focus on the constructive aspects of this representation, thus following a line of studies presented so far e.g. in [5], [6], [7] and [8].

The starting point is the Kolmogorov theorem in a notation used by Sprecher:

Theorem 1 (Sprecher, 1996). *Every continuous function* $f : I^n \to R$ *can be represented as a sum of continuous real-valued functions:*

$$f(x_1, \ldots, x_n) = \sum_{q=0}^{2n} \chi_q(y_q). \tag{1}$$

In this representation, the $x_1, \ldots, x_n$ *are the parameters of an embedding of* I_n *into* R_{2n+1}:

$$y_q = \eta_q(x) = \sum_{p=1}^{n} \lambda_p \psi(x_p + qa) \tag{2}$$

with a continuous real-valued function ψ *and suitable constants* λ_p *and* a. *This embedding is independent of* f.

J.R. Dorronsoro (Ed.): ICANN 2002, LNCS 2415, pp. 474–479, 2002.

In [7] and [8], Sprecher gives a numerical procedure for computing the "inner" function ψ and setting the "outer" function χ. However, the inner function given easily shows to be non-continuous, so the definition presented here has been corrected accordingly in order to give a continuous ψ function.

This paper is organized as follows: section 2 describes the ingredients needed for performing the construction. Then, section 3 shows how this procedure might be used to gain understanding of a given function f (an image function). The paper concludes with a discussion.

2 The Revised Sprecher Algorithm

2.1 The Inner Function

Be $n \geq 2$ the dimensionality of the function f, and $m \geq 2n$ the number of terms in eq. (1). Also, be $\gamma \geq m + 2$ a natural number, which is used as radix in the following (a good choice for $n = 2$ is $m = 5$ and $\gamma = 10$). A constant a is given by $a = (\gamma(\gamma-1))^{-1}$, and x will asign a vector $(x_1, \dots , x_n)$. For the decimal base, a has the representation $0.0111 \dots$.

One further definition: $\beta(r) = (n^r - 1)/(n - 1) = 1 + n + n^2 + \dots + n^{(r-1)}$ (e.g. for $n = 2$ the sequence $1, 3, 7, 15, 31, \dots$).

The constants λ_p in theorem 2 will be computed by the expression

$$\lambda_p = \begin{cases} 1 & p = 1 \\ \sum_{r=1}^{\infty} \gamma^{-(p-1)\beta(r)} & p > 1 \end{cases} \tag{3}$$

Using those definitions, the inner function $\psi(x)$ can be defined. The construction is given for all terminating rational numbers $d_k \in I$, which have in the decimal notation to the base γ not more than k digits (as $0.031, 0.176, 0.200$, if $\gamma = 10$ and $k = 3$). The notation $d_k = [i_1, \dots , i_k]_\gamma$ means that d_k to the base γ has the notation $0.i_1 i_2 \dots i_k$. Then, the inner function is defined as follows:

$$\psi_k(d_k) = \begin{cases} d_k & \text{for } k = 1; \\ \psi_{k-1}(d_k - \frac{i_k}{\gamma^k}) + \frac{i_k}{\gamma^{\beta(k)}} & \text{for } k > 1 \text{ and } i_k < \gamma - 1; \\ \frac{1}{2}\left(\psi_k(d_k - \frac{1}{\gamma^k}) + \psi_{k-1}(d_k + \frac{1}{\gamma^k})\right) & \text{for } k > 1 \text{ and } i_k = \gamma - 1. \end{cases} \tag{4}$$

It can be easily seen that this recursive definition always terminates.

Figure 1 gives the graph of $\psi(x)$ for $n = 2$ und $\gamma = 10$. It features some kind of structural self-similarity. There are two self scalings. The section a with increase α is replaced by γ segments, with the first $(\gamma - 2)$ segments having identical increase $\alpha_1 < \alpha$ and going to 0, and the last 2 segments having increase $\alpha_2 > \alpha$ going to ∞ for growing k. The "switching point", where the two replacement schemes change, approximates the point $d_k + \delta_k$ with

$$\delta_k = \frac{\gamma - 2}{\gamma - 1}\gamma^{-k} = (\gamma - 2)\sum_{r=k+1}^{\infty}\gamma^{-k}. \tag{5}$$

The function ψ_x has some mathematical features, as being continuous, strict monotone increasing, being flat nearly everywhere and concave.

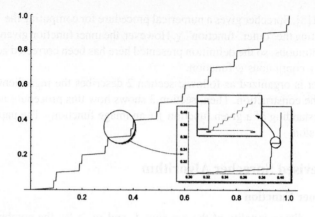

Fig. 1. Graph of the inner function ψ_x for $n = 2$ and $\gamma = 10$.

2.2 Inner Superposition

Now, we have to consider the superposition of the inner functions, as given by

$$\eta_q(\mathbf{d}_k) = \sum_{p=1}^{n} \lambda_p \psi(d_{kp} + qa) \tag{6}$$

for a fixed $0 \le q \le m = 2n$. Here, $\mathbf{d}_k$ shall be the vector composed of n rational numbers from the d_{kp}.

This superpositions has an important property: For two arbitrary vectors $\mathbf{d}_k, \mathbf{d}_k' \in I^n$, the distance of their images under $\eta()$ is never smaller than $\gamma^{-n\beta(k)}$.

2.3 The Outer Function

The construction of the outer function is basically a look-up. Each value from I_n is assigned a different "height" under the mapping η. Now, the values of f at point $x \in I_n$ gives the value of χ at the height assigned to x by η.

However, this would not suffice to gain an exact representation of f for growing k. This is, where the $m \ge 2n + 1$ shifted versions of η come into play, but the final proof can not be given here.

If such a look-up is made for all $\mathbf{d}_k$, the function χ has to extended for being defined over its full domain. This is done by the help of fuzzy singletons. Be

$$\Gamma_k = \frac{1}{2} \left[\frac{1}{\gamma^{\beta(k+1)}} - \frac{1}{\gamma^{n\beta(k+1)}} \right]. \tag{7}$$

Then, a fuzzy singleton is designed around $\eta(\mathbf{d}_k)$ by

$$\begin{aligned}
\theta(\mathbf{d}_k; x) = &\; \sigma[\Gamma_k^{-1}(x - \eta(\mathbf{d}_k)) + 1] \\
&- \sigma[\Gamma_k^{-1}(x - \eta(\mathbf{d}_k)) - (\gamma - 2)b_k)]
\end{aligned} \tag{8}$$

Here, $\sigma(x)$ is a continuous function with $\sigma(x) \equiv 0$ for $x \leq 0$ and $\sigma(x) \equiv 1$ for $x \geq 1$.

From the minimum separation of images of η, it can be seen that the fuzzy singletons around $\eta(\mathbf{d}_k)$ are not overlapping.

2.4   The Sprecher Algorithm

Based on the constructions given in the foregoing sections, the Sprecher algorithm gives an iterative procedure for approximating the representation of a given function f by the Kolmogorov theorem.

The values $\epsilon > 0$ and $\delta > 0$ being choosen such that they fulfill

$$0 < \frac{m - n - 1}{m + 1} \epsilon + \frac{2n}{m + 1} \leq \delta < 1 \tag{9}$$

Therefrom $\epsilon < 1 - (n/(m - n + 1))$ (e.g. $\epsilon < 1/3$ for $n = 2$).

The algorithm starts with $f_0 \equiv f$ and the number $k_0 = 0$. Then, for each $r \geq 1$, from f_{r-1} a function f_r is computed in four layers. The sum of all functions f_r converges to f, thus approximating a representation of f.

I.   Layer 1 computes all $\psi(x)$.

Function $f_{r-1}(x)$ with $x \in I^n$ is known. Now, a natural number $k_r > k_{r-1}$ is determined fulfilling that for all $p = 1, \ldots, n$ from $|x_p - y_p| \leq \gamma^{-k_r}$ it follows that $|f_{r-1}(x) - f_{r-1}(y)| \leq \epsilon \|f_{r-1}\|$. Via

$$\mathbf{d}_{k_r}^q = \mathbf{d}_{k_r} + q \sum_{r=2}^{k_r} \frac{1}{\gamma^r} = \mathbf{d}_{k_r} + q a_{k_r} \tag{10}$$

we obtain the values $\mathbf{d}_{k_r}^q$ for $q = 0, \ldots, m$. Now, we can compute the values $\psi(\mathbf{d}_{k_r}^q)$ from eq. (4).

II.   Layer 2 computes the linear combinations $\eta_q(x)$ of $\psi(x)$.

This is achieved by the equation

$$\eta_q(\mathbf{d}_{k_r}^q) = \sum_{p=1}^{n} \lambda_p \psi(d_{k_r p}^q + qa). \tag{11}$$

III.   In Layer 3 the values $\chi_q[\eta_q(x)]$ are computed.

The $m + 1$ functions of one variable $\chi_q^r(y)$ are given by

$$\chi_q^r(y) = \frac{1}{m + 1} \sum_{\mathbf{d}_{k_r}^q} f_{r-1}(\mathbf{d}_{k_r}) \theta(\mathbf{d}_{k_r}^q; y) \tag{12}$$

The y-values are substituted by the η_q-values of layer 2:

$$\chi_q^r[\eta_q(x)] = \frac{1}{m + 1} \sum_{\mathbf{d}_{k_r}^q} f_{r-1}(\mathbf{d}_{k_r}) \theta(\mathbf{d}_{k_r}^q; \eta_q(x)). \tag{13}$$

IV. In Layer 4 the $f_r(x)$ are computed as linear combination of the $\chi_q[\eta_q(x)]$.
This is done by computing

$$f_r(x) = f_{r-1}(x) - \sum_{q=0}^{m} \chi_q^r[\eta_q(x)]. \tag{14}$$

This ends the rth iteration step.

3 Image Function

The procedure may be applied for the representation of image functions. From the Sprecher algorithm, the following algorithm can be derived for getting the representation of an image subsampled for e.g. $k = 2$ and resampling it for a higher k, as e.g. $k = 3$:

I. **(Offline-Phase)** For all $d_{31}, d_{32} = 0.000, \dots, 0.999$ and $q = 0, \dots, 4$:

1. Compute

$$\eta_3 = \eta(\mathbf{d}_3 + qa) = \eta(d_{31} + qa, d_{32} + qa).$$

2. Find a pair $\mathbf{d}_2$ and $\mathbf{d}_2'$ of points from D_2, which are neighbors in the ranking induced by η in D_2 and fulfilling:

$$\eta(\mathbf{d}_2) \le \eta_3 \le \eta(\mathbf{d}_2').$$

3. Now, be $d^q = (\mathbf{d}_x^q, \mathbf{d}_y^q)$ the value from $\mathbf{d}_2$ and $\mathbf{d}_2'$, for which in case $\theta(\mathbf{d}_2; \eta_3)\theta(\mathbf{d}_2'; \eta_3) \ne 0$ holds that $\theta(d^q; \eta_3) \ne 0$, otherwise one of both values is arbitraily choosen .

4. Enter into a table T_{q23} at position (d_{31}, d_{32}) the triple $(d_x^q - 0.01, d_y^q - 0.01, \theta(d^q; \eta_3))$.

II. **(Online-Phase)** A result array I_r of 1000×1000 positions is initialised with values 0. For all $d_{31}, d_{32} = 0.000, \dots, 0.999$ and $q = 0, \dots, 4$ then:

1. Get the value (d_x^q, d_y^q, θ) in T_{q23} at position (d_{31}, d_{32}).
2. Add the value

$$\frac{f(d_x^q, d_y^q)}{5} \cdot \theta$$

to the value at position (d_{31}, d_{32}) in I_r.

Figure 2 shows the result of resampling the Lena image from the representation of its image function for $k = 2$ at the next level $k = 3$. The right subfigure illustrates the manner how the Sprecher algorithm for one value of k attempts to generalize the unknown parts of the image function by placing fuzzy singletons around each already looked up position of the unit square. This gives the granular structure of the resampled image.

Fig. 2. Resampling of the Lena image according to the Kolmogorov theorem for $k = 3$ at all positions from D_2.

4   Summary

The construction of a representation of functions of n variables by superposition of functions of one variable and addition was considered in this paper in more detail. Following the approach given by David Sprecher, an algorithm was given for approximating such a representation. The generalization of one approximation step to the next was considered by providing the technical procedure of such a computation for an image (the Lena image).

References

1. Kolmogorov, A.N.: On the representation of continuous functions of several variables by superpositions of continuous functions of one variable and addition. Doklady Akademii Nauk SSSR **114** (1957) 679–681 (in Russian).
2. Hecht-Nielsen, R.: Kolmogorov's mapping neural network existence theorem. In: Proceedings of the First International Conference on Neural Networks. Volume III., IEEE Press, New York (1987) 11–13
3. Kreinovich, V., Nguyen, H., Sprecher, D.: Normal forms for fuzzy logic – an application of kolmogorov's theorem. Technical Report UTEP-CS-96-8, University of Texas at El Paso (1996)
4. Nguyen, H., Kreinovich, V.: Kolmogorov's theorem and its impact on soft computing. In Yager, R.R., Kacprzyk, J., eds.: The Ordered Weighted Averaging operators. Theory and Applications. Kluwer Academic Publishers (1997) 3–17
5. Sprecher, D.: On the structure of continuous functions of several variables. Transcations of the American Mathematical Society **115** (1965) 340–355
6. Cotter, N.E., Guillerm, T.J.: The CMAC and a theorem of Kolmogorov. Neural Networks **5** (1992) 221–228
7. Sprecher, D.: A numerical implementation of Kolmogorov's superpositions I. Neural Networks 9 (1996) 765–772
8. Sprecher, D.: A numerical implementation of Kolmogorov's superpositions II. Neural Networks 10 (1997) 447–457

A New Method of Feature Extraction and Its Stability

Nojun Kwak and Chong-Ho Choi

School of Electrical Eng., ASRI, Seoul National University
San 56-1, Shinlim-dong, Kwanak-ku, Seoul 151-742, KOREA
{triplea, chchoi}@csl.snu.ac.kr

Abstract. In the classification on a high dimensional feature space such as in face recognition problems, feature extraction techniques are usually used to overcome the so called 'curse of dimensionality.' In this paper, we propose a new feature extraction method for classification problem based on the conventional independent component analysis. The local stability of the proposed method is also dealt with. The proposed algorithm makes use of the binary class labels to produce two sets of new features; one that does not carry information about the class label – these features will be discarded – and the other that does. The advantage is that general ICA algorithms become available to a task of feature extraction by maximizing the joint mutual information between class labels and new features, although only for two-class problems. Using the new features, we can greatly reduce the dimension of feature space without degrading the performance of classifying systems.

1 Introduction

Feature extraction is very important in many applications and subspace methods such as principle component analysis (PCA) and Fisher's linear discriminant analysis (FLD) have been used for this purpose. Recently, in neural networks and signal processing communities, independent component analysis (ICA), which was devised for blind source separation problems, has received a great deal of attention because of its potential applications in various areas. The advantage of ICA is that it uses higher order statistics, while PCA and FLD use second order statistics. But it leaves much room for improvement since it does not utilize output class information.

In this paper, we propose a new ICA-based feature extraction algorithm (ICA-FX) for classification problem which is an extended version of [1]. It utilizes the output class information in addition to having the advantages of the original ICA method. This method is well-suited for classification problems in the aspect of constructing new features that are strongly related to output class. By combining the proposed algorithm with existing feature selection methods, we can greatly reduce the dimension of feature space while improving classification performance.

J.R. Dorronsoro (Ed.): ICANN 2002, LNCS 2415, pp. 480–485, 2002.
© Springer-Verlag Berlin Heidelberg 2002

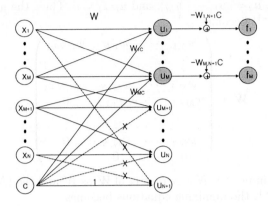

Fig. 1. Structure of ICA-FX

This paper is organized as follows. In Section 2, we propose a new feature extraction algorithm. In Section 3, local stability condition of the proposed algorithm is presented. Simulation results are presented in Section 4 and Conclusions follow in Section 5.

2   Algorithm: ICA-FX

Suppose that there are N zero-mean normalized input features $\boldsymbol{x} = [x_1, \cdots, x_N]^T$ and a binary output class $c \in \{-1, 1\}$. Our purpose of the feature extraction is to find $M(\leq N)$ new features $\boldsymbol{f}_a = [f_1, \cdots, f_M]^T$ from $\boldsymbol{x}$ containing maximal information of the class c.

Using the Fano's inequality and data processing inequality in information theory [2], if we restrict our attention on linear transformations from $\boldsymbol{x}$ to $\boldsymbol{f}_a$, the feature extraction problem can be restated as the following optimization problem:

(FX problem.) Find $K_a \in \Re^{M \times N}$ that maximize $I(c; \boldsymbol{f}_a)$, where $\boldsymbol{f}_a = K_a \boldsymbol{x}$. Here, $I(c; \boldsymbol{f}_a)$ denotes the mutual information between the class label c and the feature vector $\boldsymbol{f}_a$.

To solve this problem, we interpret feature extraction problem in the structure of the blind source separation (BSS) problem in the following.

(Mixing.) Assume that there exist N independent sources $\boldsymbol{s} = [s_1, \cdots, s_N]^T$ which are also independent of class label c and the observed features $\boldsymbol{x}$ is the linear combination of the sources $\boldsymbol{s}$ and c with the mixing matrix $A \in \Re^{N \times N}$ and $\boldsymbol{b} \in \Re^{N \times 1}$ such that

$$\boldsymbol{x} = A\boldsymbol{s} + \boldsymbol{b}c. \tag{1}$$

(Unmixing.) Our unmixing stage is a little bit different from the BSS problem and this is shown in Fig. 1. Here, $\boldsymbol{x}$ is fully connected to $\boldsymbol{u} = [u_1, \cdots, u_N]$,

c is connected to $\boldsymbol{u}_a = [u_1, \cdots, u_M]$, and $u_{N+1} = c$. Thus, the unmixing matrix $\boldsymbol{W} \in \Re^{(N+1)\times(N+1)}$ becomes

$$
\boldsymbol{W} = \begin{pmatrix}
w_{1,1} & \cdots & w_{1,N} & w_{1,N+1} \\
\vdots & & \vdots & \vdots \\
w_{M,1} & \cdots & w_{M,N} & w_{M,N+1} \\
w_{M+1,1} & \cdots & w_{M+1,N} & 0 \\
\vdots & & \vdots & \vdots \\
w_{N,1} & \cdots & w_{N,N} & 0 \\
0 & \cdots & 0 & 1
\end{pmatrix}. \tag{2}
$$

If we denote the upper left $N \times N$ matrix of $\boldsymbol{W}$ as W and the $(N+1)$th column of $\boldsymbol{W}$ as $\boldsymbol{v} \in \Re^{N \times 1}$, the unmixing equations becomes

$$
\boldsymbol{u} = W\boldsymbol{x} + \boldsymbol{v}c. \tag{3}
$$

Suppose we have made $\boldsymbol{u}$ somehow equal to the scaled and permuted version of source $\boldsymbol{s}$, i.e.,

$$
\boldsymbol{u} = \Lambda \Pi \boldsymbol{s}, \tag{4}
$$

where Λ is a diagonal matrix corresponding to appropriate scale and Π is a permutation matrix. Then, u_i's $(i = 1, \cdots, N)$ are independent of class c, and among the elements of $\boldsymbol{f} = W\boldsymbol{x}(= \boldsymbol{u} - \boldsymbol{v}c)$ $\boldsymbol{f}_a = [f_1, \cdots, f_M]^T$ will contain the same amount of the mutual information with class c as original features $\boldsymbol{x}$ have, i.e.,

$$
I(c; \boldsymbol{f}_a) = I(c; \boldsymbol{x}). \tag{5}
$$

if the unmixing matrix $\boldsymbol{W}$ be nonsingular. Thus, we can extract $M(< N)$ dimensional new feature vector $\boldsymbol{f}_a$ by a linear transformation of $\boldsymbol{x}$ containing the maximal information about the class if we can make (4) hold.

Theorem 1. *In the structure shown in Fig. 1, we can obtain independent u_i's $(i = 1, \cdots, N)$ that are also independent of c by the learning rule*

$$
\begin{aligned}
W^{(t+1)} &= W^{(t)} + \mu_1[I_N - \boldsymbol{\varphi}(\boldsymbol{u})\boldsymbol{f}^T]W^{(t)} \\
\boldsymbol{v}^{(t+1)} &= \boldsymbol{v}^{(t)} - \mu_2\boldsymbol{\varphi}(\boldsymbol{u}_a)c.
\end{aligned} \tag{6}
$$

Here $W_{N+1} = [w_{1,N+1}, \cdots, w_{M,N+1}]^T \in \Re^M$, $\boldsymbol{\varphi}(\boldsymbol{u}) = [\varphi_1(u_1), \cdots, \varphi_N(u_N)]^T$, $\boldsymbol{\varphi}(\boldsymbol{u}_a) = [\varphi_1(u_1), \cdots, \varphi_M(u_M)]^T$, where $\varphi_i(u_i) = -\frac{dp_i(u_i)}{du_i}/p_i(u_i)$, $\boldsymbol{f} = W\boldsymbol{x}$, I_N is a $N \times N$ identity matrix, and μ_1 and μ_2 are learning rates that can be set differently.

Proof. If we assume that $u_1, \cdots, u_N, u_{N+1}$ are independent of each other, the log likelihood of the given data becomes

$$
L(\boldsymbol{u}, c, \boldsymbol{W}) = \log|\det \boldsymbol{W}| + \sum_{i=1}^{N} \log p_i(u_i) + \log p(c), \tag{7}
$$

because

$$p(\boldsymbol{x}, c) = |\det \boldsymbol{W}| \, p(\boldsymbol{u}, u_c) = |\det \boldsymbol{W}| \prod_{i=1}^{N} p_i(u_i) \, p(c). \qquad (8)$$

Using the maximum likelihood estimation criterion, we are to maximize L, and this can be achieved by the steepest ascent method. Because the last term in (7) is a constant, differentiating (7) with respect to $\boldsymbol{W}$ leads to

$$\frac{\partial L}{\partial w_{i,j}} = \frac{adj(w_{j,i})}{|\det \boldsymbol{W}|} - \varphi_i(u_i) x_j \qquad i = 1, \cdots, N, \ j = 1, \cdots, N$$

$$\frac{\partial L}{\partial v_i} = -\varphi_i(u_i) c \qquad i = 1, \cdots, M \qquad (9)$$

where $adj(\cdot)$ is adjoint and $\varphi_i(u_i) = -\frac{dp_i(u_i)}{du_i}/p_i(u_i)$. Note that c has categorical values.

We can see that $|\det \boldsymbol{W}| = |\det W|$ and $\frac{adj(w_{j,i})}{|\det \boldsymbol{W}|} = W_{i,j}^{-T}$. Thus the learning rule becomes

$$\Delta W \propto W^{-T} - \boldsymbol{\varphi}(\boldsymbol{u}) \boldsymbol{x}^T$$

$$\Delta \boldsymbol{v} \propto -\boldsymbol{\varphi}(\boldsymbol{u}_a) c. \qquad (10)$$

Since the two terms in (10) have different tasks regarding the update of separate matrix W and W_{N+1}, we can divide the learning process and applying natural gradient on updating W, we get (6).

Note that the learning rule for W is the same as the original ICA learning rule [3], and also note that $\boldsymbol{f}_a$ corresponds to the first M elements of $W\boldsymbol{x}$. Therefore, we can extract the optimal features by the proposed algorithm when it find the optimal solution by (6).

3   Stability Conditions

The local stability analysis in this paper undergoes almost the same procedure as that of general ICA algorithms in [4]. In doing so, to cope with troublesome terms, we modify the notation of the weight update rule for $\boldsymbol{v}$ in (6) a little bit with the following:

$$v_i^{(t+1)} = v_i^{(t)} - \mu_i^{(t)} \varphi_i(u_i) c v_i^* v_i^{(t)}, \quad i = 1, \cdots, M \qquad (11)$$

where v_i^* is the optimal value for v_i. The learning rate $\mu_i^{(t)} (> 0)$ changes over time t and i, appropriately chosen to equalize $\mu_i^{(t)} v_i^* v_i^{(t)} = \mu_2$. This modification is justified because $v_i^* v_i^{(t)} (\simeq v_i^{*2})$ is positive when $v_i^{(t)}$ is near the optimal value v^*. In weight update rule (11) along with the update rule for W in (6), the local stability condition is obtained in the following theorem.

Theorem 2. *The stability of the stationary point of the proposed algorithm*

$$W = \Lambda \Pi A^{-1}, \quad v = -\Lambda \Pi A^{-1} b \tag{12}$$

is governed by the nonlinear moment

$$\kappa_i = E\dot{\varphi}_i(e_i) E e_i^2 - E\varphi_i(e_i)e_i, \tag{13}$$

and it is stable if

$$1 + \kappa_i > 0, \quad 1 + \kappa_j > 0, \quad (1 + \kappa_i)(1 + \kappa_j) > 1 \tag{14}$$

for all (i, j) pair. Thus the sufficient condition is

$$\kappa_i > 0, \quad 1 \le i \le N. \tag{15}$$

Here, $E(\cdot)$ is the expectation and $e = [e_1, \cdots, e_N]^T$ is the scaled and permuted version of s ($e = \Lambda \Pi s$) which is estimated by u.

The proof of the algorithm is omitted because of the space limitation. Note that this is the same as the stability condition for standard ICA in [4].

4   Simulation Results

(Simple problem.) Suppose we have two input features x_1 and x_2 uniformly distributed on [-1,1] for a binary classification, and the output class y is determined as follows:

$$y = \begin{cases} 0 & \text{if } x_1 + x_2 < 0 \\ 1 & \text{if } x_1 + x_2 \ge 0 \end{cases}$$

Here, $y = 0$ corrsponds to $c = -1$ and $y = 1$ corresponds to $c = 1$.

This problem is linearly separable, and we can easily distinguish $x_1 + x_2$ as an important feature. But feature extraction algorithms based on conventional unsupervised learning, such as the conventional PCA and ICA, cannot extract $x_1 + x_2$ as a new feature because they only consider the input distribution; i.e., they only examine (x_1, x_2) space.

For this problem, we performed ICA-FX with $M = 1$ and could get $u_1 = 43.59x_1 + 46.12x_2 + 36.78y$ from which a new feature $f_1 = 43.59x_1 + 46.12x_2$ is obtained.

(Pima indian diabetes.) This data set consists of 768 instances in which 500 are class 0 and the other 268 are class 1. It has 8 numeric features with no missing value. This dataset is used in [5] to test their feature extraction algorithm MMI. We have compared the performance of ICA-FX with those of LDA, PCA, and MMI.

As in [5], we used LVQ-PAK[1] to train the data and separated the data into 500 training data and the other 268 as test data. The meta-parameters of LVQ-PAK were set to the same as those in [5].

---

[1] http://www.cis.hut.fi/research/software.shtml.

In Table 1, classification performances are presented. The classification rates of MMI, LDA, and PCA are from [5]. As shown in the table, the performance of ICA-FX is about 3~10% better than those of the other methods. Even with one feature, the classification rate is as good as the cases where more features are used.

Table 1. Classification performance for Pima data

No. of	Classification performance (%) (LVQ-PAK)			
features	PCA	LDA	MMI	ICA-FX
1	64.4	65.8	72.0	81.3
2	73.0	–	77.5	81.7
4	74.1	–	78.5	81.5
6	74.7	–	78.3	81.0
8	74.7	–	74.7	81.0

5   Conclusions

In this paper, we have proposed an algorithm for feature extraction. The proposed algorithm is based on ICA and can generate appropriate features for classification problems.

In the proposed algorithm, we added class information in training ICA. The added class information plays a critical role in the extraction of useful features for classification. With the additional class information we can extract new features containing maximal information about the class. The number of extracted features can be arbitrarily chosen. We presented the justification of the proposed algorithm.

Since it uses the standard feed-forward structure and learning algorithm of ICA, it is easy to implement and train. Experimental results show that the proposed algorithm generates good features that outperform the original features and other features extracted from other methods for classification problems. The proposed algorithm is useful for two-class problems, and more works are needed to extend the proposed method for multi-class problems.

References

1. N. Kwak, C.-H. Choi, and C.-Y. Choi, "Feature extraction using ica," in *Proc. Int'l Conf. on Artificial Neural Networks 2001*, Vienna Austria, Aug. 2001, pp. 568–573.
2. T.M. Cover, and J.A. Thomas, *Elements of Information Theory*, John Wiley & Sons, 1991.
3. A.J. Bell and T.J. Sejnowski, "An information-maximization approach to blind separation and blind deconvolution," *Neural Computation*, vol 7, pp. 1129–1159, 1995.
4. J.-F. Cardoso, "On the stability of source separation algorithms," *Journal of VLSI Signal Processing Systems*, vol. 26, no 1/2, pp. 7–14, Aug. 2000,
5. K. Torkkola and W.M. Campbell, "Mutual information in learning feature transformations," in *Proc. Int'l Conf. Machine Learning*, Stanford, CA, 2000.

Visualization and Analysis of Web Navigation Data

Khalid Benabdeslem[1], Younes Bennani[1], and Eric Janvier[2]

[1]LIPN – CNRS, University of Paris 13
99 Av Jean Baptiste Clément - 93430 Villetaneuse, France
{kbe, younes}@lipn-univ.paris13.fr
[2]NumSight, Champ Montant 16 C 2074 Marin – Switzerland
e.janvier@numsight.com

Abstract. In this paper, we present two new approaches for the analysis of web site users behaviors. The first one is a synthetic visualization of Log file data and the second one is a coding of sequence based data. This coding allows us to carry out a vector quantization, and thus to find meaningful prototypes of the data set. For this, first the set of sessions is partitioned and then a prototype is extracted from each of the resulting classes. This analytic process allows us to categorize the different web site users behaviors interested by a set of categories of pages in a commercial site.

1 Introduction

Clustering represents a fundamental and pratical methodology for the comprehension of great sets of data. For multivariate data, we can use geometrical distance concepts in an euclidian space, but for data consisting of sequences (or images for example), there is no geometric equivalent [5].

Traditionally, two general techniques are used to deal with this problem. The first approach is to reduce all heterogeneous data to one fixed vectorial representation. For example, vectorial space representation is largely used in information extraction for the representation of textual documents having different formats and different length and histograms are used to resume web access sequences. Practically, this technique is useful and has the advantage of being relatively simple. It makes it possible to profit from clustering algorithms advantages in vector space. One second approach of clustering is to use a defined distance between each pair of individuals. Given N objects, we generate a N^2 distances matrix. There is a significant number of clustering algorithms which can use this matrix (ex: AHC). However, for a great data base, time and space complexity becomes increasingly delicate for such algorithms. In this paper, we choose the first technique by using one coding method that we propose to reorganize sequence based data.

J.R. Dorronsoro (Ed.): ICANN 2002, LNCS 2415, pp. 486–491, 2002.
© Springer-Verlag Berlin Heidelberg 2002

2   Data Base

We would like to describe a method of exploratory analysis of dynamic behaviors for web site visitors. To validate our work, we use the access log representing thousands of individuals in a commercial web site.

Therefore, our data base includes individuals which represent web site visitors, in which each one is supposed to be interested by a set of pages. Each user (session) will be represented by a succession of pages[4].

The final sentence of a table caption should end without a period

Session	Pages
S_1	2 7 4 1 5
S_2	8 4 32
S_3	3 9 23 20

3   Coding

Actually, the sites are dynamic. That means that the pages are not characterized by a fixed variables as: hierarchical address (URL), the content ....etc. They are represented by a numerical identifiers which have no meaning. For that reason, we have elaborated a novel method of coding sessions from the Log file which consists in characterizing a page by its passage importance, i.e. by its weight of precedence and succession relative to the other pages of the site which appear in the log file.

The principle is to calculate for each page its frequency of precedence and succession over all the other pages and to regroup these frequencies in one matrix : Pages × (previous Pages + next Pages) that we call: quasi-behavioral matrix. We remarked that this coding has a serious problem of size. In fact, if we don't have a enough pages in the site, we can not have a table with enough data to get a robust model. To solve this problem, we proposed to slide the matrix along the log file. In other words, we computed a quasi-behavioral matrix for each time period. This method allowed us to increase the number of samples and to have several examples for each page.

Example:

Let's suppose we have a Log file with 40 URLs over 30 days.

If we slide the matrix along the month day by day, we get a 40 × 30, ie 1200, samples base. We not only characterize the pages by behavioral variables i.e. as the web site visitors perceive the site, but we increase the information about the pages by coding their behavioral characteristics in detail, day by day.

For example, for URL_i , we can code its vector over all the others URLs as follow:

In the kth day of the file, the URL_i appeared a times as an entry page, preceded b times by the URL_j, succeeded c times by URL_l and appeared d times as an exit page.

4 Neural Clustering for the Quasi-Behavioral Visualization

We remember that the log files have an important size and contains relevant information about the web site visitors behaviors.

For the profiles extraction, we use Kohonen's topological map algorithm [8] in order
to group the similar pages[8], then we apply an ascendant hierarchical clustering to
optimize the classes. Therefore, we talk about clusters of neurons.

The obtained clusters need to be labeled. However, at the time of use of the map, each
URL can activate more than one neuron, depending on its passage importance change
day after day. Theoretically, if the map is well done, the different neurons which are
activated by the same page must be close to one another.

Practically, to label the map in this case, we proceed by a majority vote. We have
applied this process to a log file of a commercial web site (www.123credit.com) having a size of 1 Giga byte, and got the following map:

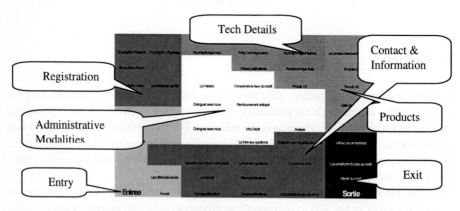

Fig. 1. Mapping of a commercial site (www.123credit.com)

5 Sessions Coding

Instead of representing the sessions by the succession of visited pages, we represent
them by the succession of the class of pages resulting from the unsupervised clustering. This coding reduces the variety of the traffic but it doesn't solve the problem of
variability of the length. In fact, several problems appear in the treatment of sequence
based data :

The presence of "holes" (missing data)

The complex variability of the length

The symbolic nature of the problem which implies a particular study about the similarity of the individuals. In order to solve this problem, we have developed a new
method which allows to code the set of sessions in a table of individuals versus variables, whose number of columns is fixed and without any missing data.

The principle of this coding, that we called "position weight" consists in awarding a weight to each category according to its position in the session : the further a category appears in a session, the more its weight increases. By the way, the coding also includes the frequency of the category in the session. The algorithmic form of this coding is the following :

Coding of "position weight" algorithm

```
// Initialization

For i = 1 to N do

  Weight(i) = 0

  Frequency(i)=0

End for

// position weight and  frequencies

For j ∈ session do

  If  j = 1st category of the session then

    Weight(j) = α //α : is an initial parameter

  End if

  If  Frequency(i) = 0 then

    Weight(j) = Weight(Previous(j)) + ΔW

  Else

    Weight(j) = [Weight(j)+(Weight(Prev(j)) + ΔW)]/2

    Frequency(i) = Frequency(i) + 1

  End if

End for
```

Therefore, to each category, is attributed, for one given session, its position weight and its frequency. The sessions are thus coded in a table whose lines represent the users (visitors) and with two columns for each category (so that the total number of columns is twice the number of categories): one for the "position weight" and the other for the frequency.

6   Vector Quantization for the Behavioral Visualization

The representation of the sessions set of the log file on the map is not easily under-stood.

The vector quantization (QV) has been introduced to create statically represented and economically stored references. The basic idea of this technique results from the fact that in the session based representation space, the coding vectors occupy only a sub spaces [6].

This quantization consists in partitioning the coded sessions in a set of classes and extracting a prototype from each of these classes.

The partitionning is done by Lind Buzo Gray (LBG) algorithm [7].

Finally, to extract a prototype "path" from a class in an uncoded (by „ position weight ") form, i.e. in a neural sequence form, the following method is then used:

In the prototypes obtained from the last iteration of LBG algorithm, we only keep neurons whose appearance frequency is higher than a empirically chosen threshold.

The previously selected categories are sorted by ascending order of their position weight.

This is done for all non empty classes and we thus get a number of paths smaller than or equal to the number of classes found by the LBG algorithm (figure 2) .

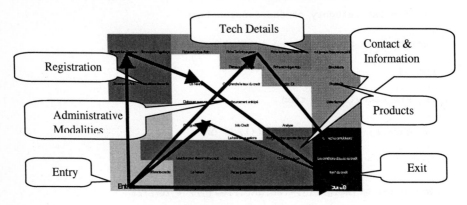

Fig. 2. Paths representation on the map.

7   Recognition

To recognize a given session in comparison with previously found prototypes, we calculate the distorsion which is simply the euclidian distance between the vectors resulting from coding (by position method) of both the session and a prototype.

This computation is done for all prototypes and the one whose distorsion with the session is the smallest is recognized as being the most similar.

8   Results

We used a Log access base of 10.000 sessions describing different behaviors over the visualization of 50 significant page categories. We applied our coding method (position weight) which reorganized the base in a $10.000 \times (50 \times 2)$ data table.
With a 16 bytes coding, we got 14 session prototypes.
For the recognition, we used a test base of 1000 sessions labeled by the 14 session prototypes previously found, and obtained an 80% success rate. We think that the mistaken 20% are due to the loss of information induced by the coding.

9   Conclusion

Thanks to a sequence analysis process based on both a new coding method and vector quantization, we could summarize the set of web site visitors navigations into categories representing significant profiles which we have represented on a synthetic map.
This work enabled us to have a first comprehension of the web site visitors whose behaviors are rather difficult to target a priori.

References

1. K. Benabdeslem and Y. Bennani, „ Classification connexionniste pour l'analyse des comportements d'internautes ", *Technical report, LIPN – CNRS,* France, 2000.
2. A. Joshi, "On Mining Web Access Logs". Technical Report, CS Department, University of Maryland, College Park, 1999.
3. K. Benabdeslem, Y. Bennani, E. Janvier „Analyse de données comportementales d'Internautes", *Journée Thématique: Exploration de données issues d'Internet, Institut Galilée, LIPN – CNRS, Villetaneuse-*France, 2 Mars 2001.
4. Benabdeslem K., Bennani Y. et Janvier E. – "Connectionnist approach for Website visitors behaviors mining", Proceedings of ACS/IEEE International Conference on Computer Systemes and Applications, Lebanon 2001.
5. Cadez I., Heckerman D., Meek C., Smyth P., White S. "Visualization of Navigation Patterns on a Web Site Using Model Based Clustering" in proceedings of the KDD 2000.
6. A. Gourinda. "Codage et reconnaissance de la parole par la quantification vectorielle"; thèse de doctorat ; Université Nancy I ; 1988.
7. Linde Y., Buzo A. et Gray R.M., "An algorithm for the VQ design" IEEE, Trans. On Communication, Vol. Com-28, 1980, pp.84-95.
8. T. Kohonen, "Self Organizing Map and Association Memories". *Springer, Vol 8, Springer Verlag,* 1984.

Missing Value Estimation
Using Mixture of PCAs

Shigeyuki Oba[1], Masa-aki Sato[2,5], Ichiro Takemasa[3], Morito Monden[3],
Ken-ichi Matsubara[4], and Shin Ishii[1,5]

[1] Nara Institute of Science and Technology, Japan.
shige-o@is.aist-nara.ac.jp
http://hawaii.aist-nara.ac.jp/~shige-o/
[2] ATR Human Information Science Laboratories, Japan.
[3] Osaka University, Japan.
[4] DNA Chip Research Institute, Japan.
[5] CREST, JST.

Abstract. We apply mixture of principal component analyzers (MPCA) to missing value estimation problems. A variational Bayes (VB) method for MPCA with missing values is developed. The missing values are regarded as hidden variables and their estimation is done simultaneously with the parameter estimation. It is found that VB method is better than maximum likelihood method by using artificial data. We also applied our method to DNA microarray data and the performance outweighed the conventional k-nearest neighbor method.

1 Introduction

Mixture of Gaussian (MG) models have been used for various purposes such as clustering, classification, vector quantization, regression and general probabilistic distribution estimation. However, their flexibility often causes the so-called "curse of dimension" for high-dimensional data. The covariance matrix of each Gaussian needs parameters of $O(d^2)$ and computational complexity of $O(d^3)$, where d is the data dimension.

In recent years, several alternative models have been proposed; they are mixture of principal component analyzers (MPCA) [1] and mixture of factor analyzers (MFA) [5]. They are latent variable models involving q-dimensional latent feature vectors and the feature dimension q is controllable by considering the model flexibility. The covariance matrix of each Gaussian needs parameters of $O(dq)$ and computational complexity of $O(dq^2)$. When $d \gg q$, the number of parameters and the computational complexity become much smaller than those in the original MG model and "curse of dimension" can be avoided.

In this article, we apply MPCA to high-dimensional data with missing values. A variational Bayes (VB) method [3, 4, 7] for MPCA with missing values is developed. The missing values are regarded as hidden variables and their estimation is done simultaneously with the parameter estimation.

J.R. Dorronsoro (Ed.): ICANN 2002, LNCS 2415, pp. 492–497, 2002.

Typical examples for such data appear in the fields of bioinformatics, e.g., in DNA microarray analyses, which measures expression level of thousands of genes under a specific condition.

It is found that VB method is better than maximum likelihood (ML) method for both of the model inference and the missing value estimation by using artificial data. We also applied our method to DNA microarray data and the performance outweighed the conventional k-nearest neighbor method [8].

2   MPCA Model

In an MG model:

$$p(\boldsymbol{y}|\boldsymbol{\theta}) = \sum_{i=1}^{m} g_i p(\boldsymbol{y}|\theta_i), \tag{1}$$

a variable vector $\boldsymbol{y}$ is assumed to be generated from one of m Gaussian distributions, $p(\boldsymbol{y}|\theta_i) = \mathcal{N}(\boldsymbol{y}|\boldsymbol{\mu}_i, \boldsymbol{\Sigma}_i)$ $(i = 1, ..., m)$. g_i is the mixing rate of the i-th Gaussian distribution (unit model) and $\boldsymbol{\theta}$ denotes the whole set of parameters $(g_1, ..., g_m, \theta_1, ..., \theta_m)$. Hereafter, $\mathcal{N}(\boldsymbol{y}|\boldsymbol{\mu}, \boldsymbol{\Sigma})$ denotes the density function of a multivariate Gaussian distribution with a mean $\boldsymbol{\mu}$ and a covariance matrix $\boldsymbol{\Sigma}$.

Mixture of principal component analyzers (MPCA) [1] is a special case of MG, in which each unit model is a probabilistic principal component analyzer (PPCA) [1]. PPCA is a probabilistic generative model of the principal component analyzer (PCA). In PPCA, a variable vector $\boldsymbol{y} \in \mathbb{R}^d$ is assumed to be generated by a noisy linear transformation of a q-dimensional hidden feature vector $\boldsymbol{x} \in \mathbb{R}^q$:

$$\boldsymbol{y} = \boldsymbol{W}\boldsymbol{x} + \boldsymbol{\mu} + \boldsymbol{\epsilon}, \tag{2}$$

where $\boldsymbol{W}$ ($d \times q$ matrix) and $\boldsymbol{\mu}$ (d vector) define the linear transformation from $\boldsymbol{x}$ to $\boldsymbol{y}$. $\boldsymbol{\epsilon}$ denotes a noise generated from a d-dimensional isotropic normal distribution $\mathcal{N}(\boldsymbol{0}, \tau^{-1}\mathbf{I}_d)$. $\tau (> 0)$ is a scalar and $\mathbf{I}_d$ denotes a $d \times d$ identity matrix. The hidden feature vector $\boldsymbol{x}$ is generated from a q-dimensional isotropic Gaussian distribution $\mathcal{N}(\boldsymbol{0}, \mathbf{I}_q)$.

The joint distribution $p(\boldsymbol{y}, \boldsymbol{x}|\theta)$ is then given by

$$p(\boldsymbol{y}, \boldsymbol{x}|\theta) = \exp\left[-\frac{\tau}{2}||\boldsymbol{y} - \boldsymbol{W}\boldsymbol{x} - \boldsymbol{\mu}||^2 - \frac{1}{2}||\boldsymbol{x}||^2 + \frac{d}{2}\ln\tau - \frac{d+q}{2}\ln 2\pi\right], \tag{3}$$

where $\theta = (\boldsymbol{W}, \boldsymbol{\mu}, \tau)$. By marginalizing over $\boldsymbol{x}$, we obtain a Gaussian distribution with mean $\boldsymbol{\mu}$ and covariance $\boldsymbol{\Sigma} = \boldsymbol{W}\boldsymbol{W}' + \tau^{-1}\mathbf{I}_d$. An ML estimation for matrix $\boldsymbol{W}$ results in a matrix consisting of principal axes of covariance $\boldsymbol{\Sigma}$. Namely, the ML estimation for PPCA is equivalent to the conventional PCA.

Then, the probabilistic generative model of MPCA can be written as

$$p(\boldsymbol{y}|\boldsymbol{\theta}) = \sum_{i=1}^{m} g_i \int d\boldsymbol{x}_i p(\boldsymbol{y}, \boldsymbol{x}_i|i, \theta_i) = \sum_{i=1}^{m} \int d\boldsymbol{x}_i p(\boldsymbol{y}, \boldsymbol{x}_i, z_i = 1|\theta_i), \tag{4}$$

$$p(\boldsymbol{y}, \boldsymbol{x}_i, z_i = 1|\theta_i) = g_i \exp\left[-\frac{\tau_i}{2}||\boldsymbol{y} - \boldsymbol{W}_i\boldsymbol{x}_i - \boldsymbol{\mu}_i||^2 - \frac{1}{2}||\boldsymbol{x}_i||^2 + \frac{d}{2}\ln\tau_i - \frac{d+q}{2}\ln 2\pi\right].$$

z_i is a discrete hidden variable indicating the ith unit model. When y is generated from the ith unit, $z_i = 1$. From its definition, $\sum_{i=1}^{m} z_i = 1$. The whole set of parameters is denoted by $\theta = \{g_i, W_i, \tau_i, \mu_i \mid i = 1, ..., m\}$.

In this study, we assume that there is a missing part in the input vector y; namely, $y = [y_o, y_h]$ where y_o and y_h are the observed part and the missing part, respectively. In this case, PPCA (equation (3)) can be described as:

$$
\begin{aligned}
p(y, x | \theta) &= p(y_o, y_h, x | W, \tau, \mu) \\
&= p(x | y_o, W_o, \tau, \mu_o) p(y_h | x, y_o, W, \tau, \mu) p(y_o | W_o, \tau, \mu_o) \\
&= \mathcal{N}(x | \overline{x}, R_x^{-1}) \mathcal{N}(y_h | \overline{y_h}, \tau^{-1} R_{y_h}) \mathcal{N}(y_o | \mu_o, \tau^{-1} R_{y_o}),
\end{aligned}
\tag{5}
$$

where $R_x = I_q + \tau W'W$, $R_{y_h}^{-1} = I_{dh} - \tau W_h R_x^{-1} W_h'$, $R_{x_o} = I_q + \tau W_o'W_o$, $R_{y_o}^{-1} = I_{do} - \tau W_o R_{x_o}^{-1} W_o'$, $\overline{x} = \tau R_{x_o}^{-1} W_o'(y_o - \mu_o)$, $\overline{y_h} = \mu_h + W_h \overline{x}$, $W = [W_o, W_h]$ and $\mu = [\mu_o, \mu_h]$. d_o and d_h are the dimensions of the observed part and the missing part, respectively.

3   Variational Bayes Method

Since MPCA is a latent variable model, expectation-maximization (EM) algorithm is used for ML estimation. A variational Bayes (VB) method [7,3,4] can be also applied to MPCA. Assuming that observed data set $\mathcal{D}$ is generated from a probabilistic generative model $p(\mathcal{D}, \mathcal{H} | \theta)$ with a hidden variable $\mathcal{H}$ and a parameter θ, the problem is to estimate the unknown $\mathcal{H}$ and θ. The VB estimation maximizes the VB free energy:

$$
\mathcal{F}[\mathcal{D}, q(\mathcal{H}, \theta)] = \int d\mathcal{H} d\theta q(\mathcal{H}, \theta) \ln \frac{p(\mathcal{D}, \mathcal{H} | \theta) p(\theta)}{q(\mathcal{H}, \theta)},
\tag{6}
$$

with respect to the trial distribution $q(\mathcal{H}, \theta)$ that approximates the posterior distribution of the unknown variables, $p(\mathcal{H}, \theta | \mathcal{D})$. $p(\theta)$ is a prior distribution of parameter θ. We also assume independence between the hidden variable and the parameter: $q(\mathcal{H}, \theta) = q(\mathcal{H})q(\theta)$. Subject to this assumption, the maximization of the free energy (6) with respect to $q(\mathcal{H})$ and $q(\theta)$ gives the iterative VB algorithm [3,4,7,5]. The missing values are regarded as a part of the hidden variable $\mathcal{H}$ and they can be estimated in the course of the VB algorithm. The free energy after the convergence approximates the log-evidence $\log p(\mathcal{D})$. The evidence, $p(\mathcal{D}) = \int d\mathcal{H} d\phi p(\mathcal{D}, \mathcal{H} | \theta) p(\theta)$, is the likelihood of a given model and it gives a criterion for model selection.

In order to do the ML estimation for MPCA, Tipping and Bishop [1] presented a two-stage EM algorithm. We extend it to a VB estimation and modify the algorithm in order to handle missing values (see Figure 1). Although the modified algorithm is different from the original VB, the increase of the free energy at each stage is guaranteed.

The VB estimation for MPCA needs a prior distribution. For parameter $\theta = \{g_i, W_i, \tau_i, \mu_i | i = 1, ..., m\}$, we assume a conjugate prior distribution. We also assume a hierarchical automatic relevance determination (ARD) prior [2] for W_i.

E-step (1) Trial distribution of hidden indicator variable $q(z_i(t))$ $\propto$ $p(z_i(t)|y_o(t))$ is calculated. Sufficient statistics (s. s.) are calculated using $q(z_i(t))$.

M-step (1) Trial distribution of parameter μ, $q(\mu)$, is updated using s.s.

E-step (2) Trial distribution of hidden feature variable and missing values, $q(x_i(t), y_h(t)) \propto p(x_i(t), y_h(t)|y_o(t), i)$, is calculated. S.s. are calculated using $q(x_i(t), y_h(t))$ and $q(z_i(t))$.

M-step (2) Trial distribution of parameters, $q(\theta)$, excluding μ, is updated using s.s.

Fig. 1. Two-stage EM algorithm

4 Model Selection

The model selection problem for MPCA involves determination of two types of model structure parameters, i.e., the number of units m, and feature dimension q, for each unit. In the ML estimation, information criteria like the minimum description length (MDL) are often used for the model selection. MDL uses the number of parameters as a model complexity measure, in which the characteristics of individual parameters are ignored. Although the ignorance is justified by the central limit theorem, it holds only for the large sample limit.

The model complexity term involved in the log-evidence, which is approximated by the VB free energy, is given by the Kullback-Leibler divergence between the posterior and the prior distributions, and it takes account of the estimation uncertainty for each parameter.

Redundant feature dimension is automatically reduced using ARD hierarchical prior that controls the length of principal axes. For PPCA, ARD for dimension reduction was first introduced by Bishop[6].

In the course of the VB algorithm, the number of effective data points for a certain unit sometimes becomes a very small value. When it becomes smaller than a specific threshold, the unit is naturally deleted.

5 Experiments

5.1 Artificial Data

In order to evaluate our method, a test data set is produced as follows. There are 200 data points, each of which is generated from an unknown MPCA model (M_{org}). The principal dimension of each unit is two, while the data dimension is 19. For each of 100 data points in the data set, two entries out of 19 dimension are randomly selected, and the corresponding entry values are deleted. Thus, an artificial data set involving missing values is prepared.

Using the above data set, we evaluate the missing value estimation and the model estimation. The missing value estimation is evaluated by normalized root mean squared error (NRMSE) and the model estimation is evaluated by the

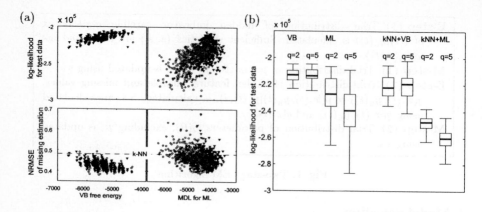

Fig. 2. (a)Comparison between VB and ML performances (b)Performance comparison between VB and ML

log-likelihood for the test data set, $\ln p(D_{10000}|M_{est})$, calculated using D_{10000} sampled from M_{org} independently of the training data set. M_{est} represents the estimated model.

The proposed method simultaneously estimates the missing values and the model parameter. The inference is done within the ML or VB framework. Initial values of parameters are set randomly for each number of clusters $m = 1, 2, ..., 20$.

For comparison, we also conduct an MPCA estimation after filling the missing values based on the k-nearest neighbor (k-NN) method.

Figure 2(a) shows the results. As can be seen in the lower figure, the average performance of the missing value estimation is better in VB than in ML or k-NN. The variation of performance is smaller in VB than in ML. These figures show that the free energy exhibits good correlation with the missing value estimation performance and the model estimation performance. Namely, when the free energy value is large, the missing value estimation and the model estimation are good. Figure 2(b) shows the generalization ability of ML and VB. ML and VB preprocessed by the k-NN missing value estimation are also evaluated. We can see that VB method with the simultaneous estimation of the missing values and the model parameters shows best performance.

5.2 Bioinformatics Data

We use here human DNA microarray data with 2029 genes and 47 samples. It consists of 2.5% missing entries due to experimental failures. The number or the location of the missing entries is different in one gene with the others. The number of genes, without missing entries, is 1251, and the other genes have at most 6 missing entries. A test data is prepared by using the data subset for the 1251 genes which has no missing data. From the data subset, 2.5% entries are randomly selected and they are regarded as missing entries. The performance of our method for the missing value estimation is compared to k-NN which has been found to be good for microarray data [8]. Figure 3 shows the results. In k-NN, various k values are tested. In MPCA, the simultaneous estimation method is used. The feature dimension in MPCA is set at $q = 10$. This figure shows that our VB MPCA method overtakes the k-NN method, and the estimation performance exhibits good correlation with the free energy.

These results imply that our method can be used for analyses of microarray data involving missing values.

6 Conclusion

We proposed a VB MPCA algorithm in order to deal with high-dimensional data including missing values. The proposed method can estimate both missing values and MPCA model parameters simultaneously, and the performance outweighs the existing methods. Our method can be a good clustering method for high-dimensional data including missing values, e.g., bioinformatics data.

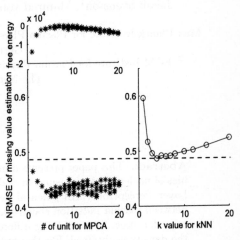

Fig. 3. Comparison between VB MPCA and k-NN using DNA microarray data

References

1. M. E. Tipping, C. M. Bishop, "Mixtures of Probabilistic Principal Component Analyzers," Neural Computation, vol. 11, pp. 443–482, 1999.
2. R. M. Neal, Bayesian learning for neural networks, Springer Verlag. ,1996.
3. H. Attias, "Learning parameters and structure of latent variable models by variational Bayes." in Proceedings of Uncertainty in Artificial Intelligence, 1999.
4. M. Sato, "On-Line Model Selection Based on the Variational Bayes," Neural Computation, Vol. 13, pp.1649-1681, 2001.
5. Z. Ghahramani and M. J. Beal. "Variational Inference for Bayesian Mixture of Factor Analyzers," Advances in Neural Information Processing Systems, vol. 12, pp.449–455, 2000.
6. C. M. Bishop, "Variational Principal Components," In IEE Conference Publication on Artificial Neural Networks, pp. 509–514, 1999.
7. S. Waterhouse, D. Mackay, T. Robinson, "Bayesian methods for mixture of experts," Advances in Neural Information Processing Systems, vol. 8, pp. 351–357, 1996
8. O. Troyanskaya, M. Cantor, G. Sherlock, P. Brown, T. Hastie, R. Tibshirani, D. Botstein and R. B. Altman, "Missing value estimation methods for DNA microarrays," Bioinformatics, Vol. 17, no. 6, 2001, pp. 520-525

High Precision Measurement of Fuel Density Profiles in Nuclear Fusion Plasmas

Jakob Svensson[1], Manfred von Hellermann[2], and Ralf König[1]

[1] Max Planck Institute for Plasma Physics, Boltzmannstrasse 2, 85748 Garching,
Germany
[2] FOM Institute for Plasma Physics, Rijnhuizen, 3430BE Nieuwegein,
The Netherlands

Abstract. This paper presents a method for deducing fuel density profiles of nuclear fusion plasmas in realtime during an experiment. A Multi Layer Perceptron (MLP) neural network is used to create a mapping between plasma radiation spectra and indirectly deduced hydrogen isotope densities. By combining different measurements a cross section of the density is obtained. For this problem, precision can be optimised by exploring the fact that both the input errors and target errors are known a priori. We show that a small adjustment of the backpropagation algorithm can take this into account during training. For subsequent predictions by the trained model, Bayesian posterior intervals will be derived, reflecting the known errors on inputs and targets both from the training set and current input pattern. The model is shown to give reliable estimates of the full fuel density profile in realtime, and could therefore be utilised for realtime feedback control of the fusion plasma.

1 Introduction

In todays and tomorrows large scale nuclear fusion experiments, two problems are associated with deducing fuel density (normally deuterium) profiles. The first is related to the large plasma core temperatures ($\sim 10^8$ K), at which all hydrogen is dissociated into free electrons and nuclei, giving no line radiation from which the density could otherwise be deduced. Line radiation is therefore enforced by charge transfer of electrons from neutral deuterium atoms in high power beams injected into the plasma core. The deexcitation and release of a photon following this charge capture can then be detected as line radiation. With knowledge about the local electron density and the local density of the beam, the intensity of this line can be directly related to the local deuterium density [1]. The second problem is related to the extraction of this line intensity and local beam density from a complex spectrum containing a very large number of overlapping features orig-

J.R. Dorronsoro (Ed.): ICANN 2002, LNCS 2415, pp. 498–503, 2002.

inating from the beam itself and from radiation at the colder plasma edge (fig 1).

It turns out that this spectrum together with an extra input for the local electron density contains the information necessary to deduce the local deuterium (fuel) density of the plasma [2]. The input for one MLP is therefore a spectrum (fig 1) and corresponding local electron density. The targets are taken from a database of indirectly inferred fuel densities from a large number of experiments. With one MLP mapping, direct extraction of a fuel density can be achieved. A combination of such mappings for different positions in the plasma gives a full reconstructed fuel density profile, available in realtime. Since the errors on both the input spectra and targets are known in advance, we have developed an algorithm that takes this into account during training and subsequent usage of the network. This approach differs from the more computationally demanding Monte Carlo based methods for taking into account known input uncertainty ([3] [4]), in that it is based on a modified cost function derived from a local linear approximation of the network. Section 2 deals with the specifics of this model and the Bayesian posterior intervals used as error bars, and in section 3 some results will be shown.

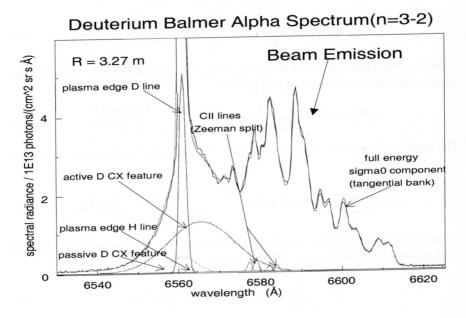

Fig. 1. Example of MLP input vector. JET pulse 40346. From the area under the feature marked 'active D CX feature' and the part of the spectrum marked 'beam emission' the local deuterium density can be deduced. Components are shown at half amplitude for clarity.

2 The MLP Model

In this model we will assume that the target value for pattern p is well approximated by a normal distributed random number with standard deviation σ_t^p. At the high signal levels of the input spectra the Poisson distributed photon count of an individual channel is well approximated by a normal distribution with standard deviation $\sigma_{x_i}^p$, where i is the channel number for input vector p. By using a local linear approximation of the MLP network mapping f, valid for small values of $\sigma_{x_i}^p$,

$$f(\mathbf{x}^p + \Delta\mathbf{x}) \approx f(\mathbf{x}^p) + \nabla f(\mathbf{x}^p)^{\mathbf{T}}\Delta\mathbf{x} \tag{1}$$

the effect of the input error on the output neuron can be calculated

$$\sigma_x^p = \sqrt{\sum_{i=1}^{N_{IN}}(\sigma_{x_i}^p)^2(\frac{\partial f}{\partial x_i^p})^2} \tag{2}$$

Assuming independence of input and target errors, gives for the standard deviation of the output neuron

$$\sigma^p(\mathbf{x}^p) = \sqrt{(\sigma_t^p)^2 + \sum_{i=1}^{N_{IN}}(\sigma_{x_i}^p)^2(\frac{\partial f}{\partial x_i^p})^2} \tag{3}$$

If this x-dependent standard deviation is now taken as the new error of the targets, the likelihood function will become

$$p(t^p|\mathbf{x}^p, \mathbf{w}) = \frac{1}{\sqrt{2\pi}\sigma^p(\mathbf{x}^p, \mathbf{w})} \exp{-\frac{(t^p - f(\mathbf{x}^p, \mathbf{w}))^2}{2\sigma^p(\mathbf{x}^p, \mathbf{w})^2}} \tag{4}$$

The negative logarithm of the full likelihood function for the training set is then given by

$$E = \frac{1}{2}\sum_{p=1}^{N_p}\{\log\sigma^p(\mathbf{x}^p, \mathbf{w})^2 + \frac{(t^p - f(\mathbf{x}^p, \mathbf{w}))^2}{\sigma^p(\mathbf{x}^p, \mathbf{w})^2}\} + \frac{1}{2}\alpha\mathbf{w}^T\mathbf{w} \tag{5}$$

where a weight decay term has also been added. In comparison with a standard least squares cost function for MLP networks

$$E = \frac{1}{2}\sum_{p=1}^{N_p}(t^p - f(\mathbf{x}^p, \mathbf{w}))^2 + \frac{1}{2}\alpha\mathbf{w}^T\mathbf{w} \tag{6}$$

the new cost function will have an extra $\mathbf{w}$-dependent term, and an extra $\mathbf{w}$-dependent denominator. The gradient of E with respect to the weights will therefore be more complicated:

$$\frac{\partial E}{\partial w_{ji}} = \sum_{p=1}^{N_p}2\frac{\partial E^p}{\partial(\sigma^p)^2}w_j\{(\sigma_{x_i}^p)^2 g'(a_j)\frac{\partial f}{\partial x_i} + x_i g''(a_j)q_j^p\} + \frac{\partial E^p}{\partial f}x_j w_j g'(a_j) + \alpha w_{ji} \tag{7}$$

$$\frac{\partial E}{\partial w_j} = \sum_{p=1}^{N_p} 2 \frac{\partial E^p}{\partial (\sigma^p)^2} g'(a_j) q_j^p + \frac{\partial E^p}{\partial f} g(a_j) + \alpha w_j \qquad (8)$$

where

$$q_j^p = \sum_{i=1}^{N_{IN}} (\sigma_{x_i}^p)^2 w_{ji} \frac{\partial f}{\partial x_i} \qquad (9)$$

$$\frac{\partial E^p}{\partial (\sigma^p)^2} = \frac{1}{2} \left(\frac{1}{(\sigma^p)^2} - \frac{(t^p - f)^2}{(\sigma^p)^4} \right) \qquad (10)$$

$$\frac{\partial E^p}{\partial f} = \frac{f - t^p}{(\sigma^p)^2} \qquad (11)$$

Double subscript ji refers to the weight from input neuron i to hidden neuron j. Single subscript j refers to hidden neuron j. The summations should include the bias neurons, with $x_{bias} = 1$, $q_{bias} = 0$ and $\sigma_{x_{bias}}^p = 0$. g is the activation function used for the hidden layer neurons. The gradient (eq. 7 and 8) can now be used in a standard gradient descent algorithm. The solution found will represent the optimum for the case of known input and output errors. In practice, because of the longer training times associated with this gradient expression, it is often possible to start with the gradient calculated from eq. 6, and then fine tune with the suggested gradient later during the training process.

In the Bayesian framework, error bars are usually associated with the width of the posterior predictive distribution for the target process given an input [5], [6]:

$$p(t|\mathbf{x}, \mathbf{D}) = \int p(t|\mathbf{x}, \mathbf{w}) p(\mathbf{w}|\mathbf{D}) d\mathbf{w} \qquad (12)$$

The width of this distribution will give a measure of the uncertainty (at $\mathbf{x}$) of the outcome of the process that generated the data. In this application though, the process that generated the target data (which here is the outcome of a manual indirect modeling process) is not interesting in itself, only the expected value of the process (the underlying 'true' value, distorted by noise), and the uncertainty related to *that expected value*. This is not the same as the width of the predictive distribution. We have therefore instead used the variance of the network output at a given $\mathbf{x}$, under the posterior distribution of the weights:

$$\sigma_{post}^2(\mathbf{x}) = \int (f(\mathbf{x}, \mathbf{w}) - m(\mathbf{x}))^2 p(\mathbf{w}|\mathbf{D}) d\mathbf{w} \qquad (13)$$

m is here the specific network function $f(\mathbf{x}, \mathbf{w}_{opt})$ found by the minimization of eq. 5. After some manipulations including Laplace's approximation [7], eq. 13 will give the following expression for the error due to the posterior distribution

$$\sigma_{post}(\mathbf{x}) = \sqrt{(\nabla_w f)^T \mathbf{A}^{-1} (\nabla_w f)} \qquad (14)$$

where $\mathbf{A}$ is the Hessian matrix of the negative logarithm of the posterior at $\mathbf{w}_{opt}$. In this Bayesian interpretation, the negative logarithm of the posterior is equivalent to eq. 5 if α is a hyperparameter for the precision of the Gaussian weight prior [8]. The elements of the Hessian matrix have to be calculated from the derivatives of eq. 7 and 8. Since the dimensionality of $\mathbf{w}$ is very large, we will use a diagonal approximation of $\mathbf{A}$, keeping only the terms on the diagonal [5]. The total error of a network prediction will thus be

$$\sigma_f(\mathbf{x}) = \sqrt{\sigma_{post}^2(\mathbf{x}) + \sigma_{\mathbf{x}}^2(\mathbf{x})} \tag{15}$$

where $\sigma_{\mathbf{x}}^2(\mathbf{x})$ is the error caused by the current input vector, calculated in the same way as for eq. 2. The result of using eq. 14 and 15 on a fuel density profile is shown in figure 2

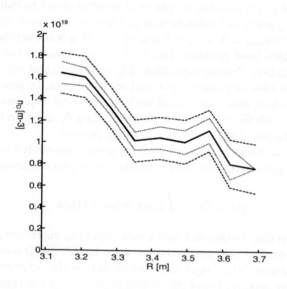

Fig. 2. Fuel density profile reconstructed from eight MLP networks, one for each spatial position at which measurements of spectra (fig 1) are made. Dashed lines are total error (eq. 15) and dotted lines the posterior error contribution (eq. 14). For clarity, 2σ error bars are shown. JET pulse 42243, t=7.83s

3   Results

In figure 3 is shown a scatter plot of the targets and MLP outputs for a test set containing 8000 spectra. The overall gain from using the model described with cost function 5 as compared to using cost function 6 was a 20% increase in accuracy. The average relative error on the test set was below 5%, which is very low considering that there is a lower limit of the accuracy (about 2%) due to the photon statistics of the input channels. The speed with which the targets are calculated (including error bars) from the MLP mapping is sufficient for a 20 Hz realtime reconstruction of the internal fuel density profile, making feedback control of the fusion plasma based on fuel density profiles possible.

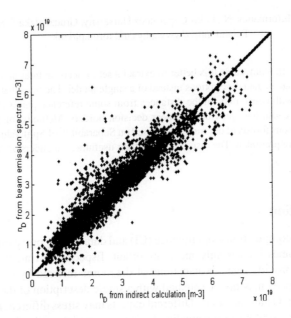

Fig. 3. Scatter plot of MLP outputs (abscissa) versus targets (ordinate) for a test set comprising 8000 spectra (fig 1).

References

1. Mandl, W.: Development of Active Balmer-Alpha Spectroscopy at JET JET-IR(92)05, PhD thesis, (1992)
2. Svensson J., von Hellermann M., König R. W. T.: Direct measurement of JET local deuteron densities by neural network modelling of Balmer alpha beam emission spectra. Plasma Physics and Controlled Fusion, **43** 389–403 (2001)
3. Wright W. A.: Neural network regression with input uncertainty. Neural networks for signal processing, IEEE workshop Cambridge UK, 284–293 (1998)
4. Tresp V., Ahamad S., Neuneier R., Training neural networks with deficient data Advances in neural information processing systems vol 6, 128–135 (1994)
5. Bishop C. M.: Neural Networks for Pattern Recognition. Oxford University Press, 1995
6. Neal R., M.: Bayesian Learning for Neural Networks. Springer Verlag, 1996
7. Gelman A., Carlin J. B., Stern H. S., Rubin D. B.: Bayesian Data Analysis. Chapman & Hall, 1996
8. MacKay D. J. C.: Bayesian methods for adaptive models. PhD thesis, California Institute of Technology, 1992

Heterogeneous Forests of Decision Trees

Krzysztof Grąbczewski and Włodzisław Duch

Department of Informatics, Nicholas Copernicus University, Grudziądzka 5, 87-100 Toruń, Poland. http://www.phys.uni.torun.pl/kmk

Abstract. In many cases it is better to extract a set of decision trees and a set of possible logical data descriptions instead of a single model. The trees that include premises with constraints on the distances from some reference points are more flexible because they provide nonlinear decision borders. Methods for creating heterogeneous forests of decision trees based on Separability of Split Value (SSV) criterion are presented. The results confirm their usefulness in understanding data structures.

1 Introduction

Recent trends in computational intelligence (CI) and data mining prove that understanding of data becomes increasingly more important. Especially in medicine CI expert systems always provide some explanations of diagnoses inferred by decision support tools. Human experts may not be satisfied with a single description of the problem they examine, even if it is quite accurate. Different experts may stress different factors in their evaluations of the problem, corresponding to different models of the data they are used to.

One of the most attractive ways to describe the classification process is to use logical rules, which can be easily extracted from decision trees. Usually data mining systems provide their users with just a single set of rules (the one that seems to be the most precise). At best several sets of rules with increasing accuracy are provided [1]. However, many sets of rules with similar complexity and accuracy may exist, using for example different feature subsets, bringing more information of interest to the domain expert. Therefore methods that are aimed at finding many different descriptions of the same data are worth investigation.

Multiple solutions to the problem may be even more informative when they use different types of rule premises. The tests usually concern the features describing data, but to make a decision tree heterogeneous they may include conditions that test the distances from the data vectors to a selected point in the feature space.

We have augmented our decision tree based on the *Separability of Split Value* (SSV) criterion [3,4] with the capability to generate heterogeneous forests of trees instead of single trees. The algorithms have been tested on several datasets coming from the UCI repository of machine learning problems [5].

2 SSV Criterion

The SSV criterion is one of the most efficient among criteria used for decision tree construction [3,4]. It's basic advantage is that it can be applied to both continuous and

J.R. Dorronsoro (Ed.): ICANN 2002, LNCS 2415, pp. 504–509, 2002.

discrete features, which means that methods based on it can operate on raw data without the need for any data preprocessing. The *split* value (or *cut-off point*) is defined differently for continuous and discrete features. In the case of continuous features the split value is a real number, in other cases it is a subset of the set of alternative values of the feature. The best split value is the one that separates the largest number of pairs of objects from different classes, so for both types of features *left side* (*LS*) and *right side* (*RS*) of a split value s of feature f for a given dataset D can be defined:

$$\mathrm{LS}(s, f, D) = \begin{cases} \{x \in D : f(x) < s\} & \text{if } f \text{ is continuous} \\ \{x \in D : f(x) \notin s\} & \text{otherwise} \end{cases} \qquad (1)$$

$$\mathrm{RS}(s, f, D) = D - -\mathrm{LS}(s, f, D)$$

where $f(x)$ is the f's feature value for the data vector x. The *separability of a split value* s is defined as:

$$\mathrm{SSV}(s) = 2 * \sum_{c \in C} |\mathrm{LS}(s, f, D) \cap D_c| * |\mathrm{RS}(s, f, D) \cap (D - D_c)| \qquad (2)$$

$$- \sum_{c \in C} \min(|\mathrm{LS}(s, f, D) \cap D_c|, |\mathrm{RS}(s, f, D) \cap D_c|) \qquad (3)$$

where C is the set of classes and D_c is the set of data vectors from D which belong to class $c \in C$. Similar criterion has been used for design of neural networks by Bobrowski et al. [6].

Among all the split values which separate the maximal number of pairs of vectors from different classes the best is the one that separates the smallest number of pairs of vectors belonging to the same class. For every dataset containing vectors which belong to at least two different classes, for each feature which has at least two different values, there exists a split value with largest separability. When the feature being examined is continuous and there are several different split values with maximal separability close to each other, the split value closest to the average of all of them is selected. To avoid such situations it is good to examine split values that are natural for a given dataset (i.e. centered between adjacent feature values that occur in the data vectors). If there are non-maximal (regarding separability) split values between two maximal points or if the feature is discrete, then the best split value is selected randomly.

Decision tree is constructed recursively by searching for best splits. At each stage when the best split is found and the subsets of data resulting from the split are not completely pure (i.e. contain data belonging to more than one class) each of the subsets is being analyzed the same way as the whole data. The decision tree built this way gives maximal possible accuracy (100% if there are no contradictory examples in the data) which usually means that the created model overfits the data. To remedy this a cross validation training is performed to find the optimal pruning parameters for the tree. Optimal pruning produces a tree capable of good generalization of the patterns used in the tree construction process.

3 Heterogeneous SSV Trees

The separability measure can be applied not only to the features supplied in the dataset. It can be calculated for other values like linear or nonlinear combinations of features or the distances from the data vectors to given points in the feature space.

Although the computational complexity of the analysis of such new features is not very high [2], the problem becomes very complex because the number of possible features to be analyzed is infinite. In the case of distance features there are some "natural" restrictions on reference points selection (for example they may be searched among the training data vectors).

Because different distance measures lead to very different decision borders, such enhancement may significantly simplify the structure of the resulting tree (i.e. discover simple class structures).

4 Construction of Forests

There are two ways to generate forests of trees, or more generally different models of the data, related to the variance due to the data sample and the variance due to the flexibility of the data model itself. The simplest way of finding multiple solutions to a problem is to apply a data mining technique to different samples of data depicting the same problem. Most computational intelligence methods, including decision trees, are not quite stable, for some data sets small perturbations of the training data may lead to remarkably different results [7]. It can be easily seen in cross validation tests, where each training pass is performed for different data selection. In section 5 we present some examples of application of this method.

Another possibility of forest generation is to explore beam search strategy which looks for the best tree (it is more efficient than multiple data resampling and adaptive model training). In each of beam search stages there is a beam of n decision trees (search states). Each stage adds a single split to the trees in current beam, so very simple trees in first beams and more complex structures in next stages are created. The usual technique to search for the best tree is to build the shortest tree with maximal accuracy possible (the first final state found), and then prune it to obtain good generalization (pruning is determined in SSV empirically by means of cross validation). This method has proven to be quite accurate, but it is justified strictly in the case of the shortest tree, other trees may require different pruning parameters. To find a forest all trees that appear in a beam at any stage of the search process are ordered by their estimated accuracy. To estimate the accuracy a cross validation on the training data is performed and a validation error is assigned to each beam position which is a pair (s,p) where s is search stage (beam index) and p is the position in that beam (the states in a beam can be sorted by the errors for the training set). The precise algorithm goes as follows:

- Perform n-fold cross validation for the training set. For each part:
  - Run beam search process (do not stop when first final state is found – stop when all the states in current beam are final);
  - for each pair (s,p) where $s \in \{1,\ldots,m\}$, m is the number of search stages (the number of beams generated), $p \in \{1,\ldots,k_s\}$, k_s is the number of states in

sth beam, calculate the validation error $E_V = E_{TRN} + nE_{TST}$, where E_{TRN} and E_{TST} are the numbers of misclassified samples in the training and test (validation) parts, respectively.

- Run beam search for the whole training data.
- Assign an average validation error to all pairs (s, p) that correspond to some states in the generated beams.
    - Discard the pairs that do not correspond to a state in any of cross validation parts (the idea is that such search states are not common and would not generalize well);
    - When calculating the average values remove the maximal and the minimal ones.
- Sort all the beam states positions with increasing average validation error.
- Present a required number of leading states as the forest.

5   Results

We have applied the above algorithm to several real life datasets from the UCI repository [5]. In all the cases we have found some sets of logical rules describing the data. The rule sets (in fact decision trees) which after sorting got to the beginning of the state list were very similar to the best known logical description of that data. Several alternative sets have been found for each analyzed dataset. The notation d(_,vnnn) used in the rules means the square of euclidean distance to the vector nnn).

The hypothyroid dataset was created from medical tests screening for hypothyroid problems. Since most people were healthy 92.5% of cases belong to the normal group, and 8% of cases belonging to the primary hypothyroid or compensated hypothyroid group. 21 medical facts were collected in most cases, with 6 continuous test values and 15 binary values. A total of 3772 cases are given for training (results from one year) and 3428 cases for testing (results from the next year). Thus from the classification point of view this is a 3 class problem with 22 attributes.

The forest search algorithm has found many rule sets of very high accuracy and confirmed that the highest results for this dataset may be obtained with methods that cut the space perpendicularly to the axes. Because of the rectangular decision borders needed in this case, the premises with distance tests occur at quite deep levels of the trees.

One of the most interesting rule sets is 99.84% accurate (6 errors) on the training set and 99.39% (21 errors) on the test set (all the values of continuous features are multiplied here by 1000):

1. if TSH > 6.05 ∧ FTI < 64.72 ∧ (T3 < 11.5 ∨ d(_,v2917) < 1.10531) then primary hypothyroid
2. if TSH > 6.05 ∧ FTI > 64.72 ∧ on_thyroxine = 0 ∧ thyroid_surgery = 0 ∧ TT4 < 150.5 then compensated hypothyroid
3. else healthy.

The Wisconsin breast cancer dataset contains 699 instances, with 458 benign (65.5%) and 241 (34.5%) malignant cases. Each instance is described by 9 attributes with integer value in the range 1-10 and a binary class label. For 16 instances one attribute is missing. The data had been standardized before the application of the algorithm.

The results for this dataset are surprisingly good – single premise rules are more accurate, than much more complicated homogeneous rule sets known so far. Some of the rules chosen for the forest are presented in Table 1.

Table 1. Classification rules for the Wisconsin breast cancer data.

Rules	Accuracy / Sensitivity / Specificity
1. if d(_,v303) < 62.7239 then malignant 2. else benign	97.3% / 97.9% / 96.9%
1. if d(_,v681) < 51.9701 then malignant 2. else benign	97.4% / 98.8% / 96.8%
1. if d(_,v613) < 65.3062 then malignant 2. else benign	97.1% / 97.9% / 96.7%
1. if d(_,v160) < 36.4661 then malignant 2. else benign	97.1% / 98.8% / 96.3%
1. if d(_,v150) < 39.5778 then malignant 2. else benign	97.1% / 98.8% / 96.3%

The **appendicitis** data contains only 106 cases, with 8 attributes (results of medical tests), and 2 classes: 88 cases with acute appendicitis (class 1) and 18 cases with other problems (class 2). For this small dataset very simple classification rules have been found by Weiss and Kapouleas [9] using their PVM (Predictive Value Maximization) approach and by us in some of our earlier approaches [1,3].

The heterogeneous forest algorithm discovered several new descriptions of this data which are simple, accurate and very sensitive. Some of them are presented in Table 2 (the distances were calculated for standardized data).

Table 2. Classification rules for the appendicitis data.

Rules	Accuracy / Sensitivity / Specificity
1. if d(_,v73) < 12.0798 ∨ MBAA > 1174.5 then class 1 2. else class 2	92.5% / 98.8% / 66.7%
1. if d(_,v22) < 16.4115 ∨ MBAP > 12 then class 1 2. else class 2	92.5% / 98.8% / 66.7%
1. if d(_,v22) < 16.4115 ∨ MBAA > 1174.5 then class 1 2. else class 2	92.5% / 98.8% / 66.7%
1. if d(_,v8) < 13.385 ∧ HNEA ∉ (9543, 9997) then class 1 2. if d(_,v8) > 13.385 ∧ MBAA > 1174.5 ∧ MNEP > 51 then class 1 3. else class 2	96.2% / 98.8% / 85.7%

6   Discussion

Medical datasets often contain small number of examples (sometimes 100 or less) with relatively large number of features. In such cases of large spaces with sparse samples it is quite likely that different logical expressions may accidently classify some data well, thus many data mining systems may find solutions which are precise but not meaningful according to the experts' knowledge. Providing experts with several alternative descriptions gives more chance that they find interesting explanations compatible with their experience and notice new regularities in the data, leading to a better understanding of the problem. Heterogeneous forests are capable of finding simple classification descriptions when nonlinear decision borders are necessary and can find reference vectors which sometimes may be used as prototypes for similarity based methods.

A forest consists of trees which very often differ in their decisions (different samples are misclassified by different trees). Such trees may successfully act as a committee giving answers interpretable as probabilities. It can also yield better classification accuracy, but even if it does not, it will provide an interesting solution to the problem of combining the comprehensibility of decisions with high accuracy and confidence.

Another possible extension of this work is to automatically detect how many trees are interesting enough to compose the forest. This can be done by means of forest committees or the evaluation of statistical significance [8] of the classification accuracies differences.

Acknowledgments.   Support by the Polish Committee for Scientific Research, grant 8 T11C 006 19, is gratefully acknowledged.

References

1. Duch W, Adamczak R. and Grąbczewski K. (2001) Methodology of extraction, optimization and application of crisp and fuzzy logical rules. *IEEE Transactions on Neural Networks* **12**: 277–306
2. Duch W, Grąbczewski K. (2002) Heterogeneous adaptive systems. *World Congress of Computational Intelligence*, Honolulu, May 2002
3. Grąbczewski K, Duch W. (1999) A general purpose separability criterion for classification systems, *4th Conference on Neural Networks and Their Applications*, Zakopane, Poland, pp. 203–208
4. Grąbczewski K. Duch W. (2000) The Separability of Split Value Criterion, *5th Conference on Neural Networks and Soft Computing*, Zakopane, Poland, pp. 201–208
5. Blake, C.L, Merz, C.J. (1998) UCI Repository of machine learning databases http://www.ics.uci.edu/ mlearn/MLRepository.html. Irvine, CA: University of California, Department of Information and Computer Science.
6. Bobrowski L, Krętowska M, Krętowski M. (1997) Design of neural classifying networks by using dipolar criterions. *3rd Conf. on Neural Networks and Their Applications*, Kule, Poland
7. Breiman L. (1998) Bias-Variance, regularization, instability and stabilization. In: Bishop, C. (Ed.) Neural Networks and Machine Learning. Springer, Berlin, Heidelberg, New York
8. Dietterich T. (1998) Approximate Statistical Tests for Comparing Supervised Classification Learning Algorithms, Neural Computation **10**, 1895–1923
9. S.M. Weiss, I. Kapouleas. "An empirical comparison of pattern recognition, neural nets and machine learning classification methods", in: *Readings in Machine Learning*, eds. J.W. Shavlik, T.G. Dietterich, Morgan Kauffman Publ, CA 1990

Independent Component Analysis for Domain Independent Watermarking

Stéphane Bounkong, David Saad, and David Lowe

Neural Computing Research Group, Aston University, Birmingham B4 7ET, UK,
{bounkons,saadd,lowed}@aston.ac.uk, http://www.ncrg.aston.ac.uk/

Abstract. A new principled domain independent watermarking framework is presented. The new approach is based on embedding the message in statistically independent sources of the covertext to mimimise covertext distortion, maximise the information embedding rate and improve the method's robustness against various attacks. Experiments comparing the performance of the new approach, on several standard attacks show the current proposed approach to be competitive with other state of the art domain-specific methods.

1 Introduction

Interest in watermarking techniques has grown significantly in the past decade, mainly due to the need to protect intellectual property rights (IPR). Research has mainly focused on digital images, audio or video data, where economic interests are more apparent, with a plethora of techniques. In spite of their common root, the techniques developed are domain specific and cannot easily be transferred across domains, making it difficult to provide a principled comprehensive theoretical approach to watermarking. The latter is a prerequisite to a methodological optimization of watermarking methods. The present paper describes a domain independent watermarking framework which aims at maximising the information embedding rate and the robustness against various attacks while mimimising the information degradation.

2 Domain Independent Watermarking

In the past few years, significant attention has been drawn to blind source separation by Independent Component Analysis (ICA) [1]. The recent discovery of efficient algorithms and the increase in computational abilities, have made it easier to extract statistically independent sources from given data.

ICA is a general purpose statistical technique which, given a set of observed data, extracts a linear transformation such that the resulting variables are as statistically independent as possible. Such separation may be applied to audio signals or digitized images [1], assuming that they constitute a sufficiently uniform class so that a statistical model can be constructed on the basis of

J.R. Dorronsoro (Ed.): ICANN 2002, LNCS 2415, pp. 510–515, 2002.

observations. Experiments conducted on a set of digitized images that we examined, show that this hypothesis holds, giving us a general domain independent framework [1].

The suggested framework can be based on various generative methods. In this paper we will focus on a particular method for identifying statistically independent sources - ICA. We now describe the ICA generative model and a simple watermarking scheme based on it. Technical details have been omitted for brevity.

2.1   ICA Generative Model

ICA describes a set of latent variables, also termed Independent Components (IC), which can be observed only through their linear combination. By definition, these variables are random and statistically mutually independent.

$$x_i = a_{i1}s_1 + a_{i2}s_2 + \ldots + a_{il}s_l, \text{ for all } i = 1, \ldots, n \tag{1}$$

where $a_{i,j}$ are real coefficients, s_i are the latent independent variables and the x_i are observed measurements. Using a matrix notation, the previous equation can be written as $\boldsymbol{x} = A\boldsymbol{s}$; and the inverse (de-mixing) process can be described by $\boldsymbol{s} = W\boldsymbol{x}$, where W is the de-mixing matrix and inverse (or pseudo-inverse if $n \neq l$) of A.

2.2   Basic Watermarking Scheme

Basic watermarking schemes can be described in three steps. Firstly, a given message m, also termed a watermark, is embedded into the covertext X (e.g. a digitized image, audio or a transformed version) providing a watermarked covertext $\hat{X}$. Then, the watermarked text may be attacked either maliciously or non-maliciously, resulting in the attacked covertext Y. Finally, a decoder tries to extract m from Y given or not side information. This is summarised in figure 1.

2.3   Domain Independent Watermarking (DIW) Scheme

In the framework studied in this paper, X may be derived from any media, such as audio signals or digitized images. The de-mixing matrix W obtained by the ICA algorithm for the different domains are different but the principle remains the same: representing the covertext through a set of IC.

Given a covertext, a set of relevant IC are chosen and modified such that they carry m. Various efficient approaches have been suggested for hiding/embedding information. We used the distortion-compensated Quantization Index Modulation (QIM) method [5], that has been shown to be close to optimal in the case

---

[1] In the case of multiple, significantly different, covertext groups, one may construct a different model for each group.

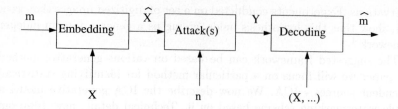

Fig. 1. A general watermarking scheme where m is the embedded message, X is the covertext, $\hat{X}$ the watermarked covertext, Y the attacked covertext and $\hat{m}$ an estimate of m.

of additive Gaussian attacks and is easy to use. It is based on quantizing the covertext real-valued IC to some central value, followed by a quantized addition/subtraction representing the binary message bit. This may also be modified by a prescribed noise template making it difficult to identify the QIM embedding process and its parameters.

The watermarked covertext $\hat{X}$ is then mixed back to the original covertext space, generating the watermarked covertext, as illustrated in Fig.2.

The decoding process proceeds in a similar way. The description of the attacked text is computed from the attacked covertext by employing the de-mixing matrix W giving us the corrupted source Y. $\hat{m}$ is computed from Y in conjunction with other available information (e.g. attack characteristics, original covertext, cryptographic key, ...; see Figure 2).

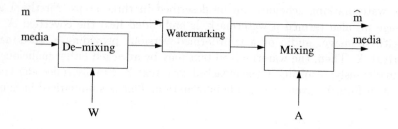

Fig. 2. This figure represents a domain independent watermarking scheme where m is the embedded message and $\hat{m}$ is an estimate of m. A and W are, respectively, the mixing and de-mixing matrices used to get the independent components.

3   Experimental Results

We carried out a few experiments, comparing the performance of our approach to other watermarking methods. The covertext used in our experiments was arbitrarily chosen to be digitized images. For the DIW approach, the latter are divided in contiguous patches. Each patch is marked independently following the method described above, see 2.3.

For comparison purposes, two other watermarking schemes have been tested under the same attacks and using the same embedding and decoding methods. Both methods operate in the discrete cosine transform (DCT) domain.

Comp1 This scheme is based on the DCT of the whole image, X, selecting a random coefficient set for the message m to be embedded in using QIM.

Comp2 In the second scheme, the image is divided into contiguous patches. The DCT of each patch is used as covertext X. A set of coefficients is selected and then quantized for embedding m.

In both schemes, $\hat{X}$ undergoes an inverse DCT, to provide the watermarked image. Notice that *local* methods such as Comp 2 and DIW are much more computationally efficient than *global* methods like Comp 1. Furthermore, watermarking parameters have been optimized in all methods, and separately for each specific attack.

3.1 Experiments

We carried out four experiments where watermarked pictures are attacked either by: a) white noise (WN) of mean zero and of various standard deviation values; b) JPEG lossy compression with different quality levels; c) resizing with various factors; d) a combination of attacks: resizing with a factor of 0.5, followed by JPEG compression with a quality factor of 70, followed by WN of zero mean and of standard deviation 15.

These attacks are, arguably, the most commonly used attacks as a benchmark in this field. The set of images used comprises eleven gray-scale pictures representing *natural*, as opposed to computer generated, scenes. The experiments are carried out ten times for each set of parameters for each picture, providing both mean performance and error bars on the measurements.

Each algorithm embeds, using a quantization method characterized by a quantization step δ, a message m of length 1024 bits with a maximum distortion of 38 dB as suggested in [3,4]. The distortion induced by the watermarking systems is measured by the peak signal to noise ratio (PSNR). A simple decoding scheme based on nearest decoding is also used for all systems. Table 3.1 summarises the parameters used in the experiments.

Table 1. Summary of the watermarking schemes parameters

Attack			Noise		JPEG		Resizing	
Scheme	Transform	Patch size	Coef. Rg.	δ	Coef. Rg.	δ	Coef. Rg.	δ
DIW	ICA	16 by 16	38-50	155	6-10	36	6-10	36
Comp1	DCT	-	101-1124	70	2081-20624	70	2-1985	70
Comp2	DCT	16 by 16	6-23	80	2-19	80	4-18	80

3.2   Results

Figure 3a, shows that all schemes are quite robust considering that the 38 dB attack distortion threshold is reached for a standard deviation of about 3. It also shows that DIW is the most robust method of those examined for a WN attack. In the case of DIW and the decoding method used, it is easy to see a direct relation between δ and the robustness of the process, since the noise in the feature space is also Gaussian. This may not be the case if other decoding methods, such as the Bayesian approach will be used. Moreover it also shows that one potential weakness of the DIW scheme, the ICA restriction of extracting only non-Gaussian sources, is not highly significant, even in the case of a Gaussian noise attack.

Figure 3b shows that all systems are quite robust against JPEG compression. However, for very low quality levels, under 15, performances decrease significantly, and are less stable as shown by the error bars. Furthermore the threshold of 38 dB distortion is reached at a quality level of about 90. DIW achieves here the best results on average.

Figure 3c shows excellent performances for Comp1 under resizing attacks. DIW and Comp2 achieve excellent results for resizing factor greater than 0.5; their performances decrease significantly for stronger attacks. Intuitively this can be explained by their patches' localised nature. Low resizing factors affect severly the capacity of these schemes and the picture quality. For a 0.25 resizing factor, the picture size is reduced by more than 93% in storage.

Figure 3d shows the results of the schemes against a combination of attacks based on a possible scenario. It appears that Comp2 performs better than DIW (which performs better than Comp1), presumably due to the resizing component.

4   Conclusions

A new principled domain independent watermarking framework is presented and examined. Experiments show highly promising performance in comparison with other state of the art methods on a limited set of attacks. The attacks include four of the most common attacks: white noise attack, JPEG lossy compression, resizing and a combination of attacks.

The main advance is that since the watermarking combines an information-theoretic embedding across a space of statistically independent sources, the same technique works across different media. Being based on local information and a linear transform, our method is economical in the computational costs required (unlike global methods relying on non-linear transforms like Comp1) and offers additional security in the use of *specific* mixing/de-mixing matrices that are not easy to obtain (in contrast to methods based on a simple transformation like Comp1 and Comp2). Further research will focus on theoretical aspects of this scheme, optimizing the decoding process and other improvements of its robustness against specific attacks.

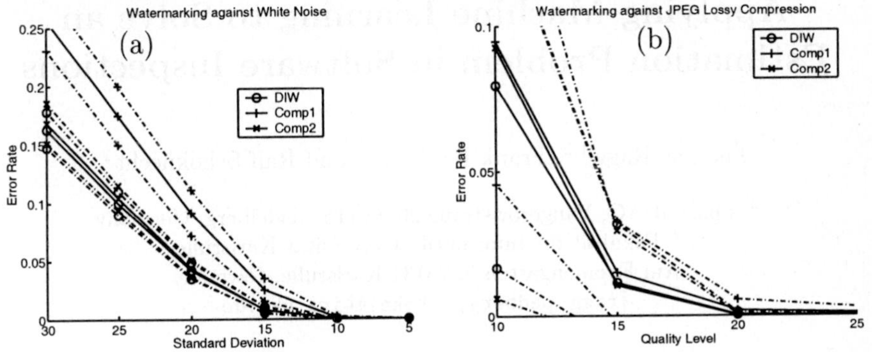

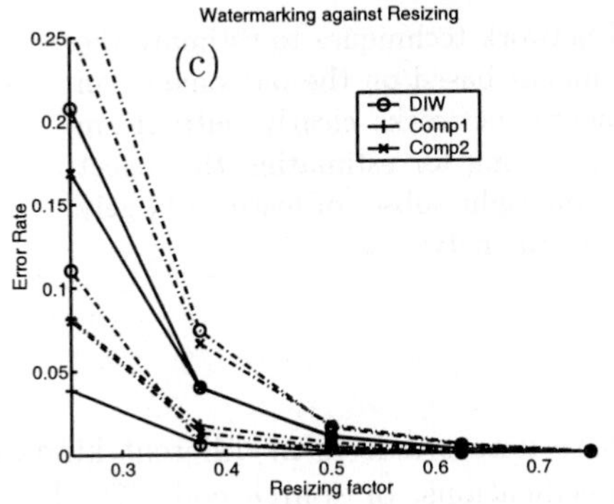

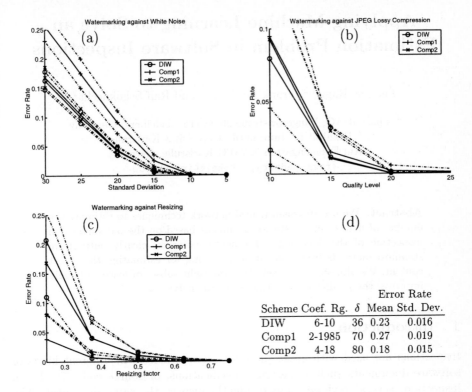

(d)

Scheme	Coef. Rg.	δ	Error Rate Mean	Error Rate Std. Dev.
DIW	6-10	36	0.23	0.016
Comp1	2-1985	70	0.27	0.019
Comp2	4-18	80	0.18	0.015

Fig. 3. The performance of the three watermarking tested: DIW, Comp1 and Comp2, against various attacks; solid lines and symbols represent the mean values; dashed lines denote error bars. (a) White noise of different standard deviation values. (b) JPEG lossy compression for different quality levels. (c) Resizing for different factor. (d) Combination of attacks: resizing 0.5, followed by JPEG 70, followed by WN 15, the attack distortion has a PSNR of about 23 dB.

References

1. Hyvärinen, A., Karhunen, J., Oja, E.: *Independent Component Analysis.* Wiley-Interscience, NY (2001).
2. Cox, I., Miller, M. L., Bloom, J. A.: *Digital Watermarking.* Principles and Practice, Morgan Kaufmann, SF (2001).
3. Petitcolas, F. A. P., Anderson, R. J.: *Evaluation of Copyright Marking Systems.* IEEE Inter. Conf. on Multimedia Computing and Systems. **1**, 574–579 (1999).
4. Petitcolas, F. A. P., Anderson, R. J., Kuhn, M. G.: *Information Hiding - a Survey.* Proceeding of IEEE Multimedia Systems 99. **87-7**, 1062–1078 (1999).
5. B. Chen and G.W. Wornell, *Quantization Index Modulation : A Class of Provably Good Methods for Digital Watermarking and Information Embedding*, IEEE Trans. Inform. Theory. **47-4**, 1423–1443 (2001).

Applying Machine Learning to Solve an Estimation Problem in Software Inspections

Thomas Ragg [1,2], Frank Padberg [2], and Ralf Schoknecht [2]

[1] phase-it AG, Vangerowstraße 20, 69115 Heidelberg, Germany
[2] Fakultät für Informatik, Universität Karlsruhe,
Am Fasanengarten 5, 76131 Karlsruhe, Germany
{ragg, padberg, schokn}@ira.uka.de

Abstract. We use Bayesian neural network techniques to estimate the number of defects in a software document based on the outcome of an inspection of the document. Our neural networks clearly outperform standard methods from software engineering for estimating the defect content. We also show that selecting the right subset of features largely improves the predictive performance of the networks.

1 Introduction

Inspections are used with great success to detect defects in different kinds of software documents such as designs, specifications, or source code [7]. In an inspection, several reviewers independently inspect the same document. The outcome of an inspection is a zero-one matrix showing which reviewer detected which defect. Some defects will be detected by more than one reviewer, but usually not all the defects contained in a document are detected. How many defects actually are contained in a document is unknown. To have a basis for management decisions such as whether to re-inspect a document or to pass it on, it is important in software engineering practice to *reliably estimate* the number of defects in a document from the outcome of an inspection.

Defect content estimation methods for software inspections currently fall into two categories: capture-recapture methods [6,12] and curve-fitting methods [15]. Both approaches use the zero-one matrix of the inspection as the only input to compute the estimate. Several studies show that the defect content estimates computed by these methods are much too unreliable to be used in practice [2,4,12,14]. Both methods show extreme outliers and a high variation in the error of the estimates. A possible explanation is that these methods do not take into account the experience made in past inspections (no learning).

In this paper, we view defect content estimation for software inspections as a *machine learning* problem: the goal is to learn from empirical data collected during *past* inspections the relationship between certain observable features of an inspection and the true number of defects in the document being inspected. A typical example of an observable feature is the total number of different defects detected in an inspection. With our approach, knowledge gained in the past is exploited in the estimation process.

J.R. Dorronsoro (Ed.): ICANN 2002, LNCS 2415, pp. 516–521, 2002.

To solve the machine learning problem, we apply the following techniques. We use feature selection based on *mutual information*. As estimation models, we train *neural networks*. For training, we take a *Bayesian* approach and use an error function with a *regularization* term. The selection of the final model is based on the *model evidence*. We have applied this framework successfully in other application domains [10].

The application of neural networks to defect content estimation in software inspections is novel, although neural networks have previously been used in software reliability. Closest to our work are Khoshgoftaar and Szabo [8], but their use of neural network techniques is improper. To estimate the number of defects in software modules, they use 10 different static code metrics as input features. Even after having reduced the number of features using principal component analysis, their training dataset is too small to avoid overfitting a non-linear model. In addition, when training a network they keep adding hidden units and layers until the network achieves a prescribed error bound on the training data, which occurs not until the network has 24 hidden units. No regularization or model selection techniques are used. As a result, the predictive performance of their networks is completely insufficient for software engineering practice.

We validate our machine learning approach applying a jackknife technique to an empirical inspection dataset. We also compare against a particular capture-recapture method (Mt (MLE) [6]) and against a particular curve-fitting method (DPM [15]). The machine learning approach achieves a mean of absolute relative errors of 5 percent. This is an improvement by a factor of 4, respectively, 7, as compared to the other approaches. In addition, no outlier estimates occur for this dataset when using machine learning.

2 Feature Selection and Bayesian Learning

2.1 Data Collection and Feature Generation

The machine learning approach requires a database containing empirical data about past inspections. For each inspection in the database, we need to know the zero-one matrix of the inspection and the true number num of defects in the inspected document. The zero-one matrix is compiled during any inspection. The true number of defects in a document can be approximated by adding up all the defects detected in the document during development and maintenance. This is sufficient for practical purposes; defects which are not detected even during deployment of the software are not relevant for its operational profile.

From the zero-one matrix of an inspection, we generate a set of five candidate features:

- the total number tdd of different defects detected in the inspection;
- the average, maximum, and minimum number ave, max, min of defects detected by a reviewer;
- the standard deviation std of the number of defects detected by a reviewer.

These features are measures for the overall success of an inspection, respectively, for the performance of the individual reviewers.

Our empirical dataset consists of 16 inspections which were conducted on different specification documents during controlled experiments [1]. For each inspection in the dataset, we know its zero-one matrix as well as the true number of defects contained in the document, because the defects had been seeded into the documents.

2.2 Feature Ranking and Subset Selection

Estimating a non-linear dependency between the input features and the target will be more robust when the input-target space is low-dimensional ("empty space phenomenon"). As a rule of thumb, we deduce from Table 4.2 in [13] that for a dataset of size 16 at most two features should be used as input.

To select the two most promising features from our five candidate features, we use a forward selection procedure based on *mutual information*. The mutual information $MI(X;T)$ of two random vectors X and T is defined as

$$MI(X;T) = \iint p(x,t) \cdot \log \frac{p(x,t)}{p(x)\,p(t)}.$$

The mutual information measures the degree of stochastic dependence between the two random vectors [5]. To compute the required densities from the empirical data, we use Epanechnikov kernel estimators [10,13].

As the first step in the selection procedure, that feature f is selected from the set of candidate features which has maximal mutual information with the target for the given dataset. For example, to maximize $MI(f;num)$ the total number tdd of different defects detected in the inspection is selected. In each subsequent step, that one of the remaining features is selected which maximizes the mutual information with the target when *added* to the already selected features. For example, to maximize $MI((tdd,f);num)$ in the second step, the standard deviation std of the reviewers' inspection results is selected. This way the features are ranked by the amount of information which they add about the target. In our example, the ranking is (tdd, std, max, min, ave). As input to the neural networks, we use the two features tdd and std.

2.3 Regularization and Bayesian Learning

The functional relationship between the input features and the target which has been learned from the empirical data must generalize to previously unseen data points. The theory of regularization shows that approximating the target as good as possible on the training data, for example, by minimizing the mean squared error E_D on the training data, is not sufficient: it is crucial to balance the training error against the model complexity [3]. Therefore, we train the neural networks to minimize the *regularized error* $E = \beta \cdot E_D + \alpha \cdot E_R$. The regularization term E_R measures the model complexity, taking into account the weights w_k in the network. We choose the weight-decay $\frac{1}{2} \sum w_k^2$ as the regularization term.

The factors α and β are additional parameters. Instead of using cross-validation, we take a Bayesian approach to determine the weights w_k and the parameters α and β during training [9,10]. This is done iteratively: we fix α and β and optimize the weights w_k using the fast gradient descent algorithm Rprop [11]. Afterwards, we update α and β; see [10] for the update rule. We alternate several times between optimizing the weights and updating α and β.

2.4 Model Evidence and Selection

To find the model which best explains the given dataset, we systematically vary the number h of hidden units in the networks from 1 to 10. Since the dataset is small, we put all hidden units in a single layer, restricting the search space to models with moderate non-linearity.

For each network topology, that is, for each number of hidden units, we train 50 networks. As the final model, we select the network which maximizes the *posterior probability* $P(\theta \mid D)$, where $\Theta = (\underline{w}, \alpha, \beta, h)$ is the parameter vector and $\underline{w}$ is the weight vector of the network. To determine the final model, we again use a Bayesian approach. We assign the same prior to each topology h, integrate out α, β, and $\underline{w}$, and estimate the posterior probability by the *model evidence* $P(D \mid h)$ for which a closed expression can be given [10]. The model evidence is known to be in good correlation with the generalization error as long as the number of hidden units is not too large [3]. Therefore, instead of choosing the network with the best evidence from all 500 trained networks, we first choose the topology which has the best *average* evidence. From the 50 networks with the best-on-average topology we select the network with the best evidence as the final model.

Table 1 shows the model selection when the second datapoint is left out from our dataset as the test pattern and the remaining 15 datapoints are used for training. For 1 to 5 hidden units, the correlation between the model evidence and the test error of the corresponding 50 networks is strongly negative, whereas for larger networks the correlation is positive (or absent). The test error grows as

Table 1. Example for model selection.

hidden units	1	2	3	4	5	6	7	8	9	10
mean evidence	-22.6	**-22.5**	-23.6	-24.0	-25.7	-34.3	-35.3	-40.2	-40.7	-40.8
best evidence	-19.5	**-20.0**	-19.2	-19.3	-18.6	-18.5	-18.5	-19.9	-15.9	-16.3
rel. test error	3.5	**3.5**	3.7	3.7	4.0	4.3	5.0	4.2	3.9	5.2
correlation	-0.46	**-0.86**	-0.67	-0.89	-0.77	0.55	0.44	0.25	0.05	-0.02

the number of hidden units increases. Since the networks with two hidden units show the best average evidence, the selection procedure chooses the best model with two hidden units as the final model. The minimal test error is reached with one or two hidden units; thus, the selection procedure yields a network with good predictive performance.

3   Validation and Results

To validate the machine learning approach, we apply a jackknife to our empirical dataset. One by one, we leave out an inspection from the dataset as the test pattern, use the remaining 15 inspections as the training patterns, and compute the relative estimation error for the test pattern. For the inspection left out, we also compute the error for the linear regression model, the capture-recapture method Mt(MLE) [6] and the curve-fitting method DPM [15]. Recall that the last two methods take as input only the zero-one matrix of the inspection which is to be estimated.

The results are given in Table 2. The neural network approach achieves a mean of absolute relative errors of 5.3 percent. The other methods show high

Table 2. Relative estimation errors for the 16 test patterns.

pattern	1	2	3	4	5	6	7	8	9	10	11	12	13	14	15	16
Mt	-20	-27	-17	-24	-22	-25	-14	-25	-39	-67	0	0	-47	-33	-7	-13
DPM	-7	-20	-3	-17	-11	-29	18	-7	-34	-61	73	27	-40	-27	113	87
linear	3	-20	3	-3	4	-14	14	-18	0	-22	33	33	-13	7	0	7
NN	-3	-3	-3	-3	4	4	4	-14	-17	-17	13	0	0	0	0	0

estimation errors and a high error variation. The mean errors are 12.1 percent for the linear model, 23.7 percent for Mt, and 35.8 percent for DPM. Clearly, the machine learning approach yields a strong improvement over the standard methods from software engineering. The improvement over the linear model shows that the function to be learned has a non-linear component.

Using more input features for fitting the unknown function *decreases* the performance of the models significantly. In Figure 1, the number of input features is varied from 1 to 5 according to the ranking of the features given in subsection 2.2. For each set of features, the model selection procedure of subsection 2.4 is applied to each of the 16 jackknife datasets; then, the average model evidence and testing error over the 16 resulting models is computed. The average model evidence is highest when only two features are used, namely, tdd and std. The average testing error is minimal with these two input features. This result experimentally justifies having selected two input features as was suggested by the rule of thumb in subsection 2.2.

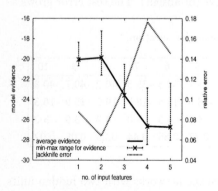

Fig. 1. The jackknife error of the machine learning approach when using input vectors of increasing dimensionality.

The feature tdd is an important input for the estimation because it gives a lower bound for the number of defects in the document. Yet, two rather different inspections can lead to the same total number of different defects detected. For example, in one inspection some reviewers might detect a large number of defects while others detect only a few defects; in some other inspection, each reviewer might detect about the same number of defects. The feature std distinguishes between two such cases, thus being an important supplement to the feature tdd.

The process we have described for building defect content estimation models for software inspections can easily be deployed in a business environment. The process can run automatically without constant interaction by a machine learning specialist. In particular, the estimation models can automatically adapt to new empirical data.

References

1. Basili, Green, Laitenberger, Lanubile, Shull, Sorumgard, Zelkowitz: "The Empirical Investigation of Perspective-Based Reading", Empirical Software Engineering 1:2 (1996) 133-164
2. Biffl, Grossmann: "Evaluating the Accuracy of Defect Estimation Models Based on Inspection Data From Two Inspection Cycles", Proceedings International Conference on Software Engineering ICSE 23 (2001) 145-154
3. Bishop: *Neural Networks for Pattern Recognition.* Oxford Press, 1995
4. Briand, El-Emam, Freimut, Laitenberger: "A Comprehensive Evaluation of Capture-Recapture Models for Estimating Software Defect Content", IEEE Transactions on Software Engineering 26:6 (2000) 518-540
5. Cover, Thomas: *Elements of Information Theory.* Wiley, 1991
6. Eick, Loader, Long, Votta, Vander Wiel: "Estimating Software Fault Content Before Coding", Proceedings International Conference on Software Engineering ICSE 14 (1992) 59-65
7. Gilb, Graham: *Software Inspection.* Addison-Wesley, 1993
8. Khoshgoftaar, Szabo: "Using Neural Networks to Predict Software Faults During Testing", IEEE Transactions on Reliability 45:3 (1996) 456-462
9. MacKay: "A practical bayesian framework for backpropagation networks", Neural Computation 4:3 (1992) 448-472
10. Ragg, Menzel, Baum, Wigbers: "Bayesian learning for sales rate prediction for thousands of retailers", Neurocomputing 43 (2002) 127-144
11. Riedmiller: "Supervised learning in multilayer perceptrons – from backpropagation to adaptive learning techniques", International Journal of Computer Standards and Interfaces 16 (1994) 265-278
12. Runeson, Wohlin: "An Experimental Evaluation of an Experience-Based Capture-Recapture Method in Software Code Inspections", Empirical Software Engineering 3:3 (1998) 381-406
13. Silverman: *Density Estimation for Statistics and Data Analysis.* Chapman and Hall, 1986
14. Vander Wiel, Votta: "Assessing Software Designs Using Capture-Recapture Methods", IEEE Transactions on Software Engineering 19:11 (1993) 1045-1054
15. Wohlin, Runeson: "Defect Content Estimations from Review Data", Proceedings International Conference on Software Engineering ICSE 20 (1998) 400-409

Clustering of Gene Expression Data by Mixture of PCA Models

Taku Yoshioka, Ryouko Morioka, Kazuo Kobayashi, Shigeyuki Oba,
Naotake Ogawsawara, and Shin Ishii

Nara Institute of Science and Technology,
8916-5 Takayama, Ikoma, Nara 630-0101, Japan

Abstract. Clustering techniques, such as hierarchical clustering, k-means algorithm and self-organizing maps, are widely used to analyze gene expression data. Results of these algorithms depend on several parameters, e.g., the number of clusters. However, there is no theoretical criterion to determine such parameters. In order to overcome this problem, we propose a method using mixture of PCA models trained by a variational Bayes (VB) estimation. In our method, good clustering results are selected based on the free energy obtained within the VB estimation. Furthermore, by taking an ensemble of estimation results, a robust clustering is achieved without any biological knowledge. Our method is applied to a clustering problem for gene expression data during a sporulation of *Bacillus subtilis* and it is able to capture characteristics of the sigma cascade.

1 Introduction

DNA microarrays provide a way of comprehensive analyses of gene expression patterns [4]. By using microarrays, we can measure expression level of thousands of genes simultaneously. Genes are characterized by their expression patterns obtained by experiments in different conditions. Based on the assumption that functionally related genes exhibit similar expression patterns, biologists infer function of unknown genes.

Clustering techniques, such as hierarchical clustering, k-means algorithm and self-organizing maps (SOM), are widely used to analyze gene expression data (see [3] for review). In general, clustering results by these algorithms significantly depend on several factors, e.g., distance measure in the hierarchical clustering, the number of clusters and initial center allocation in the k-means algorithm or SOM. However, there is almost no theoretical way to determine such factors automatically from given expression data. Then, biologists often apply these algorithms many times with various settings and pick up a reasonable result based on their biological inspection.

Mixture of probabilistic models is an alternative for data clustering. A mixture model assumes the following stochastic process from which a datum is generated. First, a cluster is probabilistically selected. Then, according to the

J.R. Dorronsoro (Ed.): ICANN 2002, LNCS 2415, pp. 522–527, 2002.

probabilistic model for the selected cluster, a datum is generated. If the parameters of the mixture model are determined from given data, the probability that a datum belongs to a cluster is calculated. This is considered as "soft" clustering rather than "hard" clustering as the k-means algorithm does. It is advantageous that a soft clustering algorithm can represent uncertainty of the clustering result.

In this article, we propose an automatic method to obtain a reasonable clustering result based on a framework of the Bayes inference. We use mixture of probabilistic PCA (MPCA) models [5] for the cluster analysis. MPCA models are trained by a variational Bayes (VB) estimation. The VB estimation introduces a free energy which gives a lower bound of the evidence (marginal likelihood). In the proposed method, the number of units (clusters) is determined based on the free energy. We also consider an ensemble of MPCA results to obtain a robust clustering result. Our method is applied to gene expression data during a sporulation of *Bacillus subtilis* [1]. Experimental results show that our method is able to capture characteristics of the sporulation process.

2   MPCA Model and Variational Bayes Method

Suppose a d-dimensional observation vector $\boldsymbol{y}$ is generated by a noisy linear transformation from a $q(< d)$-dimensional hidden variable vector $\boldsymbol{x}$:

$$\boldsymbol{y} = \boldsymbol{W}\boldsymbol{x} + \boldsymbol{\mu} + \boldsymbol{\epsilon}. \tag{1}$$

$\boldsymbol{W}$ is a d-by-q transformation matrix and $\boldsymbol{\mu}$ is a d-dimensional vector. The hidden variable vector $\boldsymbol{x}$ obeys a q-dimensional unit normal distribution $\mathcal{N}(\boldsymbol{x}|0, \boldsymbol{I}_q)$ and noise $\boldsymbol{\epsilon}$ obeys a d-dimensional normal distribution $\mathcal{N}(\boldsymbol{\epsilon}|0, \sigma^2 \boldsymbol{I}_d)$ with a variance σ^2. A generative model of the probabilistic PCA [5], $p(\boldsymbol{y}, \boldsymbol{x}|\theta)$, is defined by the above stochastic process, where θ is a parameter set: $\theta \equiv \{\boldsymbol{W}, \boldsymbol{\mu}, \sigma^2\}$. The probability of an observation is given by integrating out the hidden variable: $p(\boldsymbol{y}|\theta) = \int d\boldsymbol{x} p(\boldsymbol{y}, \boldsymbol{x}|\theta)$. If we assume a diagonal but not univariate variance for $\boldsymbol{\epsilon}$, the stochastic process (1) defines a probabilistic factor analysis.

An MPCA model is defined as a mixture of M PPCA components:

$$p(\boldsymbol{y}|\boldsymbol{g}, \{\theta_k\}_{k=1}^M) = \sum_{k=1}^M p(\boldsymbol{y}|\theta_k) g_k, \tag{2}$$

where $p(\boldsymbol{y}|\theta_k)$ is the k-th PPCA component model and $\boldsymbol{g} = (g_1, ..., g_M)$ is a mixing rate vector that satisfies $\sum_{k=1}^M g_k = 1$. Tipping and Bishop derived a maximum likelihood estimation for the MPCA model [5].

One of the objectives of the Bayes inference is to obtain the evidence (marginal likelihood) of a given dataset $Y = \{\boldsymbol{y}_i\}_{i=1}^N$:

$$p(Y|\mathcal{M}) = \int d\theta p(Y|\theta, \mathcal{M}) p_0(\theta|\mathcal{M}), \tag{3}$$

where θ is a set of model parameters and $p_0(\theta|\mathcal{M})$ is a prior distribution for the parameters. $\mathcal{M}$ denotes the model, e.g., an MPCA model with a particular number of PPCA components. Equation (3) suggests that the evidence is the likelihood of model $\mathcal{M}$ with the given dataset Y. Then, good models can be selected based on the evidence. Since the integration in equation (3) is intractable for MPCA models, however, some approximation methods are necessary. A variational Bayes (VB) estimation can deal with the intractability. In the previous study, we formulated a VB estimation for the MPCA model [2]. In the VB estimation, a free energy is introduced for a model and it gives a lower bound of the evidence. Therefore, good models can be selected based on the free energy instead of the evidence.

3   Gene Expression Data

Bacillus subtilis forms a spore in response to stresses such as starvation. This process is referred to as a sporulation. When a *B. subtilis* detects a stress, it starts a sporulation process, which can be divided into characteristic stages. Transcriptional factors called sigma factors regulate genes related to the sporulation process. Although there have been known 19 sigma factors, five of them (sigH, sigE, sigF, sigG and sigK) are most important in the sporulation. Roughly speaking, each of genes related to the sporulation process is activated by one of the sigma factors. Namely, each of the genes has a single label that corresponds to a sigma factor. However, true labels are unknown except for a small number of genes.

When a process starts, expression of sigH is activated. Then, the cell is divided into a mother cell and a prespore. During the separation, sigE and sigF are activated in the mother cell and the prespore, respectively. After the separation, the prespore becomes a forespore, and two programs in the mother cell and the forespore proceed concurrently. In the forespore, sigG is activated by sigF after sigE is activated in the mother cell. Finally, sigG activates sigK in the mother cell. The activation sequence of the sigma factors is referred to as a sigma cascade.

This study intends to capture the above-mentioned characteristics of the sigma cascade, by applying our clustering method based on MPCA models to expression data in the sporulation process. It should be noted that we do not utilize the biological knowledge on the sigma cascade in the clustering analysis.

4   Evaluation of Clustering Results

In order to evaluate a clustering result, we see the correspondence between clusters and gene groups $G = \{G_l\}_{l=1}^L$, where G_l consists of genes regulated by a sigma factor l. We can utilize the labels to evaluate a clustering result. However, the labels may be noisy due to the interpretation ambiguity of gene functions.

Let $p_i(m)$ be the cluster membership probability that gene i is in the m-th cluster. $p_i(m)$ is obtained from an MPCA result. We estimate the conditional probability that the m-th cluster corresponds to the l-th gene group by

$$p(l|m) = \frac{\sum_{\{i \in G_l\}} p_i(m) + \alpha}{\sum_{l'=1}^L \sum_{\{i \in G_{l'}\}} p_i(m) + L\alpha}, \tag{4}$$

where $\alpha(=1.0)$ is a constant. Given a prior probability of cluster indices, $p(m)$, the joint distribution $p(l,m)$ is obtained. Here, we assume $p(m) = 1/M$. A clustering result is evaluated from a biological point of view by the mutual information

$$I = -\sum_{l=1}^{L} p(l) \ln p(l) + \sum_{m=1}^{M} p(m) \sum_{l=1}^{L} p(l|m) \ln p(l|m). \tag{5}$$

5 Result

By using *B. subtilis* microarrays, expression data of 2,815 genes are obtained every 30 minutes during the sporulation that proceeds over 9 hours. Consequently, expression pattern of each gene is obtained as a time sequence represented by a 19 dimensional vector. When measuring the expression level of genes, brightness of the scanned microarray varies in one experiment to others. Since this results in the fluctuation of the average expression level, we first adjust the mean of the expression data for 2,815 genes to zero. A temporal smoothing filter is then applied to the adjusted data. For discrete time-series data x_i ($i = 1, ..., D$), the smoothed data $y(t)$ is obtained by [6]

$$y(t) = \frac{1}{F} \sum_{i=1}^{D} F_k(t_i - t) x(t_i) \qquad F = \sum_{i=1}^{D} F_k(t_i - t),$$

$$F_k(t) = S(t) \left(\frac{t}{\tau}\right)^{k-1} e^{-t/\tau} \ (k \geq 1), \tag{6}$$

where $S(t)$ is a step function. We set time constant τ equal to the experiment interval (30 minutes) and $k = 1$. t_i is the time of the i-th experiment. Although a smoothed datum $y(t)$ can be obtained for arbitrary time, we calculate the smoothed data for $\{t_i\}_{i=1}^{19}$. In the sporulation process, expression of genes that are not related to the sporulation is relatively down regulated. According to this observation, genes whose average expression level is below a certain threshold ($=0$) are removed. Then, 1,351 genes are retained.

MPCA models are applied to the 1,351 smoothed gene expression data. For each number of units between 3 and 20, 200 MPCA models are randomly initialized, and they are trained by the VB algorithm. Consequently, 3,600 clustering results are obtained. In figure 1, the horizontal and vertical axes denote the number of units and the free energy after convergence, respectively. From this figure, 7 units models are statistically plausible. In figure 2, the horizontal and the vertical axes denote the free energy and the mutual information, respectively. The two values are positively correlated. Thus, the free energy can be used as a criterion for choosing clustering results that are consistent with the biological labels.

As in figure 2, the variation of the mutual information is large where the free energy is large. Thus, it is still suspicious to infer biological functions (labels)

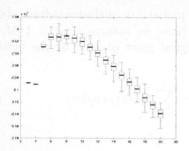
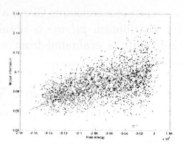

Fig. 1. The number of units and the free energy.

Fig. 2. The free energy and the mutual information.

only from the model with the largest free energy value. We consider here an ensemble of good clustering results that have the same number of clusters. The clustering result associated with the largest free energy model in the ensemble is regarded as the reference clustering result. For each cluster in another clustering result (associated with another model), the corresponding cluster in the reference clustering result is determined as follows. Let $p_i^r(m_r)$ be the cluster membership probability that the i-th gene belongs to the m_r-th reference cluster. Similarity between cluster m_r in the reference and cluster m in another clustering result is measured by

$$C(m_r, m) = \sum_{i=1}^{N} p_i^r(m_r) p_i(m). \qquad (7)$$

The cluster index in the reference, corresponding to cluster m in another clustering result, is determined as $argmax_{m_r'} C(m_r', m)$. The final clustering result is obtained by averaging the cluster membership probability over the corresponding clusters in the ensemble.

We use clustering results that have 7 clusters, because such results are plausible according to figure 1. Clustering results associated with the top 5 free energy models are combined. By averaging the cluster center vector over the corresponding clusters, representative expression pattern for each cluster is obtained (figure 3). According to this figure, genes in cluster 4 are expressed in the early stage in the sporulation process. Then, genes in cluster 2, 6 and 5 are expressed in sequence. Finally, genes in cluster 7 are expressed. This order is almost consistent with the labels of genes assigned to each cluster based on the averaged cluster membership probability (see table 1).

6   Conclusion

In this study, we proposed a clustering method for gene expression data using mixture of PCA models trained by the VB estimation. Experimental results showed that the free energy can be used as a criterion for choosing good clustering

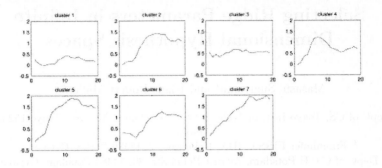

Fig. 3. Representative expression pattern of each of 7 clusters in the ensemble clustering result.

results. By taking an ensemble of good clustering results, a reasonable clustering result was obtained without *a priori* biological knowledge.

Table 1. Assignment of genes to the clusters. A number next to a gene name is the cluster index to which the gene belongs. It is determined by the maximum of the averaged cluster membership probability.

sigH	sigF	sigE		sigG		sigK
spoIIAB (4)	spoIIQ (2)	yaaH (6)	spoVE (6)	sspF (5)	gerAA (5)	cotD (7)
sigF (4)		spoVB (6)	yrbA (5)	sspE (5)	sspB (5)	ykvP (7)
spoIIAA (2)		spoIVCA (5)	spoIID (1)	spoIVB (1)	spoVT (1)	cotX (7)
		spoVD (1)	spoIIIA (2)	gdh (1)	gerBB (2)	spoVFB (7)
				spoIVB (1)	gerAC (3)	spoVFA (7)
				spoVT (1)	gerAB (7)	cotS (7)

References

1. Driks A, and Losick R.: Bacillus Subtilis spore coat. Microbiol. Mol. Biol. Rev. **63**(1) (1999) 1–20
2. Oba, S., Ishii, S. and Sato, M.: Variational Bayes method for mixture of principal component analyzers. Proceedings of 7th International Conference on Neural Information Processing (2000) 1416–1421
3. Quackenbush, J.: Computational analysis of microarray data. Nature Reviews Genetics **2**(6) (2001) 418–427
4. Schena, M., Shalon, D., Davis, R.W. and Brown, P.O.: Quantitative monitoring of gene expression patterns with a complementary DNA microarray. Science **270** (1995) 467–470
5. Tipping, M.E. and Bishop, C.M.: Mixtures of probabilistic principal component analyzers. Neural Computation **11** (1999) 443–482
6. Yoshida, W., Ishii, S. and Sato, M.: Reconstruction of chaotic dynamics using a noise-robust embedding method. IEEE International Conference on Acoustics, Speech, and Signal Processing (ICASSP) **I** (2000) 181–184

Selecting Ridge Parameters in Infinite Dimensional Hypothesis Spaces

Masashi Sugiyama[1] and Klaus-Robert Müller[2,3]

[1] Dept. of CS, Tokyo Inst. of Tech., 2-12-1, Ookayama, Meguro, Tokyo 152-8552, Japan
[2] Fraunhofer FIRST, IDA, Kekulestr. 7, 12489 Berlin, Germany
[3] Dept. of CS, U Potsdam, August-Bebel-Str. 89, 14482 Potsdam, Germany
sugi@og.cs.titech.ac.jp, klaus@first.fhg.de

Abstract. Previously, an unbiased estimator of the generalization error called the subspace information criterion (SIC) was proposed for a finite dimensional reproducing kernel Hilbert space (RKHS). In this paper, we extend SIC so that it can be applied to any RKHSs including *infinite* dimensional ones. Computer simulations show that the extended SIC works well in ridge parameter selection.

1 Introduction

Estimating the generalization capability is one of the central issues in supervised learning. So far, a large number of generalization error estimation methods have been proposed (e.g. [1,9,4,8,5]).

Typically an asymptotic limit in the number of training samples is considered [1,9,4]. However, in supervised learning, the small sample case is of high practical importance. Hence methods that work in the finite sample case as e.g. the VC-bound [8], which gives a probabilistic upper bound of the generalization error, are becoming increasingly popular.

Another generalization error estimation method that works effectively with finite samples is the subspace information criterion (SIC) [5]. Among several interesting theoretical properties, SIC is proved to be an unbiased estimator of the generalization error. The original SIC has been successfully applied to the selection of subspace models in linear regression. However, its range of applicability is limited to the case where the learning target function belongs to a specified *finite* dimensional reproducing kernel Hilbert space (RKHS).

In this paper, we therefore extend SIC so that it can be applied to any RKHSs including *infinite* dimensional ones. We further show that when the kernel matrix is invertible, SIC can be expressed in a much simpler form, making its computation highly efficient. Computer simulations underline that the extended SIC works well in ridge parameter selection.

J.R. Dorronsoro (Ed.): ICANN 2002, LNCS 2415, pp. 528–534, 2002.
© Springer-Verlag Berlin Heidelberg 2002

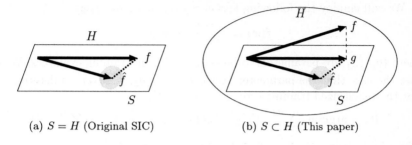

(a) $S = H$ (Original SIC)   (b) $S \subset H$ (This paper)

Fig. 1. Original SIC and extension carried out in this paper. H is a reproducing kernel Hilbert space that includes the learning target function f. S is the subspace spanned by $\{k(\boldsymbol{x}, \boldsymbol{x}_m)\}_{m=1}^{M}$. g is the orthogonal projection of f onto S. (a) Setting of the original SIC [5]. It was shown that when $S = H$, SIC is an unbiased estimator of the generalization error between $\hat{f}$ and f with finite samples. $S = H$ implies that the RKHS H whose dimension is at most M ($< \infty$) is considered. (b) Setting of this paper. We consider the case that $S \subset H$, which allows any RKHS H including infinite dimensional ones. We show that the extended SIC is an unbiased estimator of the generalization error between $\hat{f}$ and g.

2 Supervised Learning and Kernel Ridge Regression

Let us discuss the regression problem of approximating a target function from a set of M *training examples*. Let $f(\boldsymbol{x})$ be a *learning target function* of L variables defined on a subset $\mathcal{D}$ of the L-dimensional Euclidean space $\mathbb{R}^L$. The training examples consist of *sample points* $\boldsymbol{x}_m$ in $\mathcal{D}$ and corresponding *sample values* y_m in $\mathbb{R}$: $\{(\boldsymbol{x}_m, y_m) \mid y_m = f(\boldsymbol{x}_m) + \epsilon_m\}_{m=1}^{M}$, where y_m is degraded by unknown additive noise ϵ_m. We assume that ϵ_m is independently drawn from a distribution with mean zero and variance σ^2. The purpose of regression is to obtain the optimal approximation $\hat{f}(\boldsymbol{x})$ to the learning target function $f(\boldsymbol{x})$ that minimizes a *generalization error*.

In this paper, we assume that the unknown learning target function $f(\boldsymbol{x})$ belongs to a specified *reproducing kernel Hilbert space* (RKHS) H [9,8]. We denote the reproducing kernel of H by $k(\boldsymbol{x}, \boldsymbol{x}')$. In previous work [5], it was assumed that $\{k(\boldsymbol{x}, \boldsymbol{x}_m)\}_{m=1}^{M}$ span the whole RKHS H (Figure 1 (a)). This holds only if $\dim H \le M$ ($< \infty$). In contrast, we do not impose any restriction on the dimension of the RKHS H in this work. Possibly the dimension is infinity, so we can treat a rich class of function spaces such as e.g. a Gaussian RKHS (Figure 1 (b)). We measure the generalization error of $\hat{f}(\boldsymbol{x})$ by

$$J_G = \mathrm{E}_\epsilon \|\hat{f} - f\|^2, \tag{1}$$

where E_ϵ denotes the expectation over the noise and $\| \cdot \|$ is the norm in the RKHS H. This generalization measure is commonly used in the field of function approximation (e.g. [3]). Since Eq.(1) includes the unknown learning target function $f(\boldsymbol{x})$, it cannot be directly calculated. The aim of this paper is to give an estimator of Eq.(1) that can be calculated without using $f(\boldsymbol{x})$.

We will employ the following kernel regression model $\hat{f}(\boldsymbol{x})$:

$$\hat{f}(\boldsymbol{x}) = \sum_{p=1}^{M} \theta_p k(\boldsymbol{x}, \boldsymbol{x}_p), \tag{2}$$

where $\{\theta_p\}_{p=1}^{M}$ are parameters to be estimated from training examples. We consider the case that the parameter vector $\boldsymbol{\theta} = (\theta_1, \theta_2, \ldots, \theta_M)^\top$ is determined so that the regularized training error is minimized[1]:

$$\hat{\boldsymbol{\theta}}_\alpha = \operatorname{argmin}_{\boldsymbol{\theta}} (\sum_{m=1}^{M} (\sum_{p=1}^{M} \theta_p k(\boldsymbol{x}_m, \boldsymbol{x}_p) - y_m)^2 + \alpha \sum_{p=1}^{M} \theta_p^2), \tag{3}$$

where α is a positive scalar called the *ridge parameter*. Let $\boldsymbol{y} = (y_1, y_2, \ldots, y_M)^\top$, I denote the identity matrix, and K be the M-dimensional matrix with the (m, p)-th element $k(\boldsymbol{x}_m, \boldsymbol{x}_p)$. Then $\hat{\boldsymbol{\theta}}_\alpha$ is given by

$$\hat{\boldsymbol{\theta}}_\alpha = (\hat{\theta}_1, \hat{\theta}_2, \ldots, \hat{\theta}_M)^\top = X_\alpha \boldsymbol{y}, \quad \text{where} \quad X_\alpha = (K^2 + \alpha I)^{-1} K. \tag{4}$$

3   SIC for Infinite Dimensional RKHSs

First, we will briefly review the original SIC [5] that is applicable when $\{k(\boldsymbol{x}, \boldsymbol{x}_m)\}_{m=1}^{M}$ span the whole RKHS H.

When the functions $\{k(\boldsymbol{x}, \boldsymbol{x}_m)\}_{m=1}^{M}$ span the whole space H, the learning target function $f(\boldsymbol{x})$ is expressed as $f(\boldsymbol{x}) = \sum_{p=1}^{M} \theta_p^* k(\boldsymbol{x}, \boldsymbol{x}_p)$, where the true parameter vector $\boldsymbol{\theta}^* = (\theta_1^*, \theta_2^*, \ldots, \theta_M^*)^\top$ is unknown[2]. Letting $\|\boldsymbol{\theta}\|_K^2 = \langle K\boldsymbol{\theta}, \boldsymbol{\theta} \rangle$, the generalization error J_G is expressed as $J_G = E_\epsilon \|\hat{\boldsymbol{\theta}}_\alpha - \boldsymbol{\theta}^*\|_K^2$.

The key idea of SIC is to assume that a learning matrix X_u that gives an unbiased estimator $\hat{\boldsymbol{\theta}}_u$ of the unknown $\boldsymbol{\theta}^*$ is available:

$$E_\epsilon \hat{\boldsymbol{\theta}}_u = \boldsymbol{\theta}^*, \quad \text{where} \quad \hat{\boldsymbol{\theta}}_u = X_u \boldsymbol{y}. \tag{5}$$

Using $\hat{\boldsymbol{\theta}}_u$, the generalization error J_G is roughly estimated by $\|\hat{\boldsymbol{\theta}}_\alpha - \hat{\boldsymbol{\theta}}_u\|_K^2$. Performing some approximations based on this idea, the subspace information criterion (SIC) is given as follows [5]:

$$SIC = \|\hat{\boldsymbol{\theta}}_\alpha - \hat{\boldsymbol{\theta}}_u\|_K^2 - \sigma^2 \operatorname{tr}\left(K(X_\alpha - X_u)(X_\alpha - X_u)^\top\right) + \sigma^2 \operatorname{tr}\left(K X_\alpha X_\alpha^\top\right), \tag{6}$$

where $\operatorname{tr}(\cdot)$ denotes the trace of a matrix. The name *subspace information criterion* (SIC) was first introduced for selecting subspace models. It was shown in [5] that Eq.(6) is an unbiased estimator of the generalization error J_G, i.e., $E_\epsilon SIC = J_G$. When the noise variance σ^2 in Eq.(6) is unknown, one of the practical methods for estimating σ^2 is given by $\hat{\sigma}^2 = \|K X_\alpha \boldsymbol{y} - \boldsymbol{y}\|^2 / (M - \operatorname{tr}(K X_\alpha))$ [9].

SIC requires a learning matrix X_u that gives an unbiased estimate $\hat{\boldsymbol{\theta}}_u$ of the true parameter $\boldsymbol{\theta}^*$. When $\{k(\boldsymbol{x}, \boldsymbol{x}_m)\}_{m=1}^{M}$ span the whole RKHS H, such X_u surely exists and is given by

$$X_u = K^\dagger, \tag{7}$$

---

[1] Note that the discussion in this article is valid for any linear estimators.

[2] When $\{k(\boldsymbol{x}, \boldsymbol{x}_m)\}_{m=1}^{M}$ are over-complete, $\{\theta_p^*\}_{p=1}^{M}$ are not determined uniquely. In this case, we assume that $\boldsymbol{\theta}^*$ is given by $K^\dagger (f(\boldsymbol{x}_1), f(\boldsymbol{x}_2), \ldots, f(\boldsymbol{x}_M))^\top$, where † denotes the Moore-Penrose generalized inverse.

where $\dagger$ denotes the Moore-Penrose generalized inverse. However, obtaining X_u when $\{k(\boldsymbol{x}, \boldsymbol{x}_m)\}_{m=1}^M$ do not span the whole RKHS H is an open problem that we aim to solve in the following.

Now, we consider the case when $\{k(\boldsymbol{x}, \boldsymbol{x}_m)\}_{m=1}^M$ do not span the whole RKHS H, possibly $\dim H$ is infinity. Let S be a subspace spanned by $\{k(\boldsymbol{x}, \boldsymbol{x}_m)\}_{m=1}^M$. Since the learning target function $f(\boldsymbol{x})$ does not generally lie in the subspace S, $f(\boldsymbol{x})$ can be decomposed as $f(\boldsymbol{x}) = g(\boldsymbol{x}) + h(\boldsymbol{x})$, where $g(\boldsymbol{x})$ belongs to the subspace S and $h(\boldsymbol{x})$ is orthogonal to S. Then the generalization error can be expressed as $\mathrm{E}_\epsilon \|\hat{f} - f\|^2 = \mathrm{E}_\epsilon \|\hat{f} - g\|^2 + \|h\|^2$. Since the second term $\|h\|^2$ is irrelevant to $\hat{f}$, we ignore it and focus on the first term $\mathrm{E}_\epsilon \|\hat{f} - g\|^2$ (see Figure 1(b)). Let us denote the first term by J_G':

$$J_G' = \mathrm{E}_\epsilon \|\hat{f} - g\|^2. \tag{8}$$

If we regard $g(\boldsymbol{x})$ as the learning target function, then the setting is exactly the same as the original SIC. Therefore, we can apply the idea of SIC and obtain an unbiased estimator of J_G'. However, the problem is that we need a learning matrix X_u that gives an unbiased estimate $\hat{\boldsymbol{\theta}}_u$ of the true parameter[3] $\boldsymbol{\theta}^*$. The following theorem solves this problem.

Theorem 1 [4] *For an arbitrarily chosen RKHS H and the kernel regression model given by Eq.(2), a learning matrix X_u that gives an unbiased estimate $\hat{\boldsymbol{\theta}}_u$ of the true parameter $\boldsymbol{\theta}^*$ is given by*

$$X_u = K^\dagger. \tag{9}$$

Eq.(9) is equivalent to Eq.(7). Therefore, the above theorem shows that SIC is applicable irrespective of the choice of the RKHS H. If $\{k(\boldsymbol{x}, \boldsymbol{x}_m)\}_{m=1}^M$ span the whole RKHS H, SIC is an unbiased estimator of the generalization error J_G. Otherwise SIC is an unbiased estimator of J_G', which is an essential part of the generalization error J_G (see Figure 1 again):

$$\mathrm{E}_\epsilon SIC = J_G'. \tag{10}$$

Now we show that when the kernel matrix K is invertible, SIC can be computed much simpler. Substituting Eq.(9) into Eq.(6), SIC is expressed as

$$SIC = \|\hat{\boldsymbol{\theta}}_\alpha\|_K^2 - 2\langle K\hat{\boldsymbol{\theta}}_\alpha, K^\dagger \boldsymbol{y}\rangle + \|K^\dagger \boldsymbol{y}\|_K^2 + 2\sigma^2 \mathrm{tr}\left(KX_\alpha K^\dagger\right) - \sigma^2 \mathrm{tr}\left(K^\dagger\right). \tag{11}$$

Since the third and fifth terms are irrelevant to α, they can be safely ignored. When K^{-1} exists, a practical expression of SIC for kernel regression is given by

$$\begin{aligned} SIC_{practical} &= \|\hat{\boldsymbol{\theta}}_\alpha\|_K^2 - 2\langle K\hat{\boldsymbol{\theta}}_\alpha, K^{-1}\boldsymbol{y}\rangle + 2\sigma^2 \mathrm{tr}\left(KX_\alpha K^{-1}\right) \\ &= \boldsymbol{y}^\top X_\alpha^\top K X_\alpha \boldsymbol{y} - 2\boldsymbol{y}^\top X_\alpha \boldsymbol{y} + 2\sigma^2 \mathrm{tr}\left(X_\alpha\right). \end{aligned} \tag{12}$$

---

[3] When the true function $f(\boldsymbol{x})$ is not included in the model, the term 'true parameter' is used for indicating the parameter in $g(\boldsymbol{x})$, i.e., $g(\boldsymbol{x}) = \sum_{p=1}^M \theta_p^* k(\boldsymbol{x}, \boldsymbol{x}_p)$.

[4] Proof is available from 'ftp://ftp.cs.titech.ac.jp/pub/TR/01/TR01-0016.pdf'.

Here we have shown that the sequential estimation by Gaussian mixture models derives the boosting-like algorithm for the regression problems. In the next section we study the Gaussian mixture models for the regression problems and derive a sequential algorithm for regression problems which uses the information of the noise distribution more explicitly than the other boosting algorithms.

3   A New Sequential Algorithm for Regression Problems

The algorithm derived in this section uses the estimated result of the noise distribution.

The true regression function $g(x)$ is estimated by $F(x)$ which consists of the combination of the simple predictors, that is, $F(x) = \sum_{t=1}^{T} c_t f(x|\theta_t) + c_0$, where c_t's are real numbers. Furthermore we estimate the probability density of the noise ϵ by Gaussian mixture model. Here we suppose that the distribution of ϵ does not depend on the value of x.

We explain how the algorithm works. At T step the estimators for the regression function and the probability density are supposed to be obtained as $F(x)$ and $\hat{p}(\epsilon)$. Hence the estimated conditional probability is given as $p_T(y|x) = \hat{p}(y - F(x))$. At the next step the conditional probability is updated to $p_{T+1}(y|x) = (1 - \delta)\hat{p}(y - F(x) - cf(x|\theta)) + \delta\phi_\sigma(y - F(x) - cf(x|\theta) - \mu)$. In other words the update of the estimator is written as

$$\{F(x), \hat{p}(\epsilon)\} \longrightarrow \{F(x) + cf(x|\theta) + \delta\mu, \; (1 - \delta)\,\hat{p}(\epsilon + \delta\mu) + \delta\,\phi_\sigma(\epsilon - (1 - \delta)\mu)\}.$$

The problem is to find the appropriate values of $(\theta, c, \delta, \mu, \sigma)$. Here the log-likelihood is used to measure the goodness of fit to the data. If the values of c and δ are small, the log-likelihood of $p_{T+1}(y|x)$ is expanded as

$$\sum_{i=1}^{n} \log p_{T+1}(y_i|x_i) \approx \sum_{i=1}^{n} \log \hat{p}(y_i - F(x_i)) + \delta \sum_{i=1}^{n} \frac{\phi_\sigma(y_i - F(x_i) - \mu)}{\hat{p}(y_i - F(x_i))}$$

$$+ c \sum_{i=1}^{n} (-f(x_i|\theta)) \left. \frac{d}{d\epsilon} \log \hat{p}(\epsilon) \right|_{\epsilon = y_i - F(x_i)}. \tag{5}$$

When $\hat{p}(\epsilon)$ is the Gaussian mixture such as $\sum_{t=1}^{T} a_t \phi_{\sigma_t}(\epsilon - \mu_t)$, the third term of right side in (5) is written as $\sum_{i=1}^{n} v_i \tilde{\epsilon}_i f(x_i|\theta)$, where $v_i = \frac{\sum_{t=1}^{T} w_{ti}}{\hat{p}(y_i - F(x_i))}$, $\tilde{\epsilon}_i = y_i - F(x_i) - \frac{\sum_{t=1}^{T} w_{ti}\mu_t}{\sum_{t=1}^{T} w_{ti}}$ and $w_{ti} = \frac{a_t}{\sigma_t^2}\phi_{\sigma_t}(y_i - F(x_i) - \mu_k)$. This is the weighted correlation between the shifted residual $\tilde{\epsilon}$ and the simple predictor $f(x|\theta)$ by the weight v. The parameter of the simple predictor can be found by maximizing this correlation. The other parameters can be found by EM algorithm [7]. Accordingly we can construct the sequential algorithm for the regression problems as follows.

Suggested Algorithm
Step 1. Set $(F(x), \hat{p}(\epsilon)) = (0, 0)$, the initial weights $v_i = 1$ and the initial shifted residuals $\tilde{\epsilon}_i = y_i$.

Step 2. For $t = 1$ to T

1) Solve $\theta_t = \arg\min_{\theta \in \Theta} \sum_{i=1}^{n} v_i (\tilde{\epsilon}_i - f(x_i|\theta))^2$

2) Let $\hat{\epsilon}_i$ be $y_i - F(x_i)$. Find the values of $(\delta_t, \mu_t, \sigma_t, c_t)$ which maximize
$$\sum_{i=1}^{n} \log \left\{ (1 - \delta)\hat{p}(\hat{\epsilon}_i - cf(x_i|\theta_t)) + \delta\phi_\sigma(\hat{\epsilon}_i - cf(x_i|\theta_t) - \mu) \right\},$$
such that $0 \le \delta \le 1$, $\mu \in \mathbb{R}$, $0 < \sigma$, $c \in \mathbb{R}$.
The EM algorithm can be applied for this maximization.

3) $F(x) \leftarrow F(x) + c_t f(x|\theta_t) + \delta_t \mu_t$
$\hat{p}(\epsilon) \leftarrow (1 - \delta_t) \, \hat{p}(\epsilon + \delta_t \mu_t) + \delta_t \phi_{\sigma_t}(\epsilon - (1 - \delta_t)\mu_t)$

4) Calculate the values of v_i and $\tilde{\epsilon}_i$ from the updated $F(x)$ and $\hat{p}(\epsilon)$.

Step 3. Output $F(x)$.

In the suggested algorithm the mean square error is used to find the parameters of the simple predictors instead of the correlation. This is because the value of the correlation goes to infinity when the output value of the simple predictor is not bounded.

4 Simple Numerical Experiments

In this section we compare the performance of the suggested algorithm with the greedy algorithm.

In the first simulation the true regression function is defined as $g(x) = \frac{\sin(20x)}{4x}$ where x is uniformly distributed on the interval $[-1, 1]$. The distribution of the noise is chi square distribution whose freedom is five. Here chi square distribution is linearly transformed so as to have zero mean and standard deviation 0.7. The number of the training data is 200 and each algorithm is run 100 times to calculate the mean value of the test error.

In the second simulation the true regression function is Friedman's function [4], which has 7 independent variables which are uniformly distributed on $[-1, 1]^7 \in \mathbb{R}^7$. The regression function is defined as $g(z(x)) = 10\sin(\pi z_1 z_2) + 20(z_3 - 0.5)^2 + 10z_4 + 5z_5$ where $z(x) = (x + 1)/2$. The distribution of the noise is same as the first simulation but the standard deviation is 1. Note that the true function depends on only five independent variables x_1 to x_5. The number of the training data is 400 and each algorithm is run 20 times.

The simple predictor is written as $f(x|\theta) = \sum_{k=1}^{2} u_k \tanh(x \cdot w_k + h_k) + h_0$, where the dimension of w_k is same as that of x. To evaluate the performance of each algorithm we use normalized mean square error: $\sum_{i=1}^{N}(\hat{y}_i - g(x_i))^2 / \sum_{i=1}^{N}(\bar{y} - g(x_i))^2$, where $(x_i, y_i), i = 1, \dots, N$ are the test data, $\hat{y}_i$ is the estimated value on x_i and $\bar{y}$ is the average value of y_i.

The results of the simulation are shown in the Fig. 1. In these simulations the test error of the suggested algorithm is smaller than that of the greedy algorithm. Especially in the second simulation greedy algorithm exhibits more over-fitting than the suggested algorithm. We think that the rough estimation of the probability density of the noise can be efficient for the regression problems.

the blocks which are obtained, taking into account that they can be grouped into four classes: manuscript text, typed text, drawings and photographic images.

This work is structured as follows. In the following section, we give a brief explanation of the *coordinate-cluster transformation* we have used as the feature selection method. Our experimental work to classify blocks from documents is presented in Section 3 and 4. Finally, we summarize our work, draw some conclusions, and propose several lines for future extensions to the work presented here.

2   Feature Extraction

The work by Kurmyshev [10] presents a transformation of binary images that has been successfully applied to noisy texture recognition [11]. This transformation takes into account the different configurations of a binary $N \times N$ square window ($2^{N \times N}$ different configurations), and constructs a histogram to capture the number of occurrences of each window configuration throughout the image. This transformation is called *coordinate-cluster transformation* (CCT).

In order to classify the blocks into the four classes (manuscript text, typed text, drawings and photographic images), we followed the same approach, considering that blocks from each different class maintain a homogeneous-enough texture. The original work tested different values of N, but all of them obtained analogous results. Therefore, we used $N = 3$ to reduce the size of the transformation, that is, we obtained a final representation of the samples composed of $2^{3 \times 3} = 512$ values.

A set of 800 images (300 × 300 dots per inch) was used. To avoid a biased representation, all the images in the database contained the same number of pixels (1.4 million). Each image was binarized and every isolated pixel deleted in order to obtain the CCT. Some fragments of image examples (before and after binarization) and their CCT are shown in Figure 1.

3   Neural-Based Classification

Multilayer Perceptrons (MLPs) are the most common artificial neural networks used for classification. For this purpose, the number of output units is defined as the number of classes, C, and the input layer must hold the input patterns. Each unit in the (first) hidden layer forms a hyperplane in the pattern space; boundaries between classes can be approximated by hyperplanes. If a sigmoid activation function [12] is used, MLPs can form smooth decision boundaries which are suitable to perform classification tasks. The activation level of an output unit can be interpreted as an approximation of the a posteriori probability that the input pattern belongs to the corresponding class. Therefore, an input pattern can be classified in the class $i^\star$ with maximum a posteriori probability:

$$i^\star = \operatorname*{argmax}_{i \in C} \Pr(i|x) \approx \operatorname*{argmax}_{i \in C} g_i(x, \omega), \tag{1}$$

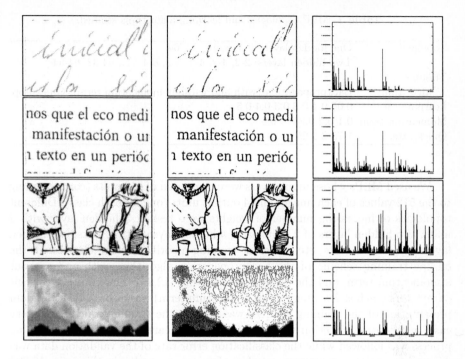

Fig. 1. Fragments of images from the database. Downwards, manuscript text, typed text, drawing and photographic image. From left to right, the original image, the binarized image and the resulting histogram are shown.

where $g_i(x, \omega)$ is the i-th output of the MLP given the input pattern, x, and the set of parameters of the MLP, ω.

3.1 Input Data

The 800 samples were randomly split into two balanced sets: 80% for training and 20% for testing. Thus, the training data was composed of 640 samples (160 samples from each of the 4 classes). From this data, we randomly selected 20% of the patterns for validation purposes (160 samples with 40 samples from each class).

The input data was normalized by dividing each of the 512 values, which made up each sample, by the maximum value of all the training samples.

3.2 Training the Neural Networks

The training of the MLPs was carried out with the neural net software package "SNNS: Stuttgart Neural Network Simulator" [13]. In order to successfully use neural networks as classifiers, a number of considerations had to be taken into account, such as the topology of the MLPs, the training algorithm, and the selection of the parameters of the training algorithm [12].

Table 1. MLP topologies and learning algorithms studied.

Topology:	One hidden layer: 2, 4, 8, 16, 32, 64
	Two hidden layers: 2-2, 4-2, 4-4, 8-2, 8-4, ..., 64-32, 64-64
Training	
algorithm:	Backpropagation (with and without momentum term), Quickprop
Learning rate:	0.05 0.1 0.2 0.3 0.4 0.5
Momentum term:	0.1 0.2 0.3 0.4 0.5
Quick rate:	1.75 2 2.25

We used MLPs whose input layers were composed of 512 units (corresponding to the 512 values of each image) and 4 output units (one for each class). Different topologies of increasing number of weights were tested: a hidden layer with 2 units, two hidden layers of 2 units each, a hidden layer with 4 units, two hidden layers of 4 and 2 units, etc. Different training algorithms were also tested: the on-line training scheme of the backpropagation algorithm, with and without a momentum term, and the quickprop algorithm [12,13]. The influence of their parameters (such as learning rate or momentum term) was also studied. Random presentation of the training samples was used in the training process. In every case, a validation criterion was used to stop the learning process: the training process was followed while the classification error rate of the validation data was decreasing (up to a maximum number of training epochs, in this case, 10,000 epochs).

During this training process, we first tested the influence of the topology of the MLP by increasing the number of weights until the classification error rate of the validation data did not improve. Under this experimental scheme, we achieved a 100% classification of the validation data for a topology of one hidden layer of 8 units. We followed our experimentation by testing different training algorithms (see Table 1), obtaining a 100% classification of the validation data for different configurations of training algorithms and parameters. For this reason, we decided to use the configuration with the quickest stopping convergence (302 epochs): an MLP with one hidden layer of 8 units and the Quickprop training algorithm with a learning rate equal to 0.05 and a quick rate equal to 2.

4   Final Experiment

Once we had selected the best combination of topology, learning algorithm and parameters for the MLP, according to the classification error rate for the validation data, we performed the final experiment. Note that we have not yet used the test data so far.

In order to obtain statistically significant results, five different balanced partitions of the data were done for this final experiment. Each partition was composed of 160 samples (20% of the data) with 40 samples from each class. This final experiment entailed five runs, using the *Leaving-One-Out* scheme [14]: training the MLP model with the data of four partitions (640 training samples with

Table 2. Classification error rates for each run, final average result and the confusion matrix for the final experiment. Classes 1 to 4 are manuscript text, typed text, drawings and photographic images.

Experiment	Error rate		class	1	2	3	4	no class
Run 1	1.88		1	200	0	0	0	0
Run 2	0		2	0	195	1	0	4
Run 3	0.62		3	0	4	195	1	0
Run 4	2.50		4	0	0	0	200	0
Run 5	1.25							
Average	1.25							

160 samples from each class[1]) and testing with the other data of only one partition (160 test samples, 40 samples from each class). Therefore, the classification error rate of the test set reported below is the average result of the five runs of the experiment. The results for each run of the experiment, the average result and the confusion matrix are shown in Table 2. We finally achieved an excellent performance on the test set: 1.25% classification error rate with a 95% confidence interval of [0.68..2.29].

5   Conclusions and Future Work

In this work, we show how useful the *coordinate-cluster transformation* is for representing the image features in the classification of blocks from documents. This is an important process in document analysis, where the reliability in the classification of the different blocks from a given document helps to classify the whole document and allows us to properly manage the information contained in it.

We considered a database of 800 samples from four different block classes (manuscript text, typed text, drawings and photographic images). We used the CCT of the samples as the input of an artificial neural network. Different configurations of topology and training algorithms were tested, obtaining an excellent classification performance (1.25% $\in$ [0.68..2.29] classification error rate with a 95% confidence interval).

As future lines of work, this approach will be useful in a document segmentation process using sequences of small fragments from the document. It could also be assumed that different classes of documents have different textures, and, therefore, this approach could be used in a multipurpose one-step document classification method.

---

[1] Out of this data, 20% was selected for validation. Therefore, 512 training samples (128 from each class) and 128 validation samples (32 from each class) were used.

References

1. R. Cattoni, T. Coianiz, S. Messelodi, and C. M. Modena. Geometric layout analysis techniques for document image understanding: a review. Technical Report 9703-09, IRST – Instituto Trentino di Cultura, 1998.
2. Y. Y. Tang, M. Cheriet, J. Liu, J. N. Said, and C. Y. Suen. *Handbook of Pattern Recognition and Computer Vision*, chapter Document analysis and recognition by computers. World Scientific Pub. Co., 1999.
3. S. W. Lee and D. S. Ryu. Parameter-free geometric document layout analysis. *IEEE Transactions on Pattern Analysis and Machine Intelligence*, 23(11):1240–1256, 2001.
4. D. Dori, D. Doermann, C. Shin, R. Haralick, I. Phillips, M. Buchman, and D. Ross. *Handbook on Optical Character Recognition and Document Image Analysis*, chapter The representation of document structure: a generic object-process analysis. World Scientific Pub. Co., 1996.
5. K. Summers. Near-wordless document structure classification. In *Proceedings of the International Conference on Document Analysis and Recognition*, pages 462–465, 1995.
6. J. Sauvola and M. Pietikäinen. Adaptive document image binarization. *Pattern Recognition*, 33:225–236, 2000.
7. Y. Yang and H. Yan. An adaptative logical method for binarization of degraded document images. *Pattern Recognition*, 33:787–807, 2000.
8. H. Noda, M. N. Shirazi, and E. Kawaguchi. MRF-based texture segmentation using wavelet decomposed images. *Pattern Recognition*, 35:771–782, 2002.
9. T. Ojala and M. Pietikäinen. Unsupervised texture segmentation using feature distributions. *Pattern Recognition*, 33:447–486, 2000.
10. E. V. Kurmyshev and M. Cervantes. A quasi-statistical approach to digital binary image representation. *Revista Mexicana de Física*, 42(1):104–116, 1996.
11. E .V. Kurmyshev and F. J. Cuevas. Reconocimiento de texturas binarias degradas por ruido aditivo usando la transformada de cúmulos coordinados. In *Proceedings of the VI Taller Iberoamericano de Reconocimiento de Patrones (TIARP'01)*, pages 179–187, 2001.
12. D. E. Rumelhart, G. E. Hinton, and R. J. Williams. *PDP: Computational models of cognition and perception, I*, chapter Learning internal representations by error propagation, pages 319–362. MIT Press, 1986.
13. A. Zell et al. *SNNS: Stuttgart Neural Network Simulator. User Manual, Version 4.2*. Institute for Parallel and Distributed High Performance Systems, Univ. of Stuttgart, 1998.
14. R. O. Duda, P. E. Hart, and G. Stork. *Pattern classification*. John Wiley, 2001.

Feature Selection via Genetic Optimization

Sancho Salcedo-Sanz, Mario Prado-Cumplido, Fernando Pérez-Cruz, and
Carlos Bousoño-Calzón *

Dpto. Teoría de la señal y Comunicaciones, Universidad Carlos III de Madrid,
28911 Leganés, Madrid, Spain.
{sancho, mprado, fernando, cbousono}@tsc.uc3m.es

Abstract. In this paper we present a novel Genetic Algorithm (GA)
for feature selection in machine learning problems. We introduce a novel
genetic operator which fixes the number of selected features. This operator, we will refer to it as *m-features* operator, reduces the size of the
search space and improves the GA performance and convergence. Simulations on synthetic and real problems have shown very good performance
of the *m-features* operator, improving the performance of other existing
approaches over the feature selection problem.

1 Introduction

Feature selection is an important task in supervised classification problems due
to irrelevant features, used as part of a classification procedure, can increase
the cost and running time of the system, and make poorer its generalization
performance [1].

There are two different approaches to the Feature Selection Problem (FSP).
The first method tries to identifying an appropriate set of features, independently
of its classification performance, which preserve most of the information provided
by the original data. This approach is known as filter method for feature selection
[1]. The second approach directly selects a subset of m features out of the n
available ones in such a way that the performance of the classifier is improved
or, at least, is not degraded. This method, known as wrapper method, is more
powerful than filter methods, but it is also computationally more demanding [2],
[3]. The search of the best feature subset can be performed by means of any
search algorithm like hill-climbing, greedy or genetic algorithms, etc.

In this paper we present a novel Genetic Algorithm (GA) as wrapper method.
We have introduced a novel operator which is able to fix the number of selected
features, reducing this way the size of the search space and improving the GA
performance on the FSP. We show that our approach has very good performance,
achieving the best solution in problems with synthetic data and improving the
results of the KDD Cup 2001 winner [4].

The article is structured as follows: Section 2 describes the FSP, Section 3
briefly reviews GAs and its application to feature selection, and propose the

---

* This work has been partially supported by a CICYT grant number: TIC-1999-0216.

J.R. Dorronsoro (Ed.): ICANN 2002, LNCS 2415, pp. 547–552, 2002.

m-features operator. In Section 4, we show, by means of computer experiments, the validity of the proposed approach for both synthetic and real data. Finally, Section 5 ends the paper with some concluding remarks.

2   Feature Selection Problem

The Feature Selection Problem (FSP) for a learning problem from samples can be addressed in the following way: Given a set of labeled data points $(\mathbf{x}_1, y_1), \ldots, (\mathbf{x}_l, y_l)$, where $\mathbf{x}_i \in \mathbb{R}^n$ and $y_i \in \{\pm 1\}$, choose a subset of m features $(m < n)$, that achieves the lowest classification error [2].

Following [2], we will define the FSP as finding the optimum n-column vector $\boldsymbol{\sigma}$, where $\sigma_i \in \{1, 0\}$, that defines the subset of selected features, which is found as:

$$\boldsymbol{\sigma}^o = arg \min_{\sigma, \alpha} \left(\int V(y, f(\mathbf{x} * \sigma, \alpha)) dP(\mathbf{x}, y) \right) \tag{1}$$

where $V(\cdot, \cdot)$ is a loss functional, $P(\mathbf{x}, y)$ is the unknown probability function the data was sampled from and we have defined $\mathbf{x} * \sigma = (x_1 \sigma_1, \ldots, x_n \sigma_n)$. The function $y = f(\mathbf{x}, \alpha)$ is the classification engine that is evaluated for each subset selection, $\boldsymbol{\sigma}$, and for each set of its hyper-parameters, α. This functional can be solved using the ERM principle [5].

3   Genetic Algorithms for Feature Selection

Genetic Algorithms (GAs) have been previously used in feature selection in both filter methods [6] and wrapper methods [7]. However, up until now, its use in wrapper methods has been restricted to small size problems, due to the high computational cost of the resulting algorithm. In this paper we propose a novel GA for the FSP, which has a better performance and a faster convergence than other GA approaches to the problem. In the next section a brief summary of the standard GA is given, and then we present the improvements introduced in the GA for the FSP.

3.1   Background

Genetic Algorithms (GAs) are a class of robust problem solving techniques based on a population of solutions which evolves through successive generations by means of the application of the genetic operators: selection, crossover and mutation [8].

Selection is the process by which individuals in the population are randomly sampled with probabilities proportional to their fitness values. An elitist strategy, consisting in passing the highest fitness string to the next generation, is applied in order to preserve the best solution encountered so far in the evolution. The selected set, of the same size of the initial population, is subjected to the crossover operation. Firstly, the binary strings are coupled at random. Secondly, for each pair of strings, an integer position along the string is selected uniformly at random. Two new strings are composed by swapping all bits between the selected position and the end of the string. This operation is applied to the couples with probability P_c less than one. By means of the mutation operation, every bit in every string of the population may be changed from 1 to 0, or vice versa, with a very small probability, P_m.

3.2   A Genetic Algorithm for the FSP

The population of the GA for the FSP is formed by k binary vectors σ, which evolves by the iterative procedure of the genetic operators described above.

In order to adapt the GA to the FSP, we introduce a novel genetic operator which fixes the number of features selected by the GA to m features, i.e fixes the number of 1s in the individuals to a given number m. We have named it *m-features* operator. The *m-features* operator works in the following way: after the crossover and mutation operators, the binary string will present p 1s, in general different from the desired number of features m. If $p < m$ the *m-features* operator adds $(m - p)$ 1s randomly; if $p > m$, the *m-features* operator randomly selects $(p - m)$ 1s and removes them from the binary string.

The *m-features* operator forces the GA to search in a reduced search space. In fact, the standard GA searchs for vector σ° in a space of size 2^n, whereas introducing the *m-features* operator, the size of the search space is reduced to $\binom{n}{m} \ll 2^n$. The application of the *m-features* operator will also obtain as a consequence a much faster convergence of the genetic algorithm.

The GA can be used with the *m-features* operator (we call it *best m-features* mode) or without it (we say then that the GA is running in *standard* mode). The GA running in *standard* mode will search for the best set of features in the whole search space, whereas the GA running in *best m-features* mode searchs for the best set of m features in the search space. Note that the *m-features* operator is an adaptation of the GA to the FSP, due to the number of 1s in the individuals is the number of selected features.

The fitness function associated the each individual is the classification error obtained classifying l training points $(\mathbf{x} * \sigma, y)$. Due to GAs maximizes the fitness function, and the objective function in the FSP is minimizing the error probability, a modified fitness function is introduced:

$$F = 100(1 - err) \tag{2}$$

where err is the probability of error given by the classifier.

4   Experiments

In order to show the performance of our algorithm, we solve two classification problems: the first one is taken from [2], and is formed by two artificial datasets, where we can test the ability of our algorithm to select a small number of features in presence of many irrelevant and redundant ones. The second is a real application in molecular bioactivity for drug design, also known as *Thrombin Binding Problem* used for the first time in the KDD (Knowledge Discovery and Data Mining) Cup in 2001 [4].

4.1   Synthetic Data

Linear Problem. Six dimensions out of 202 were relevant. The labels are equiprobable. The first three features $\{x_1, x_2, x_3\}$ were drawn form $x_i = yN(i, 1)$ and the second three features $\{x_4, x_5, x_6\}$ were drawn as $x_i = N(0, 1)$ with probability of 0.7, otherwise the first three were drawn as $x_i = N(0, 1)$ and the second three as $x_i = yN(i - 3, 1)$. The remaining features were drawn for both class from $x_i = N(0, 20)$. In this problem four of the first six features are redundant and the rest are irrelevant, being 2 the optimal number of features.

Nonlinear Problem. Two dimensions out of 52 were relevant. The labels are equiprobable. The data are drawn from the following: if $y = -1$ then $\{x_1, x_2\}$ are drawn from $N(\mu_1, \Sigma)$ or $N(\mu_2, \Sigma)$ with equal probability, $\mu_1 = \{-\frac{3}{4}, -3\}$ and $\mu_2 = \{\frac{3}{4}, 3\}$ and $\Sigma = I$. If $y = 1$ then $\{x_1, x_2\}$ are drawn again from two normal distributions with equal probability, with $\mu_1 = \{3, -3\}$ and $\mu_2 = \{-3, 3\}$ and the same Σ as before. The rest of the features are noise $x_i = N(0, 20)$.

We use the GA in *best 2-features* mode in both problems. Standard values for P_c and P_m have been used; the population size has been fixed to $k = 50$ binary strings, $P_c = 0.6$ and $P_m = 0.01$. The classifier used is a SVM machine with a linear kernel for the linear problem and a gaussian kernel for the nonlinear problem. Figure 1 shows the evolution of the GA in both problems for a training set of 50 samples, taking the average test error on 500 samples over 30 independent training sets. The GA with the *m-features* operator needs, in average, 24 and 19 generations for achieving the optimal solution in the linear and nonlinear problems respectively (see Figure 1).

4.2   Molecular Bioactivity for Drug Desing

The first step in the discovery of a new drug is to identify and isolate the receptor to which it should bind, followed by testing many molecules for their ability to bind to the target site. Thus, it is needed determining what separates the active (binding) compounds from the inactive (non-binding) ones. In this case the problem consists in predicting if a given drug binds to a target site on thrombin, a key receptor in blood. This task can be seen in a machine learning context as a FSP. Further information of the problem can be found in [4].

The data used in this paper are taken from the ones used in the KDD Cup [4]. Each example (observation) has a length of 139351 binary features, but, following the winner criteria explained in [4], we selected 100 features using mutual information. In the training set there are 1909 examples and only 42 of them bind. Hence the data are highly unbalanced (42 positive examples is only

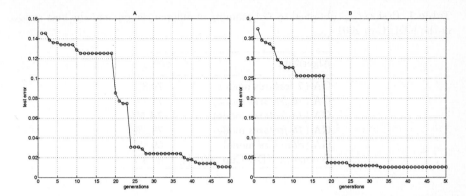

Fig. 1. GA (*best 2-features* mode) performance on A: linear problem, B: nonlinear problem. The x-axis is the GA generations, and the y-axis the test error as a fraction of test points.

2.2% of the data). The test set contains 634 additional compounds, which were generated based on the assay results recorded for the training set.

We apply a GA using the *m-features* operator for feature selection using as inductive classifier a very simple approach, the OR operator. The use of this operator is fully justified because there are very few 1s in each feature and we are looking for features with 1s highly correlated with the positive outputs. In addition the OR classifier provides a very fast way of computing the percentage of error in classification for this problem, without training or adjusting any hyper-parameter. Following the KDD rules, the error in classification is evaluated according to a *weighted accuracy criterion*, due to the unbalanced nature of the number of positive and negative examples:

$$err_{wac} = \left[\frac{1}{2} \left(\frac{\#\{\hat{y} : y = 1 \wedge \hat{y} = 1\}}{\#\{y : y = 1\}} \right) + \frac{1}{2} \left(\frac{\#\{\hat{y} : y = -1 \wedge \hat{y} = -1\}}{\#\{y : y = -1\}} \right) \right]. \quad (3)$$

Table 1 shows the results of the best solution achieved by our GA running in *best 7-features, 10-features, 13-features* and *16-features* modes, *standard* mode and the results obtained for the winner of the KDD Cup 2001 (Bayes Network) [4].

Note that our algorithm achives better results that the best existing algorithm. The best solution was obtained with the GA working in *best 13-features* mode, with a $err_{wac} = 74.5\%$, an improvement of 6% over the best solution given in the KDD Cup[1]. The best solution obtained with the GA working in *standard* mode selected 18 features, with a $err_{wac} = 69.6\%$, which is still better than the best solution in the KDD Cup. However, in general, the GA running in *standard* mode performs worse than running in *m-feature* mode.

---

[1] Only the 7% of all competitors at the KDD achieved err_{wac} higher than 60.0%.

Table 1. Best err_{wac} obtained by GA running in different modes and the best err_{wac} in the KDD Cup 2001.

Algorithm	err_{wac}
GA (*7-features* mode)	70.8%
GA (*10-features* mode)	71.1%
GA (*13-features* mode)	**74.5%**
GA (*16-features* mode)	73.5%
GA (*Standard* mode)	69.6%
Winner of KDD Cup 2001 [4]	68.4%

5 Conclusions

In this paper a novel Genetic Algorithm for Feature Selection in classification problems has been presented. We have adapted the standard GA to the FSP, introducing a novel operator (*m-features* operator) which allows selecting the best m features of the input data for a given problem. This operator also reduces the size of the search space, improving the performance and the convergence of the GA in the FSP.

Experiments in synthetic problems and in a real application, the Thrombin Binding Problem, have shown very good performance of our algorithm, achieving the correct solution in synthetic problems in a short time and better results than any other existing approach for the Thrombin Binding Problem.

References

1. Blum, A. and Langley, P.: Selection of Relevant Features and Examples in Machine Learning. Artificial Intelligence, 97, (1997) 245–271.
2. Weston, H., Mukherjee, S., Chapelle, O., Pontil, M., Poggio, T., Vapnik, V.: In: Sara A Solla, Todd K Leen, and Klaus-Robert Muller, (eds): Feature Selection for SVMs. Advances in NIPS 12, MIT Press, (2000) 526–532.
3. Kohavi, R., John, G. H.: Wrappers for Features Subset Selection. Int. J. Digit. Libr, 1, (1997) 108–121
4. Annual KDD Cup: http://www.cs.wisc.edu/~dpage/kddcup2001/.
5. Vapnik, V. N.: Statistical learning theory. John Wiley & sons, New York, 1998.
6. Chen, H. Y., Chen, T. C., Min, D. I., Fischer G. W. and Wu, Y. M: Prediction of Tacrolimus Blood Levels by Using the Neural Network with Genetic Algorithm in Liver Transplantation Patients. Therapeutic Drug Monitoring 21, (1999) 50–56.
7. Vafaie, H and De Jong, K. A.: Genetic Algorithms as a Tool for Features Selection in Machine Learning. Proc. of the 4th int. conference on tools with artificial systems, IEEE computer society press, Arlintong, VA, (1992) 200–204.
8. Goldberg, D. E: Genetic Algorithm in Search Optimization and Machine Learning. Addison-Wesley Publishing Company, Inc (1989).

Neural Networks, Clustering Techniques, and Function Approximation Problems

Jesús González, Ignacio Rojas, and Héctor Pomares

Department of Computer Architecture and Computer Technology
E.T.S. Ingeniería Informática
University of Granada
E. 18071 Granada (Spain)

Abstract. To date, clustering techniques have always been oriented to solve classification and pattern recognition problems. However, some authors have applied them unchanged to construct initial models for function approximators. Nevertheless, classification and function approximation problems present quite different objectives. Therefore it is necessary to design new clustering algorithms specialized in the problem of function approximation.

1 Introduction

Clustering algorithms have proved successful initializing models for classification problems. Nevertheless, in recent papers some authors have attempted to solve function approximation problems by means of clustering techniques without making significant changes to them [4,3]. There are, however, some particular concepts of function approximation and classification problems which should be taken into account: (1) The output variable is discrete for classification and continuous for function approximation problems. (2) In a function approximation problem, output values different from the optimum ones may be accepted if they are "sufficiently close" to them. This behavior is not desirable for a classifier. (3) Clustering techniques do not take into account the interpolation properties of the approximator system, since this is not necessary in a classification problem.

Therefore, new clustering algorithms specially fitted to the problem of function approximation, which take these differences into account, would improve the performance of the approximator.

2 CFA: A New Clustering Technique for Function Approximation

The objective of CFA is to increase the density of prototypes in the input areas where the target function presents a more variable response, rather than just in the zones where there are more input examples. Thus, CFA reveals the structure of training data in the input space, and it also preserves the homogeneity in the output responses of data belonging to the same cluster. To carry out this task,

J.R. Dorronsoro (Ed.): ICANN 2002, LNCS 2415, pp. 553–558, 2002.

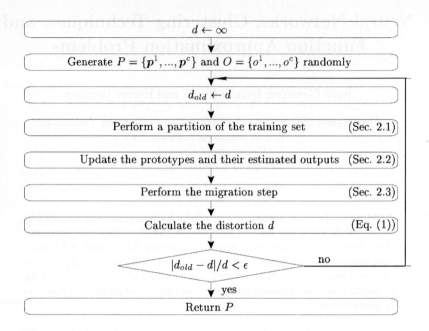

Fig. 1. General description of CFA.

CFA incorporates a set $O = \left\{o^1, ..., o^c\right\}$ representing an estimation of the output response of each cluster. The value of each o^j is calculated as a weighted average of the output responses of the training data belonging to cluster P_j, as will be explained in section 2.2. The objective function to be minimized is:

$$d = \frac{\sum_{j=1}^{c} \sum_{x^i \in P_j} \|x^i - p^j\|^2 \omega_{ij}}{\sum_{j=1}^{c} \sum_{x^i \in P_j} \omega_{ij}} \tag{1}$$

where ω_{ij} weights the influence of each training example x^i in the final position of the j-th prototype. Mathematically, ω_{ij} is defined as:

$$\omega_{ij} = \frac{|f(x^i) - o^j|}{\max_{l=1}^{n} \{f(x^l)\} - \min_{l=1}^{n} \{f(x^l)\}} + \mu_{\min} \tag{2}$$

with $\mu_{\min} > 0$. The first term of the sum calculates a normalized distance (in the range [0,1]) between $f(x^i)$ and o^j, and the second term is a minimum contribution threshold. As $\mu_{\min}$ decreases, CFA forces the prototypes to concentrate on input zones where the output variability is greater, thus preserving the homogeneity in the output space of the points x^i belonging to each cluster P_j by means of minimizing the distance of $f(x^i)$ with respect to o^j. On the contrary, if $\mu_{\min}$ increases, CFA pays more attention to the input space, acquiring the behavior of HCM for a sufficiently large value of $\mu_{\min}$.

Once c and ϵ have been fixed, the basic organization of the proposed algorithm consists, as shown in Figure 1, of the iteration of three main steps: the partition of the training examples, the updating of the prototypes and their estimated outputs, and the migration of prototypes from clusters with lower distortions to clusters with bigger distortions.

2.1   Partition of the Training Data

The partition of training data is performed as in HCM, using equation

$$\mu_j(\boldsymbol{x}^i) = \begin{cases} 1 \text{ if } \|\boldsymbol{x}^i - \boldsymbol{p}^j\|^2 < \|\boldsymbol{x}^i - \boldsymbol{p}^l\|^2, & \forall l \neq j \\ 0 \text{ otherwise} \end{cases} \tag{3}$$

2.2   Updating of the Prototypes and Their Estimated Outputs

The updating is carried out by an iterative process that updates $\boldsymbol{p}^j$ as the weighted mean of the training data belonging to each cluster, and o^j as the weighted mean of their output responses until convergence is met. The expressions used to update $\boldsymbol{p}^j$ and o^j are:

$$\boldsymbol{p}^j = \frac{\sum\limits_{\boldsymbol{x}^i \in P_j} \boldsymbol{x}^i \omega_{ij}}{\sum\limits_{\boldsymbol{x}^i \in P_j} \omega_{ij}} \tag{4}$$

$$o^j = \frac{\sum\limits_{\boldsymbol{x}^i \in P_j} f(\boldsymbol{x}^i) \omega_{ij}}{\sum\limits_{\boldsymbol{x}^i \in P_j} \omega_{ij}} \tag{5}$$

and the algorithm to perform this step is as follows:

$d' \leftarrow \infty$
repeat
    $d'_{old} \leftarrow d'$
    Update $\boldsymbol{p}^j$ using (4), $j = 1, ..., c$
    Update o^j using (5), $j = 1, ..., c$
    Update ω_{ij} using (2), $i = 1, ..., n,$ $j = 1, ..., c$
    Calculate d' using (1)
until $|d'_{old} - d'|/d' < \epsilon$

2.3   Migration of Prototypes

As the iteration of the partition and updating steps only moves prototypes locally, CFA also incorporates a random migration step to avoid local minima. This step moves prototypes allocated in input zones where the target function

Table 1. Mean and standard deviation of the approximation error after the local minimization procedure using RBFNNs from 4 to 10 prototypes to approximate f.

c	NRMSE (dev)			
	HCM	FCM	ELBG	CFA
4	0.851 (0.016)	0.835 (0.205)	0.211 (0.118)	0.176 (0.051)
5	0.818 (0.034)	0.812 (0.341)	0.202 (0.132)	0.081 (0.028)
6	0.759 (0.263)	0.783 (0.317)	0.152 (0.122)	0.090 (0.008)
7	0.344 (0.281)	0.248 (0.115)	0.111 (0.072)	0.081 (0.016)
8	0.227 (0.368)	0.150 (0.162)	0.093 (0.064)	0.053 (0.028)
9	0.252 (0.386)	0.300 (0.160)	0.073 (0.057)	0.056 (0.027)
10	0.087 (0.100)	0.285 (0.334)	0.064 (0.039)	0.047 (0.015)

each clustering algorithm, the Levenberg-Marquardt method was applied to reduce their approximation error. Table 1 shows average NRMSEs and standard deviations for the executions of each clustering technique.

4   Conclusions

This paper presents a new clustering algorithm specially designed for function approximation problems. It is shown that the proposed approach is specially useful in the approximation of functions with a high output variability, improving the performance obtained with traditional clustering algorithms that are oriented towards classification problems, and which ignore this information when initializing the model.

Acknowledgment. This work has been partially supported by the Spanish *Ministerio de Ciencia y Tecnología*, under project DPI2001-3219.

References

1. J. C. Bezdek. *Pattern Recognition with Fuzzy Objective Function Algorithms.* Plenum, New York, 1981.
2. R. O. Duda and P. E. Hart. *Pattern Classification and Scene Analysis.* Wiley, New York, 1973.
3. N. B. Karayiannis and G. W. Mi. Growing Radial Basis Neural Networks: Merging Supervised and Unsupervised Learning with Network Growth Techniques. *IDEE Trans. Neural Networks,* 8(6):1492–1506, Nov. 1997.
4. J. Moody and C. J. Darken. Fast Learning in Networks of Locally-Tuned Processing Units. *Neural Computation,* 1(2):281–294, 1989.
5. M. Russo and G. Patane. *Improving the LBG Algorithm,* volume 1606 of Lecture Notes in Computer Science, pages 621–630. ISSN: 0302-9743, Springer-Verlag, 1999.

Evolutionary Training of Neuro-fuzzy Patches for Function Approximation

Jésus González[1], Ignacio Rojas[1], Hector Pomares[1], Alberto Prieto[1], and K. Goser[2]

[1] Department of Computer Architecture and Computer Technology
University of Granada, Spain
[2] Department of Electrical Engineering
Microelectronics University of Dortmund, Germany

Abstract. This paper describes how the fundamental principles of GAs can be hybridized with classical optimization techniques for the design of an evolutive algorithm for neuro-fuzzy systems. The proposed algorithm preserves the robustness and global search capabilities of GAs and improves on their performance, adding new capabilities to fine-tune the solutions obtained.

1   Introduction

Fuzzy systems comprise one of the models best suited to function approximation problems [4], but due to the non linear dependencies between the parameters that define the antecedents of the system rules, the solution search space for this type of problem contains many local optima, so it is necessary to develop a training algorithm that is capable of reaching a good, well-fitting solution.

Evolutive algorithms [3] are robust tools for this kind of task, and are capable of performing an exploration of the whole search space and at the same time exploit and perfect the solutions obtained, with above-average results, until the best solution is achieved.

2   Description of the Neuro-fuzzy System

Let us consider the sample set X belonging to an unknown function f. A fuzzy system to approximate this function can be expressed as the following rule set R:

$$\textbf{IF } x_1 \text{ is } X_1^i \textbf{ AND ... AND } x_d \text{ is } X_d^i \textbf{ THEN } y = R_i \ \forall i = 1, ..., m \quad (1)$$

where X_k^i ($k = 1, ..., d$) is the membership function associated with the variable x_k in the i-th rule and R_i is the consequent of the rule. Each consequent R_i is a constant or *singleton* scalar value and represents the contribution to the i-th rule output. In classical fuzzy systems [4] a fixed number of membership functions are defined for each of the input variables, and the system rules are generated from a table in which all the possible combinations of such functions are established.

J.R. Dorronsoro (Ed.): ICANN 2002, LNCS 2415, pp. 559–564, 2002.
© Springer-Verlag Berlin Heidelberg 2002

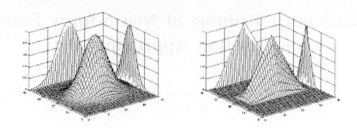

Fig. 1. Patches generated with gaussian and triangular membership functions.

This procedure means that the number of rules grows exponentially with the dimension d of the input vectors and so the problem becomes unmanageable for a large number of input variables. Another problem is that if the training examples are grouped in clusters dispersed throughout the input space, many rules are generated that will never be activated.

One way to avoid these problems is to use free rules [2] or neuro-fuzzy patches [1]. This type of rules can be located by the fuzzy system in the areas of the input space that contribute to minimizing the output error, thus producing systems that are better fitted and that have fewer parameters. Figure 1 illustrates patches generated with gaussian and triangular membership functions.

The degree of activation of the i-th patch is calculated by the equation:

$$\alpha_i(\boldsymbol{x}) = \prod_{k=1}^{d} \mu_k^i(x_k) \tag{2}$$

where $\mu_k^i(x_k)$ is the degree of membership of the variable x_k to the membership function X_k^i, and the system output is obtained by the expression: 3.

$$F(\boldsymbol{x}) = \frac{\sum_{i=1}^{m} \alpha_i(\boldsymbol{x}) R_i}{\sum_{i=1}^{m} \alpha_i(\boldsymbol{x})} \tag{3}$$

The location of these neuro-fuzzy patches within the input space (centre values c_k^i and amplitude values r_k^i) is a complex problem, one that is approached in this paper by the use of evolutive algorithm described in the following section.

3   Evolutive Algorithms

An evolutive algorithm [3] is a probabilistic search algorithm that evolves a population of individuals during a fixed number of iterations or generations. Each individual represents a potential solution to the problem that must be resolved and is implemented as a more or less complex data structure.

According to the above definition, many evolutive algorithms can be formulated for a given problem. These algorithms differ in many respects; they may use different data structures to represent individuals, different genetic operators to transform them, different ways of initializing the population, additional methods to handle restrictions and different values for the parameters of the algorithm (population size, application probabilities of the different operators, etc.) Nevertheless, all evolutive algorithms share a common principle: a population of individuals undergoes certain transformations, and during this evolution the individuals compete for survival.

3.1   Initial Population

As an initial step, a subset of individuals are generated applying some heuristics. The positions of the patches within the input space are determined by a clustering and the widths r_k^i ar estimated by the heuristic of the closest input vectors. Having established the patches that constitute each neuro-fuzzy system in the initial subset, their consequents can be calculated in an optimum way by the application of any practical method to resolve linear equations. As the covariance matrix of the system is symmetric and defined positive, it is possible to use the Cholesky decomposition, one of the most efficient methods of resolving linear equations, to find the optimum consequents [4].

The remaining individuals within the initial population are generated in cascade from the first subset by applying mutation operators. By means of this initialization procedure we are able to achieve an initial population that is sufficiently diverse so that the search process may be started.

3.2   Evaluation Function

The evaluation function is the part of the algorithm that is responsible for guiding the search towards the global optimum. For this purpose, it assigns aptitude values to each of the individuals within the population. The closer an individual is to the global optimum, the better is its aptitude. As a measure of the aptitude of a fuzzy system, we choose the *Normalized Root Mean Squared Error NRMSE)*.

3.3   Reproduction Process

Genetic operators are applied to each individual within the population, with an application probability of p_c for the crossover operators, and of p_m for the mutation operators. Due to the problems of establishing these probabilities a priori, the algorithm presented here implements a dynamic adaptation mechanism of p_c and p_m [6] which chooses the values that are appropriate at all times, based on the state of convergence of the population. Once the probabilities have been chosen, the genetic operators described below are applied.

Crossover of neuro-fuzzy systems. One of the two progenitors is chosen and a random choice is made of a number of patches to be exchanged with the other progenitor. Each patch selected from the first progenitor is exchanged with that from the second that is closest to it in the input space. In this way, the two systems that are derived from the crossover store the information from their two progenitors while avoiding the production of systems with overlapping crossovers, thanks to the locality principle imposed by the exchange process. Once the patches have been exchanged, the Cholesky decomposition is applied to the covariance matrix of each descendant.

Mutation of neuro-fuzzy systems. We applied the OLS algorithm to the neuro-fuzzy system to be mutated, as a prior step to carrying out the alteration. This algorithm has been cited in the literature as being capable of performing an optimum calculation of the consequents of the neuro-fuzzy rules. Moreover, it has another important characteristic. For each neuro-fuzzy rule, it calculates an output error reduction coefficient. The higher the coefficient, the more sensitive to movement is the rule and the greater is the error that is produced if it is altered. By using the OLS algorithm before altering the fuzzy system, we obtain a measure of the sensitivity of each patch, information which is highly useful when we have to decide which patch to modify.

After calculating the fuzzy system error reduction coefficients, each patch is assigned an alteration probability that is inversely proportional to the error reduction coefficient, and a randomly-selected patch is altered. The Cholesky decomposition is then applied to the system to update its consequents.

3.4   Selection Process

Evolutive algorithms are directed by the principle of the survival of the fittest. On this basis, in order to select the composition of the new generation, each individual is assigned a survival probability in direct proportion to its aptitude. The greater the aptitude, the higher the probability. After these probabilities have been generated, the population of the succeeding population is created by the *Roulette Wheel* method [3].

If we consider this subsequent generation, it is necessary to check that the best individual from the previous generation has survived, because although this individual has the highest survival probability, the random nature of the selection process might have led to its rejection. If this should be the case, this individual would be exchanged with the worst one of the current generation.

3.5   Fine-Tuning the Solutions

Evolutive algorithms comprise a powerful search tool, but they are incapable of reaching a solution that precisely fits the problem in the time available; nevertheless, they do achieve an intermediate solution that could serve as a starting point for a local search process that would achieve a solution with the desired degree of precision. To speed up the evolution of the system, we introduce a gradient-descent step into each generation, which is applied to the best individual in the population, but only if this individual represents an improvement on the best

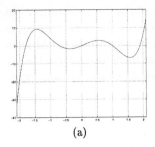 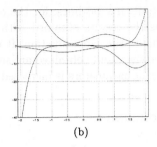

(a)   (b)

Fig. 2. (a) Function $f(x)$ (dashed line) and system output with six neuro-fuzzy patches (solid line). (b) Neuro-Fuzzy patches generated by the system to approximate $f(x)$.

Table 1. Different approaches, numbers of rules, parameters and errors for the function f.

Algorithm	Reg.	Par.	MSE	NRMSE
Dickerson [1]	6	24	3.069	
Pomares [4]	6	10	1.35	0.17
	7	12	0.46	0.10
Our approach	3	9	6.3825	0.369
	4	12	0.2469	0.072
	5	15	0.009	0.013
	6	18	0.0005	0.003

one of the previous generation. This step should not be carried out in too many iterations, as gradient descent is a very costly process. It is intended to refine the best solution within the population so that in the following generation it can be crossed with others and thus generate partially-fitted solutions; this local fit can then be transmitted to the whole population in successive generations.

When the evolutive algorithm has finished searching, the best solution found is subjected to a gradient descent, this time with a higher number of iterations.

4   Results

To test the capacity of this fuzzy system to locate the patches in the zones where they really minimize the output error, and to be obtain a graphic representation of the results to observe the exact position of each patch, and how it contributes to the output from the system, we chose the objective function $f(x)$ used by Dickerson and Kosko in [1]:

$$f(x) = 3x(x-1)(x-1.9)(x+0.7)(x+1.8) \tag{4}$$

where x moves in the interval $[-2.1, 2.1]$.

The evolutive algorithm was run with a population of 50 individuals and for 100 generations. Figure 2 shows the original function in dashed lines and the system output in solid lines, for the case of the six neuro-fuzzy patches. Table 1 compares the results obtained with those of the other two approaches used to approximate the same function. It can be seen that, for the same number of parameters, the results obtained are similar to those presented in [4], although if the number of parameters is increased then a better approximation is achieved.

5 Conclusions

This study presents a different approach to the study of fuzzy systems in which the rules are adapted to the problem by covering the zones of the input space that most contribute to reducing the global approximation error of the system.

We have also presented an evolutive algorithm to train the system and hybridize the fundamental principles of genetic algorithms with those of classical optimization algorithms, thus achieving an algorithm that provides the power of evolutive algorithms but at the same time one that fits the solutions with the desired degree of precision.

The simulations performed show that the synergy of the different paradigms and techniques used produce excellent results for the construction of fuzzy systems.

Acknowledgment. This work has been partially supported by the Spanish *Ministerio de Ciencia y Tecnología*, under project DPI2001-3219.

References

1. J. A. Dickerson and B. Kosko. Fuzzy function approximation with ellipsoidal rules. *IEEE Trans. Syst. Man and Cyber. - Part B*, 26(4):542–560, 1996.
2. J. S. R. Jang, C. T. Sun, and E. Mizutani. *Neuro-Fuzzy and Soft Computing*. Prentice-Hall, ISBN 0-13-261066-3, 1997.
3. Z. Michalewicz. *Genetic Algorithms + Data Structures = Evolution Programs*. Springer-Verlag, 3rd edition, 1996.
4. H. Pomares, I. Rojas, J. Ortega, J. González, and A. Prieto. A systematic approach to a self-generating fuzzy rule-table for function approximation. *IEEE Trans. Syst. Man and Cyber. - Part B*, 30(3):431–447, 2000.
5. M. Russo and G. Patanè. Improving the lbg algorithm. In *International Work-Conference on Artificial and Natural Neural Networks, IWANN99*, pages 621–630, 1999.
6. M. Srinivas and L. M. Patnaik. Adaptive probabilities of crossover and mutation in genetic algorithms. *IEEE Transattions on Systems, Man and Cybernetics*, 24(4):656–666, 1994.

Using Recurrent Neural Networks for Automatic Chromosome Classification*

César Martínez, Alfons Juan, and Francisco Casacuberta

Departamento de Sistemas Informáticos y Computación
Instituto Tecnológico de Informática
Universidad Politécnica de Valencia, 46020 Valencia (Spain)
{cmargal,ajuan,fcn}@iti.upv.es

Abstract. Partial recurrent connectionist models can be used for classification of objects of variable length. In this work, an Elman network has been used for chromosome classification. Experiments were carried out using the *Copenhagen* data set. Local features over normal slides to the axis of the chromosomes were calculated, which produced a type of time-varying input pattern. Results showed an overall error rate of 5.7%, which is a good performance in a task which does not take into account cell context (isolated chromosome classification).

1 Introduction

The genetic constitution of individuals at cell level is the focus of cytogenetics, where genetic material can be viewed as a number of distinct bodies –the *chromosomes*–. In computer-aided imaging systems, which are now widely used in cytogenetic laboratories, the automatic chromosome classification is an essential component of such systems, since it helps to reduce the tedium and labour-intensiveness of traditional methods of chromosome analysis.

In a normal, nucleated human cell, there are 46 chromosomes represented in the clinical routine by a structure called *the karyotype*, which shows the complete set of chromosomes organized into 22 classes (each of which consists of a matching pair of two *homologous* chromosomes) and two sex chromosomes, XX in females or XY in males.

Producing a karyotype of a cell is of practical importance since it greatly facilitates the detection of abnormalities in the chromosome structure. Suitable chromosome staining techniques produce dark bands which are perpendicular to the longitudinal axis and are characteristic of the biological class thereby allowing its recognition. Automatic classification is based on this band pattern and other kinds of information (e.g., width) along the chromosomes. This work presents a pattern recognition application to the classification of microscopic greyscale images of chromosomes by connectionist systems, the Recurrent Neural Networks (RNN).

* Work supported by the Spanish "Ministerio de Ciencia y Tecnología" under grant TIC2000-1703-CO3-01.

2   Background

2.1   Introduction to the Pattern Recognition Problem

In the early days, chromosomes were stained uniformly and as a result they could be distinguished only on the basis of size and shape, the so-called *Denver groups*. However, nowadays most routine karyotyping is carried out on Giemsa-stained chromosomes. The chromosomes appear as dark images on a light background and have a characteristic pattern of light and dark bands which is unique to each class, and which is referred to as *G-banding*. Fig. 1 shows the karyogram of a normal metaphase cell, where the characteristic band pattern can be observed. It serves as a basis for the image processing and pattern recognition algorithms.

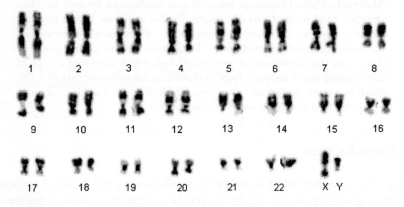

Fig. 1. Normal metaphase cell karyogram (images extracted from the Copenhagen data set).

2.2   Parameter Estimation

In 1989, Piper and Granum established a set of 30 features that were implemented in the chromosome analysis system MRC Edinburgh [1]. The features were classified according to how much a priori information is needed to measure them. Four feature *levels* were distinguished:

1. Direct measures on the image: area, density and convex hull perimeter.
2. Requirement of the axis: length, profiles of density, gradient and shape.
3. Requirement of axis plus polarity: asymetrical weight features.
4. Requirement of axis, polarity and centromere location: centromeric indices.

In this work, the parameters obtained from the images require the calculation of the longitudinal axis. The input patterns are built from measures over normal slices at unitary distance over the axis (details of the preprocessing methods are given in Section 3).

2.3   Recurrent Neural Networks

In many applications, time is inextricably bound up with many behaviors (such as language) and object representations. The question of how to represent time in connectionist models is very important, since it might seem to arise as a special problem which is unique to parallel processing models. This is because as the parallel nature of computation appears to be at odds with the serial nature of temporal events.

This work deals with the use of recurrent links in order to provide networks with a dynamic memory. In this approach, hidden unit patterns are fed back to themselves, acting as the context of prior internal states. Among the different networks proposed in the literature, the Elman network (EN) presents some advantages because of its internal time representation: the hidden units have the task of mapping both external input and the previous internal state (by means of the context units) to some desired output. This develops internal representations which are useful encodings of the temporal properties of the sequential input [7].

Fig. 2 shows an EN, where trainable connections are represented with dotted lines. Connections from the output of hidden units to the context layer are usually fixed to 1.0, and activations of hidden units are copied on a one-to-one basis. Basically, the training method involves activating the hidden units by means of input sequence segments plus prior activation of context units. Then, the output produced is compared with the teaching ouput and the backpropagation of the error algorithm is used to incrementally adjust connection strengths, where recurrent connections are not subject to adjustment [7].

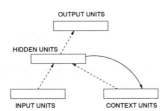

Fig. 2. A Recurrent Neural Network [7].

3   Materials and Methods

3.1   Corpus and Feature Extraction

The corpus used in the experiments was the *Cpa*, a corrected version of the large Copenhagen image data set *Cpr*. It consists of 2,804 karyotyped metaphase cells, 1,344 of which are female and 1,460 are male [4].

As a first step in the preprocessing stage, the histograms of the images are normalized and the holes are filled by means of contour chains to avoid problems in the axis calculation. Afterwards, mathematic morphology algorithms are applied (dilation-erosion) to smooth the chromosome outlines, which prevents the algorithm from producing an axis with more than two terminals.

used clustering method, the hard competitive learning as an online version of k-means (see for example [9]). Initially, we tried the procedure with 5 to 15 clusters and decided according to the error on a validation data set which number to use for the final results.

The individual cluster runs are combined using our ensemble method [7, 25], assigning to every data point a membership function. To map the clusters to classes, we use only the labeled data. We present experiments where all data are labeled (for a comparison with standard classification methods) and experiments where only 5% and 10% respectively of all the data are labeled. [2]

3.2   Data Sets - Results

Three benchmark are data sets are used for our simulations: twonorm [4], banana [18, 11, 19], and Pima Indians Diabetes [17]. Please note, that the same experimental setup is used for every comparisons made.

Twonorm is a 20-dimensional data set with 2 classes which are Gaussian clusters. The Bayes error rate is of 1.6%. 300 data points are used for the training set and 4000 as a testing one. We reach a mean error rate of 2.83% with 0.05 standard deviation. In [18, 11] a mean error of 3.2% is reported for AdaBoost, 2.7% for EM and 2.9% for RFB. Training our mixed model with 90% unlabeled points we get 2.95% misclassification (0.24 standard deviation) and 2.86% for 95% unlabeled points (0.2 st. deviation). The best reported results in [11] are 6.2%−38.5% for their MixtBoost variants, 24.8% for EM and 38.9% for AdaBoost (for 95% unlabeled points). We see that our method is very stable for the number of unlabeled data points. Even if only a few data points are labeled (15 and 30 respectively) we reach almost the same results as when all data points are labeled.

Banana is a 2-dimensional data set, with 2 non-convex classes and best reported error 10.7% by an AdaBoost variant, see [18]. For this data set 400 patterns are used as a training set and 4900 as a testing one. A mean misclassification rate of 13.5% is achieved with a standard deviation of 0.7, where papers report mean misclassification rate between 10.7% and 15.2% for Adaboosting, some of its variants, RFB, SVM, EM methods (see [18, 11]). In the case of 95% unlabeled data points, we reach a misclassification rate of 30.4% with 10.2 standard deviation, where comparison papers report 24.8% for EM, 37.5% for AdaBoost and 25.6% − 35.3% for mixed variants of AdaBoost. For this data the difference of performance between 0% unlabeled and 95% unlabeled is large, but comparable with other methods in the literature.

The Pima Indians Diabetes data set consists of 768 observations and 8 variables describing plasma glucose concentration, diastolic blood pressure, e.t.c., for the subjects. There are 2 classes (negative and positive). Our mixed method applied on the 461 (60% of the whole set) training observations, returns a mean

---

[2] All our experiments have been performed in R, a system for statistical computation and graphics, which implements to well-known S-language for statistics, available freely via CRAN, whose master site is at http://www.R-project.org.

error of 30% with 1.5 standard deviation. Results reported in [18] with different methods recover the 73%-76% of the classification rate. However, with 95% unlabeled data the classification rate turns out to be 35% (5.4 st. deviation).

Table 1. Experimental Results

	0%	90%	95%
Twonorm	2.83 ± 0.05	2.95 ± 0.24	2.86 ± 0.2
Banana	13.5 ± 0.7	18.9 ± 3.8	30.4 ± 10.2
Pima	30.0 ± 1.5	30.8 ± 2.6	35.6 ± 5.4

4   Conclusion

In this paper we present a scheme to treat the problem of the labeled-unlabeled data. This semi-supervised problem turns to be of great interest due to its duality being partially a supervised task (where the class labels of the data are known) and an unsupervised one (unlabeled data). It appears that, in many real world problems the lack of labeled data stresses the necessity of exploring methods which can maximize the advantage of this a priori information in combination to any possible gain by the use of unlabeled data. Our aim is to inject indirectly labels to the points by introducing an association between unsupervised structures and known classes. Thus, unlabeled points would learn their labels according to the behavior of the smaller entities (clusters) where they belong to. As the first part of our algorithm structures the data in an unsupervised manner and the information of the labels is only used for the second step, it can give good performance even if only a small number of data points are labeled. The use of an ensemble method for clustering yields a stable segmentation of the data set and enables to use the additional information of cluster probabilities. Another possible use of this scheme could be as a pre-processing scheme for a pre-classifying of all available data points, which can then be fitted in a real classification task, assuming that all the labels are now known.

References

1. S. Becker. JPMAX: Learning to recognize moving objects as a model-fitting problem. In *Advances in NIPS*, volume 7, pages 933–940, 1995.
2. S. Becker and G. E. Hinton. A self-organizing neural network that discovers surfaces in random-dot stereograms. *Nature*, 355:161–163, 1992.
3. A. Blum and T. Mitchell. Combining labeled and unlabeled data with Co-training. In *Proc. of COLT*, 1998.
4. L. Breiman. Bias, variance, and arcing classifiers. Technical report, Tech. Rep. 460, Statistics Department, University of California, Berkeley, CA, USA, 1996.
5. J. C. Burges. A tutorial on Support Vector Machines for pattern recognition. *Data Mining and Knowledge Discovery*, 2(2), 1998.
6. M. Collins and Y. Singer. Unsupervised models for named entity classification. In *Proc. of EMNLP*, 1999.

7. Evgenia Dimitriadou, Andreas Weingessel, and Kurt Hornik. A combination scheme for fuzzy clustering. In *Advances in Soft Computing (AFSS02)*, pages 332–338, 2002.

8. Y. Freund and R. Schapire. Experiments with a new boosting algorithm. In *Machine Learning: Proc. of the 13th international conference*, 1996.

9. Bernd Fritzke. Some competitive learning methods, April 5 1997. http://www.neuroinformatik.ruhr-uni-bochum.de/ini/VDM/research/gsn/.

10. S. Goldman and Y. Zhou. Enhancing supervised learning with unlabeled data. In *Int. Joint Conference on Machine Learning*, 2000.

11. Y. Grandvalet, F. d'Alche Buc, and C. Ambroise. Boosting mixture models for semi-supervised learning. In G. Dorffner, H. Bischof, and K. Hornik, editors, *ICANN 2001*, pages 41–48, Vienna, Austria, 2001. Springer.

12. P. J. Green and B. Silverman, editors. *Nonparametric Regression and Generalized Linear Models*. Chapman and Hall, London, 1994.

13. D. Miller and H. Ugar. A mixture of experts classifier with learning based both on labeled and unlabeled data. *Adv. in Neural Inf. Proc. Systems*, 9:571–577, 1996.

14. T. Mitchell. The role of unlabeled data in supervised learning. In *Proc. of the 6th International Colloquium on Cognitive Science*, 1999.

15. K. Nigam, A. McCallum, S. Thrun, and T. Mitchell. Text classification from labeled and unlabeled documents using EM. In *Proc. of National Conference on Artificial Intelligence*, 1998.

16. T. J. O'Neil. Normal discrimination with unclassified observations. *J. of the American Statistical Association*, 73:821–826, 1978.

17. Original owners: National Institute of Diabetes, Digestive, and Kidney Diseases. Pima indians diabetes. Donor of database: Vincent Sigillito (vgs@aplcen.apl.jhu.edu). ftp.ics.uci.edu://pub/machine-learning-databases.

18. G. Raetsch, T. Onoba, and K. R. Mueller. Soft margins for AdaBoost. *Machine Learning*, 42(3):287–320, 2001.

19. Benchmark Repository. Banana.

20. R. Schapire and Y. Singer. Improved boosting algorithms using confidence-rated predictions. In *Proc. 11th Annual Conf. on Computational Learning Theory*, 1998.

21. Matthias Seeger. Input-dependent regularization of conditional density models. Technical report, In. f. Adaptive and N. Computation, Univ. of Edinburgh, 2000.

22. Matthias Seeger. Learning with labeled and unlabeled data. Technical report, Inst. for Adaptive and Neural Computation, Univ. of Edinburgh, 2001.

23. D. Titterington, A. Smith, and U. Makov, editors. *Statistical Analysis of Finite Mixture Distributions*. Wiley Series in Probability and Mathem. Stat., 1985.

24. V. N. Vapnik. *Statistical Learning Theory*. Wiley, 1998.

25. Andreas Weingessel, Evgenia Dimitriadou, and Kurt Hornik. A voting scheme for cluster algorithms. In Gerald Krell, Bernd Michaelis, Detlef Nauck, and Rudolf Kruse, editors, *Neural Networks in Applications, Proc. 4th Int. Workshop NN'99*, pages 31–37, 1999.

26. C. K. I. Williams. Prediction with gaussian processes: From linear regression to linear prediction and beyond. In M. I. Jordan, editor, *Learning in Graphical Models*. Kluwer, 1997.

27. C. K. I. Williams and C. E. Rasmussen. Gaussian processes for regression. *Advances in Neural Information and Processing Systems*, 8, 1996.

28. T. Zhang and F. Oles. A probability analysis on the value of unlabeled data for classification problems. In *Int. Joint Conf. on Machine Learning*, 2000.

Using Perceptrons for Supervised Classification of DNA Microarray Samples: Obtaining the Optimal Level of Information and Finding Differentially Expressed Genes

Alvaro Mateos, Javier Herrero, and Joaquín Dopazo

Bioinformatics Unit, Spanish National Cancer Center (CNIO),
c/ Melchor Fernández Almagro 3, 28029, Madrid, Spain
{amateos, jherrero, jdopazo}@cnio.es
http://bioinfo.cnio.es

Abstract. The success of the application of neural networks to DNA microarray data comes from their efficiency in dealing with noisy data. Here we describe a combined approach that provides, at the same time, an accurate classification of samples in DNA microarray gene expression experiments (different cancer cell lines, in this case) and allows the extraction of the gene, or clusters of co-expressing genes, that account for these differences. Firstly we reduce the dataset of gene expression profiles to a number of non-redundant clusters of co-expressing genes. Then, the cluster's average values are used for training a perceptron, that produces an accurate classification of different classes of cell lines. The weights that connect the gene clusters to the cell lines are used to asses the relative importance of the genes in the definition of these classes. Finally, the biological role for these groups of genes is discussed.

1 Introduction

DNA microarray technology opens up the possibility of obtaining a snapshot of the expression level of thousands of genes in a single experiment, [1]. Serial experiments measuring gene expression at different times, or distinct experiments with diverse tissues, patients, etc., allows collecting gene expression profiles under the different experimental conditions studied. Unsupervised clustering of these conditions has been extensively used in the case of classification of different types of cancers, where the molecular signature of distinct tumoral tissues has demonstrated to be a valuable diagnostic tool (see for example [2,3]). Recent papers, on the other hand, have proposed the use of supervised methods like support vector machines (SVM) [4] or supervised neural networks [5] as an alternative. The neural network approach used by Khan et al. [5] uses a previous processing of the data by applying principal component analysis (PCA). Only a few of the principal components are used in the learning process of a perceptron. Here we propose a similar, combined approach, in which all the gene expression patterns are firstly reduced to a set of clusters of co-expressing genes. The clustering step is carried out using the Self Organising Tree Algorithm (SOTA) [6], a

J.R. Dorronsoro (Ed.): ICANN 2002, LNCS 2415, pp. 577–582, 2002.
© Springer-Verlag Berlin Heidelberg 2002

growing hierarchical version of SOM [7] that provides a robust and appropriate framework for the clustering of big amounts of noisy data, such as DNA microarray data [8,9]. The average expression values of the clusters so obtained are used to train a perceptron for classifying the experimental conditions. This approach provides a superior accuracy in the classification of samples than the alternative unsupervised classification. Another additional advantage is that clusters of co-expressing genes, unlike in the Khan et al. [5] approach, retain the biological meaning in the way that they constitute related biological entities that, most probably, are playing similar roles in the cell. These roles are studied using Gene Ontology (GO) terms [10].

2   Methods

2.1   General Description of the Data and the Algorithm

DNA microarray multiple experiment data are usually represented as a matrix of gene expression values. Rows represent the distinct values of expression observed for genes across conditions, and are known as gene expression profiles. Columns correspond to the expression values obtained in experiments for the genes studied. Matrices usually have thousands of rows and tens of columns. Currently, the data contains a non negligible noise component. Here we study nine different classes of cell lines analysed by Alizadeh et al. [2], which include different stages of development of lymphocytes and four distinct cancer lines.

The approach proposed here implies the following steps: (1) reduction of the whole gene expression dataset to their different representative expression profiles by clustering them with SOTA at different resolution levels, that is, producing different slices at different levels of the complete hierarchy. (2) training, at each level, a perceptron with the average values of the clusters found. (3) finding the optimal resolution level of the hierarchy based on the number of true positives in the classification. (4) detect the clusters of genes with major influence in the classification, on the basis of the weights of the perceptron.

2.2   Definition of Clusters: Finding the Optimal Information Level

SOTA is a hierarchical growing neural network [6,9]. SOTA's divisive growing allows the hierarchical clustering of gene expression profiles at different resolution levels, using for this different thresholds for clusters variability (see [9] for details). Each resolution level represents a slice of the complete hierarchy. For each resolution level, a perceptron having as many input nodes as clusters and nine output nodes, corresponding to the nine cell lines to be learned, was trained. The number of cell lines properly classified, that is, the number of true positives, was obtained using the leave-one-out cross-validation procedure. Figure 1 shows the optimal level of information, which corresponds to a tree with 223 clusters.

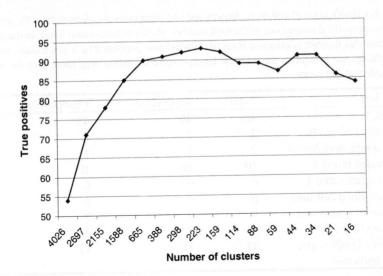

Fig. 1. Number of true positives (cell lines properly classified) for different number of clusters taken as input nodes for the perceptron. The clusters are obtained by applying SOTA with different thresholds, which produces a hierarchical classification of the gene expression profiles at different levels of resolution, that is, a tree with different number of clusters. The optimal level of resolution was found at 223 clusters.

In the dataset used, a number of clusters higher than the optimal causes a worst performance, probably as a consequence of two factors: the bias present due to the composition in the dataset and the high number of parameters that the perceptron has to learn from the data. On the other hand, a number of clusters lower than the optimal causes an excessive compression of the information, decreasing the number of true positives too.

2.3 Classification of Samples

A perceptron with a single input layer with 223 input nodes and nine nodes in the output layer, corresponding to the nine types of cell lines is able of identifying 93 out of the 96 cell lines. Table 1 shows how the perceptron, performs better or, in the worst case, equal than a unsupervised classification obtained by average linkage (as shown in the reference [2], in the figure 1). We have also studied the number of cycles necessary for optimal learning (data not shown), that resulted to be of 1000.

2.4 Weights of the Perceptron Define the Relative Importance of Different Gene Expression Patterns in the Definition of Classes

The strength of the weight connections from the clusters of genes to the classes (cell lines) is far from being homogeneous. The different clusters of co-expressing genes

Table 1. Classification of cell lines. Supervised and unsupervised columns show the number of cell lines properly classified out of the total number, shown in the column Total. In the unsupervised case this number is obtained from the cluster most populated by a given class, no matter other classes are present in this cluster. Unsupervised clustering was taken directly from the Alizadeth et al. [2] paper.

Cell line	Total	Supervised	Unsupervised
DLBCL	46	45	43
Germinal centre B	2	2	2
Nl. Lymph node/tonsil	2	1	1
Activated blood B	10	10	10
Resting/activated T	6	6	6
Transformed cell lines	6	6	6
Follicular lymphoma	9	9	7
Resting blood B	4	3	2
Chronic lymphocytic leukemia	11	11	11

are contributing at different extents to the definition of the classes. We have selected for further analysis those clusters whose weights are in the tails of the weights distribution, outside of the interval comprising the 99% of them.

2.5   Checking the Biological Meaning of the Clusters of Co-expressing Genes with Strongest Weights trough Datamining

The most strong evidence supporting the procedure for selecting genes relevant in the classification comes from their biological role. It can be inferred from the GO categories. GO is divided into molecular function, biological process and cellular component. The distribution of GO terms (taken from Compugen, Inc., http://www.cgen.com, http://www.labonweb.com) corresponding to the genes in the clusters with strongest weights has been studied. Only three among the selected clusters contained a significant number of genes to apply this simple, but efficient, datamining procedure: clusters #153 (56 genes), #154 (23 genes) and #158 (91 genes). There are 4026 clones in the whole dataset, corresponding to 3348 different genes. Only 727 of them were annotated in GO. Figure 2 shows the distribution of go terms, corresponding to biological processes, found for cluster #158.

Genes related to oncogenesis, cell proliferation and cell cycle are clearly overrepresented in the cluster. Similarly, genes related with signaling processes are underrepresented. This is the typical profile of a cell undergoing a cancer process. The molecular functions of the genes are coherent with the processes observed: *Kinase, transferase, nucleotide binding* and *enzyme inhibitor* are GO terms clearly overrepresented in this cluster, whereas, *DNA binding* and *receptor* are underrepresented (data not shown). The strongest connections of cluster #158 is with cell lines *Germinal centre B, Transformed cell lines* and *Resting blood B*. It is very reasonable to expect from a group of

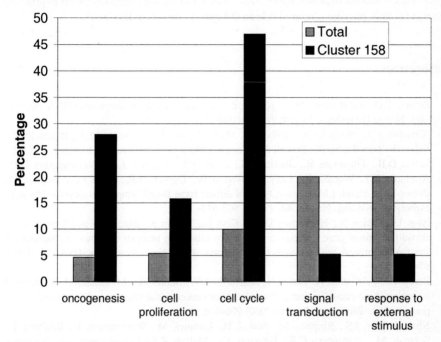

Fig. 2. Study of GO terms. Percentage of GO terms related to process found in cluster #158 (black bars) compared with the corresponding percentage observed in the whole dataset of genes (gray bars). Note that genes may be assigned to more than one function so, the percentages do not sum 100. They are obtained as genes with a term divided by the total of genes, no matter a particular gene is scored more than once.

genes like this a discrimination between cell lines that include cancer cell lines (*Transformed cell lines*, a laboratory cell line) and non cancer cell lines (*Germinal centre B* and *Resting blood B*). Similar tendencies (although not so clear) were also observed for clusters #153 and #154, probing that the genes in the clusters identified as discriminant are playing exactly the roles one may expect from them.

3   Conclusions

We have described a combined approach that provides, at the same time, an efficient classification of experimental conditions in DNA array gene expression experiments, and a method for the extraction of the gene, or clusters of co-expressing genes, responsible for these differences. Pre-clustering of the data, apart from reducing the dimensionality of the problem, avoids the effect of the different size of the clusters that may bias the result towards the cluster or clusters of largest size. An additional advantage is that, opposite to other approaches (like using PCA, see [5]), the classification thus obtained depends upon "real" biological entities (groups of co-expressing genes) probably playing a related role in the cell, as shown in this work.

References

1. Brown, P.O. and Botsein, D.: Exploring the new world of the genome with DNA microarrays. Nature Biotechnol. 14 (1999) 1675-1680.
2. Alizadeh A.A., Eisen M.B., Davis R.E., Ma C., Lossos I.S., Rosenwald A., Boldrick J.C., Sabet H., Tran T., Yu X., Powell J.I., Yang L., Marti G.E., Moore T., Hudson J. Jr, Lu L., Lewis D.B., Tibshirani R., Sherlock G., Chan W.C., Greiner T.C., Weisenburger D.D., Armitage J.O., Warnke R., Levy R., Wilson W., Grever M.R., Byrd J.C., Botstein D., Brown P.O., Staudt LM. Distinct types of diffuse large B-cell lymphoma identified by gene expression profiling. Nature. 403 (2000) 503-511
3. Alon U., Barkai N., Notterman D. A., Gish K., Ybarra S., Mack D., and Levine A. J.: Broad patterns of gene expression revealed by clustering analysis of tumor and normal colon tissues probed with oligonucleotide arrays. Proc. Natl. Acad. Sci. USA. 96 (1999) 6745-6750
4. Furey TS, Cristianini N, Duffy N, Bednarski DW, Schummer M, Haussler D.: Support vector machine classification and validation of cancer tissue samples using microarray expression data. Bioinformatics 16 (2000) 906-914.
5. Khan, J. Wei, J.S., Ringnér, M., Saal, L.H., Ladanyi, M., Westermann, F., Berthold, F., Schwab, M., Antonescu C.R., Peterson, C., Meltzer P.S.: Classification and diagnostic prediction of cancers using gene expression profiling and artificial neural networks. Nature Med. 7 (2001) 673-579.
6. Dopazo, J. and Carazo, J.M.: Phylogenetic reconstruction using a growing neural network that adopts the topology of a phylogenetic tree. J. Mol. Evol 44 (1997) 226-233
7. Kohonen, T.: Self-organizing maps. Springer-Verlag, Berlin. (1997).
8. Dopazo, J.; Zanders, E.; Dragoni, I.; Amphlett, G.; Falciani, F. Methods and approaches in the analysis of gene expression data. J. Immunol Meth 250 (2001) 93-112
9. Herrero, J., Valencia, A. and Dopazo, J.: A hierarchical unsupervised growing neural network for clustering gene expression patterns. Bioinformatics. 17 (2001) 126-136.
10. The Gene Ontology Consortium: Gene Ontology: tool for the unification of biology. Nature Genet 25 (2000) 25-29

Lower Bounds for Training and Leave-One-Out Estimates of the Generalization Error

Gerald Gavin[1] and Olivier Teytaud[2]

[1] LASS, UCBL, UMR 5823 CNRS, 43 bd du 11 nov 1918 69622 Villeurbanne Cédex
[2] ISC, UMR 5015 CNRS, 67 Bd Pinel, F-69675 Bron Cédex.

Abstract. In this paper, we compare two well-known estimates of the generalization error : the training error and the leave-one-out error. We focuse our work on lower bounds on the performance of these estimates. Contrary to the common intuition, we show that in the worst case the leave-one-out estimate is worse than the training error.

1 Introduction

A fundamental problem in machine learning is that of obtaining an accurate estimate for the generalization ability of a learning algorithm trained on a finite data set. Many estimates have been proposed in the litterature. The most famous are the training error, the hold-out, the *k-fold* cross validation, the leave-one-out and the boostrap. An interesting question is to know if estimates are better than other ones considering particular learning algorithms. In this paper we propose to compare the leave-one-out and the training error estimates for algorithm performing training error minimization over a hypothesis class H of VC dimension d. Our studies are in the worst-case framework.

In part 2, we focus on the training error estimate. Vapnik's theory provides many bounds of the accuracy of the training error. The upper and lower bound on the precision of the training error are equivalent. We propose in this paper to improve the lower bound of Anthony and Bartlett [1].

In part 3, we focus on the leave-one-out estimate. Ron and Kearns [7] have proposed an upper bounds equal (ignoring logarithm factor) to the one obtained for the training error estimate. They concluded by saying that "the leave-one-out is not worse that the training error". In this part, we present lower bounds equivalent to the upper bound.

In part 4, we prove the results. The proofs deals with a natural probability distribution defined in part 2. The tools used are very simple and intuitive. The complete proofs appear in [5].

2 Training Error

Let X be the input space, $Y = \{0; 1\}$ the output space, D a distribution over $X \times Y$ and H be a set of prediction functions (functions from X into Y). For

J.R. Dorronsoro (Ed.): ICANN 2002, LNCS 2415, pp. 583–588, 2002.
© Springer-Verlag Berlin Heidelberg 2002

$h \in H$, we denote by $er_D(h)$ the generalization error defined as the probability with respect to D that $h(x) \neq y$. A training sample z_n is a set of n observations $(x_i, y_i) \in X \times Y$. We will suppose that the observations are drawn identically and independently (iid) according to D. We denote by $er_{z_n}(h)$ the training error defined as the frequency of misclassifications on the training sample. A set H has the uniform convergence property if the training error uniformly converges in probability to the generalization error. The uniform convergence property ensures good generalization properties for any learning algorithm approximatively based upon empirical risk minimization, provided that the training sample size is large enough. This leads to the definition of sample complexity.

Vapnik and Chervonenkis introduced a combinatorial notion computed for a set of classifiers H, called the VC-dimension. It quantifies the intuitive notion of complexity of H. Its finiteness is a sufficient condition to prove that H has the uniform convergence property, without making any assumptions on the probability distribution D. Furthermore, Vapnik and Chervonenkis [10] provide an explicit sample complexity bound in their famous theorem, and show that the finiteness of d implies the uniform convergence property. The problem of the generalization is reduced to a combinatorial problem : the computation of the VC-dimension. One of the main criticism of this theorem is due to the intractability of its application. Indeed, even for small VC-dimension, it requires millions of examples to learn with usual precision and confidence. It would be interesting to know if it is possible to drastically reduce this bound on the sample complexity or if the choice of making no assumption on the distribution probability is too ambitious. Alexander [8], Talagrand [9] and Long [6] proposed a significative improvements where the logarithmic factor $\log\left(\frac{1}{\varepsilon}\right)$ is removed. In practice, for standard precision and fiability parameters, it could be less interesting than Vapnik and Chervonenkis result because the constants are higher. But in a theoretical point of view, it is very interesting because equivalent lower bounds were shown. Indeed many lower bounds on the sample complexity in $\Omega(d/\varepsilon^2)$ have been shown. As far as we know, the best result is due to Bartlett and Anthony [1].

Theorem 1 (Long - Anthony and Bartlett). *Let H a hypothesis class of dimension d. For any $n \geq d, \delta \geq 0$,*
  i) with probability at least $1 - \delta$

$$\sup_{h \in H} |er_{z_n}(h)) - er_D(h))| \leq \sqrt{\frac{2140}{n}\left(4d + \ln\frac{1}{\delta}\right)}$$

  ii) Let A be a training error minimization algorithm. If $\delta < 1/64$, then there exists a probability distribution $X \times Y$ such that with probability at least $1 - \delta$,

$$er_D(A(z_n)) - \inf_{f \in H} er_D(f) \geq \max\left(\sqrt{\frac{d}{320n}}; \sqrt{\frac{1}{n}\ln\frac{1}{\delta}}\right)$$

The gap between these two bounds is too large for practitionners. We propose to improve this lower bound in a very simple way. Given a set of classifiers H of

VC dimension d, we consider a set $x = \{x_1, x_2, ..., x_d\}$ of d points shattered by H. Let's denote by U_d the uniform discrete distribution probability over $x \times \{0, 1\}$, i.e. $P_{U_d}(\{(x_i, 0)\}) = P_{U_d}(\{(x_i, 1)\}) = 1/2d$. The Bayes error of U_d is equal to $1/2$. All the following results are shown by considering this simple distribution. The following result provides a lower bound on the precision, i.e the difference beteween the training error estimate and its expectation. (the proof is given in section 4).

Theorem 2. *Let A be a training error minimization algorithm. Over the probability distribution U_d if $n \geq \frac{25d^2 \ln(6d)}{2}$, $\delta \leq \frac{1}{20}$ and $d \geq 10$ with a probability at least δ*

$$|er_{z_n}(A(z_n)) - er_D(A(z_n))| \geq \sqrt{\frac{9d}{100n}}$$

This result significantly improves the result of Bartlett and Anthony. The gap with the upper bound is thus reduced within a factor 10. The extension of this result for the leave-one-out estimate is quite straightforward. The idea consists in noticing that for large training samples, the training error and the leave-one-out error are equal with a probability closed to 1.

3   Leave-One-Out Error

The leave-one-out estimate is computed by running the learning algorithm n times, each times removing one of the n training examples, and testing the resulting classifier on the training example that was deleted; the fraction of failed tests is the leave-one-out estimate. Using n/k examples as test set instead of one example, leads to D tests instead of n, and is called k-folds cross-validation. In practice the leave-one-out estimate is considered to be more accurate that the training error. The empirical intuition of resampling methods is to validate the classifiers on unseen data. We denote by $er^A_{cv}(z_n)$ the leave-one-out error on the training sample z_n for an algorithm A. Estimates of the expectation of K-folds cross-validation discrepancy is made in [3], but the results are trivial in the case $K = n$. [2] provide bounds as well, in the cross-validation case $K << n$ too. Results in particular cases of algorithms are provided in [4]. Kearns & Ron [7] propose bounds on the sample complexity by considering the leave-one-out estimate instead of the training error.

Theorem 3 (Kearns and Ron). *Let A be any algorithm performing training error minimization over a hypothesis class H of VC dimension d.*

i) Then for every $\delta > 0$, with probability at least $1 - \delta$

$$\left|er^A_{cv}(z_n) - er_D(A(z_n))\right| \leq \left(8\sqrt{\frac{(d+1)\left(\ln\left(\frac{9n}{d}\right) + 2\right)}{n}}\right) / \delta$$

ii) There exists a probability distribution over $X \times Y$ such that with probability $1 - \delta$ at least,

$$\left|er^A_{cv}(z_n) - er_D(A(z_n))\right| = \Omega\left(\frac{d}{n}\right)$$

These results let think that improvements are possible and that leave-one-out is a better estimate in the framework of the uniform convergence in the agnostic case. However, we get a lower bound nearly equal to the upper bound proposed by Kearns & Ron. Indeed on the probability distribution U_d, the probability that the leave-one-out error is exactly equal to the training error tends to 1 when the size n of the training set increases.

Theorem 4. *Let A be a training error minimization algorithm. For the probability distribution U_d if $n \geq 175d^3$, $\delta \leq \frac{4}{5}\left(5/66 - \frac{1}{d\sqrt{70\pi}}\right)$ with probability at least $1 - \delta$ is lower bounded by*

$$\left|er_{cv}^{A}(z_n) - er_D(A(z_n))\right| \geq \frac{\left(\frac{1}{2}\sqrt{\frac{d-\frac{1}{5}}{2}} - \frac{1}{\sqrt{11}}\right)}{\sqrt{n}}$$

As far as we know, it is the first time that such a dependence between the precision and the VC dimension and the training sample size is shown. This result shows that the leave-one-out estimate is not better than the training error in the worst case.

Theorem 5. *Let A be a training error minimization algorithm. There exists a probability distribution such that the sample complexity ensuring precision at most $\frac{2d-1}{6d^2}$ with confidence at least $1 - \delta$ is at least*

$$d^3 \times \frac{\left(\frac{\exp(-1/12)\sqrt{2\pi(1+1/(3d))} - 0.15}{16\pi(1+1/(3d))}\right)^2}{\delta^2} - 1 \text{ provided that this quantity is larger than}$$

$18d^4 ln(16d)$.

This implies in particular that:

- the risk is $\geq \min\left(\sqrt{\frac{d^3 \times 0.001035}{n+1}}, \sqrt{\frac{0.007583}{d\ln(16d)}}\right)$ if the precision is lower than $(1 - \frac{d}{2})/(3d)$.
- the sample complexity ensuring precision at most $(1 - \frac{d}{2})/(3d)$ with confidence $1 - \delta$ is at least $0.001035 \times d^3/\delta^2 - 1$, if $d\ln(16d) \geq 0.0000575/\delta^2$.

4   Proof (Sketch) of the Results

Consider X_i's iid random variables distributed according to law U_d. Consider $0 < \epsilon < \frac{1}{2}$. Define $N_i = card\{i/X_i = x_i\}$, $\Delta_i = |card\{i/X_i = x_i \wedge Y_i = 1\} - card\{i/X_i = x_i \wedge Y_i = 0\}|$. $\Delta = \sum \Delta_i$. Consider $\epsilon_1, t \geq 0$. Define $N = n/d - \epsilon_1 n$ and $N' = n/d + \epsilon_1 n$. Consider the following events:

$A_i : N_i \geq n/d - \epsilon_1 n = N$. $A = \cap A_i$.
$A_i' :$ For $y \in \{0, 1\}$, $card\{i/X_i = x_i \wedge Y_i = y\} \in [n/(2d) - \epsilon_1 n/2, n/(2d) + \epsilon_1 n/2]$. $A' = \cap A_i'$. Notice that $A' \Rightarrow A$.
$B_i : \Delta_i > 1$. $B = \cap_{i=1}^{d} B_i$.
$C : \Delta/n \geq \epsilon$
$D : \exists i/\Delta_i = 0$
$E : card\{i/N_i \text{ is even}\} \geq \lceil d/2\rceil$
$\eta: N_i = \eta_i$ (we identify a family of d-uples of integers and a family of events).

4.1 Preliminary Lemma

The proofs of the following lemmas are given in [5].

Lemma 1 (Empirical error). *The difference between the training error and the generalization after learning by the Empirical Risk Minimization algorithm is $\frac{\Delta}{2n}$.*

Lemma 2 (Leave-one-out error). *if B occurs, then the leave-one-out error is equal to the training error.*

Lemma 3 (Leave-one-out error). *If $D \cap A'$ occurs, then the difference between the error in generalization and the leave-one-out error is lower bounded by $\frac{1}{2d} - \frac{d+1}{2}\epsilon_1$.*

Lemma 4 (Many probabilities). *The followings hold: $P(A) \geq 1 - d\exp(-2n\epsilon_1^2)$, $P(C|A) \geq 1 - \frac{1}{1+\frac{t^2}{n+d\epsilon_1 n}}$ with $\epsilon = \frac{1}{2}\left(\sqrt{\frac{d-d^2\epsilon_1}{2n}} - t/n\right)$, $P(B|A) \geq \left(1 - 2\sqrt{\frac{1}{2(n/d-\epsilon_1 n)\pi}}\right)^d$, $P(D \cap A') \geq (\frac{1}{2} - 4d\exp(-n\epsilon_1^2)) \times \left(1 - \left(1 - \frac{\exp(-1/12)}{\sqrt{2N'\pi}}\right)^{\lceil d/2 \rceil}\right)$.*

4.2 Proof of Theorem 3, 4, 5

Proof (Proof of theorem 3). If C occurs, then the difference between the training error and the generalization error is lower bounded by ϵ (lemma 1). By writing $P(C) \geq P(C|A)P(A)$ and by using the results of lemma 4, we state that the difference between the training error and the generalization one is lower bounded by $\frac{1}{2}\left(\sqrt{\frac{d-d^2\epsilon_1}{2n}} - t/n\right)$ with probability at least $P(C|A)P(A)$.

By setting $\epsilon_1 = \frac{1}{5d}$ and $t = \frac{\delta}{(1-\delta)\sqrt{n}}$, the result occurs.

Note that the developed proof including lemma 1 and the necessary part of lemma 4 is very short and intuitive.

Proof (Proof of theorem 5). If B occurs, then the leave-one-out error is equal to the training error (lemma 2). By writing that $P(C \cap B) \geq P(C \cap B|A)P(A) \geq (P(B|A) + P(C|A) - 1)P(A)$ and by using the results of lemma 4, we state that the difference between the generalization error and the leave-one-out error is lower bounded by $\frac{1}{2}\left(\sqrt{\frac{d-\epsilon_1^2}{2n}} - \frac{t}{n}\right)$ with probability at least $(P(B|A) + P(C|A) - 1)P(A)$.

By setting $\epsilon_1 = \frac{1}{5d}$ and $t = \sqrt{\frac{n}{11}}$, the result occurs.

Once again, the proof is very short and intuitive.

Proof (Proof of theorem 6). This is a consequence of lemma 3 and 4. The idea is that when one point among the x_i's has the same number of positive and negative instances, then there is a trouble in leave-one-out.

5 Comments and Further Work

We consider bounds on the uniform convergence of estimates to expectations. We proved lower bounds on the difference between the estimate (leave-one-out or training estimates) *for the chosen algorithm* and its expectation. In this paper we consider the worst case and training error minimization algorithms. The results show that the leave-one-out is not better than the training error in the worst case, and sometimes much worse. It would be interesting to extend this work to practical algorithms and other estimates.

References

1. M. ANTHONY, P.L. BARTLETT, Neural Network Learning: Theoretical Foundations, Cambridge University Press, 1999.
2. M. ANTHONY, S.B. HOLDEN, Cross-Validation for Binary Classification by Real-Valued Functions: Theoretical Analysis, Proceedings of the Eleventh International Conference on Computational Learing Theory, pp. 218–229, 1998.
3. A. BLUM, A. KALAI, J. LANGFORD. Beating the hold-out: Bounds for k-fold and progressive cross-validation. Proceedings of the Twelfth International Conference on Computational Learning Theory, 1999.
4. L. DEVROYE, L. GYORFI, G. LUGOSI, A probabilistic theory of pattern recognition, 1996.
5. G. GAVIN, O. TEYTAUD, Lower bounds for Empirical, cross-validation and Leave-One-Out estimates of the generalization error, research report, 2002.
6. P.M. LONG, The Complexity of Learning According to Two Models of a Drifting Environment, Proceedings of the eleventh Annual Conference on Computational Learning Theory, 116–125, ACM press, 1998.
7. M. KEARNS, D. RON, Algorithmic Stability and Sanity-Check Bounds for Leave-One-Out Cross Validation, AT& T Labs research Murray Hill, New Jersey and MIT Cambridge, MA, 1997.
8. K. ALEXANDER, Probabilities Inequalities for Empirical Processes and a Law of the Iterated Logarithm, Annals of Probability, volume 4, pp. 1041–1067, 1984
9. M. TALAGRAND, Sharper Bounds for Gaussian and Empirical Processes, The Annals of Probability, 22(1), 28–76, 1994
10. V.N. VAPNIK, Estimation of Dependences Based on Empirical Data, 1982.

SSA, SVD, QR-cp, and RBF Model Reduction*

Moisés Salmerón, Julio Ortega, Carlos García Puntonet,
Alberto Prieto, and Ignacio Rojas

Department of Computer Architecture and Computer Technology.
University of Granada. E.T.S. Ingeniería Informática.
Campus Aynadamar s/n. E-18071 Granada (Spain). moises@ugr.es

Abstract. We propose an application of SVD model reduction to the class of RBF neural models for improving performance in contexts such as on-line prediction of time series. The SVD is coupled with QR-cp factorization. It has been found that such a coupling leads to more precise extraction of the relevant information, even when using it in an heuristic way. Singular Spectrum Analysis (SSA) and its relation to our method is also mentioned. We analize performance of the proposed on-line algorithm using a 'benchmark' chaotic time series and a difficult-to-predict, dynamically changing series.

1 Introduction

In this work we suggest a method for improving the prediction performance of RBF (*Radial Basis Function*) models. For this purpose, the SVD and QR-cp matrix decomposition operations are used in a way similar to how usually SVD linear model reduction is performed. What distinguishes our method from SVD reduction in linear model theory, is the fact that the QR-cp step reorganizes the results of the SVD computation. This serves to identify the relevant information for prediction in the input series and also the relevant nodes in the network, thus yielding parsimonious RBF models. An important characteristic of the proposed method is also its capacity for on-line operation, although this could depend to a given degree on parallelization of the matrix routines.

1.1 SVD Model Reduction for Linear Models

SVD-based model reduction can refer to a variety of algorithms that make use of such a matrix decomposition for simplifying and reducing dimensionality in model-driven forms of data adjustment. The SVD (*singular value decomposition*) of a rectangular matrix A of $\mathbb{K}^{m \times n}$ (where $\mathbb{K}$ usually denotes the real or complex number field) dates from the 1960's and 1970's decades. Its computer implementation has been thoroughly analyzed both for serial [GL96,Ste01] and parallel [BCC+97,BL85] computing architectures, although efforts in parallelization (specially for specific applications) are still being developed (e.g., [SOP+02]).

---

* Research partially supported by the Spanish MCyT Projects TIC2001-2845, TIC2000-1348 and DPI2001-3219.

J.R. Dorronsoro (Ed.): ICANN 2002, LNCS 2415, pp. 589–594, 2002.

For a linear system the SVD computation allows to determine whether the system has full or deficient rank. Normally, one looks for the *singular values* σ_i ($i = 1, 2, \ldots, n$), and values near to (or exactly) zero account for linear dependence within the column set. A reduced basis for the subspace of $\mathbb{K}^m$ spanned by the columns in matrix $\boldsymbol{A}$ is sought of dimension equal to r, the numerical rank of the argument matrix. In this case we express $\boldsymbol{A}$ as

$$\boldsymbol{A} = \sum_{i=1}^{r} \sigma_i \boldsymbol{u}_i \boldsymbol{v}_i^T \ , \tag{1}$$

where r denotes the numerical rank, and $(\boldsymbol{u}_i, \boldsymbol{v}_i)$ are the pairs of left- and right-singular vectors from the SVD computation. Solving the reduced linear system with the $\boldsymbol{A}$ in eq. (1) leads to a reduced linear model with $r < n$ parameters.

1.2   Basic Operation of Singular Spectrum Analysis

The *Singular-Spectrum Analysis* (SSA for short) methodology has been systematized recently in the 1996 monograph of Elsner and Tsonis [ET96] and the 2001 book by Golyandina et al. [GNZ01]. This is a technique that makes use of SVD for extraction of the trend, oscillatory and "structureless" noise parts in time series data. We point out that the idea had been implicitly suggested in several papers before the 1990's, e.g., by Broomhead and King [BK86]. The first step in the SSA method is to set up the *trajectory matrix* $\boldsymbol{A}$ by using "moving windows" of width W along the time series values $\{f_t : t = 0, 1, \ldots, N\}$; that is:

$$\boldsymbol{A} \equiv [\boldsymbol{A}_1^T \boldsymbol{A}_2^T \cdots \boldsymbol{A}_K^T]^T \ , \tag{2}$$

where $\boldsymbol{A}_j = (f_j, f_{j+1}, \ldots, f_{j+W-1})$. The structure of the resulting $\boldsymbol{X}$ is that of a *Hankel matrix* [Bjö96]. A total of $K = N - W + 1$ windows are needed to "cover" all of the N time series values. The second step is the SVD of this trajectory matrix. The series is then approximately reconstructed by an expansion

$$f_t = \sum_{k=1}^{m} f_t^{(k)}, \quad t = 1, 2, \ldots, N \ , \tag{3}$$

where the column index set $\{1, 2, \ldots, r\}$ is partitioned as a set of disjoint classes $\{G_k, k = 1, 2, \ldots, m\}$ such that

$$\boldsymbol{A} = \sum_{i=1}^{r} \sigma_i \boldsymbol{u}_i \boldsymbol{v}_i^T = \sum_{k=1}^{m} \sum_{s \in G_k} \boldsymbol{u}_s \boldsymbol{v}_s^T \ ; \tag{4}$$

and $f_t^{(k)}$ is a smoothed time series obtained from diagonal averaging within the outer product matrices belonging to the G_k group [GNZ01]. We notice that "traditional" SVD corresponds to the special case $G_k = \{k\}, \forall k = 1, 2, \ldots, r$. It remains a difficult problem that of determining the "optimal" partition of indexes, such that (a) excessive smoothing is avoided, and (b) the additive-efect implicit assumption in eq. (4) makes sense; that is, it leads to interpretable or more-or-less identifiable effects for every $f_t^{(k)}$ series. These are contradictory requirements. For, if a smaller number of groups are used for the partition (for the sake of identifiability), smoothing has to operate on a higher number of outer products, which in turn can render the method useless.

2 Online RBF Model That Use SVD and QR-cp

SSA is a essentially model-free technique [GNZ01]. It does not tell us how to perform adjustment of the series within each index group of $u_s v_s^T$ matrices, it just smooths out each component series $f_t^{(k)}$. Another drawback of the SSA technique is its lack of consideration of possible on-line adjustment, this being partially due to its birth within the statistics community. It has been claimed that SSA offers results for moderate N that resemble very well its asymptotically predicted behavior, but no formal proof has been given to date assuring such an hypothesis. Other "more classical" time series decomposition methods (such as those arising from the Wold Decomposition Theorem [Pol99]) are therefore most likely to be used in practice by theoretically inclined researchers. For on-line prediction, however, neither SSA nor traditional decomposition are useful. But it can be suspected that the SVD role within SSA might be incorporated in a similar way into powerful on-line models; that is exactly what we propose in the following.

In the method we describe here (thoroughly described in [SOPP01]), a RBF neural network performs on-line modelling of an input time series. A reduced model is constructed both for the input delays and for the hidden layer of locally receptive Gaussian activation units [MD89]. This is accomplished by setting up successive trajectory matrices in the way shown in eq. (2), but coupling the SVD with a QR-cp (QR with *column pivoting* [Bjö96]) phase in which we compute

$$\bar{V} P = Q[R_1 R_2] \ , \tag{5}$$

where $\bar{V} = [v_1 v_2 \cdots v_r]$ (r is still the numerical rank of A) and the v_i's are the r leftmost right singular vectors of A. Matrix Q is $r \times r$ and orthogonal, while R_1 and R_2 are upper-triangular $r \times r$ and $(n-r) \times r$ matrices, respectively. The point here is considering the P permutation matrix as indicating the relevant columns in matrix A. A justification for doing so was heuristically given in [GL96]. This has been theoretically proven correct in more detail in one of the author's PhD thesis [Sal01].

We suggest that successive matrices A be set up from the input stream of time series values, on the one hand, and from the RBF hidden nodes activation values, on the other. In the case of input matrices, SVD and QR-cp would determine the relevant input lags, which are precisely those corresponding to the select columns. In the case of RBF activations, a *pruning* routine might use SVD and QR-cp to select the relevant nodes, which are those associated with the selected columns in successive activation trajectory matrices. All this can be done on-line, provided

3 Statistical Power

To assess the quality of the proposed method compared to Genehunter's NPL-score, we created artificial data sets each consisting of ten pedigrees. Each pedigree was a family with three children of which one was married and had two children. Two disease models were used to generate the data: A recessive model with $d = 0.1$, $f_0 = 0.05$, $f_1 = 0.05$ and $f_2 = 0.9$ and a dominant model with $d = 0.02$, $f_0 = 0.05$, $f_1 = 0.9$ and $f_2 = 0.9$. For all the data the likelihood as presented in this article was computed and, using Genehunter, their NPL-score.

We assumed that for every pedigree only the marker data at the disease locus is available. Since the procedure to handle missing marker data and to make use of adjacent markers does not differ between the methods presented here, there is no need to investigate this regime. To make the single marker informative enough, we assume that it is very polymorphic: ten alleles with equal population occurance. This situation is comparable to chromosome wide available data with less polymorphic markers. After randomly assigning alleles to the founders, the marker data for the non-founders was computed according to a randomly chosen v, which is, of course, not presented to any of the methods. The index i is dropped, since we only investigate a single locus, which can be either linked or unlinked to the disease. Using the same v the number of mutated genes is computed for every individual and the affection status is set randomly using the appropriate penetrance value. A pedigree is included in the data set if at least two non-founders are affected. In this way, we construct 3,000 data sets for the recessive as well as for the dominant case.

We compare the scores found in these data sets with the scores from 3,000 artificially created null data sets. These null data sets are simply copies from the other data sets, but the marker data is generated again using a new random inheritance vector. In this way, there is no linkage between the affection status and the marker data. Therefore, we expect the scores in this data base (the null scores) to be (significantly) less than the others. As with all methods, true positives (there was linkage and we detected it) compete with true negatives (there was no linkage and we did not find anything). The higher the percentage of true possitives, the lower the percentage of true negatives.

For all data sets the NPL- and likelihood-score are computed. This results in 3,000 linked scores and 3,000 null scores for the two methods. Depending on where the threshold between 'linkage' and 'no linkage' is set, there is a higher fraction of true positives or true negatives. This is shown in figure 2a for the dominant and in figure 2b for the recessive model. The statistical power of a particular score is expressed by how close the curve lies in the upper right corner. A rectangular shape is a perfect classifier, whereas a diagonal line from the upper left to the lower right corner is the worst one can get. When we set the model parameters to the ones used to create the artificial data, the so obtained likelihood score gives an idea what the maximum achievable curve looks like. This is the solid curve in both figures. It is clear that the likelihood curve is quite close to the maximum one. In both figures, there is a clear improvement compared to the NPL-score.

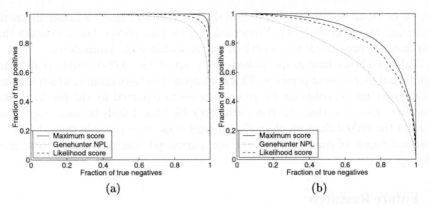

Fig. 2. The statistical power of the various methods can be seen in the plots. The solid line shows the maximum possible power, which is the likelihood score where the correct disease model is specified. The dotted line is Genehunter's NPL score, the dashed line is the non-parametric likelihood score defined in equation 1. In both cases, the latter had the most statistical power. Left is the dominant disease model, right the recessive one.

3.1 Significant Cases for Linkage

Although the previous subsection shows that we can expect an improved statistical power on the whole range, in practical situations one usually is only interested in the right most part of figure 2. This is the region where we can expect with a very high certainty that every positive sample found is really a case of linkage. In other words: the probability that a case labeled as 'linkage' is incorrect, the p-value, is very small.

We can not find the threshold needed for such small p-values by simply counting scores in the null data base, since this would require at least about $1/p$ samples to estimate the threshold reliably enough. For small p-values, this number can be unfeasibly high. Therefore, we use our 3,000 null samples to

Table 1. The number of linked cases out of 3,000 which could be found with a associated p-value less than indicated. The left table is the dominant, the right one the recessive case. The numbers can be read as: 'Given a disease with the dominant properties and a data set as described, we have a probability of 33% to detect linkage if it is present with a p-value less than 10^{-4} (and 5% for Genehunter).'

<table>
<tr><td colspan="3" align="center">Dominant</td><td colspan="3" align="center">Recessive</td></tr>
<tr><td>p-value</td><td>Likelihood</td><td>Genehunter</td><td>p-value</td><td>Likelihood</td><td>Genehunter</td></tr>
<tr><td>$< 10^{-2}$</td><td>2449 (82%)</td><td>1594 (53%)</td><td>$< 10^{-2}$</td><td>661 (22%)</td><td>171 (5.7%)</td></tr>
<tr><td>$< 10^{-3}$</td><td>1713 (57%)</td><td>632 (21%)</td><td>$< 10^{-3}$</td><td>282 (9.4%)</td><td>21 (0.7%)</td></tr>
<tr><td>$< 10^{-4}$</td><td>990 (33%)</td><td>153 (5.1%)</td><td>$< 10^{-4}$</td><td>114 (3.8%)</td><td>2 (0.1%)</td></tr>
<tr><td>$< 10^{-5}$</td><td>495 (17%)</td><td>27 (0.9%)</td><td>$< 10^{-5}$</td><td>39 (1.3%)</td><td>1 (0.0%)</td></tr>
<tr><td>$< 10^{-6}$</td><td>188 (6.3%)</td><td>3 (0.1%)</td><td>$< 10^{-6}$</td><td>18 (0.6%)</td><td>1 (0.0%)</td></tr>
<tr><td>$< 10^{-7}$</td><td>55 (1.8%)</td><td>0 (0%)</td><td></td><td></td><td></td></tr>
</table>

get an estimate of the tail of the null distribution and make a linear fit on the logarithm of the probability. Numerical studies (not shown here) indicate that this linear approximation is a perfect match within error boundaries.

Table 1 shows how many linked cases out of the 3,000 could be detected significantly for several p-values. This is compared to Genehunter, which assumes a Gaussian null distribution for the NPL score as reported by the program. It is clear from the table that our non-parametric likelihood finds far more significant cases in the linked data set than Genehunter does. We conclude that also in the practical region of small p-values the non-parametric likelihood is the best score one can use.

4   Future Research

One of the major advantages of describing the problem in terms of a Bayesian network, is the trivial extension to diseases caused by more than one locus; an aspect of linkage analysis which is more and more desired, but still in its early stages of development. In most approaches, one simply performs a single locus scan and hopes that all loci have enough effect on their own to result in a signicant score. Very few approaches were recently made to treat two loci simultaneously without neglecting the correlation between them, thus leading to more powerful tests. But these approaches are still based on comparisons between pairs of individuals instead of taking the whole pedigree.

In the Bayesian framework, one simply copies a large part of the network shown in figure 1. One copy is pointing to locus i and one to locus j. The nodes determining the affection status, A, however, are shared by both networks. Obviously, the disease model, specified by the parameters f, is more complicated now, since the conditional probability for the affection status is given by $p(A|G_iG_jf)$. In this case, a completely free to choose f contains $3^2 = 9$ parameters. In these cases, it is useful to think whether certain priors on f are appropriate. For instance, AND- or OR-like mechanisms. Also known statistics (such as the a priori probability to have the disease or the probability to have the disease given that some relative is affected) can be used to restrict the number of free parameters. These numbers are usually based on a large population (e.g. all inhabitants in a country) such that they are known very precisely.

References

1. L. Kruglyak, M.J. Daly, M.P Reeve-Daly, and E.S. Lander. Parametric and non-parametric linkage analysis: A unified multipoints approach. *American Journal of Human Genetics*, 58:1347–1363, 1996.
2. N. Friedman, D. Geiger, and N. Lotner. Likelihood computations using value abstraction. In *Proc. Sixteenth Conf. on Uncertainty in Artificial Intelligence (UAI)*, 2000.
3. R.C. Elston and J. Stewart. A general model for the genetic analysis of pedigree data. *Human Heredity*, 21:523–542, 1971.

On Linear Separability of Sequences and Structures

Alessandro Sperduti

Dipartimento di Matematica Pura ed Applicata, Università di Padova,
Via Belzoni 7, 35131 Padova, Italy

Abstract. Linear separability of sequences and structured data is studied. On the basis of a theoretical model, necessary and sufficient conditions for nonlinear separability are derived by a well known result for vectors. Examples of sufficient conditions for linear separability of both sequences and structured data are given.

1 Introduction

Historically the concept of linear separability for a set of labeled vectors has been very important since it allowed to characterize the computational limits of the Perceptron model and, consequently, it stimulated the development of more powerful neural network models, such has feed-forward networks. The introduction of neural network models for the processing of sequences (see for example, [9,2, 3]) and structured data [8,6,7,4] increased the complexity of the problems which can be faced using a neural approach. The increase in complexity of data, however, makes it more difficult to identify classification problems which are easy to solve. In fact, the standard notion of linear separability cannot be applied directly to the case of sequences and structures.

The aim of this paper is to try to redefine for these cases the concept of linear separability and to extend some of the results already known for vectors. Even if here we are able to give necessary and sufficient conditions on linear separability of sequences and structures, as well as some examples of sufficient conditions for linear separability in these contexts, much more work is still needed in order to define an efficient algorithm for learning these classes of problems.

2 Preliminaries on Data Structures

In this paper we assume that instances in the learning domain are DOAGs (directed ordered acyclic graphs) or DPAGs (directed positional acyclic graphs). A DOAG is a DAG $\mathcal{D}$ with vertex set $|(\mathcal{D})$ and edge set $\text{edg}(\mathcal{D})$, where for each vertex $v \in |(\mathcal{D})$ a total order on the edges leaving from v is defined. DPAGs are a superclass of DOAGs in which it is assumed that for each vertex v, a bijection $P : \text{edg}(\mathcal{D}) \to I\!N$ is defined on the edges leaving from v. The *indegree* of node v is the number of incoming edges to v, whereas the *outdegree* of v is the number of outgoing edges from v.

J.R. Dorronsoro (Ed.): ICANN 2002, LNCS 2415, pp. 601–606, 2002.

We shall require the DAG (either DOAG or DPAG) to possess a supersource[1], i.e. a vertex $s \in |(\mathcal{D})|$ such that every vertex in $|(\mathcal{D})|$ can be reached by a directed path starting from s. Given a DAG $\mathcal{D}$ and $v \in |(\mathcal{D})|$, we denote by $ch[v]$ the set of children of v, and by $ch_k[v]$ the k-th child of v.

We shall use lowercase bold letters to denote vectors, and calligraphic letters for representing graphs. A data structure $\mathcal{Y}$ is a DAG whose vertices are labeled by vectors of real-valued numbers which either represent numerical or categorical variables. Subscript notation will be used when referencing the labels attached to vertices in a data structure. Hence y_v denotes the vector of variables labeling vertex $v \in |(\mathcal{Y})|$.

In the following, we shall denote by $\#^{(i,c)}$ the class of DAGs with maximum indegree i and maximum outdegree c. The class of all data structures defined over the label universe domain $\mathcal{Y}$ and skeleton in $\#^{(i,c)}$ will be denoted as $\mathcal{Y}^{\#^{(i,c)}}$.

Here, we consider binary classification tasks where the training set T is defined as a set of couples $(\mathcal{U}, \xi(\mathcal{U}))$, where $\mathcal{U} \in \mathcal{U}^{\#^{(i,c)}}$ and $\xi()$ is the target function $\xi : \mathcal{U}^{\#^{(i,c)}} \to \{-1, 1\}$.

3   Theoretical Model

In accordance with several popular neural network models for processing of sequences and structures, we define the net input of a linear recursive neuron for a node v of a structure $\mathcal{U} \in \mathcal{U}^{\#^{(i,c)}}$ as:

$$net(v) = \boldsymbol{W}^t \boldsymbol{u}_v + \sum_{j=1}^{c} \hat{w}_j net(ch_j[v]), \tag{1}$$

where $\boldsymbol{u}_v \in \mathcal{U}$, $\boldsymbol{W}$ are the weights associated to the input label, and $\hat{w}_j$ are the weights on the recursive connections. Note that, the net input of the neuron for a node v is computed recursively on the net computed for all the nodes pointed by it. The output of the linear recursive neuron for $\mathcal{U}$ is then computed as

$$o(\mathcal{U}) = sgn(net(s)) = \begin{cases} 1 & \text{if } net(s) \geq 0 \\ -1 & \text{otherwise.} \end{cases}$$

where s is the *supersource* of $\mathcal{U}$.

4   Linear Separability for Sequences

When the structure at hand is a sequence $\mathcal{S} = (s_0, \ldots, s_{n-1})$ of length n, equation (1) reduces to $net(s_i) = \boldsymbol{W}^t \boldsymbol{u}_{s_i} + \hat{w}_1 net(s_{i-1})$, $i = 0, \ldots, n-1$ where $\boldsymbol{u}_{s_i}$

---

[1] If no supersource is present, a new node connected with all the nodes of the graph with null *indegree* can be added.

is the label attached to s_i, and by definition, $net(s_{-1}) = 0$. Thus the output of the neuron for $\mathcal{S}$ can be written as

$$o(\mathcal{S}) = sgn(\boldsymbol{W}^t \underbrace{\sum_{i=0}^{n-1} \boldsymbol{u}_{s_i}(\hat{w}_1)^i)}_{\boldsymbol{X}(\mathcal{S}, \hat{w}_1)} .$$

The interpretation of the above equation is straightforward: the sequence is encoded as a polynomial (in a vector field) of the recursive connection $\hat{w}_1$. The set of labels attached to the elements of the sequence are mapped, through the parameter $\hat{w}_1$, into a vector of the same dimension. Specifically, a sequence of length n is transformed by a polynomial of order $(n-1)$. The standard concept of *linear separability* of a set of vectors, can thus be restated very naturally

Definition 1 (Linear Separability for Sequences) *A set of sequences $\boldsymbol{S} = \{\mathcal{S}_1, \ldots, \mathcal{S}_m\}$ is linearly separable (l.s.) if there exists a real number $\bar{w}$ such that the set $\mathcal{X}(\boldsymbol{S}, \bar{w}) = \{\boldsymbol{X}(\mathcal{S}_1, \bar{w}), \ldots, \boldsymbol{X}(\mathcal{S}_m, \bar{w})\}$ is linearly separable.*

Using this definition, it is not hard to reformulate some well known results for set of vectors, like the theorem

Theorem 1 (Linearly Nonseparable Vectors [5]). *A set of vectors $\{Y_1, \ldots, Y_k\}$, $Y_i \in \mathbb{R}^d$, is not l.s. if and only if there exists a positive linear combination of the vectors that equals $\boldsymbol{0}$, i.e., there exists $b_i \geq 0$, $1 \leq i \leq k$, such that $\sum_{i=1}^{k} b_i Y_i = \boldsymbol{0}$, and $b_j > 0$ for some j, $1 \leq j \leq k$.*

Note that, in the above theorem, the information about the class of the vectors is retained into the set of vectors by substituting negative examples $(Y_i, -1)$ with corresponding positive examples $(-Y_i, 1)$. In our framework, assuming a similar transformation for (all labels of) negative sequences, we can state

Theorem 2 (Linearly Nonseparable Sequences). *A set of sequences $\boldsymbol{S} = \{\mathcal{S}_1, \ldots, \mathcal{S}_m\}$ is not l.s. if and only if $\forall w \in \mathbb{R}$ there exists a set of nonnegative numbers b_i, $i = 1, \ldots, m$, not all null, such that*

$$\sum_{i=1}^{m} b_i \boldsymbol{X}(\mathcal{S}_i, w) = \boldsymbol{0}, \quad \text{where} \quad \boldsymbol{X}(\mathcal{S}_i, w) = \sum_{j=0}^{l_i - 1} \boldsymbol{u}_j^{\mathcal{S}_i}(w)^j .$$

Proof: The proof follows immediately by Definition 1 and Theorem 1. $\square$

Equation (2) can be written in a more compact form as $\sum_{j=0}^{M-1}(\boldsymbol{L}_j \boldsymbol{b})(w)^j = \boldsymbol{0}$, where M is the maximum length of the sequences in $\mathcal{S}$, $\boldsymbol{b}^t = [b_1, \ldots, b_m]$, and the matrices $\boldsymbol{L}_j$, $0 \leq j < M$, are defined as

$$\boldsymbol{L}_j = \left[\boldsymbol{\lambda}_j^{\mathcal{S}_1}, \ldots, \boldsymbol{\lambda}_j^{\mathcal{S}_m} \right], \quad \text{where for } q = 1, \ldots, m, \ \boldsymbol{\lambda}_j^{\mathcal{S}_q} = \begin{cases} \boldsymbol{u}_j^{\mathcal{S}_q} & \text{if } j < l_q \\ \boldsymbol{0} & \text{if } j \geq l_q \end{cases}$$

Then we have

Theorem 3. *If $\exists b \in I\!\!R^m$ not null, and with the non null components all of the same sign, such that $\forall j$, $b \in kernel(L_j)$, then S is not l.s.*

Proof: If every $kernel(L_j)$ contains the vector b, then either b or $-b$ will satisfy the conditions of Theorem 2 regardless of the value of w. $\square$

In the following, we give some examples of sufficient conditions for l.s.

Theorem 4. *S is l.s. if one of the following conditions are satisfied:*

i) the columns of L_0 are linearly separable;

ii) the set of vectors $\sigma_{S_i} = \sum_{j=0}^{l_i-1} u_j^{S_i}$ are linearly separable;

iii) the set $Tails = \{u_{l_1-1}^{S_1}, u_{l_2-1}^{S_2}, \ldots, u_{l_m-1}^{S_m}\}$ is linearly separable;

Proof: If condition *i)* is true, then choosing $\bar{w} = 0$, for $q = 1, \ldots, m$, we have that $X(S_q, \bar{w}) = u_0^{S_q}$. Since the vectors $u_0^{S_q}$ are linearly separable, then also S is linearly separable. If condition *ii)* is satisfied, then trivially S is linearly separable by setting $w = 1$. If condition *iii)* is true, then, by definition of linear separability, there will be a vector $\bar{W}$ such that, given $\xi_{j,i} = cos(\bar{\varphi}_{j,i})$ where $\bar{\varphi}_{j,i}$ is the angle between $\bar{W}$ and $u_i^{S_j}$, for each i, $\bar{W}u_{l_i-1}^{S_i} = \|\bar{W}\| \|u_{l_i-1}^{S_i}\| \xi_{i,l_i-1} > 0$.
Moreover, for each j, assuming $w > 0$

$$|\sum_{i=0}^{l_j-2} \xi_{j,i} \|u_i^{S_j}\| (w_1)^i | < \|u_{i_j^*}^{S_j}\| \sum_{i=0}^{l_j-2} (w_1)^i = \|u_{i_j^*}^{S_j}\| \frac{w_1^{l_j-1} - 1}{w_1 - 1},$$

where i_j^* is such that for each i, $\|u_{i_j^*}^{S_j}\| \geq \|u_i^{S_j}\|$. In order for S to be linearly separable, we have to find a $w_1 = \tilde{w}$ such that for each j

$$\|u_{l_j-1}^{S_j}\| (\tilde{w})^{l_j-1} \xi_{j,l_j-1} - \|u_{i_j^*}^{S_j}\| \frac{\tilde{w}^{l_j-1} - 1}{\tilde{w} - 1} > 0. \qquad (2)$$

Equation (2) is satisfied for each j if we choose $\tilde{w}$ to be the maximum over j of the quantities

$$w_j = \frac{\|u_{l_j-1}^{S_j}\| \xi_{j,l_j-1} + \|u_{i_j^*}^{S_j}\|}{\|u_{l_j-1}^{S_j}\| \xi_{j,l_j-1}} > 1.$$

$\square$

5   Linear Separability for Structures

Let consider a structure $\mathcal{U} \in \mathcal{U}^{\#^{(i,c)}}$, then

$$o(\mathcal{U}) = o(s) = sgn(W^t \underbrace{\sum_{v \in I(\mathcal{U})} u_v \sum_{p \in Path_{\mathcal{U}}(v)} \prod_{j=1}^{n} (\hat{w}_j)^{\rho_j(p)}}_{\Xi(\mathcal{U}, \hat{w}_1, \ldots, \hat{w}_n)}),$$

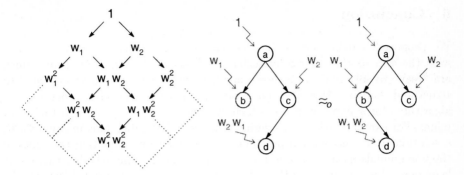

Fig. 1. Example of DAGs which are output-equivalent.

where s is the supersource of $\mathcal{U}$, $Path_\mathcal{U}(v)$ is the set of paths within $\mathcal{U}$ from s to v, and $\rho_j(p)$ is the number of occurrences within path p of the jth pointer.

From the above equation, it is clear that the structure $\mathcal{U}$ is encoded as a polynomial, in n variables, of degree $(P-1)$, where P is the maximum length of the paths from s to any node of $\mathcal{U}$. In [4] it was observed that in recursive neural networks, when assuming causality and stationarity, there exist distinct DAGs that get the same output (a *collision* occur), regardless of the value associated to the weights of the network. When considering linear recursive networks collisions are even more frequent. This concept has been formalized in [1] with the introduction of a binary relation $\approx_o$ called *output-equivalence*: two DAGs are output-equivalent if they collide (see Fig. 1). From the above considerations, it is clear that structures that are *output-equivalent* cannot be linearly separable. In the following, we assume that structures belonging to different classes cannot be *output-equivalent*.

Even in this case it is not difficult to define l.s.: it suffices to extend the existential quantifier in Def. 1 to the whole set of recursive weights. Results similar to the ones reported in Theorems 2 and 3 can trivially be obtained, as well as Theorem 4 conditions *i)* and *ii)*. Moreover, condition *iii)* of Theorem 4 can be modified in the following way

Theorem 5. *Let* $\boldsymbol{U} = \{\mathcal{U}_1, \ldots, \mathcal{U}_m\}$ *be a set of DAGs, and* $\mathcal{B}_{\boldsymbol{U}} = \{\nu^{\mathcal{U}_1}(d_j), \ldots, \nu^{\mathcal{U}_m}(d_j)\}$, *with*

$$\nu^{\mathcal{U}_j}(d_j) = \sum_{v \in \mathcal{L}_{\mathcal{U}_j}(d_j)} u_v,$$

where d_j *is the maximum depth of* $\mathcal{U}_j$, *and* $\mathcal{L}_{\mathcal{U}_j}(d_j)$ *is the set of nodes in* $|(\mathcal{U}_j)$ *at level* d_j, *then* $\boldsymbol{U}$ *is l.s. if* $\mathcal{B}_{\boldsymbol{U}}$ *is l.s.*

Proof: The proof is obtained by observing that considering $\hat{w}_1 = \hat{w}_2 = \cdots = \hat{w}_c = w$, each DAG $\mathcal{U}_j$ returns an output that is equal to the one generated by a sequence with labels $\nu^{\mathcal{U}_j}(0), \ldots, \nu^{\mathcal{U}_j}(d_j)$ and processed by a neuron for sequences with weight w. By applying Theorem 4, condition *iii)*, the Theorem is proved.

6 Conclusion

We proposed a definition of linear separability for sequences and structures with the aim to characterize classification tasks in structured domains which are "easy" to solve. We have shown how results for vectors can be extended to structured domains. However, much work need to be done in defining efficient algorithms for solving problems which are linearly separable in structured domains. For example, when considering sequences with label domain equal to $I\!R$, a solution to a classification problem can be found by computing the roots of the polynomials used to encode each sequence and checking whether there is at least one value for which all the polynomials are positive (provided that all the occurrences of labels in negative examples are changed of sign). A nice property of this algorithm is that it can decide about the linear separability of the training set and that the accuracy used in computing the roots can be adapted to the "amount" of separability of the sequences.

When considering a label domain equal to $I\!R^r$, $r > 1$, however, the problem becomes more difficult and no obvious algorithm seems to exist. Even more difficult is the situation in which structures are considered; in fact, each structure, in the proposed framework, is encoded by a multi-dimensional multi-varied polynomial.

References

1. M. Bianchini, M. Gori, and F. Scarselli. Theoretical properties of recursive neural networks with linear neurons. *IEEE Transactions on Neural Networks*, 12(5):953–967, 2001.
2. J. L. Elman. Finding structure in time. *Cognitive Science*, 14:179–211, 1990.
3. S. E. Fahlman. The recurrent cascade-correlation architecture. Technical Report CMU-CS-91-100, Carnegie Mellon, 1991.
4. P. Frasconi, M. Gori, and A. Sperduti. A framework for adaptive data structures processing. *IEEE Transactions on Neural Networks*, 9(5):768–786, 1998.
5. K.-Y. Siu, V. Roychowdhury, and T. Kailath. *Discrete Neural Computation*. Englewood Cliffs, New Jersey: Prentice Hall, 1995.
6. A. Sperduti, D. Majidi, and A. Starita. Extended cascade-correlation for syntactic and structural pattern recognition. In Petra Perner, Patrick Wang, and Azriel Rosenfeld, editors, *Advances in Structural and Syntactical Pattern Recognition*, volume 1121 of Lecture notes in Computer Science, pages 90–99. Springer-Verlag, Berlin, 1996.
7. A. Sperduti and A. Starita. Supervised neural networks for the classification of structures. *IEEE Transactions on Neural Networks*, 8(3):714–735, 1997.
8. A. Sperduti, A. Starita, and C. Goller. Learning distributed representations for the classification of terms. In *Proceedings of the International Joint Conference on Artificial Intelligence*, pages 509–515, 1995.
9. R. J. Williams and D. Zipser. A learning algorithm for continually running fully recurrent neural networks. *Neural Computation*, 1:27–280, 1989.

Stability-Based Model Order Selection in Clustering with Applications to Gene Expression Data

Volker Roth, Mikio L. Braun, Tilman Lange, and Joachim M. Buhmann

Institute of Computer Science, Dept. III, University of Bonn,
Römerstraße 164, 53117 Bonn, Germany {roth|braunm|lange|jb}@cs.uni-bonn.de

Abstract. The concept of *cluster stability* is introduced to assess the validity of data partitionings found by clustering algorithms. It allows us to explicitly quantify the quality of a clustering solution, without being dependent on external information. The principle of maximizing the cluster stability can be interpreted as choosing the most *self-consistent* data partitioning. We present an empirical estimator for the theoretically derived stability index, based on resampling. Experiments are conducted on well known gene expression data sets, re-analyzing the work by Alon *et al.* [1] and by Spellman *et al.* [8].

1 Introduction

Clustering is an unsupervised learning problem which aims at extracting hidden structure in a data set. The problem is usually split up into several steps. First, it has to be specified what kind of structure one is interested in. Then, the cluster model has to be chosen such that it extracts these structural properties. Finally, the "correct" number of clusters has to be selected. The validation schemes for this task can roughly be divided into *external* and *internal* criteria. An external criterion matches the solution to a priori information, i.e. external information not contained in the data set. An internal criterion uses only the data at hand and does not rely on further a priori information.

In this work, we address the last step, the *model order selection* problem. We assume that the modeling step has already been finished and that it remains to estimate the correct number of clusters. We try to make as few assumptions about the actual clustering method used, in order to keep our criterion as general as possible. Therefore, a clustering algorithm over some object space $\mathcal{X}$ is a function which assigns one of k labels to each element of $\mathbf{X}$, where $\mathbf{X}$ is a finite subset of $\mathcal{X}$. Specifically, we do not assume that the object space is a vector space or that the clustering provides some fitness measure, as would be the case if clustering were understood as mixture density estimation. Furthermore, our criterion is *internal* and does not rely on additional knowledge.

Our validation approach is based on the notion of *cluster stability*. The criterion measures the variability of solutions which are computed on different data sets sampled from the same source. The stability concept has a clear statistical interpretation as choosing the most self-consistent clustering algorithm (see section 2).

From a technical viewpoint, our method refines related approaches initially described in [2], and later generalized in different ways in [3] and [4]. From a conceptual viewpoint, however, it extends these heuristic approaches by providing a clear theoretical background for model selection in unsupervised clustering problems.

J.R. Dorronsoro (Ed.): ICANN 2002, LNCS 2415, pp. 607–612, 2002.

2 Stability

Requirements for the goodness of a clustering algorithm are derived from the following considerations. The clustering algorithm α is applied to some data $\mathbf{X} = (X_1, \dots, X_n)$. It is required that one gets essentially the same solution for a second data set $\tilde{\mathbf{X}} = (\tilde{X}_1, \dots, \tilde{X}_n)$ sampled from the same source. We call such algorithms *self-consistent*.

By assumption, a clustering algorithm only computes labels for its input set $\mathbf{X} = (X_1, \dots, X_n)$, and not for the whole object space $\mathcal{X}$. Therefore, we need an auxiliary construction to compare clustering solutions on different sets. We extend the labeling $\mathbf{Y} = (Y_1, \dots, Y_n) := \alpha(\mathbf{X})$ to the whole object space by a predictor g trained on $\mathbf{X}, \mathbf{Y}$, written as $g(\bullet; \mathbf{X}, \mathbf{Y})$. We set $\alpha_{\mathbf{X}}(x) := \alpha(\mathbf{X})_i = Y_i$ if $x = X_i$ and $\alpha_{\mathbf{X}}(x) := g(x; \mathbf{X}, \alpha(\mathbf{X}))$ if $x \notin \{X_i | i\}$. The self-consistency (or stability) for two i.i.d. data sets $\mathbf{X}, \tilde{\mathbf{X}}$ is then measured as the fraction of differently labeled points:

$$d_{\mathbf{X}}(\alpha_{\mathbf{X}}, \alpha_{\tilde{\mathbf{X}}}) := n^{-1} \sum_{i=1}^{n} \mathbf{1}\{\alpha_{\mathbf{X}}(X_i) \neq \alpha_{\tilde{\mathbf{X}}}(X_i)\}. \tag{1}$$

This quantity depends on the choice of the predictor g. Note that the predictor is only applied to the clustering solution of $\tilde{\mathbf{X}}$, while the solution on $\mathbf{X}$ remains fixed. Due to this asymmetry, the measured stability cannot be arbitrarily reduced by choosing a certain g. This means that even the predictor minimizing d cannot induce artifical stability. On the other hand, the measured stability value is still meaningful for other choices of g, which can lead to lower measured stability. In this case, the only possible side-effect is that intrinsically stable solutions might remain undetected. These considerations are an improvement over what Fridlyand and Dudoit proposed in [3], because the question of choosing the predictor remained un-addressed in their work.

Up to now, we have neglected the fact that the label indices might be re-ordered from $\alpha(\mathbf{X})$ to $\alpha(\tilde{\mathbf{X}})$, i.e. label 1 on $\mathbf{X}$ might correspond to 2 on $\tilde{\mathbf{X}}$. In other words, we can compare two labelings only up to permutations of the labels. Comparing two solutions should be invariant under permutations of the labels, and measure the actual disagreement between the solutions. We therefore minimize over all permutations to obtain the best possible match and define

$$d_{\mathbf{X}}^{\mathfrak{S}_k}(\alpha_{\mathbf{X}}, \alpha_{\tilde{\mathbf{X}}}) := \min_{\pi \in \mathfrak{S}_k} d_{\mathbf{X}}(\alpha_{\mathbf{X}}, \pi(\alpha_{\tilde{\mathbf{X}}})) \tag{2}$$

where k is the number of labels and $\mathfrak{S}_k$ is the set of all permutations of the k labels. Fortunately, it is not necessary to evaluate all $k!$ permutations. Instead, this problem can be solved by the *Hungarian method* for maximum weighted bipartite matchings in $O(k^3)$ [6]. It holds that $d^{\mathfrak{S}_k} \leq 1 - 1/k$. To see this, let α, β be two labelings and π be uniformly drawn from $\mathfrak{S}_k$. Then, easy computations show that $E_\pi(d^{\mathfrak{S}_k}(\alpha, \pi(\beta))) = 1 - 1/k$. Therefore, the minimum is smaller than $1 - 1/k$.

Finally, we are interested in the average instability and thus define

$$S^{\mathfrak{S}_k}(\alpha) := \mathsf{E}_{\mathbf{X}, \tilde{\mathbf{X}}}(d_{\mathbf{X}}^{\mathfrak{S}_k}(\alpha_{\mathbf{X}}, \alpha_{\tilde{\mathbf{X}}})). \tag{3}$$

In order to select a number of clusters k, we need to compare S for different values of k. In the $n \to \infty$ limit, the instability defined above can be interpreted as the probability of labeling an object differently, depending on the fact, whether it was part of the training set or not. In other words, the clustering algorithm is trying to guess its own labels on a second data set. We can use this observation in order to derive a normalizing scheme for comparing S for different values of k. Taking a game-theoretic approach, the game consists in predicting the correct label Y within k possible labels. On correct guess, one looses l_t, otherwise one looses l_f. The expected loss for guess G is given by $l_f P\{G \neq Y\} + l_t P\{G = Y\}$. The game should be fair over different values of k in the following sense: The perfect player should have expected loss 0, hence $l_t = 0$. The random player, who wins with probability $1/k$, should always have expected loss 1, so that $l_f = k/(k-1)$ and the expected loss is $k/(k-1) \times P\{G \neq Y\}$. Therefore, we set

$$S(\alpha_k) := \frac{k}{k-1} S^{\mathfrak{S}_k}(\alpha). \tag{4}$$

For practical applications, the following departures from the idealized setting discussed so far arise. First of all, n is finite. Under these conditions, the expected loss of the random player, given by the random predictor ϱ_k, which assigns k labels uniformly at random, is smaller than $1 - 1/k$. Therefore, we replace $k/(k-1)$ by the estimated instability of the random estimator at set size n and get

$$S_n(\alpha) := \frac{S^{\mathfrak{S}_k}(\alpha)}{S^{\mathfrak{S}_k}(\varrho_k)}. \tag{5}$$

Second of all, only a finite number of data is available to compute the expectation in (3). Here, we use a resampling scheme in which we repeatedly split the data set into two distinct halves and use these as samples for X and $\tilde{X}$. This is the estimator used in our experiments. We choose the number of clusters for which the estimated instability was minimal.

3 Experiments

3.1 Clustering Analysis of Tumor and Normal Colon Tissues

Alon et al. report the application of clustering methods for analyzing expression patterns of different cell types. The dataset (available at http://bio.uml.edu/LW/alonEtal.html.) used is composed of 40 colon tumor samples and 22 normal colon tissue samples, analyzed with an *Affymetrix* Oligonucleotide Array [5]. The grey-value coded expression levels are depicted in figure 2. Alon et al. focused on two different grouping tasks: (i) identifying clusters of genes, whose expression patterns are correlated; (ii) separating tissue types (e.g. cancerous and non-cancerous samples) on the basis of their gene expression similarity. The clustering results are roughly "validated" in an external fashion by a priori knowledge of gene functionality. The question of how many clusters to choose, however, remains open.

In the following we re-analyze this dataset using the *internal* clustering validation method proposed. For all data sets, cluster analysis is performed using a deterministic annealing optimization method for the k-means objective function. Prediction is done with *nearest centroid* classification. As the main result, we can identify six highly stable gene clusters. Five of these six clusters are almost non-informative with respect to

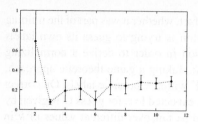 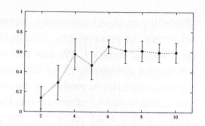

Fig. 1. In-stability curve for gene clusters (left) and for tissue clusters (right), averaged over 20 data splits. The x axis corresponds to the number of clusters k.

separating tissue types. The expression patterns of one group of genes, however, turns out to be highly discriminative.

Gene Clusters. The clustering of genes reveals groups of genes whose expression is correlated across tissue types. Following [1], the expression patterns are normalized such that the average intensity across the tissues is 0, and the standard deviation is 1. In a second step, the columns of the array are also standardized to zero mean and unit variance. Finally, the data are projected onto the first 15 leading principal components, and clustered by a deterministic annealing version of the k-means algorithm [7]. In the left panel of figure 1, the resulting in-stability curve is depicted. It shows highly stable clusterings for $k = 3$ and $k = 6$. The latter solution has the interesting property that it clearly distinguishes between genes that are sensitive to discriminating between normal and tumor tissue (cluster 1 in figure 2), and other genes that are more or less non-informative for this task (clusters 2-5).

Tissue clusters. In order to find a grouping solution for tissue types, we select those features that have been shown to be sensitive for discriminating normal and tumor tissue in the above gene clustering experiment (i.e. the set of genes collected in cluster 1). The resulting features are projected onto the five leading principle components. The right panel of figure 1 depicts the in-stability curve for grouping solutions found by minimizing the k-means criterion within a deterministic annealing framework. The most stable solution separates the tissues into $k = 2$ clusters, which agree with the "true" classes in more than 85% of all cases. We can thus conclude that both the gene and the tissue clusterings, which have been assigned a high self-consistency by our *internal* validation criterion, are also biologically meaningful.

3.2 The Yeast Cell Cycle Dataset

The yeast cell cycle dataset [8] shows the fluctuation of expression levels of yeast genes over cell cycles. Using periodicity and correlation algorithms, Spellman *et al.* identified 800 genes that meet an objective minimum criterion for cell cycle regulation. By observing the time of peak expression, Spellman *et al.* crudely classified the 800 genes into five different groups.

In our experiments, we investigated both the validity of clustering solutions and their correspondence with the classification proposed. From the complete dataset (available at http://cellcycle-www.stanford.edu), we use the 17 *cdc28* conditions for each

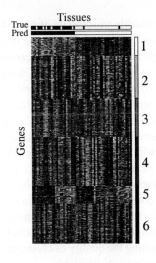

Tissues

Genes

1

2

3

4

5

6

Fig. 2. Data set of expression levels of 2000 genes in 22 normal and 40 tumor colon tissues. The rows correspond to expression patterns for individual genes, and the columns to different tissues. The rows have been re-arranged according to the 6 stable gene-clusters (see left panel of figure 1), the columns according to the two stable tissue-clusters (see right panel of figure 1). The two horizontal bar diagrams show the predicted ("Pred") and true tissue ("True") type when clustering tissues. Black indicates normal tissue, white indicates tumor tissue.

gene, after log-transforming and normalization to zero mean and unit variance. We group the 17-dimensional data by minimizing the k-means cost function using a deterministic annealing strategy. The estimated instability curve over the range of $2 \leq k \leq 20$ is shown in figure 3. For each k, we average over 20 random splits of the data. A dominant peak occurs for $k = 5$, with an estimated misclassification risk of $S(\alpha) \approx 19\%$.

In order to compare a five cluster solution with the labeling proposed by (Spellman *et al.*), we again use the bipartite matching algorithm to break the inherent permutation symmetry. The averaged agreement rate over all five groups is $\approx 52\%$. This may be viewed as a rather poor performance. However, the labels cannot be really considered as the "ground truth", but "...only as a crude classification with many disadvantages", as stated by the authors. Moreover, a closer view on the individual agreement rates per group shows, that at least two groups could be matched with more than 70% agreement.

4   Conclusions

We have introduced the concept of *cluster stability* as a means for solving the model order selection problem in unsupervised clustering. Given two independent data sets from the

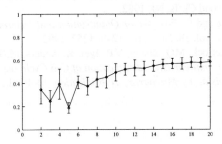

Fig. 3. Estimated instability for the Yeast Cell-Cycle dataset vs. number of classes.

same source, we measure the self-consistency as the attainable misclassification risk under the assumption that the labels produced by the clustering algorithm were the true labels. Alternatively, the stability measure can be interpreted as the disagreement of two solutions where one was extended to the whole object space by a predictor. Taking expectations over tuples of data sets, and normalizing by the stability of a random predictor, we derive a stability measure that allows us to compare solutions for different numbers of clusters in a fair and objective way. In order to estimate the cluster stability in practical applications, we use an empirical estimator that emulates independent samples by resampling. Unlike many other validation indices proposed in the literature, the estimated instability has a clear interpretation in terms of misclassification risk. The experimental results for gene expression datasets effectively demonstrate that cluster stability is a suitable measure for estimating the most self-consistent data partitioning. A comparison with biological prior knowledge shows, that for the chosen grouping model these self-consistent solutions divide the data into meaningful clusters. We conclude that our internal validation method is highly suited for assessing the quality of clustering solutions in biological applications where prior knowledge is rarely available.

Acknowledgments. This work has been supported by the German Research Foundation (DFG), grants #Buh 914/4, #Buh 914/5.

References

1. U. Alon, N. Barkai D. A. Notterman, K. Gish, S. Ybarra, D. Mack, and A. J. Levine. *Broad patterns of gene expression revealed by clustering analysis of tumor and normal colon tissues probed by oligonucleotide arrays.* Proc. Natl. Acad. Sci., 96:6745–6750, 1999.
2. J. Breckenridge. *Replicating cluster analysis: Method, consistency and validity.* Multivariate Behavioral research, 1989.
3. J. Fridlyand & S. Dudoit. *Applications of resampling methods to estimate the number of clusters and to improve the accuracy of a clustering method.* Stat. Berkeley Tech Report. No. 600, 2001.
4. E. Levine, E. Domany. *Resampling Method for Unsupervised Estimation of Cluster Validity.* Neural Computation 13: 2573–2593, 2001.
5. D. H. Mack, E. Y. Tom, M. Mahadev, H. Dong, M. Mittman, S. Dee, A. J. Levine, T. R. Gingeras, D. J. Lockhart. In: *Biology of Tumors*, eds. K. Mihich, C. Croce, (Plenum, New York), pp. 123, 1998.
6. C.H. Papadimitriou & K. Steiglitz. *Combinatorial Optimization, Algorithms and Complexity*, Prentice-Hall, Englewood Cliffs, NJ, 1982.
7. K. Rose, E. Gurewitz and G. Fox. *Vector Quantization and Deterministic Annealing*, IEEE Trans. Inform. Theory, Vol. 38, No. 4, pp. 1249–1257, 1992.
8. P.T. Spellman, G. Sherlock, MQ. Zhang, V.R. Iyer, K. Anders, M.B. Eisen, P.O. Brown, D. Botstein, B. Futcher. *Comprehensive Identification of Cell Cycle-regulated Genes of the Yeast Saccharomyces cerevisiae by Microarray Hybridization.* Molecular Biology of the Cell 9, 3273–3297, 1998.

EM-Based Radial Basis Function Training with Partial Information

Pedro J. Zufiria[1] and Carlos Rivero[2]

[1] Grupo de Redes Neuronales,
Departamento de Matemática Aplicada a las Tecnologías de la Información,
ETSI Telecomunicación, Universidad Politécnica de Madrid,
Ciudad Universitaria s/n, 28040 Madrid, Spain
pzz@mat.upm.es,
http://www.mat.upm.es
[2] Departamento de Estadística e Investigación Operativa,
Facultad de Ciencias Matemáticas, Universidad Complutense de Madrid,
28040 Madrid, Spain
crivero@eucmos.sim.ucm.es

Abstract. This work presents an EM approach for nonlinear regression with incomplete data. Radial Basis Function (RBF) Neural Networks are employed since their architecture is appropriate for an efficient parameter estimation. The training algorithm expectation (E) step takes into account the censorship over the data, and the maximization (M) step can be implemented in several ways. The results guarantee the convergence of the algorithm in the GEM (Generalized EM) framework.

1 Introduction

The expectation-maximization (EM) algorithms are being used for efficiently estimating parameters in different training contexts. These algorithms were developed for situations where the available set of data was not complete [1,7]; since then, they have been applied to situations where the data incompleteness is not evident at all [4]. This is the case when the complexity of the maximum likelihood estimation given the complete data set is reduced when compared to the maximum likelihood estimation corresponding to the observed data. This fact can be exploited via an EM perspective in order to obtain an efficient estimation algorithm with very suitable convergence properties.

Radial Basis Functions (RBFs) are quite popular neural network supervised models for nonlinear regression, due to their particular structure and available training algorithms (e.g., EM-based schemes) [4,10]. In the seminal work [2] where a general additive context is considered, an estimation procedure is presented that exploits such structure so that the observations are decomposed into the individual terms and, hence, the parameters of each term can be estimated separately. In the specific RBF setting, the linear dependency on the output layer weights can be also exploited in the maximization procedure [5]. EM-based schemes in RBF models for system identification have also being studied [3].

J.R. Dorronsoro (Ed.): ICANN 2002, LNCS 2415, pp. 613–618, 2002.
© Springer-Verlag Berlin Heidelberg 2002

All the above mentioned methods assume that the observations are fully available and the expectation-based estimation only affects internal variables which are defined for decomposing the maximization step into a set of simpler problems. Several approaches can be found in the literature for addressing the problem of missing information or data [6,8,11]. Here we are interested in the case in which the observations suffer from a specific type of censorship. This case was first considered for the RBF setting in [12]. There, a novel algorithm was proposed where the estimation procedure was focused on refining information concerning the observations. The scheme made use of a novel procedure [9] based on the idea of sequential imputation-improvement, similar to the EM scheme. The procedure did not require the Gaussian assumption and had good performance in simulation examples. Nevertheless, since it cannot be framed in the EM context, convergence of the method has not been theoretically characterized.

In this paper, an alternative method is presented which also deals with this type of censorship in the observed data. This method, very well suited for exploiting the specific architecture of the RBF network, is framed in the EM context and his convergence is theoretically characterized.

2   Notation, Model, and Algorithm

Let us consider the following general RBF model

$$y_k = \omega^t o_k(\theta) + e_k, \quad k = 1, ..., N,$$

where $o_k(\theta) = (o_{k,1}(\theta_1), ..., o_{k,M}(\theta_M))^t$, the errors terms e_k are i.i.d. upon $N(0, \sigma^2)$, and the independent variables $o_{k,i}(\theta) = o_{k,i}(\theta_i) = f(\theta_i, x_k)$, with f being a real valued real basis function.

We want to estimate the network weights $\omega = (\omega_1, ..., \omega_M)^t$ and $\theta = (\theta_1, ..., \theta_M)^t$ under the assumption that there is some censorship on the observed data y_k. For doing so we consider a partition of the real line in disjoint intervals $\mathbf{R} = \bigcup (l_k, u_k]$. It will be assumed that the dependent data y_k may be either grouped or non-grouped. If $k \in K_{ng}$ the value y_k is observed. Otherwise, when $k \in K_g$, the value y_k is missed and only a classification interval is known, $(l_k, u_k]$, which overlaps it.

Here we propose an EM-based algorithm to estimate the parameters (ω, θ) using the particular form of the model (it is a linear model when θ is known) and using the information of the grouped data.

Following [2], we define the following non-observed variables

$$y_{k,i} = \omega_i o_{k,i}(\theta_i) + e_{k,i}, \quad k = 1, ..., N \quad \text{and} \quad i = 1, ..., M,$$

where $e_{k,i}$ are i.i.d. $N(0, \sigma_i^2)$, such that $\sum_{i=1}^{M} e_{k,i} = e_k$ for all $k = 1, ..., N$.

The σ_i are chosen satisfying $\sigma^2 = \sum_{i=1}^{M} \sigma_i^2$. Note that the values $y_{k,i}$ are non-observed, regardless y_k be grouped or non-grouped. Note also that given a value for $y_{k,i}$, we can estimate the observed variables provided the partition intervals are fixed. Hence, we can then apply an EM-based approach to the problem.

Under these conditions we have that $\widetilde{y} = \left((y_{k,i})_{k=1,\ldots,N;i=1,\ldots,M} \right)^t$ has a normal distribution and $\log L_c(\omega, \theta)$ takes the form

$$\log L_c(\widetilde{y}, \omega, \theta) = \text{constant} + \sum_{k=1}^{N} \sum_{i=1}^{M} \left(-\frac{1}{2\sigma_i^2} \right) \left[(\omega_i o_{k,i}(\theta_i))^2 - 2y_{k,i}\omega_i o_{k,i}(\theta_i) \right].$$

Note that the $y_{k,i}^2$ terms, grouped in the constant, are not relevant in the log-likelihood. Hence, the expectation step of the EM algorithm only requires to compute the expectation of each $y_{k,i}$.

2.1 Expectation Step

Precisely, assuming fixed values of ω and θ, we compute $\widehat{y}_{k,i} = E_{\omega,\theta}(y_{k,i}|y_k)$ if $k \in K^{ng}$ and $\widehat{y}_{k,i} = E_{\omega,\theta}(y_{k,i}|y_k \in (l_k, u_k])$ if $k \in K^g$.

The first expectation can be computed in a similar way to [2,5] only considering the explicit formulation for the parameters ω_i. Writing the density function of $e_{k,i}$ conditioned by the sum e_k we get $e_{k,i}|e_k \equiv N\left(\frac{\sigma_i^2}{\sigma^2} e_k, \frac{\sigma_i^2}{\sigma^2} \left(1 - \frac{\sigma_i^2}{\sigma^2} \right) \right)$.

In this case $E(e_{k,i}|e_k) = \frac{\sigma_i^2}{\sigma^2} e_k$, as a consequence of normality. Since $e_k = y_k - \sum_{i=1}^{M} \omega_i o_{k,i}(\theta_i)$ and $e_{k,i} = y_{k,i} - \omega_i o_{k,i}(\theta_i)$, we get

$$E_{\omega,\theta}(y_{k,i}|y_k) = \omega_i o_{k,i}(\theta_i) + E(e_{k,i}|y_k) = \omega_i o_{k,i}(\theta_i) + E(e_{k,i}|e_k).$$

Hence we have that

$$\widehat{y}_{k,i} = E_{\omega,\theta}(y_{k,i}|y_k) = \omega_i o_{k,i}(\theta_i) + \frac{\sigma_i^2}{\sigma^2} \left(y_k - \sum_{i=1}^{M} \omega_i o_{k,i}(\theta_i) \right), \text{ if } k \in K^{ng}.$$

For the expectation corresponding to grouped data we have the following result.

Theorem 1.

$$\widehat{y}_{k,i} = E_{\omega,\theta}(y_{k,i}|y_k \in (l_k, u_k]) = E_{\omega,\theta}(y_{k,i}|y_k = E_{\omega,\theta}(y_k|y_k \in (l_k, u_k]))$$

$$= \omega_i o_{k,i}(\theta_i) + \frac{\sigma_i^2}{\sigma^2} \left(\widehat{y}_k - \sum_{i=1}^{M} \omega_i o_{k,i}(\theta_i) \right),$$

being $\widehat{y}_k = E_{\omega,\theta}(y_k|y_k \in (l_k, u_k])$.

Which means that we can first estimate the censored value y_k and then estimate $y_{k,i}$, conditioned to such value.

The proof of this theorem is based on the following result about conditional expectations. Afterwards, normality will do the rest.

Lemma 1. *Let* X, Y *be random variables over a probability space* $(\Omega, \mathcal{A}, P)$. *For each event* $A \in \mathcal{A}$, *it is*

$$E(X|Y \in A) = E(E(X|Y)|Y \in A).$$

PROOF: $E(X|Y)$ verifies

$$\int_B X dP = \int_B E(X|Y) dP, \qquad \text{for all } B \in \sigma(Y),$$

where $\sigma(Y) = \{Y \in A | A \in \mathcal{A}\}$ is the σ-field generated by variable Y. Then

$$E(X|Y \in A) = \frac{1}{P(Y \in A)} \int_{\{Y \in A\}} X dP$$

$$= \frac{1}{P(Y \in A)} \int_{\{Y \in A\}} E(X|Y) dP = E(E(X|Y)|Y \in A). \qquad \blacksquare$$

This results translates to

$$E_{\omega,\theta}(y_{k,i}|y_k \in (l_k, u_k)) = E_{\omega,\theta}(E(y_{k,i}|y_k)|y_k \in (l_k, u_k)).$$

In addition, due to normality of variables, the inner expectation is linear taking the following form

$$E(y_{k,i}|y_k) = \omega_i o_{k,i}(\theta_i) + \frac{\sigma_i^2}{\sigma^2}\left(y_k - \sum_{i=1}^{M} \omega_i o_{k,i}(\theta_i)\right),$$

which, as mentioned, is linear in the conditioning variable y_k. Then

$$E_{\omega,\theta}(E(y_{k,i}|y_k)|y_k \in (l_k, u_k))$$

$$= \omega_i o_{k,i}(\theta_i) + \frac{\sigma_i^2}{\sigma^2}\left(E(y_k|y_k \in (l_k, u_k)) - \sum_{i=1}^{M} \omega_i o_{k,i}(\theta_i)\right)$$

$$= E_{\omega,\theta}(y_{k,i}|y_k = E_{\omega,\theta}(y_k|y_k \in (l_k, u_k))),$$

as we wanted to prove. $\qquad \blacksquare$

2.2   Maximization Step

Once $\widehat{y}_{k,i}$ is computed, the maximization can be easily performed since

$$\max_{\omega,\theta} \log L_c\left((\widehat{y}_{k,i})_{k=1,\dots,N;i=1,\dots,M}, \omega, \theta\right) \qquad (1)$$

$$= \max_{\omega,\theta} \sum_{k=1}^{N} \sum_{i=1}^{M}\left(-\frac{1}{2\sigma_i^2}\right)(\widehat{y}_{k,i} - \omega_i o_{k,i}(\theta_i))^2 = \min_{\omega,\theta} \sum_{i=1}^{M} \frac{1}{2\sigma_i^2} \sum_{k=1}^{N}(\widehat{y}_{k,i} - \omega_i o_{k,i}(\theta_i))^2.$$

This minimization is equivalent to M partial minimizations. Hence, for each $i = 1, \dots, M$, we compute $\min_{\omega_i, \theta_i} \sum_{k=1}^{N}(\widehat{y}_{k,i} - \omega_i o_{k,i}(\theta_i))^2$. In addition we can exploit the linear dependency of ω_i to decompose the minimization procedure, minimizing first on ω_i (linear) and them on θ_i (nonlinear).

If we split $\min_{\omega_i, \theta_i}$ into $\min_{\omega_i} \min_{\theta_i}$ we guarantee that the objective function is reduced, although the global minimum may not be reached. Hence, the iterative procedure can be framed within the GEM schemes with similar convergence properties to the EM framework.

Further refinements can be performed on the minimization involving ω_i so that the log-likelihood of the observed data is minimized instead of (1).

3   Simulation Example

This example taken from [12], selects points from the graph of $2\sin(x) + 0.5$:
$x_k = \frac{2}{5}k$, $y_k = 2\sin x_k + 0.5$, $k = 0, \ldots, 14$, where a saturation effect
that limits the data values to the range $[-2, 2]$ is considered. This makes
$y_3 = y_4 = y_5 = 2$ a set of grouped data, being $K^g = \{3, 4, 5\}$ and $K^{ng} = \{0, 1, 2, 6, 7, 8, 9, 10, 11, 12, 13, 14\}$. Hence $y_k = 2\sin x_k + 0.5$, $k \in K^{ng}$, $y_k = 2$, $k \in K^g$.

The selected RBF has two basis functions: $\tilde{f}(x) = \sum_{i=1}^{2} w_i \exp\left(-\frac{(x-\theta_{1i})^2}{2\theta_{2i}^2}\right)$.
Concerning noise, $\sigma_i = \sigma_j, \forall i, j$ and $\sigma = 0.25$ were assumed. The initial values
for the parameters where randomly selected upon uniform distributions $w_i \in U[-1, 1]$, $\theta_{1i} \in U[2i-1, 2i+1]$ and $\theta_{2i}^2 \in U[0, 1]$, $i = 1, 2$.

The algorithm was performed so that a single iteration on w was carried
out for each gradient descent iteration, that is both iterations where performed
alternately. The results obtained after 1000 iterations are displayed in Figure
1. The resulting RBF, together with the original function used to generate the
data, as well as the data and the final estimated $\widehat{y}_k$ values for grouped data are
all displayed there.

Note that the approximation is good in the range of available data. Also,
the estimated values for y_k, $k \in K^g$ fit quite well, partially compensating the
original censorship.

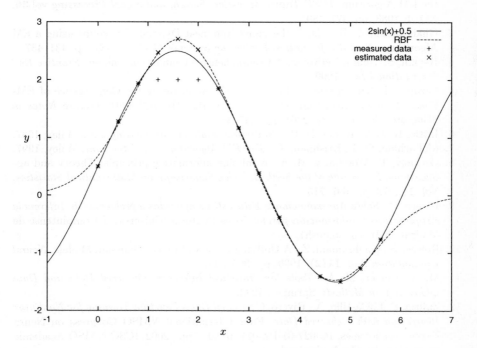

Fig. 1. Regression results for simulation example.

4 Concluding Remarks

An EM-based formulation for training RBF Neural Networks in the context of missing information has been presented. The procedure can be framed in the Generalized EM setting for convergence characterization to the maximum likelihood estimate of the observed data. Closed results for computing the expectation step have been provided. The specific structure of the RBF allows efficient algorithms for the maximization step. Simulations show that the training algorithm addresses in a straightforward manner the data censorship problem.

Acknowledgements. The authors acknowledge support from Project PB97-0566-C02-01 and BFM2000-1475 of the Programa Nacional de Promoción General del Conocimiento of the Ministerio de Educación y Cultura y Ministerio de Ciencia y Tecnología of Spain, and Project 10978 of the Universidad Politécnica de Madrid.

References

1. Dempster, A. P., Laird N. M., Rubin, D. B.: Maximum Likelihood from Incomplete Data via the EM algorithm. *Journal of the Royal Statistical Society*, Vol. **B 39**, 1977, pp. 1–38.
2. Feder, M., Weinstein, E.: Parameter Estimation of Superimposed Signals Using the EM Algorithm, *IEEE Trans. Acoustics, Speech, and Signal Processing* vol.**36**, NO. 4, 1988, pp. 477–489.
3. Ghahramani, Z., Roweis, S.: Learning Nonlinear Dynamical Systems using a EM Algorithm., *Neural Information Processing Systems 11 (NIPS'98)*, pp. 431–437.
4. Haykin, S.: Neural Networks, A Comprehensive Foundation. 2nd ed. *Prentice Hall International In.c*, 1999.
5. Lázaro, M., Santamaría, I., Pantaleón, C.: Accelerating the Convergence of EM-Based Training Algorithms for RBF Networks. *IWANN 2001, Lecture Notes in Computer Science 2084*, 2001, pp. 347–354.
6. Little, R. J. A., Rubin, D. B.: *Statistical analysis with missing data*, Wiley, 1987.
7. McLachlan, G. J., Krishnan, T.: *The EM Algorithm and Extensions*, Wiley, 1997.
8. Orchard, T., Woodbury, M. A.: A missing information principle: Theory and applications. *Proceeding of the Sixth Berkeley Symposium on Mathematical Statistics*, Vol. **1**, 1972, pp. 697–715.
9. Rivero, C.: *Sobre Aproximaciones Estocásticas aplicadas a problemas de Inferencia Estadística con información parcial*, Tesis Doctoral, Universidad Complutense de Madrid, 2001. (In spanish).
10. Roweis, S., Ghahramani, Z.: A Unifying Review of Linear Gaussian Models, *Neural Computation*, Vol. **11(2)**, 1999, pp. 305–345.
11. M. A. Tanner, M. A.: *Tools for Statistical Inference. Observed Data and Data augmentation Methods.* Springer, 1993.
12. Zufiria, P. J., Castillo, Á., Rivero, C.: *Radial Basis Function Training for Nonlinear Regression with Censored Data*, First International NAISO Congress on Neuro Fuzzy Technologies, 100027-01-PZ-071, p. 80, 7 pp., 2002. ICSC NAISO Academic Press Canada/The Netherlands.

Stochastic Supervised Learning Algorithms with Local and Adaptive Learning Rate for Recognising Hand-Written Characters

Matteo Giudici[1], Filippo Queirolo[1], and Maurizio Valle[2]

[1] Department of Production Engineering
University of Genova, Via Cadorna 2, 17100 Savona, Italy
Ph.: +39 019 264555, fax: +39 019 264558

[2] Department of Biophysical and Electronic Engineering
University of Genova, Via all'Opera Pia 11/a, 16145 Genova, Italy
Ph.: +39 010 3532775, fax: +39 010 353 2777, valle@dibe.unige.it

Abstract. Supervised learning algorithms (i.e. Back Propagation algorithms, BP) are reliable and widely adopted for real world applications. Among supervised algorithms, stochastic ones (e.g. Weight Perturbation algorithms, WP) exhibit analog VLSI hardware friendly features. Though, they have not been validated on meaningful applications. This paper presents the results of a thorough experimental validation of the parallel WP learning algorithm on the recognition of handwritten characters. We adopted a local and adaptive learning rate management to increase the efficiency. Our results demonstrate that the performance of the WP algorithm are comparable to the BP ones except that the network complexity (i.e. the number of hidden neurons) is fairly lower. The average number of iterations to reach convergence is higher than in the BP case, but this cannot be considered a heavy drawback in view of the analog parallel on-chip implementation of the learning algorithm.

1 Introduction

Neural Networks (NNs) are an efficient computational paradigm for solving many real world problems. At present there is a growing interest in applications like Optical Characters Recognition (OCR), remote-sensing images classification, industrial quality control analysis and many others in which NNs can be effectively employed [2]. Most widely used and reliable learning algorithms are gradient descent ones, namely Back Propagation (BP). Among gradient descent learning algorithms, the Weight Perturbation (WP) algorithm was formerly developed to simplify the circuit implementation [5] and, although it looks more attractive than BP for the analog VLSI implementation, its efficiency in solving real world problems has not yet been heavily investigated. Many researchers have recently proposed circuit architectures for the analog VLSI implementation of Multi Layer Perceptrons (MLPs) based networks to achieve low power consumption, small size and high speed [12]. In a previous paper, we have presented an algorithm for the local and adaptive management of the learning rate parameter [8]. The local learning rate update rule was inspired by [11] and [15] and it is similar to the one proposed by [14], though it differs from the cited ones in the way the learning rate value is updated. In [10] we

J.R. Dorronsoro (Ed.): ICANN 2002, LNCS 2415, pp. 619–624, 2002.
© Springer-Verlag Berlin Heidelberg 2002

4 Simulation Results

To evaluate the performance of the F-LR and A-LR WP algorithms, we analysed the results coming from experiments on a large number of network topologies (i.e. with different numbers of Hidden Neurons – HN) and many learning parameters sets.

From the experiments, we identified the best results in terms of VE, ATI, standard deviation of VE, standard deviation of ATI.

F-LR Simulation Results. To evaluate the best network topology and learning parameters set-up we had to investigate a four-dimension analysis space (i.e. step, η, γ, HN). Starting from the analysis and results coming from our past experience [10] we set γ =0.5. Then we exhaustively investigated the resulting three-dimensional analysis space. Best performance are summarized in Fig. 1 and Fig. 2. In Fig. 1 the VE and ATI behaviour versus the *step*, with the η (LR) as parameter and HN = 50, are shown. Best result (i.e. VE = 2.82%) is given by LR=0.001 and *step* = 0.01. In Fig. 2 the VE and ATI behaviour versus the HN number, with the step as parameter and LR= 0.001, are shown. Best result (i.e. VE = 2.82%) again is given by HN= 50 and *step* = 0.01.

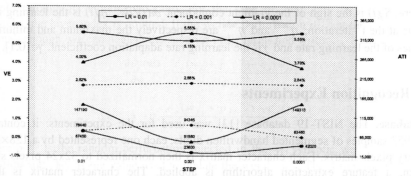

Fig. 1. F-LR: VE and ATI versus the *step* value with the η (LR) as parameter and HN = 50

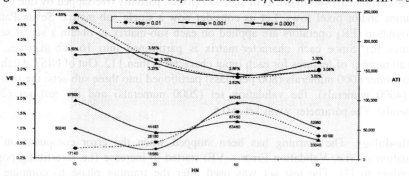

Fig. 2. F-LR: VE and ATI versus the the HN number with the step as parameter and η = 0.001

A-LR Simulation Results. Starting from the results reported in the previous subsection and in [9], we set the learning rate range as follows: $\eta_{ij} \in [0.001, 0.0001]$ and $\gamma = 0.5$. Then we had to investigate a bi-dimension analysis space (i.e. step, HN).

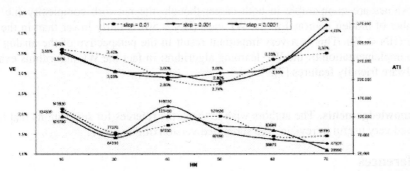

Fig. 3. A-LR: VE and ATI versus the HN number with the *step* as parameter and $\eta \in [10^{-3}, 10^{-4}]$

Comparison. In Table 1 we present the results of the comparison between the F-LR and the A-LR algorithms reporting the GE values versus the rejection rate (RR) with HN = 50, *step* = 0.01, η= 0.001 (F-LR) and $\eta \in [0.001, 0.0001]$. Please note that the Rejection Rate (RR) is the percentage of patterns that are not classified (i.e. rejected) according to a given confidence criterion (e.g., a pattern is classified if the difference between the two highest neuron outputs is higher than a confidence threshold).

Table 1. GE versus RR: comparison between the WP algorithms with F-LR and A-LR

RR	0%	1%	2%	3%	4%	5%	6%	7%	8%	9%	10%
F-LR GE	3.82%	3.18%	2.70%	2.32%	1.99%	1.63%	1.34%	1.14%	0.96%	0.77%	0.72%
A-LR GE	3.78%	3.21%	2.75%	2.30%	1.93%	1.54%	1.30%	1.05%	0.86%	0.75%	0.63%

5 Conclusions

This paper presents the results of a thorough experimental validation of a stochastic learning algorithm (i.e. the Weight Perturbation algorithm) on a fully meaningful application: the classification of hand-written characters. We adopted an adaptive and local management of the learning rate parameter in order to increase the efficiency.
We compared the results with those obtained with the Back Propagation learning algorithm (see [10]). Table 2 summarizes the comparison results.

Table 2. Comparison results between BP and WP algorithms with F-LR and A-LR

	VE	VE St. Dev.	ATI	ATI St. Dev.	GE	GE St. Dev.
BP F-LR	2.70%	0.14%	232	7.8	3.73%	0.17%
BP A-LR	2.61%	0.18%	77	1.5	3.58%	0.13%
WP F-LR	2.82%	0.15%	67450	26169	3.82%	0.33%
WP A-LR	2.74%	0.13%	121630	21362	3.78%	0.28%

The higher number of ATI in the WP case is not a drawback: it will be compensated by the very high speed of the analog on-chip parallel implementation [7]. The GE and

VE values are comparable in both cases; though, the network complexity (i.e. the number of hidden neurons) in the WP case (HN = 50) is fairly lower than in the BP case (HN = 70). This is a very important result in the perspective of the analog on-chip implementation of the WP learning algorithm: in fact the WP algorithms exhibit hardware friendly features [7].

Acknowledgements. The authors wish to thank the referees for suggestions that have proved very useful in revising the submitted version of this paper.

References

[1] Cauwenberghs, G., *A Fast Stochastic Error-Descent Algorithm for Supervised Learning and Optimization*, in Advances in Neural Information Processing Systems 5 (NIPS5), pp. 244–251, 1993

[2] Tarassenko, L.: *A Guide to Neural Computing Applications*. Arnold Pub, London (1999)

[3] Alspector, J., Meir, R., Yuhas, B., Jayakumar, A., and Lippe, D., *A Parallel Gradient Descent Method for Learning in Analog VLSI Neural Networks*, in Advances in Neural Information Processing Systems 5 (NIPS5), pp. 836–844, 1993

[4] Alspector, J., Lippe, D.: *A Study of Parallel Perturbative Gradient Descent*. Advances in Neural Information Processing Systems (i. e. NIPS96) (1996) 803–810

[5] Jabri, M. A., et al.: *Adaptive Analog VLSI Neural Systems*. Chapman & Hall, 1996

[6] G.M. Bo, D.D. Caviglia, M. Valle, *An Analog VLSI Implementation of a feature Extractor for Real Time Optical Character Recognition*, IEEE Journal of Solid State Circuits, Vol. 33, N. 4, April 1998, pp. 556–564 (ISSN: 0018 – 9200).

[7] Valle, M.: *Analog VLSI Implementation of Artificial Neural Networks with Supervised On-chip Learning*. to appear on Analog Integrated Circuits and Signal Processing, Kluwer Academic Publishers, 2002.

[8] Bo, G. M., Caviglia, D. D., Chiblè, H., Valle, M.: *Analog VLSI On-chip Learning Neural Network with Learning Rate Adaptation*. In chapter 14 of: Learning on Silicon, Adaptive VLSI Neural Systems, G. Cauwenberghs, M., Bayoumi, A. (Eds.). The Kluwer International Series In Engineering And Computer Science Volume 512. Kluwer Academic Publ., Netherlands (1999) 305–330

[9] Giudici, M., Queirolo, F., Valle, M.: *Back Propagation and Weight Perturbation Neural-Networks Classifiers: Algorithms Validation and Comparison.* submitted to: IEEE Transactions on Neural Networks

[10] Giudici, M., Queirolo, F., and Valle, M.: *Evaluation of gradient descent learning algorithms with adaptive and local learning rate for recognising hand-written numerals*. 10th European Symposium on Artificial Neural Networks, ESANN'02, Bruges (Belgium), April 24 – 26, 2002. 10th European Symposium on Artificial Neural Networks, ESANN'02, Bruges (Belgium), April 24–26, 2002, pp. 289–294

[11] Jacobs, B. A., *Increased Rates of Convergence through Learning Rate Adaptation*, Neural Networks, Vol. 4, pp. 295–307, 1988

[12] Cauwenberghs, G., Bayoumi, M. A. (ed.): *Learning on Silicon, Adaptive VLSI Neural Systems*. Kluwer Academic Pub (1999)

[13] Grother, P. J.: *NIST Special Database 19*. National Institute of ST, (1995)

[14] Almeida, L.B., Langlois, T., Amaral, J.D., *On-line step size adaptation*, Technical Report INESC RT07/97

[15] Tollenaere, T., SuperSAB: *Fast Adaptive Back propagation With Good Scaling Properties*, Neural Networks, Vol. 3, pp. 561–573, 1990

Input and Output Feature Selection

Alejandro Sierra and Fernando Corbacho

Escuela Técnica Superior de Informática
Universidad Autónoma de Madrid
28049 Madrid, Spain
{Alejandro.Sierra, Fernando.Corbacho}@ii.uam.es

Abstract. Feature selection is called wrapper whenever the classification algorithm is used in the selection procedure. Our approach makes use of linear classifiers wrapped into a genetic algorithm. As a proof of concept we check its performance against the UCI spam filtering problem showing that the wrapping of linear neural networks is the best. However, making sense of data involves not only selecting input features but also output features. Generally, this is considered too much of a human task to be addressed by computers. Only a few algorithms, such as association rules, allow the output to change. One of the advantages of our approach is that it can be easily generalized to search for outputs and relevant inputs at the same time. This is addressed at the end of the paper and it is currently being investigated.

1 Introduction

Feature selection has been investigated in pattern recognition for decades [1]. There are several interrelated reasons behind this practice although no principled foundation. First of all, irrelevant and distracting features degrade machine learning performance. Surprisingly enough, decision trees, which are designed to choose the best features, deteriorate by 5% to 10% when a random feature is added to the dataset [2]. The reason seems to be the scarcity of data down the tree which makes random features look good by chance.

Second, selecting features is a simple means of addressing overfitting by reducing the number of parameters. For instance, setting to zero some of the coefficients of least squares can often reduce the high variance of the predicted values. Alternatively, coefficients can be shrinkaged by means of a penalty term as in ridge regression [3]. The penalizing term amounts to add a diagonal term to the standard regression matrix, which makes it non-singular. It is thus equivalent to discarding correlated features. Similarly, weight decay and regularization in neural networks is another penalizing technique which generally improves prediction accuracy. Small weights favor networks that are more nearly linear [4].

Finally, reducing the number of features is the best way to end up with interpretable and simple solutions, always a very desirable trait. Even decision trees, generally acknowledged as the most interpretable learning machines, are hard to read when plagued by too many features.

J.R. Dorronsoro (Ed.): ICANN 2002, LNCS 2415, pp. 625–630, 2002.

Traditionally, it is a human expert who chooses the problem to be solved (output feature) and its hopefully predictive inputs. In general, feature selection or construction elaborates the inputs for a fixed output. However, output feature selection can also be automated. There are situations in which this is becoming worthwhile and some algorithms have been already developed. For instance, association rules try to find joint values of features that happen most frequently. It is most often applied to binary data as in market basket analysis. Here, the features are the products sold and each pattern corresponds to the set of items bought by a client [5]. It is not obvious which features should be input or output. Neither is this clear in collaborative filtering and recommender systems, which scan databases of user preferences to propose new products [6]. These algorithms are also being applied to make sense of web navigation.

The organization of the paper is as follows. In Section 2 we review our approach to feature selection and classification. The results on the UCI e-mail filtering problem are presented in Section 3. Finally, our ideas about output feature selection can be found in Section 4 together with our conclusions and future work.

2   Genetic Algorithms as Wrappers

Let us focus on pattern classification. Consider a collection of training patterns $(\mathbf{x}, t)$ where t is the class to which $\mathbf{x}$ belongs. These patterns are drawn independently from the same probability density function $p(\mathbf{x}, t)$. The quality of a classifier $y(\mathbf{x})$ can be measured by the loss $L(t, y(\mathbf{x}))$ incurred. The expected value of this loss, the so-called risk, is

$$R = \int L(t, y(\mathbf{x}))p(\mathbf{x}, t)d\mathbf{x}dt. \qquad (1)$$

For an indicator loss (0-1 loss), the risk becomes the probability of misclassification. Learning amounts to estimating the model which minimizes the risk over the set of models supported by the learning machine [8]. If we do not enforce making use of the whole set of input features, what we have to minimize is a possibly huge number of averages such as

$$R_1 = \int L(t, y(x_1))p(x_1, t)dx_1dt \qquad (2)$$

$$R_{13} = \int L(t, y(x_1, x_3))p(x_1, x_3, t)dx_1dx_3dt. \qquad (3)$$

Our inductive principles consists in breaking this task into its natural two parts:

- find the solution of each individual problem

$$R_{i_1 \ldots i_n} = \int L(t, y)p(x_{i_1}, \ldots, x_{i_n}, t)dx_{i_1} \ldots dx_{i_n}dt \qquad (4)$$

  by means of a conventional and fast linear algorithms with cross-validation complexity control and
- scan the space of subsets of features

$$\{R_1, R_2, R_3, \ldots, R_{11}, R_{12}, \ldots, R_{111}, R_{112}, \ldots\} \qquad (5)$$

by making use of the global search capabilities of a genetic algorithm [9].

2.1 Coding, Fitness, and Genetic Operators

A genetic algorithm is a population based optimization algorithm inspired by natural selection. In nature, a trait responds to selection if it varies among individuals, it is heritable and it is correlated with reproductive success [12]. These conditions can be easily simulated in a computer and applied to the evolution of subsets of features and their corresponding classifiers [10,11].

Each feature subset $R_{i_1 \ldots i_n}$ or problem is encoded by means of a binary string of length equal to the number of available features. A 1 (0) means that the corresponding feature is (not) to be used. For instance, in a problem with 10 input features:

$$R_{137} \rightarrow (1010001000). \qquad (6)$$

It is important to notice that the actual classifier is not evolved but calculated from $R_{i_1 \ldots i_n}$ by a linear optimization method. Once found, the performance of the classifier is used by the GA to calculate its fitness and thus the average reproductive success of the subset. More specifically, the fitness function for problem $R_{i_1 \ldots i_n}$ is

$$\text{Fitness}(R_{i_1 \ldots i_n}) = \epsilon(R_{i_1 \ldots i_n}) + \alpha(\%\text{features}) \qquad (7)$$

where $\epsilon(R_{i_1 \ldots i_n})$ is the sum of training and validation classification error incurred by the model which minimizes $R_{i_1 \ldots i_n}$, and the second term is a complexity penalizing term proportional to the percentage of input features used. Constant α combines numerically both objectives giving priority to classification accuracy ($\alpha = 0.01$). Risk $R_{i_1 \ldots i_n}$ is minimized over linear models $y(\mathbf{x}) = \sum_{j=1}^{n} w_j x_{i_j}$ by means of one of the following three algorithms:

- Fisher's discriminant analysis (FDA) [13]. FDA projects class means as far from each other as possible keeping the classes as compact as possible around their means.
- Functional link network (FLN) [14]. FLNs are linear networks without hidden units whose ability to learn non-linear mappings depends on their being fed with adequate extra features traditionally constructed as powers of the original attributes. They can be easily trained by means of the classical delta rule.
- Naive-Bayes (NB) [17]. NB classifies patterns by maximizing the a posteriori probabilities which can be easily calculated provided feature independence is assumed.

The methods can be turned non-linear in the input features and still linear in parameters by just invoking powers and products of the original features. We use universal stochastic sampling in the selection process and uniform crossover to avoid polynomial ordering problems. The evolution of each run is halted after 50 generations without finding better individuals.

3   Experimental Results

As a proof of concept we have conducted an experiment on the UCI spam database which consists of 4601 e-mails each characterized by 57 features and classified as spam (39.4%) or non-spam. Spam is unsolicited commercial e-mail such as web ads or chain letters [16]. Most of the features (1-48) correspond to the frequency of appearance of certain key words and key characters (49-54) in the e-mail. The other three features (55,56,57) have to do with uninterrupted sequences of capital letters. The donors of this set claim a 7-8% misclassification error rate [7]. This is in agreement with our linear results (see Table 1). Recently, a generalized additive model using a cubic smoothing spline has been reported to reach a 5.3% test error rate [15].

For each linear classifier, we have conducted 10 executions of our wrapping GA on the spam database, which has been randomly broken into training (70%), validation (15%) and test sets. The test set does not take part in the training process whatsoever and its rate is therefore an estimate of the generalization ability of the resulting classifiers. Table 1 shows the average training, validation and test error rates of the fittest classifiers. The corresponding errors for the whole set of features are also shown for comparison. The main lesson we can extract from these numbers is that generalization improves after selecting features by our global procedure. This improvement is dramatic for Naive Bayes (from 17.8% down to 8.8%) which is the algorithm most sensitive to redundant features [17].

FLNs do not seem to be much disrupted by non-informative features. Although the average evolved FLN performs worse on the training set than the FLN built with the full set of features, it does outperform the latter on the validation and test sets. There is no doubt that this improvement is due to the rejection of certain features, 40% on average. Features such as all, email, over, mail, receive, people, report, telnet, data, 1999 are always discarded. As expected, the most relevant features include address, 3d, remove, order, will, free, credit, money, George, $, etc.

As mentioned above, recent results indicate non-linear input-output relationships in this set [15]. If we add the squares of the features to the dataset, which makes a total of 114 features, the test error rate reached by an FLN trained on the full set of features is 6.95%, and 6.08% after selecting features. Moreover, Fisher's discriminant cannot be constructed on this data set due to rank deficiency and Naive Bayes yields a 22% error rate. As expected, the higher the degree the more FDA and Naive Bayes benefit from the wrapping approach.

4   Output Feature Selection and Discussion

As noted before, making sense of a database involves selecting both input and output features. This is necessary if we want to extract all possible knowledge out of data. Even the UCI repository already contains databases such as the automobile dataset with a warning to this respect: "several of the attributes in

Table 1. Average training, validation and test error rates for three different evolutionary wrappings on the UCI spam filtering problem. The corresponding errors for the whole set of features are written below. FDA and NB greatly benefit from the selection procedure. Column 5 shows the average number of features used by the evolved classifiers.

Wrapping	Training error rate %	Validation error rate %	Test error rate %	Number of features
Functional	7.35	5.31	7.56	30
link net	6.18	6.81	7.67	57
Fisher	8.18	7.03	8.09	28
discr.	9.25	10.72	10.42	57
Naive	8.57	5.94	8.83	18
Bayes	18.57	21.01	17.80	57

the database could be used as a class attribute" [7]. If we ask our algorithm to search for outputs (only one output is considered for simplicity) we have to find the model which minimizes

$$R_{i_1\ldots i_n;j} = \int L(t_j, y(x_{i_1}, \ldots, x_{i_n}))p(x_{i_1}, \ldots, x_{i_n}, t_j)dx_{i_1}\ldots dx_{i_n}dt_j \qquad (8)$$

where the nature of the problem and number of classes depend on the chosen output feature j which must not belong to the input feature set. Our coding scheme (see Section 2.1) can be easily updated for the chromosomes to encode this output feature.

We claim that our wrapper approach based in a genetic algorithm constitutes a natural means of accomplishing this *less* supervised task. After all, natural evolution distinguishes itself for the absence of objectives outside reproductive success. No living organism is taught exactly what to do to survive. The surviving organisms are those who find a niche, i.e., those who spot a new problem and solve it. This is precisely what output feature selection is all about after substituting natural environments by databases. There are different ways of simulating this niching behavior in a GA of which fitness sharing is probably the simplest [18].

Learning is called supervised whenever training patterns with class labels are available. In this respect, output feature selection algorithms are still supervised learning, although less supervised than usual. Interestingly enough, real life problems are poorly supervised and the more intelligent an agent the less supervision it requires. But we know from unsupervised learning how difficult it can become to learn. The key ingredient is designing an appropriate super-objective to lead learning. Association rules lack in general this objective and limit their output to a collection of rules ordered by coverage (number of patterns they apply to) and support or accuracy on the coverage set. This makes sense only for nominal features and we are also dealing with continuous features and constructing full fledged models. Our fitness function has proved effective for input feature selection as shown in the spam filtering problem. It systematically improves the generalization ability of the evolved models. We still have to develop smart objectives taking into account not only accuracy and complexity but also coverage, feasibility, interest, etc. All these topics are currently under research and will be reported in the near future.

References

1. Kittler, J.: Feature set search algorithms. In: Chen, C. H. (ed.): Pattern recognition and signal processing. The Netherlands: Sijthoff an Noordhoff (1978).
2. John, G. H. Enhancements to the data mining process. PhD Dissertation, Computer Science Department, Stanford University (1997).
3. Hoerl, A. E., Kennard, R.: Ridge regression: biased estimation for non-orthogonal problems. Technometrics 12 (1970) 55–67.
4. Weigend, A. S., Rumelhart, D. E., Huberman, B. A.: Generalization by weight-elimination with application to forecasting. In: Lippman, R. P., Moody, J. E., Touretzky, D. S. (eds.): Advances in Neural Information Processing Systems, Vol. 3. Morgan Kaufmann, San Mateo, CA (1991) 875–882.
5. Agrawal, R., Mannila, H., Srikant, R., Toivonen, H., Verkamo, A. I.: Fast discovery of association rules. Advances in Knowledge Discovery and Data Mining. AAAI/MIT Press, Cambridge, MA (1995).
6. Resnick, P., Varian, H.: Recommender systems. Communications of the ACM 40(3) (1997) 56–58.
7. Blake, C. L., Merz, C. J.: UCI Repository of machine learning databases [http://www.ics.uci.edu/~mlearn/MLRepository.html]. Irvine, CA: University of California, Department of Information and Computer Science (1998).
8. Cherkassky, V., Mulier, F.: Learning From Data: Concepts, Theory and Methods. John Wiley & Sons, New York (1998).
9. Goldberg, D. E.: Genetic Algorithms in Search, Optimization and Machine Learning. Addison-Wesley, Reading, MA (1989).
10. Sierra, A., Macías, J. A., Corbacho, F.: Evolution of Functional Link Networks. IEEE Transactions on Evolutionary Computation 5 (1) (2001) 54–65.
11. Sierra, A.: High order Fisher's discriminant analysis. Pattern Recognition 35 (6) (2002) 1291–1302.
12. Stearns, S. C., Hoekstra, R. F.: Evolution: an introduction. Oxford University Press, Oxford (2000).
13. Fisher, R. A.: The use of multiple measurements in taxonomic problems. Annals of Eugenics 7 (1936) 179–188.
14. Pao, Y. H., Park, G. H., Sobajic, D. J.: Learning and generalization characteristics of the random vector functional link net. Neurocomputing 6 (1994) 163–180.
15. Hastie, T., Tibshirani, R., Friedman, J.: The Elements of Statistical Learning. Data Mining, Inference and Prediction. Springer Series in Statistics, Springer, New York (2001).
16. Cranor, Lorrie F., LaMacchia, Brian A.: Spam! Communications of the ACM 41(8) (1998) 74–83.
17. Langley, P., Sage, S.: Induction of selective Bayesian classifiers. In: de Mántaras, R. L., Poole, D. (eds.): Proc. Tenth Conference on Uncertainty in Artificial Intelligence, Seattle, WA. Morgan Kaufmann, San Francisco, CA (1994) 399–406.
18. Mahfoud, S. W.: Niching methods. In: Back, T., Fogel, D. B., Michalewicz, Z. (eds.): Evolutionary Computation 2. Institute of Physics Publishing, Bristol (2000) 87–92.

Optimal Extraction of Hidden Causes

Luis F. Lago-Fernández and Fernando Corbacho

E.T.S. de Informática, Universidad Autónoma de Madrid, 28049 Madrid, Spain

Abstract. This paper presents a new framework extending previous work on multiple cause mixture models. We search for an optimal neural network codification of a given set of input patterns, which implies hidden cause extraction and redundancy elimination leading to a *factorial code*. We propose a new entropy measure whose maximization leads to both maximum information transmission and independence of internal representations for factorial input spaces in the absence of noise. No extra assumptions are needed, in contrast with previous models in which some information about the input space, such as the number of generators, must be known a priori.

1 Introduction

The generative view in machine learning takes inputs as random samples drawn from some particular, possibly hierarchical, distribution. Hence, an input scene composed of several objects might be more efficiently described using a different generator for each object rather than just one generator for the whole input (independence of objects). The goal in a multiple cause learning model is, therefore, to discover a vocabulary of *independent causes*, or generators, such that each input can be completely accounted for by the cooperative action of a few of these generators. A common approach to achieve this goal is *independent component analysis*, in which redundancy is minimized constrained to the conservation of transmitted information [1].

In this paper we will deal with the problem of optimal codification of an input space with a neural network. This optimal coding means that, given an input set, the network should be able to extract the information that is useful for the task it has to perform, eliminating useless and redundant components. Assuming all information is important, the problem reduces to finding the most compact representation of the whole input space. On one hand, the network must be able to discriminate among all the input patterns. This, in mathematical terms, means that the mutual information between input and output is maximal, and in the absence of noise reduces to the following condition:

$$H(Y) = H(X), \tag{1}$$

where $H(X)$ and $H(Y)$ are the input and output entropies respectively. On the other hand, the network must be able to eliminate all the redundancy that is present in the input, finding the minimum dimension space that satisfies the

J.R. Dorronsoro (Ed.): ICANN 2002, LNCS 2415, pp. 631–636, 2002.
© Springer-Verlag Berlin Heidelberg 2002

information preserving constraints. This, as Barlow [2] has pointed out, is equivalent to finding an internal representation in which the neurons respond to causes that are as independent as possible. The optimal solution would be to find a *factorial* code, in which the sum of individual entropies equals the output entropy:

$$S(Y) \equiv \sum_{i=1}^{n} H(Y_i) = H(Y),\tag{2}$$

where the sum extends to all the hidden neurons in the network. We will search for neural codes that are as discriminant and factorial as possible, which means $H(Y)$ maximization and $S(Y)$ minimization.

In this context, many approaches leading to redundancy elimination have been proposed. Barlow [2] gave a proof of equation 2, and applied $S(Y)$ minimization to the problem of coding keyboard characters on 7 binary outputs that occur as independently of each other as possible. Földiák [3] applied a combination of Hebbian and anti-Hebbian mechanisms to a set of input patterns consisting of random horizontal and vertical bars. These mechanisms were able to reduce statistical dependency between representation units, while preserving information. This problem has been addressed by other authors [4,5] as a paradigm for multiple cause unsupervised learning. Recently, O'Reilly [6] has used a generalized recirculation algorithm in combination with inhibitory competition and Hebbian learning for a similar problem, where it has shown to perform quite well using only a few training patterns.

Although previous approaches are able to find good solutions to the problem of cause extraction, they generally depart from assumptions that should not be known a priori. For example, Barlow [2] assumes that a code is known, and reduces the problem to re-ordering the alphabet to find the minimum sum of bit entropies; and in [3,4,5] the number of coding units is forced to be equal to the number of hidden causes, which highly reduces the problem complexity.

Here we propose a different approach that is able to find a discriminant and factorial (DF) code with the only assumption that such a factorial code exists for the problem we are dealing with. In the next section we will introduce a new entropy measure whose maximization leads to a DF code without any extra assumption or constraint. After introducing the measure, we will show an application to the above problem [3] in contrast with other approaches. In the last section we discuss the limitations and applicability of this new theoretical framework.

2   New Entropy Measure for Cause Extraction

Recalling the two previous conditions for a good codification of the input space, namely discriminability (maximum $H(Y)$) and factoriality (minimum $S(Y)$), we will combine both requirements into a single condition. We will define a new entropy measure on the output space, $M(Y)$, as the quotient between the squared output entropy and the sum of individual output entropies:

$$M(Y) = \frac{H(Y)^2}{S(Y)} = \frac{H(Y)^2}{\sum_{i=1}^{n} H(Y_i)} \tag{3}$$

As we will see, $M(Y)$ is maximal when the network is performing a DF codification of the input. In this case $M(Y)$ takes precisely the value $H(X)$. The only requirement here is that a factorial code exists. In such a case, the following theorem establishes the relation between DF codes and maximum $M(Y)$ codes:

Theorem 1. *Let X be a factorial input space, and let Y be a coding space for X (each element $x \in X$ has a unique representation $y \in Y$). Then the code is a DF code if and only if $M(Y) = H(X)$.*

We will give only a sketch of the proof here. Note that, as $H(Y) \leq S(Y)$, we have:

$$M(Y) \leq H(Y) \leq H(X) \tag{4}$$

So the maximum value for $M(Y)$ is $H(X)$, and it occurs when $H(Y) = S(Y)$ (factorial code) and $H(Y) = H(X)$ (discriminant code). If we assume factorial problems we can always find an optimal code by maximizing this new entropy measure.

3 Results

As stated in the introduction, we will test the new measure with a problem introduced by Földiák [3]. It consists of learning to discriminate among different stimuli composed of vertical and horizontal bars that appear randomly with a specific probability p. Here we will use an input space of 5×5 pixels; in figure 1A we show some of the input patterns. Note that all of them are constructed from a set of only 10 generators. Learning these generators would lead to an optimal coding.

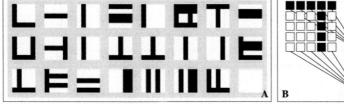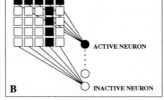

Fig. 1. A. Subset of the input patterns used to test the new entropy measure. All patterns are generated as a random combination of 5 vertical and 5 horizontal bars that appear independently with a probability $p = 0.15$. B. Neural network structure. Each of the hidden units responds to a subregion of the input space (its receptive field) determined by a 5×5 binary matrix. We show the receptive fields (only connections with value 1 are drawn) and the corresponding states of two neurons for a given input pattern.

To tackle the problem, we will use a neural network with $N = 20$ units, each one connected to the input layer with weights that are either 0 or 1. In other words, the receptive field of a given neuron is defined by a 5×5 binary matrix that determines the region in the input space it responds to. Let us remark that the number of neurons in the network exceeds the number of hidden causes. We expect the network organizes itself so that it uses the minimum number of neurons to explain the input. This is in contrast with other algorithms, such as backpropagation, that tend to use all the available resources.

The neuron dynamics is not relevant for our present analysis, since our interest is focused on finding the receptive fields that best account for the input structure. So we will simply assume that the neuron is active if its receptive field is stimulated, and silent otherwise. In figure 1B we show a scheme of the network structure.

Different receptive field configurations will produce different codes. In our quest for an optimal code, we searched the receptive field space looking for the code that best satisfies the discriminability and factoriality criteria, i.e. maximizes $M(Y)$. The search was performed using genetic algorithms with the standard PGAPack libraries [7]; results are shown in figures 2 and 3. In figures 2A and 2B we show the evolution of $H(Y)$, $M(Y)$ and $S(Y)$ as the genetic algorithm performs maximization of $H(Y)$ and $M(Y)$ respectively. In the first case, we observe a fast convergence of $H(Y)$ to its maximum allowed value ($H(X) = 4.14$). However $S(Y)$ does not decrease, which is a clear indicator of the lack of independence between neurons. In the second case, $H(Y)$ converges more slowly ($H(X) = 4.18$), but the small difference between $H(Y)$ and $S(Y)$ indicates that the network is performing an almost factorial codification.

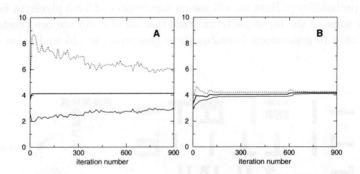

Fig. 2. $H(Y)$ (solid thick), $M(Y)$ (solid), and $S(Y)$ (dotted) versus iteration number of the genetic algorithm for $H(Y)$ maximization (A) and $M(Y)$ maximization (B). In both cases we used a neural network with $N = 20$ hidden units.

When we try to maximize $H(Y)$, the network fails to find a factorial code (final receptive fields are shown in figure 3A). This problem can be overcome by forcing the network to have $N = 10$ neurons (results not shown). In this case the only discriminant solution implies learning the input generators. When maximizing $M(Y)$, however, the network uses exactly 10 neurons (figure 3B) whose receptive fields resemble each of the hidden causes.

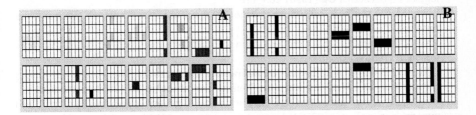

Fig. 3. A. Receptive fields after maximization of $H(Y)$. B. Receptive fields after maximization of $M(Y)$. The gray level indicates the probability of activation of a neuron with such a receptive field. Neurons with a dark receptive field are stimulated more often than neurons with a brighter one.

4 Discussion

We have introduced a new quantity, $M(Y)$, that measures the degree of discriminability and factoriality of a code. In the absence of noise, if a factorial code exists for a given input space, this quantity turns out to be equal to the input entropy $(M(Y) = H(X))$, and for any other code we have $(M(Y) < H(X))$. Maximization of $M(Y)$ leads to factorial codes without any assumption on the input space apart from the existence of such a factorial code. This approach improves previous results [2,3,4,5,6] in which some extra information about the input structure is assumed.

We assumed no noise is present in the input patterns, which led us to the discriminability condition 1. In the presence of noise a more complex expression in terms of mutual entropy can be derived.

Although this theoretical framework provides a solution to the problem of finding factorial representations, the approach we followed here involves computationally expensive entropy calculations that make it difficult to apply to most real problems. Some approximation to $M(Y)$ involving only local computations seems to be essential.

Other research approaches related to learning factorial representations [8,9, 10] and nonlinear dimensionality reduction [11,12] emphasize different aspects of the problem. Work in progress attempts to unify some of these concepts within a single framework, and at the same time develop an approximation method that avoids entropy calculations and deals with noisy problems.

References

1. Deco, G., Obradovic, D.: An information theoretic approach to neural computing (Springer-Verlag, New York, 1996) 65–107

2. Barlow, H.B., Kaushal, T.P., Mitchison, G.J.: Finding minimum entropy codes. Neural Computation **1** (1989) 412–423
3. Földiák, P.: Forming sparse representations by local anti-Hebbian learning. Biological Cybernetics **64** (1990) 165–170
4. Saund, E.: A multiple cause mixture model for unsupervised learning. Neural Computation **7** (1995) 51–71
5. Dayan, P., Zemel, R.S.: Competition and multiple cause models. Neural Computation **7** (1995) 565–579
6. O'Reilly, R.C.: Generalization in interactive networks: the benefits of inhibitory competition and Hebbian learning. Neural Computation **13** (2001) 1199–1241
7. Levine, D.: Users guide to the PGAPack parallel genetic algorithm library (1996) www-fp.mcs.anl.gov/CCST/research/reports_pre1998/comp_bio/stalk /pgapack.html
8. Hinton, G.E., Dayan, P., Frey, B.J., Neal, R.: The wake-sleep algorithm for unsupervised Neural Networks. Science **268** (1995) 1158–1161
9. Lee, D.D., Seung, H.S.: Learning the parts of objects by non-negative matrix factorization. Nature **401** (1999) 788–91
10. Olshausen, B.A., Field, D.J.: Emergence of simple-cell receptive field properties by learning a sparse code for natural images. Nature **381** (1996) 607–609
11. Roweiss, S.T., Saul, L.K.: Nonlinear dimensionality reduction by locally linear embedding. Science **290** (2000) 2323–2326
12. Tenenbaum, J.B., de Silva, V., Langford, J.C.: A global geometric framework for nonlinear dimensionality reduction. Science **290** (2000) 2319–2323

Towards a New Information Processing Measure for Neural Computation

Manuel A. Sánchez-Montañés and Fernando J. Corbacho

E.T.S. Ingeniería Informática, Universidad Autónoma de Madrid
Ctra. Colmenar Viejo, Km. 15, Madrid 28049, Spain.
{Manuel.Sanchez-Montanes, Fernando.Corbacho}@ii.uam.es

Abstract. The understanding of the relation between structure and function in the brain requires theoretical frameworks capable of dealing with a large variety of complex experimental data. Likewise neural computation strives to design structures from which complex functionality should emerge. The framework of information theory has been partially successful in explaining certain brain structures with respect to sensory transformations under restricted conditions. Yet classical measures of information have not taken an explicit account of some of the fundamental concepts in brain theory and neural computation: namely that optimal coding depends on the specific task(s) to be solved by the system, and that autonomy and goal orientedness also depend on extracting relevant information from the environment and specific knowledge from the receiver to be able to affect it in the desired way. This paper presents a general (i.e. implementation independent) new information processing measure that takes into account the previously mentioned issues. It is based on measuring the transformations required to go from the original alphabet in which the sensory messages are represented, to the objective alphabet which depends on the implicit task(s) imposed by the environment-system relation.

1 Introduction

Classical information theory schools [13] [6] [5] search for optimal ways of coding information. It is not the aim of this short paper to provide a detailed comparison of the different approaches, so we refer the interested reader to [4] for detailed expositions on this topic. More specifically, information theory has received widespread attention in the neural computation arena ([2] [7] [1] [3]) to cite a few. In this regard we fully agree with Atick [1] in the use of information theory as a basis for a first principles approach to neural computation. The relevance derives, as Atick claims, from the fact that the nervous system possesses a multitude of subsystems that acquire, process and communicate information. To bring to bear the more general problem of information we follow Weaver's [13] classification at three different levels: Technical problems: how accurately can the symbols of communication be transmitted? Semantic problems: how precisely do the transmitted symbols convey the desired meaning? Effectiveness

J.R. Dorronsoro (Ed.): ICANN 2002, LNCS 2415, pp. 637–642, 2002.
© Springer-Verlag Berlin Heidelberg 2002

problems: how effectively does the received meaning affect the receiver's conduct in the desired way?

We claim that any autonomous system (including the brain) living in an active environment must solve these three problems. Yet, as Weaver [13] already pointed out, classical information theory deals mainly with the technical problem. We claim that even nowadays a shift of view is necessary to take into proper consideration the knowledge of the receiver within the information theoretical framework. This paper provides a step towards dealing with the semantic and the effectiveness problems by making optimal coding depend on the specific task(s) to be solved by the system as well as on the specific knowledge about the environment/receiver in order to affect it in the desired way.

To address all these issues we present a new information processing measure based on measuring the transformations required to go from the original alphabet in which the sensory messages are represented, to the objective alphabet which depends on the implicit task imposed by the system-environment relation. In the classical approach the amount of information decreases when it undergoes any processing, that is, processing is passive instead of being active (i.e. elaborating the data and approaching the goal). In this regard a perfect communication channel has maximal mutual information yet minimal information processing. We search for a general information processing framework independent of the particularities of any specific implementation. As a proof of concept this paper presents a specific information processing measure that shows these properties (although it is not claimed this measure to be the only possible one).

2   Desirable Properties for a Measure of Information Processing

Next we list a set of desired properties that, we claim, the new information processing measure must have if it is to be considered for the design of any autonomous system (whether artificial or biological).

(a) It should take into account the task(s) to be solved by the system. The input to the system can be statistically rich and complex, yet it may be mostly useless if it is not related to the task.

(b) It should take into account how much the input data (number of transformations) has to be processed in order to extract the relevant information for the task. In case the output is a copy of the input, this measure should be null.

(c) It should be a compromise between reduction of spurious information and extraction of the relevant part of the information.

(d) It should account for uncertainties introduced by different means, such as: lost of meaningful information, environmental noise, etc.

(e) It should be able to deal with systems composed of stochastic elements.

3   Definition of the Information Processing Measure

To address all these issues we present a new information processing measure based on measuring the transformations required to go from the original alphabet in which the sensory messages are represented, to the objective alphabet which depends on the implicit task imposed by the system-environment relation.

3.1   Distance between Two Alphabets

We start with the definition of the equivalence between 2 alphabets and follow with the definition of the distance between two alphabets.

Definition: Two alphabets A_1 and A_2 are equivalent for a system if and only if they have the same number of symbols that are actually used for the system, and there is a bijective mapping between them.

Next we introduce a notion of distance between two alphabets. Intuitively, we expect the distance to be 0 when the alphabets are equivalent and to grow whenever any symbol in one alphabet is "represented" by more than one symbol in the other alphabet or when symbols in A_2 are not represented in A_1.

Definition: The distance (pseudo-distance) between an alphabet A and the objective alphabet G is defined by

$$d(A, G) = \beta \frac{H(G|A)}{H(G)} + \frac{H(A|G)}{H(A)} \tag{1}$$

where the first term reflects the amount of uncertainty remaining about G after a symbol in A has been observed. We call this the *uncertainty term*, since the larger the uncertainty term the more undetermined is G given A. The second term, on the other hand, corresponds to the degree of degeneration incurred. It indicates that the alphabet A has an excessive number of symbols to express what G expresses. That is, the *degeneration term* indicates the existence of spurious symbols, hence, minimizing this term leads to more compact codes.

The entropies on the denominators have a normalization role so that the measure is not biased for larger dimensionality in one of the terms and does not depend on the logarithmic base. For instance $\frac{H(G|A)}{H(G)}$ reflects the percentage of relevant information that is being lost. The free parameter β, which must be positive, reflects the relative weighting between both terms. For several problems a simplified version of (1) is enough where $\beta = 1$.

Next we list some of the properties that the previous distance measure exhibits: (a) In the general case it is positive semi definite, reflexive, and satisfies the triangular inequality with a definition of addition of two processes. It satisfies the symmetry property only when $\beta = 1$ (in this case it is a strict distance), being a pseudo-distance in the other cases (see [12] for the corresponding proofs to the different properties). (b) The performance of a system is optimal (distance to objective = 0) if and only if the output of the system is the objective or a relabeling of the objective space [12]. (c) Lastly, code redundancy, noise and lost of important information for the task make the measure larger.

3.2 Definition of the Amount of Information Processing Performed by a System

The central hypothesis is that as the distance between two alphabets is decreased the number of transformations/operations required to go from one to the other decreases.

Definition: the amount of information processing performed by a system with input space X, output space Y and objective/goal space G is defined as

$$\Delta P = d(X, G) - d(Y, G) \tag{2}$$

expanding and rearranging terms

$$\Delta P = \left(\frac{H(X|G)}{H(X)} - \frac{H(Y|G)}{H(Y)} \right) - \beta \left(\frac{H(G|Y)}{H(G)} - \frac{H(G|X)}{H(G)} \right) \tag{3}$$

The first term corresponds to *degeneration reduction*, that is, minimization of spurious information. Whereas the absolute value of the second term corresponds to *uncertainty creation*. That is, loss of relevant information. ΔP becomes larger as Y gets closer to G reflecting the fact that the system processing is taking the system closer to the goal. This can be achieved by minimizing degeneration and/or uncertainty. ΔP may take positive and negative values. The term of loss of information is always zero or positive (see [12] for proof). For a perfect communication channel $\Delta P = 0$ (see [12] for proof). To compare two processes P_1 and P_2 with the same X and G spaces, it suffices to compare $d(Y_1, G)$ and $d(Y_2, G)$.

4   Empirical Results

To validate the new information processing measure, three test cases are used as a proof of concept. The test cases describe learning the optimal structure of a network of linear classifiers in different problem sets. The output of the *ith* classifier (y_i) is 1 in case $m_i x + b_i > 0$, 0 otherwise, where x is the input pattern. The binary vector composed by all the classifiers outputs Y determines to which class the input vector belongs to. The classifiers configurations have been generated by searching the parameter space by means of a genetic algorithm due to its global search properties. The β is kept constant during the training phase and its value is tuned using the validation set.

   The dataset *gaussians* consists of three equiprobable clusters of data elements belonging to two different classes. Figure 1 displays the configuration selected when using ten linear classifiers. With the new measure, processing is equal to redundancy extraction minus lost of important information. Yet, mutual information only takes into account uncertainty minimization ignoring extraction of redundancies.

   The *cancer1* and *heart1* databases are taken from the *proben1* archive [10]. For these two databases we use simplified classifiers (the coefficients are either

0, 1 or −1) for ease of interpretation. In the case of the *heart1* database the test error can be improved by using classifiers with continuous coefficients.

In all cases we have compared the solutions that optimize the new measure with respect to the solutions that optimize mutual information. The new measure proves to be clearly superior under conditions of noise, overfitting and allocation of optimal number of resources (Table 1).

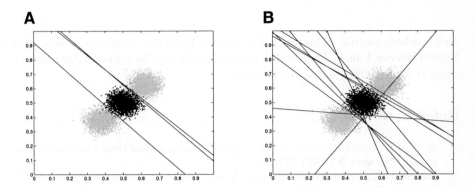

Fig. 1. A. Results when using a maximum number of 10 processing elements with the new information processing measure. Notice that the new measure needs to use only 3 out of the maximum 10. B. Similarly for mutual information.

Table 1. Test errors for the different databases. C is the number of used classifiers out of the maximum (15 for the *heart1* database, 10 for the others).

Data Set	Distance Test error	C	Mutual Information Test error	C	Proben1 Test error
gaussians	1.64	3	1.88	10	-
cancer1	0.57	2	11.49	10	1.149
cancer2	4.598	1	14.94	10	5.747
heart1	22.17	5	57.4	15	20.00

5   Conclusion

In summary many of the information based techniques are implementation dependent and local (e.g. with respect to a single attribute) whereas our measure is implementation independent (general enough to be applicable to a wide range of implementations) and global, that is, it takes into account the overall performance of the system. The entropy distance [8] can be considered as a simplified case of the proposed information processing measure. Specific instances of the entropy distance have been used in different applications, such as in the inference of decision trees [9]. Future work includes validation using more expressive languages such as general feedforward and recurrent neural networks.

Acknowledgments. This work has greatly benefited from discussions with Ramón Huerta, Luis F. Lago and Francisco Rodríguez. We acknowledge financial support from MCT grant BFI2000-0157.

References

1. Atick, J.J.: Could information theory provide an ecological theory of sensory processing? Network **3** (1992) 213-251
2. Barlow, H.B.: Unsupervised learning. Neur. Comput. **1** (1989) 295–311
3. Borst, A., Theunissen, F.E.: Information theory and neural coding. Nat. Neurosc. **2** (1999) 947–57
4. Cover, T.M., Thomas, J.A.: Elements of Information Theory. John Wiley, New York (1991)
5. Fisher, R.A.: Statistical Methods and Scientific Inference, 2nd edition. Oliver and Boyd: London (1959)
6. Kolmogorov, A.: Three approaches to the quantitative definition of information. Problems of Information Transmission **1** (1965) 1–11
7. Linsker, R.: Self-organization in a perceptual network. IEEE Computer **21** (1988) 105–117
8. MacKay, D.J.C.: Information Theory, Inference and Learning Algorithms. Textbook in preparation, to be published by Cambridge University Press. Available from http://www.inference.phy.cam.ac.uk/mackay/itprnn/ (1999)
9. López de Mántaras, R.: A Distance-based Attribute Selection Measure for Decision Tree Induction. Machine Learning Journal **6** (1991) 81–92
10. Prechelt, L.: PROBEN1: A Set of Neural Network Benchmark Problems and Benchmarking Rules, Technical Report 21/94, (1994), Fakultat fur Informatik, Universität Karlsruhe, Germany
11. Ruderman, D. L.: The statistics of natural images. Network **5** (1994) 517–548
12. Sánchez-Montañés, M.A., Corbacho, F.: Towards a new information processing measure for neural computation. Internal Report, (2002), ETS de Informática, Universidad Autñoma de Madrid, Spain. Available at http://www.ii.uam.es/~msanchez/papers (2002)
13. Shannon, C.E., Weaver, W.: The Mathematical Theory of Communication. University of Illinois Press, Urbana (1949)

A Scalable and Efficient Probabilistic Information Retrieval and Text Mining System

Magnus Stensmo

IBM Almaden Research Center
San Jose, CA 95120, USA
magnus@us.ibm.com

Abstract. A system for probabilistic information retrieval and text mining that is both scalable and efficient is presented. Separate feature extraction or stop-word lists are not needed since the system can remove unneeded parameters dynamically based on a local mutual information measure. This is shown to be as effective as using a global measure. A novel way of storing system parameters eliminates the need for a ranking step during information retrieval from queries. Probability models over word contexts provide a method to suggest related words that can be added to a query. Test results are presented on a categorization task and screen shots from a live system are shown to demonstrate its capabilities.

1 Introduction

A practical information retrieval and text mining system has to be fast, scalable and efficient both in terms of time and space usage. This applies to index building as well as information retrieval. In this paper we describe a system that has these attributes. Indexing is essential since it's impossible to build a fast system without an index. The most common form of an index is the *inverted index* in which terms occurring in documents are the index keys, and the document occurrences (and perhaps positions within a document) are the values, or *postings lists*. Many common words (eg "the", "of", etc.) are not relevant as search terms and are often removed as being part of a *stop-word list*. This list needs maintenance and is domain dependent. It is possible to build such lists automatically by selecting terms according to some method, but this has to be done in a separate pass and takes extra time. At run-time, a result list is then ranked according to some ranking function, eg Term-Frequency-inverse-Document-Frequency (TFiDF) which requires both document and term frequencies.

The following describes a system that avoid these problems with novel methods for indexing, storage and dynamic term selection using only term frequencies. We show that this can be done without significant loss of precision in a standardized categorization task. We also show how a probabilistic model over words can be used to guide the user during information retrieval by suggesting related terms.

J.R. Dorronsoro (Ed.): ICANN 2002, LNCS 2415, pp. 643–648, 2002.

2   System Description

Probabilistic information retrieval (IR) views documents as having been created by a probabilistic process. The idea is to build a reasonable model, estimate its parameters from observed data and use it to predict future events. A common problem in IR is to rank documents based on a query. Using a probabilistic model, we can obtain a probability for each document given the query, and in that way obtain ranking according to the *probability ranking principle*, described as optimal [6]. Many different probability models are possible for IR and almost all of them differ in some respects. Either the event model differs or the probability of terms and documents are assumed to be different.

The probability models presented here are based on a multinomial generative event model [3,4]. This means that individual word occurrences (terms t) in the documents are considered to be events and a document (d) is the collection of such word events. The documents in the document set consist of sequences of terms, t. Each term occurrence is an event and the document, d, is the collection of the word events. $t \epsilon T$, where T is the set of all terms. Likewise the document identifiers $d \epsilon D$ are from a set of documents, D. In the following p(a) stands for the probability of a variable $A = a$, where A is a discrete random variable and a is one of values that A can take on. Probabilities are estimated from observed counts. Some of the counts are going to be very small and thus quite unreliable since there is a huge difference in the estimated probability if the count is zero or one. Laplace aka Equivalent Sample Size m-estimates of probability [4] are used to yeild better estimates.

Single Terms. The probability of a term t, p(t) is the number of times the term has occurred in the entire corpus, C_t, divided by the total number of term occurrences, N. p(t) = $\frac{C_t}{N}$ and p(t,d) = $C_{t,d}/N$ is the joint probability. If we sum all the times the terms occur in each document we should get $\sum_{d \epsilon D} C_{t,d} = C_t$, likewise $\sum_{t \epsilon T} C_{t,d} = C_d$. This means that C_d is the total number of word occurrences in document d which is the length of the document, $|d|$.

Sequences of Terms. The probability that a particular sequence of terms, $t_1, \ldots, t_n$, comes from a particular document d can be written p($d|t_1, \ldots, t_n$). Bayes's Rule says

$$p(d|t_1, \ldots, t_n) = p(d) \frac{p(t_1, \ldots, t_n|d)}{p(t_1, \ldots, t_n).} \tag{1}$$

The probability of a sequence of terms p($t_1, \ldots, t_n$) means the number of times this sequence occurred in the corpus and is $\frac{C_{t_1, \ldots, t_n}}{N - n + 1}$, since the number of possible sequences of length n is $N - n + 1$. p($t_1, \ldots, t_n|d$) means p($d, t_1, \ldots, t_n$)/p(d) as above. The number of possible sequences of length at least one out of N elements is $2^N - 1$. The number of probability estimations are therefore very large and since there are sampling errors due to unseen word combinations, we would like to build a model that approximate the distribution we are looking for. A simplified model use a fraction of the space needed for the original model, and smooth unseen combinations of probabilities.

The simplification used is called Simple Bayes or Naive Bayes. It states that we consider the word events to be independent given a particular document so that $p(t_1, \ldots, t_n|d) = \prod_{i=1}^{n} p(t_i|d)$. The reason for doing this is to reduce the number of parameters that we need to estimate. We will also use another simplification and assume that all terms are unconditionally independent, so that $p(t_1, \ldots, t_n) = \prod_{i=1}^{n} p(t_i)$. Equation 1 becomes ($p(t_i|d) = p(t_i, d)/p(d)$)

$$p(d|t_1, \ldots, t_n) = p(d) \frac{\prod_{i=1}^{n} p(t_i|d)}{\prod_{i=1}^{n} p(t_i)} = p(d) \prod_{i=1}^{n} \frac{p(t_i, d)}{p(t_i)p(d)}$$

The result of this is that we only need to estimate probabilities for singletons t and d, and pairs (t, d). The number of such pairs is much smaller than the number of combinations of terms. Note that we know that the independence assumptions are not valid and will in general over-estimate the probabilities for sequences. It is therefore necessary to normalize the probability values after using this formula. The use of the independence assumptions leads to a probability estimation that is independent of the order of terms in the sequence. This is of course not correct but is helpful since this is often how queries are interpreted, ie "John Smith" should normally return the same results are "Smith John".

If a query terms is in a particular document we will assume that the term is statistically independent of the document, so that $p(t_i, d) = p(t_i)p(d)$. This means that this terms contributes with a factor one to the product, ie it leaves it unchanged. It is possible to further reduce the number of pairs needed. Tests have shown that a mutual information measure can be used to remove pairs based on mutual information measures. These removed terms are then treated similarly to independent terms. The actual computations in the system are done in the logarithmic space so that the products become sums.

$$\log p(d|t_1, \ldots, t_n) = \log p(d) + \sum_{i=1}^{n} \log \frac{p(t_i, d)}{p(t_i)p(d)} = B_d + \sum_{i=1}^{n} M_{t_i, d} \quad (2)$$

The logarithm of the conditional probability can be written as the sum of a bias term B_d and the terms M_{d, t_i}. Experiments have shown that it is possible to reduce both B and M values to their integer parts with little loss of precision in a classification task. Furthermore, the range of the M values is very limited and can be expressed using few bits. This results in a very reduced representation that does not degrade the performance significantly.

The parameter values are also stored interleaved with the document identifiers in the postings lists so that all parameters that are needed for the probabilistic ranking are available as soon as the correct database record has been read. The ranking is done by adding the appropriate values using Equation 2. The result is a very fast system that does lookup and ranking of results in one single step.

2.1   Additional Probability Models

Feature Extraction by "Backward Probabilities". Document retrieval is done by ordering the documents by their conditional probabilities given a certain query term t_j (or term sequences), $p(d_i|t_j)$ for all documents. We can also turn this probability "backwards" by using $p(t_j|d_i)$. This means that we are interested in terms that are relevant given that we have a certain document d_j. Such terms are features. However, the terms that have a high conditional probability will often be non-content words due to their frequency of occurrence. The following discussion shows how to avoid this. If we note that $p(t_j|d_j)$ can be interpreted as the weight of an arc from d_i to t_j, $d_i \rightarrow t_j$, we can formulate a better formula for feature extraction by also using the arrow pointing back from t_j to d_i, $d_i \leftarrow t_j$.

The following example illustrates: A common term such as "the" will have a high conditional probability with almost all documents. This means that $p(\text{the}|d_i)$ will be high for most documents. The backward arc, however, $p(d_i|\text{the})$, will have a low value since "the" does not identify one specific document but almost all of them. The product will thus be high*low=low. An uncommon term such as "elephant" will have a high conditional probability only with certain documents. This means that $p(\text{the}|d_i)$ will be high for some documents about elephants. The backward arc $p(d_i|\text{the})$, will have a high value since "elephant" does identify one or a few specific document. For these documents, the product will thus be high*high=high. This observation leads to a simple yet efficient way to get good features.

Probabilistic Categorization. To use the above model for classification (Equations 1 and 2), we associate the category labels that have been assigned to a training set to the terms in those documents. Several different categories can be given for each document in the training set which is how it is different from document retrieval. Alternatively, we can view document retrieval as classification where each document is its own class. For example, a document in the training set is labeled "wheat, grain, Asia". These labels are then associated with the document terms in the same way document identifiers were associated with the document terms in the section about document retrieval above. It is then possible to find the conditional probability of each of the labels using unlabeled text. Such a categorization model can then be used to find the conditional probabilities of the different categories given the terms in a new document of unknown categorization. If several of the predicted categories have high probability we assign several categories to the document in question.

Suggestions of Related Words. To build a probability model over words in context, a window is slid over each word in each document. The words adjacent to the word in the center are associated with it in the same manner as with document labels above. This is done for every word in every document. The result is a probability model that can be used to predict related words to a query, since it will give high values to words used in the context of the query words. For example, if the word "bank" occurs adjacent to "financial" and "river", then the conditional probability of both "financial" and "river" will be high if the query word is "bank". This will be helpful to the user since she can then disambiguate the query accordingly.

3 Results

Categorization Task. Testing of the system was done by building a categorizer on a training set and then applying it to a test set. This is a common task in text mining studies and many results have been published, for a summary of recent methods and comparative results, see [8]. The data sets were taken from Reuters-21578 [2]. The training set has 7769 documents and the test set has 3019 documents. There are 90 categories, each document has an average of 1.3 categories.

Many methods have been used for categorization. Naive Bayes is only a little bit worse than expensive methods. The best was a Support Vector Machine but it was only 7.5% better, based on $F_1 = 0.7956$ for Naive Bayes and $F_1 = 0.8599$ for a Support Vector Machine F_1 measure is from van Rijsbergen and is defined equal to $2rp/(r+p)$, where r is the recall and p is the precision [8]. A high value for both is better. The methods described in [8] require feature extraction to be done prior to the classification. This is not necessary for the present method. Training of the system was done by collecting the statistics necessary for computing all values needed for Equation 2.

Mutual information is defined to be [5] $\sum_{x \in X} \sum_{y \in Y} p(x, y) \log \frac{p(x,y)}{p(x)p(y)}$, for discrete random variables X and Y. To maximize the mutual information while removing some terms, we see that terms with a low $p(t_j, d_i) \log \frac{p(t_j,d_i)}{p(t_j)p(d_i)}$, can be removed. The pairs with a low value for this measure are removed since they contribute less to the overall measure of mutual information. This turns out to be a good method for pruning as can be seen in Figure 1. For this dataset the range of Integer values -7 to 14 for the pair values, which requires 5 bits, and -19 to -2 for the singular values, which also requires 5 bits for the Integer representation of the parameters. Note that this may be reduced further by pruning. This number of bits should be compared to the number of bits needed to store a Float value which is normally 32, or a Double which is 64 bits. The saving is thus at least 6-8 times, while the quality of the results is only marginally reduced.

A complete system has been constructed that can instantly retrieve documents, find keywords from one or several documents and find related words to any query. It was built using Java servlets and is run in a web browser and has been built with several large data sets.

4 Summary and Discussion

We have presented a system for probabilistic information retrieval and text mining that is highly scalable, fast both in terms of indexing and retrieval, and that can use a reduced amount of storage without the explicit use of stop-word lists and without a separate feature extraction pass before indexing.

648   M. Stensmo

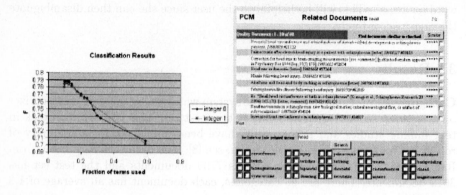

Fig. 1. Left. Results of the categorization task. The x-axis is the fraction of document-term pairs used for the classification, and the y-axis shows the resulting F value. The line marked *integer 0* indicates results when float values were used for categorization according to Equation 2 and *integer 1* indicates results when integer values were used. The results are very similar but integer values reduce the storage requirements for the coefficients by a factor of 7. **Right.** Part of the results page for the query "head" in a medical database about schizophrenia. 60 documents were found and are presented in order of $p_d(d_i|\text{Query})$, for documents d_i. To the right is a normalized probability (all values are divided by the top probability) shown as 1-5 '*'. The check boxes to the right are used to find *Similar Documents* from the checked ones when the *Similar button* is used. The bottom part shows *Related Words*. These words are shown in decreasing relevance as described in the text. They can be included (left box) or excluded (right box) to form a new query which is executed when the *Search button* is used.

References

1. Domingos, P., Pazzani, M. (1996). Beyond Independence: Conditions for Optimality of the Simple Bayes Classifier. *Proceedings of the 13th International Conference on Machine Learning*.
2. Lewis, D. (1998). The Reuters-21578 Text Categorization Test Collection Description. http://www.research.att.com/~lewis/reuters21578.html
3. McCallum, A., Nigam, K. (1998). A Comparison of Event Models for Naive Bayes Text Classification. American Association for Artificial Intelligence Workshop on Learning for Text Classification.
4. Mitchell, T. (1997). *Machine Learning.* WCB/McGraw-Hill, Boston, MA.
5. Papoulis, A. (1984). *Probability, Random Variables, and Stochastic Processes*, McGraw Hill, New York, second edition.
6. van Rijsbergen, C.J. (1979). *Information Retrieval.* 2nd Ed. Butterworths, London.
7. Robertson, S.E. & Sparck-Jones, K. (1976). Relevance weighting of search terms. Journal of the American Society for Information Science, 27:129–146.
8. Yang, Y. (1999). A re-examination of text categorization methods. *Proc. SIGIR 1999*.

Maximum and Minimum Likelihood Hebbian Learning for Exploratory Projection Pursuit

Donald MacDonald[1], Emilio Corchado[1], Colin Fyfe[1], and Erzsebet Merenyi[2]

[1] Applied Computation Intelligence Research Unit, University of Paisley, High St.
Paisley, Scotland
{mcdo-ci0, corc-ci0, fyfe-ci0}@Paisley.ac.uk
http://ces.paisley.ac.uk/
[2] Department of Electrical and Computer Engineering
Rice University, 6100 Main St.
Houston, Texas, USA
{Erzsebet}@Rice.edu

Abstract. This paper presents an extension to the learning rules of the Principal Component Analysis Network which has been derived to be optimal for a specific probability density function. We note this pdf is one of a family of pdfs and investigate the learning rules formed in order to be optimal for several members of this family. We show that the whole family of these learning rules can be viewed as methods for performing Exploratory Projection Pursuit. We show that these methods provide a simple robust method for the identification of structure in remote sensing images.

1 Introduction

One problem with the analysis of high dimensional data is identifying structure or patterns which exist across dimensional boundaries. By projecting the data onto a different basis of the space, these patterns may become visible. This presents a problem - how does one decide which basis is optimal for the visualisation of the patterns, without foreknowledge of the patterns in the data.

One solution is Principal Component Analysis (PCA), which is a statistical technique aimed at finding the orthogonal basis that maximises the variance of the projection for a given dimensionality of basis. This involves finding the direction which accounts for most of the data's variance, the first principal component; this variance is then filtered out. The next component is the direction of maximum variance from the remaining data and orthogonal to the 1st PCA basis vector.

We [7,6] have over the last few years investigated a negative feedback implementation of PCA defined by (1) - (3). Let us have an N-dimensional input vector, $\mathbf{x}$, and an M-dimensional output vector, $\mathbf{y}$, with W_{ij} being the weight linking the jth input to the ith output. The learning rate, η is a small value which will be annealed to zero over the course of training the network. The activation passing from input to output through the weights is described by (1).

J.R. Dorronsoro (Ed.): ICANN 2002, LNCS 2415, pp. 649–654, 2002.

The activation is then fed back though the weights from the outputs and the error,e calculated for each input dimension. Finally the weights are updating using simple Hebbian learning.

$$y_i = \sum_{j=1}^{N} W_{ij}x_j, \forall i \tag{1}$$

$$e_i = x_i - \sum_{i=1}^{M} W_{ij}y_i \tag{2}$$

$$\Delta W_{ij} = \eta e_j y_i \tag{3}$$

We have subsequently modified this network to perform clustering with topology preservation [8], to perform Factor Analysis [12,4] and to perform Exploratory Projection Pursuit [11,9] (EPP).

PCA is a very powerful and simple technique for investigating high dimensional data. One of the reasons for this is that structure in data is often associated with high variance. But this is not necessarily so: interesting structure can be found in low variance projections. Thus EPP tries to search for interesting directions onto which the data will be projected so that a human can investigate the projections by eye. We thus have to define what interesting means. Most projections through high dimensional data will be approximately Gaussian and so the degree of interestingness of a projection is measurable by its distance from the Gaussian distribution. We have previously introduced nonlinear functions to (3) and shown that the resulting system performs EPP [11,9] and applied it to visual [10] and to audio data [14].

This paper will deal with a new variation of the basic network which also performs Exploratory Projection Pursuit and apply it to high dimensional astronomical data.

2   Maximum Likelihood Learning

Various researchers e.g. [19,16] have shown that the above learning rules can be derived as an approximation to the best linear compression of the data. Thus we may start with the cost function

$$J = 1^T E\{(\mathbf{x} - W^T\mathbf{y})^2\} \tag{4}$$

which we minimise to get (3). In this paper, we use the convention that $.^2$ is an element-wise squaring.

We may show that the minimisation of J is equivalent to minimising the negative log probabilities of the residual, $\mathbf{e}$, if $\mathbf{e}$ is Gaussian [2] and thus is equal to maximising the probabilities of the residual. Let $p(\mathbf{e}) = \frac{1}{Z}\exp(-\mathbf{e}^2)$ Then we can denote a general cost function associated with the network as

$$J = -\log p(\mathbf{e}) = (\mathbf{e})^2 + K \tag{5}$$

where K is a constant. Therefore performing gradient descent on J we have

$$\Delta W \propto -\frac{\delta J}{\delta W} = -\frac{\delta J}{\delta e}\frac{\delta e}{\delta W} \approx \mathbf{y}(2\mathbf{e})^T \qquad (6)$$

where we have discarded a relatively unimportant term [16].

We have considered an extension of the above in [13,5] with a more general cost function

$$J = f_1(e) = f_1(\mathbf{x} - W^T\mathbf{y}) \qquad (7)$$

Let us now consider the residual after the feedback to have probability density function

$$p(\mathbf{e}) = \frac{1}{Z}\exp(-|\mathbf{e}|^p) \qquad (8)$$

The we can denote a general cost function associated with this network as

$$J = -\log p(\mathbf{e}) = (\mathbf{e})^p + K \qquad (9)$$

where K is a constant, Therefore performing gradient descent on J we have

$$\Delta W \propto -\frac{\delta J}{\delta W} = -\frac{\delta J}{\delta e}\frac{\delta e}{\delta W} \approx \mathbf{y}(p(|\mathbf{e}|^{p-1}sign(\mathbf{e})))^T \qquad (10)$$

We would expect that for leptokurtotic residuals (more kurtotic than a Gaussian distribution), values of $p < 2$ would be appropriate, while platykurtotic residuals (less kurtotic than a Gaussian), values of $p > 2$ would be appropriate. It is a common belief in the ICA community [15] that it is less important to get the exactly the correct distribution when searching for a specific source than it is to get an approximately correct distribution i.e. all supergaussian signals can be retrieved using a generic leptokurtotic distribution and all subgaussian signals can be retrieved using a generic platykurtotic distribution.

Therefore the network operation is:

Feedforward: $y_i = \sum_{j=1}^{N} W_{ij}x_j, \forall i$

Feedback: $e_j = x_j - \sum_{i=1}^{M} W_{ij}y_i$

Weight change: $\Delta W_{ij} = \eta y_i sign(e_j)|e_j|^p$

Fyfe and MacDonald [13] used a single value of p (p=1)in this rule and described this method as performing a robust PCA, but this is not strictly true since only the original ordinary Hebbian learning rule actually performs PCA. It might seem to be more appropriate to link this family of learning rules to Principal Factor Analysis since this method makes an assumption about the noise in a data set and then removes the assumed noise from the covariance structure of the data before performing a PCA. We are doing something similar here in that we are biasing our PCA-type rule on the assumed distribution of the residual. By maximising the likelihood of the residual with respect to the actual distribution, we are matching the learning rule to the pdf of the residual.

More importantly, we may also link the method to EPP. Now the nature and quantification of the interestingness is in terms of how likely the residuals are under a particular model of the pdf of the residual. As with standard EPP, we also sphere the data before applying the learning method to the sphered data.

2.1   Minimum Likelihood Hebbian Learning

Now it is equally possible to perform gradient ascent on J. In this case we find a rule which is the opposite of the above rules in that it is attempting to minimise the likelihood of the residual under the current assumptions about the residual's pdf. The operation of the network is as before but this time we have

Weight change: $\Delta W_{ij} = -\eta y_i sign(e_j)|e_j|^p$

This corresponds to the well known anti-Hebbian learning rule. Now one advantage of this formulation compared with the Maximum Likelihood Hebbian rule is that, in making the residuals as unlikely as possible, we are having the weights learn the structure corresponding to the pdf determined by the parameter p. With the Maximum rule, the weights learn to remove the projections of the data which are furthest from that determined by p. Thus if we wish to search for clusters in our data (typified by a pdf with $p > 2$), we can use Maxmimum Likelihood learning with $p < 2$ [5], which would result in weights which are removing any projections which make these residuals unlikely. Therefore the clusters would be found by projecting onto these weights. Alternatively we may use Minimum Likelihood learning with $p > 2$, which would perform the same job: the residuals have to be unlikely under this value of p and so the weights converge to remove those projections of the data which exhibit clustering.

3   Experimental Results

In this paper we use two remote sensing data sets: the first is the 65 colour spectra of 115 asteroids used by [17]. The data set is composed of a mixture of the 52-colour survey by Bell et al. [1] together with the 8-colour survey conducted by Zellner et al. [20] providing a set of asteroid spectra spanning 0.3-2.5μm. A more detailed description of the dataset is given in [17]. When this extended data set was compared by [17] to the results of Tholen [18] it was found that the additional refinement to the spectra leads to more classes than the taxonomy produced by Tholen. Figure 1 compares Maximum Likelihood Hebbian learning ($p = 0$, an improper distribution since it does not integrate to 1 but we use it to illustrate how robust the method is) with Principal Component Analysis (PCA, $p = 2$) on the Asteroid Data set. Since we wish to maximise the likelihood of the residuals coming from a leptokurtotic distribution, this method is ideal for identifying the outliers in the data set. Clearly the ML method is better at this than standard PCA (Figure 1).

The second data set is an data set constructed by [3] which consists of 16 sets of spectra in which each set contains 32x32 spectra descriptions of which are given in [17] and [3]. Figure 2 gives a comparison of the projections found by the Minimum Likelihood Hebbian learning with values of $p = 2.2$ and $p = 2.3$. We see that both show the 16 clusters in the data set very precisely (standard PCA with $p = 2$ does not perform nearly so well) but that even with this small change in the p-parameter, the projections are quite different.

We see therefore that both Maximum and Minimum Likelihood Hebbian learning can be used to identify structure in data. First results suggest that

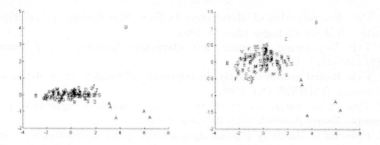

Fig. 1. Principal Component Analysis (left figure) and Maximum Maximum Likelihood Hebbian Learning (right figure) on the Asteroid Data. We used a value of p = 0. The Maximum Likelihood method identifies a projection which spreads the data out more than PCA.

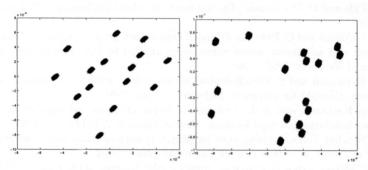

Fig. 2. The figure shows the results of the Minimum Likelihood Hebbian Learning network using $p = 2.3$ (left figure) and $p = 2.2$ (right figure).

Minimum Likelihood is more stable and accurate and that both methods outperform the previous neural implementation of EPP [11]. Future work will be based on a rigorous comparison of these methods.

References

1. J. F. Bell, P. D. Owensby, B.R. Hawke, and M.J. Gaffey. The 52 colour asteroid Survey: Final results and interpretation. *Lunalr Planet Sci. Conf, XiX,* pages 57–58, 1988.
2. C. Bishop. *Neural Networks for Pattern Recognition.* Oxford: Clarendon Press, 1995.
3. H. Cetin and D. W. Levandowski. Interactive classification and mapping of multi-dimensional remotely sensed data using n-dimensional probability density functions (npdf). *Photogrammetric Engineering and Remote Sensing,* 57(12):1579–1587, 1991.
4. D. Charles and C. Fyfe. Modelling multiple cause structure using rectification constraints. *Network: Computation in Neural Systems,* 9:167–182, May 1998.
5. E. Corchado and C. Fyfe. Maximum likelihood hebbian learning. In *Tenth European Symposium on Artificial Neural Networks, ESANN2002,* 2002.

654   D. MacDonald et al.

6. C. Fyfe. Pca properties of interneurons. In *From Neurobiology to Real World Computing, ICANN 93*, pages 183–188, 1993.
7. C. Fyfe. Introducing asymmetry into interneuron learning. *Neural Computation*, 7(6):1167–1181, 1995.
8. C. Fyfe. Radial feature mapping. In *International Conference on Artificial Neural Networks, ICANN95*, Oct. 1995.
9. C. Fyfe. A comparative study of two neural methods of exploratory projection pursuit. *Neural Network*, 10(2):257–262, 1997.
10. C. Fyfe and R. Baddeley. Finding compact and sparse distributed representations of visual images. *Network: Computation in Neural Systems*, 6(3):1–12, 1995.
11. C. Fyfe and R. Baddeley. Non-linear data structure extraction using simple hebbian networks. *Biological Cybernetics*, 72(6):533–541, 1995.
12. C. Fyfe and D. Charles. Using noise to form a minimal overcomplete basis. In *Seventh International Conference on Artificial Neural Network, ICANN99*, 1999.
13. C. Fyfe and D. MacDonald. Epsilon-insensitive hebbian learning. *Neurocomputing*, 2001.
14. M. Girolami and C. Fyfe. An extended exploratory projection pursuit network with linear and non-linear lateral connections applied to the cocktail party problem. *Neural Networks*, 1997.
15. A. Hyvarinen and E. Oja. Independent component analysis: A tutorial. Technical report, Helsinki University of Technology, April 1999.
16. Juha Karhunen and Jyrki Joutsensalo. Representation and separation of signals using nonlinear pca type learning. *Neural Network*, 7(1):113–127, 1994.
17. E. Merenyi. Self-organising anns for planetary surface composition research. *Journal of Geophysical Research*, 99(E5):10847-10865, 1994.
18. D.J. Tholen. Asteroid Taxonomy from Cluster Analysis of Photometry. PhD thesis, University of Arizona, Tucson, 1984.
19. Lei Xu. Least mean Square error reconstruction principle for self-organizing neural-nets. *Neural Network*, 6(5):627–648, 1993.
20. B. Zellner, D.J. Tholen, and E.F. Tedesco. The eight colour asteroid survey: Results from 589 minor planets. *Icarus*, pages 355–416, 1985.

Learning Context Sensitive Languages with LSTM Trained with Kalman Filters*

Felix A. Gers[1], Juan Antonio Pérez-Ortiz[2], Douglas Eck[3], and
Jürgen Schmidhuber[3]

[1] Mantik Bioinformatik GmbH, Neue Gruenstrasse 18, 10179 Berlin, Germany
[2] DLSI, Universitat d'Alacant, E-03071 Alacant, Spain
[3] IDSIA, Galleria 2, 6928 Manno, Switzerland

Abstract. Unlike traditional recurrent neural networks, the Long Short-Term Memory (LSTM) model generalizes well when presented with training sequences derived from regular and also simple nonregular languages. Our novel combination of LSTM and the decoupled extended Kalman filter, however, learns even faster and generalizes even better, requiring only the 10 shortest exemplars ($n \leq 10$) of the context sensitive language $a^n b^n c^n$ to deal correctly with values of n up to 1000 and more. Even when we consider the relatively high update complexity per timestep, in many cases the hybrid offers faster learning than LSTM by itself.

1 Introduction

Sentences of regular languages are recognizable by finite state automata having obvious recurrent neural network (RNN) implementations. Most recent work on language learning with RNNs has focused on them. Only few authors have tried to teach RNNs to extract the rules of simple context free and context sensitive languages (CFLs and CSLs) whose recognition requires the functional equivalent of a potentially unlimited stack. Some previous RNNs even failed to learn small CFL training sets [9]. Others succeeded at CFL and even small CSL training sets [8,1], but failed to extract the general rules and did not generalize well on substantially larger test sets.

The recent *Long Short-Term Memory* (LSTM) method [6] is the first network that does not suffer from such generalization problems [4]. It clearly outperforms traditional RNNs on all previous CFL and CSL benchmarks that we found in the literature. Stacks of potentially unlimited size are automatically and naturally implemented by linear units called *constant error carousels* (CECs).

In this article we focus on improving LSTM convergence time by using the decoupled extended Kalman filter (DEKF) [7] learning algorithm. We compare DEKF with the original gradient-descent algorithm when applied to the only CSL ever tried with RNNs, namely, $a^n b^n c^n$.

* Work supported by SNF grant 2100-49'144.96, Spanish Comisión Interministerial de Ciencia y Tecnología grant TIC2000-1599-C02-02, and Generalitat Valenciana grant FPI-99-14-268.

J.R. Dorronsoro (Ed.): ICANN 2002, LNCS 2415, pp. 655–660, 2002.

2   NGnet & Probabilistic Model

The NGnet [1], which transforms an N-dimensional input vector x into a D-dimensional output vector y, is defined by the following equation.

$$y = \sum_{i=1}^{M} \left(\frac{g_i \mathcal{N}_N(x|\mu_i, S_i)}{\sum_{j=1}^{M} g_j \mathcal{N}_N(x|\mu_j, S_j)} \right) W_i \tilde{x}, \tag{1}$$

where $\tilde{x} \equiv (x', 1)'$ and the prime $(')$ denotes a transpose. M is the number of units. $g \equiv (g_1, \cdots, g_M)'$ is a mixing rate vector. $\mathcal{N}_N(x|\mu_i, S_i)$ denotes an N-dimensional Gaussian function with an N-dimensional center μ_i and an $N \times N$ inverse covariance matrix[4]. W_i is a $D \times (N+1)$ linear regression matrix.

The NGnet can be interpreted as a probabilistic model, in which a pair of input and output, (x, y), is a probabilistic event [1]. Each event is assumed to be generated from a single unit, whose index is regarded as a hidden variable. The probabilistic model is defined by the joint distribution for a triplet (x, y, i):

$$P(x, y, i|\theta) = g_i \mathcal{N}_N(x|\mu_i, S_i) \mathcal{N}_D(y|W_i \tilde{x}, B_i), \tag{2}$$

where B_i is a $D \times D$ diagonal matrix denoting the inverse variance of the output. $\theta \equiv \{(g_i, \mu_i, S_i, W_i, B_i)|i = 1, \cdots, M\}$ is a set of model parameters. With this distribution, the expected output y for a given input x, $E[y|x] \equiv \int dy P(y|x, \theta) y$, is identical to equation (1). Namely, distribution (2) provides a probabilistic model for the NGnet.

3   Variational Bayes Inference

From equation (2), the distribution of an observed variable pair, (x, y), is given by the following mixture model using an indicator variable $z \equiv (z_1, \cdots, z_M)'$:

$$P(x, y|\theta) = \sum_{\{z\}} P(x, y, z|\theta); \quad P(x, y, z|\theta) = \exp \left[\sum_{i=1}^{M} z_i \log P(x, y, i|\theta) \right], \tag{3}$$

where $z_i = 1$ ($z_j = 0$ for $j \neq i$) indicates that (x, y) is generated from the i-th unit. In formulation (3), z is a hidden variable in place of i.

After observing a data set $(X, Y) \equiv \{(x(t), y(t))|t = 1, \cdots, T\}$, Bayes inference tries to obtain the posterior distribution of unknown variables, $P(Z, \theta|X, Y)$, where $Z \equiv \{z(t)|t = 1, \cdots, T\}$ denote the hidden variables corresponding to (X, Y). According to the Bayes theorem, the posterior distribution is given by

$$P(Z, \theta|X, Y) = P(X, Y, Z|\theta) P_0(\theta)/P(X, Y), \tag{4}$$

---

[4] $\mathcal{N}_N(x|\mu_i, S_i) \equiv (2\pi)^{-N/2} |S_i|^{1/2} \exp \left[-\frac{1}{2}(x - \mu_i)' S_i (x - \mu_i) \right].$

where $P(X, Y, Z|\theta) \equiv \prod_{t=1}^{T} P(x(t), y(t), z(t)|\theta)$ is the likelihood for the complete data set (X, Y, Z). $P_0(\theta)$ is a prior distribution of the model parameters θ. $P(X, Y) \equiv \sum_{\{Z\}} \int d\theta P(X, Y, Z|\theta) P_0(\theta)$ is called the marginal likelihood.

The marginal likelihood, which is the likelihood for a given model, gives a criterion for model selection. Since the marginal likelihood cannot be calculated analytically, we use a variational Bayes (VB) method developed in [3,4]. In the VB method, the posterior distribution $P(Z, \theta|X, Y)$ is approximated by a trial posterior distribution $Q(\theta, Z)$. The free energy for the trial posterior distribution $Q(\theta, Z)$ is defined by

$$F[Q] \equiv \sum_{\{Z\}} \int d\theta Q(Z, \theta) \log \frac{P(X, Y, Z|\theta) P_0(\theta)}{Q(Z, \theta)}$$

$$= \log P(X, Y) - \text{KL}(Q(Z, \theta) \| P(Z, \theta|X, Y)), \tag{5}$$

where $\text{KL}(\cdot \| \cdot)$ denotes the Kullback-Leibler (KL) divergence between two distributions. Note that the KL divergence becomes minimum (zero) when $Q(Z, \theta) = P(Z, \theta|X, Y)$. By maximizing (5) with respect to Q, the posterior distribution and the (log) marginal likelihood are approximated simultaneously.

We assume that the prior distribution of the model parameters is given by the following conjugate distribution:

$$P_0(\theta|\xi) = \prod_{i=1}^{M} P_0(g_i) P_0(\mu_i|S_i) P_0(S_i|\sigma_i) P_0(W_i|B_i, \varUpsilon_i) P_0(B_i|R_i) \tag{6}$$

$$\log P_0(g_i) = \gamma_{0,i} \log g_i + const.$$

$$\log P_0(\mu_i|S_i) = -\frac{\gamma_{0,i}}{2} (\mu_i - m_{0,i})' S_i (\mu_i - m_{0,i}) + \frac{1}{2} \log |S_i| + const.$$

$$\log P_0(S_i|\sigma_i) = -\frac{\gamma_{s0,i}}{2} \text{Tr}[\sigma_i S_i] + \frac{\gamma_{s0,i} - N - 1}{2} \log |S_i| + \frac{N \gamma_{s0,i}}{2} \log \sigma_i + const.$$

$$\log P_0(W_i|B_i, \varUpsilon_i) = -\frac{1}{2} \text{Tr}[W_i' B_i W_i \varUpsilon_i] + \frac{N+1}{2} \log |B_i| + \frac{D}{2} \log |\varUpsilon_i| + const.$$

$$\log P_0(B_i|R_i) = -\frac{\gamma_{\beta0,i}}{2} \text{Tr}[B_i R_i] + \frac{\gamma_{\beta0,i} - 2}{2} \log |B_i| + \frac{\gamma_{\beta0,i}}{2} \log |R_i| + const.,$$

where $m_{0,i}$ is an N-dimensional vector and all hyper parameters with subscript '0' are constants. $\text{Tr}[\cdot]$ denotes a matrix trace. σ_i, $\varUpsilon_i \equiv diag(v_{i,1}, \cdots, v_{i,(N+1)})$ and $R_i \equiv diag(\rho_{i,1}, \cdots, \rho_{i,D})$ are hyper parameters that parameterize the prior distribution of the model parameters. We also assume that a hierarchical prior distribution of the hyper parameters $\xi \equiv \{(\sigma_i, \varUpsilon_i, R_i)|i = 1, \cdots, M\}$, $P_0(\xi)$, is given by the product of the Gamma distribution and they are integrated out, i.e., $P_0(\theta) = \int d\xi P_0(\theta|\xi) P_0(\xi)$.

By introducing a confidence parameter κ ($\kappa > 0$), which corresponds to the multiplicity of the observed data, the free energy (5) is modified into

$$F^\kappa[Q] \equiv \sum_{\{Z\}} \int d\theta d\xi Q(Z, \theta, \xi) \log \frac{P(X, Y, Z|\theta)^\kappa P_0(\theta|\xi) P_0(\xi)}{Q(Z, \theta, \xi)}. \tag{7}$$

Moreover, we proposed a direct manipulation of functions, whereas FDA relies on a pre-processing phase. In this phase, each input function is projected on a set of smooth functions (for instance on a truncated B-Spline base). In the present paper, we show that FMLP can also be applied to pre-processed functions. The main advantage of such a solution is that computation time can be greatly reduced without impairing performances (for low dimensional input spaces). As this approach uses linear models to represent input and weight functions, in high dimensional spaces this model has reduced performances compared to the direct approach, which can represent weight functions thanks to nonlinear models.

The paper is organized as follows. We first introduce the FMLP model and our new projection based approach. Then we give important theoretical results for the new model: universal approximation and consistency of parameter estimation. We conclude with some experiments that illustrate differences between direct and projection base approaches.

2   Functional MLP

2.1   Functional Neurons

As explained in the introduction, the main idea of FMLP is to replace in a neuron a linear form on $\mathbb{R}^n$ by a continuous linear form defined on the input space of the neuron. This idea was proposed and studied on a theoretical point of view in [8]. Basically, a generalized neuron takes input in E, a normed vectorial space. It is parametrized thanks to a real number b and a weight form, i.e., an element w of E^*, the vectorial space of continuous linear forms on E. The output of the neuron is $T(w(x) + b)$.

Of course, it is quite difficult to represent arbitrary continuous linear forms. That's why we proposed in [7] to restrict ourselves to functional inputs. More precisely, we denote μ a finite positive Borel measure on $\mathbb{R}^n$ (the rational for using such a measure will be explained in section 3), and $L^p(\mu)$ the space of measurable functions from $\mathbb{R}^n$ to $\mathbb{R}$ such that $\int |f|^p d\mu < \infty$. Then we can define a neuron that maps elements of $L^p(\mu)$ to $\mathbb{R}$ thanks to an activation function T, a numerical threshold b and a weight function, i.e. a function $w \in L^q(\mu)$ (where q is the conjugate exponent of p). Such a neuron maps an input function g to $T(b + \int wg \, d\mu) \in \mathbb{R}$.

2.2   Functional MLP

As a functional neuron gives a numerical output, we can define a functional MLP by combining numerical neurons with functional neurons. The first hidden layer of the network consists exclusively in functional neurons (defined thanks to weight functions w_i), whereas subsequent layers are constructed exclusively with numerical neurons. For instance, an one hidden layer functional MLP with real output computes the following function:

$$H(g) = \sum_{i=1}^{k} a_i T \left(b_i + \int w_i g \, d\mu \right) \tag{1}$$

3   Practical Implementation

3.1   Two Problems

Unfortunately, even if equation 1 uses simplified linear forms, we still have two practical problems: it is not easy to manipulate functions (i.e., weights) and we have only incomplete data (i.e., input functions are known only thanks to finite sets of input/output pairs). Therefore, computing $\int wg\,d\mu$ is quite difficult.

3.2   Projection Based Solution

We have already proposed a direct solution to both problems in [6,7]. The main drawback of the direct solution is that computation times can be quite long.

In this paper, we propose a new solution based on projected representation commonly used in FDA (see [5]). The main idea is to use a regularized representation of each input function. In this case we assume that the input space is $L^2(\mu)$ and we consider $(\phi_i)_{i\in\mathbb{N}^*}$ a Hilbertian base of $L^2(\mu)$. We define $\Pi_n(g)$ as the projection of $g \in L^2(\mu)$ on the vectorial space spanned by $(\phi_1,\ldots,\phi_n)$ (denoted $span(\phi_1,\ldots,\phi_n)$), i.e. $\Pi_n(g) = \sum_{i=1}^{n}(\int \phi_i g\,d\mu)\phi_i$. Rather than computing $H(g)$, we try to compute $H(\Pi_n(g))$. It is important to note that in the proposed approach we consider $\Pi_n(g)$ to be the effective input function. Therefore, with the projection based approach, we do not focus anymore on computing $H(g)$.

In order to represent the weight functions, we consider another Hilbertian base of $L^2(\mu)$, $(\psi_j)_{j\in\mathbb{N}^*}$ and we choose weight functions in $span(\psi_1,\ldots,\psi_p)$. For the weight function $w = \sum_{j=1}^{p}\alpha_j\psi_j$, we have:

$$\int w\Pi_n(g)\,d\mu = \sum_{i=1}^{n}\sum_{j=1}^{p}(\int \phi_i g\,d\mu)\alpha_j\int \phi_i\psi_j\,d\mu \qquad (2)$$

For well chosen bases (e.g., B-splines and Fourier series), $\int \phi_i\psi_j\,d\mu$ can be easily computed exactly (especially if we use the same base for both input and weight functions!). Moreover, the values do not depend on actual weight and/or input functions. Efficiency of the method is deeply linked to this fact which is a direct consequence of using truncated bases.

A FMLP based on this approach uses a finite number of real parameters because each weight function is represented thanks to its p real coordinates in $span(\psi_1,\ldots,\psi_p)$. Weights can be adjusted by gradient descent based algorithms, as the actual calculation done by a functional neuron is now a weighted sum of its inputs if we consider the finite dimensional input vector $(\int \phi_1 g\,d\mu,\ldots,\int \phi_n g\,d\mu)$. Back-propagation is very easily adapted to this model and allows efficient calculation of the gradient.

3.3   Approximation of the Projection

Of course, $\Pi_n(g)$ cannot be computed exactly because our knowledge on g is limited to a finite set of input/output pairs, i.e., $(x_k, g(x_k))$. We assume that the measurement points x_k are realizations of independent identically distributed random variables $(X_k)_{k \in \mathbb{N}}$ (defined on a probability space $\mathcal{P}$) and we denote P_X the probability measure induced on $\mathbb{R}^n$ by those random variables. This observation measure weights measurement points and it seems therefore natural to consider that $\mu = P_X$.

Then we can define the random element $\widehat{\Pi}_n(g)^m$ as the function $\sum_{i=1}^n \widehat{\beta}_i \phi_i$ that minimizes $\frac{1}{m} \sum_{k=1}^m \left(g(X_k) - \sum_{i=1}^n \beta_i \phi_i(X_k) \right)^2$. This is a straightforward generalized linear problem which can be solved easily and efficiently with standard techniques. Therefore, rather than computing $H(\Pi_n(g))$ we compute a realization of the random variable $H(\widehat{\Pi}_n(g)^m)$. Following [1] we show in [4] that $\widehat{\Pi}_n(g)^m$ converges almost surely to $\Pi_n(g)$ (in $L^2(\mu)$). This proof is based on an application of theorem 1.1.1 of [3].

4   Theoretical Results

4.1   Universal Approximation

Projection based FMLP are universal approximators:

Corollary 1. *We use notations and hypotheses of subsection 3.2. Let F be a continuous function from a compact subset K of $L^2(\mu)$ to $\mathbb{R}$ and let ϵ be an arbitrary strictly positive real number. We use a continuous non polynomial activation function T. Then there is $n > 0$ and $p > 0$ and a FMLP (based on T and using weight functions in $span(\psi_1, \ldots, \psi_p)$) such that $|H(\Pi_n(g)) - F(g)| < \epsilon$ for each $g \in K$.*

Proof. Details can be found in [4]. First we apply corollary 1 of [7] (this result is based on very general theorems from [8]) to find a FMLP H based on T that approximates F with precision $\frac{\epsilon}{2}$. This is possible because $(\psi_j)_{j \in \mathbb{N}^*}$ is a Hilbertian base of $L^2(\mu)$ which is its own dual. p is obtained as the maximum number of base functions used to represent weight functions in the obtained FMLP.

H is continuous (because T is continuous) and therefore uniformly continuous on K. There is $\eta > 0$ such that for each $(f, g) \in K^2$, $\|f - g\|_2 < \eta$ implies $|H(f) - H(g)| < \frac{\epsilon}{2}$. As $(\phi_i)_{i \in \mathbb{N}^*}$ is a basis and K is compact, there is $n > 0$ such that for each $g \in K$, $\|\Pi_n(g) - g\|_2 < \eta$, which allows to conclude.

In the practical case, $\Pi_n(g)$ is replaced by $\widehat{\Pi}_n(g)^m$ which does not introduce any problem thanks to the convergence result given in 3.3.

4.2   Consistency

When we estimate optimal parameters for a Functional MLP, our knowledge of data is limited in two ways. As always in Data Analysis problems, we have a finite set of input/output pairs. Moreover (and this is specific to FDA), we have also a limited knowledge of each input function. Therefore we cannot apply classical

MLP consistency results (e.g., [9]) to our model. Nevertheless, we demonstrate a consistency result in [4], which is summarized here.

We assume that input functions are realizations of a sequence of independent identically distributed random elements $(G^j)_{j \in \mathbb{N}}$ with values in $L^2(\mu)$ and we denote $G = G^1$. In regression or discrimination problems, each studied function G^j is associated to a real value Y^j ($(Y^j)_{j \in \mathbb{N}}$ is a sequence of independent identically distributed random variables and we denote $Y = Y^1$). Our goal is that the FMLP gives a correct mapping from g to y. This is modeled thanks to a cost function l (in fact we embed in l both the cost function and the FMLP calculation). If we call w the vector of all numerical parameters of a FMLP, we have to minimize $\lambda(w) = E(l(G, Y, w))$.

As we have a finite number of functions, the theoretical cost is replaced by the random variable $\widehat{\lambda}_N(w) = \frac{1}{N} \sum_{j=1}^{N} l(G^j, Y^j, w)$. Moreover, $H(\Pi_n(g))$ is replaced by the random variable $H(\widehat{\Pi}_n(g)^m)$. Therefore, we have to replace l by an approximation which gives the random variable $\widehat{\lambda}_N(w)^m = \frac{1}{N} \sum_{j=1}^{N} l(G^j, Y^j, w)^m$.

We call $\widehat{w}_N^m$ a minimizer of $\widehat{\lambda}_N(w)^m$ and W^* the set of minimizers of $\lambda(w)$. We show in [4] that $\lim_{n \to \infty} \lim_{m \to \infty} d(\widehat{w}_n^m, W^*) = 0$.

5   Simulation Result

We have tried our model on a simple discrimination problem: we ask to a functional MLP to classify functions into classes. Examples of the first class are functions of the form $f_d(x) = \sin(2\pi(x - d))$, whereas functions of the second class have the general form $g_d(x) = \sin(4\pi(x - d))$.

For each class we generate example input functions according to the following procedure:

1. d is randomly chosen uniformly in $[0, 1]$
2. 25 measurement points are randomly chosen uniformly in $[0, 1]$
3. we add a centered Gaussian noise with 0.7 standard deviation to the corresponding outputs

Training is done thanks to a conjugate gradient algorithm and uses early stopping: we used 100 training examples (50 of each class), 100 validation examples (50 of each class) and 300 test examples (150 for each class).

We have compared on those data our direct approach ([6]) to the projection based approach. For both approaches we used a FMLP with three hidden functional neurons. Each functional neuron uses 4 cubic B-splines to represent its weight functions. For the projection based approach, input functions are represented thanks to 6 cubic B-splines. Both models use a total of 19 numerical parameters. For the direct approach, we obtain a recognition rate of 94.4 % whereas the projection based approach achieves 97% recognition rate. Moreover, each iteration of the training algorithm takes about 20 times more time for the direct approach than for the projection based approach.

We have conducted other experiments (for instance the circle based one described in [6]) which give similar results. When the dimension of the input space of each function increases, the generalized linear model used to represent input and weight functions becomes less efficient than a non linear representation ([2]). We have therefore to increase the number of parameters in order to remain competitive with the direct method which can use numerical MLP to represent weights.

6 Conclusion

We have proposed in this paper a new way to work with Functional Multi Layer Perceptrons (FMLP). The projection based approach shares with the direct one very important theoretical properties: projection based FMLP are universal approximators and parameter estimation is consistent for such model.

The main advantage of projection based FMLP is that computation time is greatly reduced compared to FMLP based on direct manipulation of input functions, whereas performances remain comparable. Additional experiments are needed on input functions with higher dimensional input spaces (especially on real world data). For this kind of functions, advantages of projection based FMLP should be reduced by the fact that they cannot use traditional MLP to represent weight and input functions. The direct approach does not have this limitation and should therefore use less parameters. Our goal is to establish practical guidelines for choosing between both approaches.

References

1. Christophe Abraham, Pierre-André Cornillon, Eric Matzner-Lober, and Nicolas Molinari. Unsupervised curve clustering using b-splines. Technical Report 00-04, ENSAM-INRA-UM 11-Montpellier, October 2001.
2. Andrew R. Barron. Universal Approximation Bounds for Superpositions of a Sigmoidal Function. *IEEE Trum. Information Theory*, 39(3):930–945, May 1993.
3. Helga Bunke and Olaf Bunke, editors. *Nonlinear Regression, Functional Relations and Robust Methods*, volume II of Series in Probability and Mathematical Statistics. Wiley, 1989.
4. Brieuc Conan-Guez and Fabrice Rossi. Projection based functional multi layer perceptrons. Technical report, LISE/CEREMADE & INRA, http://www.ceremade.dauphine.fr/, february 2002.
5. Jim Ramsay and Bernard Silverman. *Functional Data Analysis*. Springer Series in Statistics. Springer Verlag, June 1997.
6. Fabrice Rossi, Brieuc Conan-Guez, and François Fleuret. Functional data analysis with multi layer perceptrons. In *IJCNN 2002/WCCI 2002*, volume 3, pages 2843–2848. IEEE/NNS/INNS, May 2002.
7. Fabrice Rossi, Brieuc Conan-Guez, and Fraqois Fleuret. Theoretical properties of functional multi layer perceptrons. In *ESANN 2002*, April 2002.
8. Maxwell B. Stinchcombe. Neural network approximation of continuous functionals and continuous functions on compactifications. *Neural Networks*, 12(3):467–477, 1999.
9. Halbert White. Learning in Artificial Neural Networks: A Statistical Perspective. *Neural Computation*, 1(4):425–464, 1989.

Natural Gradient and Multiclass NLDA Networks

José R. Dorronsoro and Ana González*

Dpto. de Ingeniería Informática and Instituto de Ingeniería del Conocimiento
Universidad Autónoma de Madrid, 28049 Madrid, Spain

1 Introduction

Natural gradient has been recently introduced as a method to improve the convergence of Multilayer Perceptron (MLP) training [1] as well as that of other neural network type algorithms. The key idea is to recast the training process as a problem in quasi maximum log–likelihood estimation of a certain semiparametric probabilistic model. This allows the natural introduction of a riemannian metric tensor G in the probabilistic model space. Once G is computed, the "natural" gradient in this setting is $G(W)^{-1}\nabla_W e(X, y; W)$, rather than the ordinary euclidean gradient $\nabla_W e(X, y; W)$. Here $e(X, y; W)$ denotes an error function associated to a concrete pattern (X, y) and weight set W. For instance, in MLP training, $e(X, y; W) = (y - F(X, W))^2/2$, with F the MLP transfer function. Viewing $(y - F(X, W))^2/2$ as the log–likelihood of a probability density, the metric tensor is

$$G(W) = \int \int \frac{\partial \log p}{\partial W} \left(\frac{\partial \log p}{\partial W} \right)^t p(X, y; W) dX dy. \tag{1}$$

$G(W)$ is also known as the Fisher Information matrix, as it gives the variance of the Cramer–Rao bound for the optimal W estimator. In this work we shall consider a natural gradient–like training for Non Linear Discriminant Analysis (NLDA) networks, a non–linear extension of Fisher's well known Linear Discriminant Analysis introduced in [6] (more details below). Instead of following an approach along the previous lines, we observe that (1) can be viewed as the covariance $G(W) = E[\nabla_W e(X, y; W)\nabla_W e(X, y; W)^t]$ of the random vector $\nabla_W e(X, y; W)$. In turn, its expectation $E[\nabla_W e(X, y; W)]$ is just the gradient of the MLP standard error function. Thus, we shall derive a random variable expectation expression for the gradient of an NLDA criterion function $J(W)$ in the form $\nabla J(W) = E[\Psi(X, W)]$, with Ψ a random vector, which in turn will lead us to define an NLDA information–like matrix as

$$\mathcal{I} = E[(\Psi(X, W) - \overline{\Psi})(\Psi(X, W) - \overline{\Psi})^t].$$

Notice that, contrary to the situation for MLPs, this can no longer be seen as a Gauss–Newton procedure, as the Fisher criterion functions are not given

---

* With partial support of TIC 01–572 and CAM 02–18

J.R. Dorronsoro (Ed.): ICANN 2002, LNCS 2415, pp. 673–678, 2002.

by mean square errors. This approach was first proposed in [3] for the 2 class case. The Fisher's output is then 1 dimensional and there is just essentially one way of defining Fisher's criterion function as $J(W) = s_T/s_B$, where $s_T = E[(y - \overline{y})^2]$ is the total output covariance and $s_B = \pi_1(\overline{y}^1 - \overline{y})^2 + \pi_2(\overline{y}^2 - \overline{y})^2$ is the between class covariance. In the C class setting, however, Fisher's outputs are $(C-1)$ dimensional, and the previous covariances become now the matrices $s_T = E[(Y - \overline{Y})(Y - \overline{Y})^t]$ and $s_B = \sum_{c=1}^{C} \pi_c(\overline{Y}^c - \overline{Y})(\overline{Y}^c - \overline{Y})^t$. Here $\overline{Y} = E[Y]$ denotes the overall output mean, $\overline{Y}^c = (\overline{y}_1^c, \ldots, \overline{y}_{C-1}^c)^t = E[Y|c]$ denotes the class conditional output means and π_c are the class prior probabilities. Several options do exist now for the definition of a C class Fisher's criterion function and in section 2 we shall briefly review two of them. We shall derive C class NLDA gradients, define the associated natural gradients and analyze their complexity in section 3, while in section 4 a numerical illustration will be given.

2 Multiclass Fisher Discriminant Analysis

The general objective of Fisher's Discriminant Analysis is to linearly transform a original feature vector X in a new $C - 1$ dimensional vector Y so that the new features concentrate each class c, $c = 1, \ldots, C$, around its mean $\overline{Y}^c$ while keeping these means apart. For 2 classes the above defined $J(W)$ captures this objective. For more than 2 classes, however, a suitable scalar criterion has to be defined from the matrices s_T and s_B. The most straightforward extension is to define $J_1(W) = |s_T|/|s_B|$, as the determinants $|\cdot|$ somewhat measure the scatter of the points of each class. If S_T and S_B denote the total and between class input covariances, we have $s_T = W^t S_T W$, $s_B = W^t S_B W$, and it easily follows [2,4] that $\nabla J_1(W) = 2J_1 S_T \left(W s_T^{-1} s_B - S_T^{-1} S_B W\right) s_B^{-1}$. Now, if U is the matrix whose columns are the eigenvectors of $S_T^{-1} S_B$, i.e., $S_T^{-1} S_B U = U \Lambda$, with Λ diagonal, it is well known that a solution of $\nabla J_1(W) = 0$ is given by a choice $W^O = (U^1, \ldots, U^{C-1})$ of $C - 1$ eigenvectors of U; then, $J_1(W^O) = 1/\Pi_1^{C-1} \lambda_i$, with $\lambda_1, \ldots, \lambda_{C-1}$ the eigenvalues associated to the U^i. Thus, to minimize J_1 we simply have to take in W^O the eigenvectors associated to the $C - 1$ largest eigenvalues of $S_T^{-1} S_B$. We will also assume that they are arranged in descending eigenvalue order, i.e., $\lambda_1 > \ldots > \lambda_{C-1}$. However, and as we shall see, the J_1 gradient with respect to hidden weights will be quite costly to compute. To simplify this, we will consider [2] a second criterion function, $J_2(W) = tr(\tilde{\Lambda} s_T)/tr(s_B)$, where $\tilde{\Lambda} = \frac{1}{\lambda_1}\Lambda$, i.e., $\tilde{\lambda}_1 = 1$. It can then be seen that

$$\nabla J_2 = \frac{2}{tr(s_B)}\left(S_T W \tilde{\Lambda} - S_B W J_2\right) = \frac{2}{tr(s_B)}S_T\left(W \tilde{\Lambda}\frac{1}{J_2} - S_T^{-1} S_B W\right) J_2. \quad (2)$$

Setting $\nabla J_2(W) = 0$ it follows that a solution of (2) is obtained taking again as W the eigenvectors of $S_T^{-1} S_B$ with eigenvalues $\tilde{\Lambda}/J_2$, that is, $J_2 = 1/\lambda_1$. It can be easily seen that $\lambda_1 \leq 1$ and therefore $J_2 \geq 1$. In the next section we shall define J_1 and J_2 based NLDA networks, compute the hidden weight gradients of J_1 and J_2 and show how to define a natural gradient for them.

3 Natural Gradient in Multiclass NLDA Networks

We will consider the simplest possible NLDA architecture, with D input units, a single hidden layer with H units and $C - 1$ linear output units. We denote network inputs as $X = (x_1, \ldots, x_D)^t$, the weights connecting these D inputs with the hidden unit h as $W_h^H = (w_{1h}^H, \ldots, w_{Dh}^H)^t$ and the weights connecting the hidden layer to the i output by $W_i^O = (w_{1i}^O, \ldots, w_{Hi}^O)^t$. As in an standard MLP, the network transfer function $F(X, W)$ is thus given by $y = F(X, W) = (W^O)^t O$, with $O = O(X, W) = (o_1, \ldots, o_H)^t$ the outputs of the hidden layer, that is, $o_h = f((W_h^H)^t X)$, with f the sigmoidal or tanh function. The optimal weights $W_* = (W_*^O, W_*^H)$ will minimize, however, either one of the previously considered Fisher criterion functions, $J_1(W^O, W^H) = |s_T|/|s_B|$, $J_2(W^O, W^H) = tr(\tilde{A}s_T)/tr(s_B)$, where s_T and s_B are now computed over the hidden layer ouputs O. These criterion functions depend on both the linear output weights W^O and the non linear hidden weights W^H. However, once the W^H are given, the W^O can be computed from the hidden layer outputs O by the previous eigen computations. Thus, in some sense, the weights W^H determine the values of the W^O weights. This suggests the following general setting for standard gradient descent in NLDA networks: if we have at time t a given hidden weight set W_t^H, we update it to W_{t+1}^H as follows

1. We compute first the hidden outputs O and then the associated Fisher weights W_{t+1}^O.
2. We compute the new hidden weights W_{t+1}^H as $W_{t+1}^H = W_t^H - \eta_t \nabla_{W^H} J_t^H(W_t^H)$, where $J_t^H(W^H)$ represents the general criterion function $J(W_{t+1}^O, W^H)$. Here η_t denotes a possibly time varying learning rate.

To apply this procedure we need the gradient $\nabla_W J_t^H(W) = \nabla_W J(W^O, W)$. We observe that both J_1 and J_2 can be written in the form $J_i(W) = \phi(s_T)/\phi(s_B)$ and, therefore,

$$\frac{\partial J_i}{\partial w_{kl}} = \frac{1}{\phi(s_B)} \left(\frac{\partial \phi(s_T)}{\partial w_{kl}} - J_i \frac{\partial \phi(s_B)}{\partial w_{kl}} \right). \tag{3}$$

It can be shown [2] that if the entries of a matrix s depend of a parameter w, then $\partial |s|/\partial w = |s| tr(s^{-1}(\partial s/\partial w))$. Applying this to J_1, (3) can be written as

$$\frac{\partial J_1}{\partial w_{kl}} = J_1 tr \left(s_T^{-1} \frac{\partial s_T}{\partial w_{kl}} - s_B^{-1} \frac{\partial s_B}{\partial w_{kl}} \right). \tag{4}$$

Notice that the W^O weights are invariant under translations of the hidden outputs O. We can thus assume that $\overline{O} = 0$, which also implies that $\overline{Y}$ is also 0 and, hence, $s_T = E[YY^t]$ and $s_B = \sum_c \pi_c \overline{Y}^c (\overline{Y}^c)^t$. As a consequence,

$$\frac{\partial s_T}{\partial w} = E[\left(\frac{\partial Y}{\partial w} \right) Y^t + Y \left(\frac{\partial Y}{\partial w} \right)^t], \quad \frac{\partial s_B}{\partial w} = \sum_1^C \pi_c E_c[\left(\frac{\partial Y}{\partial w} \right) Y^t + Y \left(\frac{\partial Y}{\partial w} \right)^t],$$

where $E_c[z] = E[z|c]$ denotes class c conditional expectation. To finish with (4) we need the partials $\partial y_i/\partial w_{kl}$ that are easily shown to be given by $w_{li}^O f'(a_l)x_k = w_{li}^O A_{kl}$. For J_2 we can obtain a simpler expression, for we have

$$\frac{\partial tr(\tilde{A}s_T)}{\partial w_{kl}} = \sum_1^{C-1} \tilde{\lambda}_i \frac{\partial E[y_i^2]}{\partial w_{kl}} = 2\sum_1^{C-1} \tilde{\lambda}_i E\left[y_i \frac{\partial y_i}{\partial w_{kl}}\right];$$

$$\frac{\partial tr(s_B)}{\partial w_{kl}} = \sum_1^C \pi_c \sum_1^{C-1} \frac{\partial E_c[y_i]^2}{\partial w_{kl}} = 2\sum_1^C \pi_c \sum_1^{C-1} E_c[y_i]E_c[\frac{\partial y_i}{\partial w_{kl}}],$$

and therefore, writing $\tau_B = tr(s_B)$

$$\frac{\partial J_2}{\partial w_{kl}} = \frac{2}{\tau_B}\left(\sum_1^{C-1} \tilde{\lambda}_i w_{li}^O E[y_i A_{kl}] - J_2 \sum_1^C \pi_c \sum_1^{C-1} w_{li}^O E_c[y_i]E_c[A_{kl}]\right)$$

$$= \frac{2J_2}{\tau_B}\left\{E\left[(\sum_{i=1}^{C-1} \lambda_i w_{li}^O y_i)A_{kl}\right] - \sum_1^C \pi_c E_c\left[\sum_{i=1}^{C-1} w_{li}^O y_i\right] E_c[A_{kl}]\right\}. \quad (5)$$

Now, (4) and (5) can be written as the expectation of a random vector Ψ_{kl}, i.e.,

$$\frac{\partial J_i}{\partial w_{kl}}(W) = E[Z_{kl}] - \mu_{kl} = E[\Psi_{kl}(X,W)].$$

For instance, for J_1 we have

$$Z_{kl} = J_1 tr\left(s_T^{-1}\left[\left(\frac{\partial Y}{\partial w_{kl}}\right)Y^t + Y\left(\frac{\partial Y}{\partial w_{kl}}\right)^t\right]\right),$$

$$\mu_{kl} = J_1 tr\left(s_B^{-1}\left(\sum_1^C \pi_c E_c\left[\left(\frac{\partial Y}{\partial w_{kl}}\right)Y^t + Y\left(\frac{\partial Y}{\partial w_{kl}}\right)^t\right]\right)\right).$$

Similarly, for J_2 we have

$$Z_{kl} = \frac{2J_2}{\tau_B}\left(\sum_{i=1}^{C-1} \lambda_i w_{li}^O y_i\right) A_{kl} = \frac{2J_2}{\tau_B}z_l A_{kl},$$

$$\mu_{kl} = \frac{2J_2}{\tau_B}\sum_1^C \pi_c E_c\left[\sum_{i=1}^{C-1} w_{li}^O y_i\right] E_c[A_{kl}] = \frac{2J_2}{\tau_B}\sum_1^C \pi_c E_c[\zeta_l]E_c[A_{kl}],$$

where $z_l = \sum_i \lambda_i w_{li}^O y_i$ and $\zeta_l = \sum_i w_{li}^O y_i$ In particular, we can define the covariance matrix $\mathcal{I}$ of the Ψ_{kl} as

$$\mathcal{I}_{(kl)(mn)} = (\mathcal{I})_{(kl)(mn)}(W) = E[(\Psi_{kl} - \overline{\Psi}_{kl})(\Psi_{mn} - \overline{\Psi}_{mn})^t]$$

$$= E[(Z_{kl} - \overline{Z}_{kl})(Z_{mn} - \overline{Z}_{mn})^t], \quad (6)$$

and therefore $(\mathcal{I})_{(kl)(mn)} = cov(Z_{kl}Z_{mn})$. Thus, we can now define natural gradient descent for multiclass NLDA networks simply changing the previous gradient

Table 1. Percentage of epochs needed by NG to achieve the J_2 value attained by OG after 1000 epochs and final relative $J_2 - 1$ values.

Run	1	2	3	4	5	6	7	8	9	10
Epoch %	6	13	3	2	2	3	3	3	3	3
$J_2 - 1$ %	47.5	53.5	54.0	54.2	56.2	60.6	61.1	61.8	63	67
Run	11	12	13	14	15	16	17	18	19	20
Epoch %	3	7	13	3	3	4	9	4	2	4
$J_2 - 1$ %	67	67.7	74.1	75.9	75.6	77	80	80.2	81.5	91.3

descent formula to $W_{t+1}^H = W_t^H - \eta_t \mathcal{I}^{-1} \nabla_W J_t(W_t^H)$. The greater complexity of J_1 also shows in the complexity of its gradient, which is $O(NDH(C-1)^2) = O(NDHC^2)$, while that of the gradient of J_2 is just $O(NDHC)$. Gradient descent on J_1 is thus $O(C)$ times costlier than on J_2. On the other hand, natural gradient descent adds for both a complexity of $O(ND^2H^2)$, as D^2H^2 expectations have to be computed and this clearly dominates matrix inversion. In other words, we can expect for J_2 natural gradient descent to be about $DH/(C-1)$ times costlier than ordinary descent, while it will be $DH/(C-1)^2$ times costlier for J_1. For this reason, we shall numerically illustrate its use in the next section only for J_2.

4 Numerical Examples

We shall compare natural (NG) and ordinary (OG) gradient learning in the well known problem of forensic glass classification [5]. It has 6 different glass types with a total of 214 9–dimensional inputs. The original features measure different chemical glass properties. We have used a $9 \times 6 \times 5$ network architecture, which implies that J_2 natural gradient will be about $54/5 \simeq 10.8$ times costlier than ordinary gradient descent. This means that, to be worthwhile, natural gradient should arrive at a given J_2 value about 10 times faster than ordinary gradient. To test this, we have performed a 20 run simulation, starting at different initial weights selected from initial uniform values in $[-1, 1]$. On each run 1000 ordinary and natural gradient epochs were done. Table 1 gives for each run the percentage of epochs used by NG to achieve the minimum OG value (second row) and the percentage of the value $J_2 - 1$ given by NG with respect to that of OG (notice that, as pointed out before, $J_2 \geq 1$ always). It is clearly seen that NG was much faster than OG. For instance, in 15 runs NG needed to arrive to the OG minimum J_2 value less than 5% the number of OG epochs, and in 18 runs the number of epochs was below 10%. The average percentage is 4.65%. Similarly, J_2 NG values were much better than those of OG (the average percentage is 67.4 %) and usually they were achieved well before 1000 epochs. Figure (1) shows the NG and OG J_2 evolution in a typical run. In a classification problem, the sample mean error probability MEP is defined as $MEP = 1 - \sum_i N_i c_{ii}/N$, with c_{ii} the percentage of patterns in class i correctly classified, N_i the number of sample patterns in class i and $N = \sum_i N_i$. Obviously, the smaller MEP is, the better is

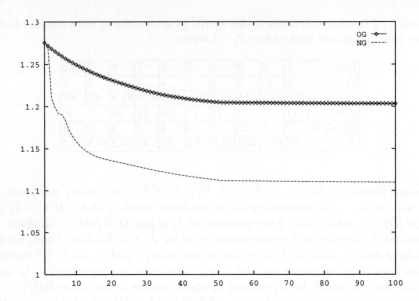

Fig. 1. NG (bottom) and OG J_2 evolution in a typical forensic glass run. NG gets below the OG minimum very fast, and its minimum is quite better than that of OG.

the corresponding classifier. Here a simple classifier is defined assigning each new pattern to the class whose output mean is closer to that pattern's NLDA output. Again, the NG MEP values were better than those achieved by OG. The average NG training MEP was 24% while that of OG was about 35%. We can conclude that NG performs better than OG both in terms of the J_2 value attained and of the MEP value of the associated classifier. Moreover, NG achieves this with considerable less computation efforts.

References

1. H. Park, S. Amari, K. Fukumizu, "Adaptive Natural Gradient Learning Algorithms for Various Stochastic Models", Neural Networks 13 (2000), 755–764.
2. P.A. Devijver, J. Kittler, Pattern Recognition: A Statistical Approach, Prentice Hall, 1982.
3. J.R. Dorronsoro, A. González, C. Santa Cruz, "Natural gradient learning in NLDA networks", Lecture Notes in Computer Science 2084, Springer Verlag 2001, 427–434.
4. K. Fukunaga, Introduction to statistical pattern recognition. Academic Press, 1972.
5. B.D. Ripley, Pattern Recognition and Neural Networks, Cambridge U. Press, 1996.
6. C. Santa Cruz, J.R. Dorronsoro, "A nonlinear discriminant algorithm for feature extraction and data classification", IEEE Transactions in Neural Networks 9 (1998), 1370–1376.

Part IV

Kernel Methods

A Greedy Training Algorithm for Sparse Least-Squares Support Vector Machines

Gavin C. Cawley and Nicola L.C. Talbot

School of Information Systems
University of East Anglia
Norwich, U.K. NR4 7TJ
gcc@sys.uea.ac.uk

Abstract. Suykens *et al.* [1] describes a form of kernel ridge regression known as the least-squares support vector machine (LS-SVM). In this paper, we present a simple, but efficient, greedy algorithm for constructing near optimal sparse approximations of least-squares support vector machines, in which at each iteration the training pattern minimising the regularised empirical risk is introduced into the kernel expansion. The proposed method demonstrates superior performance when compared with the pruning technique described by Suykens *et al.* [1], over the motorcycle and Boston housing datasets.

1 Introduction

Ridge regression [2] is a method from classical statistics that implements a regularised form of least-squares regression. Given training data,

$$\mathcal{D} = \{\boldsymbol{x}_i, y_i\}_{i=1}^{\ell}, \quad \boldsymbol{x}_i \in \mathcal{X} \subset \mathbb{R}^d, \quad y_i \in \mathcal{Y} \subset \mathbb{R},$$

ridge regression determines the parameter vector, $\boldsymbol{w} \in \mathbb{R}^d$, of a linear model, $f(\boldsymbol{x}) = \boldsymbol{w} \cdot \boldsymbol{x} + b$, by minimising the objective function

$$\mathcal{W}(\boldsymbol{w}) = \frac{1}{2}\|\boldsymbol{w}\|^2 + \frac{\gamma}{\ell}\sum_{i=1}^{\ell}(y_i - \boldsymbol{w} \cdot \boldsymbol{x}_i - b)^2. \tag{1}$$

The objective function used in ridge regression (1) implements a form of Tikhonov regularisation [3] of a sum-of-squares error metric, where γ is a regularisation parameter controlling the bias-variance trade-off [4]. This corresponds to penalised maximum likelihood estimation of $\boldsymbol{w}$, assuming the targets have been corrupted by an independent and identically distributed (i.i.d.) sample from a Gaussian noise process, with zero mean and variance σ^2, i.e.

$$y_i = \boldsymbol{w} \cdot \boldsymbol{x}_i + b + \epsilon_i, \quad \epsilon \sim \mathcal{N}(0, \sigma^2).$$

A non-linear form of ridge regression [1,5,6], the least-squares support vector machine, can be obtained via the so-called "kernel trick", whereby a linear ridge regression model is constructed in a high dimensional feature space,

J.R. Dorronsoro (Ed.): ICANN 2002, LNCS 2415, pp. 681–686, 2002.
© Springer-Verlag Berlin Heidelberg 2002

$\mathcal{F}$ ($\phi : \mathcal{X} \rightarrow \mathcal{F}$), induced by a non-linear kernel function defining the inner product $\mathcal{K}(\boldsymbol{x}, \boldsymbol{x}') = \phi(\boldsymbol{x}) \cdot \phi(\boldsymbol{x}')$. The kernel function, $\mathcal{K} : \mathcal{X} \times \mathcal{X} \rightarrow \mathbb{R}$ may be any positive definite "Mercer" kernel. The objective function minimised in constructing a least-squares support vector machine is given by

$$W_{\mathrm{LS-SVM}}(\boldsymbol{w}, b) = \frac{1}{2}\|\boldsymbol{w}\|^2 + \frac{\gamma}{\ell} \sum_{i=1}^{\ell} (y_i - \boldsymbol{w} \cdot \phi(\boldsymbol{x}_i) + b)^2.$$

The representer theorem [7] indicates that the solution of an optimisation problem of this nature can be written in the form of an expansion involving training patterns, i.e. $\boldsymbol{w} = \sum_{i=1}^{\ell} \alpha_i \phi(\boldsymbol{x}_i)$. The output of the least-squares support vector machine is then given by the kernel expansion

$$f(\boldsymbol{x}) = \sum_{i=1}^{\ell} \alpha_i \mathcal{K}(\boldsymbol{x}_i, \boldsymbol{x}) + b.$$

It can easily be shown [5,6] that the optimal coefficients of this expansion are given by the solution of a set of linear equations

$$\begin{bmatrix} \boldsymbol{\Omega} & \boldsymbol{1} \\ \boldsymbol{1}^T & 0 \end{bmatrix} \begin{bmatrix} \boldsymbol{\alpha} \\ b \end{bmatrix} = \begin{bmatrix} \boldsymbol{y} \\ 0 \end{bmatrix},$$

where $\boldsymbol{\Omega} = \boldsymbol{K} + \ell\gamma^{-1}\boldsymbol{I}$, $\boldsymbol{K} = [k_{ij} = \mathcal{K}(\boldsymbol{x}_i, \boldsymbol{x}_j)]_{i,j=1}^{\ell}$, $\boldsymbol{y} = (y_1, y_2, \ldots, y_\ell)^T$, $\boldsymbol{\alpha} = (\alpha_1, \alpha_2, \ldots, \alpha_\ell)^T$ and $\boldsymbol{1} = (1, 1, \ldots, 1)^T$.

1.1   Imposing Sparsity

Unfortunately, unlike the support vector machine, the kernel expansion implementing a least-squares support vector machine is in general fully dense, i.e. $\alpha_i \neq 0$, $\forall\, i \in \{1, 2, \ldots, \ell\}$; this, along with the $\mathcal{O}(\ell^2)$ space and $\mathcal{O}(\ell^3)$ time complexities of the training algorithm make this approach impractical for very large-scale applications. Suykens $et\ al$ [6] propose an iterative pruning procedure to obtain a sparse approximation of the full kernel expansion: A LS-SVM is trained on the entire dataset, yielding a vector of coefficients, $\boldsymbol{\alpha}$. A small fraction of the data (say 5%), associated with coefficients having the smallest magnitudes, is discarded and the LS-SVM retrained on the remaining data. This process is repeated until a sufficiently small kernel expansion is obtained. Model selection is performed at each iteration to refine values for the regularisation parameter, γ and any kernel parameters, in order to obtain adequate generalisation. In this paper we propose a constructive training algorithm for sparse approximation of least-squares support vector machines, adding terms to the kernel expansion in a greedy manner. The proposed algorithm also takes into account the residuals of all training patterns, rather than just those included in the kernel expansion, eliminating the need for further model selection.

2   Method

We begin by introducing an improved formulation of the objective function that includes the residuals for all training patterns, rather than just those patterns currently included in the kernel expansion. If the weight vector, w, can be closely approximated by a weighted sum of a limited subset of the training vectors, i.e., $w \approx \sum_{i \in S} \beta_i \phi(x_i)$, $S \subset \{1, 2, \ldots \ell\}$, then we obtain the objective function minimised by the greedy sparse least-squares support vector machine

$$\mathcal{L}(\beta, b) = \frac{1}{2} \sum_{i,j \in S} \beta_i \beta_j k_{ij} + \frac{\gamma}{\ell} \sum_{i=1}^{\ell} (y_i - \sum_{j \in S} \beta_j k_{ij} - b)^2. \tag{2}$$

Setting the partial derivatives of $\mathcal{L}$ with respect to β and b to zero, and dividing through by $2\gamma/\ell$, yields:

$$\sum_{i \in S} \beta_i \sum_{j=1}^{\ell} k_{ij} + \ell b = \sum_{j=1}^{\ell} y_j$$

and

$$\sum_{i \in S} \beta_i \left\{ \frac{\ell}{2\gamma} k_{ir} + \sum_{j=1}^{\ell} k_{jr} k_{ji} \right\} + b \sum_{i=1}^{\ell} k_{ir} = \sum_{i=1}^{\ell} y_i k_{ir}, \qquad \forall \, r \in S$$

These equations can be expressed as a system of $|S|+1$ linear equations in $|S|+1$ unknowns,

$$H \begin{bmatrix} \beta \\ b \end{bmatrix} = \begin{bmatrix} \Omega & \Phi \\ \Phi^T & \ell \end{bmatrix} \begin{bmatrix} \beta \\ b \end{bmatrix} = \begin{bmatrix} c \\ \sum_{k=1}^{\ell} y_k \end{bmatrix},$$

where $\Omega = [\frac{\ell}{2\gamma} k_{ij} + \sum_{r=1}^{\ell} k_{rj} k_{ri}]_{i,j \in S}$, $\Phi = (\sum_{j=1}^{\ell} k_{ij})_{i \in S}$, $c = (\sum_{j=1}^{\ell} y_j k_{ij})_{i \in S}$. Starting with only a bias term, b, a sparse kernel machine is iteratively constructed in a greedy manner. During each iteration, the training pattern minimising the objective function (2) is incorporated into the kernel expansion. Training can be terminated once the kernel expansion has reached a pre-determined size, or if the reduction in the objective function falls below some threshold value. Note further model selection is not generally necessary as the second summation of (2) is over all training patterns.

2.1   Efficient Implementation

At each iteration, H is extended by additional row and column. The inversion of H_i at the ith iteration can be performed efficiently given H_{i-1}^{-1} computed during the previous iteration, via the block matrix inversion identity

$$\begin{bmatrix} A & B \\ C & D \end{bmatrix}^{-1} = \begin{bmatrix} A^{-1} + A^{-1} B S^{-1} C A^{-1} & -A^{-1} B S^{-1} \\ -S^{-1} C A^{-1} & S^{-1} \end{bmatrix}, \tag{3}$$

where $S = (D - CA^{-1}B)$. In this case, C and B are row and column vectors respectively and D is a scalar, and so S is also a scalar, giving

$$\begin{bmatrix} A & b \\ b^T & c \end{bmatrix}^{-1} = \begin{bmatrix} A^{-1} + \frac{1}{k}A^{-1}bb^T A^{-1} & -\frac{1}{k}A^{-1}b \\ -\frac{1}{k}b^T A^{-1} & \frac{1}{k} \end{bmatrix}, \tag{4}$$

where $k = c - b^T A^{-1} b$. This allows the inversion of H_i with a complexity of only $\mathcal{O}(n^2)$ operations.

3   Results

The Motorcycle benchmark consists of a sequence of accelerometer readings through time following a simulated motor-cycle crash during an experiment to determine the efficacy of crash-helmets (Silverman [8]). Figure 1 shows conventional and greedy sparse support vector machine models of the motorcycle dataset, using a Gaussian radial basis function kernel,

$$\mathcal{K}(x, x') = \exp\left\{ -\sigma^{-2} \|x - x'\|^2 \right\}.$$

The greedy sparse model is functionally identical to the full least-squares support vector machine model with only 15 basis vectors comprising the sparse kernel expansion. Figure 2 compares the 10-fold root-mean-square (RMS) cross-validation error of greedy sparse and pruned least-squares support vector machines as a function of the number of training patterns included in the resulting kernel expansions. The regularisation and kernel parameters for the pruned model were determined in each trial via minimisation of the 10-fold cross-validation error. The cross-validation error is consistently lower for the greedy sparse model regardless of the number of patterns forming the kernel expansion, without the need for further model selection.

The Boston housing dataset describes the relationship between the median value of owner occupied homes in the suburbs of Boston and thirteen attributes representing environmental and social factors believed to be relevant [9]. Figure 3 compares the 10-fold root-mean-square (RMS) cross-validation error of greedy sparse and pruned least-squares support vector machines. Again the error for the greedy sparse method is consistently lower.

4   Summary

This paper presents a simple but efficient greedy training algorithm for constructing sparse approximations of least-squares support vector machines. The proposed algorithm demonstrates performance superior to that of the pruning algorithm of Suykens *et al.* [1] on two real-world benchmark tasks. The new algorithm is also considerably faster as the need for model selection in each iteration is eliminated. The method also provides a plausible approach for large-scale regression problems as it is no longer necessary to store the entire kernel matrix at any stage during training.

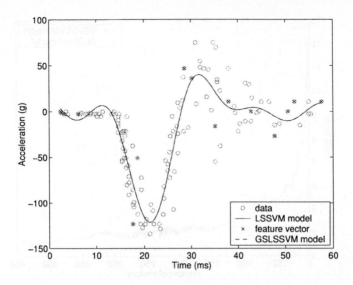

Fig. 1. Least-squares support vector machine (LS-SVM) and greedy sparse least-squares support vector machine (GSLSSVM) models of the motorcycle data set; note the standard and sparse models are essentially identical.

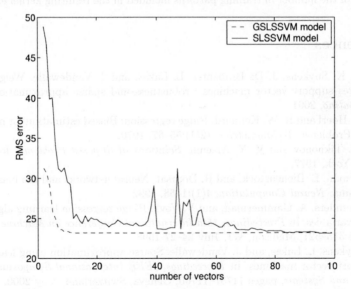

Fig. 2. Cross-validation error of greedy sparse (GSLSSVM) and sparse (SLSSVM) least-squares support vector machine models, over the motorcycle dataset, as a function of the number of training patterns included in the resulting kernel expansions.

Acknowledgements. The authors would like to thank Rob Foxall for his helpful comments on previous drafts of this manuscript. This work was supported by Royal Society research grant RSRG-22270.

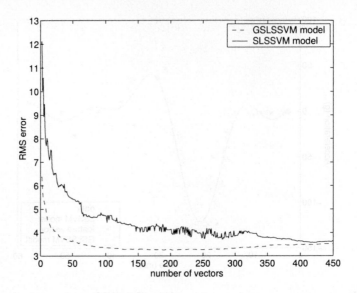

Fig. 3. Cross-validation error of greedy sparse (GSLSSVM) and sparse (SLSSVM) least-squares support vector machine models, over the Boston housing dataset, as a function of the number of training patterns included in the resulting kernel expansions.

References

[1] J. A. K. Suykens, J. De Brabanter, L. Lukas, and J. Vandewalle. Weighted least squares support vector machines : robustness and sparse approximation. *Neurocomputing*, 2001.

[2] A. E. Hoerl and R. W. Kennard. Ridge regression: Biased estimation for nonorthogonal Problems. *Technometrics*, 12(1):55–67, 1970.

[3] A. N. Tikhonov and V. Y. Arsenin. *Solutions of ill-posed Problems*. John Wiley, New York, 1977.

[4] S. Geman, E. Bienenstock, and R. Doursat. Neural networks and the bias/variance dilemma. *Neural Computation*, 4(1):1–58, 1992.

[5] C. Saunders, A. Gammerman, and V. Vovk. Ridge regression learning algorithm in dual variables. In *Proceedings, 15th International Conference on Machine Learning*, pages 515–521, Madison, WI, July 24–27 1998.

[6] J. Suykens, L. Lukas, and J. Vandewalle. Sparse approximation using least-squares support vector machines. In *Proceedings, IEEE International Symposium on Circuits and Systems*, pages 11757–11760, Geneva, Switzerland, May 2000.

[7] G. S. Kimeldorf and G. Wahba. Some results on Tchebycheffian spline functions. *J. Math. Anal. Applic.*, 33:82–95, 1971.

[8] B. W. Silverman. Some aspects of the spline smoothing approach to non-parametric regression curve fitting. *Journal of the Royal Statistical Society*, B, 47(1):1–52, 1985.

[9] D. Harrison and D. L. Rubinfeld. Hedonic prices and the demand for clean air. *Journal Environmental Economics and Management*, 5:81–102, 1978.

Selection of Meta-parameters for Support Vector Regression

Vladimir Cherkassky and Yunqian Ma

Department of Electrical and Computer Engineering
University of Minnesota
Minneapolis, MN 55455
{cherkass, myq}@ece.umn.edu

Abstract. We propose practical recommendations for selecting meta-parameters for SVM regression (that is, ε-insensitive zone and regularization parameter C). The proposed methodology advocates analytic parameter selection directly from the training data, rather than resampling approaches commonly used in SVM applications. Good generalization performance of the proposed parameter selection is demonstrated empirically using several low-dimensional and high-dimensional regression problems. In addition, we compare generalization performance of SVM regression (with proposed choice ε) with robust regression using 'least-modulus' loss function (ε=0). These comparisons indicate superior generalization performance of SVM regression.

1 Introduction

This study is motivated by growing popularity of support vector machines (SVM) for regression problems [2–9]. However, SVM regression application studies are performed by 'expert' users having good understanding of SVM methodology. Since the quality of SVM models depends on a proper setting of SVM meta-parameters, the main issue for practitioners trying to apply SVM regression is how to set these parameter values (to ensure good generalization performance) for a given data set. Whereas existing sources on SVM regression [2,3,7] give some recommendations on appropriate setting of SVM parameters, there is clearly no consensus and (some) contradictory opinions. Hence, resampling remains the method of choice for many applications. Unfortunately, using resampling for (simultaneously) tuning several SVM regression parameters is very expensive in terms of computational costs and data requirements.

In this paper, we suggest simple yet practical analytical approach to SVM regression parameter setting directly from the training data. The proposed approach (to parameter selection) combines well-known theoretical understanding of SVM regression (that provides basic analytical form of unknown dependency for parameter selection) with empirical tuning of such dependencies using several synthetic regression data sets. The practical validity of the proposed approach is demonstrated using several low-dimensional and high-dimensional regression problems. We also

J.R. Dorronsoro (Ed.): ICANN 2002, LNCS 2415, pp. 687–693, 2002.
© Springer-Verlag Berlin Heidelberg 2002

and $w_{ij} = 1$ for non-missing values, thus not discriminating between different elements of the kernel matrix $\mathbf{K}$. For the solution we used a dual-step-first interior point method[1][4]. For comparison we also performed the spectral truncation method as described at the end of Section 2. Based on the completed matrices $\mathbf{K}_{AC}$ and $\mathbf{K}_{ST}$ we trained a support vector machine and evaluated test error and number of support vectors.

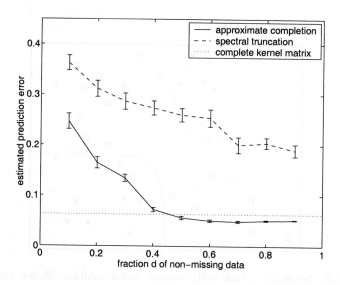

Fig. 2. Estimated prediction error for XOR classification (see Figure 1) based on AC and spectral truncation as a function of the fraction d of non-missing data. The curves show an average over 30 instantiations of $\mathbf{K}^*$, and the errorbars indicate one standard deviation of the estimated mean.

Figure 2 shows the estimated prediction error for the XOR task as a function of the fraction d of non-missing data averaged over 30 random instances of $\mathbf{K}^*$. It is observed, that the AC yields by far lower error values than the spectral truncation method. The average error based on $\mathbf{K}_{AC}$ approaches the error based on the original $\mathbf{K}$ for values of $d \geq 0.4$. In Figure 3 (left) we plotted the average number of support vectors for the same set of simulations. Again we see that $\mathbf{K}_{AC}$ performs much better than $\mathbf{K}_{ST}$, leading to fewer support vectors for values of $d > 0.2$. However, the superior performance has its price in terms of computational complexity. In Figure 3 (right) we plotted the average CPU time in seconds required for the matrix completion. While the spectral truncation methods is very fast independently of d, the AC takes orders of magnitude more time increasing with d. This behaviour is due to the dual step first strategy and would be reversed for the primal step first algorithm, indicating that in practice

---

[1] MATLAB code for the approximate completion problem can be found at
  http://orion.math.uwaterloo.ca:80/~hwolkowi/

the algorithm should be chosen depending on the value of d. Also it appears that the MATLAB implementation used could be accelerated. A more detailed discussion of the time complexity of the algorithm used can be found in [4].

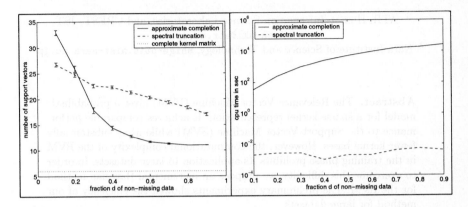

Fig. 3. (Left) Average number of support vectors as a function of the fraction d of non-missing data for the XOR task (for details see Figures 1 and 2). **(Right)** CPU time required for a single kernel matrix completion.

5 Discussion and Conclusion

We demonstrated that semidefinite programming can be used to complete kernel matrices with missing values. It appears, however, that semidefinite programming may have many more applications in data analysis and machine learning. In particular, many problems in machine learning are related to finding appropriate distance measures and Euclidean embeddings that are closely related to semidefinite matrix completion as discussed here.

References

1. D. Achlioptas, F. McSherry, and B. Schölkopf. Sampling techniques for kernel methods. In S. B. Thomas G. Dietterich and Z. Ghahramani, editors, *Advances in Neural Information Processing Systems*.
2. N. Cristianini and J. Shawe-Taylor. *An Introduction to Support Vector Machines*. Cambridge University Press, Cambridge, UK, 2000.
3. D. Haussler. Convolutional kernels on discrete structures. Technical Report UCSC-CRL- 99-10, Computer Science Department, University of California at Santa Cruz, 1999.
4. C. Johnson, B. Kroschel, and H. Wolkowicz. An interior-point method for approximate positive semidefinite completions. *Comput. Optim. Appl.*, 9(2):175–190, 1998.
5. L. Vanderberghe and S. Boyd. Semidefinite programming. *SIAM Rev.*, 38(1):49–95, 1996.

Incremental Sparse Kernel Machine

Masa-aki Sato[1] and Shigeyuki Oba[2]

[1] ATR, Human Information Science Laboratories, and CREST, JST,
masa-aki@atr.co.jp
[2] Nara Institute of Science and Technology, shige-o@is.aist-nara.ac.jp

Abstract. The Relevance Vector Machine (RVM) gives a probabilistic model for a sparse kernel representation. It achieves comparable performance to the Support Vector Machine (SVM) while using substantially fewer kernel bases. However, the computational complexity of the RVM in the training phase prohibits its application to large datasets. In order to overcome this difficulty, we propose an incremental Bayesian method for the RVM. The preliminary experiments showed the efficiency of our method for large datasets.

1 Introduction

The SVM [3] has been recognized as a state-of-the-art method for machine learning. It makes predictions based on a linear combination of kernel functions defined on a subset of the training data points called the support vectors. This sparse kernel representation avoids overfitting to the training data and increases the generalization ability. However, the main disadvantage of the SVM is its lack of probabilistic model. Therefore, it cannot estimate the uncertainty of a prediction, which gives important information in application to real-world problems. Recently, a probabilistic model for a kernel representation called the Relevance Vector Machine (RVM) [1] was proposed. The Bayes estimation of the RVM with an Automatic Relevance Determination (ARD) prior [1,2] leads to a sparse kernel representation. The RVM achieves comparable performance to the SVM while using substantially fewer kernel bases [1]. However, the principal disadvantage of the RVM is its computational complexity in the training phase. It requires $O((\# \ of \ data)^3)$ computation time. This prohibits the application of the RVM to large datasets.

In order to overcome this difficulty, we propose an incremental Bayesian method for the RVM. The proposed method can learn a large amount of data incrementally and gives a sparse kernel representation. It requires $O((\# \ of \ bases)^2 * (\# \ of \ data))$ computation time, which is much smaller than the computation time required by the RVM for a large amount of data.

The Gaussian Process (GP) [4] also provides a probabilistic model for kernel representation. An online learning method for the GP was proposed [5] recently. However, the GP has no inherent mechanism for basis selection. In order to get a sparse kernel representation, it is necessary to introduce heuristic mechanisms [5]. The advantage of our method is that the sparse kernel representation is an

J.R. Dorronsoro (Ed.): ICANN 2002, LNCS 2415, pp. 700–706, 2002.
© Springer-Verlag Berlin Heidelberg 2002

inherent property of the Bayes estimation with the ARD prior. The preliminary experiments showed the efficiency of our approach for large datasets.

2   Relevance Vector Machine

In supervised learning, a set of input-output pairs $(\boldsymbol{X}, \boldsymbol{Y}) \equiv \{\boldsymbol{x}(n), \boldsymbol{y}(n)|n = 1, \ldots, N\}$ is given to a learner, where $\boldsymbol{x}$ and $\boldsymbol{y}$ represent an L-dim. input vector and a D-dim. output vector, respectively. The task of the learner is to predict the output $\boldsymbol{y}$ for a new input data $\boldsymbol{x}$ based on the training data $(\boldsymbol{X}, \boldsymbol{Y})$ and some prior knowledge of the problem. The RVM [1] is defined as a probabilistic model for the kernel representation:

$$P(\boldsymbol{y}|\boldsymbol{x}, \boldsymbol{X}, \boldsymbol{W}, \sigma) = \mathcal{N}(\boldsymbol{y}|\boldsymbol{f}(\boldsymbol{x}, \boldsymbol{X}, \boldsymbol{W}), \sigma^{-1}), \tag{1}$$

$$\boldsymbol{f}(\boldsymbol{x}, \boldsymbol{X}, \boldsymbol{W}) = \sum_{n=0}^{N} \boldsymbol{w}_n \phi_n(\boldsymbol{x}) = \sum_{n=0}^{N} \boldsymbol{w}_n K(\boldsymbol{x}, \boldsymbol{x}(n)), \tag{2}$$

where $\boldsymbol{w}_n$ denotes a D-dim. weight vector, $\phi_n(\boldsymbol{x}) \equiv K(\boldsymbol{x}, \boldsymbol{x}(n))(n = 1, \ldots, N)$ is a kernel basis function defined on an input point $\boldsymbol{x}(n)$, and $\phi_0(\boldsymbol{x}) \equiv 1$. $\mathcal{N}(\boldsymbol{y}|\boldsymbol{\mu}, \sigma^{-1})$ denotes a normal distribution over $\boldsymbol{y}$ with mean $\boldsymbol{\mu}$ and variance σ^{-1}. Accordingly, the conditional likelihood for the training data can be written as $P(\boldsymbol{Y}|\boldsymbol{X}, \boldsymbol{W}, \sigma) = \prod_{n=1}^{N} \mathcal{N}(\boldsymbol{y}(n)|\boldsymbol{f}(\boldsymbol{x}(n), \boldsymbol{X}, \boldsymbol{W}), \sigma^{-1})$. Applying the Bayesian method together with the ARD prior [1,2], one can get a sparse kernel representation: $\boldsymbol{y} = \boldsymbol{f}(\boldsymbol{x}, \boldsymbol{X}_B, \boldsymbol{W}) = \sum_{j=0}^{J} \boldsymbol{w}_j \phi_j(\boldsymbol{x})$, where $\boldsymbol{X}_B \equiv \{\boldsymbol{x}_B(j)|j = 1, \ldots, J\} \subset \boldsymbol{X}$ represents a reduced set of input data that defines the sparse kernel basis $\phi_j(\boldsymbol{x}) \equiv K(\boldsymbol{x}, \boldsymbol{x}_B(j))$ and is called the relevance vector [1].

3   Incremental Sparse Kernel Machine

3.1   Incremental Bayesian Method

In the original RVM [1,2], the posterior parameter distribution is calculated for the entire dataset $(\boldsymbol{X}, \boldsymbol{Y})$. The incremental Bayesian method proposed here uses only a part of the dataset in one learning epoch. At the beginning of the learning, a dataset $(\boldsymbol{X}_D, \boldsymbol{Y}_D) \equiv \{\boldsymbol{x}_D(m), \boldsymbol{y}_D(m)|m = 1, \ldots, M\}$ is selected from the dataset $(\boldsymbol{X}, \boldsymbol{Y})$. The current basis set $\boldsymbol{X}_B \equiv \{\boldsymbol{x}_B(j)|j = 1, \ldots, J\}$ is also selected from the input dataset $\boldsymbol{X}$. The conditional likelihood for the current dataset $(\boldsymbol{X}_D, \boldsymbol{Y}_D)$ under the current basis set $\boldsymbol{X}_B$ is given by

$$P(\boldsymbol{Y}_D|\boldsymbol{X}_D, \boldsymbol{X}_B, \boldsymbol{W}, \sigma) = \prod_{m=1}^{M} \mathcal{N}(\boldsymbol{y}_D(m)|\boldsymbol{f}(\boldsymbol{x}_D(m), \boldsymbol{X}_B, \boldsymbol{W}), \sigma^{-1}), \tag{3}$$

where $\boldsymbol{f}(\boldsymbol{x}, \boldsymbol{X}_B, \boldsymbol{W})$ is defined in (2). We employ the hierarchical ARD prior [2] for the model parameters $\boldsymbol{W} \equiv \{\boldsymbol{w}_j|j = 0, \ldots, J\}$:

$$P_0(\boldsymbol{W}|\sigma, \boldsymbol{\alpha}) = \prod_{j=0}^{J} \mathcal{N}(\boldsymbol{w}_j|\overline{\boldsymbol{w}}_{j0}, (\sigma\alpha_j)^{-1}), \quad P_0(\boldsymbol{\alpha}) = \prod_{j=0}^{J} \Gamma(\alpha_j|a_{j0}, \bar{\alpha}_{j0}^{-1}), \tag{4}$$

epoch, i.e., $\overline{w}_{j0} = \overline{w}_j, \overline{\alpha}_{j0} = \overline{\alpha}_j$, and $a_{j0} = a_j$. An alternative method is to use the obtained posterior $Q_W(W_{old}|\sigma)$ as the new prior for W_{old}. In this case, the matrix $\overline{A}$ in (11) becomes a block diagonal matrix and the block matrices are given by $\overline{A}_{old} = \Sigma_{old}$ and $\overline{A}_{new} = \text{diag}(\overline{\alpha}_{new})$ in an obvious notation. We also have information on the inverse-variance parameter σ, so we reset the prior parameter as $\overline{\tau}_0 = \overline{\tau}, \gamma_{\sigma0} = \gamma_\sigma$, and $\gamma_{\tau0} = \gamma_\tau$ or use the obtained posterior $Q_\sigma(\sigma)$ as the new prior for σ.

We repeat the above process until all data (X, Y) are processed. The predictive distribution after the learning can be calculated by using the posterior parameter distribution: $\hat{P}(y|x, X_B) = \int dW d\sigma P(y|x, X_B, W, \sigma) Q_W(W|\sigma) Q_\sigma(\sigma) = \mathcal{T}(y|\overline{W}' \cdot \phi(x), \Delta(1 + \phi(x)'\Sigma^{-1}\phi(x)), 2\gamma_\sigma)$, where $\mathcal{T}(y|\mu, C, \gamma)$ denotes the t-distribution with mean μ, variance $C/(1 - 2\gamma^{-1})$, and degree of freedom γ. $\phi(x)$ denotes the $(J+1)$-dim. vector defined by $\phi_j(x) = K(x, x_B(j))(j = 1, \ldots, J)$ and $\phi_0(x) \equiv 1$.

4   Experiments

The applicability of our incremental Sparse Kernel Machine (SKM) was investigated by using large datasets. The Gaussian kernel is used in all experiments. The problem is a prediction task for the chaotic Mackey-Glass (M-G) model, which is defined by the differential equation

$$ds(t)/dt = -bs(t) + as(t - \tau)/(1 + s(t - \tau)^{10}), \tag{13}$$

where $a = 0.2, b = 0.1$, and $\tau = 17$ [8]. The task was to predict $y = s(t + 85)$ from the delay coordinate $x = (s(t), s(t - 6), s(t - 12), s(t - 18))$. The training data were prepared by adding the Gaussian noise with different levels to the generated time-series. The N/S (noise-to-signal) ratios of the added noise were 0.0, 0.11, and 0.22 in terms of standard deviation. Three datasets with 500, 1000, and 10000 data points were used to train the incremental SKM, the ν-SVM [9,10] and the Kernel Principal Components Regression (KPCR) method [8]. The NRMSE (normalized root mean squared error) for the test dataset and the corresponding number of bases are listed in Table 1. The SKM showed better performance for the data with noise than did the ν-SVM or the KPCR, while the ν-SVM and the KPCR showed better performance for noiseless data. It should be noted that the SKM achieved good performance with a much fewer bases than the ν-SVM. The SKM selected moderate number of bases even when a large amount of data were given. Fig. 1 shows the basis points (relevance vector) obtained by the SKM for 10000 training data together with the M-G attractor.

5   Discussion

In this paper, we formulated the incremental Sparse Kernel Machine (SKM) and showed the efficiency of this method for large datasets. The Bayes estimation using the ARD prior automatically eliminated insignificant bases and gave a sparse

Table 1. NRMSE for SKM, ν-SVM and KPCR. The number of bases are presented in parentheses.

	n/s=0.0			n/s=0.11			n/s=0.22		
	500	1000	10000	500	1000	10000	500	1000	10000
SKM	0.130	0.109	0.088	0.238	0.201	0.110	0.333	0.322	0.146
	(145)	(148)	(169)	(142)	(250)	(290)	(223)	(413)	(334)
ν-SVM	0.037	0.013	0.004	0.275	0.197	0.148	0.467	0.395	0.355
	(462)	(704)	(3177)	(333)	(507)	(2201)	(334)	(590)	(2769)
KPCR	0.038	0.008	***	0.307	0.280	***	0.443	0.414	***

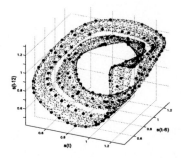

Fig. 1. Relevance vector

kernel representation for large datasets. We also examined another implementation of SKM, where the dataset in one epoch was the entire dataset and the basis set was given by $X_B = \{X_{\text{new}}, X_{\text{old}}\}$. However, this implementation did not significantly improve performance, while it required $O((\# \ of \ bases)*(\# \ of \ data)^2)$ computation time. In this paper, we only considered regression problems. However, our method could be easily extended to classification problems by using the same method developed in [2]. Application of our method to real-world problems remains a task for future research.

References

1. Tipping, M. E. (2000). The Relevance Vector Machine. *NIPS* **12**, pp. 652–658.
2. Bishop, C. M. & Tipping, M. E. (2000). Variational relevance vector machines. *Proc. of 16th Conf. UAI*, pp. 46–53.
3. V. N. Vapnik. (1995). *The Nature of Statistical Learning Theory*. Springer-Verlag.
4. Williams, C. K. I. & Rasmussen, C. E. (1996). Gaussian Process for Regression. *NIPS* **8**. pp. 514–520.
5. Csató, L. & Opper, M. (2001). Sparse Representation for Gaussian Process Models. *NIPS* **13**.
6. Attias, H. (1999). Inferring parameters and structure of latent variable models by variational Bayes. *Proc. of 15th Conf. UAI*, pp. 21–30.
7. Sato, M. (2001). On-line Model Selection Based on the Variational Bayes. *Neural Computation*, **13**. 1649-1681.

8. Rosipal, R. et al. (2001). Kernel PCA for Feature Extraction and De-Noising in Non-linear Regression. *Neural Computing & Applications*. 10(3).
9. Chang, C. and Lin, C., (2001), LIBSVM: a library for SVM. (available at http://www.csie.ntu.edu.tw/~cjlin/libsvm.)
10. Scholköpf, B. et al. (2000). New support vector algorithms. *Neural Computation*, **12**, 1207–1245.

Frame Kernels for Learning

Alain Rakotomamonjy and Stéphane Canu

P.S.I INSA de Rouen,
Avenue de l'université
76801 Saint Etienne du Rouvray France
alain.rakotomamonjy@insa-rouen.fr

Abstract. This paper deals with a way of constructing reproducing kernel Hilbert spaces and their associated kernels from frame theory. After introducing briefly frame theory, we give mild conditions on frame elements for spanning a RKHS. Examples of different kernels are then given based on wavelet frame. Thus, issues of this way of building kernel for semiparametric learning are discussed and an application example on a toy problem is described.

1 Introduction

The purpose of this paper is to present hypothesis spaces based on frame elements in which one looks for the solution of a learning from examples problem. Learning from examples can be viewed as the estimation of the functional dependency between an input $\mathbf{x}$ and an output $\mathbf{y}$ of a system given a set of examples $\{(x_i, y_i), x_i \in \mathcal{X}, y_i \in \mathcal{Y}, i = 1 \ldots \ell\}$. Here, we will assume this point of view and we suppose that the examples are independently drawn according to an unknown probability density function. We also consider that $\mathcal{X} = \mathbb{R}^d$ and $\mathcal{Y} = \mathbb{R}$.

This problem of functional estimation from sparse data is an ill-posed problem and a classical way to turn it in a well-posed one is to use regularization theory. In this context, the solution of the problem is the function f that minimizes the regularized empirical risk :

$$R[f] = \frac{1}{\ell} \sum_{i=1}^{\ell} C(y_i, f(y_i)) + \lambda \|f\|_{\mathcal{H}}^2 \qquad (1)$$

where $C(\cdot, \cdot)$ is a cost function, $\mathcal{H}$ is a Reproducing Kernel Hilbert Space and λ a regularization parameter. Depending on the cost function used, this minimization problem leads either to SVM or Regularization networks [4]. Under general conditions [7], the solution of this minimization problem is :

$$f(x) = \sum_{i=1}^{\ell} a_i K(x, x_i) + \sum_{j=1}^{m} b_j g_j(x) \qquad (2)$$

where $K(\cdot, \cdot)$ is a semi-positive definite kernel associate to the RKHS $\mathcal{H}$ and $\{g_j\}_{j=1 \ldots m}$ a set of functions spanning the null space of the functional $\|f\|_{\mathcal{H}}$.

J.R. Dorronsoro (Ed.): ICANN 2002, LNCS 2415, pp. 707–712, 2002.

Thus, RKHS plays a central role as it describes the hypothesis space where one looks for the function f. Several papers highlight the importance of using a wisely chosen RKHS as it influences largely the generalization capability of f [1,10]. Our aim in this paper is to present a way for building RKHS from set of functions with general conditions owing to the frame theory, thus allowing a practitioner to build a kernel adapted to its problem at hand. Besides, we outline a possible application of this approach on semiparametric learning. Before concluding, an example of a regression toy problem is given in section 3.

2 Regression Estimation with Frame Kernels

The frame theory was introduced by Duffin et al [2,3] in order to establish general conditions under which one can reconstruct perfectly a function f in a Hilbert space $\mathcal{H}$ from its inner product $(\langle \cdot, \cdot \rangle_{\mathcal{H}})$ with a family of vectors $\{\phi_n\}_{n \in \Gamma}$. Γ is an index set of either finite or infinite dimension.

A set of vectors $\{\phi_n\}_{n \in \Gamma}$ is a frame of a Hilbert Space $\mathcal{H}$ if there exists two constants $A > 0$ and $\infty > B \geq A > 0$ so that

$$\forall f \in \mathcal{H}, \qquad A||f||^2 \leq \sum_{n \in \Gamma} |\langle f, \phi_n \rangle|^2 \leq B||f||^2 \tag{3}$$

a tight frame is a frame which bounds A and B are equal. Any function f belonging to $\mathcal{H}$ can be decomposed in the following way :

$$f = \sum_{n \in \Gamma} \langle f, \bar{\phi}_n \rangle \phi_n = \sum_{n \in \Gamma} \langle f, \phi_n \rangle \bar{\phi}_n \tag{4}$$

where $\bar{\phi}_n$ is the dual frame element of ϕ_n. If $\{\phi_n\}_{n \in \Gamma}$ is an orthonormal basis or a tight frame of $\mathcal{H}$ then $\bar{\phi}_n$ is respectively equal to ϕ_n or $\frac{1}{A}\phi_n$. For more general cases, the dual frame elements can be recovered by means of algorithms like the frame algorithm [2] or the one proposed by Grochenig [6]. Another interesting point of frame theory is that any finite set of functions belonging to $\mathcal{H}$ is a frame of the set it spans. From this statement, it becomes simple to build RKHS as we show in the next theorem.

Theorem 1. *Let $\{\phi_n\}_{n=1...N}$ be a finite set of non-zero functions of a Hilbert Space $\mathcal{B}$ of $\mathbb{R}^\Omega$ so that :*

$$\forall n \; 1 \leq n \leq N, \qquad ||\phi_n|| < \infty$$

and

$$\exists M, \forall t \in \Omega, \; \forall n \; 1 \leq n \leq N, \qquad |\phi_n(t)| \leq M$$

Let $\mathcal{H}$ be the set of functions so that:

$$\mathcal{H} = \{f : \exists a_n \in \mathbb{R}, \quad n = 1 \ldots N, f = \sum_n^N a_n \phi_n\}$$

$(\mathcal{H}, \langle \cdot, \cdot \rangle_{\mathcal{B}})$ is a RKHS and its Reproducing Kernel is $K(s, t) = \sum_{n=1}^N \bar{\phi}_n(s) \phi_n(t)$

The proof of this theorem can be found in the technical report [9]. Gao et al. [5] have also constructed a reproducing kernel from frame theory, but their kernel is different as they build it from the frame operator, thus leading to a space with different regularization properties. Applying the above theorem on a subset of $L_2(\mathbb{R}^d)$ functions gives an easy way for obtaining reproducing kernel, as given below :

Examples :

- Consider a finite set of wavelet

$$\left\{ \psi_{j,k}(t) = \frac{1}{\sqrt{a^j}} \psi \left(\frac{t - k u_0 a^j}{a^j} \right), j \in [j_{min}, j_{max}], k \in [k_{min}, k_{max}] \right\}$$

  where $(a, u_0) \in \mathbb{R}^2$, and $(j_{min}, j_{max}, k_{min}, k_{max}) \in \mathbb{Z}^4$, then the span of these functions endowed with L_2 inner product is a RKHS. Depending on a and u_0, this frame can be either tight or not [8]. An example of one-dimensional Daubechies-2 orthogonal wavelet kernel is depicted on figure 1.
- Consider the set of functions on $\mathbb{R}$ $\left\{ \phi_n(t) = \frac{sin(\pi(t-n))}{\pi(t-n)} \right\}_{n=1...N}$. The space spanned by these frame elements associated to $L_2(\mathbb{R})$ inner product form an RKHS.

Thus, the set of the hypothesis space in which one can look for the solution of the minimizer of equation (1) can be defined in this way depending on the problem at hand.

In a classical context of learning with kernels (like SVM or regularization networks), the functions $\{g_j\}_{j=1...m}$ are usually chosen to be a constant, and thus $m = 1$. However, for some learning problem, one may have some prior information about the underlying function generating the data, in this case, as stated by Smola [11] it is reasonable to take advantage of such information by including it in the learning machine, and this is usually done in the context of semiparametric learning by using the function g_j.

Within this framework, using frame based kernels becomes very appreciable owing to the flexibility it allows. In fact, consider $G = \{g_i\}_{i=1...n}$ as a set of n linearly independent functions that satisfy theorem 1, hence, any subset of G, $\{g_i\}_{i \in \Gamma}$, Γ being an index set of size $n_o < n$ can be used for building a RKHS $\mathcal{H}_K$ while the remaining vectors can be used in the parametric part of equation 2. Hence in this case, the solution of (1) is written

$$f(\cdot) = \sum_{i=1}^{n} c_i \sum_{k \in \Gamma} \bar{g}_k(x_i) g_k(\cdot) + \sum_{j \in C_\Gamma} d_j g_j(\cdot)$$

The flexibility comes from the fact that in an approximation problem, any elements of G can be regularized (if involved in the span of $\mathcal{H}_K$) or be kept as it is (if used in the parametric part). Intuitively, one should move any vector that

comes from "good" prior knowledge, in the parametric part of the approximation while leaves in the kernel expansion the others frame elements. Notice also
that only the subset of G which is used in the parametric part has to be linearly
independent.

3   Application on a Toy Regression Problem

We apply this setting for learning in a toy regression problem where the function
to be approximated is :

$$f(x) = \sin x + sinc(\pi(x - 5)) + sinc(5\pi(x - 2)) \tag{5}$$

where $sinc(x) = \frac{\sin x}{x}$. Data used for the approximation is corrupted by an additive noise, thus $y_i = f(x_i) + \epsilon_i$ where ϵ_i is a gaussian noise of standard deviation
0.2 . Points x_i are drawn from uniform random sampling of interval $[0, 10]$. The
cost function $C(\cdot, \cdot)$ that we is used is the quadratic loss function.

In this experiment, we suppose that some additional knowledge on the approximation problem is available, and thus its exploitation using semiparametric
approximation should lead to good performance.

Basis functions and kernels that we used are:

– Gaussian kernel and the sinusoidal basis functions $\{1, \sin(x), \cos(x)\}$
– Gaussian kernel and the wavelet basis functions :

$$\left\{\psi_{j,k}(x) = \frac{1}{\sqrt{a^j}}\psi\left(\frac{x - ku_0a^j}{a^j}\right), j \in \{0, 5\}\right\}$$

– Wavelet kernel and wavelet basis functions : the latter are the same as in
  the previous case , whereas the kernel is built only with low dilation wavelet
  $(j = -10)$. In a nutshell, we can consider that the the original wavelet span
  has been splitted in two RKHS. One that leads to a hypothesis space that
  has to be regularized and another one that does not have to be controlled.
– Sinc kernel and Sin/Sinc basis functions : in this setting, the kernel is given
  from the following frame elements : $\phi_i(x) = \{sinc(j\pi(x - k)) : j \in \{3, 6\}\}$,
  and the basis functions are $\{1, \sin x, \cos x, sinc(\pi(x - k)\}$, with $k \in [0\ldots9]$.

For each kernel, model selection has been solved by cross-validation using 50
datasets. Then, after having spotted the best hyperparameters, the experiment
has been run a hundred times and the true generalization error, in a mean-
square sense, was evaluated. Table 1 summarizes all these trials and describes the
performance improvement achieved by different kernels compared to the gaussian
kernel and sin basis functions. Figure 2 represents an example of estimation
for the different basis functions and kernels and the corresponding pointwise
square errors. These results lead to the following comments. First of all, one can
note that using semiparametric learning allows one to enhance performance if a
"good" guess on the basis function are made. However, even if the knowledge of
the basis functions are not available but only an assumption on the smoothness

Table 1. Generalization performance for semiparametric regression networks for differents setting of kernel and basis functions. The number in parentheses reflects the number of trials for which the model has been the best model.

Kernel / Basis Functions	M.S.E	Improvement (%)
Gaussian / Sin	0.0216 ± 0.0083 (6)	0
Gaussian / Wavelet	0.0202 ± 0.0072 (4)	4.6
Wavelet / Wavelet	0.0195 ± 0.0077 (2)	9.7
Sinc / Sin	0.0156 ± 0.0076 (88)	27.8

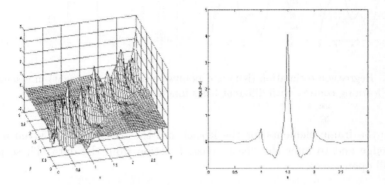

Fig. 1. Left: Example of a Daubechies-2 orthogonal wavelet kernel $K(x,y)$. Right: $K(x, 3 - x)$.

of these functions are known, this prior information can be turned into useful basis function by means of wavelet basis. In fact, one can specify the desired smoothness (or the frequency contents of these bases) by acting on the dilation of wavelet. The *sinc/sin* combination of basis function and kernels gives the best result in this experiment. This seems to empirically justify the statement that one advantage of using frame based kernel lies in the flexibility of choosing basis functions depending on the "good" (functions not regularized) or "bad" (functions used for kernel) knowledge on these functions.

4   Conclusions and Perspectives

Some simple conditions for obtaining reproducing kernel hilbert spaces from any finite set of functions has been given in this paper. The associated kernel suitable for some learning algorithm are described by frame and dual frame elements of the space. This setting enhances the flexibility of semiparametric learning as it allows to choose from a set of functions which one should be regularized (and thus used in the kernel) and which should not. The toy example illustrates these statements and gives promising results. Now, the important points to be investigated are a way for cheaply computing the kernel, searching for how to

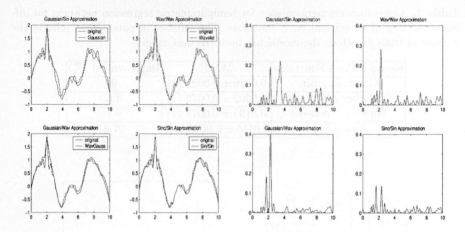

Fig. 2. Regression estimation (left) and pointwise square error (right) in a semiparametric learning context with different basis functions and kernel.

choose the frame elements for the kernel and the basis function given a prior knowledge and to infer the "best" basis function from the data. These points are still open issues.

References

1. N. Cristianini and J. Shawe-Taylor. *Introduction to Support Vector Machines*. Cambridge University Press, 2000.
2. I. Daubechies. *Ten Lectures on Wavelet*. SIAM, CBMS-NSF regional conferences edition, 1992.
3. R. Duffin and A. Schaeffer. A class of nonharmonic fourier series. *Trans. Amer. Math. Soc.*, 72:341–366, 1952.
4. T. Evgeniou, M. Pontil, and T. Poggio. Regularization networks and support vector machines. *Advances in Computational Mathematics*, 13(1):1–50, 2000.
5. J. Gao, C. Harris, and S. Gunn. On a class of a support vector kernels based on frames in function hilbert spaces. *Neural Computation*, 13(9):1975–1994, 2001.
6. K. Grochenig. Acceleration of the frame algorithm. *IEEE Trans. Signal Proc.*, 41(12):3331–3340, 1993.
7. G. Kimeldorf and G. Wahba. Some results on Tchebycheffian spline functions. *J. Math. Anal. Applic.*, 33:82–95, 1971.
8. S. Mallat. *A wavelet tour of signal processing*. Academic Press, 1998.
9. A. Rakotomamonjy and S. Canu. Learning, frame, reproducing kernel and regularization. Technical Report TR2002-01, Perception, Systèmes et Information, INSA de Rouen, 2002.
10. B. Schölkopf, P. Y. Simard, A. J. Smola, and V. Vapnik. Prior knowledge in support vector kernels. In M. I. Jordan, M. J. Kearns, and S. A. Solla, editors, *Advances in Neural information processings systems*, volume 10, pages 640–646, Cambridge, MA, 1998. MIT Press.
11. A. Smola. *Learning with Kernels*. PhD thesis, Published by: GMD, Birlinghoven, 1998.

Robust Cross-Validation Score Function for Non-linear Function Estimation

Jos De Brabanter, Kristiaan Pelckmans, Johan A.K. Suykens, and Joos Vandewalle

K.U.Leuven ESAT-SCD/SISTA, Kasteelpark Arenberg 10, B-3001 Leuven Belgium
{jos.debrabanter,johan.suykens}@esat.kuleuven.ac.be

Abstract. In this paper a new method for tuning regularisation parameters or other hyperparameters of a learning process (non-linear function estimation) is proposed, called robust cross-validation score function (CV_{S-fold}^{Robust}). CV_{S-fold}^{Robust} is effective for dealing with outliers and non-Gaussian noise distributions on the data. Illustrative simulation results are given to demonstrate that the CV_{S-fold}^{Robust} method outperforms other cross-validation methods.

Keywords. Weighted LS-SVM, Robust Cross-Validation Score function, Influence functions, Breakdown point, M-estimators and L-estimators

1 Introduction

Most efficient learning algorithms in neural networks, support vector machines and kernel based methods require the tuning of some extra learning parameters, or *hyperparameters*, denoted here by θ. For practical use, it is often preferable to have a data-driven method to select θ. For this selection process, many data-driven procedures have been discussed in the literature. Commonly used are those based on the cross-validation criterion of Stone [5] and the generalized cross-validation criterion of Craven and Wahba [1]. In recent years, results on L_2, L_1 cross-validation statistical properties have become available [9]. However, the condition $E\left[e_k^2\right] < \infty$ (respectively, $E\left[|e_k|\right] < \infty$) is necessary for establishing weak and strong consistency for L_2 (respectively, L_1) cross-validated estimators. On the other hand, when there are outliers in the y observations (or if the distribution of the random errors has a heavy tail so that $E\left[|e_k|\right] = \infty$), then it becomes very difficult to obtain good asymptotic results for the L_2 (L_1) cross-validation criterion. In order to overcome such problems, a robust cross-validation score function is proposed in this paper.

This paper is organized as follows. In section 2 the classical S-fold crossvalidation score function is analysed. In section 3 we construct a robust crossvalidation score function based on the trimmed mean. In section 4 the repeated robust S-fold crossvalidation score function is described. In section 5 we give an illustrative example.

J.R. Dorronsoro (Ed.): ICANN 2002, LNCS 2415, pp. 713–719, 2002.

2   Analysis of the S-Fold Cross-Validation Score Function

The cross-validation procedure can basically be split up into two main parts: (a) constructing and computing the cross-validation score function, and (b) finding the hyperparameters as: $\theta^* = \text{argmin}_\theta [CV_{S-fold}(\theta)]$. In this paper we focus on (a).

Let $\{z_k = (x_k, y_k)\}_{k=1}^N$ be an independent identically distributed (i.i.d.) random sample from some population with distribution function $F(z)$. Let $F_N(z)$ be the empirical estimate of $F(z)$. Our goal is to estimate a quantity of the form

$$T_N = \int L(z, F_N(z)) \, dF(z), \tag{1}$$

with $L(\cdot)$ the loss function (e.g. the L_2 or L_1 norm) and where $E[T_N]$ could be estimated by cross-validation. We begin by splitting the data randomly into S disjoint sets of nearly equal size. Let the size of the s-th group be m_s and assume that $\lfloor N/S \rfloor \leq m_s \leq \lfloor N/S \rfloor + 1$ for all s. Let $F_{(N-m_s)}(z)$ be the empirical estimate of $F(z)$ based on $(N - m_s)$ observations outside group s and let $F_{m_s}(z)$ be the empirical estimate of $F(z)$ based on m_s observations in group s. Then a general form of the S-fold cross-validated estimate of T_N is given by

$$CV_{S-fold}(\theta) = \sum_{s=1}^S \frac{m_s}{N} \int L\left(z, F_{(N-m_s)}(z)\right) dF_{m_s}(z). \tag{2}$$

Let $\hat{f}^{(-m_S)}(x; \theta)$ be the regression estimate based on the $(N - m_s)$ observation not in group s. Then the least squares S-fold cross-validated estimate of T_N is given by $CV_{S-fold}(\theta) = \sum_{s=1}^S (m_s/N) \sum_{k=1}^{m_s} (1/m_s) \left(y_k - \hat{f}^{(-m_s)}(x_k; \theta)\right)^2$.

The cross-validation score function can be written as a function of $(S + 1)$ means and estimates a location-scale parameter of the corresponding s-samples. Let $u = L(v)$ be a function of a random variable v. In the S-fold cross-validation case, a realization of the random variable v is given by $v_k = \left(y_k - \hat{f}^{(-m_S)}(x_k; \theta)\right)$ $k = 1, ..., m_s$ $\forall s$ with

$$CV_{S-fold}(\theta) = \sum_{s=1}^S \frac{m_S}{N}\left[\frac{1}{m_s}\sum_{k=1}^{m_s} L(v_k)\right] = \sum_{s=1}^S \frac{m_S}{N}\left[\frac{1}{m_s}\sum_{k=1}^{m_s} u_k\right]$$
$$= \hat{\mu}\left[\hat{\mu}_1\left(u_{11}, ..., u_{1m_1}\right), ..., \hat{\mu}_S\left(u_{S1}, ..., u_{Sm_S}\right)\right], \tag{3}$$

where u_{sj} denotes the j-th element of the s-th group, $\hat{\mu}_s(u_{s1}, ..., u_{sm_1})$ denotes the sample mean of the s-th group and $\hat{\mu}$ is the mean of all the sample group means. Consider only the random sample of the s-th group and let $F_{m_s}(u)$ be the empirical distribution function. Then $F_{m_s}(u)$ depends in a complicated way on the noise distribution $F(e)$, the θ values and the loss function $L(\cdot)$. In practice $F(e)$ is unknown (except for the assumption of symmetry around 0). Whatever the loss function would be (L_2 or L_1), the distribution $F_{m_s}(u)$ is always concentrated on the positive axis with an asymmetric distribution. Another important consequence is the difference in the tail behavior of the distribution $F_{m_s}(u)$. L_2 creates a more heavy-tailed distribution than the L_1 loss function.

3 Robust S—Fold Cross-Validation Score Function

A classical cross-validation score function with L_2 or L_1 works well in situations where many assumptions (such as $e_k \sim N(0, \sigma^2)$, $E\left[e_k^2\right] < \infty$ and no outliers) are valid. These assumptions are commonly made, but are at best approximations to reality. There exist a large variety of approaches towards the robustness problem. The approach based on influence functions [2] will be used here. Rather than assuming $F(e)$ to be known, it may be more realistic to assume a mixture noise model for representing the quantity of contamination. The distribution function for this noise model may be written as $F_\epsilon(e) = (1 - \epsilon) F(e) + \epsilon H(e)$, where ϵ and $F(e)$ are given and $H(e)$ is an arbitrary (unknown) distribution, both $F(e)$ and $H(e)$ being symmetric around 0. An important remark is that the regression estimate $\hat{f}^{(-m_s)}(x; \theta)$ must be constructed via a robust method, for example the weighted LS-SVM [6].

In order to understand why certain location-scale estimators behave the way they do, it is necessary to look at the various measures of robustness. The effect of one outlier on the location-scale estimator can be described by the influence function (IF) which (roughly speaking) formalizes the bias caused by one outlier. Another measure of robustness is how much contaminated data a location-scale estimator can tolerate before it becomes useless. This aspect is covered by the breakdown point of the location-scale estimator.

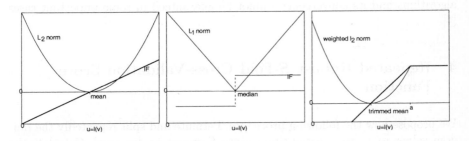

Fig. 1. Different norms with corresponding influence function (IF): (left) L_2; (Middle) L_1; (Right) weighted L_2.

The influence function of the S-fold crossvalidation score function based on the sample mean is sketched in Fig.1. We see that the IF is unbounded in $\mathbb{R}$. This means that an added observation at a large distance from the location-scale parameter (mean, median or trimmed mean) gives a large value in absolute sense for the IF. For example the breakdown point of the mean is 0%. One of the more robust location-scale estimators is the median, a special case of an L-estimator. Although the median is much more robust (breakdown point is 50 %) than the mean, its asymptotic efficiency is low. But in the asymmetric distribution case, the mean and median estimate not the same quantity.

A compromise between mean and median (trade-off between robustness and asymptotic efficiency) is the trimmed mean defined as $\hat{\mu}_{(\beta_1,\beta_2)} = \frac{1}{a}\sum_{i=g_1+1}^{N-g_2} u_{(i)}$, where β_1 (the trimming proportion at the left) and β_2 (the trimming proportion at the right) are selected so that $g_1 = \lfloor N\beta_1 \rfloor$ and $g_2 = \lfloor N\beta_2 \rfloor$ and $a = N - g_1 - g_2$. The trimmed mean is a linear combination of the order statistics given zero weight to g_1 and g_2, extreme observations at each end and equal weight $1/(N - g_1 - g_2)$ to the $(N - g_1 - g_2)$ central observations. Remark that $F_N(u)$ is asymptotic and has only a tail at the right, so we set $\beta_1 = 0$ and β_2 will be estimated from the data. The IF is sketched in figure 1.c. The S-fold cross-validation score function can be written as

$$CV_{S-fold}^{Robust}(\theta) = \hat{\mu}_{(0,\beta_2)}\left[\hat{\mu}_{(0,\beta_{2,1})}\left(u_{1(1)}, ..., u_{1(m_1)}\right), ..., \hat{\mu}_{(0,\beta_{2,s})}\left(u_{S(1)}, ..., u_{S(m_S)}\right)\right] \tag{4}$$

with ordering $u_{(1)} \leq ... \leq u_{(N)}$. It estimates a location-scale parameter of the s-samples, where $\hat{\mu}_{(0,\beta_{2,s})}\left(u_{s(1)}, ..., u_{s(m_1)}\right)$ is the sample trimmed mean of the s-th group, and $\hat{\mu}_{(0,\beta_2)}$ is the trimmed mean of all the sample group trimmed means. We select $\beta_{2,s}$ to minimize the standard deviation of the trimmed mean based on a random sample.

More sophisticated location-scale estimators (M-estimators) have been proposed in [3].With appropriate tuning constant, M-estimates are extremely efficient; their breakdown point can be made very high. However, the computation of M-estimates, for asymmetric distributions, requires rather complex iterative algorithms and its convergence cannot be guaranteed in some important cases [4].

4   Repeated Robust S-Fold Cross-Validation Score Function

We propose now the following procedure. Permute and split repeatedly the random values $u_1, ..., u_N$ - e.g. r times - into S groups as discussed. Calculate the S-fold cross-validation score function for each split and finally take the average of the r estimates

$$\text{Repeated\_}CV_{S-fold}^{Robust}(\theta) = \frac{1}{r}\sum_{j=1}^{r} CV_{S-fold,j}^{Robust}(\theta). \tag{5}$$

The sampling distributions of the estimates, based on mean or trimmed means, are asymptotically Gaussian. The sampling distributions of both standard as robust cross-validations are shown in figure 2. The repeated S-fold cross-validation score function has about the same bias as the S-fold cross-validation score function, but the average of r estimates is less variable than for one estimate.

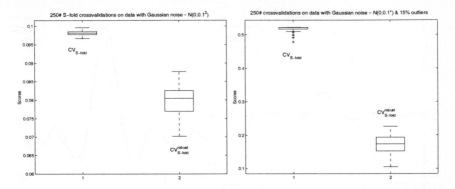

Fig. 2. Sampling distribution of the crossvalidation score function: (Left) Gaussian noise on $\sin(x)/x$; (Right) Gaussian noise & 15% outliers. Significant improvements are made by using the robust CV score function.

5   Illustrative Example

In a simulation example we use here LS-SVMs and weighted LS-SVMs with RBF kernel as the non-linear function estimator. LS-SVMs are reformulations to the standard SVMs [8]. The cost function is a regularized least squares function with equality constraints. The solution of this set of equations is found by solving a linear Karush-Kuhn-Tucker system, closely related to regularization networks, Gaussian processes and kernel ridge regression. The solution can be computed efficiently by iterative methods like the Conjugate Gradient algorithm.

Simulations were carried out to show the differences among the two criteria $\mathrm{CV}_{S-fold}^{L2}$ and $\mathrm{CV}_{S-fold}^{Robust}$. The data $(x_1, y_1), ..., (x_{200}, y_{200})$ are generated from the nonparametric regression curve $y_k = f(x_k) + e_k$, $k = 1, ..., 200$, where the true curve is $f(x_k) = \sin(x)/x$, the observation errors $e_k \sim^{i.i.d.} N(0, 0.1^2)$ and $e_k \sim^{i.i.d.} F_\epsilon(e)$. Define the mixture distribution $F_\epsilon(e) = (1-\varepsilon) F(e) + \varepsilon \Delta_e$ which yields observations from F with high probability $(1-\varepsilon)$ and from the point e with small probability ε. We take for $F(e) = N(0, 0.1^2)$, $\Delta_e = N(0, 1.5^2)$ and $\varepsilon = (1 - \Phi(1.04)) = 0.15$, where $\Phi(\cdot)$ is the standard normal distribution function. A summary of the results is given in Table 1.

Table 1. The performance on a **test** set of a weighted LS-SVM with RBF kernel and tuned hyperparameters using different criteria. Significant improvements are obtained by the robust CV score method in the case of outliers.

criteria	$e \sim \mathcal{N}(0, 0.1^2)$			$e \sim (1-\epsilon)\mathcal{N}(0, 0.1^2) + \epsilon(\mathcal{N}(0, 1.5^2))$		
	$\|\cdot\|_\infty$	$\|\cdot\|_1$	$\|\cdot\|_2$	$\|\cdot\|_\infty$	$\|\cdot\|_1$	$\|\cdot\|_2$
CV_{S-fold}	0.0470	0.0162	$3.95.10^{-4}$	0.148	0.0435	0.0030
CV_{S-fold}^{Robust}	0.0529	0.0178	$4.76.10^{-4}$	0.089	0.0280	0.0012

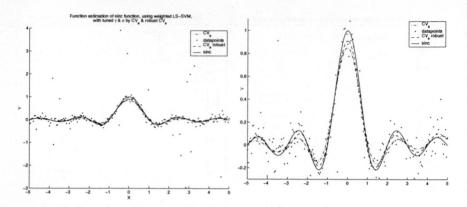

Fig. 3. Result of the weighted LS-SVM where the hyperparameters are tuned by S-fold cross-validation and robust S-fold cross-validation. The robust CV method outperforms the other method. (Left) 15% outliers; (Right) zoom on the estimate.

6   Conclusion

Cross-validation methods are frequently applied for selecting hyperparameters in neural network methods, usually by using L_2 or L_1 norms. However, due to the asymmetric and non-Gaussian nature of the score function, better location-scale parameters can be used to estimate the performance. In this paper we have introduced a robust cross-validation score function method which applies concepts of influence function to the cross-validation methodology. Simulation results suggest that this method can be very effective and promising, especially with non-Gaussian noise distributions and outliers on data. The proposed method has a good robustness/efficiency trade-off such that it performs sufficiently well where L_2 performs optimally. In addition the robust method will improve in many situations where the classical methods will fail.

Acknowledgements. This research work was carried out at the ESAT laboratory and the Interdisciplinary Center of Neural Networks ICNN of the Katholieke Universiteit Leuven, in the framework of the Belgian Program on Interuniversity Poles of Attraction, initiated by the Belgian State, Prime Minister's Office for Science, Technology and Culture (IUAP P4-02 & IUAP P4-24), the Concerted Action Project MEFISTO of the Flemish Community and the FWO project G.0407.02. JS is a postdoctoral researcher with the National Fund for Scientific Research FWO – Flanders.

References

1. Craven P. and Wahba G. (1979). Smoothing noisy data with spline functions. *Numer. Math.*, 31, 377–390.

2. Hampel F.R. (1974). The influence curve and its role in robust estimation. *J.Am.Stat.Assoc.* 69, 383–393.
3. Huber P.J. (1964). Robust estimation of a location parameter. *Ann.Math. Stat.* 35, 73–103
4. Marazzi A. and Ruffieux C. (1996). Implementing M-estimators of the Gamma Distribution. Lecture Notes in Statistics, 109, Springer Verlag, Heidelberg.
5. Stone M. (1974). Cross-validatory choice and assessment of statistical predictions. *J. Royal Statist. Soc. Ser. B* 36 111–147.
6. Suykens J.A.K., Vandewalle J. (1999). Least squares support vector machine classifiers. *Neural Processing Letters.* Vol. 9, 293–300
7. Suykens J.A.K., De Brabanter J., Lukas L., Vandewalle J. (2002). "Weighted least squares support vector machines: robustness and sparse approximation", *Neurocomputing*, in press.
8. Vapnik V. (1995) *The nature of statistical learning theory*, Springer-Verlag.
9. Yang Y. and Zheng Z. (1992). "Asymptotic properties for cross-validation nearest neighbour median estimates in non-parametric regression: the L_1-view", *Probability and statistics*, 242–257.

Compactly Supported RBF Kernels for Sparsifying the Gram Matrix in LS-SVM Regression Models

Bart Hamers, Johan A.K. Suykens, and Bart De Moor

K.U.Leuven, ESAT-SCD/SISTA, Kasteelpark Arenberg 10, B-3001 Leuven, Belgium
{bart.hamers,johan.suykens}@esat.kuleuven.ac.be

Abstract. In this paper we investigate the use of compactly supported RBF kernels for nonlinear function estimation with LS-SVMs. The choice of compact kernels, recently proposed by Genton, may lead to computational improvements and memory reduction. Examples, however, illustrate that compactly supported RBF kernels may lead to severe loss in generalization performance for some applications, e.g. in chaotic time-series prediction. As a result, the usefulness of such kernels may be much more application dependent than the use of the RBF kernel.

Keywords. Support vector machines, nonlinear function estimation, compactly supported kernels, direct and iterative methods.

1   Introduction

Recently, kernel methods for pattern recognition and nonlinear function estimation have received a lot of attention. The performances of these methods is often excellent, although one of the disadvantages is the upscaling to larger data sets. This is caused by the fact that many optimization methods demand the storage of the Gram matrix. Genton [3] recently showed an efficient method for constructing kernels with compact support without destroying the positive definiteness of the kernel. In this paper we will study the result of the use of compactly supported RBF kernels. RBF kernels are frequently used in nonlinear function estimation problems [2]. A compactified version of this kernel could be computationally attractive. In this paper we will apply this kernel to a number of toy problems and real life data sets. As a result, we observe that on certain problems, such as chaotic time series prediction, the use of compactly supported RBF kernels leads to loss in generalization performance, while for other problems (e.g. in lower dimensional problems) the quality of the results is comparable.

This paper is organized as follows. In section 2 we discuss the compactly supported RBF kernel. In section 3 we discuss methods for solving LS-SVM systems and how to exploit sparseness in the Gram matrix. In section 4 illustrations on artificial and real life data sets are given.

J.R. Dorronsoro (Ed.): ICANN 2002, LNCS 2415, pp. 720–726, 2002.
© Springer-Verlag Berlin Heidelberg 2002

2   Kernel Matrix and Compactly Supported Kernels

The kernel functions that are used in the support vector literature [1] are functions $K : \mathbb{R}^d \times \mathbb{R}^d \to \mathbb{R} : (x, z) \mapsto K(x, z)$. One works with kernels that satisfy the Mercer condition. Given a training data set $\{x_i, y_i\}_{i=1}^N$ with inputs $x_i \in \mathbb{R}^d$ and outputs $y_i \in \mathbb{R}$ the kernel matrix or Gram matrix $\Omega \in \mathbb{R}^{N \times N}$ is positive (semi-)definite, where $\Omega_{ij} = K(x_i, x_j)$.

In nonlinear function estimation, a frequently used kernel is the radial basis function (RBF) kernel $K(x, z) = \exp(-\|x - z\|^2 / \sigma^2)$. where $\sigma \in \mathbb{R}$ is a tuning parameter of the model. This Gaussian kernel is a special case of the class of Matérn type kernels [3]. An important property of this class of kernels is that they easily can be transformed into *compactly supported kernels*. This means that the kernel will be zero if $\|x - z\|$ is larger than a cut-off distance θ'. As explained in Genton [3] one can multiply the kernel by $\max\{0, (1 - \|x - z\|/\theta')^{\nu'}\}$ where $\theta' > 0$ and $\nu' \geq (d + 1)/2$ to ensure positive definiteness. The danger of cutting off a kernel in another way is that one will loose positive definiteness. In this paper we will investigate the use of *compactly supported Gaussian RBF kernel* (CS-RBF)

$$K(x, z) = \max\left\{0, \left(1 - \frac{\|x - z\|}{3\sigma}\right)^{\nu'}\right\} \exp\left(-\frac{\|x - z\|^2}{\sigma^2}\right). \tag{1}$$

In order to avoid having too many extra parameters we decided to take the *cut-off point* $\theta' = 3\sigma$, where σ denotes the bandwidth of the Gaussian RBF kernel. ν' is chosen to be equal to the dimension of the input variables for the odd cases; when the dimension is even, it is augmented by one.

3   Nonlinear Function Estimation Using LS-SVMs

We test the CS-RBF kernel in the context of LS-SVMs for nonlinear function estimation. This method is closely related to regularization networks, Gaussian processes and kernel ridge regression [1,2]. The emphasis in the LS-SVM formulation is on primal-dual interpretations as in standard SVM, but simplified to a ridge regression formulation in the primal weight space which can be infinite dimensional. In the primal weight space one has the model $y_i = w^T \varphi(x_i) + b + e_i$ with $\varphi(\cdot)$ the mapping to a high dimensional feature space as in standard SVMs. e_i denotes the error for the i-th training data point. One minimizes $\min_{w,b,e}(1/2)w^T w + \gamma \sum_{i=1}^N e_i^2$ s.t. $y_i = w^T \varphi(x_i) + b + e_i$ for $i = 1, .., N$. For this constrained optimization problem one constructs a Lagrangian. The dual problem gives the KKT system

$$\begin{bmatrix} 0 & 1_v^T \\ \hline 1_v & \Omega + I_N/\gamma \end{bmatrix} \begin{bmatrix} b \\ \alpha \end{bmatrix} = \begin{bmatrix} 0 \\ y \end{bmatrix} \tag{2}$$

with $\alpha = [\alpha_1; ...; \alpha_N]$, $y = [y_1; ...; y_N]$, $1_v = [1; ...; 1]$. This results in the model $\hat{f}(x) = \sum_{i=1}^N \alpha_i K(x, x_i) + b$ with use of the kernel trick $K(x_i, x_j) = \varphi(x_i)^T \varphi(x_j)$.

This model can be robustified and sparsified as explained in [7]. Many algorithms for solving the linear system require a positive definite matrix which is not the case here. Therefore, one can transform this system into $H\eta = 1_v$ and $H\nu = y$ with $H = \Omega + I_N/\gamma$ positive definite. From this we find that $b = \eta^T 1_v/s$ and $\alpha = \nu - b\eta$ where $s = \eta^T 1_v$.

One of the standard numerical methods for solving the linear systems with matrix H is the *Cholesky factorization* [5]. An important disadvantage is that the matrix has to be completely stored in memory. Applying the CS-RBF kernel leads to a sparse matrix. The memory requirements become proportional to the number of non-zero elements n_z. The computational cost is reduced by making efficient use of the zero elements in the matrix. There exist different permutation algorithms (column count permutation, symmetric minimum degree, reverse Cuthill-McKee,...)[4] on the elements of the sparse matrix that give a higher degree of sparseness in the Cholesky factor.

A second important class of methods to solve linear systems are Krylov methods. Such iterative methods are suitable for solving large scale problems. The *conjugate gradient* (CG) method can only be applied to positive definite matrices [6],[5]. The most demanding part in this algorithm is the matrix-vector product between H and the conjugate directions. This can also be reduced by a CS-RBF kernel. The number n_z can be exploited at this point. In the CG method the condition number $\kappa(H)$ determines the convergence (note that this also depends on σ and the regularization constant γ).

4   Examples

In this section we will investigate the use of the CS-RBF kernel on a number of artificial and real-life data sets.

4.1   Sinc Toy Problem

Here we compare CS-RBF and RBF kernels for a noisy sinc function $f(x) = \sin(x)/x$ estimated by LS-SVMs. The tuning parameters are selected as $\gamma = 1.5$ and $\sigma = 3.7$. The inputs were take between -20 and 20 with an interspacing of 0.03. We added Gaussian noise to the inputs with zero mean and standard deviation 0.1. Fig.1 shows that the performance of regression with the RBF and CS-RBF are almost the same. CS-RBF gives a slightly larger bias and less smooth results. The pointwise variance of $\hat{f}(x)$ is larger for the CS-RBF kernel.

An advantage of the CS-RBF kernel is the sparse H matrix. For large data sets this results in a memory reduction. In the example of the sinc-function with 1334 training points, the number of non-zero elements decreases from $1334^2=1779556$ to $n_z =850160$. In this one-dimensional problem the H matrix also has a very clear band structure. Notice that the H matrix is independent of the y_i values of the training set. This means, that for each regression problem with the same x_i values for the training set and hyperparameter set (γ, σ), the H matrix has this sparse band structure. This band structure, in combination

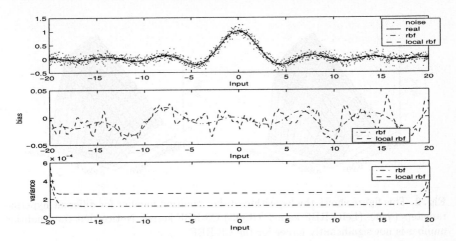

Fig. 1. LS-SVM results for nonlinear function estimation on the sinc function. The middle and bottom part show respectively the bias and variance of both estimates for both kernels.

with the sparseness in the matrix, makes that there is a speed-up in the training procedure. Depending on the used methods (Cholesky or Conjugate Gradient), the time needed to solve the two systems is the following: the Cholesky factorization needs 9.7450 sec. cpu-time to solve the two linear systems for the Gaussian RBF and 4.3270 sec for the compactly supported RBF. The conjugate gradient method needs respectively 4.1470 sec and 2.6540 sec. Hence, we typically observe that the compactly supported kernel results in a memory reduction and a speed-up of about 50%.

We also tested the influence of the localization on the condition number of the matrix H. Fig.2 shows that there is only a small difference in the condition number $\kappa(H)$ for the different values of (γ, σ). Therefore, the speed of convergence for CG with RBF or CS-RBF kernels is comparable.

4.2 Boston Housing Data

As a second example, we tested the Boston housing data set. This data set consists of 506 cases in 14 attributes. We trained LS-SVMs on 406 randomly selected training data and used 100 points as test set. We normalized the data except the binary variables. In Table 1 we show the performances of the LS-SVM for different values of the hyperparameter σ where $\gamma = 30$ is kept constant. The performances of RBF and CS-RBF kernels were comparable on all performed tests. We see that by decreasing σ, the H matrix will become more and more sparse as a result of the localization of the kernel.

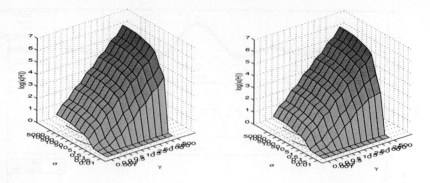

Fig. 2. This figure shows the logarithm of the condition number for different hyperparameters (γ, σ). (Left) RBF kernel; (Right) CS-RBF kernel. Notice that the condition number is not significantly larger for the CS-RBF.

Table 1. Performance for different values of bandwidth σ for CS-RBF kernels on the Boston housing data. MSEtr and MSEtest are respectively the mean squared error on the training and test set. The ratio n_z/N^2 characterizes the degree of sparseness in the Gram matrix.

	$\sigma = 1.5$	$\sigma = 2.0$	$\sigma = 10$
MSEtr	5.8e-3	1.3e-2	1.20e-1
MSEtest	1.1e-1	1.0e-1	8.45e-2
n_z/N^2	0.37	0.84	1

4.3   Santa Fe Chaotic Laser Data Time Series Prediction

In a third example, we use LS-SVM for time series prediction on the Santa Fe laser data set. The model that we use is based on a trained *one step ahead predictor* $\hat{y}_k = f(y_{k-1}, y_{k-2}, ..., y_{k-n})$ with $n = 50$, where y_k denotes the true output at discrete time instant k. In Fig.3 we see that a good iterative prediction performance is obtained for the RBF kernel with hyperparameters $(\gamma, \sigma) = (70, 4)$ found by 10-fold crossvalidation. For the same hyperparameters the CS-RBF kernel has a very bad performance as can be seen in Fig.3. For almost similar performance either the cut-off point $\theta' = 50\sigma$ or the bandwidth of CS-RBF kernel has to be increased. Unfortunately, both reduce the degree of sparseness in the Gram matrix to zero.

5   Conclusion

We have studied the use of compactly supported RBF kernel based on recent work by Genton. RBF kernels are frequently used for many applications. The use of a compactly supported version of the RBF kernel could result in a sparse Gram matrix, and thus decrease the computational cost and memory requirements. In our study we have seen that for certain problems

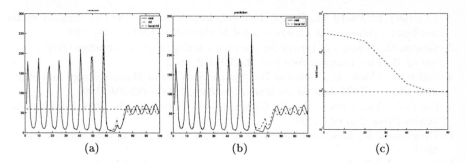

(a)   (b)   (c)

Fig. 3. Santa Fe laser data prediction: (a) (-) real data, (- .) RBF kernel (- -) CS-RBF kernel; (b) (- -) CS-RBF with cut-off point $\theta' = 50\sigma$ having no sparseness; (c) MSE on testdata with respect to cut-off point θ', showing bad results for smaller θ', i.e. sparse Gram matrix.

the generalization performance of the RBF and the compactly supported RBF remains comparable, as well as the conditioning of the matrices towards iterative methods of conjugate gradient. However, on a problem of chaotic time series prediction the compactly supported RBF kernels fails to produce good results when having a sparse Gram matrix. As a result one may conclude that compactly supported RBF kernels may be useful for some specific applications but one should be careful to use it in a general context.

Acknowledgements. Our research is supported by grants from several funding agencies and sources: Research Council KUL: Concerted Research Action GOA-Mefisto 666 (Mathematical Engineering), IDO (IOTA Oncology, Genetic networks), several PhD/postdoc & fellow grants; Flemish Government: Fund for Scientific Research Flanders (several PhD/postdoc grants, projects G.0256.97 (subspace), G.0115.01 (bio-i and microarrays), G.0240.99 (multilinear algebra), G.0197.02 (power islands), G.0407.02 (support vector machines), research communities ICCoS, ANMMM), AWI (Bil. Int. Collaboration Hungary/Poland), IWT (Soft4s (softsensors), STWW-Genprom (gene promotor prediction), GBOU McKnow (Knowledge management algorithms), Eureka-Impact (MPC-control), Eureka-FLiTE (flutter modeling), several PhD grants); Belgian Federal Government: DWTC (IUAP IV-02 (1996-2001) and IUAP V-10-29 (2002-2006): Dynamical Systems and Control: Computation, Identification & Modelling), Program Sustainable Development PODO-II (CP-TR-18: Sustainibility effects of Traffic Management Systems); Direct contract research: Verhaert, Electrabel, Elia, Data4s, IPCOS; BDM is a full professor at K.U.Leuven Belgium, JS is a professor at K.U.Leuven Belgium and a postdoctoral researcher with FWO Flanders.

References

1. Cristianini N., Shawe-Taylor J., *An Introduction to Support Vector Machines*, Cambridge University Press, 2000.

2. Evgeniou T., Pontil M., Poggio T., "Regularization networks and support vector machines", *Advances in Computational Mathematics*, **13**(1), 1–50, 2000.
3. Genton M., "Classes of kernels for machine learning: a statistics perspective", *Journal of Machine Learning Research*, **2**, 299–312, 2001.
4. Gilbert J., Moler C., Schreiber R., "Sparse matrices in Matlab: design and implementation", *SIAM Journal on Matrix Analysis*, **13**(1), 333–356, 1992.
5. Golub G., Van Loan C. *Matrix Computations*, Baltimore: The John Hopkins University Press, 2nd ed., 1990.
6. Greenbaum A., *Iterative Methods for Solving Linear Systems*, Philadelphia: SIAM, 1997.
7. Suykens J.A.K., De Brabanter J., Lukas L., Vandewalle J., "Weighted least squares support vector machines: robustness and sparse approximation", *Neurocomputing*, in press.

The Leave-One-Out Kernel

Koji Tsuda[1] and Motoaki Kawanabe[2]

[1] AIST CBRC, 2-41-6, Aomi, Koto-ku, Tokyo, 1350064 , Japan
[2] Fraunhofer FIRST, Kekuléstr. 7, 12489 Berlin, Germany
koji.tsuda@aist.go.jp, nabe@first.fraunhofer.de

Abstract. Recently, several attempts have been made for deriving *data-dependent* kernels from distribution estimates with parametric models (e.g. the Fisher kernel). In this paper, we propose a new kernel derived from any distribution estimators, parametric or nonparametric. This kernel is called the Leave-one-out kernel (i.e. LOO kernel), because the leave-one-out process plays an important role to compute this kernel. We will show that, when applied to a parametric model, the LOO kernel converges to the Fisher kernel asymptotically as the number of samples goes to infinity.

1 Introduction

In kernel-based learning algorithms [6], a kernel function has to be defined a priori. Most algorithms require the kernel function to be positive semidefinite, and such kernels are often called "Mercer kernels" [9]. The design of Mercer kernels is an important topic in the study of kernel-based learning algorithms [6]. Here, one major direction is to derive a kernel function based on the estimated input distribution (e.g. [4,8,7]). The Fisher kernel [4] is constructed by a distribution estimate with a parametric model, and is applied to many tasks successfully, e.g. protein classification [4]. One important contribution of the Fisher kernel is that it enables to apply the kernel machines to discrete data such as sequences of different lengths or graphs, which used to be hard to deal with.

When the parametric model is not known for given data, nonparametric distribution estimators are often used [2]. Typical methods are e.g. kernel density estimators, k-nearest neighbor methods, orthogonal series estimators and so on. However, in principle, the Fisher kernel method cannot be applied for nonparametric estimators.

In this paper, we propose a general method to derive a Mercer kernel from any distribution estimates, parametric or nonparametric. The leave-one-out process plays an important role for obtaining this kernel. In order to compute the kernel function between x_i and x_j, we consider the leave-one-out density estimates $\hat{p}^{(i)}$ and $\hat{p}^{(j)}$, where x_i and x_j are left out, respectively. Then, the Hellinger inner product between $\hat{p}^{(i)}$ and $\hat{p}^{(j)}$ in the space of probability distributions [1] is taken as the kernel function, which is called the *Leave-one-out kernel* (LOO kernel). By constructing the correspondence between x_i and the leave-one-out estimate $\hat{p}^{(i)}$, the inner product in the space of probability distributions is imported as

J.R. Dorronsoro (Ed.): ICANN 2002, LNCS 2415, pp. 727–732, 2002.

where $H_{ab}(\hat{\boldsymbol{\theta}}) = \frac{1}{n-1}\sum_{i\neq k}\partial_{\theta_a}\partial_{\theta_b}\log p(\boldsymbol{x}_i|\hat{\boldsymbol{\theta}})$. For notational convenience, $\partial_{\theta_a} := \partial/\partial\theta_a$. It is verified that

$$H_{ab}(\hat{\boldsymbol{\theta}}) = -G_{ab}(\hat{\boldsymbol{\theta}}) + o_p(1), \tag{5}$$

because we have $H_{ab}(\hat{\boldsymbol{\theta}}) = \int \partial_{\theta_a}\partial_{\theta_b}\log p(\boldsymbol{x}|\hat{\boldsymbol{\theta}})p(\boldsymbol{x}|\hat{\boldsymbol{\theta}})d\boldsymbol{x} + o_p(1)$ from the law of large numbers and its first term is rewritten as follows:

$$\int \partial_{\theta_a}\partial_{\theta_b}\log p(\boldsymbol{x}|\hat{\boldsymbol{\theta}})p(\boldsymbol{x}|\hat{\boldsymbol{\theta}})d\boldsymbol{x} = \int \partial_{\theta_a}\partial_{\theta_b}p(\boldsymbol{x}|\hat{\boldsymbol{\theta}})d\boldsymbol{x} - \int \frac{\partial_{\theta_a}p(\boldsymbol{x}|\hat{\boldsymbol{\theta}})\partial_{\theta_b}p(\boldsymbol{x}|\hat{\boldsymbol{\theta}})}{p(\boldsymbol{x}|\hat{\boldsymbol{\theta}})}d\boldsymbol{x}$$

$$= -\int \partial_{\theta_a}\log p(\boldsymbol{x}|\hat{\boldsymbol{\theta}})\partial_{\theta_b}\log p(\boldsymbol{x}|\hat{\boldsymbol{\theta}})p(\boldsymbol{x}|\hat{\boldsymbol{\theta}})d\boldsymbol{x}$$

$$= -G_{ab}(\hat{\boldsymbol{\theta}}).$$

The first term of (4) is described as

$$\sum_{i\neq k}\boldsymbol{u}(\boldsymbol{x},\hat{\boldsymbol{\theta}}) = -\boldsymbol{u}(\boldsymbol{x}_k,\hat{\boldsymbol{\theta}}). \tag{6}$$

By substituting (5) and (6) into (4), we have

$$\hat{\boldsymbol{\theta}}^{(k)} - \hat{\boldsymbol{\theta}} = \frac{-1}{n-1}G(\hat{\boldsymbol{\theta}})^{-1}\boldsymbol{u}(\boldsymbol{x}_k|\hat{\boldsymbol{\theta}}) + o_p(n^{-1}). \tag{7}$$

By the Taylor expansion of $p^{1/2}(\boldsymbol{x}|\hat{\boldsymbol{\theta}}^{(k)})$ around $\hat{\boldsymbol{\theta}}$, we have the following:

$$p^{1/2}(\boldsymbol{x}|\hat{\boldsymbol{\theta}}^{(k)}) - p^{1/2}(\boldsymbol{x}|\hat{\boldsymbol{\theta}}) = \frac{\nabla_{\boldsymbol{\theta}}p(\boldsymbol{x}|\hat{\boldsymbol{\theta}})}{2p^{1/2}(\boldsymbol{x}|\hat{\boldsymbol{\theta}})}(\hat{\boldsymbol{\theta}}^{(k)} - \hat{\boldsymbol{\theta}}) + o_p(n^{-1}). \tag{8}$$

By substituting (8) into K_H,

$$K_H(\boldsymbol{x}_i,\boldsymbol{x}_j) = \frac{1}{4}(\hat{\boldsymbol{\theta}}^{(i)} - \hat{\boldsymbol{\theta}})^\top [\int (\frac{\nabla_{\boldsymbol{\theta}}p(\boldsymbol{x}|\hat{\boldsymbol{\theta}})}{p(\boldsymbol{x}|\hat{\boldsymbol{\theta}})})(\frac{\nabla_{\boldsymbol{\theta}}p(\boldsymbol{x}|\hat{\boldsymbol{\theta}})}{p(\boldsymbol{x}|\hat{\boldsymbol{\theta}})})^\top p(\boldsymbol{x}|\hat{\boldsymbol{\theta}})d\boldsymbol{x}](\hat{\boldsymbol{\theta}}^{(j)} - \hat{\boldsymbol{\theta}})$$

$$= \frac{1}{4}(\hat{\boldsymbol{\theta}}^{(i)} - \hat{\boldsymbol{\theta}})^\top G(\hat{\boldsymbol{\theta}})(\hat{\boldsymbol{\theta}}^{(j)} - \hat{\boldsymbol{\theta}}) + o_p(n^{-2}). \tag{9}$$

By substituting (7) into (9), we have

$$K_H(\boldsymbol{x}_i,\boldsymbol{x}_j) = \frac{1}{4(n-1)^2}\boldsymbol{u}(\boldsymbol{x}_i,\hat{\boldsymbol{\theta}})^\top G^{-1}(\hat{\boldsymbol{\theta}})\boldsymbol{u}(\boldsymbol{x}_j,\hat{\boldsymbol{\theta}}) + o_p(n^{-2}).$$

4 Simulations

First of all, we will show the simulation result with the one-dimensional Gaussian $p(x|\mu,\eta) = \frac{\eta}{\sqrt{2\pi}}\exp(-\frac{\eta^2(x-\mu)^2}{2})$, where μ and η are the mean and the inverse of standard deviation. Fig.1 shows an example of the kernel matrices of two methods between 30 samples drawn from the Gaussian. The difference is so

The Fisher Kernel   The LOO kernel

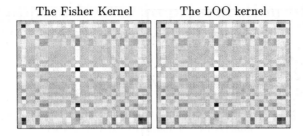

Fig. 1. The comparison between the two kernel matrices

small that they look almost the same. It shows that, in such a low dimensional problem, 30 samples are enough for asympotic approximation.

Secondly, we perform a clustering experiment. In most conventional partitional clustering methods [5] utilize the Euclidean or Mahalanobis distances, thus a cluster is assumed as spherical or ellipsoidal shaped. However, when the cluster shape is not actually spherical (e.g. Fig.2), these methods will fail. So, a proper distance measure is needed for successful clustering. In [8,7], the distance was derived from parametric density models, mostly the Gaussian mixtures, and excellent results were obtained. Here, we will show a similar result by using a nonparametric density estimate, namely the k-nearest neighbor method. The two-dimensional artificial dataset is shown in Fig.2. We performed the k-NN density estimate (k=15), and the result is shown in the lower left of Fig.2. Then, the LOO kernel matrix is derived from this estimate. The lower right of Fig.2 shows the kernel matrix, where the clear separation of two clusters is already visible. When the k-means clustering is applied in the feature space, the two non-spherical clusters are nicely separated (Fig. 2 upper left), whereas the k-means in the input space cannot separate (Fig. 2 upper right). The advantage of this method over Gaussian mixture based methods [8,7] is that the iterative EM learning is not necessary for computing kernels. Unlike the approaches based on the Gaussian mixture, our kernel does not have a local minima problem.

5 Conclusion

In this paper, we proposed the leave-one-out kernel, which is derived from parametric and nonparametric distribution estimates. Since the LOO kernel is approximated by the Fisher kernel in parametric cases, it can be considered as an extension of the Fisher kernel. The leave-one out process is commonly used for estimating generalization error (i.e. LOO cross validation). In future works, it would be interesting to explore the connections between the LOO kernel and the LOO cross validation. Notice that the LOO kernel is defined only among existing examples, not for unseen examples. So in supervised learning, we have to know the input vectors of test set in advance (i.e. the transductive setting). The generalization to unseen examples would be an interesting future topic as well.

2   The Algorithm

We consider a set of labelled training examples $(x_1, c_1), ..., (x_l, c_l)$, where x_i is an element of some $\mathbb{R}^n$ and the label c_i is an element of a finite set, the class to which x_i belongs to. Let us denote by l_c the number of training examples in class c, and by m the number of classes. We will map the data with $\phi : \mathbb{R}^n \to H$ into the Hilbert space H. For each class c we seek a vector $w_c \in H$ and a real value ρ_c representing the class in the following sense: for a data point x the sign of the function

$$d_c(x) = (w_c - \gamma \bar{w}_c) \cdot \phi(x) - \rho_c \qquad (1)$$

gives us information about the class of x. Here $\bar{w}_c = \frac{1}{m-1} \sum_{d \neq c} w_d$ and the dot denotes the scalar product in the Hilbert space H. The parameter $\gamma \in [0, 1]$ lets us adjust the influence of the other classes relative to class c. We will use the shorthand notation $(w, \rho, \xi) = (w_1, ..., w_m, \rho_1, ..., \rho_m, \xi_1, ..., \xi_l)$. For parameters $\gamma \in [0, 1]$ and $\nu \in (0, 1]$, we define a cost function

$$f(w, \rho, \xi) = \sum_c \left(\frac{1}{2} ||w_c||^2 - \frac{1}{2}\gamma w_c \cdot \bar{w}_c - \rho_c + \frac{1}{\nu l_c} \sum_{i \in c} \xi_i \right) \qquad (2)$$

and consider the quadratic program:

$$\begin{aligned}
\text{minimize} \quad & f(w, \rho, \xi) \\
\text{subject to} \quad & (w_{c_i} - \gamma \bar{w}_{c_i}) \cdot \phi(x_i) \geq \rho_{c_i} - \xi_i \\
& \xi_i \geq 0 .
\end{aligned} \qquad (3)$$

The constraints tell us that if a training example is not assigned to its class by (1), then the corresponding ξ_i is positive and this is penalised in the cost function. In addition the cost function favours vectors w_c of small norm and large values of ρ_c. With γ close to 1, the second term in the cost function moves the vectors w_c towards the most discriminant examples. We mention that for $\gamma = 0$ we have m independent problems. With the parameter ν the number of outliers can be controlled, see below. Note that there is no explicit constraint on the vectors w_c to be expressible in terms of the data of their class. We show in the appendix that the function f is convex and bounded from below on the set of feasible points. Thus instead of optimising (3) we can equivalently optimise the dual problem [7].

2.1   The Dual Problem

To determine the dual of (3), we consider the Lagrangian

$$L(w, \rho, \xi) = f(w, \rho, \xi) - \sum_i \alpha_i [(w_{c_i} - \gamma \bar{w}_{c_i}) \cdot x_i - \rho_{c_i} + \xi_i] - \sum_i \beta_i \xi_i \qquad (4)$$

with multipliers $\alpha_i, \beta_i \geq 0$. Setting the derivatives with respect to w_c, ρ_c, ξ_i equal to zero yields a system of linear equations, which is solved by

$$w_c = \sum_{i \in c} \alpha_i x_i \ , \tag{5}$$

$$\alpha_i = \frac{1}{\nu l_{c_i}} - \beta_i \le \frac{1}{\nu l_{c_i}} \ , \qquad \sum_{i \in c} \alpha_i = 1 \ . \tag{6}$$

Equation (5) shows that the vector w_c representing class c is expressible as a linear combination of examples in class c only. Note that for $\gamma = 1$ the solution for w_c is not unique. However, the cost function of all these solutions has the same value, so that we can work with the solution (5). The examples x_i with $\alpha_i \ne 0$ are called support vectors. Substitution of (5) and (6) in L and using $k(x_i, x_j) = \phi(x_i) \cdot \phi(x_j)$ yields the dual problem

$$\text{maximize} \quad -\frac{1}{2} \sum_{i,j} \alpha_i \alpha_j y_{ij} k(x_i, x_j)$$

$$\text{subject to} \quad 0 \le \alpha_i \le \frac{1}{\nu l_{c_i}} \ , \quad \sum_{i \in c_i} \alpha_i = 1 \tag{7}$$

with

$$y_{ij} = \begin{cases} 1 & \text{if } c_i = c_j \\ -\gamma \frac{1}{m-1} & \text{if } c_i \ne c_j \end{cases} \ .$$

We recall that the dual is a concave function. When solving the dual problem we find the optimal values for the multipliers α_i and thus with (5) the optimal vectors w_c. Now given the vectors w_c the optimal ρ_c and ξ_i can be found very easily, and we can proceed as in reference [6]. First we note that these optimal values can be found for each class separately. Consider a class c and set $a_i = (w_c - \gamma \bar{w}_c) \cdot x_i$ for all x_i in class c. We want to minimise $-\rho_c + \frac{1}{\nu l_c} \sum_{i \in c} \xi_i$ subject to $a_i \ge \rho_c - \xi_i$ and $\xi_i \ge 0$. For fixed ρ_c the optimal value of ξ_i is the positive part of $\rho_c - a_i$. Now when we change ρ_c in the negative direction the term $\frac{1}{\nu l_c} \sum \xi_i$ will change proportionally to the relative number of points x_i that have a nonzero ξ_i i.e. points with $a_i < \rho_c$, called outliers. Changing ρ_c in the positive direction changes the sum proportional to the number of points with $a_i \le \rho_c$. Thus the optimal value is found when there are at most νl_c points with $a_i < \rho_c$ and at least νl_c points with $a_i \le \rho_c$. Note that when νl_c is an integer the values ρ_c and ξ_i are not unique. Furthermore the Karush–Kuhn–Tucker conditions tell us that there is no support vector with $a_i > \rho_c$.

2.2 Special Cases

Example 1. Estimating the Support of a Distribution:
If we have only one class, $m = 1$, problem (3) reduces to

$$\text{minimize} \quad \frac{1}{2} ||w||^2 - \rho + \frac{1}{\nu l} \sum_i \xi_i \tag{8}$$

$$\text{subject to} \quad w \cdot \phi(x_i) \ge \rho - \xi_i, \ \xi_i \ge 0 \ ,$$

which tries to separate the data in feature space from zero [6].

Example 2. Support Vector Machine:
In the case of 2 classes and $\gamma = 1$, we define $w = w_1 - w_2$, $b = \frac{1}{2}(\rho_1 - \rho_2)$ and $\rho = \frac{1}{2}(\rho_1 + \rho_2)$. Furthermore we define labels $y_i = 1$ if the example x_i is in the first class, and $y_i = -1$ otherwise. Then (3) reduces to

$$\text{minimize} \quad \frac{1}{2}\|w\|^2 - \rho + \frac{1}{\nu}\sum_i \frac{1}{l_{c_i}}\xi_i \tag{9}$$
$$\text{subject to} \quad y_i(w \cdot \phi(x_i) - b) \geq \rho - \xi_i, \ \xi_i \geq 0 \ .$$

This is the well-known ν–SVM [8] with the following two differences: there is no restriction on ρ to be positive. Therefore we find the constraint that the Lagrange multipliers of each class sum to one. Second the penalty ξ_i for each outlier is weighted by a factor one over the number of examples in its class, instead of the total number of examples. This is necessary in our formulation because it will assure the existence of a feasible point.

3   Experiments and Discussion

3.1   Toy Experiment

We tested the algorithm on artificial data in two dimensions, Fig. 1. Each of the three clusters represents a class and we plotted the lines where the decision functions (1) are zero. One can see that for $\gamma = 1$ the algorithm focuses on the space between the classes, Fig. 1A and B, whereas for $\gamma = 0$ the parameters of each class are learned separately, C and D.

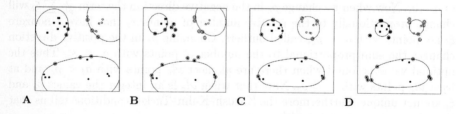

A   B   C   D

Fig. 1. A toy example with the three classes o, •, and ■. For each class the line shows the zero level of the decision function (1). We used a Gaussian kernel and maximal coupling $\gamma = 1$ for **A,B** and $\gamma = 0$ for **C,D**. Plot **A** shows the solution with $\nu = 0.1$ and **B** with $\nu = 0.5$. The support vectors are marked with a ◯. We can see how the parameter ν controls the number of outliers. For $\gamma = 0$, in **C** with $\nu = 0.1$, and in **D** with $\nu = 0.5$, the parameters of each class are optimised independently. As a result the support vectors are not necessarily the vectors closest to the other classes. The solution is simply an estimate of the support of each class as in [6]

3.2   Real-World Data

We tested the algorithm on the US postal service database of handwritten digits. The digits are images of size 16×16. For convenience we trained the algorithm on the digits 1,2 and 3. There are 2394 training examples and 628 in the testing set. We divided each data vector with values in [-1,1] by 16 and used a Gaussian kernel of variance 0.1 as in [9]. The results are summarised in Table 1. We say that an example of class c is accepted, if the decision function (1) is positive, and otherwise it is rejected. Note that one example can be falsely accepted by more than one class. We can see that the algorithm achieves very low false acceptance rates. If we don't want to reject any example, we can use a different strategy. We set the class of x to $\arg\max_c d_c(x)$. We see in Table 1 that with this strategy the error of 2.7% is close to the error achieved by a support vector machine, trained one–against–all (2.1%), but uses considerably fewer support vectors and has three times less parameters to learn. Furthermore, the error for $\gamma = 1$ is significantly smaller than the error of 4.8% in the case $\gamma = 0$, where classification is based on the estimate of the support of each class.

Table 1. Results on the USPS handwritten digits dataset: The algorithm was trained on the digits 1,2 and 3. The first two columns give the values of the parameters ν and γ. The values in the other four columns are given in percentage, with the absolute values in brackets. The third column gives the relative number of test examples whose decision function (1) has a negative value. The fourth column gives the relative number of examples for which there is a decision function of an other class with a positive value. The fifth column shows the relative error in attributing each test example the class whose decision function has maximum value. The last column shows the relative number of support vectors. The last row shows for comparison a support vector machine trained one–against–all. There, the number of support vectors is the total number of different support vectors for all three classifiers.

γ	ν	false reject	false accept	error argmax	number SV
1	0.01	13.1 (82)	1.1 (7)	2.7 (17)	7.4 (176)
0.5	0.01	15.1 (95)	1.4 (9)	4.0 (25)	8.7 (207)
0	0.01	14.2 (89)	4.9 (31)	4.8 (30)	8.2 (197)
1	0.1	19.6 (123)	1.0 (6)	3.8 (24)	13.7 (328)
SVM				2.1 (13)	11.6 (277)

3.3   Discussion

The main inspiration for our algorithm comes from the resemblance of the support vector algorithms for one and two classes. We proposed a general algorithm for the case of multiple classes and showed the applicability to a real–world problem. Even if this is just a first test the results show the potential utility of the algorithm.

Appendix

First we will show that the function (2) is bounded on the set of feasible points. For this we write $w = (w_1, ..., w_m)$ and define $g_c(w) = ||w_c||^2 - \gamma w_c \cdot \bar{w}_c$. For $m > 1$ a short calculation shows

$$\sum_c g_c(w) = \frac{m}{m-1} \left((m - 1 + \gamma) \frac{1}{m} \sum_c ||w_c||^2 - \gamma m || \frac{1}{m} \sum_c w_c ||^2 \right) .$$

With $m - 1 + \gamma \geq \gamma(m - 1) + \gamma = \gamma m$, for $\gamma \in [0, 1]$, and Jensens inequality it follows

$$\sum_c g_c(w) \geq 0 . \tag{10}$$

Thus we deduce that f in (2) is bounded on the set of feasible points. To see that f is convex choose $t \in [0, 1]$ and another vector $v = (v_1, ..., v_m)$. One calculates

$$tg_c(w) + (1 - t)g_c(v) - g(tw_c + (1 - t)v_c) = t(t - 1)g_c(w - v) , \tag{11}$$

and because of (10) the sum over all classes of (11) is greater or equal to zero, thus f is convex.

References

[1] B. E. Boser, I. M. Guyon, and V. N. Vapnik. A training algorithm for optimal margin classifiers. In *Proceedings of the 5th Annual ACM Workshop on Computational Learning theory*, 1992.
[2] V. Vapnik. *Statistical Learning Theory*. Wiley, 1998.
[3] J. Weston and C. Watkins. Support vector machines for multi-class pattern recognition. In *Proceedings of the 6th European Symposium on Artificial Neural Networks (ESANN)*, 1999.
[4] K. Cramer and Y. Singer. On the learnability and design of output codes for multiclass Problems. *Computational Learning Theory*, pages 35–46, 2000.
[5] Y. Lee, Y. Lin, and G. Wahba. Multicategory support vector machines. In *Proceedings of the 33rd Symposium on the Interface*, 2001.
[6] B. Schölkopf, J. C. Platt, J. Shawe-Taylor, A. J. Smola, and R. C. Williamson. Estimating the support of a high-dimensional distribution. *Neural Computation*, 13(7), 2001.
[7] D. P. Bertsekas. *Nonlinear Programming*. Athena Scientific, 1995.
[8] B. Schölkopf, A . J . Smola, R. C. Williamson, and P. L. Bartlett. New support vector algorithms. *Neural Computation*, 12(5), 2000.
[9] B. Schölkopf. *Support Vector Learning*. R. Oldenbourg Verlag, Munich, 1997.

Robust De-noising by Kernel PCA

Takashi Takahashi[1] and Takio Kurita[2]

[1] Department of Applied Mathematics and Informatics,
Ryukoku University, Ootsu, Shiga 520–2194, JAPAN
`takataka@math.ryukoku.ac.jp`
[2] National Institute of Advanced Industrial Science and Technology(AIST),
Tsukuba, Ibaraki 305–8568, JAPAN

Abstract. Recently, kernel Principal Component Analysis is becoming a popular technique for feature extraction. It enables us to extract non-linear features and therefore performs as a powerful preprocessing step for classification. There is one drawback, however, that extracted feature components are sensitive to outliers contained in data. This is a characteristic common to all PCA-based techniques. In this paper, we propose a method which is able to remove outliers in data vectors and replace them with the estimated values via kernel PCA. By repeating this process several times, we can get the feature components less affected with outliers. We apply this method to a set of face image data and confirm its validity for a recognition task.

1 Introduction

Principal Component Analysis is one of the most powerful techniques for feature extraction, and it has widely applied to many tasks such as face recognition[1]. Applying PCA to a set of data, one can extract the feature components accounting for as much variance of the data as possible by a linear transformation. Nonlinear variants of PCA have also investigated by many researchers. Among these, kernel PCA can be considered as a natural generalization of standard(linear) PCA[2]. It enables us to extract nonlinear features via a nonlinear map Φ specified by kernel functions $k(x, y) = \Phi(x) \cdot \Phi(y)$ which substitute dot products between pairs of data $x \cdot y$.

Both standard PCA and kernel PCA can reconstruct the data with minimal error by projecting them onto the principal subspaces and dropping the components with small variance. Consequently, we can employ these methods for de-noising the data[3]. However, the data are sometimes contaminated with outliers due to occlusion etc(see Fig. 1). Since such outliers can affect every principal component, it is difficult to remove their influence upon reconstruction. To solve this problem, we propose a method which is able to remove outliers in data vectors and replace them with the estimated values via kernel PCA. By repeating this process several times, we can get the feature components less affected with outliers. We apply this method to a set of face image data and confirm its validity for a recognition task.

J.R. Dorronsoro (Ed.): ICANN 2002, LNCS 2415, pp. 739–744, 2002.

2   Kernel PCA and De-noising

2.1   Kernel PCA

Kernel PCA can be derived using the known fact that PCA can be carried out on the dot product matrix instead of the covariance matrix[1,2]. Let $\left\{x_i \in R^M\right\}_{i=1}^N$ denote a set of data. Kernel PCA first maps the data into some feature space F by a function $\Phi : R^M \to F$, and then performs standard PCA on the mapped data. Defining the data matrix X by $X = [\Phi(x_1)\Phi(x_2)\cdots\Phi(x_N)]$, the covariance matrix in F becomes

$$C = \frac{1}{N}\sum_{i=1}^N \Phi(x_i)\Phi(x_i)^{\mathrm{T}} = \frac{1}{N}XX^{\mathrm{T}}. \tag{1}$$

For simplicity, we assume that the mapped data are centered[1]: $\frac{1}{N}\sum_{i=1}^N \Phi(x_i) = 0$. Although it is in some cases intractable to carry out the direct eigendecomposition of C, we can find the eigenvalues and eigenvectors of C via solving the eigenvalue problem

$$\lambda u = K u. \tag{2}$$

The $N \times N$ matrix K is the dot product matrix defined by

$$K = \frac{1}{N}X^{\mathrm{T}}X \quad \text{where} \quad K_{ij} = \frac{1}{N}\Phi(x_i) \cdot \Phi(x_j) = \frac{1}{N}k(x_i, x_j). \tag{3}$$

Let $\lambda_1 \geq \ldots \geq \lambda_P$ be the nonzero eigenvalues of K ($P \leq N$ and $P \leq M$), and $u^1, \ldots, u^P$ the corresponding eigenvectors. Then C has the same eigenvalues and there is a one-to-one correspondence between the nonzero eigenvectors $\{u^h\}$ of K and the nonzero eigenvectors $\{v^h\}$ of C: $v^h = \alpha_h X u^h$ where α_h is a constant for normalization[4]. If both of the eigenvectors have unit length, $\alpha_h = 1/\sqrt{\lambda_h N}$. In the following discussion, we assume $\|u^h\| = 1/\sqrt{\lambda_h N}$ so that $\alpha_h = 1$.

For a test data x, its h-th principal component y_h can be computed using kernel functions as

$$y_h = v^h \cdot \Phi(x) = \sum_{i=1}^N u_i^h k(x_i, x). \tag{4}$$

Then the Φ-image of x can be reconstructed from its projections onto the first $H(\leq P)$ principal components in F by using a projection operator P_H as

$$P_H \Phi(x) = \sum_{h=1}^H y_h v^h. \tag{5}$$

The procedure of kernel PCA is equivalent to that of standard PCA on the mapped data. Hence, kernel PCA inherits some properties of standard PCA. For instance, the reconstruction error $\sum_i \|\Phi(x_i) - P_H \Phi(x)\|^2$ is minimal among all H dimensional projection operators in F.

---

[1] This is not the case in general. However, all calculations can be reformulated to deal with centering[2].

2.2   Iterative De-noising with Gaussian Kernels

In order to perform de-noising, we need to reconstruct the data in the input space R^M rather than in F. This can be achieved by seeking a vector z satisfying $\Phi(z) = P_H\Phi(x)$. If such a z exists, it will be a good approximation of x in the input space. However, it will not always exist and it may not be unique even if it exists. Gaussian kernels $k(x, y) = \exp(-\|x - y\|^2/c)$ come under the case.

To settle the problem, Mika et. al[3] proposed to approximate z by minimizing $\rho(z) = \|\Phi(z) - P_H\Phi(x)\|^2$. For kernels satisfying $k(x, x) \equiv$ constant for all x, we can maximize the following expression instead of $\rho(z)$.

$$\tilde{\rho}(z) = (\Phi(z) \cdot P_H\Phi(x)) + \Omega = \sum_{i=1}^{N} w_i k(x_i, z) + \Omega \tag{6}$$

where $w_i = \sum_{h=1}^{H} y_h u_i^h$ and Ω denotes the terms independent of z. Employing standard gradient ascent methods for this, they derived an iteration scheme for computing optimal z:

$$z(t) = \frac{\sum_{i=1}^{N} w_i k(x_i, z(t-1)) x_i}{\sum_{i=1}^{N} w_i k(x_i, z(t-1))} \tag{7}$$

provided that $z(0) = x$.

3   Robust De-noising by Kernel PCA

When we employ kernel PCA as a feature extractor for other applications such as classification, it is desired that the extracted principal components $\{y_h\}$ do not affected with noise in the data. Although this can be achieved to some extent by discarding the components with smaller eigenvalues, it is insufficient for dealing with outliers. Mika et. al's method can not solve this problem since the principal components $\{y_h\}$ are unchanged by their method. One of the possible approaches is to modify the above iteration method so that the principal components are also updated:

$$y_h(t) = \sum_{i=1}^{N} u_i^h k(x_i, \tilde{x}(t)), \qquad w_i(t) = \sum_{h=1}^{H} u_i^h y_h(t), \quad \text{and} \tag{8}$$

$$z(t) = \frac{\sum_{i=1}^{N} w_i(t) k(x_i, \tilde{x}(t)) x_i}{\sum_{i=1}^{N} w_i(t) k(x_i, \tilde{x}(t))} \tag{9}$$

where $\tilde{x}(t)$ is given by

$$\tilde{x}(t) = \begin{cases} x & t = 0 \\ z(t-1) & \text{otherwise.} \end{cases} \tag{10}$$

However, this procedure is considered to be seeking the minimizer of $\|\Phi(z) - P_H\Phi(\tilde{x}(t))\|^2$ at each step. Therefore, it may fail during the iteration unless $\tilde{x}(t) = z(t-1)$ becomes a good approximation of x.

For avoiding this problem, we introduce a "certainty" of data and replace Equation (10) with

$$\tilde{x}(t) = B(t)x + (I - B(t))z(t-1). \tag{11}$$

for $t > 0$. The matrix I is the $M \times M$ identity matrix, and $B(t)$ is a $M \times M$ matrix defined as $B(t) = \text{diag}(\beta_1(t)\ldots,\beta_M(t))$. Each element $\beta_j(t)$ denotes the "certainty" of x_j $(j = 1,\ldots,M)$. We estimate the certainty $\beta_j(t)$ by using the difference between x_j and the corresponding reconstruction $z_j(t-1)$ as

$$\beta_j(t) = \exp\left(-(x_j - z_j(t-1))^2/(2\sigma_j^2)\right). \tag{12}$$

The parameter σ_j denotes the standard deviation of the differences and it is estimated by employing robust estimation method[5]:

$$\sigma_j = 1.4826\left(1 + 5/(N-1)\right)\underset{i}{\text{med}}\sqrt{\varepsilon_{ij}^2} \tag{13}$$

where ε_{ij}^2 denotes the squared error between the j-th component of the i-th training data and its reconstruction, and $\text{med}(x)$ means the median of x. Equations (10) and (12) mean that the input x_j itself is used for the next de-noising step if it has high certainty, otherwise the reconstructed value z_j is used as its estimate. We can get the principal components less affected with outliers by repeating the scheme several times.

4   Experiment

4.1   De-noising Experiments

To test our method, we applied it to the ARFace database of face images [6]. The training data consist of 93 images of 31 persons (three images for each person), and each image was normalized to 18×25 pixels. We performed kernel PCA with these data using Gaussian kernels[2]. Then we applied our algorithm for de-noising two types of noisy data set (Fig. 1(a)). One set(pixel-wise noise) consists of images generated by adding pixel-wise noise with probability 0.4 to the training data[3], while the other set(sunglasses) consists of face images with sunglasses. We used 31 principal components for reconstructing the pre-images, and iterated the algorithm 20 times.

Figure 1 and Table 1 show the results. Table 1 shows the mean squared error between the reconstructed images and the original(noiseless) images. For

---

[2] The parameter c was set to be $3.0M$ so that the best performance was acquired in the classification experiments described below

[3] Each pixel was flipped to black or white with probability $0.4/2$.

comparison, the results obtained by other methods are also shown: 'initial' indicates the errors of the initial reconstruction $z(1)$, the next two rows indicate the errors acquired by our method, provided that in 'w/o certainty' condition Equation (10) was used instead of Equation (11), and the last row indicates the errors acquired by applying Mika et. al's method 20 times. It is shown that our method significantly improves the resulting images. It is also noticed that the pixel-wise 'certainty' of Equation (12) has an effect. Figure 2 depicts the changes of errors during iteration in the case of pixel-wise noise.

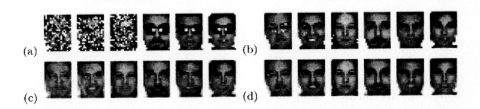

Fig. 1. Results of the de-noising experiments: (a) and (b): Noisy test data and their originals, (c) and (d): De-noising by Mika et. al's method and our method.

Table 1. Results of the de-noising experiments(MSE) and the classification experiments (error rate): MSE indicates the mean squared error between the reconstructed images and the original images.

	MSE		error rate(%)	
	pixel-wise	sunglasses	pixel-wise	sunglasses
initial	0.885	0.854	30.5	12.9
our method	0.368	0.562	3.2	3.2
w/o certainty	0.700	0.669	34.0	25.8
Mika et. al's	0.889	0.859		

4.2 Classification Experiments

In this section, we evaluate the effectiveness of our method via a classification task. The task is to classify the face images used in the previous experiments into 31 classes according to their identities. We used multinomial logit model as the classifier. This model is a special case of the generalized linear model[7], and it can be regarded as one of the simplest neural network for multi-way classification problems. Given the face images, the principal components of each image are computed via our de-noising method. Then the classifier receives the vector composed of principal components $(y_1, \ldots, y_{31})$ and determines the class

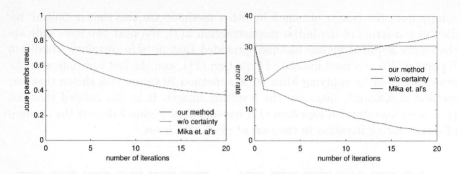

Fig. 2. Time course of MSE (left) and error rate (right).

of the image. The experiments were performed in the identical conditions as the de-noising experiments.

Table 1 shows the error rates for two types of noisy test data. The results obtained by Mika et. al's method are omitted since it produces the same results as 'initial' condition. Figure 2 depicts the changes of the error rates during iteration in the case of pixel-wise noise.

References

1. M. Kirby and L. Sirovich. Application of the Karhunen-Loeve procedure for the characterization of human faces. *IEEE Transactions on Pattern Analysis and Machine Intelligence*, 12(1):103–108, 1990.
2. B. Schölkopf, A. J. Smola, and K.-R. Müller. Nonlinear component analysis as a kernel eigenvalue Problem. *Neural Computation*, 10:1299–1319, 1998.
3. S. Mika, B. Schölkopf, A. Smola, K.-R. Müller, M. Scholz, and G. Rätsch. Kernel PCA and de-noising in feature spaces. In *Advances in Neural Information Processing Systems 11*, 1999.
4. P. Moerland. An on-line EM algorithm applied to kernel PCA. Technical report, IDIAP, 2000.
5. P. J. Huber. *Robust Statistics*. John Wiley & Sons, 1981.
6. A. M. Martinez and R. Benavente. The AR face database. *CVC Technical Report*, (24), 1998.
7. P. McCullaph and J. A. Nelder. *Generalized Linear Models*. Chapman and Hall, 1983.

Maximum Contrast Classifiers

P. Meinicke, T. Twellmann, and H. Ritter

University of Bielefeld, Germany
pmeinick@techfak.uni-bielefeld.de

Abstract. Within the Bayesian setting of classification we present a method for classifier design based on *constrained density modelling*. The approach leads to maximization of some contrast function, which measures the discriminative power of the class-conditional densities used for classification. By an upper bound on the density contrast the sensitivity of the classifiers can be increased in regions with low density differences which are usually most important for discrimination. We introduce a parametrization of the contrast in terms of modified kernel density estimators with variable mixing weights. In practice the approach shows some favourable properties: first, for fixed hyperparameters, training of the resulting *Maximum Contrast Classifier (MCC)* is achieved by linear programming for optimization of the mixing weights. Second for a certain choice of the density contrast bound and the kernel bandwidth, the *maximum contrast* solutions lead to sparse representations of the classifiers with good generalization performance, similar to the maximum margin solutions of support vector machines. Third the method is readily furnished for the general multi-class problem since training proceeds in the same way as in the binary case.

1 Introduction

In the Bayesian setting of classification with a "zero-one" loss function [2] the optimal classifier for an M-class problem is given by

$$c(\mathbf{x}) = \arg\max_{j} \pi_j p(\mathbf{x} \mid j) \tag{1}$$

where the $p(\mathbf{x} \mid j)$ are the class-conditional densities with class label $j = 1, \ldots, M$ and the π_j are the apriori probabilities of class membership while the posterior probabilities are given by $P(j \mid \mathbf{x}) = \pi_j p(\mathbf{x} \mid j) / \sum_k \pi_k p(\mathbf{x} \mid k)$. For notational simplicity, in the following we will assume equal apriori probabilities, i.e. $\pi_j = 1/M$. Because in almost all cases the class specific densities are unknown, they have to be replaced by the corresponding estimates. Then (1) is usually referred to as a "plug-in" classifier, which is consistent[1] if the estimation of the class-conditional densities is consistent [6]. Because we usually have no information about the functional form of the densities, in general nonparametric density

---

[1] consistent here means that for increasing training set sizes the risk of misclassification converges to the smallest possible value, the so-called Bayes risk.

J.R. Dorronsoro (Ed.): ICANN 2002, LNCS 2415, pp. 745–750, 2002.
© Springer-Verlag Berlin Heidelberg 2002

estimates are required to achieve consistency. This gives rise to a serious problem: because of the notoriously slow convergence of density estimation, especially in high dimensions, huge amounts of data are required to build good classifiers. Therefore Vapnik [8] argued that it might be a better idea to directly address the classification problem instead of tackling the more general problem of density estimation and he gave a well-motivated framework for practical construction of "direct" classifiers, which can be viewed as the foundation of the ongoing success story of support vector machines (SVMs).

Due to the fact that the SVM has been introduced for binary classification by optimizing a single hyperplane in a feature space [1], there is no unique extension of the SVM framework for multi-class problems. This fact is reflected by the diversity of multi-class extensions for SVMs with individual pros and cons (see e.g. [3]). In most approaches, the multi-class task is traced back to a set of binary classification problems. Additionally, by a separating hyperplane alone one cannot decide whether a data example actually belongs to any of the represented classes or if it is an "outlier".

In contrast to SVM classifiers the density-based definition (1) directly suggests solutions to the above mentioned problems: right from the definition the general multi-class problem is addressed in the same way as the binary case. Thereby the contrast provides a direct measure of confidence and outliers can be detected by a threshold on the class-conditional densities.

Therefore we may ask whether it is possible to solve the classification task within the density-based framework without giving up the task-oriented specialization of the SVM approach, which is the key to construction of good classifiers. The answer to this question is that we must give up the generality of density estimation in that we have to specify what we want to do with the resulting densities. Good density estimators on their own cannot be the objective anymore, instead, if we want good classifiers, we need *constrained* densities, tailored to the task at hand. In other words: we somehow have to revise the original plug-in concept, where density estimators and classifiers are constructed independently in separate steps. With the Quantizing Density Estimators (QDE) we recently proposed a framework for unsupervised learning of projection sets [4] where we showed how constrained density modelling can be utilized to perform PCA or vector quantization. In the following we show how constrained density modelling can be utilized for classifier construction and we start with the motivation of a suitable learning objective, which we shall refer to as *maximum contrast* and thus the corresponding classifier is termed *Maximum Contrast Classifier*.

2   Maximum Contrast

In order to specialize density modelling to the classification task we require class-conditional probability density functions (pdf) which are as discriminative as possible. For an example from a certain class j this means that in regions with a low average difference ("contrast") between $\hat{p}(\mathbf{x} \,|\, j)$ and another class-conditional density we should try to enlarge this contrast. On the contrary in regions where the contrast is large we can reduce it to a value which is still high enough for good discrimination. Thus, the idea is to enlarge (small) differences between class-conditional pdf in order to increase the discriminative power of the densities. To get into detail we now define the single bounded density contrast as

$$C_k(\mathbf{x}, j) = \min\{\hat{p}(\mathbf{x} \mid j) - \hat{p}(\mathbf{x} \mid k), \gamma_{\max}\} \tag{2}$$

where $\gamma_{\max} > 0$ is an upper bound, which should prevent the learning algorithm from enhancing contrast in regions where discrimination is already easy. Thus a suitable objective is to maximize the average contrast given by

$$C = \sum_{j=1}^{M} \sum_{k \neq j} \int C_k(\mathbf{x}, j) p(\mathbf{x}, j) d\mathbf{x}. \tag{3}$$

In order to maximize the corresponding sample average, for an M-class problem we require a set of labeled training examples $\mathcal{X} = \{\mathbf{x}_1, \dots, \mathbf{x}_N\}$. For the labeling we have M index sets I_j containing indices of examples from class j, i.e. for $i \in I_j$ correct classification yields $c(\mathbf{x}_i) = j$. Then we can state the *empirical contrast* as

$$\hat{C} = \frac{1}{N} \sum_{j=1}^{M} \sum_{k \neq j} \sum_{i \in I_j} C_k(\mathbf{x}_i, j) \tag{4}$$

which has to be maximized.

Now we have to choose a set of candidate functions from which we select the required density function. Because this set should provide some flexibility with respect to contrast maximization the usual kernel density estimator [7] isn't a quite good choice, since the only free parameter is the kernel bandwidth which doesn't allow for any local adaptation. On the other hand if we allow for local variation of the bandwidth we get a complicated contrast which is difficult to maximize due to nonlinear dependencies on the parameters. The same is true if we treat the kernel centers as free parameters. However, if we modify the kernel density estimator to have flexible mixing weights according to

$$\hat{p}(\mathbf{x} \mid j) = \sum_{i \in I_j} \alpha_{ij} K(\mathbf{x}, \mathbf{x}_i) \quad \text{with} \quad \sum_{i \in I_j} \alpha_{ij} = 1, \ \alpha_{ij} \geq 0 \tag{5}$$

we get an objective function which is linear in the parameters α_{ij} weighting the contribution of the normalized kernel functions. Thus we have class-conditional pdf with mixing weights α_{ij} which control the contribution of a single training example to the density function.

Optimization of mixing weights is now achieved by *maximizing the sample contrast* w.r.t. the α_{ij}. This can be done by solving the linear program

$$\max \sum_{j=1}^{M} \sum_{i \in I_j} \sum_{k \neq j} (\gamma_{ijk} + \epsilon \delta_{ijk}) \quad \text{subject to:} \quad \sum_{i \in I_j} \alpha_{ij} = 1 \ \forall j, \quad \alpha_{ij} \geq 0, \tag{6}$$

$$\hat{p}(\mathbf{x_i} \mid j) - \hat{p}(\mathbf{x_i} \mid k) - \gamma_{ijk} - \delta_{ijk} \geq 0, \quad \gamma_{ijk} \leq \gamma_{max}, \quad \delta_{ijk} \geq 0 \tag{6a}$$

where each slack variable γ_{ijk} measures a single contrast $C_k(\mathbf{x}_i, j)$ as defined in (2), while δ_{ijk} measures that part of the density difference which exceeds $\gamma_{\max}$. The δ_{ijk} prevent the program from singularities which may occur for the case where too many γ_{ijk} equal $\gamma_{\max}$. Since the influence of δ_{ijk} should be small as compared with γ_{ijk}, the δ_{ijk} are weighted by some small $\epsilon < \gamma_{\max}$. Given that we use unnormalized kernel functions with $K(\mathbf{x}, \mathbf{x}) = 1$ we have $\gamma_{ijk} \leq 1$ and we can choose $\epsilon = \gamma_{\max}^2$ in order to decrease the influence for decreasing $\gamma_{\max}$.

Therefore, the program has two hyperparameters which have to be treated as constants during linear optimization: the kernel bandwidth and the upper bound $\gamma_{\max}$ on a single contrast. In the following we will study the influence of these hyperparameters on the solutions in some simulations.

3 Simulations

In order to illustrate the role of the two hyperparameters and in particular to show the influence of the (single) contrast limit $\gamma_{\max}$ on the solutions of the above linear program we now consider the Maximum Contrast Classifiers (MCCs) for synthetic data.

The first kind of synthetic data consisted of two 50-point sets in 1D with equal unit distances between neighbours within a single set and with a mutual overlap of 20 points of each set (neighbouring points of different classes have distance 0.5). MCCs with Gaussian kernels were trained according to (6). For kernel bandwidths $\sigma = 10$ and 25 we trained MCCs for a series of increasing $\gamma_{\max}$, which we varied according to $\gamma_{\max} = 0.75^n$, $n = 0, 1, \ldots$. In Fig. 1 the corresponding solutions for the mixing weights α_{i1} are depicted as image rows where neighbouring rows are associated with the weights of neighbouring data points of the first training set. From the evolution of the mixing weights for decreasing $\gamma_{\max}$ we can observe an interesting behaviour for suitable bandwidths: while the probability mass is concentrated on one mixing weight at $\gamma_{\max} = 1$, it is shifted more and more towards the region of class overlap (upper region) for decreasing $\gamma_{\max}$ with only a small number of non-zero weights according to a threshold of 10^{-7}.

For further illustration of this behaviour we trained MCCs with Gaussian kernels of fixed width for different values of $\gamma_{\max}$ on a 2D toy data set. As depicted in Fig. 2 we have only a small number of non-zero mixing weights (encircled symbols). With decreasing $\gamma_{\max}$ the complexity of the decision function varies from a simple line to the final class separating boundary function. Comparable to Fig. 1 for decreasing $\gamma_{\max}$ the major part of the probability mass is concentrated near the decision boundary. Summarizing, the form of the decision boundary depends on both, the kernel bandwidth and $\gamma_{\max}$.

4 Performance Comparison

To illustrate the performance of the proposed classifier, we compare the classification results achieved on three benchmark datasets (*Pima-Indians-Diabetes*,

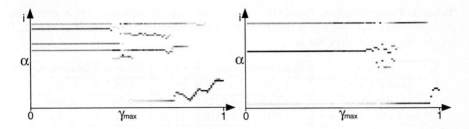

Fig. 1. Evolution of α_{i1} for growing γ_{max} (left: $\sigma=10$, right: $\sigma=25$). Each row i depicts the value of α_{i1} for increasing γ_{max}. Dark pixels correspond to high values for α_{i1}. For the purpose of illustration the values are scaled to the range of $0, .., 255$.

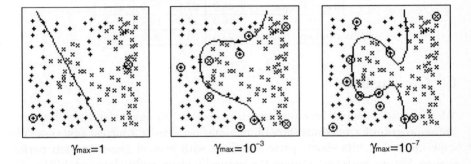

Fig. 2. Maximum Contrast classifiers for different γ_{max} on toy data in 2D with training points from two classes according to symbols + and ×; circles indicate training points with non-zero mixing weights.

Breast-Cancer, Heart) with those of a Support-Vector-Machine (SVM), a RBF-Network (RBF) and a Gaussian Kernel-Density-Classifier (KDC) based on bandwidth optimized kernel density estimators. The reference values for the SVM and RBF are taken from [5]. To guarantee comparable results the setup for the KDC and MCC was exactly the same as described in [5]: Each dataset was normalized to zero mean and standard deviation one and was partitioned 100 times in different pairs of training and test set (ratio of 60%:40%)^2.

Table 1 summarizes the optimal parametrization together with the corresponding averaged classification error and sparseness of the MCC for each dataset. A comparison of the averaged classification error of the four classifiers is given in table 2. While for the Breast-Cancer dataset the sparse solution of the MCC shows a classification error competitive with those of the SVM, it is slightly outperformed on the Heart and Diabetes dataset.

---

^{2} The partitions can be downloaded from http://ida.first.gmd.de/~raetsch/data/bench marks.htm. We thank Gunnar Rätsch for providing the datasets.

Table 1. Parametrization (bandwidth, γ_{max}), averaged classification error with standard deviation and sparseness (fraction of training data with $\alpha_i > 10^{-7}$) of the MCC for the three benchmark datasets.

Dataset	Bandwidth	γ_{max}	Error	Std.dev.	Sparseness
Breast-Cancer	1.333	0.32	25.44	4.55	5.16%
Heart	1.210	0.0742	16.80	3.70	27.35%
Pima Indian Diabetes	1.704	0.2316	24.26	2.14	2.76%

Table 2. Summary of averaged classification errors on the three benchmark datasets for SVM, RBF, KDC and MCC.

Database	SVM	RBF	KDC	MCC
Breast-Cancer	26.0	27.6	26.9	25.4
Heart	16.0	17.1	16.2	16.8
Pima Indian Diabetes	23.5	24.1	26.0	24.3

5   Conclusion

We proposed a density based approach to classifier design which is realized by a linear program for maximization of an empirical density contrast function. The experimental results show sparse solutions with a good generalization performance. Future work will address the naturally given multi-class capability.

References

1. C. Cortes and V. Vapnik. Support-vector networks. *Machine Learning*, 20(3):273–297, 1995.
2. R. O. Duda and P. E. Hart. *Pattern Classification and Scene Analysis*. Wiley, New York, 1973.
3. C.-W. Hsu and C.-J. Lin. A comparison on methods for multi-class support vector machines. *IEEE Transactions on Neural Networks*, 13:415–425, 2002.
4. P. Meinicke and H. Ritter. Quantizing density estimators. In *Advances in Neural Information Processing Systems 14*, 2001.
5. G. R. Rätsch, T. Onoda, and K.-R. Müller. Soft margins for AdaBoost. Technical Report NC-TR-1998–021, University of London, Egham, UK, 1998.
6. B. Ripley. *Pattern Recognition and Neural Networks*. Cambridge University Press, Cambridge, 1996.
7. D. W. Scott. *Multivariate Density Estimation*. Wiley, 1992.
8. V. Vapnik. *The Nature of Statistical Learning Theory*. Springer, New York, 1995.

Puncturing Multi-class Support Vector Machines

Fernando Pérez-Cruz and Antonio Artés-Rodríguez*

University Carlos III, Department of Signal Theory and Communications
Avda. Universidad 30, 28911 Leganés (Madrid) SPAIN
fernandop@ieee.org

Abstract. Non-binary classification has been usually addressed by training several binary classification when using Support Vector Machines (SVMs), because its performance does not degrade compared to the multi-class SVM and it is simpler to train and implement. In this paper we show that the binary classifiers in which the multi-classification relies are not independent from each other and using a puncturing mechanism this dependence can be pruned, obtaining much better multi-classification schemes as shown by the carried out experiments.

1 Introduction

Support Vector Machines (SVMs) have rapidly become one of the most common techniques for solving binary classification problems, due to its ease of implementation and its state-of-the-art results in large number of applications [4]. SVMs have been also used for multiple class problems in which the lables might belong to more than 2 classes. For such problems there are several approaches and none of them seems to be superior to the others. Basically, we can either solve several binary problems [1] or we can directly solve the multiclass problem by extending the SVM binary formulation [6], which it is very hard to implement and train. Among the binary settings, we can describe several approaches: the comparison of each class against all the others [5], known as one-vs-all; the comparison of each class against all the other classes individually [3], known as all-pairs, or the comparison of a subset of classes against the rest of them using error correcting codes [2], this last one presents as particular cases the previous two.

In the paper in which it was unified the binary approaches for multi-class problems [1], they propose that the best way to increase the performance of the overall system (reduce the class' mislabelling) is to increase the number of binary classifiers as much as one can, because being more classifiers to choose from it will make the appearance of mistakes harder. The problem with this approach is that the classifiers can not be regarded as independent from each other, which is a needed condition for error correcting codes to be effective [7], so increasing the number of them might not increase the system performance and it could even degrade it. To obtain a higher performance (lower probability of erroneous decisions) we propose in this paper a puncturing mechanism that will eliminate

* This work has been partially supported by CICYT grant TIC2000-0380-C03-03.

J.R. Dorronsoro (Ed.): ICANN 2002, LNCS 2415, pp. 751–756, 2002.

those classifiers that degrade the performance and obtaining, as a consequence, much less complex multi-classification schemes.

The rest of the paper is outlined as follows. In Section 2, the error correcting codes are reviewed and its application to non-binary problem resolution is presented. In Section 3, we discuss the limitation of error correcting codes for multi-class problems and how we can use puncturing to improve the system performance. Computers experiment with the vowel data set are shown in Section 4. We end this paper in Section 5 with some concluding remarks.

2   Channel Coding for SVMs

Error protection codes are a family of techniques for detecting and correcting channel errors in digital communications. In the general setting, we have several symbols, let say k, to be transmitted through a binary communication channel. Previous to transmission, the symbols have to be matched with a binary string that can be handled by the communication system. The shortest string that can be used to distinguish between k symbols is $\lceil \log_2 k \rceil$, but with a single mistake in the transmitted bits, due to a channel error, it will produce a symbol error. One can assign longer sequences to each symbol expecting that, if done right, the string of bits would present more differences with other symbol's strings and would be able to detect or correct errors produce by the digital communication channel and, as a consequence, the mistake will not be transferred to the symbol itself. The price to pay is that as the length of the string increases the bit error rate does (for a fixed symbol throughput), because the time dedicated to each bit decreases [7]. For each digital communication system there is an optimum between the length of the string and the correcting capabilities which is usually sought when designing the code.

The learning problem from samples for multiple outputs can be described as finding decision rules that separate with the least number of errors a given set of labeled samples $((\mathbf{x}_1, y_1), \ldots, (\mathbf{x}_l, y_l)$ for $\mathbf{x}_i \in \mathbb{R}^d$ and $y_i \in \{1, 2, \ldots, k\})$. If we assign a binary string to each class and train a binary classifier with all the samples for each label of the string, the analogy with the channel communication readily arises. We will slightly modify the binary string to a ternary one, following [1], the entries to these strings will be either ± 1 or 0 indicating, respectively, the label of each class in the binary problem (± 1) or that it is absent from it (0). We show in Figure 1 three different ways in which a four-class problem can be divided into binary classifiers.

In order to assign a class to a new sample, we need to compute the output for all binary classifiers and construct an n-dimensional vector to be compared to each class' string. The class selection can be done using either Hamming distance if the outputs of the classifiers are binary (ternary) or using Euclidean distance if these outputs have not been threshold previously. For the one-vs-all approach, it simplifies to finding the classifier with largest output.

	one-vs-all				all-pairs						H(7,4)						
	C1	C2	C3	C4	C1	C2	C3	C4	C5	C6	C1	C2	C3	C4	C5	C6	C7
Class 1	1	-1	-1	-1	1	1	1	0	0	0	-1	-1	-1	1	-1	1	1
Class 2	-1	1	-1	-1	-1	0	0	1	1	0	-1	-1	1	-1	1	1	-1
Class 3	-1	-1	1	-1	0	-1	0	-1	0	1	-1	1	-1	1	1	-1	-1
Class 4	-1	-1	-1	1	0	0	-1	0	-1	-1	1	-1	1	1	-1	-1	-1

Fig. 1. We show three type of transformations of a four-class problem into binary classifiers. The first on is the on-vs-all approach in which each class takes the positive label once and the rest takes the negative label. The second approach compares all pairs of two classes in each binary classifier. One class takes the positive label the other class the negative label and the rest of the classes are absent from training. The last one is a channel coding approach in which we have used some of the words of weight three of a cyclic Hamming code H(7,4), in which a set of labels is compared to the rest of them.

3 Puncturing

The question that readily arises is how do I choose the strings to obtain the best performance. In [1], the authors propose that the larger the code is the better, because they use codes much longer than the minimum necessary one and the larger ones perform best. But seldom the code with largest number of classifier is the best. The selection of large codewords can be explained by the limiting trade off parameter. In channel coding this parameter is the bit energy (probability of channel errors) and in non-binary classification is the available computational resources, which is a far less limiting factor. So under this viewpoint, one would usually tend to use as many binary classifiers as he/she can handle. But it can be seen that as the number of classifiers increases the performance of the overall system does not always improves and it can even get worse, which does not follow what would be expected from coding theory. This fact can be discussed over the independence of the binary classifiers. Coding theory in order to assert that the symbol error rate decreases with increasing code length needs to assure that the bit error rate in the channel is independent bit by bit. But we cannot assert this point over a set of binary classifiers that have been trained with the same set of samples (with different labels). We cannot expect that with an increasing number of binary classifiers the overall probability of error will be reduced and could even expect an increase in it as some classifiers might present erroneous decisions for symbols in the borderline between its true class and an incorrect decision.

In most error correcting codes we have enough redundancy, so if one of the bits are missing, we can still decode without error the transmitted symbol. This applies as well for the error correcting codes for non-binary classification, for example, if we delete the 5th classifier in the all-pair codes in Figure 1, the classes 2 and 4 are still distinguishable, although there is not a direct comparison between them. This technique is known in channel coding as puncturing and it is widely use for convolutional and block codes. The idea is that one can remove one/several of the redundancy bits that the sequence is still decodable without error, without considering the non-transmitted bits. The puncturing, when used over a communication system, removes part of the redundancy, increasing the energy of the remaining bits[1]. The advantage of this technique is that it can be used to prune the pernicious classifiers, that do to provide an improve in each class determination.

The puncturing mechanism, that we propose to provide a better performance, relies in a iterative pruning of the worst classifier until the best set of binary classifier has been found. We start with the n classifiers and measure the performance of all the combinations of $n-1$ classifiers, keeping the set of classifiers with least number of errors in each iteration. This procedure is repeated until there is not any classifier left and we have n set of binary classfiers with a decreasing number of them to choose from. Then, the best set of classifiers is selected comparing the best performance of all the remaining subset of classifiers.

4   Computer Experiments

We have taken the vowel data set from the UCI Machine Learning Repository for testing the puncturing of SVM binary classifiers for multiple classes problem. We have selected this problem because the best result reported in UCI is a 44% of classification error using nearest neighbour classifier and this data set has been solved using channel codes in [1] and the best achieved result was a 39% using an all-pairs code and binary SVMs with a polynomial kernel of fourth degree. Also, this is one of the hardest multiple class problems because one has to distinguish between 11 different vowels pronounce by 15 people only 6 times. The training and test sets divides the people in 2 sets, respectively, with 8 and 7 in each one. The difficult part of this task is to extract the characteristics that separates the different vowels instead of separating the speakers. In order to train the classfier we used the 10 features and whether the speaker was a woman or a man, all the features where preprocess to present zero mean and unit standard deviation.

We have use this data set and have used as code: BCH(7,4), BCH(15,4), BCH(31,5), BCH(63,7), one-vs-all and all-pairs. The number of binary classifiers is k for the one-vs-all, $\binom{k}{2}$ for all-pairs and for the BCH codes is the first number in its descriptor. The BCH can be easily generated as shown in [7] and for selecting the string for each vowel we have selected randomly among all the possible codewords, deleting first the all zero (-1) codeword. We have used an RBF kernel, $\kappa(\mathbf{x}_i, \mathbf{x}_j) = \exp\left(-\frac{\|\mathbf{x}_i - \mathbf{x}_j\|^2}{2\sigma^2}\right)$, and have use 8-fold cross validation (each time one of the training person was left a side) to select the hyperparameters σ and C. The hyper-parameters have been chosen in two different ways. First by using the same σ and C for all the and seeking for the best global behaviour (Equal). Second, by treating each binary classifier independently and setting the best σ and C for each one of them and afterwards

---

[1] This is necessary because the tools to construct good block codes do not allow to use an arbitrary length for them.

their outputs were joined together to form the output of the multi-class problem (Individual). We show in Table 1 the results for these two approaches, for both the outputs of the binary SVMs were considered to be a real number and we used Euclidean distance to compare between the n-dimensional output word and the valid codewords.

Table 1. We show the error rate for the six selected codes and the two ways the hyperparameters where chosen.

	on-vs-all	all pairs	BCH(7,4)	BCH(15,4)	BCH(31,5)	BCH(63,7)
Equal	43.5%	39.4%	44.4%	48.1%	45.5%	42.0%
Individual	42.4%	39.2%	42.4%	39.8%	41.8%	41.6%

From this results one can notice that it is a more wiser approach to train each classifier independently and afterwards join their outputs together than seeking for the best parameters all together, because the tuning of each codeword individually get the best classifier for each set of labels. This can be explained if one sees each binary classifier as a hard-decisor, in this case the output of one classifier is not affected by rest of them and the best will be to train each one individually and also it is widely known that using soft outputs leads to better error correcting performance [7]. It can bee seen that the Achieved results with the all-pairs and BCH(15,4) codes are as good as the best result reported in [1], where it was proposed the use of channel coding for non-binary classification with SVMs.

We have punctured the 6 proposed multi-class classifiers using the procedure described in the previous section for both ways of hyperparameters selection. We use the same 8-fold cross validation scheme and we show the result over the 7 people that were neither use for hyperparameter nor for classifier selection in Table 2.

Table 2. The error rate for the six punctured codes are shown. The number in brackets indicates the remaining binary classifiers.

	on-vs-all	all pairs	BCH(7,4)	BCH(15,4)	BCH(31,5)	BCH(63,7)
Equal	43.5% (11)	29.0% (29)	42.4% (6)	40.5% (8)	37.0% (9)	36.4% (30)
Individual	42.4% (11)	24.9% (20)	42.4% (7)	37.2% (10)	34.4% (10)	32.4% (14)

All the classifiers have provided a much better result after they were punctured. This is due to the binary classifiers with worst classification properties were pruned and the errors they induced were now corrected by the remaining classifers. The best result in this case is given by the all-pair approach in which we have a reduction of the test error over 14 percentual points and have reduced the number of binary classifiers from 55 to 21. We show in Table 3 the 21 pairs of classifiers and the used hyper-parameters.

In this paper we propose a novel approach for SVM regression in which a hyper-spherical insensitive zone is defined around the estimate, following the Euclidean norm the regression estimation problem defines [5]. This hyper-spherical insensitive zone will allow us to equally treat every sample, being them penalized by the same factor if they lie outside this insensitive zone. The SVM for multi-regression problem can not be solved using quadratic programming, and thus an iterative procedure similar to the one in [2] is used.

The rest of the paper is outlined as follows. In section 2, we state the multi-dimensional regression estimation problem and we formulate it as a constrained optimization problem. In section 3, we show that the Multidimensional Support Vector Regressor (MSVR) can be solved using an iterative procedure. We present some simulations results in Section 4. The paper ends with some concluding remarks in Section 5.

2   Multi-dimensional Support Vector Regressor

The SVR for regression estimation defines an insensitive zone around the estimate, if a sample is further than ε than the estimate is penalized by C and is not otherwise [4]. If the 1D SVR is applied over each direction of the multidimensional problem, we will have samples as close as ε that will be penalized and samples as far as $\sqrt{k}\varepsilon$, in Euclidean norm, that will not. Moreover, the samples will not suffer an equal penalty, because if the sample is further than ε in every direction will be penalized by $k \times C$ and only by C if it is only in one direction. These limitations are illustrated for a 2D problem in Figure 1a, in which the square represents the insensitive zone for the two 1D-SVR. The ideal result would be the one pictured by the circle in Figure 1a, in which the samples inside will not be penalized and the samples further that ε_2 from $r(\mathbf{x})$ will.

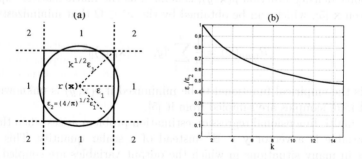

Fig. 1. A hyper-cubic and a hyper-spherical insensitive zones are shown in (a) for a bi-dimensional regression estimation problem, in which they have equal areas. The areas marked with a 1 will be penalized just once and the samples in 2 will be penalized twice. In (b) we show the ratio between ε_1 and ε_2, respectively, the values of the hyper-cubic and hyper-spherical insensitive zone, as a function of k, in order to assure equal hyper-volume.

The Multi-dimensional SVR (MSVR) with a hyper-spherical insensitive zone can then be similarly stated as the regular SVR. The MSVR, given a labeled training data set $((\mathbf{x}_i, y_i) \; \forall i = 1, \ldots, n$, where $\mathbf{x}_i \in \mathbb{R}^d$ and $y_i \in \mathbb{R})$ and a nonlinear transformation to a higher dimensional space $(\phi(\cdot), \mathbb{R}^d \xrightarrow{\phi(\cdot)} \mathbb{R}^H$ and $d \leq H)$, solves:

$$\min_{\mathbf{w}^j, b^j, \xi_i} \sum_{j=1}^{k} \|\mathbf{w}^j\|^2 + C \sum_{i=1}^{n} \xi_i \tag{2}$$

subject to

$$\|\mathbf{y}_i - \mathbf{W}\phi(\mathbf{x}_i) - \mathbf{b}\|^2 \leq \varepsilon + \xi_i \qquad \forall i = 1, \ldots, n \tag{3}$$

$$\xi_i \geq 0 \qquad \forall i = 1, \ldots, n \tag{4}$$

where $\mathbf{W} = [\mathbf{w}^1, \ldots, \mathbf{w}^k]^T$ and $\mathbf{b} = [b^1, \ldots, b^k]^T$ define the k-dimensional linear regressor in the H-dimensional feature space.

3   Resolution of the MSVR

In order to solve the MSVR, we first introduce the constrains (3) and (4) into (2), making use of the Lagrange multipliers, leading to the minimization of

$$L_P = \sum_{j=1}^{k} \|\mathbf{w}^j\|^2 + C \sum_{i=1}^{n} \xi_i - \sum_{i=1}^{n} \alpha_i(\varepsilon + \xi_i - \|\mathbf{y}_i - \mathbf{W}\phi(\mathbf{x}_i) - \mathbf{b}\|^2) - \sum_{i=1}^{n} \mu_i \xi_i \tag{5}$$

with respect to $\mathbf{w}^j$, b^j and ξ_i and its maximization with respect the Lagrange multipliers, α_i and μ_i. The solution to this problem is given by the Karush-Kuhn-Tuker Theorem [1], that states the following conditions (KKT conditions); namely (3), (4), and

$$\frac{\partial L_P}{\partial \mathbf{w}^j} = 2\mathbf{w}^j - 2\boldsymbol{\Phi}^T \mathbf{D}_\alpha [\mathbf{y}^j - \boldsymbol{\Phi}\mathbf{w}^j - \mathbf{1}b^j] = 0 \qquad \forall j = 1, \ldots, k \tag{6}$$

$$\frac{\partial L_P}{\partial b^j} = \boldsymbol{\alpha}^T [\mathbf{y}^j - \boldsymbol{\Phi}\mathbf{w}^j - \mathbf{1}b^j] = 0 \qquad \forall j = 1, \ldots, k \tag{7}$$

$$\frac{\partial L_P}{\partial \xi_i} = C - \mu_i - \alpha_i = 0 \qquad \forall i = 1, \ldots, n \tag{8}$$

$$\mu_i, \alpha_i \geq 0 \qquad \forall i = 1, \ldots, n \tag{9}$$

$$\alpha_i \{\varepsilon + \xi_i - \|\mathbf{y}_i - \mathbf{W}\phi(\mathbf{x}_i) - \mathbf{b}\|^2\} = 0 \qquad \forall i = 1, \ldots, n \tag{10}$$

$$\mu_i \xi_i = 0 \qquad \forall i = 1, \ldots, n \tag{11}$$

where $\boldsymbol{\Phi} = [\phi(\mathbf{x}_1), \ldots, \phi(\mathbf{x}_n)]^T$, $\boldsymbol{\alpha} = [\alpha_1, \alpha_2, \ldots, \alpha_n]^T$, $\mathbf{y}^j = [y_{1j}, y_{2j}, \ldots, y_{nj}]^T$ and $\mathbf{y}_i = [y_{i1}, y_{i2}, \ldots, y_{ik}]^T$. The MSVR can not be solved as the regular SVR, since the KKT conditions can not be replaced into the functional in (5) to obtain a quadratic problem linearly restricted. We can instead use an iterative algorithm, similar to the one proposed to solve the SVR in [2]. This procedure first obtains $\mathbf{w}$ and b, considering α_i fixed and, secondly, recomputes α_i from the achieved solution in the first step. The algorithm iterates until the SVR solution is reached.

In order to apply this algorithm for the MSVR, we have joined together (6) and (7) to obtain $\mathbf{w}^j$ and b^j from the linear system in which α_i must remain unchanged.

$$\begin{bmatrix} \boldsymbol{\Phi}^T \mathbf{D}_\alpha \boldsymbol{\Phi} + \mathbf{I} & \boldsymbol{\Phi}^T \alpha \\ \alpha^T \boldsymbol{\Phi} & \alpha^T \mathbf{1} \end{bmatrix} \begin{bmatrix} \mathbf{w}^j \\ b^j \end{bmatrix} = \begin{bmatrix} \boldsymbol{\Phi}^T \mathbf{D}_\alpha \mathbf{y}^j \\ \alpha^T \mathbf{y}^j \end{bmatrix} \tag{12}$$

Once, we have solved (12), we recompute the α_i values fulfilling the remaining KKT conditions, so when the algorithm stops, we will be at the MSVR solution. The value of α_i can be computed from the KKT conditions (8)-(11):

$$\alpha_i = \begin{cases} 0, & e_i < 0 \\ C, & e_i > 0 \end{cases} \tag{13}$$

where we have defined the error over each sample as $e_i = \varepsilon - \|\mathbf{y}_i - \mathbf{W}\phi(\mathbf{x}_i) - \mathbf{b}\|^2$.

3.1   Reproducing Kernels Hilbert Space

The system in (12) needs to know the nonlinear mapping $\phi(\cdot)$, which usually is not. In order to work with kernels $(\kappa(\mathbf{x}_i, \mathbf{x}_j) = \phi^T(\mathbf{x}_i)\phi(\mathbf{x}_j))$ as the SVR does, we use the Representer theorem [3], which states that the best solution can be expressed as a linear combination of the training samples in the feature space:

$$\mathbf{w}^j = \sum_{i=1}^{n} \beta_i^j \phi(\mathbf{x}_i) = \boldsymbol{\Phi}^T \beta^j \tag{14}$$

If we reorder (6), collecting the terms depending on $\mathbf{w}^j$ in one side of the equation and replace (14) in it and multiply it by the pseudo-inverse of $\boldsymbol{\Phi}\mathbf{D}_\alpha$, we obtain:

$$(\mathbf{D}_\alpha \boldsymbol{\Phi}\boldsymbol{\Phi}^T \mathbf{D}_\alpha)^{-1}\mathbf{D}_\alpha[\boldsymbol{\Phi}\boldsymbol{\Phi}^T \mathbf{D}_\alpha \boldsymbol{\Phi}\boldsymbol{\Phi}^T + \boldsymbol{\Phi}\boldsymbol{\Phi}^T]\beta^j = [\mathbf{y}^j - \mathbf{1}b^j] \tag{15}$$

which can be simplified to:

$$[\mathbf{H} + \mathbf{D}_\alpha^{-1}]\beta^j = [\mathbf{y}^j - \mathbf{1}b^j] \tag{16}$$

where we have defined the kernel matrix as $\mathbf{H} = \boldsymbol{\Phi}\boldsymbol{\Phi}^T$. We can now join together (16) and (7) to form a linear system of equations that it does not depend on the nonlinear mapping ϕ, just on its kernel.

$$\begin{bmatrix} \mathbf{H} + \mathbf{D}_\alpha^{-1} & \mathbf{1} \\ \alpha^T \mathbf{H} & \alpha^T \mathbf{1} \end{bmatrix} \begin{bmatrix} \beta^j \\ b^j \end{bmatrix} = \begin{bmatrix} \mathbf{y} \\ \alpha^T \mathbf{y}^j \end{bmatrix} \tag{17}$$

The resolution of the system in (17) for each set of β^j is equal for every dimension of $\mathbf{y}$, except for the independent term. So its resolution will be much simpler than that of k one-dimensional SVR.

4 Examples

We first illustrate the MSVR with a synthetic example and afterwards we will solve a multi-dimensional estimation with real data. For the synthetic problem, the input vector is 2D and each component were generated independently from a Gaussian distribution function of zero mean and standard deviation equal to 10. The output vector $\mathbf{y} \in \mathbb{R}^5$, described by: $y_{i1} = 4\sin(x_{i1}) - 2sinc(x_{i2}) + 5 + n_{i1}$, $y_{i2} = 3\sin(x_{i1}) - 3\cos(x_{i2}) + 2 + n_{i2}$, $y_{i3} = -5sinc(x_{i1}) + 4\sin(x_{i2}) + 1 + n_{i3}$, $y_{i4} = -2\sin(x_{i1}) - 4\sin(x_{i2}) - 5 + n_{i4}$ and $y_{i5} = 4sinc(x_{i1}) - 2\cos(x_{i2}) - 3 + n_{i5}$, where n_{ij} are random Gaussian variables with zero mean and standard deviation equal to 0.5. We have solved this problem for several values of ε_2 and have run 10 independent trials. We have used an RBF kernel with $\sigma = 0.5$ and $C = 10$. The achieved results are shown in Figure 2 in which we have plotted the Mean Square Error (MSE), $MSE = \sum_{i=1}^{N_t s} \|\mathbf{y}_i - \mathbf{W}\phi(\mathbf{x}_i) - \mathbf{b}\|^2$, and the fraction of Support Vectors for the different values of ε_2. The value of ε_1, used for the 1D SVR, is $\varepsilon_1 = 0.6974 \times \varepsilon_2$, which is the one that gives the same hyper-volume for both insensitive zones.

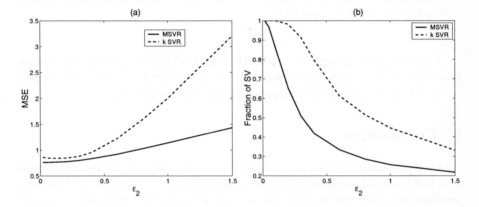

Fig. 2. The MSE is shown for the MSVR and for the k SVR in (a). In (b) we have plotted the fraction of support vector.

The MSVR gives a lower prediction error and fewer support vectors than using 5 SVRs. Furthermore, the MSVR is very robust when the value of ε is not correctly set as illustrated in Figure 2.

The second experiment deals with the simultaneous one-step ahead prediction of four clinical variables commonly used in therapeutic drug monitoring of patients who have undergone kidney transplantation. Eighteen lagged inputs formed by anthropometrical, clinical and biochemical data of patients are used to predict Cyclosporine blood concentration, the daily dosage, alkaline phosphates and the creatinine clearance. The data was split into a training set (665 samples) and a validation set (442 samples). We have used an RBF kernel with a $\sigma = 13.4$ and a $C = 35$, which were obtained using 8-fold CV over the training set. In

Figure 3 we show the MSE and the fraction of support vectors for the MSVR and 4 one-dimensional SVR. The input data were preprocessed to present zero mean and unit standard deviation.

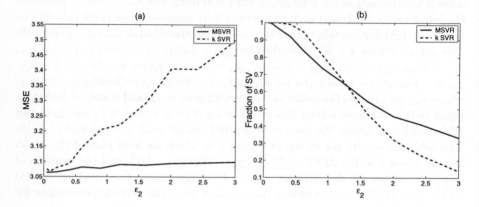

Fig. 3. The MSE is shown for the MSVR and for the k SVR in (a). In (b) we have plotted the fraction of support vectors.

As in the synthetic data we can see that the MSVR does not only achieves the best solution as ε varies but its solution is more robust and with fewer support vectors than using the SVR independently for each component.

5   Conclusions

We have presented a novel approach for solving multiple output regression problems in which we have defined a hyper-sphere insensitivity zone, that allow us to penalize only once the samples that are not placed inside the insentivity zone. The MSVR does not only achieves better predictions but it is also more robust than the use of SVR independently for each varaible to be estimated.

References

1. R. Fletcher. *Practical Methods of Optimization*. Wiley, Second Ed., 1987.
2. F. Pérez-Cruz, A. Navia-Vázquez, P. L. Alarcón-Diana, and A. Artés-Rodríguez. An IRWLS procedure for SVR. In *Proc. of the EUSIPCO'00*, Finland, Sept 2000.
3. B. Schoelkopf and A. Smola. *Learning with kernels*. M.I.T. Press, 2001.
4. V. N. Vapnik, S. Golowich, and A. Smola. Support vector method for function approximation, regression estimation, and signal processing. In M. Mozer, M. Jordan, and T. Petsche, eds, *NIPS 1997*, pages 169–184.
5. V. N. Vapnik. *Statistical Learning Theory*. Wiley, 1998.

Detecting the Number of Clusters Using a Support Vector Machine Approach

Javier M. Moguerza[1], Alberto Muñoz[2], and Manuel Martín-Merino[3]

[1] University Rey Juan Carlos, c/ Tulipán s/n, 28933 Móstoles, Spain
j.moguerza@escet.urjc.es
[2] University Carlos III of Madrid, c/ Madrid 126, 28903 Getafe, Spain
albmun@est-econ.uc3m.es
[3] University Pontificia of Salamanca, c/ Compañía 5, 37002 Salamanca, Spain
mmerino@ieee.org

Abstract. In this work we introduce a new methodology to determine the number of clusters in a data set. We use a hierarchical approach that builds upon the use of any given (user-defined) clustering algorithm to produce a decision tree that returns the number of clusters. The decision rule takes advantage of the ability of Support Vector Machines (SVM) to detect both density gaps and high-density regions in data sets. The method has been successfuly applied on a variety of artificial and real data sets, covering a broad range of structures, group densities, data dimensionalities and number of groups.

1 Introduction

It is a well known fact that there is no universal clustering algorithm valid for any given data set: some algorithms are well suited for non-overlapping clusters, some other for mixtures of normal densities, and so on (see for instance, [6]). Knowledge of the number of clusters is crucial for many algorithms, but procedures for determining it depend on the statistical structure of the data set, frequently unknown, and often these procedures are unsuccessful [7].

Recently, the Support Vector (SV) approach [4,15] has been used to detect clusters [2], where a fine SV algorithm is presented to compute a set of contours which enclose the data points and correspond to cluster boundaries. However, as the authors state, there is a degradation in the algorithm performance as dimension rises, related to the number of support vectors required to describe cluster contours.

In our approach, rather than trying to directly detect clusters with SVMs, we will show how the SV approach can help any given clustering algorithm to detect the 'correct' number of clusters, and we will also point out how SVMs can help to validate the suitability of the particular clustering method used.

To this aim, notice that for clustered data sets, there are two (often complementary) facts that are worth considering: (a) clusters are usually surrounded by

J.R. Dorronsoro (Ed.): ICANN 2002, LNCS 2415, pp. 763–768, 2002.
© Springer-Verlag Berlin Heidelberg 2002

density gaps (valleys in the probability distribution), and (b) clusters frequently correspond to high density regions in the data space.

We will show in next section how SVMs can be used to detect fact (a). Regarding fact (b), it is equivalent to detect the support of the data distribution, and has been solved in [12,14]. The detection of these two situations will serve as a basis to build a strategy to solve the problem under discussion.

2   Cluster Detection Using a Hierarchical SVM-Based Approach

Our approach dwells in recursively testing the presence of clusters in the data set under consideration. The mildest assumption when trying to detect structure is to take as null hypothesis the existence of just two clusters in the data set. If this null hypothesis is not rejected, the process is iterated on the two clusters obtained until no more structure is detected. To induce a binary partition in the data set we will use a predefined clustering algorithm. Remarks on the particular choice of the clustering procedure will be made at the end of this section.

Being overlap a problem for any clustering algorithm, we will work with the support of the data set distribution, calculated using a one-class SVM with a RBF-kernel (see [12] for details). The points outside the support correspond to low density regions, which match with overlap zones among clusters. This pre-process increases the probability of correct operation for any cluster algorithm. The next step is to train a two-class SVM on the classification induced by the clustering algorithm. The choice of a linear kernel in the SVM formulation is reasonable for convex clusters, and non-convex data sets can be put in this context using RBF-kernels. For the sake of geometric interpretation, we will concentrate on the linear case in this very first work.

After SVM training, we will have a number of SV points and a separating hyperplane to detect density gaps. Note that in this context, a low error in the trained SVM indicates that the separating hyperplane is located at a density gap, that is, between the two clusters. We could try to use the proportion $\#SV/n$ (n=size of the data set) to detect this situation (see [5] pag. 100-101). But, as [2] points out, counting support vectors causes trouble in high dimensional settings, and this could be aggravated if the data size is small. Instead, we use in this paper the average of the distances from the support vectors to the SVM hyperplane to detect density gaps. Let d_{SV} be this average distance. In case of separable clusters, no bounded support vectors will be available, and the value $2/\|w\|$ can be used to estimate the distance between the two groups, where $\mathbf{w}$ is the weight vector corresponding to the classification rule induced by the function $< \mathbf{w} \cdot \mathbf{x} > +b$ (b constant) (see [5]).

Next, for each of the two clusters we calculate the average of the distances from the points in the support of the data set to their k nearest neighbours (k small). A small k guarantees that only distances between points belonging to the same cluster are considered. Regarding to k, the robustness of the procedure is ensured by the fact that only data in the support are considered. Lets call d_{s1}

and d_{s2} these distances respectively. The use of an average distance for each cluster makes the procedure robust against the case of groups with different density distributions. Now, if $d_{SV} \leq \min(d_{s1}, d_{s2})$, it implies that density at the boundary is not smaller than density in the support of the group distributions. Therefore, there is no cluster structure in the data set. On the other hand, if $\min(d_{s1}, d_{s2}) < d_{SV}$, then there are at least two regions in the data set with unequal densities. Therefore, two clusters are detected. This procedure is recursively applied to each cluster until no more clusters are found. The sketch of the algorithm is:

Calculate the support S of the data set using a one-class SVM
Number of clusters = ncluster(S)

where *ncluster* is the following procedure:

Procedure ncluster(S)

1. *Induce on S a two-class classification problem using a clustering algorithm. Denote S_1 and S_2 the clusters obtained.*
2. *Train a SVM on the classification problem resulting from step 1. Let H be the separating hyperplane.*
3. *Calculate the quantities:*

   $d_{SV} = $ *average of distances from the SV to H*

   $d_{s1} = $ *average of distances from S_1 points to their nearest neighbours*

   $d_{s2} = $ *average of distances from S_2 points to their nearest neighbours*

4. *IF $d_{SV} \leq \min(d_{s1}, d_{s2})$ THEN $nc = 1$*
   ELSE $nc = ncluster(S_1) + ncluster(S_2)$
5. *return(nc)*

Regarding the particular clustering algorithm, in this work we use EM-type clustering algorithms like k-means [3] and mixture-based algorithms [1,10]. These algorithms are well suited for symmetrical clusters and mixtures of normal distributions, a quite usual problem, and many high-dimensional problems can be reduced to this setting via random projections [8]. Note that the partition induced on the support of the data set can be used to induce a partition on the full data set using the SVM hyperplane, nearest neighbours techniques or any other particular method suited for the chosen clustering algorithm. In relation to the data size, it is worth considering that for small data sets could be convenient to omit the support density estimation step, to avoid the risk of an excesive reduction in the data size. Regarding computing time and scalability, it is directly related to the clustering algorithm used.

3   Experiments

Next we show the performance of the proposed algorithm on two well-known data sets. For all the experiments, the chosen number of neighbours involved in distance calculations has been set to $k = 4$. Experiments using different values of k (always $k < 10$) reported similar results. We begin with the well-known Iris data set, a standard benchmark in statistical and pattern recognition literature.

3.1   The Iris Data Set

The data set is given by a 150×4 matrix of four measurements taken on iris flowers. It is well known that there are three clusters. One of the clusters (made up by setosa flowers) is linearly separable from the other two. The remaining two clusters (made up by virginica and versicolor flowers respectively) show severe overlap. This set is clustered in [2] using a SVM approach, after being reduced into the two dimensional subspace formed by the first two principal components. Our analysis is performed with the original data set in $\mathbb{R}^4$. The clustering algorithm chosen in procedure *ncluster* is k-means. Similar results are obtained when using hierarchical algorithms. Notice that the procedure *ncluster* does not need a particular set of cluster centers, but simply the cluster label for each data point, in order to train the SVM in step 2. The result of the algorithm is shown in Figure 1. Notice that the main aim of this paper is to detect the number of clusters, and therefore no confusion matrices are given. Of course, the particular clustering algorithm chosen can benefit from the procedure's output, as stated at the end of section 2. In this example, for which we have the true labels, we obtain 66 support data points, 26 from setosa class, perfectly isolated, and 40 belonging to the remaining two clases. For these 66 data points, the k-means algorithm misclassifies just three data points from the versicolor class into the virginica group. Regarding Figure 1, $\min d_{si}$ stands for $\min(d_{s1}, d_{s2})$. Notice that the large $d_{SV} = 1.33$ obtained in the first step accurately reflects the fact that setosa data points are well separated from the rest.

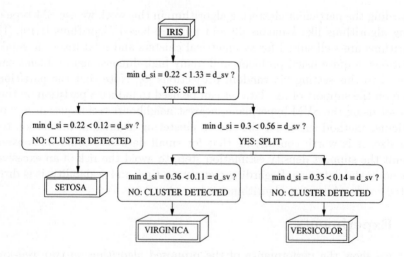

Fig. 1. Detection of clusters in iris data set using procedure ncluster

3.2   Breast Cancer Data

This data set is given by a 699×9 matrix of clinical measures taken on breast cancer patients [9]. The real class distribution is 65.5% benign, and 34.5% malignant. This data set is interesting for three reasons: relatively high dimension, overlap and different density distributions for each group.

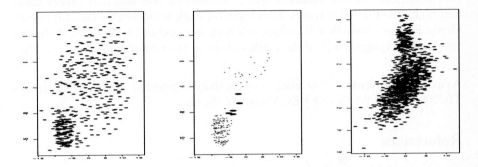

Fig. 2. Left: Sammon mapping of breast cancer data. Middle: Same mapping, only projected support points and support vectors. Right: A data set with severe overlap.

We process this data set in $\mathbb{R}^9$ with *ncluster* using the k-means algorithm to induce the binary partitions. In the first split, we obtain $d_{SV} = 11.9$, $d_{s1} = 2.25$ and $d_{s2} = 12$. As $\min(2.25, 12) < d_{SV} = 11.9$ the procedure accepts there are at least two clusters in the data set. Notice that the density is quite different for each cloud, and the sample sizes are not balanced. The next step applies the same system to each detected group. In the first group, $d_{SV} = 9.5 < \min\{d_{si}\} = \min(12.1, 12.3)$, so this group cannot be splitted. For the second group, $d_{SV} = 1.6 < \min\{d_{si}\} = \min(2.3, 2.3)$, so we attain the same conclusion. Therefore, the procedure detects two groups, and this is the correct assumption for this data set. Left side of Figure 2 shows a Sammon mapping [11] of the breast cancer data set, and gives us an idea of the structure and overlap of the set. The central picture of this figure shows the projections of points corresponding to the support of the distribution and also the projection of the (five) SVM support vectors. Note that it is just a two-dimensional projection.

We have applied the procedure *ncluster* to a variety of artificial data sets, using different number of groups, densities (mixtures of normal and uniform distributions), and different covariance structures. These experiments will be detailed in a more extensive paper. Here is only worth considering that the procedure can deal with severe overlap, due to the first step of support distribution extraction. For instance, the right picture in Figure 2 shows a mixture of two bivariate normals with strong overlap. Our procedure gave values: $d_{s1} = 0.2$, $d_{s2} = 0.25$ and $d_{SV} = 1$, therefore detecting there were two groups.

4   Conclusions

In this paper a new methodology to fix the number of clusters in a data set has been presented. The method proceeds recursively, splitting the data set under study in two groups, and relies on the use of two types of SVMs to decide when to accept or reject a given binary partition. The procedure can work together with any given clustering method and is suitable to deal with a variety of baffling situations, such as different density structures, non balanced cluster sizes, and high dimensional settings. Experimental work corroborates the statements above. Further research will include real high dimensional data, like text-mining sets, and thorough study of the methodology in the case of non linear kernels.

Acknowledgments. This work was partially supported by DGICYT grants TIC2001-0175-C03-03 and BEC2000-0167 (Spain).

References

1. J.D. Banfield and A.E. Raftery. *Model-based Gaussian and Non-Gaussian Clustering.* Biometrics, 49 (3) (1993) 803–822.
2. A. Ben-Hur, D. Horn, H.T. Siegelmann and V. Vapnik. *Support Vector Clustering.* Journal of Machine Learning Research 2:125–137, 2001.
3. L. Bottou and Y. Bengio. *Convergence Properties of the K-Means Algorithms.* Advances in Neural Information Processing Systems 7, pp. 585–592, 1995.
4. C. Cortes and V. Vapnik. *Support Vector Networks.* Machine Learning, 20:1–25, 1995.
5. N. Cristianini and J. Shawe-Taylor. *An Introduction to Support Vector Machines.* Cambridge University Press, 2000.
6. B. Everitt. *Cluster Analysis. Second Edition.* Gower, Hampshire, 1986.
7. A. Hardy. *On the Number of Clusters.* Computational Statistics & Data Analysis 23 (1996) 83–96.
8. S. Kaski. *Dimensionality Reduction by Random Mapping.* Proc. Int. Joint Conf. on Neural Networks, volume 1, pages 413–418, 1998.
9. O.L. Mangasarian and W.H. Wolberg. *Cancer diagnosis via linear programming* SIAM News,Volume 23, Number 5, 1990, 1–18.
10. C. Fraley and A.E. Raftery. *How Many Clusters? Which Clustering Method? Answers Via Model-Based Cluster Analysis.* The Computer Journal – Vol. 41, 801–823.
11. J.W. Sammon. *A nonlinear mapping for data structure analysis.* IEEE Transactions on Computers 18 (5) (1969) 401–409.
12. B. Schölkpf, J.C. Platt, J. Shawe-Taylor, A.J. Smola and R.C. Williamson. *Estimating the Support of a High Dimensional Distribution.* Proceedings of the Anual Conference on Neural Information Systems 1999 (NIPS*99). MIT Press, 2000.
13. B.W. Silverman. *Density Estimation for Statistics and Data Analysis.* Chapman & Hall, London, 1990.
14. D.M.J. Tax and R.P.W. Duin. *Support Vector Domain Description.* Pattern Recognition Letters, 20:1991–1999, 1999.
15. V. Vapnik. *The Nature of Statistical Learning Theory.* Springer, New York, 1995.

Mixtures of Probabilistic PCAs and Fisher Kernels for Word and Document Modeling

George Siolas and Florence d'Alché-Buc

Laboratoire d'Informatique, Université de Paris 6,
8, rue du capitaine Scott, 75015 Paris, France
{georges.siolas, florence.dalche}@lip6.fr

Abstract. We present a generative model for constructing continuous word representations using mixtures of probabilistic PCAs. Applied to co-occurrence data, the model performs word clustering and allows the visualization of each cluster in a reduced space. In combination with a simple document model, it permits the definition of low-dimensional Fisher scores which are used as document features. We investigate the models' potential through kernel-based methods using the corresponding Fisher kernels.

1   Introduction

One of the most challenging questions in Text Mining is to construct document representations that improve the performance and interpretability of Information Retrieval systems. Although most statistical learning methods use flat vector representations, recent results show that richer representations incorporating word semantics seem to have interesting features. Two approaches have been explored so far: methods that integrate *a priori* linguistic knowledge into document representations [7,8] and data-driven methods that generate relevant document representations from data [3,1]. Our work falls into the second category as it addresses the problem by the joint use of a generative document model and Fisher scores. Feature extraction from generative models has been introduced by Jaakola and Haussler [5] in bio-informatics for discriminative purposes, while Hofmann [4] followed a similar approach in a document retrieval context. Our work's novelty is that we do not consider words as mere tokens anymore but as continuous vectors generated by a hidden model. We show that mixtures of probabilistic PCAs for modeling produce meaningful word visualizations alongside a word clustering mechanism that captures semantic aspects of the vocabulary. We end up by pointing that Fisher scores, calculated from a simple document model incorporating the word model, provide a method for performing effective kernel-based classification from low dimensional features.

2   Word Modeling Using Probabilistic PCAs

In Machine Learning, document representation is commonly based on the assumption that words are symbolic tokens, yet linguists argue that continuous

J.R. Dorronsoro (Ed.): ICANN 2002, LNCS 2415, pp. 769–774, 2002.

word representations are more adequate for capturing semantics [6]. The standard form of a document is a vector containing the indexing words as attributes, with a measure of their (weighted) in-document frequencies as respective values. For an entire collection of documents, the gathering of all such "document-word" vectors gives the co-occurrence matrix of the data $\mathbf{N} = [n(i, j)]$ where $n(i, j)$ indicates the frequency of word j in document i. As a starting point for constructing continuous word representations we will use the transposed co-occurrence matrix $\mathbf{N}^{\mathbf{T}}$: we consider documents as attributes of words. This matrix transposition does not affect the informational content of the problem but provides the conceptual means of an extended word modeling. In an Information Extraction perspective, our goal is then twofold: infer a parametric model capable both of performing a semantic clustering of the vocabulary and of providing a low-dimensional projection of the relative word positioning. We will show next that modeling with mixtures of probabilistic PCAs introduced by Tipping & Bishop [9] is well suited for these purposes.

If $\mathbf{N}^{\mathbf{T}}$ is considered to be an $N \times d$ matrix, each word $\mathbf{w_j}$, $j = 1..N$ is a vector of dimension d. Our goal is to learn a generative model and to compute the probability densities $p(\mathbf{w_j})$ corresponding to the observed words $\mathbf{w_j}$. Probabilistic PCA is a factor analysis model and as such it considers that observed data is generated from a set of latent variables $\mathbf{z_j}$, $j = 1..N$ of lower dimensionality q through a parametric linear function:

$$\mathbf{w_j} = \mathbf{Q}\mathbf{z_j} + \boldsymbol{\mu} + \varepsilon, \ \boldsymbol{\mu}\text{: mean of } \mathbf{w}, \ \varepsilon\text{: noise} \tag{1}$$

The latent variables are independent and follow the Gaussian distribution, $\mathbf{z} \sim N(0, \mathbf{I})$. The noise model is also a Gaussian, $\varepsilon \sim N(0, \boldsymbol{\Psi})$, with $\boldsymbol{\Psi}$ diagonal. Consequently, the distribution of the data vectors will also be a Gaussian, $\mathbf{w} \sim N(\boldsymbol{\mu}, \mathbf{C})$. In the particular case of probabilistic PCA, if we choose $\boldsymbol{\Psi} = \sigma^2 \mathbf{I}$ it can be shown that the columns of matrix $\mathbf{Q}$ span the same sub-space as the one of conventional PCA. The probability distribution of the data becomes:

$$p(\mathbf{w_j}) = (2\pi)^{-\frac{d}{2}} |\mathbf{C}|^{-\frac{1}{2}} \exp\{-\frac{1}{2\sigma^2}(\mathbf{w_j} - \boldsymbol{\mu})^{\mathbf{T}}\mathbf{C}^{-1}(\mathbf{w_j} - \boldsymbol{\mu})\} \tag{2}$$

where the covariance $\mathbf{C}$ is equal to $\sigma^2 \mathbf{I} + \mathbf{Q}\mathbf{Q}^{\mathbf{T}}$ and parameters $\mathbf{Q}$ and σ^2 are calculated iteratively by the EM algorithm. However, as there is no reason why $\mathbf{w}$ should behave like a Gaussian in real world, we will use the more flexible form of a mixture model. Data is now being generated not by a singe latent variable distribution but by many, each with its own parameter set:

$$p(\mathbf{w_j}) = \sum_{c=1}^{M} \pi_c p(\mathbf{w_j}|c : \theta_c), \ \theta_c = \{\boldsymbol{\mu}_c, \mathbf{Q}_c, \sigma_c^2\} \tag{3}$$

with the constraint $\sum \pi_c = 1$. In a manner analogue to standard PCA, data is projected in an optimal way in lower dimension but this time in M different spaces instead of a single one.

The system is capable of learning the values of parameters $\{\pi_c, \boldsymbol{\mu}_c, \mathbf{Q}_c, \sigma_c^2\}$ from data but for maximum flexibility we would also like it to automatically decide how many PCAs there will be in the final model. This is achieved by using a slight variation of the Split-Merge EM algorithm [10]. The basic idea is that each time the system converges to a certain likelihood level, we split the projection c_1 that has the highest π_c among the projections (the PCA which is responsible for generating the most of the words) in two new projections c_2 and c_3 with parameters $\pi_{c_2} = \pi_{c_3} = \frac{\pi_{c_1}}{2}$ and $\theta_{c_2} = \theta_{c_1} + \epsilon$, $\theta_{c_3} = \theta_{c_1} + \epsilon$ where ϵ is some small random noise. After the splitting, the model parameters are recalculated and if the log-likelihood continues to rise the newly added projections are kept.

Once the word model becomes available it can be integrated into many different document models, ranging from the naive "bag-of-words" to elaborate sequential models, with the final goal being to extract Fisher scores which will be used as document representations. By choosing for our first tests the "bag-of-words" approach, we make the assumption that word appearances inside a document are not correlated in any way. The log-probability of observing the N-word document $\mathcal{D}_i$ of length $|\mathcal{D}_i|$ is then:

$$\mathcal{L}(\mathcal{D}_i) = \log p(\mathcal{D}_i) = \frac{1}{|\mathcal{D}_i|} \sum_{j=1}^{N} n(i,j) p(\mathbf{w_j}) \tag{4}$$

where $p(\mathbf{w_j})$ is the word probability provided by equation 3. While the independence assumption is not realistic and its mostly used for simplicity reasons, one can easily see that the suggested document model still captures word dependencies through the clustering effect of the mixture model.

3   Fisher Scores Extraction

Learning the model consists on finding the set of parameters that minimize the error function or equivalently that maximize the log-likelihood function $\mathcal{L}(\mathcal{D})$. A parametric probabilistic model defines a Riemannian manifold in which we can maximize $\mathcal{L}(\mathcal{D})$ by following the direction of the steepest ascent according to the natural gradient $\mathbf{G} = E[\nabla_\theta \mathcal{L}(\mathcal{D}) \nabla_\theta^T \mathcal{L}(\mathcal{D})]$. This quantity, the Fischer Information matrix, fully describes the geometry of the parameter space. Knowing the metrics of the model space is necessary to bridge the gap between the parametric model and many I.R. applications (such as text classification) which require a measure of similarity -and hence defined metrics- between data objects. This analysis has been further developed in [5] where the data is mapped from Euclidean space to parameter space through the natural transformation $\Phi_\mathcal{D} = \mathbf{G}^{-1} \mathbf{U}_\mathcal{D}$ with $\mathbf{U}_\mathcal{D} = \nabla_\theta \mathcal{L}(\mathcal{D})$. This choice is justified by the fact that a kernel based on this mapping (named Fisher kernel) is proportional to the inner product of the parameter space:

$$K(\mathcal{D}_1, \mathcal{D}_2) \propto \Phi_{\mathcal{D}_1}^\mathbf{T} \mathbf{G} \Phi_{\mathcal{D}_2} = \mathbf{U}_{\mathcal{D}_1}^\mathbf{T} \mathbf{G}^{-1} \mathbf{U}_{\mathcal{D}_2} \tag{5}$$

Further simplification of the Fisher kernel to $K(\mathcal{D}_1, \mathcal{D}_2) \propto \mathbf{U}_{\mathcal{D}_1}^\mathbf{T} \mathbf{U}_{\mathcal{D}_2}$ shows that we can construct similarity measures solely from Fisher scores $\mathbf{U}_\mathcal{D}$. Hence, it is possible to build various kernel functions from equation 4 by computing the Fisher score for each parameter $\{\pi_c, \boldsymbol{\mu}_c, \mathbf{W_c}, \sigma_c^2\}$. Parameter π_c is particularly interesting for clustering and classification purposes. Its corresponding Fisher score for a particular document $\mathcal{D}_i$ takes the form of an M-sized vector $\mathbf{U}_{\mathcal{D}_i}^{\pi}$ where the value of attribute c will be equal to:

$$\mathbf{U}_{\mathcal{D}_i}^{\pi}(c) = \frac{\partial \mathcal{L}(\mathcal{D}_i)}{\partial \pi_c} = \frac{1}{|\mathcal{D}_i|} \sum_{j=1}^{N} n(i,j) \frac{p(\mathbf{w_j}|c)}{p(\mathbf{w_j})} \tag{6}$$

The interpretation of equation 6 is that the more words contained in document $\mathcal{D}_i$ are generated by the same PCA projection c, the highest the value of its corresponding attribute in the Fisher score vector will be. It further reveals that the mixture model has an underlying clustering mechanism similar in operation to Latent Semantic Indexing (LSI) [2] which is not a probabilistic model but a dimensionality reduction method based on the idea that words are generated from a reduced number of latent concepts. The analogy is even clearer if we compare our model to the probabilistic version of LSI presented in [3]. Probabilistic LSI associates each element of the co-occurrence matrix to a class of latent variables $\mathbf{z} \in Z = \{\mathbf{z_1}, \ldots, \mathbf{z_M}\}$. The probability of observing word $\mathbf{w_j}$ in a given document $\mathcal{D}_i$ is given by:

$$p(\mathbf{w_j}|\mathcal{D}_i) = \sum_{\mathbf{z} \in Z} p(\mathbf{w_j}|\mathbf{z})p(\mathbf{z}|\mathcal{D}_i) \tag{7}$$

Assuming that the latent variables $\mathbf{z}$ are replaced by probabilistic PCA projections c, equation 7 becomes $p(\mathbf{w_j}|c)p(c|\mathcal{D}_i)$ where the second probability function is expressed by the Fisher score of equation 6. In consequence, by using the π_c Fisher score we can produce clustering kernel functions in a manner similar to probabilistic LSI. The use of probabilistic PCAs for word modeling goes further in the sense of I.E., providing not only a rich parametric model but also means of optimal data visualization. Moreover, dimensionality reduction by probabilistic PCA does not require hard data clustering, in contrary to an approach where Gaussian mixture modeling is followed by conventional PCA.

4   Numerical Simulation

Numerical simulations were performed with the "20 newsgroups" dataset. This dataset contains 20,000 Usenet postings equally distributed among 20 newsgroups. The newsgroups' subjects are computers, religion, politics, science and recreational issues. We split the original multi-class problem in 20 binary problems by forming 20 new datasets each containing the totality of a newsgroup (the positive class) and an equal number of randomly chosen messages belonging to all the other newsgroups (the negative class). To simplify and be compatible with previous work [8] we use a relatively small 200-word index. In each binary problem we put 2/3 of the 1,000 positive and 2/3 of the 1,000 negative documents in the training set and the rest in the test set.

In order to visualize the formation of the projections we choose the dimension q of the latent variables to be equal to 2. From an I.E. perspective, the quality of the data clustering produced by the mixture model is very important. Figure 1 shows an example projection from the "comp.windows.x" sub-problem.

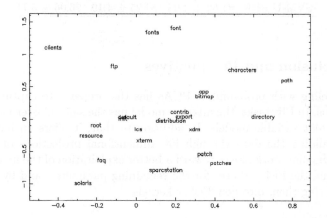

Fig. 1. Word projection from the "comp.windows.x" sub-problem

Considering the projection as a whole, we notice that the words selected are indeed related to the newsgroup's subject. Moreover, the proximity of some singular/plural word forms proves that relative word positioning inside the projection is meaningful. Since we observed similar results for the rest of the newsgroups, we infer that the proposed model efficiently captures word semantics.

Our next test was to evaluate the potential of the Fisher score document representation $\mathbf{U}_{\mathcal{D}}^{\pi}$ when it is used in discriminative tasks. We compared the performance of three kernel-based classifiers: a k-nearest-neighbors (kNN) classifier with an inner product kernel using the original 200-attribute document representation; the same kNN classifier but this time using the reduced attribute Fisher score representation $\mathbf{U}_{\mathcal{D}}^{\pi}$ and a radial basis function (RBF) as kernel (kNN-U-RBF); finally, a Support Vector Machine (SVM-U-RBF) with the same reduced representation and RB function as in the previous case. During learning we allowed models with at most 10 clusters for each binary problem and only the best performing model was held. The mean number of clusters and consequently the mean size of the $\mathbf{U}_{\mathcal{D}}^{\pi}$ vector was equal to 8. Table 1 shows the average performance and standard deviation of the classifiers in the 20 binary problems. The first and most encouraging conclusion is that the reduced features representation gives comparable results in terms of accuracy and precision to those of the full 200-attributes representation. The second is that the clustering approach seems to trade the high recall of the full features representation for superior precision, a fact more visible in the case of the SVMs. Finally, the significantly smaller deviations observed for the $\mathbf{U}_{\mathcal{D}}^{\pi}$-based methods reflect a more equable performance which is better appreciated in difficult classification tasks.

Table 1. Comparative classification results

Kernel	Accuracy	Precision	Recall
kNN	81.33 ± 6.56	77.66 ± 8.66	91.05 ± 7.90
kNN-U-RBF	78.39 ± 3.82	78.20 ± 4.51	81.25 ± 6.28
SVM-U-RBF	81.62 ± 4.08	84.25 ± 4.19	76.06 ± 5.71

5   Conclusion and Perspectives

Word modeling with probabilistic PCAs has the property to capture semantic relationships. In I.R. tasks, the mixture model has the same clustering capability as other latent variable models. In addition to that, it offers an intuitive way of understanding the data through low dimensional projections of the latent concepts. Ongoing work moves toward a better exploitation of the model's power by extracting the Fisher scores for the remaining parameters and by combining and integrating them into new Fisher kernels.

References

[1] Nello Cristianini, Huma Lodhi, and John Shawe-Taylor. Latent semantic kernels for feature selection. Technical Report NC-TR-2000-080, Royal Holloway, University of London, June 2000.

[2] S. Deerwester, S.T. Dumais, G.W. Furnas, T.K. Landauer, and R. Harshman. Indexing by latent semantic analysis. *Journal of the American Society for Information Science*, 1990.

[3] Thomas Hofmann. Probabilistic latent semantic analysis. *Uncertainity in Artificial Intelligence*, 1999.

[4] Thomas Hofmann. Learning the similarity of documents: An information-geometric approach to document retrieval and categorization. *Advances in Neural Information Processing Systems*, 12:914–920, 2000.

[5] Tommi S. Jaakola and David Haussler. Exploiting generative models in discriminative classifiers. In *Advances in Neural Information Processing Systems*. MIT Press, 1998.

[6] Sabine Ploux and Bernard Victorri. Construction d'espaces semantiques à l'aide de dictionnaires informatisés des synonymes. *Traitement Automatique des Langues*, 39(1):161–182, 1998.

[7] Sam Scott and Stan Matwin. Text classification using wordnet hypernyms. In *Proceedings of the COLING-ACL'98 Workshop*, pages 45–52. University of Montreal, Quebec, Canada, 1998.

[8] George Siolas and Florence d'Alché-Buc. Support vector machines based on a semantic kernel for text categorization. In *Proceedings of the IEEE-INNS-ENNS International Joint Conference on Neural Networks*, 2000.

[9] Michael Tipping and Christopher Bishop. Probabilistic principal component analysis. *Journal of the Royal Statistical Society*, 61(B):611–622, 1999.

[10] N. Ueda, R. Nakano, Z. Ghahramani, and G. Hinton. Smem algorithm for mixture models. In *Advances in Neural Information Processing Systems*. MIT Press, 1999.

Part V

Robotics and Control

methods [4] such as temporal-difference (TD) learning, Q-learning and actor-critic methods are not suited for training the CPG, which is a special case of recurrent neural networks. In order to deal with RL for CPG, we propose a new RL method called the CPG-actor-critic method. We applied this method to the biped robot used in [3]. The computer simulation showed that our RL method was able to train the CPG so that the biped robot walk stably.

2   CPG-Actor-Critic

In this paper, we study reinforcement learning (RL) for rhythmic movement using a central pattern generator (CPG), depicted in Fig. 1(a). The equation of motion for a physical system such as for a robot is formally written as

$$\dot{\mathbf{x}} = F(\mathbf{x}, \tau), \tag{1}$$

where $\mathbf{x}$ and $\dot{\mathbf{x}}$ denotes the physical state and its time derivative, respectively. The control signal (torque) from the CPG is denoted by τ. $F(\mathbf{x}, \tau)$ represents the vector field of the system dynamics. The equation of motion for a CPG, which is a special case of recurrent neural networks, is given by

$$c_i \dot{\nu}_i = -\nu_i + I_i, \quad y_i = G_i(\nu_i), \quad I_i = \sum_j W_{ij} y_j + I_i^{bias} + I_i^{ext}, \tag{2}$$

$$I_i^{ext} = \sum_k A_{ik} S_k, \quad \tau_\alpha = \sum_i T_{\alpha i} y_i,$$

where ν_i, y_i and c_i represent the i-th neuron state variable, its output and its time constant, respectively. I_i, I_i^{bias} and I_i^{ext} represent the total input, the bias input and the external input to the i-th neuron, respectively. The external input I_i^{ext} is a weighted sum of the sensory feedback signal S_k with the connection weight A_{ik}. The output function $G_i(\cdot)$ is assumed to be a sigmoidal function or some threshold function. The connection weight from the j-th neuron to the i-th neuron is denoted by W_{ij}. The control signal to the physical system τ_α is a weighted sum of the CPG neuron output y_i with the connection weight $T_{\alpha i}$.

When we try to apply the actor-critic method to the CPG controller system, there are several difficulties. In this method, (Fig. 1(b)), the CPG controller becomes the actor. Since the CPG output depends on its current internal state, the future reward also depends on the internal state of the CPG. Therefore, the reinforcement learning task in this method becomes a partially observable Markov decision problem (POMDP), which is much more difficult than a MDP, even if the physical system state is fully observable. Another source of difficulty comes from the fact that the CPG is a recurrent neural network. The standard actor-critic algorithm is not suited for training recurrent neural networks.

In order to overcome these difficulties, we propose a new RL method that is called the CPG-actor-critic method (Fig. 1(d)). In this method, the CPG is divided into two modules, i.e., the basic CPG and the actor. Correspondingly,

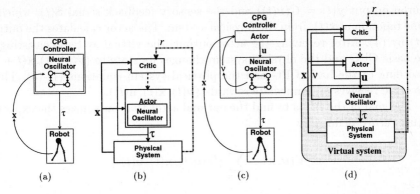

Fig. 1. (a),(c): CPG controller. (b): Actor-critic. (d): CPG-actor-critic method.

input to the CPG neuron, I_i, is divided into two parts:

$$I_i = I_i^{fix} + u_i, \quad I_i^{fix} = \sum_j W_{ij}^{fix} y_j + I_i^{bias}, \qquad (3)$$

$$u_i = \sum_j W_{ij}^{act} y_j + \sum_k A_{ik} S_k. \qquad (4)$$

The basic CPG is defined by the fixed connection weight W_{ij}^{fix} and receives the input, **u**, from the actor. The actor in this method is a linear controller and receives the basic CPG neuron output **y** and the physical system feedback signal **S**. The actor calculates its output to the basic CPG, **u**, by (4). This CPG-actor-critic method has dual aspects. From the control viewpoint, the CPG controller consists of the basic CPG and the actor (i.e., $W_{ij} = W_{ij}^{fix} + W_{ij}^{act}$) as shown in Fig. 1(c). From the RL viewpoint, the actor sends the virtual control signal, **u**, to the virtual system, which consists of the basic CPG and the physical system as shown in Fig. 1(d). Since the actor in this method is a linear controller, training of the actor can be done by using the gradient ascent method described below.

The critic receives the virtual system state, namely, it receives both the physical state and the basic CPG state, and predicts the future reward for the current state. When the physical system state is fully observable, the RL problem in this CPG-actor-critic method becomes a MDP. Consequently, we can avoid the difficulties mentioned earlier.

3 Learning Algorithm

The RL algorithm for the CPG-actor-critic method is explained in this section. For explanatory simplicity, we assume the discrete time notation, that is, it is assumed that the differential equations, (1) and (2), are discretized by an appropriate method.

The current virtual system state at time t is represented by the physical state $\mathbf{x}(t)$ and the basic CPG state $\boldsymbol{\nu}(t)$. The actor receives the basic CPG neuron output $\mathbf{y}(t) = G(\boldsymbol{\nu}(t))$ and the sensory feedback signal $\mathbf{S}(t)$, which is some function of $\mathbf{x}(t)$, from the physical system. The actor calculates the output $\mathbf{u}(t)$ by (4). After receiving the actor output, the virtual system consisting of the basic CPG and the physical system changes its state to $(\boldsymbol{\nu}(t+1), \mathbf{x}(t+1))$ according to the basic CPG and the physical system equations (1)-(3). Then, the critic receives an immediate reward $r(\boldsymbol{\nu}(t), \mathbf{x}(t), \mathbf{u}(t))$.

The goal of the RL is to find the optimal actor that maximizes the expected future return defined by

$$V(\boldsymbol{\nu}, \mathbf{x}) \equiv \sum_{t=0}^{\infty} \gamma^t r(\boldsymbol{\nu}(t), \mathbf{x}(t), \mathbf{u}(t)), \tag{5}$$

where $\nu(0) = \nu$ and $\mathbf{x}(0) = \mathbf{x}$ are assumed. $\gamma (0 < \gamma \leq 1)$ is a discounted factor. The action-value function $Q(\boldsymbol{\nu}, \mathbf{x}, \mathbf{u})$, which is called the Q-function, is defined by

$$Q(\boldsymbol{\nu}, \mathbf{x}, \mathbf{u}) = r(\boldsymbol{\nu}, \mathbf{x}, \mathbf{u}) + \gamma V(\boldsymbol{\nu}(t+1), \mathbf{x}(t+1)), \tag{6}$$

where $\nu(t) = \nu$, $\mathbf{x}(t) = \mathbf{x}$ and $\mathbf{u}(t) = \mathbf{u}$ are assumed. The Q-function, $Q(\boldsymbol{\nu}, \mathbf{x}, \mathbf{u})$, indicates the expected future return for the current state and action $(\boldsymbol{\nu}, \mathbf{x}, \mathbf{u})$, when the current actor is used for the subsequent states. From the definition, the Q-function should satisfy the following consistency condition:

$$Q(\boldsymbol{\nu}(t), \mathbf{x}(t), \mathbf{u}(t)) = r(\boldsymbol{\nu}(t), \mathbf{x}(t), \mathbf{u}(t)) + \gamma Q(\boldsymbol{\nu}(t+1), \mathbf{x}(t+1), \mathbf{u}(t+1)). \tag{7}$$

The critic in the CPG-actor-critic method approximates the Q-function based on (7) as in the Sarsa algorithm [4]. As a function approximator for the critic, we employ the normalized Gaussian neural network (NGnet). The efficient on-line EM algorithm for the NGnet has been derived in [5]. The unit deletion and creation mechanism have also been incorporated in this on-line EM algorithm. This algorithm is suited for learning dynamic environment [6] such as the CPG-actor-critic method where the target Q-function for the critic depends on the current actor that is also modified according to the critic prediction.

The learning process for the CPG-actor-critic method is as follows. In the first phase, the state of the basic CPG and the physical system, $(\boldsymbol{\nu}(t), \mathbf{x}(t))$, is updated according to (1)-(4) by using the fixed actor for a given period of time. In this period, the critic receives the immediate reward $r(\boldsymbol{\nu}(t), \mathbf{x}(t), \mathbf{u}(t))$ and is trained in on-line fashion such that the consistency condition (7) is satisfied. The input to the critic is the system state and the action $(\boldsymbol{\nu}(t), \mathbf{x}(t), \mathbf{u}(t))$. The teacher output for the critic is given by the right hand side of (7). Then, the model parameters of the NGnet are adjusted by using the on-line EM algorithm. The system state and the action trajectory $\{(\boldsymbol{\nu}(t), \mathbf{x}(t), \mathbf{u}(t)) | t = 0, 1, \ldots, t_{max}\}$ are saved.

In the second phase, the actor is trained using the saved trajectory. In order to increase the Q-function value (i.e., the expected future return), the weight parameters of the actor are updated by the gradient ascent method: $\Delta \psi \propto \frac{\partial Q(\boldsymbol{\nu}, \mathbf{x}, \mathbf{u})}{\partial \mathbf{u}} \frac{\partial \mathbf{u}}{\partial \psi}$, where ψ represents the weight parameters of the actor, W_{ij}^{act} or A_{ik}.

The above procedure defines one episode. The reinforcement learning proceeds by repeating these episodes.

4 Experiment

In the following, we apply the CPG-actor-critic method to the biped robot studied in [3]. The robot consists of five links in the sagittal plane as shown in Fig. 2(a). Each leg consists of two links and link-1 represents the remainder of the body, which is given as a point mass. The robot state is represented by $\mathbf{x} = (x_1, \dot{x}_1, h_1, \dot{h}_1, \theta_2, \dot{\theta}_2, \dots, \theta_5, \dot{\theta}_5)$, where x_1 and h_1 represent the horizontal and vertical coordinates of link-1, respectively. $\theta_i (i = 2, \dots, 5)$ represents the angle of link-i from the vertical axis.

The structure of the CPG that controls the biped robot is also adopted from [3] and shown in Fig. 2(b). There are six oscillators interacting with each other. An oscillator consists of two mutually inhibiting parent neurons, ν_{2i-1} and ν_{2i} ($i = 1, \dots, 6$), whose output functions are the threshold function defined by $y_i = G_i(\nu_i) = \max(0, \nu_i), (i = 1, \dots, 12)$. Each parent neuron, $\nu_i (i = 1, \dots, 12)$, has a daughter neuron ν_{i+12} whose output function is the identity function. The daughter neuron is solely connected to its parent neuron with excite-inhibit mutual connections.

The torque $\tau_i (i = 1, \dots, 6)$, which is applied to the i-th joint (Fig. 2(a)), is calculated from the CPG neuron output: $\tau_i = -T_i^I y_{2i-1} + T_i^E y_{2i}$ for $i = 1, \dots, 4$, and $\tau_i = (-T_i^I y_{2i-1} + T_i^E y_{2i}) \Xi_{i-1}$ for $i = 5, 6$. Ξ_i is an indicator function of the link-i ($i = 4, 5$), i.e., $\Xi_i = 1$ (0) when the foot link-i touches (is off) the ground. The values of T_i^I and T_i^E are given in [3]. In the following experiment, all connection weights between CPG neurons are fixed to the value given in [3], namely, $W_{ij}^{act} \equiv 0$ is assumed in (4). We also assume a specific form of the sensory feedback to the CPG as in [3]:

$$
\begin{aligned}
I_1^{ext} &= a_1 S_1 - a_2 S_2 + a_3 S_3 + a_4 S_6, & I_9^{ext} &= -a_6 S_3 - a_7 S_4 - a_8 S_7, \\
I_3^{ext} &= a_1 S_2 - a_2 S_1 + a_3 S_4 + a_4 S_5, & I_{11}^{ext} &= -a_6 S_4 - a_7 S_3 - a_8 S_8, \qquad (8) \\
I_5^{ext} &= a_5 S_4, \quad I_7^{ext} = a_5 S_3, & I_{2i}^{ext} &= -I_{2i-1}^{ext} \quad \text{for } i = 1, \dots, 6,
\end{aligned}
$$

where $\mathbf{S} = \{\theta_2, \theta_3, \theta_4 \Xi_4, \theta_5 \Xi_5, \Xi_4, \Xi_5, \dot{\theta}_4 \Xi_4, \dot{\theta}_5 \Xi_5\}$.

The main task of the reinforcement learning for the biped robot is to adjust the sensory feedback connections $\{a_i | i = 1, \dots, 8\}$ in (8) such that the robot is able to walk stably. In order to encourage the robot to walk, an immediate reward $r(\boldsymbol{\nu}(t), \mathbf{x}(t), \mathbf{u}(t))$ is given by $\tilde{r}(\mathbf{x}(t+1))$,

$$
\tilde{r}(\mathbf{x}) = 0.5 r_{height}(\mathbf{x}) + 0.02 r_{speed}(\mathbf{x}).
$$

The reward $r_{height}(\mathbf{x}) = h_1 - 0.8 - \min(h_4, h_5)$, encourages the hip position to stay high, where h_i ($i = 4, 5$) denotes the foot height of the link-i. The reward $r_{speed}(\mathbf{x}) = \max(-1, \min(\dot{x}_1, 1))$ encourages the robot to move toward the right direction. The maximum period length of one episode was 5 sec. If the robot tumbled before 5 sec, the episode was terminated at that point. The initial value of the sensory feedback connections were set to small random values.

The learning curve for the total return in one episode is shown in Fig. 2(c). At the beginning of the learning, the robot soon fell down (Fig. 2(d)). After about 4000 episodes, the robot started to run in place (Fig. 2(e)). After about 5800 episodes, the robot started to walk (Fig. 2(f)). The final sensory feedback weight values after learning were $\mathbf{a} = \{1.00, -0.31, 1.00, 0.65, 0.11, 0.25, 1.00, 0.49\}$ which are quite different from the hand-tuned values used in [3], $\mathbf{a} = \{0.15, 0.10, 0.15, 0.15, 0.30, 0.15, 0.30, 0.15\}$.

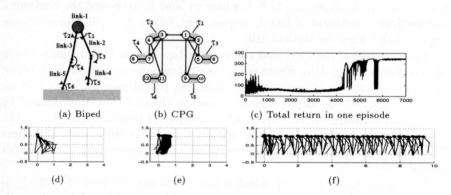

Fig. 2. (a): Biped robot. (b): CPG. (c): Learning curve. (d): Robot gait pattern before learning. (e): after 5500 episodes. (f): after 7000 episodes.

5   Discussion

In this paper, we proposed a new RL method called the CPG-actor-critic method and applied it to biped locomotion. Although the result was successful, the learning process was rather unstable. It is necessary to increase the stability of our RL method in the future. In the current simulation, the CPG internal weights were fixed and only sensory feedback connections were adjusted by RL. It remains for future study to adjust the CPG internal weights by the CPG-actor-critic method.

References

1. Hirai, K., et al. 1998. *Proceedings of ICRA* 2:1321-1326
2. Grillner, S., Wallen, P. and Brodin, L. 1991. *Annu. Rev. Neurosci.* 14:169-199
3. Taga, G., Yamaguchi, Y., and Shimizu, H. 1991. *Biol. Cybern.* 65:147-159
4. Sutton, R. S. and Barto, A. G. 1998. *Reinforcement learning.* MIT Press
5. Sato, M. and Ishii, S. 2000. *Neural Computation* 12:407-432
6. Sato, M. and Ishii, S. 1999. *NIPS 11*, 1052-1058
7. Morimoto J. and Doya K. 2001. *Robot. Auton. Syst.*, 36:37-51.

Dynamical Neural Schmitt Trigger for Robot Control

Martin Hülse and Frank Pasemann

Fraunhofer Institute for Autonomous Intelligent Systems (AIS)
Schloss Birlinghoven, 53754 Sankt Augustin, Germany
{martin.huelse, frank.pasemann}@ais.fraunhofer.de
http://www.ais.fraunhofer.de/INDY

Abstract. Structure and function of a small but effective neural network controlling the behavior of an autonomous miniatur robot is analyzed. The controller was developed with the help of an evolutionary algorithm, and it uses recurrent connectivity structure allowing non-trivial dynamical effects. The interplay of three different hysteresis elements leading to a skilled behavior of the robot in challenging environments is explicitly discussed.

1 Introduction

A modular neuro-dynamics approach to behavior control of autonomous systems starts with the basic assumption that the particular abilities of these systems are based on non-trivial internal dynamical features which are provided by neural systems with recurrent connectivity structure. As has been argued elsewhere [2], there are of course many difficult problems with such an approach. These are, for instance, related to questions like what type of dynamics and what type of recurrent structure to use for the generation of a successful behavior.

To tackle these difficulties we use an evolutionary algorithm, called ENS^3 (evolution of neural systems by stochastic synthesis), for the structural development of neural networks, which optimizes parameters with respect to a given fitness function at the same time. This evolutionary algorithm was successfully applied to benchmark control tasks [7] and to robot control tasks [8]. After having generated several examples of effective neuro-controllers, we analyze the dynamics of the resulting neural networks, the relation to the underlying connectivity structure, the relevance of specific dynamical properties for the resulting behavior or control strategy and look for differences in their performance, like differences in robustness, for example. In addition, kind of lesion techniques are used and parameters values can be varied by hand to study details of a structure-function relationship.

In this paper we want to report a simple mechanism, called a *dynamical neural Schmitt Trigger*. The context in which this mechanism became active is the following: We evolved neuro-controllers for Khepera robots [3] which should be able to move continuously (exploration behavior) in a given environment cluttered with obstacles (obstacle avoidance) [9], [4]. The robot could use eight

J.R. Dorronsoro (Ed.): ICANN 2002, LNCS 2415, pp. 783–788, 2002.
© Springer-Verlag Berlin Heidelberg 2002

proximity (infrared) sensors, 6 in front and 2 in the rear, and they where driven by two motors. There were many evolved networks solving the task, larger ones using 3 or more internal neurons, but also a few with no internal neurons at all. Noteworthy is the fact that most of the effective controllers used recurrent connections for their two output neurons (compare [8]). There are some situations in this setting, which usual robot controllers have difficulties to handle. These are situations where the robot drives into sharp corners or runs into dead ends. Then the robots usually just come to a rest (Braitenberg-like controllers [1]) or they start to oscillate left-right-left. We observed some robots which in these situations quickly turned around at large angles and then moved out of such an "unpleasant situation".

When analyzing the corresponding networks, we realized that their output units were connected and that one or both output neurons had positive self-connections. To be sure that the underlying dynamical feature is exactly that of a hysteresis phenomenon, observed for instance for single neurons with super-critical positive self-connection [5], we reduced the number of controller inputs to only two, and used the ENS^3 algorithm to generate appropriate networks for this more demanding setup.

2   Evolving a Neuro-controller

In the following experiment the mean value of the three left proximity sensors, respectively, the three right proximity sensors was calculated and used as inputs $Inp1$ and $Inp2$ for the neuro-controller. The two proximity sensors at the rear were not used. The initial neural structure for this experiment has only two input and two output neurons. The input neurons are simply buffers and the output neurons 3 and 4 are of additive type with sigmoidal transfer function $tanh$, so that the motors can turn forward and backward. Bias terms are set to zero. The fitness function used for the evaluation of the controllers says: For a given time T go straight ahead as long and as fast as possible. This is coded in terms of the two network output signals $Out3$, $Out4$ as follows:

$$F := \sum_{t=1}^{T} (M_0(t) + M_1(t))(2 - |Out4(t) - Out3(t)|), \tag{1}$$

with $0 < M_0$, $M_1 < 1$ defined by $M_0 := max\{0, Out3\}$, $M_1 := max\{0, Out4\}$. There is also a stopping condition: If the robot collides before T time steps the evaluation of the network stops and the value of the current performance is taken as the maximum performance of the individual.

Figure 1(a) shows one of the resulting networks which generates a very successful robot behavior. Both output neurons of this controller have positive self-connections, $w_{33} = 3.19$ and $w_{44} = 1.46$, respectively. Furthermore, one can find inhibitory connections $w_{34} = -7.42$ and $w_{43} = -2.94$ between the two output neurons establishing a feedback loop. In figure 1(b) a typical path of the simulated robot is plotted. It shows both, obstacle avoidance and exploration

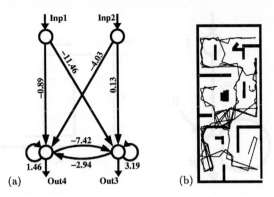

Fig. 1. (a) A small evolved neurocontroller with only two input neurons. (b) Typical paths of a simulated robot controlled by this network.

behavior. The behavior of the physical robot controlled by this network is comparable to that of the simulated one. Especially it is observed that the robots leave sharp corners as well as dead ends as it is indicated, for instance, in the upper right corner of the environment in figure 1(b).

For the considered small network with only two inputs it is easy to relate explicite input configurations with typical situations during the interaction of the robot with its environment. Furthermore, it is possible to simulate and visualize the whole dynamics of this network and to relate special dynamics to the observable behavior of the robot. Thus, we will be able to precisely explain how the recurrent network structure creates this well skilled robot behavior. With this aim in view, we summarize what will be called

3   A Dynamical Neural Schmitt Trigger Module

The *dynamical neural Schmitt Trigger module* consist of a single additive neuron with sigmoidal transfer function $tanh$ and excitatory self-connection w_s, as shown in figure 2. Let θ denote a fixed internal bias or a stationary input to this neuron. For given parameter values of θ and w_s the fixed point equation for its discrete-time dynamics reads

$$x^* = \theta + w_s \tanh(x^*) , \quad x^* \in \mathbf{R}. \tag{2}$$

The stability condition for a fixed point x^* is given by $w_s \tanh'(x) < 1$; i.e., we are looking for bifurcation points in (θ, w_s)-parameter space, for which the condition

$$w_s \tanh'(x) = 1 \tag{3}$$

is satisfied; i.e., parameter values for which a fixed point gets unstable. For this equation to hold, the self-connection must be strong enough, i.e. $w_s > 1$, since $0 < \tanh'(x) < 1$. Furthermore $\tanh'$ is symmetric, i.e. there will exist two fixed

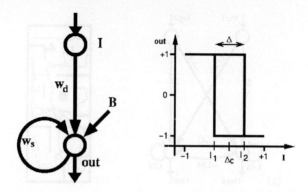

Fig. 2. Basic setup for a dynamical Schmitt trigger module with a buffered input I, weight w_d, excitatory self-connection w_s, and a modulating input B.

points $x_{1,2}^*$, with $x_1^* = -x_2^*$ satisfying condition (3). Thus, there will be two critical θ-values

$$\theta_{1,2}^* = \pm x^* \mp w_s \tanh(x^*),$$

and for the corresponding *hysteresis interval* Δ^* with center at $\Delta_c^* = 0$, we get $\Delta^* = |\theta_1^* - \theta_2^*| = 2|\theta_1^*|$.

Using the identities $\tanh'(x) = (1 - \tanh^2(x))$ and $\tanh^{-1} = \frac{1}{2}\ln\frac{1+y}{1-y}$ for fixed self-connections w_s the critical θ-values $\theta_{1,2}^*$ are calculated by solving equation (3) for x. For Δ^* we obtain

$$\Delta^* = \left| \ln\left[\frac{1+\alpha}{1-\alpha}\right] - 2 \cdot w_s \cdot \alpha \right|, \quad \alpha := \sqrt{1 - \frac{1}{w_s}}, \quad w_s > 1. \quad (4)$$

If we now feed this neuron by an external input I with weighted connection w_d then in this input space we will observe the characteristic jumps at values $I_{1,2} = \frac{\theta_{1,2}}{w_d}$. That is in input space the hysteresis interval with center at zero has length $\Delta = \frac{\Delta^*}{w_d}$. If, in addition, we have a "slowly" varying input B modulating the neuron, then the center Δ_c of the hysteresis interval Δ for the "fast" input signal I is shifted dynamically by B, i.e. $\Delta_c = -\frac{B}{w_d}$, and jumps will occur at

$$I_{1,2} = \frac{\theta_{1,2} - B}{w_d}, \quad w_d \neq 0, \quad B \in \mathbf{R}. \quad (5)$$

4   Network Dynamics and Robot Behavior

Both output neurons of the considered network (figure 1(a)) have "super-critical" self-connections. Therefore two hysteresis effects should cooperate during the control actions. But, in addition, there is a third hysteresis phenomenon involved which is associated to the 2-loop (w_{34}, w_{43}) between the output neurons. Such purely inhibitory 2-loops are known to have parameter domains where two stable fixed points co-exist with a period-2 attractor [6]. A complete overview about

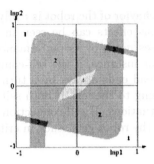

Fig. 3. Dynamical properties of the controller output configuration (compare text).

the dynamical properties of the output configuration is given in figure 3 where one can find four different domains in the $(Inp1, Inp2)$-space. For input values in domain 1 (white) there exists only one fixed point attractor, in domain 2 (light grey) there are two fixed point attractors, in domain 3 (dark grey) there are three of them, and in the eye shaped domain 4 we have two fixed point attractors co-existing with a period-2 orbit.

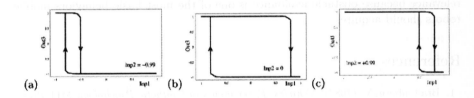

Fig. 4. Hysteresis domain of $Inp1$ for output neuron $Out3$ of the network (fig. 1(a)) with $Inp2$ fixed.

To relate this picture to the behavior of the controlled robot one may have a look at the corresponding hysteresis diagrams plotted in figure 4. These diagrams for the output $Out3$ of neuron 3 are related to typical situations of an obstacle avoidance task. For instance figure 4(a) represents a turn to the right if there is an obstacle on the left (right input $Inp2$ is low: no obstacle).

The hysteresis interval of figure 4(c) represents a situation for which both input values reach their maximum. This will only rarely occur in experiments, because for this input configuration (narrow impasses or sharp corners) the robot will always turn away. In contrast to the "a" and "c" pictures the hysteresis interval of figure 4(b) is much larger. As can be seen from figure 3 this interval $-0.6 < Inp1 < 0.6$ exists for input values $-0.6 < Inp2 < 0.6$. These input constellations are related to deadlock situations, because sharp corners or impasses usually leads to sensor values in this domain. The oscillatory mode, also existing for $(Inp1, Inp2)$-values around the origin, will be never observed because inputs usually will sweep over this domain staying in the fixed point mode.

for creating CPGs out of oscillators (and the different gait patterns produced) is described in [9].

In reference [5] Ijspeert builds on his previous work [10] and extends the evolved lamprey CPG into a locomotion controller for a salamander. This involved adding two segmental oscillators (which were previously evolved for lamprey swimming) and then evolving the connections between them so that it would produce the quadruped walking patterns for a salamander. The leaky-integrator neuron model was used for all neurons in the networks.

3   Methodology

3.1   Artificial Reflexes

The task of the muscle stretch reflex in biological systems [11] is to control simple movements by regulating the length of muscles. The reflexes implemented in [3] had a similar task: to control the position of an actuator on an animat. The actuator can be any kind of device, for example a leg or a wheel, although in this paper and [3] it was a D.C. motor (which subsequently drove a leg on the animat). The reflexes were implemented using multi-layer neural networks with forward and recurrent connections. A typical reflex ANN, evolved using an ES, is shown in Fig. 1.

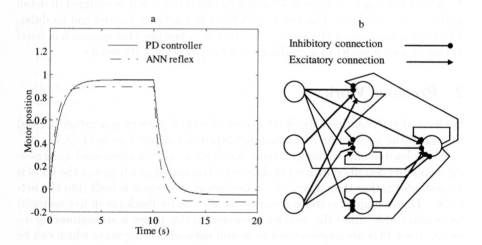

Fig. 1. A typical reflex ANN. a) Performance in controlling D.C. motor, compared to PD controller. b) Network structure (all units have sigmoid activation function)

For more complete descriptions of the methods used to implement the artificial reflexes, see [3,4].

3.2 Action Layer

The main focus of this paper is the additional work carried out since [3], specifically to evolve different gaits for a quadrupedal robot, rather than a bipedal one. The quadruped gaits will be created by modules in the action layer of the ANS presented in [2]. Each module is currently inspired from biological CPGs [12], which perform tasks such as respiration and locomotion in real animals. The outputs from the CPG control the robot's actuators via the reflexes described above.

A neuron model similar to the leaky integrator neuron model was used; this was the same as used in [3]. It was felt that the use of this neural model would give rise to a wider range of possible output patterns from the CPG, without the need for specifically patterned input signals. The activation of the neuron is given by its membrane potential:

$$mp(t+1) = mp(t) \cdot k_d + \sum_{j=1}^{n} w_j i_j \qquad (1)$$

where the summed terms are the weighted inputs to the neuron, $mp(t)$ is the membrane potential at time t and k_d is a constant decay rate. If the membrane potential is above a threshold, the neuron immediately produces an output for a period of time (from t_0 to t_1):

$$if(mp(t_0) > threshold) \Rightarrow out(t) = 1 \qquad (t_0 \leq t < t_1) \qquad (2)$$

Once t_1 has been reached, the membrane potential is reset to a start point resting potential.

The synapses in the ANN have a weight and a type. The weight may be inhibitory or excitatory and the type specifies the period of time the neuron produces an output for: either a short or a long time (this is to simulate the effects of different neurotransmitters in the nervous system [13]). The system can be extended so that the output period can be specified for other connection types. The different times are important for altering the neuron activity. This allows any rhythmic pattern to be generated by the network more effectively than using the previously available neural models. The inputs to the network are simply tonic levels, to provide a stimulus to the network. The outputs from the CPG networks (four outputs in the case of the quadruped gaits) are connected to four reflex ANNs, one for each leg of the quadruped. These reflexes directly control four legs on a simulated robot.

A $(\mu + \lambda)$ ES (Evolutionary Strategy) is used to create the action network. Any neuron can be connected to any other, with a total of sixteen neurons in the ANN. The network is evolved to produce the patterns of activity regardless of input. The fitness of each chromosome is evaluated as the distance the simulated robot moved during a set time. Each chromosome encodes the weights, synapse type and the connection information as a linear string.

4   Results

In all of the results displaying the actual leg positions of the robot the following
line styles represent a separate leg:

Front left   —————————   Front right   — — — — —
Rear left   ······················   Rear right   — · — · — · —

The gaits chosen to be evolved at this stage were the gallop, trot, pronk, walk
and pace [14]. Due to space restrictions, only gallop and trot are shown here.

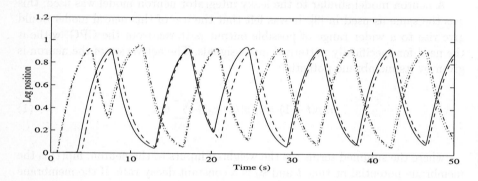

Fig. 2. Gallop leg positions

Fig. 2 shows the leg positions of the robot while under control of the fittest
CPG network. As can be seen, the positions of the legs are consistent with that
of a gallop. The two fore legs (solid and dashed lines) are moving as one pair,
and are anti-phase to the two hind legs (dotted and dash-dotted lines). The
pattern of leg positions is quite stable, with the only anomaly being in the first
and second steps. The next gait to be evolved was the trot. In this gait, the
diagonal legs move in pairs, with each diagonal pair moving in anti-phase from
each other. The results of the evolution for this type of gait are shown in Fig. 3
below.

The leg positions produced by the CPG for a trotting gait were almost perfect. The only problem occurred in the final step where one leg (the right hind
leg) was moved forward before the other leg in that pair.

5   Effect of Modularization on Ease of Evolution

It was noted that as the gaits became complex (a walk contains four phases
compared with the two phases in a pace), they became more difficult to evolve.
It was decided to partition the network into modules rather than one homogeneous network to test whether this improved the efficiency of the evolutionary
algorithm. In particular, the oscillator which produced the fundamental stepping
frequency and the part of the network which produced individual activation for
legs were allowed to evolve separately.

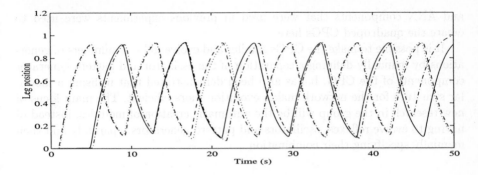

Fig. 3. Trot quadruped leg patterns

It was found that when this system was tested, not only did the system evolve more quickly, but that it could produce better approximations of the more complex gaits like the walk in fewer generations with less neurons. Fig. 4b shows a comparison between the speed of evolution of the modular and non-modular systems. This effect might be due to the different parts of the network interfering with each other in the homogenous system [15].

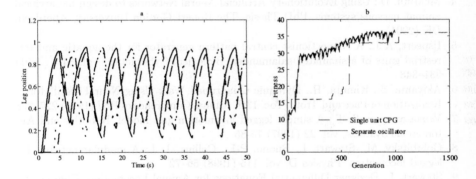

Fig. 4. a) Leg patterns for separate oscillator and pattern generator for gallop b) Best fitness for single unit CPG and separate oscillator and pattern generator for gallop

6   Conclusions and Future Work

Central pattern generators have been created using evolutionary algorithms. The task of these central pattern generators was to produce walking patterns for simulated and real quadruped robots, and this was successfully achieved.

The flexibility of the artificial nervous system model has allowed the central pattern generator networks to remain simple, as they use the services provided by the layer below (reflexes) to achieve their goal. In addition, the same techniques and ANN components that were used in previous experiments were used to create the quadruped CPGs here.

It is possible to make the CPGs smaller and evolve in lower numbers of generations by manually defining a separation of the oscillator and pattern generator components of the CPG. It has also been demonstrated that utilising a modular structure for the network makes evolution more efficient. The main focus of new research in the group will be the automatic creation of modules, instead of having to evolve reflexes, oscillators and pattern generators separately and then manually specifying their combination.

References

1. MacLeod, C.: The Synthesis of Artificial Neural Networks Using Single String Evolutionary Techniques. PhD Thesis, The Robert Gordon University, Aberdeen, UK (1999)
2. MacLeod, C., Maxwell G., McMinn, D.: A Framework for Evolution of an Animat Nervous System. EUREL European Advanced Robotics Systems Development: Mobile Robotics, Leiria, Portugal (1998)
3. McMinn, D., Maxwell, G., MacLeod, C.: An Evolutionary Artificial Nervous System for Animat Locomotion. EANN 2000: International Conference on Engineering Applications of Neural Networks, London, UK (2000)
4. McMinn, D.: Using Evolutionary Artificial Neural Networks to design hierarchical animat nervous systems, PhD Thesis, The Robert Gordon University, Aberdeen, UK (2001)
5. Ijspeert, A.J.: A connectionist central pattern generator for the aquatic and terrestrial gaits of a simulated salamander, Biological Cybernetics, Vol. 84:5 (2001) 331–348
6. Akiyama, S., Kimura, H.: Dynamic Quadruped Walk Using Neural Oscillators - Realization of Pace and Trot. Proc. 13th Annual Conference of RSJ (1995) 227–228
7. Venkataraman, S. T.: A simple legged locomotion gait model. Robotics and Autonomous Systems, vol. 22 (1997) 75-85
8. Golubitsky, M., Stewart, I., Buono, P-L., Collins, J. J.: A modular network for legged locomotion. Physica D, vol. 115 (1998) 56–72
9. Stewart, I.: Designer Differential Equations for Animal Locomotion. Complexity, vol. 5, part 2 (1999) 12–22
10. Ijspeert, A.J., Kodjabachian, J.: Evolution and development of a central pattern generator for the swimming of a lamprey, Artificial Life 5:3 (1999) 247–269
11. Kandel, E. R., Schwartz, J. H., Jessell, T. M. (eds.): Principles of Neural Science 3rd edition. Appleton & Lange (1991) 591–593
12. Haines, D. E. (ed.): Fundamental Neuroscience. Churchill Livingstone Inc. (1997) 340–341
13. Levitan, B., Kaczmarek, L. K.: The Neuron: Cell and Molecular Biology 2nd Edition. Oxford University Press (1997) 100–115
14. Gray, J.: Animal Locomotion. Weidenfeld and Nicolson (1968) 266–276
15. Cho, S., Shimohara, K.: Modular Neural Networks Evolved by Genetic Programming. Proceedings of the 1996 IEEE International Conference on Evolutionary Computation, Nagoya, Japan (1996) 681–684

Saliency Maps Operating on Stereo Images Detect Landmarks and Their Distance

Jörg Conradt[1], Pascal Simon, Michel Pescatore, and Paul F.M.J. Verschure

Institut of Neuroinformatics, ETH / University Zurich
Winterthurerstrasse 190, CH-8057 Zürich
{conradt, psimon, michelp, pfmjv}@ini.phys.ethz.ch
http://www.ini.ethz.ch

Abstract. We present a model that uses binocular visual input to detect landmarks and estimates their distance based on disparity between the two images. Feature detectors provide input to saliency maps that find landmarks as combinations of features. Interactions between feature detectors for the left and right images and between the saliency maps enables corresponding landmarks to be found. We test the model in the real world and show that it reliably detects landmarks and estimates their distances.

1 Introduction

Animals need a spatial representation of their environment to navigate and return home. There exist neurons in the hippocampus that are only active in a small part of the environment, the so-called place field [1]. These neurons provide a spatial representation, which is built on a variety of sensory information. However, the dominant stimuli are distant visual landmarks [2]. This paper proposes a model that detects landmarks in natural scenes using a stereo video system and estimates their directions and distances.

A number of biologically inspired algorithms exist that find salient objects in monocular camera images, some of which use saliency maps (SMs) [3]. While it is possible to extract a small number of distance cues from monocular images, e.g. landmark size, occlusions, etc., these approaches do not estimate distance.

In contrast, multiple images taken from different positions contain an important distance cue, the differences in the landmark's position in each image, called disparity. It turns out that detecting the same landmark in both images is a hard problem [4] due to e.g. only partially overlapping visual areas of both cameras, varying lighting conditions, noise in the signal.

The approach we explore in this paper is based on biologically inspired Saliency Maps (SMs) [5, 6] that receive preprocessed input from Feature Detectors (FDs). Interactions between both cameras' FDs and SMs support the detection of corresponding landmarks in both images and allow the estimation of their direction and distance.

---

[1] Corresponding author

J.R. Dorronsoro (Ed.): ICANN 2002, LNCS 2415, pp. 795–800, 2002.

2 The Hardware Setup

We use a mobile Koala Robot (K-Team, Switzerland) with a stereo pan-tilt camera system (fig. 1) to perform real world experiments. A standard Personal Computer (Atmel processor, 1400 MHz, RedHat Linux 7.2) controls the robot and the cameras. Both the robot's cameras (type VPC-795) are mounted on pan-tilt devices, allowing each camera to turn in 2 degrees-of-freedom with minimal translation of the camera's optical center. Both cameras can move independently, but their tilt angles are coupled in software. In the PC, two standard PCI framegrabber boards (Hauppage WinTV) acquire the images. To increase the computational efficiency, we reduce the image resolution (768x576) by a factor of 4 in both directions. The landmark detection, and the control of the pan-tilt system and the robot run in real-time on the PC. The system is implemented in MatLab, Version 6 (The Math Works, USA).

3 Description of the Model

3.1 Feature Detectors

A saliency map based approach to finding landmarks in images consists of multiple processing stages. Firstly, we extract four different channels from every image: red, blue, green and brightness information, with normalized color channels. Every channel is then convolved with a set of feature detectors (FDs) that respond best to particular patterns in the image. The FDs used are circles of different radius, and horizontal, vertical, and diagonal edges (shown in fig. 3). Convolving the FDs with all channels results in 24 independent feature selective maps. To preserve the FDs' polarity, the 24 maps are split into 48 feature maps (FMs), each containing either a map's positive stimulus or the absolute values of a map's negative stimulus.

3.2 Saliency Map Based Detection of Landmarks

All FMs are used as input to a Saliency Map (SM) that is composed of a competitive neural network as proposed in Amari and Arbib's neural fields approach for modeling cortical information processing [5, 6]. The goal of the network is to take the FDs' spatially localized stimuli as input, have them compete and finally output a winning target. The neurons' activation dynamics in the SM is expressed as

$$\tau \dot{u}(\mathbf{x}) = -u(\mathbf{x}) + S(\mathbf{x}) + h + \sum_{\mathbf{x}'} w_k(\mathbf{x}, \mathbf{x}') \cdot \sigma(u(\mathbf{x}')), \qquad (1)$$

with h being the baseline activation level and $\sigma(u)$ controlling the local threshold of activation. $S(\mathbf{x})$ is the cumulative weighted input from the FDs, defined as

$$S(\mathbf{x}) = \sum_n w_n \cdot stimulus_n(\mathbf{x}). \tag{2}$$

Depending on the choice of the parameter h and the form of σ and w_k, the activation dynamics of (1) can have various stable equilibrium points. We are interested in a solution which has uniform low activation in regions without stimuli, and which forms a peak of activity at the location of the most significant combination of features. This is achieved by using a smoothed step-function

$$\sigma(u) = 1/(e^{(-cu)} + 1) \tag{3}$$

as the transfer function and an interaction kernel with short-range excitation and a long-range inhibition term H_0

$$w_k(\mathbf{x}, \mathbf{x}') = k \cdot e^{-(\mathbf{x}-\mathbf{x}')^2/\sigma_w^2} - H_0. \tag{4}$$

The values of the constants H_0, k, c, and σ_w^2 have to be determined based on the magnitude of the stimulus $S(\mathbf{x})$, as outlined in [5]. The SM has found a target if the local activity in the network exceeds a global threshold.

3. 3 Finding Corresponding Landmarks in Stereo Images

We will use the offset between an object's positions in both images to estimate its distance from the cameras. To find landmarks in two images, we use independent FMs and SMs for visual input from a left and a right camera. Ideally, the first targets found independently in both images would correspond to each other, as would the second, the third, and so on. However, false pairing will occur due to e.g. the shifted visual fields of both cameras, noise in the signal, or varying lighting conditions.

Interactions between Left and Right Processing Streams

The detection of corresponding targets despite a potentially substantial offset is easier at the level of the FMs, where the landmarks' features are separate. To enhance potentially corresponding stimuli, we introduce a coupling between FDs on the left and right processing streams that respond to the same feature (e.g. small bright circles). Activity in one FM excites a one-sided Gaussian shaped region in the corresponding FM of the other camera, whereas the absence of activity inhibits a small region (fig. 2). Corresponding landmarks lie at similar vertical position in both images, such that the vertical extension of the interaction kernel is minimized to avoid false pairing. In contrast, the horizontal width of the interaction adapts to a landmark's expected offset in both images, which is reduced during consecutive iterations of the algorithm as described below.

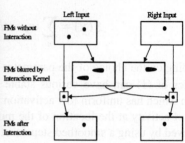

Fig. 1. The mobile Koala robot with stereo pan-tilt cameras.

Fig. 2. Interactions between corresponding FMs enhance corresponding stimuli and suppress stimuli without matches.

Additionally, both SMs interact to enhance activity that originates from corresponding stimuli but different FMs. Activity in one SM excites the other within a one-sided Gaussian-shaped region, centered on the same position.

Determining Landmarks

If neurons in both saliency maps at similar vertical locations exceed a threshold, the positions in both images define the new target. The required adjustments in camera pan and tilt angles to center the target are estimated by linear approximation. The peaks of activity in both SMs are then shifted to the center of the map to adjust for the cameras' rotation.

If a target was found outside of the image center, the SMs continue processing input to increase the accuracy in estimating the target's position. In such a case, we introduce a short-term attentional bias by increasing the connection weights w_n between the SMs and those FMs that contributed to the target. Additionally, the width of the interaction Gaussian is reduced, as the positions of the target will be nearer in both images.

If the target is already close to the images' center when detected, a new landmark is declared found. The landmarks' coordinates and the cameras' pan and tilt angles are stored. Finally, a strong negative activity is introduced in both saliency maps at the position of the landmark, generating an inhibition of return [5], allowing other targets to be detected in subsequent iterations. Additionally, the weights w_n between the FMs and the SMs are reset to initial values.

3.4 Estimating a Landmark's Direction and Distance

When the SMs have found a landmark, its direction and distance are estimated. A neural net maps the 7 given coordinates (the landmark's x and y position in both images, 2 camera pan angles, and one common camera tilt angle) to the landmark's direction and distance.

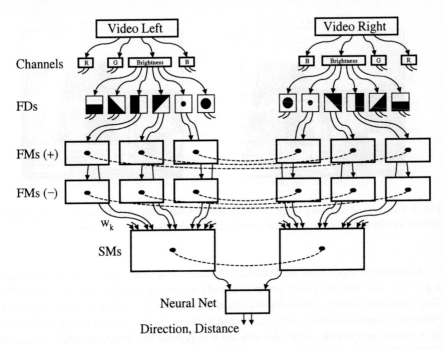

Fig. 3. From top to bottom: Each video image is split into four channels, each convolved with 6 FDs. The resulting 48 FMs on each side (only six shown here) serve as weighted stimulus for a SM that detects landmarks. The SMs provide input to a neural net estimating direction and distance to the landmark. Interactions between corresponding FMs belonging to the left and right stream and between the SMs are indicated by dotted lines.

4 Results and Discussion

The number of landmarks detected in a stereo image varies dramatically with the image content. The algorithm typically detects 16-20 corresponding targets in our office environment, using the whole pan-tilt range of 150 horizontal and 45 vertical degrees. These targets do not all represent distinct landmarks, as some targets fall on the same object. A result of the detection algorithm for a single pair of stereo images is shown in figure 4. In this scene, nine distinct targets are found. High contrasts targets are detected first, here a green key-ring mascot on the shelf, followed by a blue light bulb in the lower right-hand corner. Successively, the monitor, the computer rack, the oscilloscope, the folder, and the wooden lamp-holder are detected. Finally, the cables at the very top are found. Then, the algorithm returns to the previously detected objects, but attends to those found first more frequently. The error in estimating a landmark's distance is below 10% within six meters.

Random dot stereograms, for example, show that humans recognize depth based on disparity without recognizing objects [4]. This inspired us to implement interactions between FMs, rather than between SMs only, which turned out to it increases the reliability in matching points.

Left Image

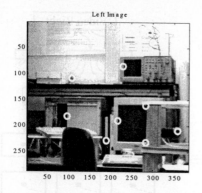

Right Image

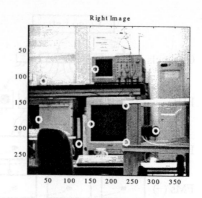

Fig. 4. Examples of matching target pairs denoted by white circles are shown. Here, the pan-tilt system was disabled and the algorithm searched for landmarks in static images.

A number of algorithms for landmark detection and matching have been proposed, e.g. [7-9], but they are either computationally expensive or biologically implausible. The algorithm presented here detects landmarks and their distance in natural environments in a computationally inexpensive and biologically inspired way. The model runs in real time on a mobile robot. We will use it to provide the directions and distances of landmarks for a model of hippocampal place fields, for use in robot navigation.

References

[1] J. O'Keefe and L. Nadel, *The Hippocampus As a Cognitive Map*: Oxford University Press, 1978.

[2] P. J. Best, A. M. White, and A. Minai, "Spatial Processing in the Brain: The Activity of Hippocampal Place Cells," *Annual Reviews of Neuroscience*, vol. 24, pp. 459–86, 2001.

[3] L. Itti and C. Koch, "A saliency-based search mechanism for overt and covert shift of visual attention," *Vision Research*, vol. 40, pp. 1489–1506, 2000.

[4] I. P. Howard and B. J. Rogers, *Binocular Vision and Stereopsis*, vol. 29. New York, Oxford: Oxford University Press, Clarendon Press, 1995.

[5] S. Amari, "Dynamics of pattern formation in lateral-inhibition type neural fields.," *Biological Cybernetics*, vol. 27, pp. 77–87, 1977.

[6] S. Amari and M. A. Arbib, "Competition and cooperation in neural nets," in *Systems Neuroscience*, Metzler, Ed. New York: Academic Press, 1977, pp. 119–165.

[7] P. E. Trahanias, S. Valissaris, and T. Garavelos, "Visual Landmark Extraction and Recognition for Autonomous Robot Navigation," presented at International Conference on Intelligent Robots and Systems, Grenoble, France, 1997.

[8] S. Livatino and B. C. Madsen, "Autonomous Robot Navigation with Automatic Learning of Visual Landmarks," presented at International Symposium on Intelligent Robotic Systems, Coimbra, Portugal, 1999.

[9] G. Bianco, A. Zelinsky, and M. Lehrer, "Visual Landmark Learning," presented at International Robots and Systems, Takamatsu, Japan, 2000.

A Novel Approach to Modelling and Exploiting Uncertainty in Stochastic Control Systems

Randa Herzallah and David Lowe

NCRG, Aston University, UK {herzarom,d.lowe}@aston.ac.uk

Abstract. We consider an inversion-based neurocontroller for solving control problems of uncertain nonlinear systems. Classical approaches do not use uncertainty information in the neural network models. In this paper we show how we can exploit knowledge of this uncertainty to our advantage by developing a novel robust inverse control method. Simulations on a nonlinear uncertain second order system illustrate the approach.

1 Introduction

Recently neural network models have become favourite candidates in the field of nonlinear system identification and control due to their ability to approximate multi-variable nonlinear mappings. Besides having nonlinear features, dynamic systems may have noise events affecting their inputs and outputs, and usually they are time-variant. Because artificial neural networks can be adapted on line [1], usually they are capable of good performance in such situations. However for most real control problems where disturbances play an important part and where a relatively big sampling interval is used, the predicted output of the neural network is inherently uncertain. However neural networks now have the ability to model general distributions rather than just producing point estimates, and in particular can produce estimates of the uncertainties involved in its own predictions [2]. The most recent research interest is now to go beyond the classical methods for identification and control by accounting for the model uncertainty. In [3] a systematic procedure that accounts for the structured uncertainty when a neural network model is integrated in an approximate feedback linearisation control scheme has been developed. The use of an adaptive critic controller when there is input uncertainty has been discussed in [4]. The application of recently developed minimal resource allocating network ($MRAN$) in a robust manner under faulty conditions has been demonstrated in [5]. But none of the recent works have considered the possibility of using the neural network's estimate for error bars. In this paper we address for the first time the use of this extra knowledge to develop a robust control method for uncertain nonlinear systems. Knowledge concerning the uncertainty involved in the control signal determines a region around the predicted value of the control signal where sampling can be used to search for a better value of the control signal. Since the true plant model is usually unknown, we also employ an approximating neural network for the forward plant to measure the effect of the sampled control values.

J.R. Dorronsoro (Ed.): ICANN 2002, LNCS 2415, pp. 801–806, 2002.

2   Adaptive Inverse Control and Error Bar Modelling

Figure 1 illustrates the classical inverse adaptive control technique, where the neural network is learning to recreate the inputs that are necessary to produce the desired process outputs. Specifically the inputs to the neural inverse model include the past values of the inputs and outputs of the plant $z(t) = [y(t - 1), ..., y(t-n), u(t-d-1), ..., u(t-m)]$ and the desired output value $y_{ref}(t)$. The network predicts the process inputs, $\hat{u}(t-d)$, (where d is the relative degree of the plant) necessary to produce the desired outputs, $y_{ref}(t)$. In this work the basic goal is to model the uncertainty of the control signal $u(t - d)$ by modelling its statistical properties, expressed in terms of the conditional distribution function $p(u(t - d)|z(t), y_{ref}(t))$. The uncertainty involved in the control signal can be estimated in several ways. Here an error bar estimate known as a predictive error bar, as reported in [2] will be used. This approach is based on the important result that for a network trained on minimum square error the optimum network output approximates the conditional mean of the target data (control signals), or $\hat{u}(t-d) = < u(t-d)|(z(t), y_{ref}(t)) >$, and that the local variance of the target data can be estimated as $\sigma^2_{\hat{u}(t-d)} = \|u(t - d) - \hat{u}(t - d)\|^2$. Thus the distribution of the target data could be described by a Gaussian function with an input-dependent mean given by $\hat{u}(t - d)$ and an input-dependent variance given by $\sigma^2_{\hat{u}(t-d)}$. In the implementation of predictive error bars two neural networks should be used. Each network shares the same input and hidden nodes, but has different final layer links which are estimated to give the approximated conditional mean of the target data in the first network, and the approximated conditional mean of the variance in the second network. Thus the second network predicts the noise variance of the predicted mean by the first network. Optimisation of the weights is a two stage process: The first stage determines the weights w_1 conditioning the regression on the mapping surface. Once these weights have been determined, the network approximations to the target values are known, and hence so are the conditional error values on the training examples. In the second stage the inputs to the networks remain exactly as before, but now the target outputs of the network are the error values. This second pass determines the weights w_2 which condition the second set of output noise to the squared error values $\sigma^2_{\hat{u}(t-d)}$.

3   Problem Formulation and Solution Development

Dynamic programming is a powerful tool in stochastic control problems [6]. However, it performs poorly when the order of the system increases. The algorithm proposed here circumvents the dynamic programming scaling problem by using the predicted neural network error bars to define an importance sampling distribution which limits the possible control solutions to consider. Accepting the inaccuracy of neural networks and assuming the output of the inverse control network can be approximated by a Gaussian distribution of control signals, the mean and variances can be obtained as discussed previously. Using just the mean estimate of the control is typically suboptimal in nonlinear systems. Even

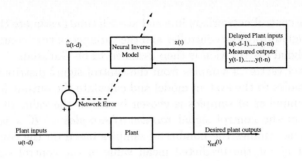

Fig. 1. Training of an inverse controller.

though the Gaussian assumption used here is an approximation, using the on-line variance estimate of the neural network determines a region around the predicted mean value where sampling can be used to obtain a better estimate of the control signal than the mean. The distribution assumption is Gaussian but the predicted mean and variance are nonlinear functions of previous states, thus allowing for good models of forward and inverse plant behaviour provided the inverse plant is a function. If this is not the case a similar approach, but using Gaussian mixtures (a mixture density network) could be employed. Based on estimates of the mean and variance of the distribution of control signal values, we can construct the following algorithm incorporating the uncertainty directly.

1. Based on the pre-collected input-output data, an accurate forward model of the process is constructed and trained off line. It is assumed to be described by the following neural network model:

$$\hat{y}(t) = N_f(y(t-1), ..., y(t-n), u(t-d), ..., u(t-m)) \qquad (1)$$

   where $y(t)$ is the measured plant output, $u(t)$ is the measured plant input, n is maximum delay of the output, m is the maximum delay of the input, d is the relative degree of the plant.

2. An accurate inverse model of the plant should also be constructed, and trained off line to approximate the conditional mean of the control vector and the conditional variance. It is assumed to be described by the following neural network

$$x(t) = g(y(t), y(t-1), .., y(t-n), u(t-d-1), .., u(t-m)) \qquad (2)$$

   where $\hat{u}(t-d) = x(t)w_1$, $var_{\hat{u}(t-d)} = x(t)w_2$, and where $x(t)$ is the predicted hidden variable from the neural network at each instant of time t, w_1 is the weight of the linear layer estimated to predict the conditioned mean of the control signal, and w_2 is the weight of the linear layer estimated to predict the variance of the predicted control signal.

3. At each instant of time t the desired output is calculated from the reference model output, which should be chosen to have the same relative degree as that of the plant.

4. Bring the control network on line and at each time t estimate the appropriate control signal from the controller and the variance of that control signal. The control signal distribution is then assumed to be Gaussian.

5. Generate a vector of samples from the control signal distribution, and feed these samples to the system model and calculate the output from each sample. The number of samples is chosen based on the value of the predicted variance of the control signal *number of samples* $= K \times var_{\hat{u}(t-d)}$. This equation determines the number of samples based on the confidence of the controller about the predicted mean value of the control signal. So more samples are generated for larger variance.

6. Based on the effect of each sample on the output of the model, the most likely control value is taken, which is assumed to be the value that minimises the following cost function.

$$J(t) = \underset{u \in U}{Min} \underset{\bar{v}}{E} [(\hat{y}(t) - y_r(t))^2] \tag{3}$$

where U is a vector containing the sampled values from the control signal distribution, E is the expected value of the cost function over the random noise variable $\bar{v}$. Because we are using a neural network to model the system, and because the neural network predicts the mean value for the output of the model averaged over the noise on the data, the above function can be optimised directly.

7. Go to step 3.

4 Simulation Study

4.1 Introduction

In order to illustrate the validity of the theoretical developments, we consider the liquid-level system described by the following second order equation:

$$\begin{aligned} y(t) = {}& 0.9722y(t-1) + 0.3578u(t-1) - 0.1295u(t-2) \\ & - 0.3103y(t-1)u(t-1) - 0.04228y^2(t-2) + 0.1663y(t-2)u(t-2) \\ & - \bar{v}y^2(t-1)y(t-2) - 0.3513y^2(t-1)u(t-2) \\ & + 0.3084y(t-1)y(t-2)u(t-2) + 0.1087y(t-2)u(t-1)u(t-2) \end{aligned} \tag{4}$$

This model has been used in [7] to illustrate theoretical development for the direct adaptive controller. Because disturbances play an important part in real world processes, a stochastic component, $\bar{v}$, has been added to this model. This component is taken to be a Gaussian random variable $\mathcal{N}(0.03259, 0.2)$. The plant has been considered to be described by equation (4). In order to identify the plant, an input-output model described by $\hat{y}(t) = N_f(y(t-1), y(t-2), u(t-1), u(t-2))$ was chosen. Where N_f is a Gaussian radial basis function network. This neural network model was trained using the scaled conjugate gradient optimisation algorithm, based on noisy input output data measurements taken from the plant with sampling time of 1s. The input to the plant and the model was a sinc function followed by a sine wave in the interval $[-1, 1]$ with additive Gaussian noise of zero mean and 0.3 variance. The single optimal structure for the neural network found by applying the cross validation method consisted of 6 Gaussian basis functions. Similarly an input output model described by $\hat{u}(t - 1) = N_c(y(t - 1), y(t - 2), y(t), u(t - 2))$ was chosen to find the inverse model of the plant, where N_c is a Gaussian radial basis function network. The training data has been the same as in the forward model. A neural network with 6 Gaussian basis functions was found to be the best model.

4.2   Classical Inverse Control Approach

After training the inverse controller off line, the control network is brought on line and the control signal is calculated at each instant of time from the control neural network and by setting the output value at time t, $y(t)$, equal to the desired value $y_r(t)$, where $y_r(t) = 0.2 * r(t - 1) + 0.8 * y_r(t - 1)$ and r is the set point. The predicted mean value from the neural network was forwarded to the plant. After running the process for about 600 time steps the output of the plant was seen to increase in an unbounded fashion, and the classical inverse controller was unable to force the plant output to follow the reference output.

4.3   Proposed Control Approach

In our new approach, both the mean and the variance of the control signal were estimated. Following the procedure presented earlier, the best control signal was found and forwarded to the plant. This control signal was obtained from a small number of samples from the Gaussian distribution, typically less than 27 samples. The overall performance of the plant under the proposed method is shown in figure 2a, where it is evident that the system outputs remain stable across the whole region, and that the proposed sampling approach managed to stabilise the plant. The control signal is shown in figure 2d, and the variance of this control law is shown in figure 2e. The error from the absolute difference between the plant output and the desired output of the classical inverse controller is shown in figure 2b, while that of the proposed sampling approach is shown in figure 2c. From these two figures we can see that the sampling approach control performs in a stable manner with smaller error values than that of the classical one.

5   Conclusions

Inverse Control can be considered as a good control strategy if the model of the plant happens to be invertible. Uncertainty around the control signal can be estimated if neural networks have been used to learn the inverse mapping from the plant output to the plant input.

The main contribution of this paper is that it provides a systematic procedure to use this uncertainty measure in order to improve the generalisation property of

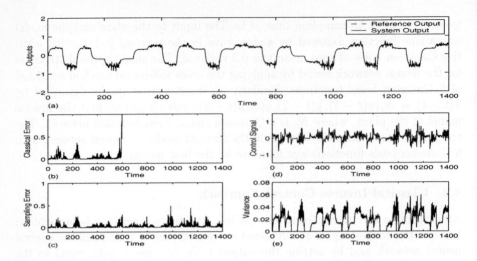

Fig. 2. System response using the sampling approach. (a) System output and the reference output. (b) Error from the classical inverse control showing unstable divergence. (c) Error from the proposed sampling algorithm demonstrating long term stability. (d) Control signal. (e) Control signal variance

the controller. Simulation experiments demonstrated the successful application of the proposed strategy to improve the controller performance for a class of nonlinear control system.

References

1. Narendra, K.S., Parthasarathy, K.: Neural Networks in Control Systems.
2. Lowe, D., Zapart, C.: Point wise Confidence Interval Estimation by Neural Networks: A Comparative study Based on Automative Engine Calibration. Neural Computation and Application. 8 (1999) 77–85 .
3. Botto, M.A., Wams B., Boom, T., Costa, J.: Robust Stability of Feedback Linearised Systems Modelled with Neural Networks: Dealing with Uncertainty. Engineering Applications of Artificial Intelligence. 13 (2000) 659–670.
4. Huang, Z., Balakrishnan, S.N.: Robust Adaptive Critic Based Neurocontrollers for Systems with Input Uncertainties. IEEE. (2000) 67–72.
5. Li, Y., Sundararajan, N., Saratchandran, P.: Neuro-Flight Controllers for Aircraft Using Minimal Resource Allocating Networks (MRAN). Neural Computing and Applications. (2001) 172–183.
6. Larson, R.E.: State Increment Dynamic Programming. American Elsevier Publishing Company, INC. New York. (1968).
7. Ahmed, M.S.: Neural-Net-Based Direct Adaptive Control for a Class of Nonlinear Plants. IEEE Transactions on Automatic Control. 45 (2000) 119–124.

Tool Wear Prediction in Milling Using Neural Networks

*Rodolfo E. Haber[1,2], A. Alique[1], and J.R. Alique[1]

[1]Instituto de Automática Industrial (CSIC).
km. 22,800 N-III, La Poveda. 28500. Madrid.
SPAIN.
rhaber@iai.csic.es
[2] School of Computer Science and Engineering.
Universidad Autónoma de Madrid.
Ciudad Universitaria de Cantoblanco
Ctra. de Colmenar Viejo,km 15. 28049 - Madrid
SPAIN

Abstract. An intelligent supervisory system, which is supported on a model-based approach, is presented herein. A model, created using Artificial Neural Networks (ANN), able to predict the process output is introduced in order to deal with the characteristics of such an ill-defined process. In order to predict tool wear, residuals errors are used as basis of a decision-making algorithm. Experimental tests are made in a professional machining center. The attained results show the suitability and potential of this supervisory system for industrial applications.

1 Introduction

Nowadays, many classical supervisory systems incorrectly assess process condition, because they lack the relevant information or because their strategies are insufficiently elaborated as to account for certain events. These shortcomings in reliability encourage the development of intelligent supervision strategies [1].

Presented in this paper is an intelligent supervisory system for tool wear prediction, created with a model-based approach. In order to deal with nonlinear process characteristics, a Neural Network Output Error (NNOE) model, able to predict online the resultant cutting force under actual cutting conditions, is proposed. In order to predict tool wear, a decision-making algorithm that uses residual errors (difference between measured and predicted cutting force) is applied.

In section 2 a brief presentation of neural networks that focuses on the type of ANN used, including training algorithm is given. In section 3, a fault detection algorithm on the basis of residual computation is described. In section 4, a short description of the machining process and its general characteristics is shown. In section 5 the experimental set-up, the design considerations, the network topology for

---

* Corresponding author

J.R. Dorronsoro (Ed.): ICANN 2002, LNCS 2415, pp. 807–812, 2002.
© Springer-Verlag Berlin Heidelberg 2002

the best training result obtained and the experimental results are presented. Finally, the authors conclusions are given.

2   Artificial Neural Networks and Dynamic Process Models

Among all modeling structures, the so-called Output Error (OE) model is one of the most widely used. This work deals with a neural network output error (NNOE) model because of the regression vector, which is similar to that one of an OE model.

$$\hat{y}(t) = g(\varphi(t), \theta). \tag{1}$$

where $\varphi(t) = \left[\hat{y}(t-1|\theta), ..., \hat{y}(t-n_A|\theta), u(t-1), ..., u(t-n_B)\right]^T$ is a regression vector, n_A is the number of past predictions used for determining the predictions, n_B is the number of past inputs, θ is the vector containing the weights and $g(\cdot)$ is the function performed by the neural network.

The class of MLP considered here consists in only one hidden layer with hyperbolic tangent activation function H, and a linear activation function, L, at the output. Such configuration is advantageous considering a foreseeable real-time implementation of the network [2].

$$\hat{y}_i(\mathbf{w}, \mathbf{W}) = L_i \left[\sum_{j=1}^{Q} W_{ij} H_j \left(\sum_{k=1}^{M} w_{jk} u_k + w_{j0} \right) + W_{i0} \right]. \tag{2}$$

where Q is the number of output neurons, M is the number of neurons in the hidden layer, $\mathbf{u}$ are the inputs, and $\hat{y}_i$ is the output of the network. The weights are specified by the matrices $\mathbf{w}$ (input-to-hidden layer weights) and $\mathbf{W}$ (hidden-to-output layer weights). Both matrices are included in θ. As the training algorithm a version of the Levenberg-Marquardt method was selected [3].

3   Supervision System

Among all the available methods for industrial processes, model-based fault detection can be considered the most popular approach [4]. The essence of this method is the use of residuals from the comparison of actual process behavior with the behavior of a reference process model.

One simple strategy for residual evaluation and fault detection in any supervisory system is the weighted sum squared residuals (WSSR) [5]

$$e_M(t) = y(t) - \hat{y}(t). \tag{3}$$

where $y(t)$ is the process's actual output, $\hat{y}(t)$ is the predicted (model) output, and t is the discrete time and $e_M(t)$ is the residual sequence.

Under normal operating conditions, the process under supervision can be considered a zero-mean white-noise process with a covariance matrix $V_R(t)$. The deviation of a certain variable η is used to detect the failure on the basis of some threshold (ε) and $V_R(t)$ both computed empirically considering the length of the chosen time window [$t\text{-}N_T\text{-}1: t$].

$$\eta = \sum_{t-N_T+1}^{t} e_M^T(t)\cdot V_R(t)\cdot e_M(t) \begin{cases} > \varepsilon \rightarrow \text{fault} \\ \leq \varepsilon \rightarrow \text{no fault} \end{cases}. \qquad (4)$$

The threshold level scaled with the size of the input vector can be used as an adaptive threshold. Therefore, the threshold value can be estimated adaptively combining the vector norm of the residuals and their derivatives. The use of the time derivative of residual vectors allows to evaluate the trend in the matching degree of model and process. The utilisation of (6) means to establish the threshold on the basis of the greater element of each vector. The bigger the infinity norm of the residual vector and its derivative the closer the threshold. The use of (5) could give slightly better results because the Euclidean norm yields a realistic balance of residuals and their derivatives.

$$\varepsilon(t) = \frac{1}{|e_M|_2 + |\dot{e}_M|_2}. \qquad (5)$$

$$\varepsilon(t) = \frac{1}{|e_M|_\infty + |\dot{e}_M|_\infty}. \qquad (6)$$

where $|\cdot|_\infty$ and $|\cdot|_2$ are the infinity norm and Euclidean norm, respectively; e_M and $\dot{e}_M$ are the residual vector and its derivative in the window [$t\text{-}N_T\text{-}1, t$].

4 The Machining Process. Experimental Set-Up

For the case study is selected the milling process since milling is one of the most complex of machining operations [6]. Tthe milling process can be formally described by a discrete nonlinear relationship

$$\hat{F}(t) = G(\mathbf{F}, \mathbf{f}, \mathbf{a}). \qquad (7)$$

where G is an unknown function to identify, F is the cutting force exerted during the removal of metal chips, a is the radial depth of cut and f is the relative feed speed between tool and worktable. Considering a representation with vectors, $\mathbf{f}, \mathbf{a}$ and $\mathbf{F}$, are the inputs and output respectively defined as $\mathbf{F} = [F(t-1) \cdots F(t-n)]$, $\mathbf{f} = [f(t-1), \cdots, f(t-m)]$, $\mathbf{a} = [a(t-1) \cdots a(t-m)]$, t, is the discrete time instant and n, $m \in Z$ A successful identification scheme should insure $\hat{F}(t)$ values as close as possible to those of $F(t)$ (actual output).

The experimental tests are conducted on a 5.8kW-4 axes milling machine equipped with CNC, which is interfaced with a personal computer by an RS-232 communication link. The architecture for milling process supervision and control is illustrated in figure 1.

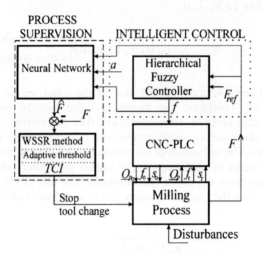

Fig. 1. Scheme for milling process supervision and control.

The main goal of this work is to implement a supervisory system (block in dashed line) composed by a neural network model (7) and a WSSR algorithm (see section 3). Another block named „hierarchical fuzzy controller" is depicted, but the rationale behind introducing and using this technique is out of scope of this paper [7]. The model objective is to predict in real-time, one step ahead, the resultant mean cutting force $\hat{F}$. The process failure detection, in this case tool wear, is made on the basis of a WSSR algorithm. If the tool is sufficiently worn, the final decision is to stop the process and replace the tool.

5   Model of the Milling Process. Tool Wear Prediction

The training algorithm was developed using the program MATLAB. The topology was initially chosen as follows: two inputs f and a, one output $\hat{F}$, a linear activation function at the output, and one hidden layer using hyperbolic tangent for the activation function. The type of model was selected starting from the *a priori* knowledge of the milling process and the types of models considered in previous works. An ANN with 6 neurons in the input, 12 neurons in the hidden layer and one in the output layer, was selected. Modifying (7) for a second order output error model, the one step prediction is evaluated with the previous model outputs

$$\hat{F}(t) = g\left(\hat{F}(t-1),\ \hat{F}(t-2),\ f(t-1),\ f(t-2),\ a(t-1),\ a(t-2)\right) \quad (8)$$

In order to evaluate the model's behaviour in actual working conditions, tailor-made simulation software was developed and real-time supervision of machine tools was employed. The application for Windows NT was programmed in Visual C++.

The libraries of controllers (i.e., different types of fuzzy controllers) and process models (i.e., neural networks) were programmed in C++ and compiled in two dynamic link libraries (DLL).

The real time behavior of the model is depicted in Fig. 2. The behavior of predicted and measured cutting forces is shown at the top of the picture. The feed-rate command signal is shown in the middle of the graph. For each case considered, the chosen profile is depicted at the bottom of the picture. The vertical line represents the actual values also shown at the bottom of the dialog in text boxes.

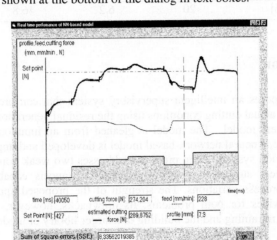

Fig. 2. Real-time running of the recurrent neural network.

The tool condition index (*TCI*), which ranges from 0% (completely new tool) to 100% (worn tool), was used for predicting the tool state. Considering that to produce one part many operations must be executed, each operation is composed by many time windows $[t-N_r-1, t]$ depending on the overall operating time. *TCI* is calculated by breaking the operation down into a certain number of zones, and calculating a threshold for each zone, defined as

$$TCI = \frac{\sum_{k=1}^{NW} C_k}{NW} \cdot 100 \quad [\%] \quad C_k = \begin{cases} 1 & \forall \alpha = \eta - \varepsilon > 0 \\ 0 & \text{otherwise} \end{cases} \quad (9)$$

where $NW = N/N_T$ is the total number of windows in one machining operation, C_k is an integer ($C_k \in [0, NW]$).

A clear transition in tool condition is reflected in the *TCI* (see Table 1). Tool condition was also visually examined by the operator at the end of the machining operation to corroborate actual tool condition. For half-worn and worn tools, tool condition was adequately estimated. Since the model does not recreate the system perfectly, the residuals will fluctuate for changing inputs, even in a new-tool situation and in spite of the strategy chosen to compute the threshold (empirical or adaptive).

Table 1. Tool condition index (TCI) for new, half-worn and worn tools.

	N_T [Samples]	TCI [%]		
		New tool	Half-worn tool	Worn tool
(ε =0.2)	20	9.2	68.3	100
$\varepsilon(t)$ [5]	20	6.2	57.6	100
$\varepsilon(t)$ [6]	20	6.2	46.8	76.8

6   Conclusion

This paper proposes an intelligent supervisory system that can predict tool wear in real time under actual cutting conditions using the residual generation approach on the basis of a process model. The model is gleaned from an input/output formulation. For this purpose, a neural network-based model is developed and implemented.

The supervision system herein proposed addresses two weak points in present-day machining-process supervision systems: lack of diagnosis capability and inconsistency with process variations. The strength of the proposed approach lies in the solution it provides for both problems. Its drawbacks, however, are the need for neural-network retraining and the building of a larger library of models.

References

1. Haber R., Peres C., Alique J.R., Ros S. Fuzzy supervisory control of end milling process. Information Sciences 89 (1996) 95-106.
2. Haykin S., Neural Networks: A comprehensive foundation. 2nd edition, IEEE Press, 1999.
3. Fletcher R., Practical Methods of Optimization. 2nd. Edition, John Wiley&Sons (2000).
4. Isermann R. Process fault detection based modelling and estimation methods – A survey. Automatica 20 (4) (1984) 387-404.
5. Tzafestas S., Watanabe K. Modern approaches to system sensor fault detection and Diagnosis. Journal A 31 (4) (1990) 42-57.
6. Grabec I., Chaos generated by the cutting process, Physics Letter 117 (1986) 384-386.
7. Haber R. E., Haber R.H., Alique A., Ros S., Application of knowledge-based systems for supervision and control of machining processes. In: S.K. Chang (ed.): Handbook of Software Engineering and Knowledge Engineering Vol. II. World Scientific Publishing (2002) 327-362.

Speeding-up Reinforcement Learning with Multi-step Actions

Ralf Schoknecht and Martin Riedmiller

Institute of Logic, Complexity and Deduction Systems
University of Karlsruhe, 76128 Karlsruhe, Germany
{schokn,riedml}@ira.uka.de

Abstract. In recent years hierarchical concepts of temporal abstraction have been integrated in the reinforcement learning framework to improve scalability. However, existing approaches are limited to domains where a decomposition into subtasks is known a priori. In this paper we propose the concept of explicitly selecting time scale related actions if no subgoal-related abstract actions are available. This is realised with multi-step actions on different time scales that are combined in one single action set. The special structure of the action set is exploited in the MSA-Q-learning algorithm. By learning on different explicitly specified time scales simultaneously, a considerable improvement of learning speed can be achieved. This is demonstrated on two benchmark problems.

1 Introduction

Standard Reinforcement Learning algorithms, like trajectory-based *Q-learning*, scale very badly with increasing problem size. Among others, one intuitive reason for this is that according to the problem size the number of decisions from the start state to the goal state increases [1]. In order to keep the number of decisions to the goal tractable hierarchical approaches based on temporal abstraction have been proposed. The main contributions are the *Option* approach [6], the *Hierarchy of Abstract Machines* (HAM) [4] and the *MAXQ* approach [1]. The two latter are based on the notion that the whole task is decomposed into subtasks each of which corresponds to a subgoal. These hierarchical RL approaches are thus able to solve problems of the following two types:

1. **Abstract actions given**: Abstract actions for achieving subgoals are given in terms of actions that are lower in the hierarchy.
2. **Subgoals given**: The concrete realisation of abstract actions in terms of subordinate actions is not known but a decomposition into subtasks is given.

The option approach is formulated in a very general way so that it is not restricted to subgoal-related abstract actions. However, the main algorithm, *Intra-option Q-learning* [6], requires that abstract actions be subgoal-related. Hence, the minimal requirement for the efficient application of existing hierarchical RL algorithms is that a decomposition of the whole problem into subproblems is

J.R. Dorronsoro (Ed.): ICANN 2002, LNCS 2415, pp. 813–818, 2002.

known. Thus, problems from technical process control, e.g. cart-pole balancer or mountain car, as well as general navigation tasks cannot profit from those approaches because subgoals are not known in advance or do not exist at all.

Even if no subgoals are available many control problems show the following characteristic. When controlling the mountain car, for example, on the one hand it is necessary to use a temporal resolution that is fine enough to provide the needed reactivity when a switch of action is required. On the other hand between those action switches the same action will be applied for several consecutive time steps. Hence, the temporal resolution could be coarser which would result in less decisions to the goal. This dilemma cannot be resolved by existing RL approaches. In this paper we propose a new hierarchical approach to RL that is suited for such problems where no decomposition into subproblems is known in advance. The main idea is to let the agent explicitly select abstract actions that correspond to fixed time scales. Such *multi-step actions (MSAs)* consist of a sequence of the same primitive action that is applied for several consecutive time steps. The MSA is executed as a whole and can be interpreted as acting on a coarser time scale. These abstract actions are *building blocks* of optimal paths and can therefore reduce the number of decisions to the goal. The idea of repeating the same primitive action is not new [2,5]. The work presented here differs from [2] in that we focus on the special class of actions with a *pure* time-dependent termination condition. In [5] only actions on one time scale are considered. We extend this work by combining different length MSAs, i.e. different time scales, in *one* action set. This creates a structure that is exploited in *MSA-Q-learning*.

2   Q-Learning with Multi-step Actions

The objective of RL is to learn how to behave optimally in unknown environments. The learning situation is modelled as a *Markov Decision Process* (MDP). An agent interacts with the environment by selecting an action a from the available action set $\mathcal{A}$ and receiving feedback about the resulting immediate reward r. As a consequence of the action the environment makes a (stochastic) transition from a state s to a state s'. Accumulated over time the obtained rewards yield an evaluation of every state concerning its long-term desirability. This value function is optimised during learning and by greedy evaluation of the value function an optimal policy can be derived.

As described we consider a discrete time RL problem with a primitive time step Δt. Primitive actions last exactly one such primitive time step. In the following, the set of primitive actions is denoted as $\mathcal{A}^{(1)}$. We define the set of all *multi-step actions* (MSAs) of *degree* n as $\mathcal{A}^{(n)} = \{a^n | a \in \mathcal{A}^{(1)}\}$ where a^n denotes the MSA that arises if action a is executed in n consecutive time steps. The next decision is only made after the whole MSA has been executed. Thus, the MSA has a time-dependent termination condition after n primitive time steps. In the general option framework defined in [6] MSAs can therefore be modelled as special semi-Markov options[1]. In order to retain the possibility of

---

[1] A Markov option would require a state-dependent termination condition.

learning optimal policies different time scales are combined in one action set. We denote such combined action sets as $\mathcal{A}^{(n_1,\ldots,n_k)} = \mathcal{A}^{(n_1)} \cup \ldots \cup \mathcal{A}^{(n_k)}$.

In the following we investigate how the concept of MSAs can be integrated in learning algorithms like Q-learning. The agent maintains Q-functions for all time scales it is acting on. When executing action a^j of degree j in a state s the agent goes to state s' after j time steps are elapsed and updates the corresponding Q-value as follows

$$Q^{k+1}(s, a^j) = (1 - \alpha)Q^k(s, a^j) + \alpha[r(s, a^j) + \gamma^j \max_{a' \in \mathcal{A}} Q^k(s', a')] \qquad (1)$$

where α is the learning rate and γ denotes the discount factor. The reward accumulated on the primitive time scale is $r(s, a^j) = \sum_{\tau=t}^{t+j-1} \gamma^{\tau-t} r(s_\tau, a)$. Note, for $j = 1$ the update rule (1) reduces to classical trajectory-based Q-learning.

Until now MSAs have been viewed as indivisible units. We looked at each action only at the time scale at which it was executed. Consider, for example, an action a^n of degree n. When this action is selected in s_t we obtain the transition $(s_t, a^n) \to s_{t+n}$ together with the reward $r(s_t, a^n)$. This information is used *only* for updating $Q(s_t, a^n)$ according to (1). The experience contained in the transition could be used more efficiently by also looking inside the MSA. When executing a^n all actions a^k, $k = 1, \ldots, n-1$, are executed implicitly. The transition $(s_t, a^n) \to s_{t+n}$ contains all information necessary to update the Q-values for those lower-level actions at all intermediate states. For a^k with $k < n$, for example, $Q(s_{t+i}, a^k)$ can be updated for $i = 0, \ldots, n-k$. It is convenient to carry out these updates in a backward manner where the index i descends from $n - k$ to 0. This ensures a faster propagation of the correct values. The modified Q-learning algorithm which includes these update rules for all lower-level actions in the action set is referred to as *MSA-Q-learning*. It enables to extract more training examples from the same experience. The idea resembles the intra-option methods introduced in [6]. There, however, the intra-option Q-learning algorithm was only applicable to Markov options. The MSA-Q-learning algorithm we propose here is applicable to a special kind of semi-Markov options, namely MSAs. In the form presented here, we refer to the intra-option method as *intra-MSA* method. It enables to learn at different time scales simultaneously.

3   Results

3.1   The Rooms Gridworld

In the rooms gridworld [6] the objective of the agent is to move from a start state to the goal in a minimum number of steps. The state transition is stochastic, e.g. the agent moves in the intended direction only with probability $\frac{2}{3}$ and otherwise with equal probability in one of the remaining directions. The reward is zero everywhere outside the goal and one in the absorbing goal state. We use a discount factor $\gamma = 0.9$, zero initialised Q-functions and an ϵ-greedy policy for training with $\epsilon = 0.1$. Experimentally, $\alpha = 0.25$ was determined to be a good value for the learning rate.

In the following we investigate how different action sets influence the speed of learning with MSA-Q-learning. The performance measure we use is the quality of the learned policy, i.e. the number of steps to the goal on a test trajectory, plotted against the gathered experience, i.e. the accumulated number of training time steps. Figure 1 shows the learning curves for MSA-Q-learning with different action sets averaged over 10000 experiments. The best result is obtained with the action set $\mathcal{A}^{(1,8)}$.

In order to compare learning with MSAs to learning with options we will state the performance of MSA-Q-learning in terms of steps to the goal plotted against number of training episodes[2]. When using only primitive actions it takes 26 episodes until a performance level of 80 steps is achieved[3]. Using options, this perfor-

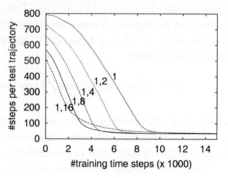

Fig. 1. Learning curves with MSA-Q-learning for the stochastic rooms gridworld with different action sets.

mance level is already reached after about 10 episodes, which is an improvement of 62%. With MSA-Q-learning the 80-steps level is reached after 16 training episodes which is an improvement of 38% compared to the primitive action set. As one might have expected, the MSA approach does not improve the option approach which involves much more prior knowledge to specify the subgoals. But this result demonstrates that MSAs can significantly speed up learning in stochastic domains compared to the primitive action set.

3.2   Mountain Car

In the mountain car benchmark the task consists in controlling a car from the bottom of a valley into a goal region on the hill by applying a thrust either to the left or to the right. The immediate reward is 5.0 inside the goal region and -0.1 outside. We formulate the task as infinite horizon task with no absorbing goal state, i.e. the control is not stopped when the goal region is reached. This makes the task more realistic and harder to learn.

In this benchmark domain there are obviously no known subgoals that the agent should try to reach on its way to the goal region. Approaches that are based on a task decomposition are therefore not applicable here. On the other hand an optimal policy will only involve about two action switches until the goal region is reached. This is a typical situation where MSAs are suited. The number of decisions to the goal region can be reduced by selecting larger time scale actions but the necessary reactivity is retained because lower time scale actions are also available.

---

[2] This performance measure that is used in [6] neglects the length of individual trajectories and cannot be directly found in Figure 1

[3] We chose this level because it can be read off reasonably well in [6].

Table 1. Learning with different action sets. The second row shows after how many primitive training time steps a 'good' policy of quality 1.152 is permanently achieved on the average. In the third row the speed-up relative to $\mathcal{A}^{(1)}$ is depicted. The fourth row indicates if an optimal policy is possible with the chosen action set.

action set	$\mathcal{A}^{(1)}$	$\mathcal{A}^{(2)}$	$\mathcal{A}^{(4)}$	$\mathcal{A}^{(1,2)}$	$\mathcal{A}^{(1,4)}$	$\mathcal{A}^{(1,8)}$
'good' policy learned after time step	224584	107244	-	75596	49565	34556
speed-up relative to $\mathcal{A}^{(1)}$	1	2.1	-	3.0	4.5	6.5
optimal policy possible	yes	yes	no	yes	yes	yes

As the state space is continuous, we use a grid-based function approximator for the Q-functions. In a trajectory-based learning approach the origin of a transition $(s, a^j) \rightarrow s'$ does not necessarily lie on a grid point. The Kaczmarz update rule [3] is suitable to adapt the Q-function in this situation. We use zero initialised Q-functions, a discount factor $\gamma = 0.9$ and an an ϵ-greedy policy for training with $\epsilon = 0.1$. Experimentally, $\alpha = 0.3$ was determined to be a good value for the learning rate[4].

In the following we investigate how different action sets influence the speed of MSA-Q-learning. Again, we assess the performance by measuring the quality of the learned policy, i.e. the accumulated discounted reward per test trajectory, plotted against the gathered experience. All trajectories start on the bottom of the valley. This is a 'difficult' starting state because the thrust is not sufficient to reach the goal directly by driving up the hill. Instead, the car has to gain energy by moving up the opposing hill. After 80 steps the trajectory is aborted and the car is set back to the starting state. Using an optimal policy the agent reaches the goal region after 29 steps and does not leave it any more. Thus, the optimal reward that can be achieved in 80 steps is about 1.392. A policy that needs one step more to reach the goal region has a reward of about 1.152. In order to determine the speed of learning we measure after how many training time steps a policy of this quality is permanently achieved on the average. Table 1 shows quantitative results for the learning speed with different action sets. If the degree of the actions

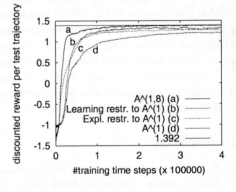

Fig. 2. Learning curves with MSA-Q-learning for the mountain car with action set $\mathcal{A}^{(1,8)}$ and exploration or learning restricted to the primitive time scale.

in a homogeneous action set becomes too large ($\mathcal{A}^{(4)}$) optimal policies cannot be learned any more. Moreover, heterogeneous action sets with MSAs lead to a

---

[4] Although the mountain car is a deterministic domain $\alpha < 1$ is required because the effects of transitions from different states in the same cell have to be averaged.

considerable speed-up of learning. With $\mathcal{A}^{(1,8)}$, for example, learning is 6.5 times faster than with $\mathcal{A}^{(1)}$.

Learning with $\mathcal{A}^{(1,n)}$ differs from learning with $\mathcal{A}^{(1)}$ in two important ways. Firstly, by allowing actions of degree n to be selected exploration steps on a larger time scale become possible. And secondly, an explicit representation of the Q-function for larger time scale actions is established. Thus, n-step actions can be selected and adapted during learning when they are optimal *and* they can be used to determine the maximum in the Q-learning rule (1). In the following experiment we show that faster learning with MSAs depends on both factors. Firstly, we restrict the use of 8-step actions to exploration steps without learning on that time scale. Secondly, we use the 8-step actions only for learning without exploration on that time scale. And thirdly, we use 8-step actions neither for exploration nor for learning which is equivalent to using $\mathcal{A}^{(1)}$. Figure 2 depicts the resulting learning curves. If only one of the two factors is missing the learning curve degrades. Thus, both exploration *and* learning on larger time scales are responsible for the success of MSA-Q-learning in the mountain car domain.

4   Conclusions

We extended Q-learning by integrating the concept of multi-step actions (MSAs) together with the intra-MSA method. The new MSA-Q-learning algorithm efficiently uses training experience from multiple explicitly specified time scales. In two benchmark domains learning could be considerably accelerated by using MSA-Q-learning together with action sets that combine primitive actions and MSAs. The success of the MSAs is due to an implicit reduction of problem size, which enables the agent to reach the goal with less decisions. The concept of MSAs can be especially applied to unstructured domains for which a decomposition is not known in advance or does not exist at all.

Our objective for future work is to choose the time scales adaptively during learning. This would lead to action sets that are tailored to the system dynamics in different regions of the state space.

References

1. T. G. Dieterich. Hierarchical reinforcement learning with the MAXQ value function decomposition. *Journal of Artificial Intelligence Research*, 13:227–303, 2000.
2. A. McGovern, R.S. Sutton, and A.H. Fagg. Roles of macro-actions in accelerating reinforcement learning. In *Grace Hopper Celebration of Women in Computing*, 1997.
3. S. Pareigis. Adaptive choice of grid and time in reinforcement learning. In *Advances in Neural Information Processing Systems*, volume 10. MIT Press, 1998.
4. R. E. Parr. *Hierarchical Control and Learning for Markov Decision Processes*. PhD thesis, University of California, Berkeley, CA, 1998.
5. T. J. Perkins and D. Precup. Using options for knowledge transfer in reinforcement learning. Technical report, University of Massachusetts, Amherst, 1999.
6. R. S. Sutton, D. Precup, and S. Singh. Between mdps and semi-mdps: A framework for temporal abstraction in reinforcement learning. *Artificial Intelligence*, 112:181–211, 1999.

Extended Kalman Filter Trained Recurrent Radial Basis Function Network in Nonlinear System Identification

Branimir Todorović[1*], Miomir Stanković[1], and Claudio Moraga[2**]

[1]Faculty of Occupational Safety, University of Niš, 18000 Niš, Yugoslavia,
bssmtod@EUnet.yu
[2]Department of Artificial Intelligence, Polytechnic University of Madrid, Spain, and
Department of Computer Science, University of Dortmund, Germany
moraga@cs.uni-dortmund.de

Abstract. We consider the recurrent radial basis function network as a model of nonlinear dynamic system. On-line parameter and structure adaptation is unified under the framework of extended Kalman filter. The ability of adaptive system to deal with high observation noise, and the generalization ability of the resulting RRBF network are demonstrated in nonlinear system identification.

1 Introduction

The system identification problem can be defined as inferring the relationship between past input-output data and future outputs of the system. In this paper we consider the Recurrent Radial Basis Function (RRBF) network, sequentially trained by Extended Kalman Filter (EKF) as a model of the unknown relationship. The architecture of the RRBF network implements the Nonlinear AutoRegresive with eXogenous inputs (NARX) model, which is chosen because of its simplicity, representational and learning capabilities [2]. The on-line structure adaptation of RRBF network is achieved by combining the growing and pruning of the hidden neurons and connections. The Kalman filter consistency test is used as the criterion for adding new hidden neurons – network growing. The on-line pruning algorithm uses the statistics estimated by EKF to determine the significance of the connections and neurons in order to decide whether to prune or not.

We have tested the ability of the proposed adaptive system to deal with the high observation noise of the unknown variance. The generalization property of the resulting network is considered as well.

---

[*] The work of B. Todorović was supported by a Scholarship of the German Academic Exchange Service (DAAD) under the Stability Pact for South East Europe.
[**] The work of C. Moraga was supported by the Spanish State Secretary of Education and Universities of the Ministry of Education, Culture and Sports (Grant SAB2000-0048), and by the Social Fund of the European Community.

J.R. Dorronsoro (Ed.): ICANN 2002, LNCS 2415, pp. 819–824, 2002.
© Springer-Verlag Berlin Heidelberg 2002

hidden layers with 20 and 10 neurons respectively, thus giving the 460 adaptable parameters in total. Networks were trained sequentially for 100 000 time steps.

Our algorithm, applied to the same problem, after a sequential training on 3000 samples produced the recurrent RBF network with 2 inputs, 12 hidden neurons and one output neuron, i.e. 61 adaptable parameters. The output of the plant and the RRBF network prediction are presented in Fig. 2(a). The growing and pruning pattern is presented in Fig. 2(b). The normalized mean square error, measured on the 200 test data samples was NMSE=1.539e-4.

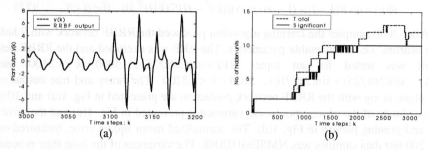

(a)   (b)

Fig. 2. (a) Outputs of the plant and the network (b) Growing and pruning pattern

7 Conclusion

We have considered the RRBF network, sequentially trained by EKF, as a model of nonlinear dynamic system. The architecture of RRBF network is based upon NARX model. The EKF is applied in estimation of the augmented state vector: parameters and outputs of RRBF network. Combining growing and pruning of the hidden units and connections we achieve the on-line structure adaptation of the network. Statistical criteria for growing and pruning were derived using Kalman filter's estimate of the state estimation error and the innovation statistics. The ability of the adaptive system to deal with high observation noise of unknown variance, and the generalization ability of the resulting RRBF network are demonstrated.

References

[1] Hassibi, B., Stork, D., Wolff, G. J.: Optimal Brain Surgeon and General Network Pruning, *IEEE Int. Conf. Neural Networks*, San Francisco (1993) 293–299.
[2] Ling, T., Horne, B. G., Tino, P., Giles, L., C., Learning long-term dependencies in NARX recurrent neural networks, *IEEE Trans. Neural Networks*, 7(6), (1996),p.1329
[3] Narendra, K. S., Parthasarathy, K., Identification and control of dynamical systems using neural networks," *IEEE Trans. Neural Networks* 1(1), (1990), 4–27
[4] Todorović, B., Stanković, M: Training recurrent radial basis function network using extended Kalman filter: parameter, state and structure estimation, In *Proc. South-Eastern Europe Workshop on Computational Intelligence and Information Technology*, Press T. University of Nis, Yugoslavia (2001)
[5] Williams, R.J.: Some observations on the use of the extended Kalman filter as a recurrent network learning algorithm, Technical Report NU_CCS_92-1. Boston: Northeastern University, College of Computer Science (1992)

Integration of Metric Place Relations in a Landmark Graph

Wolfgang Hübner and Hanspeter A. Mallot

Eberhard-Karls University, Tübingen, Germany
wolfgang.huebner@uni-tuebingen.de

Abstract. This paper describes a graph embedding procedure which extends the topologic information of a landmark graph with position estimates. The graph is used as an environment map for an autonomous agent, where the graph nodes contain information about places in two different ways: a panoramic image containing the landmark configuration and the estimated recording position. Calculation of the graph embedding is done with a modified "multidimensional scaling" algorithm, which makes use of distances and angles between nodes. It will be shown that especially graph circuits are responsible for preventing the path integration error from unbounded growth. Furthermore a heuristic for the MDS–algorithm is described, which makes this scheme applicable to the exploration of larger environments. The algorithm is tested with an agent building a map of a virtual environment.

1   Introduction

Graphs are an efficient way to code environment maps for autonomous agents. A graph $G = (V, E)$ contains information only about salient places of the environment, represented by the node set V, and the topological structure of stored places coded in the edge set E. Compared with occupancy grids, graphs are a sparse sampled representation of the environment, where the sampling density can be adapted to the environment or the agent's needs. These advantages are paid for with the need for local navigation strategies allowing travel between nodes. The edge set of the graph contains information about routes through the environment which can be traversed by the agent without taking the risk of getting lost. Therefore the agent is only able to perform topological navigation [4].

The behavior of visual homing [2] is based on the ability of the agent to identify a place by the surrounding landmark–configuration. A possible technical way to capture the landmark configuration is a conical mirror vertically mounted above a camera, providing the agent with a 360° view. It has been shown [3] that an array of 72 pixels taken from the horizon line (see figure 1) is sufficient to guide the agent back to a known place from within a certain area around the recording position.

Experiments have shown that the portion of an environment that could be represented with places defined by the landmark configuration is mainly limited

J.R. Dorronsoro (Ed.): ICANN 2002, LNCS 2415, pp. 825–830, 2002.

a) 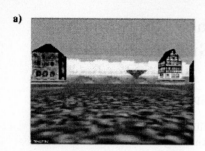 b)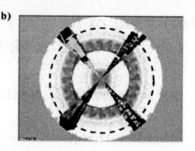

Fig. 1. a) Normal view of the environment. **b)** Panoramic view as seen from underneath the mirror. The dashed circle shows the location of the 72 pixels which are used by the visual–homing procedure.

by the ambiguity of the images. Ambiguous images are not only a result of the minimalistic visual approach. Increasing the resolution of the image may extend the size of the representable area by a certain amount. However this can't be a general solution for arbitrarily large environments.

A possible way to distinguish between similar views seen at different locations is to label the images with the recording position. Additional position information gives the agent the ability to perform survey navigation [4], i.e. the agent is able to find shortcuts apart from learned routes. This article describes a method which extends the topological graph information with metric relations between nodes. The following chapter describes the algorithm which calculates global position estimations from local movement measures obtained by path integration [1]. The last chapter shows results from an agent exploring a virtual environment and draws some conclusions about the map quality.

2   Calculation of the Metric Graph

2.1   Movement in the Graph

An agent which uses visual landmark information and metric relations simultaneously can be modeled by a state vector containing the perceived view ($I_t \in \mathbb{N}^{72}$), the instantaneous position (x_t, y_t) and the instantaneous heading (ϕ_t):

$$S_t = (I_t, x_t, y_t, \phi_t) \tag{1}$$

The change in the state vector after a rotation Θ and a translation of length d is given by:

$$S_{t+1} = (I_{t+1}, x_t + \hat{d}\sin(\phi_t + \hat{\Theta}), y_t + \hat{d}\cos(\phi_t + \hat{\Theta}), \phi_t + \hat{\Theta}) \tag{2}$$

I_{t+1} is the perceived view after the movement is completed. A detailed mathematical description of image changes generated by a moving cone mirror is given

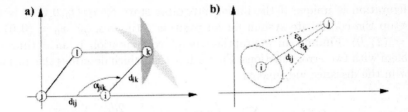

Fig. 2. a) Position estimates for k are affected by accumulated errors along the paths {j,i,k} and {j,l,k}. Calculating the position for k with respect to the whole network will bound the position error on k and consequently for all nodes linked to k. **b)** In order to calculate a reliable path from j to i, the error in the position estimate of the target node i must be limited independent from the distance d_{ij}. As a consequence the direction error ϵ_ϕ must decrease proportional to d_{ij}^{-1}

in [3]. The values $\hat{d}$ and $\hat{\Theta}$ are the intended translation and rotation plus a noise term. A node v_i in the graph contains the state vector at recording time S_i.

If the agent returns to an already known place by visual homing, a circuit in the graph is build. Considering the error in the movement measures, it is clear that a direct application of (2) will lead to erroneous position estimates along the path and to contradicting position estimates for the start node (see figure 2a). Instead of calculating path integration along single paths we use a graph–embedding procedure which takes all available routes into account and prevent the accumulation of errors.

2.2 Multidimensional Scaling

The distance between two nodes, or the length of one edge can be calculated from two successive integrator states stored for two nodes v_i and v_j:

$$d_{ij} = \sqrt{\left(x_i - x_j\right)^2 + \left(y_i - y_j\right)^2} \qquad (3)$$

The angle between two edges sharing one common node are calculated from three states:

$$\alpha_{jik} = \pi - \arctan\left(\frac{y_j - y_i}{x_j - x_i}\right) + \arctan\left(\frac{y_k - y_i}{x_k - x_i}\right) \qquad (4)$$

Graph embedding is mathematically equal to finding a function $f(V) \to \mathbb{R}^2$ which assigns a position $\tilde{x}_i$ to the node v_i. The result of the graph embedding is a configuration of points $X = (\tilde{x}_1, \tilde{x}_2, \ldots, \tilde{x}_n)$, where pairs or triples of nodes should fulfill the geometrical constraints given by (3) and (4).

Such problems can be solved by *multidimensional scaling* methods [5]. A closed form solution exists if all pairs of distances are known, i.e. if the graph is fully connected [5]. Instead of using all distances, which is impractical for moving agents, we use the additional angle information. The resulting point configuration is unique if the initial integrator state $S_0 = (I_0, 0, 0, 0)$ is used to setup the coordinate system for all position estimates, i.e. $\tilde{x}_0 = (0, 0)$ and $\tilde{x}_1 = (\tilde{x}_1, 0)$. Finding an appropriate point configuration is an optimization problem with two error functions. First a function which describes the mismatch error in the distance judgements:

$$E_d(X) = \sum_{(i,j)|d_{ij} \neq 0} \left(\| \tilde{x}_j - \tilde{x}_i \| - \frac{2d_{ij}}{\max d_{ij}} \right)^2 \tag{5}$$

and second the error of the angular match[1]:

$$E_\alpha(X) = \sum_{(j,i,k)|\alpha_{j,i,k} \neq 0} \left\| \frac{\tilde{x}_j - \tilde{x}_i}{\| \tilde{x}_j - \tilde{x}_i \|} - R(\alpha_{jik}) \frac{\tilde{x}_k - \tilde{x}_i}{\| \tilde{x}_k - \tilde{x}_i \|} \right\|^2 \tag{6}$$

Finally, both functions are combined over a weighted sum:

$$E(X) = \lambda E_d(X) + (1 - \lambda)E_a(X), \quad \lambda \epsilon [0, 1] \tag{7}$$

The weighting parameter λ could be used to compensate systematic errors in the agents path integrator. In the experiments λ is kept constant at 0.5.

Given an arbitrary configuration X, function (6) is limited by $E_\alpha(X) \leq \sum 4$. Contrary to function (6), function (5) has no upper limit, which could make the weighting factor λ obsolete and results in an overfitting of the distance values. This problem is solved by scaling down the measured edge distances by $\frac{1}{2} \max (d_{ij})$ so that the maximal distance equals 2. Selecting a start configuration randomly in a circle around the origin with a radius of $2|V|$ guarantees a solution for which function (5) and (6) are equally accounted. The resulting configuration is scaled up again by $\frac{1}{2} \max (d_{ij})$.

2.3   Application to Large Graphs

The optimization problem defined by (7) has a dimensionality of $2|V| - 3$. With a growing number of nodes a direct application of the MDS–method becomes impractical. Therefore the following heuristic is used to calculate the position estimations in real–time:

1. A subgraph around the current location v_c is selected containing all nodes which are less than ϵ edges away from the current node[2]: $G' \subseteq G$ with $V' = \{v | d(v, v_c) \leq \epsilon\}$.
2. The MDS–procedure embeds G' into a local coordinate system with $x_c' = (0, 0)$ as the origin, resulting in a point configuration X' for the subgraph.

---

[1] $R(\alpha_{jik}) \in I\!R^{2 \times 2}$ is a rotation matrix
[2] Using the graph distance as a selection method is one of many possibilities. It is e.g. also possible to use the smallest circuit containing v_c.

a)   b)

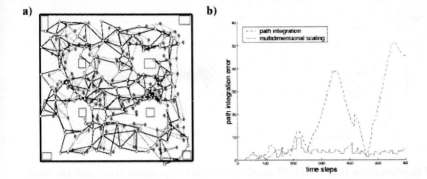

Fig. 3. a) Resulting map. Nodes marked by diamonds indicate the correct positions. Nodes marked by dots are the position estimations calculated with the MDS–method. Landmarks are shown as gray boxes. **b)** Error in the path integrator position over a time interval. Jumps in the mds–curve show points where the path integrator was recalibrated. Decreases in the path integration curve are only random.

3. G' is merged back into the global map G. The coordinate transformation is described by a rotation R and a translation T which minimizes the position difference between the corresponding point sets X and $RX' + T$. For this problem a closed form solution exists (see e.g. [6] for a detailed description).

4. The new position estimates are combined with older ones using a simple time filter in order to make the global map more stable. This becomes necessary especially as the subgraph selection could break graph circuits.

3   Results

For the experiments we use a simulation of a khepera–robot which explores an arena with a size of 180cm × 180cm. The arena contains 8 toyhouses as landmarks enclosed by a textured wall (see figure 1a). The agents path integrator was able to measure rotations and translations with a precision of ±10%. The heading of the agent is assumed to be known within the visual resolution of the panoramic image, i.e. ±2.5° with 72 pixels (see figure 1b).

The agent follows a simple exploration strategy in order to build a map of the environment. At each time step two distance measures between the instantaneous state S_t and the node states S_i are calculated. First the image similarity $d_I(I_t, I_i) = \max_i(\mathrm{corr}(I_t, I_i))$ with $\mathrm{corr}(I_t, I_i) = \max_n \sum_{m=0}^{71} I_t(n) I_i(m-n)$ and second the metric distance $d_M(x_t, x_i) = \min_i \|x_t - y_i\|$. If d_I is below a certain threshold ($d_I \leq 0.98$) and d_M is greater than 5cm a new node is added and the agent selects a new exploration direction by rotating about a fixed angle of 90°. If $d_I \geq 0.98$ and $d_M \leq 5$cm for the same node, i.e. $\arg\max(\mathrm{corr}(I_t, I_i)) = \arg\min(\|x_t - y_i\|)$ the agent tries to reach this node by visual homing. After a successful homing trail, a new edge is added. Finally, the position estimates are updated with the MDS–algorithm and the path inte-

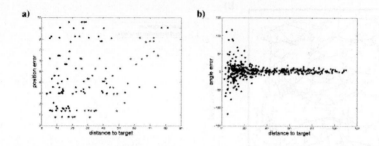

Fig. 4. The plots show that the MDS–method bounds the position error over the whole environment. Therefore the map could be used to plan paths between arbitrary pairs of nodes. **a)** Position error of the target node x_g with respect to the distance from start node x_s. **b)** Error in the calculated driving direction ϵ_α with respect to the distance.

grator is recalibrated to the improved position estimation. The resulting graph is shown in figure 3a. The thresholds for d_I and d_M are used to adjust the mesh density so that the recalibration frequency is adapted to the path integrator noise.

The map quality is measured by the agent's ability to navigate reliably over larger distances apart from learned routes. Given a start node v_s and a target node v_g the direction to the goal is calculated by $\alpha = \arctan(y_s - y_g)/(x_s - x_g)$. If the error ϵ_i in the position estimates is bounded for all nodes, i.e. $\forall_i : ||\epsilon_i|| \leq$ *const.* the error in the calculated driving direction must decrease proportional to the distance $\epsilon_\alpha \sim ||x_s - x_g||^{-1}$ (see figure 2b), which is indeed the case for the resulting map (see figure 4a and 4b).

The results show that metrically embedded landmark graphs can be established from noisy data. Such graphs form a powerful representation of space for navigation, detouring, and other tasks in spatial cognition.

References

1. Benhamou S., Séguinot V.: *How to find one's way in the labyrinth of path integration models.* J. theor. Biol. **174** (1995) 463–466
2. Cartwright B.A., Collet T.S.: *Landmark learning in bees.* Journal of Computational Physiology **A 151** (1983) 521–543
3. Franz M.O., Schölkopf B., Mallot H.A., Bülthoff H.: *Where did I take that snapshot? Scenebased homing by image matching.* Biol. Cybernetics **79(3)** (1998) 191–202
4. Kuipers B.: *The spatial semantic hierarchy.* Artifical Intelligence **119** (2000) 191–233
5. Mardia K.V., Kent J.T., Bibby J.M.: *Multivariate Analysis.* Academic Press, Inc., (1982) 413–415
6. Lu F., Milios,E.: *Robot Pose Estimation in Unknown Environments by Matching 2D Range Scans.* Journal of Intelligent and Robotic Systems **18** (1997) 249–275

Hierarchical Object Classification for Autonomous Mobile Robots

Steffen Simon, Friedhelm Schwenker, Hans A. Kestler, Gerhard Kraetzschmar, and Günther Palm

University of Ulm, Neural Information Processing Department, 89069 Ulm, Germany

Abstract. An adaptive neural 3D-object recognition architecture for mobile robot applications is presented. During training, a hierarchy of LVQ classifiers based on feature vectors with increasingly higher dimensionality is generated. The hierarchy is extended exactly in those regions of the feature space, where objects cannot be distinguished using lower-dimensional feature vectors. During recall, this system can produce object classifications in an anytime fashion with increasingly more detailed and higher confident results. Experimental data obtained from application to two real-world data sets are very encouraging. We found many of the confusion classes to represent meaningful concepts, with obvious implications for symbol grounding and integration of subsymbolic and symbolic representations.

1 Introduction

Autonomous mobile robots need the capability to recognize a wide variety of objects and classify them into an object hierarchy in order to be useful for various service tasks. For example, the service robot in our department [1] is supposed to look for and find known object instances in its environment, and to classify and localize any new instances of known classes. Given the resource constraints of typical mobile robots and demanding resource requirements of complex service robot applications, the computational costs of object classification should scale well with available processing time, even if less accurate classification results can be obtained.

All these requirements motivated the development of an adaptive hierarchical neural object classification architecture, consisting of a tree-structured system of LVQ classifiers. When moving from the tree root towards leaves, the dimensionality of the input feature vectors increases, while coverage of overall feature space decreases. The tree-structured classifier hierarchy is generated during training. On recall, the classifier hierarchy permits object classification in an anytime fashion, delivering quickly rough classification results at low cost and computing increasingly more accurate and more confident results as more computational time is alloted.

Section 2 briefly surveys image preprocessing performed prior to classification. Section 3 presents the hierarchical object classification approach and its

J.R. Dorronsoro (Ed.): ICANN 2002, LNCS 2415, pp. 831–836, 2002.

```
V := confusion-matrix(S, C)
calculate set K of confusion classes
IF k == 0 THEN EXIT
FOREACH i in k DO
    train-HC(S_i, K_i, l+1)
```

In the first step of this training procedure a LVQ network is trained with a subset of the training set $S(l)$, where $S(l)$ is the set of samples at level of detail l. S_i denotes the set of training data contained in the Voronoi cells of the prototypes in K_i. The classification performance of the LVQ network is tested on the whole training set. The result of this classification is represented by the matrix V. In every recursion step the number of prototypes for each class may be decreased in order to adopt to the actual number of remaining training patterns. Standard deviation σ is calculated from all samples correctly recognized during training by the respective prototype. It is the standard deviation of the distance between samples and prototype. The training is aborted either if there is only one class left or no more training data is available at a higher level of detail.

3.4   Recall

On recall, the nearest prototype either maps the input pattern x^μ to a confusion class K_ν or to its class label $z \in C.$, i.e. a leaf of the classification hierarchy. The recursive recall phase with an input pattern x, a set of confusion classes K (initially the root of the classification tree), and a level of detail l contains the following steps:

```
recall-HC(x, K, l):
    IF x(l) is not available THEN EXIT
    p = recall-LVQ(x(l))
    IF p in K THEN
            return recall-HC(x, K_p, l+1)
    ELSE
            return (class(p),believe(b))
```

The recall algorithm shows some properties which allow for a trade-off between processing time and classification accuracy: First, the number of levels has not to be fixed; subsequent calls to the classifier may differ in resolution. Second, the recall phase may be aborted after an arbitrary recursion step resulting in a less accurate classification result with a smaller believe value b. Finally, the stepwise refinement of the feature data allows for a kind of lazy evaluation of the feature extraction process: Features at a higher level of detail only have to be calculated, if the respective hierarchy level is really reached.

4   Experiments

Two different real-world data sets were used in order to evaluate the performance of the classifier. Features were calculated at five levels of detail from concatenated orientation histograms from 1×1 to 5×5 image segments and $m = 8$ bins, yielding feature space dimensions from 8 to 200. Classification results are based on one five-fold cross-validation run. Table 1 shows the detailed cross-validation errors.

Data set A: COIL 100

The classifier was tested on the COIL-100 set, containing 7200 images of one hundred objects recorded at 72 different angles (see [7] for details). The top-level network started with eight prototypes per class. This number was decreased by 25% in each recursion step, until a minimum number of two prototypes per class was reached. The resulting network consists of about 6000 subclassifiers and 60000 neurons with a hierarchy depth of five. The mean number of neurons traversed during recall was 1439 or 2.4%, mean computation time was 1 msec/classification on top-level and 9 msec/classification for the whole hierarchy (800 MHz Athlon running Linux). Using a sufficiently large belief threshold, the false acceptance rate (FAR) can be reduced to zero in this experiment. However, this leads to high false rejection rates (FRR): for example, using a threshold of 0.8 at the first level resulted in FAR close to 0 and FRR of about 45%.

Data set B: Robot driving sequence

Camera images where recorded for seven different 3-D objects (water bottle, blue bottle, milk box, ice tea box, cylinder, stacked cylinder, coffee machine) with an resolution of 384×288 pixels. 2100 test scenes were acquired by the robot during a run through the office environment, with mixed natural and artificial lighting conditions. Images containing objects were labeled by hand, yielding 1370 images for training. Due to the small number of classes, only four neurons per class are used. A classification accuracy of 99.512% ±0.76% (100 % on the training set) was attained.

Table 1. Results of one five-fold cross-validation run on the data sets. The mean classification error and standard deviation of the root classifier respective the whole hierarchy are given. In the third column all classification results with a believe value <0.7 were rejected, the mean classification error of the remaining classification results is given.

COIL-100	1-step LVQ	HC	HC with threshold = 0.7
training set	44.99% ± 1.33%	0.64% ± 0.08%	0.05% ± 0.02%
test set	51.24% ± 1.88%	8.44% ± 1.28%	0.97% ± 0.29%
robot sequence	1-step LVQ	HC	HC with threshold = 0.7
training set	6.11% ± 0.61%	0% ± 0%	0% ± 0%
test set	7.15% ± 0.76%	0.488%± 0.76%	0% ± 0%

5   Conclusions

We presented an approach for hierarchical neural object classification based on a set of LVQ classifiers organized in a tree structure and an appropriate training algorithm. The method proved to provide very good classification results. Its

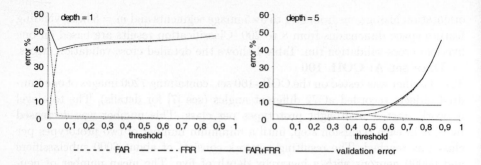

Fig. 1. False acception rate (FAR) and false rejection rate (FRR) based on the COIL-100 cross-validation run. Results for a hierarchy depth of 1 and 5 are given. The unthresholded validation error is plotted for comparison.

inherent resource adaptivity permits its use in an anytime fashion on autonomous mobile robots, or similar situations involving resource constraints.

Another interesting result is that the confusion classes derived by the algorithm in most cases represent meaningful concepts. Typically, prototypes in confusion classes represent similar objects, or objects that look similar from particular points of views. Thus, a natural object hierarchy is falling out of the classification process, which could be at least a partial solution for the symbol grounding problem. The relationship between subsymbolically derived object hierarchies and pre-defined symbolic object taxonomies remains a subject of further study.

References

1. Kestler, H.A., Sablatnög, S., Simon, S., Enderle, S., Baune, A., Kraetschmar, G.K., Schwenker, F., Palm, G.: Concurrent Object Identification and Localization for a Mobile Robot. Number 4 in Künstliche Intelligenz, Springer Verlag (2000) 23–29
2. Simon, S., Kestler, H., Baune, A., Schwenker, F., Palm, G.: Object classification with simple visual attention and a hierarchical neural network for subsymbolic-symbolic coupling. In: Proceedings of the 1999 IEEE International Symposium on Computational Intelligence in Robotics and Automation. (1999) 244–249
3. Roth, M., Freeman, W.: Orientation histograms for hand gesture recognition. Technical Report 94-03, Mitsubishi Electric Research Laboratories, Cambridge Research Center (1995)
4. Mel, B.: Seemore: Combining colour, shape, and texture histogramming in a neurally-inspired approach to visual object recognition. Neural Computation 9 (1997) 777-804
5. Kohonen, T.: Self Organizing Maps. Springer-Verlag (1995)
6. Schwenker, F., Kestler, H., Simon, S., Palm, G.: 3D Object Recognition for Autonomous Mobile Robots Utilizing Support Vector Machines. In: Proceedings of the 2001 IEEE International Symposium on Computational Intelligence in Robotics and Automation. (2001) 344–349
7. Nene, S.A., Nayar, S.K., Murase, H.: Columbia Object Image Library (COIL-100). Technical Report CUCS-006-96, Columbia University (1996)

Self Pruning Gaussian Synapse Networks for Behavior Based Robots

J.A. Becerra, R.J. Duro, and J. Santos

Grupo de Sistemas Autónomos, Universidade da Coruña, Spain
ronin@mail2.udc.es, {richard, santos }@udc.es

Abstract. The ability to obtain the minimal network that allows a robot to perform a given behavior without having to determine what sensors the behavior requires and to what extent each must be considered is one of the objectives of behavior based robotics. In this paper we propose Gaussian Synapse Networks as a very efficient structure for obtaining behavior based controllers that verify these conditions. We present some results on the evolution of controllers using Gaussian Synapse Networks and discuss the way in which they improve the evolution through their ability to smoothly select to what extent each signal and interval is considered within the internal processing of the network. In fact, the main result presented here is the way in which these networks provide a very efficient mechanism to prune the networks, allowing the construction of minimal networks that only make use of the signal intervals required.

1 Introduction

In the late eighties and early nineties, artificial evolution was proposed as a means to automate the design procedure of behavior based robot controllers [1][2][6]. Many authors have taken up this issue and have developed different evolutionary mechanisms and strategies in order to obtain robotic controllers, mostly applied to single behavior modules in the form of Artificial Neural Networks (ANNs).

When considering more complex structures, mechanisms are required for reducing the participation of the designer so that coordinated sets of behaviors can evolve more freely. In addition, an attempt must be made to reduce unnecessary complexity. The larger the number of modules that are reused the more economical the design process. One would like to reduce as much as possible the undesired effects of interferences between modules or due to the limitations induced by the finite time requirements of simulation and evolution processes, which do not allow the controllers to face all of the possible world configurations during training or evolution.

Some authors have addressed the problem of deciding if it is possible to evolve controllers for real robots through simulations that are not as complex as the real world. Jakobi [8] calls this "behavior transference" and obtains a minimal set of conditions for this transference from a controller evolved in a simulated setting to one in the real world to be successful. These conditions may be summarized as:

- Both systems must be able to support the behavior.
- There must be a mapping from the base set of external variables of the real world onto the base set of external variables of the simulation.

J.R. Dorronsoro (Ed.): ICANN 2002, LNCS 2415, pp. 837–843, 2002.

- Two sets of functions must exist such that the controller is base set exclusive and base set robust.

He contends that base set exclusive controllers are those whose internal state depends only on those features of the environment described by the base set of external variables. Base set robust implies a controller operating reliably not only in the agent environment system, but also in a whole set of associated agent environment systems.

Now, we contend that to reliably build an efficient multi-behavior robot controller, able to operate even in environments the robot has never seen before or under different robot architectures, it is not sufficient to have a base set exclusive simulation, but a global controller that is base set exclusive for each behavior within the global architecture. That is, each subcontroller must be base set exclusive, which means that the controller must only consider the information that is relevant for its objectives and completely ignore the rest. In this article we show how the network used as controller permits discovering this minimal base set of necessary external information. In addition, it must use the smallest possible architecture so as to be able to completely determine its state space through a limited simulation process.

In general, we do not know how many nodes will be required to solve a problem with an ANN. If we consider generalization we know that, if we contemplate too many nodes, low errors obtained during training will not lead to good generalization due to a possible overtraining. The same may happen with a neural robotic controller. Within the ANN field, pruning methods have been employed to reduce the computing elements of a network, eliminating noncontributory units and weights [11]. A method commonly used is to introduce a "weight decay" [7][10] that subtracts a small portion from each weight during each update. This decay toward zero is at a rate proportional to the connection´s magnitude. Thus connections disappear unless reinforced. These mechanisms can use or ignore a connection, but cannot decide to use particular intervals of the signals that circulate through it. Consequently, the resulting networks must compensate for the signal intervals that are not necessary through the use of other connections and nodes.

We would like to provide a mechanism within the controller architecture that is being evolved or trained to permit an easy way to ignore information or processing paths totally or partially in the form of ignoring certain values of the signals. At the same time, one would like to provide a search space that is as smooth as possible towards the optimum. This is the main objective of this paper, to present and test an artificial neural network that structurally and smoothly permits deciding the appropriate inputs and/or architecture (in terms of number of nodes in each layer and connections between them). The network is based on the non-linearities induced in the synapses when these become Gaussian functions.

2   Structure of the Gaussian Synapse Based ANNs and Evolution

Most neural networks employed in behavior based robotics have been based on multilayer feedforward architectures or recurrent networks [1][5][6][8]. In these structures the neurons in each layer are usually completely connected to those of the next, being a synaptic weight the element that determines the importance of each connection. In general, this weight presents a fixed value whatever the input to the synapse. These types of networks have been proven capable of great plasticity and adaptation to different classes of problems. However, due to the limited processing capacity of their nodes and connections, their scalability to complex problems has usually implied a significant increase in their sizes and thus made the process of obtaining them more time consuming and liable to local minima.

To prevent these drawbacks, one approach of great interest is to seek structures that correspond in a more direct manner to the context in which they are employed. Obviously, if we increase the processing order of nodes and/or connections appropriately, we may obtain a larger processing capability with an architecturally simpler structure.

Thus, one possible strategy is to employ activation functions that differ from the usual step or sigmoid. Another option is to increase the processing power of synapses. This is the approach we have followed. The behavior architectures we propose here use Gaussian Synapse Artificial Neural Network modules. They are evolved in simulated environments and then transferred to real robots. The networks consist of a set of nodes interconnected as a MLP, where instead of using real valued weights, we have resorted to a very simple non-linear function: the Gaussian function.

In effect, the synapses have become a sort of variable gain filtering elements; through their three parameters (center, amplitude and variance) they decide what intervals of their inputs are relevant for processing and how relevant they are. These networks can be trained, using the GSBP algorithm [3] or they can be evolved. In this work, as we are focusing on the evolution of multi-level behavior controllers for autonomous robots, we will consider the second option, that is, evolution.

Regarding the evolutionary process, we have selected a Macro Evolutionary Algorithm (MA) [9]. In our case, the original structure of MAs has been adapted to the development of ANN based robot controllers subdividing it into races to prevent the clustering that arises when a random initial distribution of the low fitness individuals of small populations is employed. Evolution, as well as the simulation of the environment and robots, is carried out using SEVEN [4], a distributed multi-behavior architecture evolver developed by our group.

3 Results and Comparison

As a demonstrative example of the effects of using Gaussian Synapse ANNs we have chosen one of the typical behaviors in robot literature: wall following. In this behavior the robot (a Pioneer II with sonar sensors in our case) has to move near the walls avoiding collisions. In figure 1 we display the evolution of fitness for 6 different experiments: using numerical synapses with 4, 8 and 12 neurons in the hidden layer and the same 3 evolutions but using Gaussian Synapses instead of numerical synapses. The other parameters of the evolutions are as follows: 1 input layer with 8 nodes (sonar sensors), 1 hidden layer, 1 output layer with 2 nodes (linear velocity and angular velocity), 1600 individuals clustered into 8 races, 2000 generations of evolution, local migration every 40 generations, global migration every 80 generations. The fitness of each individual is evaluated as its average fitness for 8 runs, each consisting of 1000 steps.

We can see that the fitness when using Gaussian Synapses is better in every case. In addition, if we plot the final ANNs obtained for the case of using Gaussian Synapses, a lot of the synapses have disappeared leading to very similar ANNs in the three test cases, which started from networks of different sizes. The results correspond to ANNs with only 3 inputs and 2 or 3 hidden neurons (example in figure 2). In the cases where Gaussian Synapses were not used, even if we want to prune, it

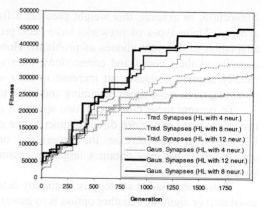

Fig. 1. Fitness of best individual for each experiment.

is not obvious how to do it without a complex analysis of the network because, with very few exceptions, connections do not present values of 0 or near 0.

Something that is interesting is that pruning was not actively valued in any way so, why does the evolutionary algorithm do it? And why is the fitness consistently better? The answer lies in the fact that using as few connections and nodes as possible is an advantage from an evolutionary or, in general, from a training point of view. In terms of evolution, the particular case of ANNs is a problem with a high degree of epistasis as there is a large dependence between genes (ANN parameters) and, therefore, the goodness of a gene depends not only on its value, but also on the value of other genes. We have to bear in mind that we seldom know the optimal size of the network, so we must start from an overdimensioned structure. Also, as it is evident, the smaller the number of connections (as long as they are enough) the smaller the search space one is really considering. So the problem of obtaining an optimal network can be solved by the evolutionary (or training) algorithm in two ways. It either eliminates connections or it compensates their values to obtain the final function. Normal networks with numerical weights present only two ways of eliminating nodes or connections: to make a synaptic weight 0 or adjust neuron bias. Epistasis makes this process even harder.

On the other hand, when we use Gaussian Synapses, the evolutionary or training algorithm has additional ways of pruning. It can very easily move the center of a Gaussian in a synapse well away from the input interval, thus making the value it transmits effectively 0. It can make the variance 0, or near 0, and adjust the amplitude so as to make the synapse value a constant. In addition, Gaussian synapse networks present the freedom for the evolutionary or training algorithm to adjust the three values to make a synapse only active for a subset of incoming values. In figure 2 we can see some examples of this: a synapse that carries always a constant value, a synapse that filters an input to activate a node only when objects are near and how the output neurons activate in a different way with different intervals of the same hidden neuron output.

With the possibilities that Gaussian Synapses offer, the evolutionary algorithm may effectively prune a network. This is useful to eliminate unnecessary elements in the network to achieve a higher fitness shuffling fewer parameters. As a consequence, the controller is better because fitness is higher. It is faster because the ANN is smaller. And it is more robust because the ANN only uses the sensors it needs and in the necessary ranges, avoiding interferences from other sensors, or even the same sensors, in ranges not contemplated during evolution. We can see this in figure 3 where the robot, which evolved to follow walls from the left, has to deal with a "never seen during evolution" wall on the right. The Gaussians Synapses ANN based controller is able to carry out the wall following task in the new world, but the traditional one is not and the robot crashes when it detects walls on the right nearer than found during the evolutionary process. As shown on the picture on the right of fig. 3, the controllers were tested on the real robot under different conditions obtaining the same results.

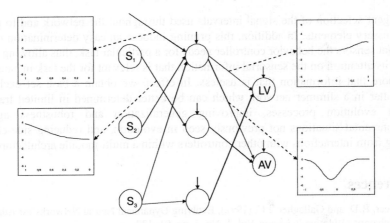

Fig. 2. Neural network resulting from evolution. We display the Gaussians obtained for some of the synapses.

If we look at the fitness again, we see how the evolution problems grow with the network size when not using Gaussian Synapses because the evolutionary algorithm has to adjust many more parameters. In the case of Gaussian synapses it just deletes those it does not require. If fact, the fitness is better when starting from larger networks, probably because when starting from smaller networks it has to activate some synapses that where inactive, due to the random nature of genes at the beginning of the evolution process, and which are necessary.

4 Conclusions

We have seen in one example how Gaussian Synapses are very well suited for automatically pruning an ANN. In our example, the resulting smaller ANN translates into a better, faster and more robust controller without any kind of penalty. The smooth modification of the parameters of the Gaussian functions permits an

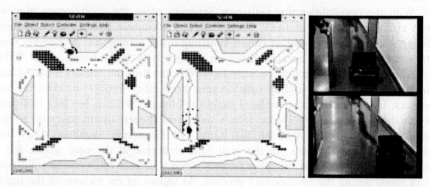

Fig. 3. Simulated robot with non Gaussians Synapse ANN (left. The robot collided with a corner), simulated robot with Gaussian Synapse ANN (center) and real robot with Gaussian Synapse ANN (right).

intelligent selection of the signal intervals used throughout the network and to prune unnecessary elements. In addition, this pruning induces an easy determination of the relevant sensors the behavior controller needs for a particular task, thus allowing it to focus its attention on the sensorial information that is relevant for the task in hand and to ignore that information that is useless. In effect, we obtain a base set exclusive controller in a slimmer network which can be better determined in limited training and/or evolution processes, improving generalization and robustness against environmental situations not previously seen in evolution, and reducing side-effects arising from interactions with other controllers within a multi-module architecture.

References

[1] Beer, R.D. and Gallagher, J.C. (1992), Evolving Dynamical Neural Networks for Adaptive Behavior, *Adaptive Behavior*, Vol. 1, No. 1, pp. 91–122.

[2] Cliff, D., Harvey, I. and Husbands, P. (1992), *Incremental Evolution of Neural Network Architectures for Adaptive Behaviour*, Tech. Rep. No. CSRP256, Brighton, School of Cognitive and Computing Sciences, Univ. Sussex, UK.

[3] Duro, R.J., Crespo, J.L., and Santos, J. (1999), Training Higher Order Gaussian Synapses, *Foundations and Tools for Neural Modeling*, J. Mira & J.V. Sánchez-Andrés (Eds.), *LNCS*, pp. 537–545, Vol. 1606, Springer-Verlag, Berlín.

[4] Duro, R.J., Becerra, J.A., and Santos, J. (2000), Evolving ANN Controllers for Smart Mobile Robots, *Future Directions for Intelligent Information Systems and Information Sciences*, Nikola Kasabov (Ed.), pp. 34–64, Physica Verlag.

[5] Floreano, D. and Mondada, F. (1998), Evolutionary Neurocontrollers for Autonomous Mobile Robots, *Neural Networks*, Vol. 11, pp. 1461–1478.

[6] Harvey, I., Husbands, P. and Cliff, D. (1993), Issues in Evolutionary Robotics, J-A. Meyer, H. Roitblat, and S. Wilson (Eds.), *From Animals to Animats 2. Proceedings of the Second International Conference on Simulation of Adaptive Behavior (SAB92)*, MIT Press, Cambridge, MA, pp. 364–373.

[7] Hinton, G.E. (1986), Learning Distributed Representations of Concepts, *Proc. of the 8th Annual Conference of the Cognitive Science Society*, pp. 1–12, Erlbaum, Hillsdale (Ed), Amherst.

[8] Jakobi, N. (1997), Evolutionary Robotics and the Radical Envelope of Noise Hypothesis, Adaptive Behavior, Vol. 6, No. 2, pp. 325–368.

[9] Marín, J. and Solé, R.V. (1999), Macroevolutionary Algorithms: A New Optimization Method on Fitness Landscapes, *IEEE Transactions on Evolutionary Computation*, Vol. 3, No. 4, pp. 272–286.

[10] Plaut, D.S., Nowlan, S., and Hinton G.E. (1986), *Experiments on Learning by Backpropagation*, Tech Rep CMU-CS-86-126, Dept CS, Carnegie-Mellon Univ.

[11] Reed, R. (1993), Pruning Algorithms - A Survey, *IEEE Transactions on Neural Networks*, Vol. 4, No. 5, pp. 740–747.

Second-Order Conditioning in Mobile Robots

Samuel Benzaquen and Carolina Chang

Grupo de Inteligencia Artificial, Departamento de Computación y TI
Universidad Simón Bolívar
Apartado Postal 89000, Caracas 1080, Venezuela
{96-28234,cchang}@ldc.usb.ve   http://www.gia.usb.ve

Abstract. We have proposed a neural network that learns to control avoidance behaviors of a physical mobile robot through classical conditioning and operant conditioning. In this article we test whether our network can acquire second-order conditioning. During training we first associate the activation of the robot's infrared sensors with collisions. Then, the activation of a visual sensor is repeatedly paired with the activation of the infrared sensors. Results show that the robot learns to elicit avoidance responses whenever the visual sensor becomes active.

1 Introduction

Over the past years we have been interested in mimicking biological mechanisms of learning in physical mobile robots. We have proposed and tested a neural network model for classical conditioning and operant conditioning [3], added mechanisms of habituation to the network [2], explored sequence learning [7], and compared our network with related work, such as [1,8]. In this paper we focus on the phenomenon of classical conditioning known as second-order conditioning.

In the classical or Pavlovian conditioning paradigm [5], learning occurs by repeated association of a Conditioned Stimulus (CS), which normally has no particular significance for an animal, with an Unconditioned Stimulus (UCS), which has significance and always gives rise to an Unconditioned Response (UCR). The response that comes to be elicited by the CS after classical conditioning is known as the Conditioned Response (CR), which is frequently very similar to the UCR. For example, every time a dog eats some food (UCS), it salivates (UCR). If the dog repeatedly hears a bell (CS) before being fed it will eventually begin to salivate (CR) soon after the bell is heard. This type of learning allows an animal to predict the events that take place in its environment.

Pavlov also described the second-order conditioning, which is a two-step process. In the first step, the CS is paired with the UCS (*i.e.* classical conditioning). In the second step, another neutral stimulus CS2 is presented repeatedly before the CS. After enough training, presentation of the CS2 alone produces the CR. Learning takes place even in the absence of the UCS, because the CS acquired UCS-like properties in step 1. Second-order conditioning is a complex phenomenon of animal learning. However, it seems feasible that it is acquired through the same mechanisms of classical (*i.e.*, first-order) conditioning.

J.R. Dorronsoro (Ed.): ICANN 2002, LNCS 2415, pp. 844–849, 2002.

In this paper we test the ability of our neural network for acquiring second-order conditioning. As we have done in the past, we let a robot learn through classical conditioning to avoid obstacles in a closed environment. Learning is achieved by repeated association of the robot's infrared sensor activation (CSs) with a collision signal (UCS). After training the robot can manage to avoid obstacles (UR) before a collision takes place. In a second step of conditioning, we use the robot's camera to detect color bricks (CS2) on the walls of the environment. Detection of the color bricks is often followed by the activation of the robot's short range infrared sensors. Hence, if our neural network can explain second-order conditioning, then the robot should learn to avoid impending collisions whenever the color bricks are detected.

2   The Neural Network Model

In recent years we have developed a neural network model that learns simultaneously to produce obstacle avoidance behaviors and light approach behaviors in a mobile robot [3]. The neural network is based on a detailed theory of learning proposed by Grossberg, which was designed to account for a variety of behavioral data on learning in vertebrates [4]. In this paper we have used a simplified version of our network, which focuses only on its mechanisms of classical conditioning. A description of the complete neural network can be found in [3].

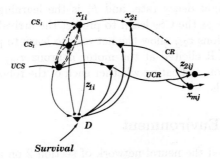

Fig. 1. The neural network model

Figure 1 depicts the neural network. Sensory nodes correspond to the conditioned stimuli, whose activation x_{1i} is given by:

$$x_{1i}(t) = \frac{I_i(t)}{\sum_j I_j(t)} \tag{1}$$

Here I_i represents a sensor value which codes proximal objects with large values, and distal objects with small values.

The drive node D codes the need to release an avoidance behavior. It becomes active when a homeostatic signal (*e.g.* survival instinct) is active in combination with a UCS, or with a CS that has learned to predict the UCS. The activation of drive node D is determined by the weighted sum of all the CS inputs, plus the UCS input, which is presumed to have a large, fixed connection strength:

$$y(t) = \sum_i x_{1i} z_{1i}(t) - T_y + UCS(t) \tag{2}$$

where $y(t)$ is the activation of the drive node D, z_{1i} is the adaptive weight connecting the sensory node x_{1i} to the drive node, T_y is a threshold that controls how easily the drive node is activated, and UCS(t) represents the collision status at time t ($UCS = 1$ if a collision just occurred, and $UCS = 0$ otherwise).

The activation of the drive node and of the sensory nodes converge upon a population of polyvalent cells. Each polyvalent cell receives input from only one sensory node, and all polyvalent cells also receive input from the drive node. We denote by x_{2i} the activation of the ith polyvalent cell:

$$x_{2i}(t) = x_{1i}(t) f(y(t)) \tag{3}$$

where $f(y(t)) = 1$ if $y(t) > 0$, and $f(y(t)) = 0$ otherwise. The multiplication of $x_{1i}(t)$ and $f(y(t))$ in Eq. 3 codes the need for joint activation from the sensory nodes (CS) and the drive node, in order for the polyvalent cells to become active.

The learning laws are given by:

$$z_{1i}(t) = E_1 z_{1i}(t-1) + F_1 x_{1i}(t) f(y(t)) \tag{4}$$

$$z_{2ij}(t) = E_2 z_{2i}(t-1) + F_2 x_{2i}(t) x_{mj}(t) \tag{5}$$

where E_i is the weight decay rate, and F_i is the learning rate. The adaptive connections z_{1i} grow as the CSs learn to predict the arrival of the UCS. On the other hand, connections z_{2ij} allow the network to learn to produce a CR, which is similar to the UCR elicited at the motor neurons x_{mj}. The motor neurons have binary activation, indicating whether each of the robot's motors is moving forward or backwards.

3   Robot and Environment

We have implemented the neural network of section 2 on a Khepera miniature mobile robot (K-Team SA, Préverenges, Switzerland). Khepera is a 55mm diameter differential drive robot (see figure 2(a)). It has eight infrared proximity sensors, and a color video turret. We situate the robot in a 48x26cm rectangular environment made out of white LEGO bricks, as shown in figure 2(b).Panel (d) is an image of the environment from the robot's view. Note that the walls occasionally have some color bricks, which are the neutral stimuli (CS2) used during second-order conditioning.

We use the *Khepera Integrated Testing Environment* (KITE) [6] to process the visual input. KITE is a tool for the evaluation of navigation algorithms for mobile robots. In particular, it features a segmentation module that detects and segments colored objects, automatically and in real time. For instance, in figure 2(c) KITE has segmented a blue brick from the background. The figure shows the bounding box of the segmented object.

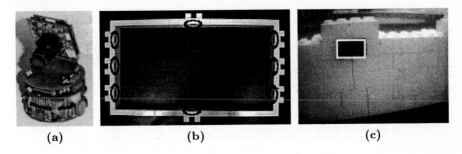

Fig. 2. The Khepera miniature robot. (a) Khepera has 8 infrared sensors and a CCD color camera. (b) The environment is a 48x26cm rectangle made out of LEGO bricks. The picture shows the location of the color bricks on the walls. (c) The environment as seen from the robot's camera.

4   Experiments and Results

The goal of this work is to achieve second-order conditioning in our mobile robot. To this end, we designed the following experiment. By default the robot always moves forward. We use information from the robot's encoders to detect collisions (UCS). The UCS always elicits an avoidance response (UCR). Whenever the robot collides with a wall, it turns to a side and then continues moving forward.

Training is done in two steps. In the first step the walls of the environment are made out of white LEGO bricks only. Readings from the robot's six front infrared sensors are neutral stimuli (CSs) that initially elicit no responses. The Khepera's infrared sensors can detect the proximity of objects within a range of 2.5cm. Learning through classical conditioning takes place as the robot navigates its environment because activation of any of the infrared sensors is often followed by a collision with a wall. Once the first-order conditioning is well established, activation of the infrared sensors alone elicits avoidance responses (CR). Therefore, after learning no more collisions take place.

In the second step of training we add some color bricks to the walls of the environment, as shown in figure 2(b). Initially the color bricks are regarded by the robot as neutral stimuli (CS2). The color bricks are detected by the robot's camera from an average distance of 8cm. Second-order conditioning develops when the robot navigates, due to the repeated association of the detection of the color bricks with the activation of the infrared sensors.

We trained the robot 10 times. In all trials the robot learned to avoid obstacles using its infrared sensors and camera. Table 1 shows the number of collisions required to learn to avoid obstacles through classical conditioning. It also shows the distance from the wall at which the responses are elicited, after first- and second-order conditioning. Second-order conditioning allowed the robot to predict impending collisions more ahead of time, resulting in a smoother navigation.

Figure 3 shows the trajectories followed by the robot after first-order and second-order conditioning. The images are taken from a top-mounted camera. A tracking algorithm depicts the trajectory of the robot, identified by a pink mark.

Table 1. Results of 10 training trials.

	Minimum	Average	Maximum
Number of collisions	34	45	57
Distance from wall at which the CR is elicited by the IRs	1.5cm	1.9cm	2.25cm
Distance from wall at which the CR is elicited by the visual input	3.2cm	5.1cm	7cm

We have used white floor only to improve visualization of the pictures. The top row of the figure (panels (a), (b), and (c)) shows some trajectories after classical conditioning. The bottom row (panels (d), (e), and (f)) shows the trajectories after second-order conditioning, when navigating from the same initial positions and directions as above. The walls are completely white in panels (a), (b), and (c), while color bricks have been added to the walls of panels (d), (e), and (f).

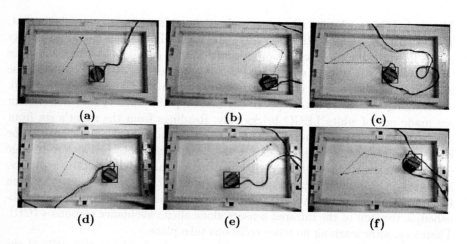

(a)   (b)   (c)

(d)   (e)   (f)

Fig. 3. First- and second-order conditioning. Panels (a), (b) and (c) show robot's trajectories after the infrared sensors have learned to predict impending collisions. Panels (d), (e) and (f) show the trajectories from the same initial points as before but after second-order conditioning. Avoidance behaviors are produced with more anticipation.

In figure 3(a), the infrared sensors become active when the robot gets close to the wall, which elicits a conditioned avoidance response. In (d) the response is generated at a farther distance because the robot has detected the color brick on the wall. Panel (b) shows the trajectory at a corner. The robot detects the walls twice, producing two turns to the right. When the robot detects the first wall in (e), it elicits a turn to the right as in (b). However, while the robot is turning, the camera detects a color brick on the other wall. Hence, the robot does not move forward but keeps turning. Panels (c) and (f) show once again that the responses are elicited with more anticipation after second-order conditioning.

5   Discussion

In this paper we have shown that our neural network can acquire second-order conditioning. The unsupervised network was trained on real time to control the reactive navigation of a physical mobile robot. In the past we have used this network to let a robot learn through first-order conditioning. Now we did not need to introduced any new neural mechanisms in order to account for second-order conditioning. This is one of the properties of the conditioning circuit [4], which is the theoretical foundation of this network. Our neural network could, in principle, acquire third- and higher-order conditioning.

Acknowledgments. We would like to thank Dr. Paolo Gaudiano, who donated the Khepera robot that we have used in this project.

References

1. Baloch, A., Waxman, A. M.: Visual learning, adaptive expectations, and behavioral conditioning of the mobile robot. *Neural Networks, 4*, (1991) 271–302.
2. Chang, C.: Improving hallway navigation in mobile robots with sensor habituation. Proceedings of the 2000 IEEE International Joint Conference on Neural Networks, (2000) 143–147.
3. Chang, C., Gaudiano, P.: Application of biological learning theories to mobile robot avoidance and approach behaviors. *J. of Complex Systems, 1*(1), (1998) 79–114.
4. Grossberg, S., Levine, D. Neural dynamics of attentionally modulated Pavlovian conditioning: blocking, interstimulus interval, and secondary reinforcement. *Applied Optics, 26*, (1987) 5015–5030.
5. Pavlov, I. P. Conditioned Reflexes. Oxford University Press (1927).
6. Şahin E., and Gaudiano P. : KITE: The Khepera Integrated Testing Environment. Proceedings of the First International Khepera Workshop. Paderborn, Germany. (1999) 199–208.
7. Quero, G., and Chang, C.:Sequence Learning in Mobile Robots using Avalanche Neural Networks. *Lecture Notes in Computer Science 2085*, 508–515.
8. Verschure, P. F. M. J., Kröse, Ben J. A., Pfeifer, R.: Distributed adaptive control: The self-organization of structured behavior *Robotics and Autonomous Systems, 9*, (1992) 181–196.

An Optimal Sensor Morphology Improves Adaptability of Neural Network Controllers

Lukas Lichtensteiger and Rolf Pfeifer

Artificial Intelligence Lab, Computer Science Department, University of Zurich,
Winterthurerstr. 190, CH-8057 Zürich, Switzerland, llicht@ifi.unizh.ch

Abstract. Animals show an abundance of different sensor morphologies, for example in insect compound eyes. However, the advantages of having highly specific sensor morphologies still remain unclear. In this paper we show that an appropriate sensor morphology can improve the learning performance of an agent's neural controller significantly. Using a sensor morphology that is "optimised" for a given task environment the agent is able to learn faster and to adapt more quickly to changes.

1 Introduction

The behavior of an autonomous agent is thoroughly affected by its morphology, i.e., the construction of its body, the placement of sensors and actuators, their specific properties, the materials used, etc [1]. An appropriate morphology can in some cases simplify the control problem for a given task, and there is evidence that it can also make the system more stable with respect to environmental changes [2]. In biological systems morphology is widely exploited [3,1]. An animal's body has always evolved together with the neural controller to survive in a specific econiche. For example, there is evidence that compound eyes of insects feature special morphologies (facet distributions) that are tailored for optical-flow based flight guidance and control [4,5,6]. There are strong morphological differences between different species and sometimes even within a single species (e.g., sex differences), and it is suspected that these differences relate to differences in the respective task environments of the animals; however, so far very little is known about this correspondence. In this paper we propose that an important function of a "good" sensor morphology is to enhance the adaptability of the neural controller, i.e., its capability to quickly cope with changes in its task or its environment. We compare three different sensor morphologies and study how the choice of a particular sensor morphology influences the learning performance and adaptability of a simple neural network controller.

2 Method

We consider a flat, 2-dimensional world where an agent is moving at constant velocity v with respect to an obstacle. We assume that the agent has to always

J.R. Dorronsoro (Ed.): ICANN 2002, LNCS 2415, pp. 850–855, 2002.

keep a sufficiently large lateral distance S_x to the obstacle by only using visual information from a single circular array of light sensors. In [7] it was shown that

$$S_x = \frac{v}{u}\rho(\alpha)\sin^2\alpha \qquad (1)$$

where α is the viewing angle, $\rho(\alpha)$ is the morphology (angular sensor distribution) of the sensor array, and u is the speed of the *image* of the obstacle over the robot's "retina" (the optical flow). In previous experiments both in simulation and on a real-world robot [7,8,9] we were evolving the sensor morphology $\rho(\alpha)$ for a *fixed, homogeneous* neural controller. There the optimal sensor morphology for the task of estimating lateral distance was found to be

$$\rho(\alpha) = k\frac{1}{\sin^2\alpha} \qquad (2)$$

where k is an arbitrary constant defining the image resolution. The most important property of this solution is the fact that it eliminates the dependency of S_x on α yielding $S_x = \frac{kv}{u}$. In [10] we were discussing potential benefits of this "optimal" sensor morphology, like reduced computational requirements, higher noise-immunity, sampling rate matched to magnitude of optical flow, etc. In this paper we will investigate how the choice of a particular sensor morphology affects the learning performance of a neural network used for processing the sensory data. Apart from the morphology given by (2) we consider

$$\text{all sensors positioned at equal angular intervals:} \quad \rho(\alpha) = k\alpha \qquad (3)$$

$$\text{a standard, (1-dimensional) pinhole-type camera:} \quad \rho(\alpha) = k\frac{1}{\cos^2\alpha}. \qquad (4)$$

The agent's task was simply to learn to distinguish between objects whose lateral distance S_x was bigger than a given threshold and others that were too close, given the input data from a certain sensor morphology. Equation (1) yields S_x as a function of the one-dimensional optical flow u which could be determined using the spatial and temporal light intensity gradients from the sensor array. However, for the present experiments we were not interested in the details of measuring optical flow, so we instead used equation (1) to simulate "ideal" flow values u for a certain sensor (more precisely, for a pair of neighboring sensors) at angle α and given a specific morphology $\rho(\alpha)$ (either (2), (3) or (4)) and a certain lateral distance S_x (the agent's speed v remained always constant, we used $v = 1$ throughout).

These simulated sensor readings were then directly input into a standard feed-forward neural network with one hidden layer, one output neuron and tanh activation function. The network was trained with optical flow values for objects at 10 different lateral distances S_x, where 5 of the distances were smaller than an arbitrarily chosen critical distance and 5 were larger. For every lateral distance considered every input unit was set either to the corresponding flow value or to a "background value" that would correspond to the flow generated by an object very far away, and all 2^N possible input combinations were presented to the network (with N input units). This amounts to training the network with objects of every possible size and shape (as perceived by the network, neglecting boundary effects) but without having two or more objects at different lateral distances at the same time (this restriction was chosen for simplicity). The network had to learn to produce a positive value at the output unit whenever there was at least one input value corresponding to the optical flow of an object closer than the critical distance, and to produce a negative value for all other cases. Training was stopped as soon as the sign of the output unit value was correct for all (10×2^N) training samples, and the number of training epochs required was recorded. Then the generalisation ability of the trained network was verified on data from a test set generated in the same way as the training set but using objects at different lateral distances (but with the same value for the critical distance). Next, the already trained network had to relearn, i.e., learn a different task that slightly contradicted the previous one: A new training set was generated where the value for the critical distance was about 30% larger than the original one (this kind of task change could happen for example when the agent's body size changes). Again, 5 of the 10 lateral distances used to generate the training set where chosen above and the other 5 below the threshold. The number of epochs required to learn the new task was recorded and the network's generalisation ability was tested (for the new critical distance). All data sets were generated separately for each of the three morphologies (using the corresponding one of equation (2), (3), or (4)) but with the same values for lateral distances and critical distances for all morphologies. One run of the experiment consisted of training, testing, re-training, and retesting the neural network separately on the corresponding data sets for each of the three morphologies. The initial values for the network's weights (before the first training) were chosen randomly from a uniform distribution on the interval $[-1.0, 1.0]$ (different ranges were also tested); within one run the same initial weights were used for all three morphologies. For learning we used standard backpropagation in batch learning mode with fixed learning rate and a momentum term.

3   Results

Figure 1 shows the number of training epochs needed for learning (graphs in left column) and re-learning a new task (graphs in right column). All graphs compare the results obtained for the three different training sets generated by simulating the three particular sensor morphologies: "std" refers to the standard pinhole camera (4), "hom" to the homogeneous circular array (3), and "opt" designates the special morphology (2). The graph at the top left shows the number of training epochs needed to successfully train a neural network with 3 input and 5 hidden units for the task described above. The backprop momentum parameter was set to 0.5 while the learning rate (normalised by the size of the training set) was varied from 0.3 to 3.0 in steps of 0.3. Each point represents an average over 10 independent runs with different random initial values for the weights; errorbars indicate the standard deviation. Runs not converged within 5×10^5 epochs were discarded from the mean calculation since their convergence

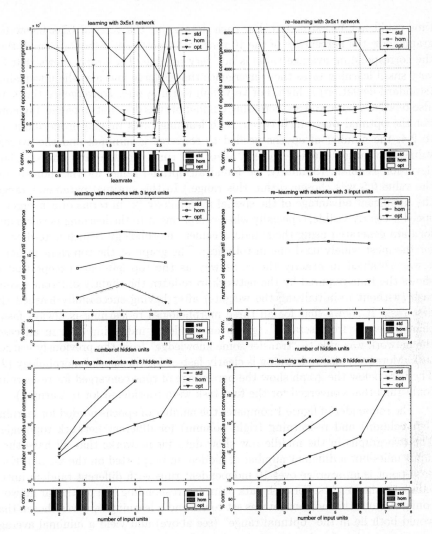

Fig. 1. Number of training epochs needed for learning (graphs in left column) and re-learning a changed task (graphs in right column). Top row: Comparison of the number of epochs needed to train a network with 3 input and 5 hidden units with input data generated by three different sensor morphologies. Each data point is an average over 10 independent runs with errorbars indicating standard deviation. For each morphology the learning rate (x-axis) was varied from 0.3 to 3.0. The bars below the graphs indicate for each data point the percentage of runs that converged to a solution. The faster convergence for the "opt"-morphology is clearly visible. Middle row: Learning performance depending on the number of hidden units (training epochs vs. number of hidden units) for networks having 3 input units. Bottom row: Learning performance depending on the number of input units (training epochs vs. number of input units) for networks having 8 hidden units. Details see text.

time could not be estimated. For each data point the vertical bars below the graph show the percentage of the 10 runs that converged; if they are small then the corresponding average value is not very reliable. The graph shows that for very small learning rates the convergence time is large and also the uncertainty (standard deviation) is high. For very high learning rates convergence times are also relatively large and an increasing number of runs does not converge anymore. For the optimal range in between, with learning rates ranging from about 1.2 to 2.4, the number of epochs needed to converge is minimal and the average values are quite reliable – nearly all of the runs converge and the standard deviation is small. (Similarly, we found an optimal range for the momentum; the value 0.5 was chosen within this range.) In this optimal parameter range, the significant advantage of the special morphology (2) in terms of convergence speed is clearly visible, especially when comparing it to the learning performance for data generated using the standard camera morphology (which is to-date by far the most widely used one in robotics). The graph at the top right in figure 1 was obtained in exactly the same way as the top left one, except that it shows the time required by the network to re-learn the slightly different second task (without re-initialising the weights) after having successfully learned the first task. It shows that for all three morphologies re-learning is always faster than to learn "from scratch" (graph on the left), indicating that in all cases the network was able to "transfer some knowledge" from the previously learned task. Moreover, also re-learning is clearly fastest for the special morphology (2). The bars below the graph show the percentage of runs converged for re-learning (only runs that converged for the first task were considered for re-learning).

The remainder of figure 1 compares the number of epochs needed for learning (left column) and re-learning (right column) for different network topologies. The two graphs in the middle row show data for networks that all have three input units but a different number of hidden units (plotted on the x-axis). Each data point is an average over 10 independent runs with different random initial values for the weights. For each data point (corresponding to a particular network configuration) the momentum was set to 0.5 and a learning rate was chosen that would both lie in the "optimal range" (see above) and yield a minimal average convergence time for the given network configuration. The two graphs show that the learning times measured by training epochs do not depend very much on the number of hidden units and that, again, re-learning is faster than learning. But most importantly, for all network configurations considered both learning and re-learning are much faster on the data sets generated by the special morphology (2) as compared to the other two morphologies (note the log-scale on the y-axis).

Finally, the bottom row of figure 1 shows the number of epochs needed for learning (left) and re-learning (right) for networks that all have 8 hidden units but a different number of input units (plotted on the x-axis). The data was obtained in the same way as for the graphs in the middle row. For a constant number of hidden units the convergence times increase with the number of input units N since the task becomes more difficult (the network has to learn 10×2^N different training vectors; when N became larger than the number of hidden units none of the networks was able to learn the task anymore). However, for all configurations the learning performance on the data sets generated for the special morphology (2) is generally again significantly better than for the other two morphologies. We were investigating all possible network topologies having between 2 and 23 hidden units (in steps of 3) and between 2 and 6 input units (in steps of 1). All these network configurations generally were much faster in learning and re-learning input data generated for the special morphology (2) than for the other two morphologies. Verifying the generalisation ability of the networks showed no significant dependency on the particular morphology used.

4   Conclusion

We have shown that choosing an appropriate sensor morphology for an autonomous agent can improve the learning performance of its neural controller significantly. Using a specific sensor morphology that is "optimised" for a given task environment the agent can learn faster and adapt quicker to changes. Maybe this increased adaptability is one of the reasons why biological evolution has come up with so many highly specific sensor morphologies.

References

1. Pfeifer, R. (2000). On the role of morphology and materials in adaptive behavior. In: *From Animals to Animats: Proceedings of the Sixth International Conference on the Simulation of Adaptive Behaviour*, J.-A. Meyer et al (eds.)
2. Panerai, F. and Sandini, G. (1998). Oculo-motor stabilization reflexes: integration of inertial and visual information. *Neural Networks*, **11**, pp. 1191–1204
3. Lehrer, M. (1998). Looking all around: Honeybees use different cues in different eye regions. *The Journal of Experimental Biology*, **201**, 3275–3292
4. Franceschini, N., Pichon, J., and Blanes, C. (1992). From insect vision to robot vision. *Philosophical Transactions of the Royal Society of London B* **337**:283–294.
5. Horridge, G.A. (1978). Insects which turn and look. *Endeavour* **1**:7–17.
6. Collett, T.S. (1978). Peering – a locust behavior pattern for obtaining motion parallax. *Journal of Experimental Biology* **76**:237–241.
7. Lichtensteiger, L. and Eggenberger, P. (1999). Evolving the Morphology of a Compound Eye on a Robot. In: *Proceedings of the Third European Workshop on Advanced Mobile Robots (Eurobot '99)*. IEEE, Piscataway, NJ, USA; 1999; p.127–134.
8. Lichtensteiger, L. (2000). Towards optimal sensor morphology for specific tasks: Evolution of an artificial compound eye for estimating time to contact. In: *Sensor Fusion and Decentralized Control in Robotic Systems III*, Gerard T. McKee, Paul S. Schenker (eds.), Proceedings of SPIE Vol. 4196, pp. 138–146
9. Lichtensteiger, L. and Salomon, R. (2000). The Evolution of an Artificial Compound Eye by Using Adaptive Hardware. In: *Proceedings of the 2000 Congress on Evolutionary Computation*, 16-19 July 2000, La Jolla Marriott, San Diego, CA, USA, pp. 1144–1151
10. Lichtensteiger, L. (2002). Evolving Task Specific Optimal Morphologies for an Artificial Insect Eye. To appear in: *Proceedings of the First International Workshop on Morpho-functional Machines*

where $\Delta X(.)$ is a well-defined function, which for each end-effector orientation provides the relative position of X with respect to X^*.

Note that X^* is not moved by varying ν, and thus it depends only on θ. We now like to obtain the inverse function, namely θ as a function of X^*. This is straightforward, except in the cases where the forward kinematics mapping is not one-to-one. If solutions are discrete, this can be handled by keeping track of the different solution branches. In the case of redundant robots, the situation is affordable with a learning architecture able to associate X^* with a subspace of θ, but one should restrict the workspace to (X, Ω) such that for every solution θ available for X^*, the compatible Ω are the same.

Either way we get independence of θ with respect to Ω, and we can write

$$\theta = \tau(X^*).$$

Using (1), we obtain:

$$\theta = \tau(X - \Delta X(\Omega)), \tag{2}$$

where both $\tau(.)$ and $\Delta X(.)$ are 3D functions.

2.2   Calculus of ν

To simplify the exposition in this section, we will consider by now that Ω is a rotation matrix.

First we define a fixed configuration of the first three joints, θ_0, to be used as reference. Then, we define a new function $\Omega_0(\cdot)$ such that $\Omega_0(\theta)$ is the rotation that transforms the orientation of the end-effector at a configuration (θ, ν) to the orientation it would have at (θ_0, ν):

$$\Omega_0(\theta)\ \Omega(\theta, \nu) = \Omega(\theta_0, \nu). \tag{3}$$

Note that $\Omega_0(.)$ is independent of ν and the only requirements are that the range for ν at every θ is contained in that at θ_0, and that the last links and joints in ν are not flexible.

We shall now define the function $\phi_0(.)$ such that $\phi_0(\Omega)$ is the ν value which at θ_0 yields the orientation Ω.

We can apply $\phi_0(\cdot)$ to both members of the equality (3), leading to:

$$\phi_0(\Omega_0(\theta)\ \Omega(\theta, \nu)) = \phi_0(\Omega(\theta_0, \nu)),$$

and thus,

$$\phi_0(\Omega_0(\theta)\ \Omega) = \nu. \tag{4}$$

2.3   The Target Decomposition

Supposing we are able to learn $\tau(.)$, $\Delta X(.)$, $\phi_0(.)$ and $\Omega_0(.)$, the inverse kinematics can be calculated in two phases. First we obtain θ following equation 2, and then we calculate ν according to equation 4.

We have obtained expressions for θ and ν as a composition of functions, each having as domain a part of the input, X or Ω. However, ϕ_0 and Ω_0 have respectively as input and output a rotation matrix, which could be compacted as Eulerian angles. Thus, by redefining adequately ϕ_0 and Ω_0, we can transform (4) into:

$$\phi_0(Eul[\Omega_0(\theta)\ \Omega]) = \nu, \tag{5}$$

$Eul[R]$ being the eulerian expression of rotation matrix R.

Now, not only $\tau(.)$ and $\Delta X(.)$, but also $\phi_0(.)$ and $\Omega_0(.)$ are 3D functions. Thus, their learning can be expected to require a number of samples orders of magnitude lower than that needed to learn the whole IKM directly.

3   Learning

The function $\Delta X(.)$ is a special case because of its simplicity, and will be considered separately from the other three functions. If, through external sensors, the set-up permits acquiring the position X^* at which the last three axes cross, then this function is not even needed: it suffices to consider (X^*, Ω) directly as input. If, on the contrary, X^* needs to be derived from (X, Ω), then a simple procedure entailing only three observations and the motion of the last two joints can be applied.

The remaining functions $\tau(.)$, $\Omega_0(.)$ and $\phi_0(.)$ are inverse functions, in the sense that we cannot generate the output for a given input. Their learning can be accomplished by means of the following two algorithms:

Learning of ϕ_0
Repeat for $i = 1$ to whatever
   Select $\nu^{i)}$
   Move to $(\theta_0, \nu^{i)})$. Observe $\Omega^{i)}$
   Learn with $Eul[\Omega^{i)}]$ as input and $\nu^{i)}$ as
   output

Learning of Ω_0 and τ
   Select ν' arbitrarily
   Move to (θ_0, ν'). Observe $\Omega^{0)}$
   Repeat for $i = 1$ to whatever
   Select $\theta^{i)}$
   Move to $(\theta^{i)}, \nu')$. Observe $(X^{i)}, \Omega^{i)})$
   Learn Ω_0 with $\theta^{i)}$ as input and $Eul[\Omega^{0)}(\Omega^{i)})^T]$
   as output
   Learn τ with $X^{i)} - \Delta X(\Omega^{i)})$ as input and $\theta^{i)}$
   as output

A similar strategy permits to perform on-line learning, i.e., learning that is integrated in normal working operation. This requires access to the inverse $\phi_0^{-1}(\nu)$, which gives the orientation that the argument ν produces when $\theta = \theta_0$. By using PSOMs [7], the learning of a function automatically makes available a proper estimation of its inverse and, therefore, a separate estimator for $\phi_0^{-1}(\nu)$ is not required.

4   Experimental Results

We have used the PUMA robot as a testbed to validate our procedure in a controlled setting. The ranges allowed for the six joints [1] are as follows: $[-150, -10]$, $[-215, -100]$, $[-35, 80]$, $[-110, 170]$, $[-100, 100]$, $[-100, 100]$. Note that this defines a very large workspace.

Local Parametrized Self-Organizing Maps (PSOM) [7] with a subgrid size of 4 knots per axis have been chosen as an appropriate neural model to experiment with the decomposition approach.

For the control experiment (labeled "standard") we simply generate grids in the joint space covering the workspace above. Then we move the robot to the different configurations represented in the grid to obtain the associated positions and orientations. Thus, each knot in the grid requires one movement.

In the experiment to test our decomposition approach, we generate a grid for θ and move the first three robot joints to traverse each of its knots in order to get simultaneously points for τ and Ω_0. In the same way, a grid for ν is generated and movements are carried out accordingly to get ϕ_0 points.

Orientations and rotations are represented with five elements of the corresponding rotation matrix that determine it univocally.

Tables 1 and 2 show the precisions attained with an increasing number of movements. Units are millimeters for position and radians for orientation. The precision was evaluated by querying for 400 random position-orientation configurations inside the workspace. The tables only cover numbers of movements that seem reasonable. It was impossible with our computer memory resources (allowing grids of up to 262,144 points) to reach precisions under 1 mm and .01 radians with the standard procedure, whereas the decomposition procedure only needed 686 and 1024 movements to get these precisions, respectively.

A final and important remark is that the time to obtain good precisions was also orders of magnitude faster with the decomposition approach. This is due to lower searching times to get the closer knot in the grids and to lower complexity in the optimizations performed in the PSOMs.

5   Concluding Remarks

The purpose of this paper is to propose a method to learn the IK mapping with a reasonable number of movements when a high accuracy is required.

In addition to learning efficiency, our decomposition procedure has other advantages over classic learning of IK in some contexts. For example, in [5] we tackled IK learning for a REIS robot placed in a Space Station mock-up, whose

Table 1. Position (in millimeters) and orientation (in radians) precisions obtained with different numbers of movements using the standard procedure.

number of movements	position mean error	position stdev. error	orientation mean error	orientation stdev. error
64	591	234	2.145	0.655
729	195	132	0.316	0.253
4096	38	50	0.138	0.163

Table 2. Position (in millimeters) and orientation (in radians) precisions obtained with different numbers of movements using the new decomposition procedure.

number of movements	position mean error	position stdev. error	orientation mean error	orientation stdev. error
54	57.7	27.2	0.420	0.255
128	9.7	6.0	0.158	0.150
250	3.0	2.7	0.068	0.134
432	1.0	0.8	0.020	0.029
686	0.6	1.0	0.015	0.032
1024	0.2	0.2	0.006	0.028

mission was to insert and extract cards from a rack. If, due to launching stress or tear-and-wear, the IK mapping would strongly deviate from the nominal one, the movements required for relearning could damage the rack (or further damage the robot). With the procedure here proposed, it is possible to learn to move in the complete workspace without actually moving everywhere, and only approach risk zones after learning has been successfully completed.

References

1. Fu K.S., González R.C. and Lee C.S.G., 1987: *Robotics: Control, Sensing, Vision, and Intelligence*, New York: McGraw-Hill.
2. Kröse B.J.A. and van der Smagt P.P., 1993: *An Introduction to Neural Networks* (5th edition), Chapter 7: "Robot Control". University of Amsterdam.
3. Martinetz T.M., Ritter H.J. and Schulten K.J, 1990: Three-dimensional neural net for learning visuomotor coordination of a robot arm. *IEEE Trans. on Neural Networks*, **1**(1): 131–136.
4. Ritter H., Martinetz T. and Schulten K.J. , 1992: *Neural Computation and Self-Organizing Maps*. New York: Addison Wesley.
5. Ruiz de Angulo V. and Torras C., 1997: Self-calibration of a space robot. *IEEE Trans. on Neural Networks*, **8**(4): 951-963.
6. Walter J.A. and Schulten K.J., 1993: Implementation of self-organizing neural networks for visuo-motor control of an industrial arm. *IEEE Trans. on Neural Networks*, **4**(1).
7. Walter J. and Ritter H., 1996: Rapid learning with parametrized self-organizing maps. *Neurocomputing*, **12**: 131-153.

Part VI

Selforganization

The Principal Components Analysis Self-Organizing Map

Ezequiel López-Rubio, José Muñoz-Pérez, and José Antonio Gómez-Ruiz

E.T.S.I. Informática
Universidad de Málaga
Campus de Teatinos, s/n. 29071-Málaga (SPAIN)
{ezeqlr,munozp,janto}@lcc.uma.es

Abstract. We propose a new self-organizing neural model that performs Principal Components Analysis (PCA). It is also related to the ASSOM network, but its training equations are simpler. Furthermore, it does not need any grouping of the input samples by episodes. Experimental results are reported, which show that the new model has better performance than the ASSOM network in a number of benchmark problems.

1 Introduction

The adaptive subspace self-organizing map (ASSOM) is an evolution of the self-organizing feature map (SOFM) proposed by Kohonen [1]. The ASSOM was first presented as an invariant feature detector [2]. This property has been further studied in [3], and its relations with wavelets and Gabor filters have been reported in [4]. This network is an alternative to the standard principal component analysis (PCA) algorithms, as it looks for the most relevant features of the input data. The ASSOM has been successfully applied to the handwritten digit recognition problem [5] which many neural network researchers have addressed. Also, it has been used for texture segmentation [6].

In Section 2 we propose our model. The differences between the original Kohonen's approach and our proposal are considered in Section 3. Finally, sections 4 and 5 deal with experimental results and conclusions, respectively.

2 The PCASOM Model

2.1 Neuron Weights Updating

The ASSOM network stores a vector basis in each unit. The weight update involves a rather complicated transformation to obtain a vector basis that is similar to the older one, but closer to the present input samples. Here we propose an alternative way to

J.R. Dorronsoro (Ed.): ICANN 2002, LNCS 2415, pp. 865–870, 2002.
© Springer-Verlag Berlin Heidelberg 2002

store the information: we use the *covariance matrix*. The covariance matrix of an input vector **x** is defined as

$$\mathbf{R} = E\left[(\mathbf{x} - E[\mathbf{x}])(\mathbf{x} - E[\mathbf{x}])^T\right] \tag{1}$$

where $E[\cdot]$ is the mathematical expectation operator and we suppose that all the components of **x** are real random variables.

If we have M input samples, $\mathbf{x}_1, ..., \mathbf{x}_M$, we can make the following approximation:

$$\mathbf{R} \approx \mathbf{R}^I = \frac{1}{M-1} \sum_{i=1}^{M} (\mathbf{x}_i - E[\mathbf{x}])(\mathbf{x}_i - E[\mathbf{x}])^T \tag{2}$$

Note that this is the best approximation that we can obtain with this information (it is an unbiased estimator with minimum variance). Now, if we obtain N new input samples, $\mathbf{x}_{M+1}, ..., \mathbf{x}_{M+N}$, we may write:

$$\mathbf{M} = \frac{1}{N} \sum_{i=M+1}^{M+N} (\mathbf{x}_i - E[\mathbf{x}])(\mathbf{x}_i - E[\mathbf{x}])^T \tag{3}$$

$$\mathbf{R} \approx \mathbf{R}^{II} = \frac{1}{M+N-1}\left((M-1)\mathbf{R}^I + N\mathbf{M}\right) \tag{4}$$

Both $\mathbf{R}^I$ and $\mathbf{R}^{II}$ are approximations of $\mathbf{R}$, but $\mathbf{R}^{II}$ is more accurate because it takes into account the $N+M$ input samples. Equation (4) is a method to accumulate the new information (the last N input samples) to the old information (the first M input samples). Now we need to approximate the expectation of the input vector $E[\mathbf{x}]$. We may use a similar approach:

$$E[\mathbf{x}] \approx \mathbf{e}^I = \frac{1}{M} \sum_{i=1}^{M} \mathbf{x}_i , \quad \mathbf{e}^{II} = \frac{1}{N} \sum_{i=M+1}^{M+N} \mathbf{x}_i \tag{5}$$

$$E[\mathbf{x}] \approx \mathbf{e}^{III} = \frac{1}{M+N}\left(\sum_{i=1}^{M} \mathbf{x}_i + \sum_{i=M+1}^{M+N} \mathbf{x}_i\right) = \frac{1}{M+N}\left(M\mathbf{e}^I + N\mathbf{e}^{II}\right) \tag{6}$$

Analogously, $\mathbf{e}^I$, $\mathbf{e}^{II}$ and $\mathbf{e}^{III}$ are approximations of $E[\mathbf{x}]$, but $\mathbf{e}^{III}$ is more accurate because it takes into account all the input samples. Note that if we suppose $N \ll M$, we have $\mathbf{e}^I \approx \mathbf{e}^{III}$. We use $\mathbf{e}^I$ to approximate $E[\mathbf{x}]$ in (2), and $\mathbf{e}^{III}$ to approximate $E[\mathbf{x}]$ in (4). Note that we do not use $\mathbf{e}^{II}$ in (4) because it is a poor approximation of $E[\mathbf{x}]$ while $\mathbf{e}^{III}$ is better, and both of them can be calculated only when the last N input samples are known.

Every processing unit of our model stores approximations to the matrix **R** and the vector $E[\mathbf{x}]$. They will use the above equations to update these approximations. So, in the time instant t the unit i stores the matrix $\mathbf{R}_i(t)$ and the vector $\mathbf{e}_i(t)$.

Let the *learning rate for the covariance matrix* η_R and the *learning rate for the mean* η_e be

$$\eta_R = \frac{N}{N+M-1}, \quad \eta_e = \frac{N}{N+M} \tag{7}$$

Then we may rewrite (6) and (4) as

$$\mathbf{e}^{III} = \left(1-\eta_e\right)\mathbf{e}^I + \eta_e\mathbf{e}^{II} \tag{8}$$

$$\mathbf{R}^{II} = \left(1-\eta_R\right)\mathbf{R}^I + \eta_R\mathbf{M} \tag{9}$$

Note that these expressions are analogous to the weight update equations of the competitive learning and the self-organizing feature map (SOFM).

2.2 Competition among Neurons

When an input sample is presented to a self-organizing map, a competition is held among the neurons. Here we use an adaptation of the ASSOM approach. We note the *orthogonal projection* of a vector $\mathbf{x}$ on an orthonormal vector basis $B=\{\mathbf{b}_h \mid h=1,...,K\}$ as $\hat{\mathbf{x}} = Orth(\mathbf{x}, B)$. The vector $\mathbf{x}$ can be decomposed in two vectors, the orthogonal projection and the *projection error*, i. e., $\mathbf{x} = \hat{\mathbf{x}} + \tilde{\mathbf{x}}$. Every unit (say, i) of our network has an associated vector basis $B^i(t)$ at every time instant t. It is formed by the K eigenvectors corresponding to the K largest eigenvalues of $\mathbf{R}_i(t)$. This is rooted in the Principal Components Analysis (PCA). Note that $B^i(t)$ must be orthonormalized in order for the system to operate correctly. The difference among input vectors $\mathbf{x}^i(t)$, $i=1,...,N$ and estimated means are projected onto the vector bases of all the neurons. The neuron c that has the minimum sum of projection errors is the winner:

$$c = \arg\min_j \left(\sum_{i=1}^{N} \left\| \mathbf{x}^i(t) - \mathbf{e}_j(t) - Orth\left(\mathbf{x}^i(t) - \mathbf{e}_j(t), B^j(t)\right) \right\|^2 \right) \tag{10}$$

2.3 Network Topology

We consider a topology that defines which neurons are neighbors. This means that our model is a *computational map*. Typically the neurons form a rectangular lattice. When a neuron c wins the competition it is updated. Its neighbors are also updated, according to the *degree of neighborhood* π_{ic} between winning neuron c and its neighbor i. The update equations for neuron i are the following:

$$\mathbf{e}_i(t+1) = \mathbf{e}_i(t) + \eta_e(t)\pi_{i,c}(t)\left[\frac{1}{N}\left(\sum_{j=1}^{N}\mathbf{x}_j\right) - \mathbf{e}_i(t)\right] \tag{11}$$

$$R_i^*(t+1) = \frac{1}{N} \sum_{j=1}^{N} (x_j - e_i(t+1))(x_j - e_i(t+1))^T \tag{12}$$

$$R_i(t+1) = R_i(t) + \eta_R(t)\pi_{i,c}(t)\left(R_i^*(t+1) - R_i(t)\right) \tag{13}$$

The learning process is divided into two phases, like the standard self-organizing map algorithms: the *ordering phase* and the *convergence phase*. It is during the ordering phase when the topological ordering of the neurons takes place.

3 Comparison with ASSOM

The PCASOM algorithm is quite different from ASSOM. The PCASOM model has the following advantages:

a) It is solidly rooted on statistics (KL transform, PCA), because the model equations have been derived directly from statistical considerations.

b) Update equation is more stable, because it is based on matrix sums, and not on quotients and matrix products. Furthermore, the learning rate has a direct justification.

c) Does not need episodes. This means that it is no longer needed a criterion to group the samples that belong to the same subspace into episodes. The *a priori* knowledge that the ASSOM required to build episodes is not available in a number of applications. PCASOM supports batch mode, but it is optional.

d) Does not need dissipation of spurious components. The need for this step in the ASSOM model suggests that there is an undiscovered instability in this model. Hence, the fact that PCASOM does not need it is an advantage, both in computation time and reliability of the results.

e) It has a wider capability to represent the input distribution. In the ASSOM, each neuron is able to adapt to a subset of the input of the form $\alpha_l b_l + ... + \alpha_K b_K$, where K is the number of basis vectors. On the other hand, each PCASOM neuron adapts to a subset of the form $e + \alpha_l b_l + ... + \alpha_K b_K$, where e is the estimated mean. If we desired to reduce the representation capability of the PCASOM to match that of the ASSOM, we would only have to fix $e_i(t)=0$ $\forall i \forall t$, i. e., to deactivate the learning of the means.

4 Experimental Results

Our set of experiments is devoted to the comparison of the classification performance of ASSOM and PCASOM. We have selected several standard benchmark databases from the *UCI Repository of Machine Learning Databases and Domain Theories* [7].

We have split the complete sample set into two disjoint subsets. The training subset has been presented to the networks in the training phase, while the test subset has been used to measure the classification performance of the trained networks. Each of the two subsets have 50% of the samples.

All the samples of the training subset have been presented to a unique network (non supervised learning). When the training has finished, we have computed the winning neuron for all the samples of the training subset. For every neuron, we have computed its *receptive field*, i. e., the set of training samples for which this neuron is the winner. Each neuron has been assigned to the class with the most training samples in its receptive field. Finally, we have presented the test samples to the network, one by one. If the winning neuron corresponds with the class of the test sample, we count it as a successful classification. Otherwise, we count it as a classification failure.

We have use the same network architecture with both ASSOM and PCASOM: a 4x4 rectangular lattice, with 2 basis vectors per neuron. The neighbourhood function has been always a Gaussian. We have selected a linear decay rate of the neighbourhood width σ in the ordering phase. In the convergence phase we fix $\sigma=0.04$. In the experiments with ASSOM, it is mandatory that the samples are grouped into episodes. As one could expect, the benchmark databases include no information about how this grouping should be done, so we have formed each episode with 10 or 20 consecutive samples (according to the original database order). On the other hand, the PCASOM does not need any grouping, as explained, so we have used only one sample per episode, i. e., no grouping.

The parameters of both models have been optimized in order to make a useful comparison. The optimal values for the parameters that we have used have been the following. For the ASSOM model: initial learning rate $\lambda(0)=0.45$, initial neighbourhood width $\sigma(0)=4$, dissipation parameter $\delta=0.2$. For the PCASOM model: initial learning rates $\eta_e=1$ and $\eta_R=1$, initial neighbourhood width $\sigma(0)=3.5$.

The experiments have been performed in an Origin 2000 computer. The 'CPU time' results correspond to monoprocessor MATLAB 6.0 processes, running on a MIPS R10000 microprocessor (196 MHz clock).

We can see that the PCASOM model outperforms ASSOM in the BalanceScale, Ionosphere, Segmentation and Yeast databases. It must be noted that these results are achieved even with 20000 epochs (PCASOM), when the computational effort is much less than ASSOM. There are also two cases (Contraceptive and Haberman) where the classification performance is similar, but the PCASOM has less computations.

Table 1. Correct classification results on the test subset

	ASSOM		PCASOM	
# Episodes * Ep. size	10,000 * 10	20,000 * 20	20,000 * 1	60,000 * 1
BalanceScale	59.935%	57.692%	66.346%	68.269%
Contraceptive	44.761%	46.394%	42.721%	42.585%
Glass	52.380%	52.427%	36.190%	36.215%
Haberman	73.514%	73.842%	73.684%	71.710%
Ionosphere	79.428%	74.857%	76.071%	84.213%
PimaIndiansDiabetes	66.406%	60.156%	65.104%	63.541%
Segmentation	31.428%	35.238%	50.476%	56.190%
Yeast	39.783%	39.648%	38.430%	45.331%

Table 2. CPU time results, in tens of thousands of seconds

	ASSOM		PCASOM	
# Episodes * Ep. size	10,000 * 10	20,000 * 20	20,000 * 1	60,000 * 1
BalanceScale	0.87768	3.17866	0.36624	1.03283
Contraceptive	0.93265	3.30942	0.45345	1.28803
Glass	2.17091	7.68927	1.04644	3.00307
Haberman	0.58045	2.12949	0.23423	0.68150
Ionosphere	0.88435	2.92071	1.57546	4.84513
PimaIndiansDiabetes	0.62082	2.24226	0.28243	0.84441
Segmentation	2.44941	8.48273	2.00363	5.15219
Yeast	3.07094	11.09871	1.43532	4.39100

5 Conclusions

We have proposed a new self-organizing neural model that performs Principal Components Analysis (PCA). It is also related to the ASSOM network, but its training equations are much simpler, and its input representation capability is broader. Furthermore, the grouping of the input samples into episodes and the dissipation step are no longer needed. Experimental results have been reported, with an application to classification. Experiments show that the new model has better performance than the ASSOM network with less computations in a number of benchmark problems.

References

[1] Kohonen, T.: The self-organizing map. Proceedings of the IEEE 78 (1990) 1464–1480.
[2] Kohonen, T.: The adaptive-subspace SOM (ASSOM) and its use for the implementation of invariant feature detection. In F. Fogelman-Soulié & P. Galniari (Eds.), Proceedings of the ICANN'95, International Conference on Artificial Neural Networks 1, 3–10. Paris: EC2 & Cie (1995).
[3] Kohonen, T.: Emergence of invariant-feature detectors in the adaptive-subspace SOM. Biological Cybernetics 75 (1996), 281–291.
[4] Okajima, K.: Two-dimensional Gabor-type receptive field as derived by mutual information maximization. Neural Networks 11(3) (1998) 441–447.
[5] Zhang, B., Fu, M., Yan, H., Jabri, M. A.: Handwritten digit recognition by adaptive-subspace self-organizing map (ASSOM). IEEE Transactions on Neural Networks 10(4) (1999) 939–945.
[6] Ruiz del Solar, J.: TEXSOM: Texture segmentation using self-organizing maps. Neurocomputing 21 (1–3) (1998) 7–18.
[7] Murphy, P. M.: UCI Repository of Machine Learning Databases and Domain Theories [online]. Available: http://www.ics.uci.edu/~mlearn/MLRepository.html. Date of access: March 2001.

Using Smoothed Data Histograms for Cluster Visualization in Self-Organizing Maps[*]

Elias Pampalk[1], Andreas Rauber[2,**], and Dieter Merkl[2]

[1] Austrian Research Institute for Artificial Intelligence (OeFAI)
Schottengasse 3, A-1010 Vienna, Austria
elias@oefai.at
[2] Department of Software Technology and Interactive Systems
Vienna University of Technology
Favoritenstr. 9-11/188, A-1040 Vienna, Austria
{andi, dieter}@ifs.tuwien.ac.at

Abstract. Several methods to visualize clusters in high-dimensional data sets using the Self-Organizing Map (SOM) have been proposed. However, most of these methods only focus on the information extracted from the model vectors of the SOM. This paper introduces a novel method to visualize the clusters of a SOM based on smoothed data histograms. The method is illustrated using a simple 2-dimensional data set and similarities to other SOM based visualizations and to the posterior probability distribution of the Generative Topographic Mapping are discussed. Furthermore, the method is evaluated on a real world data set consisting of pieces of music.

1 Introduction

The Self-Organizing Map (SOM) [1] is frequently employed in exploratory data analysis to project multivariate data onto a 2-dimensional map in such a way that data items close to each other in the high-dimensional data space are close to each other on the map. In the interactive process of data mining such maps can be utilized to visualize clusters in the data set to support the user in understanding the inherent structure of the data.

Several methods to visualize clusters based on the SOM can be found in the literature. The most commonly used method, i.e. the U-Matrix [2], visualizes the *distances between the model vectors* of units which are immediate neighbors. Alternatively, the *model vectors can be clustered* [3] using techniques such as k-means or hierarchical agglomerative clustering. The clusters or dendrograms can be visualized on top of the map. Another possibility to analyze the cluster structure of the SOM is to *project the model vectors* into low-dimensional spaces and use a color coding to link the projections with the original map [4,5]. A

---

[*] Part of this work was supported by the Austrian FWF under grant Y99-INF.
[**] Part of the work was performed while the author was an ERCIM Research Fellow at IEI, Consiglio Nazionale delle Ricerche (CNR), Pisa, Italy.

J.R. Dorronsoro (Ed.): ICANN 2002, LNCS 2415, pp. 871–876, 2002.
© Springer-Verlag Berlin Heidelberg 2002

similar approach is to *mirror the movement of the model vectors* during SOM training in a two-dimensional output space using Adaptive Coordinates [6]. A rather different approach to visualize clusters are *data histograms* which count how many data items are best represented by a specific unit. For an overview of different possibilities to visualize the data histograms see e.g. [7]. However, the problem with data histograms is that considering only the best matching unit for each data item ignores the fact that data items are usually represented well by more than just one unit.

The visualization method presented in this paper is based on smoothed data histograms (SDH). The underlying idea is, that clusters are areas in the data space with a high density of data items. The results are related to the posterior probability distribution obtained by e.g. the Generative Topographic Mapping (GTM) [8], however the technique itself is very simple and computationally not heavier than the calculation of the standard data histogram.

The remainder of this paper is organized as follows. Section 2 presents the calculation of SDHs and discusses the similarities to other methods. In Section 3 the method is evaluated using a data set consisting of 359 pieces of music, and in Section 4 some conclusions are drawn.

2   Smoothed Data Histograms

2.1   Principles

The SOM consists of units which are usually ordered on a rectangular 2-dimensional grid which is referred to as map. A model vector in the high-dimensional data space is assigned to each of the units. During the training process the model vectors are fitted to the data in such a way that the distances between the data items and the corresponding closest model vectors are minimized under the constraint that model vectors which belong to units close to each other on the map, are also close to each other in the data space.

The objective of the SDH is to visualize the clusters in the data set through estimation of the probability density of the high-dimensional data on the map. This is achieved by using the SOM as basis for a smoothed data histogram. The map units are interpreted as bins. The bin centers in the data space are defined by the model vectors and the varying bin widths are defined through the distances between the model vectors. The membership degree of a data item to a specific bin is governed by the smoothing parameter s and calculated based on the rank of the distances between the data item and all bin centers. In particular, the membership degree is s/c_s to the closest bin, $(s-1)/c_s$ to the second, $(s-2)/c_s$ to the third, and so forth. The membership to all but the closest s bins is 0. The constant $c_s = \sum_{i=0}^{s-1} s - i$ ensures that the total membership of each data item adds up to 1.

The SDH is illustrated and the similarities to other methods are discussed using the 2-dimensional data set presented in Figures 1(a) and 1(b). The data set consists of 5000 samples, randomly drawn from a probability distribution that is a mixture of 5 Gaussians. The SOM consists of 10×10 units which are

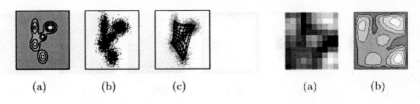

Fig. 1. The 2-dimensional data space. (a) The probability distribution from which (b) the sample was drawn and (c) the model vectors of the SOM.

Fig. 2. The SDH ($s = 8$) visualized using (a) gray shadings and (b) contours.

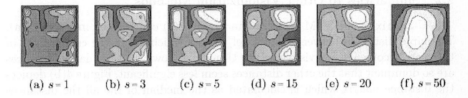

(a) $s = 1$ (b) $s = 3$ (c) $s = 5$ (d) $s = 15$ (e) $s = 20$ (f) $s = 50$

Fig. 3. Different values for the smoothing parameters s and their effects on the SDH.

arranged on a rectangular grid. The model vectors in the data space, where the immediate neighbors are connected by a line, are shown in Figure 1(c).

Figure 2(a) illustrates the SDH for $s = 8$. The SOM has been rotated so that the orientation of the latent space corresponds to the orientation of the data space. The gray shadings are chosen so that darker shadings correspond to lower values, while areas shown in lighter shadings correspond to higher values of the SDH. To further simplify the SDH visualization a 5-level contour plot is presented in Figure 2(b), where the values between units are interpolated. A detailed comparison of the SDH and the model vectors in the data space reveals, that the cluster centers found by the SDH correspond well to the cluster centers of the data set.

The influence of the parameter s can be seen in Figure 3. For $s = 1$ the results are identical with those of the data histogram, where clusters cannot easily be identified. For example, the cluster in the upper left corner is represented by three different peaks. For increasing values of s the general characteristics of the data space become more obvious until levelling out into a coarser cluster representation. For extremely high values of s there is only one big cluster which has its peak approximately in the center of the map.

The different cluster shapes reflect the hierarchical structure of the clusters in the data. In particular, $s \geq 50$ resembles the top level in the hierarchy, where the whole data set appears as unity. On the next level, with $s \approx 20$, the difference between the upper right cluster and the rest of the data set becomes noticeable. In the range $5 \leq s \leq 15$ the 5 clusters can easily be identified, and $s < 5$ depicts the noise within the data. The appropriate hierarchical level and thus the optimal value for the smoothing parameter s depends on the respective application and in particular on the noise level contained in the data.

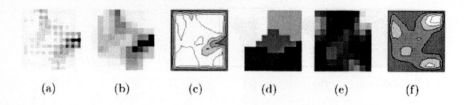

(a)   (b)   (c)   (d)   (e)   (f)

Fig. 4. Alternatives to the SDH, in particular: (a) U-matrix, (b) distance matrix, (c) as contour plot, (d) k-means, (e) posterior distribution of GTM, (f) as contour plot.

2.2   Similarities to Other Visualization Methods

The U-matrix of the SOM presented in Figure 1(c) can be seen in Figure 4(a). The large distances between the model vectors which are in the center right of the map become clearly visible in the U-matrix. However, these large distances are so dominant that the other distances seem less significant. Figure 4(b) depicts the distance matrix which is calculated as the median of the all the distances of the model vectors between a unit and its immediate neighbors. Although the big gap in the center right of the map is still very dominating, the contour plot depicts that there are 5 clusters (cf. Figure 4(c)).

Generally, there is a close relationship between the distance of the model vectors and the probability density of the data due to an important characteristic of the SOM algorithm known as *magnification factors*. Areas with a high density of data items are represented by more model vectors and thus with more detail than sparse areas. Another advantage of visualizing the distances between the model vectors is that there is no need for further parameters. Note that the cluster centers found by the SDH with $s = 8$ and with the distance matrix are approximately identical. Yet, as we will see in Section 3, the distance matrix provides sub-optimal results in many real-world data mining applications.

Figure 4(d) illustrates the k-means (with $k = 5$) clustering of the model vectors where all map units belonging to one cluster have the same gray shading. Although the 5 clusters can clearly be identified, it is necessary to know how many clusters there are beforehand. The Davies-Bouldin index [9], for example, indicates that $k = 2$ would be a better choice, resulting only in the separation of the upper right cluster from the other four clusters.

The main advantage of the posterior probability of the GTM (cf. Figures 4(e - f)) is the precisely defined statistical interpretation. The GTM was trained using 10×10 latent points, 4 basis functions, and $\sigma = 3$. Although this paper does not deal with visualizing the GTM, it is interesting to note the similarities between the posterior probability and the SDH.

3   Visualizing a Music Collection

The SDH was evaluated using a data set consisting of 359 pieces of music which are represented by 1200 features based on Fourier-transformed frequency spectra. The main goal of this experiment is to obtain a clustering of music according

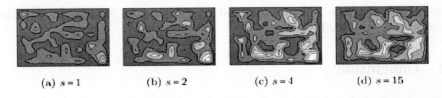

(a) $s = 1$　　　　(b) $s = 2$　　　　(c) $s = 4$　　　　(d) $s = 15$

Fig. 5. Influence of the parameter s on the SDH visualization of the music collection.

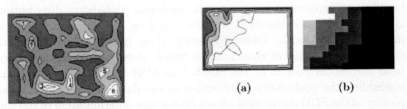

(a)　　　　(b)

Fig. 6. Music collection with SDH, $s = 3$.

Fig. 7. Alternative visualizations: (a) distance matrix, (b) k-means, $k = 6$.

to perceived acoustic similarities similar to the results presented in [10]. For full details on the feature extraction and experiments see [11]. Figure 5 depicts the effects of different s values for the SDH of the 14×10 SOM trained with the music collection. The best results were obtained with $s = 3$ (cf. Figure 6). Using, for example, $s = 15$ the cluster in the upper left of the map disappears. Using values below $s = 3$ the cluster in the lower right becomes too dominant. Yet, in all cases the cluster structure is clearly visible.

We will now take a closer look at the interpretation of some of the clusters. Cluster 1 in Figure 6 represents music with very strong beats. In particular, several songs of the group *Bomfunk MCs* are located there, but also songs such as *Blue* by *Eiffel 65* or *Let's get loud* by *Jennifer Lopez*. Cluster 2 represents music mainly by the rock band *Red Hot Chili Peppers* including songs like *Californication* and *Otherside*. Cluster 3 represents more aggressive music by bands such as *Limp Bizkit, Papa Roaches, Korn*. Cluster 4 represents slightly less aggressive music by groups such as *Guano Apes* and *K's Choice*. Cluster 5 represents concert music and classical music used for films, e.g., the well known *Starwars* theme. Cluster 6 represents peaceful, classical pieces such as *Für Elise* by *Beethoven* or *Eine kleine Nachtmusik* by *Mozart*.

Figure 7 depicts the distance matrix and the k-means visualization of the same SOM. While the former clearly identifies Cluster 1 in the upper left corner, all other clusters are not revealed, mainly because of the general similarity of data in that area and the dominance of the difference to the cluster in the upper left. The features extracted represent the dynamic patterns of the music pieces and emphasize in particular beats reoccurring in fixed time intervals. Thus, for example, the values of songs by *Bomfunk MCs* are much higher than those of *Für Elise*. The general tendency of the increasing strength of the beats can also be guessed from the k-means visualization for $k = 6$ in Figure 7(b), yet, the cluster structure itself is not visible.

4   Conclusion

The smoothed data histogram is a simple, robust, and computationally light cluster visualization method for self-organizing maps. We illustrated SDHs using a simple 2-dimensional data set and discussed the similarities to other SOM based visualizations as well as to the posterior probability distribution of the GTM. Furthermore, the SDH was evaluated on a real world data set consisting of pieces of music. Observed results show improved cluster identification compared to alternative visualizations based solely on the information extracted from the model vectors of the SOM. The SDH is able to identify clusters by resembling the probability distribution of the data on the map. The low complexity of the SDH calculation allows interactive determination of the smoothing parameter s, and thus is well suited for interactive data analysis. We provide a Matlab® toolbox to visualize SDHs together with several demonstrations at *http://www.oefai.at/~elias/sdh*.

References

1. Kohonen, T.: Self-Organizing Maps. 3rd edn. Springer-Verlag, Berlin (2001)
2. Ultsch, A., Siemon, H.: Kohonen's self-organizing feature maps for exploratory data analysis. In: Proc Int'l Neural Network Conference (INNC'90), Dordrecht, Netherlands, Kluwer (1990) 305–308
3. Vesanto, J., Alhoniemi, E.: Clustering of the Self-Organizing Map. IEEE Transactions on Neural Networks **11** (2000) 586–600
4. Himberg, J.: Enhancing the SOM based data visualization by linking different data projections. In: Proc Int'l Symp on Intelligent Data Engineering and Learning (IDEAL'98), Hong Kong (1998) 427–434
5. Kaski, S., Kohonen, T., Venna, J.: Tips for SOM processing and colorcoding of maps. In: Visual Explorations in Finance. Springer Verlag, Berlin, Germany (1998)
6. Merkl, D., Rauber, A.: Alternative ways for cluster visualization in self-organizing maps. In: Proc Workshop on Self-Organizing Maps (WSOM97), Espoo, Finland
7. Vesanto, J.: SOM-Based data visualization methods. Intelligent Data Analysis **3** (1999) 111–126
8. Bishop, C., Svensen, M., Williams, C.: GTM: The generative topographic mapping. Neural Computation **10** (1998) 215–235
9. Davies, D., Bouldin, D.: A cluster separation measure. IEEE Transactions on Pattern Analysis and Machine Intelligence **1** (1979) 224–227
10. Rauber, A., Frühwirth, M.: Automatically analyzing and organizing music archives. In: Proc 5. Europ Conf on Research and Advanced Technology for Digital Libraries (ECDL 2001). Darmstadt, Germany, Springer (2001)
11. Pampalk, E.: Islands of Music: Analysis, Organization, and Visualization of Music Archives. Master's thesis, Vienna University of Technology, Austria (2001)
    http://www.oefai.at/~elias/music

Rule Extraction from Self-Organizing Networks

Barbara Hammer[1], Andreas Rechtien[1], Marc Strickert[1], and Thomas Villmann[2]

[1] Department of Mathematics/Computer Science,
University of Osnabrück, D-49069 Osnabrück, Germany
[2] Clinic for Psychotherapy and Psychosomatic Medicine, University of Leipzig,
Karl-Tauchnitz-Straße 25, D-04107 Leipzig, Germany

Abstract. Generalized relevance learning vector quantization (GRLVQ) [4] constitutes a prototype based clustering algorithm based on LVQ [5] with energy function and adaptive metric. We propose a method for extracting logical rules from a trained GRLVQ-network. Real valued attributes are automatically transformed to symbolic values. The rules are given in the form of a decision tree yielding several advantages: hybrid symbolic/subsymbolic descriptions can be obtained as an alternative and the complexity of the rules can be controlled.

1 Introduction

For more than a decade, rule extraction from neural networks has been an important though not yet satisfactorily solved problem in machine learning [1]. A basic difficulty is caused by a property of neural networks which, in practice, is the strength of neural networks: they implement highly nonlinear systems with real-valued distributed internal representations. Hence, their capacity is different from the capacity of rule sets. They are often more powerful compared to simple logical rules. Nevertheless, a conversion of networks to rules is desirable, if possible: rules are more easily understood by humans; they can be verified and erroneous regions be identified; they can be integrated in expert systems, to name just a few advantages.

Various approaches tackle multi-layer perceptrons, the most common network type (see e.g. [1,3,7,12]). Some of them treat the networks as black boxes and try to learn appropriate rules using the network as an oracle. Others approximate the behavior of single neurons by logical rules. The result of such approximation is often a huge number of rules due to the nonlinearities in the network. Therefore, several approaches restrict the possibly complex behavior of network parts in a way that allows their approximation with logical rules. Fuzzy networks, specific choices of the weights, or factorizing activation functions are some of the above simplification methods. Rule extraction mechanisms for other network architectures such as recurrent networks exist, too [2].

We will focus on self-organizing networks where behavior is based on prototyping and metric information. It is commonly argued that self-organizing networks can be easily understood by humans. Two-dimensional self-organizing feature maps [5] which preserve the metric structure of possibly high dimensional data, for example, are often used for visualization of data. Learning vector quantization [5] provides prototypes for several classes which can give an idea of typical attributes for the classes. Nevertheless, the algorithms do not yield precise logical rules which could be verified or embedded

J.R. Dorronsoro (Ed.): ICANN 2002, LNCS 2415, pp. 877–883, 2002.
© Springer-Verlag Berlin Heidelberg 2002

in expert systems, and they allow humans only a partial insight into their behavior. Therefore, rule extraction from self-organizing networks is an interesting and due to the simplicity of the involved networks a promising task. So far, only few algorithms for rule extraction in self-organizing networks have been proposed [10,11,13]. [10,13] introduce a pipeline for data mining: huge data sets are processed using an appropriately trained and postprocessed unsupervised self-organizing feature map and rule extraction.

Here we will focus on GRLVQ, a supervised self-organizing algorithm used for classification [4,9]. We will present a method to extract decision trees from trained GRLVQ-networks including automatic discretization of possibly real-valued features. The method naturally allows hybridization, modularization, and further training of the classifier. We present results for benchmarks from the UCI repository [6].

2   GRLVQ

GRLVQ as proposed in [4] constitutes a generalization of LVQ and GLVQ [5,9] with the advantages that training can be performed with an adaptive metric and the dynamics obey a stochastic gradient descent. Assume a finite set $X = \{(x^i, y^i) \in \mathbb{R}^n \times \{1, \ldots, C\} \mid i = 1, \ldots, m\}$ of training data is given and a clustering of the data into C classes is to be learned. We denote the components of a vector $x \in \mathbb{R}^n$ by $(x_1, \ldots, x_n)$ in the following. GRLVQ chooses a fixed number of vectors in $\mathbb{R}^n$ for each class, so called prototypes. Denote the set of prototypes by $W = \{w^1, \ldots, w^K\}$ and assign the label $c^i = c$ to w^i iff w^i belongs to the cth class, $c \in \{1, \ldots, C\}$. The receptive field of w^i is defined by $R^i_\lambda = \{x \in X \mid \forall w^j \, (j \neq i \rightarrow |x - w^i|_\lambda \leq |x - w^j|_\lambda)\}$. Here $\lambda_i \geq 0$ are scaling factors with $\sum_i \lambda_i^2 = 1$. $|x - y|_\lambda = \left(\sum_i \lambda_i (x_i - y_i)^2\right)^{1/2}$ denotes the weighted Euclidian metric. The training algorithm adapts the prototypes w^i and the factors λ_i such that for each class $c \in \{1, \ldots, C\}$, the corresponding prototypes represent the class as accurately as possible. That means, the difference of the points belonging to the cth class, $\{x^i \in X \mid y^i = c\}$, and the receptive fields of the corresponding prototypes, $\bigcup_{c^i = c} R^i_\lambda$, should be as small as possible. This is achieved through a stochastic gradient descent on the cost function

$$C := \sum_{i=1}^m \text{sgd} \left(\frac{d^+_\lambda(x^i) - d^-_\lambda(x^i)}{d^+_\lambda(x^i) + d^-_\lambda(x^i)} \right)$$

where $\text{sgd}(x) = (1 + \exp(-x))^{-1}$ denotes the logistic function, $d^+_\lambda(x^i) = |x^i - w^{i+}|^2_\lambda$ is the squared weighted Euclidian distance of x^i to the nearest prototype w^{i+} of the same class as x^i, and $d^-_\lambda(x^i) = |x^i - w^{i-}|^2_\lambda$ is the squared weighted Euclidian distance of x^i to the nearest prototype w^{i-} of a different class than x^i. Taking the derivatives with respect to the prototypes and weighting factors, respectively, and adding normalization of the factors λ_i yields the very stable and efficient learning rule of GRLVQ [4].

3   Extraction of Rules

We assume in the following that the input dimensions of the training data are linearly transformed before training so that they occupy approximately the same range. After training GRLVQ includes information about the importance of the input dimensions through the magnitudes of the weighting factors. We would like to extract a decision tree from a trained GRLVQ-network of the following specific form: each node N of the tree may possess an arbitrary number C^N of children. An interior node N is labeled by an index $I^N \leq n$ and real values $W_1^N < \ldots < W_{C^N-1}^N$. Each leaf L is labeled with a class number $C_L \leq C$. We call such a tree BB-tree for short.[1] A data point x is classified as follows by a BB-tree: an interior node N corresponds to the decision in which of the C^N intervals $(-\infty, W_1^N], (-W_1^N, W_2^N], \ldots, (-W_{C^N-2}^N, W_{C^N-1}^N], (-W_{C^N-1}^N, \infty)$ the component x_{I^N} of x lies. I^N is the index attached to node N. This setup corresponds to the decision which of the C^N children should be considered next. Starting at the root, a path in the tree is specified for the input x in this way. The class label at the leaf of this path corresponds to the output for x. Obviously, a BB-tree can be transformed to a set of IF-THEN-rules where each path of the tree denotes a rule, and the number of conditions in the rule is given by the length of the path. The intervals used for the decisions can be identified with appropriate linguistic attributes if the components of the data are real valued.

In order to extract a BB-tree of a trained GRLVQ-network, we need to define a ranking of the input dimensions to determine the indices I^N attached to the interior nodes, and we need to define the interval borders W_i^j. A GRLVQ-network provides both in a natural way: a ranking of the dimensions is induced by the weighting factors λ_i. Interval borders are induced by the prototypes; if dimension i is considered, we project the prototypes to the i th coordinate and use the mid-points between neighbored projections as interval borders. Assume Λ is the list of indices i sorted according to the magnitude of the weighting factors λ_i. $first(\Lambda)$ denotes the first entry, $rest(\Lambda)$ the rest of the list. X denotes the training set and W the set of prototypes. We assume that all dimensions i which, due to a small weighting factor λ_i, do not contribute to the classification are deleted. Moreover, all idle prototypes, i.e. prototypes with empty receptive fields are deleted, too. We propose the following recursive tree-extraction procedure:

BB-Tree (X, W, Λ):

  delete all idle prototypes from W,

  if STOP: output a leaf with class $argmax_c|\{x^i \mid y^i = c, (x^i, y^i) \in X\}|$

  else: output an interior node N with $|W|$ children,

  choose $I^N := first(\Lambda)$,

  denote by $[a_1, \ldots, a_{|W|}]$ a sorted list of $\{x_{I^N}^i \mid (x^i, y^i) \in X\}$

  choose $W_i^N := (a_i + a_{i+1})/2, i = 1, \ldots, |W| - 1$ (∗)

  choose the i th child of N, $i = 1, \ldots, |W|$, as the output of

  BB-Tree $(\{x \in X \mid x_{I^N} \in (W_{i-1}^N, W_i^N]\}, W, rest(\Lambda) \bullet [first(\Lambda)])$

Here $\bullet$ denotes the concatenation of lists. $W_0^N := -\infty$ and $W_{|W|}^N := \infty$. *STOP* denotes a stopping criterion, e.g. the remaining data points from X are contained in one class or a maximum height of the tree is reached. The procedure recursively splits the training set using guillotine cuts in the different dimensions. The location of the cuts is induced by the prototypes. Obviously, this procedure yields precisely the classification

---

[1] BB stands for Babsi-Bäumchen; the name is due to historical reasons.

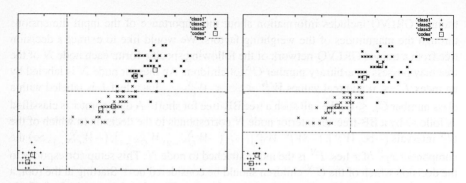

Fig. 1. Classification of BB-trees extracted for the Iris data set. The maximum height is restricted to 2 corresponding to the weighting factors λ_i. Class 2 corresponds to two leafs in the right tree.

of the GRLVQ-network for one-dimensional data. If $n > 1$, we can limit the number of points which might increase the classification error of the extracted tree in comparison to the GRLVQ-network: a point x can be misclassified only if the distance $|w_i - x_i|$ to its respective prototype is larger than $d/(2n)$ for some dimension i. d denotes the minimum pairwise distance between prototypes measured in the L_1-norm. Hence the error does not increase if the distance between prototypes is large compared to the distance of data from their respective prototypes. We will see that the classification obtained by a BB-Tree is sometimes even better than the GRLVQ-network.

Note that usually n recursive steps are sufficient because all interval borders in further recursive steps coincide with already used interval borders. However, separation in $(*)$ between *all* neighbored prototypes yields trees with a large fan-out, particularly at the root, hence a large number of rules is produced. Therefore, we alter this step and only choose interval borders between prototypes in such a way that the distance of their projections is larger or equal to the mean distance of all projections in $(*)$. Then, a depth of the trees of more than n may be obtained. Naturally, one can often simplify the resulting set of rules of the BB-tree, e.g. by deleting nodes with only one child or by merging preconditions of the rules. Limiting the maximum height of the tree and the number of prototypes limits the complexity of the rule set. Moreover, further training at the recursive steps could be added until training data are perfectly fitted.

	Iris	Monks1	Monks2	Monks3	Wisconsin
Decision tree (ID3/C4.5)	96	98.6	67.9	94.4	96
Sigmoidal neural network	98	100	100	93.1	96.7
Rules from neural networks	95.7-100	100	80.6-100	97.2-100	95-99
GRLVQ	98	100	60	97	96.5
BB-Tree	95-98	96-100	70	96	95-97

Fig. 2. Results on test sets for various classifiers. Rule extraction covers typical results for several algorithms [3]. The accuracy of BB-trees depends on the run and the tree's height.

4 Experiments

Since GRLVQ, unlike commonly sequential decision-tree algorithms, explores appropriate rankings in parallel it is easy to construct examples where the BB-tree algorithm performs superior, e.g. XOR expanded with an additional input dimension which contains pure noise. However, our focus here lies on simple standard benchmarks which show that the extracted rules correctly approximate GRLVQ behavior. We tested the algorithm on UCI-benchmarks [6] for which results for decision-trees [8] and neural networks are available [3]. Training patterns with missing values have been deleted. The reported results have been verified in several runs separating data into training set and test set. Data are linearly scaled to zero mean and unit variance before training. Learning rates are 0.5 for the prototypes and 0.05 for the factors λ_i until convergence.

Iris: The task is to predict three classes of plants based on four numerical values for 150 instances. It is well known that the projection to the last two dimensions allows a very good accuracy of 98%, and classification based on the last dimension still allows an accuracy of 95%, though the classification is less noise tolerant. Taking more dimensions into account allows 100% accuracy, although overfitting is usually inevitable. GRLVQ with two prototypes for each class results in an accuracy of 98% and an accuracy of about 97% for the extracted BB-trees. The weighting $\lambda = (0.02, 0.08, 0.74, 0.66)$ emphasizes the last two dimensions. Rule sets for BB-trees of height 2 yield (for transformed data)

$$\{x_3 \le -0.55 \to \text{class1}, \ (x_3 > -0.55 \wedge x_4 \le 0.53) \to \text{class2}, \text{ else class3}\}$$
or $\quad \{x_4 \le -0.5 \to \text{class1}, \ (x_3 \le 0.55 \wedge -0.5 < x_4 \le 0.53) \to \text{class2}, \text{ else class3}\}$

where the difference of rule sets between several runs is due to large regions without data points (see Fig. 1). Training with one prototype for each class yields a more distinct weighting $(0, 0, 0.95, 0.8)$ and a rule set which separates only in one dimension $\{x_4 < -0.95 \to 1, \ -0.95 < x_4 < 0.58 \to 2, \text{ else } 3\}$ with an accuracy of 95%. Training with more prototypes puts some emphasis on the other dimensions, e.g. 10 prototypes for each class yield the weighting $\lambda = (0.12, 0.11, 0.68, 0.71)$ which results in a larger tree with preconditions on x_1 and the better accuracy of 98%.

Monks: Monks1-3 are artificial binary classification tasks with 6 discrete input dimensions and 556, 601, or 554 instances, respectively, created according to the rules for class 1: $\{x_1 = x_2 \vee x_5 = 1\}$, {precisely two of $x_1, \dots, x_6$ are 1}, and $\{(x_2 \ne 3 \wedge x_5 \ne 4) \vee (x_4 = 1 \wedge x_5 = 3)\}$, respectively, where the last set is disrupted by 5% noise.

For Monks1, training with 10 prototypes for each class yields a classification accuracy of $96 - 100\%$ on a test set where the weighting factors $\lambda \approx (0.3, 0.2, 0, 0, 0.9, 0)$ clearly extract the relevant attributes. The extracted trees are usually high with $18 - 30$ rules. One of the rules corresponds to a very short path in the tree and implements the decision $x_5 \le 1$; about 26.6% of the patterns is covered by this rule. The other rules cover only $1 - 7\%$ of the data set and implement the test $x_1 = x_2$. Obviously, this requires guillotine cuts between all possible values of x_1 and x_2 resulting in a large number of rules if rules for both possible outputs are taken into account.

Monks2 is a task fitted to feedforward neural networks which can represent m-of-n rules with few neurons, whereas algorithms based on IF-THEN rules or prototyping need more resources for the representation. Correspondingly, we trained 30 prototypes for each class obtaining an accuracy of 60% for GRLVQ and about 70% for the extracted BB-trees on a test set. The trees are usually large with $50 - 60$ rules. Moreover, strong overfitting can be observed for both, GRLVQ and the BB-trees, with a training accuracy of 70% or 80%, respectively. However, BB-trees give better results than GRLVQ and are comparable to the performance of decision trees.

For Monks3, training with 2 prototypes for each class yields 97% accuracy for GRLVQ and 96% accuracy for the extracted BB-trees. The weighting emphasizes the second and fifth attribute $\lambda \approx (0, 0.8, 0, 0, 0.6, 0)$. Extracted rules for trees of height 2 are $(x_2 = 0 \land x_5 \notin \{3, 4\}) \to 1$, $(x_2 = 1 \land x_5 \neq 4) \to 1$, $x_2 \in \{3, 4\} \to 1$, else 0. Note that the intervals resulting from BB-trees are converted to the discrete inputs in the above rules. Important attributes have been identified and the above rules overlap with the original ones. They yield better accuracy if the noise of 5% is taken into account.

Wisconsin breast cancer: The set consists of 699 cases with 9 discrete attributes and 2 classes. Training with 2 codebooks for each class yields an accuracy of about 96%. The weighting equals $\lambda \approx (0, 0.9, 0, 0, 0, 0.4, 0, 0, 0)$. An extracted rule set for trees of height 2 and discrete attributes is $\{(x_2 \leq 4 \land x_6 \leq 6) \to 0, \text{else } 1\}$. These are very simple rules which overlap with results reported in [3] and have a very low complexity.

5 Conclusions

We have proposed an algorithm for rule extraction from trained GRLVQ-networks, an efficient and stable generalization of LVQ with adaptive weighting factors. Unlike decision tree algorithms, GRLVQ evolves an appropriate ranking in parallel. The accuracy of the achieved rule sets of the BB-tree algorithm is similar to GRLVQ and to corresponding results for other rule-extraction mechanisms or symbolic decision tree extraction algorithms reported in literature. The obtained rule sets in the BB-tree algorithm are usually small; moreover, the complexity of the rule sets can be controlled limiting the height of the extracted trees. Hybrid solutions can naturally be integrated if necessary, attaching sets of prototypes to the leafs instead of explicit class labels.

The reported results have been obtained for comparably small data sets; experiments for larger data sets are forthcoming. It is natural that the obtained rules can often be further simplified depending on the number of prototypes used for training for larger data sets. Cuts between prototypes of the same class can be eliminated, unifying the respective intervals. A large number of leafs may correspond to regions without training points; hence no decision is performed within these regions and the corresponding rules can be discarded. An automatic simplification procedure based on these heuristics seems promising though it has not yet been implemented. Another possibility to improve the classification accuracy is to drop the a priori fixed choice of the prototypes and weighting factors: further training at each node (or after n recursive steps) would allow to use adapted prototypes and weighting factors for the remaining data points at each node for optimal further splitting. Though being computationally more expensive, the resulting rule sets and accuracy might be considerably better.

References

1. R. Andrews, J. Diederich, A.B. Tickle, A survey and critique of techniques for extracting rules from trained artificial neural networks. *Knowledge Based Systems* 8: 373–389, 1995.
2. M. Casey, The dynamics of discrete-time computation, with application to recurrent neural networks and finite state machine extraction. *Neural Computation*, 8(6): 1135–1178, 1996.
3. W. Duch, R. Adamczak, K. Grabczewski, A new method of extraction, optimization and application of crisp and fuzzy logical rules. *IDEE TNN* 12: 277–306, 2001.
4. B. Hammer, T. Villmann, Estimating relevant input dimensions for self-organizing algorithms. In: N. Allison, H. Yin, L. Allinson, J. Slack (eds.), *Advances in Self-Organizing Maps*, pp.173–180, Springer, 2001.
5. T. Kohonen. *Self-Organizing Maps*. Springer, 1997.
6. C.J. Mertz, P.M. Murphy. *UCI repository of machine learning databases*, http://www.ics.uci.edu/~mlearn/MLSummary.html
7. D. Nauck, F. Klawonn, R. Kruse. *Foundations on Neural-Fuzzy Systems*, Wiley, 1997.
8. J.R. Quinlan. *C4.5: programs for machine learning*. Morgan Kaufmann, 1993.
9. A. S. Sato, K. Yamada. Generalized learning vector quantization. In G. Tesauro, D. Touretzky, and T. Leen, editors, *Advances in NIPS*, volume 7, pp.423–429. MIT Press, 1995.
10. M. Siponen, J. Vesanto, O. Simula, and P. Vasara. An approach to automated interpretation of SOM. In: *Advances in Self-Organizing Maps*, pp.89–94, Springer, 2001.
11. A.-H. Tan. Rule learning and extraction with self-organizing neural networks. In: *Proceedings of the 1993 Connectionist Models Summer School*, Hillsdale, pp.192–199, Lawrence Erlbaum Associates, 1994.
12. A.B. Tickle, R. Andrews, M. Golea, and J. Diederich. The truth will come to light: directions and challenges in extracting knowledge embedded within trained artificial neural networks. *IEEE TNN* 9:1057–1068, 1998.
13. A. Ultsch. Knowledge extraction from self-organizing neural networks. In: O. Opitz, B. Lausen, R. Klar (eds.), *Information und Classification*, pp.301–306, Springer, 1993.

Predictive Self-Organizing Map for Vector Quantization of Migratory Signals

Akira Hirose and Tomoyuki Nagashima

Dept. Frontier Informatics / RCAST, University of Tokyo, Tokyo 153-8904, Japan

Abstract. Predictive self-organizing map (P-SOM) that performs an adaptive vector quantization of migratory signals is proposed. The P-SOM separates continuously varying components of the signal from random noise, resulting in a better performance of the adaptive vector quantization. An application to a communication system is presented.

1 Introduction

The self-organizing map (SOM) is an excellent method to perform adaptive vector quantizations [1]. It usually quantizes signals in a statistically stationary state. As an extension of the conventional SOM, we propose a method to quantize signals whose statistical properties such as average values and velocities are varying in time. The proposed method can be called predictive SOM (P-SOM) since it quantizes migratory signals by predicting future positions of signal clusters in information space.

Assume that we construct a mobile communication receiver using the quasi-coherent detection scheme (a digital demodulation method employing asynchronous local oscillators without phase-locked loops (PLL's) for parallel signal processing) that is capable of receiving signals containing phase rotations caused by the Doppler effect. When a transmitter, receiver and/or reflecting objects are moving in the communication environment, the phase detected by the quasi-coherent detector is always rotating.

There have been several proposals of neural-network applications to communications. Among the neural networks, the SOM is effective in particular to solve the problems related to nonlinearity in transmitters and receivers [2],[3]. However, the SOM receivers, in spite of the robustness against nonlinearity, cannot follow steady motion of signals such as the Doppler phase rotation, but yields residual deviation in letting the reference (code-book) vectors self-organize. Therefore, they need to be combined with another adaptive linear filters such as the Wiener filter.

To solve such a problem, we propose the P-SOM that possesses not only the weights corresponding to the signal values themselves but also those expressing time-derivative values such as velocity and acceleration. By making all of the weights self-organize, the P-SOM generates a set of predictive reference vectors that can track the migratory signal without deviation. The predictive reference vectors enable a separation of continuously varying statistical properties from

J.R. Dorronsoro (Ed.): ICANN 2002, LNCS 2415, pp. 884–889, 2002.

random noise components. The P-SOM quantizes signals adaptively based on the predicted reference vectors generated by the time-derivative quantities.

The P-SOM method is applicable to adaptive vector quantizations of various dynamical signals such as motion images. At the same time, the method is considered increasingly important in the advanced communications where the orthogonal frequency-domain multiplexing (OFDM) technique is employed to cope with multipath environments [4]. Because the parallel construction of detectors in the OFDM systems has a difficulty in being equipped with PLL's, the receivers should rely on digital signal processing concerning the phase tracking and detection. Based on our proposal, the single P-SOM framework can treat all of the nonlinearity, random noise and the phase rotation.

2 Construction, Dynamics, and Noise Evaluation

P-SOM in general: Figure 1(a) shows the construction of the P-SOM we proposed here. The P-SOM selects a winner and generates an output $c(t) = i_{\text{winner}}(t)$ among the *predictive* reference vectors $\tilde{w}_1$ mentioned below.

$$c(t) = \arg \min_{1 \leq i \leq i_{\text{max}}} \left| x(t) - \tilde{w}_i(t) \right| \tag{1}$$

In the P-SOM, we prepare the reference vectors $\tilde{w}_i(t)$ and neural weights $w_i(t)$ separately. The former takes charge of prediction, whereas the latter is in charge of self-organization. We define the reference vectors $\tilde{w}_i(t)$ corresponding to the signal values themselves as well as time-derivative ones, e.g., $\dot{\tilde{w}}_i(t)$ expressing the velocity of a cluster moving in the information space, by using the unit-time earlier signal-equivalent (ordinary) weights $w_i(t - \Delta t)$ and time-derivative-equivalent weights $\dot{w}_i(t - \Delta t)$ (first order), $\ddot{w}_i(t - \Delta t)$ (second order), etc., as

$$\tilde{w}_i(t) = w_i(t - \Delta t) + \dot{w}_i(t - \Delta t)\Delta t + \frac{1}{2!}\ddot{w}_i(t - \Delta t)(\Delta t)^2 + \cdots \tag{2}$$

$$\dot{\tilde{w}}_i(t) = \dot{w}_i(t - \Delta t) + \ddot{w}_i(t - \Delta t)\Delta t + \cdots \tag{3}$$

$$\vdots$$

The signal-equivalent weights $w_i(t)$ and the time-derivative weights such as $\dot{w}_i(t)$ and $\ddot{w}_i(t)$ self-organize according to the SOM dynamics as

$$w_i(t) \leftarrow \tilde{w}_i(t) + h_{0(i;c)}\{x(t) - \tilde{w}_i(t)\} \tag{4}$$

$$\dot{w}_i(t) \leftarrow \dot{\tilde{w}}_i(t) + h_{1(i;c)}\{\dot{x}(t) - \dot{\tilde{w}}_i(t)\} \tag{5}$$

$$\ddot{w}_i(t) \leftarrow \ddot{\tilde{w}}_i(t) + h_{2(i;c)}\{\ddot{x}(t) - \ddot{\tilde{w}}_i(t)\} \tag{6}$$

$$\vdots$$

where $h_{k(i;c)}$ determines the self-organization speed and k means that the coefficient is related to the k-th order time-derivative. As is the case of the conventional SOM, the values are dependent on the topological relations between the

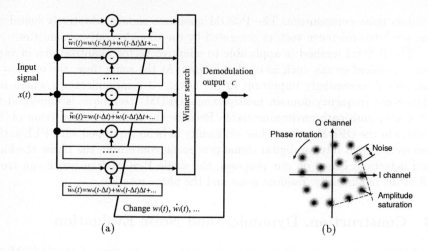

Fig. 1. (a) Construction of P-SOM and (b) constellation of received and quasi-coherently detected signal.

weights to self-organize and the winner. That is, the value for the winner $h_{k(c;c)}$ is large, that for the neighbors $h_{k(i\in\text{neighbors};c)}$ is small, while that for others is zero $(h_{k(i\in\text{others};c)} = 0)$.

The P-SOM is constructed as mentioned above. Both the reference vectors and the weights are categorized into those corresponding to the signal values themselves and those related to the k-th time-derivative quantities. Assuming that the average value of the input signal $x(t)$ in each cluster varies continuously, that is, its motion is characterized by velocity, acceleration, etc., we connect those references and weights to one another consistently as expressed by (2)–(6). The P-SOM can then distinguish continuously migrating components from random noise components of signal variation. Hence we expect that the system performs an appropriate adaptive vector quantization of the temporally varying signals.

As an example, the follows are considered for a time-sequential signal $x(t)$ that migrates uniformly in information space. The conventional SOM always has a residual deviation inevitably between the reference winner $w_c(t)$ and the moving signal $x(t)$. Contrarily, the P-SOM predicts an appropriate position of the winning reference vector $w_c(t)$ by taking into account the first and second terms of (2), with the velocity weight $\dot{w}_i(t)$ converging at a constant (average migration velocity of $x(t)$ in each cluster). Therefore, it can obtain the winning reference without any residual deviation.

P-SOM demodulator: In this section, we consider a particular P-SOM system for adaptive vector quantization of complex-valued quasi-coherently detected signals that have almost steady Doppler phase shift. Figure 1(b) shows the signal constellations of received signals in the complex domain for a 16-ary QAM (quadrature amplitude modulation) communication system. The received signal contains noise, nonlinear distortion such as amplitude saturation and harmonic distortion, and phase rotation. The Doppler effect is brought about by relative motion of transmitter, receiver and reflectors. Because such a motion is accompanied with substantial motion with mass, which is not the case of virtual motion in virtual space, we have only to take the velocity (first-order derivative) into consideration, neglecting acceleration and higher-order derivatives that are expected to be small.

We can also assume usually that the signal migration caused by the nonlinearity is negligibly small or short in duration so that the Doppler phase rotation is the dominant steady variation. Moreover, in this case, all the reference vectors have identical phase rotation speed. Therefore, we need only two derivative quantities for the prediction, i.e., the phase speed reference vector $\tilde{\theta}$ and the phase speed weight $\dot{\theta}$ owned jointly by all the references $\tilde{w}_i$ ($\forall i$). According to the above consideration, we determine the neural dynamics as

$$\tilde{w}_i(t) = w_i(t - \Delta t) \, e^{i\dot{\theta}(t)\Delta t} \tag{7}$$

$$\tilde{\theta}(t) = \dot{\theta}(t - \Delta t) \tag{8}$$

$$w_i(t) \leftarrow \tilde{w}_i(t) + h_{0(i;c)}\{x(t) - \tilde{w}_i(t)\} \tag{9}$$

$$\dot{\theta}(t) \leftarrow \tilde{\theta}(t) + h^\theta_{1(c;c)} \left\{ \left(\arg[x(t)] - \arg\left[\tilde{w}_i(t - \Delta t)\Big|_{i=c}\right] \right) / \Delta t \;\; - \tilde{\theta}(t) \right\} \tag{10}$$

where $h^\theta_{1(c;c)}$ denotes the self-organization coefficient for the phase speed $\dot{\theta}$ whose role is the same as that of $h_{1(i;c)}$. When $\overline{|w_c|^2}=1$ for 16QAM, they are related on average as $h_{1(i;c)} \approx 0.95 h^\theta_{1(c;c)}$. Note that the speed reference vector $\tilde{\theta}$ is always the winner and used for the predictions of all the reference vectors $\tilde{w}_i(t)$. This fact simplifies the construction and reduces the calculation cost.

Noise evaluation: The noise increase in the conventional-SOM and P-SOM demodulators is caused by the fluctuation of the reference vectors w_i or $\tilde{w}_i$ affected by the noise contained in the signal $x(t)$. A series of calculations yields an equivalent noise power (variance) σ^2 expressed as

$$\sigma^2 = \frac{2}{2 - h_{0(c;c)}} \frac{2}{2 - h_{1(c;c)}} \sigma_0^2 \quad \text{or} \quad \frac{2}{2 - h_{0(c;c)}} \frac{2 - \frac{1}{2}h^\theta_{1(c;c)}}{2 - h^\theta_{1(c;c)}} \sigma_0^2 \tag{11}$$

where σ_0^2 denotes the noise power in $x(t)$. The details will be given elsewhere. When $h_{1(c;c)}$ (or $h^\theta_{1(c;c)}$) $= 0$, σ becomes the noise in the conventional SOM. The result (11) suggests that the noise increase caused by the time-derivative terms is less than 0.1[dB] when the self-organization coefficient is $h^\theta_{1(c;c)} \leq 0.1$.

3   Simulation Experiment

We evaluate the P-SOM demodulation performance in a simulation experiment based on the dynamics (7)–(10). The signal fed to the system is shown in Fig. 1(b). We assume that the signal includes additive noise, nonlinearity and phase rotation. Specifically speaking, a 16-QAM signal $x(t)$, whose average squared amplitude is normalized to unity, is saturated such that $|x_{\text{sat}}| =$

Categorical Topological Map

Mustapha Lebbah[1,2], Christian Chabanon[1], Fouad Badran[3], and
Sylvie Thiria[2]

[1] RENAULT, Direction de la recherche, TCR RUC T 62,
1 avenue du golf F 78288 Guyancourt Cedex
{Mustapha.Lebbah, Christian.Chabanon}@Renault.com
[2] Laboratoire LODYC, Université Paris 6, Tour 26-4^{e}étage,
4 place Jussieu 75252 Paris cedex 05 France.
{lebbah, thiria}@lodyc.jussieu.fr
[3] CEDRIC, Conservatoire National des Arts et Métiers,
292 rue Saint Martin, 75003 Paris, France
{badran}@cnam.fr

Abstract. This paper introduces a topological map dedicated to an automatic classification categorical data. Usually, topological maps uses a numerical (or binary) coding of the categorical data during the learning process. In the present paper, we propose a probabilistic formalism where the neurons now represent probability tables. Two examples using actual and synthetic data allow to validate the approach. The results show the good quality of the topological order obtained as well as its performances in classification.

1 Introduction

The algorithm of topological maps proposed by Kohonen [6] is a self-organization algorithm which provides quantification and clustering of the observation space. In the paper [3], we presented an algorithm dedicated to binary data using topological maps. In this paper, we generalize this approach to categorical data using a probabilistic formalism. In section 2 we present the Categorial Topological Map (CTM) and its learning procedure based on the EM algorithm which maximizes the likelihood of the data. In section 3 we present two examples on real and synthetic data which allow to validate this approach.

2 Categorical Topological Map (CTM)

Let $A = \{\mathbf{z}_i, i = 1..N\}$ be the learning data set. We assume that a given observation $\mathbf{z}_i$ is a M dimensional vector $\mathbf{z}_i = (z_i^1, z_i^2, ..., z_i^k, ..., z_i^M)$ where the kth component z_i^k is a categorical variable with n_k modalities taking its value in $\mathcal{A}_k = \{\xi_1^k, \xi_2^k, ..., \xi_{n_k}^k\}$, such as each observation $\mathbf{z}_i$ is a realisation of a random variable which belongs to $\mathcal{A}_1 \times \mathcal{A}_2 \times ... \times \mathcal{A}_M$. We assume also that the M categorical variables characterizing a given observation $\mathbf{z}$ are independent. Under this assumption $p(\mathbf{z}) = \prod_{k=1}^M p(z^k)$ where each $p(z^k)$ is a one dimensional table representing the probabilities of the n_k modalities of $\mathcal{A}_k$.

J.R. Dorronsoro (Ed.): ICANN 2002, LNCS 2415, pp. 890–895, 2002.
© Springer-Verlag Berlin Heidelberg 2002

The goal of this paper is to present a new topological map dedicated to these categorical data. As for the traditional topological maps, the lattice $\mathcal{C}$ has a discrete topology defined by an undirect graph. Usually this graph is a regular grid in one or two dimensions. We denote N_{neuron} the number of neurons in $\mathcal{C}$. For each pair of neurons (c,r) on the map, the distance $\delta(c,r)$ is defined as the shortest path between c and r on the graph. In the following, we introduce a Kernel positive function K ($\lim_{|\delta| \to \infty} K(\delta) = 0$) and its associated family K_T parametrized by T : $K_T(\delta) = [1/T]K(\delta/T)$, which controls the size of the neighborhood. Following Luttrel [5] and Anouar et al [1], we introduce a probabilistic formalism to deal with topological map dedicated to categorical data. We assume that the lattice $\mathcal{C}$ is duplicated in two similar maps $\mathcal{C}_1$ and $\mathcal{C}_2$ provided with the same topology as $\mathcal{C}$. At each neuron $c_1 \in \mathcal{C}_1$, we associate M probability tables denoted by $p(z^k/c_1)$ where $(k = 1 \ldots M)$. The kth probability table is defined by the n_k values of $p(z^k = \xi_j^k/c_1)$, $(j = 1..n_k)$. The probability $p(\mathbf{z}/c_1)$ can be expressed using the independency hypothesis of its component as : $p(\mathbf{z}/c_1) = \prod_{k=1}^{M} p(z^k/c_1)$ which describes the data generated by c_1. We assume that each neuron c_1 of $\mathcal{C}_1$ is a distortion of a neuron c_2 of $\mathcal{C}_2$ and this distortion is described by the probability $p(c_1/c_2)$. In order to introduce a topological order we assume that $p(c_1/c_2) = [1/T_{c_2}]K_T(\delta(c_1,c_2))$ where $T_{c_2} = \sum_{r \in \mathcal{C}_2} K_T(\delta(c_2,r))$. Under the "Markov" property : $p(\mathbf{z}/c_1,c_2) = p(\mathbf{z}/c_1)$, we can see that each neuron of $\mathcal{C}_2$ can generate data by $p(\mathbf{z}/c_2) = \sum_{c_1 \in \mathcal{C}_1} p(c_1/c_2)p(\mathbf{z}/c_1)$. The probability $p(\mathbf{z}/c_2)$ is a mixture of probabilities completely defined from the map and the probability tables $p(\mathbf{z}/c_1)$, the probability tables which describe $p(\mathbf{z})$ is a mixture of probabilities defined on $\mathcal{C}_2$: $p(\mathbf{z}) = \sum_{c_2 \in \mathcal{C}_2} p(c_2)p(\mathbf{z}/c_2)$. In this expression, $p(c_2)$ represents the a priori probability of the neuron c_2. The aim of the learning algorithm is to estimate all the parameters of the model. These parameters are: the N_{neuron} parameters $\theta^{c_2} = p(c_2)$ and for each neuron c_1 the different values $\theta_j^{k,c_1} = p(z^k = \xi_j^k/c_1)$. In the following we denote : $\theta^{k,c_1} = \{\theta_j^{k,c_1}, j = 1 \ldots n_k\}$ and $\theta^{c_1} = \cup_k \theta^{k,c_1}$ the parameters which define neuron c_1. We assume that the N observations of the learning set $\mathcal{A}$ are independent observations and generated according to $p(\mathbf{z})$. The learning procedure estimates these parameters by maximizing the likelihood of the observations: $p(\mathbf{z}_1, \mathbf{z}_2, \ldots \mathbf{z}_N) = \prod_i p(\mathbf{z}_i)$. Since our model suggests that the observations are generated by two cells c_1 and c_2, we introduce the boolean variables $\chi_i^{c_1,c_2} = 1$ if $\mathbf{z}_i$ is generated by c_1 and c_2 and is equal to 0 otherwise. So each observation $\mathbf{z}_i$ is associated the hidden variable χ_i whose components are the $\chi_i^{c_1,c_2}$ and use the EM algorithm to maximize the likelihood of the observations, [2]. The application of EM for the maximization gives rise to the iterative algorithm of CTM. In order to present this algorithm, we define the conditional probability $p(c_1, c_2/\mathbf{z}_i)$, which can be expressed as :

$$p(c_1, c_2/\mathbf{z}_i) = \frac{\theta^{c_2} K_T(\delta(c_1, c_2))p(\mathbf{z}_i/c_1)}{\sum_{r \in \mathcal{C}_2} \theta^r p(\mathbf{z}_i/r)} \tag{1}$$

The learning procedure is expressed as follows:

- **Initialisation Step :** Choose an initial values $\theta_0^{c_1}$ and $\theta_0^{c_2}$ for the parameters.
- **The iteration step :** Compute the current values of $p(c_1, c_2/\mathbf{z}_i)$ by applying equation (1) and the new parameters using the equations (2) and (3):

$$\theta^{c_2} = \frac{\sum_{i \in A} \sum_{c_1 \in C_1} p(c_1, c_2/z_i)}{\sum_{c_2 \in C_2} \sum_{i \in A} \sum_{c_1 \in C_1} p(c_1, c_2/z_i)} \qquad (2)$$

$$\theta_{j_0}^{k,c_1} = \frac{\sum_{i \in \tau_{k,j_0}} \sum_{c_2 \in C_2} p(c_1, c_2/z_i)}{\sum_{j=1..n_k} \sum_{i \in \tau_{k,j}} \sum_{c_2 \in C_2} p(c_1, c_2/z_i)} \qquad (3)$$

In the last equation $\tau_{k,j} = \{i$ such that $z_i^k = \xi_j^k\}$ represents the set of observations $\mathbf{z}_i$ for whom the kth component takes the modality ξ_j^k.

- **Repeat** the iterative procedure until convergence.

As for traditional topological map, the topological order is introduced by repeating the procedure using decreasing values of T.

3 Experiments

In the following we used Categorical Topological Maps (CTM) for an automatic classification of two distinct sample surveys. The first experiment we present, deals with real data and prove the adequacy and the accuracy of CTM for market segmentation. We used this problem in order to give an example of the progress of the method and the accuracy it achieves. The second experiment deals with artificial data, as these data have been created for comparison purposes [4], the performances of CTM can be compared to those of several cluster algorithms dedicated to binary data.

   In the first experiment the learning data base $\mathcal{A}$ is extracted from a survey resulting from Belgium insurances. It tries to characterize the insurants according to various criteria and to classify them in two groups according to their ease to create accidents. In this experiment we consider 9 categorical variables with two or three modalities: **Use** (**Private, Professional**), **Sex**(**Male,Female,Society car**), **Language** (**French, Other**), **Age**(**Old Men, Middle Age, Young**), **Localisation** (**Capital, Province**), **Bonus**(1,2), **Police** (**86,Other**), **Power**(**Big, Small**), **Car Age** (**Old, New**). The 1106 insurants are labeled according to the number of accidents for which they are responsible: classe "1" is for those who never had any accident, class "2" characterizes those responsible for at least one accident. A first CTM map was trained, using the global data base of 1106 categorical vectors and a 2-D map of 5×5 neurons. At the end of this learning phase, each neuron is thus associated with 9 probability tables with two or three values corresponding to the categorical variables. The Figure 1.a presents the probability tables estimated by CTM for the first neuron (left-top corner of the map). It is seen that some modalities are very likely: this neuron represents persons who are most often **Professional** (0.99), living in

Province (0.85) and having an **Old** car (0.81).

At the end of the learning phase, each insurant is assigned to the neuron having the highest probability when using the parameters estimated during the learning process; the neuron takes the label ("1" or "2") which is the most frequent in its subsample (majority vote rule). Figure 1.*b* shows the map after the labeling process. In this map, the neurons corresponding to empty subsets are not labeled. Clearly, this map exhibit a topological order, Class "1" and "2" being represented by distinct connected area of neurons. Figure 1.*c* shows the 25 posterior probabilities $p(c/\mathbf{z})$ computed from the map for the insurant $\mathbf{z} = (\mathbf{Pf}, \mathbf{M}, \mathbf{Fr}, \mathbf{OM}, \mathbf{Pr}, 1, 86, \mathbf{S}, \mathbf{N})$. It can be seen that the most likely neuron is surrounded with neurons with high probability too. This property is general and can be observed for every insurant of $\mathcal{A}$.

In figure 2, we show the CTM map corresponding to four among the nine probability tables (**Sex**, **Age**, **Power** of car, **Age** of car) estimated during the learning process. In figure 2.*a*, each neuron is represented by the histograms of the four variables: for a given neuron, the upper right histogram is dedicated to the variable **Age** of the insurant, the upper left to the variable **Sex**, the bottom right to the variable **Age** of **car** and the bottom left to the variable **Power** of car. One can see that, for the six neurons without label, CTM computes very coherent parameters. These parameters allow to define insurant types which do not appear in the learning data base $\mathcal{A}$, but could exist.

Another view of this topological order is shown in figure 2.*b*. We ploted in place of histogram, the most probable modality, provided that this probability was greater than 0.8 if the categorical variable has two modalities and 0.6 if it has three modalities. Different regions on the map can be interpreted: for example the right region at the top of the map is dedicated to **Young** people, **Old Men** are localized at the top left corner and **Middle Age** at the bottom. The mixing of the different modalities can be seen looking at figure 3, where we displayed the probability for the three modalities of variable **Age** : the left map shows the probability of the modality **Old Man**, the middle is for the **Middle Age** and the right for **Young**. The various regions present some covering, this is due to the fact that nine categorical variables are used and are active during the classification.

- CTM can be used for visualization purpose and allows a fine analysis of symbolic data.
- The 9 probability tables of a given neuron characterize a particular small subset of insurants.
- Each class is represented by its different neurons, looking at the set of probability tables for one class provides symbolic information.

In order to validate the performances of the CTM map, we compute the classification error rate on the 1106 insurants of the data base and found 15%. We test the generalization using cross validation. The total data base was divided into three equal subsets, the learning phase was repeated three times using at

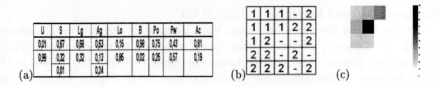

(a)   (b)   (c)

Fig. 1. (a).The 9 probability tables associated with the neuron located at the left-upper corner of the map with a probability of 0.99 that the insurant represented by this neuron has **Professional Use** of their car. (b).CTM classifier, neuron without label represent empty subset of data. (c).Probability density of assignment for a given example (The more intense the gray, the higher the probability)

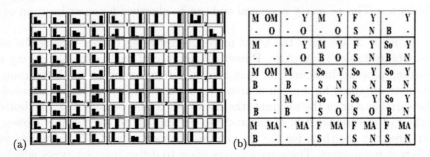

(a)   (b)

Fig. 2. 5 × 5 Neuron CTM. (a). We have displayed four among the nine probability tables associated with each neuron : the upper right corner cell corresponds to: the **Age** of insurant, the upper left: the **Sex**, the bottom right: the **Age** of car and the bottom left: **Power** of car. (b). We have displayed the highest probability associated with the modality of the fourth variables described below. (M:Male, F:Female, So:Society car; OM:Old Men, MA:Middle Age, Y: Young people;B:Big power, S:Small power; O:Old car, N:New car)

each time 2 subsets for learning and the classification error rate was computed on the third one: the mean of the 3 classification error rates for the test is 18.3%. showing the ability of the method to generalize.

    In the second experiment, in order to compare the performances of CTM with other clustering algorithms we use binary data distributed on the web site www.wu-wien.ac.at/am, [4]. We extract from the benchmark two different data bases made of 6000 individuals. These data are artificial data simulated for comparison purposes, they mimic typical situations from tourism marketing. The tourists are classified in 6 classes according to their answer("Yes" or "No") to twelve questions. Six different synthetic data bases with increasing difficulties are available; we select the two difficult problems according to the performances given by the computed Bayes classifier (scenario 5ind, scenario 5dep). Two 2-D topological maps with 5 × 5 neurons were trained using CTM. The comparison with seven different clustering algorithms (Hard Competitive Learning with Euclidian or Absolute Distance HCL-ED or HCL-AD, Neural Gas NGAS-ED or NGAS-AD, k-Means, Self Organizing Map SOM, and a

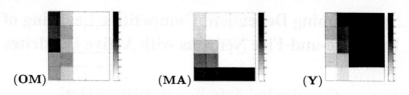

Fig. 3. Topological Map describing the probability distribution of the 3 Modalities of Age variable,(**OM**: Old Men,**MA**: Middle Age, **Y**: Young) (The more intense the gray, the higher the probability)

variant of NGAS Topology Representing Networks TRN) are presented in table 1. This comparison shows that CTM allows to approach the theoretical limit provided by the Bayes classifier.

Table 1. Comparison of the classification performances reached by CTM and seven clustering algorithms on the two simulated data sets (scenario 5ind and scenario 5dep). TBR:Theoretical Bayes Rate

	HCL-ED	HCL-AD	k-means	NGAS-ED	NGAS-AD	SOM	TRN	CTM	TBR
5ind	71%	83%	51%	71%	74%	51%	72%	85%	89%
5dep	49%	58%	48%	52%	59%	49%	48%	71%	79%

4   Conclusion

In this paper we showed that a probabilistic formalism adapted to categorical topological map allows to provide accurate visualisation and classification of sample surveys. Introducing probabilities on the topological maps allows us to make different analysis on the data. CTM provides a pertinent partition of the observations and the topological order of the map gives useful insight of the existing topological relationship between those subsets.

References

1. Anouar, F. Badran, F. Thiria, S. Probabilistic self-organizing map and radial basis function networks. Neurocomputing 20, 83-96. (1998)
2. Dempster,A. P. Laird, N. M. Rubin, D. B. Maximum likelihood from incomplete data via the EM algorithm. Journal of royal Statistic Society, Series B, 39, 1-38
3. Lebbah, M. Badran, F. Thiria, S. Topological Map for Binary Data, ESANN 2000, Bruges, April 26-27-28, 2000, Proceedings
4. Leich, F. Weingessel, A. Dimitriadou, E. Competitive Learning for Binary Data. Proc of ICANN'98, septembre 2-4. Springer Verlag. (1998)
5. Luttrel S. P, 1994. A Bayesian Analysis of Self-Organizing Maps, Neural Computing vol 6
6. Kohonen, T. Self-Organizing Map. Springer, Berlin.(1994)

Spike-Timing Dependent Competitive Learning of Integrate-and-Fire Neurons with Active Dendrites

Christo Panchev*, Stefan Wermter, and Huixin Chen

The Informatics Centre, School of Computing and Technology, University of Sunderland
St. Peter's Campus, Sunderland SR6 0DD, United Kingdom
*http://www.his.sunderland.ac.uk/christo/

Abstract. Presented is a model of an integrate-and-fire neuron with active dendrites and a spike-timing dependent Hebbian learning rule. The learning algorithm effectively trains the neuron when responding to several types of temporal encoding schemes: temporal code with single spikes, spike bursts and phase coding. The neuron model and learning algorithm are tested on a neural network with a self-organizing map of competitive neurons. The goal of the presented work is to develop computationally efficient models rather than approximating the real neurons. The approach described in this paper demonstrates the potential advantages of using the processing functionalities of active dendrites as a novel paradigm of computing with networks of artificial spiking neurons.

1 Introduction

For a long time, dendrites have been thought to be the structures where complex neuronal computation takes place, but only recently we have begun to understand how they operate. Dendrites do not simply collect and pass synaptic inputs to the soma, but in most cases they shape and integrate these signals in complex ways [1]. With our growing knowledge of such processing, there is a stronger argument for taking advantage of the processing power and active properties of the dendrites, and integrating their functionality into artificial neuro-computing models [2]. The features of the models presented here are a computationally optimized interpretation of processing functionalities observed in real neurons.

2 Spike Processing with Active Dendrites

Real neurons show a passive response only under very limited conditions. In many brain areas, such as the cerebellar cortex and neocortex, a reduction of ongoing synaptic activity has been shown to increase the membrane time constant and input resistance, suggesting that synaptic activity can reduce both parameters [3,4]. The model of a neuron with active dendrites presented in this paper is based on such observations. It builds upon the leaky integrate-and-fire neuron. The developed model of an artificial neuron has a set of new active dendrites. In the equations describing the model, the s, d and m indices indicate that the variable or parameter belongs to a synapse, dendrite or the membrane

J.R. Dorronsoro (Ed.): ICANN 2002, LNCS 2415, pp. 896–901, 2002.

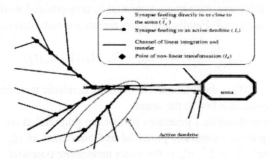

Fig. 1. Model of a neuron with active dendrites.

respectively. For the active dendrite i with a set of $\mathcal{S}^i$ synapses, the total post-synaptic current I_s^i is described by:

$$\tau_s \frac{d}{dt} I_s^i(t) = -I_s^i + \sum_{j \in \mathcal{S}^i} c^{ij} \sum_{t^{(f)} \in \mathcal{F}^j} \delta(t - t^{(f)})$$

where synaptic connection j at dendrite i has weight c^{ij}, $\mathcal{F}^j$ is the set of pre-synaptic spike times filtered as Dirac δ-pulses, and τ_s is the synaptic time constant. In addition, the neuron has a number of synapses feeding close to or directly to the soma. The same as the above equation holds for the total current $\overline{I}_s$ from these synapses.

Further, the current passing through the dendrite into the soma is described by:

$$\tau_d^i \frac{d}{dt} I_d^i(t) = -I_d^i + R_d^i I_s^i(t)$$

Here, the time constant τ_d^i and resistance R_d^i define the active properties of the artificial dendrite as they depend on the incoming post-synaptic current. They are defined as functions of I_s^{i*} which is the maximum of $I_s^i(t)$ since the last pre-synaptic spike:

$$\tau_d^i \triangleq \tau_d^i(I_s^{i*}) = \tau_m - \frac{\tau_m - \tau_s}{1 + e^{-\frac{10}{\tau_s}(\tau_s^2 I_s^{i*} + 1)}} \qquad \text{and} \qquad R_d^i \triangleq R_d^i(\tau_d^i) = \frac{\theta}{BE}$$

$$\text{with} \quad A = \frac{1}{\tau_s - \tau_d^i}, \quad B = A \frac{R_m}{\tau_m}, \quad C = \frac{\tau_m \tau_s}{\tau_m - \tau_s}, \quad D = \frac{\tau_m \tau_d^i}{\tau_m - \tau_d^i} \quad \text{and}$$

$$E = min\left(-Ce^{-\frac{t}{\tau_s}} + De^{-\frac{t}{\tau_d^i}} + (C - D)e^{-\frac{t}{\tau_m}}\right), \ t > 0$$

For low synaptic input, this leads to values of τ_d^i approaching the time constant of the soma τ_m, and for high inputs τ_d^i approaches the time constant of the synapse τ_s which is usually much faster than τ_m. The effect is that a dendrite receiving strong post-synaptic input generates a sharp earlier increase of the membrane potential at the soma, whereas the potential generated from a lower input signal will be delayed. Furthermore, R_d^i is defined such that for a single spike at a synapse with strength c^{ij}, the value of the maximum of the soma membrane potential is proportional to the neuron's firing threshold, i.e. it equals $c^{ij}\theta$. Finally, the soma membrane potential u_m is:

$$\tau_m \frac{d}{dt} u_m(t) = -u_m + R_m(I_d(t) + \overline{I_s}(t))$$

where $I_d(t) = \sum_i I_d^i(t)$ is the total current from the dendritic tree, and $\overline{I_s}(t)$ is the total current from synapses attached to the soma.

The current from dendrite i generates part of the potential at the soma, which will be referred to as *partial membrane potential* and annotated as u_m^i. The *total partial membrane potential* $u_m^d = \sum_i u_m^i$ is the soma membrane potential generated from all dendrites.

The introduced active properties of the dendrites are the basis for the development presented in the next section, where the ability to control the *time* and *value* of the maxima of the membrane potentials plays a critical role in the learning algorithm.

3 Spike-Timing-Dependent Hebbian Learning

The spike-timing dependent Hebbian learning algorithm developed here adjusts the synaptic weight c^{ij} of synapse j at dendrite i, so that a post-synaptic spike occurs at the time when the partial membrane potential u_m^i is at its maximum. Immediately following a post-synaptic spike at time t' in a simulation with time step Δt, the synapse receives two weight correction signals, from the dendrite Δc_d^i and from the soma Δc_m:

$$\Delta c_d^i = \frac{2}{\pi} \arcsin\left(\frac{\Delta I_d(t')}{\sqrt{\Delta t^2 + \Delta I_d^2(t')}}\right) , \ \Delta c_m = -\frac{2}{\pi} \arcsin\left(\frac{\Delta u_m^d(t')}{\sqrt{\Delta t^2 + \Delta u_m^{d\,2}(t')}}\right)$$

where $\Delta I_d(t')$ and $\Delta u_m^d(t')$ are the changes in the post-dendritic current and partial soma membrane potential just before the post-synaptic spike. The correction signal Δc_d^i sent from the dendrite follows the rule: if a post-synaptic spike occurs in the rising phase of the partial membrane potential u_m^i, i.e. before it reaches its maximum, the synaptic strength will be increased so that next time the maximum will occur earlier. Respectively, the synaptic strength will be decreased if a post-synaptic spike occurs after the maximum (Figure 2 (A)). The role of the correction signal Δc_m sent from the soma is to prevent the weights of the synapses from reaching high values simultaneously, or to prevent a total decay in the synaptic strength. Its rule is opposite to the one for the dendrite. Based on the two signals, the total correction signal for the synapse is:

$$\Delta c^{ij} = \begin{cases} \frac{\Delta c_d^i + \Delta c_m}{2} & \text{if } |\Delta c_m| > \varepsilon, \\ \Delta c_d^i & \text{if } |\Delta c_m| \le \varepsilon. \end{cases}$$

where the constant ε allows the neuron to fire without generating a correction signal from the soma when the potential is sufficiently close to the maximum.

Finally, following a post-synaptic spike, the synaptic weights are updated with learning rate η according to:

$$c_{new}^{ij} = \begin{cases} c_{old}^{ij} + \eta \Delta c^{ij}(1 - c_{old}^{ij}) & \text{if } \Delta c^{ij} > 0, \\ c_{old}^{ij} + \eta \Delta c^{ij} c_{old}^{ij} & \text{if } \Delta c^{ij} < 0. \end{cases}$$

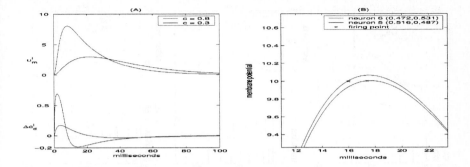

Fig. 2. (A) The correction signal Δc_d^i that would be sent from the dendrite to a synapse with weight 0.8 or 0.3 in the the event of a post-synaptic spike. $\Delta c_d^i = 0$, i.e. no change in the weight occurs, if the post-synaptic spike is at the point of maximum of u_m^i; (B) Soma membrane potential of two neurons with different weights, receiving two input spikes at different dendrites. The second spike is delayed 2 ms. If the maxima of partial membrane potentials coincide or are close in time, the neuron will reach the firing threshold earlier (neuron 6). If the maxima are not close in time, the neuron will reach the threshold later (neuron 5) or not reach it at all.

There have been several suggestions for spike-timing dependent and Hebbian learning algorithms [5,6,7,8]. The learning algorithm presented in this paper achieves very precise tuning of the synapses in response to input spikes representing information with different temporal encoding schemes. The algorithm leads to a normal distribution and intrinsic normalization of the synaptic weights, which allows competitive behaviour of the neurons with dynamic synapses in a network (Figure 2 (B)).

4   Experiments

4.1   Learning to Respond to Different Temporal Codes

The next three examples demonstrate the responses of neurons trained on temporal encoding with single spikes, spike bursts and phase coding. The neuron model and the learning algorithm are able to detect the temporal properties of the input independently on the encoding scheme being applied. The neuron in the first example receives single spikes at three synapses each belonging to a different dendrite (Figure 3 (A)). In the second example, the neuron receives two decaying spike bursts with fixed onset times (Figure 3 (B)). The third example presents a neuron responding to the phase of an input spike with respect to a global oscillation (Figure 3 (C)). The trained neurons fire near the maximum of the partial membrane potentials.

4.2   Competitive Learning

This section demonstrates an application of the neuron with active dendrites and its learning algorithm in a network of self-organizing competitive neurons. The network consist of 2 input and 10 competitive neurons. The input is encoded in the relative spike timing for the two input neurons. Each competitive neuron receives feedforward signals

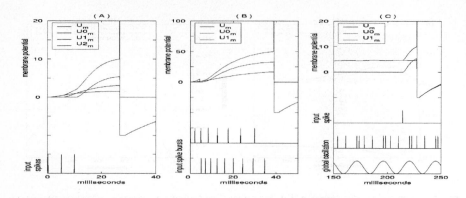

Fig. 3. (A) Response of a trained neuron receiving three input spikes at 0, 5 and 10 ms. u_m is the membrane potential at the soma, u_m^0 is the partial membrane potential generated from the dendrite receiving the first spike, u_m^1 and u_m^2 are for the dendrites receiving the second and the third spikes respectively. (B) Response of a trained neuron receiving two input spike bursts with onset times at 0 and 5 ms. u_m^0 is the partial potential generated from the dendrite receiving the first spike burst, and u_m^1 is for the dendrite receiving the second spike burst. (C) Response of a trained neuron receiving a single spike 8 ms before a peak of an oscillation with a period of 24 ms. u_m^0 is the partial membrane potential generated from the dendrite receiving as input the oscillation spike train and u_m^1 is the partial membrane potential for the dendrite receiving the single spike.

from the input neurons via excitatory synapses at different active dendrites. Furthermore, each competitive neuron receives lateral connections from all other competitive neurons via synapses attached to the soma. These synapses have a fast and strong direct influence on the soma membrane potential and are very efficient for lateral connections.

Figure 4 (A) shows the beginning of the formation of a self-organizing map after 50 epochs. After full training, a well formed self-organizing map is observed (Figure 4 (B)). Each competitive neuron responds only to a particular interval of input values. Since the competitive neurons are relatively fine-tuned to respond only to a particular interval of input values, the feedforward connections are sensitive to noise in the weights. Such noise will destroy the map. On the other hand, due to the fine tuning, the competitive neurons are very robust to noise in the lateral connections. The network was tested with the lateral inhibition removed, and showed relatively little overlap of the responses of the different neurons in the map (Figure 4 (C)). The responses of the trained neurons exhibit clear selectivity to the input. A zoomed-in example of the response of neurons 5 and 6 without lateral connections is shown in Figure 2 (B).

5   Conclusions

The developed new model of a neuron with active dendrites and spike-timing dependent Hebbian learning algorithm are viewed as a contribution towards novel efficient computing models of networks with artificial spiking neurons. The introduction of the active dendrites plays a critical role in achieving a learning algorithm which goes beyond the relative timing of the pre- and post-synaptic spikes to incorporate functions of the membrane potential at the dendrite and at the soma, and the synaptic strength.

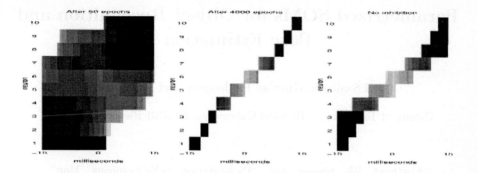

Fig. 4. Self-organization of competitive neurons with active dendrites. The three graphs show the response of the network to two input spikes with relative delays from the interval $[-15, 15]$ ms. The darker color indicates faster response of the competitive neuron to the particular input. Lighter colors indicate later post-synaptic spikes. White areas indicate no post-synaptic response. Left (A): Early self-organized formation after 50 training epochs; Middle (B): A well formed self-organizing map where each competitive neuron responds only to a particular interval of input values; Right (C): Response of the trained network with all lateral connections removed.

The algorithm trains the neurons independently on the temporal code being used at the input and achieves precise selective responses. The presented experiments show details of the functionalities of the neuron, the learning algorithm and their application in training a network of competitive neurons. Further work will build upon these encouraging results and concentrate on applying the model of a neuron with active dendrites and the spike-timing dependent learning algorithm in the development of more complex neural structures such as cell assemblies and synfire chains.

References

1. Stuart, G., Spruston, N., Häusser, M., eds.: Dendrites. Oxford University Press (2001)
2. Wermter, S., Panchev, C.: Hybrid preference machines based on inspiration from neuroscience. Cognitive Science Research (2002) to appear.
3. Häusser, M., Clark, B.: Tonic synaptic inhibition modulates neuronal output pattern and spatiotemporal synaptic integration. Neuron **19** (1997) 665–678
4. Paré D., Shink, E., Gaudreau, H., Destexhe, A., Lang, E.: Impact of spontaneous synaptic activity on the resting properties of cat neocortial pyramidal neurons in vivo. Journal of Neurophysiology **79** (1998) 1450–1460
5. Song, S., Miller, K.D., Abbott, L.F.: Competitive hebbian learning though spike-timing dependent synaptic plasticity. Nature Neuroscience **3** (2000) 919–926
6. Panchev, C., Wermter, S.: Hebbian spike-timing dependent self-organization in pulsed neural networks. In: Proceedings of World Congress on Neuroinformatics. Vienna, Austria (2001)
7. Storck, J., Jäkel, F., Deco, G.: Temporal clustering with spiking neurons and dynamic synapses: towards technological applications. Neural Networks **14** (2001) 275–285
8. Natschläger, T., Ruf, B.: Pattern analysis with spiking neurons using delay coding. Neurocomputing **26–27** (1999) 463–469

Parametrized SOMs for Object Recognition and Pose Estimation

Axel Saalbach, Gunther Heidemann, and Helge Ritter

Faculty of Technology – Bielefeld University – D-33501 Bielefeld, Germany

Abstract. We present the "Parametrized Self-Organizing Map" (PSOM) as a method for 3D object recognition and pose estimation. The PSOM can be seen as a continuous extension of the standard Self-Organizing Map which generalizes the discrete set of reference vectors to a continuous manifold. In the context of visual learning, manifolds based on PSOMs can be used to represent the appearance of various objects. We demonstrate this approach and its merits in an application example.

1 Introduction

Automatic recognition and pose estimation of unknown 3D objects requires suitable representations of the objects which come into question. Among the numerous approaches regarding the representation of objects, *view based* techniques which are based either directly on the object images themselves or on eigenvector decomposition are favorable, since they can be obtained from machine learning algorithms — in contrast to other approaches, e.g. those based on syntactic models.

The appearance of an object in a scene depends typically not only on its physical properties but also on additional degrees of freedom like object pose or lighting conditions. In a controlled environment the possible appearances of an object can be described by a variation of these parameters. All possible images of an object form an appearance manifold M in a representation space X. The continuity of this manifold is thereby influenced by the visual complexity of the scenario [1].

In this paper we investigate the representation of such appearance manifolds based on PSOMs with respect to classification and pose estimation performance. We demonstrate the superiority of PSOMs compared to a standard nearest neighbor classifier (NN-classifier) and how the PSOM approach performs with different interpolation strategies.

The concept of a parametrized representation of appearance manifolds is a general approach which has been proposed for pose estimation of common household items [3] as well as for mobile robots [1] and active vision [7]. The main difference of these approaches is in the generation of the manifold and the parameter estimation. In the following we will focus on the PSOM since it is applicable to tasks with arbitrary degrees of freedom and the parameter estimation is based on numerical standard techniques.

J.R. Dorronsoro (Ed.): ICANN 2002, LNCS 2415, pp. 902–907, 2002.

2   Extending the SOM

A standard SOM [2] usually consists of a low dimensional grid A where each element in the grid is associated with a single neuron or *reference vector* $\mathbf{w_a} \in M$ where $M \subset X \subseteq \mathbb{R}^n$. The position of the reference vector $\mathbf{w_a}$ in the grid is determined by its index $\mathbf{a} \in A$.

During an iterative learning process the reference vectors become topologically ordered according to the grid. The response to an input $\mathbf{x}$ can be seen as the best-matching node $\mathbf{a}^* = \mathrm{argmin} \parallel \mathbf{w_a} - \mathbf{x} \parallel$ with the reference vector $\mathbf{w_{a^*}}$. Thus the discrete approximation of the $\mathbb{R}^n$ by the SOM can be seen as an ordered, nonlinear projection from $\mathbb{R}^n$ to A.

Given a pre-structured SOM, a PSOM [8,9] can be obtained by an extension of the discrete SOM grid to continuous grid positions $\mathbf{s} \in S \subseteq \mathbb{R}^m$, while the reference vectors $\mathbf{w_a}$ are replaced by a vector valued function $\mathbf{w}(\cdot)$: $\mathbf{s} \to \mathbf{w(s)}$.

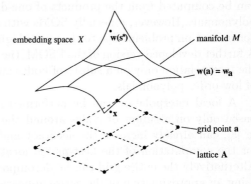

The extension to a PSOM also affects the response to an input. While the discrete grid is replaced by a contin-

Fig. 1. Training setup from a 3x3 PSOM [5].

uous variable, the best-match node becomes a best-match location $\mathbf{s}^*$ according to a distance function $dist(\cdot)$ leading to a minimization problem for $\mathbf{s}^* = \mathrm{argmin}\ dist(\mathbf{w(s)} - \mathbf{x})$ and the corresponding location $\mathbf{w(s^*)}$ on the manifold M. The calculation of $\mathbf{s}^*$ can be performed in an iterative procedure using numerical standard techniques like gradient descent or the Levenberg-Marquardt algorithm and holds similarities to the dynamics of a recurrent network [9].

The distance function $dist(\cdot)$ is usually defined as the Euclidian metric, but the distance calculation can be restricted to certain components as will be described later:

$$dist(\mathbf{x}, \mathbf{x}')^2 = \sum_{k=1}^{n} p_k (\mathbf{x} - \mathbf{x}')^2, \qquad p_k \in \{0, 1\} \tag{1}$$

One way to construct the manifold $\mathbf{w(s)}$ in close analogy to the SOM is as a weighted sum of the reference vectors $\mathbf{w_a}$ and a set of basis functions $H(\mathbf{a}, \mathbf{s})$:

$$\mathbf{w(s)} = \sum_{\mathbf{a} \in A} H(\mathbf{a}, \mathbf{s}) \mathbf{w_a} \tag{2}$$

whereby the basis functions should meet the condition of ortho-normality (Eq.3) and partition of unity (Eq.4).

$$H(\mathbf{a}_i, \mathbf{a}_j) = \delta_{i,j}, \forall \mathbf{a}_i, \mathbf{a}_j \in A \qquad \text{(ortho-normality)} \qquad (3)$$

$$\sum_{\mathbf{a} \in A} H(\mathbf{a}, s) = 1, \forall s \qquad \text{(partition of unity)} \qquad (4)$$

Ortho-normality ensures that the manifold passes through all support points, while the partition of unity allows the mapping of constant functions [9].

By the topological organization of the reference vectors the approach contains additional information about the curvature of the data distribution. In case of a rectangular multidimensional SOM topology, the basis functions $H(\mathbf{a}, s)$ can be computed from the products of one-dimensional Lagrange interpolation polynomials. However, especially SOMs with a large number of nodes can cause generalization problems due to the use of high degree interpolation polynomials. A further developed version of the PSOM, the so called *local PSOM* [8], performs the interpolation only on a subset of nodes and avoids such problems by the use of low-order polynomials.

A local interpolation can be performed in several ways. An interpolation, based only on a subset of nodes around the best-matching node, as proposed in [8], can already increase the accuracy and reduces the computational effort for the computation of the best-match location compared to a global PSOM. Alternatively the entire grid can be decomposed into competing (local) PSOMs using an approximation of the entire appearance manifold for the estimation of the best-match location, leading to a further increased classification performance.

3   PSOMs for Classification and Pose Estimation

So far, the PSOM has been applied successfully in combination with other neural classifiers for hand posture recognition [5] and robotics [9]. Here the PSOM is applied directly for object recognition and pose estimation.

The approximation of an appearance manifold M requires a suitable set of reference vectors D. A set of topologically ordered vectors that approximates the "true" appearance manifold of an object can be either obtained from orderless data by a SOM to which the reference vectors of the PSOM are attached after training, or by a reasonably spaced sampling over the degrees of freedom that determine the object appearance in the image.

In a classification application unknown input can be classified by a matching against competing object manifolds based on PSOMs, in contrast to e.g. a NN-classifier which is directly based on the reference vectors. Figure 2 illustrates the approach of a parametric appearance "trajectory" (1D-manifold) in a three dimensional eigenspace. The circular SOM is based on a data set which consists of 72 images from the *Columbia Object Image Library* (COIL-20) and shows an object in steps of 5 degrees.

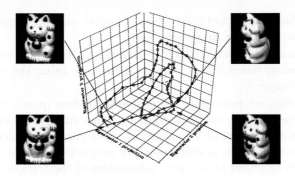

Fig. 2. Illustration of an eigenspace trajectory of COIL-20 object views in an eigenspace based on the three most important eigenvectors.

Obviously the closest point on the "winner-manifold" regarding a given input comprises not only the information about the class of the input but also supplementary image parameters. Another interesting characteristics of the PSOM that can be used for the estimation of these parameters is the ability for *associative completition* of fragmentary data.

To estimate additional image parameters the representation space X that describes the object in terms of features is extended to the Cartesian product X_P of X and the space of these image parameters. Given a PSOM that is based on X_P and an input with no parameter information, the modified Euclidian metric described in Eq.1 can be used for the estimation of the closest point on the manifold while the original features act as a constraint in the calculation of the closest point $\mathbf{w}(\mathbf{s}^*)$ on the manifold. The resulting vector can be seen as an associative completion of the input which holds the requested parameter information.

4   Results for Classification and Pose Estimation (COIL-20)

For an evaluation of object recognition and pose estimation with appearance manifolds based on PSOMs, we used the processed version of the well known *Columbia Object Image Library* (COIL-20) that was developed in the context of the recognition system used by Murase and Nayar [3]. The COIL-20 database consists of 1440 grey-scale images of 20 different objects and is available in several versions via Internet [4].

The images show common household items in steps of 5 degrees, covering a range of 360 degrees using a motorized turntable.

In order to apply the PSOM approach on this data set we performed a feature extraction on the data, which lead to a more compact image representation. As a first step, the images were further preprocessed, mainly the energy of all images has been normalized and the mean subtracted. For the feature extraction the images were finaly projected onto the eigenvectors with the largest eigenvalues, computed from the entire image set.

While an analytical computation of eigenvectors is often impractical, we used an approximation based on an iterative procedure using the neural network proposed by Sanger [6].

Since the data set holds only one degree of freedom (pose), the corresponding object trajectory can be represented by a one-dimensional circular PSOM. Therefore feature vectors of two adjacent views can be connected by a local PSOM by means of linear interpolation. For pose estimation the representation space has been extended by the pose information of the training data.

To evaluate classification and pose estimation performance of PSOMs using different interpolation strategies, the COIL-20 data set was divided in two disjoint subsets, each consisting of images with 10 degree spacing. Both subsets have been used in a two-fold-cross validation for classification and pose estimation. For comparison we give the results of a NN-classifier. For pose estimation only the results for correctly classified views have been used.

Two direct interpolation strategies are the already described decomposition of the circular SOMs into several local PSOMs and the dynamic interpolation based on the best matching node. Figure. 3 shows the average recognition rate of both approaches and of the NN-classifier as well as the pose estimation performance. All tests were repeated with an increasing number of eigenvectors used for the representation until all approaches reach a perfect classification accuracy. Even while a classification accuracy of 100% can be achieved on this data set, all approaches show a considerably different behavior depending on the dimension of the eigenspace.

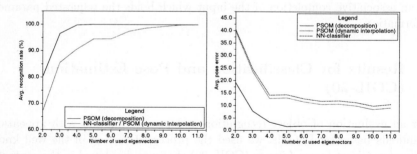

Fig. 3. Comparison of the PSOM approaches and an NN-classifier. Average recognition rate (Left). Average error in the pose estimation (Right).

As expected, classification performance of the dynamic interpolation is identical to that of a NN-classifier because both approaches rely on the best matching node. However, the PSOM decomposition that uses manifolds for classification as well as for pose estimation performs remarkably better, especially in the low dimensional eigenspace.

According to the average pose error the NN-classifier performs rather bad while the PSOM decomposition achieves much better results than the theoretical optimum of 5 degrees of a NN-classifier. The reason for the high pose error of the NN-classifier are a large number of correctly classified objects which cause pose errors of up to 180 degrees. By contrast, the representation by PSOMs can capture the quite subtle appearance variations, demonstrating the advantages of the approximation of the appearance manifolds compared to the direct use of the training data in a NN-classifier.

5 Conclusion

In this paper we introduced the PSOM for appearance-based object representation. The PSOM extends a set of topologically ordered training vectors to a parametrized appearance manifold that can be used for classification as well as for pose estimation by means of associative completion of partial input.

The PSOM showed promising results and can be used as rapid learning technique for the generation of object representations based on small structured data sets. Its main advantages compared to other approaches are the flexible interpolation scheme, its extendibility to tasks with arbitrary degrees of freedom, and the high accuracy in approximation of the object manifolds, which leads to good pose estimation results even for poorly structured objects. The parameter estimation of the response to a given input is based on numerical standard techniques.

References

[1] J.L. Crowley and F. Pourraz. Continuity properties of the appearance manifold formobile robot position estimation. *Image and Vision Computing*, 2001.

[2] Teuvo Kohonen. *Self-Organizing maps*. Springer series in information science. Springer, Berlin, Heidelberg, New York, third edition, 2001.

[3] H. Murase and S.K. Nayar. Visual learning and recognition of 3-d objects from appearance. *Int'l J. Computer Vision*, 14:5–24, 1995.

[4] S. A. Nene, S. K. Nayar, and H. Murase. Columbia Object Image Library (COIL-20). Technical Report CUCUS-006-96, Dept. Computer Science, Columbia Univ. New York, N.Y. 10027, 1996.

[5] C. Nölker and H. Ritter. Parametrized SOMs for hand posture reconstruction. In S.-I. Amari, C.L. Giles, M. Gori, and V. Piuri, editors, *Proc. IEEE-INNS-ENNS Int'l Joint Conf. on Neural Networks IJCNNY'2000*, 2000.

[6] T.D. Sanger. Optimal unsupervised learning in a single-layer linear feedforward neural network. *Neural Networks*, 2:459–473, 1989.

[7] Michael A. Sipe and David Casasent. Global feature space neural network for active object recognition. In *IJCNN'99*, Washington, D.C., 1999.

[8] Jörg Walter and Helge Ritter. Local PSOMs and Chebyshev PSOMs – improving the parametrised self-organizing maps. In *Proc. ICANN*, Paris, volume 1, pages 95–102, October 1995.

[9] Jörg Walter and Helge Ritter. Rapid learning with parametrized self-organizing maps. *Neurocomputing*, 12:131–153, 1996.

An Effective Traveling Salesman Problem Solver Based on Self-Organizing Map

Alessio Plebe

University of Messina Department of Cognitive Science v. Concezione 6–8, Messina, Italy alessio@ylem.njit.edu

Abstract. Combinatorial optimization seems to be a harsh field for Artificial Neural Networks (ANN), and in particular the Traveling Salesman Problem (TSP) is an exemplar benchmark where ANN today are not competitive with the best heuristics from the operations research literature. The thesis upheld in this work is that the Self-Organizing feature Map (SOM) paradigm can be an effective solving method for the TSP, if combined with appropriate mechanisms improving the efficiency and the accuracy. An original TSP-solver based on the SOM is tested over the largest TSP benchmarks, on which other ANN typically fail.

1 Introduction

In the last two decades ANN have been shown to be very successful in a wide range of applications, where the mathematical problem to be solved is in terms of statistical classification or approximation of continuous functions. Neural Networks have not been as successful when applied to combinatorial optimization problems, a recent and comprehensive review is in [15]. The TSP is probably the most famous topic in the field of combinatorial optimization, for a general overview see [13]. Neural networks for the TSP has been studied since the early work of Hopfield & Tank [6] and Durbin & Willshaw [5] (see [10] for a survey). The Hopfield approach has an intrinsic limitation in the quadratic non-convex formulation of the TSP objective function with dramatic effects on large sized problems [14]. Most recently, several algorithms have relied on the more promising Self-Organizing feature Map (SOM) [7]. The straight implementation of the SOM is also quite inefficient, but several researcher have applied it into more refined algorithms: [1,4,3].

Angéniol et al. [1] first addressed a drawback of the pure SOM: the mismatch between the relative spatial distribution of neurons and the location of the targets. In our view this is the most critical problem, and also in the algorithm here presented one of the key feature has been introduced in order to tackle neuron/target mismatch situations, but with a different and more refined approach. This strategy, together with other variation on the SOM, has been first introduced by Plebe and Anile [9] in a double-TSP solver, a problem where two tours has to be optimized, related to the planning of a two-arms robot [11]. The proposed modifications improved dramatically the performances of the

J.R. Dorronsoro (Ed.): ICANN 2002, LNCS 2415, pp. 908–913, 2002.

SOM network. Most of the basic principle of this method has been here adapted to the standard TSP, and a suitable strategy has been developed for the initial placement of the neurons.

2 The Regeneration Process

The first key component of the algorithm is a mechanism for deleting and creating neurons, which breaks the smooth self-organization process every time an excessive density of neurons in an area without targets (or an excess of targets in an area with low neural density) occurs. Given a set of N targets $\mathcal{T}$, the initial network is a set of N neurons $\mathcal{X}$, and both targets T and neurons X have an associated coordinate vector t and x.

The normal process essentially update a neuron X_w, winner of the competition on a target T_j, towards the target itself:

$$x_w \leftarrow x_w + \eta(t_j - x_w) \tag{1}$$

where η is the learning rate, which decreases during the process. If X_w is also the winner in past competitions on T_j (typically at least three times), is accepted as node assigned to the target T_j:

$$x_w \leftarrow t_j \tag{2}$$

The regeneration is essentially a create-and-delete process:

$$\mathcal{X} \leftarrow \mathcal{X} - \{\widetilde{X}\} + \{\widehat{X}\} \tag{3}$$

where $\widetilde{X}$ is the *worst* neuron, $\widehat{X}$ is the newly generated neuron, placed nearby the current winner X_w. In each iteration of the algorithm, two parameters are checked to determine if the neural adaptation is still effective, or the neural regeneration should be issued. These two parameters are called *frustration* (F) and *knowledge* (K). The *frustration* F is increased when the result of the competitions is not progressing the construction of the tours: every time the winner is already assigned to a different target, or is different from the previous winner. On the contrary, it is decreased by successful neural assignments. The *knowledge* K is simply increased by any competition involving a new target, and decreased by an assignment of a neuron to a target. The regeneration state occurs when the following condition is met:

$$F K > \gamma U \tag{4}$$

where U is the number of target not yet assigned, and γ is a constant. When the neural adaptation process resumes, frustration F and knowledge K are reset to zero. Fig. 1 illustrates the effect of the regeneration.

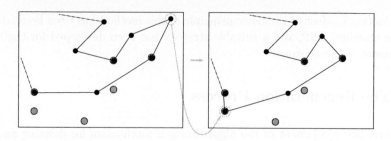

Fig. 1. Example of regeneration: before (left) and after (right) the process.

3 A Metrics for the Competition

In the conventional SOM competition process, the winner X_w of equations (1) and (2) is computed by:

$$X_w = \arg \min_{X_i \in \mathcal{X}} \{D(X_i, T_j)\} \tag{5}$$

where D is the Euclidean metrics. In the new algorithm a different metrics is used

$$X_w = \arg \min_{X_i \in \mathcal{X}} \left\{ \overset{*}{D}(X_i, T_j) \right\} \tag{6}$$

defined as:

$$\overset{*}{D}(X_i, T_j) = \begin{cases} D(X_i, T_j) & \text{if } \frac{\partial D}{\partial s}\big|_{X_i^+} \geq 0 \wedge \frac{\partial D}{\partial s}\big|_{X_i^-} \leq 0 \\ D^\perp(X_i, X_{i+1}, T_j) & \text{if } \frac{\partial D}{\partial s}\big|_{X_i^+} \leq 0 \wedge \frac{\partial D}{\partial s}\big|_{X_{i+1}^-} \geq 0 \\ \infty & \text{otherwise} \end{cases} \tag{7}$$

where D is the usual Euclidean distance between two points, and $D^\perp$ is the distance from a segment joining two points to a third point, and s is a linear coordinate along the tour trajectory. This competition rule has simple geometrical motivations: as long as the tour trajectory is oriented towards the target, the next neuron is closer to the target, and there is no need to compute the distance. It is only necessary to compute the distance when a change in this orientation occurs. A change in orientation can occur either at a neuron, or along a part of the trajectory connecting two neurons. In the first case it is more appropriate to use the point to point distance as a score in the competition for the winner neuron. In the second case the distance is measured to the segment, and if the segment is closest than the nearest of the two neurons connected by the segment, it is selected as the winner. The effect of including the segment distance in the competition is illustrated in an example in Figure 2.

4 The Initial Layout

In [9] the initial tour has a special setup suitable for the mechanics of the two-armed robot for which the TSP application was developed. In this work, in

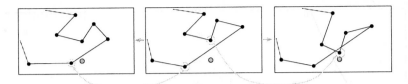

Fig. 2. Example of advantage of the new competition rule. Starting from the situation of the mid panel, the conventional SOM competition will evolve in the right situation, while the left panel is the result of the new rule.

order to compute a proper starting tour, independent from the specific problem, the targets are first checked for clusters. The first identification of clusters is made with a very simple quantization technique. The number N_C of clusters is assumed as the maximum tessellation of the working space in $N_C \times N_C$ equally spaced boxes, such that there are at least N_C non-empty boxes. This procedure also suggest the initial guess of the vector c_i for each cluster C, which is refined recursively in a *kmean*-like manner.

Then, a tour is computed across the mean points of the clusters, just using exactly the same SOM-algorithm, on a N_C-sized TSP problem. Now the starting tour for the complete problem is built, placing all N neurons along the edge of the solved N_C TSP. Being N_{C_i} the number of targets in the cluster C_i, on every edge $\langle C_i, C_{i+1} \rangle$ there will be $\frac{N_{C_i} + N_{C_{i+1}}}{2}$ neurons, equally spaced. It has to be noted that clustering has been frequently used to divide the TSP into smaller sub-problems and combine the sub-solutions separately [8], while in this work the process is global, and clustering is only used to efficiently define the starting tour. An example is given in Fig. 4.

5   Application to Large TSP Instances

It is well known that serious limitations of ANN approaches to the TSP arise from large problem instances. There are no published result of neural network attempt for benchmark problems over 30000 targets, to the best of our knowledge, SOM-based algorithms typically experience convergence problems on such large instances. In most of the recent publications on neural approaches to TSP, the computation speed is not reported, however from direct experience time performances become typically unacceptable for large problems.

The present algorithm has been tested over all Euclidean problems in the TSPLIB [12], which is the benchmark set used in the TSP community. As a "optimum" comparison, the same problem have been solved with the `concorde` code [2], known as the current state-of-the-art for large TSP, based on the *Chained-Lin-Kernighan* method. The `concorde`, which is also able to compute the optimum tour, has been executed with a reduced quality, no more then 10% of gap from the neural solution, for faster performance.

Figure 3 reports the resulting computation time, on a 1.2GHz Athlon PC running Linux, measured with the `getrusage` system call. The `angeniol` algo-

Coordinating Principal Component Analyzers

Jakob J. Verbeek, Nikos Vlassis, and Ben Kröse

Informatics Institute, University of Amsterdam
Kruislaan 403, 1098 SJ Amsterdam, The Netherlands

Abstract. Mixtures of Principal Component Analyzers can be used to model high dimensional data that lie on or near a low dimensional manifold. By linearly mapping the PCA subspaces to one global low dimensional space, we obtain a 'global' low dimensional coordinate system for the data. As shown by Roweis et al., ensuring consistent global low-dimensional coordinates for the data can be expressed as a penalized likelihood optimization problem. We show that a restricted form of the Mixtures of Probabilistic PCA model allows for a more efficient algorithm. Experimental results are provided to illustrate the viability method.

1 Introduction

With increasing sensor capabilities, powerful feature extraction methods are becoming increasingly important. Consider a robot sensing its environment with a camera yielding a stream of 100×100 pixel images. The observations made by the robot often have a much lower intrinsic dimension than the 10.000 dimensions the pixels provide. If we assume a fixed environment, and a robot that can rotate around its axis and translate through a room, the intrinsic dimension is only three. Linear feature extraction techniques are able to do a fair compression of the signal by mapping it to a much lower dimensional space. However, only in very few special cases the manifold on which the signal is generated is a linear subspace of the sensor space. This clearly limits the use of linear techniques and suggests to use non-linear feature extraction techniques.

Mixtures of Factor Analyzers (MFA) [2] can be used to model such non-linear data manifolds. This model provides local linear mappings between local latent spaces and the data space. However, the local latent spaces are not compatible with each other, i.e. the coordinate systems of neighboring factor analyzers might be completely differently oriented. Hence, if we move through the data space from one factor analyzer to the next we cannot predict the latent coordinate of a data point on the one factor analyzer from the latent coordinate on the other factor analyzer.

Recently, a model was proposed that integrates the local linear models into a global latent space, allowing for mapping back and forth between the global latent and the data-space [5]. The idea is that there is a linear map for each factor analyzer between the data-space and the global latent space. The model, which is fitted by maximizing penalized log-likelihood with an algorithm closely

J.R. Dorronsoro (Ed.): ICANN 2002, LNCS 2415, pp. 914–919, 2002.
© Springer-Verlag Berlin Heidelberg 2002

related to the Expectation-Maximization (EM) algorithm [4], is discussed in the next section. In Section 3, we show how we can reduce the number of parameters to be estimated and simplify the algorithm of [5]. These simplifications remove the iterative procedure from the M-step and remove all matrix inversions from the algorithm. The price we pay is that the covariance matrices of the Gaussian densities we use are more restricted, as discussed in the same section. Experimental results are given in Section 4. A discussion and conclusions are provided in Section 5.

2   The Density Model

To model the data density in the high dimensional space we use mixtures of a restricted type of Gaussian densities. The mixture is formed as a weighted sum of its component densities, which are indexed by s. The mixing weight and mean of each component are given by respectively p_s and $\boldsymbol{\mu}_s$. The covariance matrices of the Gaussian densities are constrained to be of the form:

$$\mathbf{C} = \sigma^2(\mathbf{I}_D + \rho\boldsymbol{\Lambda}\boldsymbol{\Lambda}^\top), \qquad \boldsymbol{\Lambda}^\top\boldsymbol{\Lambda} = \mathbf{I}_d, \qquad \rho > 0 \qquad (1)$$

where D and d are respectively the dimension of the high-dimensional/data-space and the low-dimensional/latent space. We use $\mathbf{I}_d$ to denote the d-dimensional identity matrix. The d columns of $\boldsymbol{\Lambda}$, in factor analysis known as the *loading matrix*, are D-dimensional vectors spanning the local PCA. Directions within the PCA subspace have variance $\sigma^2(1+\rho)$, other directions have σ^2 variance. This is as the Mixture of Probabilistic Principal Component Analyzers (MPPCA) model [7], with the difference that here we do not only have isotropic noise outside the subspaces but also isotropic variance inside the subspaces. We use this density model to allow for convenient solutions later. In [9] we provide a Generalized EM algorithm to find maximum likelihood solutions for this model.

The same model can be rephrased using hidden variables $\mathbf{z}$, which we use to denote 'internal' coordinates of the subspaces. We scale the coordinates $\mathbf{z}$ such that: $p(\mathbf{z} \mid s) = \mathcal{N}(\mathbf{z}; 0, \mathbf{I}_d)$. The internal coordinates allow us to clearly express the link to the global latent space, for which we denote coordinates with $\mathbf{g}$. All mixture components have their own linear mapping to the global space, given by a translation $\boldsymbol{\kappa}$ and a matrix $\mathbf{A}$, i.e. $p(\mathbf{g} \mid \mathbf{z}, s) = \delta(\boldsymbol{\kappa}_s + \mathbf{A}_s\mathbf{z})$, where $\delta(\cdot)$ denotes the distribution with mass 1 at the argument. The generative model reads:

$$p(\mathbf{x} \mid \mathbf{z}, s) = \mathcal{N}(\mathbf{x}; \boldsymbol{\mu}_s + \sqrt{\rho_s}\sigma_s\boldsymbol{\Lambda}_s\mathbf{z}, \sigma_s^2\mathbf{I}_D), \quad (2)$$

$$p(\mathbf{x}) = \sum_s p_s\mathcal{N}(\mathbf{x}; \boldsymbol{\mu}_s, \sigma_s^2(\mathbf{I}_D + \rho_s\boldsymbol{\Lambda}_s\boldsymbol{\Lambda}_s^\top)), \quad p(\mathbf{g}) = \sum_s p_s\mathcal{N}(\mathbf{g}; \boldsymbol{\kappa}_s, \mathbf{A}_s\mathbf{A}_s^\top). \quad (3)$$

We put an extra constraint on the projection matrices:

$$\mathbf{A}_s = \alpha_s\sigma_s\sqrt{\rho_s}\mathbf{R}_s, \qquad \mathbf{R}_s^\top\mathbf{R}_s = \mathbf{I}_d, \qquad \alpha_s > 0, \qquad (4)$$

hence $\mathbf{R}_s$ implements only rotations plus reflections.

Note that the model assumes that locally there is a linear correspondence between the data space and the latent space. It follows that the densities $p(\mathbf{g} \mid \mathbf{x}, s)$ and $p(\mathbf{x} \mid \mathbf{g}, s)$ are Gaussian and hence both $p(\mathbf{g} \mid \mathbf{x})$ and $p(\mathbf{x} \mid \mathbf{g})$ are mixtures of Gaussian densities. In the next section we discuss how this density model allows for an efficient learning scheme, as compared to the expressive but expensive MFA model proposed in [5].

3   The Learning Algorithm

The goal is, given observable data $\{\mathbf{x}_n\}$, to find a good density model in the data-space *and* mappings $\{\mathbf{A}_s, \boldsymbol{\kappa}_s\}$ that give rise to 'consistent' estimates for the hidden $\{\mathbf{g}_n\}$. With consistent we mean that if a point $\mathbf{x}$ in the data-space is well modeled by two PCA's, then the corresponding estimates for its latent coordinate $\mathbf{g}$ should be close to each other, i.e. the subspaces should 'agree' on the corresponding $\mathbf{g}$.

Objective Function: To measure the level of agreement, one can consider for all data points how uni-modal the distribution $p(\mathbf{g} \mid \mathbf{x})$ is. This idea was also used in [10]. There the goal was to find a global linear low-dimensional projection of supervised data, that preserves the manifold structure of the data. In [5] it is shown how the double objective of likelihood and uni-modality can be implemented as a penalized log-likelihood optimization problem. Let $Q(\mathbf{g} \mid \mathbf{x}_n) = \mathcal{N}(\mathbf{g}; \mathbf{g}_n, \boldsymbol{\Sigma}_n)$ a Gaussian approximation of the mixture $p(\mathbf{g} \mid \mathbf{x}_n)$ and $Q(s \mid \mathbf{x}_n) = q_{ns}$. We define:

$$Q(\mathbf{g}, s \mid \mathbf{x}_n) = Q(s \mid \mathbf{x}_n) Q(\mathbf{g} \mid \mathbf{x}_n). \tag{5}$$

As a measure of uni-modality we can use a sum of Kullback-Leibler divergences:

$$\sum_{ns} \int d\mathbf{g} \, Q(\mathbf{g}, s \mid \mathbf{x}_n) \log \left[\frac{Q(\mathbf{g}, s \mid \mathbf{x}_n)}{p(\mathbf{g}, s \mid \mathbf{x}_n)} \right] = \sum_n D_{KL}(q_{ns} \| p_{ns}) + \sum_{ns} q_{ns} \mathcal{D}_{ns}$$

where $\mathcal{D}_{ns} = D_{KL}(Q(\mathbf{g} \mid \mathbf{x}_n) \| p(\mathbf{g} \mid \mathbf{x}_n, s))$ and $p_{ns} = p(s \mid \mathbf{x}_n)$. The total objective function, combining log-likelihood and the penalty term, then becomes:

$$\Phi = \sum_n \log p(\mathbf{x}_n) - D_{KL}(\{q_{ns}\} \| \{p_{ns}\}) - \sum_s q_{ns} \mathcal{D}_{ns} \tag{6}$$

$$= \sum_{ns} \int d\mathbf{g} \, Q(\mathbf{g}, s \mid \mathbf{x}_n) \left[-\log Q(\mathbf{g}, s \mid \mathbf{x}_n) + \log p(\mathbf{x}_n, \mathbf{g}, s) \right]. \tag{7}$$

The objective corresponds to a constrained EM procedure, c.f. [8] where the same idea is used to derive a probabilistic version of Kohonen's Self-Organizing Map [3]. Our density model differs with that of [5] in two aspects: (i) we use an isotropic noise model outside the subspaces (as opposed to diagonal covariance matrix) and (ii) we use isotropic variance inside the subspace (as opposed to general Gaussian). Also, using our density model it turns out that to optimize

$\mathbf{\Phi}$ with respect to Σ_n, it should be of the form[1] $\Sigma_n = \beta_n^{-1}\mathbf{I}_d$. Therefore we work with β_n from now on. Using our density model and $\mathbf{g}_{ns} = \mathbf{g}_n - \boldsymbol{\kappa}_s$ and $\mathbf{x}_{ns} = \mathbf{x}_n - \boldsymbol{\mu}_s$ we can write (7) as:

$$\mathbf{\Phi} = \sum_{ns} q_{ns} \left[-\frac{d}{2}\log\beta_n - \log q_{ns} - \frac{e_{ns}}{2\sigma_s^2} - \frac{v_s}{2}\left[d\beta_n^{-1} + \frac{\mathbf{g}_{ns}^\top\mathbf{g}_{ns}}{\rho_s+1}\right]\right. \tag{8}$$

$$\left. -D\log\sigma_s + \frac{d}{2}\log\frac{v_s}{\rho_s+1} + \log p_s\right] + const. \qquad \text{with}$$

$$e_{ns} = ||\mathbf{x}_{ns} - \alpha_s^{-1}\mathbf{\Lambda}_s\mathbf{R}_s^\top\mathbf{g}_{ns}||^2 \quad \text{and} \quad v_s = \frac{\rho_s+1}{\sigma_s^2\rho_s\alpha_s^2}, \tag{9}$$

where v_s is the inverse variance of $p(\mathbf{g} \mid s, \mathbf{x})$ and e_{ns} is the squared distance between $\mathbf{x}_n$ and $\mathbf{g}_n$ mapped into the data space by component s.

Optimization: To optimize $\mathbf{\Phi}$ we use an EM-style algorithm, a simplified version of the algorithm provided in [5]. The simplifications are: (i) the iterative process to solve for the $\mathbf{\Lambda}_s, \mathbf{A}_s$ is no longer needed; an exact update is possible and (ii) the algorithm no longer involves matrix inversions. The same manner of computation is used: in the E-step, we compute the uni-modal distributions $Q(s, \mathbf{g} \mid \mathbf{x}_n)$, parameterized by $\beta_n, \mathbf{g}_n$ and q_{ns}. Let $\langle \mathbf{g}_n \rangle_s = E_{p(\mathbf{g}\mid\mathbf{x}_n,s)}[\mathbf{g}]$ denote the expected value of $\mathbf{g}$ given $\mathbf{x}_n$ and s. We use the following identities:

$$\langle \mathbf{g}_n \rangle_s = \boldsymbol{\kappa}_s + \mathbf{R}_s\mathbf{\Lambda}_s^\top\mathbf{x}_{ns}\alpha_s\rho_s/(\rho_s+1), \tag{10}$$

$$\mathcal{D}_{ns} = \frac{v_s}{2}\left[d\beta_n^{-1} + ||\mathbf{g}_n - \langle\mathbf{g}_n\rangle_s||^2\right] + \frac{d}{2}[\log\beta_n - \log v_s]. \tag{11}$$

The distributions Q can be found by iterating the fixed-point equations:

$$\beta_n = \sum_s q_{ns}v_s, \qquad \mathbf{g}_n = \beta_n^{-1}\sum_s q_{ns}v_s\langle\mathbf{g}_n\rangle_s, \qquad q_{ns} = \frac{p_{ns}\exp{-\mathcal{D}_{ns}}}{\sum_{s'}p_{ns'}\exp{-\mathcal{D}_{ns'}}},$$

where we used $p_{ns} = p(s \mid \mathbf{x}_n)$. In the M-step, we update the parameters of the mixture model. Using notation:

$$C_s = \sum_n q_{ns}||\mathbf{g}_{ns}||^2, \qquad E_s = \sum_n q_{ns}e_{ns}, \qquad G_s = d\sum_n q_{ns}\beta_n^{-1}, \tag{12}$$

the update equations are:

$$\boldsymbol{\kappa}_s = \frac{\sum_n q_{ns}\mathbf{g}_n}{\sum_n q_{ns}}, \qquad \boldsymbol{\mu}_s = \frac{\sum_n q_{ns}\mathbf{x}_n}{\sum_n q_{ns}}, \qquad \alpha_s = \frac{C_s + G_s}{\sum_n q_{ns}(\mathbf{g}_{ns}^\top\mathbf{R}_s\mathbf{\Lambda}_s^\top\mathbf{x}_{ns})},$$

$$\rho_s = \frac{D(C_s + G_s)}{d(\alpha_s^2 E_s + G_s)}, \qquad \sigma_s^2 = \frac{E_s + \rho_s^{-1}\alpha_s^{-2}[C_s + (\rho_s+1)G_s]}{(D+d)\sum_n q_{ns}}, \qquad p_s = \frac{\sum_n q_{ns}}{\sum_{ns'}q_{ns'}}.$$

Note that the above equations require E_s which in turn requires $\mathbf{\Lambda}_s\mathbf{R}_s^\top$ via equations (9) and (12). To find $\mathbf{\Lambda}_s\mathbf{R}_s^\top$ we have to minimize:

$$\sum_n q_{ns}e_{ns} = \sum_n q_{ns}||\mathbf{x}_{ns} - \alpha_s^{-1}\mathbf{\Lambda}_s\mathbf{R}_s^\top\mathbf{g}_{ns}||^2 = -\sum_n q_{ns}\mathbf{x}_{ns}^\top(\mathbf{\Lambda}_s\mathbf{R}_s^\top)\mathbf{g}_{ns} + const.$$

[1] Once we realize that the matrices $\mathbf{V}_s$ in [5] are of the form $c\mathbf{I}_d$ with our density model, it can be seen easily by setting $\partial\mathbf{\Phi}/\partial\Sigma_n = 0$ that $\Sigma_n = \beta^{-1}\mathbf{I}_d$.

This problem is known as the 'weighted Procrustes rotation' [1]. Let

$$C = [\sqrt{q_{1s}}\mathbf{x}_{1s} \cdots \sqrt{q_{ns}}\mathbf{x}_{ns}][\sqrt{q_{1s}}\mathbf{g}_{1s} \cdots \sqrt{q_{ns}}\mathbf{g}_{ns}]^\top, \quad \text{with SVD:} \quad C = \mathbf{U}\mathbf{L}\mathbf{\Gamma}^\top,$$

where the $\mathbf{g}_{ns}$ have been padded with zeros to form D-dimensional vectors, then the optimal $\mathbf{\Lambda}_s \mathbf{R}_s^\top$ is given by the first d columns of $\mathbf{U}\mathbf{\Gamma}^\top$.

4   Experimental Illustration

To demonstrate the method, we captured 40×40 pixel gray valued images of a face with a camera. The face has two degrees of freedom, namely looking up-down and left-right. We learned a coordinated mixture model with 1000 images. We used a global PCA projection to 22 dimensions, preserving over 70% of the variance in the data set. We used a latent dimensionality of two and 20 mixture components. We initialized the coordinated mixture model by clamping the latent coordinates $\mathbf{g}_n$ at coordinates found by Isomap [6] and clamping the β_n at small values for the first 50 iterations. The q_{ns} were initialized uniformly random, and updated from the start. The obtained coordinated mixture model was used to map 1000 'test' images. For each test image $\mathbf{x}_n$ we approximated $p(\mathbf{g} \mid \mathbf{x}_n)$ with a single Gaussian $Q_n = \arg\min_Q D_{KL}(Q\|p(\mathbf{g} \mid \mathbf{x}_n))$ with a certain mean and standard deviation. In Figure 1 we show these means (location of circle) and standard deviations (radius). To illustrate the discovered parametrization further, two examples of linear traversal of the latent space are given.

5   Conclusions and Discussion

We showed how a special case of the density model used in [5] leads to a more efficient algorithm to coordinate probabilistic local linear descriptions of a data manifold. The M-step can be computed at once, the iterative procedure to find solutions for a Riccati equation is no longer needed. Furthermore, the update equations do not involve matrix inversion anymore. However, still d singular values and vectors of a $D \times D$ matrix have to be found.

The application of this method to partially supervised data sets is an interesting possibility and a topic of future research. Another important issue, not addressed here, is that often when we collect data from a system with limited degrees of freedom we actually observe *sequences* of data. If we assume that the system can vary its state only in continuous manner, these sequences should correspond to paths on the manifold of observable data. This fact might be exploited to find a low dimensional embedding of the manifold. In [9] we report on promising results of experiments where we used this model to map omni-directional camera images, recorded through an office, to a 2d latent space (the location in the office), where the data was 'supervised' in the sense that 2d workfloor coordinates are known.

Acknowledgment. This research is supported by the Technology Foundation STW (project nr. AIF4997) applied science division of NWO and the technology program of the Dutch Ministry of Economic Affairs.

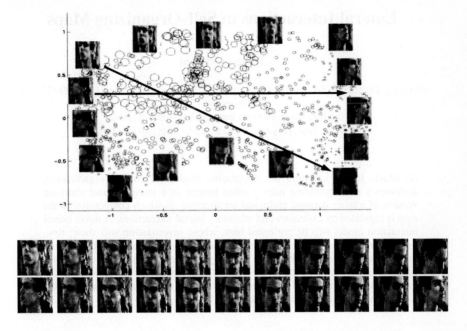

Fig. 1. Latent coordinates and linear trajectories in the latent space.

References

1. T.F. Cox and M.A.A. Cox. *Multidimensional Scaling*. Number 59 in Monographs on statistics and applied probability. Chapman & Hall, 1994.
2. Z. Ghahramani and G.E. Hinton. The EM Algorithm for Mixtures of Factor Analyzers. Technical Report CRG-TR-96-1, University of Toronto, Canada, 1996.
3. T. Kohonen. *Self-Organizing Maps*. Springer Series in Information Sciences. Springer-Verlag, Heidelberg, Germany, 2001.
4. R.M. Neal and G.E. Hinton. A view of the EM algorithm that justifies incremental, sparse, and other variants. In M.I. Jordan, editor, *Learning in Graphical Models*, pages 355–368. Kluwer Academic Publishers, Dordrecht, The Netherlands, 1998.
5. S.T. Roweis, L.K. Saul, and G.E. Hinton. Global coordination of local linear models. In T.G. Dietterich, S. Becker, and Z. Ghahramani, editors, *Advances in Neural Information Processing Systems 14*. MIT Press, 2002.
6. J.B. Tenenbaum, V. de Silva, and J.C. Langford. A global geometric framework for nonlinear dimensionality reduction. *Science*, 290(5500):2319–2323, 2000.
7. M.E. Tipping and C.M. Bishop. Mixtures of probabilistic principal component analysers.*Neural Computation*, 11(2):443–482, 1999.
8. J.J. Verbeek, N. Vlassis, and B. Kröse. The Generative Self-Organizing Map: A Probabilistic Generalization of Kohonen's SOM. Technical Report IAS-UVA-02-03, Informatics Institute, University of Amsterdam, The Netherlands, May 2002.
9. J.J. Verbeek, N. Vlassis, and B. Kröse. Procrustes Analysis to Coordinate Mixtures of Probabilistic Principal Component Analyzers. Technical report , Informatics Institute, University of Amsterdam, The Netherlands, February 2002.
10. N. Vlassis, Y. Motomura, and B. Kröse. Supervised dimension reduction of intrinsically low-dimensional data.*Neural Computation*, 14(1):191–215, January 2002.

Lateral Interactions in Self-Organizing Maps

Roberto Viviani

Abteilung Psychiatrie III, Universitätsklinikum Ulm, Leimgrubenweg 12, D-89075 Ulm,
Germany
roberto.viviani@medizin.uni-ulm.de
http://www.psychiatrie.uni-ulm.de/~Viviani

Abstract. In the literature on topographic models of cortical organization, Kohonen's self-organizing map is often treated as a computational short-cut version of a more detailed biological architecture, in which competition in the map is regulated by excitatory and inhibitory lateral interactions. A novel lateral interaction model will be presented here, whose investigation will show: first, that the behavior of the two models is not identical; and second, that the lateral interaction architecture behaves similarly to non-topographic algorithms, constructing representations of the input at intermediate levels of detail in the initial phases of training. This observation supports a novel interpretation of the topographic organization of the cerebral cortex.

1 Introduction

Topographic self-organizing algorithms have been used to model the topographic organization of cortical areas. The first studies that addressed this issue used Kohonen's self-organizing map [1], while subsequent studies used biologically more realistic models of the cortex [2, 3] (see [4] for a review). An essential constituent of these algorithm is the definition of a halo of excitation around a 'winner' unit in the cortical map, and a larger area of inhibition further away ('Mexican hat'). In Kohonen's self-organizing map [5] this center-surround combination of excitation and inhibition is defined explicitly by an exponential function. In more biologically realistic models, the center-surround effect is achieved by a combination of lateral intra-layer excitatory and inhibitory connections. Since the intra-layer dynamic that ensues from the existence of these lateral connections demonstrably leads to a blob of activation in the map that is analogous to the Mexican hat, these algorithms have been considered roughly equivalent [6].

More generally, self-organizing algorithms may be viewed as on-line learning strategies for the training of quantizers. Unlike standard quantization algorithms such as LBG [7], many of these algorithms update not only the nearest neighbor of the current input at each step, but also a subset of the codebook (in the case of Kohonen's self-organizing map, this extended neighborhood is defined by the Mexican hat). The progression of the algorithm goes hand in hand with the gradual reduction of the size of the extended neighborhood.

The individual algorithms are characterized by the way in which the neighborhood is defined. Topographic algorithms, such as Kohonen's self-organizing map, define a truncated lattice, whose topology determines neighborhood membership as a function of the distance of the code vectors in the lattice. Other algorithms use a function of the

J.R. Dorronsoro (Ed.): ICANN 2002, LNCS 2415, pp. 920–926, 2002.

distortion between the input and the code vectors to determine neighborhood membership [8, 9]; these algorithms are not topographic, and have not been used to model biological networks.

The literature on soft competition quantizers in the signal processing community often views such algorithms as estimators of a probability density that incorporate a form of smoothing in their estimate [10–12]. By gradually reducing the effect of smoothing through a stochastic relaxation schedule, a sequence of estimates is produced, which avoid local minima and gradually construct a more detailed representation of the input distribution. From this perspective, the prototype of this family of algorithms is deterministic annealing, which implements this learning strategy consequently [8]. However, such views are not directly applicable to a topographic quantizer such as Kohonen's self-organizing map. In a smoothed estimator, the smoothing term is a function of the estimate, which leads to a non-topographic algorithm. By contrast, in Kohonen's self-organizing map it is largely a function of the position of the code vectors in the map lattice, which is fixed and essentially unrelated to the estimate. Hence, the intermediate and final results of these algorithms usually differ. Qualitatively, it is well known that the behavior of Kohonen's self-organizing map may be described as a progressive stretching of the output topology over the input space; only exceptionally does this stretching correspond to intermediate phases of approximation. This is especially the case if there are areas of zero-probability space in the input distribution. By contrast, a theoretically motivated soft quantizer such as the deterministic annealing algorithm approximates the input distribution at different degrees of smoothing at all stages.

In this paper we investigate the properties of the neighborhood in a novel, biologically realistic model, in which the lateral connections determine the location and form of the Mexican hat. The stochastic relaxation schedule is implemented by a progressive pruning of the excitatory lateral connections. We will show that, while the quantizers obtained by our lateral connections model are still topographically organized, they also construct intermediate representations of the input distribution like those of non-topographic algorithms. This discovery is interesting because it justifies the use of this algorithm to model psychological development, which is characterized by a gradual refinement of initially gross representations [13, 14].

2 Network Architecture

The network is composed of an input and an output layer or map. The N units in the output map are arranged in a one-dimensional lattice structure defining the distances between them. To avoid edge effects, units at the border of the lattice are considered adjacent. We will denote a unit in the output map as y_j, $j = 1, \ldots N$ to refer to its activation, j to refer to its position in the output lattice. Also, j_i will be the code assigned by the quantizer to the current input i.

The weights connecting the input to the output layer constitute the set of code vectors $\mathbf{w}_j$, $j = 1, \ldots N$ and are the only weights that learn adaptively during training.

The intrinsic weights of the horizontal connections in the output layer w^{intr} are initialized to an exponential function of the distance on the output map between the originating and destination units, j and j':

$$w_{j'j}^{\text{intr}} = \frac{\alpha}{1 + \exp\left[-\beta \cdot \left(\rho - |j - j'|\right)\right]} \, ,$$

where the parameters of the function are such as to create a narrower, positive halo of excitation around j for the one group of weights, and a larger area of inhibition for the other.

The update of the network proceeds in two steps. In the first step, the units in the output map behave as receptive fields of width ρ_I tuned to the Euclidean distance between each code vector and the input vector $\mathbf{x}$:

$$y_j = \exp\left(\frac{-\|\mathbf{w}_j - \mathbf{x}\|^2}{\rho_I^2}\right).$$

In the second step, each unit in the output map is updated according to the activation it receives from the horizontal intrinsic connections:

$$y_{j'} = \frac{2}{1 + \exp\left[-\gamma \cdot \sum_{j=0}^{N} y_j \left(w_j^{\text{exc}} + w_j^{\text{inh}}\right)\right]} - 1 \, ,$$

which is the logistic function with a range between -1 and 1 and gain γ. The originality of our model consists in the fact that the units are not updated in random order, as in a Hopfield network [15], but that the order of their activation is followed, starting from the least active unit. There are two reasons to proceed in this way. First, the model captures the fact that more active units reach their discharge threshold before less active units. Second, the net remains functional at subsequent phases of pruning of the lateral connections. If the units are updated in random order, the network becomes dysfunctional at moderate degrees of pruning.

During training, the code vectors are updated using a soft competitive learning rule:

$$\Delta \mathbf{w}_j = \eta \sigma(j, j_{\text{win}})(\mathbf{x} - \mathbf{w}_j)$$

$$j_{\text{win}} = \arg \max_{j=1\ldots N} y_j$$

where η is the learning rate, and $\sigma(j, j_{\text{win}})$ is the neighborhood membership function. In Kohonen's self organizing map, this function takes the form

$$\sigma_T(j, j_{\text{win}}) = \exp\left(-\frac{\|j - j_{\text{win}}\|^2}{\rho_T^2}\right),$$

while in the deterministic annealing algorithm, the membership function is

$$\sigma_A(j, j_{\text{win}}) = \frac{\exp\left(-\dfrac{\|\mathbf{x} - \mathbf{w}_j\|^2}{\rho_A^2}\right)}{\displaystyle\sum_{j'=1}^{N} \exp\left(-\dfrac{\|\mathbf{x} - \mathbf{w}_{j'}\|^2}{\rho_A^2}\right)} .$$

In these algorithms, ρ_T and ρ_A determine the size of the neighborhood and are gradually reduced during training. Because we let the horizontal connections determine the update code vectors subset, in our network $\sigma(j, j_{\text{win}}) = y_j$, and the learning rule becomes

$$\Delta \mathbf{w}_j = \eta\ y_j(\mathbf{x} - \mathbf{w}_j) .$$

To reduce the size of the udpate codebook subset, a pruning schedule was instituted for the excitatory lateral weights at each training step t:

$$w_{t+1}^{\text{exc}} = \begin{cases} w_t^{\text{exc}}, & \text{if } w > \kappa \\ 0, & \text{if rnd} < \dfrac{1}{E \cdot N} \end{cases}$$

where κ is a pruning factor that was gradually increased from 0.4 to 1.8, E the number of steps in the train epoch, and rnd a random value between 0 and 1.

3 Simulations

The purpose of the first simulation is to illustrate the characteristic allocation of code vectors of the lateral interaction model at intermediate stages of training (Fig. 1). The input is constituted by a mixture of eight Gaussian sources. The position of the centroids is such that the input can be grouped into two larger clusters of four sources or into four clusters of two. A one-dimensional map of 30 units was used. The lateral interactions model builds dense clusters of code vectors representing input clusters of larger size. At different stages of training, these clusters represent prototypes of the input at different levels of detail. By contrast, because of its strong topographic

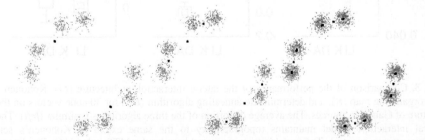

Fig. 1. Allocation of 30 code vectors (*black dots*) in the lateral interaction model at intermediate stages of training (*from left to right*). The small gray dots are the inputs. Not all code vectors can be distinguished due to their close location

prescription, Kohonen's self-organizing map allocates code vectors to the empty areas between the clusters, corresponding to the stretching process of the map topology (Fig. 2). These code vectors do not contribute to the reduction of the quantization error, and do not correspond to any meaningful psychological 'prototype'.

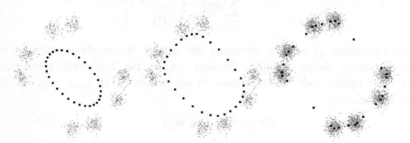

Fig. 2. The same simulation as in Fig. 1 carried out with Kohonen's self-organizing map

In the second simulation, we carried out 15 trials of training using the lateral interaction architecture, Kohonen's self-organizing map, and the deterministic annealing algorithm on a mixture of eight eight-dimensional Gaussian sources located at isotropic random locations on the unit hypersphere. Kohonen's self-organizing map maintains code vectors in the empty space between the squares even in the final stages of training. By contrast, the lateral connection model employs almost all code vectors to encode parts of the input distribution (Fig. 3, right).

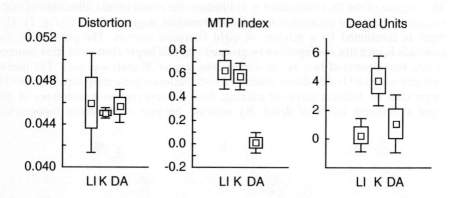

Fig. 3. Comparison of the performance of the lateral interaction architecture (*LI*), Kohonen's self-organizing map (*K*), and deterministic annealing algorithm (*DA*) for 30 code vectors on the mixture of Gaussian sources. The average distortion of the three algorithms is similar (*left*). The lateral interaction model maintains topographicity to the same extent as Kohonen's self organizing map (*center*). Topographicity is measured here by the MTP index [16]. The lateral interaction model differs from Kohonen's self organizing map in the location of its code vectors, as exemplified by the number of dead units at the end of training (*right*)

4 Discussion

The lateral interaction models presented here qualitatively differs from Kohonen's self-organizing map because the attractor in the map is not necessarily centered on the winner, determining the update coefficient symmetrically around it. Rather, the attractor tends to be located in a position such as to match the activation area in the map caused by the current input as much as possible (the winner may not be in the center of this area). This shift, attempting to include the most activated (least-distorting) units, resembles the encoding prescription for a quantizer under noisy conditions [11, 17]. As a consequence of the shift, the neighborhood is no longer 'fixed' around the winner.

The topographicity constraint has a different interpretation in Kohonen's topographic map and in the lateral interaction model. In this latter, the amount of lateral interaction translates directly into the extent to which the input should be approximated. Topographicity in Kohonen's topographic map, while a useful instrument for the visualization of data, appears to have no computational role.

References

1. Obermayer, K., Ritter, H., Schulten, K.: A principle for the formation of the spatial structure of cortical feature maps. Proceedings of the National Academy for Science USA **87** (1990) 8345-8349
2. Ben-Yishai, R., Bar-Or, R. L., Sompolinsky, H.: Theory of orientation tuning in visual cortex. Proceedings of the National Academy for Science USA **92** (1995) 3844-3848
3. Somers, D. C., Nelson, S. B., Sur, M.: An emergent model of orientation selectivity in cat visual cortical simple cells. The Journal of Neuroscience **15** (1995) 5448-5465
4. Ferster, D., Miller, K. D.: Neural mechanisms of orientation selectivity in the visual cortex. Annual Review of Neuroscience **23** (2000) 441-471
5. Kohonen, T.: Self-organized formation of topologically correct feature maps. Biological Cybernetics **43** (1982) 59-69
6. Ritter, H., Martinetz, T. M., Schulten, K.: Neural Computation and Self-Organizing Maps. Wiley, Reading (Mass., 1992)
7. Linde, Y., Buzo, A., Gray, R. M.: An algorithm for vector quantiser design. IEEE Transactions on Communications **28** (1980) 84-95
8. Rose, K., Gurewitz, E., Fox, G. C.: Vector quantization by deterministic annealing. IEEE Transactions on Information Theory **38** (1992) 1249-1257
9. Martinetz, T. M., Berkovich, S. G., Schulten, K. L.: 'Neural-gas' network for vector quantization and its application to time-series prediction. IEEE Transactions on Neural Networks **4** (1993) 558-569
10. Yair, E., Zeger, K., Gersho, A.: Competitive learning and soft competition for vector quantizer design. IEEE Transactions on Signal Processing **40** (1992) 294-308
11. Luttrell, S.P.: Derivation of a class of training algorithms. IEEE Transactions on Neural Networks **1** (1990) 229-232
12. Cherkassky, V., Mulier, F.: Learning from Data. Concepts, Theory, and Methods. John Wiley & Sons, New York (1998)
13. Keil, F.C.: On the emergence of semantic and conceptual distinctions. Journal of Experimental Psychology: General **112** (1983) 357-385
14. Mandler, J. M., Bauer, P. J., McDonough, L.: Separating the sheep from the goats: Differentiating global categories. Cognitive Psychology **23** (1991) 263-298

15. Hopfield, J. J.: Neurons with graded response have collective computational properties like those of two state neurons. Proceedings of the National Academy for Science USA **81** (1984) 3088-3092

16. Bezdek, J. C., Pal, N. R.: An index of topological preservation for feature extraction. Pattern Recognition **28** (1995) 381-391

17. Luttrell, S.P.: A Bayesian analysis of self-organizing maps. Neural Computation **6** (1994) 767-794

Complexity Selection of the Self-Organizing Map

Anssi Lensu* and Pasi Koikkalainen

University of Jyväskylä, Laboratory of Data Analysis
P.O.Box 35 (MaD), FIN-40351 Jyväskylä, Finland
anssi@mit.jyu.fi and pako@mit.jyu.fi

Abstract. This paper describes how the complexity of the Self-Organizing Map can be selected using the Minimum Message Length principle. The use of the method in textual data analysis is also demonstrated.

1 Introduction

The Self-Organizing Map [6,7] (SOM) is a projection and clustering method, which is frequently applied in data analysis. The SOM representation of data can be characterized as a nonlinear manifold, or as a principal surface [4], whose dimension, d_v, is usually lower than the dimension, d_x, of original data. In order to represent multivariate data accurately the lower dimensional SOM surface tends to fold inside the higher dimensional data space. In practice, we expect that the SOM representation is a principal regression surface, whose points $v \in \mathbb{R}^{d_v}$ go through data $x \in \mathbb{R}^{d_x}$ such that $x(v) = \mathbb{E}[X \mid v'(X) = v]$, and that there shall be zero mean residual error $\epsilon = X - x(v)$. Here $\mathbb{E}$ denotes expectation and $v'(x)$ is the projection $v' = \arg\min_v \|x(v) - x\|^2$. We may interpret ϵ as noise, while surface $x(v)$ is the model of our interest.

In this paper we select the SOM model $x(v)$ using an information theoretic criterion that minimizes the joint complexity, Comp, of $x(v)$ and residual ϵ:

$$\min_{x(v)}[\text{Comp}(x(v), \epsilon)] = \min_{x(v)}[\text{Comp}(\epsilon \mid x(v)) \, \text{Comp}(x(v))] \,. \tag{1}$$

The best known information theoretic model selection methods are *Minimum Message Length* (MML) by Wallace and Boulton [15] and *Minimum Description Length* (MDL) by Rissanen [13]. MML uses a message length approach for selecting models and does not require the data to have a predefined distribution, which makes it suitable for our implementation. MDL uses penalized likelihood and is therefore distribution dependent. In MML (and MDL) literature the complexity is measured as the length, Len, of the compressed binary description of data set Ω using model $\mathcal{M}$. To locate the simplest sufficient model we minimize:

$$\min_{\mathcal{M}} \text{L}(\Omega) = \min_{\mathcal{M}} \underbrace{[\text{Len}(\mathcal{M}) + \text{Len}(\Omega \mid \mathcal{M})]}_{Data\ description\ length} = \min_{x(v)} [\underbrace{\text{Len}(x(v))}_{=\,Model} + \underbrace{\text{Len}(\epsilon \mid x(v))}_{+\,Residual}] \,.$$
$$\tag{2}$$

The question is then, how to measure $\text{Len}(x(v))$ and $\text{Len}(\epsilon \mid x(v))$ in practice?

* This work is supported by the Academy of Finland under project #37190, CATO/LAMDA.

J.R. Dorronsoro (Ed.): ICANN 2002, LNCS 2415, pp. 927–932, 2002.

1.1   The Self-Organizing Map and Its Complexity

In this context we describe the Self-Organizing Map only briefly, for more details see [7]. In Self-Organizing Maps the principal surface $x(v)$ is approximated with a finite lattice of nodes (*neurons*) $\hat{v}_k, k = 1, 2, \ldots, N_v$. In most applications the dimension of the SOM, d_v, is one (curve) or two (surface).

Let $x_j = [x_j^1, x_j^2, \ldots, x_j^{d_x}]^T$ denote a data vector in $\mathbb{R}^{d_x}$, and let vectors x_j, $j = 1, 2, \ldots, N_x$ form the set of data samples, Ω. In a SOM with N_v neurons, each neuron $\hat{v}_k$ is associated with a prototype vector w_k, which contains d_x weight values, which correspond to the elements of the data vectors $x_j \in \mathbb{R}^{d_x}$.

There are several training algorithms to fit the SOM representation to data. A good review of the methods can be found from [2]. The purpose of training is to find positions w_k of the neurons $\hat{v}_k$ in the data space $\mathbb{R}^{d_x}$. In this paper we expect[1] that the outcome of the training is a discrete representation of a smooth surface or curve, which satisfies $x(\hat{v}_k) = \mathbb{E}[X \mid v'_{bmu}(x) = \hat{v}_k]$, where $v'_{bmu} = \arg\min_{\hat{v}_r} \|x(\hat{v}_r) - x\|^2$ (*bmu* = the *best matching unit*).

In our case the smoothness of the SOM representation is in a key role. To control the complexity of the SOM, we use an implicit smoother, which is defined by the number of neurons N_v. A SOM with only a few neurons corresponds to a principal curve with strong smoothing, while a large number of neurons makes the SOM more flexible. This is implemented via a kernel smoother (see [2]) and a tree-structured, multiresolution training algorithm, TS-SOM [9,10], where several SOMs are built, starting from simple models and advancing to more complex ones. Each SOM model is called a TS-SOM *layer*. Figure 1 illustrates how the number of neurons affects the smoothness of the SOM.

Fig. 1. 1-D TS-SOM layers in $\mathbb{R}^2$, $N_x = 500$. Increasing N_v makes the curve more flexible and the SOM starts to represent the noise. Layers 3 and 4 are rather smooth.

1.2   An Example Problem

In one of our applications the problem is to select an appropriate SOM complexity for textual data mining. We have developed a data mining method [11], which is similar to WEBSOM [8], but more hierarchical. In our method, delimited character sequences are first grouped to build a map of words. Then sequences of identified words are grouped to build a map of sentences, and finally the distribution of sentences is used to build a map of documents. We have

---

[1] Note that in the literature there are also different opinions about the objective of the SOM.

found that in this kind of application, where SOMs are used in a sequence, the selection of SOM complexity is very important, as demonstrated in Sect. 3.

Traditionally the complexity of SOM, the number of neurons, has been chosen either intuitively or after a thorough visual or statistical inspection of the data distribution on the latent SOM representation. With TS-SOM this is quite practical, because several SOM resolutions coexist.

2   Model Selection for Self-Organizing Maps

Several attempts have been made to measure the outcome of the Self-Organizing Map (for examples see the references in [7]). Some of the methods evaluate the convergence of the training, while some others try to make sure that the topology of training data is preserved in modeling. Typically these methods can be used for selecting good training parameters or for finding the most suitable topological configuration for a dynamically growing SOM grid. However, they usually cannot be used to compare SOMs with differing amounts of neurons.

There are two exceptions: Cottrell, et al. [3] suggest a coefficient of variation $CV(\theta) = 100\,\sigma_\theta/\mu_\theta$ calculated from bootstrap samples of intra-class sum of squares, θ. However, the results in the paper are not easily interpretable to make it easy to choose the right number of units in complex problems. Hyötyniemi [5] used the Minimum Description Length for SOM training, and also suggested a method for choosing the number of neurons. In this work, there is a strong assumption that the data clusters around the neurons can be modeled as Gaussian probability densities, and the example only proves applicability for cases where the training data has been sampled from four separable normal distributions.

Our goal was to develop a method, which does not make any assumptions about the distribution of the training data nor the distributions of the clusters. It still has the ability to compare SOM models with different numbers of neurons using a universal yardstick for model comparison: complexity. Although our method does not guarantee that we are able to find the true model $v(x)$, it prevents us from making bad subjective choices in automated applications.

2.1   The MML Principle

The basic idea of using Minimum Message Length for model selection [12] is to find model m_i from some model set $\mathcal{M} = \{m_1, m_2, \ldots\}$, which is able to represent data set $\Omega = \{x_1, x_2, \ldots, x_{N_x}\}$, $x_j \in \mathbb{R}^{d_x}$, using an encoded binary string S_i, whose length (calculated in bits) is the smallest. The binary string S_i for model m_i consists of the description of the model m_i and the description of the original data using m_i. In principle, the length of the coded string is calculated as a sum: $\text{Len}(m_i) + \text{Len}(\Omega \mid m_i)$, but in practice the coding requires several code strings, which are catenated together.

We assume that the variable values are continuous, and that the model family (TS-SOM), the dimensionality of the SOM, d_v, and the amount of data (d_x and N_x) are known by the decoder. Therefore, these parameters need not be included in the message. In this case the MML description of data set Ω consists of

- C^{par}: parameters ℓ, a, b and γ, where ℓ is the TS-SOM layer, which defines the number of neurons ($N_v = 2^{d_v \ell}$), and a, b and γ are the coding parameters of real numbers (see Sect. 2.2). Len(C^{par}) is constant for all models, m_i.
- C^{book}: the codebook for quantized real numbers (see Sect. 2.2)
- $\{s_k\}$: the set of sizes of data subsets classified to neurons k (see Sect. 2.2)
- $\{w_k\}$: the weight vectors w_k of the TS-SOM layer ℓ (coded using C^{book})
- $\{\epsilon_j\}$: the residual vectors, indicating the directions of data points from the closest neurons: $\epsilon_j = x_j - w_{bmu}$, where bmu is the index of the closest neuron for x_j (coded using C^{book})

Our "best" model (accurate, but still smooth), TS-SOM layer ℓ^{best}, minimizes the true description length. Due to coding inefficiencies, using MML we can only compute the upper and lower bounds of the true (but unknown) description length L(Ω), but we know that $\min_\ell L^{low}(\Omega) \leq \min_\ell L(\Omega) \leq \min_\ell L^{up}(\Omega)$, where

$$L^{up} = \mathrm{Len}(C^{par}) + \mathrm{Len}(C^{book}) + \mathrm{Len}(\{s_k\}) + \mathrm{Len}(\{w_k\}) + \mathrm{Len}(\{\epsilon_j\}) \quad (3)$$

$$L^{low} = \mathrm{Len}(\{w_k\}) + \mathrm{Len}(\{\epsilon_j\}) . \quad (4)$$

There are several methods for the actual calculation of MML (see [1]). In MML an *Accuracy of Parameter Value* (AOPV) is chosen probabilistically, and all real numbers are coded using this accuracy. Because the formulas for AOPV presented in the MML literature assume a known data distribution, we decided to use a nonparametric approach for the optimization of accuracy. This idea was presented by Rissanen [13] for roughly uniform distributions, while our variation (see Sect. 2.3) can also be used in the non-uniform case.

2.2 The Coding of Real Numbers

In our situation both the SOM weights and the residuals are real numbers, there are a total of $n = d_x (N_v + N_x)$ of them. A commonly used approach for coding the floating point variable values [13,12] is to divide the real axis into 2^γ subintervals, identified by integer values, which are then coded using a *codebook* [12] approach. If the ranges of the training variables are similar, it is sufficient to use the same quantization parameters: γ, $a = \min(\{x_j^u, w_k^u\}_{u=1}^{d_x})$ and $b = \max(\{x_j^u, w_k^u\}_{u=1}^{d_x})$, and a common codebook for all real numbers.

To make coding more efficient, Huffman codes are used for the integer indices. If the number of times when index g occurs within the numbers is n_g, and their sum $n = \sum_{g=1}^{2^\gamma} n_g$, coding of reals requires (in fractions of bits; see [14,13,12])

$$\mathrm{Len}(\{w_k\}) + \mathrm{Len}(\{\epsilon_j\}) = \sum_{g=1}^{2^\gamma} n_g c_g \text{ bits, where } c_g = \begin{cases} -\log_2 \frac{n_g}{n} & \text{if } n_g > 0 \\ 0 & \text{otherwise.} \end{cases} \quad (5)$$

Chaitin's prefix code (see [13]) was used for the amounts of data points, $\{s_k\}$, and thus $\mathrm{Len}(\{s_k\}) = \sum_{k=1}^{N_v}(\lfloor \log_2 2s_k \rfloor + 2\lfloor \log_2 2 \lfloor \log_2 2s_k \rfloor \rfloor)$. A fractional version of Chaitin's code was used for the codebook, and thus the total length of the codebook, C^{book}, is

$$\text{Len}(C^{book}) = \sum_{g=1,c_g \geq 1}^{2^\gamma} (c_g + 2\log_2(2c_g)) + \sum_{g=1,0<c_g<1}^{2^\gamma} 3c_g + \sum_{g=1,c_g=0}^{2^\gamma} 1 . \quad (6)$$

2.3   Optimization of the Number of Subintervals

The original real numbers have been measured with some *precision*, $2^{-\delta}$, while our discrete coding is using an accuracy of $2^{-\gamma}(b-a)$. Using a small γ introduces error in the quantization, and a large γ makes the coding inefficient. The goal for the optimization of accuracy is to find a good balance between these two. To achieve this, we use again the MML principle. We calculate the length of the description of all real numbers using our accuracy, and then estimate the number of bits needed to represent the exact numerical values by coding the errors. The optimal number of subintervals minimizes the sum of these two quantities:

$$\min_{\gamma}[\text{Len}(data + cbook \,|\, acc = 2^{-\gamma}(b-a)) + \text{Len}(errors + cbooks \,|\, acc = 2^{-\delta})] .$$
$$(7)$$

A suitable approximation of optimal precision can be obtained by calculating the *integer* values $\gamma = 1, 2, 3, 4, \ldots$ and choosing the minimum. The quantization accuracy has to be the same for all models, m_i, to make the results comparable.

3   Example Results

Figure 1 demonstrates the method for an artificial data, where $d_x = 2$ and $d_v = 1$. The "best" 1-D SOM model for the data is TS-SOM layer 4, because its complexity value is the smallest both for L^{up} and L^{low}:

TS-SOM layer	1	2	3	4	5	6	7
L^{up} (in bits)	10 504	9 072	7 394	6 373	6 642	7 502	8 938
L^{low} (in bits)	7 212	6 704	5 901	5 276	5 279	5 604	6 439

In another example we examine how well MML type of coding corresponds to word categories of human written documents. We analyzed $N_D = 1\,285$ natural language survey answers written by Finnish school children. There were $N_w = 7\,328$ words and we were interested to know if our 2-D TS-SOM, using a slightly improved version of our word coding [11] where $d_x = 24$, is able to represent similar words with SOM neurons properly. Figure 2 a) depicts portions of the word TS-SOM layers. On layer 4 (too simple model) completely different words have been classified together, and on layer 6 (too complex model) different forms of the same word end up in different groups.

To evaluate our method, we performed a laborious task of verifying the clustering results of layers 4 to 6 by hand. We calculated Shannon's entropy [14] for two things: $\sum_{k=1}^{N_v} \text{H}(neuron_k)$ and $\sum_{i=1}^{N_w} \text{H}(word_i)$. The former measures the entropies of word distributions on the neurons, and the latter how the different forms of the same word are distributed on the SOM. Figure 2 b) depicts how our description length compares to these measures. According to our complexity measures, L^{up} and L^{low}, the entropy measures, and also visual examinations, the best layer is 5. The results have been good with other data sets, as well.

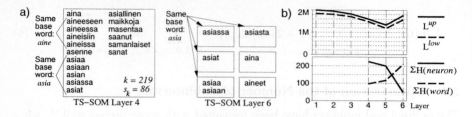

Fig. 2. a) Portions of word TS-SOM layers. Layer 4 classifies completely different words together and layer 6 separates different forms of the same word. b) Graphs illustrating the behavior of our description length and the entropy measures.

4   Conclusions

Our model selection method is able to choose the proper TS-SOM layer both for low and high dimensional data sets. The method does not make strong assumptions about the data distribution, and it could also be used with other SOM algorithms for comparing candidate models with different amounts of neurons.

References

1. Baxter, R. and Oliver, J. *MDL and MML: Similarities and Differences*. Technical Report TR94/207, Dept. of Computer Science, Monash University, 1994.
2. Cherkassky, V. and Mulier, F. *Learning from Data*. John Wiley and Sons, 1998.
3. Cottrell, M., et al. 'A Statistical Tool to Assess the Reliability of Self-Organizing Maps'. In *Advances in Self-Organizing Maps*. Pages 7–14. Springer, 2001.
4. Hastie, T., et al. *The Elements of Statistical Learning*. Springer, 2001.
5. Hyötyniemi, H. 'Minimum Description Length (MDL) Principle and Self-Organizing Maps'. In Proc. *WSOM'97*. Pages 124–129. Libella, Espoo, Finland, 1997.
6. Kohonen, T. 'Self-Organized Formation of Topologically Correct Feature Maps'. *Biological Cybernetics*, 43. Pages 59–69. 1982.
7. Kohonen, T. *Self-Organizing Maps – Third Edition*. Springer, 2001.
8. Kohonen, T., et al. 'Self Organization of a Massive Text Document Collection'. In *Kohonen Maps*. Pages 171–182. Elsevier Science, 1999.
9. Koikkalainen, P. and Oja, E. 'Self-Organizing Hierarchical Feature Maps'. In Proc. *IJCNN'90: Int'l Joint Conf. on Neural Networks*. Pages 279–284. IEEE Press, 1990.
10. Koikkalainen, P. 'Tree-Structured Self-Organizing Maps'. In *Kohonen Maps*. Pages 121–130. Elsevier Science, 1999.
11. Lensu, A. and Koikkalainen, P. 'Similar Document Detection using Self-Organizing Maps'. In Proc. *KES'99: 3rd Int'l Conf. on Knowledge-Based Intelligent Information Engineering Systems*. Pages 174–177. IEEE Press, 1999.
12. Oliver, J. and Hand, D. *Introduction to Minimum Encoding Inference*. Technical Report TR94/205, Dept. of Computer Science, Monash University, 1994.
13. Rissanen, J. *Stochastic Complexity in Statistical Inference*. World Scientific, 1989.
14. Shannon, C.E. 'A Mathematical Theory of Communication'. *Bell Systems Technical Journal*, 47. Pages 143–157. 1948.
15. Wallace, C.S. and Boulton, D.M. 'An Information Measure for Classification'. *Computer Journal*, Vol. 11. Pages 185–194. 1968.

Nonlinear Projection with the Isotop Method

John A. Lee* and Michel Verleysen**

Université catholique de Louvain
Laboratoire de Microélectronique
Place du Levant, 3, B-1348 Louvain-la-Neuve
verleysen@dice.ucl.ac.be

Abstract. Isotop is a new neural method for nonlinear projection of high-dimensional data. Isotop builds the mapping between the data space and a projection space by means of topology preservation. Actually, the topology of the data to be projected is approximated by the use of neighborhoods between the neural units. Isotop is provided with a piecewise linear interpolator for the projection of generalization data after learning. Experiments on artificial and real data sets show the advantages of Isotop.

1 Introduction

Often the analysis of numerical data raises some difficulties because of their high dimensionality. This problem can be attenuated by projection techniques such as the well-known Principal Component Analysis (PCA, [6]). However, PCA is a strictly linear method that is unable to detect nonlinear dependencies between variables. Numerous nonlinear projection methods have been created to address this issue. For example, the nonmetric Multidimensional Scaling (MDS, [12]) and Sammon's nonlinear mapping (NLM [11]) are based on the preservation of either pairwise dissimilarities or Euclidean distances. Neural versions of the NLM, like Curvilinear Component Analysis (CCA, [3,4]), generally show better performance, particularly when they do not use the traditional Euclidean metrics [9,13]. Finally, nonlinear projection can be achieved by the Self-Organizing Map (SOM, [8,14,10]), that works with true topology preservation rather than the more constraining distance reproduction. In this framework, Isotop is a new nonlinear projection algorithm combining the advantages of the SOM and the distance preserving algorithms like Sammon's NLM.

The following of this paper describes how Isotop works (Sect. 2) and shows some results of experiments (Sect. 3). Finally, Sect. 4 draws the conclusions and sketches some perspectives for future developments.

---

* This work was realized with the support of the 'Ministère de la Région wallonne', under the 'Programme de Formation et d'Impulsion à la Recherche Scientifique et Technologique'.
** M.V. works as a senior research associate of the Belgian FNRS.

J.R. Dorronsoro (Ed.): ICANN 2002, LNCS 2415, pp. 933–938, 2002.
© Springer-Verlag Berlin Heidelberg 2002

2   Description of Isotop

Isotop proceeds in three stages: vector quantization of the raw data, linking of the neighboring prototypes and mapping to the projection space.

Assuming that N data patterns are stored in matrix X (one row x_i per pattern), Isotop first proceeds with a vector quantization (VQ) step. For example, the well-known Competitive Learning (CL, [1]) may be used. This neural algorithm transforms the raw data into a set of n representative units called prototypes. These are stored as the rows p_j of matrix P and may be seen as neurons. Formally, P is initialized with randomly selected rows in the data set and is then modified adaptively in several epochs (sweeps of the data set). For each data row x_i the closest prototype p_* (best matching unit) is modified according to the rule:

$$p_* \leftarrow \alpha^t(x_i - p_*) \tag{1}$$

where α_t is a learning rate with values between 0 and 1, decreasing as epochs go by.

The second step of Isotop consists in defining neighborhood relations between the prototypes. Actually, this task is realized by linking prototypes that are close to each other. For example, each prototype can be linked with the k closest ones, with k being a predetermined constant. Another possible method links each prototypes with the ones lying closer than a fixed radius ϵ. In both cases, the result is a connected structure, where each link can be characterized by its Euclidean length.

The third and last step of Isotop builds the mapping from the d-dimensional data space to the p-dimensional projection space. It uses only the neighborhoods defined by the links. The link lengths define distances $\delta_{j,k}$ between direct neighbors j and k. These distances can be extended to any pair (k, l) of prototypes by summing the lengths associated with the shortest path [5] walking from k to l. Such distances help to build matrix M whose rows m_j correspond to those of P and contain the coordinates of the neurons in the projection space. Matrix M is initialized randomly around zero. Next, the twisted structure of links has to be unfolded in the projection space, in order to retrieve the same neighborhoods as in the data space. This goal is reached by randomly stimulating the mapped prototypes. At time t, stimulus $g(t)$ is drawn from a zero-mean, unit-variance, p-dimensional Gaussian distribution. Defining the best matching unit (BMU) as the closest mapped prototype m_* from the stimulus, all prototypes m_j are then moved towards the stimulus. The movement of each mapped prototype becomes smaller and smaller as its neighborhood distance from the BMU grows. Formally, adjustments are made according to:

$$m_j \leftarrow \alpha^t \nu_j^t(g^t - m_j) \tag{2}$$

where α^t is a learning rate with time-decreasing values between 0 and 1. The neighborhood factor ν_j is defined as:

$$\nu_j^t = \exp^{\frac{1}{2}\left(\frac{\delta_{*,j}}{\lambda^t}\right)^2} \tag{3}$$

where $\delta_{*,j}$ is the neighborhood distance from the BMU p_* to the prototype p_j and where λ^t is a time-decreasing neighborhood width.

The choice of a Gaussian distribution for the network stimulation is arbitrary in the absence of a priori information about what would be the best distribution of the mapped prototypes. The Gaussian distribution is just an 'average' choice as is the choice of a uniformly distributed rectangular grid when using a SOM. An advantage of the Gaussian pdf is its smoothness, by comparison to the sharp edges of a SOM grid.

Once the three learning stages are completed, Isotop has build a mapping between the data space and the projection space, resulting from the correspondence between the rows of P and the rows of M. Starting from this discrete representation of the mapping, Isotop can work in conjunction with a piecewise linear interpolator. Such device generalizes the mapping and projects new data.

3 Experiments

Experiments on artificial and real data sets have been conducted in order to compare Isotop with the SOM, which is the most used neighborhood preserving mapping algorithm for nonlinear projection.

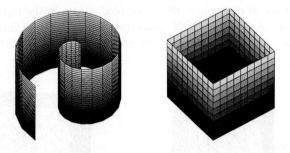

Fig. 1. Artificial data sets: Swiss roll (left) and open box (right)

The implemented SOM algorithm uses rectangularly shaped grids, with a hexagonal neighborhood structure. The neighborhood factor is an exponentially decreasing function of the grid distance from the best matching unit, as described in [7]. Isotop uses Competitive Learning for the VQ step.

As an illustrative example, the 'Swiss roll' data set (see left of Fig. 1) contains 10000 samples. It can be unfolded by Isotop (300 prototypes) and by the SOM (30 × 10 prototypes). The projections are shown in Fig. 2. Despite of a careful parameterization, a grid shaped SOM poorly unfolds the data because its learning process occurs in the data space, whose dimensionality (3D) is higher than the one of the grid (2D). The result is a map that jumps from one spire to the following one in the Swiss roll, as shown in Fig. 2. Isotop does not suffer

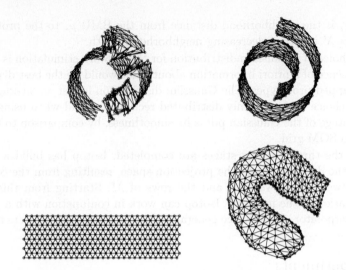

Fig. 2. Swiss roll unfolded by a SOM (left) and by Isotop (right), both shown in the 3D data space (top) and the 2D projection shown (bottom)

from this shortcoming since it builds its mapping in the projection space, whose dimensionality is ideally equal to the one of the linked structure (2D).

Another artificial example is the 'open box' shown in the right part of Fig. 1 (20000 samples). A 30 × 10 SOM converges easily. But this time, the problem comes from the rectangular shape of the grid that difficultly fits to the topology

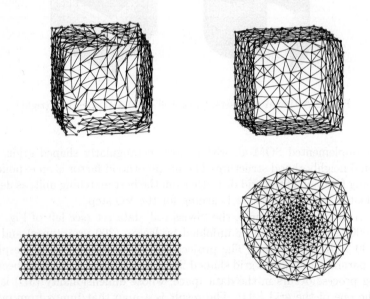

Fig. 3. Open box unfolded by a SOM (left) and by Isotop (right), both shown in the 3D data space (top) and the 2D projection shown (bottom)

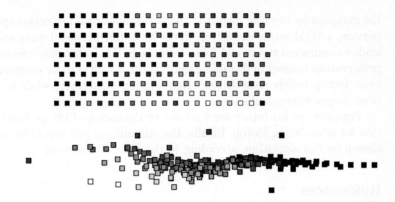

Fig. 4. Abalone dataset: nonlinear projection from 7 to 2 dimensions by a SOM (above) and by Isotop (bottom); the gray level of each prototype is proportional to the mean age of the shells it represents (black is young and white is old)

of the box. Indeed, Fig. 4 shows that some neighborhoods are not preserved: faces of the box are cut out into two parts that are not directly contiguous on the grid. With the help of its data-driven linking step, Isotop (300 prototypes) works with no difficulties and perfectly reproduces the neighborhoods.

Finally, Isotop and the SOM have been applied to a real database, namely the 'Abalone' set from the UCI machine learning repository [2]. This set gathers various attributes from 4177 abalone shells in order to determine their age. Among the nine given attributes, the sex and the age (given by the number of rings in the shell) are eliminated because they are respectively nominal and integer-valued. The seven kept attributes are real values related to the size and weight of the shells. Each attribute is normalized to have zero mean and unit variance. Next, the dimensionality is reduced from 7 to 2 by a SOM and Isotop. After some preliminary analysis, it appears that the data cloud is quite elongated. Therefore, the above-mentioned SOM algorithm give good results when used with 20×10 prototypes. In the same way, the stimulation distribution of Isotop is modified in order to have standard deviations equal to 4 and 1. Like the SOM, Isotop works with 200 prototypes. The results are shown in Fig. 4, where the gray level of each prototype is proportional to the mean age of the shells it represents. Both projection method converges well. At first sight, the regular grid of the SOM is visually pleasant, compared to the irregular cloud given by Isotop. However, a careful examination shows that Isotop preserves more information about the data than the SOM: the global shape of the data cloud is well reproduced and the outliers are still visible.

4 Conclusion

Isotop has been shown as an effective nonlinear projection method. Isotop combines advantages of different projection methods. Indeed, in the same way as many distance preserving algorithms do (e.g. Sammon's mapping), Isotop builds the mapping by working mainly in the low-dimensional projection space; by comparison, a SOM learns exclusively in the high-dimensional data space, what often leads to undesired twists and folds. Like the SOM, Isotop also uses neighborhood preservation instead of distance preservation, which is more constraining. Moreover, Isotop builds data-driven neighborhood structures, while a SOM suffers from its predetermined shape.

Perspectives for future work relates to the choice of the probability distribution for stimulating Isotop. Ideally, the stimulation pdf should be automatically chosen by the algorithm, according to the data distribution.

References

1. A. Ahalt, A. K. Krishnamurthy, P. Chen, and D. E. Melton. Competitive learning algorithms for vector quantization. *Neural Network*, 3:277–290, 1990.
2. C. L. Blake and C. J. Merz. UCI repository of machine learning databases, 1998.
3. P. Demartines and J. Hérault. Vector Quantization and Projection Neural Network. In A. Prieto, J. Mira, and J. Cabestany, editors, *Lecture Notes in Computer Science*, volume 686, pages 328–333. Springer-Verlag, 1993.
4. P. Demartines and J. Hérault. Curvilinear Component Analysis: A self-organizing neural network for nonlinear mapping of data sets.*IEEE Transaction on Neural Networks*, 8(1):148–154, January 1997.
5. E. W. Dijkstra. A note on two problems in connection with graphs.*Numerical Mathematics*, (1):269–271, 1959.
6. I. T. Jollife. *Principal Component Analysis*. Springer-Verlag, New York, 1986.
7. T. Kohonen. Self-organization of topologically correct feature maps.*Biological Cybernetics*, 43:59–69, 1982.
8. T. Kohonen. *Self-Organizing Maps*. Springer, Heidelberg, 2nd edition, 1995.
9. J. A. Lee, A. Lendasse, N. Donckers, and M. Verleysen. A Robust Nonlinear Projection Method. In M. Verleysen, editor,*Proceedings of ESANN'2000, 8th European Symposium on Artificial Neural Networks*, pages 13–20. D-Facto public., Bruges (Belgium), April 2000.
10. H. Ritter, T. Martinetz, and K. Schulten. *Neural Computation and Self-Organizing Maps*. Addison-Wesley, 1992.
11. W. Sammon, J. A nonlinear mapping algorithm for data structure analysis.*IEEE Transactions on Computers*, CC-18(5):401–409, 1969.
12. R. N. Shepard. The analysis of proximities: Multidimensional Scaling with an unknown distance function. *Psychometrika*, 27:125–140, 1962.
13. J. B. Tenenbaum, V. de Silva, and J. C. Langford. A global geometric framework for nonlinear dimensionality reduction. *Science*, 290(5500):2319–2323, 2000.
14. C. Von Der Malsburg. Self-organization of orientation sensitive cells in the striate cortex.*Kybernetik*, 14:85–100, 1973.

Asymptotic Level Density of the Elastic Net Self-Organizing Feature Map

Jens Christian Claussen and Heinz Georg Schuster

Institut für Theoretische Physik und Astrophysik
Leibnizstr. 15, 24098 Christian-Albrechts-Universität zu Kiel, Germany
claussen@theo-physik.uni-kiel.de
http://www.theo-physik.uni-kiel.de/~claussen/

Abstract. Whileas the Kohonen Self Organizing Map shows an asymptotic level density following a power law with a magnification exponent 2/3, it would be desired to have an exponent 1 in order to provide optimal mapping in the sense of information theory. In this paper, we study analytically and numerically the magnification behaviour of the Elastic Net algorithm as a model for self-organizing feature maps. In contrast to the Kohonen map the Elastic Net shows no power law, but for onedimensional maps nevertheless the density follows an universal magnification law, i.e. depends on the local stimulus density only and is independent on position and decouples from the stimulus density at other positions.

Self Organizing Feature Maps map an input space, such as the retina or skin receptor fields, into a neural layer by feedforward structures with lateral inhibition. Biological maps show as defining properties topology preservation, error tolerance, plasticity (the ability of adaptation to changes in input space), and self-organized formation by a local process, since the global structure cannot be coded genetically. The self-organizing feature map algorithm proposed by Kohonen [1] has become a successful model for topology preserving primary sensory processing in the cortex [2], and an useful tool in technical applications [3].

The Kohonen algorithm for Self Organizing Feature Maps is defined as follows: Every stimulus $\mathbf{v}$ of an euclidian input space V is mapped to the neuron with the position $\mathbf{s}$ in the neural layer R with the highest neural activity, given by the condition

$$|\mathbf{w_s} - \mathbf{v}| = \min_{\mathbf{r} \in R} |\mathbf{w_r} - \mathbf{v}| \tag{1}$$

where $|.|$ denotes the euclidian distance in input space. In the Kohonen model the learning rule for each synaptic weight vector $\mathbf{w_r}$ is given by

$$\mathbf{w_r^{new}} = \mathbf{w_r^{old}} + \eta \cdot g_{\mathbf{rs}} \cdot (\mathbf{v} - \mathbf{w_r^{old}}) \tag{2}$$

with $g_{\mathbf{rs}}$ as a gaussian function of euclidian distance $|\mathbf{r} - \mathbf{s}|$ in the neural layer. The function $g_{\mathbf{rs}}$ describes the topology in the neural layer. The parameter η determines the speed of learning and can be adjusted during the learning process. Topology preservation is enforced by the common update of all weight vectors whose neuron $\mathbf{r}$ is adjacent to the center of excitation $\mathbf{s}$.

J.R. Dorronsoro (Ed.): ICANN 2002, LNCS 2415, pp. 939–944, 2002.
© Springer-Verlag Berlin Heidelberg 2002

1   The Elastic Net Feature Map

The Elastic Net [4] was proposed for solving optimization problems like the famous Travelling Salesman Problem. Here we apply this concept to feature maps. The Elastic Net is defined as a gradient descent in the energy landscape

$$E = -\sigma^2 \sum_{\mu} \ln \sum_{r} e^{-(\mathbf{v}^{\mu}-\mathbf{w}_r)^2/2\sigma^2} + \frac{\tilde{\kappa}}{2} \sum_{r} |\mathbf{w}_{r+1} - \mathbf{w}_r|^2 \tag{3}$$

with the input vectors denoted by $\mathbf{v}^{\mu}$. Here r is the index of the neurons in an one-dimensional array (for the TSP: with periodic boundary conditions), and $\mathbf{w}_r$ is the synaptic weight vector of that neuron. For $\sigma \to 0$ (3) becomes

$$\lim_{\sigma \to 0} E = \frac{1}{2} \sum_{\mu} (\mathbf{v}^{\mu} - \mathbf{w}_{s(\mathbf{v}^{\mu})})^2 + \frac{\tilde{\kappa}}{2} \sum_{r} |\mathbf{w}_{r+1} - \mathbf{w}_r|^2. \tag{4}$$

Here $s(\mathbf{v}^{\mu})$ denotes the neuron with the smallest distance to the stimulus, the winning neuron, which is assumed to be nondegenerate. A gradient descent in the first term (which can be interpreted as an entropy term [5]) leads for sufficiently small σ to the condensation of (at least) one weight vector to each input vector, if the input space is discrete. The second term is the potential energy of an elastic string between the weight vectors, and gradient descent in this term leads to a minimization of the (squared!) distances between the weight vectors.

Depending on parameter adjustment [6, 7] a gradient descent in E can provide near-optimal solutions to the TSP within polynomial processing time [8], similar as the Kohonen algorithm [3]. We remark that in the Travelling Salesman application (if the numbers of neurons and cities are chosen to be equal) both the Elastic Net and the Kohonen algorithm share the same zero [3] and first [9] order terms and are therefore related for the final state of convergence, although their initial ordering process is different.

The update rule of the Elastic Net Algorithm is the gradient descent in (3):

$$\frac{1}{\eta} \delta \mathbf{w}_r = \sum_{\mu} (\mathbf{v}^{\mu} - \mathbf{w}_r) \frac{e^{-(\mathbf{v}^{\mu}-\mathbf{w}_r)^2/2\sigma^2}}{\int d\mathbf{r}' \, e^{-(\mathbf{v}^{\mu}-\mathbf{w}_{r'})^2/2\sigma^2}} + \tilde{\kappa} \triangle \mathbf{w}_r, \tag{5}$$

where $\triangle w_r = w_{r-1} - 2w_r + w_{r+1}$ denotes the discrete Laplacian.

If we apply this concept to feature maps, we have to replace the sum over all input vectors by an integral over $\int p(\mathbf{v}) d\mathbf{v}$, i.e. a probability density. If we interpret (5) as a neural feature mapping algorithm, it is a pattern parallel learning rule, or batch update rule, where contributions of all patterns are summed up to one update term. In the brain, hovever, patterns are presented serially in a stochastic sequence. Therefore we generalize this algorithm to serial presentation:

$$\frac{1}{\eta} \delta \mathbf{w}_r = (\mathbf{v} - \mathbf{w}_r) \frac{e^{-(\mathbf{v}-\mathbf{w}_r)^2/2\sigma^2}}{\int d\mathbf{r}' \, e^{-(\mathbf{v}-\mathbf{w}_{r'})^2/2\sigma^2}} + \kappa \triangle \mathbf{w}_r. \tag{6}$$

In Monte Carlo simulations of this model, one chooses input vectors $\mathbf{v}$ according to the probablility density function $p(\mathbf{v})$ and updates $\mathbf{w_r}$ for every neuron $\mathbf{r}$ in the neural layer according to (6). The algorithm can be viewed as a stochastic approximation algorithm that converges if the conditions $\sum_{t=0}^{\infty} \eta^2(t) < \infty$ and $\sum_{t=0}^{\infty} \eta(t) = \infty$ for the time development of parameter η are fulfilled [10] The simultaneous adjustment of κ and σ has been discussed in [6, 7] for the special case of the TSP optimization problem. For the TSP it appears necessary to adjust κ/σ to a system-size-dependent value to avoid 'spike defects' for small κ/σ and 'frozen bead defects' for large κ/σ when annealing $\sigma \rightarrow 0$ [7]. Both 'defects' are no defects in feature maps, the 'spike defects' can only occur for delta-peaked stimuli (cities) together with a dimension-reduction.

The aim in feature maps is different. Using the Kohonen algorithm, one tries to start with large-ranged interaction in the neural layer to avoid global topological defects. This is not directly possible for the Elastic Net, as its learning cooperation is restricted to next-neighbour. Only the strength of the elastic spring κ can be initialized with a high value and decreased after global ordering. The parameter σ is to be interpreted as a resolution length in feature space, e. g. the distance between two receptors in skin or retina. For selectivity of the winner-take-all mechanism, one would choose σ smaller or alike the average or minimal distance between adjacent weight vectors.

2 Asymptotic Density and the Magnification Factor

In this paper we consider the case of continuously distributed input spaces with same dimensionality as the neural layer, so there is no reduction of dimension.

The magnification factor is defined as the density of neurons $\mathbf{r}$ (i. e. the density of synaptic weight vectors $\mathbf{w_r}$) per unit volume of input space, and therefore is given by the inverse Jacobian of the mapping from input space to neural layer: $M = |J|^{-1} = |\det(d\mathbf{w}/d\mathbf{r})|^{-1}$. (In the following we consider the case of noninverting mappings, where J is positive.) The magnification factor is a property of the networks' response to a given probability density of stimuli $P(\mathbf{v})$. To evaluate M in higher dimensions, one in general has to compute the equilibrium state of the whole network using global knowledge on $P(\mathbf{v})$.

For one-dimensional mappings (and possibly for special geometric cases in higher dimensions) the magnification factor may follow an universal magnification law, i.e. $M(\bar{\mathbf{w}}(\mathbf{r}))$ is a function only of the local probability density P and independent of both location $\mathbf{r}$ in the neural layer and $\bar{\mathbf{w}}(\mathbf{r})$ in input space.

An optimal map from the view of information theory would reproduce the input probability exactly ($M \sim P(\mathbf{v})^\rho$ with $\rho = 1$), according to a power law with exponent 1, equivalent to all neurons in the layer fire with same probability. An algorithm of maximizing mutual information has been given by Linsker [11].

For the classical Kohonen algorithm the magnification law (for one-dimensional mappings) is a power law $M(\bar{\mathbf{w}}(\mathbf{r})) \propto P(\bar{\mathbf{w}}(\mathbf{r}))^\rho$ with exponent $\rho = \frac{2}{3}$ [12]. For a discrete neural layer and especially for neighborhood kernels with different shape and range there are corrections to the magnification law [3, 13, 14].

3   Magnification Exponent of the Elastic Net

The necessary condition for the final state of algorithm (6) is that for all neurons r the expectation value of the learning step vanishes:

$$\forall_{r \in R} \quad 0 = \int d\mathbf{v}\, p(\mathbf{v}) \delta \mathbf{w_r}(\mathbf{v}). \tag{7}$$

Since this expectation value is equal to the learning step of the pattern parallel rule (6), equation (7) is the stationary state condition for *both* serial and parallel updating. Inserting the learning rule (6) to condition (7), we obtain for the invariant density $\bar{w}_r$ in the one-dimensional case:

$$0 = \int \left((v - \bar{w}_r) \frac{e^{-(v-\bar{w}_r)^2/2\sigma^2}}{\int dr'\, e^{-(v-\bar{w}_{r'})^2/2\sigma^2}} + \kappa \triangle \bar{w}_r \right) P(v) dv.$$

In the limit of a continuous neural layer for every stimulus v there exists one unique center of excitation s with $v = w_s$. Thus we can substitute integration over dv by integration over ds. Using the Jacobian $J(s) := d\bar{w}(s)/ds$, we have

$$0 = \int \left((\bar{w}(s) - \bar{w}(r)) \frac{e^{-(\bar{w}(s)-\bar{w}(r))^2/2\sigma^2}}{\int dr'\, e^{-(\bar{w}(s)-\bar{w}(r'))^2/2\sigma^2}} + \kappa \triangle \bar{w}(r) \right) P(\bar{w}(s)) J(s) ds.$$

The second term becomes $\kappa \frac{dJ}{dr}$. The normalization integral is ($p := s - r'$):

$$\int e^{-(\bar{w}(s)-\bar{w}(r'))^2/2\sigma^2} dr' = \int e^{-p^2/2(\sigma/J(s))^2} dp + o(\sigma^3) = \sqrt{2\pi} \cdot \frac{\sigma}{J(s)} + o(\sigma^3).$$

For the following equations, we define the abbreviation $\bar{P}(r) := P(\bar{w}(r))$. Using parametric differentiation, substitution $ds = dw_s/(dw_s/ds) = dw_s/J(s)$, and saddlepoint expansion (method of steepest descent) for $\sigma \to 0$, the first integral becomes (after Simic [15]):

$$\frac{1}{\sqrt{2\pi}\,\sigma} \cdot \int (\bar{w}(s) - \bar{w}(r)) e^{-(\bar{w}(s)-\bar{w}(r))^2/2\sigma^2} P(\bar{w}(s)) J(s)^2 ds$$

$$= \frac{\sigma}{\sqrt{2\pi}} \frac{1}{J(r)} \frac{d}{dr} \int e^{-(\bar{w}(s)-\bar{w}(r))^2/2\sigma^2} P(\bar{w}(s)) J(s)^2 ds$$

$$= \frac{\sigma}{\sqrt{2\pi}} \frac{1}{J(r)} \frac{d}{dr} \int e^{-(\bar{w}(s)-\bar{w}(r))^2/2\sigma^2} P(\bar{w}(s)) J(w(s)) dw(s)$$

$$= \sigma^2 \frac{1}{J(r)} \frac{d}{dr} (\bar{P}(r) J(r)) + o(\sigma^4) = \sigma^2 \left(\frac{d\bar{P}}{dr} + \frac{\bar{P}}{J} \frac{dJ}{dr} \right) + o(\sigma^4). \tag{8}$$

Neglecting higher orders of σ, we obtain

$$0 = \frac{\sigma^2}{J(r)} \cdot \frac{d}{dr} \left(\bar{P} J + \kappa \frac{dJ}{dr} \right). \tag{9}$$

This is a first-order nonlinear differential equation for $J(r)$ to a given input density $P(\bar{r})$. However, this can be expressed explicitly only if (additional to $P(v)$) the complete equilibrium state $\bar{w}(r)$ is known, and then one obtains $J(r)$ directly by evaluating the first derivative. Thus the differential equation (9) gives further insight only if $J(r)$ follows an universal scaling law without explicit dependence on the location r, that is, J is a function of $\bar{P}$ only.

The ansatz $J(r) = J(\bar{P}(r))$ leads for all r, where $d\bar{P}/dr \neq 0$, to the differential equation for the invariant state of the one-dimensional Elastic Net Algorithm

$$\frac{dJ}{d\bar{P}} = -\frac{J}{\bar{P}} \cdot \left(1 + \frac{\kappa}{\sigma^2} \frac{J}{\bar{P}}\right)^{-1}. \tag{10}$$

The first derivative depends only on $J/\bar{P}$. The gradient field of (10) has two regimes: For $\kappa/\sigma^2 \to 0$ ('soft string tension') $dJ/d\bar{P} = -J/\bar{P}$, therefore $M = J^{-1} \sim P(v)^1$. The magnification exponent is asymptotically 1 and cortical representation is near to the optimum given by information theory. For $\kappa/\sigma^2 \to \infty$ ('hard string tension') $dJ/d\bar{P} \to 0$, therefore $M = J^{-1}$ has a constant value. Here all adaptation to the stimuli vanishes, equivalent to a magnification exponent of zero.

Substituting $X := \ln P$, $Y := -\ln J$ and $Z := X + Y$, (10) can be solved exactly (see Fig. 1)

$$\ln M = \frac{1}{2} \left(\ln(PM) + \ln\left(1 + \frac{1}{2}\frac{\kappa}{\sigma^2}\frac{1}{PM}\right)\right) + \text{const.} \tag{11}$$

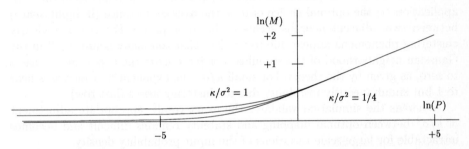

Fig. 1. Solutions of equation (10) for $\kappa/\sigma^2 = 1$, $1/2$ (middle) and $1/4$.

Thus the magnification exponent depends only on the local input probability density $M \sim P^{\rho(P)}$, and we have $\rho_q = \frac{dY}{dX} = \rho + X\frac{d\rho(X)}{dX}$, where $\rho = \rho_q$ for limiting cases with $d\rho(X)/dX \to 0$. For $\kappa \to 0$ the magnification exponent shifts from 1 to zero according to equation (10), rewritten as

$$\frac{1}{\rho_q} = \frac{dX}{dY} = \left(1 + \frac{\kappa}{\sigma^2}e^{-Z}\right) = \left(1 + \frac{\kappa}{\sigma^2}\frac{1}{PM}\right). \tag{12}$$

Finally we remark that the decomposition (6) of the parallel update rule to update responses to the stimuli is not unique. Especially the elastic term can be decomposed in a siutable stimulus-dependent manner so that elasticity is appended only in vicinity of the stimulus. This Local Elastic Net reads

$$\delta\mathbf{w_r} = \eta \cdot \{A^\sigma(\mathbf{v}, \mathbf{w_r}) \cdot (\mathbf{v} - \mathbf{w_r}) + \kappa((1 - \nu) \cdot A^{(\alpha\sigma)}(\mathbf{v}, \mathbf{w_r}) + \nu) \cdot \triangle\mathbf{w_r}\},$$

where A is a normalized gaussian function of distance, $\alpha \simeq 1$ and $0 \leq \nu \leq 1$. A small global elasticity (e.g. $\nu = 0.05$) smoothes fluctuations, but the "forgetting" due to global relaxation is reduced which improves convergence. The Magnification law of the Local Elastic Net is similar as for the Elastic Net [9].

4   Numerical Verification of the Magnification Law

To calculate the asymptotic level density numerically, we considered the map of the unit interval to a onedimensional neural chain of 100 neurons with fixed first and last neuron. The learning rate was 0.5. The stimulus probability density was chosen exponentially as $\exp(-\beta w)$ with $\beta = 4$. After an adaptation process of $5 \cdot 10^7$ steps further 10% of learning steps were used to calculate average slope and its fluctuation (shown in brackets) of $\log J$ as a function of $\log P$. (The first and last 10% of neurons were excluded to eliminate boundary effects). The (local) magnification exponents were obtained as

$\downarrow \sigma \quad \kappa \rightarrow$	0.24	0.024	0.0024	0.00024
0.0003	0.00 (0.01)	0.03 (0.01)	0.15 (0.02)	0.29 (0.03)
0.001	0.03 (0.01)	0.03 (0.01)	0.15 (0.02)	0.28 (0.02)
0.003	0.04 (0.01)	0.03 (0.01)	0.23 (0.01)	0.49 (0.01)
0.01	0.03 (0.01)	0.25 (0.01)	0.77 (0.02)	0.96 (0.06)
0.03	0.23 (0.01)	0.70 (0.03)		

For the Elastic Net the parameter choice appeared crucial: Same as in the TSP application [7] the optimal choice of σ as the average distance (in input space) between two adjacent neurons seems to be appropriate. For larger σ clearly clustering phenomena appear due to the fact that too many neurons fall in the Gaussian neighborhood of the stimulus. For large κ/σ^2 the exponent decreases to zero, as given by the theory. For small κ/σ^2 the exponent first increases near to 1 but simultaneously instability due to clustering arises (last row).

Whereas the simulation validates the exact result, appropriate adjustment of κ/σ^2 between optimal mapping and stability remains difficult and becomes intractable for large-scale variations of the input probability density.

References

1. T. Kohonen 1982. *Biological Cybernetics* **43**, 59-69.
2. K. Obermayer, G. G. Blasdel, and K. Schulten 1992. *Phys. Rev. A* **45**, 7568-7589.
3. H. Ritter, T. Martinetz, and K. Schulten 1992. *Neural Computation and Self-Organizing Maps.* Addison-Wesley.
4. R. Durbin and D. Willshaw 1987. *Nature* **326**, 689-691.
5. P. D. Simic 1990. *Network* **1**, 89-103.
6. R. Durbin, R. Szeliski, and A. Yuille 1989. *Neural Computation* **1**, 348-358.
7. M. W. Simmen 1991. *Neural Computation* **3**, 363-374.
8. J. Hertz, A. Krogh, and R. G. Palmer 1991. *Intr. to the Theory of Neural Comp.* Addison-Wesley, Reading, MA.
9. J. C. Claussen (born Gruel) 1992. Diploma thesis, Kiel,
   J. C. Claussen (born Gruel) and H. G. Schuster 1994. Preprint.
10. T. Kohonen 1991. in: *Artificial Neural Networks*, ed. T. Kohonen et al. North-Holland, Amsterdam.
11. R. Linsker 1989. *Neural Computation* **1**, 402-411.
12. H. Ritter and K. Schulten 1986. *Biological Cybernetics* **54**, 99-106.
13. D. R. Dersch and P. Tavan 1995. *IEEE Trans. Neur. Netw.* **6**, 230-236.
14. H. Ritter 1991. *IEEE Transactions on Neural Networks* **2**, 173-175.
15. P. D. Simic 1994. Private communication.

Local Modeling Using Self-Organizing Maps and Single Layer Neural Networks

Oscar Fontenla-Romero[1], Amparo Alonso-Betanzos[1] Enrique Castillo[2],
Jose C. Principe[3], and Bertha Guijarro-Berdiñas[1]

[1] Laboratory for Research and Development in Artificial Intelligence,
Department of Computer Science, University of A Coruña,
Campus de Elviña s/n, 15071 A Coruña, Spain
{oscarfon@mail2.udc.es,ciamparo@udc.es,cibertha@udc.es},
http://www.dc.fi.udc.es/lidia
[2] Department of Applied Mathematics and Computer Sciences,
University of Cantabria, Avda de los Castros s/n, 39005 Santander, Spain
[3] Computational NeuroEngineering Laboratory,
Electrical and Computer Engineering Department, University of Florida,
Gainesville, FL 32611, USA

Abstract. The paper presents a method for time series prediction using local dynamic modeling. After embedding the input data in a reconstruction space using a memory structure, a self-organizing map (SOM)
derives a set of local models from these data. Afterwards, a set of single
layer neural networks, trained optimally with a system of linear equations, is applied at the SOM's output. The goal of the last network is
to fit a local model from the winning neuron and a set of neighbours
of the SOM map. Finally, the performance of the proposed method was
validated using two chaotic time series.

1 Introduction

The problem of time series prediction can be viewed as a function approximation
problem, because the system tries to find the mapping that outputs the next
instance of the time series based on previous data and desired responses. The
aim of function approximation is to estimate a complex function $f(x)$ by means
of another function $\hat{f}(x)$ composed of basic functions,

$$\hat{f}(x) = \sum_{i=1}^{n} \alpha_i \phi_i(x) \tag{1}$$

where α_i are the coefficients, and $\phi_i(x); i = 1, 2, \cdots, n$ are the basic functions.
Many methods have been proposed to select the number and type of these functions and to obtain the corresponding coefficients. In particular, neural networks
have been employed widely to solve this problem. Most of the neural approaches
are based on the global model of the data presented in (1). An alternative to a

J.R. Dorronsoro (Ed.): ICANN 2002, LNCS 2415, pp. 945–950, 2002.

single global predictor is local modeling where the overall predictive function is the union of several local estimators [1], i.e.,

$$\hat{f}(x) = \bigcup_{i=1}^{m} \hat{f}_i(x) \qquad (2)$$

where the union sign must be understood as the function $f(x)$ to be defined on different support spaces. It was experimentally shown in [2] that local models are very suitable for function approximation. Moreover, functions which may be too complex for a given neural network to approximate them globally may be more easily estimated with local models. In this paper, an efficient and fast neural system for local dynamic modeling, composed of a self-organizing map and a set of single layer neural networks, is presented. This method follows essentially the topology proposed by Principe et al. [3] but in this paper a weighted least squares estimation to fit linear local models was used.

2   Dynamic Modeling

Takens [4] proved that an observable sample $x(n)$ of a dynamic system and its delayed versions $x(n) = [x(n), x(n-\tau), \ldots, x(n-(N-1)\tau)]$, where τ is a specific time delay, can be used to create a trajectory in a Euclidean space of size N, which preserves the dynamical invariants (correlation dimension and Lyapunov exponents) of the original dynamical system. The dimension N of the space must be larger than $2D$, where D is the dimension of the attractor.

Dynamic modeling implies a two step procedure [5]. The first step is to transform the observed time series into a trajectory in a reconstruction space by using an embedding technique [6]. The most common is a time delay embedding, which can be implemented with a delay line with a size specified by Takens' embedding theorem. The second step is to build the predictive model from the trajectory in the reconstruction space. Many techniques have been developed for the purpose of non-linear dynamical modeling. They can be categorised into local and global models. Global approximations can be made with many different function representations, e.g., polynomials, multilayer perceptrons and radial basis function extended with memory structures [8]. Unlike global models, which are based on the data from the entire attractor, local models are established using data from local regions. The state space is partitioned into small regions and a local model is established to describe the dynamics in each region. Some previous neural approaches, based on SOMs, for local dynamic modeling have been proposed, see for example [7,8]. However, the work presented in this paper has several differences, the main being that the fitting of local models is done using the weights of the SOM.

3   Description of the System

The proposed method is composed of three blocks, as is schematically shown in Fig. 1:

1. *An embedding layer* implemented by a time delay line. This stage transforms the original space of the time series, $x(n)$, into a reconstruction space. The output of this layer is a sequence of N-dimensional state vectors, $\boldsymbol{x}(n) = [x(n), x(n-\tau), \ldots, x(n-(N-1)\tau)]^T$, created from the input signal.

2. *A self-organizing map* trained using Kohonen's learning [9]. The input of this network is the pair $(d(n), \boldsymbol{x}(n))$, where $d(n)$ is the desired response. If a time series prediction scenario is employed then $d(n) = x(n + \gamma\tau)$, where γ is a predetermined prediction step. The SOM, formed by $P \times Q$ neurons, will represent the system dynamics in the discrete output lattice, like a codebook, but enhanced with a neighborhood relationships, i.e., states that are adjacent in the reconstruction space will be represented by neurons that are neighbours in the output space. The SOM has been selected because its neighborhood relations assure the continuity among the local models that is a desirable property in local modeling [1].

3. *A set of single layer neural networks.* The goal of this subsystem is to fit the local models from the weights of the self-organizing map, $\boldsymbol{w}_k^{(1)} = [w_{0k}^{(1)}, w_{1k}^{(1)}, \ldots, w_{Nk}^{(1)}]^T; k = 1, \ldots, P \times Q$. Weight $w_{0k}^{(1)}$ is associated with the input $d(n)$ and the other N correspond to the $\boldsymbol{x}(n)$ vector. For the training of each one of the $P \times Q$ single layer networks a new powerful and fast learning algorithm proposed in [10] was used, which always obtains the global optimum in a direct (non iterative) manner. It was demonstrated in [10] that the minimization problem based on the mean squared error (MSE) can be rewritten equivalently (up to first order) as follows:

$$\min_{\boldsymbol{W}, \boldsymbol{b}} E[(\boldsymbol{d} - \boldsymbol{y})^T (\boldsymbol{d} - \boldsymbol{y})] \approx \min_{\boldsymbol{W}, \boldsymbol{b}} E[(\boldsymbol{f}'(\bar{\boldsymbol{d}}) * \bar{\varepsilon})^T (\boldsymbol{f}'(\bar{\boldsymbol{d}}) * \bar{\varepsilon})] \tag{3}$$

where $\boldsymbol{W}$ is the weight matrix, $\boldsymbol{b}$ the bias vector, $\boldsymbol{y}$ the output, $\boldsymbol{f}$ the neural activation function, $\bar{\boldsymbol{d}} = \boldsymbol{f}^{-1}(\boldsymbol{d})$ and '$*$' denotes the element-wise Hadamard product of the vectors $\boldsymbol{f}'(\bar{\boldsymbol{d}})$ and $\bar{\varepsilon} = \bar{\boldsymbol{d}} - (\boldsymbol{Wx} + \boldsymbol{b})$. If the alternative cost function on the right hand side of (3) is used, the weights of the network do not appear inside the nonlinear function. Therefore, the optimal weights can be obtained taking the derivatives of this alternative cost function with respect to the weights, and for the output of each single layer network, we have a system of $N + 1$ linear equations with $N + 1$ unknowns:

$$\left. \begin{array}{l} \sum_{i=1}^{N} \left(\sum_{s \in T} w_{is}^{(1)} w_{ps}^{(1)} f'^2(w_{0s}^{(1)}) \right) w_i^{(2)} = \sum_{s \in T} \left(f^{-1}(w_{0s}^{(1)}) - b^{(2)} \right) f'^2(w_{0s}^{(1)}) w_{ps}^{(1)}; \\ p = 1, 2, \ldots, N, \\ \sum_{i=1}^{N} \left(\sum_{s \in T} w_{is}^{(1)} f'^2(w_{0s}^{(1)}) \right) w_i^{(2)} \quad = \sum_{s \in T} \left(f^{-1}(w_{0s}^{(1)}) - b^{(2)} \right) f'^2(w_{0s}^{(1)}), \end{array} \right\} \tag{4}$$

where $w_i^{(2)}$ and $b^{(2)}$ are, respectively, the weights and the bias of a single layer neural network and T is the set formed by the winning neuron in the SOM and N_L neighbours. The best value for N_L must be determined experimentally.

The training of the system is carried out in two stages, first, the SOM is trained over the overall data set, and second, all the single neural networks are trained using the weights of the SOM. The aim of the proposed method is to obtain a finite set of local models to approximate the global dynamics of the data.

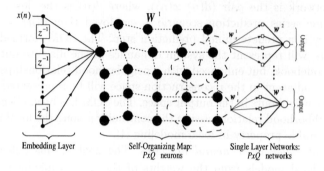

Embedding Layer   Self-Organizing Map: Single Layer Networks:
                              PxQ neurons   PxQ networks

Fig. 1. Structure of the proposed neural system.

Once training is completed, the system works as follows: (1) given a input vector $x(n)$, supplied by the first block of the system, the winner neuron of the SOM is obtained. In this process, only $x(n)$ is used and not the desired output ($d(n)$) which is unknown at this time. (2) The single layer network, associated with the winner neuron, is activated and supplies the output of the system.

4   Experimental Results

The proposed system was validated using synthetic and real-world time series. In both cases the goal was to predict the $x(n + 1)$ sample ($\tau = 1, \gamma = 1$) using a reduced number of previous samples. The synthetic data used was the Lorenz chaotic time series. In this case, a grid of 25×25 neurons in the SOM was used. The embedding dimension, N, was chosen as 4. In order to train the single layer neural networks, the winning neuron and 41 neighbours were used. The train data contained 3000 samples and the test data 1000. Fig. 2 shows the results obtained for this data. Fig. 2(a) depicts the two-dimensional attractor of the chaotic time series and Fig. 2(b) shows, for the train data, the sequence of winning neurons connected with lines. As it can be seen, the trajectory of the chaotic signal have been captured by the SOM. Fig. 2(c) shows, for the test data, a plot of the real versus the desired output and Fig. 2(d) the real time series (solid line) and the generated time series (dashed line). The normalised MSE obtained by the system for the test data was 1.58e-3.

The real-world data set employed in the simulations was the laser time series from the Santa Fe Time Series Competition [6]. The embedding dimension was chosen as 5. In this case a 30×30 map and 52 neighbours were used. The train data contained 2000 samples and the test data 1000. Fig. 3 presents the results

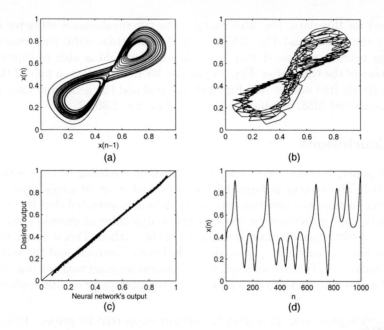

Fig. 2. Results for Lorenz data: (a) two-dimensional attractor, (b) trajectory of winning neurons in the SOM, (c) real versus desired output and (d) observed (solid line) and predicted (dashed line) time series.

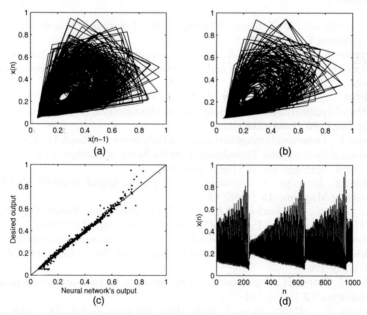

Fig. 3. Results for Laser data: (a) two-dimensional attractor, (b) trajectory of winning neurons in the SOM, (c) real versus desired output and (d) observed (solid line) and predicted (dashed line) time series.

obtained for this data. Fig. 3(a) contains the two-dimensional attractor of the chaotic time series and Fig. 3(b) depicts, for the train data, the sequence of winning neurons connected with lines. Again, the SOM is able to capture the trajectory of the time series. Fig. 3(c) shows, for the test data, a plot of the real versus the desired output and Fig. 3(d) the real and the generated time series. The normalised MSE obtained for the test data was 1.36e-2.

5   Conclusions

In this paper, a neural system for local dynamic modeling of time series was presented. The system is formed by a SOM and a set of single layer neural networks. Simulations over two benchmark data demonstrated that this method is able to capture accurately the underlying dynamics of chaotic signals. An additional advantage obtained is the speed of the method. This is due to the two efficient learning methods employed, i.e., Kohonen's learning and the system of linear equations. This allows a quick incremental learning without degradation in performance on previous data that is very appropriate for on-line applications.

Acknowledgements. This work is partially supported by project PGIDT01-PXI10503PR.

References

1. Crutchfield, J.P., McNamara, B.S.: Equations of motion from a data series. Complex Systems **1** (1987) 417
2. Lawrence, S., Tsoi, A., Back, A.: Function approximation with neural networks and local methods: Bias, variance and smoothness. Australian Conference on Neural Networks (1996) 16–21
3. Principe, J., Wang, L., Motter, M.: Local dynamic modeling with self-organizing maps and applications to nonlinear system identification and control. Proceedings of the IEEE **86** (1998) 2240–2258
4. Takens, F.: Detecting strange attractors in turbulence. In Rand, D., Young, L., eds.: Dynamical Systems and Turbulence (Lecture Notes in Mathematics). Volume 898, New York, Springer-Verlag (1980) 365-381
5. Haykin, S., Principe, J.: Dynamic modeling with neural networks. IEEE Signal Processing Magazine **15** (1988) 66
6. Weigend, A.S., Gershenfeld, N.A.: Time Series Prediction: Forecasting the Future and Understanding the Past. Addison-Wesley, Reading, MA (1994)
7. Moshou, D., Ramon, H.: Extended self-organizing maps with local linear mappings for function approximation and system identification. Proceedings of the WSOM'97 (1997) 181–186
8. Walter, J., Ritter, H.: Rapid learning with parametrized self-organizing maps. Neurocomputing **12** (1996) 131–153
9. Kohonen, T.: Self-Organizing Feature Maps. Springer-Verlag, New York (1995)
10. Castillo, E., Fontenla-Romero, O., Alonso-Betanzos, A., Guijarro-Berdiñas, B.: A global optimum approach for one-layer neural networks. Neural Computation **14** (2002)

Distance Matrix Based Clustering of the Self-Organizing Map

Juha Vesanto and Mika Sulkava

Neural Networks Research Centre
Helsinki University of Technology
P.O.Box 5400, FIN-02015 HUT, Finland
Juha.Vesanto@hut.fi, Mika.Sulkava@hut.fi

Abstract. Clustering of data is one of the main applications of the Self-Organizing Map (SOM). U-matrix is a commonly used technique to cluster the SOM visually. However, in order to be really useful, clustering needs to be an automated process. There are several techniques which can be used to cluster the SOM autonomously, but the results they provide do not follow the results of U-matrix very well. In this paper, a clustering approach based on distance matrices is introduced which produces results which are very similar to the U-matrix. It is compared to other SOM-based clustering approaches.

1 Introduction

The Self-Organizing Map (SOM) [2] quantizes the training data set with a representative set of prototype vectors. The quantization process is regularized by the neighborhood relations defined between map units so that the topology of the data set is preserved. These properties make the SOM an important tool in data mining with main applications in visualization and clustering [8].

One of the most frequently used methods in clustering using SOM is the U-matrix [6] which visualizes distances between each map unit and its neighbors (see Fig. 1b). Unfortunately, when clusters are identified by visual inspection only, the procedure requires human intervention, and the results may be different when performed by different people. In this paper, automated clustering of the SOM based on the distances between neighbors is investigated. The introduced algorithm is based on the one proposed by Vellido *et al.* [7]. It is significantly improved by additional constraints and a pruning phase.

2 Base Clusters

The SOM consists of a regular, usually two-dimensional, grid of map units. Each unit i is represented by a prototype vector, $\mathbf{m}_i = [m_{i1}, \ldots, m_{id}]$ where d is input vector dimension. The units are connected to adjacent ones by a neighborhood relation. The set of neighboring map units of the map unit i is denoted by N_i. The prototype vectors define a tessellation of the input space into a set of Voronoi

J.R. Dorronsoro (Ed.): ICANN 2002, LNCS 2415, pp. 951–956, 2002.

sets $V_i = \{\mathbf{x} \mid \|\mathbf{x} - \mathbf{m}_i\| < \|\mathbf{x} - \mathbf{m}_j\| \, \forall j \neq i\}$. In effect each data vector belongs to the Voronoi set of the prototype to which it is closest.

Using SOM for clustering is a two-level approach: first the data is partitioned into a few hundred Voronoi sets, each corresponding to one map unit, and then the map units are clustered. All data vectors in a Voronoi set belong to the same cluster as the corresponding map unit. By clustering the SOM rather than the data directly, significant gains in the speed of clustering can be obtained [8].

Several approaches for the second level — clustering the map units — have been suggested[1]: for example a second-level SOM, k-means or some agglomerative clustering scheme [4,5,8]. Recently, Vellido $et\ al.$ proposed a clustering algorithm based on distance matrices [7]. Distance matrices are visualization techniques based on showing the distances between neighboring map units [3,6]. Since the point density of the map prototypes roughly follows the probability density function of the data, these distances are approximately inversely proportional to the density of the data, and thus distance matrices are a mode-seeking approach to clustering.

In Vellido's approach, the distance matrix is used to identify cluster centers from the SOM. The rest of the map units are then assigned to the cluster whose center is closest. This algorithm is simple and fast, but it also makes the implicit assumption that the border between two clusters lies on the middle-point between their cluster centers. In this paper, an enhanced version based on region-growing is used, which divides the map into a set of base clusters:

1. Local minima of the distance matrix are found. This is done by finding the set of map units i for which: $f(\mathbf{m}_i, N_i) \leq f(\mathbf{m}_j, N_j)$, $\forall j \in N_i$, where $f(\mathbf{m}_i, N_i)$ is some function of the set of neighborhood distances $\{\|\mathbf{m}_i - \mathbf{m}_j\| \mid j \in N_i\}$, associated with map unit i. In the experiments, median distance was used. The set of local minima may have units which are neighbors of each other. Only one minimum from each such group is retained.
2. Initialization. Let each local minimum be one cluster: $C_i = \{\mathbf{m}_i\}$. All other map units j are left unassigned.
3. Calculate distance $d(C_i, \{\mathbf{m}_j\})$ from each cluster C_i to (the cluster formed by) each unassigned map unit j.
4. Find the unassigned map unit with smallest distance and assign it to the corresponding cluster. Two optional constraints can be used to limit the growth of the clusters:
   - The continuity constraint: only those unassigned map units are considered for merging which are neighbors of the units in the clusters [5]. This ensures that the clusters form continuous areas on the map.
   - Cluster border constraint: map units on borders between clusters may have been identified beforehand using, for example, presence of empty

---

[1] The simplest approach is to handle the map units themselves as the final clusters. However, this is not very sensible. Because of the neighborhood relations, neighboring map units reflect the properties of the *same* rather than different clusters. To remove this overlap, the neighborhood relations can be removed ($N_i = \emptyset$, $\forall i$) in which case the SOM reduces to the k-means algorithm.

map units, as proposed in [9]. Connections to such border units can be removed, thus creating barriers for the region-growing procedure.

5. Repeat from step 3 until no more connections can be made.
6. If there are any unassigned map units, for example unconnected map units due to the cluster border constraint, they are assigned to the same cluster as the closest (neighboring) map unit.

This procedure provides a partitioning of the map into a set of base clusters, the number of which is equal to the number of local minima on the distance matrix. A problem is that the distance matrix may have some local minima which are a product of random variations in the data rather than real local maxima of the probability density function. Such base clusters should be pruned out of the clustering. In the following, this is done in a hierarchical fashion.

3   Cluster Hierarchy

In cluster analysis, constructing a cluster hierarchy is often beneficial. Apart from the need for pruning above, a cluster hierarchy may represent the true structure of the data set better than a single-level partitioning: some clusters can be considered super-clusters, consisting of several sub-clusters. In data mining this allows the data to be investigated at several levels of detail.

Agglomerative clustering algorithms can be used to construct the cluster hierarchy starting from any set of base clusters (Fig. 1c-d). However, most agglomerative algorithms produce binary trees which may not be representative of the true structure. If in reality, a super-cluster consists of three (or more) sub-clusters, the binary tree will have one (or more) extra intermediate clusters which should be pruned out, for example as follows:

1. Start from root (top level) cluster.
2. For the cluster c under investigation, generate different sub-cluster sets. A sub-cluster set may contain either sub-clusters of cluster c or sub-clusters of c's sub-clusters (sub-sub-clusters).
3. Each sub-cluster set defines a partitioning of the data in the investigated cluster. Investigate each partitioning using some clustering validity measure, for example Davies-Bouldin index [1]. However, the Davies-Bouldin index depends mainly on the cluster centroids, and thus is not sensitive to local density of the data. In its stead, a measure of the gap between the two clusters can be used:

$$I_{gap} = \frac{1}{C} \sum_{i=1}^{C} \max_{j} \{ \frac{S_i + S_j}{d_{ij}} \} \tag{1}$$

$$S_i = E\{\|\mathbf{m}_k - \mathbf{m}_l\| \mid \mathbf{m}_k, \mathbf{m}_l \in C_i, k \in N_l, V_k, V_l \neq \emptyset\} \tag{2}$$

$$d_{ij} = E\{a\|\mathbf{m}_k - \mathbf{m}_l\| \mid \mathbf{m}_k \in C_i, \mathbf{m}_l \in C_j, k \in N_l\}. \tag{3}$$

where C is the number of clusters, E is the average and $a = 2$ iff $V_k = \emptyset \vee V_l = \emptyset$, and $a = 1$ otherwise. The coefficient a is used to reflect the fact that an empty map unit between two clusters does not really belong to either, and thus the distance between the clusters is approximately twice the distance from either cluster to the empty map unit.

4. Select the best sub-cluster set (for example the one with minimum I_{gap}), and prune the corresponding intermediate clusters.
5. Select an uninvestigated cluster (if any), and continue from step 2.

This procedure gives a pruned cluster tree (Fig. 1e) together with goodness measures of the clustering quality of the sub-cluster sets of each node in the tree. A particular partitioning is obtained from this tree by starting from the top with all data in a single cluster, and traversing the tree downwards by always splitting the intermediate node with best clustering validity index, until a predetermined number of clusters has been obtained. Alternatively, some validity index can be used to select the best number of clusters.

4   Comparisons

In this section, three different ways to cluster the SOM are compared to each other in a 2-dimensional clustering problem. The data set consists of 2200 data points divided to 7 clusters, plus 20 outlier points (Fig. 1a). The data was normalized to have unit variance in each dimension, and then a SOM with approximately 200 map units was trained[2] using the batch training algorithm and final neighborhood width of 1. After training the SOM (see Fig. 1b), the following three algorithms were used to cluster the map units from 2 upto 20 clusters[3]:

1. k-means clustering of the map prototypes
2. forming base clusters by clustering the map prototypes with region-growing, followed by agglomerative clustering of the base clusters, pruning the tree using I_{gap} and selecting the final partition as discussed above (cluster distances are calculated as distances between their centroids)
3. as previous item, but using Vellido's algorithm to form the base clusters and Davies-Bouldin index in the pruning phase

Fig. 1f-h show examples of the clustering results. These were evaluated using mutual information as a similarity measure between the true clusters $\{C_i\}$ and the acquired clusters $\{\hat{C}_i\}$:

$$m(C, \hat{C}) = -\sum_i p_i \log(p_i) - \sum_j \hat{p}_j \log(\hat{p}_j) + \sum_{i,j} p(i,j) \log(p(i,j)) \qquad (4)$$

where p_i is the probability of $\mathbf{x} \in C_i$, $\hat{p}_j$ is the probability of $\mathbf{x} \in \hat{C}_j$ and $p(i,j)$ is the probability of $\mathbf{x} \in C_i \wedge \mathbf{x} \in \hat{C}_j$. The tests were repeated 20 times such that each time the data set was regenerated. The averages and standard deviations

---

[2] Using SOM Toolbox: http://www.cis.hut.fi/projects/somtoolbox
[3] Notice that the two latter algorithms are limited by the number local minima in the distance matrix, and therefore did not reach the full maxima of 20 clusters.

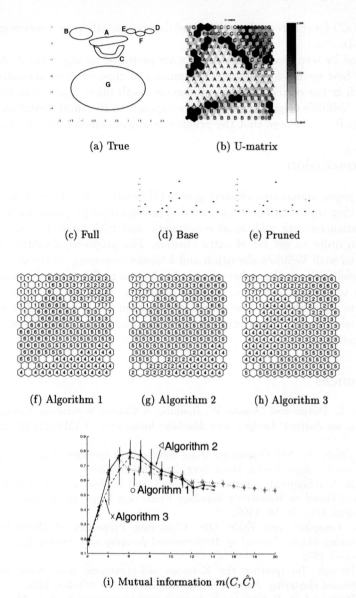

(a) True   (b) U-matrix

(c) Full   (d) Base   (e) Pruned

(f) Algorithm 1   (g) Algorithm 2   (h) Algorithm 3

(i) Mutual information $m(C, \hat{C})$

Fig. 1. (a) Data set with true clusters indicated with encircled areas and the letters. (b) U-matrix of the data with empty map units shown as black, and correct clustering result (obtained by majority voting in each map unit) with the letters. (c) Dendrogram starting from each map unit. (d) Dendrogram starting from base clusters. (e) Pruned dendrogram, encircled clusters are those selected for (g). (f-h) Results of tested algorithms for 7 clusters, respectively. (i) Average mutual information of true and acquired clusters, and associated standard deviations (vertical lines).

of $m(C, \hat{C})$ for different algorithms and different numbers of clusters are shown in Fig. 1i.

It can be seen that the best results are produced by algorithm 2. Algorithm 1 gives best results for high ($\geq$ 12) number of clusters, but especially around 7, which is the correct number of clusters, both other algorithms give better results. Vellido's original algorithm corresponds to the third algorithm without pruning. It can be seen that the pruning produces considerable benefits.

5 Conclusion

In this paper, automated clustering of SOM based on distances between neighboring map units has been considered. The algorithm proposed by Vellido has been enhanced with additional constraints, and with an hierarchical pruning phase in order to get rid of extra clusters. The proposed algorithm has been compared with Vellido's algorithm and k-means clustering. In the tests k-means works well with small number of clusters, but its performance decreases as the number of clusters increases. Vellido's algorithm, on the other hand, produces quite too many clusters. The proposed algorithm combines Vellido's algorithm with a local density based pruning procedure and produced the best results in our experiments.

References

1. David L. Davies and Donald W. Bouldin. A Cluster Separation Measure. *IEEE Trans. on Pattern Analysis and Machine Intelligence*, PAMI-1(2):224–227, April 1979.
2. Teuvo Kohonen. *Self-Organizing Maps*, volume 30 of *Springer Series in Information Sciences*. Springer, Berlin, Heidelberg, 2nd edition, 1995.
3. Martin A. Kraaijveld, Jianchang Mao, and Anil K. Jain. A Nonlinear Projection Method Based on Kohonen's Topology Preserving Maps. *IEEE Trans. on Neural Networks*, 6(3):548–59, 1995.
4. Jouko Lampinen and Erkki Oja. Clustering Properties of Hierarchical Self-Organizing Maps. *Journal of Mathematical Imaging and Vision*, 2(2-3):261–272, November 1992.
5. F. Murtagh. Interpreting the Kohonen self-organizing map using contiguity-constrained clustering. *Pattern Recognition Letters*, 16:399–408, 1995.
6. A. Ultsch and H. P. Siemon. Kohonen's Self Organizing Feature Maps for Exploratory Data Analysis. In *Proceedings of International Neural Network Conference (INNC'90)*, pages 305–308, Dordrecht, The Netherlands, 1990. Kluwer.
7. A. Vellido, P.J.G Lisboa, and K. Meehan. Segmentation of the on-line shopping market using neural networks. *Expert Systems with Applications*, 17:303–314, 1999.
8. Juha Vesanto and Esa Alhoniemi. Clustering of the Self-Organizing Map. *IEEE Transactions on Neural Networks*, 11(2):586–600, March 2000.
9. Xuegong Zhang and Yanda Li. Self-Organizing Map as a New Method for Clustering and Data Analysis. In *Proceedings of International Joint Conference on Neural Networks (IJCNN'93)*, pages 2448–2451, 1993.

Mapping the Growing Neural Gas to Situation Calculus

Dimitrios Vogiatzis and Andreas Stafylopatis

National Technical University of Athens, Dept. of Electrical and Computer
Engineering, 157 73 Zografou, Athens, Greece,
dimitrv@central.ntua.gr, andreas@cs.ntua.gr

Abstract. We propose a dynamic mapping of the operation of the
Growing Neural Gas model to Situation Calculus, with the purpose of
grounding the relatively higher level concepts of Situation Calculus to
lower level signals. Since both the Situation Calculus and the Growing
Neural Gas model were conceived with the express purpose of describing dynamic phenomena, this transformation is natural. We believe that
the transformation will also be useful in data mining tasks. Finally, we
present experimental results as an early evaluation of our method.

1 Introduction

Let us assume that there is a domain where a distribution varies with time
and there are no category labels. Because of the non-stationary nature of the
distribution we can deploy the *Growing Neural Gas* (GNG) network [4]. The
GNG model performs dimensionality reduction and vector quantisation, at any
time instance there is one or more connected components which represent the
data distribution. We aim to transform the nodes, weights and maps of the GNG
into the higher level structures of *Situation Calculus* (SC), which is a very good
way to formally describe change in a domain [10].

The motivation behind the proposed method is to overcome the knowledge
acquisition problem for systems of symbolic Artificial Intelligence (AI). It has
been observed that the most expensive and time consuming task is the construction and the debugging of the knowledge base [6]. We also aim to associate
higher level concepts to their referents (the signals), which in turn is a way of
grounding SC symbols [5].

In dynamic domains there have also been interesting methods to transform
the relatively lower lever signals to higher level concepts. For instance, *Clarion*
is a hybrid system which comprises a lower level reinforcement learning component and a higher level rule extraction component [11]. Another interesting
approach is the *Symbolic Dynamic Programming for First-Order MDPs* [3]. The
authors devised a method for symbolic learning and symbolic representation of
the value function and policies. The symbolic form that has been chosen is the
Situation Calculus, which has been enhanced with a special notation to handle
probabilities.

J.R. Dorronsoro (Ed.): ICANN 2002, LNCS 2415, pp. 957–962, 2002.

Our method differs from the aforementioned methods in the sense that it is based on unsupervised learning, whereas the other two methods employ reinforcement learning. Furthermore, there is a method to extract finite state automata from trained recurrent neural networks [8]. Our algorithm presents two differences; first it works on-line so the extracted knowledge is subject to constant revision, and second the expressive power of a calculus such as SC is more extended than a finite state automaton.

In Sect. 2 we present an introduction to the features of SC which are useful in method. In Sect. 3 we present the mapping algorithm from GNG to SC; the experiments are in Sect. 3.1 and Sect. 3.2; finally, the conclusions and future extensions are presented in Sect. 4.

2   Situation Calculus

Situation Calculus is a sorted second order language that can represent dynamic phenomena in domains such as planning, simulation, robotics and decision theory. Informally speaking, there is a transition from one situation to next because of actions. In every situation, there are a number of *fluents* that represent properties of *domain objects*. In the following we shall be using a version of SC which is loosely based on [10]. Formally stated, in SC there are the following sorts: $\mathcal{A}, \mathcal{S}, \mathcal{F}, \mathcal{D}$, for actions, situations, fluents, domain objects respectively. Next, we provide definitions of sorts with examples from the blocks' world.

Situations For the situation sort there are the following symbols:
- sit_1, which denotes the initial situation
- $\sqsubseteq$, which denotes an ordering of situations; e.g. $sit_1 \sqsubseteq sit_2$, denotes that situation sit_1 is an ancestor of situation sit_2

Domain objects Constants and variables in SC are referred to as domain objects; e.g. *robot, block1, block2, block3*.

Actions The functional symbol which denotes actions is defined as: $(\mathcal{A} \cup \mathcal{D}) \rightarrow \mathcal{A}$; e.g. *drop(block1)*

Fluents The functional symbol for relational fluents is defined as: $(\mathcal{A} \cup \mathcal{D}) \times \mathcal{S}$. The interpretation is that a property is true for a domain object in a specific situation; e.g. *is_carrying(block1, sit₁)*

As the world passes from one situation to the next one, most of the world remains unchanged apart from a small number of entities; the question is how to capture the dynamics of the world succinctly, this is an expression of the perennial *frame problem* as it was stated in [7]. In the domain of situation calculus the frame axioms are expressed by actions which have no effect on certain fluents.

3   Mapping GNG to SC

A GNG network is made of one or more connected components which have nodes. Each node has a neighbour and a weight vector w. It is necessary to introduce two concepts, the *component position* and the *component identity*. The component position is the centroid of the nodes' weight, i.e. $\sum_{i=1}^{n_{comp}} w_i/n_{comp}$. The component identity is defined as a two dimensional vector, where the first dimension is the average internode distance, i.e $\sum_{i,j} \frac{\|w_i-w_j\|_2}{n_{comp}}$; the second dimension is the average number of neighbours per node, i.e. $\frac{\sum_i num\_neigh(n_i)}{n_{comp}}$, where n_{comp} is the number of nodes in a component. Furthermore, we assume that a non stationary distribution nst, is made of a finite number of stationary distributions st_i (we also refer to st_i as time frame), i.e. $nst = st_1 \cup st_2 \ldots st_n$. We also assume that there is enough time in each st_i for the GNG to adapt. We define adaptation as the minimisation of the *quantisation error* [1], naturally we look for the case where the quantisation error is below a certain threshold. In such a case, the distribution is accurately represented by the GNG network and it is exactly at this phase that the mapping from GNG to SC takes place. An abrupt increase of the quantisation error marks the end of the current st_i and the transition to st_{i+1}.

The algorithm is presented in Table 1, where there are basically two notions: The first is that a domain object corresponds to a cluster of components' positions, the second is that a fluent corresponds to a cluster of components' identities (we have used the fuzzy c-means [2] clustering algorithm). In other words, fluents and domain objects at a lower level are actually spaciously extented entities. Another way to see a fluent is as a data distribution and objects are the 'positions' a fluent may assume, having said that what actually happens is the identification of fluents across the same or different situations. This has a consequence, that the knowledge base of SC is not monotonic. For instance, let us say that at time frame st_i we apply the mapping from GNG to SC, which involves situations st_1 to st_i. After that, at situation st_{i+1} the mapping will involve situations st_1 to st_{i+1}, which means that because of clustering the number and labels of fluents and domain objects might be different, the same holds for actions.

The transition from st_i to st_{i+1} is represented as an action. In SC, actions are deterministic, furthermore we do not need a boundless proliferation of actions. Thus, we always assign the same action labels to transitions, apart from the case where the same fluent(s) on the same object(s) leads to different results.

Finally, the components that remain identical as far as their position and identity is concerned, correspond to fluents that have been unaffected by actions, consequently they are frame axioms.

3.1   Artificial Data Experiment

We examine a data distribution, where the GNG maps are depicted for three different time frames st_1, st_2, st_3 (see Fig. 1). Based on the quantisation error (see Fig. 2) we obtain three situations, with $sit_1 \sqsubset sit_2 \sqsubset sit_3$. The domain objects correspond to the positions of the rectangle and the curved components, let us call them d_1, d_2, d_3 for the rectangle and d_4, d_5, d_6 for the three curves. We were able to identify the rectangle and the curves across time based on the identity criterion (the application was trivial in this case), the same holds for

Table 1. The GNG to SC algorithm

---

1. Apply the GNG algorithm on a set of data vectors
2. Wait till the current quantisation error is below a certain threshold.
3. Detect the connected components for all time frames.
   (a) Each time frame corresponds to a situation.
   (b) For every component, record average inter-node distance and the average number of neighbours per node, i.e. component identity. Cluster components, based on their identity and assign a fluent to each component.
   (c) Record positions of the components and cluster them. Each cluster corresponds to a domain object.
4. Assign the minimum number of actions so that each transition is deterministic.
5. Go to step 1

---

the positions of the curves. Furthermore, the curves have not been displaced and have remain unaffected, therefore they represent frame axioms. Thus there is the following result: $rectangle.d_i \xrightarrow{a} rectangle.d_{i+1}, \forall i \in \{1,2\}$, which means that fluent $rectangle$ is applicable on domain objects d_1, d_2, d_3 and action a permits the transition between different situations.

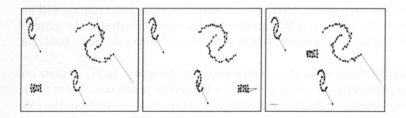

Fig. 1. The GNG maps for the artificial data distribution at 3 different time frames

3.2   Sleep Data Experiment

The data set concerns various measurements (heart rate, chest volume and blood oxygen concentration) of a sleeping subject. In total, there are 34000 samples (see http://reylab.bidmc.harvard.edu/download/competition/DataSetB.html).

There are 16 time frames, which correspond to 16 situations (with $sit_1 \sqsubset sit_2 \sqsubset \ldots sit_{16}$) as it can be detected from the quantisation error shown in Fig. 3. In each time frame there is one GNG component, apart from the sixteenth where

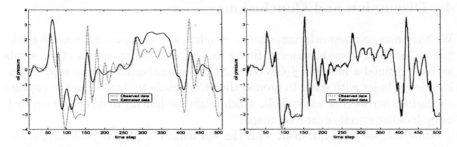

Fig. 2. The quantisation error for the arti-**Fig. 3.** The quantisation error for the sleep
ficial data (3 spikes)   data (16 spikes)

there are two. After clustering the components and the components positions we
obtain 4 fluents (a, b, c, d), 5 domain objects $(cl_1, cl_2, cl_3, cl_4, cl_5)$ and 3 actions
(act_1, act_2, act_3). The results are summarised in Table 2, where the table on the
left depicts the fluents that are applicable in each situation, as well as the objects
which they refer to. The table on the right denotes the actions that permit the
transition from a pair of (fluent, domain object), to a new pair. The conclusions
that can be drawn are that from sit_1 to sit_9, fluents concern only the domain
objects cl_1 and cl_2. Then fluents are displaced to cl_3 for the next 6 situations.
Finally, a novel fluent which corresponds to a novel data distribution appears in
sit_{16}.

Table 2. Fluents, objects and actions for the sleep data

situations	domain objects				
	cl_1	cl_2	cl_3	cl_4	cl_5
1	a				
2		b			
3		a			
4	b				
5		a			
6	a				
7	a				
8	c				
9		c			
10			b		
11			a		
12			b		
13			b		
14			a		
15			b		
16				b	d

$$a.cl_1 \overset{act_1}{\to} b.cl_2$$
$$a.cl_1 \overset{act_2}{\to} a.cl_1$$
$$a.cl_1 \overset{act_3}{\to} c.cl_1$$
$$a.cl_2 \overset{act_1}{\to} b.cl_1$$
$$a.cl_3 \overset{act_1}{\to} b.cl_3$$

$$b.cl_1 \overset{act_1}{\to} a.cl_1$$
$$b.cl_2 \overset{act_1}{\to} a.cl_2$$
$$b.cl_3 \overset{act_1}{\to} a.cl_3$$
$$b.cl_3 \overset{act_2}{\to} b.cl_3$$
$$b.cl_3 \overset{act_3}{\to} b.cl_2 \wedge d.cl_5$$
$$c.cl_1 \overset{act_1}{\to} c.cl_2$$
$$c.cl_2 \overset{act_2}{\to} b.cl_2$$

4   Discussion and Conclusions

We have mapped network components, weights, and the quantisation error into fluents, actions, domain objects, frame axioms and situations; in other words we have defined a function $f(GNG) \to SC$. The motivation was to supplement knowledge bases with data; to ground the SC symbols to signals; and to capture similarities across time as symbols, which might be difficult to do otherwise, such as by looking at self-organising maps.

However, we have not dealt with the full length of SC like reasoning about time, concurrent actions, action durations and a sophisticated solution of the frame problem. On the other hand, we have ignored certain aspects of the GNG model, like the time if takes for the GNG to reach a low quantisation error. Furthermore, the clustering of components and components' positions with the fuzzy c-means provides a degree of membership; this could be mapped as the degree or intensity of a fluent.

An essential part of our algorithm is the comparison of network components, i.e. graphs. That is the reason behind the introduction of the components' identity. An alternative idea that can be explored is based on the recognition of an unknown tree X based on the noisy subtree Y of the original tree. This has been made possible with the introduction of some tree-editing operations, like insertion deletion or substitution of one node by another [9].

References

1. H. U. Bauer, M. Herrmann, and T. Villmann. Neural Maps and Topographic Vector Quantization. *Neural Networks*, 12:659–767, 1999.
2. J.C. Bezdek. *Pattern Recognition with Fuzzy Objective Function algorithms*. Plenum, New York, 1981.
3. C. Boutilier, R. Reiter, and B. Price. Symbolic Dynamic Programming for First-Order MDPs. In *IJCAI, 690-697.*, 2001.
4. B. Fritzke. A self-organizing network that can follow non-stationary distributions. In *Proceedings of ICANN*, pages 613–618. Springer, 1997.
5. S. Harnad. The Symbol Grounding Problem. *Physica*, D(42):335–346, 1990.
6. J. Hendler and E. A. Feigenbaum. Knowledge Is Power: The Semantic Web Vision. In N. Zhong, Y. Yao, J. Liu, and S. Ohsuga, editors, *Web Intelligence Research and Development*, pages 18–29. Springer, 2001.
7. J. McCarthy and P. Hayes. Some philosophical problems from the standpoint of artificial intelligence. In B. Meltzer and D. Mitchie, editors, *Machine Intelligence*, number 4. Edinburgh University Press, 1969.
8. C. Omlin and C.L. Giles. Extraction of rules from discrete-time recurrent neural networks. *Neural Networks*, 9(1):41–52, 1996.
9. B. J. Oommen and R. K. S. Loke. On the pattern recognition of noisy subsequence trees. *Pattern analysis and machine Intelligence*, 23(9):929–946, 2002.
10. R. Reiter. *Knowledge in Action, Logical Foundations for Specifying and Implementing Dynamical Systems*. MIT press, 2001.
11. R. Sun, T. Peterson, and C. Sessions. Beyond simple rule extraction: Acquiring planning knowledge from neural networks. In *IEEE-INNS-ENNS IJCNN*, 2000.

Robust Unsupervised Competitive Neural Network by Local Competitive Signals

Ernesto Chiarantoni[1], Giuseppe Acciani[1], Girolamo Fornarelli[1], and
Silvano Vergura[1]

[1] Dipartimento di Elettrotecnica ed Elettronica
Politecnico di Bari, Via E. Orabona, 4
70125 Bari, Italy
{chiarantoni, acciani}@poliba.it
{fornarelli, vergura}@deemail.poliba.it

Abstract. Unsupervised competitive neural networks have been recognized as a powerful tool for pattern analysis, feature extraction and clustering analysis. The global competitive structures tend to critically depend on the number of elements in the networks and on the noise property of the space. In order to overcome these problems in this work is presented an unsupervised competitive neural network characterized by units with an adaptive threshold and local inhibitory interactions among its cells. Each neural unit is based on a modified competitive learning law in which the threshold changes in learning stage. It is shown that the proposed neuron is able, during the learning stage, to perform an automatic selection of patterns that belong to a cluster, moving towards its centroid. The properties of this network, are examined in a set of simulations adopting a data set composed of Gaussian mixtures.

1 Introduction

In recent years Unsupervised Competitive Neural Networks (UCNNs) have been recognized as a powerful tool for pattern analysis, feature extraction and clustering analysis [1–6]. These networks are mainly based on the winner-take-all (WTA) mechanism, where competitive signals acting among all the elements of the network are required [7–8]. Therefore, the learning space of each unit in a competitive neural network is marked by the position of the neighboring units and by the local properties of the input space. These limitations could be removed if global competition is replaced by a locally generated signal. The idea of the new neural network is to limit the influence of the neighboring units adopting a local measure of the density to condition the learning of each unit. The proposed learning algorithm is based on a modify version of the winner signals for each single cell. Utilizing local signals to condition the learning, it is possible to adopt simplified version of global competitive signals. This architecture is able to find center of cluster by exploiting only localized information. In [9] has been proposed an application of this model of unsupervised neural network to obtain features from environmental data.

J.R. Dorronsoro (Ed.): ICANN 2002, LNCS 2415, pp. 963–968, 2002.
© Springer-Verlag Berlin Heidelberg 2002

2 Learning Algorithm

Consider N processing elements, where each unit receives an input signal $x = \begin{pmatrix} x_1 & x_2 & \dots & x_N \end{pmatrix}$ from an external data-base and is characterized by the weight vector $w = \begin{pmatrix} w_1 & w_2 & \dots & w_N \end{pmatrix}$. When the input is received, the unit computes the Euclidean distance $d(x, w)$ between the input and the weight. Then, the following competitive learning algorithm is proposed:

Step 1

$$y(k) = f\big(\sigma(k) - d(x(k), w(k))\big) \tag{1}$$

Step 2

$$\Delta w(k) = \big(\alpha(k) + \beta(k) y(k)\big)\big(x - w(k)\big) \qquad w(k+1) = w(k) + \Delta w(k) \tag{2}$$

Step 3

$$\Delta \sigma(k) = \big(\alpha(k) + \beta(k) y(k)\big)\big(d(x, w(k)) - \sigma(k)\big) \tag{3a}$$

$$\sigma(k+1) = \sigma(k) + \Delta \sigma(k) \tag{3b}$$

Step 4:

$$\Delta \alpha(k) = -\alpha(k) y(k) \delta \qquad \alpha(k+1) = \alpha(k) + \Delta \alpha(k) \tag{4a}$$

$$\Delta \beta(k) = -\Delta \alpha(k) \qquad \beta(k+1) = \beta(k) + \Delta \beta(k) \tag{4b}$$

Step 5:   Iterate to Step 1 until

$$\Delta \sigma(k) < \varepsilon \tag{5}$$

where y is the neuron output, $f(\cdot)$ is the logistic function with $f(u)=0$ if $u \leq 0$, $0 < f(u) < 1$ if $u > 0$, $\alpha(k) \in [0,1[$ is the learning coefficient [1], $\beta(k) \in [0,1[$ is a coefficient that changes with the same rate of α, $0 \leq \delta < 1$ is a constant value that governs the dependence of α on y, whereas $\sigma(k)$ is an adaptive threshold which will be clarified later on. The behavior of the algorithm can be described as follows. From equation (1) it follows that the activation of the neuron ($y > 0$) occurs only if the extreme of the input vector x is within the hyper-sphere with radius given by σ and center given by the extreme of w. When $y = 0$, equation (2) is driven by the value of α. On the other hand, when α goes to zero, the value of y predominates and becomes a "winner" indicator. In this way, it is possible to implement a learning law that generate a transition from a phase of "weak" undifferentiated learning ($y = 0$), where each datum influences the learning, to a phase of "locally" selective learning, where the competitive signal βy enables the unit to learn only in the local "winner region". In order to adapt the 'winner region' to the feature of the input space, an adaptive threshold is introduced to assure that when the learning process starts, the neuron is unable to find a maximum of density without dependence on the local property of the input space. Namely, the idea is to increase σ when there is no activation, and vice-versa [10]. During a phase of undifferentiated learning (y =0), equation (3a) is driven by α. On the other hand, while w moves toward its center, the selective learning (y >0) takes place, α goes to zero, σ decreases, the neighborhood size of w decreases and, consequently, input vectors of a reduced neighbor of w contribute to the learning.

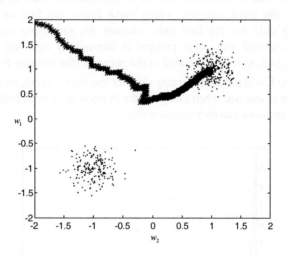

Fig. 1. Trajectory of the weight vector in bi-dimensional case obtained for $\sigma(0) = 10$

3 Network with Local Connections

Differently from WTA paradigms, the network proposed herein is an array of N units, which are characterized by *local* inhibitory connections. In particular, the proposed network architecture is described by the following equations:

$$\Delta w_1(k) = \big(\alpha(k) + \beta(k)y_1(k)\big)\big(x - w_1(k)\big)$$

$$\Delta w_2(k) = \big[\big(\alpha(k) + \beta(k)y_2(k)\big)\big(x - w_2(k)\big)\big]\,\theta\,(y_1)$$

$$\cdots \quad \cdots \quad \cdots$$

$$\Delta w_k(k) = \big[\big(\alpha(k) + \beta(k)y_k(k)\big)\big(x - w_k(k)\big)\big]\,\big[\theta\,(y_1)\theta\,(y_2)...\theta\,(y_{k-1})\big]$$

$$\cdots \quad \cdots \quad \cdots$$

$$\Delta w_N(k) = \big[\big(\alpha(k) + \beta(k)y_N(k)\big)\big(x - w_N(k)\big)\big]\,\big[\theta\,(y_1)\theta\,(y_2)...\theta\,(y_{N-1})\big]$$

(6)

where $\theta\,(\cdot)$ describes the action of the inhibitory links, with $\theta\,(y_k)=0$ if $y_k>0.5$ and $\theta\,(y_k)=1$ if $y_k<0.5$. The behavior of the network can be described as follows.

At the beginning each unit receives samples from the input space. However, only the first unit learns, since it has no input inhibitory links, whereas all the other units are inhibited, due to the presence of input inhibitory links. Mathematically, this is because $\theta\,(y_1)=0$ when the first unit has $y_1>0.5$. As a consequence, from (6) it follows that the units from 2 to N are inhibited. Successively, when the first unit has found a cluster center, its threshold value decreases, y_1 decreases and $\theta\,(y_1)=1$. Therefore, the first inhibitory link is turned off and the second unit can start its learning phase. This means that the input samples, which make active the first unit, continue to produce learning only for the first unit, whereas the remaining samples produce learning for the second unit. This process is iterated by applying the following strategy. If $\theta\,(y_k)=1$, unit $k+1$ is added to the network (the samples that make active the units from 1 to k will produce learning only for the first k units), on the other hand, the learning stage is stopped when either all the N units have been utilized or the last unit added to the network satisfies equation (5).

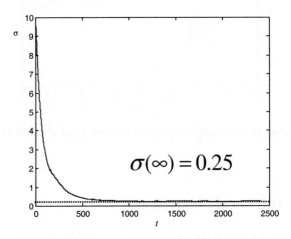

Fig. 2. Time-behavior of the thresholds for a network composed of 8 units $\sigma(0)=10$.

Notice that the proposed network architecture combines the advantages of both sequential cluster search and competitive networks. This feature leads to a major advantage over the performances of UCNNs. Namely, due to the presence of inhibitory links, lower priority units cannot affect higher priority units. As a consequence, if two networks are composed of N and $(N+M)$ units, the common N units will behave in the same way in both the networks and will found exactly the same centroids. This means that it is always possible to add some extra units (in order to check if some cluster in the data structure has been missed) without modifying the behavior of the previous units. Notice that UCNNs lack of this property, that is, they suffer of the problem of the dependence of the final partitions of the data on the number of the network elements.

In order to show the behavior of a single NUSD is been used an input space constituted by two groups of bi-dimensional Gaussian distributed pattern elements and parameter values: $\alpha(0) = 0,01$, $\beta(0) = 0,01$, $\delta = 0,001$, $\sigma(0) = 10$. In this simulation has been used a network composed of 8 elements. In figure 1 is reported the trajectory of weight vector obtained for first unit. The extreme of the weight vector (stars in the figure), starting from its initial position, tends initially to fluctuate towards the center of the whole input space, because all the input vectors participate in the learning in the same way, then converges towards the centroid of the data set located at (1, 1). Figure 2 reports the progress of the threshold $\sigma(t)$ as function of the number t of the presented samples.

In order to show the behavior of the proposed architecture and the ability to select the proper number of units, has been used the previous network with the same parameter values and an input space constituted by four groups of bi-dimensional Gaussian distributed pattern elements. In this case 3 elements have their thresholds unchanged to their initial values, 5 elements have been involved in the learning process and of these only 4 elements converge towards a dense region of the input space, that is, they have found the corresponding cluster center (the trajectories of weigth vector are shown in Figure 3).

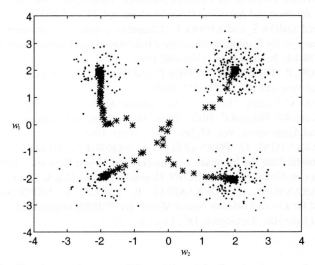

Fig. 3. Network of 8 units: convergence of the weights of the first 4 units towards the centroids.

4 Conclusions

In this paper a new model of unsupervised neural network as well as an innovative competitive learning algorithm have been described. The proposed network belongs to the class of unsupervised networks and is suitable for the applications as pattern recognition and feature extraction. The network is based on a model of neural unit able to self-tune to a single centroid of a cluster avoiding problems relevant to global search of the winner as in competitive "winner take-all" paradigms. Even though the proposed unit requires the computation of adaptive threshold for each learning step and the network structure imposes a penalty in the learning stage, because it becomes pseudo-sequential in the first stage of learning.

References

1.   KOHONEN, T.: "Self-organizing maps", *Springer-Verlag, Third Edition* 2001.
2.   MAO J., JAIN ANIL. K., "Artificial Neural Networks for Feature extraction and multivariate data Projection", *IEEE Trans. Neural Networks, vol. 6, pp. 296-317, Mach. 1995*
3.   McNEILL D., CARD C. H., MURRAY A. F., "An Investigation of Competitive Learning for Autonomous Cluster Identification in Embedded Systems", *International Journal of Neural Systems, vol. 11, n. 4, (2001), pp. 389-398*
4.   BISHOP C. M., "Neural Networks for Pattern Recognition", *Oxford University Press (1995)*
5.   XU L., "Best Harmony, Unified RPCL and Automated Model Selection for Unsupervised and Supervised Learning on Gaussian Mixtures, Three-Layer Nets and ME-RBF-SVM Models", *International Journal of Neural Systems, vol. 11, n. 1, (2001), pp. 43-69*
6.   ITO R. and SHIDA T. and KINDO T. "Competitive models for unsupervised clustering" Transactions of the Institute of Electronics Information and Communication Engineers D-II, Vol. J79D-II, Number 8, pp. 1390-400, 1996
7.   J MEADOR, P. HYLANDER, "A Pulse Coded Winner-Take-All Circuit*" Proc. Int. Conf. On Neural Networks Vol. II, 299-302, 1996
8.   PEDRONI A. Volnei, "Inhibitory Mechanism Analysis of Complexity O(N) MOS Winner-Take-All Networks" IEEE Trans. On Circuits and Systems – 1 Fundamental Theory and Applications, Vol. 42, No. 3, pp. , March 1995.
9.   E. CHIARANTONI, G. FORNARELLI, S. VERGURA, "Redundancy Reduction in Environmental Data Set by means of Unsupervised Neural Network" International Joint Conference on Neural Networks-IJCNN, Honolulu, Hawaii, U.S.A. 12-17 May 2002
10. E. CHIARANTONI, G. FORNARELLI, F. VACCA, S. VERGURA "Dynamical Threshold for a feature detector Neural Model" INNS-IEEE International Joint Conference on Neural Networks Washington, DC, 14-19 July 2001

Goal Sequencing for Construction Agents in a Simulated Environment

Anand Panangadan and Michael G. Dyer⋆

Computer Science Department, University of California at Los Angeles,
Los Angeles CA 90095, USA
{anand,dyer}@cs.ucla.edu

Abstract. A connectionist architecture enables a society of agents to efficiently construct 2D structures. The agents use internal spatial maps to compute a sequence of construction actions that reduces total distance traveled. All computations are done over grids of neurons interacting locally. Simulation results are presented.

1 Introduction

Here, a group of autonomous agents efficiently construct structures in a simulated 2-D continuous environment. Construction involves arranging same-sized discs to form a given 2-D pattern. [2] describes an architecture that performs this task; however, its greedy approach moves a disc to its closest drop-site and this can prove inefficient. For example in figure 1, the fastest way to build the three walls is to build the right-most wall first, then the middle wall and finally the wall closest to the source of discs - if the nearest wall is built first, this would block direct paths from the discs to the other two walls. The agent(s) should also use the upper source of discs to build the upper parts of the walls and the lower source for the lower parts of the walls. This minimizes the distance between the initial and final positions of the discs. Here, we extend the architecture in [2] to enable the agents to place discs in an order that reduces the total distance traveled. Interference between agents is also reduced without communication. The algorithms are computed over girds of neurons using only local interactions.

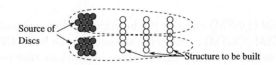

Fig. 1. Arrangement of discs and drop-sites where a greedy approach is inefficient.

⋆ This work supported in part by an Intel University Research Program grant to the second author.

J.R. Dorronsoro (Ed.): ICANN 2002, LNCS 2415, pp. 969–974, 2002.

2   Architecture

Agents can sense discs and other agents near them within a limited range. Disc locations are encoded via a grid of neurons (the *Sensor grid*) that divides the sensing area into 30×30 unit squares. Sensor neuron activation indicates a disc at the corresponding square. Construction involves moving toward a disc, grabbing it when nearby, moving toward a drop-site and then dropping it. An agent can move forward and turn smoothly in any direction.

The architecture is shown in figure 2. Each agent uses its own Egocentric Spatial Maps (ESMs) [2] to represent the spatial relationship between itself and the discs. An ESM contains neurons arranged in a uniform grid that maintains an egocentric view of the world (the center node always represents the current location of the agent). Each ESM neuron corresponds to a small square area of the world and its activation indicates a disc being present in the corresponding area. Thus each agent maintains a "birds-eye" view of the world around it. As the agent moves, activations are passed to neighboring neurons to maintain egocentricity. New sensor input is integrated into the ESM from the Sensor Grid.

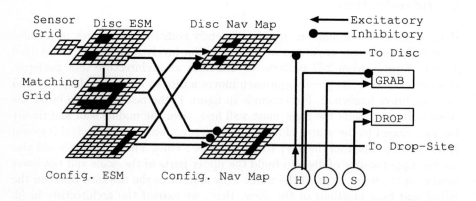

Fig. 2. The architecture: Activations (dark cells) on the Matching grid show which discs (marked on the Disc ESM) should be moved to drop-sites (on the CESM). H, D, and S represent the internal state nodes Have-Disc, At-Disc and At-Drop-Site respectively.

The *Disc ESM* (DESM) encodes the locations of discs around the agent. The *Configuration ESM* (CESM) encodes the structure to be built. CESM activations are also shifted as the agent moves, but those activations that encode the structure to be built are set a priori and are not updated by the sensors. Each DESM neuron is connected to a corresponding *Disc Navigation Map* (DNM) neuron. Active DESM neurons initiate spreading activation on corresponding DNM neurons. The CESM active neurons encode drop-off sites and thus discs, once placed at these sites, should not be picked up. Hence these neurons represent obstacles in the path and inhibit their corresponding neurons in the DNM.

Let n denote an arbitrary neuron in a Navigation map and $nb(n)$ the set of eight neighboring neurons of n. Let $a_n(t)$ be the activation of n at time t.

$$a_n(0) = \begin{cases} 1, & \text{if } n \text{ represents a drop-off site} \\ -1, & \text{if } n \text{ represents an obstacle} \\ 0, & \text{otherwise} \end{cases}$$

$$a_n(t+1) = \max_{m \in nb(n)} (a_m(t) - d(n,m)), a_n(t) \geq 0$$

where $d(m,n)$ is proportional to the distance between the locations represented by nodes m and n. This is a parallel implementation of Dijkstra's shortest path algorithm.The above equations are iterated until the activations stop changing: $a_n(t+1) = a_n(t) \forall n$. The gradient created by the spreading activation is the planned path to the nearest disc that is not already at a construction location. Similarly, the Configuration Navigation Map (CNM) is used to compute a path to locations where discs should be dropped. Every node is activated by the corresponding node from the CESM and inhibited by the nodes of the DESM.

There are three *Internal State Nodes* that indicate the current stage of the construction task: *Have-Disc* indicates if the agent is holding a disc; *At-Disc* and *At-Drop-Site* are active if the agent is near a disc or drop-site respectively. If Have-Disc is active, the agent moves toward a drop site else it moves toward a disc by following the gradient on the corresponding Navigation map. At-Disc and At-Drop-Site determine if a disc has to be picked up or dropped.

3   Sequence of Dropping Discs

With this architecture, the agents drop discs at target sites nearest to disc sources. This greedy approach is a distributed neural implementation of the "matrix scan" heuristic for the minimum weighted perfect matching (MWPM) or assignment problem for bipartite graphs [8]. The vertices of one partition are the initial locations of discs and the vertices of the other partition are the drop sites. The edges represent path distances between two vertices. The MWPM problem is to compute a one-to-one matching between disc locations and drop sites such that the sum of the paths is a minimum. Modeling construction as a MWPM problem has the property that in an optimal solution, paths between matched pairs do not cross, reducing interference between agents.

This approach assumes that path length between two locations will not change during construction. However, in the general case, the order in which discs are placed affects the distance traveled by agents to place the remaining discs as illustrated in figure 1. Thus, for efficient construction, the agents have to plan the temporal sequence of goal locations such that the path from every disc to its drop site is a straight line (not blocked by any previously dropped disc). The algorithm described below determines such a sequence in two steps. Each agent first computes a matching between discs and drop sites represented in its ESMs using a greedy heuristic. It then determines which of the drop sites are not on any of the paths (between a disc and its matching drop site) and proceeds to fill these first. The algorithm is implemented using only a grid of neurons (called the *Matching grid*) that can spread activations among neighboring cells. At the end of the algorithm, match$[n] := 1$ for all nodes n that are on a path between a matched disc and drop-site location. Let N be the set of all nodes on the Matching grid and S, T be the set of nodes representing disc locations and drop-sites respectively. If every node n has an activation a_n, let n^+ be the neighboring node of n along the strongest gradient of activation: $n^+ = \text{argmax}_{n' \in nb(n)}(a'_n)$.

```
calculate_matching (S, T)
    returns   match[n] ∈ {0,1}, ∀ n ∈ N
    1:  match[n] := 0 ∀ n ∈ N
    2:  spread activation a from all s ∈ S (aₛ = 1)
    3:  while there are unmatched nodes in T
    4:      for all unmatched nodes t ∈ T
    5:          {set bₙ := 1 for all nodes n on path from t to nearest disc}
    6:          bₜ₊ := 1; if (bₙ = 1) then bₙ₊ = 1
    7:          if (bₛ = 1) ∧ (aₛ = 1), s ∈ S
    8:              {match s to t, where t originated b activation that reached s}
    9:              bₛ := 2; if (bₙ = 2) then set bₙ' := 2, ∀ n' ∈ nb(n) ∧ bₙ' = 1
    10:             set match[n] := 1 for all nodes n on path from s to t
    11:             aₛ := 0
```

In lines 4-6, every drop-site location t spreads activation b *against* the gradient of the spreading activation initiated (from discs) in line 2. Thus, $b_n = 1$ identifies paths between discs and drop sites. When b activation reaches a disc location s (line 7), s and t are matched by setting match$[n] = 1$ for all nodes between them. The nodes on this path are identified by spreading $b_n = 2$ from the disc location (line 9). Activation from more than one drop site might reach s, but only the closest drop site (whose activation reaches s first) is matched since a_s is reset to 0 in line 10 after b activation reaches s.

If the path between a disc and a drop site crosses another drop-site t, then match$[t] = 1$. Thus, drop-sites whose match value is 0 are chosen first with inhibitory connections from the nodes in the Matching grid to the corresponding neurons on the CNM. To identify the discs that are to be moved to the selected drop-sites, activation is spread along those nodes n with match$[n] = 1$ from the selected drop-sites. If this activation reaches a node representing a disc location, then the corresponding neuron in the DESM is excited.

4   Results and Discussion

The performance of the algorithm is studied in both structured (discs and drop-sites grouped into piles/walls) and unstructured (discs and drop-sites randomly placed) environments. The number of agents is varied from 1 to 10 and they are initially distributed randomly within a square of 100×100 units. match$[n]$ values are shown for structured environments that consists of three parallel walls and one source of discs (figure 3a) and two sources of discs ((figure 3c). The positions of discs while construction is in progress are shown in figures 3b, d. With one source of discs, the sequencing algorithm fills the farthest wall first, then the

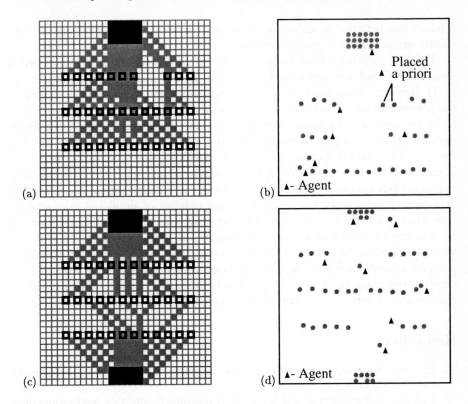

Fig. 3. Match values (gray squares) for 3 walls (dark unfilled squares) and (a) one source (dark filled sqaures) and (c) two sources of discs. (b, d) Corresponding environment with 7 agents at an intermediate stage of construction.

middle wall and finally the nearest wall (while the greedy algorithm fills them in the reverse order). Note that two discs were already in place and activation flowed around them. With two sources of discs, the agents assign the outer walls to the top and bottom piles. The middle wall is filled from both the piles and is the first to be built. The greedy approach builds the outer walls first and agents have to move around these to build the middle wall. The sequencing algorithm has no advantage in unstructured environments since the probability that three or more discs/sites are collinear is small.

5 Conclusions and Related Work

The simulation results show that planning the sequence of disc placement reduces the total distance traveled to complete the construction task. Implementing good heuristic solutions to the MWPM problem directly on the spatial representation is a convenient way of reducing interference between agents without the need for communication. The algorithm also takes into account obstacles on paths between discs and drop-sites. All computations are performed using only local interactions between neighboring nodes and hence this algorithm can be efficiently implemented in parallel systems.

Hopfield networks have been used for solving combinatorial optimization problems [5] and a formulation to solve the perfect matching problem is given in [7]. However, these methods do not always give good solutions [12] because of the presence of hard constraints [4]. Moreover, the weights in such a network have to be pre-computed by hand. Solving path computation problems using spreading activations is an idea borrowed from the working of ant colonies [3] and has been used for data dispersion in sensor networks [6]. In these works, activation is spread in the forward direction from a single source node and a reverse activation is then directed against the gradient of the forward activation. Spreading activation has been used to control a large number of simulated micro-robots that destroy a brain tumour [9] and to indicate paths that have been discovered by a group of robots to a human situated away from the robots [11]. In our work, the activation is spread on an internal spatial representation to plan a path for a single mobile agent. Goal sequencing strategies that do not explicitly consider the blocking of paths by previously dropped discs are studied in [10]. A comparison of average-case bounds for greedy MWPM heuristics is given in [1].

References

1. D. Avis. A survey of heuristics for the weighted matthing problem. *Networks*, 13:475–493, 1983.
2. G. Chao, A. Panangadan, and M. G. Dyer. Learning to integrate reactive and planning behaviors for construction. In *Proceedings of the 6th International Conference On the Simulation Of Adaptive Behavior*, 2000.
3. G. Dorigo, M. Di Caro. Antnet: Distributed stigmergetic control for communications networks. *Journal of Artificial Intelligence Research*, 9:317–265, 1998.
4. A. H. Gee, S. V. B. Aiyer, and R. Prager. An analytical framework for optimizing neural networks. *Neural Networks*, 6:79–97, 1993.
5. J. Hopfield and D. Tank. Neural computation of decisions in optimization problems. *Biological Cybernetics*, 52:141–152, 1985.
6. C. Intanagonwiwat, R. Govindan, and D. Estrin. Directed diffusion: A scalable and robust communication paradigm for sensor networks. In *Proc of the 6th Annual Intl. Conf. on Mobile Computing and Networks (MobiCOM 2000)*, 2000.
7. S. Y. Kung. *Digital Neural Network*, chapter 9. PTR Prentice Hall, 1993.
8. J. M. Kurtzberg. On approximation methods for the assignment problems. *J. Assoc. Comput. Mach.*, 9:419–439, 1962.
9. M. A. Lewis and G. A. Bekey. The behavioral self-organization of nanorobots using local rules. In *Proc. of the 1992 IEEE/RSJ Intl. Conf. on Intelligent Robots and Systems*, 1992.
10. A. Panangadan and M. G. Dyer. Construction by autonomous agents in a simulated environment. In *Proc. of the Intl. Conf. on Artificial Neural Networks*, 2001.
11. D. Payton, M. Daily, R. Estkowski, M. Howard, and C. Lee. Pheromone robotics. *Autonomous Robots*, 11(3), 2001.
12. G. V. Wilson and G. S. Pawley. On the stability of the travelling salesman problem algorithm of Hopfield and Tank. *Biological Cybernetics*, 58:63–70, 1988.

Nonlinear Modeling of Dynamic Systems with the Self-Organizing Map

Guilherme de A. Barreto and Aluizio F.R. Araújo

Department of Electrical Engineering, University of São Paulo (USP)
Av. Trabalhador Sancarlense, 400, CEP. 13560-590, São Carlos-SP, Brazil
{gbarreto, aluizioa}@sel.eesc.sc.usp.br
http://www.sel.eesc.sc.usp.br/lasi.htm

Abstract. In this paper we propose an unsupervised neural modeling technique, called *Vector-Quantized Temporal Associative Memory* (VQ-TAM). Using VQTAM, the Kohonen's self-organizing map (SOM) becomes capable of approximating dynamical nonlinear mappings from time series of measured input-output data. The SOM produces modeling results as accurate as those produced by multilayer perceptron (MLP) networks, and better than those produced by radial basis functions (RBF) networks, both the MLP and the RBF based on supervised training. In addition, the SOM is less sensitive to weight initialization than MLP networks. The three networks are evaluated through simulations and compared with the linear ARX model in the forward modeling of a hydraulic actuator.

1 Introduction

Training a neural network using input-output data from a nonlinear plant can be considered as a nonlinear function approximation problem. In this paper, it is proposed a new neural modeling technique which uses unsupervised neural networks for function approximation, instead of the usual supervised ones (MLP and RBF) [1]. By means of this technique, called *Vector-Quantized Temporal Associative Memory* (VQTAM), it is shown that the Self-Organizing Map (SOM) [2] can be successfully used to approximate nonlinear input-output mappings. Computer simulations illustrate this approximation ability of the SOM using the VQTAM approach and compare the obtained results with those produced by MLP and RBF networks and linear models.

The remainder of the paper is organized as follows. Section 2 introduces the VQTAM technique and its main properties. It is also shown how the VQTAM can be used together with SOM in the identification of forward dynamics of a hydraulic actuator. In Section 3, the approximation results of the SOM algorithm are compared with those produced by other linear and nonlinear methods. The paper is concluded in Section 4.

J.R. Dorronsoro (Ed.): ICANN 2002, LNCS 2415, pp. 975–980, 2002.

2   Temporal Associative Memory Using the SOM

The Self-Organizing Map (SOM) is an unsupervised neural algorithm designed to represent neighborhood (spatial) relationships among vectors of an unlabelled data set [2]. The neurons in the SOM are put together in an output layer, $\mathcal{A}$, in one-, two- or even three-dimensional arrays. Each neuron $i \in \mathcal{A}$ has an weight vector $\mathbf{w}_i \in \Re^n$ with the same dimension as the input vector $\mathbf{x} \in \Re^n$. An important characteristic of the SOM algorithm is that it can learn only *static* input-output mappings [3], which are described mathematically by $\mathbf{y}(t) = \mathbf{f}(\mathbf{u}(t))$, where $\mathbf{u}(t) \in \Re^n$ and $\mathbf{y}(t) \in \Re^m$ denote the input and output vectors, respectively. This paper however is interested in showing that, with slight modifications, the SOM can also be used to approximate nonlinear dynamic mappings, such as those described as follows [4]:

$$\mathbf{y}(t+1) = \mathbf{f}[\mathbf{y}(t), \ldots, \mathbf{y}(t - n_y + 1); \mathbf{u}(t), \ldots, \mathbf{u}(t - n_u + 1)] \qquad (1)$$

where n_u and n_y are the maximum delays (orders) of the input and output memories, respectively. For the SOM to be able to learn dynamic mappings, it must have some type of *short-term memory* (STM) mechanism [5]. That is, the SOM should be capable of temporarily storing past information about the system's input and output vectors.

By using STM mechanisms, such as delay lines, the SOM algorithm is used to approximate the nonlinear function $\mathbf{f}(\cdot)$. To achieve this, the input vector to the SOM, $\mathbf{x}(t)$, is slightly modified so that it has from now on two parts. The first part, denoted as $\mathbf{x}^{in}(t)$, carries data about the input of the dynamic mapping being learned. The second part, denoted $\mathbf{x}^{out}(t)$, contains data related to the output of this mapping. The weight vector $\mathbf{x}(t)$ of neuron i has its dimension increased accordingly. These changes in $\mathbf{x}(t)$ and $\mathbf{w}_i(t)$ are written as:

$$\mathbf{x}(t) = \begin{pmatrix} \mathbf{x}^{in}(t) \\ \mathbf{x}^{out}(t) \end{pmatrix} \qquad \text{and} \qquad \mathbf{w}_i(t) = \begin{pmatrix} \mathbf{w}_i^{in}(t) \\ \mathbf{w}_i^{out}(t) \end{pmatrix} \qquad (2)$$

It is worth noting the difference between this unsupervised strategy and that used in training supervised networks. In MLP and RBF networks, the vector $\mathbf{x}^{in}(t)$ is presented to the network input, while the $\mathbf{x}^{out}(t)$ is used at the network output to compute the error signal that guides network training. When using the SOM with the definitions in (2), the vector $\mathbf{x}^{out}(t)$ is presented to the network input together with the vector $\mathbf{x}^{in}(t)$ and no error signal is explicitly computed.

Depending on the variables chosen to build the vectors $\mathbf{x}^{in}(t)$ and $\mathbf{x}^{out}(t)$ one can use the SOM algorithm to learn the forward or the inverse dynamics of a nonlinear plant. For example, if one wishes to approximate the forward dynamics as defined in (1) the following definitions apply:

$$\mathbf{x}^{in}(t) = [\mathbf{y}(t), \ldots, \mathbf{y}(t - n_y + 1); \mathbf{u}(t), \ldots, \mathbf{u}(t - n_u + 1)] \qquad (3)$$

$$\mathbf{x}^{out}(t) = \mathbf{y}(t+1) \qquad (4)$$

During training, the winning neurons are found using only the portion corresponding to $\mathbf{x}^{in}(t)$:

$$i^*(t) = \arg \min_{i \in \mathcal{A}} \{ \| \mathbf{x}^{in}(t) - \mathbf{w}_i^{in} \| \} \qquad (5)$$

In updating the weights both, $\mathbf{x}^{in}(t)$ and $\mathbf{x}^{out}(t)$, are used:

$$\Delta\mathbf{w}_i^{in}(t) = \eta(t)h(i^*,i;t)[\mathbf{x}^{in}(t) - \mathbf{w}_i^{in}(t)] \tag{6}$$

$$\Delta\mathbf{w}_i^{out}(t) = \eta(t)h(i^*,i;t)[\mathbf{x}^{out}(t) - \mathbf{w}_i^{out}(t)] \tag{7}$$

where $\eta(t)$ is the learning rate and $h(i^*,i;t)$ is a Gaussian-type neighborhood function given by $h(i^*,i;t) = \exp\left(-\|\mathbf{r}_i(t) - \mathbf{r}_{i^*}(t)\|^2/\sigma^2(t)\right)$, where $\mathbf{r}_i(t)$ and $\mathbf{r}_{i^*}(t)$ are, respectively, the locations of neurons i and i^* in the output array. The variables $\eta(t)$ and $\sigma(t)$ decay exponentially with time according to $\eta(t) = \eta_0\,(\eta_T/\eta_0)^{(t/T)}$ and $\sigma(t) = \sigma_0\,(\sigma_T/\sigma_0)^{(t/T)}$, where η_0 and σ_0 denote the initial values of $\eta(t)$ and $\sigma(t)$, while η_T and σ_T are the final ones, after T training iterations. As training proceeds, the SOM algorithm learns to associate the outputs, $\mathbf{x}^{out}(t)$, of the dynamic mapping being learned with the corresponding inputs, $\mathbf{x}^{in}(t)$. At the same time, the SOM performs vector quantization on both the input and output spaces of the mapping. Thus, this technique will be referred to as *Vector-Quantized Temporal Associative Memory* (VQTAM).

For validation, the vector $\mathbf{y}(t+1)$ is unknown, and hence, the vector $\mathbf{x}^{out}(t)$ is not built. A trained SOM is then used to obtain estimates for this output vector from the output portion of the weight vector, $\mathbf{w}_i^{out}(t)$. For the case of estimating the forward dynamics, one gets:

$$\hat{\mathbf{y}}(t+1) \equiv \mathbf{w}_{i^*}^{out}(t) \tag{8}$$

where the winning neuron, $i^*(t)$, is found as defined in (5). The estimation process continues for M steps until an entirely new time series is built from the estimated values of $\mathbf{y}(t+1)$.

3   Computer Simulations

Figure 1 shows measured values of the valve size (input variable), $u \in \Re$, and the oil pressure (output variable), $y \in \Re$, of a hydraulic actuator. As can be seen in the oil pressure time series, there are very oscillative behaviors caused by mechanical resonances. These data have been used in benchmarking studies on nonlinear system identification [1]. The SOM, MLP and RBF networks are used as nonlinear identification models to approximate the forward nonlinear dynamics of the hydraulic actuator. These three neural networks are also compared with the usual linear model, known as the *Autoregressive model with Exogenous Inputs* (ARX):

$$\hat{y}(t+1) = \sum_{i=0}^{n_y-1} a_i y(t-i) + \sum_{j=0}^{n_u-1} b_j u(t-j) \tag{9}$$

where a_i and b_j are the coefficients of the model and $\hat{y}(t+1)$ is the estimated value for the plant output at time step $t+1$. The coefficients are computed by the Least-Squares Method [6]. The approximation accuracy is evaluated through the *Root Mean Square Error* (RMSE):

$$RMSE = \sqrt{\frac{1}{M} \sum_{t=0}^{M-1} (y(t+1) - \hat{y}(t+1))^2} \qquad (10)$$

where M is the length of the estimated series. The data are presented to the four models without any preprocessing stage. A total number of $N = 1024$ samples are available for both the input and output variables. The first 512 samples are used to train the three networks and to compute the coefficients of the linear ARX model, while the remaining 512 samples are used to validate the four models. A training epoch is defined as one presentation of the training samples. For all the simulations, it is assumed $n_y = 3$ and $n_u = 2$, as suggested in [1].

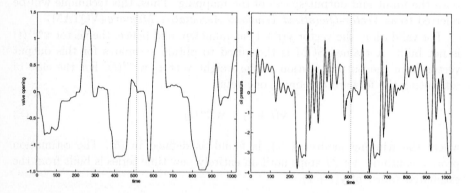

Fig. 1. Measured values of valve position (left) and oil pressure (right).

The SOM network has six input units, since $\dim(\mathbf{x}^{in}) + \dim(\mathbf{x}^{out}) = 5+1 = 6$, and an output layer with 500 neurons. The weights are randomly initialized between 0 and 1, and adjusted for 600 epochs. The training parameters are the following: $\eta_0 = 1.0$, $\eta_T = 10^{-5}$, $\sigma_i = 250$, $\sigma_T = 10^{-3}$ and $T = 600 \times 512 = 3 \times 10^5$. The time series generated by the ARX and the SOM model during validation are shown in Figure 2. In these Figures one can note that the estimated values provided by the linear ARX model were not very accurate ($RMSE = 1.0133$), while the accuracy of the SOM was much better ($RMSE = 0.2051$).

The MLP network has five input units, since $\dim(\mathbf{x}^{in}) = 5$, one hidden layer with ten neurons and one output neuron. The neurons in the hidden layer have hyperbolic tangent transfer functions, while the output neuron has a linear transfer function. The MLP network is trained with backpropagation algorithm with momentum. The values for the learning rate and the momentum factor are set to 0.2 and 0.9, respectively. The training is stopped if RMSE ≤ 0.001 or a maximum number of 600 training epochs is reached. The RBF network also has five input units, an intermediate layer with neurons with Gaussian basis

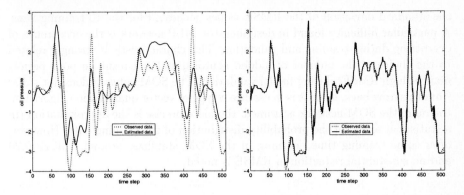

Fig. 2. Simulation of the ARX (left) and the SOM (right) models on validation data. Solid line: estimated signal. Dotted line: observed (true) data.

function, and one output neuron. Following the RBF design in [7], the number of neurons in the intermediate layer is the same as the number of training samples, and hence there is a Gaussian kernel centered at every training vector. The intermediate-to-output weights are just the target values, so the output is simply a weighted average of the target values of training cases close to the given input case. The only weights that need to be learned are the radii of the Gaussian kernels. However, in this paper, this parameter was deterministically varied to evaluate its effect in the accuracy of the approximation. The statistics describing the results produced by the three networks are shown in Table 1. For the SOM and the MLP were performed ten training runs, and for the RBF network the radius of the Gaussian kernels was varied from 0.1 (minimum RMSE) to 1.0 (maximum RMSE) in increments of 0.1.

Table 1. RMSE values for the SOM, MLP and RBF networks in the identification of the forward dynamics of the hydraulic actuator.

Forward Modeling			
RMSE	Min	Max	Mean	$\sqrt{var}$
MLP	0.1162	0.2493	0.1554	0.0457
SOM	0.2051	0.2665	0.2259	0.0215
RBF	0.2067	0.4103	0.2994	0.0774

In this table, one can note that the MLP network provides the best results in general. The SOM algorithm, in its turn, produces better results than a RBF network with approximately the same number of neurons. An interesting result is that the SOM is less sensitive to weight initialization than the MLP network, as can be seen in the fifth column of Table 1. This sensitivity is measured through the standard deviation of the RMSE values generated for the 10 training runs. A particular difficulty found in designing the MLP network is the occurrence of overfitting during training and validation. This characteristic is strongly related to the choice of the number of hidden neurons, and can result in poor performance (high RMSE) during model validation. The SOM network does not suffer from this drawback, since it is in essence a type of vector quantization algorithm: the more the SOM network is trained, the more precise is the approximation (in a statistical sense) of the probability distribution of the training data. However, after some training time, learning in the SOM stabilizes around RMSE=0.20 and no substantial reduction in RMSE is noted.

4   Conclusion

The simulations shown in this paper illustrate the potential of the VQTAM technique. Additional tests should be performed, such as residual analysis of the estimation error, noise and fault tolerance, etc., to demonstrate effectively the viability of using the SOM algorithm in identification and control of nonlinear dynamic systems. Currently, research is being conducted with the aim of designing a predictive nonlinear controller using the SOM and the VQTAM approach. Also, a theoretical analysis intends to show that the SOM algorithm, using the VQTAM scheme, can be used for universal function approximation.

Acknowledgements. The authors thank FAPESP for the financial support (grant Nr. 98/12699-7).

References

1. Sjöberg, J., Zhang, Q., Ljung, L., Benveniste, A., Deylon, B., Glorennec, P.-Y., Hjalmarsson, H., Juditsky, A.: Nonlinear Black-Box Modeling in System Identification: A Unified Overview. Automatica **31:12** (1995) 1691–1724
2. Kohonen, T.: Self-Organizing Maps. 2nd. edn. Springer-Verlag, Berlin Heidelberg New York (1997)
3. Walter, J., Ritter, H.: Rapid Learning with Parametrized Self-Organizing Maps. Neurocomputing **12** (1996) 131–153
4. Norgaard, M., Ravn, O., Poulsen, N. K., Hansen L. K.: Neural Networks for Modelling and Control of Dynamic Systems. Springer-Verlag, Berlin Heidelberg New York (2000)
5. Araújo, A. F. R., Barreto, G. A.: Context in Temporal Sequence Processing: A Self-Organizing Approach and Its Application to Robotics. IEEE Trans. Neural Nets. **13:1** (2002) 45–57
6. Ljung, L., Glad, T.: Modeling of Dynamic Systems. Prentice-Hall, Englewood Cliffs NJ (1994)
7. Specht, D. F.: A Generalized Regression Neural Network. IEEE Trans. Neural Nets. **2:5** (1991) 568–576

Implementing Relevance Feedback as Convolutions of Local Neighborhoods on Self-Organizing Maps

Markus Koskela, Jorma Laaksonen, and Erkki Oja

Laboratory of Computer and Information Science, Helsinki University of Technology
P.O.BOX 5400, 02015 HUT, Finland
{markus.koskela,jorma.laaksonen,erkki.oja}@hut.fi

Abstract. The Self-Organizing Map (SOM) can be used in implementing relevance feedback in an information retrieval system. In our approach, the map surface is convolved with a window function in order to spread the responses given by a human user for the seen data items. In this paper, a number of window functions with different sizes are compared in spreading positive and negative relevance information on the SOM surfaces in an image retrieval application. In addition, a novel method for incorporating location-dependent information on the relative distances of the map units in the window function is presented.

1 Introduction

The data organization provided by the Self-Organizing Map (SOM) [1] can be utilized in searching for interesting data items. Due to the topology-preservation property of the SOM, neighboring map units contain similar feature vectors. If we already know that certain map units contain data items which are in some manner similar to the item we are interested in, a natural strategy is to focus the further search in the neighborhoods of these map units. This kind of setting arises, e.g. in iterative multi-round information retrieval where, on each query round, the user marks the retrieved items as relevant or nonrelevant to the query. The system then uses this information in estimating what the user is looking for. This kind of iterative refinement of a query is known as *relevance feedback* in information retrieval literature [2]. Content-based image retrieval (CBIR) has been a subject of recent intensive research effort. It differs considerably from textual information retrieval as, unlike text that consists of words, images do not consist of such basic building blocks which could directly be utilized in retrieval applications. Instead, the retrieval is based on visual features extracted from the images and alternative retrieval paradigms must be used. One common approach is *query by example*, where the user specifies her object of interest by giving or pointing out examples of interesting or relevant images. Relevance feedback is essential here, as the systems are normally not capable of returning the desired image on the first query round [3]. A CBIR system implementing

J.R. Dorronsoro (Ed.): ICANN 2002, LNCS 2415, pp. 981–986, 2002.

relevance feedback tries to learn the optimal correspondence between the high-level concepts people use and the low-level features obtained from the images. The user thus does not need to explicitly specify weights for different features as the weights are formed implicitly by the system. This is desirable, as it is generally a difficult task to give low-level features such weights which would coincide with human perception of images.

2   PicSOM

The PicSOM [4,5] image retrieval system is a framework for research on methods for content-based image retrieval. The methodological novelty of PicSOM is to use several parallel Self-Organizing Maps trained with separate data sets. After training the SOMs, their map units are connected with the images of the database. This is done by locating the best-matching map unit (BMU) for each image. Also, among the images which have a common BMU, the best-matching one is used as a visual label for that unit. As a result, the different SOMs impose different similarity relations on the images and the system is able to adapt to different kinds of retrieval tasks. The spreading of the positive and negative responses on the SOMs has been an integral part of the system from the beginning, but it has not been thoroughly examined until now. Instead of the standard SOM version, PicSOM uses a special form of the algorithm, the Tree Structured Self-Organizing Map (TS-SOM) [6]. The hierarchical TS-SOM structure is useful for two purposes. First, it reduces the complexity of training large SOMs by exploiting the hierarchy in finding the BMU for an input vector. Second, the hierarchical representation of the image database produced by a TS-SOM can be utilized in browsing the images in the database. The PicSOM home page including a working demonstration of the system for public access is located at *http://www.cis.hut.fi/picsom.*

3   Relevance Feedback with Self-Organizing Maps

The basic assumption in the PicSOM method is that images similar according to specific visual features are located near each other on the SOM surfaces. Therefore, we are motivated to spread the relevance information given by the user to the shown images also to the neighboring units. This is done as follows. All relevant images are first given equal positive weight inversely proportional to the number of relevant images. Likewise, nonrelevant images receive negative weights that are inversely proportional to their total number. The overall sum of these relevance values is thus zero. For each SOM layer, the values are then mapped from the images to their BMUs where they are summed. Finally, the resulting sparse value fields on the SOM surfaces are low-pass filtered to produce qualification values for each SOM unit and its associated images. This process is illustrated in Figure 1. Content descriptors that fail to coincide with the user's conceptions mix positive and negative values in nearby map units. Therefore, they produce lower qualification values than those descriptors that match the

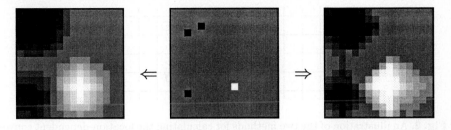

Fig. 1. An example of how positive and negative map units, shown with white and black marks on the middle figure, are convolved with shift-invariant (leftmost figure) and location-dependent (rightmost figure) window functions.

user's expectations and impression of image similarity. As a consequence, the different content descriptors and the TS-SOMs formed from them do not need to be explicitly weighted as the system automatically takes care of weighting their opinions. On each SOM, we first search for a fixed number, say 100, of unseen images with the highest qualification values. After removing duplicates, the second stage of processing is carried out. Now, the qualification values of all images in this combined set are summed up on all SOMs. 20 images with the highest total qualification values have then been used in the experiments as the result of the query round.

3.1   Shift-Invariant Window Functions

Spreading of the response values can be performed by convolving the sparse value fields with a tapered (or rectangular) window or kernel function. The one-dimensional convolution of signal $x[n]$ and window $w[n]$ is a basic signal processing operation defined as

$$y[n] = x[n] * w[n] = \sum_{k=-M}^{M} x[n-k]w[k] \; . \tag{1}$$

On SOM surfaces the convolutions have to be two-dimensional. Due to computational reasons this has been implemented as one-dimensional horizontal convolution followed by one-dimensional vertical convolution. This can be done because the convolution kernels we have used have been separable and shift-invariant. The following window functions have been used in the experiments:

$$w_r[n] = 1 \qquad\qquad \text{(rectangular)} \tag{2}$$

$$w_t[n] = \frac{M - |n|}{M} \qquad\qquad \text{(triangular)} \tag{3}$$

$$w_g[n] = e^{-(\frac{n}{\alpha})^2} \qquad\qquad \text{(truncated Gaussian)} \tag{4}$$

$$w_x[n] = e^{-\frac{|n|}{\beta}} \qquad\qquad \text{(truncated exponential)} \tag{5}$$

The truncated Gaussian and exponential windows require a parameter controlling the decay of the window. Here, α and β have been selected so that

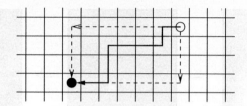

Fig. 2. An illustration of the two methods for calculating the location-dependent convolutions on the SOM grid. In the "path" method, the minimum path ○ → ● is solved with dynamic programming. In the "sum" method, horizontal and vertical one-dimensional location-dependent convolutions are calculated in both orders and then averaged.

$w_g[\pm\frac{M}{2}] = w_x[\pm\frac{M}{4}] = \frac{1}{2}$. The length of the window, $N = 2M + 1$, is the predominant parameter of any window function. With small N, the search expands only to the immediate neighbors of the relevant items. As N grows the search area widens. As the computational complexity of the convolution is linearly dependent on the window length, it is beneficial to be able to use as small windows as possible.

3.2   Location-Dependent Window Functions

Information on the distances between neighboring SOM codebook vectors in the feature space has earlier been used mainly in visualization [7,1]. If the relative distance of two SOM units is small, they can be regarded as belonging to the same cluster and, therefore, the relevance response should easily spread between the neighboring map units. Cluster borders, on the other hand, are characterized by large distances and the spreading of responses should be less intensive. For each neighboring pair of map units according to 4-neighborhood, say i and j, the distance in the original feature space is calculated and then scaled so that the average neighbor distance is equal to one. The normalized distances d_{ij} are then used for calculating location-dependent convolutions with two alternative methods, illustrated in Figure 2. The "path" method uses dynamic programming to solve the minimum path length along the 4-neighborhood grid between two arbitrary map units i and j. Given a maximum allowed distance M, we can calculate and tabularize the between-node distances d_{ij} for non-neighboring map units. Then the two-dimensional convolution functions were formed from equations (2–5) by setting $n = d_{ij}$. In the alternative "sum" method, a computationally faster solution is obtained by performing one-dimensional location-dependent convolution first horizontally with kernel values obtained again from equations (2–5) with $n = d_{ij}$. The result of the horizontal convolution was then similarly convolved with vertical one-dimensional location-dependent kernels. As the order of the successive one-dimensional convolutions now matters, the original impulse-valued SOM surface was convolved again, now first vertically and then horizontally, and the two slightly-different convolution results were averaged. In preliminary experiments it was observed that the difference of the two methods is not substantial in this setting and that the "path" method is computationally

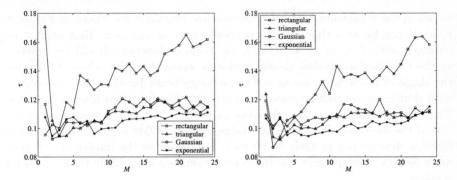

Fig. 3. Average τ values with different shift-invariant (left figure) and location-dependent (right figure) window functions of varying width M.

not feasible with large window sizes. Therefore, only the "sum" method was used in the experiments.

4   Experiments

We used an image database containing 59 995 images from the Corel Gallery 1 000 000 product. From the database images, we have created manually five ground truth image classes: **faces** (1115 images, *a priori* probability 1.85%), **cars** (864 images, 1.44%), **planes** (292 images, 0.49%), **sunsets**, (663 images, 1.11%), and **horses**, (486 images, 0.81%). As image features, we used a subset of MPEG-7 [8] content descriptors for still images, viz. *Scalable Color, Dominant Color, Color Structure, Color Layout, Edge Histogram*, and *Region Shape*. The image queries are always started with one reference image that belongs to the image class in question. Therefore, initial browsing is not needed and we can limit the search exclusively to the 256×256-sized bottommost TS-SOM levels. From each of the above classes, 20 random images were selected to the set of reference images and an image query was then run for each of these images. Then we get the final results by averaging the results of the 100 individual runs. For performance evaluation, we used the τ measure [5] which coincides with the question "how large portion of the whole database needs to be browsed through until, on the average, the searched image will be found". The τ measure can be obtained for a relevance feedback system by simulating the responses of a human user. This can be done by examining each output of the system and marking the returned images either as relevant or non-relevant according to whether they belong to a ground truth image class $\mathcal{C}$. From this data, we calculate the average number of shown images needed before a hit occurs. The τ measure for $\mathcal{C}$ is then obtained by dividing the average number of shown images by the size of the database. The measure yields a value in the range $\tau \in [\frac{\rho_C}{2}, 1 - \frac{\rho_C}{2}]$ where ρ_C is the *a priori* probability of the class $\mathcal{C}$. For values $\tau < 0.5$, the performance of the system is thus better than random picking of images and, in general, the smaller the τ value the better the performance. The resulting values of the τ measure with different window functions are shown in Figure 3. First, it can be seen that a small window length is sufficient. Best results are obtained with $2 \leq M \leq 4$. Second, the window function should be tapered, as the rectangular window clearly performs worse than the others. Otherwise the shape of the window seems not to be a significant factor. Third, the results with shift-invariant and location-dependent window functions are quite similar. This is probably due to the relatively large size of the maps compared to the size of the database. In many cases using smaller SOMs is preferable and then location dependency is likely to be more important as the images are mapped more densely to map units and the convolutions are more likely to cross cluster borders.

5   Conclusions

In this paper, experiments on implementing relevance feedback on multiple SOMs were presented. The SOM surfaces are convolved with a window function in order to spread the relevance feedback responses provided by a human user. Different shapes and sizes of the convolution kernel have been studied. In addition, a method for combining location-dependent distance information of the map units in the window function was presented. Here, using the location-dependent window functions did not improve the results. Still, they may prove out to be useful with smaller SOMs. Also, it was seen that using small window sizes suffices, which is computationally advantageous in actual implementations.

References

1. Kohonen, T.: Self-Organizing Maps. Third edn. Volume 30 of Springer Series in Information Sciences. Springer-Verlag (2001)
2. Salton, G., McGill, M.J.: Introduction to Modern Information Retrieval. Computer Science Series. McGraw-Hill (1983)
3. Lew, M.S., ed.: Principles of Visual Information Retrieval. Springer-Verlag (2000)
4. Laaksonen, J.T., Koskela, J.M., Laakso, S.P., Oje, E.: PicSOM – Content-based image retrieval with self-organizing maps. Pattern Recognition Letters **21** (2000) 1199-1207
5. Laaksonen, J., Koskela, M., Laakso, S., Oje, E.: Self-organizing maps as a relevance feedback technique in content-based image retrieval. Pattern Analysis & Applications **4** (2001) 140–152
6. Koikkalainen, P., Oje, E.: Self-organizing hierarchical feature maps. In: Proc. IJCNN-90, International Joint Conference on Neural Networks, Washington, DC. Volume II., Piscataway, NJ, IEEE Service Center (1990) 279–285
7. Ultsch, A., Siemon, H.P.: Kohonen's self organizing feature maps for exploratory data analysis. In: Proc. INNC'90, Int. Neural Network Conf., Dordrecht, The Netherlands, Kluwer (1990) 305–308
8. MPEG: Overview of the MPEG-7 Standard (version 5.0) (2001) ISO/IEC JTC1/SC29/WG11 N4031.

A Pareto Self-Organizing Map

Andrew Hunter[1] and Richard Lee Kennedy[2]

[1] Department of Computer Science,
University of Durham
Science Labs, Durham, UK
andrew1.hunter@durham.ac.uk
http://www.durham.ac.uk/andrew1.hunter/index.html
[2] Department of Science,
University of Sunderland,
Sunderland, Tyne and Wear, UK
Lee.Kennedy1@sunderland.ac.uk

Abstract. Self Organizing Features Maps are used for a variety of tasks in visualization and clustering, acting to transform data from a high-dimensional original feature space to a (usually) two-dimensional grid. SOFMs use a similarity metric in the input space, and this composes individual feature differences in a way that is not always desirable. This paper introduces the concept of a Pareto SOFM, which partitions features into groups, defines separate metrics in each partition, and retrieves a set of prototypes that trade off matches in different partitions. It is suitable for a wide range of exploratory tasks, including visualization and clustering. . . .

1 Introduction

The Self-Organizing Feature Map [6] is a ubiquitous form of neural network. It has two major features: neurons that act as prototype vectors for input vectors, and that reflect clustering in the training data; and a topological ordering property, which ensures that vectors closely situated in input space are likely to be assigned to nearby units on the topological map. Uses include cluster analysis, visualisation and novelty detection.

All forms of SOFM use a metric in the input space to determine the similarity of the input vector, $\mathbf{x}$, to the prototypes, $\mathbf{p}^{(i)}$, both during training (when the prototypes are adjusted to reflect clustering in the training cases) and during execution (when the closest prototype to the input vector is selected, or the metric is used to rate the "closeness" to each prototype). The most common metric is the Euclidean distance, $||\mathbf{x} - \mathbf{p}^{(i)}||$.

It is questionable whether it always makes sense to combine features in this way. Features may have very different semantics, be differently scaled, and be of different types (e.g. intermixed binary, nominal and real valued features). It is then far from clear how to compose them sensibly into a single metric, and an active research area considers how similarity should be measured [2] [9].

J.R. Dorronsoro (Ed.): ICANN 2002, LNCS 2415, pp. 987–992, 2002.

In this paper, we show how to modify the SOFM architecture so that the requirement for *a priori* construction of a similarity metric is reduced. The variables are partitioned, and a metric applied in each partition. A *Pareto set* of solutions *mutually non-dominating* in distance across partitions is generated. Each such solution represents a possible trade-off between matching in the different partitioned feature spaces. The user may then judge the value of the matched set *a posteriori*. The modified architecture is particularly useful in exploratory analysis, as in exemplar case-based retrieval systems, and clustering.

2   Multi-objective Optimization

Many neural processes, including SOFM training and execution, involve minimization of some objective measure (error function value, distance to prototype). In there are multiple objectives we look for a solution that simultaneously optimises them all. The simplest approach is to combine them into a single function using weighting parameters that reflect the relative importance of the objectives; for example, in Weigend weight elimination [8], the error function contains two terms, $E = E_T + \lambda E_W$, where E_T is the error on the training data, E_W is a regularization term based on the weight magnitudes, and λ is the weighting parameter. In SOFM prototype matching, the features are routinely scaled, both to correct implicit weighting imposed by difference of scale between variables, and to explicitly weight the contributions to reflect differences in semantic relevance.

In general, the weighting approach to multiobjective optimisation is unsatisfactory as the weighting coefficient must be arrived at *a priori*, whereas we may actually lack a clear understanding of the extent to which deterioration in one objective can be tolerated in exchange for improvements in others.

Multiobjective algorithms avoid *a priori* weighting by identifying a set of solutions which represent different trade-offs between objectives. The key concept is non-domination [1]. Given two models, A and B, we say that $A \succ B$ (A dominates B) if A has at least one objective value strictly better than B, and all objective values at least equal to B:

$$A \succ B \iff \forall i, \, m_i(A) >= m_i(B); \, \exists i, \, m_i(A) > m_i(B) \tag{1}$$

The *Pareto-optimal set* consists of all non-dominated solutions (i.e. those that are not dominated by any other solution). Multiobjective algorithms search along a *Pareto front* — the non-dominated subset of the models found during the search. Multiobjective algorithms are currently generating a great deal of interest in the Genetic Algorithms community (e.g. [3] [4]).

3   Similarity Metrics and Non-domination

Feature Maps are based on the application of a similarity metric. During both learning and execution, input vectors are compared against the prototype vectors held in the feature map, and the activations of the prototypes are the similarity.

A standard implementation of the SOFM uses the Euclidean metric, so that the distance between input vector $\mathbf{x}$ and the ith prototype, $\mathbf{p}^{(i)}$, is given by:

$$\mathcal{D}(\mathbf{x}, \mathbf{p}^{(i)}) = \sqrt{\sum_j (x_j - p_j^{(i)})^2} \qquad (2)$$

It may be reasonable to combine some subsets of features into single metrics, while measuring similarity differently for others. We introduce a partition of the feature set $\mathcal{F}$ into disjoint subsets, $\mathcal{F}_i$, such that $\mathcal{F} = \bigcup_i \mathcal{F}_i, \forall i, j : \mathcal{F}_i \cap \mathcal{F}_j = \varnothing$; define separate metrics for each partition; and define the activation of topological neurons as a *vector*, $\mathcal{D}(\mathbf{x}, \mathbf{p}^{(i)}))$, of similarity metrics in each feature partition:

$$\mathcal{D}(\mathbf{x}, \mathbf{p}^{(i)}) = [\mathcal{D}_{\mathcal{F}_1}(\mathbf{x}, \mathbf{p}^{(i)}), \mathcal{D}_{\mathcal{F}_2}(\mathbf{x}, \mathbf{p}^{(i)}), ..., \mathcal{D}_{\mathcal{F}_N}(\mathbf{x}, \mathbf{p}^{(i)})] \qquad (3)$$

We then define the *proximity set*, $\mathcal{P}$, of an input vector as the set of prototype vectors that are non-dominated in separation from the input vector:

$$\mathcal{P}(\mathbf{x}) = \{\mathbf{p}^{(i)} : \nexists j, \mathcal{D}(\mathbf{x}, \mathbf{p}^{(j)}) \succ \mathcal{D}(\mathbf{x}, \mathbf{p}^{(i)})\}, \qquad (4)$$

The proximity set always contains at least one prototype, and in principle may contain any number up to the size of the topological layer, even if this exceeds the number of partitions. An extreme version of this algorithm would be to treat each feature as a partition. Unfortunately, our analysis shows that this is likely to lead to many spurious matches, and unacceptable bloating of the proximity set – the number of partitions does have to be kept quite small. Approaches such as "lazy learning" [9] may help to reduce composition dependencies in highly multivariate domains.

4   Training the Network

The standard approach to training a feature map locates the single winning prototype, $\mathbf{p}^{(w)}$, which maximizes similarity to the input vector, then updates this prototype and the members of its topological neighbourhood, $\mathcal{N}_\tau(\mathbf{p}^w)$, using:

$$\mathbf{p}_{\tau+1}^{(i)} = \mathbf{p}_\tau^{(i)} + \eta_\tau(\mathbf{x} - \mathbf{p}_\tau^{(i)}), \ \forall i \in \mathcal{N}_\tau(\mathbf{p}^{(w)}) \qquad (5)$$

To modify the algorithm for multiobjective optimisation, we locate and update members of the proximity set and their neighbours. Once prototypes have been selected for update, the algorithm proceeds exactly as the standard Kohonen training algorithm. Similarly, one can easily modify Learned Vector Quantization [7] algorithms so that the members of the proximity set, rather than a single prototype, are updated.

$$\mathbf{p}_{\tau+1}^{(i)} = \mathbf{p}_\tau^{(i)} + \eta_\tau(\mathbf{x} - \mathbf{p}_\tau^{(i)}), \ \forall i \in \bigcup_j \{\mathcal{N}_\tau(\mathbf{p}^{(j)}) : \mathbf{p}^{(j)} \in \mathcal{P}(\mathbf{x})\} \qquad (6)$$

If observed output classes, $C(\mathbf{x}_i)$, are available in the data set, we apply a class label to a prototype, $\mathcal{C}(\mathbf{p}^{(i)})$, by finding the cases, $\mathbf{x}_i$, whose proximity sets

contain the prototype, and choosing the best represented class, provided that
the number of such cases exceeds a threshold proportion, ϑ.

$$m_{i,j} = |\{\mathbf{x}_k : \mathbf{p}^{(i)} \in \mathcal{P}(\mathbf{x}_k) \wedge C(\mathbf{x_k}) = j\}| \tag{7}$$

$$C(\mathbf{p}^{(i)}) = \begin{cases} \arg\max_i(m_{i,j}) & \max_j(m_{i,j}) > \vartheta \\ d & otherwise \end{cases} \tag{8}$$

where d is the "doubt" (unlabelled prototype) indicator.

5   Novelty Detection and Classification

In novelty detection we identify inputs that are "unlike" the prototype vectors.
An input vector is considered "unlike" a prototype if the similarity metrics exceed
threshold values in any partitions. If all members of the proximity set are unlike
the input, it is considered novel. Formally, we define a threshold vector, θ, and
check for domination of the input / prototype similarity vector by the threshold
vector. If the similarity vector is non-dominated, the vector is novel:

$$\text{novel}(\mathbf{x}) \iff \forall \mathbf{p}^{(i)} \in \mathcal{P}(\mathbf{x}) : \theta \nprec \mathcal{D}(\mathbf{x}, \mathbf{p}^{(i)}) \tag{9}$$

If the prototype vectors have class labels, we can perform classification by
voting among the members of the proximity set. As the size of the set is not
known *a priori*, we may assign a threshold proportion, Θ, which must be exceeded by the most populous class label in the proximity set. If the proportion is
not exceeded, we issue a "doubt" classification. We may also elect to exclude from
the proximity set any prototypes that exceed the novelty detection threshold.

Let $\mathcal{Q}_i(\mathcal{P}(\mathbf{x}))$ be the number of prototypes with class label i in the proximity
set, $|\mathcal{P}(\mathbf{x})|$ the cardinality of the proximity set. Then the classification assigned
by the network to input $\mathbf{x}$ is given by:

$$\mathcal{Q}_i(\mathcal{P}(\mathbf{x})) = |\{\mathbf{p}^{(j)} : \mathbf{p}^{(j)} \in \mathcal{P}(\mathbf{x}) \wedge C(\mathbf{p}^{(j)}) = i\}| \tag{10}$$

$$C(\mathbf{x}) = \begin{cases} d, & \max_i(\mathcal{Q}_i(\mathcal{P}(\mathbf{x}))) < \Theta|\mathcal{P}(\mathbf{x})| \\ \arg\max_i(\mathcal{Q}_i(\mathcal{P}(\mathbf{x}))) & otherwise \end{cases} \tag{11}$$

6   Example of Usage, Trauma Data

In this section, we illustrate the algorithm's performance on a specific problem.
The STAG (Scottish Trauma Audit Group) data set illustrates the influence of
patient condition on survival chance on admission to Accident and Emergency
Departments [5]. There are two major semantically distinct feature categories:
injury assessment and physiological indicators. Injury assessment features include: Type of injury (Blunt or Penetrating), and scores A1-A6 reflecting a
visual assessment of physical trauma in: Head, Face, Chest, Abdomen, Extremi-
ties, Overall. Physiological indicators include systolic blood pressure, respiratory

rate, and response to tests of voluntary motor control, verbal response and eye movement. In addition, age has a significant bearing on survival prognosis. It is appropriate to partition the variables so that similarity in physiological indicators is not unduly confounded with injury assessment. Age presents a problem in being related to both groups – we therefore include a copy in both partitions.

Table 1. Algorithm Settings

Phase	One	Two
Epochs	10	90
Learning Rate	0.1–0.01	0.01–0.001
Neighbourhood	1–0	0

We trained a Pareto SOFM, dimensions 6x8 on 1500 cases, of whom 750 died and 750 survived. Two stages of iterated unsupervised training were followed by class labelling. The algorithm parameters are given in table 1. A standard SOFM with the same training parameters correctly classified 1300 cases, with 172 wrong and 28 classified as "doubt" (three nodes were left unlabelled); the Pareto SOFM correctly classified 1274, with 226 wrong and no "doubt" (no nodes were left unlabelled). Adding a threshold proportion of 0.7 in classification yields 1231 correct, 186 wrong, 83 doubt. Classification performance therefore appears close to, but perhaps a little inferior to, that of the standard SOFM.

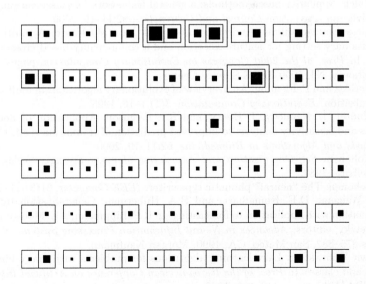

Fig. 1. Pareto Topological Map

The proximity set averaged 3.17 neurons (maximum, 11). Figure 1 illustrates the topological map for one particular case. Each neuron is represented by two squares, with the filled area representing the activation level of the neuron in each metric (an entirely filled area indicates zero separation). The three members of the proximity set are outlined; two represent maximal matches in one metric, and the other a compromise between the two. This conveys richer information about similarity to prototypes than a standard SOFM, showing quite clearly the alternative matches, and possible trade-off in similarities between the two partitions. The approach does therefore appear to be useful for visualisation and exploratory analysis applications, despite the lower classification performance.

7   Conclusion

Topological maps rely on similarity metrics which compose differences in multiple features, which may lead us to confound semantically different features. This paper introduces a novel method to partition features, and to detect similarity in a non-dominated fashion across a small number of partitions, locating a range of prototypes that are "similar in different ways" to the input. The technique is useful in performing cluster analysis on data with a natural feature partition structure. It is suitable for any application of similarity metrics to semantically decomposable feature spaces, including case based reasoning and clustering.

References

1. A. Ben-Tal. Characterization of pareto and lexicographic optimal solutions. In Fandel and Gal, editors, *Multiple Criteria Decision Making Theory and Application*, pages 1–11. Springer-Verlag, 1979.
2. W. Duch. Similarity based methods: a general framework for classification, approximation and association.*Control and Cybernetics*, 62:11–19, 2000.
3. Christos Emmanouilidis, Andrew Hunter, and John MacIntyre. A multiobjective evolutionary setting for feature selection and a commonality-based crossover operator. In *Proc. of the 2000 Congress on Evolutionary Computation*, pages 309–316, Piscataway, NJ, 2000. IEEE Service Center.
4. C. Fonseca and P. Fleming. An overview of evolutionary algorithms in multiobjective optimization. *Evolutionary Computation*, 3(1):1–16, 1995.
5. A. Hunter, L. Kennedy, J. Henry, and R.I. Ferguson. Application of neural networks and sensitivity analysis to improved prediction of trauma survival. *Computer Methods and Algorithms in Biomedicine*, 62:11–19, 2000.
6. T. Kohonen. *Self-Organisation and Associative Memory*. Springer-Verlag, Berlin-Heidelberg, 1987.
7. T. Kohonen. The "neural" phonetic typewriter. *IEEE Computer*, 21(3):11–22, 1988.
8. A.S. Weigend, D.E. Rumelhart, and B.A. Huberman. Generalization by weight-elimination with application to forecasting. In R.P. Lippmann, J.E. Moody, and D.S. Touretzky, editors, *Advances in Neural Information Processing Systems*, volume 3, pages 875–882, San Mateo, CA, 1990. Morgan Kaufmann.
9. Werner Winiwarter. Lazy learning algorithms for problems with many binary features and classes. In *Proc. of the Iberoamerican Conference on Artificial Intelligence (IBERAMIA)*, pages 112–123, 1998.

A SOM Variant Based on the Wilcoxon Test for Document Organization and Retrieval

Apostolos Georgakis, Costas Kotropoulos, and Ioannis Pitas

Department of Informatics,
Aristotle University of Thessaloniki,
Thessaloniki 54006, Greece.
{apostolos, costas}@zeus.csd.auth.gr

Abstract. A variant of the self-organizing maps algorithm is proposed in this paper for document organization and retrieval. Bigrams are used to encode the available documents and signed ranks are assigned to these bigrams according to their frequencies. A novel metric which is based on the Wilcoxon signed-rank test exploits these ranks in assessing the contextual similarity between documents. This metric replaces the Euclidean distance employed by the self-organizing maps algorithm in identifying the winner neuron. Experiments performed using both algorithms demonstrates a superior performance of the proposed variant against the self-organizing map algorithm regarding the average recall-precision curves.

1 Introduction

Document organization and retrieval has been a vivid research and development area for the past 30 years with goals spanning from: *indexing* and *retrieval* to *representation* and *categorization* [1]. A fundamental problem in the area is the evaluation of the contextual similarity between documents. This paper describes a method for evaluating the contextual similarity between documents by addressing the issue of a proper distance between texts. In doing so we assume that the contextual similarity between documents exists also in their vectorial representation. Subsequently, the above mentioned similarity can be assessed through the use of a vector norm. For this purpose, the available textual data are represented by vectors using the *vector space* model [1,2].

A plethora of document organization and retrieval systems are based on the vector space model. One system capable of clustering documents according to their contextual similarity is the well-known *Self-Organizing Map* (SOM) or *Kohonen* algorithm [3,4]. In this paper, a variant of the SOM algorithm will be presented which is based on a novel vector norm.

A modeling technique based on the vector space model is used in order to effectively encode the documents. Subsequently, a new metric based on the aforementioned modeling technique and the Wilcoxon *signed-rank* test is introduced in order to assess the above mentioned similarity. Finally, the modeling method and the norm proposed are used in constructing a document organization system.

J.R. Dorronsoro (Ed.): ICANN 2002, LNCS 2415, pp. 993–998, 2002.

In what follows, Section 2 provides a brief description of the language modeling method. Section 3 provides a detailed description of the proposed norm, whereas, the next Section contains a brief description of the SOM algorithm. Section 5 describes the variant under consideration. Finally, in Section 6 we assess the effectiveness of both algorithms by using document-based queries.

2   Vector Construction

Let us suppose that we have a training corpus. The documents of the corpus are encoded into numerical vectors using the well-know *bigram* model [2]. For this purpose the maximum likelihood estimates of the conditional probabilities for the bigrams are computed as follows: $x_{lm} = n_{lm}/N_l, \forall\, l, m \in \{1, 2, \ldots, N\}$, where n_{lm} is the number of times the bigram (lth word stem, mth word stem) occurred in the corpus, N_l is the number of times the lth word stem occurred in the corpus and N is the number of word stems in the corpus [2]. The *feature vector* corresponding to the ith document is given by:

$$\tilde{\mathbf{x}}_i = \sum_{l=1}^{N} \sum_{m=1}^{N} x_{lm} \mathbf{e}_{lm}, \tag{1}$$

where $\mathbf{e}_{lm}$ denotes the $(N^2 \times 1)$ unit norm vector having one in the $(l \times N + m)$th position and zero elsewhere. Let $\mathbf{b}_i$ denote the *indicator vector* that contains the bigrams of the ith document.

To reduce the dimensionality in both vectorial types, the elements of the feature vectors are sorted into descending order and the same permutations are performed on the elements of the indicator vectors. Afterwards, a threshold, which is denoted by n, is used to divide both vectors into two parts. The first part in both vectors contains the most significant elements and is preserved while the second part contains the non-significant elements of the vectors and is rejected.

3   Document Distance

The proposed modeling method is based on the following assumption: two documents, that are contextually similar, with high probability, contain the same set of bigrams. To assess the degree of similarity between documents a new metric is introduced which is based on the distance between the entities of the same bigrams inside the indicator vectors corresponding to the documents.

Let us denote by $\mathbf{b}_i$ and $\mathbf{b}_k$ the indicator vectors corresponding to the ith and kth documents, respectively. The distance between two entities of the same bigram in the indicator vectors is given by:

$$d_{ik}(j) = \begin{cases} j - l, & \text{if } \mathbf{b}_{i(j)} = \mathbf{b}_{k(l)} \\ n^2, & \text{else}, \end{cases} \tag{2}$$

where $\mathbf{b}_{i(j)}$ corresponds to the jth bigram (in the order of their frequencies) in the ith document.

Subsequently, the distances obtained for all the bigram pairs are transformed into their absolute values, that is, $d_{ik}^*(j) = |d_{ik}(j)|, \forall j \in \{1, 2, \ldots, n\}$. The absolute values are ranked into descending order, with tied ranks included where appropriate. Let $r_{ik}^*(j)$ denote the rank corresponding to the absolute value $d_{ik}^*(j)$ and $n_z(ik)$ denote the number of zeros encountered when evaluating Eq. (2) for the ith and jth documents respectively. No ranks are assigned to the zeros resulted from Eq. (2). Let also $N_z(ik)$ denote the total number of non-zero unsigned ranks ($i.e.$, $N_z(ik) = n - n_z(ik)$).

The last step is the computation of the signed ranks, defined by:

$$r_{ik}(j) = \begin{cases} r_{ik}^*(j), & \text{if } d_{ik}^*(j) = d_{ik}(j) \\ -r_{ik}^*(j), & \text{if } d_{ik}^*(j) = -d_{ik}(j) \\ 0, & \text{if } d_{ik}(j) = 0. \end{cases} \tag{3}$$

Let us denote by W_{ik}^+ and W_{ik}^- the sum of the positive and negative signed ranks, respectively. The distance between two documents is defined as $W_{ik} = \min\{W_{ik}^-, W_{ik}^+\}$ and will be called the *Wilcoxon* distance, henceforth.

The Wilcoxon distance is the proposed metric and is employed in a hypothesis testing to assess whether the ith and the jth documents are contextually similar or not.

The null hypothesis is true (the documents are similar) when both indicator vectors consist of the same set of bigrams. In that case, due to the association between the bigrams and their frequencies, the identical bigrams are expected to be located at the same positions in the indicator vectors and the value of W_{ik} is expected to be near zero. Subsequently, the null hypothesis is accepted if the absolute value of W_{ik} is bounded by the so-called *critical values*. In that case, the documents are contextually *relevant*, otherwise, they are *irrelevant*.

For $N_z(ik) < 25$ the critical values of the Wilcoxon test can be found in any statistical book, whereas, when the number of non-zero unsigned ranks exceeds 25 the Wilcoxon test is approximated by the normal distribution. The parameters of the distribution are: $\mu(ik) = (N_z(ik) \times (N_z(ik)+1))/4$ and $\sigma^2(ik) = (N_z(ik) \times (N_z(ik)+1) \times (2N_z(ik)+1))/24$. In that case the critical values are derived from the table of the cumulative distribution function of the normal distribution.

4   Self-Organizing Maps

The SOMs are feedforward, artificial neural networks (ANN). Each neuron is equipped with a *reference* vector which is updated every time a new *feature* vector is assigned to that particular neuron. Let W denote the set of reference vectors $\{\mathbf{w}_l(p) \in \mathbb{R}^n, l = 1, 2, \ldots, Q\}$, where the parameter p denotes discrete time and the notion Q corresponds to the total number of neurons. During the training phase, the algorithm tries to identify the *winning* reference vector $\mathbf{w}_s(p)$ to a specific feature vector $\tilde{\mathbf{x}}_h$. The index of the winning reference vector is given by: $s = \min \|\tilde{\mathbf{x}}_h - \mathbf{w}_l(p)\|$, $l = 1, 2, \ldots, Q$, where $\|\cdot\|$ denotes the Euclidean distance.

The reference vector of the winner as well as the reference vectors of the neurons in its neighborhood are modified towards $\tilde{\mathbf{x}}_h$ using the following equation:

$\mathbf{w}_i(p+1) = \mathbf{w}_i(p) + a(p) \times [\tilde{\mathbf{x}}_h - \mathbf{w}_i(p)]$, where $a(p)$ corresponds to the *learning rate* which is a monotonically decreasing parameter.

5 Wilcoxon Variant

In the proposed variant we replace the Euclidean norm with the Wilcoxon distance in identifying the winner neuron. Let $S_q(p)$ denote the set of indicator vectors that have been assigned to the qth neuron until the pth iteration. Let also $\mathbf{b}_{q_{vmed}} \in S_q(p)$ denote the so-called *vector median* corresponding to the set $S_q(p)$ [5]. The vector median corresponds to the indicator vector that minimizes the L_1 norm over the set $S_q(p)$:

$$\sum_{i=1}^{|S_q(p)|} W_{iq_{vmed}} \leq \sum_{i=1}^{|S_q(p)|} W_{ij} \ \forall \ j = 1, 2, \ldots, |S_q(p)|, \tag{4}$$

where $|S_q(p)|$ denotes the *cardinality* of the set $S_q(p)$. The vector median corresponding to the set $S_q(p)$ stands for the reference vector of the qth neuron $(\mathbf{w}_q(p) \equiv \mathbf{b}_{q_{vmed}})$.

In identifying the index of the winner neuron with respect to a specific, randomly selected, indicator vector $\mathbf{b}_h$, the Wilcoxon distances between all the reference vectors of the ANN and the indicator vector under consideration are assessed. If the null hypothesis related to the Wilcoxon distance is validated to be "true", then the reference vector corresponds to the winner neuron. It must be noted that it is possible multiple winners to stem out from the above procedure.

Let s denote the index of a winner neuron that stemmed from the above procedure. The set $S_s(p)$ is updated with the vector $\mathbf{b}_h$, that is, $S_s(p) = S_s(p) \bigcup \mathbf{b}_h$ and the corresponding vector median, $\mathbf{b}_{s_{vmed}}$, is also updated. At the completion of each iteration, that is, when all the indicator vectors have been presented to the network, the sets of reference vectors are updated as follows: $S_q(p+1) = \{\mathbf{w}_s(p)\}, \forall q = 1, 2, \ldots, Q$. Finally, prior to the completion of the training phase and in order to fine tune the map, the identification of the winning neuron is achieved by: $s = \min\{W_{hq}\}, \forall q = 1, 2, \ldots, Q$. The above equation results in assigning the bigram vector $\mathbf{b}_h$ to only one neuron during each iteration.

6 Document Organization and Retrieval

To test the proposed Wilcoxon variant against the SOM algorithm the Reuters-21578 corpus was used, which is an annotated corpus [6]. The SGML tags, the URLs, the email addresses, and the punctuation marks were removed. Subsequently, some common words and frequent terms were removed also and *stemming* was performed. Finally, the documents were encoded into numerical vectors.

These vectors are presented iteratively an adequate number of times to each one of the NNs in an effort to construct clusters containing semantically related documents. This process yields the so-called *document map* (DM) [3]. The DM corresponding to the Reuters-21578 corpus using the Wilcoxon variant can be

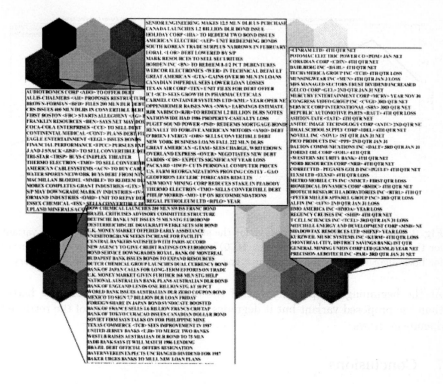

Fig. 1. The document map constructed for the Reuters-21578 for a 9×9 neural network using the Wilcoxon variant. The highlighted neurons correspond to document clusters related to "financial debts" (top middle and left), "bonds" (bottom left) and "corporate economic results" (bottom right).

seen in Fig. 1. Each hexagon on the DM corresponds to one document category and the levels of grey correspond to different document densities. Hexagons with grey levels near 255 imply that fewer documents have been assigned to these neurons, whereas, grey levels near 0 imply higher document densities.

To evaluate the performance of the algorithms, with respect to their document organization capabilities, document-queries are used. For each query, the algorithms identify the winning neurons on the computed DMs and retrieve the documents of the training corpus associated with the winners. Subsequently, the retrieved documents are ranked according to their distance from the queries using either the Euclidean or the Wilcoxon distance. Finally, the retrieved documents are labeled as either relevant or not to the document-query, with respect to the annotation category they bear.

The relevance between the retrieved documents and the queries leads to the partitioning of the training corpus into two sets, one containing the relevant documents and another with the non-relevant documents. The effectiveness of the algorithms is assessed using the *average recall-precision* curve [7]. Figure 2a and Fig. 2b depict the *eleven-point* average recall-precision curves for the standard SOM and the Wilcoxon variant for the topics with the highest frequencies.

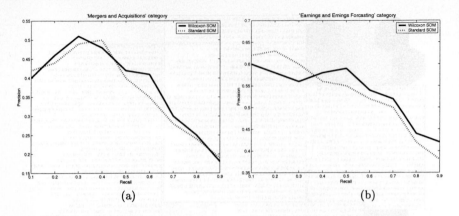

(a)   (b)

Fig. 2. (a) The recall-precision curve for the standard SOM and the Wilcoxon variant for the "Mergers and Acquisitions (ACQ)" category. (b) The recall-precision curves for both algorithms for the "Earnings and Earnings Forecasts (EARN)" category.

At the beginning the performance of the SOM algorithm was slightly better than the proposed variant but it degrades rapidly as the volume of the retrieved document grows.

7   Conclusions

A variant of the SOM algorithm for document organization and retrieval has been presented in this paper. The Euclidean distance used by the SOM algorithm in identifying the winning neuron is replaced by a novel metric which exploits the correlation between the words formulating the documents. The performance of the proposed variant with respect to the average recall-precision curves have been demonstrated to be superior than the SOM algorithm. Further investigations will be made towards the enhancement of the suggest algorithm in exploiting the latent textual information.

References

1. R. B. Yates and B. R. Neto, *Modern Information Retrieval*, ACM Press, 1999.
2. D. Manning and H. Schütze, *Foundations of Statistical Natural Language Processing*, Cambridge, MA: MIT Press, 1999.
3. T. Kohonen, *Self Organizing Maps*, Germany: Springer-Verlag, 1997.
4. T. Kohonen, S. Kaski, K. Lagus, J. Salojärvi, V. Paatero, and A. Saarela, "Organization of a massive document collection," *IEEE Trans. on Neural Networks*, vol. 11, no. 3, pp. 574–585, May 2000.
5. J. Astola, P. Haavisto, and Y. Neuro, "Vector median filters," *Proceedings of the IEEE*, vol. 78, no. 4, pp. 678–689, April 1990.
6. D. D. Lewis, "Reuters-21578 text categorization test collection, distribution 1.0," Sep. 1997, http://kdd.ics.uci.edu/databases/reuters21578/reuters21578.html.
7. R. R. Korfhage, *Information Storage and Retrieval*, New York: J. Wiley, 1997.

Learning More Accurate Metrics for Self-Organizing Maps

Jaakko Peltonen, Arto Klami, and Samuel Kaski

Neural Networks Research Centre, Helsinki University of Technology,
P.O. Box 9800, FIN-02015 HUT, Finland
{Jaakko.Peltonen, Arto.Klami, Samuel.Kaski}@hut.fi
http://www.cis.hut.fi

Abstract. Improved methods are presented for learning metrics that measure only important distances. It is assumed that changes in primary data are relevant only to the extent that they cause changes in auxiliary data, available paired with the primary data. The metrics are here derived from estimators of the conditional density of the auxiliary data. More accurate estimators are compared, and a more accurate approximation to the distances is introduced. The new methods improved the quality of Self-Organizing Maps (SOMs) significantly for four of the five studied data sets.

1   Introduction

Variable selection or feature extraction is a burning problem especially for exploratory (descriptive) data analysis. The quality of the results is determined by the selection since there is no other supervision. Poor features may emphasize uninteresting properties of the data.

An alternative view to feature extraction is that the topology and the metric of the data space need be chosen. We study the choice of the metric; if the topology need be changed it can be done as a preprocessing step.

Assume that there exists auxiliary data c paired with the primary data $\mathbf{x}$. Here $\mathbf{x}$ is vector-valued and c categorical (finite number of possible values). Assume further that the goal is to study the $\mathbf{x}$, explore or describe them, but that changes in $\mathbf{x}$ are only relevant to the extent they cause changes in c. An example is analysis of the causes of bankruptcy, where the $\mathbf{x}$ contains features of the financial state of a company and c denotes whether the company goes bankrupt or not.

In the learning metrics principle ([5,9]; see [4,5] for more detailed discussion) the distance d between two close-by points $\mathbf{x}$ and $\mathbf{x} + d\mathbf{x}$ of the primary data space is measured by approximations to the distance between the important things, the distribution of c:

$$d^2(\mathbf{x}, \mathbf{x} + d\mathbf{x}) \equiv D_{KL}(p(c|\mathbf{x}), p(c|\mathbf{x} + d\mathbf{x})) = d\mathbf{x}^T \mathbf{J}(\mathbf{x}) d\mathbf{x} . \qquad (1)$$

J.R. Dorronsoro (Ed.): ICANN 2002, LNCS 2415, pp. 999–1004, 2002.
© Springer-Verlag Berlin Heidelberg 2002

Here D_{KL} is the Kullback-Leibler divergence and $\mathbf{J}(\mathbf{x})$ is the Fisher information matrix having $\mathbf{x}$ as its parameters,

$$\mathbf{J}(\mathbf{x}) = E_{p(c|\mathbf{x})} \left[\left(\nabla_{\mathbf{x}} \log p(c|\mathbf{x}) \right) \left(\nabla_{\mathbf{x}} \log p(c|\mathbf{x}) \right)^T \right] . \tag{2}$$

In earlier studies the metric has either been incorporated into the cost function of a method [9], or the density $p(c|\mathbf{x})$ has been estimated and the Fisher information matrix computed from the estimate [5]. Here we extend the latter approach by more accurate density estimators and distance approximation. Since different kinds of data sets may require different kinds of estimators, we suggest choosing the estimator using a validation set.

2   Self-Organizing Maps in Learning Metrics

We will apply the learning metrics to Self-Organizing Maps (SOMs) [7] to improve our earlier results [5].

A SOM is a regular lattice of units i. Each unit contains a model $\mathbf{m}_i$, a representation of particular kinds of data in the data space. The model vectors are adapted with an iterative training algorithm to follow the distribution of the training data. For brevity, we call a SOM trained in learning metrics SOM-L and a SOM in Euclidean metrics SOM-E.

The training algorithm repeats two steps, winner search and adaptation. At each iteration t, an input sample $\mathbf{x}(t)$ is picked randomly from the data, and a winner SOM unit $w(t)$ is selected by

$$w(t) = \arg \min_i d^2(\mathbf{x}(t), \mathbf{m}_i(t)) , \tag{3}$$

where d^2 is the distance function. Here the distance is not in the traditional Euclidean metric but the learning metric (1) derived from the auxiliary data.

When the winner has been selected, the model vectors are all adapted towards the input sample in the steepest descent direction. For learning metrics the direction is given by the natural gradient. For the local approximation (1) this leads to the familiar update rule

$$\mathbf{m}_i(t + 1) = \mathbf{m}_i(t) + \alpha(t) h_{wi}(t)(\mathbf{x}(t) - \mathbf{m}_i(t)) \tag{4}$$

which we have used in this paper. Here $\alpha(t)$ is the learning rate and $h_{wi}(t)$ is the neighborhood function, a decreasing function of the distance between i and $w(t)$ on the SOM lattice.

In practical SOM-L training we use two approximations for calculating learning metric distances. Firstly, the matrix $\mathbf{J}(\mathbf{x})$ is computed from an estimate of the conditional density $p(c|\mathbf{x})$. The investigated alternatives are introduced in Section 3. Secondly, the global distance between two points $\mathbf{x}$ and $\mathbf{m}$ is actually defined as the minimal path integral of the local distances, where the minimum is taken over all paths between $\mathbf{x}$ and $\mathbf{m}$. Finding exact minimal path integrals is computationally prohibitive, so in Section 4 we consider several approximations. The approximations are compared empirically in Section 5.

3 Estimating the Auxiliary Distribution

Learning of the metric is based on the Fisher information matrix of a conditional density estimate. Here we discuss alternative kernel estimators.

We have previously derived the conditional densities from estimators of the joint density of x and c. Two standard estimators, the nonparametric Parzen kernel estimate and a version of Mixture Discriminant Analysis (MDA2) were used. Here we compare other estimators to MDA2; Parzen was too computationally intensive to be included as such.

Since only the conditional densities $p(c|\mathbf{x})$ are needed here, directly estimating them should improve the results. We consider two alternatives. The first is a kind of a mixture of experts (see [3]):

$$\hat{p}_{MoE}(c_i|\mathbf{x}) = \sum_{j=1}^{N_U} y_j(\mathbf{x})\psi_{ji} \,. \tag{5}$$

Here N_U is the number of mixture components. The ψ_{ji} are the parameters of the multinomial distribution generated by the expert j. Their sum is fixed to unity by softmax-reparameterization (not shown). The $y_j(\mathbf{x})$ form the gating network; we used Gaussians normalized to sum to unity for each $\mathbf{x}$.

The second method for conditional density estimation is a product of experts [2], here

$$\hat{p}_{PoE}(c_i|\mathbf{x}) = \frac{1}{Z(\mathbf{x})} \sum_{j=1}^{N_U} \exp\left(y_j(\mathbf{x})\log\psi_{ji}\right) \tag{6}$$

where $Z(\mathbf{x})$ normalizes the density to sum to one.

The MDA2 is fitted to data by maximizing the joint log-likelihood with respect to the ψ_{ji} and the parameters of the gating network by the EM algorithm. For the other models the mean conditional log-likelihood of the auxiliary data is maximized by conjugate gradient algorithms. All estimators include a free dispersion parameter, the variance of the Gaussians. This parameter is chosen to maximize the conditional likelihood of auxiliary data on a validation set.

4 Distance Approximations

The true learning metric distances are minimal path integrals which must be approximated. A simple approximation $\hat{d}_1^2$ of the squared distance, used earlier in e.g. [5], is to evaluate the metric at the input sample $\mathbf{x}$ and to extend the local distance to the whole space. In winner search the distance becomes

$$\hat{d}_1^2(\mathbf{x}, \mathbf{m}) = (\mathbf{m} - \mathbf{x})^T \mathbf{J}(\mathbf{x})(\mathbf{m} - \mathbf{x}) \,. \tag{7}$$

We call this the '1-point approximation'. It is accurate when the model vectors are close to $\mathbf{x}$. Note that for winner selection we need not know the exact distances but only which one is the smallest.

Table 1. The Data Sets

Data set	Dimensions	Classes	Samples
Landsat Satellite Data *	36	6	6435
Letter Recognition Data *	16	26	20000
Phoneme Data from LVQ_PAK [8]	20	14	3656
TIMIT Data from [10]	12	41	14994
Bankruptcy Data used in [5]	23	2	6195
* from UCI Machine Learning Repository [1]			

Our earlier conditional density estimates have been smooth, obtained with a small number of wide kernels. Such estimates may fail to notice some detail in the density but the simple local approximation (7) may be reasonably accurate because of the smoothness. However, for more accurate estimators that potentially change more rapidly the local approximation may hold only very locally.

A more accurate but still computable approximation is obtained by assuming that the minimal path is a line but that the metric may change along the line. When the metric along the line connecting $\mathbf{x}$ and $\mathbf{m}$ is evaluated at T points, the distance becomes

$$\hat{d}_T^2(\mathbf{x}, \mathbf{m}) = \frac{1}{T^2} \left(\sum_{t=1}^{T} \left((\mathbf{m} - \mathbf{x})^T \mathbf{J} \left(\mathbf{x} + \frac{t-1}{T}(\mathbf{m} - \mathbf{x}) \right) (\mathbf{m} - \mathbf{x}) \right)^{1/2} \right)^2 . \quad (8)$$

We call the above the 'T-point approximation'.

The T-point approximations involve more computation. The computational complexity of a single SOM-L training iteration becomes $\mathcal{O}(N_{DIM} N_C N_U N_{SOM} T)$ for N_{SOM} model vectors with dimensionality N_{DIM}, N_C classes, and N_U mixture components. By comparison, the complexity of the 1-point approximation is $\mathcal{O}(N_{DIM} N_C (N_U + N_{SOM}))$.

The T-point winner search may be speeded up by first using 1-point distances to winnow the set of winner candidates; e.g. the W model vectors that are closest according to $\hat{d}_1^2$ are selected and the winner is chosen from these by $\hat{d}_T^2$. In the empirical tests of Section 5 we have used $T = 10$ evaluation points and $W = 10$ winner candidates, resulting in a 20-fold speed-up compared to the unwinnowed T-point approximation, but computational time compared to the 1-point approximation was still about 100-fold.

5   Empirical Testing

The methods were compared on five different data sets (Table 1). The class labels were used as the auxiliary data and the data sets were preprocessed by removing the classes with only a few samples.

The metric was estimated from training data with the methods presented in Sections 3 and 4. The number of mixture components (10, 30, or 100) and the

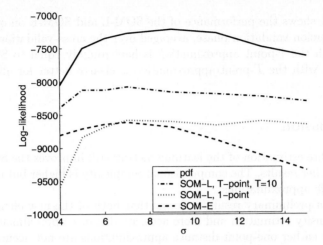

Fig. 1. Average accuracy of SOM-L vs. SOM-E for the TIMIT data over the 10-fold validation sets. The likelihood given by the best pdf estimate (mixture of 100 experts) is included for reference; it is the approximate upper limit.

dispersion parameter were selected using a validation set. The SOM-E and the SOM-L were trained in the resulting metrics, using both the 1-point and T-point ($T = 10$, $W = 10$) distance approximations for SOM-L.

The accuracy of the resulting SOMs in representing the important auxiliary data was measured by the conditional likelihood evaluated at the winner SOM units [5]. The quality of the SOM visualizations was monitored visually.

The significance of the difference between the best SOM-L and SOM-E was tested using 10-fold cross-validation. The dispersion (Gaussian variance) was validated anew in each fold to maximize map accuracy, using part of the training set for validation. The accuracy (likelihood) for the best map was then calculated for the test set.

To reduce the consumption of computational resources, we selected a suboptimal density estimator for the SOM-L having T-point distances: the estimator optimal for 1-point SOM-L was chosen with 30 components.

6   Results

The learning metrics improved the accuracy of the SOMs on all data sets; the improvements were significant by t-test ($p < 0.05$) between SOM-E and SOM-L with the T-point approximation, except for the Bankruptcy data ($p = 0.07$).

The mixture of experts (5) was best for SOM-L in three sets, the product of experts in one and MDA2 in one.

The more accurate distance approximation (8) is crucial. The SOM-L with the 1-point approximation was only comparable or worse than SOM-E on two data sets, while SOM-Ls trained with the improved approximation are on average better on all sets.

Figure 1 shows the performance of the SOM-L and SOM-E on one data set in the dispersion validation phase, averaged over the cross-validation folds. The SOM-L with the 1-point approximation is here roughly equal to SOM-E, but the SOM-L with the T-point approximation is clearly better for all dispersion values.

7 Discussion

More accurate estimation of the learning metrics still improves the SOM results from the earlier results. The computational complexity is higher but manageable with suitable approximations.

Based on preliminary results it seems that both of the new elements, more accurate density estimation and more accurate distance approximation, are required. The earlier one-point distance approximations are not accurate enough for the new density estimators capable of following more accurately the details of the conditional density.

Acknowledgment. This work was supported by the Academy of Finland, in part by the grant 52123.

References

1. Blake, C.L., and Merz C.J.   UCI Repository of machine learning databases. http://www.ics.uci.edu/~mlearn/MLRepository.html, 1998.
2. Hinton, G.E. Products of Experts. In *Proceedings of ICANN'99, the Ninth International Conference on Artificial Neural Networks*, 1–6, IEE, London, 1999.
3. Jordan, M., and Jacobs, R. Hierarchical Mixtures of Experts and the EM Algorithm. *Neural Computation*, 6:181–214, 1994.
4. Kaski, S., Sinkkonen, J. Principle of learning metrics for exploratory data analysis. Submitted to a journal.
5. Kaski, S., Sinkkonen, J., and Peltonen, J.   Bankruptcy Analysis with Self-Organizing Maps in Learning Metrics. *IEEE Transactions on Neural Networks*, 12:936–947, 2001.
6. Kaski, S., and Venna, J. Neighborhood preservation in nonlinear projection methods: An experimental study. In G. Dorffner, H. Bischof, and K. Hornik, editors, *Artificial Neural Networks–ICANN 2001*, 458–491, Springer, Berlin, 2001.
7. Kohonen, T. Self-Organizing Maps. Springer, Berlin, 1995 (Third, extended edition 2001).
8. Kohonen T, Kangas J, Laaksonen J, and Torkkola K.   LVQ_PAK: A program package for the correct application of Learning Vector Quantization algorithms. In *Proceedings of IJCNN'92, International Joint Conference on Neural Networks*, I:725-730, 1992.
9. Sinkkonen, J., and Kaski, S. Clustering based on conditional distributions in an auxiliary space. *Neural Computation*, 14:217–239, 2002.
10. TIMIT 1998. CD-ROM prototype version of the DARPA TIMIT acoustic-phonetic speech database.

Correlation Visualization of High Dimensional Data Using Topographic Maps

Ignacio Díaz Blanco, Abel A. Cuadrado Vega, and Alberto B. Diez González

Área de Ingeniería de Sistemas y Automática
Universidad de Oviedo
Campus de Viesques s/n, 33204, Gijón, Spain
{idiaz,cuadrado,alberto}@isa.uniovi.es

Abstract. Correlation analysis has always been a key technique for understanding data. However, traditional methods are only applicable on the whole data set, providing only global information on correlations. Correlations usually have a local nature and two variables can be directly and inversely correlated at different points in the same data set. This situation arises typically in nonlinear processes. In this paper we propose a method to visualize the distribution of local correlations along the whole data set using dimension reduction mappings. The ideas are illustrated through an artificial data example.

1 Introduction

Visualization and dimension reduction techniques have received considerable attention in recent years for the analysis of large sets of multidimensional data [1, 2,3] and particularly for supervision and condition monitoring of complex industrial processes [4,5,6]. These techniques allow to discover unknown features and relationships of high dimensional data in a visual manner by means of a mapping from a data space D (also *input space*) onto a low dimensional visualization space V where complex relationships among input variables can be easily represented and visualized while preserving information significant to a given problem.

Another very useful technique when dealing with high dimensional data is correlation analysis. Correlation analysis is concerned with finding how components $x_1, \cdots, x_p$ of the sample data vectors $\{\mathbf{x}_i\}_{i=1,\cdots,n}$ are mutually related. The standard way to cope with this problem is through the analysis of second order statistics such as the *correlation matrix* $\mathbf{R}$ whose coefficients $r_{ij} \in [-1, 1]$ provide a description of how variables x_i and x_j are related. These coefficients are the result of a normalized inner product –the cosine– between vectors formed by the values of x_i and x_j for the whole data set and, in consequence, they provide a correlation information of a global nature. However, in many cases data variables can be correlated in different ways for different regions of the data space. This is the case, for instance, of multimodal or nonlinear processes, which behave locally in different ways depending on the working point. Thus, we need a local description of correlation.

J.R. Dorronsoro (Ed.): ICANN 2002, LNCS 2415, pp. 1005–1010, 2002.

In this paper, we suggest a method to combine correlation analysis with the power of dimension reduction visualization methods, such as the Self-Organizing Map (SOM) [7] or the Generative Topographic Map (GTM) [8], allowing to visualize local correlations for each pair of variables x_i, x_j through the so called *correlation maps* defined in the visualization space. The paper is organized as follows. In section 2 the ideas of *local covariance* and *local correlation* are introduced, and a method to display the information provided by local second order statistics in the visualization space is proposed. In section 3 the proposed ideas are illustrated through a simple example. Finally, in section 4 some concluding remarks and future research lines are outlined.

2 Correlation Maps

2.1 Local Covariance Matrix

Let $\psi(\mathbf{y}) : \mathbb{R}^2 \to \mathbb{R}^n$ a continuous mapping which takes a point $\mathbf{y}$ of the visualization space $V \subset \mathbb{R}^2$ and obtains a point $\psi(\mathbf{y})$ pertaining to the manifold which approximates the distribution of the input data points $\mathbf{x}_i$ in the data space $D \subset \mathbb{R}^n$. Let's define the following *neighborhood function* $w_i(\mathbf{y}) = e^{-\frac{1}{2}\|\mathbf{x}_i - \psi(\mathbf{y})\|^2/\sigma^2}$, which describes the degree of locality or proximity of sample $\mathbf{x}_i$ with respect to $\psi(\mathbf{y})$ in the data space D. We define the *local mean vector* $\mathbf{m}(\mathbf{y})$ and the *local covariance matrix* $\mathbf{C}(\mathbf{y})$ associated to a point $\mathbf{y}$ in the visualization space V as

$$\mathbf{m}(\mathbf{y}) = \frac{\sum_i \mathbf{x}_i \cdot w_i(\mathbf{y})}{\sum_i w_i(\mathbf{y})} \tag{1}$$

$$\mathbf{C}(\mathbf{y}) = (c_{ij}) = \frac{\sum_i [\mathbf{x}_i - \mathbf{m}(\mathbf{y})][\mathbf{x}_i - \mathbf{m}(\mathbf{y})]^T \cdot w_i(\mathbf{y})}{\sum_i w_i(\mathbf{y})} \tag{2}$$

Taken independently, the $n \times n$ components $c_{ij}(\mathbf{y})$ of the covariance matrix $\mathbf{C}(\mathbf{y})$, can be regarded as local covariance values which describe the local dependency between variables x_i and x_j. Expressions (1) and (2) represent local versions of the sample first and second order moments of the input data distribution around the image of point $\mathbf{y}$ in the visualization space, i.e., $\psi(\mathbf{y})$, where the width factor σ is a design parameter related the degree of locality to be taken into account, allowing to establish a tradeoff between global and local correlations.

The local covariance $\mathbf{C}(\mathbf{y})$ described in (2) defines in V a field of covariance matrices from D each of which provides a local description of second order statistical features of data in D lying in the vicinity of $\psi(\mathbf{y})$.

2.2 Local Correlation Matrix

The previously defined covariance matrix provides insight in the approach of local description of second order statistics. However, in looking for correlations, correlation coefficients are preferred as they provide a normalized description of correlations in the interval $[-1, +1]$. The *local correlation matrix* around $\mathbf{y}$ can be defined as

$$\mathbf{R}(\mathbf{y}) = (r_{ij}) \quad \text{where,} \quad r_{ij} = \frac{c_{ij}}{\sqrt{c_{ii}c_{jj}}} \tag{3}$$

The local correlation matrix $\mathbf{R}(\mathbf{y})$ has $n \times n$ components $r_{ij}(\mathbf{y})$ which represent the local correlation coefficient between variable x_i and variable x_j and lie always in the interval $[-1, +1]$, where $+1$ denotes full direct correlation, 0 denotes incorrelation, and -1 denotes full inverse correlation.

2.3 Visualization of Second Order Statistical Features

Both the *covariance matrix* $\mathbf{C}(\mathbf{y})$ and *correlation matrix* $\mathbf{R}(\mathbf{y})$ are defined for each point $\mathbf{y}$ of V. In addition to this, all powerful geometrical and statistical interpretations underlying both matrices can be represented in V using scalar quantities. Thus, for instance, each component $c_{ij}(\mathbf{y})$ or $r_{ij}(\mathbf{y})$ defines a scalar quantity susceptible to be represented in the same way as SOM planes, using a color code for each pixel $\mathbf{y}$. In the same way, the principal values of the covariance matrix $\lambda_i(\mathbf{y})$ or the components of the principal vectors $\mathbf{u}_i(\mathbf{y})$ can be represented as SOM planes.

This representation provides a unified visualization of the underlying correlations and second order statistical properties in general. Moreover, it is coherent with other SOM representations such as SOM planes or the *u-matrix* providing insight in the pattern of correlation dependencies among variables or revealing the most important features describing the behavior of the underlying process for each data region.

3 Application to Artificial Data

All these ideas are illustrated in figures 1, 2 and 3. A simple 2D data set was used to train both a 1D-SOM and a 2D-SOM. Local covariances were obtained for the 2D-SOM using (1) and (2) and then plotted in both the data space D and the visualization space V. Local correlations were also obtained using (3) to build the correlation maps of $r_{xx}, r_{xy}, r_{yx}, r_{yy}$ shown in figure 2. A set of points with negative local correlations (corresponding to the right part of the "arc" in the data) can be discovered by looking at the upper left corner in the r_{xy} plane. Similarly, moderately high correlations appear in the upper right corner of the map, showing up the positive local correlations existing in the left part of the "arc" in the data space. It can also be observed that the graphical information provided by correlation maps in figure 2 is consistent with that shown in the SOM planes in figure 3, because both are descriptions in the same visualization space V. Finally, as we should expect, planes r_{xx} and r_{yy} are equal to 1, and $r_{xy} = r_{yx}$ due to the symmetry properties of correlation matrices.

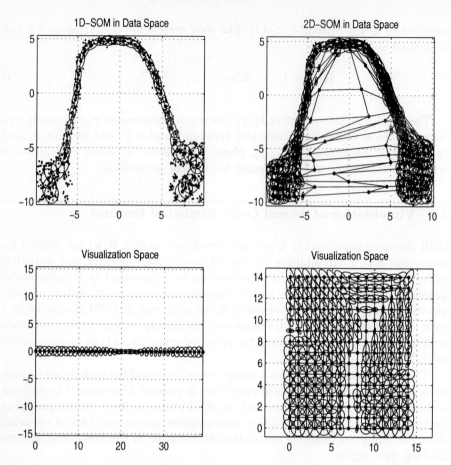

Fig. 1. Local covariances in D (top) and in V (bottom) obtained for both a 1D-SOM (left) and a 2D-SOM (right). In the thick areas (low correlations), the covariances are nearly spherical, while in thin areas (high correlations) the covariances become low rank, and oriented, showing up in V the nature of local correlations.

4 Concluding Remarks

We have proposed here a method for the visualization of local second order statistical properties using dimension reduction mappings like –but not restricted to– the SOM. The proposed idea has strong connections with local model approaches, such as [9], where local linear PCA projections are proposed to capture the nonlinear structure of data.

We showed here through an artificial data example how local second order statistical properties can be revealed by means of correlation maps, which, in addition, are consistent with other standard representations in the visualization space such as the component planes or the distance matrix. This provides an

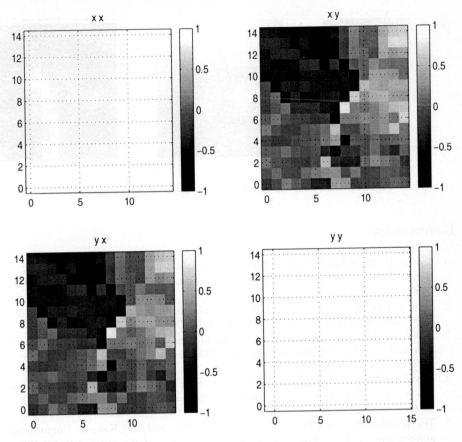

Fig. 2. Correlation Maps for the 2D-SOM show a region in V (up-left) related to highly negative local correlations and another region (up-right) revealing positive local correlations.

alternative way for high dimensional data visualization to the standard methods based on SOM $-u$-$matrix$, SOM planes, response surfaces or SOM planes rearrangement [10], as well as SOM clustering methods [11]– which combines the classical correlation analysis techniques (correlation matrix) with the power of SOM data visualization.

As a matter of further study, the idea of local second order moments is not restricted to correlation analysis or even to second order moments. Eigenvalues $\lambda_i(\mathbf{y})$ or the components of eigenvectors $\mathbf{u}_i(\mathbf{y})$ of the local covariance matrix can lead to meaningful maps, which can be derived in a straightforward manner from the ideas described here. In a similar way, higher order statistics (cumulants) can be obtained in a local fashion opening new exciting research lines in data visualization.

The ideas proposed in this paper are currently being tested in the steel industry to investigate the effects of several dozens of process variables in several quality factors of the processed coils in a tandem mill with encouraging results.

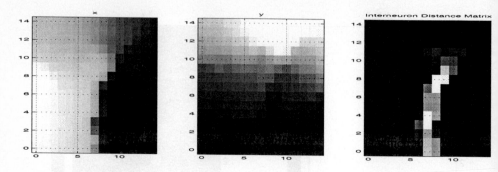

Fig. 3. SOM planes of variables x and y and distance matrix.

References

1. Joshua B. Tenenbaum, Vin de Silva, and John C. Langford. A global geometric framework for nonlinear dimensionality reduction. *Science*, 290:2319–2323, Dec, 22 2000.
2. Sam T. Roweis and Lawrence K. Saul. Nonlinear dimensionality reduction by locally linear embedding.*Science*, 290:2323–2326, Dec., 22 2000.
3. Jianchang Mao and Anil K. Jain. Artificial neural networks for feature extraction and multivariate data projection. *IEEE Transactions on Neural Networks*, 6(2):296–316, March 1995.
4. David J. H. Wilson and George W. Irwin. RBF principal manifolds for process monitoring. *IEEE Transactions on Neural Networks*, 10(6):1424–1434, November 1999.
5. Teuvo Kohonen, Erkki Oja, Olli Simula, Ari Visa, and Jari Kangas. Engineering applications of the self-organizing map. *Proceedings of the IEEE*, 84(10):1358–1384, October 1996.
6. Esa Alhoniemi, Jaakko Hollmèn, Olli Simula, and Juha Vesanto. Process monitoring and modeling using the self-organizing map. *Integrated Computer Aided Engineering*, 6(1):3–14, 1999.
7. Teuvo Kohonen. *Self-Organizing Maps*. Springer-Verlag, 1995.
8. Christopher M. Bishop, Markus Svensen, and Christopher K. I. Williams. GTM: The generative topographic mapping.*Neural Computation*, 10(1):215–234, 1998.
9. M. Tipping and C. Bishop. Mixtures of probabilistic principal component analyzers. *Neural Computation*, 11(2):443–482, 1999.
10. Juha Vesanto. Som-based data visualization methods.*Intelligent Data Analysis*, 3(2):111–126, 1999.
11. Juha Vesanto and Esa Alhoniemi. Clustering of the self-organizing map.*IEEE Transactions on Neural Networks*, 11(3):586–600, May 2000.

Part VII

Signal and Time Series Analysis

Continuous Unsupervised Sleep Staging Based on a Single EEG Signal

Arthur Flexer, Georg Gruber, and Georg Dorffner

The Austrian Research Institute for Artificial Intelligence
Schottengasse 3, A-1010 Vienna, Austria
arthur@ai.univie.ac.at

Abstract. We report improvements on automatic continuous sleep staging using Hidden Markov Models (HMM). Our totally unsupervised approach detects the cornerstones of human sleep (wakefulness, deep and rem sleep) with around 80% accuracy based on data from a single EEG channel. Contrary to our previous efforts we trained the HMM on data from a single sleep lab instead of generalizing to data from diverse sleep labs. This solved our previous problem of detecting rem sleep.

1 Introduction

Sleep staging is one of the most important steps in sleep analysis and is usually done using the traditional Rechtschaffen & Kales [7] (R&K) rules. It is a very time consuming task consisting of classifying all 30 second pieces of an approximately 8 hour recording into one of six sleep stages (wake, rem (rapid eye movement) sleep, S1, S2, S3, S4 (deep sleep)). A sleep recording is made with a minimum setting of four channels: electro-encephalogram (EEG) from electrodes C3 and C4, electro-myogram (EMG) and electro-oculogram (EOG). In order to classify each 30 second segment of sleep, the human scorer looks for defined patterns of waveforms in the EEG, for rapid eye movements in the EOG and for EMG level.

There is however a considerable dissatisfaction within the sleep research community concerning the very basics of R&K sleep staging: R&K is based on a predefined set of rules leaving much room for subjective interpretation; it is a very time consuming and tedious task; it is designed for young normal subjects only; it has a low 30 second temporal resolution; it is defined in terms of six stages neglecting the micro-structure of sleep; it cannot be automized reliably due to the large inter-scorer variability and insufficient rules for staging.

Our aim is to built an automatic continuous sleep stager, based on probabilistic principles which overcomes the known drawbacks of traditional R&K sleep staging. In previous efforts [1] we tried to find a new description of human sleep which is based on the comparably unambiguous "extreme" cornerstones of traditional sleep staging rather than merely automating and replicating R&K sleep staging. We used a Hidden Markov Model (HMM) to produce three continuous probability traces P(wake), P(deep) and P(rem) with a one second resolution.

J.R. Dorronsoro (Ed.): ICANN 2002, LNCS 2415, pp. 1013–1018, 2002.

The newly obtained continuous sleep profiles were compared to traditional R&K scoring. The two "extreme" R&K stages "wake" and "deep" could be detected very satisfactory with an accuracy of above 80%. However, we had great problems discriminating rem sleep from wakefulness and stages 1, 2 and even 3. The mean accuracy for detection of rem sleep was as low as 26%.

This paper reports about some improvements in continuous sleep staging which allow us to detect all three cornerstones of human sleep with satisfactory accuracy (around 80%). Reviewing our old results lead us to the hypothesis that the problems we encountered so far might be due to problems in the data base. Concentrating on data from a single sleep lab solved the problem of detecting rem sleep.

2   Data

In our previous efforts [1] our data base consisted of nine whole night sleep recordings from a group of healthy adults (total sleep time = 70.5h, age ranges from 20 to 60, 5 females and 4 males). We used reflection coefficients and stochastic complexity computed for EEG channels C3 and C4 and a measure of EMG level (altogether five features) for analysis with an HMM. The nine recordings had been recorded in five different European sleep laboratories during the SIESTA project [3]. The results were as described in Sec. 1: satisfactory detection of wakefulness and deep sleep, but accuracy as low as 26% for rem sleep.

The SIESTA project resulted in yet another sleep stager [9], which is based on a semi-supervised approach using Gaussian kernels plus sensor fusion to fuse information from different channels. Without giving any further detail it should suffice to say that its outputs are again three continuous probability traces. This sleep stager has been evaluated on data from eight different sleep labs. Plotting the mean entropy of the probability traces per sleep lab reveals clear lab effects (see Fig. 1). High entropy values indicate that the probability traces show little variation, i.e. that all three of them stay around .33 and therefore contain only little information. Obviously, the sleep stager seems to work quite well for some of the sleep labs and quite bad for others.

To gather further evidence for this hypothesis, we decided to concentrate our analysis on data from a single sleep lab (the one with the best results according to the entropy plot). Our new data base consists of 40 whole night sleep recordings from a group of healthy adults (total sleep time = 326.4h, age ranges from 20 to 80, 22 females and 18 males). We use only EEG channel C3 for further analysis. Twenty recordings are used to train our automatic sleep stager (training set), twenty are set aside to evaluate it (test set). Both sets are matched for sex and age.

3   Methods

HMMs [6] allow analysis of non-stationary multi-variate time series by modeling, both, the probability density functions of locally stationary multi-variate data

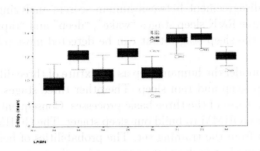

Fig. 1. Mean of the entropy of the probability traces given per sleep lab. Depicted are the means and the 25% and 75% percentile per sleep lab.

and the transition probabilities between these stable states. In the context of sleep analysis, the locally stable states can be thought of as sleep stages.

Following the classical text by Rabiner and Juang [6], an HMM can be characterized as (i) having a finite number of N states; (ii) a new state is entered based upon a transition probability distribution A which depends on the previous state (the Markovian property); (iii) after each transition an observation output symbol is produced according to a probability distribution B which depends on the current state. Although the classical HMM uses a set of discrete symbols as observation output, Rabiner and Juang [6] already discuss the extension to continuous observation symbols. Such a Gaussian Observation HMM (GOHMM) [5] has already been proposed as a model for EEG analysis and the model we now describe is the same we used for our previous work on sleep staging [1]. We use a GOHMM where the observation symbol probability distribution for state j is given by $B = \{b_j(x)\}, b_j(x) = \mathcal{N}[x, \mu_j, U_j]$, where $\mathcal{N}$ is the normal density and μ_j and U_j are the mean vector and covariance matrix associated with state j. Please note that this a simple version of the Gaussian M-component mixture given in [6] with M equal one. The Expectation-Maximization (EM) algorithm is used to train the GOHMM thereby estimating the parameter sets A and B as well as the μ_j and U_j. Viterbi decoding is used to identify most likely state sequences corresponding to a particular time series and enables the computation of the probabilities of being in any of the N states at each point in time. Full details of the algorithms can be found in [6].

A GOHMM is defined over the first reflection coefficient of the EEG channel at C3. Reflection coefficients are the coefficients of the order recursive representation of autoregressive (AR) processes [4]. We used a lattice filter representation of an AR process. The inferred a-posteriori distribution over model coefficients are the reflection coefficients (see [9] for full detail). The reflection coefficient is computed with a one second resolution.

Our aim is not to replicate R&K scoring but to find a new description of human sleep which is based on the comparably unambiguous "extreme" cornerstones of traditional sleep staging. Since R&K sleep staging is based on a predefined set of rules which leave much room for subjective interpretation there can ous description of human sleep which is based on probabilistic principles. It is therefore in line with previous recommendations [2] and work [8] on continuous sleep staging. The approach is based on Hidden Markov Models, is totally unsupervised and uses only a single channel of EEG. It is superior in performance compared to our own previous results [1]. The output in the form of three continuous probability traces captures the three main processes in human sleep: wakefulness, deep sleep and rem sleep.

The improvement in performance has been made possible by realizing that there exist clear lab effects in our data base of sleep recordings. Although a lot of effort had been put into harmonization of sleep labs and recording protocol, there seem to be differences in hardware and maybe also in filter settings which are still visible in EEG and other signals.

Acknowledgements. The recordings for this work were done within the BIOMED-2 BMH4-CT97-2040 project SIESTA, funded by the EC DG XII. The Austrian Research Institute for Artificial Intelligence is supported by the Austrian Federal Ministry of Education, Science and Culture.

References

1. Flexer A., Dorffner G., Sykacek P., Rezek I.: An automatic, continuous and probabilistic sleep stager based on a Hidden Markov Model, Applied Artificial Intelligence, Vol. 16, Num. 3, pp.199–207, 2002.
2. Kemp B.: A proposal for computer-based sleep/wake analysis, Journal of Sleep Research, 2, 179–185, 1993.
3. Kloesch G., Kemp B., Penzel T., Schloegl A., Rappelsberger P., Trenker E., Gruber G., Zeitlhofer J., Saletu B., Herrmann W.M., Himanen S.-L., Kunz D., Barbanoj M., Roeschke J., Vaerri A., Dorffner G.: The SIESTA Project Polygraphic and Clinical Database, IEEE Eng. in Medicine & Biology Magazine, 20(3)51–57, 2001.
4. Ljung L.: System Identification, Theory for the User, Prentice-Hall, Englewood Cliffs, New Jersey, 1999.
5. Penny W.D., Roberts S.J.: Gaussian Observation Hidden Markov Models for EEG analysis, Technical Report, Imperial College, London, TR-98-12, 1998.
6. Rabiner L.R., Juang B.H.: An Introduction To Hidden Markov Models, IEEE ASSP Magazine, 3(1):4–16, 1986.
7. Rechtschaffen A., Kales A.: A Manual of Standardized Terminology, Techniques and Scoring System for Sleep Stages of Human Subjects, U.S. Dept. Health, Education and Welfare, National Institute of Health Publ. No.204, Washington, 1968.
8. Roberts S., Tarassenko L.: New Method of Automated Sleep Quantification, Medical and Biological Engineering and Computing, (5), 509–517, 1992.
9. Sykacek P., Roberts S., Rezek I., Flexer A., Dorffner G.: A Probabilistic Approach to High-Resolution Sleep Analysis, in Dorffner G. et al. (eds.), Artificial Neural Networks – ICANN 2001, International Conference, Vienna, Austria, Lecture Notes In Computer Science 2130, Springer, pp. 617–624, 2001.

Financial APT-Based Gaussian TFA Learning for Adaptive Portfolio Management

Kai Chun Chiu and Lei Xu

Department of Computer Science and Engineering
The Chinese University of Hong Kong, Shatin, N.T., Hong Kong, P.R. China
{kcchiu,lxu}@cse.cuhk.edu.hk

Abstract. Adaptive portfolio management has been studied in the literature of neural nets and machine learning. The recently developed Temporal Factor Analysis (TFA) model mainly targeted for further study of the Arbitrage Pricing Theory (APT) is found to have potential applications in portfolio management. In this paper, we aim to illustrate the superiority of APT-based portfolio management over return-based portfolio management.

1 Introduction

In view of the rapid expansion of today's capital markets, quantitative analysis of financial data has been studied in the context of neural networks and machine learning. Adaptive portfolio management [1,2,3] usually refers to the study of the traditional Markowitz's portfolio theory [8] in the context of artificial neural networks.

In literature, adaptive portfolio management via maximizing the well-known Sharpe ratio [4] was studied in [1,2]. However, such approaches either treat the weights as constants or depend directly on the security returns.

Recently, a new technique called Temporal Factor Analysis (TFA) was proposed by [5] with an aim to provide an alternative way for implementing the classical financial APT model. In this paper, we consider how the APT-based Gaussian TFA model can be used for adaptive portfolio management. Comparisons with another similar, previously adopted technique is shown.

The rest of the paper is organized in the following way. Sections 2 and 3 briefly review the APT and the Gaussian TFA model respectively. Section 4 gives an algorithm for implementing the APT-based Gaussian TFA learning for adaptive portfolio management. Comparisons with the other approach by way of an empirical study is shown in section 5. Section 6 concludes the paper.

2 Review on Arbitrage Pricing Theory

APT begins with the assumption that the $n \times 1$ vector of asset returns, R_t, is generated by a linear stochastic process with k factors [6]:

$$R_t = \bar{R} + Af_t + e_t \tag{1}$$

J.R. Dorronsoro (Ed.): ICANN 2002, LNCS 2415, pp. 1019–1024, 2002.

Table 2. Risk-return statistics of the portfolio under test

Component Name	Expected Return (μ)	Risk (σ)	S_p Estimated
Risk-free Security	0.0148%	0.0018%	–
HSI	0.18%	1.48%	–
HSCCI	0.03%	2.51%	–
HSCEI	-0.20%	2.55%	–
Return-based (short-selling disallowed)	0.08%	0.61%	0.13
APT-based (short-selling disallowed)	0.19%	1.04%	0.18
APT-based (short-selling allowed)	0.33%	1.62%	0.20

5.3 Performance Evaluation

The better performance of APT-based portfolio over return-based portfolio may be attributed to the contribution of independent hidden factors in controlling portfolio weights. Two advantages are evident. The first benefit arises from dimensionality reduction as there are usually only a few hidden factors for even a large number of securities. Second, the portfolio weights may be controlled more appropriately by hidden factors rather than security returns, considering the proposition of classical APT [6] that returns are generated by several hidden factors.

6   Conclusion

In this paper, we introduce how the Gaussian TFA model can be appropriately applied to adaptive portfolio management. We find that the APT-based portfolio management demonstrates superior performance over return-based portfolio management.

References

1. Choey, M., Weigend, A.S.: Nonlinear trading models through Sharpe ratio optimization. International Journal of Neural Systems **8** (1997) 417–431
2. Moody, J., Wu, L.: Optimization of trading systems and portfolios. Proc. of Computational Intelligence for Financial Engineering (CIFEr'97) (1997) 300–307
3. Xu, L., Cheung, Y.M.: Adaptive supervised learning decision networks for traders and portfolios. Proc. of Computational Intelligence for Fin. Eng. (CIFEr'97) (1997) 206–212
4. Sharpe, W.F.: The Sharpe ratio – properly used, it can improve investment. Journal of Portfolio Management (1994) 49–58
5. Xu, L.: Temporal BYY learning for state space approach, hidden Markov model and blind source separation. IEEE Trans. on Signal Processing **48** (2000) 2132–2144
6. ROSS, S.: The arbitrage theory of capital asset pricing. J. Econ. Theory **13** (1976) 341–360
7. Xu, L.: BYY harmony learning, independent state space and generalized APT financial analyses. IEEE Trans. on Neural Networks **12** (2001) 822–849
8. Markowitz, H.: Portfolio Selection: Efficient Diversification of Investments. NY Wiley (1959)
9. Xu, L.: RBF nets, mixture experts, and Bayesian Ying-Yang learning. Neurocomputing **19** (1998) 223–257

On Convergence of an Iterative Factor Estimate Algorithm for the NFA Model

Zhiyong Liu and Lei Xu

Department of Computer Science and Engineering,
The Chinese University of Hong Kong, Shatin, N.T. Hong Kong, P.R. China

Abstract. The *iterative fixed posteriori approximation* (iterative FPA) has been empirically shown to be an efficient approach for the MAP factor estimate in the Non-Gaussian Factor Analysis (NFA) model. In this paper we further prove that it is exactly an EM algorithm for the MAP factor estimate problem. Thus its convergence can be guaranteed. We also empirically show that NFA has better generalization ability than Independent Factor Analysis (IFA) on data with small sample size.

1 Introduction

Conventional factor analysis (FA) model

$$x_t = Ay_t + e_t \tag{1}$$

tries to estimate the transformation matrix $A \in \mathbb{R}^{n \times m}$, factor $y_t \in \mathbb{R}^m$, and noise e_t just based on the observation $x_t \in \mathbb{R}^n$. However, the rotation and additive indeterminacies make FA model have many solutions [1]. Although Independent Component Analysis (ICA) is not affected by these indeterminacies, it makes an unrealistic *noise-free* assumption, i.e., ignoring e_t in (1).

To make up for ICA's *noise-free* deficiency, various research on the *noisy* ICA model has been done. Typically, we have the Non-Gaussian Factor Analysis (NFA) model first proposed in [2], and then further developed with a more efficient algorithm in [1]. NFA is of the same form as (1), but y is no longer Gaussian. NFA assumes that each element $y^{(j)}$ of y is component-wise independent from a Gaussian mixture (non-Gaussian) as:

$$p(y) = \prod_j p(y^{(j)}), p(y^{(j)}) = \sum_r \alpha_{j,r} G(y^{(j)} | m_{j,r}, \sigma_{j,r}^2) \tag{2}$$

where $G(y^{(j)} | m_{j,r}, \sigma_{j,r}^2)$ refers to a Gaussian with mean $m_{j,r}$ and variance $\sigma_{j,r}^2$.

NFA not only can estimate A, y_t and e_t, but also possesses the automatical model selection ability and two forms of regularizations, namely *data smoothing* and *normalization*, via the BYY *harmony learning*. In the iterative algorithm [1] for NFA, the first step is the maximum a posteriori (MAP) problem as $\hat{y}_t = \arg\max_y \ln[p(x_t|y_t)p(y_t)]$, where $p(x_t|y_t) = G(x_t|Ay_t, \Sigma_e)$, and Σ_e denotes the covariance matrix of noise e_t. How to tackle this problem is important.

J.R. Dorronsoro (Ed.): ICANN 2002, LNCS 2415, pp. 1025–1030, 2002.

Although the so-called *iterative fixed posteriori approximation (iterative FPA)* [3] was empirically shown as an efficient approach, no theoretical proof on its convergence was given. In this paper we aim to prove that the *iterative FPA* is exactly the EM algorithm, and thus its convergence can always be guaranteed [4,5,6]. Also we empirically show that NFA outperforms IFA, which is another typical *noisy* ICA algorithm, on the generalization ability on small data sample.

In the sequel, the algorithms for NFA and *iterative FPA* are introduced in section 2. Section 3 establishes the theoretic relationship between the *iterative FPA* and EM algorithm. Section 4 empirically compares NFA and IFA. Section 5 concludes this paper.

2   Non-Gaussian Factor Analysis and Iterative FPA for MAP Estimate

2.1   The NFA Algorithm and MAP Problem

Based on BYY *harmony learning*, a four-step iterative algorithm was proposed in [1] to estimate the NFA model, with a MAP factor estimate problem in the first step as:

$$\hat{y}_t = \arg \max_y \ln[p(x_t|y_t)p(y_t)] \tag{3}$$

This MAP factor estimate problem is important since it provides the basis for the next 3 steps which are the updating rules for other parameters, as well as the *normalization* and *data smoothing* regularizations. Details can be referred to [1]. For the MAP factor estimate problem, we have already introduced the *iterative FPA* algorithm [3] which works quite well in practice.

2.2   Iterative FPA

The so-called *fixed posteriori approximation* (FPA) approach was proposed in Tab. 2 of [1] as

$$\hat{y} = (A^T \Sigma_e^{-1} A + diag[b_1, ..., b_k])^{-1} [A^T \Sigma_e^{-1} x + d] \tag{4}$$

where $b_j = \sum_r \frac{h_{j,r}}{\sigma_{j,r}^2}$, $d^{(j)} = \sum_r \frac{h_{j,r} m_{j,r}}{\sigma_{j,r}^2}$, and the posteriori

$$h_{j,r} = \frac{\alpha_{j,r} G(y^{(j)}|m_{j,r}, \sigma_{j,r}^2)}{\sum_r \alpha_{j,r} G(y^{(j)}|m_{j,r}, \sigma_{j,r}^2)} \tag{5}$$

is approximately regarded as irrelevant to y.

Based on it an iterative algorithm can be formed, and thus we name it *iterative FPA* as follows:

step 1: calculate h based on a properly initialized y.
step 2: fix h, calculate y according to eq 4.
step 3: fix y, calculate h. If it converges, stop; otherwise, go to step (2).

In practice the *iterative FPA* performs quite well for the MAP factor estimate problem in terms of both accuracy and efficiency [3]. Below it can be proved to be exactly the EM algorithm for this problem and thus its convergence can always be guaranteed.

3   EM Algorithm and Its Equivalence to Iterative FPA for MAP Estimate

The so-called Expectation-Maximization (EM) algorithm [7] was usually introduced to tackle the incomplete-data problem. In E-step, one needs to find the expectation of the complete (including observed and "missing") data based cost function, which is typically the log-likelihood function, and M-step is to maximize the expectation obtained in E-step.

On the other hand, the MAP factor estimate problem is as shown below.

$$\hat{y} = \arg\max_{y} \ln[p(x|y)p(y)] \triangleq \arg\max_{y} \ln J(y|\Theta)$$

$$= \arg\max_{y} [\sum_{j} \ln \sum_{r} \alpha_{j,r} G(y^{(j)}|m_{j,r}, \sigma_{j,r}^2) + \ln G(x|Ay, \Sigma_e)] \tag{6}$$

where $\Theta = \{\alpha_{j,r}, m_{j,r}, \sigma_{j,r}^2, A, \Sigma_e\}$ in the defined function $J(y|\Theta)$ denotes all the currently **known** parameters. To take advantage of the EM algorithm to solve it, we first design a "missing" data $z \in \mathbb{R}^m$ to explain $y^{(j)}$ being generated by the respective component, i.e.,

$$z = [z^{(1)}, z^{(2)}, ..., z^{(m)}]^T \tag{7}$$

where $z^{(j)} \in \{1, 2, ..., k\}$, $j = 1, 2, ..., m$, and $z^{(j)} = i$ denotes $y^{(j)}$ is generated by the ith component. For simplicity we assume the number of components of all mixture models being equal to k, but later it would be shown inessential. Consequently, the J function can be obtained as:

$$J(y|\Theta, z) = [\prod_{j} \alpha_{j, z^{(j)}} G(y^{(j)}|m_{j, z^{(j)}}, \sigma_{j, z^{(j)}}^2)] G(x|Ay, \Sigma_e) \tag{8}$$

The density function $p(z^{(j)}|y^{(j)}, \Theta)$ of the "missing" data $z^{(j)}$ is with the following posterior probability:

$$p(z^{(j)}|y^{(j)}, \Theta) = \frac{\alpha_{j, z^{(j)}} G(y^{(j)}|m_{j, z^{(j)}}, \sigma_{j, z^{(j)}}^2)}{\sum_r \alpha_{j,r} G(y^{(j)}|m_{j,r}, \sigma_{j,r}^2)} \tag{9}$$

Because Θ is the **known** parameters, the distribution of $z^{(j)}$ is only relevant to, or determined by the mutually independent variable $y^{(j)}$, and so $z^{(j)}$, $j = 1, 2, ..., m$ are also mutually independent. Hence, the conditional pdf $p(z|y, \Theta)$ can be computed by:

$$p(z|y, \Theta) = \prod_{j} p(z^{(j)}|y^{(j)}, \Theta) = \prod_{j} \frac{\alpha_{j, z^{(j)}} G(y^{(j)}|m_{j, z^{(j)}}, \sigma_{j, z^{(j)}}^2)}{\sum_r \alpha_{j,r} G(y^{(j)}|m_{j,r}, \sigma_{j,r}^2)} \tag{10}$$

The expectation of the complete-data based log-J function in E-step can be then obtained as:

$$
Q(y, y^*) = \int \ln(J(y|\Theta, z)) p(z|y^*, \Theta) dz = \sum_z \ln(J(y|\Theta, z)) p(z|y^*, \Theta)
$$

$$
= \sum_{z^{(1)}=1}^{k} \cdots \sum_{z^{(m)}=1}^{k} \sum_{r=1}^{k} \delta_{r, z^{(j)}} \Big[\sum_{j=1}^{m} \ln \alpha_{j,r} G(y^{(j)}|m_{j,r}, \sigma_{j,r}^2)
$$

$$
+ \ln G(x|Ay, \Sigma_e) \Big] \prod_j p(z^{(j)}|y^{(j)*}, \Theta)
$$

$$
= \sum_{r=1}^{k} \Big[\sum_{j=1}^{m} \ln \alpha_{j,r} G(y^{(j)}|m_{j,r}, \sigma_{j,r}^2)
$$

$$
+ \ln G(x|Ay, \Sigma_e) \Big] \sum_{z^{(1)}=1}^{k} \cdots \sum_{z^{(m)}=1}^{k} \delta_{r, z^{(j)}} \prod_j p(z^{(j)}|y^{(j)*}, \Theta)
$$

$$
= \sum_{r=1}^{k} \Big[\sum_{j=1}^{m} \ln \alpha_{j,r} G(y^{(j)}|m_{j,r}, \sigma_{j,r}^2) + \ln G(x|Ay, \Sigma_e) \Big] p(r|y^{(j)*}, \Theta)
$$

$$
= \sum_{j=1}^{m} \sum_{r=1}^{k} \ln[\alpha_{j,r} G(y^{(j)}|m_{j,r}, \sigma_{j,r}^2)] p(r|y^{(j)*}, \Theta) + \ln G(x|Ay, \Sigma_e) \quad (11)
$$

where $\delta_{r, z^{(j)}} = 1$ for $r = z^{(j)}$ and otherwise, $= 0$, and y^* denotes the current y.

Then M-step requires finding y^{new} to maximize the above expectation $Q(y, y^*)$. Differentiate $Q(y, y^*)$ with respect to y we get

$$
\frac{\partial Q(y, y^*)}{\partial y} = \frac{\partial \sum_{j=1}^{m} \sum_{r=1}^{k} \ln G(y^{(j)}|m_{j,r}, \sigma_{j,r}^2) p(r|y^{(j)*}, \Theta)}{\partial y} + \frac{\partial \ln G(x|Ay, \Sigma_e)}{\partial y}
$$

$$
= \begin{pmatrix} \sum_{r=1}^{k} \frac{m_{1,r}}{\sigma_{1,r}^2} p(r|y^{(1)*}, \Theta) \\ \cdots \\ \sum_{r=1}^{k} \frac{m_{m,r}}{\sigma_{m,r}^2} p(r|y^{(m)*}, \Theta) \end{pmatrix} + y \cdot diag[b_1, ..., b_k] + A^T \Sigma_e^{-1}(x - Ay)
$$

where $b_j = \sum_{r=1}^{k} \frac{p(r|y^{(j)*}, \Theta)}{\sigma_{j,r}^2} = \sum_{r=1}^{k} \frac{h_{j,r}}{\sigma_{j,r}^2}$. Thus, $\frac{\partial Q(y, y^*)}{\partial y} = 0$ makes

$$
y^{new} = (A^T \Sigma_e^{-1} A + diag[b_1, ..., b_k])^{-1} [A^T \Sigma_e^{-1} x + d] \quad (12)
$$

where $d \in \mathbb{R}^m$ is with element $d^{(j)} = \sum_{r=1}^{k} \frac{m_{j,r}}{\sigma_{j,r}^2} p(r|y^{(j)*}, \Theta) = \sum_{r=1}^{k} \frac{h_{j,r} m_{j,r}}{\sigma_{j,r}^2}$, where $h_{j,r}$ is defined in (5). So (12) is exactly (4). Previously we assume the numbers of components of different mixture models all equal to k. However, (12) (including b and d) can be extended to the case with different component number without any further consideration.

So in E-step, we just need to calculate $p(r|y^{(j)}, \Theta) = h_{j,r}$ according to (5), and in M-step, update y as (12) or 4. It is exactly the *iterative FPA* for the MAP factor estimate problem.

4   Experimental Comparison between NFA and IFA on Generalization Performance

Independent Factor Analysis (IFA) [?] is also a *noisy* ICA model, which consists of two steps: The first step estimates the factor distribution model via *maximum likelihood learning*, and the second estimates the factor and noise via Least Mean Square (LMS) or MAP. Compared with NFA, IFA is less efficient due to its two-step implementation. Furthermore, it has no model selection ability for determining the number of factors. Moreover, its generalization ability on small data sample is unsatisfactory due to lack of regularization which is common with *maximum likelihood learning*. In this experiment we only compare their performance on generalization.

4.1   Experiment and Data Description

For comparison, we choose to implement NFA with and without *data smoothing* regularization. The basic idea of *data smoothing* is to replace a datum by a Parzen window, whose *width* can be automatically determined during the learning process.

The 50 four-dimensional sample data were generated via (1) with four source signals. The first three are generated from sub-Gaussian distribution and the last one from super-Gaussian. The noise e_t is generated with pdf $G(e_t|0, \Sigma)$, where Σ is a 4×4 matrix with diagonal elements $0.01, 0.0025, 0.01, 0.0025$ and off-diagonal elements $\sigma_{ij} = 0$. The source and observed signals are shown in figure 1. We choose the first 25 samples as training data, and the remaining 25 as test data. Both NFA and IFA are first estimated via the training data, and then in the test phase, the source signal model, i.e., $p(y)$ in NFA and its counterpart in IFA, is fixed to estimate the other parameters using the test data. In this way, the generalization ability of the two models can be compared.

4.2   Experimental Results

For the sake of brevity, only numerical results of the training data are presented and shown in table 1. The results of the test data are shown in figure 2, with means and variances of the three signals normalized as 0 and 1 respectively. The numerical results on the test data are shown in Table 1, where MSE means Mean Square Error between the normalized original and recovered signals.

As witnessed by test data, NFA with *data smoothing* regularization has better generalization ability than IFA and NFA without *data smoothing*. As a tradeoff, NFA with *data smoothing* performs not so well on the training data as compared with the other two. This is within our expectation as the regularization in fact tries to avoid over-fitting the training data with small sample size.

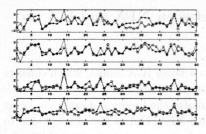

Fig. 1. Sources signals and observations, "o" denotes sources, "*" observations.

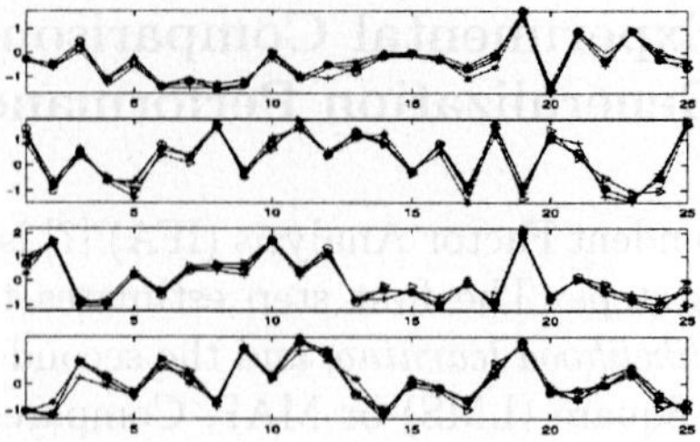

Fig. 2. "o" denotes source, "▷" NFA without DS, "+" NFA with DA, "*" IFA.

Table 1. Comparative experiments results of IFA and NFA

	MSE of training data				MSE of test data			
	signal 1	signal 2	signal 3	signal 4	signal 1	signal 2	signal 3	signal 4
NFA (with DS)	0.0588	0.0453	0.0564	0.0605	0.0757	0.0933	0.0736	0.0681
NFA (without DS)	0.0332	0.0487	0.0457	0.0333	0.0854	0.1126	0.0787	0.0974
IFA	0.0498	0.0417	0.0369	0.0533	0.0765	0.0976	0.1151	0.1013

5 Conclusions

In this paper, the *iterative FPA* is proved to be exactly an EM algorithm for the MAP factor estimate in NFA, and thus its convergence can always be guaranteed. Moreover, NFA is experimentally shown to have better generalization ability than IFA on small data sample.

References

1. Xu, L.: Byy harmony learning, independent state space and generalized apt financial analysis. IEEE Transaction on Neural Network 12 (2001) 822–849
2. Xu, L.: Bayesian kullback ying-yang dependence reduction theory. Neurocomputing 19 (1998) 223–257
3. Liu, Z.Y., Xu, L.: A comparative study on three map factor estimate approaches in non-gaussian factor analysis model. International Conference on Intelligent Data Engineering and Automated Learning (IDEAL'02) (Manchester, UK, 2002)
4. Wu, C.: On the convergence properties of the em algorithm. The Annals of Statistics 11 (1983) 95–103
5. Ma, J., Xu, L., Jordan, M.: Asymptotic convergence rate of the em algorithm for gaussian mixtures. Neural Computation 12 (2000) 2881–2907
6. Xu, L., Jordan, M.: On convergence properties of the em algorithm for gaussian mixture. Neural Computation 8 (1996) 129–151
7. Dempster, A., Laird, N., Rubin, D.: Maximum-likelihood from incomplete data via the em algorithm. J. Royal Statist. Soc. Ser. B. 39 (1977) 1–38
8. Attias, H.: Independent factor analysis. Neural Computation 11 (1999) 803–851

Error Functions for Prediction of Episodes of Poor Air Quality

Robert J. Foxall[1], Gavin C. Cawley[1], Stephen R. Dorling[2], and
Danilo P. Mandic[3]

[1] School of Information Systems
University of East Anglia
Norwich NR4 7TJ, U.K.
[2] School of Environmental Sciences
University of East Anglia
Norwich NR4 7TJ, U.K.
{rjf,gcc}@sys.uea.ac.uk
[3] Dept Electrical and Electronic Engineering
Imperial College of Science,
Technology and Medicine
London SW7 2BT, U.K.
d.mandic@ic.ac.uk

Abstract. Prediction of episodes of poor air quality using artificial neural networks is investigated. Logistic regression, conventional sum-of-squares regression and heteroscedastic sum-of-squares regression are employed for the task of predicting real-life episodes of poor air quality in urban Belfast due to SO_2. In each case, a Bayesian regularisation scheme is used to prevent over-fitting of the training data and to provide pruning of redundant model parameters. Non-linear models assuming a heteroscedastic Gaussian noise process are shown to provide the best predictors of pollutant concentration of the methods investigated.

1 Introduction

Belfast is unusual within the U.K. in that a significant fraction of the city's domestic heating is derived from coal burning, resulting from limited availability of natural gas. This leads to relatively low-level SO_2 emission, the efficient dispersion of which is highly dependent on meteorological conditions. Episodes of high ground-level SO_2 concentrations caused by emissions from tall stacks are mostly short-lived, however the longevity of SO_2 episodes caused by low-level emission may be more extended; meteorological conditions which are unconducive to efficient dispersion may persist for a period of hours to days. Given the health implications of exposure to high concentrations of SO_2, it is important to develop accurate forecast models for both the occurrence and severity of episodes of poor air quality. An ideal model should therefore produce and accurate forecast of the expected concentration of a given pollutant *and* some

J.R. Dorronsoro (Ed.): ICANN 2002, LNCS 2415, pp. 1031–1036, 2002.

means of estimating the probability that the observed concentration will exceed a preset statutory threshold level. In this paper, we compare three error functions for training multi-layer perceptron neural networks models of atmospheric pollution that attempt to address these requirements.

2   Neural Models of Air Pollution Time-Series

The parameters of a neural network model, w, are normally determined by some form of gradient descent optimisation of an appropriate error function, $E_{\mathcal{D}}$, over a set of labelled training examples,

$$\mathcal{D} = \{(\boldsymbol{x}_i, t_i)\}_{i=1}^{\ell}, \quad \boldsymbol{x}_i \in \mathcal{X} \subset \mathbb{R}^d, \quad t_i \in \mathcal{T} \subset \mathbb{R},$$

where $t_i \in (0, 1)$ for prediction of exceedences of statutory thresholds and $t_i \in \mathbb{R}^+$ for prediction of pollutant concentrations based on a vector of meteorological and other input variables $\boldsymbol{x}_i$. It has often been observed that simple maximum-likelihood estimates for the parameters of complex models often lead to severe over-fitting of the training data. In order to overcome this problem, we use instead a regularised error function, adding a term $E_{\mathcal{W}}$ penalising overly-complex models,

$$M = \alpha E_{\mathcal{W}} + \beta E_{\mathcal{D}},$$

where α and β are regularisation constants controlling the bias-variance trade-off. In this study we adopt the Bayesian regularisation scheme due to Williams [1], using a Laplace prior, i.e.

$$E_{\mathcal{W}} = \sum_{i=1}^{N} |w_i|,$$

in which the regularisation parameters α and β are integrated out analytically in the style of Buntine and Weigend [2]. An added advantage of the Laplace prior, rather than the usual Gaussian weight decay, is that redundant weights are set exactly to zero and can be pruned from the network. In the remainder of this section, we consider three data misfit terms, $E_{\mathcal{D}}$, for use in predicting episodes of poor air quality.

2.1   Logistic Regression

Logistic regression provides perhaps the most straight forward approach to predicting exceedences of statutory threshold concentrations. Assuming the target patterns, t_i, are an independent and identically distributed (i.i.d) sample drawn from a Bernoulli distribution ($t_i = 1$ indicates an exceedence, $t_i = 0$ indicates no exceedence), conditioned on the corresponding input vectors, $\boldsymbol{x}_i$, minimisation of the familiar cross-entropy error metric given by

$$E_{\mathcal{D}} = - \sum_{i=1}^{\ell} \{t_i \log y_i + (1 - t_i) \log(1 - y_i)\} \tag{1}$$

corresponds to maximisation of the likelihood of the data $\mathcal{D}$. The output layer activation function is taken to be the logistic function, $g(a) = 1/(1 + \exp\{-a\})$, restricting the output of the model to lie in the range (0, 1). Under these conditions the output of the model is a penalised maximum likelihood estimate of the Bayesian *a-posteriori* probability of an exceedence. Unfortunately, this error metric cannot be used to obtain a direct forecast of the concentration of a given pollutant, but only an indication of the likelihood this this concentration exceeds a fixed threshold.

2.2   Conventional Sum-of-Squares Regression

The sum-of-squares metric, $E_{\mathcal{D}} = \sum_{i=1}^{\ell}(t_i - y_i)^2$, with a linear output layer activation function corresponds to penalised maximum likelihood estimation of the conditional mean of the target values, assuming a Gaussian noise process with constant variance. This model can be simply extended to give the probability of an exceedence: The maximum likelihood estimate for the variance of the (Gaussian) target distribution is given by

$$\sigma^2 = \frac{1}{\ell} \sum_{i=1}^{\ell} (t_i - y_i)^2.$$

The probability that the observed concentration, c, exceeds a given threshold level, C, is then given by integrating the upper tail of the Gaussian probability density function, i.e.

$$p(c > C \mid \boldsymbol{x}) = \int_C^{\infty} \frac{1}{\sqrt{2\pi\sigma^2}} \exp\left\{ \frac{[z - y(\boldsymbol{x})]^2}{2\sigma^2} \right\} dz.$$

2.3   Heteroscedastic Sum-of-Squares Regression

A heteroscedastic regression model relaxes the assumption that the variance of the noise process is constant, and so attempts to estimate both the conditional mean and variance of the target distribution (e.g. Nix and Weigend [3], Williams [4]). For a Gaussian noise process, the network then has two output units, y^{μ}, estimating the conditional mean of the target distribution and y^{σ} estimating the conditional standard deviation. The negative logarithm of the likelihood is then given by

$$E_{\mathcal{D}} = -p(\mathcal{D} \mid \boldsymbol{w}) = \frac{1}{2} \sum_{i=1}^{N} \left\{ \frac{[t_i - y_i^{\mu}]^2}{(y_i^{\sigma})^2} + \log(y_i^{\sigma})^2 + \log 2\pi \right\},$$

which can be used to form a penalised maximum likelihood error metric as before. Again, the probability of an exceedence is given by the integral of the probability density function of the noise process above the exceedence threshold, replacing the constant variance σ by an input-dependent variance $\sigma(\boldsymbol{x})$.

3   Results

The error metrics given in the previous section are applied to the task of prediction of episodes of poor air quality in urban Belfast due to SO_2. The target data provide hourly measurements of SO_2 taken from a Belfast monitoring station over the years 1993-1996. For any prediction of SO_2 concentration to be of practical use it must be made at least a day in advance, hence the explanatory variables include an autoregressive component beginning no later than $(T - 24)$ hours, where T denotes the time in hours at which a prediction is required. Other input variables include meteorological variables such as temperature, wind speed/direction and visibility, and day of the week, Julian day, and hour of the day. Since a Gaussian distribution is inconsistent with the observed data (being strictly positive), the final target values are the logarithm of the observed SO_2 concentrations. Hence the mean prediction models of non-linear heteroscedastic Gaussian (NLG), non-linear sum-of-squares (NLS) and linear sum-of-squares (LS) are fitting maximum likelihood estimates for a log-normal distribution, while the the non-linear logistic (NLG) and linear logistic (LL) are fitting maximum likelihood estimates for a Bernoulli distribution. An exceedence is said to have occurred if the hourly mean SO_2 concentration is greater than $350 \mu g m^{-3}$. In each case, the performance statistics are computed using a four-fold cross-validation procedure, where the disjoint test partitions used in each trial are defined by the year in which the observations were made.

3.1   ROC Analysis

The *Receiver operating characteristic* (ROC) of a classifier graphically displays the trade-off between false negative $(1-$ true positive) and false positive rates obtained by varying some parameter of the model. In this case the parameter varied is the threshold probability above which an exceedence is predicted. The area under the ROC curve gives an indication of the effectiveness of a classifier, assuming that nothing is known about the optimal ratio of misclassification costs, unity being optimal. Table 1 gives the area under the ROC curve and rankings for each model. Note the linear sum-of-squares and linear logistic regression models are both poorly calibrated in that both consistently underestimate the probability of an exceedence, however this shortcoming is not revealed by the ROC diagram.

3.2   Log-Likelihood Analysis

Table 2 shows the log-likelihood computed over cross-validation test partitions for the models considered. The likelihood for the task of predicting the occurrence of an exceedence, given in the second column, are calculated using the the cross-entropy metric to evaluate the accuracy of estimates of probability of an exceedence. The likelihoods for the prediction of the concentration of SO_2 are computed using the error metrics given in the previous section.

As expected, the NLL and LL models out perform the other models in prediction of exceedences, being free of distributional assumptions regarding the

Table 1. Area under the ROC curve for models considered.

Model	Area under ROC	Rank
NLG	0.9405	5
NLS	0.9511	4
NLL	0.9605	2
LS	0.9576	3
LL	0.9625	1

Table 2. Log-likelihoods for considered models.

Model	log-likelihood (occurrence)	Rank	log-likelihood (prediction)	Rank
NLG	-1016.2 (-951.9)	5 (4)	-213607 (-33274)	3 (1)
NLS	-992.0	4 (5)	-34597	1 (2)
NLL	-809.1	1 (1)	*	*
LS	-916.0	3 (3)	-36316	2 (3)
LL	-841.7	2 (2)	*	*

noise process contaminating observations of SO_2 concentrations. Another interesting feature is the relatively poor performance of the NLG model compared to the less flexible NLS and LS models, however if the heteroscedastic variance structure is ignored and the predicted mean values used along with usual sum-of-squares estimate for the homoscedastic variance, the NLG model provides a significantly improved log-likelihood (shown in parentheses in Table 2). This is likely to be due to the observation that maximum-likelihood estimates of variance are biased since over-fitting in the model of the conditional mean reduces the apparent variance of the noise process.

3.3 McNemar's Test

Given two classifiers A and B, which classify each data point either correctly or incorrectly, McNemar's test [5] decides whether the the number of occasions that A is correct and B is incorrect is essentially the same as the number of occasions on which A is incorrect and B is correct. Table 3 gives the probabilities of the paired classifiers being essentially the same for each of the possible pairings. The lower triangle of the table gives the better classifier by this system for each pair. A (conservative) Bonferroni adjusted significance level of 0.005 is used to ensure a final significance level of 0.05 over all tests, and so there is no evidence that any of the models predict exceedences more accurately than any other.

4   Summary

In this paper, non-linear logistic regression models have demonstrated the best performance for the task of predicting episodes of poor air quality in Belfast due

Table 3. McNemar's test for considered models.

	non-linear Gaussian (NLG)	non-linear sse (NLS)	non-linear logistic (NLL)	linear sse (LS)	linear logistic (LL)
NLG	1	0.0535	0.0614	0.8750	0.2190
NLS	NLG	1	1.0000	0.1010	0.6450
NLL	NLG	NLS	1	0.0966	0.4890
LS	NLG	LS	LS	1	0.2120
LL	NLG	LL	LL	LS	1

to SO_2, although the differences in performance between classifiers are not statistically significant. None of the methods investigated however, provide a reliable predictor for exceedances of the statutory threshold concentration, assuming that the costs of false-positive and false-negative errors are equal. The non-linear heteroscedastic regression model provides the best estimate of the conditional mean of the concentration of SO_2. A further advantage of these models is that they can be doubly calibrated; not only is it possible to determine the probability that a pollutant exceeds a fixed threshold, accommodating changes in the costs of false-positive and false-negative errors, but also these models can still be used following a change in threshold level, due perhaps to the introduction of more stringent legislation.

Acknowledgements. This work was supported by the European Commission, grant number IST-99-11764, as part of its Framework V IST programme.

References

[1] Peter M. Williams. Bayesian regularisation and pruning using a Laplace Prior. *Neural Computation*, 7(1):117–143, 1995.

[2] Wray L. Buntine and Andreas S. Weigend. Bayesian back-propagation. *Complex Systems*, 5:603–643, 1991.

[3] D. A. Nix and A. S. Weigand. Learning local error bars for nonlinear regression. In G. Tesauro, D. Touretzky, and T. Leen, editors, *Advances in Neural Information Processing*, volume 7, pages 489–496. MIT Press, 1995.

[4] Peter M. Williams. Using neural networks to model conditional multivariate densities. *Neural Computation*, 8:843–854, 1996.

[5] I. McNemar. Note on the sampling error of the difference between correlated proportions or percentages. *Psychometrika*, 12:153–157, 1947.

[6] D. P. Mandic and J.A. Chambers. *Recurrent neural networks for prediction – learning algorithms, architectures and stability.* Wiley series on adaptive and learning Systems for Signal processing, communications and control. John Wiley & Sons Ltd., Chichester, 2001.

Adaptive Importance Sampling Technique for Neural Detector Training

José L. Sanz-González and Francisco Álvarez-Vaquero

Universidad Politécnica de Madrid (Dpto. SSR), ETSI de Telecomunicación-UPM,
Ciudad Universitaria, 28040 Madrid, Spain
{jlsanz, fav}@gc.ssr.upm.es

Abstract. In this paper, we develop the use of an adaptive Importance Sampling (IS) technique in neural network training, for applications to detection in communication systems. Some topics are reconsidered, such as modifications of the error probability objective function (P_e), optimal and suboptimal IS probability density functions (biasing density functions), and adaptive importance sampling. A genetic algorithm was used for the neural network training, having utilized an adaptive IS technique for improving P_e estimations in each iteration of the training. Also, some simulation results of the training process are included in this paper.

1 Introduction and Preliminaries

Importance Sampling is a modified Monte Carlo technique commonly applied to the performance analysis of radar and communication detectors [1-5]. In communications detectors, the error probability (P_e) is estimated by Importance Sampling (IS) techniques for very low P_e (e.g. $P_e < 10^{-5}$). In [6], we have proposed the use of IS techniques in neural detector training for applications in communications.

Now, in this paper, we propose an adaptive importance sampling technique in order to improve P_e-estimations in each iteration of the training, by finding a suboptimal probability density function for sampling.

Throughout the paper, we shall refer to Fig. 1, where $\mathbf{x} = (x_1, x_2, ..., x_n)$ is the input vector of the R^n-space, $y=g(\mathbf{x})$ is the scalar output, $g(\cdot)$ is a nonlinear system (e. g. a neural network), T_0 is the detection threshold, and $z=u(g(\mathbf{x})-T_0)$ is the detector output, where $u(\cdot)$ is the unit-step function (i.e. $u(t)=1$ if $t>0$ and $u(t)=0$ if $t<0$). We denote $\mathbf{X}=(X_1, X_2, ..., X_n)$ as a random vector and $f_{\mathbf{X}}(\mathbf{x} \mid H_i)$ as the probability density function (pdf) of $\mathbf{X}$ under a hypothesis H_i, $i=1, 0$ (binary hypotheses), where H_0 is the null hypothesis or symbol „0" and H_1 is the alternative hypothesis or symbol „1". $P(H_i)$ is the „a priori" probability of the hypothesis H_i, $i= 1, 0$, and $P(D_j|H_i)$ is the conditional probability of deciding H_j, $j=1, 0$, under the true hypothesis H_i, $i=1, 0$. If $g(\mathbf{x}) > T_0$ (or $z=1$), the decision is H_1; if $g(\mathbf{x}) < T_0$ (or $z=0$), the decision is H_0. Finally, $\mathrm{E}\{Z \mid H_i\}$ is the expectation of the random variable Z conditioned by H_i, $i=1, 0$, and $\mathscr{E}\{g(\mathbf{X})\}$ is the expectation of $g(\bullet)$ with respect to the pdf of $\mathbf{X}$ (i.e. $f_{\mathbf{X}}(\mathbf{x})$).

J.R. Dorronsoro (Ed.): ICANN 2002, LNCS 2415, pp. 1037–1042, 2002.

2   Error-Probability Estimations for Training

Following [6], consider the error probability as the objective function for detection in communications [7] (or the misclassification probability for applications in classifications). According to the notation given above, the error probability (P_e) can be expressed as follows

$$P_e = P(H_0)P(D_1 \mid H_0) + P(H_1)P(D_0 \mid H_1) \tag{1}$$

Also, in order to save space, let us define [6]

$$h(\mathbf{x}) = P(H_0)f_{\mathbf{X}}(\mathbf{x} \mid H_0)u(g(\mathbf{x}) - T_0) + P(H_1)f_{\mathbf{X}}(\mathbf{x} \mid H_1)u(T_0 - g(\mathbf{x})), \quad \mathbf{x} \in R^n \tag{2}$$

If we consider a new probability density function (pdf) $f_{\mathbf{X}}^*(\mathbf{x})$, such that $f_{\mathbf{X}}^*(\mathbf{x}) \neq 0$ wherever $h(\mathbf{x}) \neq 0$, $\mathbf{x} \in R^n$, then from (1) and (2) we have

$$P_e = \int_{R^n} h(\mathbf{x})\,d\mathbf{x} = \int_{R^n} \frac{h(\mathbf{x})}{f_{\mathbf{X}}^*(\mathbf{x})}f_{\mathbf{X}}^*(\mathbf{x})\,d\mathbf{x} = \mathrm{E}^*\left\{\frac{h(\mathbf{X})}{f_{\mathbf{X}}^*(\mathbf{X})}\right\} \tag{3}$$

where $\mathcal{E}^*\{\cdot\}$ means expectation with respect to $f_{\mathbf{X}}^*(\mathbf{x})$ (the Importance Sampling pdf).

The last equality in (3) is the key of the Importance Sampling technique. From the statistical inference theory applied to (3), an estimator of P_e is given by

$$\hat{P}_e = \frac{1}{N}\sum_{k=1}^{N}\frac{h(\mathbf{x}_k^*)}{f_{\mathbf{X}}^*(\mathbf{x}_k^*)} \tag{4}$$

where $\mathbf{x}_k^*$, $k=1, 2, ..., N$, are independent sample vectors whose pdf is $f_{\mathbf{X}}^*(\mathbf{x})$. Note that $\hat{P}_e$ is an unbiased and consistent estimator of P_e [6] if $f_{\mathbf{X}}^*(\mathbf{x}) \neq 0$ wherever $h(\mathbf{x}) \neq 0$, $\mathbf{x} \in R^n$.

Estimator $\hat{P}_e$, given in (4), must be computed in order to perform the neural network training (i.e. to find g(•) for minimum $\hat{P}_e$).

The unconstrained optimal solution ($\hat{P}_e$-variance is zero) for $f_{\mathbf{X}}^*(\mathbf{x})$ is given by [6]

$$f_{\mathbf{X}}^*(\mathbf{x}) = \frac{1}{P_e}[P(H_0)f_{\mathbf{X}}(\mathbf{x} \mid H_0)u(g(\mathbf{x}) - T_0) + P(H_1)f_{\mathbf{X}}(\mathbf{x} \mid H_1)u(T_0 - g(\mathbf{x}))], \mathbf{x} \in R^n \tag{5}$$

The optimal solution for $f_{\mathbf{X}}^*(\mathbf{x})$ given in (5) is not realistic, because P_e is not known "a priori" (it has to be estimated by (4)). Furthermore, in the training phase, $g(\cdot)$ is changing from one iteration to the other.

A suboptimal solution for $f_{\mathbf{X}}^*(\mathbf{x})$ was proposed in [6] for the uniparametric case. Commonly $f_{\mathbf{X}}(\mathbf{x} \mid H_i)$, $i = 1, 0$ depends on a parameter vector θ that can be related to the signal-to-noise ratio in communication applications [7]. Then we can write $f_{\mathbf{X}}(\mathbf{x}; \theta \mid H_i)$, $i = 1, 0$, where $\theta = (\theta_1, \theta_2, ..., \theta_m) \in \Theta \subset R^m$. Taking into account (5), a suboptimal multiparametric $f_{\mathbf{X}}^*(\mathbf{x})$ is as follows

$$f_{\mathbf{X}}^*(\mathbf{x}) = P(H_0)f_{\mathbf{X}}(\mathbf{x}; \theta^* \mid H_0) + P(H_1)f_{\mathbf{X}}(\mathbf{x}; \theta^* \mid H_1), \quad \mathbf{x} \in R^n \tag{6}$$

where θ^* is the θ-value that minimizes the variance of the estimator $\hat{P}_e$ (supposed unbiased) or the relative error of $\hat{P}_e$. An estimator ($\hat{\varepsilon}_{\hat{P}_e}$) of the relative error is

$$\hat{\varepsilon}_{\hat{P}_e} = \frac{\hat{\sigma}_{\hat{P}_e}}{\hat{\mu}_{\hat{P}_e}} = \sqrt{\frac{1}{N}\left[\frac{\frac{1}{N}\sum_{k=1}^{N}\left(\frac{h(\mathbf{x}_k^*)}{f_{\mathbf{X}}^*(\mathbf{x}_k^*)}\right)^2}{\left(\hat{P}_e\right)^2} - 1\right]} \tag{7}$$

where $h(\mathbf{x})$, $f_{\mathbf{X}}^*(\mathbf{x})$ and $\hat{P}_e$ are in (2), (6) and (4), respectively.

Expressions (4) and (7) are computed at each iteration of the training for a given IS-parameter vector θ (fixed or adaptive during the training), N (fixed or adjustable in each iteration), and $g(\cdot)$ which changes from one training iteration to the other. At the end of the training $\hat{P}_e$ and $\hat{\varepsilon}_{\hat{P}_e}$ are good estimations of P_e and $\varepsilon_{\hat{P}_e}$, respectively.

We have to point out that because it is not guaranteed in (6) that $f_{\mathbf{X}}^*(\mathbf{x}) \neq 0$, wherever $h(\mathbf{x}) \neq 0$, $\mathbf{x} \in R^n$, we shall have $E^*\{\hat{P}_e\} \leq P_e$, i.e. $\hat{P}_e$ is an underestimation of P_e. Note that $\hat{P}_e$ is also an underestimator of P_e if $f_{\mathbf{X}}^*(\mathbf{x}) \approx 0$ and $f_{\mathbf{X}}^*(\mathbf{x}) \square h(\mathbf{x})$ for some $\mathbf{x} \in R^n$, i.e. $\hat{P}_e < P_e$ with high probability. Therefore, we have to assure that the solution θ^* produces minimum bias in $\hat{P}_e$. The parameter vector $(\theta^*)_{opt}$ that minimizes the bias and the variance of the estimator $\hat{P}_e$ can be obtained from

$$(\theta^*)_{opt} = \arg\min_{\theta \in \Theta}\left\{\ln(\hat{\varepsilon}_{\hat{P}_e}) - \alpha \cdot \ln(\hat{P}_e)\right\} \tag{8}$$

where the first and the second terms of (8) account for the variance and the bias, respectively, of the estimator $\hat{P}_e$, and α is a positive real number that balance the importance of each term (usually $\alpha=1$). The minimization of (8) have to be performed by an adequate optimization algorithm like stochastic gradient descent algorithms [3,4] or, alternatively, by a genetic algorithm as we shall show after.

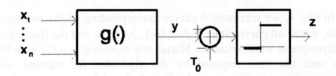

Fig. 1. The binary detector structure, where $g(\cdot)$ is a neural network.

3   Computer Simulation Results

Referring to Fig. 1, a Multi-Layer Perceptron (MLP) is the Neural Network (NN) used as nonlinear system $g(\mathbf{x})$. The parameters for the MLP are 5×5×1 (i.e. number of input nodes: $n=5$, number of hidden layer nodes: 5, number of outputs: 1) and the threshold $T_0=0.5$. A genetic algorithm is used for training the MLP.

Although, our genetic algorithm (with elitism and real-number genes in the chromosomes [6]) is not the subject of this paper (in fact, here it is considered only as a tool), we supply the parameters used in our training. These are the number of MLP's in genetic set: 20, the mutation probability: 0.1, the crossover probability: 0.1, the fitness function: $-\log(\hat{P}_e)$, and the number of iterations less than 200. We use $N=1000$ input sample vectors (patterns) in order to have enough precision in the estimations.

In order to show how Importance Sampling works in neural detector training, let us consider a detection of binary symbols in Gaussian noise. The hypotheses are $H_1 : \mathbf{x} = \mathbf{\eta} + \mathbf{a}$ and $H_0 : \mathbf{x} = \mathbf{\eta} - \mathbf{a}$, where $\mathbf{a} = (a_1, a_2, \ ... \ , a_n)$, $a_i = \mu$, $i = 1, 2, ..., n$ and μ is a real constant (for simulations $\mu=2$), $\mathbf{\eta} = (\eta_1, \eta_2, ..., \eta_n)$ is a Gaussian noise vector of independent and identically distributed zero-mean samples of unit variance. Also, we suppose for simulations that $P(H_1) = P(H_0) = 1/2$ (the symbols are equally likely). We consider the adaptive importance sampling with a parameter vector $\mathbf{\theta}^*$ in $f_{\mathbf{x}}^*(\mathbf{x})$, so the IS-pdf family is given by

$$f_{\mathbf{x}}^*(\mathbf{x}) = \frac{1}{2} \cdot (2\pi)^{-n/2} \cdot \left[\exp\left(-\frac{1}{2}\sum_{i=1}^{n}(x_i - \theta_i^*)^2 \right) + \exp\left(-\frac{1}{2}\sum_{i=1}^{n}(x_i + \theta_i^*)^2 \right) \right] \tag{9}$$

where $\mathbf{\theta}^* = (\theta_1^*, \theta_2^*, ..., \theta_n^*) \in R^n$ is the parameter vector, and $\mathbf{\theta}^* = \mathbf{a} = (a_1, a_2, \ ... \ , a_n)$ is the optimum for the first iterations of the training and $\mathbf{\theta}^*=0$ (zero vector) is the optimum at the last iterations. The optimization algorithm has to compute (8) for finding the optimum parameter vector $\mathbf{\theta}^*$ in each iteration of the training. First, the IS algorithm starts with $\mathbf{\theta}^*=\mathbf{\theta}_0$ and computes (4) and (7) for the initial set of the NNs proposed by the genetic algorithm of training. Then, the optimization algorithm (realized also by a genetic algorithm without crossover operator) finds the best $\mathbf{\theta}^*$ from (8) to apply in the next iteration of the training, and so on until the last iteration.

In Fig.2, we show the corresponding results of the error probability estimations ($\hat{P}_e$) versus iteration number with parameter $\mathbf{\theta}^*$ adaptively modified; in Fig. 3, we present the relative error $\hat{\varepsilon}_{\hat{P}_e}$, computed from (7) at the same time that $\hat{P}_e$ is computed from (4). In Fig. 4, we represent 5 curves corresponding to the evolution of the θ^*-components, where all curves start at $\theta_i=2$ ($i=1, 2, ..., 5$), and the final values are close to zero. All programs were written in Matlab and executed in a PC with Pentium III at 800 MHz, and the time required by the algorithm for training with adaptive importance sampling was 35 minutes for Figs. 2, 3 and 4.

Similar results can be obtained for other MLP structures (e.g. 8×8×1 or 8×4×1, 10×5×1, etc.), trained under the conditions given above. For example, in the case 8×8×1, we have $\hat{P}_e \approx 9 \cdot 10^{-9}$; in the case 8×4×1, we have and $\hat{P}_e \approx 10^{-8}$; then both are quasi-optimum, because $(P_e)_{\text{Bayes}} = 7.7 \cdot 10^{-9}$ (the optimal Bayes detector [7]).

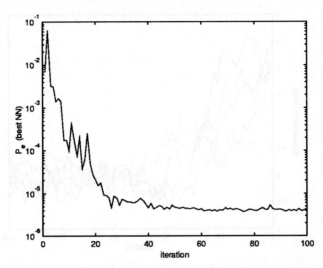

Fig. 2. Error Probability ($\hat{P}_e$) versus iteration number for the neural network training. Adaptive Importance Sampling with 1000 input sample vectors (patterns) to estimate $\hat{P}_e$, and a parameter vector $\theta^* = (\theta_1^*, \theta_2^*, \dots, \theta_5^*)$, $\theta^*_{initial} = (2,2,2,2,2)$, $\theta^*_{final} \approx 0$. An MLP of 5×5×1 nodes, and a genetic algorithm for training. (See also Figs. 3 and 4).

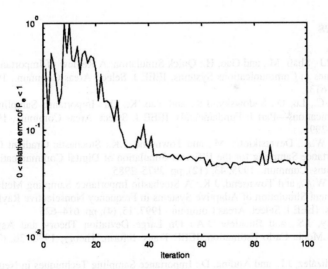

Fig. 3. Relative Error of $\hat{P}_e$ ($\hat{\varepsilon}_{\hat{P}_e}$) versus iteration number for the neural network training, under the same conditions of Fig. 2 and Fig. 4.

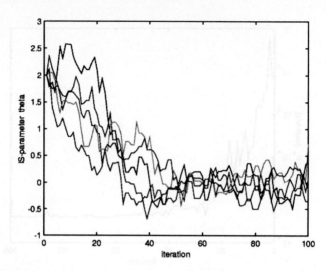

Fig. 4. IS-parameter vector theta (θ^*) versus iteration number for the neural network training, under the same conditions of Fig. 2 and Fig. 3. The parameter vector θ^* is adaptively (optimally) modified as the training progresses. Each curve corresponds to θ_i^*, $i=1, 2, \ldots, 5$, and $\theta^*=(\theta_1^*, \theta_2^*, \ldots, \theta_5^*)$.

References

1. Smith, P.J., Shafi, M., and Gao, H.: Quick Simulation: A Review of Importance Sampling Techniques in Communications Systems. IEEE J. Select. Areas Commun., 1997, 15, (4), pp. 597–613.
2. Chen, J.-C., Lu, D., Sadowsky, J.S., and Yao, K.: On Importance Sampling in Digital Communications—Part I: Fundamentals. IEEE J. Select. Areas Commun., 1993, 11, (3), pp. 289–299.
3. Al-Qaq, W.A., Devetsikiotis, M., and Townsend, J.K.: Stochastic Gradient Optimization of Importance Sampling for the Efficient Simulation of Digital Communication Systems. IEEE Trans. Commun., 1995, 43, (12), pp. 2975–2985.
4. Al-Qaq, W.A., and Townsend, J.K.: A Stochastic Importance Sampling Methodology for the Efficient Simulation of Adaptive Systems in Frequency Nonselective Rayleigh Fading Channels. IEEE J. Select. Areas Commun., 1997, 15, (4), pp. 614–625.
5. Sadowsky, J.S., and Bucklew, J.A.: On Large Deviation Theory and Asymptotically Efficient Monte Carlo Estimation. IEEE Trans. Inform. Theory, 1990, 36, (3), pp. 579–588.
6. Sanz-González, J.L. and Andina, D.: Importance Sampling Techniques in Neural Detector Training. Lectures Notes in Artificial Intelligence, LNAI 2167 (Machine Learning: ECML'01), pp. 431–441, Springer-Verlag, 2001.
7. Poor, H.V.: An Introduction to Signal Detection and Estimation, (Second Edition) Springer-Verlag, Berlin (1994).

State Space Neural Networks for Freeway Travel Time Prediction

Hans van Lint, Serge P. Hoogendoorn, and Henk J. van Zuylen

Delft University of Technology
Faculty of Civil Engineering and Geosciences
Traffic Engineering and Transportation Planning Department
Stevinweg 1, P.O. Box 5048, 2600 GA Delft, The Netherlands
H.vanLint@citg.tudelft.nl

Abstract. The highly non-linear characteristics of the freeway travel time prediction problem require a modeling approach that is capable of dealing with complex non-linear spatio-temporal relationships between the observable traffic quantities. Based on a state-space formulation of the travel time prediction problem, we derived a recurrent state-space neural network (SSNN) topology. The SSNN model is capable of accurately predicting experienced travel times - outperforming current practice by far - producing approximately zero mean normally distributed residuals, generally not outside a range of 10% of the real expected travel times. Furthermore, analyses of the internal states and the weight configurations revealed that the SSNN developed an internal models closely related to the underlying traffic processes. This allowed us to rationally eliminate the insignificant parameters, resulting in a Reduced SSNN topology, with just 63 adjustable weights, yielding a 72% reduction in model-size, without loss of predictive performance.

1 Introduction

One of the most complex and challenging problems in engineering is predicting future states of non-linear spatio-temporal processes, such as tidal flows in estuaries, deformation of steel under dynamic loads, or traffic flow propagation on large road networks. Formulating mathematical models for these problems requires either sound theory or advanced data driven models. For some problems, however, the available theory might not be sound or comprehensive enough, while data-driven approaches might lead to models that have too many adjustable parameters or require too many a-priori design choices, such as input- and model-selection. In this article we will investigate a typical example of such a problem within the field of traffic engineering: travel time prediction on road networks.

2 Travel Time Prediction on Road Networks

Freeway travel time prediction for Advanced Traffic Information Systems (ATIS) is a typical example of a complex spatio-temporal modeling problem. ATIS aim to provide road users (cars, trucks) with up to date traffic information (such as expected travel time), by means of pre-trip (websites, telephone services) or en-route (variable

J.R. Dorronsoro (Ed.): ICANN 2002, LNCS 2415, pp. 1043–1048, 2002.
© Springer-Verlag Berlin Heidelberg 2002

message signs, in-car devices) information services. The assumption is that *informed* drivers are able to make more efficient travel decisions (e.g. route choice and departure time choice), yielding improved usage of the capacity of the road network.

2.1   Basic Notions on Freeway Travel Time Prediction

Predicting the travel time on a route requires knowledge on the future traffic conditions on that particular route. To predict the expected travel time of vehicles starting a particular route R at departure time t_0, we need to know whether or not they will encounter delays (congestion) during the period $[t_0, t_0+T]$ along their route. The problem, however, is that it is exactly *that* quantity, (T = time spent on the route) that we want to predict.

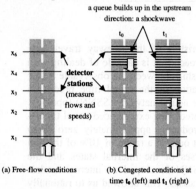

a queue builds up in the upstream direction: a shockwave

(a) Free-flow conditions   (b) Congested conditions at time t_0 (left) and t_1 (right)

Fig. 1. Free-flow versus Congested traffic conditions. Travel time prediction requires different information from different locations in congested conditions than in free-flow conditions.

From traffic theory (see e.g. [1] for an overview) we know that in free flow conditions, *information*[1] flows in the same direction as traffic does. In congested traffic conditions however, information may also flow in the opposite direction. So-called shockwaves occur (Fig. 1), as traffic from upstream is forced to slow down due to slower traffic downstream. If the differences in speed and speed-variations of the two "colliding" traffic regimes are large enough, the resulting shockwaves will move in the upstream direction, causing the queue to spill back, and vehicles to be delayed. This implicates that the spatial distribution of measurements we want to feed into a travel time prediction model also depends on the models' output.

The current travel time prediction procedure (NLTT) incorporated by the Dutch Highway authorities is a so-called instantaneous Network Level Travel Time estimator (NLTT):

$$TT_R(p) = \sum_{k \in R} TT_k(p) = \sum_{k \in R} \frac{L_k}{\overline{V}_k(p)} \tag{2.1}$$

where $TT_R(P)$ denotes the mean experienced travel time on route R for vehicles starting at time period p, $TT_k(p)$ denotes the mean experienced travel time on link k ∈ R during p, and L_k and $V_k(p)$ denote the length of link k, and the mean speed measured on link k during time period p respectively. Clearly, the lack of temporal dynamics in this model makes it unsuitable for travel time prediction, but since it is current practice we will compare our model against it in the next sections.

---

[1] *Information* here refers to the descriptive variables that we can actually measure (traffic flow, density and speeds), in this case at discrete locations (detector stations) along the route

2.2   Neural Network Approaches to Travel Time Prediction

A number of successful applications have been reported on feed-forward neural networks (FNN's) for travel time or traffic flow prediction (e.g. [3], [4]). In most that the FNN models generally outperform other tested models, such as historical profiles, ARIMA and dynamic regression models etc. The principal drawback of applying FNN's to traffic prediction, however, is that they are essentially static function approximators. Time series of speed and flow measurements are fed into the FNN model as if they were spatial patterns. The optimal look-back intervals and the spatial distribution of the detector stations used need to be determined a-priori by the FNN designer.

As is pointed out in [2], the optimal resolution and composition of the input time-series are correlated to the prediction horizon, which is a-priori unknown. Although the Recurrent and Time Delayed Neural Networks (TDNN's) in this paper clearly outperform static FNN's, predicting traffic volume at a single location for a fixed prediction horizon, is a different problem than predicting travel time on a route. To this end, we propose a different ANN topology, which *implicitly* takes into account the temporal and spatial dynamics of the travel time prediction problem.

2.3   A State-Space Approach to Travel Time Prediction: State-Space Neural Networks

Our efforts are based on a so-called discrete state-space model, which is in line with macroscopic traffic flow theory, where the link states (speeds, flows and densities) are a function of their previous states and inputs only. We generalize this concept by describing the state dynamics at route level.

$$x(t+1) = f(x(t), \{...,u_k(t),...\}) \tag{2.2}$$

$$y(t+1) = g(x(t+1), w) \tag{2.3}$$

As is depicted in eq (2.2), the state $x(t+1)$ of the links on time instant $t+1$, is uniquely defined by the state-vector $x(t)$ and the link-specific input-vectors $u_k(t)$ of the previous time period, which contain speeds and flows from measurement locations on the links, and in- and outflows from on- and off-ramps, connecting to the links. The generic function $y(t) = g(w,x(t))$, depicted in eq (2.3), calculates the experienced travel time for a vehicle starting the route at time instant $t+1$, and takes a vector $x(t+1)$ of all link states at time instant $t+1$ as inputs. It incorporates a parameter vector w, which is adjusted during calibration.

This general state-space model shows much resemblance with the Recurrent Neural Network (RNN) proposed in [5]. If we allow the input-layer to be *partially connected* to the hidden layer, and interpret the hidden layer as a state-vector at time $t+1$, the Elman network exactly represents the model of equations (2.2) and (2.3). We will refer to it from hereon as the State-Space Neural Network (SSNN). The neurons in both hidden and output-layer calculate their outputs as a weighted sum of their inputs and a bias, squashed by the well-known sigmoid transfer-function. Summing up all inputs (speeds, flows) implicates that we cannot assign any physical quantity to the hidden neuron outputs (the *internal states*). We can, however, interpret each hidden neuron's output $x_i(t)$ as a metric (scalar) *representative* for the traffic conditions (speeds and flows) on link i at time (>) t.

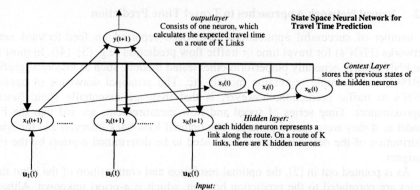

Fig. 2. The State-space representation of the travel time prediction problem naturally leads to a slightly modified version of the Elman network: the State-Space Neural Network (SSNN).

3   Results

3.1   Data

We have used synthetic data obtained from a traffic micro-simulation model (FOSIM – developed at the Delft University of Technology, see e.g. [7]) to train and test the performance of the SSNN travel time prediction model. We have set up a network in FOSIM for the A13 highway between The Hague and Rotterdam. It contains four on- and four off ramps, and two weaving sections. The route of length 7,3 km consists of 12 adjacent links, each equipped with two detectors, measuring 1 minute aggregate flows (veh/min) and one-minute averaged speeds (km/h).

Five (different) seven-hour simulation runs were used to compile all the minute-aggregate data. For each simulation run, the traffic demand patterns and the random seed generator were different, resulting in different but realistic travel time patterns. Two of the five runs (834 records) were used for testing, the rest (1251) for training. All data (both inputs and outputs) were linearly scaled to the interval [+0.1, +0.9], based on the rule-of-thumb that this leads to faster and more stable learning. The SSNN's were developed with the Neural Network Toolbox of Matlab 5, and were trained by means of Levenbergh-Marquardt optimization and Bayesian regularization (see [6] for details and references).

3.2   Results on Synthetic Data

Table 1 shows the performance of the SSNN model on test-set 2 on the following performance indicators.

MSE   (Mean Squared Error) $\dfrac{1}{P}\sum_{P}(\hat{T}_p - T_p)^2$

MAE   (Mean Absolute Error) $\dfrac{1}{P}\sum_p (\hat{T}_p - T_p)$

SAE   (Standard Deviation of Absolute Error) $\dfrac{1}{P-1}\sum_p \sqrt{(\hat{T}_p - MAE)^2}$

where $\hat{T}_p$ and T_p denote the predicted and experienced travel time in period p, subscript p denotes the time period, and capital P the total number of time periods. The SSNN model outperforms the current NLTT procedure by far, producing approximately zero mean normally distributed residuals, with a standard deviation of 17 seconds, which is quite accurate, given a free-flow travel time of 240 seconds (4 minutes) and a maximum travel time of 700 seconds (11 minutes and 20 seconds).

Table 1: Performance of State Space Neural Network vs Instantaneous Travel Time Predictor

	MSE (10^3 seconds^2)	MAE (seconds)	SAE (seconds)
SSNN	0.283	1.11	16.8
NLTT	3.60	-24.0	55.1

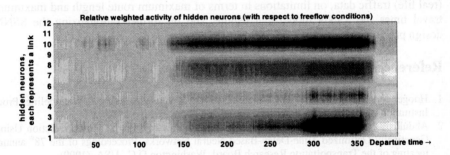

Fig. 3. The *relative internal states* (relative weighted hidden neuron activities) of the SSNN reveal the model has adapted a weight configuration closely related to the actual traffic processes (see text below for explanation).

The most interesting result however, we obtained by analyzing the *internal states* (i.e. the output of the hidden-layer neurons). Since every hidden neuron represents a link, the *weighted activity* of a hidden neuron can be interpreted as the contribution of that particular link to the output, i.e. the *expected travel time*. If we subtract the weighted activity in free-flow conditions we obtain *relative internal states*, which can be interpreted as the contribution of each hidden neuron (and its associated link) to the *expected delay* due to congestion. It is this quantity that is plotted in Fig 3, in which dark areas depict high *relative* neuron activity. The first observation is that only neurons representing links associated with traffic demand (links connected to on- and off-ramps) contribute to the expected delay: links 2,5,6,8 and 10. All other link-states only contribute (sum up to) to a constant value, which is in fact the maximum travel time the network has learned from the data. Since in all simulation runs congestion *initially* sets in at link 10, the SSNN "considers" this link as the sole bottleneck. Obviously, for the data from the simulation runs this approach suffices. To develop SSNN's with more generic congestion detection capabilities they need to be fed with a greater variety of simulation runs, in which congestion sets in at various locations and in various degrees of severity. Finally, we can now – rationally – eliminate those neurons that do not contribute to expected delay. Removing the "insignificant" from the SSNN results in a Reduced SSNN topology, with just 63 adjustable weights, yielding a 72% reduction in model-size, without loss of predictive performance!

4 Conclusions and Recommendations

The proposed state-space neural network (SSNN) topology proved capable of accurately predicting experienced freeway travel times, producing approximately zero mean normally distributed residuals, generally not outside a range of 10% of the real expected travel times. Moreover, analyses of the internal states and the weight configurations revealed that the SSNN developed an internal model closely related to the underlying traffic processes. This insight allowed us to rationally eliminate the insignificant parameters, resulting in a Reduced SSNN topology, with just 63 adjustable weights, yielding a 72% reduction in model-size, without loss of predictive performance. Future research should emphasize on training with a broader variety of (real life) traffic data, on limitations in terms of maximum route length and maximum travel times the SSNN model can handle, and on further rationalizing the SSNN design process.

References

1. Hoogendoorn, S.P., Bovy, P.H.L., State-of-the-art of vehicular traffic flow modeling, Proc. Institution of Mechanical Engineers, (2001), Vol 215 Part 1, pp 283–303
2. Abdulhai, B., Porwal, H., Recker, W., Short Term Freeway Traffic Flow Prediction Using Genetically-Optimized Time-Delay-Based Neural Networks, Proceedings of the 78th annual meeting of the Transportation Research Board, Washington D.C., USA, (1999)
3. Park, D., Rilett, L., Forecasting freeway link travel times with a multilayer feed-forward neural network, Computer-Aided-Civil-and-Infrastructure-Engineering. (1999) vol 14 n 5, p 357–367
4. Cheu, Ruey-Long, Freeway traffic prediction using neural networks, Proceedings of the International Conference on Applications of Advanced Technologies in Transportation Engineering, 1998, ASCE, Reston, VA, USA. p 247–254
5. Elman, J, *Finding structure in Time*, Cognitive Science 14, pp. 179–211 (1990)
6. Demuth Howard, Beale Mark, 1998, *Neural Network Toolbox for Use with Matlab*, The MathWorks Inc., USA
7. Vermijs, R.G.M.M., Schuurman, H., Evaluating capacity of freeway weaving sections and on-ramps using the microscopic simulation model FOSIM, Proceedings of the second international symposium on highway capacity (1994), Vol 2., pp 651–670

Overcomplete ICA with a Geometric Algorithm

Fabian J. Theis[1], Elmar W. Lang[1], Tobias Westenhuber[1], and Carlos G. Puntonet[2]

[1] Institute of Biophysics
University of Regensburg, D-93040 Regensburg, Germany
fabian.theis@mathematik.uni-regensburg.de
[2] Dept. of Architecture and Computer Technology
University of Granada, Spain

Abstract. We present an independent component analysis (ICA) algorithm based on geometric considerations [10] [11] to decompose a linear mixture of more sources than sensor signals. Bofill and Zibulevsky [2] recently proposed a two-step approach for the separation: first learn the mixing matrix, then recover the sources using a maximum-likelihood approach. We present an efficient method for the matrix-recovery step mimicking the standard geometric algorithm thus generalizing Bofill and Zibulevsky's method.

1 Introduction

In independent component analysis (ICA), given a random vector, the goal is to find its statistically independent components. This can be used to solve the blind source separation (BSS) problem which is, given only the mixtures of some underlying independent sources, to separate the mixed signals thus recovering the original sources. Recently, geometric ICA algorithms have received further attention due to their relative ease of implementation [10].

Standard ICA algorithms generally assume that at least as many sensor signals as there are underlying source signals are provided. In overcomplete ICA however, more sources are mixed to less signals. Olshausen and Fields first put the idea of 'overcomplete representations' from coding theory into an information theoretic context decomposing natural images into an overcomplete basis [8]. Later, Harpur and Prager [5] and, independently, Olshausen [9] presented a connection between sparse coding and ICA in the quadratic case.

2 Basics

For $m, n \in \mathbb{N}$ let $\mathrm{Mat}(m \times n)$ be the $\mathbb{R}$−vectorspace of real $m \times n$ matrices, and $\mathrm{Gl}(n) := \{W \in \mathrm{Mat}(n \times n) \mid \det(W) \neq 0\}$ be the general linear group of $\mathbb{R}^n$.

In the general case of linear blind source separation (BSS), a random vector $X : \Omega \to \mathbb{R}^m$ composed of **sensor signals** is given; it originates from an independent random vector $S : \Omega \to \mathbb{R}^n$, which is composed of **source signals**, by mixing with a **mixing matrix** $A = (a_1, ..., a_n)^T \in \mathrm{Mat}(m \times n)$, i.e. $X = AS$.

J.R. Dorronsoro (Ed.): ICANN 2002, LNCS 2415, pp. 1049–1054, 2002.

Here Ω denotes a fixed probability space. Only the sensor signals are known, and the task is to recover both the mixing matrix A and the source signals S. Let $a_i := Ae_i$ denote the columns of A, where e_i are the unit vectors. We will assume that the mixing matrix A has full rank and any two different columns a_i, a_j are linearly independent, $i \neq j$.

Note that in the **quadratic** case $(m = n)$ A is invertible ie. $A \in \mathrm{Gl}(n)$. Then we can recover S from A by $S = A^{-1}X$. For less sources than sensors $(m > n)$ the BSS problem is said to be **undercomplete**, and it is easily reduced to a quadratic BSS problem by selecting m sensors or applying some more sophisticated preprocessing like PCA. The separation problem we are mainly interested here is the **overcomplete** case where less sensors than sources are given $(m < n)$. The problem stated like this is ill-posed, hence further restrictions will have to be made.

3   A Two Step Approach to the Separation

In the quadratic case it is sufficient to recover the mixing matrix A in order to solve the separation problem, because the sources can be reconstructed from A and X by inverting A. For the overcomplete case as presented here, however, after finding A in a similar fashion as in quadratic ICA (**matrix-recovery step**), the sources will be chosen from the $n - m$-dimensional affine vector space of the solutions of $AS = X$ using a suitable boundary condition (**source-recovery step**). Hence with this algorithm we follow a two step approach to the separation of more sources than mixtures; this two-step approach has been proposed recently by Bofill and Zibulevsky [2] for delta distributions. It contrasts to the single step separation algorithm by Lewicki and Sejnowski [7], where both steps have been fused together into the minimization of a single complex energy function. We show that our approach resolves the convergence problem induced by the complicated energy function, and, moreover, it reflects the quadratic case as special case in a very obvious way.

4   Matrix-Recovery Step

In the first step, given only the mixtures X, the goal is to find a matrix $A' \in \mathrm{Mat}(m \times n)$ with full rank and pairwise linearly independent columns such that there exists an independent random vector S' with $X = A'S'$. If we assume that at most one component of S and S' is Gaussian, then for $m = n$ it is known [4] that A' is equivalent to A; here we consider two matrices $B, C \in \mathrm{Mat}(m \times n)$ **equivalent** if C can be written as $C = BPL$ with an invertible diagonal matrix (scaling matrix) $L \in \mathrm{Gl}(n)$ and an invertible matrix with unit vectors in each row (permutation matrix) $P \in \mathrm{Gl}(n)$. In the overcomplete case, however, no such uniqueness theorem is known, and we believe that without further restrictions it would not be true anyway.

For geometric matrix-recovery, we use a generalization of the geometric ICA algorithm [10]. Later on, we will restrict ourselves to the case of two-dimensional mixture spaces for illustrative purposes mainly. With high dimensional problems, however, geometrical algorithms need very many samples [6], hence seem less practical; but for now let $m > 1$ be arbitrary.

Let $S : \Omega \longrightarrow \mathbb{R}^2$ be an independent n-dimensional Lebesgue-continuous random vector describing the source pattern distribution; its density function is denoted by $\rho : \mathbb{R}^n \longrightarrow \mathbb{R}$. As S is independent, ρ factorizes into $\rho(x_1, \ldots, x_n) = \rho_1(x_1) \ldots \rho_n(x_n)$, with the marginal source density functions $\rho_i : \mathbb{R} \longrightarrow \mathbb{R}$. As above, let X denote the vector of sensor signals and A the mixing matrix such that $X = AS$. A is assumed to be of full rank and to have pairwise linearly independent columns. Since we are not interested in dealing with scaling factors, we can assume that the columns in A have Euclidean norm 1. If $a_1, \ldots, a_n \in \mathbb{R}^m$ denote the columns of A, we often write $A = (a_1 | \ldots | a_n)$. The **geometric learning algorithm** for symmetric distributions in its simplest form then goes as follows:

Pick $2n$ starting elements $w_1, w_1', \ldots, w_n, w_n'$ on the unit sphere $S^{m-1} \subset \mathbb{R}^m$ such that w_i and w_i' are opposite each other, i.e. $w_i = -w_i'$ for $i = 1, \ldots, n$, and such that the w_i are pairwise linearly independent vectors in $\mathbb{R}^m$. Often, these w_i are called **neurons** because they resemble the neurons used in clustering algorithms and in Kohonen's self-organizing maps. If $m = 2$, one usually takes the unit roots $w_i = \exp(\frac{n-1}{n}\pi i)$. Furthermore fix a learning rate $\eta : \mathbb{N} \longrightarrow \mathbb{R}$ such that $\eta(n) > 0$, $\sum_{n \in \mathbb{N}} \eta(n) = \infty$ and $\sum_{n \in \mathbb{N}} \eta(n)^2 < \infty$. Then iterate the following step until an appropriate abort condition has been met:

Choose a sample $x(t) \in \mathbb{R}^m$ according to the distribution of X. If $x(t) = 0$ pick a new one – note that this case happens with probability zero since the probability density function ρ_X of X is assumed to be continuous. Project $x(t)$ onto the unit sphere to yield $y(t) := \frac{x(t)}{|x(t)|}$. Let i be in $\{1, \ldots, n\}$ such that w_i or w_i' is the neuron closest to y with respect to the Euclidean metric. Then set $w_i(t + 1) := \pi(w_i(t) + \eta(t) \operatorname{sgn}(y(t) - w_i(t)))$, where $\pi : \mathbb{R}^m \setminus \{0\} \longrightarrow S^{(m-1)}$ denotes the projection onto the $(m-1)$-dimensional unit sphere $S^{(m-1)}$ in $\mathbb{R}^m$, and $w_i'(t + 1) := -w_i(t + 1)$. All other neurons are not moved in this iteration.

Similar to the quadratic case, this algorithm may be called **absolute winner-takes-all learning**. It resembles Kohonen's competitive learning algorithm for self-organizing maps with a trivial neighbourhood function (0-**neighbour algorithm**) but with the modification that the step size along the direction of a sample does not depend on distance, and that the learning process takes place on $S^{(m-1)}$ not in $\mathbb{R}^{(m-1)}$.

5   Source-Recovery Step

Using the results given above, we can assume that an estimate of the original mixing matrix A has been found. We are therefore left with the problem of reconstructing the sources using the sensor signals X and the estimated matrix A. Since A has full rank, the equation $x = As$ yields the $n - m$-dimensional affine vectorspace $A^{-1}\{x\}$ as solution space for s. Hence, if $n > m$ the source-recovery problem is ill-posed without further assumptions. An often used [7] [2] assumption can be derived using a maximum likelihood approach, as will be shown next.

The problem of the source-recovery step can be formulated as follows: Given a random vector $X : \Omega \longrightarrow \mathbb{R}^m$ and a matrix A as above, find an independent vector $S : \Omega \longrightarrow \mathbb{R}^n$ satisfying an assumption yet to be found such that $X = AS$. Considering $X = AS$, i.e. neglecting any additional noise, X can be imagined to be determined by A and S. Hence the probability of observing X given A and S can be writen as $P(X|S,A)$. Using Bayes Theorem the **posterior probability** of S is then

$$P(S|X,A) = \frac{P(X|S,A)P(S)}{P(X)},$$

the probability of an event of S after knowing X and A. Given some samples of X, a standard approach for reconstructing S is the **maximum-likelihood algorithm** which means maximizing this posterior probability after knowing the **prior probability** $P(S)$ of S. Using the samples of X one can then find the most probable S such that $X = AS$. In terms of representing the observed sensor signals X in a basis $\{a_i\}$ this is called the most probable decomposition of X in terms of the overcomplete basis of $\mathbb{R}^m$ given by the columns of A.

Using the posterior of the sources $P(S|X,A)$, we can obtain an estimate of the unknown sources by solving the following relation

$$\begin{aligned} S &= \arg\max_{X=AS} P(S|X,A) \\ &= \arg\max_{X=AS} P(X|S,A)P(S) \\ &= \arg\max_{X=AS} P(S). \end{aligned}$$

In the last equation, we use that X is fully determined by S and A, and hence $P(X|S,A)$ is trivial. Note that of course the maximum under the constraint $X = AS$ is not necessarily unique.

If $P(S)$ is assumed to be Laplacian that is $P(S_i)(t) = a\exp(-|t|)$, then we get $S = \arg\min_{X=AS} |S|_1$ where $|v|_1 := \sum_i |v_i|$ denotes the 1-norm. We can show that the solution S is unique. Note that the S may not be unique for other norms; for example considering the supremum norm $|x|_\infty$, the perpendicular from 0 onto an affine vectorspace is not unique.

The general algorithm for the source-recovery step therefore is the maximization of $P(S)$ under the constraint $X = AS$. This is a linear optimization problem which can be tackled using various optimization algorithms [3].

In the following we will assume a Laplacian prior distribution of S which is characteristic of a sparse coding of the observed sensor signals. In this case, the minimization has a nice visual interpretation, which suggests an easy to perform algorithm: The source-recovery step consists of minimizing the 1-norm $|s_\lambda|_1$ under the constraint $As_\lambda = x_\lambda$ for all samples x_λ. Since the 1-norm of a vector can be pictured as the length of a path with parallel steps to the axes, Bofill and Zibulevsky call this search **shortest-path decomposition** — indeed, one can show that s_λ represents the shortest path to x_λ in $\mathbb{R}^m$ along the lines given by the matrix columns $a_i = Ae_i$ of A.

6   Experimental Results

In this section, we give a demonstration of the algorithm. The calculations have been performed on a AMD Athlon 1 GHz computer using Matlab and took no more than one minute at most.

We mixed three speech signals to two sensor signals as shown in figure 1, left side and middle. After 10^5 iterations, we found a mixing matrix with satisfactorily small minimal column distance 0.1952 to the original matrix, and after source-recovery, we calculate a correlation of estimated and original source signals with a crosstalking error [1] $E_1(\mathrm{Cor}(S, S')) = 3.7559$. In figure 1 to the right, the estimated source signals are shown. One can see a good resemblance to the original sources, but the crosstalking error is still rather high.

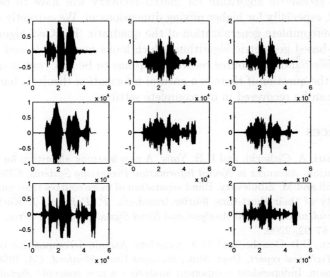

Fig. 1. Example: The three sources to the left, the two mixtures in the middle, and the recovered signals to the right. The speech texts were 'californication', 'peace and love' and 'to be or not to be that'. The signal kurtosis were 8.9, 7.9 and 7.4.

We suggest that this is a fundamental problem of the source-recovery step, which, to our knowledge, using the above probabilistic approach cannot be improved any further. To explore this aspect further, we performed an experiment using the source recovery algorithm to recover three Laplacian signals mixed with

$$A_\alpha = \begin{pmatrix} 1 & \cos(\alpha) & \cos(2\alpha) \\ 0 & \sin(\alpha) & \sin(2\alpha) \end{pmatrix},$$

where we started the algorithm already with the correct mixing matrix. We then compared the crosstalking error $E(\mathrm{Cor}(S, S'))$ of the correlation matrix of the recovered signals S' with the original ones (S) for different angles $\alpha \in [0, \frac{\pi}{2}]$.

We found that the result is nearly independent of the angle, which makes sense because one can show that the shortest-path-algorithm is invariant under coordinate transformations like A_α. This experiment indicates that there might be a general border on how good sources can be recovered in overcomplete settings.

7   Conclusion

We have presented a two-step approach to overcomplete blind source separation. First, the original mixing matrix is approximated using the geometry of the mixture space in a similar fashion as geometric algorithms do this in the quadratic case. Then the sources are recovered by the usual maximum-likelihood approach with a Laplacian prior.

For further research, two issues will have to be dealt with. On the one hand, the geometric algorithm for matrix-recovery will have to be improved and tested, especially for higher mixing dimensions m. We currently experiment with an overcomplete generalization of the quadratic 'FastGeo' algorithm [6], a histogram-based geometric algorithm, which looks more stable and also faster. On the other hand, the source-recovery step has to be analyzed in more detail, especially the question if there is a natural information theoretic barrier of how well data can be recovered in overcomplete settings.

References

[1] S. Amari, A. Cichocki, and H.H. Yang. A new learning algorithm for blind signal separation. *Advances in Neural Information Processing Systems*, 8:757–763, 1996.

[2] P. Bofill and M. Zibulevsky. Blind separation of more sources than mixtures using sparsity of their short-time fourier transform. *P. Pajunen, J. Karhunen, eds., Independent Component Analysis and Blind Signal Separation (Proc. ICA '2000)*, pages 87–92, 2000.

[3] S. Chen, D.L. Donoho, and M.A. Saunders. Atomic decomposition by basis pursuit. *Technical report, Dept. Stat., Stanford Univ, Stanford, CA*, 1996.

[4] P. Comon. Independent component analysis – a new concept? *Signal Processing*, 36:287–314, 1994.

[5] G.F. Harpur and R.W. Prager. Development of low-entropy coding in a recurrent network. *Network*, 7:277–284, 1996.

[6] A. Jung, F.J. Theis, C.G. Puntonet, and E.W. Lang. Fastgeo – a histogram based approach to linear geometric ica. *ICA2001 accepted*, 2001.

[7] M.S. Lewicki and T.J. Sejnowski. Learning overcomplete representations. *Neural Computation*, 1998.

[8] B. Olshausen and D. Field. Sparse coding of natural images produces localized, 1995.

[9] B.A. Olshausen. Learning linear, sparse, factorial codes. *Technical Report AIM-1580*, 1996.

[10] C.G. Puntonet and A. Prieto. An adaptive geometrical procedure for blind separation of sources. *Neural Processing Letters*, 2, 1995.

[11] C.G. Puntonet, A. Prieto, C. Jutten, M.R. Alvarez, and J. Ortega. Separation of sources: a geometry-based procedure for reconstruction of n-valued signals. *Signal Processing*, 46:267–284, 1995.

Improving Long-Term Online Prediction with Decoupled Extended Kalman Filters*

Juan A. Pérez-Ortiz[1], Jürgen Schmidhuber[2], Felix A. Gers[3], and Douglas Eck[2]

[1] DLSI, Universitat d'Alacant, E-03071 Alacant, Spain
[2] IDSIA, Galleria 2, 6928 Manno, Switzerland
[3] Mantik Bioinformatik GmbH, Neue Gruenstrasse 18, 10179 Berlin, Germany

Abstract. Long Short-Term Memory (LSTM) recurrent neural networks (RNNs) outperform traditional RNNs when dealing with sequences involving not only short-term but also long-term dependencies. The decoupled extended Kalman filter learning algorithm (DEKF) works well in online environments and reduces significantly the number of training steps when compared to the standard gradient-descent algorithms. Previous work on LSTM, however, has always used a form of gradient descent and has not focused on true online situations. Here we combine LSTM with DEKF and show that this new hybrid improves upon the original learning algorithm when applied to online processing.

1 Introduction

The decoupled extended Kalman filter (DEKF) [4,8] has been used successfully to optimize the training of recurrent neural networks (RNNs). In such a framework, DEKF considers learning as a filtering problem in which the optimum weights of the network are estimated efficiently in a recursive fashion. The algorithm is especially suitable for online learning situations, where weights are adjusted in a continuous fashion.

With DEKF it should be possible for a RNN to learn optimal weights for many difficult problems. However, RNNs in general [1,7,9] are hampered by *vanishing gradients* [5] that make networks unable to deal correctly with long-term dependencies. A recent novel RNN called *Long Short-Term Memory* (LSTM) [6] overcomes this problem and learns previously unlearnable solutions to numerous tasks [6,2,3], including tasks that require storing relevant events for more than 1000 subsequent discrete time steps without the help of any short training sequences.

In this study we use LSTM with forget gates [2] to predict subsequent symbols of a continual input stream (not segmented a priori into subsequences with

---

* Work supported by the Generalitat Valenciana through grant FPI-99-14-268, by the Spanish Comisión Interministerial de Ciencia y Tecnología through grant TIC2000-1599-C02-02, and by the Swiss National Foundation through grant 2100-49'144.96.

J.R. Dorronsoro (Ed.): ICANN 2002, LNCS 2415, pp. 1055–1060, 2002.

clearly defined ends) with long-term dependencies. Thus, unlike previous approaches with LSTM, the focus is on true *online* processing.

Gers et al. [2] studied a similar problem using a different setup. In their simulations, weight updates were performed after each new symbol presentation, a strategy from online learning. However, when network error became too high, a reset was done (clearing activations, states and partial derivatives) and the symbol stream was started over, a heuristic that presents information to the network in a way similar to batch learning. Thus, the previous attempt can be considered as half-way between online and offline learning. Here we apply the same LSTM architecture to the same kind of sequences, but with a *pure* online approach: there is only one single input stream; learning continues even when the network makes mistakes; and training and testing are not divided into separate phases.

All previous LSTM implementations have used a form [6] of gradient descent to adjust the weights of the network. In this paper we apply the DEKF training algorithm to the LSTM architecture for the first time; we compare experimental results obtained with the gradient descent algorithm to those of DEKF, and also comment on much worse results obtained with traditional RNNs.

2 LSTM Networks Trained by DEKF

Gradient descent algorithms, such as the original LSTM training algorithm, are usually slow when applied to time series because they depend on *instantaneous* estimations of the gradient: the derivatives of the error function with respect to the weights to be adjusted only take into account the distance between the current output and the corresponding target, using no history information for weight updating.

DEKF [8,4] overcomes this limitation. It considers training as an optimal filtering problem, recursively and efficiently computing a solution to the least-squares problem. At any given time step, all the information supplied to the network up until now is used, including all derivatives computed since the first iteration of the learning process. However, computation is done such that only the results from the previous step need to be stored.

Lack of space prohibits a complete description of DEKF; we refer the reader to previous citations for details. The extended Kalman filter is used for training neural networks (recurrent or not) by assuming that the optimum setting of the weights is stationary. However, when considering all the weights of the network together, the resulting matrices become so unmanageable (even for networks with moderate sizes) that a *node-decoupled* version of the algorithm is usually used instead to make the problem computationally tractable. The decoupled approach applies the extended Kalman filter independently to each neuron in order to estimate the optimum weights feeding it. By proceeding this way, only local interdependences are considered. The equations for iteration t of a DEKF minimizing the typical quadratic error measure can be formulated as follows:

$$G_i(t) = K_i(t-1)C_i^T(t) \left[\sum_{i=1}^{n_g} C_i(t)K_i(t-1)C_i^T(t) + R(t) \right]^{-1} \qquad (1)$$

$$w_i(t) = w_i(t-1) + G_i(t)\left[d(t) - y(t)\right] \qquad (2)$$

$$K_i(t) = K_i(t-1) - G_i(t)C_i(t)K_i(t-1) + Q_i(t) \qquad (3)$$

where n_g is the number of neurons, i defines a particular neuron (with $1 \leq i \leq n_g$), w_i is a vector with all the weights leading to neuron i, $d(t)$ is the desired response, and $y(t)$ is the actual output of the network.

Let n_i denote the number of weights leading to neuron i, and n_Y the number of output neurons of the network. The Jacobian $C_i(t)$ is an $n_Y \times n_i$ matrix containing the partial derivatives of the function defining the output $y(t)$ of the network with respect to each weight leading to neuron i. Matrices G_i, K_i, Q_i and R are initialized in a problem-specific manner and denote, respectively, the Kalman *gain*, the *error* covariance matrix, the covariance matrix of *artificial process noise*, and the covariance matrix of the *measurement noise*.

Combining DEKF with the LSTM architecture is straightforward. We consider a group of weights for each neuron in LSTM, that is, a group for each different gate, cell and output neuron (see previous references on LSTM for a detailed description of the architecture and how error derivatives are computed). At time step t we calculate the derivatives required for matrix $C_i(t)$ exactly the same way as the original LSTM gradient-descent training algorithm does, and then apply equations (1)–(3) in order to update weights $w_i(t)$.

It should be noted that DEKF's time complexity [4, p. 771] is much larger than that of gradient descent because DEKF not only has to compute the same derivatives, but also many matrix operations at every time step.

3   Experiments

We use LSTM with forget gates to predict subsequent symbols in a sequence generated by the continual embedded Reber automaton (or grammar) [10] shown in Fig. 1. Due to existence of long-term dependencies, this task is suitably difficult to show the power of both LSTM and DEKF. The learning process is completely online. The network is trained to give in real-time an output as correct as possible for the input supplied at each time step; after normalization, $y_i(t)$ is interpreted as the probability of the next symbol being the i-th symbol of the alphabet; symbols, when considered as inputs or targets, are coded with unary vectors by means of *local* or *exclusive* coding.

Network Topology and Parameters. The LSTM network has 4 memory blocks with 2 cells each. The size of the alphabet of the automaton is 7, so we consider 7 neurons in the input and output layers. Bias weights to input and output gates are initialized block-wise: -0.5 for the first block, -1 for the second, -1.5 for the third, and so on. Forget gate biases are initialized with symmetric

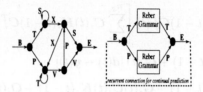

Fig. 1. Transition diagrams for standard (left) and embedded (right) Reber grammars

values: 0.5 for the first block, 1 for the second, and so on. The rest of the weights are randomly taken from a uniform distribution in $[-0.2, 0.2]$. The squashing function g is set to $\tanh(x)$, and h is set to the identity function — see [2,6] for details on the LSTM architecture. For the gradient-descent algorithm we set the learning rate to 0.5. In case of DEKF, the parameters of the algorithm suggested by Haykin [4, p. 771] turned out to be adequate for this task as well.

Training and Testing. We count the number of symbols needed by LSTM to attain error-free predictions for at least 1000 subsequent symbols (a large period of time); here "error-free" means that the symbol corresponding to the winner neuron in the network output is one of the possible transition symbols, given the current state of the Reber automaton.

Gers et al. [2] considered longer error-free sequences, but learning was not truly online, and the networks were tested with frozen weights. Therefore, although the criterion for sustainable prediction was stringent, the learning was easier in principle. On the other hand, when working online, the recurring presence of particular subsequences usually makes the network forget past history and trust more recent observations instead. This is what one would expect from an online model, which is supposed to deal with non-stationary environments.

After an initial training period, LSTM usually makes only few mistakes and tends to keep making correct predictions. To obtain a tolerant measure of prediction quality we measure the time at which the N-th error takes place after the first 1000 subsequent error-free predictions: here we consider two possible values for N, namely, 1 and 10.

4   Results and Analysis

Gradient-Descent LSTM Results. Table 1 shows the results for 9 different sequences with 9 independently initialized LSTM networks trained by the original LSTM training algorithm.[1] In one case (row 5) no correct prediction sequences (for 1000 symbols in a row) are found before the 1000000-th sequence symbol; this is indicated in the table by 1000000+.

---

[1] The average number of symbols required for learning to predict accurately in real-time (thousands of symbols) is much smaller than the number of symbols required in the offline set-up (millions). This deserves a more profound study.

Table 1. Time steps required by online LSTM (trained with gradient descent) to achieve 1000 subsequent correct predictions

Net	Sustainable prediction	Next 10 errors	Next error
1	39229	178229	143563
2	102812	144846	111442
3	53730	141801	104163
4	54565	75666	58936
5	1000000[+]	–	–
6	111483	136038	113715
7	197748	235387	199445
8	54629	123595	123565
9	85707	92312	86742

Table 2. Time steps required by online LSTM (trained by DEKF) to achieve 1000 subsequent correct predictions

Net	Sustainable prediction	Next 10 errors	Next error
1	29304	30953	30347
2	19758	322980	25488
3	20487	24106	22235
4	26175	33253	27542
5	18015	22241	19365
6	16667	1000000[+]	29826
7	23277	26664	24796
8	1000000[+]	–	–
9	29742	594117	31535

LSTM with DEKF Results. With the DEKF training algorithm, the number of symbols needed for correct prediction is even lower — compare Table 2. Although the time required to achieve 1000 error-free predictions in a row is generally lower than with the original algorithm, the number of symbols before the 10-th error is also smaller. DEKF seems to reduce the long-term memory capabilities of LSTM while increasing its online learning speed. There are 3 remarkable cases (rows 2, 6 and 9 in Table 2), however, where a very long subsequence is necessary for the 10-th error to appear; row 6 shows a particularly good result: only 3 errors occur before the 1000000-th symbol.

LSTM Analysis. Study of the evolution of gate and state activations revealed that online LSTM learns a behavior similar to the one observed in previous non-online experiments [2], that is, one memory block specializes in bridging long-time information, while the others exhibit short-term behavior only.

RTRL-RNNs Results. Experiments with traditional RTRL-trained [11] RNNs (such as the simple recurrent net [1] or the recurrent error propagation network [9]) demonstrated that they are unable to obtain sustainable error-free predictions for 1000 subsequent symbols, even after extremely long training times. Even as few as 100 subsequent correct predictions were extremely rare. DEKF applied to these architectures, however, did allow for sustainable error-free predictions. But it always required many more than 100000 symbols.

5 Conclusion

LSTM variants are applicable to true online learning situations with never-ending continual input streams. On the difficult extended Reber automaton, online LSTM yields results comparable to that of previous LSTM applications involving offline learning, and clearly outperforms traditional RNNs.

For the first time the DEKF algorithm was applied to LSTM. This led to results even better than those obtained with the original training algorithm. The DEKF-based approach reduces significantly the number of training steps necessary for error-free prediction when compared to the standard algorithm. However, it forgets more easily than original LSTM, and requires more operations per time step and weight.

References

1. Elman, J. L.: Finding structure in time. Cognitive Science **14** (1990) 179–211.
2. Gers, F. A., Schmidhuber, J., Cummins, F.: Learning to forget: continual prediction with LSTM. Neural Computation **12**, 10 (2000) 2451–2471.
3. Gers, F. A., Schmidhuber, J.: LSTM recurrent networks learn simple context free and context sensitive languages. IEEE Transactions on Neural Networks **12**, 6 (2001) 1333–1340.
4. Haykin, S.: Neural networks: a comprehensive foundation. Prentice-Hall (1999).
5. Hochreiter, S., Bengio, Y., Frasconi, P., Schmidhuber, J.: Gradient flow in recurrent nets: the difficulty of learning long-term dependencies. Kremer, S. C., Kolen, J. F. (eds.): A field guide to dynamical recurrent neural networks (2001). IEEE Press.
6. Hochreiter, S., Schmidhuber, J.: Long short-term memory. Neural Computation **9**, 8 (1997) 1735–1780.
7. Pearlmutter, B. A.: Gradient calculations for dynamic recurrent neural networks: a survey. IEEE Transactions on Neural Networks **6**, 5 (1995) 1212–1228.
8. Puskorius, G. V., Feldkamp, L. A.: Neurocontrol of nonlinear dynamical systems with Kalman filter trained recurrent networks. IEEE Transactions on Neural Networks **5**, 2 (1994) 279–297.
9. Robinson, A. J., Fallside, F.: A recurrent error propagation speech recognition system. Computer Speech and Language **5** (1991) 259–274.
10. Smith, A. W., Zipser, D.: Learning sequential structures with the real-time recurrent learning algorithm. Intl. Journal of Neural Systems **1**, 2 (1989) 125–131.
11. Williams, R. J., Zipser, D.: A learning algorithm for continually training recurrent neural networks. Neural Computation **1** (1989) 270–280.

Market Modeling Based on Cognitive Agents

Georg Zimmermann, Ralph Grothmann, Christoph Tietz, and Ralph Neuneier

Siemens AG, Otto-Hahn-Ring 6, 81730 Munich, Germany
Georg.Zimmermann@mchp.siemens.de

Abstract. In this paper, we present an explanatory multi-agent model. The agents decision making is based on cognitive systems with three basic features (perception, internal processing and action). The interaction of the agents allows us to capture the market dynamics.
The three features are derived deductively from the assumption of homeostasis and constitute necessary conditions of a cognitive system. Given a changing environment, homeostasis can be seen as the attempt of a cognitive agent to maintain an internal equilibrium. We model the cognitive system with a time-delay recurrent neural network.
We apply our approach to the DEM / USD FX-Market. Fitting real-world data, our approach is superior to a preset benchmark (MLP).

1 Introduction

A market consists of a large number of interacting agents. Thus, a natural way of explaining and predicting market prices is to analyze the decision making of the agents on the microeconomic level and to study their interaction on the macroeconomic side of the market (i. e. price formation mechanism). Recent developments in this research field are summarized in LeBaron (1999) [5].

We present a new approach of multi-agent market modeling. The decision making of an agent is modeled with a homeostatic dynamic system. Homestasis means, that the system maintains an internal equilibrium. On this basis, we derive the properties perception, internal processing and action of a cognitive system (sec. 2). As a structural representation of homeostasis, we propose so-called zero-neurons within a time-delay recurrent neural network (sec. 3).

The aggregation of agents decisions leads to a multi-agent based approach of market modeling (sec. 4). Fitting real-world data (DEM / USD FX-Market), our approach is superior to a traditional feedforward neural network (sec. 5).

2 Homeostatic Cognitive Systems

Let us define homeostasis as a dynamically stable state, in which external influences (perception) are balanced with internal expectations such that an internal equilibrium is maintained. Perception may disturb the internal stable state. In order to keep the internal equilibrium, the cognitive system is forced to develop an internal model and to initiate actions, which counterbalance the disturbances.

J.R. Dorronsoro (Ed.): ICANN 2002, LNCS 2415, pp. 1061–1067, 2002.

The three properties perception, internal processing and action constitute necessary conditions for a cognitive system [1,9]. the cognitive system. More conceptionally, in a changing environment, homeostasis is the attempt of an internal neuron to perpetuate in a dynamical stable state.

We distinguish conscious from unconscious perception. In conscious perception, the system compares selective observations to an internal expectation. Only the resulting difference has an impact on the system. Unconscious perception enters the system directly, i. e. it corresponds to stimulus-response. Unconscious perception is therefore not balanced with internal expectations [6,7,9].

The internal processing mainly consists of an internal model of the external world. In order to maintain the internal equilibrium, homeostasis implies the construction of this internal model, which balances conscious perception with internal expectations. Without this balance, the system would be driven by stimulus-response. For construction of internal expectations, an internal memory is required. This memory can be seen as merging of all relevant past information from which the internal expectation is formed [9].

To initiate actions, the internal expectation is evaluated by an objective function (e. g. profit or utility maximization). Hence, actions are always goal-oriented. In the homeostatic cognitive system (see Fig. 1) perception and action are decoupled: incoming stimuli do not have to lead to actions if they do not disturb the internal equilibrium [7]. Due to this decoupling, the action of a homeostatic cognitive system are driven by the abstract objective function rather than by a simple stimulus-response mechanism [9]. Through the interaction with the environment, the system consciously perceives the impact of its own actions. Note, that this is also a necessary condition for self-consciousness [7].

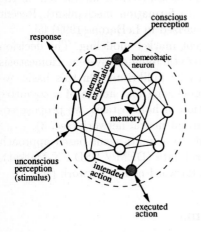

Fig. 1. A homeostatic cognitive system. Conscious perception forces the system to build an internal model, which generates internal expectations as a compensation. Unconscious perception influences the cognitive system directly in the form of stimulus-response. The system memory, which is due to the recurrent formulation, is a merging of all relevant past information. Memory is crucial for a homeostatic system, unless there is another way to build an internal dynamics.

If most of the inputs are processed in a conscious way, the cognitive system has a predominantly analytical picture of the world, because most of the input is evaluated against the internal model. In the dual case of mainly unconscious perception, the cognitive system has a more 'emotional' picture of the world.

Concerning the behavior of agents, one may argue that a more analytical description of the world corresponds to the behavior of a financial analyst, while the more 'emotional' picture can be attributed to a trader, who often relies on market rumors or moods. The more inputs are handled consciously, the more complete the 'world model' of system has to be [11].

A more detailed *inductive* description of the perception, internal processing and action can be found in Zimmermann et al. (2001) [11].

3 Modeling Homeostatic Cognitive Systems

We suggest to model homeostasis by so-called zero-neurons in a recurrent neural network framework. Zero-neurons are input-output neurons with constant, task invariant target values of 0 and input weights fixed at -1. Zero-neurons can be equipped with any common squashing (e.g. tanh) and cost function (e.g. x^2) [8].

Conscious perceptions y_τ^d are always connected to a zero-neuron. Due to the fixed input weights of the zero-neuron, a trivial solution of homeostasis by disconnecting the input y_τ^d is avoided. In fact, the input y_τ^d induces an error signal resp. flow, which forces the internal model to construct an expectation y_τ as a compensation. An error flow can only be avoided, if the internal expectation y_τ is equal to the input y_τ^d. Only the difference $y_\tau - y_\tau^d$ has an impact on the system. This matches the above definition of conscious perception [11].

Unconscious perceptions u_τ are directly connected to hidden neurons of the recurrent network. Hence, an error flow is not induced and the system is not forced to build internal expectations as a compensation. Nevertheless, unconscious perception u_τ has a direct impact on the cognitive system. This corresponds to the principle of stimulus-response [4].

The internal processing is structurally represented by a recurrent neural network similar to the one depicted in Fig. 2 [10]. The connections among the neurons are the only tunable parameters. Due to the recurrent construction, the system is able to accumulate a superposition of all past information, i. e. a memory, which is required to achieve a dynamically stable state [6].

In addition to the internal expectations y_τ, the cognitive system suggests actions α_τ. The cognitive system perceives the resulting state of the environment y_τ^d and the actually executed α_τ. If the intended action α_τ cannot be executed, the observed action α_τ^d and the intention α_τ differ [1,9].

4 Multi-agent Models by Homeostatic Cognitive Systems

We now focus on the modeling of a financial market. Each agent $i = 1, \ldots, N$ in our model is represented by a homeostatic cognitive system. The agents consciously perceive changes of the market price p_τ^d and observe external influences u_τ unconsciously. A market order of an agent i corresponds to an intended action $\alpha_{i,\tau}$ suggested by his internal model. The objective function of the agents is profit maximization. For simplicity, we neglect transaction costs and assume that the agents have unlimited credit [2].

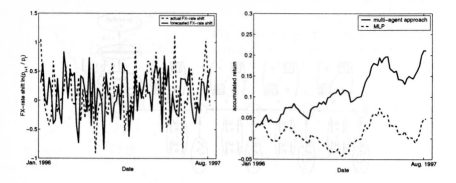

Fig. 4. Comparison of multi-agent model forecasts and actual DEM / USD FX-rate shifts, left, and accumulated return of investment calculated on the test set, right.

6   Conclusion

We developed a multi-agent approach of market modeling based on the idea of homeostatic cognitive systems. Our model explains market prices on the basis of the decision processes of multiple agents. Hence, our model emphasizes semantic specifications instead of ad-hoc functional relationships. We utilized recurrent neural networks, to merge multi-agent models, cognitive systems and econometrics into an integrated market model.

The decision making of the agents is modeled as a cognitive process with the three basic features: perception, internal processing and action. We derived these features deductively from the assumption of homeostasis. Homeostasis is an internal equilibrium, in which external influences are compensated with internal expectations. Thus, the cognitive system overcomes poor stimulus-response mechanisms. As a structural representation of homeostasis, we introduced zero-neurons within a time-delay recurrent neural network.

As a remarkable property, our multi-agent approach allows the fitting of real world data (DEM / USD FX-market). The empirical study indicated, that based on common performance measures, the proposed multi-agent model is superior to an ad-hoc MLP in an out-of-sample period of one and a half years.

We have lots of extensions and further analyses under consideration. For example, one may think of the inclusion of utility functions to guide the agent's behavior. Further more, different price formation mechanisms may be considered [2,5]. Technically, the underlying networks may be extended by *unfolding in space and time* in order to smooth the trajectories of the market dynamics [10].

References

1. Chalmers, D.: *Facing up the problem of consciousness, in: Explaining Consciousness: The Hard Problem*, ed. J. Shear, MIT Press, 1997.
2. Farmer, J.D.: *Market Force, Ecology and Evolution*, Working Paper, Santa Fe Institute, No. 98-12-116, 1998.

3. Haykin S.: *Neural Networks. A Comprehensive Foundation.*, Macmillan College Publishing, New York, 1994, 2nd edition 1998.
4. Juarrero, A.: *Dynamics in Action*, Intentional Behavior as a Complex System, MIT Press, Cambridge, MA., London 1999.
5. LeBaron, B.: *Agent-based computational finance: Suggested readings and early research*, Journal of Economic Dynamics and Control, Vol. 24, 2000, pp. 679 - 702.
6. MacPhail, E.: *The Evolution of Consciousness*, Oxford Uni. Pr., New York, 1998.
7. Metzinger, Th.: *Neural Correlates of Consciousness*, Empirical and Conceptual Questions, MIT Press, Cambridge, Massachusetts, London 2000.
8. Neuneier R. and Zimmermann H. G.: *How to Train Neural Networks.* in: *Neural Networks: Tricks of the Trade*, pages 373–423. Springer Verlag, Berlin, 1998.
9. Perlovsky, L.I.: *Neural Networks and intellect: using model-based concepts.* Oxford University Press, New York, 2001.
10. Zimmermann, H.G. and Neuneier, R.:*Neural Network Architectures for the Modeling of Dynamical Systems*, in: A Field Guide to Dynamical Recurrent Networks, Eds. Kolen, J.F., Kremer, St., IEEE Press 2001.
11. Zimmermann, H.G., Neuneier, R. and Grothmann, R.: *An Approach of Multi-Agent FX-Market Modeling based on Cognitive Systems.*, in: Proceedings of ICANN 2001.

An Efficiently Focusing Large Vocabulary Language Model

Mikko Kurimo and Krista Lagus

Helsinki University of Technology, Neural Networks Research Centre
P.O.Box 5400, FIN-02015 HUT, Finland
Mikko.Kurimo@hut.fi, Krista.Lagus@hut.fi

Abstract. Accurate statistical language models are needed, for example, for large vocabulary speech recognition. The construction of models that are computationally efficient and able to utilize long-term dependencies in the data is a challenging task. In this article we describe how a topical clustering obtained by ordered maps of document collections can be utilized for the construction of efficiently focusing statistical language models. Experiments on Finnish and English texts demonstrate that considerable improvements are obtained in perplexity compared to a general n-gram model and to manually classified topic categories. In the speech recognition task the recognition history and the current hypothesis can be utilized to focus the model towards the current discourse or topic, and then apply the focused model to re-rank the hypothesis.

1 Introduction

The estimation of complex statistical language models has recently become possible due to the large data sets now available. A statistical language model provides estimates of probabilities of word sequences. The estimates can be employed, e.g., in speech recognition for selecting the most likely word or sequence of words among candidates provided by an acoustic speech recognizer.

Bi- and trigram models, or more generally, n-gram models, have long been the standard method in statistical language modeling[1]. However, the model has several well-known drawbacks: (1) an observation of a word sequence does not affect the prediction of the same words in a different order, (2) long-term dependencys between words do not affect predictions, and (3) very large vocabularies pose a computational challenge. In languages with syntactically less strict word order and a rich inflectional morphology, such as Finnish, these problems are particularly severe.

Information regarding long-term dependencies in language can be incorporated into language models in several ways. For example, in word caches [1] the probabilities of words seen recently are increased. In word trigger models [2] probabilities of word pairs are modeled regardless of their exact relative positions.

---

[1] n-gram models estimate $P(w_t|w_{t-n+1}w_{t-n+2}\ldots w_{t-1})$, the probability of nth word given the sequence of the previous $n-1$ words. The probability of a word sequence is then the product of probabilities of each word.

J.R. Dorronsoro (Ed.): ICANN 2002, LNCS 2415, pp. 1068–1073, 2002.
© Springer-Verlag Berlin Heidelberg 2002

Mixtures of sentence-level topic-specific models have been applied together with dynamic n-gram cache models with some perplexity reductions [3]. In [4] and [5] EM and SVD algorithms are employed to define topic mixtures, but there the topic models only provide good estimates for the content word unigrams which are not very powerful language models as such. Nevertheless, perplexity improvements have been achieved when these methods are applied together with the general trigram models.

The modeling approach we propose is founded on the following notions. Regardless of language, the size of the active vocabulary of a speaker in a context is rather small. Instead of modeling all possible uses of language in a general, monolithic language model, it may be fruitful to *focus* the language model to smaller, topically or stylistically coherent subsets of language. In the absence of prior knowledge of topics, such subsets can be computed based on content words that identify a specific discourse with its own topics, active vocabulary, and even favored sentence structures.

Our objective was to create a language model suitable for large vocabulary continuous speech recognition in Finnish, which has not yet been extensively studied. In this paper a focusing language model is proposed that is efficient enough to be interesting for the speech recognition task and that alleviates some of the problems discussed above.

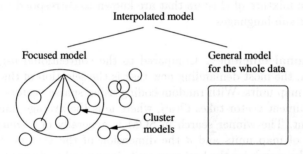

Fig. 1. A focusing language model obtained as an interpolation between topical cluster models and a general model.

2   A Topically Focusing Language Model

The model is created as follows:

1. Divide the text collection into topically coherent text 'documents', such as paragraphs or short articles.
2. Cluster the passages topically.
3. For each cluster, calculate a small n-gram model.

For the efficient calculation of topically coherent clusters we apply methods developed in the WEBSOM project for exploration of very large document collections [6][2]. The method utilizes the Self-Organizing Map (SOM) algorithm [7] for clustering document vectors onto topically organized document maps. The document vectors, in turn, are weighted word histograms where the weighting is based on idf or entropy to emphasize content words. Stopwords (e.g., function words), and very rare words are excluded, inflected words are returned to base forms. Sparse random coding is applied to the vectors for efficiency. In addition to the success of the method in text exploration, an improvement in information retrieval when compared to the standard tf.idf retrieval has been obtained by utilizing a subset of the best map units [8].

The utilization of the model in text prediction comprises the following steps:

1. Represent recent history as a document vector, and select the clusters most similar to it.
2. Combine the cluster-specific language models of the selected clusters to obtain the focused model.
3. Calculate the probability of the predicted sequence using the model and interpolate the probability with the corresponding one given by a general n-gram language model.

For the structure of the combined model, see Fig. 1. When regarded as a generative model for text, the present model is different from the topical mixture models proposed by others (e.g. [4]) in that here a text passage is generated by a very sparse mixture of clusters that are known to correspond to discourse- or topic-specific sub-languages.

Computational efficiency. Compared to the conventional n-grams or mixtures of such, the most demanding new task is the selection of the best clusters, i.e. the best map units. With random coding using sparse vectors [6] the encoding as a document vector takes $\mathcal{O}(w)$, where w is the average number of words per document. The winner search in SOM is generally of $\mathcal{O}(md)$, where m is the number of map units and d the dimension of the vectors. Due to sparse documents the search for the best map units is reduced to $\mathcal{O}(mw)$. In our experiments ($m = 2560$, $w = 100$, see Section 3.) running on a 250 MHz SGI Origin a single full search among the units took about 0.028 seconds and with additional speedup approximations that benefit from the ordering of the map, only 0.004 seconds. Moreover, when applied to rescoring the n best hypotheses or the lattice output in two-pass recognition, the topic selection need not be performed very often. Even in single-pass recognition, augmenting the partial hypothesis (and thus the document vectors) with new words requires only a local search on the map. The speed of the n-gram models depends mainly on n and the vocabulary size; a reduction in both results in a considerably faster model. The combining, essentially a weighted sum, is likewise very fast for small models. Also preliminary experiments on offline speech recognition indicate that the relative increase of the recognition time due to the focusing language model and its use in lattice rescoring is negligible.

---

[2] The WEBSOM project kindly provided the means for creating document maps.

3   Experiments and Results

Experiments on two languages, Finnish and English, were conducted to evaluate the proposed unsupervised focusing language model. The corpora were selected so that each contained a prior (manual) categorization for each article. The categorization provided a supervised topic model against which the unsupervised focusing cluster model was compared. For comparison we implemented also another topical model where full mixtures of topics are used, calculated with the EM-algorithm [4]. Furthermore, as a clustering method in the proposed focusing model we examined the use of K-means instead of the SOM.

The models were evaluated using perplexity[3] on independent test data averaged over documents. Each test document was split into two parts, the first of which was used to focus the model and the second to compute the perplexity.

To reduce the vocabulary (especially for Finnish) all inflected word forms were transformed into base forms. Probabilities for the inflected forms can then be re-generated e.g. as in [9]. Moreover, even when base forms are used for focusing the model, the cluster-specific n-gram models can, naturally, be estimated on inflected forms. To estimate probabilities of unseen words, standard discounting and back-off methods were applied, as implemented in the CMU/Cambridge Toolkit [10].

Finnish corpus. The Finnish data[4] consisted of 63 000 articles of average length 200 words from the following categories: Domestic, foreign, sport, politics, economics, foreign economics, culture, and entertainment. The number of different base forms was 373 000. For general trigram model a frequency cutoff of 10 was utilized (i.e. words occurring fewer than ten times were excluded), resulting in a vocabulary of 40 000 words. For the category and cluster specific bigram models, a cutoff of two was utilized (the vocabulary naturally varies according to topic). For the focused model, the size of the document map was 192 units and only the best cluster (map unit) was included in the focus. The results on a test data of 400 articles are presented in Fig. 2.

English corpus. The English data consisted of patent abstracts from eight subcategories of the EPO collection: A01–Agriculture; A21–Foodstuffs, tobacco; A41–Personal or domestic articles; A61–Health, amusement; B01–Separating, mixing; B21–Shaping; B41–Printing; B60–Transporting. Experiments were carried out using two data sets: pat1 including 80 000 and pat2 with 648 000 abstracts, with an average length of 100 words. The total vocabulary for pat1 was nearly 120 000 base forms, the frequency cutoff for the general trigram model 3 words resulting in vocabulary size 16 000. For pat2 these figures were 810 000, 5, and 38 000, respectively. For the category and cluster specific bigram models a cutoff of two was applied. The size of the document map was 2560 units in

---

[3] Perplexity is the inverse predictive probability for all the words in the test document.
[4] The Finnish corpus was provided by the Finnish News Agency STT.

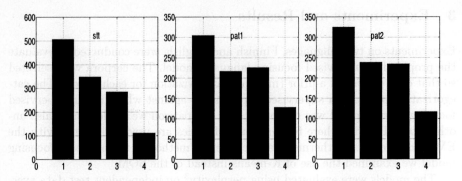

Fig. 2. The perplexities of test data using each language model for the Finnish news corpus (stt) on the left, for the smaller English patent abstract corpus (pat1) in the middle, and for the larger English patent abstract corpus (pat2) on the right. The language models in each graph from left fo right are: 1. General 3-gram model for the whole corpus, 2. Topic factor model using mixtures trained by EM, 3. Category-specific model using prior text categories, and 4. Focusing model using unsupervised text clustering. The models 2–4 were here all interpolated with the baseline model 1. The best results are obtained with the focusing model (4).

both experiments. For pat2 only the best cluster was employed for the focused model, but for pat1, with significantly fewer documents per cluster, the amount of best map units chosen was 10. The results on the independent test data of 800 abstracts (500 for pat2) are presented in Fig. 2.

Results. The experiments on both corpora indicate that when combined with the focusing model the perplexity of the general 'monolithic' trigram model improves considerably. This result is, as well, significantly better than the combination of the general model and topic category specific models where the correct topic model was chosen based on manual class label on the data. When K-means was utilized for clustering the training data instead of SOM, the perplexity did not differ significantly. However, the clustering was considerably slower (for an explanation, see Sec.2 or [6]). When applying the topic factor model suggested by Gildea and Hofmann [4] with each corpus we used 50 normal EM iterations and 50 topic factors. The first part of a test article was used to determine the mixing proportions of the factors and the second part to compute the perplexity (see results in Fig. 2).

Discussion. The results for both corpora and both languages show similar trends, although for Finnish the advantage of a topic-specific model seems more pronounced. One advantage of unsupervised topic modeling over a topic model based on fixed categories is that the unsupervised model can achieve an arbitrary granularity and a combination of several sub-topics.

The obtained clear improvement in language modeling accuracy can benefit many kinds of language applications. In speech recognition, however, it is central to discriminate between the acoustically confusable word candidates, and the average perplexity is not an ideal measure for this [11,4]. Therefore, a topic for future research (as soon as a speech data and a text corpus of related kind can be obtained for Finnish), is to examine how well the improvements in modeling translate to advancing speech recognition accuracy.

4   Conclusions

We have proposed a topically focusing language model that utilizes document maps to focus on a topically and stylistically coherent sub-language. The longer-term dependencies are embedded in the vector space representation of the word sequences, and the local dependencies of the active vocabulary within the sub-language can then be modeled using n-gram models of small n. Initially, we aimed at improving statistical language modeling in Finnish, where the vocabulary growth and flexible word order offer severe problems for the conventional n-grams. However, the experiments indicate improvements for modeling English, as well.

References

1. P. Clarkson and A. Robinson, "Language model adaptation using mixtures and an exponentially decaying cache," In *Proc. ICASSP*, pp. 799–802, 1997.
2. R. Lau, R. Rosenfeld, and S. Roukos, "Trigger-based language models: A maximum entropy approach," In *Proc. ICASSP*, pp. 45–48, 1993.
3. R.M. Iyer and M. Ostendorf, "Modelling long distance dependencies in language: Topic mixtures versus dynamic Cache model," *IEEE Trans. Speech and Audio Processing*, 7, 1999.
4. D. Gildea and T. Hofmann, "Topic-based language modeling using EM," In *Proc. Eurospeech*, pp. 2167–2170, 1999.
5. J. Bellegarda. "Exploiting latent semantic information in statistical language modeling," *Proc. IEEE*, 88(8):1279–1296, 2000.
6. T. Kohonen, S. Kaski, K. Lagus, J. Salojärvi, V. Paatero, and A. Saarela. "Organization of a massive document collection," *IEEE Transactions on Neural Networks*, 11(3):574–585, May 2000.
7. T. Kohonen. *Self-Organizing Maps*. Springer, Berlin, 2001. 3rd ed.
8. K. Lagus, "Text retrieval using self-organized document maps," *Neural Processing Letters*, 2002. In press.
9. V. Siivola, M. Kurimo, and K. Lagus. "Large vocabulary statistical language modeling for continuous speech recognition," In *Proc. Eurospeech*, 2001.
10. P. Clarkson and R. Rosenfeld, "Statistical language modeling using CMU-Cambridge toolkit," in *Proc. Eurospeech*, pp. 2707–2710, 1997.
11. P. Clarkson and T. Robinson. "Improved language modelling through better language model evaluation measures," *Computer Speech and Language*, 15(1):39–53, 2001.

Neuro-classification of Bill Fatigue Levels Based on Acoustic Wavelet Components

Masaru Teranishi[1], Sigeru Omatu[2], and Toshihisa Kosaka[3]

[1] Dept. of Electrical Eng., Nara National College of Technology
[2] Dept. of Computer and Systems Sciences, College of Eng., Osaka Prefecture University
[3] Glory Ltd.

Abstract. This paper proposes a new method to classify bills(paper moneys) into different fatigue levels due to the extent of their damage. While a bill passing through a banking machine, a characteristic acoustic signal is emitted from the bill. To classify the acoustic signal into three bill fatigue levels, we calculate the acoustic wavelet power pattern as the input to a competitive neural network with the Learning Vector Quantization(LVQ) algorithm. The experimental results show that the proposed method can obtain better classification performance than the best of conventional acoustic signal based classification methods. It is, consequently, the LVQ algorithm demonstrates a good classification.

1 Introduction

On the practical use of an Automatic Teller Machine(ATM), fatigued bills cause serious troubles. Therefore, such bills must be picked up, and should be exchanged with new ones at banks. To do the selection automatically, the development of a fatigued bills classification method that can be implemented to bank machines is desired.

This paper proposes a new method that classifies the fatigue levels of bill using the acoustic signal generated by passing a bill through a banking machine. The proposed method introduces the wavelet power pattern of the acoustic signal for classification to extract explicit feature of a fatigued bill. Furthermore, the proposed method uses a competitive neural network trained with the LVQ[1] algorithm to obtain good classification performance.

In this paper, we classify bills into three fatigue levels ; no fatigue level (level 0), average fatigued level (level 1), and much fatigued level (level 2). We carry out a classification experiment with the following procedures. First, we implement an acoustic signal acquisition system to a commercial banking machine. Some experimental acoustic signals of bills with three fatigue levels have been taken by the banking machine. Next, acoustic wavelet power patterns for each sampled signal are calculated. Then we feed the wavelet power pattern to the competitive neural network and train it by the LVQ algorithm. Finally, the classification results are shown to see the effectiveness of the proposed method.

J.R. Dorronsoro (Ed.): ICANN 2002, LNCS 2415, pp. 1074–1079, 2002.

2   Acquisition of Acoustic Signal from a Banking Machine

The block diagram of the acoustic signal acquisition system is shown in Fig.1.

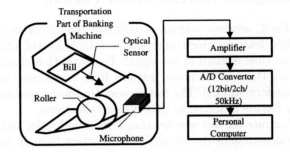

Fig. 1. Sketch of the acoustic data acquisition system.

A significant acoustic signal is generated when a bill passes through the transportation part of the banking machine. Therefore, we locate the microphone at the cover of the transportation part of the banking machine. The time series data of an acoustic signal from a bill are measured and digitized with the microphone and the Analog-Digital converter , and then stored into a personal computer. An acquisition sequence is triggered by an output of the optical sensor which detects the passing of a bill.

Examples of acoustic signals for different fatigue levels of bills are shown in Fig.2.

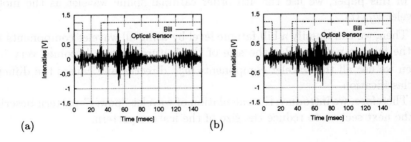

Fig. 2. Examples of acoustic signals of bills : (a) fatigue level 0, (b) fatigue level 2.

From Fig.2, the difference between fatigue levels is not appeared in the time series data of acoustic signals without being slightly changed in amplitude. Therefore, it is difficult to classify the bill into a fatigue level by directly using the acoustic signal. Hence, the proposed method transforms the acoustic signal into the wavelet power pattern as the preprocessing for the neuro-classification. The wavelet power pattern is calculated by using fast wavelet transformation described in the following section.

3   Calculating the Wavelet Power Pattern

As shown in Fig.2, acoustic signals of bills are considered as non-stationary signals. Therefore, features which represent the fatigue level of a bill are localized in frequency and time domain of the signal. To obtain those features explicitly, we now calculate the wavelet power pattern from an acoustic signal by using fast wavelet transformation[2]. The fast wavelet transformation can obtain short time frequency feature localized in time domain effectively.

First, we illustrate the fast wavelet transformation briefly, and show the procedure of the calculation of the wavelet power pattern from an acoustic signal.

3.1   Fast Wavelet Transformation of Acoustic Signal

We denote the acoustic signal by $x(t)$ where t is the discrete time. By using the fast wavelet transformation, $x(t)$ is represented as the series of "mother wavelets" that has single frequency component and localized in time domain as follows:

$$x(t) = \sum_j \sum_k w_k^{(j)} \psi_{j,k}(t) \tag{1}$$

where $\psi_{j,k}(t)$ denotes mother wavelet which has own level(i.e.,frequency) j and located in discrete time shifted position k , and $w_k^{(j)}$ denotes wavelet coefficients of $\psi_{j,k}(t)$. by using Eq.1, we can obtain the intensities of the mother wavelets, that is, wavelet components, which have level j and located in the time k among the source signal $x(t)$ as the magnitude of each $w_k^{(j)}$.

In this paper, we use the 4th order cardinal spline wavelet as the mother wavelet $\psi_{j,k}(t)$.

Then, we can classify a bill fatigue level by using all wavelet components $w_k^{(j)}$ as the feature pattern. But the size of the feature pattern becomes very large when we use all the wavelet components $w_k^{(j)}$ directly, and causes the difficulty of classification.

Therefore, we introduce the calculation of wavelet "power" pattern described in the next section to reduce the size of the feature pattern.

3.2   Calculation Procedure of a Wavelet Power Pattern

To reduce the size of the feature pattern, first, we divide entire time region of acoustic signal $x(t)$ into N sections, called "frame" $l_i (i = 1, \cdots, N)$. Next, we calculate the wavelet power $pw(j, l_i)$ for each frame l_i and each wavelet level j as follows:

$$pw(j, l_i) = \sum_{k \in l_i} |w_k^{(j)}| \tag{2}$$

Finally, we combine all $pw(j, l_i)$s into a 1-dimensional vector $\boldsymbol{p}$ as follows:

$$\boldsymbol{p} = (pw(1, l_1), pw(2, l_1), \cdots, pw(1, l_N), pw(2, l_1), \cdots, pw(M, l_N)) \qquad (3)$$

The vector $\boldsymbol{p}$ is now called "wavelet power pattern".

Examples of wavelet power pattern for different fatigue level of bill in case of $N = 8, M = 9$ are shown in Fig.3. It shows the difference of power pattern in fatigue levels more remarkably, especially on wavelet levels 1, 2, and 5.

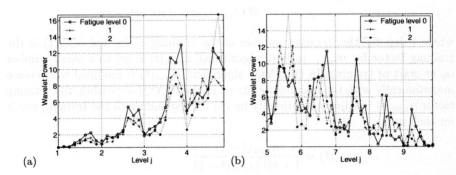

Fig. 3. Examples of wavelet power pattern for each fatigue level of bill: (a) from level 1 to 4, (b) from level 5 to 9.

4   Classification of Fatigue Level by a Competitive Neural Network

In the proposed method, wavelet power patterns are classified into three categories each of which corresponds to a fatigue level of a bill with a competitive neural network. The competitive neural network is trained by a LVQ algorithm.

The structure of the competitive neural network is two-layered; one is the input layer, the other is the competitive layer. The input layer consists of MN neuron units, and each element of the wavelet power pattern $\boldsymbol{p}$ is fed to each unit. In the competitive layer, i-th neuron unit has the predefined category C_i, and has a code vector $\boldsymbol{w}_i$. The unit whose code vector is in the nearest distance to the training vector is called "winner unit", and has the code vector $\boldsymbol{w_{win}}$. The classification of the fatigue level from the wavelet power pattern is carried out by obtaining the category of the winner unit C_{win}.

To perform the classification task appropriately, the competitive neural network is trained by the LVQ algorithm[1]. In the proposed method, the OLVQ1[1] algorithm is used in the training procedure.

4.1 OLVQ1 Algorithm

The OLVQ1 algorithm[1] is an optimized extension of the LVQ1 algorithm[1] which is a basic training algorithm for competitive neural networks. In the LVQ1 and OLVQ1 algorithms, the code vector of the winner unit w_{win} is modified by the following equation:

$$w_{win}(t+1) = w_{win}(t) + \alpha_{win}(t)(p - w_{win}(t))$$
$$\text{if } C_{win} \text{ is correct category} \qquad (4)$$
$$w_{win}(t+1) = w_{win}(t) - \alpha_{win}(t)(p - w_{win}(t))$$
$$\text{if } C_{win} \text{ is incorrect category}$$

where t denotes the iteration number of the training cycle, and $\alpha_{win}(t)$ is the training factor for w_{win}. In both algorithms, $\alpha_{win}(t)$ is set to a small number α_0 at start of training. In the LVQ1 algorithm, $\alpha_{win}(t)$ is modified to decrease monotonically with the training iteration. In the OLVQ1 algorithm, the training factor $\alpha_{win}(t)$ is adjusted optimally by (5) corresponding to the training result for each unit.

$$\alpha_{win}(t) = \frac{\alpha_{win}(t-1)}{1 + \delta(t)\alpha_{win}(t-1)}$$

$$\delta(t) = \begin{cases} +1, \text{ if } C_{win} \text{ is correct category} \\ -1, \text{ if } C_{win} \text{ is incorrect category} \end{cases} \qquad (5)$$

5   Classification Experiment

Under these configurations described above, a classification experiment for US 1 dollar bill has been carried out. In this experiment, ten bills are used for each fatigue level. We measured fifteen acoustic signals per bill. Totally, 150 acoustic signals were taken for each fatigue level. Then we calculate the wavelet power patterns for all measured acoustic signal, and used them for training of the competitive neural network. In the training, 50 wavelet power patterns for each fatigue level are used. After training, the remaining 100 patterns for each fatigue level are used as untrained patterns to obtain classification performance. The result of the experiment is indicated in Table 1 where values mean the correctly classification ratios for untrained patterns. These values are the average performance from ten classification experiments by using different training/untrained pattern combinations to reduce the sensitivity of classification performance.

For comparison, the classification performance by using the acoustic frequency band energy patterns ,which is the best performance of conventional acoustic signal based classification methods[3,4,5] is also shown in Table 1.

From the result, the classification performances are improved entirely by the proposed method. Comparing the results of the best conventional methods[3, 4,5], the classification performance for fatigue level 1 is improved remarkably. The proposed method could obtain better classification performance only use

Table 1. Classification performance

Fatigue	Level 0	Level 1	Level 2	Entire
Proposed(Wavelet)	99.0%	95.0%	89.0%	94.3%
Conventional(Band-Energy)	95.0%	83.0%	92.6%	89.2%

OLVQ1 training algorithm, compared with the best conventional method which uses the combination of two kinds of LVQ algorithms, OLVQ1 and LVQ2. The fact implies that the wavelet power patterns are more feasible feature patterns for classification method based on the competitive neural network. In the Table.1, the classification ratio for fatigue level 2 of the proposed method is lower than the conventional method. But in the conventional method, 17.0% of fatigue level 1 bills are misclassified as fatigue level 2. It means that the conventional method disposes reusable bill(fatigue level 1). This misclassification of the conventional method causes serious problem for practical implementation. At this point, the proposed method is able to improve misclassification problem.

6   Conclusion

A new method has been proposed to classify the fatigue levels of a bill based on the wavelet power pattern by a competitive neural network. The result of classification experiment shows the effectiveness using the wavelet power pattern and the advantage using of the LVQ algorithm for the neuro-classification.

References

1. Kohonen,T.: Self-Organizing Maps, Springer(1997)
2. Daubechies,I.: Ten Lectures on Wavelets, SIAM(1992)
3. Teranishi,M.,Omatu,S.,Kosaka,T.: New and Used Bill Money Classification Using Spectral Information Based on Acoustic Data of Banking Machine, Trans. IEE of Japan, **Vol.117-C, 11** (1997) 1677–1681
4. Teranishi,M., Omatu,S., Kosaka,T.: Classification of New and Used Bills Using Acoustic Cepstrum of a Banking Machine by Neural Networks, Trans. IEE of Japan, **Vol.119-C, 8/9** (1999) 955–961
5. Teranishi,M., Omatu,S., Kosaka,T.: Classification of Three Fatigue Levels for Bills Using Acoustic Frequency Band Energy Patterns, Trans. IEE of Japan, **Vol.120-C, 11**(2000) 1602–1608

Robust Estimator for the Learning Process in Neural Networks Applied in Time Series[*]

Héctor Allende[1,4], Claudio Moraga[2,3], and Rodrigo Salas[1]

[1] Universidad Técnica Federico Santa María; Dept. de Informática;
Casilla 110-V; Valparaíso-Chile; {hallende,rsalas}@inf.utfsm.cl
[2] Technical University of Madrid; Dept. Artificial Intelligence.
E-28660 Boadilla del Monte Madrid; Spain.
[3] University of Dortmund; Department of Computer Science;
D-44221 Dortmund; Germany; moraga@cs.uni-dortmund.de
[4] Universidad Adolfo Ibañez; Facultad de Ciencia y Tecnología.

Abstract. Artificial Neural Networks (ANN) have been used to model non-linear time series as an alternative of the ARIMA models. In this paper Feedforward Neural Networks (FANN) are used as non-linear autoregressive (NAR) models. NAR models are shown to lack robustness to innovative and additive outliers. A single outlier can ruin an entire neural network fit. Neural networks are shown to model well in regions far from outliers, this is in contrast to linear models where the entire fit is ruined. We propose a robust algorithm for NAR models that is robust to innovative and additive outliers. This algorithm is based on the generalized maximum likelihood (GM) type estimators, which shows advantages over conventional least squares methods. This sensitivity to outliers is demonstrated based on a synthetic data set.

Keywords: Feedforward ANN; Nonlinear Time Series; Robust Learning.

1 Introduction

FANN are very good universal approximators of functions first used in the field of engineering. Typical applications involve the analysis of spatio-temporal data called time series analysis in the field of statistics.

Recently, applications of ANN have received a huge impact on the research of nonlinear time series. Researchers in neurocomputing use feedforward networks, e.g. [6], [11], to predict future values of time series only by knowledge from the past. Since ANN are universal approximators for the unknown functions (See [12]), no assumptions need to be hypothesized for the data set. That is, no prior models are built for the unknown functions. This characteristic is quite

[*] This work was supported in part by Research Grant Fondecyt 1010101, in part by Research Grant BMBF CHL-99/023 (Germany) and in part by Research Grant DGIP-UTFSM. Work of C. Moraga was supported by Grant SAB2000-0048 of the Ministry of Education, Culture and Sport (Spain) and the Social Fund of the European Union.

J.R. Dorronsoro (Ed.): ICANN 2002, LNCS 2415, pp. 1080–1086, 2002.

different from the time series which is assumed to be represented by a linear and stationary fractional ARIMA model.

The goal of this paper is to propose a neurocomputing technique to perform robust forecasting for some non-linear time series.

2 Problem Formulation

Time series analysis is often perturbed by occasional unpredictable events that generate aberrant observations. The analyst has to rely on the data to detect which points in time are outliers to estimate the appropriate corrective actions.

The case of unknown location and type of the outlying observations has been considered extensively in the literature for outliers isolation. Fox in [8] introduced the additive and innovation types of outliers. The impact of the outliers on the parameters estimates has been studied by [2] and that on forecasting by [4].

Theory and Practice are mostly concerned with linear methods, such as ARIMA models (See [3]). However, many series exhibit features which cannot be explained in a linear framework. More recently there has been increasing interest in non-linear models. Many types of non-linear models have been proposed in the literature, see for example bilinear models [7] and non-linear ARMA models (NARMA) [5].

There is a large statistical Literature on the topic of robustness toward outliers. A robust statistical method consists of a model, that is not much affected by outliers. [2], [7] and [6] show that least square (LS) methods are quite non-robust. Furthermore, it is important to know that the LS method lacks of robustness not only for classical statistical linear models, but also for non-linear times series neural network predictor model. For LS procedure the ANN modelling shows that the outliers have a local and a semiglobal impact (See [5]).

In this work we focus on robust neural network modelling of non-linear time series which contain outliers. We present some results on the lack of robustness of LS fitting of ANN models for time series. Section (6) introduces a new robust learning process for fitting feedforward predictive models for time series. The proposed method uses a robust algorithm that limits the influence of the gross outliers upon the learning process (parameter estimation). Synthetic data studies are considered.

3 Time Series Processing

In formal terms, a time series is a sequence of vectors, depending on time t: $\underline{x}_t$, $t = 0, 1...$, theoretically, $\underline{x}_t$ can be seen as a function of the time variable t. For practical purpose, however, time is usually viewed in terms of discrete time steps, leading to an instance of x at every end point of a usually fixed size - time interval.

The problem of forecasting is stated as follows: Find a function $h : \Re^{d \times n + l} \rightarrow \Re^d$, where d is the dimension of the sampling space, n is the size of the sampling and l gives the number of exogenous variables. To obtain an estimate $\underline{x}(t +$

k) of the vector $\underline{x}$ at time $t + k$ given the values of $\underline{x}$ up to time t, plus a number of additional time-independent variables (exogenous features) π_i: $\underline{x}_{t+k} = h(x_t, x_{t-1}, ..., \pi_1, ..., \pi_l)$, where k is called the lag for prediction. Typically, $k = 1$, meaning that the subsequent vector should be estimated, but can take any value longer than 1, as well. For the sake of simplicity, we will neglect the additional variables π_i throughout this paper.

Viewed in this way, forecasting becomes a problem of function approximation, where the chosen method is to approximate the continuous-valued function h as closely as possible. In this sense, it can be compared to function approximation or regression problems involving static data vectors, and many methods from this domain can be applied here, as well (See [1] and [5]). This observation will turn out to be important when discussing the use of ANN for forecasting.

Usually the evaluation of forecasting performance is done by computing an error measure E, over a number of time series elements, such as a validation on a test set: $E = \sum_{i=0}^{N} e(\hat{x}(t - i); x(t - i))$ where e is a function measuring a single error between the estimated forecast and actual sequence element. Typically, a distance measure is used here, but depending on the problem, other functions can be used.

4   Neural Networks

ANN are computational models that consist of elementary processing elements called neurons that are connected to others by weighted links. The elementary processor is a rough mathematical representation of the neuron which receives an excitation signal as an input and then it processes it by applying a transfer function to obtain an output signal. Depending how these neurons are connected to others, we obtain different types of ANN architecture or topology. The neurons are organized in three kinds of layers, input or sensory, hidden and output layers.

ANN have received a great attention because they are able to learn from data, and then generalize when they are facing unknown new data. Suppose we have the sample $D_n = \{x_1, ..., x_t, ..., x_n\}$ belonging to some sample space $\chi \subset \Re^d$ and generated by an unknown function $h(\underline{x})$ with the addition of a stochastic component ε.

The task of the "neural learning" is to construct an estimator $g_{ANN}(\underline{x}_t, \underline{w}, D_n) \equiv \hat{h}(\underline{x}_t)$ of $h(\underline{x}_t)$, where $\underline{w} = (w_1, ..., w_p)^T$ is a set of free parameters (known as "connection weights") obtained from the parameter space $\Theta \subset \Re^p$, $\underline{x}_t = (x_{t-1}, ..., x_{t-q})$, and D_n is a finite set of observations. Since no a priori assumptions are made regarding the functional form of $h(\underline{x}_t)$, the neural model $g_{ANN}(x_{t-1}, ..., x_{t-q}, \underline{w})$ is a non-parametric estimator of the conditional density $E[h(x_{t-1}, ..., x_{t-q})/x_{t-1}, ..., x_{t-q}]$.

5   ANN for Time Series

In this paper we will deal with a non-linear Autoregressive (NAR) time series model. The central problem is to construct a function, $h : \Re^q \to \Re$ in a dynamical system with the form: $x_t = h(x_{t-1}, x_{t-2}, ..., x_{t-q}) + \varepsilon_t$, where h is an unknown smooth function and ε_t denotes noise.

A FANN provides a nonlinear approximation to h given by

$$\hat{x}_t = h(x_{t-1}, x_{t-2}, ..., x_{t-q}) = \sum_{j=1}^{\lambda} w_j^{[2]} \gamma_1 (\sum_{i=1}^{p} w_{ij}^{[1]} x_{t-i} + w_{p+1,j}^{[1]}) \qquad (1)$$

where the function $\gamma_1(\cdot)$ is a smooth bounded monotic function.

The estimated parameter $\hat{w}$ is obtained by minimizing iteratively a cost functional $L_n(\underline{w})$ i.e., $\hat{w} = arg\ min\{L_n(\underline{w}) : \underline{w} \in \Theta\}$, $\quad \Theta \subset \Re^p$, where $L_n(\underline{w})$ is for example the ordinary least squares function i.e.

$$L_n(\underline{w}) = \frac{1}{2n} \sum_{i=q+1}^{n} (x_i - g(\underline{x}_{i-1}, \underline{w}))^2 \qquad (2)$$

where $\underline{x}_i = (x_{i-1}, ..., x_{i-q})$

6   Robust Neural Networks Training

In this section we develop a robust method of fitting feedforward neural networks for NAR type models. The robust procedure we use is an adaptation to the neural network setting of a procedure known to be highly robust for fitting linear AR and ARMA models in the presence of outliers.

The loss function disclosed in (2) is very sensitive in the presence of outliers, causing that the error produced by the outlier give a large value compared to the other estimated errors caused by the training algorithm.

We propose a Robust method of fitting NAR models for Time Series with innovative or additive outliers as proposed by [2] for the AR and ARMA models. This method consists in a measure to estimate the fitting of the ANN to the data other than the traditional measure.

The M estimator of the parameter $\hat{w}$ is obtained by minimizing iteratively a cost functional $RL_n(\underline{w}) = \frac{1}{n} \sum_{t=q+1}^{n} \rho(\frac{r_t}{s_t})$, i.e., $\hat{w} = arg\ min\{RL_n(\underline{w}) : \underline{w} \in W \subseteq \Re^p\}$, where ρ is a robustifying function that introduces a bound influence of the outlier on the loss function, and $s_t(r)$ is a data-dependent robust scale estimate whose objective is to make the parameters invariant to scale transformation.

Alternatively, the estimated parameter can be obtained by solving the first order equation

$$\frac{1}{n} \sum_{t=q+1}^{n} \psi(x, r_t/s_t) D_t(w) = 0 \qquad (3)$$

where $D_i(w) = (\frac{\partial}{\partial w_1} g_{ann}\ (\underline{x}_i, \underline{w}_1), \ ..., \ \frac{\partial}{\partial w_1} g_{ann}\ (\underline{x}_i, \ \underline{w}_p))$, $r = x_t - g_{ANN}(x_{t-1}, ..., x_{t-q}, \underline{w})$ and $\psi\ (x, r) := \partial\rho(r)/\partial r$,

[6,7] and present an improved algorithm that takes third- and fourth-order cumulants into account simultaneously and, at the same time, is simpler and faster than Comon's algorithm [1], which our algorithm is based upon.

2   Improved ICA Algorithm

2.1   Cumulants and Independence

Statistical properties of the output data set $\mathbf{u}$ can be described by its moments or, more conveniently, by its cumulants $C^{(\mathbf{u})}$. Cumulants of a given order form a tensor, e.g. third-order cumulants are defined by $C_{ijk}^{(\mathbf{u})} := \langle u_i u_j u_k \rangle$ with $\langle \cdot \rangle$ indicating the mean over all data points. The diagonal elements characterize the distribution of single components. For example, $C_i^{(\mathbf{u})}$, $C_{ii}^{(\mathbf{u})}$, $C_{iii}^{(\mathbf{u})}$, and $C_{iiii}^{(\mathbf{u})}$ are the mean, variance, skewness, and kurtosis of u_i, respectively. The off-diagonal elements (all cumulants with $ijkl \neq iiii$) characterize the statistical dependencies between components. If and only if all components u_i are statistically independent, the off-diagonal elements vanish and the cumulant tensors (of all orders) are diagonal (assuming infinite amount of data).

Thus, ICA is equivalent to finding an unmixing matrix that diagonalizes the cumulant tensors $C^{(\mathbf{u})}$ of the output data u_i, at least approximately. The first order cumulant tensor is a vector and doesn't have off-diagonal elements. The second order cumulant tensor can be diagonalized easily by whitening the input data $\mathbf{x}$ with an appropriate matrix $\mathbf{W}$, yielding $\mathbf{y} = \mathbf{W}\mathbf{x}$. It can be shown that this exact diagonalization of the second order cumulant tensor of the whitened data $\mathbf{y}$ is preserved if and only if the final matrix $\mathbf{Q}$ generating the output data $\mathbf{u} = \mathbf{Q}\mathbf{y}$ is orthogonal, i.e. a pure rotation possibly plus reflections [8]. In general there doesn't exist an orthogonal matrix that would diagonalize also the third- or fourth-order cumulant tensor, thus the diagonalization of these tensors can only be approximate and we need to define an optimization criterion for this approximate diagonalization, which is done in the next section.

2.2   Contrast Function

Since the goal is to diagonalize the cumulant tensors, a suitable contrast function or cost function to be minimized would be the square sum over all third- and fourth-order off-diagonal elements. However, because the square sum over all elements of a cumulant tensor is preserved under any orthogonal transformation $\mathbf{Q}$ of the underlying data $\mathbf{y}$ [9], one can equally well maximize the sum over the diagonal elements

$$\Psi_{34}(\mathbf{u}) = \frac{1}{3!} \sum_\alpha \left(C_{\alpha\alpha\alpha}^{(\mathbf{u})} \right)^2 + \frac{1}{4!} \sum_\alpha \left(C_{\alpha\alpha\alpha\alpha}^{(\mathbf{u})} \right)^2 . \tag{3}$$

The factors $\frac{1}{3!}$ and $\frac{1}{4!}$ arise from an expansion of the Kullback-Leibler divergence of $\mathbf{u}$ and $\mathbf{y}$, which provides a principled derivation of this contrast function [7,

10]. Due to the multilinearity of the cumulants $C^{(\mathbf{u})}_{\ldots}$ in $C^{(\mathbf{y})}_{\ldots}$, (3) can be rewritten as

$$\Psi_{34}(\mathbf{Q}) = \frac{1}{3!} \sum_{\alpha} \underbrace{\left(\sum_{\beta\gamma\delta} Q_{\alpha\beta} Q_{\alpha\gamma} Q_{\alpha\delta} C^{(\mathbf{y})}_{\beta\gamma\delta} \right)^2}_{C^{(\mathbf{u})}_{\alpha\alpha\alpha}} + \frac{1}{4!} \sum_{\alpha} \underbrace{\left(\sum_{\beta\gamma\delta\epsilon} Q_{\alpha\beta} Q_{\alpha\gamma} Q_{\alpha\delta} Q_{\alpha\epsilon} C^{(\mathbf{y})}_{\beta\gamma\delta\epsilon} \right)^2}_{C^{(\mathbf{u})}_{\alpha\alpha\alpha\alpha}},$$

(4)

which is now subject to an optimization procedure to find the orthogonal matrix $\mathbf{Q}$ that maximizes it.

2.3   Givens Rotations

Any orthogonal $N \times N$ matrix such as $\mathbf{Q}$ can be written as a product of $\frac{N(N-1)}{2}$ (or more) Givens rotation matrices $\mathbf{Q}^r$ (for the rotation part) and a diagonal matrix with diagonal elements ± 1 (for the reflection part). Since reflections don't matter in our case, we only consider the Givens rotations.

A Givens rotation is a plane rotation around the origin within a subspace spanned by two selected components. For simplicity and without loss of generality we consider only $N = 2$, so that the Givens rotation matrix reads

$$\mathbf{Q}^r = \begin{pmatrix} \cos(\phi) & \sin(\phi) \\ -\sin(\phi) & \cos(\phi) \end{pmatrix}.$$

(5)

With (5) the contrast function (4) can be rewritten as $\Psi_{34}(\phi) = \Psi_3(\phi) + \Psi_4(\phi)$ with

$$\Psi_n(\phi) := \frac{1}{n!} \sum_{i=0}^{n} d_{ni} \left(\cos(\phi)^{(2n-i)} \sin(\phi)^i + \cos(\phi)^i (-\sin(\phi))^{(2n-i)} \right)$$

(6)

with some constants d_{ni} that depend only on the cumulants before rotation. The next step is to find the angle of rotation ϕ that maximizes Ψ_{34}. For $N > 2$ one has to perform several Givens rotations for each pair of components in order to find the global maximum.

To simplify this equation Comon [1] defined some auxiliary variables $\theta := \tan(\phi)$ and $\xi := \theta - \frac{1}{\theta}$ and derived

$$\Psi_3(\theta) = \frac{1}{3!} \left(\theta + \frac{1}{\theta} \right)^{-3} \sum_{i=1}^{3} a_i \left(\theta^i - (-\theta)^{-i} \right),$$

(7)

$$\Psi_4(\xi) = \frac{1}{4!} (\xi^2 + 4)^{-2} \sum_{i=0}^{4} b_i \xi^i$$

(8)

for (6), with some constants a_i and b_i depending on the cumulants before rotation. To maximize (7) or (8) one has to take their derivative and find the root giving the largest value for $\Psi_{3,4}$. With this formulation only either the third-order or the fourth-order diagonal cumulants can be maximized but not both simultaneously.

In a more direct approach and after some quite involved calculations using various trigonometric addition theorems, we were able to derive a contrast function that combines third- and fourth-order cumulants, is mathematically much simpler, has a more intuitive interpretation, and is therefore easier to optimize and approximate. We found

$$\Psi_{34}\left(\phi\right) = A_0 + A_4 \cos\left(4\phi + \phi_4\right) + A_8 \cos\left(8\phi + \phi_8\right) \,, \tag{9}$$

with some constants A_0, A_4, A_8 and ϕ_4, ϕ_8 that depend only on the cumulants before rotation (see the appendix). The third term comes from the fourth order cumulants only while the first two terms incorporate information from the third- and the fourth-order cumulants.

We found empirically that A_8 is usually small compared to A_4 by a factor of about $1/8$. In the simulations we will therefore also consider an approximate contrast function $\tilde{\Psi}_{34}\left(\phi\right) := A_0 + A_4 \cos\left(4\phi + \phi_4\right)$ which is trivial to maximize. Note that $\tilde{\Psi}_{34}$ still takes third- *and* fourth-order cumulants into account. Contrast functions for third- or fourth-order cumulants only, i.e. Ψ_3, Ψ_4, $\tilde{\Psi}_3$, or $\tilde{\Psi}_4$, can be easily obtained by setting all fourth- or third-order cumulants to zero, respectively.

3   Simulations

We compared four variants of the improved algorithm, namely those based on $\Psi_{34}(\phi)$, $\tilde{\Psi}_{34}(\phi)$, $\Psi_4(\phi)$, and $\tilde{\Psi}_4(\phi)$, with Comon's original algorithm based on $\Psi_4(\xi)$ [1], with the JADE algorithm [2], that diagonalizes 4th order cumulant matrices and the FastICA package using a fixed-point algorithm with different nonlinearities[1] [3].

We assembled three different data sets of length 44218, each mixed by a randomly chosen mixing matrix. To quantify the performances we used the error measure E proposed by Amari et al. [11], which indicates good unmixing by low values and vanishes for perfect unmixing. Table 1 shows the performance of the different algorithms in Matlab implementation[2].

For underlying symmetric distributions all algorithms perform similarly well. If the source distributions are skewed, the additional third order information is crucial and the new method and the FastICA algorithm using a corresponding nonlinearity clearly unmix better. In the case where the sources are both symmetric and asymmetric, only the new algorithm can discriminate between the different distributions. Furthermore the algorithms with approximate contrast work as well as those with the exact contrast.

4   Conclusion

We have proposed an improved cumulant based method for independent component analysis. In contrast to Comon's method [1] it diagonalizes third- and

---

[1] We have always chosen the nonlinearity yielding best performance.

[2] Code is available from http://itb.biologie.hu-berlin.de/~blaschke

Table 1. Unmixing performance for different algorithms and data sets: (i) 5 real acoustic sources [12] (mean kurtosis: 3.808) + 1 normal distributed source ($N(0, 1)$), (ii) 5 skew-normal distributed sources (mean skewness: 0.078) + 1 normal distributed source, (iii) 3 music sources [13] (mean kurtosis: 1.636) + 3 skew-normal distributed sources + 1 normal distributed source. Low values of E indicate good performance. (Simulations have been carried out on a 1.4 GHz AMD-Athlon PC using Matlab implementation).

contrast function / method		unmixing performance (E)			elapsed time/s		
		(i)	(ii)	(iii)	(i)	(ii)	(iii)
$\Psi_{34}(\phi)$	3rd+4th order	0.012	0.053	0.18	1.9	1.9	3.6
$\bar{\Psi}_{34}(\phi)$	3rd+4th order (approx.)	0.013	0.054	0.18	1.9	1.9	3.6
$\Psi_4(\phi)$	4th order	0.011	5.81	1.74	1.7	1.7	6.0
$\bar{\Psi}_4(\phi)$	4th order (approx.)	0.012	6.27	1.93	1.9	1.9	3.6
$\Psi_4(\xi)$	Comon [1]	0.012	4.51	2.11	4.9	4.9	9.2
JADE [2]		0.011	5.11	1.62	1.3	1.3	2.0
FastICA [3]		0.010	0.058	1.00	1.9	0.49	0.7

fourth-order cumulants simultaneously and is thus able to handle linear mixtures of symmetric and skew distributed source signals. Due to its mathematically simple formulation an approximate algorithm can be derived, which shows equal unmixing performance. It can handle also sources with symmetric and asymmteric components.

5   Appendix

From (6) one can derive

$$\Psi_n(\phi) = a_{n0} + s_{n4}\sin(4\phi) + c_{n4}\cos(4\phi) + s_{n8}\sin(8\phi) + c_{n8}\cos(8\phi) \ , \ \text{with}$$

$$a_{30} := \frac{1}{3!}\frac{1}{8}\left[1 \cdot 5\left(C_{111}^{(y)^2} + C_{222}^{(y)^2}\right)\right.$$
$$\left. + 9\left(C_{112}^{(y)^2} + C_{122}^{(y)^2}\right) + 6\left(C_{111}^{(y)}C_{122}^{(y)} + C_{112}^{(y)}C_{222}^{(y)}\right)\right] \ ,$$

$$a_{40} := \frac{1}{4!}\frac{1}{64}\left[1 \cdot 35\left(C_{1111}^{(y)^2} + C_{2222}^{(y)^2}\right)\right.$$
$$+ 16 \cdot 5\left(C_{1112}^{(y)^2} + C_{1222}^{(y)^2}\right) + 12 \cdot 5\left(C_{1111}^{(y)}C_{1122}^{(y)} + C_{1122}^{(y)}C_{2222}^{(y)}\right)$$
$$\left. + 36 \cdot 3\,C_{1122}^{(y)^2} + 32 \cdot 3\,C_{1112}^{(y)}C_{1222}^{(y)} + 2 \cdot 3\,C_{1111}^{(y)}C_{2222}^{(y)}\right] \ ,$$

$$s_{34} := \frac{1}{3!}\frac{1}{4}\left[6\left(C_{111}^{(y)}C_{112}^{(y)} - C_{122}^{(y)}C_{222}^{(y)}\right)\right] \ ,$$

$$c_{34} := \frac{1}{3!}\frac{1}{8}\left[1 \cdot 3\left(C_{111}^{(y)^2} + C_{222}^{(y)^2}\right)\right.$$
$$\left. - 9\left(C_{112}^{(y)^2} + C_{122}^{(y)^2}\right) - 6\left(C_{111}^{(y)}C_{122}^{(y)} + C_{112}^{(y)}C_{222}^{(y)}\right)\right] \ ,$$

$$s_{44} := \frac{1}{4!}\frac{1}{32}\left[8 \cdot 7\left(C_{1111}^{(y)}C_{1112}^{(y)} - C_{1222}^{(y)}C_{2222}^{(y)}\right)\right.$$

$$+ 48 \left(C_{1112}^{(y)} C_{1122}^{(y)} - C_{1122}^{(y)} C_{1222}^{(y)} \right) + 8 \left(C_{1111}^{(y)} C_{1222}^{(y)} - C_{1112}^{(y)} C_{2222}^{(y)} \right) \Big] ,$$

$$c_{44} := \frac{1}{4!} \frac{1}{16} \left[1 \cdot 7 \left({C_{1111}^{(y)}}^2 + {C_{2222}^{(y)}}^2 \right) \right.$$

$$- 16 \left({C_{1112}^{(y)}}^2 + {C_{1222}^{(y)}}^2 \right) - 12 \left(C_{1111}^{(y)} C_{1122}^{(y)} + C_{1122}^{(y)} C_{2222}^{(y)} \right)$$

$$\left. - 36 {C_{1122}^{(y)}}^2 - 32 C_{1112}^{(y)} C_{1222}^{(y)} - 2 C_{1111}^{(y)} C_{2222}^{(y)} \right] ,$$

$$s_{48} := \frac{1}{4!} \frac{1}{64} \left[8 \left(C_{1111}^{(y)} C_{1112}^{(y)} - C_{1222}^{(y)} C_{2222}^{(y)} \right) \right.$$

$$\left. - 48 \left(C_{1112}^{(y)} C_{1122}^{(y)} - C_{1122}^{(y)} C_{1222}^{(y)} \right) - 8 \left(C_{1111}^{(y)} C_{1222}^{(y)} - C_{1112}^{(y)} C_{2222}^{(y)} \right) \right] ,$$

$$c_{48} := \frac{1}{4!} \frac{1}{64} \left[1 \left({C_{1111}^{(y)}}^2 + {C_{2222}^{(y)}}^2 \right) \right.$$

$$- 16 \left({C_{1112}^{(y)}}^2 + {C_{1222}^{(y)}}^2 \right) - 12 \left(C_{1111}^{(y)} C_{1122}^{(y)} + C_{1122}^{(y)} C_{2222}^{(y)} \right)$$

$$\left. + 36 {C_{1122}^{(y)}}^2 + 32 C_{1112}^{(y)} C_{1222}^{(y)} + 2 C_{1111}^{(y)} C_{2222}^{(y)} \right] ,$$

and $s_{38} := c_{38} := 0$. With this it is trivial to determine the constants for $\Psi_{34}(\phi) = \Psi_3(\phi) + \Psi_4(\phi)$ in the form given in (9). We find: $A_0 := a_{30} + a_{40}$, $A_4 := \sqrt{(c_{34} + c_{44})^2 + (s_{34} + s_{44})^2}$, $A_8 := \sqrt{c_{48}^2 + s_{48}^2}$, $\tan(\phi_4) := -\frac{s_{34} + s_{44}}{c_{34} + c_{44}}$, $\tan(\phi_8) := -\frac{s_{48}}{c_{48}}$. The coefficients A_0, A_8, A_4 and ϕ_8, ϕ_4 are functions of the cumulants of 3rd and 4th order of the centered and whitened signals y.

References

1. Comon, P.: Tensor diagonalization, a useful tool in signal processing. In Blanke, M., Soderstrom, T., eds.: IFAC-SYSID, 10th IFAC Symposium on System Identification. Volume 1., Copenhagen, Denmark (1994) 77–82
2. Cardoso, J.F., Souloumiac, A.: Blind beamforming for non Gaussian signals. IEE Proceedings-F **140** (1993) 362–370
3. Hyvärinen, A.: Fast and robust fixed-point algorithms for independent component analysis. IEEE Transactions on Neural Networks **10** (1999) 626–634
4. Hyvärinen, A., Karhunen, J., Oja, E.: Independent Component Analysis. Wiley Series on Adaptive and Learning Systems for Signal Processing, Communications, and Control. John Wiley&Sons (2001)
5. Lee, T.W., Girolami, M., Bell, A., Sejnowski, T.: A unifying framework for independent component analysis. International Journal of Computers and Mathematics with Applications **39** (2000) 2–21
6. Comon, P.: Independent component analysis. In Lacoume, J., ed.: Proc. Int. Sig. Proc. Workshop on Higher-Order Statistics, Chamrousse, France (1991) 29–38
7. Cardoso, J.F.: High-order contrasts for independent component analysis. Neural Computation **11** (1999) 157–192
8. Hyvärinen, A.: Survey on independent component analysis. Neural Computing Surveys **2** (1999) 94–128
9. Deco, G., Obradovic, D.: An Information-Theoretic Approach to Neural Computing. Springer Series in Perspectives in Neural Computing. Springer (1996)
10. McCullagh, P.: Tensor Methods in Statistics. Monographs on Statistics and Applied Probability. Chapmann and Hall (1987)

11. Amari, S., Cichocki, A., Yang, H.: Recurrent neural networks for blind separation of sources. In: Proc. of the Int. Symposium on Nonlinear Theory and its Applications (NOLTA-95), Las Vegas, USA (1995) 37–42
12. John F. Kennedy Library, Boston: Sound excerpts from the speeches of president John F. Kennedy. (http://www.jfklibrary.org)
13. Pearlmutter, B.: 16 clips sampled from audio CDs. (http://sweat.cs.unm.edu/~bap)

Finding the Optimal Continuous Model for Discrete Data by Neural Network Interpolation of Fractional Iteration

Lars Kindermann[1], Achim Lewandowski[2], and Peter Protzel[2]

[1] RIKEN Brain Science Institute, Lab for Mathematical Neuroscience
Wako-shi, Saitama 351-0106, Japan
kindermann@brain.riken.go.jp
www.mns.brain.riken.go.jp/~kinderma

[2] Dept. of Electrical Engineering and Information Technology
Chemnitz University of Technology, Germany
peter.protzel@e-technik.tu-chemnitz.de, lewandowski@alewand.de
www.infotech.tu-chemnitz.de~proaut

Abstract. Given the complete knowledge of the state variables of a dynamical system at fixed intervals, it is possible to construct a mapping, which is a perfect discrete time model of the system. To embed this into a continuum, the translation equation has to be solved for this mapping. However, in general, neither existence nor uniqueness of solutions can be guaranteed, but fractional iterates of the mapping computed by a neural network can provide regularized solutions that exactly comply with the laws of physics for several examples. Here we extend this method to continuous embeddings which represent the true trajectories of the dynamical system.

1 Introduction

Data very often is sampled at fixed time intervals. If a model is derived from this data it is in general valid only on this grid. Making predictions for other input values means interpolation. Interpolation in general introduces a model: linear, bicubic or spline interpolation are most commonly used. Neural networks often assume to be model-free which just means that one does not really care about the model. Even if the model can represent the training data perfectly, there will be in general some interpolation error.

Think of having time series data sampled monthly and you have constructed a very good model to predict the next month. But now you are requested to make a prediction for next week. Or you have measured extensively how a one meter thick concrete wall blocks some radiation passing through it. Now you are asked how a half meter wall will perform instead. Having a mathematical model with "physical" meaning will usually allow you to address these questions in an intelligent way, but just based on a black-box style model like a neural network this is not a trivial task.

We will investigate this question in an otherwise ideal case and assume the following conditions fulfilled. For simplicity only discrete time-sampling is considered but the arguments can be transferred to space accordingly. Without loss of generality we will set the sampling interval Δt to "1" which will result in a nice formalism later. X_t denotes a vector, describing the state of the system at time t completely. We assume

J.R. Dorronsoro (Ed.): ICANN 2002, LNCS 2415, pp. 1094–1099, 2002.

the system is deterministic and autonomous, there is no external influence and the system is translation invariant, just adding a constant to every time will not change any experiment.

We also assume that we have a very large set of very exact data samples $(X_t, X_{t+\Delta t})$, which represents the *time-one mapping* function f of the system

$$X_{t+1} = f(X_t) \tag{1}$$

arbitrarily well, so that we can derive from it some (neural network) model, which approximates f arbitrarily exact. The discrete-time dynamics of the system is thus totally defined by f and a discrete trajectory of some initial state X_0 can be computed by repeatedly *iterating* the function f:

$$X_t = f^t(X_0) \quad t = 0, 1, \ldots \tag{2}$$

Is it theoretically possible to compute the state at other (non-integer) times from this knowledge and if so, how to do it? Mathematically this would mean to find a function $F(X, t)$ for real t, that solves the translation equation

$$F(F(X, s), t) = F(X, s + t)$$

for *real* s and t under the condition $F(X, n) = f^n(X)$ for *integer* n. If we formally define $f^t(X) == F(X, t))$ for real t, we extend the concept of iteration to non-integer iteration counts.

2 A Historical View on This Problem

This questions belongs to the mathematical area of *functional equations*. The problem to find the state at time $t = 1/2$ is equivalent to solve the functional equation

$$\varphi(\varphi(x)) = f(x) \tag{3}$$

for the unknown function φ, which is usually called an *iterative root* of f.

This problem dates back to Charles Babbage and his *Essay towards an calculus of functions* from 1815 [1], where he investigates the solutions of $\varphi(\varphi(x)) = x$, the so called *roots of identity*. The widely used notation $\varphi = f^{1/2}$ suggests the idea to allow any exponent for the iteration: $f^{1/n}$ is called the n-th iterative root of f and $f^{m/n}$ a *fractional iterate*, also defined by a functional equation

$$\varphi^m(x) = f^n(x). \tag{4}$$

Abel found 1826 a method to solve some iterative roots, based on linearization of the now so called Abel functional equation $\varphi(f(x)) = \varphi(x + c)$ [2].

In his attempt to find closed expressions h for a given function f under iteration $f^n(x) = h(x, n)$, a problem today usually investigated by chaos theory, Schröder [3] extended the concept of iteration towards real iteration indices and showed how to find solutions of eq. (3) at least near fixpoints of f by solving the now famous Schroeder equation $\varphi(f(x)) = c\varphi(x)$, which represents the eigenvalue problem of functional algebra.

Of special interest e.g. in the theory of computational complexity are iterative roots of the exponential function, i.e. solutions of

$$\varphi(\varphi(x)) = e^x. \tag{5}$$

Hardy showed 1924 [4] that there are infinitely many solutions of eq. 5, but every member of a monotonically rising subset of these solutions has the remarkable property to grow faster than any polynomial but slower than any power function. All solutions depend on some arbitrary chosen function on an interval, so the question came up for a "natural" solution. Kneser found 1938 an analytical solution to this problem in response to a "request from industry" [5].

While Isaacs solved eq. 3 in 1950 [6] and Zimmermann eq. 4 in 1978 [9] for arbitrary self mappings of abstract sets based on the concepts of orbits, these results are hard to apply to concrete questions.

Even more complicated is the task to construct continuos embeddings which take the form of iteration semigroups [8], there exists a close relationship to dynamical systems and chaos theory.

Targonsky related this mathematical questions to a philosophical discussion about the concept of time: If we ever find a physical process represented by a mapping function which is not embeddable or does not have iterative roots of every order, this suggests a minimal time interval, the *chronon* [7].

The current standard opus on this topic is the book "iterative functional equations" by Kuczma et.al. from 1990 which summarizes most known facts about these topics with great detail [11]. But despite this long history the problem of finding iterative roots is still not yet solved. A recent (2001) survey article on the current research on iterative roots states *"...one should not expect results on iterative roots in a general situation. In fact, even roots of polynomials are not described. Even worse: we do not know whether every complex cubic polynomial has a square root..."* [12]

For polynomials of order two at least the question of existence had been solved by Rice et.al. in "When is $f(f(z)) = az^2 + bz + c$ for all complex z?" [11] in 1980.

For linear self-mappings, i.e. square matrices, a *matrix power* operator A^t for real t based on eigenvalue decomposition is well known and implemented in numerical packages like Matlab, but for most other cases there is no numerical method published to compute iterative roots of given mappings.

3 Computing Fractional Iterates with a Neural Network

Confronted with an engineering problem from steel industry, we were required to compute iterative roots of a process we only knew by a dataset. The input-output mapping of the whole process could be well modelled by a neural network, but the process consisted of a succession of 7 identical processing steps in a row which should be separated. Thus it was necessary to compute the 7th iterative root of this model. We found some straight forward extensions to the MLP model that can be used to compute approximations of iterative roots and fractional iterates [13].

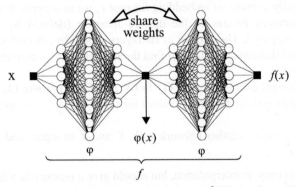

Fig. 1. A MLP computes an iterative root of the function f if both subnets which represent φ posses identical weights.

The basic idea is to use a structured network like Figure 1 to approximate f and additionally force the weights of the two subnets to be identical. Once both goals are reached, each subnet will be a model for the iterative root of f. Different algorithms for training such networks were presented by us recently [14].

This method can be extended to n-th iterative roots, where a composition of n subnetworks is trained towards f. A fractional iterate $f^{m/n}$ can then be constructed by m of these sub-networks. This is similar to the backpropagation through time formalism except in BPTT the intermediate values are usually known. Another method is to use a recurrent network which is trained in n loops and later recalled in m loops.

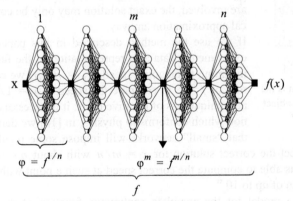

Fig. 2. Finding a fractional iterate $f^{m/n}(x)$.

In our paper [16] we could show that several basic experiments from physics can be modelled perfectly by this method for fractions of the measured time interval. But the number of sublayers is practically limited to about 10 do to learning restrictions imposed by the layer structure [14], so at most the 10th root could be calculated.

4 Continuous Embeddings

To construct a really continuous embedding for *real* t, we now apply three steps:
1. Compute fractional iterates of the desired function (defined by a dataset) as described in chapter 3. The number of sublayers should be selected as high as possible. This will theoretically allow to find the mathematically correct values of the fractional iterates.
2. Extend the original dataset $(X \rightarrow f(X))$ to $(X, t \rightarrow f^t(X))$ where $(X, 1 \rightarrow f(X))$ is the original data and the numerical values for other $t = \frac{1}{n}, \frac{2}{n}, ..., 1$ are derived from step 1.
3. Use this data to train another network with X and t as inputs and $f^t(X)$ as the desired output.

Of course step 3 is only an interpolation, but should give a reasonable interpolation for any value of t when f is relatively smooth in t. This method will also make a good model for short term extrapolation, i.e. $t > 1$ [15].

5 Example

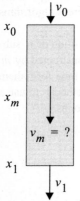

Fig. 3. What is the speed of the object at point x_m?

An object falls down within a closed tube of length $\Delta x = 1$. We inject it with different speeds v_0 and measure it's velocity v_1 after passing the tube. The goal is to compute from this data only the speed at any other point within the tube. In case when there is friction involved this is analytically a rather difficult task that involves solving complicated differential equations, even if we know the exact physics of the process. If the exact differential equation is not know, e.g. we do not know which types of frictions are involved, the exact solution may only be computed by numerical approximation anyway.

If we use the method described in this paper and think of the experimental data as a representation of the function f that maps incoming to outgoing speed, $v_1 = f(v_0)$, we could use a iterative root of $f^{1/2}$ to compute the speed at $x = 1/2$. Because of the non-uniqueness of iterative roots, it is necessary to select the solution which conforms to physics, in [16] we demonstrated recently that "small" networks will impose some regularization that will *exactly* select the correct solution for $x = m/n$ with about $n < 10$ and $m \leq n$. The network was able to compute the correct speed at such a point with a relative numerical precision of up to 10^{-6}.

To get a model for the complete continuous function of the speed vs. height, $v = f(v_0, x)$, we used another MLP and trained it with this data. Because it interpolates between this exact points, the mean error increased to about 10^{-5} but it is still a very good model for the actual physical process, especially when considering, that we used no physical laws at all to compute these results, just experimental data.

6 Conclusions

The mathematically as interesting as difficult problem of iterative roots and continuous iteration remains a pretty much unsolved problem with even no standard numerical methods available yet. However, neural networks have successfully proven to give sufficient approximations to some problems from industrial system identification [17] and also showed their capability to model several fundamental experiments from physics exactly. This may open the door to other applications of this mathematical theory.

References

1. C. Babbage: Essay towards the Calculus of functions. Phil. trans. Royal Soc. London 105 (1815) 389-424
2. N. H. Abel: Determination d'une function au moyen d'une equation qui ne contient qu'une seule variable. Manuscript 1824, in: Oeuvres compl'etes, Christiania (1881)
3. E. Schröder: Über iterierte Funktionen. Math. Ann. 3 (1871) 296-322
4. G.H. Hardy, E. Cunningham: Orders of infinity. Cambridge Tracts in Mathematics 12 (1924)
5. H. Kneser: Reelle analytische Lösungen der Gleichung $f(f(x))$ = e^x und verwandter Funktionalgleichungen. J. reine angew. Math. 187 (1950) 56-67
6. R. Isaacs: Iterates of fractional Order. Canad J. Math. 2 (1950) 409-416
7. G. Targonski: An Iteration theoretical approach to the concept of time. Colloques Internationaux du C.N.R.S. 229, Transformations ponctuelles et leurs applications, Toulouse (1973) 259-271
8. M.C. Zdun: Continuous iteration semigroups. Boll. Un. Mat. Ital. 14 A (1977) 65-70
9. G. Zimmermann: Über die Existenz iterativer Wurzeln von Abbildungen. Dissertation, Marburg/Lahn (1978)
10. R.E. Rice, B. Schweizer & A. Sklar: When is $f(f(z))$ = az^2+bz+c for all complex z? Amer. Math. Monthly 87 (1980) 252-263
11. M. Kuczma, B. Choczewski & R. Ger: Iterative Functional Equations. Cambridge University Press, Cambridge (1990)
12. K. Baron & W. Jarczyk: Recent results on functional equations in a single variable, perspectives and open problems. Aequationes Math. 61. (2001), 1-48
13. L. Kindermann: Computing Iterative Roots with Neural Networks. Proc. Fifth Conf. Neural Information Processing, ICONIP (1998) Vol. 2:713-715
14. L. Kindermann & A. Lewandowski: A Comparison of Different Neural Methods for Solving Iterative Roots. Proc. Seventh Int'l Conf. on Neural Information Processing, ICONIP, Taejon (2000) 565-569
15. L. Kindermann & T.P. Trappenberg, Modeling time-varying processes by unfolding the time domain. Proc Int'l Joint Conf. on Neural Networks, IJCNN, Washington DC (1999)
16. L. Kindermann & P. Protzel: Physics without laws - Making exact predictions with data based methods. Proc. Int'l Joint Conf on Neural Networks, Honolulu (2002)
17. L. Kindermann, P. Protzel, F. Schmid & O. Gramckow, (SIEMENS AG): Process and device for determining an intermediate section of a metal strip. International Patent WO9942232 (1999)

Support Vector Robust Algorithms for Non-parametric Spectral Analysis

José Luis Rojo-Álvarez[1], Arcadi García-Alberola[2], Manel Martínez-Ramón[1], Mariano Valdés[2], Aníbal R. Figueiras-Vidal[1], and Antonio Artés-Rodríguez[1]

[1] Dept. Signal Theory and Communications, Univ. Carlos III de Madrid, 28911 Leganés, Madrid, Spain***
[2] Dept. of Cardiology, Hospital GU Virgen de la Arrixaca, 30150 El Palmar, Murcia, Spain
{jlrojo,manel,arfv,antonio}@tsc.uc3m.es; algamur@teleline.es
http://www.tsc.uc3m.es

Abstract. A new approach to the non-parametric spectral estimation on the basis of the Support Vector (SV) framework is presented. Two algorithms are derived for both uniform and non-uniform sampling. The relationship between the SV free parameters and the underlying process statistics is discussed. The application in two real data examples, the sunspot numbers and the Heart Rate Variability, shows the higher resolution and robustness in SV spectral analysis algorithms.

1 Introduction

Non-parametric spectral analysis of time series is a widely scrutinized framework. The most relevant of the classical spectral estimators, the Welch periodogram and the Blackman-Tukey correlogram, are based on Fourier-Transform representations, either for the observed time series or for its estimated autocorrelation function [1], so their main advantages are the low computational burden required and their simplicity. On the other hand, their spectral resolution is limited due to the effect of windowing. The spectral analysis of non-uniform sampled series has also been suggested by means of the Lomb periodogram [2,3]. In both cases, the Periodogram exhibits a high sensitivity to outliers.

An alternative approach to the classical non-parametric spectral analysis can be drawn from the Support Vector (SV) framework, which was first suggested to obtain maximum margin separating hyperplanes in classification problems [4,5, 6]. In this work, we propose to modify the standard SV regression algorithm and the cost function to provide an adequate approach to non-parametric spectral analysis problems.

SV algorithms for the Welch periodogram and the Lomb periodogram are first derived. Two application examples, the sunspot numbers for the SV-Welch and the Heart Rate Variability (HRV) for the SV-Lomb algorithms are shown, to support the potential of this new approach.

*** This work has been partially supported by the project 07T/0046/2000 of the CAM

J.R. Dorronsoro (Ed.): ICANN 2002, LNCS 2415, pp. 1100–1105, 2002.
© Springer-Verlag Berlin Heidelberg 2002

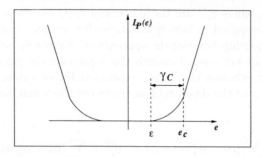

Fig. 1. Robust SV cost function.

2 SV-Welch Periodogram

The model for the harmonic decomposition of a sequence $\{y_n\}$ can be expressed in terms of the discrete-time harmonic Fourier series. The spectral analysis of the observed signal can be denoted as follows,

$$y_n = \sum_{i=1}^{N_\omega} A_i \cos(\omega_i n + \phi_i) \tag{1}$$

where the unknown parameters are amplitudes A_i, phases ϕ_i and frequencies ω_i for a number N_ω of sinusoidal components. This is a non-linear relationship, except in the case where frequencies are known, say $\omega_i = 2\pi i/N$. In this case, Eq. (1) can be linearly expressed by switching to Cartesian coordinates $c_i = A_i \cos(\phi_i)$ and $d_i = A_i \sin(\phi_i)$, and the model can be written down as

$$y_n = \sum_{i=1}^{N_\omega} [c_i \cos(\omega_i n) + d_i \sin(\omega_i n)]. \tag{2}$$

Several robust cost functions have been used in SV regression, as the Vapnik's loss function [4], Huber's robust cost [7], or the ridge regression approach [8]. We propose here a more general cost function, which has the above mentioned ones as particular cases, we force the minimization of

$$\frac{1}{2}\left(\|\bar{c}\|^2 + \|\bar{d}\|^2\right) + \frac{1}{2\gamma}\sum_{k \in I_1}\left(\xi_k^2 + \xi_k^{*2}\right) + C\sum_{k \in I_2}\left(\xi_k + \xi_k^*\right) \tag{3}$$

constrained to

$$y_k - \sum_{i=1}^{N_\omega} c_i \cos(\omega_i k) - \sum_{i=1}^{N_\omega} d_i \sin(\omega_i k) \leq \varepsilon + \xi_k \tag{4}$$

$$-y_k + \sum_{i=1}^{N_\omega} c_i \cos(\omega_i k) + \sum_{i=1}^{N_\omega} d_i \sin(\omega_i k) \leq \varepsilon + \xi_k^* \tag{5}$$

$$\xi_k, \xi_k^* \geq 0 \tag{6}$$

for $k = 1, \ldots, N$, where $\xi_k^{(*)}$ are the losses, and I_1, I_2 are the sets of samples for which losses are required to have quadratic or linear cost, respectively. Figure 1 depicts the relationship between the approximation error e_n and its corresponding loss. Note that for γ small enough this represents the regularized Vapnik's ε-insensitive cost, whereas for $\varepsilon = 0$ it represents Huber's robust cost.

The derivation of the dual problem shows the Cartesian components can be expressed as

$$c_l^W = \sum_{k=1}^{N} (\alpha_k - \alpha_k^*) \cos(\omega_l k); \quad d_l^W = \sum_{k=1}^{N} (\alpha_k - \alpha_k^*) \sin(\omega_l k) \qquad (7)$$

where $\alpha_k^{(*)}$ are the Lagrange multipliers for the constraints in (4) and (5). Therefore, the lth coefficients are the cross-correlation of the Lagrange multipliers and the sinusoid with frequency ω_l. These conditions are introduced into the Lagrange functional in order to remove the primal variables, and it is easy to show that if we write down

$$R_{cos}^W(m, k) = \sum_{i=1}^{N_\omega} \cos(\omega_i m) \cos(\omega_i k); \quad R_{sin}^W(m, k) = \sum_{i=1}^{N_\omega} \sin(\omega_i m) \sin(\omega_i k) \quad (8)$$

then the regularized L_D dual problem, to be maximized with respect to the dual variables only, can be solved by maximizing

$$\begin{aligned} L_D = -\frac{1}{2} (\bar{\alpha} - \bar{\alpha}^*)^T \left[\mathbf{R_{cos}^W} + \mathbf{R_{sin}^W} \right] (\bar{\alpha} - \bar{\alpha}^*) + (\bar{\alpha} - \bar{\alpha}^*)^T \bar{y} - \\ - \varepsilon \bar{1}^T (\bar{\alpha} + \bar{\alpha}^*) - \frac{\gamma}{2} \left(\bar{\alpha}^T \mathbf{I} \bar{\alpha} + \bar{\alpha}^{*T} \mathbf{I} \bar{\alpha}^* \right) \end{aligned} \qquad (9)$$

constrained to $0 \leq \alpha, \alpha^* \leq C$.

It can be easily shown that the SV approach leads to the non-orthogonal, regularized representation of the observation data upon the signal subspace which is generated by the Cartesian representation of the sinusoidal components in the model. The implications of this interpretation are currently being studied.

3   SV-Lomb Periodogram

For series of data samples $\{y_{t_n}\}$ which have been non-uniformly sampled in the corresponding time instants $\{t_1, \cdots, t_N\}$, the model is

$$y_{t_n} = \sum_{k=1}^{N_\omega} A_i \cos(\omega_i t_n + \phi_i) = \sum_{k=1}^{N_\omega} [c_i \cos(\omega_i t_n) + d_i \sin(\omega_i t_n)] \qquad (10)$$

and in this case the solution coefficients are given by

$$c_l^L = \sum_{k=1}^{N} (\alpha_k - \alpha_k^*) \cos(\omega_l t_k); \quad d_l^L = \sum_{k=1}^{N} (\alpha_k - \alpha_k^*) \sin(\omega_l t_k). \qquad (11)$$

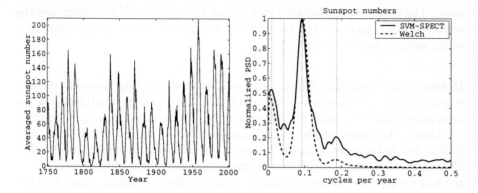

Fig. 2. Sunspot numbers. **Left:** Averaged and decimated sunspot number series. **Right:** Normalized Welch and SV spectral estimators (natural units).

Taking into account

$$R^L_{cos}(m, k) = \sum_{i=1}^{N_w} \cos(w_i t_m) \cos(w_i t_k); \quad R^L_{sin}(m, k) = \sum_{i=1}^{N_w} \sin(w_i t_m) \sin(w_i t_k)$$

(12)

the problem is now equivalent to maximize the dual functional

$$L_D = -\frac{1}{2} (\bar{\alpha} - \bar{\alpha}^*)^T \left[\mathbf{R^W_{cos}} + \mathbf{R^W_{sin}} \right] (\bar{\alpha} - \bar{\alpha}^*) + (\bar{\alpha} - \bar{\alpha}^*)^T \bar{y} -$$
$$- \varepsilon \bar{1}^T (\bar{\alpha} + \bar{\alpha}^*) - \frac{\gamma}{2} (\bar{\alpha}^T \mathbf{I} \bar{\alpha} + \bar{\alpha}^{*T} \mathbf{I} \bar{\alpha}^*)$$

(13)

constrained to $0 \le \alpha, \alpha^* \le C$.

4   Application Examples

Periodogram and sunspot numbers. A classical application of the classical spectral estimation is the search of periodicity in sunspot numbers. This experiment has been taken from [1, pp. 161-4], where preprocessing is detailed. The updated series of 3036 monthly sunspot numbers for the years from 1750 to 2001 were analyzed, aiming to study the details in the band between 0 and 2 cycles per year. The averaged Welch and SV-Welch periodograms were calculated. The free parameters in SV were $\varepsilon = 0$, $C = 100$ and $\gamma = 0.01$ as this set was previously shown to give an enhanced harmonic spectral structure.

Figure 2 shows the normalized periodograms. The Welch estimator points the 10.5 year dominant component, one sub-harmonic and one harmonic. For the same frequency resolution, the SV estimator marks 4 components, which is coherent with the analysis performed in [1], where different non-parametric methods highlighted different components (0.95, 0.188, 0.25 and 0.45 cycles per year). The peak frequencies were very similar in both the SV estimator and the several classical analysis, but the first one was able to mark all the peaks in a single spectral representation.

Robust HRV analysis. The natural oscillations of the time between consecutive heart beats are known as HRV, and it is related with the modulation of the sympathetic and the vagal nervous system on the heart rhythm [9]. In healthy conditions, the power of the oscillations observed in the LF band (from 0.04 to 0.15 Hz) is balanced with the power in the HF band (from 0.15 to 0.4 Hz). The association of the time between two consecutive beats to the time where the first couple beat happens leads to an non-uniform sampled series. Besides, ectopic beats frequently appear, but they are not related with the modulation of the autonomic system and they should be excluded from the analysis.

The Lomb periodogram has allowed to overcome the non-uniform sampling problem. Nevertheless, the detection of ectopic beats has to be manually validated by a medical expert, and this is a much time-consuming task. This fact has been a main limitation for the application of the HRV into the habitual clinical practice, and it suggests the introduction of a robust, ectopic-insensitive spectral analysis.

The SV-Lomb can be adapted by tuning its free parameters according to the nature of the HRV signal. The impact of the ectopic beats can be limited by constraining the amplitude of the Lagrange multipliers (γC) to a low enough level. Additionally, the SV-Lomb method has been observed to model the high-frequency noise from the high-frequency sinusoidal components which are non-uniform sampled, hence overestimating the HF power. This effect can be limited by choosing an insensitivity level ε and a moderate frequency resolution.

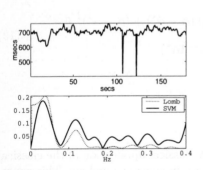

Figure 3 shows an application example to a HRV series. The free parameters were chosen as $C = 4$, $\gamma = 1$, $\epsilon = 3$ and the frequency resolution $\Delta f = 0.001Hz$. Note the SV-Lomb estimators are less affected by the inclusion of ectopic beats in the window than Lomb periodogram is. In the first case, the spectral estimator holds the power in the LF and HF bands, whereas the Lomb periodogram spectrum is flatter

Fig. 3. Heart Rate Variability. **Up:** time signal (dotted) and SV-Lomb aproximation (continuous). **Down:** Lomb periodogram (dotted) and SV-Lomb (continuous) normalized spectral estimators.

due to the influence of the wide-band spectral contamination of the ectopics. The overestimation of the frequency components near to 0.4 Hz in the SV-Lomb estimator can be observed. Further and deeper study is to be done in order to explore the possibilities of this method in HRV series analysis.

5 Conclusion

The application of the SV framework to classical, non-parametric spectral analysis is a promising framework. For uniform sampling applications, the spectral resolution of SV is higher than the corresponding classical methods. For non-uniform sampling, a more enhanced harmonic spectral structure can be captured by the SV method when compared to the Lomb periodogram. The SV spectral analysis seems to be a good approach for the robust measurement of the HRV in a clinical environment. Further work is to be done in several directions. First, the theoretical and statistical analysis of the SV spectral algorithms have to be completed. Second, the robust approach to the HRV problem has to be delimited and performed. Third, other application fields should be proposed and explored, where the advantages of the SV framework can be of interest, such as voice processing or astronomy data analysis. Finally, other spectral SV algorithms, such as signal-and-noise subspace decomposition, ARMA modeling and deconvolution, will be developed.

References

1. Marple, S.L.: Digital Spectral Analysis with Applications. Prentice-Hall, NJ, USA, 1987
2. Laguna, P., Moody, G.B., Mark, R.G.: Power spectral density of unevenly sampled data by least-square analysis: performance and application to heart rate signals. IEEE Trans. Biomed. Eng. Vol. 45, pp. 698-715. 1998.
3. Press, W.H., Vetterling, W.T., Teukolsky, S.A., Flannery, B.P.: Numerical Recipes in C. The Art of Scientific Computing. Cambridge, NY, USA. 1997
4. Vapnik, V.: The Nature of Statistical Learning Theory Springer–Verlag, NY, 1995.
5. Schölkopf, B., Sung, K.: Comparing Support Vector Machines with Gaussian Kernels to Radial Basis Function Classifiers IEEE Trans. on Signal Proc. Vol 45, n 11. 1997.
6. Pontil, M., Verri, A.: Support Vector Machines for 3D Object Recognition. IEEE Trans. on Pattern Anal. and Mach. Intell. Vol. 20, n 6, 1998.
7. Müller, K.R., Smola, A., Rätsch, G.R., Schölkopf, B., Kohlmorgen, J., Vapnik, V.: Predicting Time Series with Support Vector Machines. In Advances in Kernel Methods. Support Vector Learning. MIT Press, MA, USA. 1999.
8. Cristianini, N., Shawe-Taylor, J.: An Introduction to Support Vector Machines and Other Kernel-based Learning Methods. Cambridge Un. Press, 2000.
9. Malik, M., Camm, A.J.: Heart Rate Variability Futura Pub Co. 1995.

Support Vector Method for ARMA System Identification: A Robust Cost Interpretation

José Luis Rojo-Álvarez, Manel Martínez-Ramón, Aníbal R. Figueiras-Vidal, Mario de Prado-Cumplido, and Antonio Artés-Rodríguez

Dept. Signal Theory and Communications, Universidad Carlos III de Madrid, 28911 Leganés, Madrid, Spain
{jlrojo,manel,arfv,mprado,antonio}@tsc.uc3m.es
http://www.tsc.uc3m.es

Abstract. This paper deals with the application of the Support Vector Method (SVM) methodology to the Auto Regressive and Moving Average (ARMA) linear-system identification problem. The SVM-ARMA algorithm for a single-input single-output transfer function is formulated. The relationship between the SVM coefficients and the residuals, together with the embedded estimation of the autocorrelation function, are presented. Also, the effect of the numerical regularization is used to highlight the robust cost character of this approach. A clinical example is presented for qualitative comparison with the classical Least Squares (LS) methods.

1 Introduction

The ARMA modeling of a linear time invariant (LTI) system relating two simultaneously observed discrete-time processes is a wide framework, its applications ranging from digital communications (channel identification) to plant identification and biological signal analysis [1]. The two families of ARMA techniques, the Prediction Error Methods (such as Least Squares, LS) and the Correlation Methods, still present a number of limitations such as sensitivity to outliers, noise and order selection [2].

A robust solution can be drawn from the SVM framework [3,4,5]. Among the potential advantages of the SVM are: first, its single-minimum solution; second, it is an strongly regularized method, appropriate to ill-posed problems; and finally, it extracts the maximum information from the available samples whenever the statistical distribution is unknown. Therefore, the SVM appears as an attractive robust approach to the system identification problem.

The SVM has already been used for AR modeling in time series prediction [6], by the straightforward approach of using the embedded time series as input for the common SVM regression algorithm. Nevertheless, ARMA system identification is a troublesome framework which makes it worth to extend the regression algorithm to this particular problem structure. Moreover, little attention has been paid to the role of the numerical regularization on the cost function, and

J.R. Dorronsoro (Ed.): ICANN 2002, LNCS 2415, pp. 1106–1111, 2002.
© Springer-Verlag Berlin Heidelberg 2002

in fact it allows a very useful statistical interpretation of the different terms of the SVM algorithm for the identification problem.

In the next section, the formulation of the SVM-ARMA algorithm is presented. In Section 3, the effect of the numerical regularization on the empirical cost function is analyzed. A clinical application example is introduced in Section 4. Finally, conclusions are drawn in Section 5.

2   The SVM-ARMA Formulation

A LTI system relating N observations of two discrete-time sequences, $\{x_n\}$ and $\{y_n\}$, can be approximated by an ARMA system, which is completely defined by the following difference equation [2],

$$y_n = \sum_{i=1}^{p} a_i y_{n-i} + \sum_{j=1}^{q} b_j x_{n-j+1} + e_n \tag{1}$$

where a_n (b_n) is the p-length (q-length) sequence determining the AR (MA) coefficients, and $\{e_n\}$ stands for measurement errors. This equation is used to obtain a constraint for the output samples observed in the time-lags $n = k_o, \ldots, N$ (where $k_o = \max{(p+1, q)}$ in order to take into account initial conditions).

The SVM approach for the linear regression problem uses the L_2 norm for the structural cost and the L_ϵ-insensitive function for the empirical cost [8], the later given by

$$L_\varepsilon(v) = \begin{cases} |v| - \varepsilon, & \text{if } |v| >= \varepsilon, \\ 0, & \text{if } |v| < \varepsilon. \end{cases} \tag{2}$$

Then, by following the usual methodology for deriving SVM algorithms [3], the SVM-ARMA problem can be stated as minimizing

$$L_P\left(\xi_k, \xi_k^*, a_i, b_j\right) = \frac{1}{2}\sum_{i=1}^{p} a_i^2 + \frac{1}{2}\sum_{j=1}^{q} b_j^2 + C\sum_{k=k_o}^{N}\left(\xi_k + \xi_k^*\right) \tag{3}$$

constrained to

$$y_k - \sum_{i=1}^{p} a_i y_{k-i} - \sum_{j=1}^{q} b_j x_{k-j+1} \leq \varepsilon + \xi_k \tag{4}$$

$$-y_k + \sum_{i=1}^{p} a_i y_{k-i} + \sum_{j=1}^{q} b_j x_{k-j+1} \leq \varepsilon + \xi_k^* \tag{5}$$

and to $\xi_k, \xi_k^* \geq 0$, for $k = k_o, \ldots, N$, where ξ_k, ξ_k^* are the *slack variables* or losses. The Lagrange functional, L_{PD}, has to be minimized with respect to the primal variables a_i, b_i, ξ_k, ξ_k^* and maximized with respect to the dual variables $\alpha_k, \alpha_k^*, \beta_k, \beta_k^*$. By deriving L_{PD} with respect to the primal variables and equaling to zero, two main consequences are drawn. First, dual variables are shown to be constrained to an upper and a lower bound,

$$0 \leq \alpha_k, \alpha_k^* \leq C \tag{6}$$

for $k = k_0, \ldots, N$. Second, the following conditions can be traced,

$$a_l = \sum_{k=k_o}^{N} (\alpha_k - \alpha_k^*) \, y_{k-l} \tag{7}$$

$$b_l = \sum_{k=k_o}^{N} (\alpha_k - \alpha_k^*) \, x_{k-l+1} \tag{8}$$

showing the relationship between model coefficients, dual coefficients and the observations. These conditions are introduced into the L_{PD} functional, thus maximizing the dual counterpart,

$$L_D = -\frac{1}{2} (\bar{\alpha} - \bar{\alpha}^*)^T \left[\mathbf{R_x^q} + \mathbf{R_y^p} \right] (\bar{\alpha} - \bar{\alpha}^*) + (\bar{\alpha} - \bar{\alpha}^*)^T \bar{y} - \varepsilon \bar{1}^T (\bar{\alpha} + \bar{\alpha}^*) \tag{9}$$

where $\bar{\alpha}^{(*)} = \{\alpha_{k_o}^{(*)}, \ldots, \alpha_N^{(*)}\}^T$, $\bar{y} = \{y_{k_o}, \ldots, y_N\}^T$, and

$$R_x^q(m, k) = \sum_{j=1}^{q} x_{m-j+1} x_{k-j+1} \tag{10}$$

$$R_y^p(m, k) = \sum_{j=1}^{p} y_{m-j} y_{k-j}. \tag{11}$$

These equalities can be seen as the local-lime p^{th} and q^{th} order estimators of the m, k lags of the data autocorrelation functions.

3   On the Numerical Regularization

The numerical procedure used for solving (9) is the maximization of a quadratic problem of the form $\mathbf{x^T H x} + \mathbf{b^T x}$ with respect to $\mathbf{x}$ and with some linear constraints. In the framework of the SV regression, the matrix $\mathbf{H}$ is not invertible and the problem is usually *ad hoc* regularized by adding a small-element diagonal matrix, this is, substituting the square matrix by $\mathbf{H}' = \mathbf{H} + \gamma \mathbf{I}$ (for instance, see [8] in the SVM regression problem). The numerical regularization is in fact a L_2 norm regularization on the dual coefficients, and it can be included into the dual problem in 9 by an additional term, thus leading to the maximization of

$$L_D = -\frac{1}{2} (\bar{\alpha} - \bar{\alpha}^*)^T \left[\mathbf{R_x^q} + \mathbf{R_y^p} \right] (\bar{\alpha} - \bar{\alpha}^*) + (\bar{\alpha} - \bar{\alpha}^*)^T \bar{y} -$$
$$- \varepsilon \bar{1}^T (\bar{\alpha} + \bar{\alpha}^*) - \frac{\gamma}{2} (\bar{\alpha}^T \mathbf{I} \bar{\alpha} + \bar{\alpha}^{*T} \mathbf{I} \bar{\alpha}^*) \tag{12}$$

constrained to (6). The inclusion of this term raises the following result. Let $\bar{\alpha}^o, \bar{\alpha}^{*o}$ be the solution of the problem defined by (12). Then, the relationship between the estimated measurement error e_m and the coefficients at the saddle point $\alpha_m^o, \alpha_m^{o*}$ is given by

$$(\alpha_m^o - \alpha_m^{o*}) = f(e_m) = \begin{cases} -C, & -e_C \geq e_m \\ \frac{1}{\gamma}(e_m + \varepsilon), & -\varepsilon \geq e_m \geq -c \\ 0, & -\varepsilon \leq e_m \leq \varepsilon \\ \frac{1}{\gamma}(e_m - \varepsilon), & \varepsilon \leq e_m \leq e_C \\ C, & e_C \leq e_m. \end{cases} \qquad (13)$$

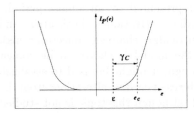

Fig. 1. Robust cost function for the SVM-ARMA algorithm.

The result can be easily shown by analyzing the Karush-Kuhn-Tucker (KKT) conditions in the solution [9]. This functional relationship is a straightforward nexus between the Lagrange multipliers and the residuals.

As the numerical regularization modifies the dual formulation, it necessarily modifies the primal formulation. Specifically, the consideration of the following cost function

$$L^C(e) = \begin{cases} 0, & |e| \leq \varepsilon \\ \frac{1}{2\gamma}(|e| - \varepsilon)^2, & \varepsilon \leq |e| \leq \gamma C + \varepsilon \\ C(|e| - \varepsilon) - \frac{1}{2}\gamma C^2, & |e| \geq \gamma C + \varepsilon \end{cases} \qquad (14)$$

leads to the following function to be minimized:

$$L_P^C(\xi_k, \xi_k^*, a_i, b_j) = \frac{1}{2}\sum_{i=1}^p a_i^2 + \frac{1}{2}\sum_{j=1}^q b_j^2 + \frac{1}{2\gamma}\sum_{k \in I_1}(\xi_k^2 + \xi_k^{*2}) + C\sum_{k \in I_2}(\xi_k + \xi_k^*) \qquad (15)$$

constrained to (4) and (5), where I_1 is the set of samples for which $\varepsilon \leq |\xi_k| \leq \xi_C$, and I_2 is the set of samples for which $|\xi_k| > e_C$ (Figure 1). The derivation of the dual problem leads to a functional to be maximized which is slightly different from (12), but it has the same solution.

Therefore, the primal function we are actually using is not the one suggested in (3), but rather (15). It can be seen as a robust cost function that allows different error penalties (insensitive, quadratic or linear cost) which can be straightly related to the residuals due to the analytical relationship (13), and the free parameters of the SVM can be tuned according to the statistical nature of the problem. A number of results can be derived from the signal analysis point of view, which are being currently developed.

4   Clinical Application Example

A kind of application where the characteristics of SVM-ARMA are appropriate and potentially useful is the digital, signal-based clinical featuring. If clinical features are to be derived after a signal processing procedure, this method should be as robust as possible. In the present section, an example of clinical featuring is presented for qualitative comparison of the SVM-ARMA with the LS solution.

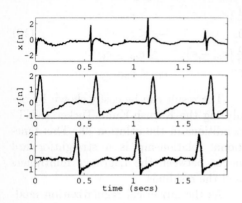

Fig. 2. Robust cost function for the SVM-ARMA algorithm.

The modeling problem is to establish an LTI system representation for the atrial and the ventricular cardiac electrical functions, from signals that are recorded in implantable defibrillators. The usual healthy cardiac rhythm is known as Sinus Rhythm (SR), and it is represented by an almost periodic cardiac activation in both atrial and ventricular channels (electrograms). If the model is shown to be robust enough, it might be used to establish indexes of cardiac alterations from the SR, although this topic is not studied here.

The digitized atrial and ventricular signals were obtained from an implantable defibrillator in a patient during SR (Figure 2). The sample rate was 128 Hz for both channels.

(a) **The LS criterion.** Samples were divided into two non overlapping subsets for fitting the model and testing, with $N = 250$ samples (three cardiac cycles) each. The Akaike Information Criterion was used to select the best model order, which was ARMA(1,8) [2](Model 1).

(b) **The SVM-ARMA algorithm.** The following procedure was followed for estimating the free parameters. First, $\gamma = 1$ was fixed. An initial SVM-ARMA(40,40) was previously obtained, as this order was a 25% of the cardiac cycle, and it was considered high enough to embed all the coefficients. With $\varepsilon = 0$ and $C = +\infty$, the coefficients were estimated (Model 2). The wide order was shortened by determining the first coefficients containing the 95% of the coefficients energy, leading to an ARMA(2,13) (Model 3). Then, the optimal value for the free parameters was found by sweeping a range for each of them according to the prediction error on the training samples. A range of values was found proper for ε and C, and a possible set of values were $\varepsilon = 0.2$ and $C = 2$.

Figure 3 shows the impulse responses obtained for the models. The estimated system in (b) exhibits oscillation, due to the extremely high number of coefficients. This situation improves after the moderation of the order in Model 3. The agreement between the SVM-ARMA(2,13) before and after parameter tuning (Model 3 and 4) is evident.

The spectral analysis of atrial and ventricular channels, together with the spectral representation of systems in Model 1 and Model 4, are shown in Figure 3. Note that: (a) the SVM spectral response provides an enhanced energy compensation in the band where it is required (4 to 18 Hz); (b) the SVM-ARMA does not distort the low-frequency components (0 to 4 Hz), while LS does.

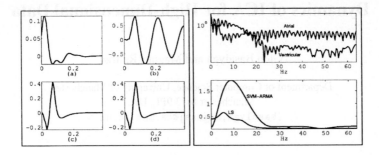

Fig. 3. Left: Impulse responses: (a) Model 1; (b) Model 2; (c) Model 3; (d) Model 4. **Right:** Frequency domain representation. Up: power spectral densities of atrial and ventricular channels. Down: modulus for Model 1 and Model 4.

Therefore, a first approach to this problem shows the SVM-ARMA exhibiting higher reproducibility and robustness than classical LS estimators. Further, specific analysis is being done on the basis of these findings.

5   Conclusions

A first approach to the SVM-ARMA system identification has been suggested. The role of the numerical regularization has been examined leading to a robust cost functional interpretation. A clinical application example has been developed in order to test the algorithm in comparison with the LS approach. Future work includes the extension to the non-linear system identification problem by using Mercer's kernels and non-uniform sampling. Also, robust modeling of cardiovascular and hemodynamic biomedical signals is being currently developed.

References

1. Proakis, J.G., Rader, C.M., Ling, F., Nikias, C.L.: Advanced Digital Signal Processing. Macmillan Publishing Company, NY, US, 1992.
2. Ljung, L.: System Identification. Theory for the User. Prentice Hall, NJ, US, 1987.
3. Vapnik, V.: The Nature of Statistical Learning Theory Springer–Verlag, NY, 1995.
4. Schölkopf, B., Sung, K.: Comparing Support Vector Machines with Gaussian Kernels to Radial Basis Function Classifiers IEEE Trans. on Signal Proc. Vol 45, n 11. 1997.
5. Pontil, M., Verri, A.: Support Vector Machines for 3D Object Recognition. IEEE Trans. on Pattern Anal. and Mach. Intell. Vol. 20, n 6, 1998.
6. Müller, K.R., Smola, A., Rätsch, G.R., Schölkopf, B., Kohlmorgen, J., Vapnik, V.: Predicting Time Series with Support Vector Machines. In Advances in Kernel Methods. Support Vector Learning. MIT Press, MA, USA. 1999.
7. Tikhonov, A.N., Arsenen, V.Y.: Solution to Ill-Posed Problems. V.H. Winston & Sons. Washington, US, 1977.
8. Smola, A.J., Schölkopf, B.: A Tutorial on Support Vector Regression. NeuroCOLT2 NC2-TR-1998-030,1998.
9. Luenberguer, D.G.: Linear and Nonlinear Programming. Addison–Wesley Pub Co, Reading, MA, 1984

Dynamics of ICA for High-Dimensional Data

Gleb Basalyga and Magnus Rattray

Department of Computer Science, University of Manchester,
Manchester M13 9PL, UK.
{basalyga,magnus}@cs.man.ac.uk

Abstract. The learning dynamics close to the initial conditions of an on-line Hebbian ICA algorithm has been studied. For large input dimension the dynamics can be described by a diffusion equation. A surprisingly large number of examples and unusually low initial learning rate are required to avoid a stochastic trapping state near the initial conditions. Escape from this state results in symmetry breaking and the algorithm therefore avoids trapping in plateau-like fixed points which have been observed in other learning algorithms.

1 Introduction

We study the dynamics of an on-line Hebbian learning algorithm for a popular statistical modelling technique, Independent Component Analysis (ICA). The goal of ICA is to find a representation of data in terms of a combination of statistically independent variables. This technique has a number of useful applications such as blind source separation, feature extraction and blind deconvolution. A review of current approaches to ICA can be found in [1].

In particular we focus on the initial transient dynamics of Hebbian ICA and its influence on the symmetry breaking processes which are often critical to performance of the learning process. For example, there is a well known permutation symmetry in multi-layer perceptrons. It is possible to permute all the connections of a neuron (inputs and outputs) with those of another neuron in the same layer without changing the network function. Symmetries give an equal chance for a learning algorithm to move in several directions and therefore lead to indecisiveness and slowing down in the learning process. A good example of the influence of symmetries is the quasi-stationary plateau in the learning dynamics that appears in backpropagation algorithms of multilayer perceptrons [2] and Sanger's PCA algorithm [3].

It has been suggested that increasing randomness of the learning process will help break such symmetry, since stochastic fluctuations should help the system to escape from unstable sub-optimal states. However in the case of Hebbian ICA considered here we find that stochastic fluctuations near the initial conditions are very large and stabilise a special class of otherwise unstable sub-optimal states (see [4,5]). In this case one must choose the learning rate carefully in order to escape from trapping in these sub-optimal states. Here we show that escaping from these sub-optimal states will typically coincide with the symmetry breaking required to learn effectively. It appears that plateau-like states

J.R. Dorronsoro (Ed.): ICANN 2002, LNCS 2415, pp. 1112–1118, 2002.

seen in other learning algorithms {2,3] are therefore unlikely to affect performance of this algorithm.

In this paper we provide a solution to the dynamics of Hebbian ICA close to the initial conditions in the limit of large input dimension. This generalises on previous results [4, 5] which were limited to the simplest single source case. We find that the dynamics can be described as a multidimensional diffusion in polynomial potential. By solving the dynamics of the multi-source case we can characterise the symmetry breaking process required for effective learning.

2   Online Hebbian Learning Rule

In order to use a statistical mechanics approach, an idealised ICA model was introduced in [4,5]. The N-dimensional data x is generated from a noiseless linear mixture of a small number M of non-Gaussian sources s and a large number $N - M$ of uncorrelated Gaussian components, $n \sim \mathcal{N}(0, I_{N-M})$,

$$x = A \begin{bmatrix} s \\ n \end{bmatrix} = A_s s + A_n n, \tag{1}$$

where $A = [A_s \ A_n]$ is the mixing matrix. To simplify the analysis we assume here that the data is already sphered, ie. the data has zero mean and an identity covariance matrix. Without loss of generality it can also be assumed that the sources each have unit variance.

The goal of ICA is to find the de-mixing matrix W such that the projection $y \equiv W^T x$ will coincide with the non-Gaussian sources s up to scaling and permutations. We consider a simple Hebbian learning rule [6] which extracts non-Gaussian sources from this mixture by maximising the non-Gaussianity of projections. The change of the de-mixing matrix W in one iteration is given by,

$$\Delta W = \eta \sigma \, x \phi(y)^T + \alpha W (I - W^T W) . \tag{2}$$

Here, η is the learning rate and σ is diagonal with $\sigma_{ii} = \text{sign}\left(E_{s_i}\{s_i \phi(s_i) - \phi'(s_i)\}\right)$ which ensures stability of the correct solution. The function $\phi(y)$ is some smooth non-linear function which is applied to every component of the vector y. An even non-linearity, e.g. $\phi(\mu) = \mu^2$, is usually used to detect asymmetric non-Gaussian signals, while an odd non-linearity, e.g. $\phi(\mu) = \mu^3$ or $\phi(\mu) = \tanh(\mu)$, is used to detect symmetric non-Gaussian signals. The first term on the right of (2) maximises some measure of non-Gaussianity of the projections. The second term provides orthogonalisation of the de-mixing matrix. The parameter α is less critical and we set $\alpha = 0.5$ in simulations.

Defining the overlaps $R \equiv W^T A_s$ and $Q \equiv W^T W$, we obtain for projections

$$y = W^T (A_s s + A_n n) = Rs + z , \quad \text{where} \quad z \sim \mathcal{N}(0, Q - RR^T) . \tag{3}$$

In order to calculate averages over Gaussian sources in section 3 , we can write $z = L\mu$, where $\mu \sim \mathcal{N}(0, I)$ and the matrix L always can be found by special factorisation, e.g. by Cholesky decomposition: $Q - RR^T = LL^T$. From equation (2) one can calculate the change in R and Q after a single learning step,

$$\Delta R = \eta \sigma \phi(y) s^{\mathrm{T}} + \alpha (I - Q) R , \tag{4}$$

$$\Delta Q = \eta \sigma (I + \alpha (I - Q)) \phi(y) y^{\mathrm{T}} + \eta \sigma y \phi(y)^{\mathrm{T}} (I + \alpha (I - Q))$$
$$+ 2\alpha (I - Q) Q + \alpha^2 (I - Q)^2 Q + \eta^2 \phi(y) x^{\mathrm{T}} x \phi(y)^{\mathrm{T}} . \tag{5}$$

The dynamics is not very sensitive to the exact value of α as long as $\alpha \gg \eta$. As α increases, Q approaches I. If one sets $Q - I \equiv q/\alpha$ and take $\alpha \to \infty$ then equation (5) converges to,

$$q = \frac{1}{2} \left(\eta \sigma \left(\phi(y) y^{\mathrm{T}} + y \phi(y)^{\mathrm{T}} \right) + \eta^2 N \phi(y) \phi(y)^{\mathrm{T}} \right) , \tag{6}$$

where we have dropped terms lower then $O(\eta^2 N)$ and $O(\eta)$. Substituting this result into equation (4) leads to

$$\Delta R = \eta \sigma \left(\phi(y) s^{\mathrm{T}} - \frac{1}{2} \left(\phi(y) y^{\mathrm{T}} + y \phi(y)^{\mathrm{T}} \right) R \right) - \frac{1}{2} \eta^2 N \phi(y) \phi(y)^{\mathrm{T}} R . \tag{7}$$

This is an example of adiabatic elimination of fast variables [7]. The informal discussion here can be put on a more rigorous framework which will be described in future work.

Due to the fluctuation term ($\sim O(\eta^2)$) in equation (7), the algorithm has a special class of sub-optimal fixed points near $R = 0$ which causes the presence of a stochastic trapping state near the initial conditions. The discussion below is restricted to describing the escape dynamics from this sub-optimal trapping state. For an account of the dynamics far from initial conditions, see [4].

3   Dynamics Close to the Initial Conditions

A random and uncorrelated choice for A and the initial entries W always leads us to expect $R = O(N^{-\frac{1}{2}})$. We set $r \equiv R \sqrt{N}$ in the following discussion, where r is assumed to be an $O(1)$ quantity. Usually, macroscopic quantities like the overlap r have a "self-averaging" property when the variance of these macroscopic quantities tends to zero in the limit $N \to \infty$ (see [2,3]). In the present case, as we will see below, the mean and variance of the change in r at each iteration are the same order. That means that the overlap r does not self-average and the fluctuations have to be considered even in the limit. In this case it is more natural to model the on-line learning dynamics as a diffusion process (see, for example [7,8]).

3.1   Odd Non-linearity

It is most common to use an odd non-linearity. This is appropriate when we need to extract a symmetrical signal. In this case the appropriate scaling for the learning rate will be $\eta = \nu/N^2$ where ν is an $O(1)$ scaled learning rate parameter. After expanding equation (7) near $r = 0$ we obtain the following expressions for the mean and covariance of the change in r at each iteration,

$$\mathrm{E}[\Delta r_{ij}] \simeq \left(-\frac{1}{2} \langle \phi^2(\mu) \rangle \nu^2 r_{ij} + \frac{1}{6} \kappa_4^j \langle \phi'''(\mu) \rangle \sigma_{ii} \nu r_{ij}^3 \right) N^{-3} ,$$
$$\mathrm{Cov}[\Delta r_{ij}, \ \Delta r_{kl}] \simeq \langle \phi^2(\mu) \rangle \nu^2 \, \delta_{ik} \delta_{jl} \, N^{-3} , \tag{8}$$

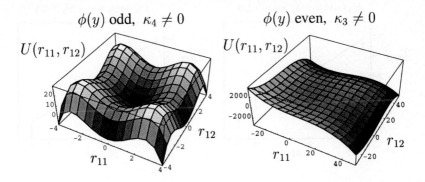

Fig. 1. Close to the initial conditions the learning dynamics is equivalent to diffusion in a polynomial potential. For symmetrical source distributions we should use an odd non-linearity in which case the potential is quartic, as shown on the left (for $K = 1$, $M = 2$). For asymmetrical source distributions we can use an even non-linearity in which case the potential is cubic, as shown on the right. Escaping over the minimal barriers results in symmetry breaking, ie. specialisation to a particular source.

where κ_4^j is the fourth cumulant of the j-th source distribution (measuring kurtosis) and brackets denote averages over Gaussian variables $\mu \sim \mathcal{N}(0, I)$. In this case the system can be described by a Fokker-Planck equation for large N with a characteristic timescale $(\delta t)^{-1} \sim O(N^3)$. The system is locally equivalent to diffusion in the following potential

$$U(r) = \sum_{i=1}^{K} \sum_{j=1}^{M} \left(\tfrac{1}{4} \langle \phi^2(\mu) \rangle \, \nu^2 r_{ij}^2 - \tfrac{1}{24} \kappa_4^j \langle \phi'''(\mu) \rangle \sigma_{ii} \nu \, r_{ij}^4 \right) \qquad (9)$$

with a diagonal diffusion matrix of magnitude $D = \langle \phi^2(\mu) \rangle \nu^2$. The shape of this potential for an example with two sources (with equal kurtosis) and one projection ($K = 1$, $M = 2$) is shown on the left of Fig. 1. In this case we have four minimal potential barriers which system has to overcome. Escaping over one barrier results in symmetry breaking, ie. the projection will specialise to a particular source. The escape time (the mean first passage time) from this initial trapping (when at least one of source signals is learned) is mainly determined by the effective size of the minimal potential barrier ΔU (see, for example, [7,8]): $T_{\text{escape}} \propto (\delta t)^{-1} \exp(\Delta U / D)$. For the case when all non-Gaussian sources have the same kurtosis $\kappa_4^i = \kappa_4$ ($i = 1, 2, ..., M$) we have:

$$T_{\text{escape}}^{\text{odd}} \propto \frac{N^3}{4MK} \exp \left[\frac{3 \langle \phi^2(\mu) \rangle \nu}{8 \, |\kappa_4 \langle \phi'''(\mu) \rangle|} \right] \qquad (10)$$

where we have assumed $\sigma_{ii} = \text{sign}(\kappa_4)$ which is a necessary condition for successful learning. We see that the time-scale for escape diverges with learning rate. The numerical simulations for the dependence of the escape time on the number of non-Gaussian sources for $N = 50$ are shown on the left in Fig. 2, on a log-log scale, showing good agreement with the theory. We note that if the sources have different kurtosis, then the algorithm will be most likely to learn those with highest kurtosis first.

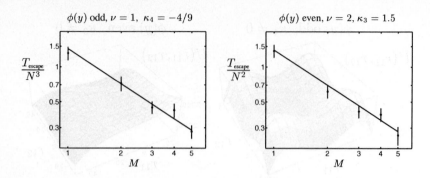

Fig. 2. Dependence of the escape time (time to learn at least one source signal) on the number of non-Gaussian sources M for Hebbian ICA with one projection ($K = 1$). The solid line shows the theoretical results. The points with errorbars denote the simulation results for $N = 50$ averaged over 50 experiments.

3.2   Even Non-linearity

In the case of even non-linearity $\phi(y)$ the appropriate scaling for the learning rate will be $\eta = \nu/N^{\frac{3}{2}}$ where ν is an $O(1)$ scaled learning rate parameter. After expanding equation (7) near $r = 0$ we find that the mean and covariance of the change in r at each iteration are given by (to leading order in N^{-1}),

$$\mathrm{E}[\Delta r_{ij}] \simeq (-\tfrac{1}{2}\langle \phi^2(\mu)\rangle \nu^2 r_{ij} + \tfrac{1}{2}\kappa_3^j \langle \phi''(\mu)\rangle \sigma_{ii}\nu r_{ij}^2 - \tfrac{1}{2}\langle \phi(\mu)^2\rangle \nu^2 \sum_{l\neq i}^{K} r_{lj})N^{-2} \,,$$

$$\mathrm{Cov}[\Delta r_{ij}, \ \Delta r_{kl}] \simeq \langle \phi^2(\mu)\rangle \nu^2 \, \delta_{ik}\delta_{jl}N^{-2} \,, \tag{11}$$

where κ_3^j is the third cumulant of the j-th source distribution (third central moment), which measures skewness, and brackets denote averages over Gaussian variables $\mu \sim \mathcal{N}(0, I)$. Again the system can be described by a Fokker-Planck equation for large N but now with shorter characteristic timescale $(\delta t)^{-1} \sim O(N^2)$. The system is locally equivalent to a diffusion process in the cubic potential

$$U(r) = \sum_{i=1}^{K}\sum_{j=1}^{M}(\tfrac{1}{4}\langle \phi^2(\mu)\rangle \nu^2 r_{ij}^2 - \tfrac{1}{6}\kappa_3^j \langle \phi''(\mu)\rangle \sigma_{ii}\nu \, r_{ij}^3 + \tfrac{1}{2}\langle \phi(\mu)^2\rangle \nu^2 \sum_{l\neq i}^{K} r_{lj}r_{ij}) \,,$$

with a diagonal diffusion matrix of magnitude $D = \langle \phi^2(\mu)\rangle \nu^2$. The shape of this potential for an example with two sources and one projection ($K = 1, M = 2$) is shown on the right of Fig. 1. In this case we have a ledge in the potential with two points of minimum height ΔU (escape points). In the general case we have for the effective size of the minimal barriers:

$$\frac{\Delta U}{D} = \frac{f(K) \, \nu^2}{\langle \phi^2(\mu)\rangle \, \kappa_3^{i\,2}} \,, \tag{12}$$

where the function $f(K)$ depends on the choice of non-linear function $\phi(\mu)$. For example, the shape of this function for $\phi(\mu) = \mu^2$ is shown on the left in Fig. 3. We see that the size of potential barrier decreases with increasing number of projections K. These

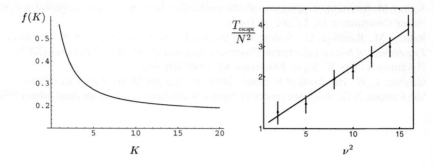

Fig. 3. Dependance of effective size of barrier of number of projections K for the specific choice of the even non-linear function $\phi(\mu) = \mu^2$ is shown on the left. Dependence of the escape time on learning rate ν for the one-dimensional Hebbian ICA ($K = M = 1$) is shown on the right. The solid line shows the theoretical results. The points with errorbars denote the simulation results for $\kappa_3 = 1.5$ and $N = 50$ averaged over 50 simulations.

results suggest that parallel algorithms for extracting asymmetrical signals may prove more efficient than deflationary ones which separate one signal at a time. The numerical simulations for the escape time for $N = 50$ are shown on the right of Fig. 2 and Fig. 3, and are consistent with the theory.

4   Conclusion

Analysis of the multivariate diffusion equation describing the dynamics of Hebbian ICA near the initial conditions shows that the fluctuations at the beginning of learning are very large compared to the average dynamical trajectory. Due to this highly stochastic nature of the initial dynamics, the plateau configurations in the learning dynamics caused by symmetries do not appear here in practice. However, in other respects the situation is worse. The strong fluctuations at the beginning not only break any permutation symmetries but also stabilise another special class of unstable sub-optimal states which leads to stochastic trapping state near initial conditions if one chooses the initial learning rate too large. The learning rate must be less than $O(N^{-1})$ initially in order to avoid trapping. The number of iterations required to escape the initial transient should be greater than $O(N)$, resulting in an extremely slow initial stage of learning for large N.

Acknowledgements. This work was supported by an EPSRC award (ref. GR/M48123).

References

1. Hyvärinen, A.: Survey on independent component analysis. Neural Computing Surveys **2** (1999) 94–128
2. Saad, D. and Solla, S. A.: Exact solution for on-line learning in multilayer neural networks. Physical Review Letters E **74** (1995) 4337–4340.
3. Biehl, M., Schlosser, E.: The dynamics of on–line principal component analysis. Journal of Physics **A31** (1998) L97

4. Rattray, M.: Stochastic trapping in a solvable model of on-line independent component analysis. Neural Computation **14**, 2 (2002) 421–435
5. Rattray, M., Basalyga, G.: Scaling laws and local minima in Hebbian ICA. To appear in *Proceedings of Neural Information Processing Systems (NIPS*2001)*, Vancouver (2002)
6. Hyvärinen A., Oja, E.: Signal Processing, **64** (1998) 301–313
7. Gardiner, C. W.: Handbook of Stochastic Methods. Springer-Verlag, New York (1985)
8. Van Kampen, N.G.: Stochastic processes in physics and chemistry. Elsevier, Amsterdam (1992)

Beyond Comon's Identifiability Theorem for Independent Component Analysis

Riccardo Boscolo[1], Hong Pan[2], and Vwani P. Roychowdhury[1]

[1] Electrical Engineering Department,
University of California, Los Angeles
Los Angeles, CA 90095
{riccardo,vwani}@ee.ucla.edu

[2] Functional Neuroimaging Laboratory,
Department of Psychiatry,
Weill Medical College,
Cornell University,
New York, NY 10021
hop2001@med.cornell.edu

Abstract. In this paper, Comon's conventional identifiability theorem for Independent Component Analysis (ICA) is extended to the case of mixtures where several gaussian sources are present. We show, in an original and constructive proof, that using the conventional mutual information minimization framework, the separation of all the *non-gaussian* sources is always achievable (up to scaling factors and permutations). In particular, we prove that a suitably designed optimization framework is capable of seamlessly handling both the case of one single gaussian source being present in the mixture (separation of all sources achievable), as well as the case of multiple gaussian signals being mixed together with non-gaussian signals (only the non-gaussian sources can be extracted).

1 Introduction

In his fundamental work [1], Comon showed that the separation of a set of stationary signals, instantaneously and linearly mixed, is always possible, as long as the mixing matrix has full rank, and at the most *one* of the original signals is gaussian distributed. This result is often cited in the literature as Comon's *identifiability theorem* for ICA, and it represents a well-known and widely mentioned result in the blind signal separation field. Although the theorem holds strictly only when a functional of the probability density function of the reconstructed signals is used as contrast function, most ICA algorithms are based on contrast functions of this type, such as the mutual information between the reconstructed signals, or its equivalent counterparts, i.e. the InfoMax principle, or the maximum likelihood (ML) principle.

In recent years, Cruces et al. [2][3][4] investigated several criteria for the extraction of a subset of sources from a linear mixture, both in the instantaneous case, and in the case of convolutive mixtures. In particular, it was shown in [3],

J.R. Dorronsoro (Ed.): ICANN 2002, LNCS 2415, pp. 1119–1124, 2002.
© Springer-Verlag Berlin Heidelberg 2002

that a suitably designed entropy minimizing framework can be used to extract
the non-gaussian sources, from mixtures containing an arbitrary number of gaussian distributed signals. The authors also introduced a moment-based iterative
algorithm that minimizes an approximation of the contrast function derived from
this principle.

In this paper, we derive a novel proof of Comon's identifiability theorem, and
extend the theorem to the case of multiple gaussian sources being mixed with
non-gaussian sources. All the results are derived from investigating the properties
of the optimization problem associated with minimizing the mutual information
between the reconstructed signals. In particular, we prove that, regardless of
the number of gaussian sources in the mixture, the resulting objective function
always has extrema that yield the separation of the non-gaussian sources (up
to scaling and permutations), and the gaussian components are *irrelevant* in
determining such extrema.

2   Separation Principle and Objective Function Definition

We make the conventional assumption that N independent and stationary source
signals $(s_1, \ldots, s_N)$ are mixed by an unknown, full-rank mixing matrix A, resulting in a set of mixtures given by $\mathbf{x} = A\mathbf{s}$. The reconstruction of the original
sources is attempted from the mixture data through a linear projection of the
type $\mathbf{y} = B\mathbf{x}$. Following the mutual information minimization principle, common to most ICA frameworks, we seek the matrix B, solution of the optimization
problem [3]:

$$B_{opt} = \arg\min_{B} I(y_1, \ldots, y_N) \tag{1}$$

Using basic information theory equalities, (1) becomes:

$$\min_{B} \sum_{i=1}^{N} H(y_i) - \log|\det B| - H(\mathbf{x}), \tag{2}$$

where $H(a) = -\int p_a(u) \log p_a(u) du$. If we assume that the mixture data has
been sphered, i.e. $Cov(\mathbf{x}\mathbf{x}^T) = I$, we can restrict the search space for the unmixing matrix B to the manifold of orthogonal matrices [5]. The problem can
be simplified as:

$$\min_{B} \sum_{i=1}^{N} H(y_i) \tag{3}$$

$$\text{s.t. } BB^T = I, \tag{4}$$

since $\log|\det(B)| \equiv 1$, and $H(\mathbf{x})$ is a constant with respect to B. The equality
constraints (4) define a sub-group of the Stiefel manifold for the case of square

---

[3] $I(y_1, \ldots, y_N) \triangleq \int p_y(\mathbf{y}) \log \dfrac{p_y(\mathbf{y})}{\prod_{i=1}^{N} p_{y_i}(y_i)} d\mathbf{y}$

matrices. If we define $F(B) \triangleq \sum_{i=1}^{N} H(y_i)$, then the gradient of the cost function defined on such manifold is given by [6]:

$$\nabla_m F(B) \triangleq \nabla F(B) - B\nabla F(B)^T B. \tag{5}$$

where $\nabla F(B)$ is the conventional gradient of $F(B)$ in the Euclidean space:

$$\nabla F(B) \triangleq \left[\frac{\partial F(B)}{\partial b_{ij}} \right] = \begin{bmatrix} \nabla H(y_1) \\ \vdots \\ \nabla H(y_N) \end{bmatrix}. \tag{6}$$

The extrema of the optimization problem (3) are given by all the matrices that satisfy the condition:

$$\nabla_m F(B) = 0 \quad \Rightarrow \quad \nabla F(B)B^T = B\nabla F(B)^T, \tag{7}$$

since $BB^T = I$.

3 Extending Comon's Identifiability Theorem

In this section, an alternative proof of Comon's well-known theorem on ICA identifiability [1] is derived, and it is extended to the case where more than one gaussian source is present in the mixture. Under the modeling assumption of Section 2, we consider mixtures of N independent sources $s_1, \cdots, s_N$, with probability density function $f_{s_1}, \cdots, f_{s_N}$, M of which are gaussian distributed. We make the further assumption that the mixing matrix A is the $N \times N$ identity matrix. This is not a restrictive assumption, since, if the mixture data is sphered, the solution spaces associated to any two full rank mixing matrices simply map to each other through an orthogonal transformation [7]. The generic reconstructed signal can be written as:

$$y_i = b_{i1}s_1 + b_{i2}s_2 + \ldots + b_{iN}s_N \qquad i = 1, \ldots, N, \tag{8}$$

and its differential entropy is given by:

$$H(y_i) = -\int_{-\infty}^{\infty} f_{y_i}(u) \log f_{y_i}(u) du, \tag{9}$$

where, because of the independence between the sources:

$$f_{y_i}(u) = \frac{1}{|b_{i1}|} f_{s_1}\left(\frac{u}{b_{i1}}\right) * \frac{1}{|b_{i2}|} f_{s_2}\left(\frac{u}{b_{i2}}\right) * \cdots * \frac{1}{|b_{iN}|} f_{s_N}\left(\frac{u}{b_{iN}}\right). \tag{10}$$

The components of the gradient of $H(y_i)$ with respect to $\mathbf{b}_i$ (ith row of B) can be computed as:

$$\frac{\partial H(y_i)}{\partial b_{ij}} = -\int_{-\infty}^{\infty} (1 + \log f_{y_i}(u)) \frac{\partial f_{y_i}(u)}{\partial b_{ij}} du \tag{11}$$

To make explicit the dependence of the entropy $H(y_i)$ on $\mathbf{b}_i$, define $h(\mathbf{b}_i) \triangleq H(y_i)$. In order to satisfy the first-order conditions given by (7), we must have that:

$$\begin{bmatrix} \nabla h(\mathbf{b}_1) \\ \vdots \\ \nabla h(\mathbf{b}_N) \end{bmatrix} [\mathbf{b}_1^T \cdots \mathbf{b}_N^T] = \begin{bmatrix} \mathbf{b}_1 \\ \vdots \\ \mathbf{b}_N \end{bmatrix} [\nabla h(\mathbf{b}_1)^T \cdots \nabla h(\mathbf{b}_N)^T]. \tag{12}$$

The resulting set of equations is equivalent to the following set of $N(N-1)$ equalities:

$$\nabla h(\mathbf{b}_k) \mathbf{b}_l^T = \nabla h(\mathbf{b}_l) \mathbf{b}_k^T \qquad k, l = 1, \ldots, N \ (k \neq l). \tag{13}$$

Using expression (11), we get:

$$\int_{-\infty}^{\infty} \log f_{y_k}(u) \left[b_{l1} \frac{\partial f_{y_k}(u)}{\partial b_{k1}} + \cdots + b_{lN} \frac{\partial f_{y_k}(u)}{\partial b_{kN}} \right] du = \tag{14}$$

$$= \int_{-\infty}^{\infty} \log f_{y_l}(u) \left[b_{k1} \frac{\partial f_{y_l}(u)}{\partial b_{l1}} + \cdots + b_{kN} \frac{\partial f_{y_l}(u)}{\partial b_{lN}} \right] du.$$

The computation of $\partial f_{y_i}(u)/\partial b_{ij}$ can be efficiently carried out in the frequency domain. Using the conventional definition of *characteristic function* of a random variable [8]:

$$\Phi_X(\omega) \triangleq \mathcal{F}\{f_X(x)\} = \int_{-\infty}^{\infty} f_X(x) e^{-j\omega x} dx, \tag{15}$$

we have from (10), using the convolution theorem:

$$\Phi_{y_i}(\omega) = \Phi_{s_1}(b_{i1}\omega)\Phi_{s_2}(b_{i2}\omega) \cdots \Phi_{s_N}(b_{iN}\omega) \qquad i = 1, \ldots, N. \tag{16}$$

If we assume that the pdfs f_{s_i} are continuous functions, with continuous derivatives almost everywhere, we can exchange the order of the integral and the derivative, and compute $\partial f_{y_i}(u)/\partial b_{ij}$ as follows:

$$\frac{\partial f_{y_i}(u)}{\partial b_{ij}} = \mathcal{F}^{-1}\left\{ \omega \Phi_{s_1}(b_{i1}\omega) \cdots \Phi_{s_i}'(b_{ij}\omega) \cdots \Phi_{s_N}(b_{iN}\omega) \right\} \tag{17}$$

where $\mathcal{F}^{-1}$ denotes the inverse fourier transform operator. The conditions imposed by (14) are satisfied, in particular, when:

$$b_{l1} \frac{\partial f_{y_k}(u)}{\partial b_{k1}} + \cdots + b_{lN} \frac{\partial f_{y_k}(u)}{\partial b_{kN}} = 0 \qquad k, l = 1, \ldots, N \ (k \neq l). \tag{18}$$

If we substitute (17) into (18), and, under the assumption that all the characteristic functions are non-zero for every ω, we divide by $\Phi_{s_1}(b_{k1}\omega) \cdots \Phi_{s_N}(b_{kN}\omega)$ the resulting expression, we obtain:

$$\frac{\omega b_{l1} \Phi'_{s_1}(b_{k1}\omega)}{\Phi_{s_1}(b_{k1}\omega)} + \ldots + \frac{\omega b_{lN} \Phi'_{s_N}(b_{kN}\omega)}{\Phi_{s_N}(b_{lN}\omega)} = 0 \qquad k,l = 1,\ldots,N \ (k \neq l). \quad (19)$$

Notice that if and only if f_{s_i} is a gaussian pdf it holds that:

$$\Phi'_{s_i}(\alpha\omega) = -\alpha\omega\Phi_{s_i}(\alpha\omega). \quad (20)$$

Therefore in the special case where $M = N$, i.e. all the original sources have a gaussian distribution, (19) simplifies as:

$$-(b_{k1}b_{l1} + \ldots + b_{kN}b_{lN})\omega^2 = -\mathbf{b}_k^T \mathbf{b}_l \omega^2 = 0 \qquad k,l = 1,\ldots,N \ (k \neq l), \quad (21)$$

which are always satisfied because of the orthogonality constraints. Therefore, *if all sources are gaussian, the resulting objective is a constant with respect to the elements of an arbitrary orthogonal unmixing matrix, and the separation is not possible.*

When M is strictly less than N, in order to simplify the notation, we can assume that the first M sources, $(s_1,\ldots,s_M)$, are gaussian distributed. The equations in (19) can be simplified as:

$$-\omega^2(b_{l1}b_{k1} + \cdots + b_{lM}b_{kM}) + \frac{\omega b_{lM+1} \Phi'_{s_{M+1}}(b_{kM+1}\omega)}{\Phi_{s_{M+1}}(b_{lM+1}\omega)} \cdots + \frac{\omega b_{lN} \Phi'_{s_N}(b_{kN}\omega)}{\Phi_{s_N}(b_{lN}\omega)} = 0$$

$$k,l = 1,\ldots,N \ (k \neq l) \quad (22)$$

The subset of orthogonal matrices that satisfy this set of equalities is given by:

$$B = \left[\begin{array}{c|c} Q & 0 \\ \hline 0 & P \end{array}\right], \quad (23)$$

where Q is an arbitrary $M \times M$ orthogonal matrix, and P is a generalized permutation matrix. Notice, in fact, that:

$$\Phi'_{s_i}(b_{ij}\omega)\big|_{b_{ij}=0} = -\jmath E[s_i] = 0 \qquad i = 1,\ldots,N, \quad (24)$$

if the sources are zero-mean[4]. This result shows that minima of the optimization problem, that was derived from the separation principle (1), appear in correspondence of matrices B that result in separation of the non-gaussian sources. Therefore, we proved the following theorem:

---

[4] In general this is not a restriction because the mean can always be removed during pre-processing of the mixtures.

Theorem 1 (Extended ICA Identifiability Theorem). *Given N independent and stationary signals $s_1, \ldots, s_N$, $M < N$ of which are gaussian distributed, the $N - M$ non-gaussian distributed signals can be reconstructed, up to scaling and permutations, from any linear mixture of the type $x = As$, where A is a full-rank $N \times N$ matrix, solving the following optimization problem:*

$$\min_{B} \sum_{i=1}^{N} H(y_i) \tag{25}$$
$$s.t. \ BB^T = I.$$

Notice that in the summation (25), the index is up to N since *the number of non-gaussian sources is not assumed to be known a-priori*, thus preserving the "blindness" of the approach to the underlying distribution of the mixed signals.

4 Conclusions

An extension to the conventional identifiability theorem for ICA is introduced and rigorously proved. We show that, even when an arbitrary number of gaussian sources is included in the set of independent signals, the conventional mutual information minimization framework is still capable of separating all the non-gaussian signals, without requiring an ad-hoc ICA implementation. In particular, the main result of this paper is shown by investigating the properties of the extrema of the optimization problem derived from the separation principle.

References

1. Comon, P.: Independent component analysis, a new concept? Signal Processing **36** (1994) 287–314
2. Cruces, S., Cichocki, A., Castedo, L.: An iterative inversion approach to blind source separation. IEEE Trans. Neural Networks **11** (2000) 1423–1437
3. Cruces, S., Cichocki, A., Amari, S.: The minimum entropy and cumulant based contrast functions for blind source extraction. In Mira, J., Prieto, A., eds.: Bio-Inspired Applications of Connectionism, Lecture Notes in Computer Science, Springer-Verlag. [6th International Work-Conference on Artificial and Natural Neural Networks (IWANN'2001)]. Volume II., Granada, Spain (2001) 786–793
4. Cruces, S., Cichocki, A., i. Amari, S.: Criteria for the simultaneous blind extraction of arbitrary groups of sources. In Lee, T.W., Jung, T.W., Makeig, S., Sejnowski, T.J., eds.: Proceedings of the 3rd International Conference on Independent Component Analysis and Blind Signal Separation, San Diego, California, USA (2001) 740–745
5. Cardoso, J.F.: Blind signal separation: statistical principles. Proceedings of the IEEE. Special issue on blind identification and estimation **9** (1998) 2009–2025
6. Edelman, A., Arias, T.A., Smith, S.T.: The geometry of algorithms with orthogonality constraints. SIAM J. Matrix Anal. Appl. **20** (1999) 303–353
7. Obradovic, D., Deco, G.: Information maximization and independent component analysis: Is there a difference? Neural Computation **10** (1998) 2085–2101
8. Papoulis, A.: Probability, Random Variables, and Stochastic Processes. WCB/McGraw-Hill (1991)

Temporal Processing of Brain Activity for the Recognition of EEG Patterns

Alexandre Hauser[1], Pierre-Edouard Sottas[1], and José del R. Millán[2,1]

[1] Swiss Federal Institute of Technology, Lab of Computational Neuroscience
CH-1015 Lausanne, Switzerland
jose.millan@epfl.ch

[2] Joint Research Centre of the European Commission, 21020 Ispra (VA), Italy
http://sta.jrc.it/abi

Abstract. This paper discusses three common strategies to incorporate temporal dynamics of brain activity to recognize 3 mental tasks from spontaneous EEG signals. The networks have been tested in a hard experimental setup; namely, generalization over different recording sessions while analyzing short time windows. It turns out that the simple local neural classifier currently embedded in our BCI, which averages the response to 8 consecutive EEG samples, is to be preferred to more complex time-processing networks such as TDNN and Elman-like. With this local classifier, users with some hours of training are able to operate several brain-actuated applications.

1 Introduction

There is a growing interest in the use of physiological signals for communication and operation of devices for the severely motor disabled as well as for able-bodied people. Over the last years evidence has accumulated to show the possibility to analyze brain-waves on-line to derive information about the subjects' mental state that is then mapped into some external action such as selecting a letter from a virtual keyboard or moving a robotics device [1], [2], [3], [4], [5], [6]. This alternative communication and control channel is called a *brain-computer interface (BCI)*.

Most BCIs are based on the analysis of electroencephalogram signals (EEG) associated to spontaneous mental activity. Thus, [3] measures slow cortical potentials over the top of the scalp, which indicate the overall preparatory excitation level of a cortical network. Other groups look at local variations of EEG rhythms. The most used of such rhythms are related to the imagination of movements and are recorded from the central region of the scalp overlying the sensorimotor cortex [2], [5]. But, in addition to motor-related rhythms, other cognitive mental tasks are being explored [4], [6], [7] as a number of neurocognitive studies have found that different mental tasks—such as imagination of movements, arithmetic operations, or language—activate local cortical areas at different extents. In this latter case, rather than looking for predefined EEG phenomena as in the previous paradigms, the approach aims at discovering EEG patterns embedded in the continuous EEG signal associated with different mental states.

J.R. Dorronsoro (Ed.): ICANN 2002, LNCS 2415, pp. 1125–1130, 2002.
© Springer-Verlag Berlin Heidelberg 2002

[2] and [3] have demonstrated that some subjects can learn to control their brain activity through appropriate, but lengthy, training in order to generate fixed EEG patterns that the BCI transforms into external actions. Other groups follow machine-learning approaches to train the classifier embedded in the BCI [4], [5], [6], [7]. Most of these approaches are based on a mutual learning process where the user and the brain interface are coupled together and adapt to each other [4], [5], [6]. This should accelerate the training time. Thus, [6] has allowed subjects to achieve good performances in just a few hours of training in the presence of feedback.

Most of these works deal with the recognition of just 2 mental states [2], [3], [4], [5] or report classification errors bigger than 15% for 3 or more tasks [5], [7]. An exception is the approach called *Adaptive Brain Interface (ABI)* [6] that achieves error rates below 5% for 3 mental tasks, while correct recognition is 70%. Contrarily to almost all other BCIs, ABI relies upon an *asynchronous* protocol where the subject makes self-paced decisions on when to stop doing a mental task and start immediately the next one[1]. This makes the system very flexible and natural to operate, and yields rapid response times—the system tries to recognize what mental task the subject is concentrated on every 1/2 second. ABI is being used to operate several brain-actuated devices; namely, a virtual keyboard, a computer game and a wheelchair [1], [8]. The description of these demonstrators is beyond the scope of this paper.

Currently, ABI uses a *static* local neural classifier where every RBF unit represents a prototype of one of the mental tasks to be recognized [6]. It comes then natural to incorporate the temporal dynamics of brain activity in order to improve the recognition rates. In this paper we explore three common strategies for doing so; namely time-delay neural networks (TDNN) [9], Elman-like recurrent networks [10], and averaging multiple responses of the neural classifier. One of these classifiers is a SVM [11], considered to be the state-of-the-art in what concerns static neural networks. We also report averaging results for the TDNN, Elman network and the current local classifier. The performance of the different networks have been measured in a hard experimental setup; namely, generalization over different sessions while analyzing short time windows. The difficulty lies in that brain activity changes from a session (with which data the classifier is trained) to the next (where the classifier is applied).

2 Experimental Protocol

In the experiments reported in this paper, subjects concentrate on 3 mental tasks out of a set of 5 possible. These are: "relax", imagination of "left" and "right" hand movements, "cube rotation", and "subtraction". The tasks consist on getting relaxed, imagining repetitive movements of the hand, visualizing a cube rotating around one of its

---

[1] In the case of *synchronous* protocols, the subject must follow a fixed repetitive scheme to switch from a mental task to the next [2], [3], [5]. A trial consists of two parts. A first cue warns the subject to get ready and, after a fixed period of several seconds, a second cue tells the subject to undertake the desired mental task for a predefined time. The EEG phenomena to be recognized are time-locked to the last cue and the BCI responds with the average decision over the second period of time. In these synchronous BCI systems, a trial lasts from 4 to 10 or more seconds.

axis, and performing successive elementary subtractions by a fixed number (e.g., 64–3=61, 61–3=58, 58–3=55, etc.). In a recording session, the subject performs the selected task during 10 to 15 seconds, and he/she chooses when to stop doing it and the next to be undertaken[2]. For the training and testing of the classifier, the user tells an operator which task is going to perform so that the operator can label the corresponding sequence of EEG samples. Each recording session lasts about 5 minutes.

During the sessions users receive feedback as follows. There are three buttons on the computer screen, each of a different color and associated to one of the mental tasks to be recognized. A button lights up when an arriving EEG sample is classified as belonging to the corresponding mental task.

EEG potentials are recorded at the 8 standard fronto-centro-parietal locations F3, F4, C3, Cz, C4, P3, Pz, and P4. The sampling rate is 128 Hz. We use the *Welch* periodogram algorithm to estimate the power spectrum of each channel over the last second. Epochs are 0.5 seconds long, what gives a frequency resolution of 2 Hz. The values in the frequency band 8-30 Hz are normalized according to the total energy in that band. Thus an EEG sample has 96 features (8 channels times 12 components each). The periodogram, and hence an EEG sample, is computed every 62.5 ms (i.e., 16 times per second). Therefore, a session has 4800 samples approximately.

3   Experimental Results

In this section we report the performance of the neural classifiers for five subjects in the recognition of 3 mental tasks from spontaneous EEG signals. MJ, MJR and CGS are advanced users of the interface, while FM and MC are beginners. The performance is measured by the *accuracy*, defined as the number of correct classifications divided by the total number of samples, and the *error*, defined as the number of incorrect classifications divided by the total number of samples. It is worth noting that accuracy and error do not always sum to 100% because the networks may give "unknown" responses to uncertain samples. The incorporation of rejection criteria to avoid making risky decisions is an important concern in BCI. From a practical point of view, a low classification error is a critical performance criterion for a BCI, for otherwise users would frustrate and stop utilizing the interface.

Each of the subjects carried out four training sessions at different times in the same day. For each subject, the different classifiers we have explored were independently trained on sessions 1, 2 and 3, and tested on sessions 2, 3 and 4 respectively. In a given session X, 75% of the samples were used for training and the remaining 25% for validation. Then, the resulting network was tested on the whole session X+1. The tables below report the performances of the classifiers over the last iteration (training on session 3 and testing on session 4), which corresponds to the end of training.

In order to compare time-processing neural networks against the (potentially) best static neural classifier, we have applied SVM with linear kernels. We have also tried SVM with polynomial and gaussian kernels with similar or even worse results. Re-

---

[2] While operating a brain-actuated application, the user does essentially the same as during the training phase. The only difference is that in the former case he/she switches to the next mental task as soon as the desired action has been performed.

garding the TDNN and Elman networks, we have explored architectures with different numbers of hidden units. It seems that networks with in between 10 and 50 units perform equally well. In the case of TDNN it has been used a time delay of 8 samples, for compatibility with the averaging technique explained below.

Table 1 gives, for each of the subjects, the accuracy of SVM with linear kernels, TDNN with a time delay of 8 samples, and Elman networks. In this case we have not used any rejection criteria and so all the EEG samples have been classified. Later in this section we will discuss the incorporation of rejection criteria to reduce errors. For each kind of network there are two results: the normal one (output of the classifier to each sample), and averaging the outputs for 8 consecutive EEG samples so as to respond every 1/2 second as in the current system. Averaging is a simple method to combine consecutive responses and, in our case, yields similar or better results than other techniques (e.g., product combination). Space limitations prevent the discussion of these alternatives. [4] and [7] report significant improvements in the classification of EEG signals when averaging over several seconds (from 2.5 to 5 seconds).

From this table we can make a number of observations. First, except for subject MC, SVM performs always worse than the time-processing networks. Taking into account the performances of the different training iterations, a Z-test across subjects on 2 independent proportions shows significant differences between SVM and time-processing networks. The fact that SVM is arguably the state-of-the-art highlights how hard is the classification task being tackled. Second, averaging leads to some improvements of the accuracy, but they are not significant statistically. Third, TDNN and Elman networks give statistically similar results.

Table 1. Accuracy of SVM with linear kernels, TDNN with a time delay of 8 samples, and Elman networks for the five subjects (no rejection criteria)

Subject	SVM		TDNN		Elman	
	Normal	Average	Normal	Average	Normal	Average
MJ	63%	65%	69%	69%	69%	70%
MJR	78%	77%	80%	80%	80%	80%
CGS	68%	67%	72%	74%	74%	75%
FM	59%	59%	60%	63%	58%	61%
MC	65%	66%	54%	55%	55%	57%

Table 2 illustrates the effects of incorporating a rejection criterion on the performance of TDNN, Elman and local networks, all of them averaging the response to 8 consecutive EEG samples. In this case, we use a confidence probability threshold, as the output of these networks is the posterior probability distribution for a sample to belong to the different classes. The local networks have a small number of prototypes per mental task, namely 4 units. Thus the classifier consists of just 12 units. For each subject, the first row gives the accuracy while the second row reports the error, for the corresponding probability threshold (from 0.75 to 0.95).

These experiments illustrate that neither the advanced users nor the beginners achieve high recognition rates. However, the modest accuracies achieved by advanced users (MJ, MJR and CGS) are compensated by the low percentages of errors. For the best networks and probability thresholds, the accuracy is 60% for MJR, 52% for CGS, 40% for MJ, while the error is always below 4%. Thus, the classification error is between 9 and 20 times smaller than the classification accuracy. Beginner users (FM and MC) also achieved similar absolute and relative classification errors, but they reached lower accuracies (33% and 17%, respectively). In addition to the appealing property of low classification errors, the system exhibits another key feature. Since it makes decisions every 1/2, a modest classification accuracy (in combination with low errors) does not preclude practical operation. In fact, recognition of a desired mental task takes in between 1 and 1.5 seconds on average. It is worth noting that 1 second is the shortest time necessary for recognition as EEG samples are derived from sequences that are 1-second long—and so subjects must stay concentrated on the task during that time to obtain a good codification.

Table 2. Performances of TDNN, Elman and local networks for different probability thresholds. The response of the networks is the average of 8 consecutive EEG samples. For each subject, the first row gives the accuracy and the second reports the error. Figures in bold indicate the best performance for each network where the error is below 5%

Sub-ject	TDNN					Elman					Local				
	0.75	0.80	0.85	0.90	0.95	0.75	0.80	0.85	0.90	0.95	0.75	0.80	0.85	0.90	0.95
MJ	47.6	41.8	**35.3**	30.5	16.8	50.3	41.4	36.3	**29.8**	18.2	61.6	57.8	53.7	48.6	**40.1**
	08.6	06.8	**02.4**	01.0	00.7	08.2	07.5	04.5	**02.7**	00.7	09.6	08.2	06.2	05.5	**02.7**
MJR	71.5	65.6	58.9	**52.2**	43.3	69.6	65.2	60.7	**54.4**	44.4	67.8	64.1	**60.4**	55.9	51.5
	12.6	09.6	06.3	**04.1**	02.6	11.8	09.3	05.5	**03.7**	02.2	07.1	05.5	**03.7**	02.6	02.2
CGS	54.0	**45.0**	33.7	19.7	05.5	54.0	47.9	36.9	26.5	**17.2**	56.8	54.2	**52.3**	50.7	48.7
	06.5	**04.2**	02.3	00.6	00.0	11.3	09.1	07.8	04.9	**01.3**	06.9	04.7	**03.0**	02.1	01.5
FM	19.8	15.9	**14.3**	08.7	06.3	22.2	16.7	**13.5**	11.1	03.2	51.6	47.6	43.7	39.7	**33.3**
	11.1	06.3	**03.2**	00.8	00.0	11.1	07.1	**01.6**	00.0	00.0	08.7	06.3	06.3	04.8	**02.4**
MC	**09.8**	05.5	04.5	03.4	01.8	13.7	**08.4**	04.7	03.7	01.6	38.3	33.5	28.2	21.9	**16.6**
	04.2	00.8	00.3	00.3	00.0	05.8	**01.8**	00.8	00.3	00.0	17.4	15.6	11.3	06.3	**03.4**

Interestingly, Table 1 points out that these simple local networks achieve better performances than TDNN and Elman networks. A Z-test across subjects on 2 independent proportions shows that the differences are statistically significant. For the statistical analysis, we have computed, for each probability threshold, a cost function that combines the error and non-response percentages.

Regarding the choice of the best time-processing network, it seems that a TDNN is slightly better for advanced users while an Elman network is more appropriate for beginners. This should be confirmed by a more systematic statistical comparison.

4 Conclusions

In this paper we have explored different neural networks for the classification of 3 mental tasks from spontaneous EEG signals. From these signals we extract simple power spectral features, and the neural classifiers make decisions every 1/2 second. Contrarily to our expectations, standard SVMs do not perform well. Their poor performance is even clearer when incorporating rejection criteria, which cannot be discussed here. These results highlight the difficulty and challenge of the problem. On the contrary, common time-processing neural networks achieve satisfactory performances. In particular, it turns out that the simple local neural classifier currently embedded in our BCI, which averages the response to 8 consecutive EEG samples, is to be preferred to more complex time-processing networks such as TDNN and Elman-like. With this local classifier, users with some hours of training are able to operate several brain-actuated applications [1], [8].

We have tested the above networks in a hard experimental setup; namely, generalization over sessions while analyzing short time windows. The difficulty lies in that brain activity changes naturally over time, especially from a session (with which data the classifier is trained) to the next (where the classifier is applied). A current area of research is to adapt on-line the classifier while the subject operates a brain-actuated application. In this respect, local neural classifiers are better suited than other methods due to their robustness against catastrophic interference and simple learning rules.

References

1. Millán, J. del R.: Brain-Computer Interfaces. In Arbib, M.A. (ed.): Handbook of Brain Theory and Neural Networks, 2nd edn. MIT Press, Cambridge (2002)
2. Wolpaw, J.R., McFarland, D.J.: Multichannel EEG-based Brain-Computer Communication. Electroenceph. Clin. Neurophysiol. 90 (1994) 444–449
3. Birbaumer, N., Ghanayim, N., Hinterberger, T., Iversen, I., Kotchoubey, B., Kübler, A., Perelmouter, J., Taub, E., Flor, H.: A Spelling Device for the Paralysed. Nature 398 (1999) 297–298
4. Roberts, S.J., Penny, W.D.: Real-Time Brain-Computer Interfacing: A Preliminary Study using Bayesian Learning. Med. Biol. Eng. Computing 38 (2000) 56–61
5. Pfurtscheller, G., Neuper, C.: Motor Imagery and Direct Brain-Computer Communication. Proc. of the IEEE 89 (2001) 1123–1134
6. Millán, J. del R., Mouriño, J., Franzé, M., Cincotti, F., Varsta, M., Heikkonen, J., Babiloni, F.: A Local Neural Classifier for the Recognition of EEG Patterns Associated to Mental Tasks. IEEE Trans. Neural Networks 13 (2002) 678–686
7. Anderson, C.W.: Effects of Variations in Neural Network Topology and Output Averaging on the Discrimination of Mental Tasks from Spontaneous EEG. J. Int. Systems 7 (1997) 165–190
8. Renkens, F., Millán, J. del R.: Brain-Actuated Control of a Mobile Platform. 7th Int. Conf. Simulation of Adaptive Behavior, Workshop Motor Control in Humans and Robots (2002)
9. Waibel, A., Hanazawa, T., Hinton, G., Shikano, K., Lang, K.J.: Phoneme Recognition using Time-Delay Neural Networks. IEEE Trans. Acoustics, Speech, Signal Proc. 37 (1989) 328–339
10. Elman, J.L.: Finding Structure in Time. Cog. Scien. 14 (1990) 179–211
11. Vapnik, V.: Statistical Learning Theory. John Wiley, New York (1998)

Critical Assessment of Option Pricing Methods Using Artificial Neural Networks

Panayiotis Ch. Andreou, Chris Charalambous, and Spiros H. Martzoukos

University of Cyprus, Department of Public and Business Administration,
Kallipoleos 75 Str., P.O. Box 20537, CY 1678 Lefkosia, Cyprus
bachris@ucy.ac.cy

Abstract. In this paper we compare the predictive ability of the Black-Scholes Formula (BSF) and Artificial Neural Networks (ANNs) to price call options by exploiting historical volatility measures. We use daily data for the S&P 500 European call options and the underlying asset and furthermore, we employ nonlinearly interpolated risk-free interest rate from the Federal Reserve board for the period 1998 to 2000. Using the best models in each sub-period tested, our preliminary results demonstrate that by using historical measures of volatility, ANNs outperform the BSF. In addition, the ANNs performance improves even more when a hybrid ANN model is utilized. Our results are significant and differ from previous literature. Finally, we are currently extending the research in order to: a) incorporate appropriate implied volatility per contract with the BSF and ANNs and b) investigate the applicability of the models using trading strategies.

1 Introduction and Contributions

Black and Scholes introduced in 1973 their milestone Options Pricing Model (OPM) that is nowadays known as the Black-Scholes Formula (BSF). Despite the fact that the BSF and its variants are considered as the most prominent achievements in financial theory in the last three decades, empirical research [1, 2] has shown that the formula suffers from systematic biases when compared to options market prices. To avoid the parametric models deficiencies we can address our attention to market-data driven models and not to depend on models that spring from the theoretical concepts of the options pricing field. Nonparametric techniques such as Artificial Neural Networks (ANNs) are the latest and most promising alternatives, in respect to unbiasedness and pricing accuracy, relative to the parametric OPM. The reasons that ANNs consist a promising method for developing an unbiased, market driven, accurate and robust option pricing tool are: a) Unlike BSF, ANNs are independent of any restrictive assumptions and hypotheses that must be relaxed before handling a certain financial problem; b) Option pricing functions are multivariate and highly nonlinear. ANNs are appropriate tools for approximating the empirical option pricing function [3, 4] since they can be used for nonlinear regression; c) The BSF's dynamics are

J.R. Dorronsoro (Ed.): ICANN 2002, LNCS 2415, pp. 1131–1136, 2002.

diachronically stationary and unchanged. Since it is known [1] that market participants change their option pricing attitudes from time to time, a stationary model will repeatedly fail to adjust to rapidly changing market behavior. ANNs if frequently trained, are able to diachronically adjust to market driving forces and the underling assets price dynamics by inductively learning using market data; d) Although BSF is widely used, its pricing ability varies according to options characteristics (see Yao et al. [5]). ANNs if properly developed, can be trained to price options in areas were the BSF is most biased [6].

The first contribution of our study is that it is concentrated on the most recent possible and examinable data spanning from May 1998 to December 2000. In addition, our study examines an extremely large number of data, much larger than any of the other similar studies implying that our results would be more accurate and more representative. The second contribution of our study is that it examines more explanatory variables compared to the published literature. We further examine the hybrid ANN target function suggested by Watson and Gupta [7] and used for options pricing by Lajbcygier et al. [4]. In this case, ANNs are trained to map the difference between actual market and BSF options prices instead of the actual market price of option. Moreover, instead of constant maturity risk-free interest rate (e.g. yield of the 3-month Treasury bill), we use nonlinear interpolation for extracting a continuous risk-free interest rate according to each option's time to maturity. In addition we very carefully discriminate the comparison between ANNs and BSF in two distinct, and very precise stages. In the first stage, BSF and ANNs are conceived as rival models whilst in the second stage they are conceived as supplementary models. In Stage 1, by using only historical or weighted implied volatilities we compare ANNs and BSF. Both types of volatility measures in this stage are conceived as forecasts of a unique volatility. In Stage 2, ANNs that use information derived by BSF are developed. Previous literature contradicts ANNs and BSF results without any specific rationale.

2   Option Pricing by Using BSF and ANNs

The modified option pricing formula for European call options for dividend-paying underlying asset is:

$$c^{bs} = S \, e^{\delta T} \, N(d_1) - X e^{r_f T} N(d_2) \tag{1}$$

where
$$d_1 = \frac{\ln\left(\frac{S}{X}\right) + (r_f - \delta)\,T + \frac{(\sigma\sqrt{T})^2}{2}}{\sigma\sqrt{T}}, \qquad d_2 = d_1 - \sigma\sqrt{T} \tag{2}$$

and c^{bs} stands for the premium paid for the European call option, S is the spot price of the underlying asset, X is the exercise price of the option, r_f is the continuously compounded risk free interest rate, d is the continuous dividend yield paid by the underlying asset, T is the time left until the option expiration date and σ^2 represents the yearly variance of the continuous rate of return for the underlying asset. Moreover, $N(.)$ stands for the standard normal cumulative distribution. Variance rate is the only variable that cannot be directly observed from the market. There are various kinds of volatility measurements, but the two most popular and the mostly widely used are the historical volatility and implied volatility.

On the contrary, nonparametric methods such as ANNs do not depend on any restrictive financial properties. A Neural Network is a collection of interconnected simple processing elements structured in successive layers and can be depicted as a network of arcs/connection and nodes/neurons. The ANN architecture used in this study is the feedforward network with three layers: an input layer with N input variables, a hidden layer with H neurons, and a single-neuron output layer. Each neuron is connected with all neurons in the previous and the forward layer. Each connection is associated with a weight and bias. A particular neuron node is composed by: 1) the vector of input signals, 2) the vector weights and the associated bias, 3) the neuron itself that is operating as a summer and finally, 4) the transfer function. In addition the outputs of the hidden layer are the inputs for the output layer. It can be shown (Cybenko [8]) that a two layer network having a sigmoid first layer and a linear second layer is operating as a nonlinear regression model and can be trained to approximate most functions arbitrarily well. The arbitrary accuracy is obtained by including enough processing nodes in the hidden layer.

The training of ANNs is consisted of several configurations. Regarding the learning algorithm used to train the ANNs, we utilized the modified Levenberg-Marquardt algorithm as this is described in [9]. We have scaled the input and output variables using the mean-variance scaling. In addition, before using this transformation, we have forgone a scaling on S, and c. Different number of hidden neurons up to 20 has been tested on various studies but most of them came up with ANNs with less than 10 hidden neurons. In this study for each input variable set of each training sample, all the available networks having two to ten hidden neurons were validated. Moreover, for a specific number of hidden neurons the network was initialized, trained and tested four times.

3   Input and Output Variables, and Data Filtering

We have applied various filtering rules in our datasets in order to eliminate undue bias. Before filtering, more than 80,000 observations were included for the period May 1998 - December 2000 and after filtering the dataset reduced to 63,825 datapoints, which is much larger than the ones used by previous researchers.

Before this study, a thorough pilot study was carried out. By using a sliding window of four months, we created six training periods of eight months duration titled as Tr1 to Tr6 and six associated testing periods of four moths duration titled as Ts1 to Ts6. For example, the first training period (Tr1) spans from May to December of 1998, the first consecutive testing period (Ts1) spans from January to April 1999, the second training (Tr2) spans from September 1998 to April 1999, etc. By using ANNs we examined 12 different explanatory variables. Based on these explanatory variables, we examined 18 different input variable combinations. For each of the above 18 input variable combinations we adopted a time consuming cross validation technique in order to define the optimal number of hidden neurons to be used with ANNs. The combination of all the above resulted in a total of 972 different ANN structures. In what follows, we present a description of the input and output variables extracted or computed by the collected data. Moreover only a subset of all models examined is presented.

The moneyness ratio $(Se^{-\delta T}/X)$ is calculated similar to [3]. Trading days and calendar days until expiration of each option were calculated. The trading days were computed assuming 252 days while the calendar days were computed assuming 365 days in a year. Almost all studies use the 252 days scheme and do not check for any other time to expiration. It is not clear which time to maturity should be fed to the networks and to BSF since volatility is annualized using 252 days (T_{252}) and interest rates and dividends are annualized using 365 days (T_{365}). Moreover, most of the studies use an approximation of the interest rate (e.g., 90 day bank bill) but an inaccurate interest rate when used may cause significant deviation of BSF from c. In this study we instead use cubic spline interpolation in deriving a continuous risk free interest rate, r_f, to be used in either case. Furthermore, three different volatility measures are presented. The first volatility considered is the 60 days annualised historical volatility (σ_{60}) that was calculated using all the past 60 log-relative index returns starting from the immediately previous trading day and spanning backwards with all the available log-returns up to sixty. In the same manner, the 30 days historical volatility (σ_{30}) was calculated. The third volatility used is the CBOE VIX Volatility Index (σ_{VIX}). VIX is one measure of the level of implied volatility and was developed by the CBOE in 1993 and is a measure of the volatility of the S&P 100 Index but it can be used for the S&P 500 due to high correlation of the two indexes. The call market value standardized by the striking price c/X is used and in addition we implement the hybrid model suggested by [4] where the target function is transformed in order to represent the pricing error between the options' market price and the BSF pricing $(c/X - c^{bs}/X)$.

The training and testing samples are chosen in such a way so that the six testing datasets cover the last two years of the complete dataset. By doing this we can have comparable results in this two-year period. Although we have used MSE to optimize the ANN, the 50th (median), the 90th and the 95th percentiles of the absolute errors were considered. In addition, the coefficient of determination (R^2) was computed and reported. In here we present some descriptive results based only on the median of the absolute errors (MdAE).

From our earlier work we concluded that the BSF with T_{252} was producing better results. So, with BSF we have used: S, X, δ, T_{252}, r_f and all three volatility forecasts σ_{60}, σ_{30}, and σ_{VIX} (BSF 1, 2, and 3). For ANNs, in total eight different input and output variable combinations are presented here. Six variable combinations are formed by using $Se^{-\delta T}$, r_f and T_{365} in conjunction with one of σ_{60}, σ_{30}, and σ_{VIX} to map c/X (ANN 1.1, 1.2, and 1.3) and $c/X - c^{bs}/X$ (ANN 2.1, 2.2, and 2.3) respectively. In addition, two more combinations are formed using S/X and T_{252} for mapping c/X (ANN 1.4-like in Hutchison et al., 1994) and $c/X - c^{bs}/X$ (ANN 2.4).

4   Numerical Results and Discussion

Stage 1: Table 1 summarizes the out of sample performance of BSF 1, 2 and 3 and ANNs 1.1, 1.2, 1.3 and 1.4. In general we can conclude that ANN models are superior to BSF models but there is no input variable combination that renders either a specific BSF or a specific ANN model as superior. From Table 1 we see that the BSF 3 (uses σ_{VIX}) is qualified only one out of six times as the lowest MdAE parametric model whereas the BSF 1 (uses σ_{60}) is qualified in three out of six times. A conclusion is that many studies, which compare ANNs with the BSF, are a priori biased against the BSF pricing accuracy by exploiting only one volatility measure (usually 60 days historical volatility).

From Table 1, we can notice that ANN 1.4, which is a volatility free model similar to the one implemented by Hutchison et al. [3] is prevailing as the lower MdAE in one out of six testing periods and in general it exhibits quite good results (given that it exploits only two input variables). By examining the behavior of σ_{VIX}, σ_{60} and σ_{30} it was obvious that for period Tr4 the S&P 500 Index exhibited relatively stable/stationary volatility levels, and in that period the ANN 1.4 performed better. So, we can conclude that during periods of unusual and unexpected volatility, ANN models that are volatility free *cannot* inductively learn the volatility pattern from the data, so it is recommended that an ANN model that incorporates a volatility input should be implemented (e.g. ANN 1.3). During periods when security market's volatility is stable and does not perturb, volatility free ANN *can* be utilized (e.g. ANN 1.4).

Table 1. Out of sample Median Absolute Errors (MdAE) of BSF and ANNs

Period	BSF 1	BSF 2	BSF 3	ANN 1.1	ANN 1.2	ANN 1.3	ANN 1.4	ANN 2.1	ANN 2.2	ANN 2.3	ANN 2.4
Ts1	9.020	9.000	7.286	7.750	6.425	6.301	6.396	5.955	5.659	6.735	5.777
Ts2	5.223	5.590	6.170	5.453	4.154	3.657	5.052	5.334	3.328	3.037	2.800
Ts3	3.501	3.407	4.993	2.758	3.356	4.224	2.895	2.654	1.941	4.393	2.657
Ts4	5.497	5.986	7.579	5.583	9.020	6.766	3.443	6.221	5.404	7.574	7.903
Ts5	4.956	5.092	5.704	3.372	2.839	2.659	3.128	4.535	2.890	4.014	5.166
Ts6	4.655	4.105	5.376	3.825	2.919	1.796	2.831	2.827	3.001	3.849	3.500

What we conclude from this section is that when BSF is compared vis-à-vis ANNs by using historical or weighted implied volatility measures then, ANNs dominate BSF in the option-pricing problem. Even if only one ANN model is implemented (e.g. ANN 1.3), it is very simple to realize that on average, the pricing errors of either BSF (1, 2 or 3) are larger (in terms of MdAE).

Stage 2: In this phase, BSF and ANN are perceived as supplementary techniques, since our effort is to find an optimal way to combine them in order to improve (if possible) the options pricing precision compared in Stage 1. Table 1 summarizes ANN 2.1 to ANN 2.4 that are relevant to this stage. In this phase of our analysis, we have changed our target function from c/X to the hybrid

1.1   Neurophysiological Background

The ERP after an error trial is characterized by two components: a negative wave called error negativity (N_E) and a following broader positive peak labeled as error positivity (P_E), [4]. Recent studies revealed that the P_E is more specific to errors while the N_E can also be observed in correct trials, cf. [4]. Although both amplitude and latency depend on the specific task, the N_E occurs delayed and less intense in correct trials than in error trials. The N_E has a fronto-central maximum, the P_E a centro-parietal maximum. At present, there is not yet a final consensus about the underlying cognitive functions. N_E seems to reflect some kind of comparison process. Due to the localization of the origin in the *anterior cingulate cortex* [4] it might be an emotional and/or attentional component. In contrast, P_E seems to be connected to conscious error detection [5].

The one study reporting error potentials in a BCI context [6] is solely based on P_E, but the neurophysiological findings indicate that also the N_E component might be useful to some degree.

2   Aims and Methods

2.1   Response Verification for BCIs

Most BCI systems allow the user to select one out of several choices. At present there are often only two classes ([1], [3], [7]), but there are also multi-class BCIs, e.g. [2], [8]. For such BCI systems an error detection algorithm can provide a useful add-on. If there are just two classes, detecting an error allows to correct the BCI classification (response-correction), for more than two classes at least wrong classifications can be rejected (response-verification).

While the idea of correcting BCI misclassifications is tempting, one has to be careful: as the detection method will not work perfectly, some correct BCI classifications can potentially get "corrected" towards a wrong choice. If the proportion of such miscorrections is non-negligible the subject will become irritated. Even if the mechanisms works well enough to theoretically increase the information transfer rate it may be unfavorable in a psychological sense. This implies the need to strictly bound the rate of false positives (FP-rate: the fraction of acual correct trials which is misclassified as an error), where we use the nomenclature that "positive" events are the ones that are to be detected, i.e., trials where the BCI algorithm missed to detect the subject's intention.

2.2   Experiments

At this stage we investigated EEG data from an attention test, while BCI feedback experiments are planned for the next step. Eight healthy subjects took part in one EEG measurement each, in which they had to perform a variant of the "d2-test", [9]. After a computer screen displayed visual stimuli, subjects had to respond to targets by pressing a key with the right index finger and to non-targets with the left index finger. Targets in the d2-test are compound symbols consisting of the letter "d" and exactly two horizontal bars that may occur in four possible positions each. Non-targets either show the letter "b" and an arbitrary number of bars (0–4) or the letter "d" and a number of bars that differs from two, see Fig. 1 for some examples. After the subject's keystroke the reaction time was displayed on the screen, either in green if the response was correct, or in red if it was erroneous. The next trial began 1.5±0.25 s later. A summary of the experiments with reaction times and error rates is given in Table 1.

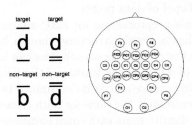

Fig. 1. Examples of targets and non-targets in the d2-test on the left. The electrode montage is shown on the right.

Table 1. Summary of the experiments with "d2-paradigm".

subject [code]	reac. [ms]	trials [#]	err. [#]	err. [%]
aa	539	977	101	10.3
ab	434	893	41	4.6
ac	556	896	8	0.9
ad	477	893	46	5.2
ah	551	884	19	2.1
ae	504	894	39	4.4
af	497	893	35	3.9
ag	529	892	42	4.7

Brain activity was recorded with 28 Ag/AgCl electrodes, cf. Fig. 1, referenced to nasion, with a broad band-pass filter. Besides EEG we recorded a horizontal and vertical electrooculogram (EOG). In an event channel timing and types of stimuli and keystrokes were stored along with the EEG signal.

No (!) trials were rejected due to artifacts, but all trials in which the subject hit two keys (simultaneously or sequentially) were sorted out.

2.3   A Pattern Matching Method

The two components that are observed in the EEG related to errors are slow cortical potentials (SCPs). In [7] we presented a successful method for classifying single trial EEG based on SCPs which can be used here with some appropriate modifications. In [7] the key for good results was the combination of high-dimensional features and robust learning machines for classification. The features we use in this study are subsampled versions of the relevant channels (marked labels in Fig. 1). Subsampling from Hz 100 to Hz 20 was done by calculating the mean of consecutive 5-tupel of data points.

The advantage of this preprocessing for ERP analysis is that the resulting classification problem has a simple structure, though being high-dimensional. The distributions of the feature vectors of each class can be modelled by a normal distribution, the mean of which is the feature of the ERP of the corresponding condition, cf. [10]. The covariance matrix is determined by non-task related brain activity. As this is approximately the same for both classes, the classification has to separate two normal distributions with equal covariance matrices. The Bayes-optimal classifier for this task is the Fisher Discriminant (FD). Dealing with high noise cases in a high-dimensional space typically requires regularization in order to obtain stable estimates of the covariance matrices.

But in the present situation we are looking for the classifier which is optimal under the constraint that the FP-rate attains a predefined value (on the training set). For linear classifiers in a separating hyperplane formulation ($w^\top x + b = 0$) this can be accomplished by adjusting the threshold b. This procedure is indeed optimal in conjunction with the FD under the forementioned assumptions, which can be seen using the Neyman-Pearson Lemma (we thank Marina Meila for this remark), or by the following direct proof.

Let $(X, Y) \in \mathbb{R}^n \times \{N, P\}$ be random variables such that the conditionals $P(X \mid Y = N)$ and $P(X \mid Y = P)$ are $\mathcal{N}(\mu_N, \Sigma)$ resp. $\mathcal{N}(\mu_P, \Sigma)$ distributed (i.e., normal distribution with mean $\mu_\sim$ and covariance matrix Σ). The problem is to maximize $P(w^\top X + b > 0 \mid Y = P)$ subject to $P(w^\top X + b > 0 \mid Y = N) = \delta$ for some fixed $\delta \in (0, 1)$. Denoting the distribution function of $\mathcal{N}(\mu, \sigma)$ by $F_{\mathcal{N}(\mu,\sigma)}$ we have $\delta = P(w^\top X + b > 0 \mid Y = N) = 1 - F_{\mathcal{N}(w^\top \mu_N + b,\, w^\top \Sigma w)}(0) = 1 - F_{\mathcal{N}(0,1)}(-\frac{(w^\top \mu_N + b)}{\sqrt{w^\top \Sigma w}})$. Hence for $\beta := F_{\mathcal{N}(0,1)}^{-1}(1 - \delta)$ we obtain the threshold $b = -\beta \sqrt{w^\top \Sigma w} - w^\top \mu_N$ in dependence from the optimal w. Substituting this term for b one can see that $P(w^\top X + b > 0 \mid Y = P) = \cdots = \int_0^\infty \frac{1}{\sqrt{2\pi}} \exp(-\frac{1}{2}[t + \beta - w^\top(\mu_P - \mu_N)(w^\top \Sigma w)^{-\frac{1}{2}}]^2)dt$, and this expression is maximized if $w^\top(\mu_P - \mu_N)(w^\top \Sigma w)^{-\frac{1}{2}}$ is maximized. Since the last term is the square root of the Rayleigh coefficient we get the same w as from FD (qed).

2.4   Amplitude Threshold Criterion

For comparision we also implemented the absolute amplitude criterion, the only algorithm for the detection of the error potential in a BCI context published so far, [6]. In this method all trials for which the amplitude of the Cz channel

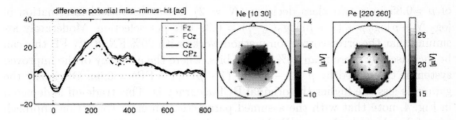

Fig. 2. Average miss-minus-hit EEG-traces at electrodes along the vertex for subject *ad*. Regions of N_E and P_E are shaded and scalp topographies for that regions are shown in the two subplots at the right.

averaged over a predefined time period exceeds a predefined threshold are classified as errors. To make the comparison fair we extracted amplitude peaks for all channels that were used in our method, and the optimal hyperplane threshold was determined by the same learning algorithm.

3   Results

In the average difference potential "miss-minus-hit" the two discussed components N_E and P_E emerge very pronounced, cf. the ERP for subject *ad* in Fig. 2. The other ERPs show the same characteristics. For all subjects an early negative and a later positive component is clearly observable. The main intersubject differences concern the latency of the two components and the fall-off after the P_E. For three subjects (*ab, ac, ad*) the peak of the N_E shows up very early about ms 10–30, but even in all other subjects it is not later than ms 60. That peak is too early to be a reaction to the visual feedback at ms 0. This observation is in agreement with the subjects' reports that in erroneous trials they often knew they were going to make a mistake while they initiated the movement but they could not withhold anymore.

As was pointed out in Section 2.1 a special demand on the error detection is the ability to strictly bound false positive classifications. The first question that arises here is: how well does the bound for FPs that we enforce on the training set carry over to the test set. Training a classifier for FP=2% resulted for all 8 data sets in FP-rates between 2.3% and 3.5%. In order to make the comparison between different data sets and parameter choices easier, we used a cross-validation in which (the thresholds of the linear) classifiers were adapted after training to obtain a predefined FP-rate so that the detection performance is reflected by the FN-value only. In Fig. 3 our pattern matching method was evaluated for FP-bounds at 1%, 2% and 3%. Subjects were sorted according to the FN-rate at FP=2%: the performance is "very good" (FN⩽11%) for 4 subjects, "good" (FN⩽22%) for 3 subjects and "not good" only for subject *aa*. White bars show the corresponding error rates for the amplitude criteron.

To assess the potential value of the proposed error detection method for improving BCI transmission rates, we take a look at an example. A BCI accuracy of $p = 0.85$ in a two class decision ($N = 2$) has a theoretical information of $\log_2 N + p \log_2 p + (1 - p) \log_2(\frac{1-p}{N-1}) = 0.39$ bits per selection. Moderately assuming that the error-correction method works with 20% FN at 3% FP this can be increased by more than 75% to 0.69 bits, where the accuracy of the improved system is calculated by $p \cdot (1 - \text{FN}) + (1 - p) \cdot (1 - \text{FP}) = 0.94$. Obviously the gain gets less the higher the original BCI accuracy is. This trade-off is depicted in Fig. 4, note that with the assumed parameters an error correction approach is useful as long as the pure BCI accuracy is lower than 96%.

Ocular artifacts. In [6] it was reported that the end of many trials contained eyeblinks, an effect that is also present in our data. So is has to be made sure that classification success is not based on ocular contamination of the EEG. Therefore we also tried a classification of errors based on the EOG signals in the same way as we did EEG-based classification. The resulting FN-rate was >95% for most subjects, and only for subject *ae* it was 77% which is still more than 8 times higher than in EEG-based detection with 9%.

4   Discussion

Our pattern recognition approach to single trial detection of the error potential provided a substantial improvement in comparison to a simple amplitude threshold criterion, and the expected gain for BCI classification is promising. The important next step is to conduct BCI experiments with real-time feedback to check whether the error potentials are of the same type in such a scenario.

Acknowledgement. We would like to thank Sebastian Mika and Klaus-Robert Müller for valuable discussions. The studies were supported by a grant of the *Bundesministerium für Bildung und Forschung* (BMBF), FKZ 01IBB02A.

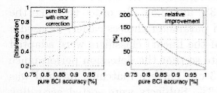
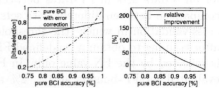

Fig. 3. Rate of FNs for detection at ms 300 with fixed FP-rate. White bars show the corresponding FP-rates for the amplitude criteron, cf. Section 2.4.

Fig. 4. The plot shows the improvement of the BCI accuracy by the proposed error correction method, when an FP-rate of 3% and an FN-rate of 20% is assumed.

References

1. Birbaumer, N., Ghanayim, N., Hinterberger, T., Iversen, I., Kotchoubey, B., Kübler, A., Perelmouter, J., Taub, E., Flor, H.: A spelling device for the paralysed. Nature **398** (1999) 297–298
2. Obermaier, B., Neuper, C., Guger, C., Pfurtscheller, G.: Information transfer rate in a five-classes brain-computer interface. IEEE Trans. Neural Sys. Rehab. Eng. **9** (2001) 283–288
3. McFarland, D.J., Miner, L.A., Vaughan, T.M., Wolpaw, J.R.: Mu and beta rhythm topographies during motor imagery and actual movements. Brain Topogr. **12** (2000) 177–186
4. Falkenstein, M., Hoormann, J., Christ, S., Hohnsbein, J.: ERP components on reaction errors and their functional significance: a tutorial. Biol. Psychol. **51** (2000) 87–107
5. Nieuwenhuis, S., Ridderinkhof, K., Blom, J., Band, G., Kok, A.: Error-related brain potentials are differentially related to awareness of response errors: evidence from an antisaccade task. Psychophysiology **38** (2001) 752–760
6. Schalk, G., Wolpaw, J.R., McFarland, D.J., Pfurtscheller, G.: EEG-based communication: presence of an error potential. Clin. Neurophysiol. **111** (2000) 2138–2144
7. Blankertz, B., Curio, G., Müller, K.R.: Classifying single trial EEG: Towards brain computer interfacing. In Diettrich, T.G., Becker, S., Ghahramani, Z., eds.: Advances in Neural Inf. Proc. Systems (NIPS 01). Volume 14. (2002) to appear.
8. Donchin, E., Spencer, K.M., Wijesinghe, R.: The mental prosthesis: Assessing the speed of a P300-based BCI. IEEE Trans. Rehab. Eng. **8** (2000) 174–179
9. Brickenkamp, R., Zillmer, E.: D2 Test of Attention. Hogrefe & Huber (1998)
10. Kohlmorgen, J., Blankertz, B.: A simple generative model for single-trial EEG classification. In: Artificial Neural Networks – ICANN. (2002)

Dynamic Noise Annealing for Learning Temporal Sequences with Recurrent Neural Networks

Pierre-Edouard Sottas and Wulfram Gerstner

Laboratory of Computational Neuroscience, EPFL, 1015 Lausanne, Switzerland

Abstract. We present an algorithm inspired by diffusion networks for learning the input/output mapping of temporal sequences with recurrent neural networks. Noise is added to the activation dynamics of the neurons of the hidden layer and annealed during learning of an output path probability distribution. Noise therefore plays the role of a learning parameter. We compare some results obtained on 2 temporal tasks with this "dynamic noise annealing" algorithm with other learning algorithms. Finally we discuss why adding noise to the state space variables can be better than adding stochasticity in the weight space.

1 Gradient Based or Stochastic Algorithms?

Learning algorithms for artificial neural networks can be roughly divided in two classes: gradient based and stochastic algorithms. Gradient based algorithms have the advantage that they learn much faster because, by definition, they try to move the parameters directly towards the optimum. However, gradient based algorithms can get stuck on a plateau or a local minimum of the error function. In contrast, some stochastic algorithms, such as simulated annealing [1], have the nice property to be global optimizers. The price to pay is generally a lot of computer time demand so that global search methods are used in practice when only the final performance is critical. The input/output mapping of temporal sequences with recurrent neural networks (RNNs) is especially faced to this problem as emphasized by the "vanishing gradients problem" [2,3]. One needs a fast global search algorithm for learning in RNNs to benefit of their universal approximator property.

We present in this paper a semi-stochastic algorithm for learning in dynamically driven RNNs. The stochastic component is combined with optimization methods that move the parameters towards the optimum. In the next section, we present this algorithm which is inspired by the formalism developed for diffusion neural networks [4,5]. Then, we compare some results obtained by a RNN of Elman type [6] optimized by this algorithm with several local and global optimizers on two time-dependent tasks, one being related to the vanishing gradients problem. Finally we give a brief conceptual comparison between several algorithms available for learning in dynamically driven RNNs.

J.R. Dorronsoro (Ed.): ICANN 2002, LNCS 2415, pp. 1144–1149, 2002.
© Springer-Verlag Berlin Heidelberg 2002

2   Dynamic Noise Annealing

Movellan et al [4,5] have developed a very interesting framework regarding diffusion neural networks. Diffusion networks can be seen as continuous RNNs with an intrinsic probabilistic component described by a set of stochastic differential equations. In particular, a Monte-Carlo approach has been successfully undertaken for learning to approximate continuous path probability distributions with diffusion networks [5].

In this paper, we do not see noise as being conceptually an integral part of the network, but rather as a learning tool for RNNs. This is achieved by annealing progressively the level of noise so that we retrieve a deterministic path (or sequence in the discrete case) after learning. Therefore the combination of the formalism developed for diffusion networks and an annealing scheme gives a powerful semi-stochastic algorithm for learning in RNNs.

The continuous RNN consists in N fully connected neurons driven by P inputs $v^I(t)$. The equations of the system are:

$$\frac{dv_j^h(t)}{dt} = -\frac{v_j^h(t)}{\tau_j} + \sum_{i=1}^{N} w_{ji}^h \, \varphi(v_i^h(t)) + I_j(t) \tag{1}$$

where the induced local fields $v_j^h(t)$ of the $j = 1, ..., N$ individual neurons constitute the state vector, τ_j are time constants, $\varphi(\cdot)$ is the activation function and $I_j(t)$ are functions of the driving input signals :

$$I_j(t) = \sum_{k=1}^{P} w_{jk}^I \, v_k^I(t) \tag{2}$$

The N neurons defining the state of the network can be interpreted as a hidden layer. A readout process is accomplished with O output neurons fed by the N neurons:

$$\frac{dv_l^o(t)}{dt} = \sum_{j=1}^{N} w_{lj}^o \, \varphi(v_j^h(t)) \tag{3}$$

Learning is accomplished by adding noise to the activation dynamics of the N neurons:

$$\frac{dv_j^h(t)}{dt} = \mu_j(v^{(\cdot)}(t)) + \sigma \, W_j(t) \tag{4}$$

where $\mu_j(.)$ is deterministic and given by the right-hand side of Eq.(1), $W_j(t)$ are independent Gaussian white noise variables and σ is the learning parameter.

Each input sequence $v^I(t)$ is applied repeatedly to the network so as to generate m hidden sequences $v_i^h(t)$ in the network. Each of these m sequences can potentially drive different output sequences. Similarly to diffusion networks, credit or blame is attached to each of the $v_i^h(t)$ by computing an importance function $p(v_i^h)$ derived from the Girsanov change of measure [7]:

$$p(v_i^h) = \exp \left\{ \sum_{l=1}^{O} \left(\frac{1}{\sigma^2} \int_0^T \mu_l^o(v_i^h(t), w) \, dv_l^o(t) \right. \right.$$
$$\left. \left. -\frac{1}{2\sigma^2} \int_0^T [\mu_l^o(v_i^h(t), w)]^2 \, dt \right) \right\} \tag{5}$$

where the first integral is an Ito stochastic integral, μ_i^o is the right part of Eq.(3), $i = 1, ..., m$ counts the sample sequences and $\sum_{i=1}^{m} p(v_i^h) = 1$. Intuitively, this can be seen as a comparison between each of the potential output sequences with the desired output sequence $v^o(t)$. The interesting point of this importance sampling is that the $p(v_i^h)$ depend in two ways on the diffusion/learning parameter σ. First, there is an implicit dependence on σ by the generated noisy sequences v_i^h. Secondly, the importance function is an explicit function of σ as given in Eq.(5). Large values of noise decrease the amplitude of the importance while very small values tend to select only the best sequence because of the finite number m of sequences v_i^h. But this is exactly what we need for an annealing scheme on σ. In the former case, differentiation between sequences is weakened, while in the latter case the importance becomes a function of only the best generated sequence. This annealed importance sampling is at the basis of the algorithm. This can be intuitively seen as the search of the best sequence generated by the network (best in a likelihood sense) as compared with the search of the minimal energy in simulated annealing.

For the update of the weights at a given amount of noise, we have adapted the EM-type algorithm of Movellan to our network. In particular we let the reader refer to [5] for the derivation of the likelihood function. In our case, we have to take care of the acyclic topology of the network. The update of the weights at each level of noise is fast because it is accomplished with a direct matrix inversion. Fig.(1.A) shows a typical evolution of the likelihood of the generated sequences during a learning run with a cooling strategy, $\sigma(l+1) = \alpha \sigma(l)$, $0 < \alpha < 1$. During the first ~ 30 learning steps, the level of noise is too large and the selection process accomplished by the importance function is not effective, all the paths having about the same importance. Therefore the mean value of the likelihood does not increase. Above this value, credit assignment and selection become effective. This is first seen by an increase of the spread of the likelihood of the v_i^h and then by an increasing mean likelihood with decreasing noise. Progressively the spread is reduced until the noise level reaches a so small value so that only one path is selected. This corresponds to the deterministic case and learning is stopped. We have tried different annealing schedules: best performances in the terms of convergence and training time were obtained with an initial σ equal to the value at which the selection process begins and a final σ chosen so that only one path contributes significantly in the importance function.

3   Results

We demonstrate the potential of our approach by applying it to a RNN that has to solve 2 input/output mapping tasks. The first is related to the vanishing gradients problem [2,3]. This problem appears in the presence of long-term dependencies between relevant inputs of the surrounding environment and target events. The RNN must classify 4 different sets of 2-dimensional sequences of length T. For each sequence, the class depends only on the first L values of the input signal. L is fixed and $T \gg L$. The system should provide an answer after

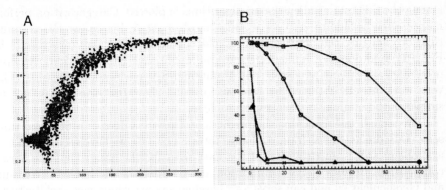

Fig. 1. A. Evolution during learning of the likelihood of the generated sequences v_i^h. B. Plot of the categorization performance versus the time lag for the dynamic noise annealing (□), simulated annealing (O), random weight guessing (∗) and a gradient based algorithm (△) under the same training time constraint. The mean value of 5 independent simulations is shown.

T, i.e. at the end of each sequence. Thus the problem can be solved only if the dynamic system is able to store information about the initial input during the interval $T - L$. This task is similar to but more difficult than the "two-sequence problem" [2,3]. Each 2-dimensional input sequence is generated as follows: the first L values are chosen from a normal distribution of given mean and variance. The value of the mean gives the criterion for the class and is chosen as either $+1$ or -1 for both inputs. The variance is 0.1. This gives 4 classes in the output. Each class is represented by an output unit that is activated only at the end of the sequence. The output sequence is therefore zero during the first $(T - Z)$ values and increases linearly until 1 during the last Z values. During $[L+1, T-Z]$, the input values are chosen as independent, zero-mean Gaussian noise with variance equal to 0.1. We fixed L and Z to 5 for all simulations. A test-sequence is considered as correctly categorized if only the correct output neuron is activated at the end of the sequence and no output neuron is activated before $(T - Z)$. A neuron l is said to be activated if its instantaneous field $v_l^o(t)$ is larger than 0.5. The number N of neurons is fixed to 10 giving a number of 160 weights. φ is taken as a logistic activation function.

We compare dynamic noise annealing with several gradient based methods and two global optimization procedures: traditional SA and random weight guessing. Random weight guessing has been included because such a procedure is known to be the fastest on the related "two-sequence problem" [3]. The comparison is accomplished with the same network, same weight initialization procedure, same training/test bases and with the same amount of training time. Traditional speed-up methods were used for all methods. Dynamic noise annealing was performed by generating $m = 20$ hidden sequences for each input sequence. Results for time lags between 1 and 100 (corresponding respectively to $T = 11$ and $T = 110$) are summarized in Fig.(1.B). Only the result obtained by the best among the different gradient based algorithms is plotted. Categorization performance already falls to zero for a time lag of about 5. Increasing the training time or changing the activation threshold condition scales only the results in Fig.(1.B) and does not influence the overall qualitative interpretation: dynamic noise annealing remains faster than traditional SA and random guessing. The very weak results obtained by random guessing show that the task is not as simple as the original two-sequence problem.

The second task concerns the prediction of future values of a financial time serie (U.S. dollar - Swiss franc exchange rates). The data we have used is publicly available [1] and is an extension of the data used for the Santa Fe Time Series competition. Results of the competition are described in the book by Weigend and Gershenfeld [8] and can serve as a baseline against which new architectures and training methods can be compared for time series analysis. In particular, we want to know whether our algorithm can improve the prediction obtained by different RNNs architectures learned by backpropagation through time (BPTT) as given by the contribution of Mozer in this book. Following the results obtained on the vanishing gradient related task, we hope that our network can extract from the series a possible long-term structure that BPTT cannot find. We applied the same preprocessing of the financial data as described in Mozer on the same training/test sets. We let the reader refer to this article for details. The task is to predict the value of an exchange rates X minutes in the future[2]. The 3-dimensional input sequences consists in the day of week, the time of day and the change in the daily exchange rates resampled at 1 minute interval. The 1-dimensional output sequences are formed by the original daily exchange rate sequences delayed by X minutes. This mimics a X minutes prediction task. Performance is given either by the normalized mean squared error (NMSE)[3] or by the percentage of the correct predicted change in direction of the series. All the results are shown in Table 1 in comparison with the best results obtained by Mozer.

Table 1. 1-minute prediction of USD-CHF exchange rate (NMSE or % of correct predicted change in direction)

Corpus	data points	results	Mozer results
extended competition set	31840	0.9991	0.998
1985/1987 set	96450	0.9934	0.985
1985/1987 set (3 clas.)	96450	71.3%	63.2%
1985/1987 set (2 cond. clas.)	25570	62.4%	58.5%

Our results are averages over 3 independent runs. An interesting point is that the final performance is very robust with respect to the weights initialization.

---

[1] http://www-psych.stanford.edu/~andreas/Time-Series/SantaFe.html

[2] We focused on the $X = 1$ prediction task, because the high average sampling rate at 1 minute would need a high-depth memory.

[3] Normalization indicates that a value of 1.0 is not better than a random walk.

Astonishingly, our results are significantly worse if measured by the NMSE, whereas we can predict significantly better the direction of change. For instance, on the 1985/1987 corpus, we can predict at 71.3% the change in direction [4]. This performance can even be increased to 75.9% with different class criterions. It still remains unclear why our network learns better the direction of change than "how much" it will change, as compared with Mozer's networks. In financial applications, traders are more interested to know the change of direction.

4   Stochasticity in the State Space

The recurrent connections enable the network to acquire state representations. Very often, people have claimed the benefits of such abstract representations. Some have even shown that such RNNs are universal approximators of dynamical systems and can simulate any Turing machine. However these results do not say how a RNN should be approximated. Global search algorithms like simulated annealing and random weights guessing add stochasticity in the weight space, a process known to be slow because of its inherent high dimensionality in RNNs. Here we add stochasticity to the state space variables instead to the weights themselves. Combined with the Movellan algorithm, that allows us to define a likelihood function which is differentiable with respect to the weights. Associated with traditional optimization methods this allows to move the parameters towards the optimum. In other words, noise, when added to the state space variables, allows to define a gradient in the weight space. The price we have to pay for this property is that we have to run the network many times (here 20). Gradient based algorithms typically run only once the network and define directly a gradient with respect to the parameters. At the opposite, most of stochastic algorithms typically move "blindly" in the weight space. Our semi-stochastic algorithm lies in between.

References

1. Kirkpatrick S., Gelatt C.D., Vecchi M.P. Optimization by Simulated Annealing. Science **220** (1983) 671–680
2. Bengio Y., Simard P., Frasconi P. Learning Long-Term Dependencies with Gradient Descent is Difficult. IEEE T. Neural Networor. **5-2** (1994) 157–166
3. Hochreiter S., Schmidhuber J. Long Short-Term Memory. Neural Comput. **9-8** (1997) 1735–1780
4. Movellan J. R.,and J.L. McClelland J. L. Learning continuous probability distributions with symmetric diffusion networks. Cognitive Sci. **17** (1992) 463–496
5. Movellan J. R., Mineiro P., Williams R. J. Modeling Path Distributions Using Partially Observable Diffusion Networks. TechReport, CogSci, UCSD (1999)
6. Elman J. L. Finding Structure in Time. Cognitive Sci. **14** (1990) 179–211
7. Oksendal B. Stochastic differential equations. Springer-Verlag (1992)
8. Weigend A. S., Gershenfeld, N. A. Times Series Prediction: Forecasting the future and understanding the past. Addison-Wesley (1994)

---

[4] with 3 classes (down, no change, up) and criterions defined as in Mozer[8]

Convolutional Neural Networks for Radar Detection

Gustavo López-Risueño[1], Jesús Grajal[1], Simon Haykin[2], and Rosa Díaz-Oliver[1]

[1] Universidad Politécnica de Madrid,
E.T.S.I. Telecomunicación, Avd. Complutense s/n, 28040 Madrid, Spain
{risueno,jesus}@gmr.ssr.upm.es,
[2] McMaster University, Adaptive System Laboratory,
1280 Main Street West, Hamilton, ON L8S 4K1, Canada
haykin@mcmaster.ca

Abstract. The use of convolutional neural networks (CNN's) for radar detection is evaluated. The detector includes a time-frequency block that has been implemented by the Wigner-Ville distribution and the Short-Time Fourier Transform to test the suitability of both techniques. The CNN detectors are compared with the classic multilayer perceptron and with several traditional non-neural detectors. Preliminary results are shown using non-correlated and correlated Rayleigh-envelope clutter.

1 Introduction

Radar detection is a rather complex problem in real-life applications since, in most of the occasions, neither the target nor the clutter have known statistics. Moreover, they depend on many factors, such as atmospheric conditions, environment, radar range resolution and target dynamics [1], and can exhibit time-varying and even chaotic behaviour [2]. Therefore, in practical situations, it is difficult to find a suitable statistical model to design the radar detector and accurately predict its performance. Neural networks can be used due to their learning ability. Moreover, several authors [3] have suggested the use of time-frequency analysis and neural networks to improve the detection performance as time-frequency representations provide the Doppler frequency information of target and clutter (see Fig. 1).

In this paper, the suitability of the former approach is evaluated using a convolutional neural network (CNN) [4,5], which is intended for image classification and has been successfully applied to handwriting recognition by Lecun et al. [4]. CNN's comprise a special class of multilayer perceptrons with high invariance to translation, scaling and distortion. Unlike previous approaches [3], CNN's avoid the need of feature extraction, since they jointly perform feature extraction and classification [5]. Along with the CNN, two time-frequency representations [6] have been tested: The Wigner-Ville Distribution (WVD), which is quadratic, and the Short-Time Fourier Transform (STFT), which is linear. The designed detector is coherent and mono-cell [1], i.e. it processes the complex envelope of the echoes from only a range cell.

J.R. Dorronsoro (Ed.): ICANN 2002, LNCS 2415, pp. 1150–1155, 2002.

Section 2 is devoted to the description of the CNN evaluated in this paper. Section 3 shows the performance of the WVD- and STFT- CNN detectors trained by back-propagation (BP). The comparison with a neural detector based on the multilayer perceptron (MLP) without a previous time-frequency block and several non-neural detectors is performed in section 4. Finally, a number of conclusions are drawn in section 5.

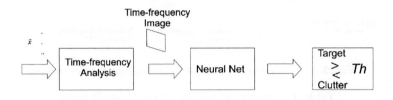

Fig. 1. Block diagram of the neural detector.

2   Convolutional Neural Network Configuration

The CNN structure is the combination of three ideas: 1) Local receptive fields, 2) shared weights, and 3) subsampling. Local receptive fields force CNN's to extract features since each particular layer comprises several feature maps. In these maps, weight sharing provides shift invariance and reduction in the number of free parameters to adjust. CNN's are mainly composed of two different types of alternating layers: Convolutional and subsampling layers. In a convolutional layer, the feature maps are obtained by convolving the previous layer outputs with a small kernel followed by a sigmoidal function. In the subsampling layers, local averaging and subsampling of the outputs of the convolutional-layer neurons are performed. Every feature map in a subsampling layer is connected only to a feature map of the previous layer.

The configuration of the CNN evaluated in this paper is shown in Fig. 2 for a 16x16-pixel WVD image as input. It follows the "bipyramidal" criterion [5][1]. If the STFT is used, a 16x16-pixel image is also obtained that is split into two 16x16-pixel input maps containing the real and imaginary part of the STFT, respectively. To construct them, 30 and 31 radar echoes are required if WVD and STFT are respectively used avoiding border effects. These are a reasonable amount of pulses integrated in conventional radars [1]. The network comprises 4 hidden layers and an full-connected ouput layer, F5. The connection, in this particular CNN, among the maps of convolutional layer C3 and the maps of subsampling layer S2 is described in Table 1. The total amount of free weights is 116 and 143 for the WVD- and STFT- CNN detectors, respectively (bias coefficients are included).

---

[1] The number of the feature maps increases while the feature map size decreases.

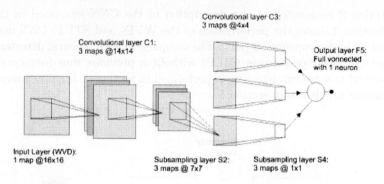

Fig. 2. Configuration of the Convolutional Neural Network.

Table 1. Connections among C3 maps and S2 maps

	S2 map ♯1	S2 map ♯2	S2 map ♯3
C3 map ♯1	X	X	
C3 map ♯2		X	X
C3 map ♯3	X	X	X

3   CNN Detector

The WVD- and STFT- CNN detectors are next evaluated using the CNN described in section 2. First of all, some aspects about the clutter and the target modelling as well as the training and testing procedure are remarked.

3.1   Clutter and Target Modelling

In this preliminary evaluation of the detector, a simple model for both clutter and target has been used. The clutter sequences have been generated as a complex white Gaussian noise and the target is a complex value with random phase, and amplitude depending on the desired signal-to-clutter ratio (SCR). For the WVD-CNN, correlated Gaussian clutter with correlation-law ρ^n has also been studied. It is important to note that WVD-CNN preprocesses the time-frequency image by removing its mean and normalizing its maximum absolute value to unity. This results in a constant false alarm rate (CFAR) regarding the clutter power. As for the STFT-CNN detector, the preprocessing cannot be used since the network is unable to learn with it. Therefore, the detector does not have CFAR performance.

3.2   Training and Testing Methods

On-line BP with momentum is used to minimize the mean square error. The learning behaviour and the performance of the neural detectors are analysed by means of cross-validation. In every epoch, the number of required examples of the estimation subset is 1000. The size of the validation subset is 200 examples. In both subsets, the probability of an example to be only clutter or target-plus-clutter is 0.5. The learning parameters that can be controlled are the training signal-to-clutter ratio (TSCR), learning rate (η) and momentum (α). For the target-plus-clutter examples in both subsets, the SCR is always the TSCR. After the training, the receiver operation characteristic (ROC) of the detector is established by the test set. It comprises 10000 examples of only clutter and 500 examples of target-plus-clutter per each analysed SCR. An example is randomly generated whenever it is needed. Therefore, in every epoch the examples of the estimation and validation subsets are different as well as the test set every time it is used.

3.3   Detector Performance

For the WVD-CNN detector, there is an optimum TSCR depending on the correlation coefficient. Fig. 3a shows the detector sensitivity regarding the TSCR for several correlation coefficients ($\eta = 0.01$ and $\alpha = 0.5$). For a given correlation, the optimum TSCR is the one that exhibits better sensitivity. Thereafter, the sensitivity is defined at 10^{-3} false alarm probability and 90% detection probability. The ROC for the uncorrelated clutter case is depicted in Fig. 3b using the optimum TSCR (0 dB). The sensitivity is SCR ≈ -0.5 dB. Also for the uncorrelated clutter case, the STFT-CNN detector's sensitivity is worse: SCR ≈ 2 dB. Its optimum TSCR is 15 dB, $\eta = 0.01$, and $\alpha = 0.5$.[2]

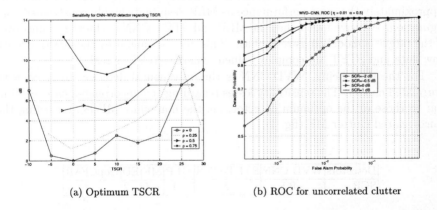

(a) Optimum TSCR   (b) ROC for uncorrelated clutter

Fig. 3. WVD-CNN detector: Sensitivity (10^{-3} false alarm probability and 90% detection probability) regarding the TSCR for several correlation coefficients and ROC for the uncorrelated case.

---

[2] STFT-CNN with correlated clutter is still under study.

4   Comparison with Other Detectors

A neural detector based on the multilayer perceptron (MLP) and other three non-neural ones have been also tested for comparison purposes. To establish a fair comparison, the MLP also requires 30 echoes, which are fed into the network without time-frequency pre-processing. Then, the input layer have 60 real-valued elements. The MLP has a hidden layer of 30 neurons and an output layer with one neuron. This configuration provides a good performance in terms of detection according to [7]. The number of weights to adapt is 1861, and 10000 examples per epoch are used by the on-line BP method. Only MLP with uncorrelated clutter is shown in this paper. Its learning parameters are TSCR = 15 dB, $\eta = 0.001$, and $\alpha = 0.25$, and its sensitivity is SCR = -1 dB.

As for the three analysed non-neural detectors, these are the matched filter (MF), the energy (ED), and the Doppler CFAR (DopCFAR) detectors. The Doppler CFAR detector [3], DopCFAR, is applied after the Discrete Fourier Transform (DFT) of the N available samples and assumes unknown clutter power. To estimate the clutter power, the energy of the DFT bins are used removing the bin of maximum value and its adjacent bins.

Table 2 summarizes the sensitivity of the evaluated detectors for the uncorrelated clutter case. The best one is the MF, which is the optimum detector for the clutter and target models. It assumes known clutter power. The DopCFAR exhibits 2 dB CFAR-losses since it estimates the clutter power. Under knowledge of the clutter power it would perform approximately as the MF. ED looses also 2 dB, but it necessarily requires the knowledge of the clutter power. Likewise, STFT-CNN and MLP detectors assume known clutter power. However, the WVD-CNN does not since it is CFAR (section 3.1). Besides, WVD-CNN presents approximately the same performance as MLP despite its highly constrained configuration. We have found that the MLP and STFT-CNN performance improves if the target is modelled as zero-phase. This may be caused by the fact that both raw data and STFT are complex-valued, whereas WVD is real.

For the time being, WVD-CNN is the only neural detector analysed with correlated clutter. The sensitivity losses regarding the uncorrelated case are similar to those of the optimum detector, i.e. the non-white MF [8], regarding the uncorrelated-clutter MF (fig. 4).

Detectors	WVD-CNN	STFT-CNN	MLP	MF	ED	DopCFAR
SCR	−0.5	2	−1	−4	−2	−2

Table 2. Sensitivity (dB) of several detectors for uncorrelated clutter. 10^{-3} false alarm and 90% detection probability.

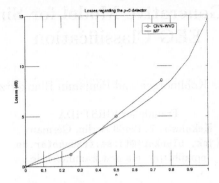

Fig. 4. Sensitivity losses for the WVD-CNN and non-white MF detectors with respect to the uncorrelated-clutter case.

5   Conclusion

A neural detector is described that is based on a time-frequency block and a CNN using less free weights than the conventional MLP scheme. Preliminary results based on simple clutter and target models show that WVD is more suitable than STFT. Besides, WVD-CNN has the same performance as MLP and has CFAR. WVD-CNN is slightly worse than traditional detectors, although further investigations suggest CNN detectors outperform traditional ones under other clutter statistics.

References

[1] J.L. Eaves and E.K. Reedy, editors. *Principles of Modern Radar*. Van Nostrand Reinhold Company, 1987.

[2] S. Haykin and J. Principe. Making sense of a complex world. *IEEE Signal Processing Magazine*, 15(3):66–81, March 1998.

[3] S. Haykin and T. K. Bhattacharya. Modular Learning Strategy for Signal Detection in a Nonstationary Environment. *IEEE Trans. on Signal Processing*, 45(6):1619–1637, June 1997.

[4] Y. Lecun, L. Bottou, Y. Bengio, and P. Haffner. Gradient-Based Learning Applied to Document Recognition. *Proceedings of the IEEE*, 86(11):2278–2324, November 1998.

[5] S. Haykin. *Neural Networks. A Comprehensive Foundation*. Prentice Hall, 2nd edition, 1999.

[6] F. Hlawatsch and G. F. Boudreaux-Bartels. Linear and Quadratic Time-Frequency Signal Representations. *IEEE Signal Processing Magazine*, pages 21–67, April 1992.

[7] D. Andina and J. L. Sanz-González. Comparison of a Neural Network Detector vs Neyman-Pearson Optimal Detector. In *IEEE International Conference on Acoustic, Speech and Signal Processing*, pages 3573–3576, 1996.

[8] M. I. Skolnik. *Introduction to Radar Systems*. McGraw-Hill, 2nd edition, 1980.

A Simple Generative Model for Single-Trial EEG Classification

Jens Kohlmorgen and Benjamin Blankertz

Fraunhofer FIRST.IDA
Kekuléstr. 7, 12489 Berlin, Germany
{jek, blanker}@first.fraunhofer.de
émailhttp://ida.first.fraunhofer.de

Abstract. In this paper we present a simple and straightforward approach to the problem of single-trial classification of event-related potentials (ERP) in EEG. We exploit the well-known fact that event-related drifts in EEG potentials can well be observed if averaged over a sufficiently large number of trials. We propose to use the average signal and its variance as a generative model for each event class and use Bayes decision rule for the classification of new, unlabeled data. The method is successfully applied to a data set from the NIPS*2001 Brain-Computer Interface post-workshop competition.

1 Introduction

Automating the analysis of EEG (electro-encephalogram) is one of the most challenging problems in signal processing and machine learning research. A particularly difficult task is the analysis of event-related potentials (ERP) from individual events ('single-trial'), which recently gained increasing attention for building brain-computer interfaces. The problem is in the high inter-trial variability of the EEG signal, where the interesting quantity, e.g. a slow shift of the cortical potential, is largely hidden in the 'background' activity and only becomes evident by averaging over a large number of trials.

To approach the problem we use an EEG data set from the NIPS*2001 Brain-Computer Interface (BCI) post-workshop competition.[1] The data set consists of 516 single trials of pressing a key on a computer keyboard with fingers of either the left or right hand in a self-chosen order and timing ('self-paced key typing'). A detailed description of the experiment can be found in [1]. For each trial, the measurements from 27 Ag/AgCl electrodes are given in the interval from 1620 ms to 120 ms *before* the actual key press. The sampling rate of the chosen data set is 100 Hz, so each trial consists of a sequence of $N = 151$ data points. The task is to predict if the upcoming key press is from the left or right hand, given only the respective EEG sequence. A total of 416 trials are labeled (219 'left' events, 194 'right' events, and 3 rejected trials due to artifacts) and can be

---

[1] publicly available at http://newton.bme.columbia.edu/competition.htm

J.R. Dorronsoro (Ed.): ICANN 2002, LNCS 2415, pp. 1156–1161, 2002.

used for building a binary classifier. One hundred trials are unlabeled and make up the evaluation test set for the competition. It should be noted here, that we construct our classifier under the conditions of the competition, i.e. without using the test set, but since we actually have access to the true test set labels, we do not participate in the competition. In this way, however, we are able to report the test set error of our classifier in this contribution.

2   Classifier Design

As outlined in [1], the experimental set-up aims at detecting lateralized slow negative shifts of cortical potential, known as 'Bereitschaftspotential' (BP), which have been found to precede the initiation of the movement [2,3]. These shifts are typically most prominent at the lateral scalp positions C3 and C4 of the international 10-20 system, which are located over the left and right hemispherical primary motor cortex.

Fig. 1 illustrates this for the given training data set. The left panel in Fig. 1 shows the measurements from each of the two channels, C3 and C4, *averaged* over all trials for *left* finger movements, and the right panel depicts the respective averages for *right* finger movements. Respective plots are also shown for channel C2, which is located next to C4. It can be seen that, on the average, a right finger movement clearly corresponds to a preceding negative shift of the potential over the left motor cortex (C3), and a left finger movement corresponds to a negative shift of the potential over the right motor cortex (C2, C4), which in this case is even more prominent in C2 than in C4 (left panel). The crux is that this effect is largely obscured in the individual trials due to the high variance of the signal, which makes the classification of individual trials so difficult. Therefore, instead of training a classifier on the individual trials [1], we here propose to exploit the above (prior) knowledge straight away and use the averages directly as the underlying model for left and right movements.

It can be seen from Fig. 1 that the difference between the average signals C4 and C3, and likewise between C2 and C3, is decreasing for left events, but is increasing for right events. We can therefore merge the relevant information from both hemispheres into only one scalar signal by using the difference of either C4 and C3 or C2 and C3. In fact, it turned out that the best (leave-one-out) performance can be achieved when subtracting C3 from the mean of C4 and C2. That is, as a first step of pre-processing/variable selection, we just use the scalar EEG signal, $y = (C2 + C4)/2 - C3$, for our further analysis.

The respective averages, $y_L(t)$ and $y_R(t)$, of the signal $y(t)$ for all left and right events in the training set, together with the standard deviations at each time step, $\sigma_L(t)$ and $\sigma_R(t)$, are shown in Fig. 2. A scatter plot of all the training data points underlies the graphs to illustrate the high variance of the data in comparison to the feature of interest: the drift of the mean.

We now use the left and right averages and the corresponding standard deviations directly as generative models for the left or right trials. Under a Gaussian assumption, the probability of observing y at time t given the left model, $M_L = (y_L, \sigma_L)$, can be expressed as

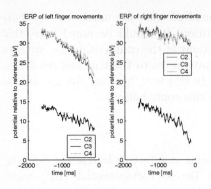
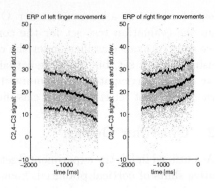

Fig. 1. Averaged EEG recordings at positions C2, C3, and C4, separately for left and right finger movements. The averaging was done over all training set trials of the BCI competition data set.

Fig. 2. Mean and standard deviation of the difference signal, $y = (C2 + C4)/2 - C3$, over a scatter plot of all data points. Clearly, there is a high variance in comparison to the drift of the mean.

$$p(y(t) \mid M_L) = \frac{1}{\sqrt{2\pi}\,\sigma_L(t)} \exp\left(-\frac{(y(t) - y_L(t))^2}{2\sigma_L(t)^2}\right). \tag{1}$$

The probability $p(y(t) \mid M_R)$ for the right model can be expressed accordingly. Assuming a Gaussian distribution is indeed justified for this data set: we estimated the distribution of the data at each time step with a kernel density estimator and consistently found a distribution very close to a Gaussian. To keep the approach tractable, we further assume that the observations $y(t)$ only depend on the mean and variance of the respective model at time t, but not on the other observations before or after time t.[2] Then, the probability of observing a complete data sequence, $\mathbf{y} = (y(1), \ldots, y(N))$, by one of the models, is given by

$$p(\mathbf{y}|M) = \prod_{t=1}^{N} p(y(t) \mid M). \tag{2}$$

Finally, the posterior probability of the model given a data sequence can be expressed by Bayes' rule,

$$p(M|\mathbf{y}) = \frac{p(\mathbf{y}|M)\,p(M)}{p(\mathbf{y})}. \tag{3}$$

We then use Bayes' decision rule, $p(M_L|\mathbf{y}) > p(M_R|\mathbf{y})$, to decide which model to choose. According to eq. (3), this can be written as

$$p(\mathbf{y}|M_L)\,p(M_L) > p(\mathbf{y}|M_R)\,p(M_R). \tag{4}$$

The evidence $p(\mathbf{y})$ vanishes in eq. (4) and we are left with the determination of the prior probabilities of the models, $p(M_L)$ and $p(M_R)$. Since there is no a

---

[2] In fact, almost all off-diagonal elements of the covariance matrix are close to zero, except for the direct neighbors, i.e. $y(t)$ and $y(t+1)$.

priori preference for left or right finger movements in the key typing task, we can set $p(M_L) = p(M_R)$ and the decision rule simplifies to a comparison of the likelihoods $p(\mathbf{y}|M)$. If we furthermore perform the comparison in terms of the negative log-likelihood, $-\log(p)$, and neglect the leading normalization factor in eq. (1) – which turns out to not diminish the classification performance, also because the left and right standard deviations, $\sigma_L(t)$ and $\sigma_R(t)$, are very similar for this data set (cf. Fig. 2) – then the decision rule can be rewritten as

$$\sum_{t=1}^{N} \frac{(y(t) - y_L(t))^2}{\sigma_L(t)^2} \quad < \quad \sum_{t=1}^{N} \frac{(y(t) - y_R(t))^2}{\sigma_R(t)^2}. \tag{5}$$

The terms on both sides are now simply the squared distances of the respective left or right mean sequence to a given input sequence, normalized by the estimated variance of each component.

3   Results and Refinements

The above approach can readily be applied to our selected quantity y from the competition data set. The result without any further pre-processing of the signal is 20.10% misclassifications (errors) on the training set, 21.07% leave-one-out (LOO) cross-validation error (on the training set), and 16% error on the test set. Next, as a first step of pre-processing, we normalized the data of each trial to zero-mean, which significantly improved the results, yielding 14.04%/15.98%/7% training/LOO/test set error.[3] A further normalization of the data to unit-variance did not enhance the result (13.56%/15.98%/7% training/LOO/test set error).

The next improvement can easily be understood from Fig. 2. Clearly, the data points at the end of the sequence have more discriminatory power than the points at the beginning. Moreover, we presume that the points at the beginning mainly introduce undesirable noise into the decision rule. We therefore successively reduced the length of the sequence that enters into the decision rule via a new parameter D,

$$\sum_{t=D}^{N} \frac{(y(t) - y_L(t))^2}{\sigma_L(t)^2} \quad < \quad \sum_{t=D}^{N} \frac{(y(t) - y_R(t))^2}{\sigma_R(t)^2}. \tag{6}$$

Fig. 3 shows the classification result for $D = 1, \ldots, N$, ($N = 151$), on the zero-mean data. Surprisingly, using only the last 11 or 12 data points of the EEG sequence yields the best LOO performance: the LOO error minimum (9.69%) is at $D = 140$ and 141, with a corresponding test set error of 6% in both cases.

A further, somehow related improvement can be achieved by excluding a number of data points from the end of the sequence when computing the mean

---

[3] The unusual result that the test set error is just half as large as the training set error was consistently found throughout our experiments and is apparently due to a larger fraction of easy trials in the test set.

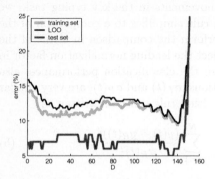

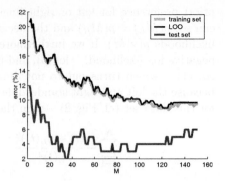

Fig. 3. Training, leave-one-out (LOO), and test set error in dependence of the starting point D of the observation window.

Fig. 4. Training, leave-one-out (LOO), and test set error in dependence of the size M of the zero-mean window (results for $D = 141$).

for the zero-mean normalization. Fig. 4 depicts the classification results for using only the first $M = 2, \ldots, N$ data points of each sequence for computing the mean for the normalization. The normalization then results in sequences that have a zero mean only for the first M data points. The LOO minimum when using $D = 141$ is at $M = 121, \ldots, 125$ (Fig. 4). The respective LOO and training set error is 8.96%. We found that this is indeed the optimal LOO error for all possible combinations of M and D. At this optimum we get a test set error of 4% for $M = 121, 122, 123$, and of 5% for $M = 124, 125$. Fig. 5 shows the respective distances (eq. (6)) of all trials to the left and right model (for $M = 121$). In Fig. 4, the test set error even reaches a minimum of 2% at $M = 34, 35, 36$, however, this solution can not be found given only the training set.

We considered other types of pre-processing or feature selection, like normalization to unit-variance with respect to a certain window, other choices of windows for zero-mean normalization, or using the bivariate C3/C4 signal instead of the difference signal. However, these variants did not result in better classification performance. Also a smoothing of the models, i.e. a smoothing of the mean and standard deviation sequences, did not yield a further improvement.

4   Summary and Discussion

We presented a simple generative model approach to the problem of single-trial classification of event-related potentials in EEG. The method requires only 2 or 3 EEG channels and the classification process is easily interpretable as a comparison with the average signal of each class. The application to a data set from the NIPS*2001 BCI competition led to further improvements of the algorithm, which finally resulted in 95–96% correct classifications on the test set (without using the test set for improving the model). We demonstrated how problem-specific prior knowledge can be incorporated into the classifier design.

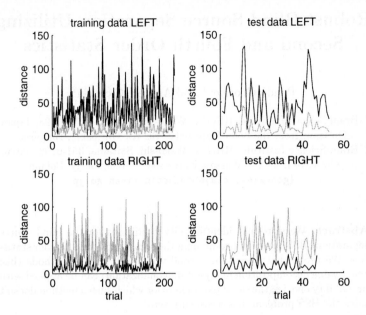

Fig. 5. Distances (cf. eq.(6)) from all trials to the finally chosen models for left and right event-related potentials ($D = 141$, $M = 121$). In almost all cases of misclassifications both models exhibit a small distance to the input. (left model: grey, right model: black)

As a result, we obtained a relatively simple classification scheme that can be used, for example, as a reference for evaluating the performance of more sophisticated, future approaches to the problem of EEG classification, in particular those from the BCI competition.

Acknowledgements. We would like to thank S. Lemm, P. Laskov, and K.-R. Müller for helpful discussions. This work was supported by grant 01IBB02A from the BMBF.

References

1. Blankertz, B., Curio, G., Müller, K.R.: Classifying Single trial EEG: Towards brain Computer interfacing. In Dietterich, T.G., Becker, S., Ghahramani, Z., eds.: Advances in Neural Information Processing Systems 14 (NIPS*01), Cambridge, MA, MIT Press (2002) to appear.
2. Lang, W., Zilch, O., Koska, C., Lindinger, G., Deecke, L.: Negative cortical DC shifts preceding and accompanying simple and complex sequential movements. Exp. Brain Res. **74** (1989) 99–104
3. Cui, R.Q., Huter, D., Lang, W., Deecke, L.: Neuroimage of voluntary movement: topography of the Bereitschaftspotential, a 64-channel DC current source density study. Neuroimage **9** (1999) 124–134

Robust Blind Source Separation Utilizing Second and Fourth Order Statistics

Pando Georgiev[1] and Andrzej Cichocki[2]

[1] Brain Science Institute, RIKEN, Wako-shi, Saitama, 351-0198, Japan,
On leave from the Sofia University "St. Kl. Ohridski", Bulgaria
[2] Brain Science Institute, RIKEN, Wako-shi, Saitama, 351-0198, Japan,
On leave from the Warsaw University of Technology, Poland
{georgiev, cia}@bsp.brain.riken.go.jp

Abstract. We introduce identifiability conditions for the blind source separation (BSS) problem, combining the second and fourth order statistics. We prove that under these conditions, well known methods (like eigen-value decomposition and joint diagonalization) can be applied with probability one, i.e. the set of parameters for which such a method doesn't solve the BSS problem, has a measure zero.

1 Introduction

The interest of blind signal processing, especially, independent component analysis (ICA) has been increased recently, due to its potential applications in many areas, including brain signal processing and other biomedical signal processing, speech enhancement, wireless communication, geophysical data processing, data mining, etc. (see e.g. the recent books [6], [11] and references therein).

The problem of blind source separation (BSS) is formulated as follows: we can observe sensor signals $\mathbf{x}(k) = [x_1(k), \ldots, x_n(k)]^T$ which are modelled as

$$\mathbf{x}(k) = \mathbf{H}\mathbf{s}(k) + \mathbf{n}(k), \tag{1}$$

where $\mathbf{H}$ is $n \times n$ non-singular unknown mixing matrix, $\mathbf{s}(k) = [s_1(k), \ldots, s_n(k)]^T$ is a vector of unknown zero mean source signals and $\mathbf{n}(k)$ is a vector of additive noise. Assume that $\mathbf{n}$ has independent components (with zero means), which are independent also with $s_i, i = 1, ..., n$.

Our objective is to estimate the mixing matrix $\mathbf{H}$ and/or a separating matrix $\mathbf{W} = \mathbf{H}^{-1}$ and source signals simultaneously.

In this paper we consider an unified model of source signals and additive noise, which is white of order 2 and 4. We assume that all source signals are uncorrelated of order 2 and 4, as some of the source signals (we don't know which) are white of order 4 but colored of order 2 (for instance colored Gaussian signals) and the rest are white of order 2 and colored of order 4. We introduce a new sufficient conditions for separation (see condition $(\mathbf{DCF}(P))$ below) stating that the sources have different autocorrelation functions or different cumulant functions of fourth order (depending on time delay). These conditions can be

J.R. Dorronsoro (Ed.): ICANN 2002, LNCS 2415, pp. 1162–1167, 2002.

considered as a generalization of those ones described in [5] and [13] (for second order statistics) and used in [4].

The second and fourth order statistics is used in [8] by a joint diagonalization procedure of covariance and cumulant matrices.

Define a covariance matrix of the sensor (resp. source) signals by

$$\mathbf{R_x}(p) = E\{\mathbf{xx}_p^T\}, \text{ (resp. } \mathbf{R_s}(p) = E\{\mathbf{ss}_p^T\}), \tag{2}$$

where E is the mathematical expectation, $\mathbf{x}_p = \mathbf{x}(k - p), \mathbf{x} = \mathbf{x}(k), \mathbf{s}_p = \mathbf{s}(k - p), \mathbf{s} = \mathbf{s}(k)$ and define the symmetric matrix $\tilde{\mathbf{R}}_\mathbf{x}(p) = \frac{1}{2}(\mathbf{R_x}(p) + \mathbf{R_x}(p)^T)$.

Define a fourth order cumulant matrix $\mathbf{C}_{\mathbf{x},\mathbf{x}_p}^{2,2}$ of the sensor signals as follows:

$$\mathbf{C}_{\mathbf{x},\mathbf{x}_p}^{2,2} = E\{\mathbf{xx}^T(\mathbf{x}_p^T\mathbf{x}_p)\} - E\{\mathbf{xx}^T\}\text{tr}E\{\mathbf{x}_p\mathbf{x}_p^T\} - 2E\{\mathbf{xx}_p^T\}E\{\mathbf{x}_p\mathbf{x}^T\}$$

and define the symmetric matrix $\tilde{\mathbf{C}}_{\mathbf{x},\mathbf{x}_p}^{2,2} = \frac{1}{2}(\mathbf{C}_{\mathbf{x},\mathbf{x}_p}^{2,2} + (\mathbf{C}_{\mathbf{x},\mathbf{x}_p}^{2,2})^T)$[1].

It is easy to see that the (i, j)-th element of $\mathbf{C}_{\mathbf{x},\mathbf{x}_p}^{2,2}$ is

$$C_{\mathbf{x},\mathbf{x}_p}^{2,2}(i, j) = \sum_{l=1}^{n} \mathbf{cum}\{x_i(k), x_j(k), x_l(k - p), x_l(k - p)\}$$

(see [3] for more general cumulant matrices).

Similarly we define analogous matrices $\mathbf{C}_{\mathbf{s},\mathbf{s}_p}^{2,2}$ and $\tilde{\mathbf{C}}_{\mathbf{s},\mathbf{s}_p}^{2,2}$ for the source signals $\mathbf{s}(k)$. Recall that a signal s is white of order 2 (resp. white of order 4) if

$$E\{s(k)s(k-p)\}=0, \forall p \geq 1 \text{ (resp. } \mathbf{cum}\{s(k-p_1), s(k-p_2), s(k-p_3), s(k-p_4)\}=0$$

for every $p_i \geq 1, i = 1, ..., 4)$ (see [12]).

In a linear data model (1), if the noise $\mathbf{n}$ is white of order 4, and the mixing matrix H is orthogonal, then the time-delayed cumulant matrices of the observation vector $\mathbf{x}(k)$ for any $p \neq 0$ satisfy $\tilde{\mathbf{C}}_{\mathbf{x},\mathbf{x}_p}^{2,2} = H\tilde{\mathbf{C}}_{\mathbf{s},\mathbf{s}_p}^{2,2}H^T$. If $\mathbf{n}$ is white of order 2, then $\tilde{\mathbf{R}}_\mathbf{x}(p) = H\tilde{\mathbf{R}}_\mathbf{s}(p)H^T$.

2 Robust Orthogonalization

In our method below we need the global mixing matrix to be orthogonal. The standard whitening procedure is not acceptable, since it enhances the noise, especially when the number of sensors is equal to the number of sources and the problem is ill conditioned. We use a preprocessing procedure, which is not sensitive to additive noise. This orthogonalization procedure allows us to define a new orthogonal mixing matrix for the preprocesed data. The idea is to use time-delayed cumulant (resp. covariance) matrices which are not sensitive to additive white noise of order 4 (resp. of order 2) and construct a positive definite

---

[1] Note that $\mathbf{C}_{\mathbf{x},\mathbf{x}_p}^{2,2}$ (resp. $\mathbf{R_x}(p)$) is symmetric, if $\mathbf{C}_{\mathbf{s},\mathbf{s}_p}^{2,2}$ (resp. $\mathbf{R_s}(p)$) is a diagonal matrix, but in order to avoid the effect of computational errors (which could destroy the symmetricity), we use $\tilde{\mathbf{C}}_{\mathbf{x},\mathbf{x}_p}^{2,2}$ (resp. $\tilde{\mathbf{R}}_\mathbf{x}(p)$).

matrix from their linear combination (for sufficiently large number of samples). Such a problem for white noise of order 2 is solved in [1] by a finite-step global convergence algorithm [14].

Let $P = \{p_1, ..., p_L\}$ be a set of positive integers with L elements. Denote

$$\mathbf{cum}_{s_i}(p) = \mathbf{cum}\{s_i(k), s_i(k), s_i(k-p), s_i(k-p)\}$$

and assume that the vectors $\left\{\left(\mathbf{cum}_{s_i}(p_1), ..., \mathbf{cum}_{s_i}(p_L)\right)\right\}_{i=1}^n$ (resp. vectors $\left(E\{ss_{p_1}\}, ..., E\{ss_{p_L}\}\right)_{i=1}^n$) are linearly independent. These conditions are necessary in order to be realized the finite-step global convergence algorithm [14] in Step 1 of the algorithm below.

The robust orthogonalization algorithm can be summarized as follows.

Algorithm Outline: Robust Orthogonalization

1. Find by the finite-step global convergence algorithm [14] a set of parameters $\{\alpha_i\}_{i=1}^L$ such that the matrix $\mathbf{C_x}(\boldsymbol{\alpha}) = \sum_{i=1}^L \alpha_i \tilde{\mathbf{C}}_{\mathbf{x},\mathbf{x}_{p_i}}^{2,2}$ (resp. $\mathbf{C_x}(\boldsymbol{\alpha}) = \sum_{i=1}^L \alpha_i \tilde{\mathbf{R}}_{\mathbf{x}}(p_i)$) is positive definite.

2. Perform an eigenvalue-decomposition of $\mathbf{C_x}(\boldsymbol{\alpha})$, $\mathbf{C_x}(\boldsymbol{\alpha}) = \mathbf{U_x}\boldsymbol{\Lambda_x}\mathbf{U_x}^T$, where the entries of diagonal matrix $\boldsymbol{\Lambda_x}$ are the positive eigenvalues of $\mathbf{C_x}(\boldsymbol{\alpha})$ and compute the preprocessing matrix $\mathbf{Q} = \boldsymbol{\Lambda_x}^{-\frac{1}{2}}\mathbf{U_x}^T$.

3. Compute the preprocesed data $\mathbf{z}(k) = \mathbf{Q}\mathbf{x}(k) = \mathbf{Q}\mathbf{H}\mathbf{s}(k)$.

Remark 1. By defining a new mixing matrix as $\mathbf{A} = \mathbf{QHD}^{\frac{1}{2}}$, where $\mathbf{D} = \sum_{i=1}^K \alpha_i \tilde{\mathbf{C}}_{\mathbf{s},\mathbf{s}_{p_i}}^{2,2}$ (resp. $\mathbf{D} = \sum_{i=1}^K \alpha_i \tilde{\mathbf{R}}_{\mathbf{s}}(p_i)$) is a diagonal (scaling) matrix with positive entries, we see that $\mathbf{C_z}(\boldsymbol{\alpha}) = \mathbf{A}\mathbf{A}^T = \mathbf{I}_n((n \times n)$ unit matrix), so, the matrix $\mathbf{A}$ is orthogonal. This orthogonality condition is necessary for performing separation of signals using either symmetric EVD, or joint diagonalization. It should be noted that in contrast to the standard prewhithening procedure for our robust orthogonalization generally $E\{\mathbf{z}\mathbf{z}^T\} \neq \mathbf{I}_m$, but we have $\sum_{i=1}^K \alpha_i \tilde{\mathbf{C}}_{\tilde{\mathbf{s}},\tilde{\mathbf{s}}_{p_i}}^{2,2} = \mathbf{I}_n$ and $\sum_{i=1}^K \alpha_i \tilde{\mathbf{C}}_{\mathbf{z},\mathbf{z}_{p_i}}^{2,2} = \mathbf{I}_n$. So, our model is $\mathbf{z} = \mathbf{A}\tilde{\mathbf{s}} + \mathbf{Q}\mathbf{n}$, where $\tilde{\mathbf{s}} = \mathbf{D}^{-\frac{1}{2}}\mathbf{s}$.

3   Sufficient Conditions for Simultaneous Blind Source Separation

We introduce the following conditions, called (**DCF**(P)) (different cumulant functions):

$$\forall i, j \neq i \; \exists l_{i,j} \in \{1, ..., L\} : \quad \text{either } E\{s_i(t)s_i(t - p_{l_{i,j}})\} \neq E\{s_j(t)s_j(t - p_{l_{i,j}})\}$$
$$\text{or } \mathbf{cum}_{s_i}(p_{l_{i,j}}) \neq \mathbf{cum}_{s_j}(p_{l_{i,j}}),$$

i.e. the sources have different autocorrelation or cumulant functions of fourth order on the set P.

Define the following matrices for the chosen set P of time delays for $\mathbf{z}$ and $\mathbf{s}$ respectively, where $\mathbf{b}, \mathbf{c} \in \mathbb{R}^L$:

$$\mathbf{Z}(\mathbf{b}, \mathbf{c}) = \sum_{i=1}^{L} \left(b_i \tilde{\mathbf{R}}_{\mathbf{z}}(p_i) + c_i \tilde{\mathbf{C}}_{\mathbf{z},\mathbf{z}_{p_i}}^{2,2} \right), \quad \mathbf{S}(\mathbf{b}, \mathbf{c}) = \sum_{i=1}^{L} \left(b_i \tilde{\mathbf{R}}_{\mathbf{s}}(p_i) + c_i \tilde{\mathbf{C}}_{\mathbf{s},\mathbf{s}_{p_i}}^{2,2} \right). (3)$$

We recall that the source signals are *uncorrelated*, if $\mathbf{R_s}(p)$ are diagonal matrices for every $p \geq 1$. If the source signals are statistically independent, then this condition is satisfied, but the converse assertion is not always true. Note that the diagonal elements of $\mathbf{R_s}(p)$ are $E\{s_i(k)s_i(k - p)\}$. We say that the source signals are *colored*, if for some $p_0 \geq 1$ the matrix $\mathbf{R_s}(p_0)$ has a nonzero diagonal element. We shall say that the source signals are *uncorrelated of order* 4, if $\mathbf{C}_{\mathbf{s},\mathbf{s}_p}^{2,2}$ are diagonal matrices for every $p \geq 1$ with diagonal elements $\mathbf{cum}_{s_i}(p)$. If the source signals are statistically independent, then this condition is satisfied, but the converse assertion is not always true. We shall say that the sources are *colored of order* 4, if for some $p_0 \geq 1$, $\mathbf{cum}_{s_i}(p_0)$ is nonzero. So, if $s_i, i = 1, ..., n$ are uncorrelated of order 4 and colored of order 4, then for some $p_0 \geq 1$, the matrix $\mathbf{C}_{\mathbf{s},\mathbf{s}_{p_0}}^{2,2}$ is a nonzero diagonal matrix.

Theorem 1. *Assume that the mixing matrix $\mathbf{A}$ is orthogonal, the source signals are uncorrelated of order 2 and 4, condition $(\mathbf{DCF}(P))$ is satisfied and the additive noise $\mathbf{n}$ is white of order 2 and 4. Then:*

(a) the matrix $\mathbf{Z}(\mathbf{b}, \mathbf{c})$ is symmetrical and can be decomposed as $\mathbf{Z}(\mathbf{b}, \mathbf{c}) = \mathbf{A}\mathbf{S}(\mathbf{b}, \mathbf{c})\mathbf{A}^T$;

(b) there exist vectors $\mathbf{b}, \mathbf{c} \in \mathbb{R}^L$ such that the matrix $\mathbf{Z}(\mathbf{b}, \mathbf{c})$ has distinct eigenvalues. Furthermore, the set $B(L)$ of all vectors $(\mathbf{b}, \mathbf{c}) \in \mathbb{R}^{2L}$ with this property form an open subset of $\mathbb{R}^{2L}$, whose complement has a measure zero;

(b) if $\mathbf{U}$ is given from an EVD of the matrix $\mathbf{Z}(\mathbf{b}, \mathbf{c})$ for some $\mathbf{b}, \mathbf{c} \in B(L)$, i.e. $\mathbf{Z}(\mathbf{b}, \mathbf{c}) = \mathbf{U}\mathbf{\Lambda}\mathbf{U}^T$, then the estimating mixing matrix is $\hat{\mathbf{A}} = \mathbf{U}$ and the separating matrix is $\mathbf{W} = \hat{\mathbf{A}}^T = \mathbf{U}^T$ (up to multiplication with arbitrary permutation and diagonal nonsingular scaling matrices).

Proof. (a) The assertion follows from the properties of the cumulants.

(b) Since $s_i, i = 1, ..., n$ are uncorrelated, $\mathbf{S}(\mathbf{b}, \mathbf{c})$ is a diagonal matrix and by (a), $\mathbf{Z}(\mathbf{b}, \mathbf{c}) = \mathbf{A}\mathbf{S}(\mathbf{b}, \mathbf{c})\mathbf{A}^T$. Observe that the matrices $\mathbf{Z}(\mathbf{b}, \mathbf{c})$ and $\mathbf{S}(\mathbf{b}, \mathbf{c})$ have the same eigenvalues. It is easy to see that the complement of $B(L)$ is a finite union of subspaces of $\mathbb{R}^{2L}$. If we prove that $B(L)$ is nonempty, then every of these subspaces must be proper (i.e. different from $\mathbb{R}^{2L}$), consequently, with a measure zero (with respect to $\mathbb{R}^{2L}$), therefore the complement of $B(L)$ must have a measure zero too.

Choose $\mathbf{b}, \mathbf{c} \in \mathbb{R}^{2L}$ arbitrary. Let $\{\sigma_i(\mathbf{b}, \mathbf{c})\}_{i=1}^n$ be the diagonal elements of the matrix $\mathbf{S}(\mathbf{b}, \mathbf{c})$. Assume that two diagonal elements of the matrix $\mathbf{S}(\mathbf{b}, \mathbf{c})$ are equal, for example $\sigma_1(\mathbf{b}, \mathbf{c}) = \sigma_2(\mathbf{b}, \mathbf{c})$. Let $l_{1,2}$ be the index defined by condition $(\mathbf{DCF}(P))$. If $E\{s_1(t)s_1(t - p_{l_{1,2}})\} \neq E\{s_2(t)s_2(t - p_{l_{1,2}})\}$ then choose a vector $\mathbf{b}(1, 2)$, which is different from $\mathbf{b}$ only in the component $b_{l_{1,2}}$. Otherwise, by condition $\mathbf{DCF}(P)$, we must have $\mathbf{cum}_{s_i}(p_{l_{i,j}}) \neq \mathbf{cum}_{s_j}(p_{l_{i,j}})$ and then we choose a vector $\mathbf{c}(1,2)$, which is different from $\mathbf{c}$ only in the component $c_{l_{1,2}}$, and put $\mathbf{b}(1,2) = \mathbf{b}$. Then $\sigma_1(\mathbf{b}(1,2), \mathbf{c}(1,2)) \neq \sigma_2(\mathbf{b}(1,2), \mathbf{c}(1,2))$, because of the condition $(\mathbf{DCF}(\mathrm{P}))$. If all diagonal elements of $\mathbf{S}(\mathbf{b}(1,2), \mathbf{c}(1,2))$ are different, we finish the proof. If not, suppose that $\sigma_i(\mathbf{b}(1,2), \mathbf{c}(1,2)) = \sigma_j(\mathbf{b}(1,2), \mathbf{c}(1,2))$ for some indexes i and j. We can change a little either the component $b_{l_{i,j}}$ of the vector $\mathbf{b}(1,2)$ or $c_{l_{i,j}}$ of the vector $\mathbf{c}(1,2)$ (keeping the other components the same) such that for the new vector $(\mathbf{b}(i,j), \mathbf{c}(i,j))$ the condition $\sigma_i(\mathbf{b}(i,j), \mathbf{c}(i,j)) \neq \sigma_j(\mathbf{b}(i,j), \mathbf{c}(i,j))$ to be satisfied (because of condition $(\mathbf{DCF}(\mathrm{P}))$ and since $\sigma_1(\mathbf{b}(i,j), \mathbf{c}(i,j)) \neq \sigma_2(\mathbf{b}(i,j), \mathbf{c}(i,j))$. Continuing in such a way, for any couple $(k,r), k \neq r$ for which $\sigma_k(\mathbf{b}(k',r'), \mathbf{c}(k',r')) = \sigma_r(\mathbf{b}(k',r'), \mathbf{c}(k',r'))$ (where $(\mathbf{b}(k',r'), \mathbf{c}(k',r'))$ is the vector considered in the previous step), we make small change either of $b_{l_{k,r}}$ or $c_{l_{k,r}}$ keeping the pair-wise difference of the diagonal elements considered in the previous steps and obtain a vector $(\mathbf{b}(k,r), \mathbf{c}(k,r))$ for which $\sigma_k(\mathbf{b}(k,r), \mathbf{c}(k,r)) \neq \sigma_r(\mathbf{b}(k,r), \mathbf{c}(k,r))$. So, after finite number of steps we obtain a vector $(\mathbf{b}^*, \mathbf{c}^*)$ for which the diagonal elements of $\mathbf{S}(\mathbf{b}^*, \mathbf{c}^*)$ are distinct. This proves the non-emptiness of the set $B(L)$ and finishes the proof of (b).

(c) This follows from the well known facts of linear algebra [10]. ∎

Corollary 1. *Under assumption (ii) of Theorem 1, an estimation of the mixing matrix is possible from the EVD of the cumulant matrix* $\mathbf{C}_{\mathbf{z},\mathbf{z}_p}^{2,2}$*, if the source signals have different cumulants of fourth order for a fixed time delay* p*, i.e., if*

$$\mathbf{cum}(s_i(k), s_i(k), s_i(k-p), s_i(k-p)) \neq \mathbf{cum}(s_j(k), s_j(k), s_j(k-p), s_j(k-p))$$

for every $i \neq j$*. When* $p = 0$*, the above condition means that the source signals have different kurtosis; in this case the conclusion is also true, if in addition, the noise is Gaussian.*

Remark 2. When the mixing matrix is not orthogonal, the condition for separation can be obtained having in mind the orthogonalization procedure: so, condition $\mathbf{DCF}(\mathrm{P})$ should be satisfied for the signals $\tilde{\mathbf{s}} = \mathbf{D}^{-\frac{1}{2}}\mathbf{s}$ (see Remark 1). In case of standard pre-whitening, the matrix $\mathbf{D}$ is equal to $E\{\mathbf{ss}^T\}$ and, when all source signals are colored of order 2, we recover identifiability conditions for second order statistics presented in [13], Theorem 2, stating that the source signals have distinct normalized autocorrelation functions. When all the source signals are colored of order four, our new identifiability conditions for four order statistics states that their normalized cumulant functions of order four should be distinct.

Remark 3. Theorem 1 gives, in particular, mathematical justification of the method in [7] and another proof of the mathematical foundation of the SOBI algorithm [2].

4   Conclusion

We develop an unified approach by second and high order statistics to BSS problem and presented new identifiability conditions. This approach gives justification of EVD and joint diagonalization procedures for solving BSS problem. It has some advantages, among which robustness to additive noise and possibility to implement efficient algorithms with orthogonality constraints [9] for symmetric eigenvalue problems and joint diagonalization of covariance and cumulant matrices.

References

1. A. Belouchrani and A. Cichocki A. "Robust whitening procedure in blind separation context". *Electronics Letters*, Vol. 36 No.24, pp. 2050–2051, 2000.
2. A. Belouchrani, K. A. Meraim, J.-P. Cardoso and E. Moulines "A blind source separation technique using second order statistics." *IEEE Trans. on Signal Processing*, Vol. 45, no. 2, pp. 434–44, Feb. 1997
3. J.-F. Cardoso, "High-order contrasts for independent component analysis" Neural Computation, vol. 11, no 1, pp. 157–192, Jan. 1999.
4. C. Chang, Z. Ding, S. F. Yau and F. H. Y. Chan "A matrix-pencil approach to blind separation of colored nonstationary signals". *IEEE Trans. Signal Proc.*, Vol. 48, No. 3, pp. 900–907, Mar. 2000.
5. C. Chang, S.F. Yau, P. Kwok, F.H.Y. Chan, F.K. Lam, *Uncorrelated and component analysis for blind source separation*, Circuits Systems and Signal Processsing, Vol. 18, No.3, pp. 225–239, 1999.
6. A. Cichocki and S. Amari. *Adaptive Blind Signal and Image Processing.* John Wiley, Chichester, 2002.
7. A. Cichocki, and R. Thawonmas " On-line algorithm for blind signal extraction of arbitrarily distributed, but temporally correlated sources using second order statistics". *Neural Processing Letters* Vol. 12, pp. 91–98, Aug. 2000.
8. I. Gorodnitsky and A. Belouchrani, "Joint cumulant and correlation based signal separation with application of EEG data analysis", Proc. 3-rd Int. Conf. on Independent Component Analysis and Signal Separation, San Diego, California, Dec. 9-13, 2001, pp.475–480.
9. A. Edelman, T. A. Arias and A. T. Smith "The geometry of algorithms with orthogonality constraints". *SIAM J. Matrix Anal. Appl.*, Vo. 20, No.2, pp. 303–353, 1998.
10. G. H. Golub and C. F. Van Loan *Matrix Computation.* J. Hopkins Univ. Press, 1989.
11. A. Hyvarinen, J. Karhunen and E. Oja, "Independent Component Analysis", John Wiley & Sons, 2001.
12. C. L. Nikias and A. Petropulu "Higher-order spectra analysis. A nonlinear signal processing framework", Prentice Hall Signal Processing Series, 1993.
13. L. Tong, R. Liu, V.C. Soon and Y.F. Huang, *Indeterminacy and Identifiability of Blind Identification*, IEEE Transactions on Circuits and Systems, Vol. 38, No.5, pp. 499–509, May 1991.
14. L. Tong, Y. Inouye and R. Liu " A finite-step global algorithm for the parameter estimation of multichannel MA processes". *IEEE Trans. Signal Proc.*, 40(10): 2547–2558, 1992.

Adaptive Differential Decorrelation: A Natural Gradient Algorithm

Seungjin Choi

Department of Computer Science and Engineering
POSTECH
San 31 Hyoja-dong, Nam-gu
Pohang 790-784, Korea
seungjin@postech.ac.kr

Abstract. In this paper, I introduce a concept of *differential decorrelation* which finds a linear mapping that minimizes the concurrent change of variables. Motivated by the differential anti-Hebbian rule [1], I develop a natural gradient algorithm for differential decorrelation and present its local stability analysis. The algorithm is successfully applied to the task of nonstationary source separation

1 Introduction

The Hebbian rule has been widely used in the domain of unsupervised learning where no target value is available. It is a correlation learning that is based on the hypothesis of Hebb [2] which is simply that concurrent activation of neurons increases the strength of connection between them. The Hebbian rule was shown to be an output variance maximizer and is closely related to principal component analysis (PCA). In contrast to the Hebbian rule, the anti-Hebbian rule updates the synaptic weights in such a way that cross-correlations between associated nodes are minimized. Hence, it is an output variance minimizer and decorrelates associated output variables. The anti-Hebbian rule was used for lateral decorrelation in a PCA neural network [3].

The differential Hebbian rule was studied as an alternative to the conventional Hebbian rule for updating synaptic weights in neural networks. The motivation of the differential Hebbian rule is that concurrent change, rather than just concurrent activation, more accurately captures the *concomitant variation* that is central to inductively inferred functional relationships [4]. The differential anti-Hebbian rule was proposed to find a linear mapping that minimizes the concurrent change of neurons. The differential anti-Hebbian rule performs differential decorrelation and was applied to the problem of independent component analysis [1].

In this paper, I consider a fully connected linear feedback network and present a natural gradient algorithm for differential decorrelation. Moreover I analyze its local stability and apply it to the problem of nonstationary source separation.

J.R. Dorronsoro (Ed.): ICANN 2002, LNCS 2415, pp. 1168–1173, 2002.

2   Differential Decorrelation

Let us consider a linear feedback network (without self-feedback connections) whose ith output node $y_i(t)$ is described by

$$y_i(t) = x_i(t) + \sum_{j \neq i} w_{ij} y_j(t). \tag{1}$$

The concurrent change of two output neurons $y_i(t)$ and $y_j(t)$ is measured by the differential correlation defined by $E\{\overset{\bullet}{y}_i(t)\, \overset{\bullet}{y}_j(t)\}$ where

$$\overset{\bullet}{y}_i(t) = \frac{dy_i(t)}{dt}, \tag{2}$$

or its discrete-time counterpart is $\overset{\bullet}{y}_i(t) = y_i(t) - y_i(t-1)$ which is its first-order approximation. The task of differential decorrelation is to learn synaptic weights $\{w_{ij}\}$ in such a way that the differential correlation between $y_i(t)$ and $y_j(t)$ (for $i \neq j$) is minimized.

The differential anti-Hebbian learning proposed in [1] has the form

$$w_{ij}(t+1) = w_{ij}(t) - \eta_t \overset{\bullet}{y}_i(t)\, \overset{\bullet}{y}_j(t), \quad \text{for } i \neq j, \tag{3}$$

where $\eta_t > 0$ is the learning rate.

3   Natural Gradient Learning

The differential anti-Hebbian rule (3) is a straightforward extension of the anti-Hebbian rule. It is a gradient descent learning for the minimization of differential variance.

The gradient descent learning is a popular method for the minimization of a given objective function. It finds a steepest descent direction when a parameter space is a Euclidean space with an orthogonal coordinate system. However, if a parameter space is a curved manifold, then an orthogonal linear coordinate system does not exist. Hence the conventional gradient (which is simply first order derivative) does not correspond to the steepest descent direction. The natural gradient learning method proposed by Amari [5], allows us to calculate the steepest descent direction when a parameter space belongs to the Riemannian manifold. In this section, I present a natural gradient algorithm for differential decorrelation and also analyze its local stability.

3.1   Algorithm

The fully connected linear feedback network is considered. The network output vector $\boldsymbol{y}(t) \in \mathbb{R}^n$ is described by

$$\begin{aligned} \boldsymbol{y}(t) &= \boldsymbol{x}(t) + \boldsymbol{W}\boldsymbol{y}(t) \\ &= (\boldsymbol{I} - \boldsymbol{W})^{-1}\boldsymbol{x}(t), \end{aligned} \tag{4}$$

where $x(t) \in \mathbb{R}^n$ is the input vector to the network and $W \in \mathbb{R}^{n \times n}$ is the synaptic weight matrix.

The objective function that is considered here is give by

$$J(W) = \tfrac{1}{2} \left\{ \sum_{i=1}^{n} \log E\{\overset{\bullet}{y}_i^2(t)\} - \log \det \left(E\left\{ \overset{\bullet}{y}(t)\overset{\bullet}{y}^T(t) \right\} \right) \right\}. \tag{5}$$

The objective function (5) is a non-negative function which takes minima if and only if $E\{\overset{\bullet}{y}_i(t)\overset{\bullet}{y}_j(t)\} = 0$, for $i, j = 1, \ldots, n$, $i \neq j$. It is a direct consequence of the Hadamard's inequality.

I calculate the total differential $dJ(W)$ due to the change dW

$$dJ(W) = J(W + dW) - J(W)$$

$$= \frac{1}{2} d \left\{ \sum_{i=1}^{n} \log E\{\overset{\bullet}{y}_i^2(t)\} \right\} - \frac{1}{2} d \left\{ \log \det \left(E\{\overset{\bullet}{y}(t)\overset{\bullet}{y}^T(t)\} \right) \right\},$$

$$= \sum_{i=1}^{n} \frac{E\{\overset{\bullet}{y}_i(t)d\overset{\bullet}{y}_i(t)\}}{E\{\overset{\bullet}{y}_i^2(t)\}} - \operatorname{tr}\{(I - W)^{-1}dW\} - \frac{1}{2} d \left\{ \log \det C_x(t) \right\}, \tag{6}$$

where $C_x(t)$ is the differential correlation matrix of $x(t)$ defined by

$$C_x(t) = E\left\{ \overset{\bullet}{x}(t)\overset{\bullet}{x}^T(t) \right\}. \tag{7}$$

Define a modified differential matrix dV as

$$dV = (I - W)^{-1}dW. \tag{8}$$

Then we have

$$dJ(W) = E\{\overset{\bullet}{y}^T(t)\Lambda^{-1}(t)dV\overset{\bullet}{y}(t)\} + \operatorname{tr}\{dV\} + d\left\{ \log \det C_x(t) \right\}, \tag{9}$$

where $\Lambda(t)$ is a diagonal matrix whose ith diagonal element is $E\{\overset{\bullet}{y}_i^2(t)\}$ and its ith diagonal element $\lambda_i(t)$ is estimated by

$$\lambda_i(t) = (1 - \delta)\lambda_i(t - 1) + \delta\overset{\bullet}{y}_i^2(t), \tag{10}$$

for some small δ (say, $\delta = 0.01$).

Hence, the gradient of the objective function (5) with respect to the modified differential matrix dV is given by

$$\frac{dJ(W)}{dV} = E\left\{ \Lambda^{-1}(t)\overset{\bullet}{y}(t)\overset{\bullet}{y}^T(t) \right\} - I \tag{11}$$

The stochastic gradient descent method leads to the updating rule for V that has the form

$$V(t + 1) = V(t) + \eta_t \left\{ I - \Lambda^{-1}(t)\overset{\bullet}{y}(t)\overset{\bullet}{y}^T(t) \right\}, \tag{12}$$

where $\eta_t > 0$ is the learning rate. It follows from the definition (8) that the learning algorithm for $\boldsymbol{W}$ is given by

$$\boldsymbol{W}(t+1) = \boldsymbol{W}(t) + \eta_t \left\{ \boldsymbol{I} - \boldsymbol{W}(t) \right\} \left\{ \boldsymbol{I} - \boldsymbol{\Lambda}^{-1}(t)\dot{\boldsymbol{y}}(t)\dot{\boldsymbol{y}}^T(t) \right\}, \qquad (13)$$

which is a natural gradient algorithm for adaptive differential decorrelation. The algorithm (13) is a differential version of the equivariant nonstationary source separation algorithm in [6].

3.2   Local Stability Analysis

Stationary points of the algorithm (13) satisfy

$$E\left\{ \boldsymbol{I} - \boldsymbol{\Lambda}^{-1}(t)\dot{\boldsymbol{y}}(t)\dot{\boldsymbol{y}}^T(t) \right\} = 0, \qquad (14)$$

which implies that $E\{\dot{y}_i(t)\dot{y}_j(t)\} = 0$ for $i, j = 1, \ldots, n$, $i \neq j$. In order to show that stationary points of (13) are locally stable, we need to show that the Hessian $d^2 \mathcal{J}$ is positive. Following suggestions in [7,6], I calculate the Hessian $d^2 \mathcal{J}$ in terms of the modified differential matrix.

For shorthand notation, we omit the time index t in the following analysis. The Hessian $d^2 \mathcal{J}$ is given by

$$\begin{aligned}
d^2 \mathcal{J} &= E\left\{ \dot{\boldsymbol{y}}^T d\boldsymbol{V}^T \boldsymbol{\Lambda}^{-1} d\boldsymbol{V}\dot{\boldsymbol{y}} + \dot{\boldsymbol{y}}^T \boldsymbol{\Lambda}^{-1} d\boldsymbol{V} d\boldsymbol{V}\dot{\boldsymbol{y}} \right\} \\
&= E\left\{ \dot{\boldsymbol{y}}^T d\boldsymbol{V}^T \boldsymbol{\Lambda}^{-1} d\dot{\boldsymbol{y}} \right\} + E\left\{ \dot{\boldsymbol{y}}^T \boldsymbol{\Lambda}^{-1} d\boldsymbol{V} d\dot{\boldsymbol{y}} \right\} \\
&= \sum_{i,j} \frac{\lambda_i}{\lambda_j} (dv_{ji})^2 + \sum_{i,j} dv_{ij} dv_{ji}, \qquad (15)
\end{aligned}$$

where the statistical expectation is taken at the solution which satisfies the condition $E\{\dot{y}_i\dot{y}_j\} = 0$ for $i \neq j$. For a pair (i, j), $i \neq j$, the summand in the first term in (15) can be rewritten as

$$\begin{aligned}
\frac{\lambda_i}{\lambda_j} (dv_{ji})^2 &+ \frac{\lambda_j}{\lambda_i} (dv_{ij})^2 + 2 dv_{ij} dv_{ji} \\
&= \begin{bmatrix} dv_{ij} & dv_{ji} \end{bmatrix} \begin{bmatrix} \frac{\lambda_j}{\lambda_i} & 1 \\ 1 & \frac{\lambda_i}{\lambda_j} \end{bmatrix} \begin{bmatrix} dv_{ij} \\ dv_{ji} \end{bmatrix}, \qquad (16)
\end{aligned}$$

which is always non-negative. Hence $d^2 \mathcal{J}$ is always positive. Therefore the algorithm (13) is locally stable around the solutions.

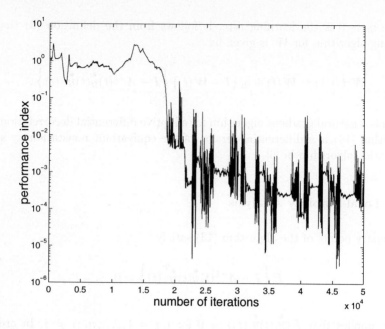

Fig. 1. The evolution of performance index.

4   Application: Nonstationary Source Separation

Source separation aims at recovering unknown original sources $s(t)$ from their linear instantaneous mixture $x(t) = As(t)$ without any knowledge of the mixing matrix $A \in \mathbb{R}^{n \times n}$. The nonstationary source separation deals with statistically independent sources whose variances are time-varying (i.e., second-order nonstationarity) [6,8]. It was shown in [9] that nonstationary source separation could be achieved by decorrelation. A natural gradient based decorrelation algorithm was proposed in [6].

Here we demonstrate the algorithm (13) can be also successfully applied to the problem of nonstationary source separation. The differential correlation matrix of $x(t)$, $C_x(t)$ and the differential correlation matrix of $s(t)$, $C_s(t)$ satisfies the decomposition $C_x(t) = AC_s(t)A^T$ where $C_s(t)$ is a diagonal matrix. Since the algorithm (13) minimizes the differential correlation between output nodes for $\forall t$, we have $(I - W) = A$.

Three digitized voice signals (sampled at 8kHz) were used as sources. The synaptic weight matrix W was initially set to be a zero matrix. The learning rate $\eta_t = .0001$ and $\delta = .07$ were used. Figure 1 shows the evolution of performance index (PI) which is defined by

$$\text{PI} = \frac{1}{n(n-1)} \sum_{i=1}^{n} \left\{ \left(\sum_{k=1}^{n} \frac{|g_{ik}|}{\max_j |g_{ij}|} - 1 \right) + \left(\sum_{k=1}^{n} \frac{|g_{ki}|}{\max_j |g_{ji}|} - 1 \right) \right\},$$

where g_{ij} is the (i,j)th element of the global system matrix $G = (I - W)^{-1} A$.

5   Conclusion

I have introduced a concept of differential decorrelation and have presented a natural gradient algorithm and its local stability analysis. The algorithm was successfully applied to the problem of nonstationary source separation. The differential correlation matrix can be defined in somewhat different manner and an algebraic method was proposed in [10]. In addition, differential kurtosis could be also employed in source separation [11].

Acknowledgments. This work was supported by KOSEF-ARIEL, by Korea Ministry of Science and Technology under Brain Science and Engineering Research Program and an International Cooperative Research Project, by Korea Ministry of Information and Communication under Advanced backbone IT technology development project, by ETRI, and by Ministry of Education of Korea for its financial support toward the Electrical and Computer Engineering Division at POSTECH through its BK21 program.

References

1. Choi, S.: Differential Hebbian-type learning algorithms for decorrelation and independent component analysis. Electronics Letters **34** (1998) 900–901
2. Hebb, D.O.: The Organization of Behavior. Wiley, New York (1949)
3. Földiák, P.: Adaptive network for optimal linear feature extraction. In: Proc. Int. Joint Conf. Neural Networks. (1989) 401–405
4. Kosko, B.: Differential Hebbian learning. In: Proc. American Institute of Physics: Neural Networks for Computing. (1986) 277–282
5. Amari, S.: Natural gradient works efficiently in learning. Neural Computation **10** (1998) 251–276
6. Choi, S., Cichocki, A., Amari, S.: Equivariant nonstationary source separation. Neural Networks **15** (2002) 121–130
7. Amari, S., Chen, T.P., Cichocki, A.: Stability analysis of learning algorithms for blind source separation. Neural Networks **10** (1997) 1345–1351
8. Choi, S., Cichocki, A., Belouchrani, A.: Second order nonstationary source separation. Journal of VLSI Signal Processing (2002) to appear.
9. Matsuoka, K., Ohya, M., Kawamoto, M.: A neural net for blind Separation of nonstationary signals. Neural Networks **8** (1995) 411–419
10. Choi, S., Cichocki, A., Deville, Y.: Differential decorrelation for nonstationary source separation. In: Proc. ICA, San Diego, CA (2001)
11. Deville, Y., Benali, M.: Differential source separation: Concept and application to a criterion based on differential normalized kurtosis. In: Proc. EUSIPCO. (2000)

An Application of SVM to Lost Packets Reconstruction in Voice-Enabled Services

Carmen Peláez-Moreno, Emilio Parrado-Hernández,
Ascensión Gallardo-Antolín, Adrián Zambrano-Miranda, and
Fernando Díaz-de-María

Dpto. Teoría de la Señal y Comunicaciones, Escuela Politécnica Superior-Univ.
Carlos III de Madrid, Avda. Universidad, 30, 28911 Leganés, Madrid, Spain.
carmen@tsc.uc3m.es,
http://www.tsc.uc3m.es

Abstract. Voice over IP (VoIP) is becoming very popular due to the huge range of services that can be implemented by integrating different media (voice, audio, data, etc.). Besides, voice-enabled interfaces for those services are being very actively researched. Nevertheless the impoverishment of voice quality due to packet losses severely affects the speech recognizers supporting those interfaces ([8]). In this paper, we have compared the usual lost packets reconstruction method with an SVM-based one that outperforms previous results.

1 Introduction

With the consolidation of IP networks as the preferred and most commonly implemented packet-based networks, the interest on the transmission of voice over these networks has considerably grown. One of its attractiveness is the possibility of using it as a vehicle for providing voice-enabled interfaces for WWW services.

However, to attain this goal, several problems need to be addressed, since the quality of voice over IP is considerably worse than the obtained using the traditional and voice-oriented telephone network. The main causes of this impoverishment of the quality are well known: the need of a severe compression of the speech signal, the unpredictable delay of the packets that carry the individual voice frames and the packet losses that occur due to congestions in the network nodes.

In this paper we focus on packet losses and its influence on both speech quality and subsequent Automatic Speech Recognition (ASR) in voiced-enable services. In this context, assuming that each packet bears a coded speech frame and simulating packet losses according to actual measurements of the Internet behavior ([2]), we asses the performance of usual reconstruction methods and propose a new one, based on a Support Vector Regressor (SVR).

When packet losses occur, current speech codecs just repeat the last available parameters. This simple solution, as stated by previous works, leads to an impoverishment of recognition performance (e.g. [8], [3]) and consequently to less reliable voice-enabled interfaces.

J.R. Dorronsoro (Ed.): ICANN 2002, LNCS 2415, pp. 1174–1179, 2002.

The paper is organized as follows: in section 2 we present the problem of packet losses in IP networks. Next, section 3 is devoted to the explanation of the SVM-based procedure we propose for parameter reconstruction. In section 4, we describe the experiments we have conducted for assessing our proposal. Finally, we draw some conclusions and outline some further work.

2   ASR Over the WWW: Facing the Packet Losses

To gain insight on the actual causes of this loss of recognition performance we have conducted the following experiment. On the one hand, we have obtained the recognition parameters from the original speech. On the other hand, we have simulated packet losses –according to real measurements due to Borella [2]- and generated another set of parameters in which the lost parameters are substituted by the previous ones. Finally, we have computed the mean square reconstruction error.

Surprisingly, as will be shown in section 4, the average errors are very low. However, the recognition performance significantly decreases. In fact, repeating the last parameters is actually a good solution since, as evidenced by Nadeu et al. [5], the bandwidth of the spectral parameters is extremely low, i.e., its time evolution is very slow.

The previous considerations allow us to conclude that slight improvements on reconstruction errors are likely going to lead to significant improvements on ASR performance.

We have tried to improve repetition results by linear low-pass filtering (according to the bandwidth estimated by Nadeu et al.), but reconstruction errors slightly increased. Consequently, we have decided to turn towards nonlinear methods; in particular, we have tried an SVM-approach.

3   An SVM-Based Reconstruction Technique

3.1   SVM

For the reconstruction of the missing frames, we have used a Gaussian Kernel Support Vector Regressor (SVR) [10]. These machines seek to determine a function $f(\mathbf{x})$, that for each data point $\mathbf{x}_i$ verifies $|f(\mathbf{x}_i) - y_i| < \epsilon$. So to speak, the algorithm fits a hosepipe of radius ϵ to the data. The smoothness of the estimated function is controlled by allowing some of the data point to remain outside the pipe.

The regressor is the result of an optimization problem applied to an RBF Network architecture whose nodes are some critical input data named Support Vectors (SVs).

$$f(\mathbf{x}) = \sum_{i=1}^{M} \alpha_i k(\mathbf{x}, \mathbf{x}_i) + b \qquad (1)$$

where $\mathbf{x}_i$ are the SVs, M is the number of SVs, k is a Gaussian kernel and α_i and b are the coefficients of the linear combination result of the optimization problem. The SVR automatically determines the SVs, so that there is no need to fix *a priori* the architecture of the RBF network.

The experiments carried out in this paper have been run with the MySVM implementation of SVR, available in [9]. The parameters of the Gaussian kernel have been determined by exploration through a cross-validation procedure.

3.2   The Proposed Approach

The application scenario entails the following steps: the speech is encoded, packetized and transmitted over an IP network. Once the bit stream reaches the application interface, some parameters are extracted to feed the automatic speech recognizer. In particular, we extract 10 LSPs –Line Spectral Pairs– and the energy of the corresponding frame. Finally, to proceed with the recognition the LSPs coefficients should be transformed into cepstral parameters; however, as this work just asses the reconstruction performance, we have omitted this last stage.

We consider the evolution of each recognition parameter as a time series. A SMV is trained for each parameter to predict the following value from previous ones. It is important to notice that the computational cost of SVM-based reconstruction is not very significant in the ASR context.

Input selection. The feature selection is one of the key points to reach successful results when working with ANN. In this case, we have recovered some prescriptions from the dynamical systems area. In particular, each one of the considered time series is seen as generated by a non-linear dynamical system defined by a (low-dimensional) state-space vector and its evolution through a state space.

As established by the "embedding theorem" [7], it is possible to reconstruct a state space equivalent to the original one. Furthermore, a state-space vector formed by time-delayed samples of the observation (in our case, the speech samples) could be an appropriate choice:

$$\mathbf{s}_n = [s(n), s(n-T), \ldots, s(n-(d-1)T)]^t \tag{2}$$

where $s(n)$ is the time series, d is the dimension of the state-space vector, T is a time delay and t means transpose.

Finally, the reconstructed state-space vector dynamic, $\mathbf{s}_{n+1} = F(\mathbf{s}_n)$, can be learned through either local or global models, which in turn will be polynomial mappings, neural networks, etc.

Considering the reconstructed state-space vector $\mathbf{s}_n$ two questions naturally arise: What should be the embedding dimension of the (reconstructed) state-space vector, d? And what should be the time delay, T? Most of the researchers who have recently proposed non-linear speech predictors have assumed $T = 1$ (following the linear case).

Recently Abarbanel et al. [1] reviewed the state of art concerning the techniques to deal with non-linear deterministic systems. In this paper we propose to apply to our particular problem the analysis techniques described by Abarbanel to determine the time delay and the embedding dimension. It follows a brief summary of these methods.

Average Mutual Information. When seeking the best value for T, the fundamental issue is to establish a right balance between a too small value (samples in the reconstructed state-vector exhibit a lot of common information) and a too large one (samples are independent). Abarbanel et al. suggest the following prescription: choose the value corresponding to the first minimum of the average mutual information

$$I(T) = \sum_{s(n),s(n+T)} P(s(n), s(n+T)) \log_2 \left[\frac{P(s(n), s(n+T))}{P(s(n))P(s(n+T))} \right] \tag{3}$$

where $P(\cdot)$ represents a probability which is estimated through a histogram.

False Nearest Neighbors. Now the issue is to determine the embedding dimension. For that purpose, Abarbanel et al. suggest the false nearest neighbors algorithm which is based on the following reasoning. For any point, we can ask whether its nearest neighbor is there due to the dynamics itself or is instead projected due to a too small reconstructed state-space vector dimension. Thus, the algorithm will compute the percentage of false nearest neighbors (those that disappear when the dimension is increased) for each of the candidate dimensions and will decide that the suitable dimension will be that for which the percentage of false nearest neighbors becomes zero (the dimension is then high enough).

4   Experiments and Results

4.1   Input Selection Results

From the Average Mutual Information, $I(T)$, we have obtained that values of T around $T = 23$ could be a good selection for every LSP and around $T = 37$ could be an appropriate choice for the energy. Nevertheless, the first minimum is not well-defined in any case. On the contrary the first valley is quite smooth. Consequently, we consider that other values in a wide neighborhood of these ones can work properly. Even more, for the energy, although the first minimum is reached at $T = 37$, the curve $I(T)$ is extremely smooth and we have decided to use also $T = 23$ for simplicity reasons.

With respect to d, the dimension of the input vector, the results are much more conclusive: $d = 2$ is the best choice in any case.

4.2   Reconstruction Experiments and Results

Database. The database which we have used in our speaker-independent continuous speech recognition experiments is the well-known Resource Management RM1 Database [6], which has a 991 words vocabulary. The speaker-independent training corpus consists of 3,990 sentences pronounced by 109 speakers. The test set contains 1,200 sentences from 40 different speakers, which corresponds to a compilation of the first four official test sets. Originally, RM1 was recorded at 16 kHz and in clean conditions; however, our experiments were performed using a (down-sampled) version at 8 kHz.

Feature extraction. The feature extraction is carried out analyzing the speech signal once every 10 ms employing a 20 ms analysis Hamming window using the HTK package [11]. Ten Linear Prediction (LP) coefficients and an energy parameter are subsequently computed for each of these analysis windows. Finally, the LSP coefficients are obtained from the LP coefficients (see, for example [4]).

Reconstruction results. With the purpose of testing the performance of the SVM-based predictor, we have chosen two subsets from the RM1 database training corpus for training and validation of the SVM, respectively. Thus, the SVM training set consists of 109 sentences, each of which belongs to one of the 109 speakers, yielding a total of 32,232 examples. Similarly, the SVM validation set provides 36,043 examples.

Table 1. Mean square prediction errors

$Parameters$	$Substitution Method (Repetition)$	$SVM prediction$
$LSP1$	$7.68 \cdot 10^{-5}$	$7.37 \cdot 10^{-5}$
$LSP2$	$12.20 \cdot 10^{-5}$	$11.75 \cdot 10^{-5}$
$LSP3$	$12.58 \cdot 10^{-5}$	$12.10 \cdot 10^{-5}$
$LSP4$	$12.82 \cdot 10^{-5}$	$12.35 \cdot 10^{-5}$
$LSP5$	$15.07 \cdot 10^{-5}$	$14.52 \cdot 10^{-5}$
$LSP6$	$13.86 \cdot 10^{-5}$	$13.42 \cdot 10^{-5}$
$LSP7$	$12.43 \cdot 10^{-5}$	$12.02 \cdot 10^{-5}$
$LSP8$	$13.64 \cdot 10^{-5}$	$13.05 \cdot 10^{-5}$
$LSP9$	$10.99 \cdot 10^{-5}$	$10.49 \cdot 10^{-5}$
$LSP10$	$9.46 \cdot 10^{-5}$	$8.94 \cdot 10^{-5}$
$Energy$	$10.20 \cdot 10^{-3}$	$9.41 \cdot 10^{-3}$

Table 1 shows the results obtained with the conventional and the SVM approaches. As we can see just slight improvements are obtained for every evaluated parameter. Nevertheless, as we have previously indicated, it is expected that slight reconstruction improvements may lead to significant increments of ASR rates.

5   Conclusions and Further Work

In this paper we have compared the usual reconstruction method used by speech codecs to circumvent the problems caused by packet losses in IP networks with an SVM-based one aiming at predicting the temporal series described by the codec parameters used by stream-based ASR approaches, obtaining encouraging reconstruction results that would likely improve the recognizer performance.

These preliminary results, however, can be refined exploring different T and d values and from the SVM predictor point of view. We expect improvements using multi-output SVM capable of exploiting the correlations among different LSP parameters corresponding to a particular frame.

Acknowledgments. This work has been partially supported by Spain CICYT grant TIC-1999-0216 and Spain CAM-07T-0018-2000

References

1. Abarbanel, H.D.I., Frison, T.W. and Tsimring, L.S.: Obtaining Order in a World of Chaos; IEEE Signal Processing Magazine, vol. 15, no. 3, pp. 49–65 (1998)
2. Borella, M. S.: Measurement and Interpretation of Internet Packet Loss, Journal of Communications and Networking, vol. 2, no. 2, pp. 93–102, (2000)
3. Kim, H. K., Cox, V.: A bitstream-based front-end for wireless speech recognition on IS-136 communications system, IEEE Transactions on Speech and Audio Processing, vol. 9, no. 5 (2001)
4. Kondoz, A. M.: Digital speech: coding for low bit rate communication systems, Ed. John Wiley & Sons, (1996)
5. Nadeu, C, Pachès-Leal, P. and Juuang, B.-H.: Filtering the time sequences of spectral parameters for speech recognition, Speech Communication 22, pp. 315–322 (1997)
6. National Institute of Standards and Technology (NIST) (distributor): The Resource Management corpus part 1 (RM1) (1992)
7. Ott, E.: Chaos in Dynamical Systems. Cambridge: Cambridge University Press (1993)
8. Peláez-Moreno, C., Gallardo-Antolín, A., Díaz-de-María, F.: Recognizing Voice over IP networks: a Robust Front-End for Speech Recognition on the WWW, IEEE Trans. on Multimedia, vol. 3, no. 2, pp. 209–18 (2001)
9. Rüping, S.: mySVM-Manual. University of Dortmund, Lehrstuhl Informatik 8, http://www-ai.cs.uni-dortmund.de/SOFTWARE/MYSVM/, (2000)
10. Schölkopf, B. and Smola, A.J.: Learning with Kernels. MIT Press, Cambridge MA, (2002)
11. Young, S. et al: HTK-Hidden Markov Model Toolkit (ver. 3.0), Cambridge University, 2000.

Baum-Welch Learning in Discrete Hidden Markov Models with Linear Factorial Constraints

Jens R. Otterpohl

Institute of Theoretical Physics, University of Bremen
28334 Bremen, Germany

Abstract. Here, I introduce a transformation-based method for extending the Baum-Welch algorithm to the training of discrete Hidden Markov Models subject to *constraints* on the parameters. A class of certain *linear factorial* constraints is described and shown to lead to exact reestimation formulas. Applying these constraints to the hidden state transitions allows to estimate processes that are cartesian products of multiple subprocesses on differing timescales. The applicability of the method has been demonstrated previously using constraints on both hidden and observation processes. The potential benefit of the approach is discussed in qualitative comparison to factorial Hidden Markov Model architectures.

1 Introduction

Traditional Hidden Markov models (HMMs) – and their derivates – have been widely applied to speech recognition (e.g.[1], [4]) and other pattern recognition problems. Another area of application of HMMs is modeling and prediction of time series.

In the context of time series modeling, one is interested in an understanding of the stochastic dynamical process generating the data. In these cases, the architecture of the HMM should be designed such that it allows for a physical interpretation of the estimated parameters. Also, in most cases, the choice of the basic architecture critically affects the success of training. If the number of free parameters in a model exceeds the dimensionality in the data at hand, one is likely to run into overfitting problems. The most direct way to reduce the degrees of freedom in a model is to impose constraints on the parameters. If one has some pre-knowledge or some assumptions about the physical characteristics of the studied system, it is therefore advisable to use this information to incorporate reasonable constraints into the training procedure.

HMMs with constraints. The strength of HMMs relies on the availability of efficient training procedures for the optimization of model parameters, such as Baum-Welch learning. It has been noted that the Baum-Welch algorithm does not generalize to work with arbitrary linear constraints [4]. Consequently,

J.R. Dorronsoro (Ed.): ICANN 2002, LNCS 2415, pp. 1180–1185, 2002.

most of the literature considers only very simple types of constraints on the parameters of a HMM, e. g. setting certain transition probabilities equal to zero (forbidden transitions) ([4], [6]) or imposing a lower bound on certain transition probabilities [4]. Another common type of constraints is *parameter tying* [1], where equivalent states, i. e. states with identical associated probabilities, are introduced. Nevertheless, I demonstrate here that Baum-Welch learning *can* be extended to work with a pretty useful class of certain 'linear factorial' constraints as introduced in the next section.

2   Linear Factorial Constraints

A standard HMM is given by a $N \times N$ transition matrix, a $N \times M$ observation matrix, and a N-dimensional vector of initial probabilities.

My objective is to modify the standard architecture to allow *deterministic* dependencies among parameters. More exactly, I introduce a coupling of both transition and observation probabilities such that observation parameters remain *independent* of transition parameters. This means that one can write b_{jk} ($j \in \{1, ..., N\}$, $k \in \{1, ..., M\}$) for representing either the observation probabilities or the transition probabilities[1] (in the latter case $M = N$). In addition to the N ordinary normalization constraints (see Eqn. 14 below), I now assume the parameters b_{jk} to satisfy K more constraints. Thus, the number of remaining free parameters is $L = N(M-1) - K$, which I denote by a vector $\overrightarrow{b^*} = (b_1^*, ..., b_L^*)$ in a space $\mathcal{L}$. The relation of these unrestricted parameters to the original b_{jk} is given by a transform f, i. e. :

$$b_{jk} = f_{jk}(\overrightarrow{b^*}) \tag{1}$$

Now I assume the transitions to be due to S subprocesses, each characterized by L_i ($i \in \{1, ..., S\}$) *free* transition (observation, resp.) parameters $b_{q+L_{i-1}}^* \in [0, 1]$ (with $L_0 = 0$, $q \in \{1, ..., L_i\}$). Thus, I have a decomposition $\mathcal{L} = \mathcal{L}_1 \times ... \times \mathcal{L}_S$ such that $\dim \mathcal{L}_i = L_i$. Further, I allow any transition f_{jk} to depend on any combination of the subprocesses. I express this dependence by $\kappa_{jkl} \in \{0, 1, ..., L_l, L_l + 1\}$, where each κ_{jkl} indicates the unique transition that contributes to f_{jk} ($\kappa_{jkl} = 0$ indicates that the lth subprocess does not contribute at all to the transition; $\kappa_{jkl} = L_l + 1$ indicates that the transition with probability $(1 - \sum_{r=1}^{L_l} b_{r+L_{l-1}}^*)$ – as given by the normalization constraint – contributes). Then, I define *linear factorial constraints* by restricting f to:

$$\forall_{j, k} : \qquad f_{jk}(\overrightarrow{b^*}) = \frac{g_{jk}}{\sum_{m=1}^{M} g_{jm}} \tag{2}$$

$$g_{jk} := \prod_{\substack{p = 1 \\ p \, : \, \kappa_{jkp} = (L_p + 1)}}^{S} \left(1 - \sum_{r=1}^{L_p} b_{r+L_{p-1}}^* \right) \prod_{\substack{p = 1 \\ p \, : \, \kappa_{jkp} \in \{1, ..., L_p\}}}^{S} b_{\kappa_{jkp}+L_{p-1}}^* \tag{3}$$

---

[1] constraints on the initial probabilities are dismissed in this discussion

$$\forall : \atop l \in \{1, ..., S\} \qquad \forall : \atop q \in \{1, ..., L_l\} \qquad \sum_{k=1}^{M} \frac{\partial g_{jk}}{\partial b^*_{q+L_l-1}} = 0 \qquad (4)$$

In order for the maximization step to yield exact reestimation formulas, the κ_{jkp} in Eqn. 3 have to be chosen such that Eqn. 4 is satisfied. This is always possible since:

$$\frac{\partial g_{jk}}{\partial b^*_{q+L_l-1}} = \chi^{jk}_{lq} \prod_{\substack{p \neq l \\ p : \kappa_{jkp} = (L_p+1)}}^{S} \left(1 - \sum_{q=1}^{L_p} b^*_{q+L_p-1}\right) \prod_{\substack{p \neq l \\ p : \kappa_{jkp} \in \{1,...,L_p\}}}^{S} \left(b^*_{\kappa_{jkp}+L_p-1}\right) \qquad (5)$$

$$\chi^{jk}_{lq} := \begin{cases} 1 \text{ if } \kappa_{jkl} = q \\ 0 \text{ if } \kappa_{jkl} = 0 \\ -1 \text{ if } \kappa_{jkl} = L_l + 1 \end{cases} \qquad (6)$$

Derivation of reestimation formulas. In the following, it is shown how to modify the standard reestimation formulas in order to incorporate the linear factorial constraints according to Eqns. 1–4.

Let $A = (a_{ij})$ be the $N \times N$ transition matrix containing the probabilities associated with the hidden states S_i, S_j, and let $B = (b_{jk})$ be the $N \times M$ observation matrix containing the probabilities to observe V_k in the hidden state S_j. Further, let $\pi = (\pi_i)$ denote the N-dimensional vector of initial probabilities of being in hidden state S_i.

One aims to find a reestimation procedure that iteratively computes an updated model $\lambda' = (A', B', \pi')$ from a current model $\lambda = (A, B, \pi)$ such that the likelihood of the observation sequence $O = \{o_1, ..., o_T\}$ is augmented, i. e. $P(O|\lambda') \geq P(O|\lambda)$. According to Baum et al. [2], this amounts to maximization of the auxiliary function $Q(\lambda, \lambda')$ (often called *quasiloglikelihood*)[2]:

$$Q(\lambda, \lambda') = \sum_{Y:P(O,Y|\lambda')>0} P(O,Y|\lambda) \ln P(O,Y|\lambda') \qquad (7)$$

$$= \sum_{i=1}^{N} \sum_{j=1}^{N} c_{ij} \ln a'_{ij} + \sum_{j=1}^{N} \sum_{k=1}^{M} d_{jk} \ln b'_{jk} + \sum_{j=1}^{N} e_j \ln \pi'_j \qquad (8)$$

where the 'expectations' c_{ij}, d_{jk} and e_j depend only on O and the current model λ:

$$c_{ij} = P(O|\lambda) \sum_{t=1}^{T-1} \zeta_t(i, j) \qquad (9)$$

$$d_{jk} = P(O|\lambda) \sum_{\substack{t=1 \\ t : o_t = V_k}}^{T} \gamma_t(j) \qquad (10)$$

---

[2] The sum in Eqn. 7 extends over all hidden state paths $Y = \{y_1, ..., y_T\}$ that lead to non-zero joint probabilities $P(O, Y|\lambda')$, $P(O, Y|\lambda)$.

$$e_j = P(O|\lambda)\gamma_1(j) \tag{11}$$

with $\zeta_t(i,j) = P(y_t = S_i, y_{t+1} = S_j|O,\lambda)$ and $\gamma_t(j) = P(y_t = S_j|O,\lambda)$
Since the three sums in Eqn. 8 are independent, maximizing $Q(\lambda, \lambda')$ is the same as maximizing each of these terms separately. Moreover, they share all the same structure. Thus, it suffices to consider one of them, say $\sum_{j=1}^{N} \sum_{k=1}^{M} d_{jk} \ln b'_{jk}$.

One has to determine the optimal update b'_{jk} such that all the parameter dependencies remain satisfied. This can be achieved naturally by updating instead the free parameters. I denote these updates by a vector $\vec{b}^{*\prime} = (b_1^{*\prime}, ..., b_L^{*\prime}) \in \mathcal{L}$. Again, the constraints are ensured implicitly through the relation :

$$b'_{jk} = f_{jk}(\vec{b}^{*\prime}) \tag{12}$$

Setting up a scheme of Lagrangian multipliers l_j ($j \in \{1, ..., N\}$) leads to the following $L + N$ optimization conditions:

$$\forall l \in \{1, ..., L\} : \sum_{j=1}^{N} \sum_{k=1}^{M} \frac{d_{jk}}{f_{jk}(\vec{b}^{*\prime})} \frac{\partial}{\partial b_l^{*\prime}} f_{jk}(\vec{b}^{*\prime}) = \sum_{j=1}^{N} l_j \sum_{k=1}^{M} \frac{\partial}{\partial b_l^{*\prime}} f_{jk}(\vec{b}^{*\prime}) \tag{13}$$

$$\forall j \in \{1, ..., N\} : \sum_{k=1}^{M} f_{jk}(\vec{b}^{*\prime}) = 1 \tag{14}$$

Evidently, all types f of constraints that reduce Eqn. 13 to a set of linear equations therefore induce exactly solvable reestimation formulas.

The special choice of f according to Eqns. 2–4 implicitly satisfies Eqn. 14, and moreover it leads to S sets of L_i linear equations:

$$\left(\sum_{\substack{j=1 \\ j,k\,:\,\kappa_{jkl} = L_l+1}}^{N} \sum_{k=1}^{M} d_{jk} \right) b_{q+L_{l-1}}^{*\prime} + \left(\sum_{\substack{j=1 \\ j,k\,:\,\kappa_{jkl} = q}}^{N} \sum_{k=1}^{M} d_{jk} \right) \sum_{r=1}^{L_l} b_{r+L_{l-1}}^{*\prime} = \sum_{\substack{j=1 \\ j,k\,:\,\kappa_{jkl} = q}}^{N} \sum_{k=1}^{M} d_{jk} \tag{15}$$

One obtains an exact solution for the updates by solving Eqns. 15 for $\vec{b}^{*\prime}$.

Reestimation procedure. In summary, I propose a reestimation procedure modified for the linear factorial constraints as follows: (1) given the unconstrained $b_l^* \in \mathcal{L}$, compute the current parameters b_{jk} from Eqn. 1; (2) evaluate the right hand side of Eqn. 10 (9, 11 resp.) via the standard forward and backward variables (see e. g. [4]); (3) determine the updates $b_l^{*\prime}$ from the solution of Eqns. 15; (4) reiterate from step (1).

From a practical point of view, this scheme has the advantage that no modification of the usual forward-backward procedure is needed. Merely the update step (3) has to be adapted in order to enforce the constraints. The additional step (1) is computationally inexpensive. However, the amount of necessary computation can be drastically reduced by modifying step (2) such that only independent 'expectations' d_{jk} are computed via the forward-backward procedure. The remaining d_{jk} can easily obtained using their dependencies.

In order to ensure that the reestimation procedure converges to a local optimum, one has to show that the above reestimation formulas induce a growth transformation, i.e. in each iteration step $P(O|\lambda') > P(O|\lambda)$ until the local maximum is reached. The proof follows the lines of the original proof by Baum et al. [2] and is omitted here.

Factorization with constrained HMMs. The class of constraints as expressed in Eqns. 2–4 – although still very restricted – proves extremely useful in practical applications. In particular, with regard to the hidden state transitions, it contains the type of constraints inherent in the structure of the well known *factorial HMM* (FHMM) as introduced in [3] in order to represent data displaying features on different timescales. Such data can be thought of as generated by a combination of independent processes, each controlling different features of the dynamics. It is evident that such a process can be mapped to a *linear-factorially constrained HMM* (LFCHMM) by assigning to each subprocess an independent state space with associated transition probabilities. The full state space is then given as the cartesian product of these spaces.

In addition to constraints on the hidden transitions, LFCHMMs also allow to specify linear factorial constraints on the observation probabilities, which proves useful in order to further restrict the number of parameters and to incorporate prior knowledge into the parameter estimation, as in the following example.

3   An Example: Application to Modeling of Behavioral Data

The method outlined above has previously been applied to an architecture specifically designed for modeling and analysis of behavioral data from pigeons [5].The pigeons' behavior has been modeled as generated by a HMM where the hidden process is the cartesian product of a renewal process (characterizing pecking dynamics) and a 2-state Markovian perceptual switching process. In addition to the embedding constraints of the hidden state transitions, the observation probabilities have been constrained to be symmetric [5].

The training algorithm for the resulting LFCHMM is able to separately extract the timescales of the percept-driven switching process and of the renewal dynamics of the pecking response (details in [5]). Testing with artificial data sets has shown that the resulting parameter estimates can be used to reconstruct the hidden switching process quite well. Notably, the constraints *both* in the transition probabilities *and* in the observation probabilities were indispensable in order to prevent severe overfitting.

4   Discussion

Using the approach of Baum-Welch learning, I have derived a sufficient condition for the existence of exact reestimation formulas for parameter estimation in constrained discrete HMMs. Moreover, I have shown that this condition is easily satisfied by a class of *linear-factorially constrained HMMs* (LFCHMMs). This type of constraints can be used to form cartesian products of independent subprocesses without incurring a parsimony of additional parameters.

It is interesting to compare the proposed LFCHMM to the well known FHMM. An obvious difference is that FHMMs assume continuous, Gaussian observation densities as opposed to the discrete observation probabilities considered here. Although the embedding of the multiple processes and their associated state spaces into a cartesian product space appears as the most natural solution to the problem of multiple timescales, in case of the FHMM it has been dismissed in favour of a compact representation of the state space. Since FHMMs do not represent the full product space, they are, on the one hand, computationally more efficient than the LFCHMMs considered here. On the other hand, this leads to the need for FHMMs to make specific assumptions regarding the way the independent subprocesses are combined into observation densities. Hence, as soon as discrete observation probabilities are considered, LFCHMMs offer a much more flexible solution, since they allow independent specification of observation probabilities over the full space of hidden states. LFCHMMs are most suited to situations where unconstrained transition probabilities ought to be combined with linear-factorially constrained observation probabilities, since in this case the FHMM approach is inappropriate. In summary, LFCHMMs seem to have a higher flexibility compared to FHMMs.

Acknowledgements. The author benefitted from interesting discussions with Frank Emmert-Streib, Udo Ernst and Klaus Pawelzik. This work was supported by the DFG Graduiertenkolleg *Komplexe Dynamische Systeme*.

References

1. Bahl, L. R., Jelinek, F. & Mercer, R. L. (1983) A Maximum Likelihood Approach to Continuous Speech Recognition *IEEE Trans. Pattern Anal. Machine Intell.* **5**, 179–190.
2. Baum, L. E., Petrie, T ., Soules, G. & Weiss, N. (1970) A Maximization Techniques Ocurring in the Statistical Analysis of Probabilistic Functions of Markov Chains. *The Annals of Mathematical Statistics* **41** (1):164–171.
3. Ghahramani, Z. & Jordan, M. I. (1997) Factorial Hidden Markov Models. *Machine Learning* **29**: 245–273.
4. Levinson, S. E., Rabiner, L. R. & Sondhi, M. M. (1983) An Introduction to the Application of the Theory of Probabilistic Functions of a Markov Process to Automatic Speech Recognition. *The Bell Systems Technical Journal* **62**(4):1035–1074.
5. Otterpohl, J. R., Haynes, J. D., Emmert-Streib, F., Vetter, G. & Pawelzik, K. (2000). Extracting the dynamics of perceptual switching from 'noisy' behaviour: An application of hidden Markov modelling to pecking data from pigeons. *J. Physiol. (Paris)*, **94**, 555–567.
6. Roweis, S. Constrained Hidden Markov Models (2000). *Advances in Neural Information Processing Systems* **12**, Solla, S. A., Leen, T. K., Müller, K.-R., eds., MIT Press.

Mixtures of Autoregressive Models for Financial Risk Analysis

Alberto Suárez

ETS de Informática, Universidad Autónoma de Madrid,
Campus Cantoblanco, Madrid-28049 Spain

Abstract. The structure of the time-series of returns for the IBEX35 stock index is analyzed by means of a class of non-linear models that involve probabilistic mixtures of autoregressive processes. In particular, a specification and implementation of probabilistic mixtures of GARCH processes is presented. These mixture models assume that the time series is generated by one of a set of alternative autoregressive models whose probabilities are produced by a gating network. The ultimate goal is to provide an adequate framework for the estimation of conditional risk measures, which can account for non-linearities, heteroskedastic structure and extreme events in financial time series. Mixture models are sufficiently flexible to provide an adequate description of these features and can be used as an effective tool in financial risk analysis.

1 Introduction

Arguments of market efficiency suggest that changes in the prices of a financial asset (for instance, a stock index) should behave in a rather unpredictable manner. In a liquid market, all rational expectations about the future evolution of an asset ought to be rapidly incorporated into the actual market price of the product. In practice, this means that the evolution of financial asset prices is very difficult to predict from the time series of past prices (weak form of market efficiency): If a reliable prediction method for the price of a given asset were available to most market agents, the information obtained from its predictions would be used to guide trading and, eventually, through the pressure exerted on the asset price by the asymmetry of offer and demand, all trends detected would be cancelled. Given the market efficiency hypothesis, which is believed to be valid at least in its weak form, it is not possible to learn how to predict prices from the identifications of patterns in historical time series. Nonetheless, there are tasks in financial analysis, other that price prediction, that involve identifying patterns which ought not to be annulled upon their detection by market agents. One of these tasks is to provide accurate and reliable quantitative measures of the risk associated to holding a financial portfolio. These measures can be used to reallocate investments in order to avert, or at least minimize, losses in the value of a given portfolio caused by fluctuations in market conditions.

In this work we propose to use mixtures of GARCH models (MixGARCH) to carry out this risk analysis. MixGARCH is an instantiation of the mixture

J.R. Dorronsoro (Ed.): ICANN 2002, LNCS 2415, pp. 1186–1191, 2002.
© Springer-Verlag Berlin Heidelberg 2002

of experts paradigm introduced by Jordan and Jacobs in the machine learning community [1]. Mixture models have have been previously introduced in the literature to analyze financial time series [2,3,4]. Our final goal is to induce from historical data models that are useful to characterize the magnitude of fluctuations in prices of financial products. These mixture models are be flexible enough to correctly capture the statistics of rare extreme fluctuations (heavy tails) and the heteroskedastic structure of the time series, both of which are highly relevant in risk analysis.

2   Heavy Tails and Correlations in Financial Time Series

Time series of prices of financial products (e.g., stock prices) are usually analyzed in terms of log-returns: Assuming that the time series is $\{S_t; t = 0, 1, 2...T\}$, the logarithmic transformation into returns

$$X_t = log\frac{S_t}{S_{t-1}} \approx \frac{S_t - S_{t-1}}{S_{t-1}}; \quad t = 1, 2, \ldots, T \tag{1}$$

is generally sufficient to render the process stationary. It is common to use relative returns instead of log-returns. The differences between the two possible definitions for the returns are negligible provided that $X_t \ll 1$.

It is an empirical observation that the tails of the unconditional and conditional distributions of returns on financial products are usually larger than those predicted by a normal distribution, which is the usual assumption made in financial models. This implies that the frequency of extreme events is severely underestimated in the usual normal models. The risks associated to a portfolio can then be severely underrated and lead to unsound risk management practices.

The assumption of market efficiency implies that all information about market expectations of a given asset should be reflected in the prices of products involving that asset in an almost instantaneous manner. This hypothesis translates in returns being independent of each other. However, markets are not perfectly efficient, and a certain hysteresis can be observed in their response. Autocorrelations in the time series of returns are generally small and short-lived. If transaction costs are considered, these memory terms cannot be translated into strategies yielding systematic profits. Nonetheless, there exist statistically significant long-term correlations in the absolute value of the returns. Intuitively, this means that large relative changes in the value of a portfolio tend to be followed by changes which are also large, but which can be of either sign (*volatility clustering*).

3   Autoregressive Models for Financial Time Series

Classical autoregressive models for conditional heteroskedasticity assume that the structure of a given time series can be reproduced by a linear model with normal innovations. In order to extend the expressive capacity of the models,

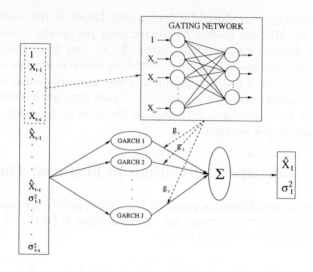

Fig. 1. MixGARCH architecture.

nonlinear predictors have been considered [5,6]. The normality hypothesis can also be abandoned in order to account for heavy tails in financial data [7].

As an alternative to the strategy where models are progressively made more involved, our proposal consists in articulating a combination of simple models that is flexible enough to capture the complexities (non-linearities, non-normal innovations, heteroskedasticity) of the time series. The probability of each of this models is generated by a gating network whose inputs are also the delayed values of the time series. In particular, the mixture of autoregressive models introduced in the present work consists of linear models with normal innovations that are combined to yield a piecewise linear model where innovations are no longer necessarily normal. A mixture of J autoregressive models has the following structure

$$X_t = f_\xi(\mathbf{X}_{(t-1)}^{(t-m)}) + u_\xi(t), \tag{2}$$

where ξ is a discrete random variable taking the values $\{1, 2, \ldots J\}$ with probability $g_\xi(\mathbf{X}_{(t-1)}^{(t-r)})$. These probabilities are be generated by a gating network. The MixGARCH architecture is displayed as in Figure 1.

In each of the models of the mixture X_t is expressed as a function of the vector of delayed values of the time series, $[\mathbf{X}_{(t-1)}^{(t-m)}]^\dagger = (X_{t-1} X_{t-2} \ldots X_{t-m})$, and of the innovations, $u_\xi(t)$. The heteroskedasticity of the time series can also be captured with a model for the innovations

$$u_\xi(t) = \sigma_\xi(t)Z_t; \quad \sigma_\xi^2(t) = h_\xi([\boldsymbol{\sigma}_\xi{}^2]_{(t-1)}^{(t-d)}, [\boldsymbol{u}_\xi{}^2]_{(t-1)}^{(t-d)}), \tag{3}$$

with the obvious definitions for the delayed vectors. Note that Eq. (3) introduces a recurrent structure in the network. The residuals Z_t, are assumed to be zero mean independent random variables, with a time-independent probability distribution $P(Z_t)$. With these assumptions, the objective is to find the functions $\{g_\xi, f_\xi, h_\xi; \xi = 1, 2, \ldots J\}$ and P that maximize the likelihood function

$$\mathcal{L} = \prod_{t=1+\max\{m,p,q\}}^{T} \sum_{\xi=1}^{J} g_\xi(\mathbf{X}_{(t-1)}^{(t-r)}) \, P \left(\frac{X_t - f_\xi(\mathbf{X}_{(t-1)}^{(t-m)})}{\sqrt{h_\xi([\boldsymbol{\sigma_\xi}^2]_{(t-1)}^{(t-d)}, [\boldsymbol{u_\xi}^2]_{(t-1)}^{(t-d)})}} \right) \tag{4}$$

Most autoregressive models posit a parametric form for these functions and then fix the various parameters by optimization of the likelihood function. Restrictions in the values of the parameters may be needed in order that the models be meaningful. For instance, it is usually required that the process be covariance-stationary. The mixture paradigm includes as particular cases the usual autoregressive models, which have $M = 1$, and assume normal innovations, possibly with a time-dependent variance.

4   Analysis of the IBEX35 Data

The mixture models described in the previous section are now applied to the analysis of the time series of relative daily returns on the IBEX35 stock index for a 7-year time span from 04/01/1993 till 09/01/2002. These 2300 trading days are divided into a training set, consisting in the series of initial 1300 prices, and a test set with the remaining data. The experts used in the fit are linear autoregressive models with AR(1)/GARCH(1,1) structure, with a gating network with sigmoidal probability functions and a $N(0,1)$ distribution for the residuals, Z_t.

A fit by likelihood maximization to a single AR(1) / GARCH(1,1) process yields the model

$$X_t = 0.0973 + 0.1338 X_{t-1} + \sigma_t Z_t$$
$$\sigma_t^2 = 0.0472 + 0.0744(X_{t-1} - (0.0973 + 0.1338 X_{t-2}))^2 + 0.8810\sigma_{t-1}^2. \tag{5}$$

With this model the correlations between the residuals are negligible, as shown the right-hand side plots in Figure 2. In contrast, the normality hypothesis is less convincingly supported with the training data (see Figure 2). In particular, extreme events are more frequent than what a Gaussian model for the innovations would predict.

A much better description of the data can be obtained with a probabilistic mixture of two AR(1)/GARCH(1,1) models

Model 1: $X_t = 0.1054 + 0.1365 X_{t-1} + \sigma_t Z_t$
$$\sigma_t^2 = 0.0098 + 0.0470(X_{t-1} - (0.1054 + 0.1365 X_{t-2}))^2 + 0.9338\sigma_{t-1}^2$$
Model 2: $X_t = -0.5450 - 0.2155 X_{t-1} + \sigma_t \epsilon_t$
$$\sigma_t^2 = 1.5041 + 0.2726(X_{t-1} - (-0.5450 - 0.2155 X_{t-2}))^2 + 0.3623\sigma_{t-1}^2 \tag{6}$$

The probabilities for the mixture are

$$g_{[1]}(X_{t-1}) = \frac{1}{1 + e^{-2.3363 - 1.3010 X_{t-1}}}; \qquad g_{[2]}(X_{t-1}) = 1 - g_{[1]}(X_{t-1}) \tag{7}$$

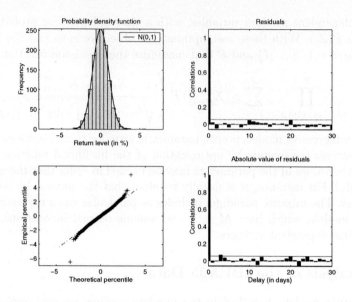

Fig. 2. Performance on training data of GARCH model . Left: Probability distribution for the residuals. Comparison with normality hypothesis. Right: Normalized autocorrelations for the IBEX35 returns time series. The range between the horizontal continuous lines corresponds to the 95% fluctuation band from random sampling.

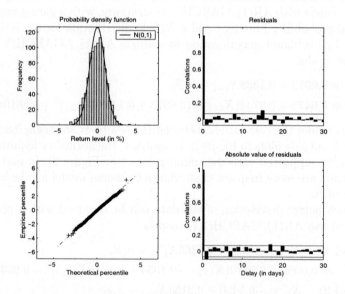

Fig. 3. Performance on test data of MixGARCH model.

This MixGARCH model accounts well for both the correlations and for heavy tails in the data. In particular, when applied in the test set, the hypothesis of normal uncorrelated residuals appears to be valid (see Figure 3).

5   Perspectives

The problem of time series modeling can be viewed as a learning task, where the input for the learning device are the vector of delays of the time series itself. The problem is then to select from the data a model that is both simple and has high predictive power. The mixture paradigm is an attempt to solve the full complex problem by means of a combination of simple models. The application of this paradigm to the modeling of financial time series can account for traits (frequency of extreme events and heteroskedasticity), which are important in risk analysis.

In this work we have restricted our studies to mixtures that consist of a maximum of two autoregressive models. With mixtures of more than two models, it is possible to articulate the models in a non-hierarchical mixture, or to structure them in a hierarchical fashion [8]. We expect this latter procedure to be more sensible: On one hand it limits the number of model parameters. On the other, it defines different levels of decision at which the complex problem is successively divided into simpler ones defined on separate approximation regions.

Acknowledgments. This work has been supported by CICyT grant TIC2001-0572-C02-02.

References

1. R. Jacobs and M. Tanner. Mixtures of X. In A. J. C. Sharkey, editor, *Combining Artificial Neural Nets*, pages 267–296, London, 1999. Springer.
2. A. S. Weigend, M. Mangeas, and A. N. Srivastava. Nonlinear gated experts for time series: Discovering regimes and avoiding overfitting. *International Journal of Neural Systems*, 6(4):373–399, 1995.
3. A. J. Zeevi, R. Meir, and R. J. Adler. Non-linear models for time series using mixtures of autorregressive models. *Preprint*, 1999.
4. C. S. Wong and W. K. Li. On a mixture autoregressive model. *Journal of the Royal Statistical Society B*, 62:95–115, 2000.
5. G. González-Rivera. Smooth transition GARCH models. *Studies in Nonlinear Dynamics and Econometrics*, 3:61–78, 1998.
6. S. Lundbergh and T. Tersvirta. Modelling economic high-frequency time series with STAR-GARCH models. *Working Paper, No. 291, Stockholm School of Economics*, 1998.
7. J. D. Hamilton. *Time Series Analysis*. Princeton University Press, Princeton, NJ, 1994.
8. M. I. Jordan and R. A. Jacobs. Hierarchical mixtures of experts and the EM algorithm. *Neural Computation*, 6:181–214, 1994.

Part VIII

Vision and Image Processing

Kernel-Based 3D Object Representation

Annalisa Barla and Francesca Odone

INFM-DISI, Università di Genova,
Via Dodecaneso 35, 16146 Genova, Italy
{barla,odone}@disi.unige.it,

Abstract. In this paper we describe how kernel-based novelty detection can be used effectively to model 3D objects from unconstrained image sequences, in order to deal with object identification and recognition. In this framework, we introduce a similarity measure based on the Hausdorff distance, well suited to represent, identify, and recognize 3D objects from grey-level images. The effectiveness of the method is shown on the representation and identification of rigid 3D objects in cluttered environments.

1 Introduction

In the *learning from examples* paradigm the solution to a problem, such as object detection and recognition (see [9,10,11] for example), is usually learned from a set of positive and negative examples. Initially this set of *training* examples is used to learn a discriminant function which will be used in the test stage to decide whether a novel example is a different instance of one of the two classes. In image-based problems, while positive examples are defined as images or portion of images containing the object of interest, negative examples are somewhat ill defined and difficult to characterize. For this reason, a representative list of informative negative examples is usually obtained by carefully selecting the most difficult negative examples or most likely false positives [10].

In this paper we show that the design of an appropriate image similarity measure makes it possible to effectively solve a multiclass classification problem even in the absence of negative examples — with the so-called *novelty detection*.

Kernel methods, which gained an increasing amount of attention in the last years after the influential work of Vapnik [13,14], reduce a learning problem of classification or regression to a multivariate function approximation problem in which the solution is found as a linear combination of certain positive definite functions named kernels, centered at the examples [3,4,15]. If the examples belong to only one class, the idea is that of determining the spatial support of the available data by finding the smallest sphere in feature space enclosing the examples [1]. The feature mapping, or the choice of the norm, is crucial.

The paper is organized as follows. The kernel method used in this paper is summarized in Section 2. Section 3 introduces and discusses a similarity measure based on the Hausdorff distance. The experiments on 3D object representation are reported in Section 4, while Section 5 is left to conclusions.

J.R. Dorronsoro (Ed.): ICANN 2002, LNCS 2415, pp. 1195–1200, 2002.

2   Kernel-Based Approach to Novelty Detection

In this section we review the method described in [1] which shares strong similarities with Support Vector Machines (SVMs) [13,14] for binary classification. The main idea behind this approach is to find the sphere in feature space of minimum radius which contains most of the data of the training set. The presence of outliers is countered by using slack variables ξ_i which allow for data points outside the sphere. This approach was first suggested by Vapnik [13] and interpreted and used as a novelty detector in [12]. The dual formulation requires the solution of a QP problem, which, as for SVMs, has the interesting property that training examples never appear isolated but they are always in pairs in the form of inner products. Thus, one can introduce a positive definite function K [14], called a *kernel*, which defines an inner product in some new space, the *feature space*, and solve the problem

$$\max_{\alpha_i} \ -\sum_i \alpha_i K(\mathbf{x}_i, \mathbf{x}_i) + \sum_i \sum_j \alpha_i \alpha_j K(\mathbf{x}_i, \mathbf{x}_j) \tag{1}$$

$$\text{subject to } \sum_i \alpha_i = 1 \quad \text{and} \quad 0 \leq \alpha_i \leq C.$$

A kernel function K is a function satisfying certain mathematical constraints [2, 14] and implicitly defining a mapping ϕ from the input space to the feature space. Like in the case of SVMs, the training points for which $\alpha_i > 0$ are the *support vectors* for this learning problem. The constraints on α_i define the *feasible* region of the QP problem. In the classification stage a new data point is considered positive if it lies sufficiently close to the sphere center. If $K(\mathbf{x}, \mathbf{x})$ is constant over the domain X, a novelty (or negative example) is detected if the inequality

$$\sum_{i=1}^{\ell} \alpha_i K(\mathbf{x}, \mathbf{x}_i) \geq \tau \tag{2}$$

is violated for some fixed value of the threshold parameter $\tau > 0$, otherwise $\mathbf{x}$ is classified as positive example. The uniqueness of the solution for a QP problem as (1) is guaranteed by the convexity of its functional. Typically, to obtain a convex functional, the positive definiteness of the function K is required. A closer look to Prob. (1) reveals that the uniqueness of the solution is ensured by the convexity of the functional in the feasible region, that is (ignoring a common scale factor):

$$\sum_{i=1}^{\ell} \sum_{j=1}^{\ell} \alpha_i \alpha_j \mathbf{K}_{ij} \geq 0 \tag{3}$$

$$\text{subject to } \alpha_i \geq 0, \text{ for } i = 1, ..., \ell,$$

where $\mathbf{K}_{ij} = K(\mathbf{x}_i, \mathbf{x}_j)$ is the associated Hessian matrix. This condition holds true for each function K that takes only non-negative values. This weaker condition allows for functions which are not Mercer's kernel to be used as kernels for novelty detection.

3   The Hausdorff Kernel

In this section we first describe a similarity measure for images inspired by the notion of Hausdorff distance (more details on how this similarity measure relates to other measures and, in particular, on previously proposed Hausdorff-based similarity measures can be found in [6,8]).Then, we discuss the conditions under which this function defines a legitimate kernel function.

Hausdorff Distances

Given two finite point sets A and B (both subsets of $\mathbb{R}^N$), the *directed Hausdorff distance* h can be written as $h(A, B) = \max_{a \in A} \min_{b \in B} ||a - b||$. The *Hausdorff distance*, a symmetric measure derived from $h(A, B)$, is defined as $H(A, B) = \max\{h(A, B), h(B, A)\}$. A way to gain intuition on Hausdorff measures which is very important in relation to the similarity method we are about to define, is to think in terms of set inclusion. Let B_ρ be the set obtained by replacing each point of B with a disk of radius ρ, and taking the union of all of these disks; effectively, B_ρ is obtained by dilating B by ρ. Then the following holds:

Prop. 1. $h(A, B) \leq \rho$ *if and only if* $A \subseteq B_\rho$.

Hausdorff-Based Measure for Grey-Level Images

Suppose to have two grey-level images, I_1 and I_2, of which we want to compute the degree of similarity; ideally, we would like to use this measure as a basis to decide whether the two images contain the same object, maybe observed from two slightly different views, or under different illumination conditions. In order to allow for grey level changes within a fixed interval or small local transformations (for instance small scale variations or affine transformations), a possibility is to evaluate the following function [6]

$$k(I_1, I_2) = \sum_p \theta(\epsilon - \min_{q \in N_p} |I_1[p] - I_2[q]|) \tag{4}$$

where θ is the unit step function. The function k counts the number of pixels p in I_1 which are within a distance ϵ (on the grey levels) from at least one pixel q of I_2 in the neighborhood N_p of p. Unless N_p coincides with p, k is not symmetric, but symmetry can be restored by taking, for example, the average:

$$K_H = \frac{1}{2}[k(I_1, I_2) + k(I_2, I_1)] \tag{5}$$

Equation (4) can be interpreted in terms of set dilation and inclusion, leading to an efficient implementation described in [6], which was inspired by Prop. 1. The similarity measure k is closely related to the directed Hausdorff distance h: computing k is equivalent to fix a maximum distance ρ_{max} (by choosing ϵ and N_p) allowed between two sets, and see if the sets we are comparing, or subsets of them, are within that distance. As a final remark, we notice that, by construction (Equation (4), $\mathbf{K}_{ij} \geq 0$ for all image pairs; as pointed out in Section 2, this means that function K_H generates matrices that satisfy (3). Therefore it can always be used as a kernel for novelty detection, even if it is not a Mercer's kernel [7].

4   Experiments on 3D Object Modeling and Recognition

In this section we present results on 3D object modeling and recognition from image sequences. As a training set we acquire an unconstrained image sequence by moving the camera around the object of interest. We do not compute image registration on the images of the training set, relying on the richness of the training set and on the fact that our similarity measure takes care of spatial misalignments. Also, since our mapping in feature space allows for some degree of deformation both in the grey-levels and in space, the effects of small illumination and pose changes are attenuated.

Fig. 1. Samples from the training set of statue A (above) and statue B (below).

Fig. 2. A few examples of negative examples, with respect to both statues of Figure 1.

Here we present results on modeling details of a static and stable scene, six marble statues of Genova Cathedral, all acquired under similar illumination conditions, thus generating similar brightness patterns (see Fig. 1 and 2), and making the recognition task more difficult. We choose not to perform any segmentation of the background, since, because of the stability of the scene, the background itself is representative of the scene detail. Fig. 1 shows samples of the training sets used for two of the six statues (statue A and statue B). After training the system with positive examples only, the performances are evaluated on a test set of both positive and negative examples, and described by means of Receiver Operating Characteristic (ROC) curves [5]. Each point of a ROC curve represents a pair consisting of the false-alarm rate and the hit-rate of the system, for some value of the threshold τ of Equation (2). The system efficiency can be evaluated by the growth rate of its ROC curve, and for a given false-alarm rate, the better system will be the one with the higher hit probability. The optimal τ for a certain recognition task is usually chosen as the $\hat{\tau}$ corresponding to the so called *equal error rate (e.e.r.)*, obtained when the percentage of false positives rate equals the percentage of false negatives rate (miss rate).

Fig. 2 illustrates samples of the negative examples with respect to both statues A and B, used in the testing phase. The performance of the Hausdorff kernel for various dilations Δ has been compared with classical kernels: linear, polynomials of various degrees d, Gaussian at different σ. Gaussian kernels produced unsatisfactory results (in the best case, the $e.e.r.$ is above 38%); polynomial kernels ($d = 2, 3$) produced the best results among classical kernels, with $e.e.r.$ stable around 30%. Hausdorff kernels with dilations on the grey-levels brought the $e.e.r.$ down to 15% ($\Delta = (1, 0, 0)$) and 11% ($\Delta = (3, 0, 0)$), but, most interestingly, a dilation on all the three directions dropped the $e.e.r.$ down to the 4%. The best performing Hausdorff kernel ($\Delta = (4, 3, 3)$) is compared with the best performing classical kernels (polynomials, with $d = 2, 3$) in the ROC curves of Fig. 3.

From the point of view of classification we sought the six statues in various images acquired from different view points and distances, using a multiscale search. Figure 4 shows the results on a few frames of a panning sequence of the identification of statue A and B at different scales and from different viewing points. The appropriately resized white and grey rectangles are the scores of statue A and B, respectively.

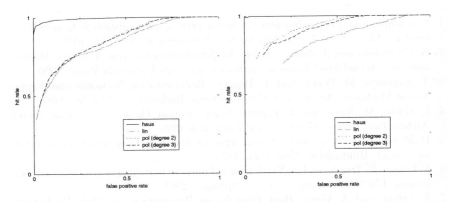

Fig. 3. Comparison between the Hausdorff kernel and the linear and polynomial ones. On the left, the ROC curve relative to statue A, with a training set of 96 elements, and a test set of 846 (pos) 3593 (neg). On the right, the ROC curve relative to statue B, with a training set of 97 elements, and a test set of 577 (pos) 3545 (neg).

5   Conclusions

This paper investigated the possibility of acquiring a representation for 3D objects, suitable for detection and identification tasks, based on a one class learning method and on a suitable similarity function. The representation or acquisition stage is based on the computation the smallest sphere in feature space containing the training data; this sphere becomes the decision surface to determine

The set $\Omega = \{\omega_1, \omega_2, \ldots, \omega_p, c_p\}$, where c_p is a residual, represents the wavelet transform of the data.

3.2 The Multiresolution Support

The multiresolution support of an image describes in a logical or Boolean way if an image g contains information at a given scale s and at a given position (x,y). if $M^l(s,x,y)=1$ (or $= true$), then g contains information at scale s and at the position (x,y). It is based on the detection at each scale of the significant wavelet coefficients. It can be defined as: [3-5]

$$M(s,x,y) = \begin{cases} 1 & \text{if } \omega_s \text{ is significant} \\ 0 & \text{if } \omega_s \text{ is not significant} \end{cases} \qquad (5)$$

Given stationary Gaussian noise, to define if ω_s is significant, it suffices to compare $\omega_s(x,y)$ to $t\sigma_s$, where σ_s is the noise standard deviation at scale s. Often t is chosen as 3. If $\omega_s(x,y)$ is small, it is not significant and could be due to noise. If $\omega_s(x,y)$ is large, it is significant. σ_s as in [3,5].

Once the support M is obtained, all the not significant wavelets coefficients, which represent inevitably the noise, are smoothed, whereas the significant ones are restored using the MHNN combined with an adaptive regularization that is determined from the M as shown in the next section.

4 Adaptive Regularization

In order to elaborate an adaptive regularization scheme in the MHNN, the weights are changed to implement different values of the regularization parameter according to different spatial activities in the image. To do it, we use the multiresolution support. For every resolution representing a specific description of the image, the pixels constituting it are associated by one particular regularization parameter according to their spatial activity.

For calculation way of the different λ associated to the different scales, we take back the method used in [7,8], in which the spatial activities of pixels are measured by means a statistics analysis formulated by a following log-linear function: $\lambda_s = \alpha \log(R_s) + \beta$, where R_s corresponds to a variance at different scales.

We can resume the different steps of the algorithm established in the present paper, as follows:

1- Wavelets decomposition of g using the à trous algorithm;
2- Scheming the multiresolution support M;
3- Smoothing of the not significant ω_s ;
4- Calculation of the regularization for each resolution basing on M;
5- Restoration of the significant ω_s , by the MHNN.

5   Results and Discussion

The algorithm is tested on *128×128* image of space (Fig. 1.a), degraded with a gaussian blur and an additive noise of 28dB SNR (Fig. 1.b). To restore the image (Fig. 1.b), the following tests are made:

Test 1:   Restoration without wavelets decomposition, and with one λ, which is calculated to adapt the global image structure, and the processing filters [12].

Test 2:   Restoration with the adaptive regularization, which is defined from the support M after wavelets decomposition of the image for $p=3$.

During the tests, the SNR improvement is calculated, and the execution is terminated when no pixel is updated.

The decompositions obtained for the three scales are reconstructed separately, and illustrated in Figures 1.c, 1.d and 1.e. It is apparent that in proportion as the scale increase the resolution contains less and less of a thin details.

The support M is shown in Figure 1.f, and the different regularization parameters are determined from it. The white areas represent flat parts of image and there are associated by a weak λ, while the dark areas correspond to edges and textures in image, and there are assigned to a high λ.

The improvement in SNR reached in *Test 1*, is of 5.82dB, while in *Test 2* it attains 8.90dB. In addition, from the Figure 1.b, we can observe the remarkable enhancement in the distorted image. Considering these results, we can conclude that the multiresolution support contributes favorably to the restoration process.

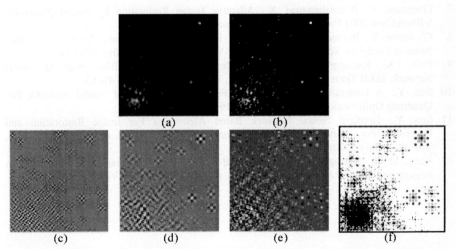

Fig. 1. (a) Degraded and noisy image (b) Restored image (*Test 2*). (c), (d) and (e) represent the reconstructed wavelets decompositions of image (a), respectively for scale *s=1, s=2* and *s=3*. (f) A multiresolution support representation of image (a).

6   Conclusion

In this paper, we try to give a solution to the image restoration problem. By using the modified Hoplfield neural network in the inversion process, we carry out an adaptive regularization scheme based on the multiresolution support obtained from the wavelets decomposition by means the *à trous* algorithm. This support is also employed to smooth the noise, which corrupt image. The results obtained are better that when it is used one single regularization for all the image structures. This improvement is due essentially to the wavelets basis, in which we can effectively track all the image non-stationarities, therefore the edges and ridges are not corrupted, as is the case in the most of the works dealing with the image restoration problem.

References

1. Gonzalez, R.C.,Woods, R.E.: Digital Image Processing. Addisson-Wesley (1992)
2. Banham, M.R., Katsaggelos, A.K.: Digital Image Restoration. IEEE Signal Processing Magazine. (March 1997) 24–41
3. Strack, J.L., Murtagh, F., Bijaoui, A.: Multiresolution Suppport Applied to Image Filtering and Restoration. Graphical Models and Image Processing. **57** (1995) 420–431
4. Murtagh, F., Strack, J.L., Bijaoui, A.: Image Restoration with Noise Suppression Using a Multiresolution Support. Astronomy and Astrophysics Supplement Series. (1995) 179–189
5. Murtagh, F., Strack, J.L.: Image Processing through Multiscale Analysis and Measurement Noise Modeling. Statistics and Computing. **10** (2000) 95–103
6. Murtagh, F., Strack, J.L., Berry, M.W.: Overcoming the Curse of Dimensionality in Clustering by Means of the Wavelet Transform. The Computer Journal, Vol. 43. **3** (2000) 107–120
7. Ghennam, S., Benmahammed, K.: Adaptive Image Restoration by Neural Networks. VIPromCom-2001 Proceedings, Zagreb, Croatia. (2001) 237–240
8. Ghennam, S., Benmahammed, K.: Image Restoration using Neural Networks. Lecture Notes in Computer Sciences, from Springer-Verlag, Vol. 2085. (2001) 227–234
9. Paik, J.K., Katsaggelos, A.K.: Image Restoration Using Modified Hopfield Neural Network. IEEE Trans. on Image Processing, Vol. 1. (January 1992) 49–63
10. Sun, Y.: A Generalized Updating Rule For Modified Hopfield Neural Network For Quadratic Optimization. Neurocomputing. **19** (1998) 133–143
11. Sun, Y.: Hopfield Neural Network Based Algorithms For Image Restoration and Reconstruction - Part I: Algorithms and Simulation. IEEE Trans. on Signal Processing, Vol. 48. **7** (July 2000) 2119-2131
12. Kang, M. G., Katsaggelos, A. K.: General Choice of the Regularization Functional in Regularized Image Restoration. IEEE Trans. on Image Processing, Vol. 4. **5** (May 1995)

Audio-Visual Speech Recognition One Pass Learning with Spiking Neurons

Renaud Séguier and David Mercier

Supélec, Team ETSN
Avenue de la Boulaie, BP28
35511 Cesson Sévigné, France
{Renaud.Seguier, David.Mercier}@supelec.fr
http://www.supelec-rennes.fr/ren/rd/etsn/

Abstract. We present a new application in the field of impulse neurons: audio-visual speech recognition. The features extracted from the audio (cepstral coefficients) and the video (height, width of the mouth, percentage of black and white pixels in the mouth) are sufficiently simple to consider a real time integration of the complete system. A generic preprocessing makes it possible to convert these features into an impulse sequence treated by the neural network which carries out the classification. The training is done in one pass: the user pronounces once all the words of the dictionary. The tests on the European M2VTS Data Base shows the interest of such a system in audio-visual speech recognition. In the presence of noise in particular, the audio-visual recognition is much better than the recognition based on the audio modality only.

1 Audio-Visual Speech Recognition

Speech recognition in noisy environments is useful in many applications, for example in car computer vocal interface or automatic ticket sale in stations and airports. The significant contribution of information contained in the movement of the lips makes it possible to improve the audio recognition rates. Audio-visual speech recognition systems use mainly HMM [3]. We propose in this article such a system exploiting impulse neurons (STAN Spatio-Temporal Artificial Neurons [15] [8]) and allowing a light training since it is only carried out on one pass. The Audio-visual features are sufficiently simple and robust to consider a real time implementation on low quality audio and video signals such as those provided by usual webcams.

2 Proposed System

The system is illustrated in Figure 1. After a specific preprocessing performed separately on the audio and video signals, a generic preprocessing makes it possible to produce impulse sequences taken into account by the STAN's which operates the classification.

J.R. Dorronsoro (Ed.): ICANN 2002, LNCS 2415, pp. 1207–1212, 2002.

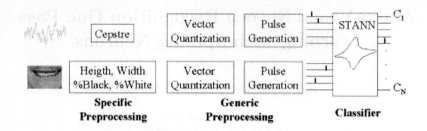

Fig. 1. Audio-Visual Speech Recognition System.

2.1 Specific Preprocessing

Audio. Cepstral coefficients are often used to characterize the sound in this type of application [6] [4]. On a 40ms sliding window, we calculate as [7] the first 12 coefficients of the cepstrum, the logarithm of the signal energy in the window and the temporal derivate of those thirteen parameters. Each one of them is normalised taking into account the values which are observed all along the sequence (from 0 to 9).

Video. Certain teams carry out a PCA (Principal Component Analysis) [1] or a DCT (Discrete Cosine Transform) [10] but most of the time dynamic contours [2] or deformable models [14] are used to characterize the shape of the mouth. Our objective is to make a real time system, thus we use features which are much simpler to extract. The height and the width of the mouth are evaluated with the method presented in [11]. For the height evaluation of the mouth, instead of working on grey levels, we use V values (from YUV color coordinate system): in these coordinates, the teeth and the dark interior of the mouth are confused, which enables us to locate the upper and lower lips more precisely.

We define in addition a sub-mouth area (see Fig. 3) centered on the mouth, in which we calculate an eight-bit grey level histogram of the pixels. The ratio between the number of pixels whose values are lower than 50 and the total pixel number gives us the percentage of dark pixels, that between the number of pixels whose values are superior to 150 and the total pixel number gives us the percentage of light pixels. We add to these four parameters (height, width, dark percentage, light percentage), the temporal derivate of the height and the width. As for the audio, we normalize each parameter over the sequence from 0 to 9.

2.2 Generic Preprocessing

We use a generic preprocessing [13] in order to convert the temporal series of features into an impulse sequence. This stage consists in applying a vector quantization (K-means [5]) separately on the audio and video features in order to extract vectors codes. At each instant, the impulse generation module compares the Euclidian distance between signal (cepstral coefficients or mouth features) and these code vectors. Each output of the impulse generation module characterizes a vector code. An impulse is then generated on the output associated with the code vector which is closest to the input signal.

2.3   Classifier

The STAN (Spatio-Temporal Artificial Neuron) works in the complex domain. An impulse sequence is converted into a vector X with complex values in the following way (Fig. 2).

The impulse of amplitude η_1 emitted at time t_1 on component j is coded at current time t by the complex number:

$$x_j(t) = \eta_1 \, exp[-\mu_S \tau_1] \, exp[-i \arctan \mu_T \tau_1] \tag{1}$$

with $i = \sqrt{-1}$, $\tau_1 = t - t_1$ and $\mu_S = \mu_T = \frac{1}{TW}$

TW depends on the application and represents the size of the temporal window inside which impulse sequence must be identified. When a new impulse η_2 is emitted at time t_2 on the same component, it is accumulated in the component j of the vector X:

$$\begin{aligned} x_j(t_2) &= \eta_1 \, exp[-\mu_S(t_2 - t_1)] \, exp[-i \arctan \mu_T(t_2 - t_1)] + \eta_2 \\ &= \rho e^{i\phi} \end{aligned} \tag{2}$$

and later:

$$x_j(t) = \rho \, exp[-\mu_S(t - t_2)] \, exp[-i \arctan (tan\phi + \mu_T(t - t_2))] \tag{3}$$

Each component of the vector X is thus reactualized as soon as an impulse is presented to the input. The comparison between X and the weight W of the neuron itself characterized by a complex vector is done here by the means of a Hermitian distance D:

$$D(X, W) = \sqrt{\sum_{j=1}^{N}(x_j - w_j)\overline{(x_j - w_j)}} \tag{4}$$

knowing that $\overline{x}$ is the complex conjugate of x.

Learning Phase. There are as many neurons in the output layer as words to be recognized. Each neuron is characterized by a weight vector. The training is done in one step only. It consists in presenting as input the audio-video sequence corresponding to the word which we wish to recognize. An impulse sequence is then generated and converted into a complex vector X (see Fig. 2) which characterizes the presented word. We carry out this procedure on the whole dictionary in order to evaluate each vector having to characterize each word of the dictionary. These vectors define each weight vector W of the STAN's used during the classification.

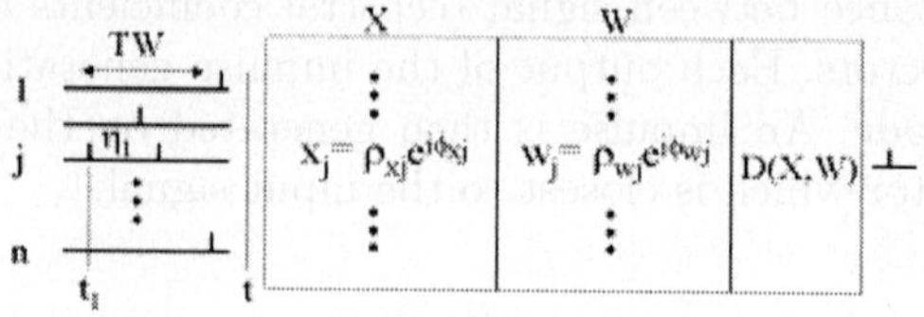

Fig. 2. The Spatio-Temporal Artificial Neuron (STAN)

Testing Phase. When an unknown audio-video sequence is presented at the input, it is translated in the form of an impulse sequence, converted into a complex vector and compared to each weight by the means of the Hermitian distance. The neuron producing the minimal distance then emits an impulse at the output: it signals the recognized word.

3   Results and Conclusion

3.1   Tests

M2VTS. Within the framework of separated word recognition, we tested our system on the first ten persons of the European Data Base M2VTS [9] (Multi Modal Checking for Teleservices and Security applications). This base is dedicated to Audio-visual recognition and identification. Each person pronounces four times (at one week interval) the digits from 0 to 9. We chose this base because it characterizes well the conditions of use in which the real time implementation of our system will have to function. The images were acquired at 25Hz with a weak resolution (288x360pixels in 4:2:2), the sound was sampled at 48kHz on 16 bits. Some people smile sometimes during acquisition, which considerably harms the performance of lipsreading.

System Parameters. We use a face detector [12] in order to locate and follow the face during the sequence. An evaluation of the motion inside the face enables us to locate the mouth rather precisely as shown in Figure 3. In the interior of the mouth, we delimit a zone of 11 pixels height in which the percentages of dark and light pixels are evaluated. With regard to the parameter setting of the STAN's, we used the same value of TW (ten units which correspond to an 400ms observation window). Thirty vectors codes were extracted from the sound, eighty from the video.

Results. Let us recall that we tested our system in the framework of separated word monospeaker recognition. For each person, we have four sequences during which the digits from 0 to 9 are pronounced. Four evaluations were thus carried out according to the number of the sequence used for the training, tests being performed on the three other sequences.

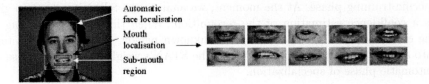

Fig. 3. Automatic Face and mouth localisation.

As one can notice on Figure 4 the performances of the combined audio-visual system (86%) are 10% higher than that of the audio system alone (76%) although the recognition capacities of the video system are definitely lower (61%). But it is in the presence of noise (white noise added to the sound signal) that the audio-visual recognition system takes all its interest. For a signal to noise ratio of 10dB for example, the performances of the combined audio-visual system (68%) are better by almost 20% compared with that of the audio system alone (49%).

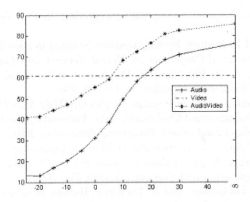

Fig. 4. Pourcentage of correct classification versus Signal to Noise ratio

3.2   Discussion

This work shows that a one pass learning system can extract sufficient information from a learning base to carry out a rather relevant monospeaker speech recognition.

The following stage will consist in conceiving a system recognition of same type as that proposed by [6] who makes a training on the whole (but one) of the people present in M2VTS Data Base and tests on the unused person.

Our final objective is to conceive an "unknown speaker" recognition system which could specialize itself on a particular person, without having to define a specific training phase. At the moment, we analyse the STAN's output to give us a confidence estimation of the recognition. When this output is strong in the case of a word pronounced by an unknown person, we would like to take into account the input signal to modify the STAN's weights and thus realise an automatic phase of specialization.

References

1. P. de Cuetos, N. Chalapathy, and W. Andrew. Audio-visual intent-to-speak detection for human-computer interaction. In *ICASSP*, 2000.
2. P. Delmas, P.Y. Coulon, and V. Fristot. Automatic snakes for robust lip boudaries extraction. In *ICASSP*, 1999.
3. S. Dupont and J. Luettin. Audio-visual Speech modeling for continuous speech recognition. *IEEE Transactions on multimedia*, 2000.
4. S. Durand and F. Alexandre. Learning Speech as acoustic sequences with the unsupervised model, tom. In *NEURAP, 8th International Conference on Neural Networks and their Applications*, 1995.
5. A. Gersho and R. M. Gray. *Vector Quantization and Signal Compression*. Kluwer Acad. Pub., 1991.
6. J. Luettin. Visual Speech and speaker recognition. In *PhD Dissertation, Univ. of Sheffield*, 1997.
7. D. Mercier and R. Séguier. Spiking neurons (stanns) in speech recognition. In *3rd WSEAS International Conference on Neural Network and Applications*, Feb 2002.
8. N. Mozayyani, A. R. Baig, and G. Vaucher. A fully neural solution for on-line handwritten Character recognition. In *IJCNN*, 1998.
9. S. Pigeon. M2vts. In *www.tele.zacl.ac.be/PROJECTS/ M2VTS/m2fdb.html*, 1996.
10. Gerasimos Potamianos and Chalapathy Neti. Automatic speechreading of impaired Speech. In *Audio-Visual Speech Processing*, September 2001.
11. R. Séguier, N. Cladel, C. Foucher, and D. Mercier. Lipreading with spiking neurons: One pass learning. In *International Conference in Central Europe on Computer Graphits, Visualization and Computer Vision*, Feb 2002.
12. R. Séguier, A. LeGlaunec, and B. Loriferne. Human faces detection and tracking in video sequence. In *Proc. 7th Portuguese Conf. on Pattern Recognition*, 1995.
13. R. Séguier and David Mercier. A generic pretreatment for spiking-neuron. Application on lipreading with stann (spatio-temporal artificial neural networks). In *International Conference on Artificial Neural Networks and Genetic Algorithms*, 2001.
14. Y. Tian, T. Panade, and J. F. Cohn. Recognizing action units for facial expression analysis. IEEE Trans. ora Patterra Analysis and Machine Iatelligence, 2001.
15. G. Vaucher. An algebraic interpretation of psp composition. In *BioSystems*, Vol 48, 1998.

An Algorithm for Image Representation as Independent Levels of Resolution

Antonio Turiel[1], Jean-Pierre Nadal[2], and Néstor Parga[3]*

[1] Air Project - INRIA. Domaine de Voluceau BP105
78153 Le Chesnay CEDEX, France
[2] Laboratoire de Physique Statistique, Ecole Normale Supérieure
24 rue Lhomond. 75231 Paris CEDEX 05. France
[3] Departamento de Física Teórica. Universidad Autónoma de Madrid
28049 Cantoblanco, Madrid. Spain

Abstract. Recently it has been shown that natural images possess a special type of scale invariant statistics (multiscaling). In this paper, we will show how the multiscaling properties of images can be used to derive a redundancy-reducing oriented wavelet basis. This kind of representation can be learnt from the data and is optimally adapted for image coding; besides, it shows some features found in the visual pathway.

Keywords. Wavelets, statistical analysis, coding

1 Introduction

In the recent years there has been much work at the boundary between the modelling of visual systems in mammals and computer vision. On one hand a better understanding of natural systems may lead to new image processing algorithms, and on the other hand analysis and modelling of images may help modelling the primary layers in visual systems.

Based on early works of Barlow [1], many works have focussed on the use of information theoretic concepts in order to address the question of efficiency of neural coding. A possible efficiency criterium is maximizing the information transfer. As shown in [2], the code which maximizes information transfer minimizes redundancy, that is, it extracts the independent components of the signal. Several theoretical studies of the primary visual system have been done, based on these ideas of information transfer and redundancy reduction [3], [4]. Any representation should arrive to a compromise between scale and translationnal invariances (that is, a multiscale representation) as they cannot be exactly fulfilled at the same time [5]. Direct statistical analysis of natural images leads also naturally to a multiscale analysis, see [6], [7].

In previous studies, a multifractal analysis of natural images has been performed on a wide variety of ensembles of natural images [8]. It has been shown

---

* To whom correspondence should be addressed

J.R. Dorronsoro (Ed.): ICANN 2002, LNCS 2415, pp. 1213–1218, 2002.

that an optimal wavelet [9] can be constructed (learned) from a set of images. It is however necessary to introduce oriented wavelets (with exactly two orientations) to provide a complete representation [10]. The representation so obtained achieves both whitening and edge detection. More importantly, the dyadic expansion on this wavelet splits the image in statistically independent components, one per level of resolution. This representation has thus several important features shared by the neural representation in mammals.

2   Dyadic Wavelet Bases

We will represent a particular image by its intensity $I(x)$ (i.e., graylevel in digitized images) at every point x in the screen. For normalization convenience, we will work over the contrast $c(x) \equiv I(x) - I_0$ where the normalization constant I_0 is chosen such that the average of c over the screen vanishes.

The starting point of our approach is the so called dyadic wavelet expansion. In this type of expansion, the signal is represented in successive levels of detail, from the coarsest (large scales) to the finest (small scales), which are obtained by resizing and translating a family of functions (wavelets) $\{\phi^r\}_r$. It is called "dyadic" because from one level of resolution to the following, the scale is divided by a factor two. The largest scale is fixed as the unity, $1 = 2^0$, and then the j-th scale is 2^{-j}. Assuming that the scale is of the same order as the dispersion of the wavelet, it is possible to distinguish up to 2^j different blocks along each dimension (2^{2j} blocks in our case, as the images are bi-dimensional). Then, a dyadic wavelet expansion for $c(x)$ corresponds to the following mathematical expression:

$$c(x) = \sum_{r=0}^{N-1} \sum_{j=0}^{\infty} \sum_{k \in (Z_{2j})^2} \alpha_{jk}^r \, \phi_{jk}^r(x) \tag{1}$$

where $\phi_{jk}^r(x) \equiv 2^j \phi^r(2^j x - k)$. For any particular family of wavelets $\{\phi^r\}_r$ for which the dyadic representation eq. (1) holds, the coefficients α_{jk}^r must satisfy the following statistical relation, known as "multiplicative cascade relation" [7]:

$$\alpha_{jk}^r \doteq \eta_{jk}^r \, \alpha_{j-1\left[\frac{k}{2}\right]}^r \tag{2}$$

where "$\doteq$" means that both sides have the same distribution, but they are not necesarily equal for any image and location j, k. The variables η_{jk}^r are independent from the $\alpha_{j-1\left[\frac{k}{2}\right]}^r$ and have the same distribution for all the wavelet indices r, resolution levels j and spatial locations k. This statistical property is a consequence of the fact that natural images are multiscaling signals [7].

3   The Optimal Wavelet Family

It is possible to determine an optimal family of wavelets under some hypothesis. The optimal family is determined by requiring the statistical equality in eq. (2)

to hold *point-by-point*, that is, the equality is true for any image, resolution and location. This is a very strong statement that will allow us to compute Ψ, which is a linear combination of the elements in the family. Under additional constraints on the rotational character of the family, the whole family can be calculated. It also allow us to extract the coefficients η_{jk}^r from the wavelet coefficients α_{jk}

$$\eta_{jk}^r = \alpha_{jk}^r / \alpha_{j-1[\frac{k}{2}]}^r \tag{3}$$

These variables η_{jk}^r provide a representation in which each level of resolution is independent of the others: η_{jk}^r and $\eta_{j'k'}^r$ are statistically independent for $j \neq j'$ [9]. It is possible then to propose a neural architecture able to extract the independent components of images (see figure 1).

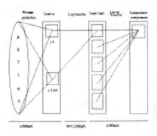

Fig. 1. A possible neural architecture to extract independent features using wavelet projection. The image is projected over ϕ_{jk}^r to produce layer α, that is, the activity of each unit is proportional to the corresponding wavelet coefficient α_{jk}^r. Then, a logarithmic transformation is applied to produce layer $\log \alpha$. Finally, those activities are linearly combined to produce the independent components $\log \eta_{jk}^r = \log \alpha_{jk}^r - \log \alpha_{j-1,[\frac{k}{2}]}^r$

The optimal wavelet family can be obtained from a set of images by means of an average of the contrast over all of them, $C(\boldsymbol{x})$. That function $C(\boldsymbol{x})$ does not give direct access to the family, but to an average wavelet Ψ, according to the following relation [9]:

$$\Psi(\boldsymbol{x}) = \frac{1}{\mathcal{N}} \left[C(\boldsymbol{x}) - \frac{1}{2} \sum_{l_1,l_2=0,1} C(2\boldsymbol{x} - \boldsymbol{l}) \right] \tag{4}$$

where $\mathcal{N}$ is a normalization constant which is obtained by requiring that $\int d\boldsymbol{x} \; \Psi^2(\boldsymbol{x}) = 1$. In the case of having a one-wavelet family, Ψ equals the mother wavelet ϕ^0. In the more general case, as it is shown in [10], the general expression is:

$$\Psi(\boldsymbol{x}) = \sum_{r=0}^{N-1} p_r \; \phi^r(\boldsymbol{x}) \tag{5}$$

for some unknown weights p_r. Assuming that the different wavelets ϕ^r are rotated versions of the same function $\phi \equiv \phi^0$, it can be proven (see [10] for details) that a possible solution of eq. (5) is $p_r \approx \delta_{r0}$, that is, $\phi \approx \Psi$. This solution holds for different values of N, the number of different orientations. So, eq. (4) can be used to obtain the non-rotated wavelet $\phi \approx \Psi$ and to extract the whole family from it. In practice we will only consider the case $N = 2$, which was shown to be very close to an orthonormal basis.

4   Properties of the Optimal Wavelet

By construction, the wavelet family splits images in independent levels of resolution (the variables η_{jk}^r). It can be learnt on-line by simple addition of new images in the averages $C(x)$ appearing in eq. (4), which in turn is equivalent to average the learnt wavelets. The experimental wavelet ϕ is shown in Figure 2; it was learnt over an ensemble of 1000 images from H. van Hateren's web database (see [11] for details on their callibration).

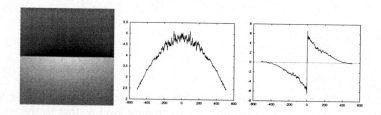

Fig. 2. Left: Gray level representation of the optimal wavelet ϕ (white: positive values, black: negative values); Middle: Horizontal cut; Right: vertical cut

The experimentally obtained ϕ defines an almost ortoghonal pyramid, that is, the autoprojections $\langle \phi | \phi_{jk} \rangle$ are negligible for $j \neq 0$, $\forall k$. The values of those autoprojections are very small, about 1%, except those for $j = 1$, which are about 10%. We think that the projections would become smaller using larger training ensembles. For $N = 2$ rotations, the two pyramids are mutually orthogonal, $\langle \phi_{jk}^r | \phi_{j'k'}^{r'} \rangle \approx 0$ if $r \neq r'$ (error less than 1% in any instance, [10]). Hence, the basis with two oriented wavelets (horizontal and vertical) acts as an orthonomal basis. The coefficients α_{jk}^r of any image c are then easily extracted by simple projection, $\alpha_{jk}^r = \langle \phi_{jk}^r | c \rangle$

5   Representation

Figure 3 shows the reconstruction at different levels of resolution of three example images. The coefficients α_{jk}^r were extracted (assuming orthonormality) just projecting on ϕ_{jk}^r. This introduces a significant error (the PSNRs for the reconstructions are 22.02 dB, 25.43 dB and 22.63 dB), although the main features are well described. From the figure it can be observed that the image is regenerated by the succesive addition of horizontal and vertical small lines. So, the wavelet representation codes the image as edges and contours, and the wavelet coefficients (which spawn the independent resolution levels) measure the relative illumination of such edges.

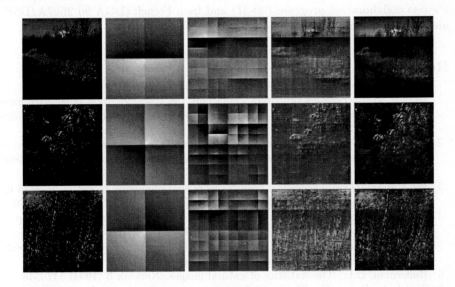

Fig. 3. *From left to right:* Original image and $\sum_{rjk} \alpha_{jk}^r \Psi_{jk}^r(x)$ for $j = 0$, $j \leq 2$, $j \leq 6$ and $j \leq 8$ with $n = 2$ orientations, for imk00480.imc (top), imk02000.imc (middle) and imk03236.imc (bottom)

6   Discussion

In this work we have seen that natural images possess an important property: they can be canonically represented in an optimal dyadic wavelet basis at two orientations. We have seen that such a basis can be expanded from a wavelet that can be calculated given a large enough learning set of images. The advantage of this particular representation lies in the fact that it naturally splits images in independent levels of resolution.

The proposed description of images as entities formed by independent levels of resolution has two main implications. First, it gives a justification of several features in the human visual system (learning capability, power spectrum extraction, edge detection), which are also present in this wavelet expansion. Our wavelet expansion could then be used to emulate a biological coding network. Second, the independency of the resolution levels is very useful to represent images compactly (independency reduces coding cost with respect to other wavelet based codings, as the one presented in [12]). We think that the approach proposed here can be useful both to understand early areas in the visual cortex and to design good coding algorithms.

Acknowledgements. A. Turiel is financially supported by a post-doctoral fellowship from INRIA. This work has been partially funded by the French-Spanish Picasso collaboration program (00-37) and by a French (DGA 96 2557A/DSP) and a Spanish grant (BMF2000-0011).

References

1. Barlow, H.B.: Possible principles underlying the transformation of sensory messages. In Rosenblith, W., ed.: Sensory Communication. M.I.T. Press, Cambridge MA (1961) 217
2. Nadal, J.P., Parga, N.: Nonlinear neurons in the low-noise limit: a factorial code maximizes information transfer. Network: Computation in Neural Systems **5** (1994) 565–581
3. van Hateren, J.H.: Theoretical predictions of spatiotemporal receptive fields of fly lmcs, and experimental validation. J. Comp. Physiology A **171** (1992) 157–170
4. Atick, J.J.: Could information theory provide an ecological theory of sensory processing? Network: Comput. Neural Syst. **3** (1992) 213–251
5. Li, X.G., Somogyi, P., Ylinen, A., Buzsáki, G.: The hippocampal CA3 network: an in vivo intracellular labelling study. J. Comp. Neurol. **339** (1994) 181–208
6. Ruderman, D.L.: The statistics of natural images. Network **5** (1994) 517–548
7. Turiel, A., Parga, N.: The multi-fractal structure of contrast changes in natural images: from sharp edges to softer textures. Neural Computation **12** (2000) 763–793
8. Nevado, A., Turiel, A., Parga, N.: Scene dependence of the non-gaussian scaling properties of natural images. Network **11** (2000) 131–152
9. Turiel, A., Parga, N.: Multifractal wavelet filter of natural images. Physical Review Letters **85** (2000) 3325–3328
10. A. Turiel, J.P.N., Parga, N.: Orientational minimal redundancy wavelets: from edge detection to perception. Submitted to Vision Research (2002)
11. van Hateren, J.H., van der Schaaf, A.: Independent component filters of natural images compared with simple cells in primary visual cortex. Proc. R. Soc. Lond. **B265** (1998) 359–366
12. Buccigrossi, R.W., Simoncelli, E.P.: Image compression via joint statistical characterization in the wavelet domain. IEEE Transactions in Image Processing **8** (1999) 1688–1701

Circular Back-Propagation Networks for Measuring Displayed Image Quality

Paolo Gastaldo[1], Rodolfo Zunino[1], Ingrid Heynderickx[2], and Elena Vicario[2]

[1] DIBE - University of Genoa - Italy
{gastaldo, zunino}@dibe.unige.it
[2] Philips Research Laboratories – Eindhoven – NL
ingrid.heynderickx@philips.com

Abstract. A system based on a neural-network estimates the perceived quality of digital pictures that had previously undergone image-enhancement algorithms. The objective system exploits the ability of feed-forward networks to handle multidimensional data with non-linear relationships. A Circular Back-Propagation network maps feature vectors into the associated quality ratings, thus estimating perceived quality. Feature vectors characterize the image at a global level by exploiting statistical properties of objective features, which are extracted on a block-by-block basis. A feature-selection procedure based on statistical analysis drives the composition of the objective metric set. Experimental results confirm the approach effectiveness, as the system provides a satisfactory approximation of subjective tests involving human voters.

1 Introduction

The improvement of displayed quality is a crucial issue for advanced digital-image enhancement algorithms. As such, reliable methods that can assess perceived quality are required. Subjective testing is the conventional approach to image quality evaluation [1]: the perceived quality is measured by asking human assessors to score the overall quality of a set of test images. Although subjective methods yield accurate results, they are very difficult to model in a reliable, deterministic way. On the other hand, objective methods [2] aim to estimate perceived quality bypassing human assessors. These techniques process numerical quantities ("objective features") extracted from images; consistency and effectiveness require them to be coherent with subjective opinions of quality perception.

This work applies neural networks (NN) to digital-image quality estimation. A Circular Back-Propagation (CBP) feedforward network [3] processes objective features worked out from each processed image, and returns the associated quality score. The overall goal is to develop a method to mimic quality perception mechanisms; to this end, the present approach exploits the ability of CBP networks to support a general paradigm, which can deal with complex mathematical models.

J.R. Dorronsoro (Ed.): ICANN 2002, LNCS 2415, pp. 1219–1224, 2002.

2 CBP Architecture

In the problem formulation adopted in this research, the neural network maps the feature vector characterizing images into the associated quality assessments. During the subjective tests, quality scores were represented by discrete values in the range [−5, +5]; thus, NN outputs are prompted as scalar quantities in that interval.

MultiLayer Perceptrons (MLPs) can efficiently tackle target-mapping problems; in MLPs, few units with global scope are encoded by the sigmoid functions within hidden units. Conversely, Radial-Basis Function (RBF) networks typically perform more efficiently if the mapping is best expressed as a superposition of locally-tuned components. Hence, the unknown characteristics of the function that maps features into scores further complicate the problem of configuring the proper NN.

Previous research showed that CBP networks [3] benefit from the advantages of both MLP and RBF paradigms, as the empirical training process drives the network to the best configuration for the problem at hand. The CBP model can be described as it follows. The input layer connects the n_i input values x_k (features) to each neuron of the "hidden layer". The u-th "hidden" neuron computes a weighted combination of input values, with coefficients ($w_{u,k}$; $u=1,\ldots,n_h$; $k=1,\ldots,n_i$):

$$r_u = w_{u,0} + \sum_{k=1}^{n_i} w_{u,k} x_k + w_{u,n_i+1} \sum_{k=1}^{n_i} x_k^2 . \tag{1}$$

Each hidden neuron finally performs a non-linear transformation of the result:

$$a_u = \sigma(r_u) ; \qquad u = 1,\ldots, n_h; \tag{2}$$

where $\sigma(x) = (1 + e^{-x})^{-1}$, r_u is the *stimulus* and a_u is the *activation*. Likewise, the *output* layer provides the actual network responses, y_v, ($v = 1,\ldots, n_o$):

$$r_v = w_{v,0} + \sum_{u=1}^{n_h} w_{v,u} a_u ; \qquad y_v = \sigma(r_v) . \tag{3}$$

The quadratic term in expression (1) sets the difference between CBP and conventional MLPs. Such augmentation is attained by one additional input, which just sums the squared values of all other inputs. The additional unit allows the overall network to adopt the standard sigmoidal behaviour, or to drift smoothly to a bell-shaped radial function approximating a Gaussian. In addition to such flexibility, conventional Back-Propagation can drive the CBP training process, thus ensuring computational efficiency. At the same time, theory proves [4] that the limited increase in the network parameters does not affect its expected generalization performance.

3 Objective Quality Assessment by Using CBP

The proposed system addresses the objective assessment of perceived quality when an enhancement algorithm is applied on a digital image (Fig. 1). The NN operates on

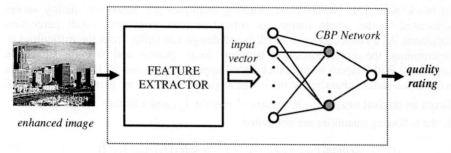

Fig. 1. The objective assessment system

objective measures worked out from each enhanced image, and directly yields the quality assessments associated with input features. The function mapping feature vectors into quality ratings is learned empirically from examples.

The effectiveness of the NN approach lies in the ability to decouple the problem of feature-selection from the design of an explicit mathematical model. Therefore, the design of the objective metric set is not involved in the set-up of the mapping function. The ability of CBP networks to render multidimensional data characterized by complex relationships is crucial to the overall approach effectiveness.

3.1   Objective Features

The present model estimates the image quality by extracting objective measures on a block-by-block basis. Thus, the picture is divided in non-overlapping blocks (of 32x32 pixel) using either a fixed sampling grid or a random sampling procedure.

The objective metric set characterizing each block is the result of a feature-selection criterion based on statistical analysis. First, a quite large set of features characterizing an image has been selected (see the Appendix). Then, the feature-selection algorithm presented in [6] has been applied. The algorithm uses skewness and kurtosis as a paradigm to characterize the statistical activity of the features. The ultimate goal is to sort out a subset of statistically relevant features. Given a library of test images $I = \{I_n; n = 1,..,n_p\}$ the feature-selection algorithm [4] processes the sets Φ_k defined as follows:

$$\Phi_k = \left\{f_{k11},..., f_{k1q},..., f_{kn_p1},..., f_{kn_pq}\right\} \quad k = 1,..,n_f , \qquad (4)$$

where f_{knb} is the value of the feature f_k for the block b of the image n, n_f is the number of features, and q is the number of blocks extracted for each test image. The algorithm yields the subset Z of $n_s (<=n_f)$ features that are classified as statistically relevant.

3.2   Input Vector Set-Up

The present approach models the image as a set of non-overlapping blocks. In principle one can design the neural network to estimate perceived quality on a block-by-block basis [5]. Nevertheless, training examples are based on quality scores associated to the whole image, as subjective tests yield an overall perception judgment. As a consequence, a block-based design can suffer from the difficulties in determining the correspondence between local quality and overall subjective judgment. With respect to the block-based approach proposed in [5], in the present work input vectors characterize the image at global level and are generated as follows.

Given an original image I_n, its enhanced version $\tilde{I}_n$, and a feature $f_k \in Z$,

1. the following quantities are computed:

$$f_k^M = median\left(f_{k\tilde{I}1},...,f_{k\tilde{I}q}\right), \quad f_k^S = stdev\left(f_{k\tilde{I}1},...,f_{k\tilde{I}q}\right); \tag{5}$$

where q is the number of blocks extracted from $\tilde{I}_n$;

2. the input vector $\vec{x}_{\tilde{I}}$ for the image $\tilde{I}$ is composed as:

$$\vec{x}_{\tilde{I}} = \left\{f_k^M, f_k^S; k = 1,..., n_S\right\}. \tag{6}$$

3.3   Neural Network Set-Up

In the present approach, CBP networks map feature vectors into quality ratings. The mapping function is learned from examples by means of an accelerated variant of the classical back-propagation algorithm, and a single output neuron in the NN yields the quality assessment for a given input vector. The network cost is expressed as:

$$E = \frac{1}{n_o n_p} \sum_{l=1}^{n_p} \sum_{v=1}^{n_o} \left(t_v^{(l)} - y_v^{(l)}\right)^2, \tag{7}$$

where n_p is the number of training patterns, and t_v are the desired training outputs. In the present application, $v=1$ and the expected output is given by the quality assessment (score) measured experimentally from a human panel.

4   Experimental Results

The quality assessment system has been tested by using an image enhancement algorithm provided by Philips. The algorithm assumes that high detailed regions having luminance variation below a certain threshold has to be regarded as noise; thus, pixels that belong to a region containing perceivable details are enhanced as follows:

$$L = (L - L_m)\gamma + L_m, \tag{8}$$

where L_m is the mean luminance over the 5x5 pixel region surrounding the pixel with luminance L, and γ can take two different values (γ_1 and γ_2) as a function of three parameters: an estimator for high spatial frequencies in the region around the pixel to be processed (σ_T), and two thresholds on the Weber curve [7] (θ_L and θ_H). In the present experiment, γ_1, σ_T and θ_H were varied, whereas γ_2 and θ_L were set to default values.

The algorithm processed a library of sixteen original grey level images (252x189 pixel). Image contents included both natural and texture-like patterns. Eventually, thirty enhanced versions for each original image were generated. Subjective data have been collected by using a double-stimulus method; human assessors were asked to score the quality of the processed image against the original one using a 11-points numerical scale ranging between –5 (i.e., the processed image is much worse than the original one) and +5 (i.e., the processed image is much better than the original one). This scale has been subsequently re-scaled in the interval [-1,1]. Eighty subjects participated in the experiment.

4.1   Test Results

The neural-network training process involved the objective metric set selected using the procedure describe in sec. 3. The eventual feature space covered the quantities *absolute value*, *contrast*, *difference variance*, *difference entropy*, extracted from the co-occurrence matrix [5] with $\theta=0$. The training and test set were obtained by dividing the data into two subsets of 360 and 180 patterns, respectively. The number of nodes in the hidden layer ($n_h=10$) was designed by use of a specific initialization technique that exploits the equivalence of the CBP model to Vector-Quantization (VQ) paradigms [4].

Fig. 2(a) shows a scatter plot of the test results with the actual subjective score as the x-axis and the correspondent NN quality estimation on the y-axis. Pearson's correlation coefficient for the test results takes the value 0.9 and confirms the effectiveness of the approach. The CBP network achieves an average prediction error $\hat{\mu}_{err} = -0.01$ over the test set. Fig. 2(b) shows the Q-Q plot comparing the prediction

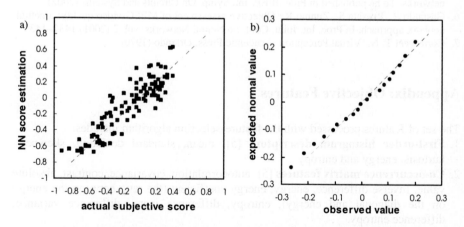

Fig. 2. Test results. Fig. 2(a) shows the correlation between actual subjective score and NN outputs. Fig. 2(b) presents the Q-Q plot that compares the quantiles of the Gaussian distribution $N(0; 0.14)$ (the dashed line) with the quantiles of the error distribution obtained for the test set

error versus the related best-fitting Gaussian approximation $N(0; 0.14)$; as the points almost fall along the dashed line representing the ideal Gaussian distribution, it can be asserted that the prediction error follows a normal distribution.

5   Conclusions

The paper presented an automated system for objective assessment of image quality. Objective features are extracted from the image and processed by a CBP network that yields as output the correspondent quality estimate. The objective metric is chosen through a feature-selection procedure based on statistical analysis, thus avoiding a priori hypotheses on the significance of the features.

The major result of the proposed model is the possibility of assessing human perception consistently by using a Circular Back-Propagation NN. Experimental evidence confirmed that the system provides a satisfactory approximation of the actual subjective scores.

References

1. Engeldrum P.: Psychometric Scaling: a Toolkit for Imaging Systems Development. Imcotek Press, Winchester (2000)
2. Ahumada A.: Computational image quality metrics: a review. SID Digest, vol. 24 (1993) 305–308
3. Ridella S., Rovetta S., Zunino R.: Circular back-propagation networks for classification. IEEE Trans. on Neural Networks, vol. 8, no. 1 (1997) 84–97
4. Ridella S., Rovetta S., Zunino R.: Circular Backpropagation networks embed Vector Quantization. IEEE Trans. on Neural Networks, vol. 10, no. 4 (1999) 972–975
5. Carrai P., Gastaldo P., Heynderickx I., Zunino R.: Image quality assessment by using neural networks. To be published in Proc. IEEE. Int. Symp. On Circuits and Systems (2002)
6. Gastaldo P., Rovetta S., Zunino R.: Objective assessment of MPEG-video quality: a neural-network approach. In Proc. Int. Joint. Conf. on Neural Networks, vol. 2 (2001) 1432–1437
7. Cornsweet T. N.: Visual Perception. Academic Press, Orlando (1970)

Appendix: Objective Features

The set of features processed with the feature-selection algorithm includes:
1. **First-order histogram descriptors** [5]: mean, standard deviation, skewness, kurtosis, energy and entropy.
2. **Co-occurrence matrix features** [5]: autocorrelation, covariance, contrast, absolute value, inverse difference, energy, energy coefficient (the ratio between the energy on the diagonal and energy), entropy, difference mean, difference variance, difference entropy.

Co-occurrence-matrix features were computed for the four principal directions ($\theta=0$, 45, 90, 135 degree) and with radius $\rho=2$.

Unsupervised Learning of Combination Features for Hierarchical Recognition Models

Heiko Wersing and Edgar Körner

Honda R&D Europe (Deutschland) GmbH
Carl-Legien-Str.30, 63073 Offenbach/Main, Germany
{heiko.wersing,edgar.koerner}@hre-ftr.f.rd.honda.co.jp

Abstract. We propose a cortically inspired hierarchical feedforward model for recognition and investigate a new method for learning optimal combination-coding cells in intermediate stages of the hierarchical network. The model architecture is characterized by weight-sharing, pooling, and Winner-Take-All nonlinearities. We show that an unsupervised sparse coding learning rule can be used to obtain a recognition architecture that is competitive with other more formally abstracted recognition approaches based on supervised learning. We evaluate the performance on object and face databases.

1 Introduction

The concept of convergent hierarchical coding assumes that sensory processing in the brain is organized in hierarchical stages of neural representations capturing increasingly complex feature combinations [1]. This concept has been criticized as leading to a combinatorial explosion of representative feature combinations under changing object views. Therefore, approaches were formulated to avoid this problem by combining hierarchical feature detection with pooling to achieve gradual invariance of response under transformations of the stimulus [2,8,11]. There is substantial experimental evidence in favor of the notion of hierarchical processing in the visual cortex, where an increase in receptive field size and stimulus complexity from initial to later processing stages can be observed [11]. Since rapid feedforward processing in recognition tasks has been shown experimentally [13], Körner et al. [5] proposed a bidirectional model for cortical processing, where an initial hypothesis on the stimulus is facilitated through a feed-forward latency encoding in relation to an oscillatory reference frame.

Methods of supervised feature optimisation which require class information were proposed such as greedy search [8] or gradient-based adaptation [7]. Nevertheless, especially unsupervised learning of features in higher hierarchical stages is still an issue of major interest. Redundancy reduction was proposed as an efficient hierarchical coding strategy [1] and has been applied to model the wavelet-like receptive fields of V1 cells by imposing sparse overcomplete representations [10]. Recently this principle and related concepts of independent component analysis were also applied to models of complex cells and higher-level contour

J.R. Dorronsoro (Ed.): ICANN 2002, LNCS 2415, pp. 1225–1230, 2002.

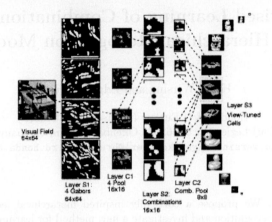

Fig. 1. Sketch of the hierarchical network. The input image is presented as a 64×64 pixel image. The S1 layer consists of 4 Gabor feature planes at 4 orientations with a dimension of 64×64 each. The C1 layer subsamples by pooling down to a resolution of 16×16 for each of the 4 S1 planes. The S2 layer contains combination coding cells with possible local connections to all of the C1 cells. The C2 layer pools the S2 planes down to a resolution of 8 × 8. The final S3 cells are tuned to particular views, which are represented as the activity pattern of the C2 planes for an input image.

coding combination cells (see [4] and references therein). In this contribution we show that a nonnegative sparse coding learning rule [4] can also be used to obtain optimized combination features in a recognition hierarchy. The resulting recognition architecture, with weights entirely based on unsupervised learning, is competitive with other recognition approaches trained by supervised learning. In Section 2 we describe our model setup with its feedforward nonlinearities and learning methods. The results on different recognition benchmarks are given in Section 3 and discussed in the concluding Section 4.

2   A Hierarchical Model of Invariant Recognition

Architecture. Our hierarchical model is based on a feedforward architecture with weight-sharing and a succession of feature-sensitive and pooling stages (see Fig. 1). The first feature-matching stage consists of an initial linear sign-insensitive receptive field summation, a Winner-Take-All mechanism between features at the same position and a final threshold function. We adopt the notation, that vector indices run over the set of neurons within a particular feature plane of a particular layer. To compute the response $q_1^l(x,y)$ of a simple cell in the first layer S1, responsive to feature type l at position (x,y) , first the image vector $\mathbf{I}$ is multiplied with a weight vector $\mathbf{w}_1^l(x,y)$ (e.g. Gabor filter) characterizing the receptive field profile:

$$q_1^l(x,y) = |\mathbf{w}_1^l(x,y) * \mathbf{I}|. \tag{1}$$

The inner product is denoted by $*$, i.e. for a 10×10 pixel image $\mathbf{I}$ and $\mathbf{w}_1^l(x, y)$ are 100-dimensional vectors. The weights $\mathbf{w}_1^l$ are normalized and characterize a localized receptive field in the visual field input layer. All cells in a feature plane l have the same receptive field structure, given by $\mathbf{w}_1^l(x, y)$, but shifted receptive field centers, like in a classical weight-sharing or convolutional architecture [2, 7]. In a second step a soft Winner-Take-All mechanism is performed with

$$r_1^l(x, y) = \begin{cases} 0 & \text{if } \frac{q_1^l(x,y)}{M} < \gamma_1 \text{ or } M = 0, \\ \frac{q_1^l(x,y) - M\gamma_1}{1 - \gamma_1} & \text{else,} \end{cases} \tag{2}$$

where $M = \max_k q_1^k(x, y)$ and $r_1^l(x, y)$ is the response after the WTA mechanism which suppresses sub-maximal responses and provides a model of latency-based competition [5]. The parameter $0 < \gamma_1 < 1$ controls the strength of the competition. The activity is then passed through a simple threshold function with a common threshold θ_1 for all cells in layer S1:

$$s_1^l(x, y) = H\big(r_1^l(x, y) - \theta_1\big), \tag{3}$$

where $H(x) = 1$ if $x \geq 0$ and $H(x) = 0$ else and $s_1^l(x, y)$ is the final activity of the neuron sensitive to feature l at position (x, y) in the S1 layer. The activities of the first layer of pooling C1-cells are given by

$$c_1^l(x, y) = \tanh\big(\mathbf{g}_1(x, y) * \mathbf{s}_1^l\big), \tag{4}$$

where $\mathbf{g}_1(x, y)$ is a normalized Gaussian pooling kernel with width σ_1, identical for all features l, and tanh is the hyperbolic tangent function. The features in the intermediate layer S2 are sensitive to local combinations of the features in the planes of the previous layer, and are thus called *combination cells* in the following. We introduce the layer activation vectors as $\bar{\mathbf{c}}_1 = (\mathbf{c}_1^1, \ldots, \mathbf{c}_1^K)$, $\bar{\mathbf{w}}_2^l = (\mathbf{w}_2^{l1}, \ldots, \mathbf{w}_2^{lK})$ with K=4. Here $\mathbf{w}_2^{lk}(x, y)$ is the receptive field vector of the S2 cell of feature l at position (x, y), describing connections to the plane k of the previous C1 cells. The combined linear summation over previous planes is then given by $q_2^l(x, y) = \bar{\mathbf{w}}_2^l(x, y) * \bar{\mathbf{c}}_1$. After the same WTA procedure with strength γ_2 as in (2), the activity in the S2 layer is given by $s_2^l(x, y) = H(r_2^l(x, y) - \theta_2)$ after thresholding with a common threshold θ_2. The step from S2 to C2 is identical to (4) and given by $c_2^l(x, y) = \tanh(\mathbf{g}_2(x, y) * \mathbf{s}_2^l)$, with Gaussian spatial pooling kernel $\mathbf{g}_2(x, y)$ with range σ_2.

Classification of an input image with C2 output $\bar{\mathbf{c}}_2$ is done by nearest neighbor match to previously stored template activations $\bar{\mathbf{c}}_2^v$ for each training view v. This can be realized e.g. by view-tuned units (VTU) [11] in an additional S3 layer with a radial basis function characteristics [11] according to $s_3^v = \exp(-||\bar{\mathbf{w}}_3^v - \mathbf{c}_2||^2)$ where $\bar{\mathbf{w}}_3^v = \mathbf{c}_2^v$ is tuned to the training C2 output of pattern v. Classification can then be performed by detecting the maximally activated VTU.

Parameter adjustment and learning. We adjust the nonlinearity parameters of the visual processing hierarchy in an incremental way. We first choose the processing nonlinearities in the initial layers to provide an optimal output for a nearest neighbor classifier based on the C1 layer activations. We then keep the initial processing layer fixed and use a sparse coding learning rule to obtain optimized combination features.

The receptive fields of the S1 layer were set as first-order Gabor filters of 4 orientations. To adjust the WTA selectivity γ_1, threshold θ_1, and pooling range σ_1 of the initial layers S1 and C1, we considered a nearest neighbor classification setup of the COIL100 images [9] based on the C1 layer activations. For each of the 100 objects there are 72 views available at subsequent rotations of 5°. We take four views at angles 0°, 80°, 160°, and 240° and store the corresponding C1 activation as a template. We can then classify a new test view by finding the template with lowest Euclidean distance in the C1 activation vector. By performing grid-like search over parameters we found an optimal classification performance at $\gamma_1 = 0.9$, $\theta_1 = 0.1$, and $\sigma_1 = 1.5$, giving a correct recognition rate of 72%. This particular parameter setting implies a certain coding strategy: The first layer of simple edge detectors combines a rather low threshold with a strong local competition between orientations. The result is a kind of "segmentation" of the input into one of the four different orientation categories (see also Figure 1). These features are pooled within a range that is comparable to the size of the Gabor S1 receptive fields.

To apply the sparse coding learning rule [4] we generated with the above S1,C1 setting an ensemble of C1 activity vectors for 1000 COIL images and extracted 10000 local 5×5 patches, each with a dimension $5 \times 5 \times 4 = 100$ and indexed by p. The learning rule is defined as the minimization of

$$E = \sum_p ||\bar{\mathbf{c}}_1^{(p)} - \sum_k s_k^{(p)} \bar{\mathbf{w}}_2^k||^2 + \lambda \sum_p \sum_k s_k^{(p)}, \qquad (5)$$

jointly in the combination features $\bar{\mathbf{w}}_2^k$ and coefficients $s_k^{(p)}$, subject to the non-negativity of both the components of $\bar{\mathbf{w}}_2^k$ and the $s_k^{(p)}$. The left part of (5) measures the error of reconstructing the input patch $\bar{\mathbf{c}}_1^{(p)}$ from a set of (nonorthogonal) basis features $\bar{\mathbf{w}}_2^k$, while the right part enforces sparse activation of the coefficients $s_k^{(p)}$. After random initialization of the $\bar{\mathbf{w}}_2^k$, the optimization is performed as a two-stage gradient descent process [10,4]: First the $\bar{\mathbf{w}}_2^k$ are fixed, and a local minimum of (5) is found in $s_k^{(p)}$ for each patch p, using an asynchronous fast fixed-point search as suggested in [14]. In the second step an average gradient step in $\bar{\mathbf{w}}^k$ is performed with $s_k^{(p)}$ set from the first step. Both steps are repeated till convergence. We chose a sparsity factor of $\lambda = 0.1$ and $k = 1, \ldots, 100$ basis features. Based on classification performance on the COIL dataset, nonlinearity parameters were set to $\gamma_2 = 0.0$, $\theta_2 = 1.7$, and $\sigma_2 = 1.0$.

For the setup of an initial quadrature pair nonlinearity without spatial pooling , Hoyer & Hyvärinen [4] obtained only collinear features of different lengths. Contrary to that, our feature set is more diverse and also contains local corner-like and more complex local combinations. This is caused by the different nature of both the input data, man-made objects here compared to texture-rich natural scenes [4], and the different initial processing nonlinearities.

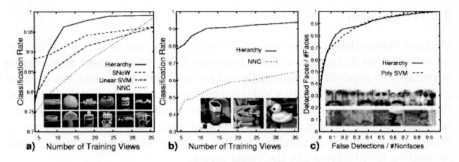

Fig. 2. Comparison of classification rates. a) compares the classification rates on the COIL100 dataset for our hierarchy to results obtained by Roth et al. using their SNoW model, a linear support vector machine, and direct image nearest neighbor classifier (NNC). In a wide regime of sufficient recognition task difficulty (compare NNC), our feature hierarchy achieves best results with high generalization. b) shows performance in a more difficult scenario of 20 objects with 8 pixel-wide position variance and cluttered surround for both training and testing data. The feature hierarchy offers strong robustness compared to NNC. c) shows an ROC plot comparison of face detection performance using a single optimized VTU, but with identical setting for the earlier hierarchical stages. The plot shows the combined rate of correctly identified faces over the rate of misclassifications as a fraction of all non-face images for the hierarchy and a polynomial kernel support vector machine classifier [3].

3   Results

In Figure 2a we compare the classification performance of our model, using C2 activity nearest neighbor matching, to the results published in [12] using the SNoW recognition approach and applying a linear support vector machine on the COIL-100 dataset. To show the application to another classification scenario using the same hierarchy, we used the ORL face image dataset (copyright AT & T Research Labs, Cambridge), which contains 10 images each of 40 people with variability in expression and pose. Without any parameter or feature modification we obtain a classification rate of 96% using 5 training views, compared to 96.5% [6] using gradient-based supervised learning on higher hierarchical stages.

Another central ability for visual recognition is the rejection of unknown stimuli. With an identical setting as described above, however, using a single sigmoidal output VTU with $s_3^1 = \tanh(\bar{\mathbf{w}}_3^1 * \bar{\mathbf{c}}_2)$, we performed gradient-based supervised optimization of $\bar{\mathbf{w}}_3^1$ with target outputs of -0.9 and 0.9 for nonfaces and faces respectively. The training ensemble (data from [3]) consists of 2429 19×19 pixel face images and 4548 non-face images. A threshold criterion was used to decide the presence or non-presence of a face for a different test set of 472 faces and 23573 non-faces. The non-face images consist of a subset of all non-face images that were found to be most difficult to reject for the support vector machine classifier considered by Heisele et al. As is shown in an ROC-plot in Figure 2b, which shows the performance depending on the variation of the detection threshold, the architecture is competitive with a high performance nonlinear SVM classifier [3].

4 Discussion

We have shown that a sparse coding learning rule allows to derive efficient local combination features in a visual hierarchy in an unsupervised way, offering advantages to supervised optimization of features through greedy search [8] or gradient-based adaptation [7], which require class information. For the databases that we considered here, we could show that the resulting representation can also be applied to face recognition and classification with good results. This generalization across domains is a highly desirable property on the way to more general recognition architectures like the visual cortex.

Acknowledgments. We thank T. Poggio, C. Goerick, J. Eggert, T. Rodemann and U. Körner for stimulating discussions and B. Heisele for providing the face image data. This work was in part supported by BMBF grant 01IB001E (LOKI project).

References

1. H. B. Barlow. The twelfth Bartlett memorial lecture: The role of single neurons in the psychology of perception. *Quart. J. Exp. Psychol.*, 37:121–145, 1985.
2. K. Fukushima. Neocognitron: A hierarchical neural network capable of visual pattern recognition. *Neural Networks*, 1:119–130, 1988.
3. B. Heisele, T. Poggio, and M. Pontil. Face detection in still gray images. Technical report, MIT A.I. Memo 1687, 2000.
4. P. O. Hoyer and A. Hyvärinen. A multi-layer sparse coding network learns contour coding from natural images. *Vision Research*, 2002. to appear.
5. E. Körner, M.-O. Gewaltig, U. Körner, A. Richter, and T. Rodemann. A model of computation in neocortical architecture. *Neural Network*, 12(7–8):989–1005, 1999.
6. S. Lawrence, C. L. Giles, A. C. Tsoi, and A. D. Back. Face recognition: A convolutional neural-network approach. *IEEE Trans. Neur. Netw.*, 8(1):98–113, 1997.
7. Y. LeCun, L. Bottorr, Y. Bengio, and P. Haffner. Gradient-based learning applied to document recognition. *Proceedings of the IEEE*, 86:2278–2324, 1998.
8. B. W. Mel and Jozsef Fiser. Minimizing binding errors using learned conjunctive features. *Neural Computation*, 12(4):731–762, 2000.
9. S. K. Nayar, S. A. Nene, and H. Murase. Real-time 100 object recognition system. In *Proc. of ARPA Image Understanding Workshop*, Palm Springs, 1996.
10. B. A. Olshausen and D. J. Field. Sparse coding with an overcomplete basis set: A strategy employed by V1? *Vision Research*, 37:3311–3325, 1997.
11. M. Riesenhuber and T. Poggio. Hierarchical models of Object recognition in cortex. *Nature Neuroscience*, 2(11):1019–1025, 1999.
12. D. Roth, M.-H. Yang, and N. Ahuja. Learning to recognize objects. In *Proc. of the Conf. on Pattern Recognition and Computer Vision*, 2000.
13. S. Thorpe, D. Fize, and C. Marlot. Speed of processing in the visual System. *Nature*, 381:520–522, 1996.
14. H. Wersing, J. J. Steil, and H. Ritter. A competitive layer model for feature binding and sensory segmentation. *Neural Computation*, 13(2):357–387, 2001.

Type of Blur and Blur Parameters Identification Using Neural Network and Its Application to Image Restoration

Igor Aizenberg[1]*, Taras Bregin[1], Constantine Butakoff[1],
Victor Karnaukhov[2], Nickolay Merzlyakov[2], and Olga Milukova[2]

[1]company "Neurotechnologies" (Ukraine)
[2]Institute for Information Transmission Problems of the Russian Academy of
Sciences (Russia)

Abstract. The original solution of the blur and blur parameters identification problem is presented in this paper. A neural network based on multi-valued neurons is used for the blur and blur parameters identification. It is shown that using simple single-layered neural network it is possible to identify the type of the distorting operator. Four types of blur are considered: defocus, rectangular, motion and Gaussian ones. The parameters of the corresponding operator are identified using a similar neural network. After a type of blur and its parameters identification the image can be restored using several kinds of methods.

Keywords. Neural network, image restoration, frequency domain.

1 Introduction

As a rule, blur is a form of bandwidth reduction of an ideal image owing to the imperfect image formation process. It can be caused by relative motion between the camera and the original scene, or by an optical system that is out of focus. Today there are different techniques available for solving of the restoration problem including Fourier domain techniques, regularization methods, recursive and iterative filters [1, 2]. All of the existing techniques are directed to the obtaining of a solution for the deconvolution problem. A problem is that without a good estimation of the blur parameters these filters show poor results. If incorrect blur model is chosen then the image will be rather distorted more than restored. Many of different algorithms for blur identification and identification of its parameters exist today, for example, the maximum likelihood blur estimation or regularization approach [3]. The disadvantage of these algorithms is their computing complexity and relatively high level of the misidentification.

In this paper we would like to present an original solution of this problem. The background for our solution is based on the learning of the specific distortions that are

* Correspondence: (IA): E-mail: igora@netvision.net.il ; (TB): E-mail: dragon@360.com.ua
(CB): E-mail: cbutakoff@ukr.net ;
(IA), (TB), (CB): E-mail: NNT Ltd., 155 Bialik str., Ramat-Gan, 52523 Israel
(VK): E-mail: vnk@iitp.ru ; (NM): E-mail: nick@iitp.ru; (OM): E-mail: milukova@iitp.ru;
(VK, NM, OM): Bolshoi Karetnyi per. 19, Moscow, 101447, Russia

J.R. Dorronsoro (Ed.): ICANN 2002, LNCS 2415, pp. 1231–1236, 2002.
© Springer-Verlag Berlin Heidelberg 2002

caused by the distorting operator in the Fourier amplitude spectrum. To identify the distorting operator, its mathematical model and its parameters, we will consider this problem as a problem of pattern recognition.

To solve the classification problem, we will use a neural network based on multi-valued neurons (MVN) [4]. It will be used for the recognition (identification) of the distorting operator or kind of blur. A similar MVN-based neural network will be used to recognize the corresponding distorting operator parameters. The multi-valued neurons have many wonderful properties. The most important of them are a high functionality and simplicity of learning.

We will consider here the classification for the four types of blur: defocus, rectangular, motion and Gaussian ones. The preliminary results for the blur and type of blur identification problem have been presented in [5], but just motion and Gaussian blur have been considered. In this paper we present the results of a significant development of the approach presented in [5].

After a type of blur and its parameters identification the image can be restored using several kinds of methods. Some fundamentals of image restoration will be considered. The image restoration (using the information obtained by the neural network) by Tikhonov regularization will be presented.

2 Multi-valued Neuron and Its Learning

As it was mentioned above, we will use a neural network based on multi-valued neurons (MVN) for the blur and blur parameters identification. Let us consider some fundamentals of MVN, its learning and networks based on it.

MVN has been deeply considered in [4]. MVN performs a mapping between n inputs and a single output. The mapping is described by multiple-valued (k-valued) function of n variables $f(x_1, ..., x_n)$ via their representation through $n+1$ complex-valued weights $w_0, w_1, ..., w_n$:

$$f(x_1, ..., x_n) = P(w_0 + w_1 x_1 + ... + w_n x_n) \qquad (1)$$

where $x_1, ..., x_n$ are variables, on which the performed function depends. Values of the function and of the variables are k^{th} roots of unity: $\varepsilon^j = \exp(i 2\pi j/k)$, $j \in \{0, k\text{-}1\}$, i is an imaginary unity. P is the activation function of the neuron:

$$P(z) = \exp(i 2\pi j/k), \text{ if } 2\pi j/k \leq \arg(z) < 2\pi (j+1)/k \qquad (2)$$

where $j=0, 1, ..., k\text{-}1$ are values of the k-valued logic, $z = w_0 + w_1 x_1 + ... + w_n x_n$ is the weighted sum , $arg(z)$ is the argument of the complex number z. The equation (2) is illustrated in Fig. 1.

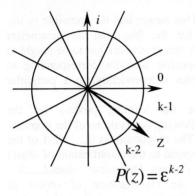

$$P(z) = \varepsilon^{k-2}$$

Fig. 1. Definition of the MVN activation function. If the weighted sum is equal to z then the output is equal to ε^{k-2}

For MVN, which performs a mapping described by the k-valued function, we have exactly k domains. Geometrically they are the sectors on the complex plane (Fig. 1).

The MVN learning is based on the same background as the perceptron learning. It means that if the weighted sum is going to the "incorrect" domain then the weights might be corrected in some way to direct the weighted sum into the correct domain. Let us consider this process in the details. If the desired output of MVN on some element from the learning set is equal to ε^q then the weighted sum should belong exactly to the sector number q. If the actual output is equal to ε^s, it means that the weighted sum belongs to the sector number s. A learning rule should correct the weights to move the weighted sum from the sector number s to the sector number q. The following correction rule for learning of the MVN has been proposed [4]:

$$W_{m+1} = W_m + C_m (\varepsilon^q - \varepsilon^s)\, \overline{X}, \qquad (3)$$

where W_m and W_{m+1} are the current and the next weighting vectors, $\overline{X}$ is the complex-conjugated vector of the neuron's input signals, C_m is the scale coefficient.

Learning algorithm, which is based on the rule (3) is very quickly converging.

3 MVN-Based Neural Network and Its Application to the Blur Identification

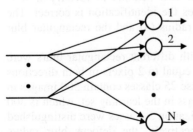

Fig. 2. MVN based neural network for pattern recognition

We will use here a single-layered MVN-based neural network, which contains the same number of neurons as a number of classes we have to classify (Fig. 2). An idea of this network has been proposed in [4]. Each neuron has to recognize pattern belonging to its class and to reject any pattern from any other class.

Any blur leads to the specific distortion of the image Fourier amplitude spectrum. The amplitude "disappears" for some frequencies. A character of this disappearance is very specific for the different types of blur (Fig. 3 illustrates this important property). Thus the Fourier spectrum amplitude contains the important information about the signal properties (existence and character of the blur,

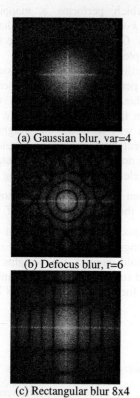

(a) Gaussian blur, var=4

(b) Defocus blur, r=6

(c) Rectangular blur 8x4

Fig. 3. Fourier spectrum amplitude's distortion implied by the different blurs

in particular). This means that it is possible to use this property for the blur and its parameters identification. A neural network has to be taught to distinguish a specific behavior corresponding to each type of blur independently on a particular image.

To organize the learning process for the network, the following reservations of the domains has been used. The output values $0,..., k/2-1$ of the ith neuron correspond to the classification of object as belonging to i^{th} class. The output values $l,..., k-1$ correspond to the classification of object as rejected for the given neuron and class (k is taken from (2)). To prepare the data for the learning and further recognition we used the normalization procedure based on the logarithmic quantization.

To test abilities of the blur and blur parameters identification, we used four kinds of blur: Gaussian, Defocus, 1D horizontal motion and rectangular ones. The images of different nature have been used: landscapes, satellite optical images, face images. The images were blurred by the mentioned blurs with the different parameters. To make our model more realistic, we corrupted the images by zero-mean Gaussian noise with the dispersion 0.3σ.

For the blur identification 140 images have been taken. Since a spectrum amplitude behavior for the rectangular and defocus blurs is often very similar, we use two-stage blur identification. Three classes have been considered on the first stage: Gaussian, 1D motion, Rectangular/Defocus. The testing results are very good: 95-97% of the correct identification. If the corresponding blur was classified as "Rectangular/Defocus", we used additional single-layered neural network of two neurons to identify if it is rectangular or defocus. For the 94% of the images the identification is correct. The rest of 6% corresponds to defocus with a small radius (4) and the rectangular blur 1x1.

For blur parameters recognition 25 classes with different rectangular blurs were created. Steps for rectangular blur parameters are equal to 2 pixels in both directions (i.e. 1x1, 1x3, ..., 1x5, 3x1, ..., 3x5, ... 5x5). These 25 classes contain 500 images in the testing set (20 per class), and 12 images per class in the learning set, which is 300 images. To test the Gaussian blur variance identification 5 classes were distinguished (variance 1 to 5, 200 images per class). Respectively, the defocus blur radius identification has been tested for the 15 classes (radius from 2 to 16, 500 images per class), 1D horizontal motion blur has been tested for the 16 classes (motion from 3 up to 18 pixels).

As the result, a recognition rate for the parameters identification is: for the Gaussian Blur 93.5% successful, for the defocus blur 94.1% successful, for the motion blur 98.1% successful and for the rectangular blur 95.6% successful.

4   General Approach to the Restoration Problem

After the blur identification, it is possible to use the following technique for the image restoration. The image restoration problem is usually formulated the obtaining of the non-distorted image $z(\zeta, \eta)$ from the given equation:

$$Az + n = u(x, y) + n(x, y) = \tilde{u}(x, y), \qquad (x, y) \in \text{W}, \qquad (4)$$

where $A : Z \to U$ (Z, U –metric spaces) is a given linear or nonlinear operator, $z \in Z$, $u \in U$, $n(x, y)$ is a noise, $\tilde{u}(x, y)$ is an output distorted image.

It is evident that whatever method we use to obtain the restored image, it must comply in a certain way with the basic equation (4). So the most general formulation of the restoration problem can be reduced to the following functional's minimization:

$$z^* = \inf_{z \in Z} \rho_U (Az, \tilde{u}), \qquad (5)$$

where ρ_U is a certain metric in U. In general it is possible to use different definitions of a distance ρ_U between two images.

The simplest way to guarantee the uniqueness and stability of the solution is to formulate "a priory" information about the original image using a functional $\Omega(z)$ that possesses stabilizing properties [6]. In this case the image restoration problem can be reduced to the conditional or unconditional optimization problem, in particular to the Tikhonov minimization [2]:

$$z^* = \inf_{z \in Z} \{ \rho_U (Az, \tilde{u}) + \alpha \Omega(z) \}, \qquad (6)$$

where α is the parameter of the regularization. Usually it is assumed that the original image is a smooth function with respect to Sobolev space, and stabilization functional in (6) is $\Omega(z) = \| z \|_{W_q^p}^q$. It is necessary to point out that the opportunity of obtaining a family of solutions that depend on a parameter α is very important. This allows us to control the visual quality of the image restoration interactively in the absence of a mathematical criterion of visual image quality.

5   Simulation Results

To identify a type of blur and then to identify its parameters we used the neural network and the algorithm presented here. Then the images have been restored using the Tikhonov's regularization algorithm (6). Let us consider the restoration example of the originally (optically) blurred image taken by the digital camera on the street (Fig.3). This image didn't participate in the learning process of the neural network.

The blur on the image has been identified as motion with parameter 6 (in every color channel). Tikhonov's filter has been used for the restoration

(a) The original blurred image   (b) The restoration result.

Fig. 3. Restoration of the originally (optically) blurred image. Motion blur has been detected.

6 Conclusions and Future Work

The main result of the presented work is the effective neural solution for the recognition of the blur and for the identification of its parameters. The results of this identification can be used for the image restoration using the Tikhnov's filter, for example. The future work will be directed to the consideration of more blurs, including the combinations of several different blurs and to the further development of the restoration technique.

References

1. W.K.Pratt, Digital Image Processing, Second Edition, N.Y.: Wiley, 1992.
2. A.N.Tikhonov, V.Y.Arsenin, Solutions of ILL-Posed Problems, N.Y.: Wiley, 1977.
3. Y.L. You and M. Kaveh, "A Regularization Approach to Joint Blur Identification and Image Restoration", *IEEE Trans. on Image Processing*, vol. 5, pp. 416–428, 1996.
4. I.N..Aizenberg, N.N.Aizenberg, J.Vandewalle *"Multi-valued and universal binary neurons: theory, learning, applications"*, Kluwer Academic Publishers, Boston/ Dordrecht/London, 2000.
5. Aizenberg., N.Aizenberg, T.Bregin, C.Butakov, E.Farberov, N.Merzlyakov, O.Milukova "Blur Recognition on the Neural Network based on Multi-Valued Neurons", *Journal of Image and Graphics*. Vol.5, 2000; Tianjin, China, *Proceedings of the First International Conference on Image and Graphics (ICIG'2000)*, Tianjin, China, August 16-18, 2000, pp.127–130.
6. O.P.Milukova " On Justification of Image Model", *SPIE Proceedings*, vol.3348, pp283–289, 1998.

Using Neural Field Dynamics in the Context of Attentional Control*

Gerriet Backer and Bärbel Mertsching

IMA-Lab, Dept. of Computer Science, University of Hamburg
Vogt-Koelln-Str. 30, 22527 Hamburg, Germany
`<surname>@informatik.uni-hamburg.de`

Abstract. We present an application of dynamic neural fields for selection and tracking in the attentional control part of an active vision system. We propose a novel two-stage selection mechanism, in which the fields are used for the first selection stage. We discuss different variants, introducing 3D neural fields and systems of interconnected fields. The dynamics can be shown to achieve important goals in active vision like robust selection, multi-object tracking, and spatiotemporal integration.

Keywords. Attentional mechanisms, neural dynamics, active vision.

1 Introduction

The purpose of intelligent vision systems is not to invert the vision process, but rather to compute a meaningful symbolic description of the most relevant aspects of an environment, thus enabling appropriate intelligent behaviour. Two of the main targets to be achieved are, therefore, the selection of relevant aspects of the environment and the transformation of the subsymbolic input signal into a symbolic representation of the selected items for memorizing, manipulation, and behaviour specification. We will show how dynamic neural fields (DNF) can be employed for the simultaneous solution of both tasks in a system of attentional control, and thereby contribute to important abilities of active vision systems.

2 Related Work

2.1 Visual Attention

The important attention model introduced by Koch and Ullman [1], was mainly influenced by human visual attention. It used parallel computation of feature maps, integrated into a master map of attention describing the overall local saliency. A WTA-process determined the location of the focus of attention (FOA). Moving attention was achieved by marking the FOA in a so-called inhibition map. The orginal model is still being improved and extended [2,3].

---

* We thank the DFG for their support on this research (Project ESAB-2 Me1289/3-2).

J.R. Dorronsoro (Ed.): ICANN 2002, LNCS 2415, pp. 1237–1242, 2002.

Connectionist models are often centered around the task of transferring a selected image area into a constant frame. Olshausen [4] introduced such a model using resolution pyramids with gating neurons directing the flow of information, while Tsotsos [5] used a so-called "inhibitory beam" to propagate the results of WTA-processes along different resolution levels. Other neural models [6,7] try to integrate location and identification processes into a single mechanism.

2.2 Dynamic Neural Fields

Dynamic neural fields have been proposed by Amari [8] as a model of cortical neural arrays. Their properties in the context of selection algorithms operating on noisy input data include hysteresis, tracking, bifurcation, and spatiotemporal integration. The dynamics of the field (for two dimensions) are given by (where u denotes the activity of the neuron at location x and time t, τ is a time constant, S a sigmoid function, h describes the (negative) resting value, w gives the weights, and i the input into the field):

$$\tau \frac{d}{dt} u(x,t) = -u(x,t) + h + \int w(x - x')S[u(x,t)]d^2x' + i(x,t) .$$

The activity change of a neuron at position x and time t depends on its actual activity level, the resting value, the weighted input from the other neurons, and the external input at this position. The connection weights for a DNF are homogenous and depend only on the distance between the neurons. In a local neighborhood, these weights are positive. The known types of neural fields include global and local inhibition. While for local inhibition the inhibitive weights are restrained to a local neighborhood and are zero for larger distances, for global inhibition the weights stay inhibitive even for large distances. Global inhibition can be achieved by defining the weights through a constant value subtracted from a normal distribution; a DoG-function is most often used for local inhibition. The main difference is that local inhibition types allow stable states with multiple local activity clusters, whereas in global inhibition at most one activity cluster is possible. In the context of attentional control, [9,10] were the first to use a global inhibition type DNF for selecting the most salient object. The achieved behaviour was limited to tracking the selected object until another object became much more salient or the object left the sensor area. In [11] a resolution pyramid with constant target masks was used to model the depth of objects for a global inhibition 3D DNF in a cue integration and selection task.

3 Architecture

Our attentional control architecture is centered around the idea of a two-stage selection mechanism; it first selects a small number of salient visual items and then selects a single focus of attention among these items. Selecting and tracking multiple objects is essential for any system which serializes high-level-computations in a dynamic environment in order to bind the extracted information to moving

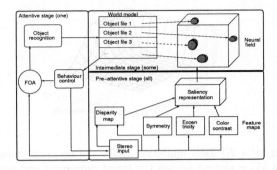

Fig. 1. Overview of the attention model divided in three computation stages

objects - in contrast to static locations. This avoids outdating of information and inefficient reallocation of attention to the same object. By using two selection stages, we come up with three computation stages (see fig. 1):

- Pre-attentive: simple feature computations for the *complete image*
- Intermediate: tracking and integration of a *small number of items*
- Attentive: complex operations (object recognition) for a *singular item*

The two selection stages are rather different. The first is mainly data-driven, selecting a few items according to their saliency. The second stage selects a single focus from the result of the first stage. The focus can be specified as a target for the saccade. This stage operates on purely symbolic data and is subject to a selected behaviour and model-driven influences (see [12] for the operation of this stage). The most important aspect of the first computation stage is the calculation of features for determining the local saliency. We have implemented area-based, edge-based, color-based, and stereo-based features (described in [13, 12]). This paper focuses on the first selection stage, part of the second computation stage, using DNF. For this stage, hysteresis, tracking, robustness against noise, spatiotemporal integration, and the selection of multiple items are desirable properties. Global inhibition type DNF in principal show these properties - except the last one. In the following, we will discuss three variants of neural fields for selecting multiple items that we examined for incorporation in our model. They differ in the expressed concept of visual objects.

3.1   2-D Local Inhibition Neural Field

Local inhibition type DNF are the natural choice in order to allow multiple activity clusters. A two-dimensional mastermap of attention containing a normalized measurement of saliency is the input into the field. The weights of the field are defined by a DoG-function. For identifying the tracking quality, we used a standard-distributed input cluster in an environment of normal distributed noise moving at different speeds. Successful tracking was denoted by the presence of exactly one activity cluster overlapping more than half of the input. Fig.

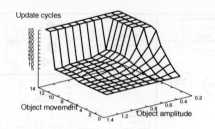

Fig. 2. Tracking by a 2D local inhibition DNF: the necessary number of updating cycles (clipped at 55) is plotted against the target movement and amplitude.

2 shows the computations necessary to keep up successful tracking for different input amplitudes and speeds. Note how the tracking performs well for a wide range of signal-noise-ratios up to a certain limiting speed. A disadvantage of this approach is the simple representation of items which are just clusters of high saliency in 2-D space. Thus the base of information used for tracking is restricted and this results in problems handling temporary occlusion situations. An additional tracking mechanism has to be used to establish a constant relation between the objects and the activity clusters based on feature information.

3.2 System of 2-D Global Inhibition Neural Fields

Alternatively, we used a number of global inhibition DNF. Each field was intended to select and track a single object, whereby the maximal number of selected objects was defined by the number of fields used. These DNF had to be connected inhibitory locally to avoid the selection of the same location by multiple fields. An advantage of this approach is that it provides the possibility of differentiating the inputs to the fields in order to improve tracking performance and the handling of temporary occlusions. We thus modified the input to each field to adapt to the features of the item selected by that specific field. The feature weights were computed by a signal-noise ratio of each feature for the active area of the field in comparison to the mean feature activity for an individual feature profile selected and tracked by each DNF. Here, selected items are clusters of high saliency in 2-D space, where the saliency comes from a particular feature weighting. The computational complexity rises nearly linearly with the number of fields used because most time is spent on the internal updating of the 2-D fields compared to the local in-between connectivity. In our experiments we used four fields. The modified updating rule is given by (u_j denotes the activity in the j-th field, $w_{k,j}$ the profile similarity between the fields k and j, and i_j the weighted feature input for the j-th field):

$$\tau \frac{d}{dt} u_j(x,t) = -u_j(x,t) + h + \int w(x-x')S[u_j(x,t)]d^2x' + \sum_k w_{k,j}S[u_k(x,t)] + i_j(x,t)$$

3.3   3-D Local Inhibition Neural Field

As depth is an important feature for attention, we extended our system to make use of a three-dimensional saliency representation. This representation is more suitable for inclusion into robotic systems where the distance to an object is necessary for navigation and manipulation purposes. An earlier version has been published in [13], where we described the computation of a 3D master map. In short, we assign saliency values to positions in 3D space according to the 2D saliency and a probability distribution for the depth of each 2D position, thereby allowing the coexistence of different depth-hypotheses. We decided to use global inhibition in the z-direction, thereby vastly reducing the computational effort in comparison to a true 3D DoG-kernel. The resulting weights are given by:

$$
w(x - x') = \begin{cases} k * \exp(\frac{x - x'}{\sigma^2}) - k_2 * \exp(\frac{x - x'}{\sigma_2^2}) & , \; x_z = x'_z \\ -H_1 & , \; x_y = x'_y, \, x_x = x'_x, \, |x_z - x'_z| > 1 \\ -H_2 & , \; else \end{cases}
$$

4   Experimental Results

We compared different approaches in temporary occlusion situations (see fig. 3) by examining arbitrary trajectories of a different number of objects with at least one temporary occlusion, each for 10 frames. We measured the mean tracking duration. The 3D local inhibition version performs best but is also associated with the highest computational effort. If three-dimensional information is not available, the 2D global fields perform better than the local version until the number of objects surpasses the number of fields used.

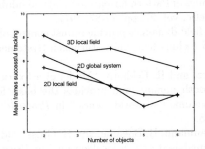

Fig. 3. Handling of temporary occlusions by DNF. The mean tracking duration by the three types of DNF is shown for temporary occlusion scenes.

5   Conclusion

We have shown that dynamic neural fields can handle a number of different tasks at once in the context of visual attention. Our architectures are all suited for the integration of robust selection and tracking of multiple objects. The known properties of DNF have thus been transferred to the domain of multiple item selection. In temporary occlusion situations, the single 2D DNF performance was improved by using either a system of 2D fields or a single 3D field. Their choice depends on the availability of stereo data and the stability of the features. In the future, we intend to integrate feature similarity and the 3D representation into a single system.

References

1. C. Koch and S. Ullman, "Shifts in selective visual attention: Towards the underlying neural circuitry," *Human Neurobiology*, vol. 4, pp. 219–227, 1985.
2. L. Itti and C. Koch, "(Feature combination strategies for saliency-based visual attention Systems," *Journal of Electronic Imaging*, vol. 10, no. 1, 2001.
3. F. Miau and L. Itti, "A neural model combining attentional orienting to object recognition: Preliminary explorations on the interplay between where and what," in *Proceedings IEEE Engineering in Medicine and Biology Society (EMBS)*, 2001.
4. B. Olshausen, C. Anderson, and D. V. Essen, "A multiscale dynamic routing circuit for forming size- and position-invariant object representations," *Journal of Computational Neuroscience*, vol. 2, no. 1, pp. 45–62, 1995.
5. J. K. Tsotsos, S. Culhane, W. Wai, Y. Lai, N. Davis, and F. Nuflo, "Modelling visual attention via selective tuning," *Artificial Intelligence*, vol. 78, pp. 507–545, 10 1995.
6. F. H. Hamker, "Distributed competition in directed attention," in *Dynamische Perzeption* (G. Baratoff and H. Neumann, eds.), pp. 39–44, 2000.
7. G. Deco, *Emergent Neural Computation Architectures*, ch. Biased Competition Mechanism for Visual Attention in a Multimodular Neurodynamical System, pp. 114–126. 2001.
8. S.-I. Amari, "Dynamics of Pattern formation in lateral inhibition type neural field," BioZogicaZ Cybernetic, vol. 27, pp. 77-87, 1977.
9. K. Kopecz, "Neural field dynamics provide robust control of attentional resources," in Aktives Sehen in technischen und natürlichen Systemen (B. Mertsching, ed.), pp. 137-144, 1996.
10. M. Pauly, K. Kopecz, and R. Eckhorn, "Model of a fixation control network performs saccades, smooth pursuit, and provides the basis for segmentation of unknown objects in dynamic real-world scenes," in *Proceedings Workshop Dynamische Perzeption*, 1998.
11. A. Corradini, U. Braumann, H. Boehme, and H. Gross, "3d neural fields and steerable filters for tontour-based person localization," in *Proceedings of the Workshop of Virtual Intelligence – Dynamic Neural Networks (VI-DYNN'98)*, (Stockholm, Sweden), June 1998.
12. G. Backer, B. Mertsching, and M. Bollmann, '(Data- and model-driven gaze control for an active vision system," *IEEE Transactions on Pattern Analysis and Machine Intelligence*, vol. 23, no. 12, pp. 1415–1429, 2001.
13. G. Backer and B. Mertsching, "Integrating time and depth into the attentional control of an active vision system," in *Dynamische Perzeption. Workshop der GI-Fachgruppe 1.0.4 Bildverstehen, Ulm, November 2000* (G. Baratoff and H. Neumann, eds.), pp. 69–74, 2000.

A Component Association Architecture for Image Understanding

Jens Teichert and Rainer Malaka

European Media Laboratory
Villa Bosch
Schloss-Wolfsbrunnenweg 33
D-69118 Heidelberg
jens.teichert@eml.org
http://www.eml.org

Abstract. A constructive approach for the detection of objects with topological variances is introduced. The architecture enables shift and scale invariant detection and is trained through supervised learning. Representations of the input data are built by combining association elements in a hierarchical grid. This leads to a big flexibility of representation employing only few element types. Simulation results are given for the task of detecting windows of buildings in real world images.

1 Introduction

One of the challenges in computer vision is the efficient and invariant recognition of objects in real world scenes. This is already done respectably for objects with fixed topology e.g., in the face recognition domain [2]. So far, very few attempts using hierarchical representations have been done for the recognition of objects with varying topologies [1,4,5]. But this seems to be very promising for the recognition of e.g. facades of buildings with varying window positions and windows with varying pane distributions. The approach presented here is motivated by hierarchical biological representations as described by e.g. Oram & Perret [3]. It accomplishes shift and scale invariance, is capable of learning invariances for objects with changing topologies and features robust recognition.

2 Proposed Architecture

The hierachical architecture of the system consists of a set of rectangular layers. On each of the L layers, a grid of association elements is positioned. The layers represent a pyramidal structure where higher layers have less elements, i.e. the element at position (l, i, j) is positioned on top of the coordinates i/c, j/c on layer $l-1$, where $c < 1$. At each discrete position (l, i, j) of the pyramid, an association element can receive an input $\mathbf{x}(l, i, j) \in \mathbb{R}^M$, can calculate an activation $\alpha(\mathbf{x}(l, i, j))$ and an output vector $\mathbf{y}(l, i, j) \in \mathbb{R}^N$. Input images are preprocessed

J.R. Dorronsoro (Ed.): ICANN 2002, LNCS 2415, pp. 1243–1248, 2002.

on different resolutions. Each preprocessed image resolution is set to a respective layer of the pyramid as external input.

At each raster position, a vector $\mathbf{y}_{sup} \in \mathbb{R}^N$ can be calculated. It contains superpositioned output from the layer below and thus reduce output information in direction of higher layers. It is calculated by a two-dimensional lowpass filter with a zero-centered gaussian kernel $f(i,j) : \mathbb{R}^2 \to [0,1]$:

$$\mathbf{y}_{sup}(l,i,j) = \sum_p \sum_q f\left(p - i/c,\, q - j/c\right) \mathbf{y}(l-1, p, q). \tag{1}$$

The spread of the centered kernel depends on the scaling factor c and has to be chosen appropriately.

The superposition allows for some direct feed-through of output vectors from layer $l-1$ to layer l. In general however, association elements on layer l will use a spacial composition of output vectors $\mathbf{y}(l-1, p, q)$ in order to compute their output. The layer $l-1$ is segmented below position (l, i, j) in R rings, each with Φ radial bounds (Fig. 1 shows the distribution used for experiments). For each segment (r, ϕ) an average output vector is calculated by:

Fig. 1. See text

$$\mathbf{y}_{r,\varphi}(l-1, i, j) = \sum_p \sum_q \mathbf{y}(l-1, p, q)\, \omega_{r,\varphi}(p - i/c,\, q - j/c) \text{ with} \tag{2}$$

$$\omega_{r,\varphi}(p,q) = \begin{cases} 1/(\#\text{raster-points in segment}(r,\varphi)) \mid p, q \in \text{segment}(r, \varphi) \\ 0 \text{ else.} \end{cases}$$

The input vector $\mathbf{x}(l, i, j)$ on the higher layer is a concatenation of those segment averages with the dimension $M = R \cdot \Phi \cdot N$:

$$\mathbf{x}(l, i, j) = (\mathbf{y}_{0,0}^T, \mathbf{y}_{0,1}^T, \ldots, \mathbf{y}_{0,\Phi-1}^T, \mathbf{y}_{1,0}^T, \ldots, \mathbf{y}_{R-1,\Phi-1}^T)^T. \tag{3}$$

2.1   Association Elements

Association elements comprise of a set $\mathcal{S}$ of adaptive detector elements and a special $s_{transport}$ element. Copies of association elements can be inserted at raster positions for performing local calculation. All elements can operate in a feed-forward-mode for pattern analysis and a suspend-mode, which will be used during learning.

A detector element s possesses an adaptive input weight $\mathbf{v}_s \in \mathbb{R}^M$ and fixed output weight $\mathbf{w}_s \in \mathbb{R}^N$ which is equipped with random values at system start. Output vectors allow each association element to respond with a characteristic signature instead of a scalar value only. Elements on higher levels can then respond to specific combinations of such signatures in the lower levels. In feed-forward-mode, the local activation α and a local output vector $\mathbf{y}_{elem}$ are calculated for input vector $\mathbf{x}$ as:

$$\alpha(\mathbf{x}) = \left| b(\mathbf{x}^T \mathbf{v}_s) \right|, \quad \mathbf{y}_{elem}(\mathbf{x}) = \alpha(\mathbf{x})\mathbf{w}_s. \tag{4}$$

The function $b(a) = tanh(a)$ is a limiting function and serves for stabilization of the activation. The weights are normalized. This yields a separation of information content and activation.

The transport element $s_{transport}$ is doing its feed-forward-mode calculation by copying the local superposition to the output: $\mathbf{y}_{elem} = \mathbf{y}_{sup}$.

2.2 Input Interface

In general, this architecture can handle any kind of pixel map-based preprocessing. The structure of the pyramid is especially advantageous, if this input can be produced for different resolution scales. For the application shown in sect. 3, gabor-filters are used, which are parameterized by an orientation d, the respective layer l and a variation σ. At each raster point a set of filter responses are calculated for D orientations. A wave vector $\mathbf{k}_0$ is scaled that the filters show one period between two adjacent raster points on the respective layer (c_0 is the resolution ratio between first layer and the input-image). The distribution of the filters in fourier space (given by $\mathcal{F}$ transformation) is indexed by $\mathbf{k}$ and defined by (see for detail and DC-value supression [2]):

$$\mathcal{F}(G^{l,d})(\mathbf{k}) = \exp(-\pi|\mathbf{k} - \mathbf{k}_0^{l,d}|^2\sigma^2) \text{ with } \mathbf{k}_0^{l,d} = \text{scale}(l)\,\text{orientation}(d) \quad (5)$$

$$\text{scale}(l) = c_0\,c^l, \quad \text{orientation}(d) = \begin{pmatrix} \cos(\frac{d\pi}{D}) \\ \sin(\frac{d\pi}{D}) \end{pmatrix} \quad (6)$$

For each orientation a rigid output-weight $\mathbf{w}^d$ is given ($w_n^d = 1$ if $n = d$, 0 else). At every raster-point of the pyramid, the final output $\mathbf{y}$ is calculated by the sum of $\mathbf{y}_{elem}$ and the output vectors $\mathbf{w}^d$ scaled by the absolute values of the respective filter reponses.

$$\mathbf{y} = \mathbf{y}_{elem} + \sum_d \left\| G^{l(\mathbf{y}),d}(\text{position}(\mathbf{y})) \right\| \mathbf{w}^d \quad (7)$$

Input information is used with (1) and (3) on the following layer. Layer 0 will receive input information only. First associations start on layer 1.

2.3 Association

The association procedure equips raster positions with association elements. After an element has been chosen and inserted in the pyramid, the local activation and output vector are calculated by its feed-forward mode. After this has been done for a complete layer, the output is used as input information for the association of the next layer and so on.

At each position (l, i, j), a winner-take-all mechanism is applied in order to select an appropriate detector element $s_{insert} \in S$ using $\mathbf{x}$ at position (l, i, j):

$$s_{insert} := \begin{cases} \arg\min_{s \in S} \left\| \frac{\mathbf{x}}{||\mathbf{x}||} - \mathbf{v}_s \right\| & \text{for } \left\| \frac{\mathbf{x}}{||\mathbf{x}||} - \mathbf{v}_s \right\| < \vartheta,\ s \neq \{\} \\ s_{transport} & \text{else.} \end{cases} \quad (8)$$

An inserted detector element has to show the minimum distance between the normalized input $\mathbf{x}$ and its weight $\mathbf{v}_{s_{insert}}$. This selection characterises the systems substantial non-linearity. Further, the distance has to be under a certain threshold ϑ. If no detector applies, a transport element is inserted, which passes the superposition signal on to the output. This is important to allow later associations on higher levels on information not used, yet. This element is most frequently used in the pyramid.

Note that an association element can be inserted at multiple positions on various layers of the pyramid. This assures shift and scale invariance and the ability of detecting an arbitrary number of patterns. Also note that for each input image a new assignment of elements to positions has to be done. Representations of scene objects consist of element combinations. Every representation is build in dependence on the actual input image and the trained element weights.

2.4 Learning

The system starts with an empty set of detector elements $\mathcal{S} = \{\}$ and iteratively learns new concepts (lines, corners, rectangles, ...). For each new scene-concept h a new detector element s_h is introduced with random weights. This element is then trained by a set $\mathcal{Z}_h$ of manually given (circular) regions z on one or more input images.

For each pattern in $\mathcal{Z}_h$ the pyramid is first equipped with the association elements as described by (8). Then the respective detector element s_h is put (clamped) to a raster position (l_c, i_c, j_c), which is determined by the size of the region z and its position in the input image. Below the element s_h (from the clamp layer l_c to the lowest layer) the shape of z forms a cone-like region $\mathcal{C}_z$ of raster positions to which the training will be applied.

After the association all elements operate in suspend-mode which prevents recursive weight changes. Then the learning rule (9) calculates new weights $\mathbf{v}^*$ for each detector element in $\mathcal{C}_z$ (if it is inserted at multiple positions, all respective weight changes are accumulated). The goal is to minimize the differences between the normalized input $\mathbf{x}$ and the input weights $\mathbf{v}$. This winner-take-all learning rule enhances the response of the element s_h to similar combinations of predecessing elements.

$$\mathbf{v}^* = \mathbf{v} + \eta\left(\frac{\mathbf{x}}{||\mathbf{x}||} - \mathbf{v}\right) \tag{9}$$

After a learning iteration for all elements in $\mathcal{C}_z$, the weights of the modified detector elements have to be normalized.

The learning rate η is calculated for each individual position in $\mathcal{C}_z$ as

$$\eta = \alpha\,\mu(\Delta_l)\,\nu(\zeta_s)\,\rho_z \quad \text{with} \quad \mu(l) = e^{-\frac{l_c - l}{\tau_{layer}}}, \quad \nu(\zeta) = e^{-\frac{\zeta}{\tau_{conf}}}. \tag{10}$$

This learning rate consists of four factors. The local activation α, a layer distance function $\mu(l)$ (elements closer to the top have a higher learning rate), a function $\nu(\zeta_s)$, which is decreasing exponentially with the number ζ_s of learning steps already taken (from system start) for the local element s. This is important for convergence and sustainment of established elements.

The last factor ρ_z is a local relevance factor. Starting from s_h downwards, a relevance value ρ is calculated recursively for each position in C_z.

$$\rho_z(l,p,q) \;=\; \min\{1, \sum_{i,j|(l,i,j)\in C_z} f(p - i/c, q - j/c)\, \rho_z(l+1,i,j)\, \alpha(l+1,i,j)\}, \quad (11)$$

$$\rho_z(position(s_h)) \;=\; 1 \qquad\qquad (12)$$

The relevance of the top element is set to the maximal value 1, elements under it integrate relevance values of the layer above scaled by their activations and the gaussian kernel f.

Fig. 2. Association-results for pane- and window-concepts trained from samples of the left image (upper row: panes, lower row: windows).

3 Experimental Results

A pyramid-geometry with 131x131 points on the lowest layer leads to ≈ 100.000 raster points with $L = 32, c = 0.91$. Further relevant parameters were $N = 10$, $D = 4$, $R = 3$ and $\Phi = 8$. Eight concepts were trained: two lines with eight samples (each), four corners with six samples (each), a pane with six samples and a window with three samples. Results in Fig. 2 show associated pane and window detector elements for a variety of images. All training samples are taken from the leftmost image. Further associations from other images are shown to illustrate the systems capability to handle shift, scale and topographical variances.

The circles shown have proportional size to the layer of the respective detectors. A full association takes a minutes on a 800Mhz Pentium III system.

4 Conclusion

In this paper, we presented a new architecture that is able to detect objects with topological variances. The system allows for shift and scale invariant, robust recognition of image feaures and builds an efficient scene representation. Since the system can also link image components together that are not represented in neighboring regions, it can also intrinsically solves the so-called binding problem.

The basic idea of our system can be further enhanced in various ways. Rotation or projective invariances could be addressed using spacial rotations of the segment masks for the input vectors in the pyramid. Object boundaries are reconstructable by tracking relevant information contributions back from the concept detector of the object down to the lowest level on the pixel map. Such a feed-backward mode could also be used for image reconstruction from fragmentary input.

So far, new detector elements are introduced with given concepts. Additional ways of crating new detector elements are reasonable. For instance established detectors could be fixed and new ones could be added during training. The total amount of detectors can be limited to a fixed number. When this "population" is completely used, new detector instantiations will overwrite detectors which are less established.

The hirarchical approach presented here is not bound to applications in image recognition. Whereever highly structured data with varying topologies are to be analyzed, this approach could be suitable. In speech analysis, for instance, certain concepts are to be detected that appear in highly variable grammatical structures. Here, single concepts in our system could relate to speech acts, pronomina, etc. Other application domains could lie in the analysis of geographical patterns or user preferences.

Acknowledgements. Parts of this work was funded by the Klaus Tschira Foundation and by the German Ministry of Education and Research (01IL905C, 01IRA12C, 01IL904D2).

References

1. K. Fukushima. Neocognitron: a neural network model for a mechanism of visual pattern recognition. *IEEE Transactions on Systems, Man and Cybernetics*, pages 826–834, 1983.
2. M. Lades, J. C. Vorbrüggen, J. Buhmann, J. Lange, C. v.d. Malsburg, R. P. Würtz, and W. Konen. Distortion invariant object recognition in the dynamic link architecture. *IEEE Transactions on Computers*, 42(3):300–311, 1993.
3. M. W. Oram and D. I. Perret. Modeling visual recognition from neurobiological constraints. *Neural Networks*, 7(6/7):945–972, 1994.
4. G. Wallis and E. T. Rolls. Invariant face and object recognition in the visual system. *Progress in Neurobiology*, pages 167–194, 1997.
5. J. Weng. Cresceptron and SHOSLIF toward comprehensive visual learning. In *Early Visual Learning*, pages 183–214. Oxford University Press, New York, 1996.

Novelty Detection in Video Surveillance Using Hierarchical Neural Networks

Jonathan Owens[1], Andrew Hunter[2], and Eric Fletcher[1]

[1] Centre for Adaptive Systems, University of Sunderland, UK
{jonathan.owens,eric.fletcher}@sunderland.ac.uk
[2] Department of Computer Science, University of Durham, UK
andrew1.hunter@durham.ac.uk

Abstract. A hierarchical self-organising neural network is described for the detection of unusual pedestrian behaviour in video-based surveillance systems. The system is trained on a normal data set, with no prior information about the scene under surveillance, thereby requiring minimal user input. Nodes use a trace activation rule and feedforward connections that are modified so higher layer nodes are sensitive to trajectory segments traced across the previous layer. Top layer nodes have binary lateral connections and corresponding "novelty accumulator" nodes. Lateral connections are set between co-occurring nodes, generating a signal to prevent accumulation of the novelty measure along normal sequences. In abnormal sequences the novelty accumulator nodes are allowed to increase their activity, generating an alarm state.

1 Introduction

To help CCTV surveillance operators maintain adequate attention levels, the next generation of automated visual surveillance systems will be attention-focussing filters that restrict events presented to the operator to those that fall outside of some definition of normality [1]. The novelty of an observed trajectory can be measured by building probabilistic descriptions of the relationship between trajectory segments [2], [3]. Neural networks have been used to create distributions of full length trajectories, or trajectory segments, where trajectory elements (e.g. flow vectors) are weighted relative to their temporal position in the sequence [4], [5].

The system described in this paper uses a hierarchical network of self-organising layers to represent trajectory segments, and a top layer of nodes with binary lateral connections and novelty accumulator nodes to measure the novelty of sequences of trajectory segments. Rather than requiring the complete trajectory to be submitted to the network, novelty is measured as trajectories are traced across the input layer, so an alarm can be generated while a novel event is occurring.

The neural network is one component of a hybrid system, which uses a temporal low-pass filter to construct a background image, and segments moving objects by background differencing. Existing objects are assigned to the segmented connected components by a minimum-distance match across feature vectors. Short-term novelty such as local erratic motion is detected by a self-organising map [6], that operates alongside the network described in this paper, which detects long-term novelty.

J.R. Dorronsoro (Ed.): ICANN 2002, LNCS 2415, pp. 1249–1254, 2002.

2 Hierarchical Self-Organising Network Structure

Self-organisation in a hierarchical structure is the basis for many powerful models, from the strictly computational [7], [8], to those that model specific biological networks [9], [10]. Our basic architecture is shown in figure 1.

Each node uses a trace activation rule, which was first hypothesised as a means by which the higher visual cortices may learn object invariance [11], and was used to that effect in [10] for learning invariant recognition of objects subject to spatial transformations. The feedforward nodes have overlapping receptive fields, except in layer 1, where the nodes simply partition the 2D image plane into fixed width cells. The final layer nodes have modifiable binary lateral connections and novelty accumulator nodes.

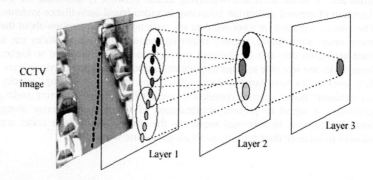

Fig. 1. General architecture, with trace activation illustrated as a decrease in greyscale value of participating nodes (for clarity, only a subset of nodes are shown)

2.1 Learning

Training is partitioned such that modification of feedforward connections to layer k is completed before training begins at layer $k+1$. Subsequently, lateral weights between top layer nodes are modified to reflect normal co-occurrences of top layer nodes.

The object tracker passes the centroid of the tracked object to the first layer of the hierarchical network. The activity of layer 1 nodes signals the recent presence of the tracked object within their receptive fields by the following activation rule:

$$x_i(t) = \begin{cases} 1 & [r(t),c(t)] \in R_i \\ x_i(t-1) * \gamma & \text{otherwise,} \quad (\gamma < 1) \end{cases} \tag{1}$$

Where x_i is the activity of first layer node i, γ is the trace rule decay constant, $[r(t),c(t)]$ is the centroid of the tracked object at time t and R_i is the receptive field of node i. Activity of nodes in higher layers is calculated as follows:

$$a_i(t) = \sum_{j \in R_i} w_{ij}(t) x_j(t) \tag{2}$$

$$x_i(t) = \begin{cases} 1 & a_i(t) = \max_n(a_n(t)) \\ x_i(t-1) * \gamma \text{ otherwise} & (\gamma < 1) \end{cases} \tag{3}$$

Where a_i is the weighted sum of the inputs to node i, R_i is the receptive field of node i, x_j is the activity of input node j, w_{ij} is the synaptic weight between node j and i, x_i is the trace activity of node i and γ is the decay constant. The winning node has its activity set to 1, while all other node activities decay according to the trace rule (1). The localised receptive fields and trace activation allow nodes to integrate inputs across a small spatio-temporal window.

Learning only occurs at the winning node at each iteration, and follows the general Hebb rule for synaptic modification [11]. In general,

$$\Delta w_{ij} = \alpha x_j x_i \big|_{j \in R_i} \tag{4}$$

Where x_i is the post-synaptic activity, x_j is the presynaptic activity and α is a learning rate. The sum of the weights impinging on a post-synaptic node is normalised to 1. Learning "chunks" normal trajectory segments into progressively larger segments at higher layers. Synaptic modification is completed in layer 2, before learning begins in layer 3.

Each layer 3 feedforward node has a corresponding "novelty accumulator" (NA) node, which is silent during learning, and is used in novelty detection. Each layer 3 node has binary lateral connections to all NA nodes in the layer, as shown in figure 2.

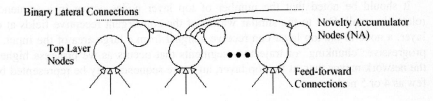

Fig. 2. Top layer nodes, novelty accumulator nodes and binary lateral connections

The lateral binary connections are modified when feedforward learning has been completed. With lateral weights set initially to zero, the training set is submitted to the feedforward section of the network, which produces sequences of winning nodes at the top layer. Binary lateral connections are set, subject to an ordered co-occurrence rule, shown below:

$$l_{ij}(t) = \begin{cases} 1 & [\tau_i > \tau_j] \\ l_{ij}(t-1) \text{ otherwise} \end{cases} \tag{5}$$

Where l_{ij} is the binary weight between pre-synaptic node j and post-synaptic node i, τ_j and τ_i are the onset times of pre- and post-synaptic activity. A connection is set if the onset time of the post-synaptic node activity occurs after the onset of pre-synaptic node activity, during a normal trajectory. Therefore, the binary weights will be set between a given node and any other nodes that have followed it in any trajectory in the training set. The binary connections give a simple indication of the co-occurrence of top layer nodes subject to an ordering constraint.

2.2 Novelty Detection

To classify the observed trajectory, the NA nodes at the top layer are switched on and activity traces are propagated through the network. The winning node at the top layer has its activity set to 1, and the non-winning node and NA node activities are calculated as follows:

$$a_k = a_i l_{ki}\big|_{k \neq i} \tag{6}$$

$$n_k(t) = \begin{cases} n_k(t-1) + \Delta n & [a_k = 0] \\ n_k(t-1) & [a_k \neq 0] \end{cases} \tag{7}$$

Where a_k is the activity of top layer node k, l_{ki} is the binary weight between node i and post-synaptic node k, n_k is the corresponding NA node, i is the index of the winning node and Δn is an arbitrary non-zero value. The NA nodes are reset at the beginning of every new trajectory.

The winning node i sends a signal through the binary lateral connections to retard the increase of NA node activity at nodes that are expected to follow in the sequence. An unusual trajectory will produce combinations of nodes dissimilar to sequences in the normal training set, and some NA nodes will not receive a retardation signal through the lateral connections, allowing them to increment their activity (eq. 7). Hence, at each time step, the value of the NA node corresponding to the winning node gives a novelty measure of the current point in the sequence.

It should be noted that the number of top layer nodes active in a sequence is relatively small. Due to the similar widths of the convergent receptive fields at each layer, a node in the top layer can receive a signal from a large area of the input. The progressive "chunking" of trajectory segments that occurs as we progress higher up the network means that, at the top layer, an entire sequence may be represented by as few as 4 or 5 nodes.

3 Training and Results

The network structure and list of learning parameters are shown in table 1. The network was trained with 311 normal trajectories, recorded from a CCTV scene over a period of five days. A human operator examined the data set to remove any unusual trajectories generated by tracking failure or genuinely novel behaviour. A set of 20 normal trajectories and 16 unusual trajectories were obtained for testing.

The classification results are shown in table 2 and a selection of trajectories are shown in figure 3. All unusual trajectories were correctly classified, three examples of which are shown in figs. 3a-c. Figure 3a illustrates the unusual occurrence of a pedestrian leaving one vehicle and entering another, while figs. 3b and 3c are novel due to the co-occurrence of trajectory segments never observed in the same sequence during learning.

Figs. 3d-f show the three false positives. Upon examination of the normal test set, the trajectory in fig. 3d was found to have no similar trajectories among the training data, due to pedestrians typically favouring a more direct route out of the car-park (towards the bottom left of the surveillance image). Figs. 3e and 3f show trajectories assigned to the normal data set, but which nevertheless show high curvature not found in any other normal examples.

Table 1. Network parameters. Note: Training only occurs at the fan-in connections to layers 2 and 3, layer 3 does not need an activity trace and layer 1 receptive fields are non-overlapping

Layer, n	Layer Size	Receptive field width	Trace decay, γ_n	Learning rate, α_n	Training epochs
1	72 x 54	9 (pixels)	0.8	—	—
2	23 x 32	11 (nodes)	0.8	0.01	100
3	7 x 12	11 (nodes)	—	0.008	100

Table 2. Classification results. Note: The 17 correctly classified normal trajectories are not shown, i.e. trajectories with no novel points

	Classification	
Data	Normal	Unusual
Normal	17	3
Unusual	0	16

(a)   (b)   (c)

(d)   (e)   (f)

Fig. 3. Trajectories with points classified as normal (white) or unusual (black). White arrows indicate direction of motion

4 Discussion

As described in section 2.2, a "retardation" signal through the binary lateral connections prevents the NA nodes which correspond to normal sequences of winning nodes, from accumulating non-zero activity, which is the measure used to classify trajectories as novel. Due to the converging feedforward receptive fields and relative sparsity of top layer activation, the generality of the "expected" sequences expressed by the binary lateral connections is quite broad. In fact, the system was able to correctly classify a normal test set, previously unseen by the network.

The main advantage of this system compared with other "global" novelty detectors is the continuous nature of the classification, it is not necessary to wait for an entire trajectory, or a significant portion, to be completed before it is submitted for classification. This is important for surveillance applications where crime prevention requires alarm reporting as the event is developing. The NA nodes effectively express the consistency of the current node with preceding nodes in the sequence, with the binary lateral connections giving an indication of an "expected" forward path that may be taken across the top layer.

The misclassified normal trajectories shown in table 2 highlight one of the main problems with self-organising systems, that of sensitivity to the sample distribution of the training data. When such false positives are encountered, the trajectory may simply be added to the normal training set and the system retrained. Future work will address such issues, particularly the matter of incremental learning, so that previously unseen normal behaviour can be learned without needing to retrain the entire network.

References

1. Foresti, G.L., Mähönen, P., Regazzoni, C.S. (eds): Multimedia Video-Based Surveillance Systems: Requirements, Issues and Solutions. Kluwer Academic Publishers
2. Stauffer, C., Grimson, W.E.L.: Learning Patterns of Activity Using Real-Time Tracking. IEEE Trans. PAMI, Vol. 22, No. 8 (2000)
3. Mattone, R., Glaeser, A., Buman, B.: A New Solution Philosophy for Complex Pattern Recognition Problems. In: Foresti, G.L., Mähönen, P., Regazzoni, C.S. (eds): Multimedia Video-Based Surveillance Systems: Requirements. Kluwer Academic Publishers
4. Johnson, N., Hogg, D.: Learning the Distribution of Object Trajectories for Event Recognition. Proc. BMVC, Vol. 2 (1995)
5. Srinivasa, N., Narendra, A.: A Topological and Temporal Correlator Network for Spatiotemporal Pattern Learning, Recognition and Recall. IEEE Trans. Neural Networks, Vol. 10, No. 2 (1999)
6. Owens, J., Hunter, A.: Application of the Self-Organising Map to Trajectory Classification. IEEE Third International Workshop on Visual Surveillance (2000)
7. Fukushima, K., Miyake, S.: Neocognitron: A New Algorithm for Pattern Recognition Tolerant of Deformation and Shifts in Position. Pattern Recognition, Vol. 15, No. 6 (1982)
8. Wang, R.: A Hybrid Learning Network for Shift-Invariant Recognition. Neural Network, Vol. 14 (2001)
9. Liaw, J.-S., King, I.K., Arbib, M.A.: Visual Perception of Translational and Rotational Motion. In: Mohan, R. (ed.): Progress in Neural Networks, Vol. 4: Machine Vision (1997)
10. Stringer, S.M. and Rolls, E.T.: Position Invariant Recognition in the Visual System with Cluttered Environments. Neural Networks, Vol. 13 (2000)
11. Földiák, P.: Learning Invariance from Transformation Sequences. Neural Computation, Vol. 3 (1991)
12. Hebb, D.: The Organisation of Behaviour. Wiley, New York (1949)

Vergence Control and Disparity Estimation with Energy Neurons: Theory and Implementation

Wolfgang Stürzl, Ulrich Hoffmann, and Hanspeter A. Mallot

Universität Tübingen, Zoologisches Institut, Kognitive Neurowissenschaften,
72076 Tübingen, Germany
wolfgang.stuerzl@uni-tuebingen.de
http://www.uni-tuebingen.de/cog/

Abstract. The responses of disparity-tuned neurons computed according to the energy model are used for reliable vergence control of a stereo camera head and for disparity estimation. Adjustment of symmetric vergence is driven by minimization of global image disparity resulting in greatly reduced residual disparities. To estimate disparities, cell activities of four frequency channels are pooled and normalized. In contrast to previous active stereo systems based on Gabor filters, our approach uses the responses of simulated neurons which model complex cells in the vertebrate visual cortex.

1 Introduction

In the literature a variety of stereo algorithms has been proposed for estimating depth from disparities, i.e. local image shifts caused by the different positions of the two eyes or cameras, see for example [6], [9]. Findings from psychophysical studies of human stereo vision and vergence adjustment were used in active stereo camera systems. Based on physiological recordings from binocular neurons in the visual cortex of cats and monkeys, so-called energy models for stereo processing have been suggested. In this paper we will describe how the responses of disparity-tuned energy neurons can be used for vergence control of a stereo camera head and for disparity estimation on images taken by two monochrome video cameras.

2 Binocular Energy Model

We give a short overview on energy neurons modeling disparity-tuned cells found in visual cortex of mammals (e.g. [3], [7]).

Linear Stage: In this paper, we consider only vertically oriented receptive fields since they are best suited to deal with horizontal disparities. The receptive field of a quadrature pair of monocular linear neurons is modeled (e.g. [1], [8]) as

$$f_\nu(x, y, \varphi) = \exp\left(-\frac{x^2}{2\sigma_x^2} - \frac{y^2}{2\sigma_y^2}\right)\left(\cos(2\pi\nu x + \varphi) + i\sin(2\pi\nu x + \varphi)\right) \ . \quad (1)$$

J.R. Dorronsoro (Ed.): ICANN 2002, LNCS 2415, pp. 1255–1260, 2002.

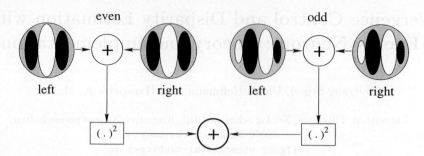

Fig. 1. Illustration of (3): Complex cells combine the output of linear neurons modeled as convolution of left and right images with even/odd (cosine/sine) Gabor kernels including a quadratic nonlinearity.

The output of the linear neurons is formulated mathematically as a convolution of image I with the receptive field function f_ν,

$$Q_\nu(x, y, \varphi) = |Q_\nu(x, y)|e^{i(\phi(x, y)+\varphi)} = \int f_\nu(x - \xi, y - \eta, \varphi)I(\xi, \eta)\, d\xi d\eta \quad . \quad (2)$$

Complex Cells: Binocular complex cells combine the output of the linear filters applied to both images as shown in Fig. 1,

$$C_\nu = |Q_{l\nu} + Q_{r\nu}|^2 = \left(\text{Re}[Q_{l\nu}] + \text{Re}[Q_{r\nu}]\right)^2 + \left(\text{Im}[Q_{l\nu}] + \text{Im}[Q_{r\nu}]\right)^2 \quad . \quad (3)$$

Two different models have been proposed for complex cells tuned to disparity D, see for example [1]:

- Phase-shift: corresponding left and right receptive fields have different phases, $C_{D\nu}(x, y, \varphi) = |Q_{l\nu}(x, y, \varphi + 2\pi\nu\frac{D}{2}) + Q_{r\nu}(x, y, \varphi - 2\pi\nu\frac{D}{2})|^2$.
- Position-shift: corresponding left and right receptive fields are centered at shifted positions, $C_{D\nu}(x, y, \varphi) = |Q_{l\nu}(x + \frac{D}{2}, y, \varphi) + Q_{r\nu}(x - \frac{D}{2}, y, \varphi)|^2$.

As discussed e.g. in [8], complex cells of the phase-shift type have limited range of preferred disparities, $D \in [-\frac{\pi}{\nu}, \frac{\pi}{\nu}]$. Consequently, only neurons with low central frequencies (small ν) can code large disparities. Since there is evidence that in humans high frequencies also contribute to perception of large disparities without utilizing a coarse-to-fine strategy [4], we use energy neurons of the pure position-shift type. Phase φ is set to zero.

3   Implementation

We use four frequency channels ($k = 1, 2, 3, 4$) with bandwidth of two octaves, i.e. $\frac{\sigma_\nu}{\nu} = \frac{3}{5}$, and center frequencies $\nu_k \in \{5\nu_0, 10\nu_0, 20\nu_0, 40\nu_0\}$, where $\nu_0 = \frac{1}{L_x}$ is the minimal frequency determined by image width L_x (see Fig. 2 b). The Gabor functions were sampled and shifted to yield zero DC component. The resulting convolution kernels for $5\nu_0$ are shown in Fig. 2 a.

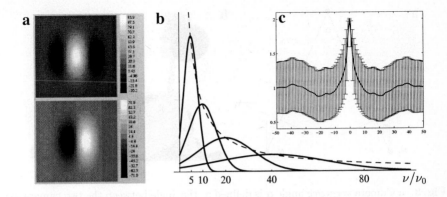

Fig. 2. a Gabor filters with bandwidth of two octaves: convolution kernel (size 49×49 pixels) for the lowest frequency $\nu_1 = 5\nu_0$. **b** In the frequency domain, complex Gabor functions are Gaussians of width $\sigma_\nu = \sigma_x^{-1}$ centered at $\nu/\nu_0 = 5, 10, 20, 40$. The inset (**c**) shows a tuning curve of a normalized complex cell (5) tuned to zero disparity (for each stimulus disparity 100,000 responses were evaluated, error bars represent standard deviation).

Monocular normalization: To reduce the influence of "inter-ocular" contrast differences the filter outputs are normalized according to

$$\hat{Q}_k(x_i, y_i) = \frac{Q_k(x_i, y_i)}{(L_x L_y)^{-1} \sum_j |Q_k(x_j, y_j)|} \tag{4}$$

(L_x and L_y are image width and height respectively).

Combination of frequency channels and binocular normalization: Complex cells of the different frequency channels, but tuned to same disparity are combined and normalized to reduce influence of local image contrast, resulting in a normalized complex cell,

$$\hat{C}_D(x_i, y_i) = \frac{\sum_k C_{Dk}(x_i, y_i)}{\epsilon + \sum_k |\hat{Q}_{lk}(x_i + \frac{D}{2}, y_i)|^2 + |\hat{Q}_{rk}(x_i - \frac{D}{2}, y_i)|^2} . \tag{5}$$

ϵ is a constant avoiding high complex cell activity in case of very low contrast which causes $\sum_k |\hat{Q}_{lk}|^2 + |\hat{Q}_{rk}|^2 \approx 0$ (in the current implementation ϵ is set to 4.0). We can rewrite (5) using (3) and dropping index k,

$$\hat{C}_D = \frac{\sum_k |\hat{Q}_{lk} + \hat{Q}_{rk}|^2}{\epsilon + \sum_k |\hat{Q}_{lk}|^2 + |\hat{Q}_{rk}|^2} = \frac{\sum 2|\hat{Q}_l|^2 + 2|\hat{Q}_r|^2 - |\hat{Q}_l - \hat{Q}_r|^2}{\epsilon + \sum |\hat{Q}_l|^2 + |\hat{Q}_r|^2}$$

$$= \frac{2 \sum |\hat{Q}_l|^2 + |\hat{Q}_r|^2}{\epsilon + \sum |\hat{Q}_l|^2 + |\hat{Q}_r|^2} \left(1 - \frac{\sum |\hat{Q}_l - \hat{Q}_r|^2}{2 \sum |\hat{Q}_l|^2 + |\hat{Q}_r|^2} \right) \tag{6}$$

$$\approx 2 \left(1 - \frac{\sum |\hat{Q}_l - \hat{Q}_r|^2}{2 \sum |\hat{Q}_l|^2 + |\hat{Q}_r|^2} \right) , \quad \text{if} \quad \epsilon \ll \sum |\hat{Q}_l|^2 + |\hat{Q}_r|^2 . \tag{7}$$

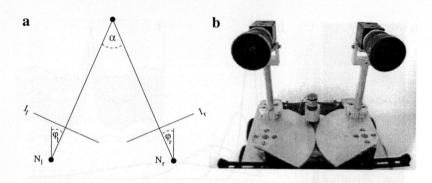

Fig. 3. a Camera vergence angle α is defined as the angle between the two camera axes ($N_{l/r}$: left/right camera nodal point, $I_{l/r}$: left/right image). **b** Stereo camera head with symmetrical vergence adjustment ($\varphi_l = \varphi_r = \frac{1}{2}\alpha$) ensured by two cogwheels driven by a single stepper motor. Each of the two monochrome cameras has a field of view of approx. 80°. Nodal point separation is 14.5 cm.

From (6) and (7) we see that $\hat{C}_D$ reaches its maximum value if $\hat{Q}_l$ equals $\hat{Q}_r$, i.e. if the left and right receptive field "look" at corresponding image parts. Using the triangle inequality it can be shown that $\hat{C}_D \in [0, 2]$:

Verging Camera Head: Taking the mean of $\hat{C}_D$ for each disparity over all positions, we compute responses of vergence controller cells C_D^V. From their activity global image disparity is estimated according to

$$D_{\text{glob}}^{\text{est}} = \arg\max_D C_D^V \quad , \tag{8}$$

$$C_D^V = (L_x L_y)^{-1} \sum_i \hat{C}_D(x_i, y_i) \quad . \tag{9}$$

We use a large range of (global) disparities with inhomogeneous resolution (highest at zero disparity), i.e. $D \in \{0, \pm 1, \pm 2, \pm 3, \pm 5, \pm 7, \pm 10, \pm 14, \pm 19, \pm 25, \pm 32, \pm 40, \pm 49, \pm 59 \text{ pixels}\}$.

As long as $D_{\text{glob}}^{\text{est}} \neq 0$, camera vergence is changed symmetrically ($\varphi_l = \varphi_r = \frac{1}{2}\alpha$) using a stepper motor which drives two cogwheels (see Fig. 3). The vergence angle α is approximately proportional to $D_{\text{glob}}^{\text{est}}$ (for our stereo system we use $\alpha = 0.5° D_{\text{glob}}^{\text{est}}$).

Disparity Estimation: After adjustment of vergence angle, the range of residual space dependent disparities is usually greatly reduced. Residual disparities are analysed with neurons tuned to disparities $D \in \{0, \pm 1, \pm 2, \ldots, \pm 10\}$. The population activity of these neurons is a rich code of stereo information that will be useful for many tasks. If disparity maps are sought, they can be obtained as the preferred disparity of the locally most active neuron,

$$D_{\text{loc}}^{\text{est}}(x_i, y_i) = \arg\max_D \hat{C}_D(x_i, y_i) \quad . \tag{10}$$

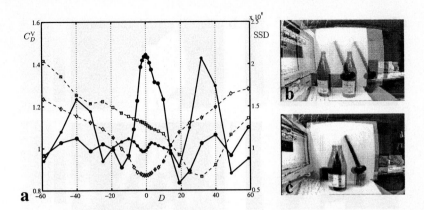

Fig. 4. a Responses of vergence controller neurons (small dots and circles) are compared to SSD (rectangles and diamonds, dashed curves) of shifted left and right images (shifts were applied according to the preferred disparity of the neurons, note different scaling on y-axes). Initially when camera axes are aligned corresponding to vergence angle 0° (see upper superimposed image, **b**) peak response is at $D = 32$ (small dots). After vergence adjustment global disparity is minimized corresponding to maximum activity of C_0^V (circles). At the resulting vergence angle of approx. 16° the residual disparities are much smaller and better fitted to the range of the local disparity detectors (**c**).

We also compute a confidence value for each pixel using

$$c(x_i, y_i) = \frac{\max_D \hat{C}_D(x_i, y_i)}{N_D^{-1} \sum_D \hat{C}_D(x_i, y_i) + \epsilon_D} \quad , \tag{11}$$

where ϵ_D was set to $1/4$ of maximum cell response, i.e. 0.5 and $N_D = 21$ is the number of preferred neuron disparities used. Locations where true disparity is out of the considered range or where disparity estimation is simply impossible, e.g. due to occlusions, will receive low confidence.

4   Results

In Fig. 4 the response of the vergence controller neurons is compared to the sum of squared differences (SSD) between left and right image. At the optimal vergence angle (approx. 16° in this example), i.e. highest response of C_0^V, the SSD has its minimum at zero image shift corresponding to maximal image correlation. This is in agreement with studies on vergence adjustment in humans, e.g. [5].

In order to check the estimated disparity map calculated after vergence adjustment according to (10) we compute a "cyclopean" view [2], i.e. we fuse left and right image using the disparity map,

$$I_{\text{fused}}(x, y) = \frac{1}{2}\Big(I_l\big(x + \tfrac{1}{2}D_{\text{loc}}^{\text{est}}(x, y), y\big) + I_r\big(x - \tfrac{1}{2}D_{\text{loc}}^{\text{est}}(x, y), y\big)\Big) \quad . \tag{12}$$

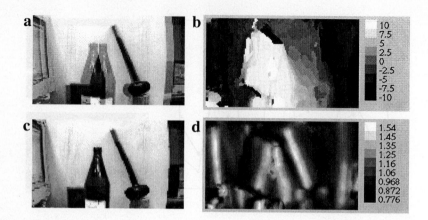

Fig. 5. Disparity estimation on stereo image: Superimposed image after vergence adjustment (**a**). The left and right images can be fused using the disparity map (**b**) and (12) eliminating all double vision (**c**). The confidence map (**d**) calculated according (11) shows high values at contours of high contrast.

By comparing the resulting fused image (Fig. 5 c) with the superposition of left and right image (Fig. 5 b) one can see that double vision has almost completely vanished.

Further evaluation of the proposed stereo algorithm will be done on a mobile robot using the activities of the disparity-tuned cells as representation of places.

References

1. Fleet, D., Heeger, D., Wagner, H.: Neural encoding of binocular disparity: Energy model, position shifts and phase shifts. Vision Research **36** (1996) 1839–1857
2. Henkel, R.D.: Fast stereovision by coherence detection. In G. Sommer, K.D., Pauli, J., eds.: Computer Analysis of Images and Patterns, LCNS 1296 (1997) 297–30
3. Hubel, D., Wiesel, T.: Receptive fields, binocular interaction and functional architecture in the cat's visual cortex. Journal of Physiology **160** (1962) 106–154
4. Mallot, H., Gillner, S., Arndt, P.: Is correspondence search in human stereo vision a coarse-to-fine process? Biological Cybernetics **74** (1996) 95–106
5. Mallot, H., Roll, A., Arndt, P.: Disparity-evoked vergence is driven by inter-ocular correlation. Vision Research **36** (1996) 2925–2937
6. Marr, D., Poggio, T.: A cooperative computation of stereo disparity. Science **199** (1976) 283–287
7. Ohzawa, I., DeAngelis, G., Freeman, R.: Stereoscopic depth discrimination in the visual cortex: Neurons ideally suited as disparity detectors. Science **249** (1990) 1037–1041
8. Qian, N.: Computing stereo disparity and motion with known binocular cell properties. Neural Computation **6** (1994) 390–404
9. Sanger, T.: Stereo disparity computation using gabor filters. Biological Cybernetics **59** (1988) 405–418

Population Coding of Multiple Edge Orientation

Niklas Lüdtke, Richard C. Wilson, and Edwin R. Hancock

University of York, Heslington, York YO10 5DD, UK,
{niklu, wilson, erh}@cs.york.ac.uk

Abstract. We present a probabilistic population coding model of Gabor filter responses. Based on the analytically derived orientation tuning function and a von Mises mixture model of the filter responses, a probability density function of the local orientation in a given point can be extracted through a parameter estimation procedure. The probability density captures angular information at edges, corners or T-junctions and also yields a contrast invariant description of the certainty of each orientation estimate, which can be characterized in terms of the entropy of the corresponding mixture component.

1 Introduction

Over the past two decades, population coding has emerged as an important model for the representation of sensory as well as motor variables in the brain [1, 3,4]. Typically, a sensory population code consists of a number of neurons broadly tuned to one or several stimulus parameters. "Classical" decoding schemes such as the population vector [1] are able to extract a unique value of an encoded quantity at a higher accuracy than individual neural tuning properties would allow for.

More recently, probabilistic decoding techniques have been developed which extract an entire probability density function (pdf) from the ensemble activities. Here, substantial contributions have been made by Sanger [3] as well as Zemel, Dayan and colleagues [4,5], whose work is mainly concerned with modelling biological information processing, particularly the perception of motion and orientation. The advantage of a probabilistic approach over "reductionist" concepts, such as the population vector, is the superior representational capacity due to the possibility of decoding multimodal distributions. Thus, a probabilistic decoding method can recover multiple measurements and, in addition, characterize the certainty of these estimates in terms of the "width" or concentration of the pdf around its modes.

In this paper, probabilistic population coding is applied to a computer vision task, namely the detection and representation of local edge orientation in computer vision. The ability to represent ambiguous inputs is used to extract multiple orientations at corner points and T-junctions, where classical edge detectors usually fail. Unlike in biological models of population coding, in which neural firing rates are considered as random variables following Poisson statistics [4],

J.R. Dorronsoro (Ed.): ICANN 2002, LNCS 2415, pp. 1261–1267, 2002.
© Springer-Verlag Berlin Heidelberg 2002

the filters are deterministic operators acting on a stochastic visual input. Even though the filtering process is deterministic per se, the responses themselves become stochastic, since they are functions of a random variable, the local contour orientation θ.

In order to recover the whole pdf of the local orientation, $p(\theta)$, representing the uncertainty and multiplicity of θ, the pdf parameters have to be estimated from the given set of filter responses, which can be achieved indirectly via the pdf of the responses. Based on a theoretical model of the tuning function of Gabor filters and the assumption of a von Mises mixture distribution of the angular input variable (local contour orientation), the corresponding pdf of the *responses* can be derived, and the mixture parameters can be determined through maximum likelihood estimation using the EM-algorithm. Thus, the pdf of the input variable is decoded from the population activities (filter outputs).

2   Orientation Tuning of Gabor Filters

The tuning function of an odd-symmetric Gabor filter can be derived analytically for a sinusoidal grating of arbitrary orientation. The result is of general relevance, since any fully anisotropic input can be expanded into a Fourier series of sinusoids with different wavelengths but equal orientation. Real images are likely to contain edges subject to some degree of blur that therefore have a dominant spatial ground frequency in their spectrum, while higher frequency components are comparatively weak. For curved edges, the aforementioned still holds approximately within the effective range of the filter mask. The characteristic spatial frequency of the edge structure at a particular location influences the tuning width of the filter responses. For practical reasons, the corresponding wavelength λ_s (in pixels) will be used. We consider a sinusoid of orientation θ:

$$S(x, y, \theta) = \sin\left[k_s\left(x \cos \theta + y \sin \theta\right)\right] ; \quad k_s = 2\pi/\lambda_s. \tag{1}$$

For simplicity, we choose a Gabor filter with aspect ratio one (radial symmetry), wavelength λ_f and vertical orientation, and we define analogously $k_f = 2\pi/\lambda_f$. Hence:

$$\mathcal{G}_{odd}(x, y) = \exp\left(-\tfrac{x^2+y^2}{2\sigma^2}\right) \sin(k_f x). \tag{2}$$

The orientation tuning function, $f(\theta)$, is the convolution of the filter with the sinusoid at the origin $(0,0)$:

$$f(\theta) = (\mathcal{G}_{odd} * S) = \int\limits_{-\infty}^{\infty} \int\limits_{-\infty}^{\infty} \mathcal{G}_{odd}(x, y) S(x, y, \theta) \, dx \, dy \propto \sinh\left(\underbrace{k_f k_s \sigma^2}_{\kappa_0} \cos\theta\right). \tag{3}$$

The parameter κ_0 is a so-called *concentration parameter*. Its reciprocal value is related to the angular variance and therefore controls the orientation tuning width. κ_0 depends on the two known filter properties $k_f = 2\pi/\lambda_f$ and σ, as well as on the unknown quantity $k_s = 2\pi/\lambda_s$. In the following, it will be assumed that the tuning functions for all filters are identical apart from an angular shift ψ indicating the preferred orientation:

$$f(\theta - \psi) = C \sinh[\kappa_0 \cos(\theta - \psi)] \, . \tag{4}$$

Here, $C = 1/\sinh(\kappa_0)$ is a normalization constant, so that $f(\theta; \psi) \in [-1, 1]$.

3 Population Coding in the "Gabor-Ensemble"

While a straight edge has a single well defined orientation, problems with conventional edge detectors occur, when either the edge is curved or multiple orientations exist within the filter mask. In addition, noise in the image can create uncertainties even in the angle of a straight edge. We can accommodate all these possibilities by adopting a stochastic model of edge orientation where the edge angle θ is governed by a probability density function $p(\theta)$.

A Gabor filter bank is an ensemble of linear orientation selective units. Therefore, the *principle of superposition* holds: the response profile for complex intensity structure, such as in corner points, where several edges coincide, is a linear combination of the response activities for the individual edge components. This assumption is crucial for the following derivations.

3.1 Expectation Values of Filter Responses

Let ψ be the preferred orientation of a filter. Assuming that superposition holds, the expectation value of the response profile is given by the convolution of the tuning function $f(\theta)$ and the pdf of the the the stimulus orientation $p(\theta)$ [4]:

$$\bar{r}(\psi) = \int_0^{2\pi} f(\theta - \psi) p(\theta) \, d\theta \, . \tag{5}$$

This equation describes the process of encoding of the pdf $p(\theta)$ in the expected response function $\bar{r}$ which is continuous in ψ. The average ensemble activities (the average response profile) are a sample of this function for a discrete set of filter orientations ψ_ν.

Thus, the decoding of the pdf $p(\theta)$ is a *deconvolution*, which is an ill-posed problem likely to require some kind of regularization. In the following, a parametric model of the expected filter responses is derived, based on the orientation tuning function and a mixture model of $p(\theta)$, whereby regularization is achieved *implicitly*.

3.2 A Mixture Model of Local Orientation

The encoding equation (5) is general in the sense that $p(\theta)$ can take any form and is not restricted to unimodality. We now choose a particular type for the pdf of edge orientation that we expect to see in the image. The probability density of the stimulus orientation is modelled as a mixture of von Mises distributions[2], where each principal edge orientation is represented by a mixture component with a mean value $\bar{\theta}_i$ and a concentration parameter κ_i:

$$p(\theta) = \frac{1}{2\pi} \sum_{i=1}^{m} \frac{P(i)}{I_0(\kappa_i)} e^{\kappa_i \cos(\theta - \bar{\theta}_i)}, \quad \text{with} \sum_{i=1}^{m} P(i) = 1. \tag{6}$$

Here, I_0 is the modified Bessel function of first kind and order zero and the term $1/2\pi I_0(\kappa_i)$ serves as a normalization factor of the i-th mixture component. Eqn.(6) can be considered a circular analogue of the Gaussian mixture density. The κ_i correspond to the $1/\sigma_i$ and the $\bar{\theta}_i$ to the μ_i in a Gaussian mixture component. The $P(i)$ are the mixing coefficients. The number of mixture components, m, will be limited to two or at most three, which incorporates the essential cases of multiple edge orientation, i.e., corners and junctions.

4   The Probability Distribution of Responses

In addition to eqn.(5), the expected response profile can also be thought of as an average over the filter responses themselves. Let $p(r; \psi)$ be the probability density over the response value of a filter of preferred orientation ψ. Then the expected (continuous) response profile is given by

$$\bar{r}(\psi) = \int_{-1}^{1} r \, p(r; \psi) \, dr. \tag{7}$$

The filter responses r_ν obtained at a particular location in an image are instances of stochastic variables, even though the filtering is, per se, a deterministic operation. Unlike standard biological models of population coding, in which neural firing rates are usually considered random variables following Poisson statistics [4], the randomness of the filter responses is created solely by the stochastic nature of the input variable.

One way of estimating the parameters of the mixture model $p(\theta)$ is through a maximum likelihood estimation. Let $\Theta = \{P(1) \ldots P(m), \bar{\theta}_1 \ldots \bar{\theta}_m, \kappa_1 \ldots \kappa_m\}$ denote the set of mixture parameters. It is essential to know the likelihood of the individual filter responses given their preferred orientations ψ_ν, the parameters Θ and κ_0, i.e. $p(r_\nu; \psi_\nu, \Theta, \kappa_0)$, in order to calculate the total likelihood of a given response profile,

$$\mathcal{L}\{r_1 \ldots r_n | \Theta, \kappa_0\} = \prod_{\nu=1}^{n} p(r_\nu; \psi_\nu, \Theta, \kappa_0).$$

The likelihood $\mathcal{L}$ depends on the parameters of the mixture pdf $p(\theta)$, together with the parameter κ_0 specifying the tuning function. Maximum likelihood estimation of the mixture parameters can be performed using standard techniques such as the EM algorithm. The remaining tuning parameter κ_0 can be obtained from previous measurement(s) using test stimuli (straight edges).

The task is now to find the pdf of a *function of a random variable*, since the filters transform the edge orientation θ via their tuning function given by (4). There is insufficient space to fully derive the resulting distribution, so we limit ourselves to quoting the result:

$$p(r;\psi) = \frac{1}{\pi C \kappa_0} \sum_{i=1}^{m} \frac{P(i)}{I_0(\kappa_i)} \frac{\exp\left[\frac{\kappa_i}{\kappa_0} \sinh^{-1}\left(\frac{r}{C}\right) \cos(\psi - \bar{\theta}_i)\right]}{\sqrt{1 + \left(\frac{r}{C}\right)^2} \sqrt{1 - \left[\frac{1}{\kappa_0}\sinh^{-1}\left(\frac{r}{C}\right)\right]^2}} \times$$

$$\cosh\left(\kappa_i \sqrt{1 - \left[\frac{1}{\kappa_0}\sinh^{-1}\left(\frac{r}{C}\right)\right]^2} \sin(\psi - \bar{\theta}_i)\right) \quad (8)$$

This density is again a mixture model. It is then possible to estimate the model parameters with the expectation maximization (EM) algorithm. Due to the complexity of the response pdf, the resultant update equations for each mixture component form a transcendental system for the pair of parameters $(\bar{\theta}_i, \kappa_i)$. However, one can obtain closed form approximations with excellent accuracy.

5   Measuring Certainty

The probabilistic approach not only yields an estimate for the different edge orientations present in the neighbourhood of the considered point (x, y) but also provides information about the certainty of these measurements through the concentration parameters, the κ_i. Zemel and colleagues [5] proposed the *resultant length* [2], which is given by

$$\rho = E\{\cos(\theta - \bar{\theta})\} = \frac{I_1(\kappa)}{I_0(\kappa)}, \quad (9)$$

as a measure of certainty within the framework of the "directional-unit Boltzmann machine" ($\bar{\theta}$ is the mean direction; $E\{.\}$ denotes the expectation value). For practical purposes, we argue that it is more suitable to use the Kullback-Leibler divergence of a given mixture component $p_i(\theta)$ from the distribution of maximum entropy, i.e. the uniform distribution $q(\theta) = 1/2\pi$.

$$K(p_i, q) = \int_0^{2\pi} p_i(\theta) \ln\left[\frac{p_i(\theta)}{q(\theta)}\right] d\theta = \frac{I_1(\kappa)}{I_0(\kappa)}\kappa - \ln[I_0(\kappa)] \quad (10)$$

This quantity, also called *relative entropy*, is, unlike the entropy itself, always positive. For convenience, the relative entropy can be normalized through a sigmoid function $g(x) = \frac{2}{1+e^{-x}} - 1$, which yields:

$$\gamma = g[K(p_i, \tfrac{1}{2\pi})] = \frac{2}{1 + \frac{\exp(-\kappa I_1(\kappa)/I_0(\kappa))}{I_0(\kappa)}} - 1. \quad (11)$$

A comparison of the certainty measures is shown in Fig. 1 (right). Our measure, γ, allows better distinction of certainties for concentration parameters of middle range ($10 < \kappa < 40$), which is important for real images.

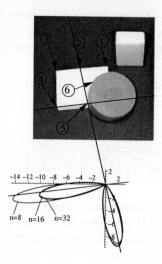
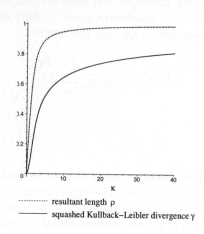

·········· resultant length ρ

———— squashed Kullback–Leibler divergence γ

Fig. 1. Left: Polar plot of the pdf $p(\theta)$ extracted at a T-junction (point 5) for 8,16, and 32 filters with overlaid ground truth (solid lines). Overall, the accuracy of the orientation estimate does not depend critically on the number of filters. **Right:** Comparison of the resultant length ρ (c.f. eqn. 9; dashed curve), proposed by Zemel and colleagues [5], with our certainty measure γ (solid curve) based on relative entropy (c.f. eqn. 11). The resultant length "saturates" too quickly, whereas γ is more suitable to discriminate between different certainties in the range of $10 < \kappa < 40$ which is important for real images.

6   Results

Fig. 1 (left) shows the pdf of orientation extracted at a T-junction (point 5) with three filter banks (8, 16 and 32 filters). As expected, two mixture components have been found, corresponding to the two principal orientations. Comparison to the ground truth shows that the orientation estimates are fairly robust with respect to the size of the population. However, with 8 filters the certainty is overestimated.

7   Conclusions

We have presented a probabilistic population coding model of Gabor filter responses. By treating the edge orientation in each point as a stochastic variable governed by a mixture of von Mises distributions, we are able to model multiple orientations and their uncertainty. We have used the EM-algorithm to estimate the model parameters on various edge configurations in a real image.

References

1. A.P. Georgopoulos, A.B. Schwarz, and R.E. Kettner. Neural population coding of movement direction. *Science*, 233:1416–1419, 1986.

2. K.V. Mardia. *Statistics of Directional Data*. Academic Press, London and New York, 1972.
3. D.T. Sanger. Probability density estimation for the interpretation of neural population Codes. *Journal of Neurophysiology*, 76(4):2790–2793, 1996.
4. R. Zemel, P. Dayan, and A. Pouget. Probabilistic interpretation of population codes. *Neural Computation*, 10(2):403–430, 1998.
5. R. Zemel, C.K.I. Williams, and M.C. Mozer. Lending direction to neural networks. *Neural Networks*, 8(4):503–512, 1995.

A Neural Model of the Fly Visual System Applied to Navigational Tasks

Cyrill Planta, Jörg Conradt, Adrian Jencik, and Paul Verschure

Institute of Neuroinformatics, Winterhurerstrasse 190, 8057 Zuerich, Switzerland

Abstract. We investigate how elementary motion detectors (EMDs) can be used to control behavior. We have developed a model of the fly visual system which operates in real time under real world conditions and was tested in course and altitude stabilization tasks using a flying robot. While the robot could stabilize gaze i.e. orientation, we found that stabilizing translational movements requires more elaborate preprocessing of the visual input and fine tuning of the EMDs. Our results show that in order to control gaze and altitude EMD information needs to be computed in different processing streams.

1   Introduction

Flies possess an elaborated visual system that supports their fast and accurate flying maneuvers. Whenever the relative position or orientation of the fly to the world changes, a global change of the illuminance distribution occurs, called optic flow. Optic flow fields of different movements such as rotation (yaw), translation, or moving upwards or downwards (lift) may resemble each other within limited areas of the visual field, but never across the entire field of view. Flies are believed to exploit optical flow by calculating the local image movements among their visual field with EMDs and integrate these signals. This processing takes place in the wide-field integrating (WF) neurons, cells with large dendritic trees whose receptive fields match certain optic flow fields ([4]).

In order to understand how EMD information is used in motor control, robots have been constructed which use EMDs and WF neurons to detect rotational egomotion and avoid obstacles (e.g. [2], [7] ,[3]). However very few researchers (e.g. [3]) have started to investigate whether the same concepts also work for robots moving in three dimensions and what the requirements are when movements other than rotational egomotions need to be detected. To address these questions we have built a flying robot capable of moving in three dimensions. The robot was controlled by a neural network which is a model of the fly visual system. We analyzed the robot's behavior in a course and altitude stabilization task, during which the trajectory, the WF neurons's responses, and the optomotor signals were recorded and analyzed (See figure 1A).

J.R. Dorronsoro (Ed.): ICANN 2002, LNCS 2415, pp. 1268–1274, 2002.

2 Methods

Robot: The robot was constructed using a blimp with a payload of 120 g. We attachedtwo b/w cameras laterally on the surface of the blimp with a field of view of 56° x 42° (Fig. 1A(d)). We mounted four motors with propellers on the ventral side of the blimp (Fig. 1A(c)). Two propellers were aligned in the horizontal plane and two in the vertical plane. The propellers could drive the robot forward at a speed of up to 3m/s and rotate it with an angular velocity of up to 112°/s. We used three radio links to exchange the onboard robot data with the neural network. Each camera had a separate video transmitter and the propellers were controlled by a third radio communication module (Fig. 1A(e)).

3D Tracking: In order to analyze the behavior of the robot accurately we tracked both the position and the orientation of the robot online. Two ceiling cameras (Fig. 1A(a)) were tracking two IR emitting diode arrays mounted on top of the blimp (Fig. 1A(b)). The transition from the two camera pixel coordinates into a pair of three dimensional coordinates was achieved by a multilayer perceptron. The tracking resolution yielded an accuracy of 5cm.

Software: We used the simulation software IQR421 [6] to implement and control the neural model, the 3D tracking and the robot. For data analysis we used both IQR421 and Matlab.

3 Model

Figure 1B gives an overview of how the fly visual system was modeled. Initially the camera image was downscaled to a resolution of 10x10 pixels. This image was the input for cells with center-surround receptive fields whose basic function was to enrich contrasts in the image similar to the function of laminar processing (Fig.1 Ba). The resulting image was then relayed to four distinct groups of EMDs which computed the four elementary motions using the Reichhardt correlation model [8]. In the fly this processing is thought to take place presynaptically to the lobular cells (Fig.1 B(b)). Like the large lobular cells the WF neurons had distinct large dendrites and integrated the responses of the EMDs (Fig.1 B(c)). There were two WF cells sensitive to left and right rotation and two that were sensitive to up or down translation. Finally all the WF neurons responses from the two processing streams were integrated and transduced linearly into a motor signal (Fig.1 B(d)). This motor signal was directed against the dominating type of movement, i.e. when a left rotation was sensed the robot rotates right and vice versa. If the WF neurons responses were below a certain threshold the robot flew forward. 2647 linear threshold neurons and 3800 synapses were needed to build the model.

4 Results

4.1 General Behavior

In order to understand whether the model could control flight behavior we analyzed in detail the relationship between neuronal responses and motor actions

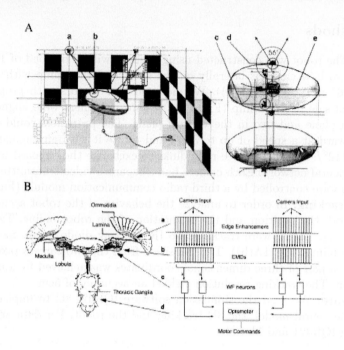

Fig. 1. A: The experimental setup and the robot. The robot moves freely in 3D being tracked by ceiling cameras (a). Visual input stems from two cameras atached laterally on the blimp (d). Checkerboard patterns are attached on the walls and the ground in order to enhance contrast. The neural network steering the robot resides on a PC and communicates via radio links with the robot (see text for further details). B: Schematic of the model. See text for explanation (Left part is adapted from [1]).

in figure 2A. To begin with we wanted to know how and how often the robot makes use of its neural network. Figure 2A describes the relevant processes in the robot and the trajectory of the robot when moving through the experimental room (the whole trajectory can be seen in inset (a)). Starting at the beginning of the big arrow in figure 2A(b) we see that the robot senses left rotation and compensates with right rotation commands. This is followed with a moment of no rotation being sensed. After that the robot senses a tiny right rotation (too small to be visible) and overcompensates until it finds itself in a state of left rotation. Again the robot compensates to the right; at the end of the big arrow no rotation is sensed by the robot. Over a distance of one meter the direction of rotation and the compensating motor command changes four times, showing that the neural network continuously controls the behavior of the robot which results in trajectories that form straight lines.

4.2   Yaw Compensation

In a first experiment we examined whether the wide-field integration of EMDs provides useful information for the robot to compensate for unintended course

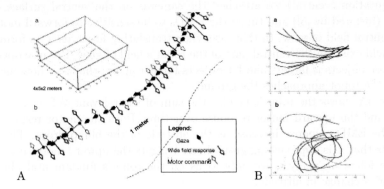

Fig. 2. A: Typical 3D trajectory of the robot. (a): overall trajectory. (b): Close up on (a). Filled black arrows visualize gaze and velocity (length of arrow); thick perpendicular arrows show the directioned sum of the WF neurons responses, where an arrow on the right of the gaze arrow means that the robot senses right rotation and vice versa. Thin arrows denote direction and strength of the current motor command (when no motor command is visible, the robot is flying forwards). These arrows point in the opposite direction of the WF arrows in order to compensate for the rotation. B: Overlaid 2D trajectories of course stabilization experiments. (a) Model enabled, (b) model disabled.

deviations while traveling through the experimental room. The performance was measured as the average angular speed of the robot, its average deviation from course. In this experiments the two cameras were placed laterally on the blimp and only horizontal optical flow was processed. In the control condition the robot would simply fly with both horizontal motors turned on, relying on its aerodynamics.

The robot with the EMD model enabled performed significantly better than the control (an average angular velocity of $1.7(+/- 0.4)°/s$, versus $16.2(+/- 0.6)°/s$ in the control condition). Figure 2B shows the overlaid two dimensional trajectories of experiments with the model enabled (a) and disabled (b). Although the robot cannot maintain its initial gaze it can still pursue a stable course (Fig. 2 B(a)) whereas in the control condition, once the robot has rotational momentum it does not recover from such deviations (Fig. 2 B(b)). This shows that the model can reliably compensate for course deviations.

4.3   Altitude Control

The next step was to test whether the robot could control its altitude and whether the EMD and WF neuron concept could be extended to altitude stabilization. We analyzed two configurations of how such a task could be implemented in our model. In the first configuration (lateral) the cameras were attached laterally on the blimp and the responses of vertically tuned EMDs's were integrated and transformed into a compensating motor response. In the other configuration (ventral) we attached the cameras on the ventral surface of the robot (inspired by [5]) and tuned the EMDs to be sensitive to forward motion in the ventral field of view. In this case when translating forwards a uniform optic flow field exists in the ventral part of the robot's field of view. Because optic flow for close objects is larger than for objects further apart, the optic flow increases when the robot approaches the ground.

Figure 3A shows the robot's height, the sum of the relevant WF neuron activities, and the upward motor response over time. Both the motor response and and the EMD response increase as the altitude of the blimp drops. The robot detects that it is approaching ground and triggers the upward motor command. In the following analysis we only focus on the robot's fundamental ability to "sense" a change in altitude.

After testing different delay values of the EMDs in both configurations in order to find the maximum sensitivity of the EMDs for lift we evaluated whether the lateral configuration was suitable for altitude control. To answer this question we calculated the ratio of correct "upwards" decisions versus wrong upwards "decisions". We also determined whether and how this ratio improves if the robot only makes decisions when the WF neuron signals are above a certain signal threshold. Finally we wanted to see how much information has to be thrown away when such a threshold is used. Figure 3B(a) summarizes this data. The maximum correct/wrong ratio which has the value of 1 in figure 3B(a) was 28. However in this case the robot would only exploit 1 percent of the information available. When no signal threshold is used the correct/wrong ratio lies around chance level. Figure 3 B(b) shows the same ratios for the ventral configuration. The ratio between correct and wrong decisions was lower than in the lateral configuration whereby the maximum in the plot stand for a correct/wrong ratio of 9.

This data suggests that the accuracy of altitude control using EMDs depends on their specific tuning and this accuracy can be increased by tresholding the signal strength.

5   Discussion

We have investigated a model of the fly visual system with the aim to understand how EMD signals can be integrated in order to achieve reliable 3D control of a flying robot. We have analyzed the behavior of the robot in a rotational course compensation task in several runs. Furthermore we evaluated the responses of WF neurons in two configurations of an altitude stabilization task in open loop experiments.

The robot performed well in the course compensation task. This is in accordance with the performance of various insectoid robots ([2], [3]) and also with the performance of other flying robots ([3]). In contrast to these robots our solution used a smaller field of view (two times 56°x42° in our case, versus nearly surround views in [2] and [3]) and also a simpler preprocessing of the image before the EMDs. This emphasizes the robustness of the EMD concept whenever there

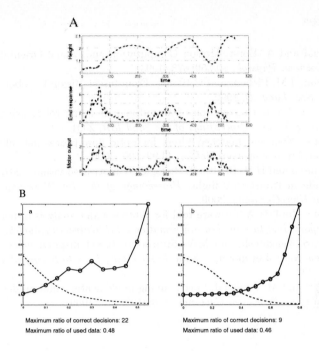

Fig. 3. A: Time course of Height, EMD response, and motor output. B: All curves are normalized in respect to their maximum value which is shown below the plots. (a) The solid curve shows the normalized ratio of correct up judgment to wrong up judgments of the robot in the lateral configuration. The dashed curve shows the ratio of data which was used for a judgement. The x-axis values are the signal tresholds. (b) Correct/wrong ratios and ratio of unused data for the ventral configuration.

is a high amount of optical flow to process, as is the case when we want to detect rotational egomotion.

On the perceptual level we saw that the EMD concept is stretched to its limits when we tried to estimate lift, which is a translational motion and elicits less optical flow than rotation. The fine tuning of EMD's and the application of a signal threshold yielded better results at the cost of discarding potentially useful information. The performance in judging its lift was better in the lateral than in the ventral configuration. One would have assumed that the configuration wich sees more optic flow would be more reliable, namely the ventral configuration. Hence the underlying cause of these results requires further investigation. Nevertheless we can conclude that if EMDs are used for the detection of translatory motion they need a specific tuning and signal thresholds need to be used for accurate judgements. Hence different processing streams are required to solve both rotational and translational compensation tasks.

References

1. M. Egelhaaf and A. Borst. Motion Computation and Visual Orientation in Flies. *Camp. Biochem. Physiol.*, 10:659–673, 1993.
2. Franchescini, J.M. Pichen, and C Blanes. From insect Vision to robot Vision. *Phil. Trans. R. Soc. Lond. B*, 1992.
3. F. Iida. Goal-directed navigation of an automonus flying robot using biologically inspired cheap Vision. *Proceedings of the 32nd ISR*, 2001.
4. H.G. Krapp. Neuronal matched filters for optic flow processing inflying insects. *International review of neurobiology*, 44:93–120, 2000.
5. T.R. Neumann and H.H. Buelthoff. Biologically motivated visual control of attitude and altitude in translatory flight. *Proceedings of the 3rd Workshop Dynamische Perzeption Ulm/Germany*, 2000.
6. P. Verschure. Iqr421: A Software tool for synthesis and analysis of neural systems. *Techn. Report, Institute of Neuroinformatics ETH/University Zurich*, 1997.
7. K. Weber, S. Venkatesh, and M.V. Srinivasan. Insect inspired behaviours for the automonous control of mobile robots. *From Living Eyes to Seeing Machines*, pages 227–248, 1997.
8. J.M. Zanker and M. Egelhaaf. Speed tuning in elementary motion detectors of the correlation type. *Biological Cybernetics*, 80:109–116, 1999.

A Neural Network Model for Pattern Recognition Based on Hypothesis and Verification with Moving Region of Attention

Masao Shimomura[1], Shunji Satoh[2], Shogo Miyake[2], and Hirotomo Aso[1]

[1] Graduate School of Engineering, Tohoku University
Aoba 05, Aramaki, Aoba-ku, Sendai-shi, 980-8579 Japan
{masao,aso}@aso.ecei.tohoku.ac.jp
[2] Graduate School of Engineering, Tohoku University
Aoba 04, Aramaki, Aoba-ku, Sendai-shi, 980-8579 Japan
{shun,miyake}@nlap.apph.tohoku.ac.jp

Abstract. We present a neural network model for pattern recognition which works successfully even if only a part of a pattern is presented on the retina. During the recognition process, (i) local features are extracted, (ii) a hypothesis for a partial pattern is generated using shift-invariant features, (iii) the hypothesis is verified by collating with the real positions of features. The verification process gradually corrects positional displacement of the presented partial pattern while the processes (i)–(iii) are executed. Computer simulations show that the model is tolerant for vast amounts of shift, deformation and noise.

1 Introduction

We human beings comprehend locations and categories of patterns in a visual scene by moving view points one after another. This mechanism is called "visual attention" or "selective attention" and it plays an important role in a human visual system [1][2]. Some neural network models have been proposed in this scheme [3][4]. These networks, however, work successfully under a restricted condition in which the whole of a target image should be projected on retina, whereas in practical cases, only a part of the image is presented in the beginning of recognition processes. Few attempts, however, have been made for such general situation.

In this paper, we propose a neural network model which can recognize patterns even if a part of a target pattern is projected on the retina at the first time. In that case, the location and the category for every pattern should be calculated, but the both cannot be solved at the same time. To overcome this problem, we adopt "Hypothesis and Verification" method, which iteratively solves the both unknown factors by generating a hypotheses and verifying the hypotheses for each factor [5]. It has been reported that this technique is effective to recognize combined mixture of patterns with identification [6] and rotated patterns [7]. Now we apply the method to categories and locations of patterns in the whole image.

J.R. Dorronsoro (Ed.): ICANN 2002, LNCS 2415, pp. 1275–1280, 2002.

2   The Structures of the Model

This model is composed of five modules: (i)retina module, (ii)feature extracting module, (iii)feature categorizing module, (iv)shift estimation module and (v)hypothesis generation/verification module. Figure 1 shows a schematic structure of the model. Feature extracting module, feature categorizing module and shift estimation module are self-organized in the learning phase. At the recognition, visual information are conveyed to the modules iteratively.

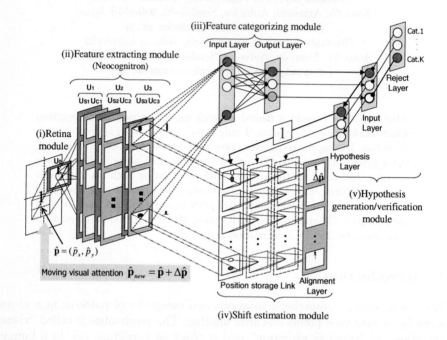

Fig. 1. A structure of the proposed model

Retina module. Retina module cuts out an partial image centered on the attention point $\hat{p} = (\hat{p}_x, \hat{p}_y)$ from a whole of the world. The output is denoted by $U_0(n)$, where $n \in R^2$ is a position in retina. The brightness $U_0(n)$ is normalized as $0(\text{black}) \le U_0 \le 1(\text{white})$.

Feature extracting module. This module is based on Fukushima's neocognitron [8]. The module has a layered structure, each layer has cell planes in which cells are arranged in two dimensions. Cells in the lowest layer U_1 detect oriented line segments, and cells in a higher layer respond to relatively complex features composed of line segments detected in U_1. In learning phase, we adopt an algorithm proposed by Fukushima et al. in [9] to extract the characteristic features of training patterns by self organization.

The properties of cell-planes and cells are that

- different features such as ⊔ and ⋁ are detected in the different cell planes, *i.e*, every cell in a cell-plane detects a same specific feature. The kind of feature is identified by a serial number of the cell-plane, k,
- a firing cell whose position in the k-th cell-plane is n indicates existence of the feature k at the position n, because cells in the cell plane have local receptive fields.

The output of feature extracting module is denoted by $u_{C3}(n, k)$. Let the number of cell-planes and the size in U_{C3} be K_{C3} and R_3, respectively. The details are in [8].

Feature categorizing module. This module generates a hypothesis (a tentative recognition result) for a partial pattern appeared in U_0. The structure is same as a single layer perceptron.

This module has K_{C3} cells in the input and the output layer has cells corresponding to pattern categories. The output of i-th cell in the input layer is given by

$$u_{fc}^{in}(i) = \sum_{n \in R_3} u_{C3}(n, i), \tag{1}$$

and the output of j-th cell in the output layer is given by

$$u_{fc}^{out}(j) = \varphi[\sum_{i=1}^{K_{C3}} w_{ij} \cdot u_{fc}^{in}(i)], \tag{2}$$

where $\varphi[\,\cdot\,]$ is a ramp function and $R_3 \in R^2$ denotes the connected area with cells in U_{C3}. In learning phase, weights w_{ij} are reinforced by competitive Hebb rule so that the value of $u_{fc}^{out}(j)$ is higher than ones of other cells if a pattern of a category j is presented. A tentative recognition result for a partial pattern is defined as $\text{argmax}_j u_{fc}^{out}(j)$.

Hypothesis generation/verification module. This module decides whether a tentative recognition result given by the feature categorizing module should be accepted or be rejected. There are three layers in this module, each layer has cells corresponding to pattern categories.

The reject layer stores a history of rejected categories, the output $u_{hgv}^{rej}(j) = 1$ means that a category "j" has been rejected. The condition $u_{hgv}^{rej}(j) = 0$ means that a category "j" is not yet rejected.

Cells in the input layer $u_{hgv}^{in}(j)$ calculate a current hypothesis. The output is given by

$$u_{hgv}^{in}(j) = u_{fc}^{out}(j) \cdot (1 - u_{hgv}^{rej}(j)). \tag{3}$$

The hypothesis layer, $u_{\mathrm{hgv}}^{\mathrm{hyp}}(j)$, receives signals from the input layer and stores a hypothesis to be verified. Namely the output of a cell gives a hypothesis as

$$u_{\mathrm{hgv}}^{\mathrm{hyp}}(j) = \begin{cases} 1 \ (j = \mathrm{argmax}_t \, u_{\mathrm{hgv}}^{\mathrm{in}}(t)), \\ 0 \ (\text{otherwise}). \end{cases} \tag{4}$$

Receiving the outputs of the hypothesis layer, the shift estimation module calculates a positional shift $\Delta\hat{p}$, then the center position of retina moves to $\hat{p}_{\mathrm{new}} = \hat{p} + \Delta\hat{p}$. In turn, a new current hypothesis for a pattern centered on $\hat{p}_{\mathrm{new}}$ is recalculated by eqn.(3). If the condition

$$\underset{j}{\mathrm{argmax}} \, u_{\mathrm{hgv}}^{\mathrm{in}}(j) = \underset{j}{\mathrm{argmax}} \, u_{\mathrm{hgv}}^{\mathrm{hyp}}(j), \tag{5}$$

is satisfied, the hypothesis is accepted and the verification process, moving attention and comparing the tentative result, are executed recursively until $\Delta\hat{p} = 0$. If eqn. (5) is satisfied and $\Delta\hat{p} = 0$, the hypothesis is adopted as a final recognition result for the input pattern in retina at the center $\hat{p}$.

If eqn. (5) is not satisfied, the hypothesis is rejected and the reject layer updates the state as

$$u_{\mathrm{hgv}}^{\mathrm{rej}}(j) = \begin{cases} 1 \ (j = \mathrm{argmax}_t \, u_{\mathrm{hgv}}^{\mathrm{hyp}}(t)), \\ 0 \ (\text{otherwise}), \end{cases} \tag{6}$$

and recreate a new hypothesis by eqn. (3), (4).

Shift estimation module. This module estimates a positional shift. The module consists of a alignment layer with position storage links. The alignment layer, $u_{\mathrm{se}}(n, k)$, has K_{C3} cell-planes of which size is R_3 and position storage links, $d(\nu, k, j)$, memorize the firing position on the k-th cell-plane of U_{C3} for patterns of category j. Vectors $\nu, n \in R_3$ denotes the location in a cell-plane.

In the learning phase, all position storage links are initially set to be zero and the links are reinforced as follows,

$$d(\nu, k, j) = \sum_m u_{\mathrm{C3}}^{j,m}(\nu, k) \ \ (k = 1 \ldots K_{\mathrm{C3}}, \ \nu \in R_3), \tag{7}$$

where $u_{\mathrm{C3}}^{j,m}$ means the response of cells in U_{C3} for the m-th training pattern belonging to category j.

In the recognition phase, a cell in the alignment layer receives signals via the links and the output is calculated as

$$u_{\mathrm{se}}(n, k) = \sum_{\nu \in R_3} \left[\sum_j [u_{\mathrm{hgv}}^{\mathrm{hyp}}(j) \cdot d(\nu, k, j)] \cdot u_{\mathrm{C3}}(n + \nu, k) \right]. \tag{8}$$

The shift $\Delta\hat{p}$ between the pattern on a retinal image and a trained pattern corresponding to the current hypothesis is calculated as follows,

$$\Delta\hat{p} = R \cdot \frac{1}{K_{\mathrm{cmp}}} \cdot \sum_{k=1}^{K_{\mathrm{C3}}} \underset{n}{\mathrm{argmax}} \, u_{\mathrm{se}}(n, k), \tag{9}$$

where R is a resolution ratio of retina to U_{C3} layer and K_{cmp} is the number of cell-planes in which one or more cells are firing in the alignment layer.

3   Computer Simulations

We evaluate the proposed model by use of ten numeric letter images as shown in Fig. 2. The size of each image is equal to retinal size: 65×65 pixels. Test patterns are shifted patterns, multiple patterns, and noisy patterns. Figure 3 shows typical examples of recognition processes. The upper image of (a)–(f) in Fig. 3 indicates the retinal image, and the lower means a hypothesis at each moment. A left side retinal image is at starting position, a right side image is a result in termination phase in which a hypothesis is verified and $\Delta \hat{p} = 0$. In these simulations, averaged computation times for every verifications are about one second by use of 800MHz Pentium III processor.

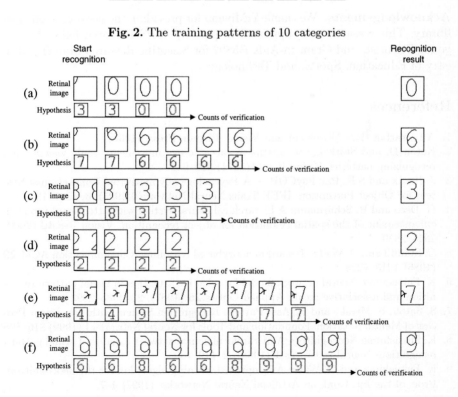

Fig. 2. The training patterns of 10 categories

Fig. 3. Examples of recognition processes for various situations

Figure 3(a), (b) are examples in which the incomplete patterns are presented at first. Verification with moving attention point makes correct result. Figure 3(c), (d) shows results for multiple patterns. Attention points converge at the one of the patterns during verification process, and the model gets the correct result. Figure 3(e), (f) show examples of deformed or noisy patterns. Even under such conditions, the proposed model works successfully by rejecting wrong hypotheses. The result of Fig. 3(f) gives an anticipation of the ability of the proposed model for recognizing cursive style.

We evaluate the robustness of the proposed model by the amount of maximal recognizable displacement. The result is that the average of ten patterns is 32.8 pixels in our model, whereas that in original neocognitron is 10.9 pixels.

4   Conclusions

We proposed a model for recognizing a pattern even if incomplete patterns are presented on the retina. By use of "Hypothesis and verification", the problem of identifying and locating of a pattern is solved. We will deal with other transformation such as scaling and apply this method to the application for cursive style.

Acknowledgements. We thank Y.Miyano for providing the neocognitron class library. This research was supported in part by JSPS Research fellowships of young Scientists and Grant-in-Aids #5892 for Scientific Research from the Ministry of Education, Sports, and Technology.

References

1. A.L. Yarbus: Eye Movement and Vision. Plenum Press. (1967)
2. Noton D. and Stark L.: Scanpaths in saccadic eye movements while viewing and recognizing patterns. Vision Res.11(1971) 929–942
3. J. Basak and S.K. Pal: PsyCOP — A Psychologically motivated Connectionist System for Object Perception. IEEE Trans.Neural Networks6(1995), 6, 1337–1354
4. G. Deco and B. Schürmann: A hierarchical neural system with attentional topdown enhancement of the spatial resolution for object recognition. Vision Res.40 (2000) 2845–2859
5. C. Stanfill and D. Waltz: Toward memorybased reasoning. Communication ACM.29 (1986) 1213–1228
6. K. Fukushima: Neural network model for selective attention in visual pattern recognition and associative recall. Applied Optics.26 (1987), 23, 4985–2992
7. S. Satoh, S. Miyake and H. Aso: Pattern Recognition System with Top-Down Process of Mental Rotation. Foundation and Tools for Neural Networks.I (1999) 816–825
8. K. Fukushima: Neocognitron:A hierarchical neural network capable of visual pattern recognition. Neural Networks.1 (1988), 2, 119–130
9. K. Fukushima and N. Wake: An improved learning algorithm for the neocognitron. Proc. of the Int. Conf. on Artificial Neural Networks. (1992) 4–7

Automatic Fingerprint Verification Using Neural Networks

Anna Ceguerra and Irena Koprinska

School of Information Technologies, University of Sydney, Sydney, Australia
{anna, irena}@it.usyd.edu.au

Abstract. This paper presents an application of Learning Vector Quantization (LVQ) neural network (NN) to Automatic Fingerprint Verification (AFV). The new approach is based on both local (minutiae) and global image features (shape signatures). The matched minutiae are used as reference axis for generating shape signatures which are then digitized to form a feature vector describing the fingerprint. A LVQ NN is trained to match the fingerprints using the difference of a pair of feature vectors. The results show that the integrated system significantly outperforms the minutiae-based system alone in terms of classification accuracy. It also confirms the ability of the trained NN to have consistent performance on unseen databases.

1 Introduction

Fingerprint verification involves matching two fingerprint images, in order to verify a person's claimed identity. The most popular approaches match local features, such as minutiae, using point-pattern matching [3,8], or graph matching [2]. These methods require extensive preprocessing to reliably extract the minutiae and are also very sensitive to noise due to image acquisition and feature extraction. Global recognition approaches, on the other hand, match features characterizing the entire image, typically extracted by filtering or transform operations [9]. They require less preprocessing than the minutiae-based approaches, but are effective when the representation is invariant to translation, rotation and scale. The invariants are addressed by registering the images with respect to a reference axis [4], which can be consistently detected in the different instances of the fingerprint. Singular points were used as reference points but their detection is not precise enough for matching.

To overcome the above limitations, this paper presents an approach which integrates local and global features and uses a NN for the final recognition. The matched local features are used as the reference axis for generating global features. In our specific implementation, minutiae and shape signatures were combined. Minutiae are first matched by a point-pattern matching. Shape signatures are then generated by using the matched minutiae as their frame of reference and are digitised to form a feature vector. Finally, a LVQ NN was trained to learn to distinguish between matching and non-matching fingerprints based on the difference of a pair of feature vectors.

J.R. Dorronsoro (Ed.): ICANN 2002, LNCS 2415, pp. 1281–1286, 2002.

2 AFV System

The AFV system consists of three modules: minutiae-based, shape signature-based and neural network module.

2.1 Minutiae-Based Module

Pre-processing and Minutiae Extraction. To remove noise and enhance the fingerprint ridge pattern, the images were first pre-processed using Gabor filters. Segmentation then limited the image to those areas that contain ridge information. Black and white image were created from a grey-scale using binarization. Finally, skeletonisation of the binarised image was performed, where ridges with a width of multiple pixels were simplified to ridges of single pixel width. Minutiae were detected by examining the neighbourhood around each pixel of the thinned ridges [3]. The position of a pixel was recorded as a *ridge termination*, if its neighbourhood contained only one other pixel, or *ridge bifurcation* if it contained three other pixels.

Minutiae Matching. Following [8], each minutia p_i is represented in terms of the counterclockwise sequence of its adjacent minutiae $\{p_{i_k} : k = 1, \ldots, K\}$, where two minutiae are adjacent if their Voronoi cells share a border. The position of the adjacent minutia p_{i_k} is expressed in terms of the minutia p_i, by the distance d_k between p_{i_k} and p_i, and the polar angle α_k between p_{i_k} and $p_{i_{k-1}}$ with respect to p_i. The respective distances and angles are combined and normalised, to form a sequence of complex numbers, and the Discrete Fourier Transform (DFT) is applied:

$$f(p_i) = \mathrm{DFT}\{u_k : k = 1, \ldots, K\},\ u_k = d_k \exp\{j\alpha_k\} / \sum_{l=1}^{K} d_l \tag{1}$$

The matching algorithm in [8] was also implemented. It uses a function quantifying the similarity between two minutiae p_m and q_o. Our implementation extended this, by incorporating a check of the minutiae type (termination or bifurcation):

$$s(p_m, q_o) = \begin{cases} 0,\ if\ \dim\{f(p_m)\} \neq \dim\{f(q_0)\}\ or\ if\ \mathrm{type}(p_m) \neq \mathrm{type}(q_0) \\ \exp\{-(\|f(p_m) - f(q_0)\|^2)/\sigma\}, \mathrm{otherwise} \end{cases} \tag{2}$$

The set of matching pairs of minutiae C is found by determining, for each minutia p_m in one fingerprint, the best matching minutia q_o in the other fingerprint. Another constraint for this pair to be in C is that they must be sufficiently similar, that is the similarity value of the two minutiae must occur above a threshold T.

2.2 Shape Signature Module

Reference Axis Generation. *Each* matched pair of minutiae (p_m, q_o) forms a reference point between the two fingerprints. Another pair of matched minutiae $(p_{m_j},$

q_{o_k}) is found by finding correspondences between the adjacent minutiae of p_m and q_o, and choosing the closest adjacent minutia p_{m_j} and its match q_{o_k}. The two points in each image form the reference line. The matching reference axes are $(\text{ref}_{p_m}, \text{ref}_{q_o})$.

Shape Signature Generation. The shape signature represents a two-dimensional boundary as a one-dimensional function [7]. It is generated in terms of the reference axis ref_{p_m} corresponding to that fingerprint. For each pixel l_u on the thinned ridges, the distance r_u from the reference point, the clockwise angle θ_u from the reference line, and the average tangent angle t_u are calculated. To calculate t_u, the ridge pixels $\{l_{u_k} : k = 1, ..., K\}$ in the neighbourhood of l_u are first found and then averaged. The tangent angle (t_{u_k}) made by each of the eight neighbouring pixels with respect to l_u is calculated as $t_{u_k} = (\arg[(r_u - r_{u_k}) + j(\theta_u - \theta_{u_k})]) \bmod \pi$.

2.3 Neural Network Module

Feature Vector Generation. An n-dimensional feature vector is calculated, by digitising the shape signature. Digitisation compresses the amount of information in the shape signature, while remembering the general shape as the clockwise angle changes. This is done using a hybrid of the compression approaches presented by [4,6,1]. The entire set of ridge pixels $\{l_u : u = 1, ..., M\}$ is divided into n subsets, where l_u belongs to the ith subset if θ_u is within the range of $[(i-1)/2n\pi, i/2n\pi)$. The average of all t_u for the ith subset is then calculated, forming the ith entry in the n-dimensional feature vector. Figure 1 is a visual summary of this step.

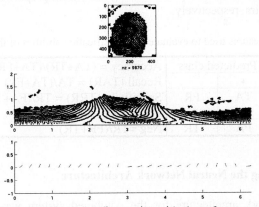

Fig. 1. Generating the feature vector: a fingerprint with the reference axis (top), its shape signature (middle), its digitised shape signature (bottom), with the direction of each feature

Determining the Similarity. Once the feature vectors for both fingerprints are generated, taking the *difference* between the two feature vectors captures the concept of their similarity. This difference is an input to a LVQ NN [5] that was trained to distinguish between matching and non-matching fingerprints.

3 Experimental Results

The goal of the experiments was two-fold: a) to find the best NN architecture for the combined minutiae/shape signature & NN-based system (AFV-combined), b) to compare the combined system with the minutiae-based module (AFV-local) alone.

3.1 Data and Performance Measures

Fingerprint Databases. Three databases were used: 'hi-optical', 'synthetic' and 'db-ipl'. The first comprises of Set B of DB3 which was captured by an optical scanner, and the second is Set B of the synthetically generated database (DB4) from the FVC2000 competition [10]. Both contain 10 distinct fingers, each finger having 8 impressions, creating two databases with 80 images each. These images had a size of 448x478 pixels. There were 3160 possible pairings from each of these databases, excluding self-pairings, and 280 (8.86%) of these were matching pairs. The Image Processing Laboratory, University of Trieste, Italy captured the images for the last database. This contains 16 distinct fingers of 8 impressions each, with 448 (5.51%) matching pairs out of the 8128 possible pairings. The images in the third database had a size of 240x320 pixels. All images have resolution of 500dpi.

Performance Measures. Five measures were used (Table 1). In addition to the most commonly used *Accuracy*, *Recall (True Acceptance Rate, TAR)* and *Specificity (True Rejection Rate, TRR)*, *Predictive value positive (Pos)*, and *Predictive value negative (Neg)* were calculated as they measure the trustworthiness of AFV system, and are most useful when compared to the ground truth proportions of matching and non-matching pairs, respectively.

Table 1. Measures used to evaluate the classification abilities of the AFV system

		Predicted class		Accuracy = (TA+TR)/(TA+FR+FA+TR)
		+	-	Recall (TAR) = TA/(TA+FR)
Actual	+	TA	FR	Specificity (TRR) = TR/(FA+TR)
class				Pos = TA/(TA+FA)
	-	FA	TR	Neg =TR/(FR+TR)

3.2 Determining the Neural Network Architecture

The LVQ network architecture for the combined system was determined using stratified 10-fold cross validation, by comparing the performance of different combinations of feature vector size and number of neurons within the competitive layer. Only the hi-optical database was used for this experiment.

The output of the minutiae-matching module ($T = 0.9$, $\sigma = 0.1$) was used to generate shape signatures for each reference axis found in each fingerprint. This information was stored, with each file containing the shape signatures for one pair of fingerprints and if they match or not. Feature vectors of size 9, 10, 11, 12, 13, 14, 15 were created from each pair of shape signatures within each file. Vectors of the same size, along with their target outputs, were bundled together defining the data for the NN with the same number of input features. Creating different combinations of input size and 8, 16, 24, 32, 40 competitive neurons gained 35 different NN architectures. Of the two output neurons, the first represents a match and the other a non-match. Stratified ten-fold cross validation was performed for each of the NNs, and the performance was measured (Fig. 2). A neural architecture with 24 input and 12 hidden units was selected, as for these parameters the performance of *Accuracy*, *TRR* and *Pos* is in the best region, and at its second best for *TAR* and *Neg*.

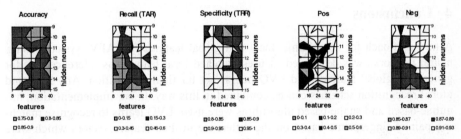

Fig. 2. Performance of different combinations of hidden neurons (9-15) and features (8-40)

3.3 Comparing AFV-local and AFV-combined

The combined system used the minutiae-matching module with $T = 0.9$ and $\sigma = 0.1$, a neural network module with 12 input, 24 competitive and 2 output neurons trained on the hi-optical database *only*. Two fingerprints were considered to match by the combined system if: (1) the minutiae-matching module found at least one reference axis, and (2) the NN module matched at least one pair of shape signatures from all possible signatures generated by the corresponding different reference axes.

Table 2. Comparison of clasification performance [%]

database	AFV system	Accuracy	TAR	TRR	Pos	Neg
hi-optical	AFV-local	85.54	12.86	92.60	14.46	91.62
	AFV-combined	90.32	3.57	98.75	21.74	91.33
synthetic	AFV-local	88.83	3.93	97.08	11.58	91.22
	AFV-combined	90.79	1.07	99.51	17.65	91.19
db-ipl	AFV-local	93.05	4.91	98.19	13.66	94.65
	AFV-combined	94.45	4.24	99.71	46.34	94.70

As it can be seen from Table 2, the complete AFV system has equal or better performance on most measures, with the most significant improvement on *Pos*, or the trustworthiness of the match decision. The only measure with no improvement is *TAR*, since this approach cannot increase the number of matches – it only refines the match decision of the minutia-based module. It can also be seen that this improvement is consistent over all databases, regardless of whether the NN has been trained on the dataset or not, which confirms the high generalization ability of LVQ.

It should be noted that the LVQ training data was of small size and highly skewed towards the non-match examples. By providing a larger training set with even proportion of matching examples, the NN performance can be further improved.

Another interesting observation is that AFV-local has trustworthiness values that are close to the actual proportions within the databases (see Sec. 3.1), whereas for AFV-combined, these values are higher than the actual proportions. This suggests that the trustworthiness values of the combined system are not dependent on the proportions within the database, while AFV-local shows this dependency.

4 Conclusions

A new approach for combining local and global features for AFV systems using neural networks was developed. It uses matched local features as reference axis for generating global features and LVQ neural net for final recognition. Any local and global recognition schemes can be combined in this way. In our implementation, minutiae-based and shape-based algorithms were used. LVQ learns to recognise matching pair of fingerprints based on the difference in their feature vectors which were extracted from the shape signatures. Experiments show that the integrated system outperforms the local system on various accuracy measures, with significant improvement in the trustworthiness. They also confirm the ability of the trained NN to have consistent performance on unseen databases. Therefore, this approach has clear potential in AFV systems that require high reliability and excellent accuracy.

Acknowledgements

We are very grateful to Piero Calucci from Tender S.p.A., Trieste, Italy and Fabio Vitali from the University of Trieste, who provided the implementations for the preprocessing and minutiae extraction. Thanks are also due to Dr Marius Tico, Tampere Univ. of Technology, Finland, for his helpful comments on minutiae matching.

References

1. Blue, J.L., Candela, G.T., Grother, P.J., Chellappa, R., Wilson, C.L.: Evaluation of Pattern Classifiers for Fingerprint and OCR Applications. Pattern Recognition 27 (1994) 485-501
2. Hollingum, J.:Automated Fingerprint Anal. Offers Fast Verif. Sensor Rev 3 (1992) 12-15
3. Jain, A.K., Hong, L., Bolle, R.: On-line Fingerprint Verif. IEEE PAMI 19 (1997) 302-314
4. Jain, A.K., Prabakhar, S., Hong, L., Pankanti, S.: FingerCode: A Filterbank for Fingerprint Representation and Matching. In Proc. CV&PR Conf., Fort Collins (1999)
5. Kohonen, T.: Self-Organizing Maps. Springer (2001)
6. Logan, B., Salomon, A.: A Content-Based Music Similarity Function. TR CRL 02 (2001)
7. Loncaric, S.: A survey of shape analysis techniques. Pattern Recogn. 31 (1998) 983-1001
8. Tico, M., Rusu, C., Kuosmanen, P.: A Geometric Invariant Representation for the Identification of Corresponding Points. In Proc. ICIP (1999) 462-466
9. Tico, M., Immonen, E., Ramo, P., Kuosmanen, P., Saarinen, J.: Fingerprint Recognition Using Wavelet Features. In Proc. ISCAS (2001) 21-24
10. www2.csr.unibo.it/research/biolab/

Fusing Images with Multiple Focuses Using Support Vector Machines

Shutao Li[1,2], James T. Kwok[1], and Yaonan Wang[2]

[1] Department of Computer Science
Hong Kong University of Science and Technology
Hong Kong, People's Republic of China
[2] College of Electrical and Information Engineering
Hunan University
Changsha, People's Republic of China

Abstract. Optical lenses, particularly those with long focal lengths, suffer from the problem of limited depth of field. Consequently, it is often difficult to obtain good focus for all the objects in the scene. One approach to address this problem is by performing image fusion, i.e., several pictures with different focus points are combined to a single image. This paper proposes a multifocus image fusion method based on the discrete wavelet frame transform and support vector machines. Experimental results show that the proposed method outperforms the conventional approach based on the discrete wavelet transform and maximum selection rule, particularly when there is slight camera/object movement or mis-registration of the source images.

1 Introduction

Optical lenses suffer from the problem of limited depth of field. According to the lens formula, only objects at one particular depth will be truly in focus for any fixed focus setting. Consequently, if one object in the scene is in focus, another object at a different distance from the lens will be out of focus and thus blurred. In general, many vision-related processing tasks, such as edge detection, image segmentation and stereo matching, can be more easily performed on focused images than on defocused ones. Hence, it is often advantageous if an everywhere-in-focus image can be recovered.

One approach to address this problem is by first estimating from the image the distance between the sensor and the objects in the scene. However, this requires a knowledge of the values of the camera parameters and/or a model of the camera point spread function, and thus involves a lot of mundane camera calibration. In this paper, we follow another approach by performing image fusion [5], in which several pictures with different focus points are combined to form a single image. If each relevant object is in good focus in at least one of the source images, the fused image will then hopefully have all these objects in focus.

The simplest fusion method just takes the pixel-by-pixel average of the source images. This, however, often produces undesirable side effects such as reduced

J.R. Dorronsoro (Ed.): ICANN 2002, LNCS 2415, pp. 1287–1292, 2002.
© Springer-Verlag Berlin Heidelberg 2002

contrast. To alleviate this problem, methods based on multiscale transforms (such as the popular discrete wavelet transform (DWT)) have been employed. The basic idea is to perform a multiresolution decomposition on each source image, then integrate all these decompositions to produce one composite representation, from which the fused image can be recovered by performing an inverse transform. However, one problem with these transforms is that their multiresolution decompositions, and consequently the fusion results, are not shift-invariant because of an underlying down-sampling process. Hence, their performance quickly deteriorates when there is slight camera/object movement or mis-registration of the source images.

Moreover, the *maximum selection* rule is usually used to fuse the multiresolution decompositions. Consider DWT as an example. After obtaining the sets of wavelet coefficients from the source images, corresponding coefficients are compared and the one with the largest absolute value is selected for use in the composite representation [2]. The rationale is that large absolute coefficients often correspond to salient features in the images. However, obviously this simple selection rule does not always work.

In this paper, we use the discrete wavelet frame transform (DWFT) [1] (also called the stationary wavelet transform [3]) to perform multiresolution decomposition, and then use the support vector machine (SVM) [4] to fuse the wavelet frame coefficients. DWFT is closely related to DWT but avoids down-sampling by using an overcomplete wavelet decomposition. Its resultant signal representation is thus both aliasing free and shift-invariant. The SVM, on the other hand, has outperformed many conventional pattern recognition approaches. The rest of this paper is organized as follows. Section 2 describes our proposed fusion scheme. Experimental results are presented in Section 3, and the last section gives some concluding remarks.

2   Proposed Fusion Scheme

Figure 1 shows a schematic diagram of our proposed multifocus image fusion method. Here, we consider the processing of just two source images (A and B), but the algorithm can be straightforwardly extended to handle more than two images. Moreover, the source images are assumed to have been registered. The algorithm consists of the following steps:

1. Decompose each source image by DWFT to d levels, resulting in a total of $3d$ details subbands and one approximation subband. Recall that the approximation subband is a low-pass filtered version of the original image. Hence, with a sufficient number of decompositions, the resultant approximation subband contains little information and so will not be used in constructing the feature vector in Step 2 (though it will still be used in reconstructing the fused image in Step 6).
2. For each details subband, compute at each pixel location a window-based activity level based on the wavelet coefficients. Specifically, consider the $b_1 b_2$

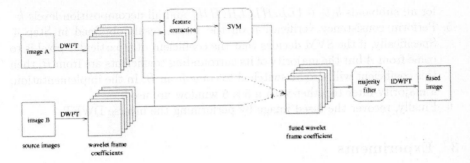

Fig. 1. Schematic diagram of the proposed fusion method.

details subband (where b_1b_2 can be either HL, LH or HH) at decomposition level k, and with the wavelet coefficient at position (i,j) denoted by $D_{b_1b_2,k}(i,j)$. Define a $p \times p$ window W_{ij} centered at (i,j), with weighting coefficients $w(u,v)$ satisfying $\sum_{(u,v)\in W_{ij}} w(u,v) = 1$. In Section 3, the following 3×3 window will be used:

$$\begin{bmatrix} \frac{1}{16} & \frac{1}{16} & \frac{1}{16} \\ \frac{1}{16} & \frac{1}{2} & \frac{1}{16} \\ \frac{1}{16} & \frac{1}{16} & \frac{1}{16} \end{bmatrix}.$$

The corresponding feature value is then computed as

$$A_{b_1b_2,k}(i,j) = \sum_{(u,v)\in W_{ij}} w(u,v)|D_{b_1b_2,k}(i,j)|. \tag{1}$$

Notice that this can be defined at *every* pixel location because the details subbands of DWFT are of the same size as the input image[1].

3. Train a SVM to determine whether the coefficients from A or B should be used in position (i,j) of the fused image. Here, we use the difference vector

$$\{A_{HL,k}(i,j) - B_{HL,k}(i,j), A_{LH,k}(i,j) - B_{LH,k}(i,j), A_{HH,k}(i,j) - B_{HH,k}(i,j)\}_k$$

as input, and the output is labeled according to

$$target(i,j) = \begin{cases} 1 & \text{if } A(i,j) \text{ is clearer than } B(i,j), \\ -1 & \text{otherwise.} \end{cases}$$

4. Perform testing of the trained SVM over the whole image. If the output of the SVM at a pixel location is positive, coefficients for all the details and approximation subbands of the fused image at this particular location will come from A, and vice versa. Denote the SVM's output at (i,j) by $out(i,j)$. Then, in other words, the fused coefficient at (i,j) is given by

$$C_{b_1b_2,k}(i,j) = \begin{cases} A_{b_1b_2,k}(i,j) & \text{if } out(i,j) > 0, \\ B_{b_1b_2,k}(i,j) & \text{otherwise,} \end{cases}$$

---

[1] On the contrary, the sizes of the subbands in DWT decrease as the decomposition goes on, and so (1) would not be well-defined if DWT were used.

for all subbands $b_1 b_2 \in \{LL, HL, LH, HH\}$ and all decomposition levels k.

5. Perform consistency verification on the fusion result obtained in Step 4. Specifically, if the SVM decides that the coefficient of a particular pixel is to come from A but the majority of its surrounding coefficients are from B, then this coefficient will also be switched to come from B. In the implementation, a majority filter together with a 5×5 window are used.

6. Finally, recover the fused image by performing the inverse DWFT.

3   Experiments

In this experiment, a reference image with good focus everywhere is used. We then produce a pair of out-of-focus images by blurring this reference image, and perform restoration using various fusion schemes. Because of the availability of the reference image, quantitative comparison of the fusion schemes is possible. In the following, two evaluative criteria, namely,

1. the root mean squared error (RMSE) between the reference image R and the fused image F, and

2. the mutual information $MI = \sum_{i_1=1}^{L} \sum_{i_2=2}^{L} h_{R,F}(i_1, i_2) \log_2 \frac{h_{R,F}(i_1,i_2)}{h_R(i_1)h_F(i_2)}$
($h_{R,F}$ is the normalized joint gray level histogram of images R and F, h_R, h_F are the normalized marginal histograms of the two images, and L is the number of gray levels),

will be used. Notice that MI measures the reduction in uncertainty about the reference image due to the knowledge of the fused image, and so a larger MI is preferred.

A linear SVM with $C = 1000$ is used. Preliminary studies show that nonlinear SVMs do not perform better here. For training, two pairs of 30×30 regions are selected from the two out-of-focus images. In one pair of these regions, the first image is clearer than the second image, and the reverse is true for the second pair. Each pixel generates one training sample for the SVM, and so the training set has a total of 1800 training samples. For the DWFT, we use the biorthogonal B-spline wavelet basis with 3 decomposition levels.

3.1   Registered Images

The first experiment is performed on an 256-level, 480×640 image (Figure 2(a)), with good focus everywhere. We blur the left and right halves separately to obtain the images in Figures 2(b) and 2(c). For comparison purposes, we also perform fusion using DWT with the maximum selection rule (DWT+max_slct), and DWFT with the maximum selection rule (DWFT+max_slct).

Table 1 compares the various fusion results quantitatively, and a visual comparison by examining the differences between the fused images and the reference image is shown in Figure 2. As can be seen, the fused image produced by DWFT+SVM is basically a combination of the focused parts in the source images. Results based on using maximum selection are much inferior.

Table 1. Performance on fusing the two images in Figures 2(b) and 2(c).

	DWT+max_slct	DWFT+max_slct	DWFT+SVM
RMSE	0.7634	0.7073	0.3433
MI	6.3607	6.4488	6.8980

3.2 Effect of Mis-registration

In this Section, we artificially shift the source image in Figure 2(b) by two pixels to the right, and the source images are consequently not perfectly registered. Figure 3 shows the differences between the fused images and the source images (Figure 2(c) and the shifted version of Figure 2(b)). Again, the fused image produced by DWFT+SVM is much superior. Also, notice that mis-registration has caused significant performance deterioration for the DWT-based method.

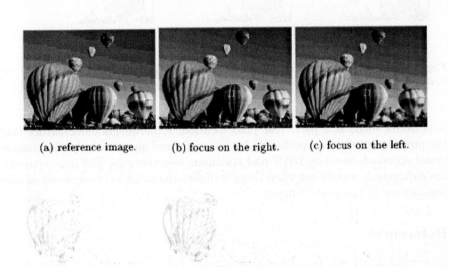

(a) reference image.   (b) focus on the right.   (c) focus on the left.

(d) DWT+max slct.   (e) DWFT+max slct.   (f) DWFT+SVM.

Fig. 2. Source images ((a)–(c)) and difference images ((d)–(f)) for the experiment in Section 3.1. The training set is selected from regions marked by the rectangles in (b) and (c).

4   Conclusion

In this paper, we use DWFT and SVM for multifocus image fusion. DWFT is advantageous over DWT in that the decomposed signal representation, and consequently the fusion result, is shift-invariant. SVM, on the other hand, outperforms the simple maximum selection rule in determining which source image

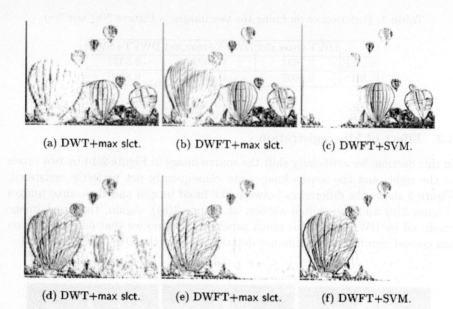

(a) DWT+max slct.   (b) DWFT+max slct.   (c) DWFT+SVM.

(d) DWT+max slct.   (e) DWFT+max slct.   (f) DWFT+SVM.

Fig. 3. Difference images ((a)–(c) are matches with Figure 2(c), while (d)–(f) are matches with the shifted version of Figure 2(b)) for the experiment in Section 3.2.

is in better focus at a particular pixel location. Experiments demonstrate that the proposed method outperforms, both visually and quantitatively, the conventional approach based on DWT and maximum selection rule. The improvements are particularly significant when there is slight camera/object movement or misregistration of the source images.

References

1. A. Laine and J. Fan Frame representations for texture segmentation. *IEEE Transactions on Image Processing* 5(5):771–780, 1996.
2. H. Li, B. S. Manjunath, and S. K. Mitra. Multisensor image fusion using the wavelet transform. *Graphical Models and Image Processing*, 57(3):235–245, May 1995.
3. G. P. Nason and B. W. Silverman. The stationary wavelet transform and some statistical applications. In A. Antoniadis and G. Oppenheim, editors, *Lecture Notes in Statistics*, volume 103, pages 281–300. Springer-Verlag, New York, 1995.
4. V. Vapnik. *Statistical Learning Theory*. Wiley, New York, 1998.
5. Z. Zhang and R. S. Blum. A categorization of multiscale decomposition based image fusion schemes with a performance study for a digital camera application. *Proceedings of the IEEE*, 87(8):1315–1326, August 1999.

An Analog VLSI Pulsed Neural Network for Image Segmentation Using Adaptive Connection Weights

Arne Heittmann[1] and Ulrich Ramacher[1], Daniel Matolin[2], Jörg Schreiter[2], and Rene Schüffny[2]

[1] Infineon Technologies AG Munich, 81739 Munich, Germany,
{Arne.Heittmann, Ulrich.Ramacher}@infineon.com

[2] Dresden University of Technology, Department of Electrical Engineering and Information Technology, 01069 Dresden, Germany
{schreite, schueffn, matolin}@iee.et.tu-dresden.de

Abstract. An analog VLSI pulsed neural network for image segmentation using adaptive connection weights is presented. The network marks segments in the image through synchronous firing patterns. The synchronization is achieved through adaption of connection weights. The adaption uses only local signals in a data-driven and self-organizing way. It is shown that for the proposed adaption rules a simple analog VLSI implementation is feasible due to the required local connections and the data-driven self-organizing approach.

1   Introduction

Perception and recognition of objects in a given scene is one of the basic tasks in image processing. A very basic scene description is the partitioning of an image into a set of coherent regions, the so-called segments [1]. Although animals and humans perform this task effortless and very reliably under realtime conditions, scene segmentation is an unsolved task for machine vision, though. Since biological systems show a excellent performance in the domain of image segmentation it seems to be promising to exploit principles that have naturally evolved in that area and to use these to build artificial systems.

It has been experimentally observed that biological systems show oscillatory firing patterns of participating neurons in the vision system. These firing pattern are used to encode information in that way, that clusters of neurons respond to individual features in the scene by synchronized firing patterns [2]. To solve the problem of image segmentation by a biologically-inspired system it is a natural way to use artificial pulsed neural networks. By assigning to each neuron a grey-valued pixel of the image those neurons that belong to the same segment should fire with the same frequency and fixed phase difference. Neurons that belong to different segments should fire asynchronously with different frequencies and varying phase difference.

J.R. Dorronsoro (Ed.): ICANN 2002, LNCS 2415, pp. 1293–1298, 2002.

To utilize these insights in technical vision applications realtime capabilities of the segmentation algorithm are required. Since many applications exhibit a large number of pixels and a large amount of neurons, realtime capabilities can be only obtained if the network can be implemented in a massive parallel way increasing the processing speed significantly to the appropriate level.

Analog circuit techniques are the most suitable way of achieving the necessary degree of parallelism and processing speed. For many models and systems found in literature for pulsed neural networks [3] [4], an analog implementation fails due to the required device precision, the area requirements, the necessity of global signals or the required network connectivity. The model we propose in this paper is based on a very simple neuron model, the integrate-and-fire neuron (IAF). Each neuron is connected to its nearest neighbors via a connection weight. To achieve synchronization the connection weights are adaptive, i.e. change their value according to a self-organizing process, that takes only local signals into account. For this approach a very area efficient solution in analog VLSI can be found and implemented very robustly with small area overhead.

2   System Dynamics

In this section the basic equations that determine the system dynamics will be presented. The neuron used here is a simple IAF neuron. The IAF neuron has two states. In the first state the neuron sends a pulse of fixed duration t_d (pulse width) and constant amplitude. The output pulse of neuron k is described by the signal X_k. If the pulse is completed the neuron resets its membrane potential a_k to an initial value and turns into a second state where the synaptic input current is continuously integrated to the membrane potential a_k:

$$a_k(t) = \int \left[\sum_{s \in N_k} W_{ks}(t) \cdot X_s(t) + W_{K0} \cdot i_k \right] dt \qquad (1)$$

W_{ks} represents the connection weight from neuron s to neuron k. N_k is a set of neurons that are connected to neuron k where neuron k is a postsynaptic neuron. i_k represents an continuous signal and represents the grey-value of the pixel k of the image (Fig. 1a). If the membrane potential reaches the threshold θ the neuron toggles its state again and sends its pulse X_k and so on. The network structure used is composed of a 2-dimensional array of neurons that are locally connected to its nearest neighbors (Fig. 1b).

2.1   Weight Adaption Rules

To achieve synchronisation the connection weights between the neurons have to be adapted. For the adaption we tried two versions that work on different parameters but with similar results.

Adaption rule 1: If the postsynaptic neuron k is in its receiving state and the presynaptic neuron l in its sending state the weight W_{kl} changes according to the following rule:

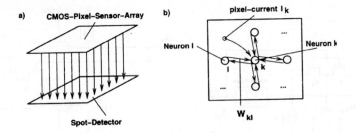

Fig. 1. a) connection of pixel to spot-detector. b) Network structure of the spot detector. Each neuron is connected locally to its nearest neighbors

$$\dot{W}_{kl} = -\gamma W_{kl} + \mu \cdot \left(a_k - \frac{\theta}{2} \right) \tag{2}$$

For all other combinations of states the weight relaxes:

$$\dot{W}_{kl} = -\gamma W_{kl} \tag{3}$$

Adaption rule 2: If the postsynaptic neuron k is in its receiving state and the presynaptic neuron l in its receiving state the weight W_{kl} changes according to the following rule:

$$\dot{W}_{kl} = -\gamma W_{kl} + \mu \cdot (a_l - a_k) \qquad \|a_l - a_k\| < \theta_a \tag{4}$$

In all other combinations of states the weight relaxes like in Eq. 3: Note, that for both rules the weight W_{kl} changes according to local available signals a_k, a_l, X_k and X_l only. Aim of both rules is to achieve either a positive feedback or a negative feedback of coupled neurons. If Eq. 2 holds, i.e. $X_l = 1, X_k = 0$, and a_k is close to the threshold θ W_{kl} increases and hence a_k rises exponentially to reach the threshold very fast. Otherwise, if a_k is far from θ W_{kl} decreases and neuron k and neuron l will be decoupled. Similar considerations can be applied for rule 2. Note that both rules produce wave-fronts of pulses. These run over the spot and stop at its boundary.

3   Analog VLSI

The proposed network dynamics are well suited for VLSI implementation. To keep area needs small we suggest certain approximations of the dynamic equations that have only small impact on the performance. To keep power consumption low the analog building blocks have been realized in subthreshold circuit technique.

IAF neuron implementations are well known. A simple neuron is build with a Schmitt-trigger, two switches T_1 and T_2, a current source I_d and a capacity, realized with the gate of T_3 (Fig. 2a).

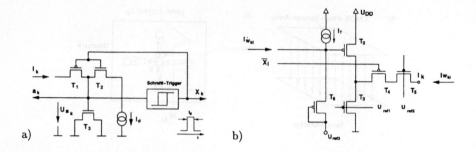

Fig. 2. a) Circuit diagram of the IAF-neuron, b) Circuit diagram of the adaptive weight

The continuous input to the neuron is represented by the current I_k. During the receiving phase the switch T_1 is closed and the current I_k charges the gate capacitance of T_3. Its voltage U_{a_k} represents the membrane potential a_k. If U_{a_k} reaches the fixed threshold U_θ of the Schmitt-trigger the output X_k rises to U_{DD} and T_1 cuts off I_k. Simultaneously, T_2 closes and I_d discharges T_3. When U_{a_k} reaches again the integrator's initial value U_{a_0} the Schmitt-trigger toggles its output, T_1 closes and the integration of the membrane potential starts again. By appropriately choosing the hysteresis $U_H = U_\theta - U_{a_0}$ of the Schmitt-trigger T_3 remains in strong-inversion and hence the gate capacity of T_3 is nearly independent of the gate voltage [7]. Further, the pulse width t_d is determined by I_d, the hysteresis of the Schmitt-trigger and the gate capacitance of T_3.

3.1   Adaptive Weights

The IAF neuron l is locally connected to the neuron k through the weight W_{kl}. For the weight a pulsed synaptic multiplier controlled by the cell's output X_l is used (Fig. 2b). T_2 and T_3 form a nearly linear voltage to current converter [5], whereas T_3 and T_5 form a current divider. Its ratio is determined through reference levels and is used to adjust the operating range of T_2. The weight W_{kl} is represented by the gate voltage of T_2 with respect to U_{DD} and hence the proportional drain currents of T_3 and T_5.

Adaptation is achieved by continuous integration of the adaption current $I_{\dot{W}_{kl}}$ on the gate capacitance of T_2. Weight relaxation as defined by (2) and (4) would require a constant conductivity. Its complex circuitry has been replaced by a simple current source I_γ discharging the gate T_2 up to a clamp voltage set by U_{ref3}. Simulations showed that the impact of this approximation has minor effects on the segmenatation quality.

The block diagram of implementations of both adaptation rules is shown in fig. 3. According to (2) the weight adaptation current is generated by a voltage-to-current converter as $I_{\dot{W}_{kl}} = f(U_{a_k} - U_{\theta/2})$. Its output is gated by a switching signal, because the adaptation is not always active.

The implementation of adaptation rule 2 is essentially the same as for rule 1. However, both membrane potentials a_k and a_l influence the adaption current.

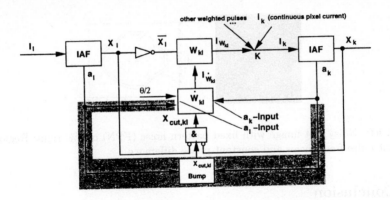

Fig. 3. Block diagram of the adaption rules. The shaded parts are used for rule 2 only, whereas the $\theta/2$-input to the adaptation circuit is supplied for rule 1

Additionally, the cut-off condition (right side of (4)) is realized in an additional building block. As that adaptation rule is symmetric for two neurons, the circuitry can be used for the two synapses in both directions.

For the voltage-to-current converter a simple differential pair is sufficient. Its tanh-characteristic slows down the synchronization process a bit without disturbing the steady-state synchrony. The cut-off condition of (4) is detected by a bump circuit [6].

4   Experimental Results

Experimentally, a test circuit for the segmentation network was designed in a $0.6\mu m$ process and its performance evaluated by simulation. The network has $11 \cdot 11$ neurons and is fed with a noisy grey-level image. The pulse width was chosen to $10\mu s$ and for typical input currents the pulse frequency was chosen to be in the range of approximately $10kHz$. The test chip implements adaption rule 1. Taking W_{kl} dimensionless γ has a value of $\gamma = 10ms^{-1}$ and μ takes a value of $\mu = 3000ms^{-2}$. After $1ms$ simulation time the pulses were analyzed. Fig. 4 shows on the left side the input image and on the right side the regions of identical pulse frequency. Region A represents all neurons that belong to the spot and region B represents all neurons that belong to the margin. As can be seen, the neurons within the spots synchronize to the same frequency. The VLSI implementation exhibit propagation phenomena of firing-waves within a spot as described above. Extensive simulations show that synchronization within a spot is stable after $10 - 20$ pulses. For low grey-level regions noise has a increasing impact on synchronization quality in that way that for lower grey-level values and fixed FPN values synchronization will decay below a crucial grey-level value.

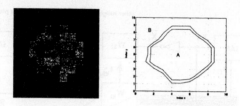

Fig. 4. left: Noisy test image with fixed pattern noise (FPN) of 5%. right: Regions of constant pulse frequencies and constant phase difference

5   Conclusion

In this paper the circuits for an analog VLSI implementation of a pulsed neural network with adaptive weights were presented. The implementations are very compact in terms of transistors used to realize one of the proposed adaption rules. The IAF neuron needs 13 transistors. For the weight adaption rule 1 14 transistors per synapse and for weight adaption rule 2 28 transistors per synapse pair W_{kl}, W_{lk} are necessary.

For both adaption rules and chosen parameters the ability to partition grey-level images into segments up to an fixed pattern noise of 5% was successfully demonstrated. For the future work a complete chip with more than $128 \cdot 128$ pixels will be implemented to study the impact of parameter variations on the segmentation quality. Since the proposed networks copes with noisy grey-level images we are very confident that with this architecture very robust image segmentation systems can be implemented successfully.

References

1. Pal, N.R., Pal, S.K.: A review on image segmentation techniques In: Pattern Recognition, Vol. 26, pp.1277–1294 (1993)
2. Recce, Michael: Encoding Information in Neuronal Activity In: Maas, W., Bishop, C.M. (eds.): Pulsed Neural Networks MIT Press, Cambridge, Massachusetts (1998)
3. Cosp, J, Madrenas, J., Moreno, J.M., Cabestany, J.: Analog VLSI Implementation of a Relaxation Oscillator for Neuromorphic Networks. In: Neuromorphic Systems, Engineering Silicon from Neurobiology, Advances in Neural Information 10, 197–208 (1998)
4. Edwards, T.E., Andreou, A.G.: VLSI Phase Locking Architectures for Feature Linking in Multiple Target Tracking Systems In: Advances in Neural Information Processing Systems, 6 (1994)
5. Vittoz, Eric A.: Analog VLSI Implementation of Neural Networks. In: Handbook of Neural Computation. Oxford University Press (1997)
6. Delbrueck, Tobias: Bump Circuits. CNS Memo 26. California Institute of Technology (1993)
7. Enz, Christian: MOS Modeling for LV-LI Circuit Design. CMOS & BICMOS IC Design. Lausanne (1995)

Kohonen Maps Applied to Fast Image Vector Quantization

Christophe Foucher, Daniel Le Guennec, and Gilles Vaucher

Supélec/ETSN, avenue de la Boulaie, BP 81127, 35511 Cesson Sévigné Cédex, France,
gilles.vaucher@supelec.fr,
http://www.supelec-rennes.fr/ren/rd/etsn/

Abstract. Vector Quantization (VQ) is a powerful technique for image compression but its coding complexity may be an important drawback. Self-Organizing Maps (SOM) are well suited for topologically ordered codebook design. We propose to use that topology for reducing image coding time. Using inter-block correlations, the nearest neighbor search is restricted to the neighborhood of the precedingly used code vector instead of the entire codebook. We obtained a reduction of up to 84% in the coding time compared to full search.

1 Introduction

Classical notions will be briefly recalled here for notational convenience.

In lossy image compression, a quantization stage is generally found. Quantization is an application from E to $F \subset E$ where F is of finite size and is called the *codebook*. If E is a vector space, the operation is called *Vector Quantization*. The elements of the codebook are called *code vectors*.

If N is the number of code vectors, we note $r = \log_2(N)$ the vector quantiser's *resolution* (in bits per vector). Let k be the vector's dimension. When coding an 8 bits per pixel image, segmented into rectangular blocks of k pixels, the compression ratio is then[1]: $CR = \frac{8k}{r}$.

The coder generally performs a so-called *Nearest Neighbour Search* (NNS). For each input vector $\mathbf{x}$, the chosen code vector minimizes the distortion d which is generally the squared euclidian distance. Basically, this implies computing N distortions per input vector. That approach is called *Full Search*.

This is computationally intensive, and several algorithms have been proposed to circumvent it. Many of them integrate *Partial Distance Search*[2] because it is more efficient than full search and has become classical[3,4,11]. We will also follow that approach here.

The code vectors must be defined so as to maximize mean quality for a given application. Thus, the codebook design algorithms use a *training set* containing vectors representative of the given application.

Among classical algorithms are the Generalized Lloyd Algorithm and the Splitting algorithm[8]. Neural competitive learning algorithms are also well

---

[1] without further entropy coding

J.R. Dorronsoro (Ed.): ICANN 2002, LNCS 2415, pp. 1299–1304, 2002.

suited for codebook design. We can cite for example plain competition, frequency sensitive competition, neural gas or self-organizing maps[12]. For a given resolution, their reconstruction quality is quite close[7].

We will use here one-dimensional SOMs (open loop). During codebook design, for each learning vector $\mathbf{x}$, the nearest code vector is searched for, let i^* be its index. Then, that code vector, and its neighbours are moved towards the learning vector following the formula:

$$\forall i \in [1, N], \quad \mathbf{y}_i \leftarrow \mathbf{y}_i + \nu \mathcal{N}(|i^* - i|)(\mathbf{x} - \mathbf{y}_i)$$

where ν is the learning coefficient and $\mathcal{N}$ is the neighbourhood function[2].

The following section will describe the proposed algorithm. It will then be tested and we will present the results obtained. The last section contains a discussion about the algorithm's performance and its potential generalisation.

2 Algorithm Description

Our algorithm uses correlations between successive blocks in an image and the topological ordering of self-organizing maps so as to reduce coding time.

2.1 Correlations between Successive Blocks

One reason for efficiency of vector quantization is the correlations between adjacent pixels in a block. Those are due to the smoothness of luminance variations in natural images. So, the larger are the blocks, the better is the compression ratio/quality trade-off. But computing complexity limits the vector size. As a consequence, there remains correlations between adjacent image blocks[8].

The block space is partitioned into cells (Voronoï domains) corresponding to all the input vectors coded by the same code vector. Because of the correlations between adjacent image blocks, entire vector sequences will be coded by the same code vector. In fact, when processing an image, the input vector follows a trajectory in the block space, and when it crosses a cell boundary, the quantization index changes.

When the codebook is not topologically ordered, there is no relation between input space and index space. So, the correlation between input blocks will be lost, except for the input vectors contained in the same cell. On the contrary, in a topologically ordered codebook, like those obtained with self-organizing maps, neighbouring cells tend to have neighbouring indices and part of the input correlation is retained in the index sequence.

2.2 Using the Topology for Fast VQ

Ways of using the index correlations may be to improve compression ratio by predictive coding[1] or to make the code more robust[10]. We propose here to use them to reduce coding time.

---

[2] For more information about Vector Quantization and image compression, see [8,12].

Effectively, the use of vector quantization for image compression is hindered by computing complexity. To code one vector, the full search requires the computation of $N = 2^{rk}$ distortion measures and so N products by pixel. The complexity is exponential in both the resolution and the vector dimension.

Different approaches where proposed to reduce coding time[3,6,11], among which is the *Partial Distance Search*(PDS)[2], often cited as a reference. During NNS, for each code vector, the squared euclidian distance is computed as a sum of squared terms. With PDS, summation is interrupted as soon as it exceeds the minimum distortion encountered so far. We propose here to improve one step further that algorithm.

Because of correlations, the current code vector will be in the previous one's neighbourhood with high probability. So, the search may be restricted to a limited neighbourhood in the SOM instead of the entire codebook. The search is then called *non-exhaustive*.

Nevertheless, when there's an important change in block content (crossing an object boundary in the image for example), it is necessary to search the entire codebook. Effectively, if the input vector is far away from its predecessor, the move in the map is probably high. In non-exhaustive mode, the search space is limited and may get stuck in an inadequate portion of the codebook. So not only the current input vector would be badly reproduced, but also many following vectors.

The transitions must thus be detected. We propose to use the distortion measure between the current input vector and the previous code vector as an indication of the continuity in block content. If that distortion exceeds a given threshold, the search is exhaustive.

The procedure for coding the input vector $\mathbf{x}(i)$ is thus:

if $d(\mathbf{y}_{\alpha(i-1)}, \mathbf{x}(i)) < T$
then NNS($\mathbf{x}(i)$, C, max($\alpha(i-1) - N'$, 1), min($\alpha(i-1) + N'$, N))
else NNS($\mathbf{x}(i)$, C, 1, N)

where $\alpha(i)$ is the quantization indices sequence, C is the codebook containing N code vectors $\mathbf{y}_1, \ldots, \mathbf{y}_N$ and NNS($\mathbf{x}$, C, i_0, i_1) is the nearest neighbour search for $\mathbf{x}$ in C restricted to the index interval $[i_0, i_1]$.

There are two coding modes: exhaustive and non-exhaustive. This algorithm uses two parameters: the distortion threshold T and the neighbourhood size N'. In that paper, NNS is the PDS but it can be any other NNS procedure.

That approach may be compared to the *probing algorithm*[9], which also uses the topological ordering to reduce NNS complexity. But the probing algorithm, not specifically designed for image compression, does not use the index correlation resulting from the inter-block correlations.

3   Experiments and Results

3.1   Experimental Protocol

Codebooks were designed with $k = 16$, and $N = 256(r = 8)$. Without entropy coding, this results in a compression ratio of 16. The learning set contains 25600 vectors extracted from several natural images. The test set is the Lena image.

A first codebook, C_C, was designed with simple competitive learning. It is initialized by randomly picking N learning vectors[3].

A second codebook, noted C_S, is constructed with a SOM following two stages. During the first one, the *ordering* phase, the neighbourhood width is decreased from N to 1. $\mathcal{N} = 1$ inside the neighbourhood, and 0 outside.

During the second phase, the *fine tuning* phase, the neighbourhood width is set to zero. So, the algorithm may be viewed as simple competitive learning initialized with a SOM, as described in [5].

3.2   Results

The coding algorithm was run with various values for N' (0, 0.01, 0.05, 0.1, 0.5 and 1 relatively to the codebook size N) and T (0.1, 0.5, 1, 5, 10, 50 and 100, relatively to the mean distortion on the learning set). The two codebooks were used to code the test image Lena. The resulting coding time and PSNR are plotted in the figure 1. Similar results were observed with other values for k (9, 4) and r (6).

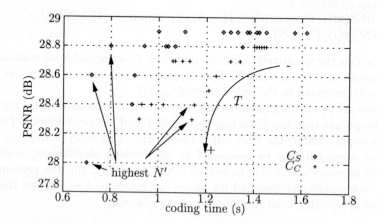

Fig. 1. Performance Measure with $r = 8$ and $k = 16$ and Test Image Lena

---

[3] it is robust enough to avoid initialization problems sometimes encountered with that algorithm in other contexts

With some values of distortion threshold and neighbourhood size, the quality much degrades. With exhaustive coding and a 128 elements codebook ($r = 7$), the PSNR is 28.1dB. So, we retained the parameter values giving a PSNR equal to or above 28dB.

Because C_C is not topologically ordered, only low values of distortion threshold and neighbourhood size give acceptable quality. In particular, a zero value for the neighbourhood size means vectors are coded by the previously used code vector in non-exhaustive mode. This is suitable for sequences of vectors which would be coded by the same code vector even with exhaustive search. This doesn't need any particular order for the codebook.

On the contrary, with C_S, good results where obtained with a relative value of 0.1 for neighbourhood size: 20% of the codebook is searched in non-exhaustive mode.

Search order has high impact on PDS efficiency. At least one of the firstly examined code vectors must be close to the input vector. If so, its distortion will be small and, used as an elimination threshold for other code vectors, it will be very discriminating.

With topological order, in exhaustive mode, the previous code vector is not a good candidate (that's why the coding is exhaustive), and all vectors in its neighbourhood are also bad candidates. Thus, to improve the algorithm, another location in the map should be searched first.

Because the image blocks are extracted from natural images, the main variation direction in the input space is $(1, \ldots, 1)'$, and the 1D SOM will naturally be mainly organized along that direction. An approximation of the location of the input vector in the map can thus be predicted by the mean component, whose computation is equivalent to projection on the main direction. That prediction is computed as: $1 + \lfloor (N - 1) \frac{\sum_{j=1}^{k} x_j}{255k} \rfloor$ where $\lfloor \cdot \rfloor$ is the integer rounding from below operator. That prediction is used in exhaustive coding mode in our experiments.

As a comparison, the same codebooks were used to code the test image with full search and plain PDS. In plain PDS, exhaustive search was conducted in the quantization indices order (from 1 to N). With full search, the encoding time was 4.88s. With 0.1 as relative neighbourhood size, and 5 as relative distortion threshold, the proposed algorithm produced an encoding time of 0.80s, with 28.8 for the PSNR. A coding time reduction of 84% was thus obtained with nearly the same quality as that of full search (0.1 dB less). Similar results (84% coding time gain relatively to full search, with PSNR lower by 0.1 dB) were also obtained with a larger test base containing 418,000 blocks from the 27 learning images.

4   Discussion

This paper has shown that a topologically ordered codebook may be used to reduce coding time in image vector quantization. The resulting algorithm was up to 6.8 times faster than full search and 2.2 times faster than plain PDS.

Two coding modes are chosen according to a threshold used to compare the distortion between the current input vector and the previous code vector. In non exhaustive mode, the NNS is limited to the neighbourhood in the SOM of the previous code vector. In exhaustive mode, all the codebook is searched.

PDS is used in both modes because it is yet faster than complete computation of each distortion. In exhaustive coding mode, due to topological ordering, it is important to begin the search away from the previous code vector. Here, a priori knowledge of input statistics allows to use mean vector component for computing the starting quantization index in the search. In a more general context, a predefined index permutation may be used instead as the search order.

That algorithm may be applied to speed up sequential implementation of self-organizing maps in other contexts than image compression. Effectively, NNS is a central concept in all competitive learning algorithms, in both system design and relaxation, for quantization, classification or representation.

References

1. C. Amerijckx, M. Verleysen, P. Thissen, and J.D. Legat. Image compression by self organized kohonen map. *IEEE Transactions on Neural Networks*, 9(3):503–507, May 1998.
2. Chang-Da Bei and Robert M. Gray. An improvement of the minimum distortion encoding algorithm for vector quantization. *IEEE Transactions on Communications*, COM-33(10):1132–1133, October 1985.
3. J. Cardinal. A fast full search equivalent for mean-shape-gain vector quantizers. In *20th Symposium on Information Theory in the Benelux*, pages 39–46, 1999.
4. Chok-Kwan Cheung and Lai-Man Po. Normalized partial distortion search algorithm for block motion estimation. *IEEE Transactions on Circuits snd Systems for Video Technology*, 10(2):417–422, April 2000.
5. Eric de Bodt, Marie Cottrell, and Michel Verleysen. Using the kohonen algorithm for quick initialisation of simple competitive learning algorithm. In *european symposium on artificial neural networks*, 1999.
6. C. Foucher, F. Durbin, D. Le Guennec, P. Leray, A. Tissot, G. Vaucher, and J. Weiss. Coding time reduction in image vector quantization by linear transforms and partial distorsion evaluation. In *IMVIP, Irish Machine Vision & Image Processing Conference*, 2001.
7. C. Foucher, D. Le Guennec, P. Leray, and G. Vaucher. Algorithmes neuronaux et non neuronaux de construction de dictionnaire pour la quantification vectorielle en traitement d'images. In *Journées Neurosciences et Sciences de l'Ingenieur (NSI)*, pages 165–168, 2000.
8. Allen Gersho and Robert M. Gray. *Vector quantization and signal compression*. Kluwer Academic, 1992.
9. J. Lampinen and E. Oja. Fast self-organization by the probing algorithm. In *International Joint Conference on Neural Networks*, volume 2, pages 503–507, 1989.
10. Dominique Martinez and Woodward Yang. Competitive learning algorithms for channel optimized vector quantizers. In *IEEE International Conference on Neural Networks*, volume 3, pages 1462–1467, 1996.
11. James McNames. Rotated partial distance search for faster vector quantization encoding. *IEEE Signal Processing Letters*, 7(9), September 2000.
12. Syed A. Rizvi and Nasser M. Nasrabadi. Neural networks for image coding: A Survey. In *IS&T/SPIE Conference on Applications of Artificial Neural Networks in Image Processing*, pages 46–57, 1999.

Unsupervised – Neural Network Approach for Efficient Video Description

Giuseppe Acciani, Ernesto Chiarantoni, Daniela Girimonte, and
Cataldo Guaragnella

D.E.E. – Electro-technology and Electronics Department
Politecnico di Bari, Via E. Orabona, 4
I-70125, Bari – Italy
{acciani,chiarantoni,guaragnella}@poliba.it

Abstract. MPEG-4 object oriented video codec implementations are rapidly emerging as a solution to compress audio-video information in an efficient way, suitable for narrowband applications.

A different view is proposed in this paper: several images in a video sequence result very close to each other. Each image of the sequence can be seen as a vector in a hyperspace and the whole video can be considered as a curve described by the image-vector at a given time instant.

The curve can be sampled to represent the whole video, and its evolution along the video space can be reconstructed from its video-samples. Any image in the hyperspace can be obtained by means of a reconstruction algorithm, in analogy with the reconstruction of an analog signal from its samples; anyway, here the multi-dimensional nature of the problem asks for the knowledge of the position in the space and a suitable interpolating kernel function.

The definition of an appropriate Video Key-frames Codebook is introduced to simplify video reproduction; a good quality of the predicted image of the sequence might be obtained with a few information parameters. Once created and stored the VKC, the generic image in the video sequence can be referred to the selected key-frames in the codebook and reconstructed in the hyperspace from its samples.

Focus of this paper is on the analysis phase of a give video sequence. Preliminary results seem promising.

Keywords. MPEG-4, video coding, video streaming, internet TV, video analysis.

1 Introduction

MPEG-4 is increasingly becoming important in the video coding framework for its ability of dealing with high to very low bit rate applications ([1-5]).

The general structure of a MPEG based codec tries to exploit the temporal redundancy of several images in a sequence to greatly reduce the required information to be efficiently transmitted to the receiving end.

Its fortune depends on the object oriented video coding structure, allowing several bit streams to be multiplexed together to form a single video frame structure to be transmitted. Each of the separate streams contains information about a single Video Object (VO), defined by means of arbitrary shape, motion and residual error coding.

J.R. Dorronsoro (Ed.): ICANN 2002, LNCS 2415, pp. 1305–1311, 2002.

Depending on the applications, the time varying available bandwidth can cause freezes of images and/or degradation of the received image quality, so that research very often addresses very low bit rate coding and/or bandwidth adaptive coding algorithms to overcome such annoying situations.

In this paper a novel technique for video reproduction is presented, trying to look at the coding problem in a sub-space description of a video sequence.

The goal the transmission of a given video should reach is to reproduce, in a end user given metrics, good perceived quality.

The video to be reproduced is segmented into shots, each one being described as a video sequence; for each shot, an appropriate, reduced in size, number of key frames are properly selected to constitute the skeleton of the image sequence; each frame of the sequence is referred to the selected frames (key frames) to obtain a good prediction of the image at hand to be coded.

The coder adapts its coding structure to the peculiarities of the video at hand, creating a Video Key-frames Codebook (VKC), and then reduces the coding requirements to the transmission of very poor information on the network.

Only a few images of the sequence should be chosen as Key-frames to represent video, and any other frame can be someway related to the Key-frames selected, to reproduce the true images along the whole video sequence.

In this way, the image reproduction of a generic video sequence can be considered as the problem of the analog signal reconstruction from its discrete time samples.

The VKC creation is based on the video analysis in a vector space; the generic image of the video sequence is firstly segmented into color coherent zones by means of an unsupervised neural network approach; subsequently, the image feature-vector is used to represent the image in a vector space, and clustering of all the images of the video sequence in the feature-space is performed, in order to select the smallest set of Video Key-frames to be used in the definition of the VKC.

The selection of the sequence Key-frame in each cluster is obtained on the basis of the minimal distance form the centroids of the obtained clusters.

Once the VKC has been created, the video reproduction can be obtained by a proper interpolation of the VKC images to obtain the generic image of the sequence.

The paper is so structured: section 2 describes the used unsupervised NN for image feature extraction and subsequent time clustering of images in the feature space; section 3 addresses the video description application, and the video reconstruction with the "nearest neighbor" application is presented, together with quantifications of the minimal bit rate for the sequence reproduction is given; section 4 describes the simulation environment and preliminary results. Conclusions and future work close the paper.

2   The Unsupervised Neural Network Approach

Several algorithms have been proposed in literature, in the field of video reordering, skimming, summarization and storyboarding: all such techniques require to select a proper subset of images of the video sequence being representative of the complete video.

This procedure requires the selection of the images in the sequence someway creating a "video spot", meaning that all the images in a time-neighborhood of such images (wherever located in the video) could be very well described by such images. This procedure requires the definition of the minimal subset of the maximally distant images, presenting such a characteristic.

The problem of selecting a metric is not easily addressable, depending mainly on the subjective definition of the obtained storyboard of the complete video sequence, so that an unsupervised approach has been used, based on the use of a neural network, to obtain the clusters of the sequence and as a second step, to extract the video key-frames from the cluster definition.

Similarly to standard competitive neural networks, the used network ([6]) is composed by M processing elements, where each unit receives an input signal $x = (x_1 \; x_2 \; ... \; x_N)$ from an external data-base and is characterized by the weight vector $w = (w_1 \; w_2 \; ... \; w_N)$. When the input is received, the unit computes the Euclidean distance $d(x,w)$ between the input and the weight. Then, for a single element of the network, the following competitive learning algorithm is proposed in table 1.

In table 1 y is the neuron output, is the logistic function with $f(u)=0$ if $u \leq 0$, $0 < f(u) < 1$ if $u > 0$, $\alpha(k)$ is the learning coefficient, $\beta(k)$ is a coefficient that changes with the same rate of α, δ is a constant value whereas $\sigma(k)$ is an adaptive threshold, the behavior of which will be clarified later on.

From first equation of table 1 it is clear that the activation of the neuron ($y > 0$) occurs only if the extreme of the input vector x is within the hyper-sphere with radius given by σ and center given by the extreme of w.

When $y = 0$, equation (2) is driven by the value of α. On the other hand, when α goes to zero, the value of y predominates and becomes a "winner" indicator. In this way, it is possible to implement a learning law that generates a transition from a phase of "weak" undifferentiated learning ($y = 0$), where each datum influences the learning, to a phase of "locally" selective learning, where the competitive signal βy enables the unit to learn only in the local "winner region". In order to adapt the 'winner region' to the feature of the input space, an adaptive threshold is introduced. Namely, the idea is to increase σ when there is no activation, and vice-versa.

During a phase of undifferentiated learning ($y = 0$), equation (3) is driven by α. On the other hand, while w moves toward its center, the selective learning ($y > 0$) takes place, α goes to zero, σ decreases, the neighborhood size decreases and, consequently, input vectors of a reduced neighbor of w contribute to the learning.

Network with Local Connections: Differently from WTA (Winner Take All) paradigms, the proposed network herein is an array of N units, which are characterized by *local* inhibitory connections.

At the beginning each unit receives samples from the input space. However, only the first unit learns, since it has no input inhibitory links, whereas all the other units are inhibited, due to the presence of input inhibitory links.

Therefore, the first inhibitory link is turned off and the second unit can start its learning phase. This means that the input samples, which make active the first unit, continue to produce learning only for the first unit, whereas the remaining samples produce learning for the second unit.

Table 1.

$$y(k) = f\big(\sigma(k) - d(x(k), w(k))\big)$$

$$\Delta w(k) = \big(\alpha(k) + \beta(k)y(k)\big)\big(x - w(k)\big) \tag{2a}$$

$$w(k+1) = w(k) + \Delta w(k) \tag{2b}$$

$$\Delta\sigma(k) = \big(\alpha(k) + \beta(k)y(k)\big)\big(d(x, w(k)) - \sigma(k)\big) \tag{3a}$$

$$\sigma(k+1) = \sigma(k) + \Delta\sigma(k) \tag{3b}$$

$$\Delta\alpha(k) = \alpha(k)y(k)\delta$$

$$\alpha(k+1) = \alpha(k) + \Delta\alpha(k) \tag{4a}$$

$$\Delta\beta(k) = -\Delta\alpha(k)$$

$$\beta(k+1) = \beta(k) + \Delta\beta(k) \tag{4b}$$

Iterate to Step 1 until $\Delta\sigma(k) < \varepsilon$

This network architecture combines the advantages of both sequential cluster search and unsupervised competitive networks (UCNN). This feature leads to a major advantage over the performances of unsupervised competitive neural networks: due to the presence of inhibitory links, lower priority units cannot affect higher priority units. As a consequence, if two networks are composed of N and $(N+M)$ units, the common N units will behave in the same way in both the networks and will found exactly the same centroids. This means that it is always possible to add some extra units (in order to check if some cluster in the data structure has been missed) without modifying the behavior of the first unit.

Notice that UCNN's lack of this property, that is, they suffer of the problem of the dependence of the final partitions of the data on the number of the network elements.

Fig. 1. The obtained clusters for the image simplified description

This NN structure has been used to simplify image description to obtain in a second step the selection of the Video Key-frames.

Presented results refer to the application to the sequence "Akiyo".

For each image of the sequence only image luminance has been considered in the clustering procedure for image simplification (feature extraction).

Seven clusters per image have been considered a good trade- off between order of the hyperspace of the images and a good segmentation into coherent regions of the image.

Figure 1 represents a typical clustering of each image in the sequence.

Once the feature description phase has been carried over all the images in the sequence, a time clustering procedure in a simplified 7-D hyperspace has been carried on.

The goal of this procedure is the detection of clusters inside the sequence; each cluster prototype, i.e. the cluster centroids represents a typical image but doesn't coincide with a given image in the sequence, so that key-frames have been chosen as the closer

Fig. 2. The five prototypes images of the Akiyo sequence.

cluster image to the detected centroid; such images can well represent the video sequence summary report.

We chose to indicate the closest images to the devised centroids after the time clustering procedure as the VKC, the Video Key-frames Codebook.

Fig. 2 reports the obtained key-frames; it can be seen, by observing the video, that the obtained key frames well represent the whole video sequence.

In a video coding framework, the predicted image construction at the receiving end is based in predictive coding techniques on the knowledge of the motion field the images is experiencing; motion fields are transmitted as side information to the receiver. In the proposed coding scheme, the side information to obtain the predicted image is substituted, when the VKC has been defined, by the transmission of the pointer (the LUT entry in the VKC to the stored Key-frame assumed as a prediction), if a zero order interpolating kernel is assumed, and on a vector of distances from the key-frames in a more general case.

Loss-less coding requires also an efficient technique for the transmission of the prediction error, i.e. the DFD (Displaced Frame Difference).

Substantially, the clustering procedure creates a look up table (LUT) to be used in the coding phase instead of the usual prediction phase.

3   The Video Coding Experimental Set-Up

Performances evaluation of the proposed video reproduction technique refers to the assumed loss-less entropy coding of the difference between the "interpolated" image and the true one.

Any image can be approximated by its key frame in the VKC, in a raw description. This means the use of the zero order interpolator in the hyperspace, which means to substitute the nearest neighbor key frame to the generic frame of the sequence.

The required bit rate to code the video in this way is really negligible if the initialization phase of the connection is neglected. If N key-frames have been selected to represent the whole video, the required bit rate might be computed as $log_2(N)$ b/frame, a very poor bit rate indeed.

Table 2.

Key-frames number	Time average squared error	Time average pixel mean entropy (b/pixel)	Time average bit rate (kb/frame)
5	21.42	2.16	54.7
10	12.29	1.91	48.1

The use of a zero order interpolating function produces high prediction errors; the choice of a better interpolating kernel function might allow lower requirements information.

In order to test the validity of our assumptions, fig. 3 reports the entropy time evolution in the case of a zero order interpolating function.

In this work, only preliminary results of the proposed idea are presented. Experimental tests have been conduced on Akiyo standard video sequence in QCIF (144×176, 300 frames) luminance only video format. As a measure of the required bit rate we chose the entropy of the prediction error corresponding to substituting the mostly similar image in the video sequence with the true one. The prediction error entropy is used as a measure of the required bit rate to obtain the perfectly reconstructed video sequence.

Coding results are reported both in visual and quantitative form respectively as a bit rate time evolution diagram and in table 2 as average values for two distinct cases of 5 and 10 selected key frames.

As notable, the bit rate per frame are very high values, but it should be noted that:

→ They refer to a zero order interpolation kernel (the nearest neighbor is used as a prediction);

→ Loss-less coding is assumed, as the entropy of the prediction error has been used as a measure of the required bit rate.

With this kind of coding structure, we can tune algorithms handles to select the coding performances: handles are represented by the VKCs size, the required received image-quality and the required information to describe the image to be reproduced.

4   Conclusion and Discussion

The creation of a Video Key-frame Codebook is here introduced to simplify the coding required information to be transmitted on very narrowband channels as those experienced in traffic congestion situations.

The proposed space approach can be considered a straightforward generalization of the forward-backward motion compensation mechanism included in MPEG-4 standard, with the fundamental difference that no time dependency exists between the image to be predicted and its preceding and following ones.

The use of a zero order multidimensional interpolating function produces high prediction errors and thus high bit rate for the difference coding; the choice of a better

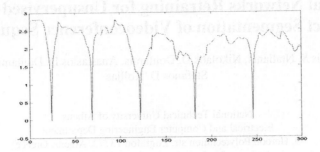

Fig. 3. Time evolution of the frame-difference entropy obtained for a zero order interpolating function. It obviously nulls on selected key-frames.

interpolating kernel function might allow lower requirements in bandwidth, on one hand, even if higher requirements in the receiver hardware performances.

Further research is still involved in the definition of a good quality multi-dimensional interpolating kernel, being able in simply reproducing the generic frame of a given sequence by interpolation carried on the video key frames.

References

[1] P Salembier, F. Marques, Region based representation of image and video: Segmentation tool for multimedia services, IEEE Trans. Circuits and systems for Video Technology, invited paper, vol. 9 no. 8, dec. 1999

[2] ISO/IEC DIS 13818–2, Information Technology – Generic Coding of Moving Pictures and Associated Audio Information – Part 2: Video, ISO, 1994.

[3] MPEG Video Group, MPEG–4 Video Verification Model Version 4.0, ISO/IEC JTC1/SC29/WG11/M1380, Proceedings of Chicago meeting, October 1996

[4] CCITT SG XV, Recommendation H.261 – Video Codec for Audiovisual Services at px64 kbit/s, COM XV–R37–E, Int. Telecommunication Union, August 1990.

[5] ITU–T Draft, Recommendation H.263 – Video Coding for low bit rate communication, Int. Telecommunication Union, November 1995.

[6] G. Acciani, E. Chiarantoni, and M. Minenna "A new non Competitive Unsupervised Neural Network for Clustering" Proc. of Intern. Symp. On Circuits and Systems, Vol. 6, pp. 273–276, London May 1994.

[7] C. Guaragnella, E. Di Sciascio, Object Oriented Motion Estimation by Sliced-Block Matching Algorithm, Proc. IEEE 15th Intl. Conf. On Pattern Recognition, Vol. 3, Image, speech and signal processing, pp. 865-869, Barcelona, Sept. 3–7, 2000.

[8] C. Cafforio, E. Di Sciascio, C. Guaragnella, Motion estimation and Modeling for Video Sequences, Proc. of EUSIPCO 98, 8–11, Rhodes, GR.

[9] A.Guerriero and V.Di Lecce, An Evaluation of the Effectiveness of image Features for Image Retrieval, J. Visual Communication and Image Representation 10,351–362 (1999).

[10] A.Del Bimbo and P. Pala, Visual image retrieval by elastic matching of user sketches, IEEE Trans. Pattern Anal. Mach. Intell. 19(2), 1997.

Neural Networks Retraining for Unsupervised Video Object Segmentation of Videoconference Sequences[*]

Klimis S. Ntalianis, Nikolaos D. Doulamis, Anastasios D. Doulamis, and Stefanos D. Kollias

National Technical University of Athens
Electrical and Computer Engineering Department
9, Heroon Polytechniou str. Zografou 15773, Athens, Greece
(kntal, adoulam, ndoulam)@image.ntua.gr

Abstract. In this paper efficient performance generalization of neural network classifiers is accomplished, for unsupervised video object segmentation in videoconference/videophone sequences. Each time conditions change, a retraining phase is activated and the neural network classifier is adapted to the new environment. During retraining both the former and current knowledge are utilized so that good network generalization is achieved. The retraining algorithm results in the minimization of a convex function subject to linear constraints, leading to very fast network weight adaptation. Current knowledge is unsupervisedly extracted using a face-body detector, based on Gaussian p.d.f models. A binary template matching technique is also incorporated, which imposes shape constraints to candidate face regions. Finally the retrained network performs video object segmentation to the new environment. Several experiments on real sequences indicate the promising performance of the proposed adaptive neural network as efficient video object segmentation tool.

1 Introduction

Video object segmentation algorithms can significantly benefit emerging multimedia applications, such as object-based video transmission and content-based image retrieval [1]. For this reason the notion of Video Objects (VOs) has been introduced by the MPEG-4 standard [2]. A VO is an arbitrarily shaped region, consisting of multiple color, texture or motion characteristics, such as a human, a ship etc. Nevertheless, no methods are suggested within the standard to segment VOs, a problem that is generally very difficult and still remains unsolved. Furthermore very little work has been presented in literature, considering neural networks as VO segmentation tools. Nevertheless, neural networks with their powerful classification capabilities can help in VO segmentation and play a significant role in the development of the new multimedia oriented standards, such as MPEG-4/7.

Color segmentation is a first attempt towards video content representation. Typical works include the morphological watershed [3] and the RSST algorithm [4], which however oversegment video objects into multiple color regions. Other schemes that use motion homogeneity criteria, like [5] and [6], cannot provide accurate VO contours, due to erroneous estimation of motion vectors. On the other hand some more sophisticated schemes have been recently proposed. In [7] color segments with

[*] This research is funded by the Institute of Governmental Scholarships of Greece.

J.R. Dorronsoro (Ed.): ICANN 2002, LNCS 2415, pp. 1312–1318, 2002.

coherent motion are merged to form the video object, while in [8], a model of the region of interest is derived using the Canny operator. The aforementioned techniques produce sufficient results when 1) adjacent video objects have different motion characteristics, and 2) color segments of each video object have coherent motion. However in real life sequences, there are several cases where little or no motion exists or when parts of the object do not have coherent motion.

In this paper, an adaptive neural network architecture is proposed for unsupervised segmentation of videoconference/videophone VOs. Each time conditions change, a retraining phase is activated and a retraining set is extracted by a face detector, based on a Gausssian probability density function (pdf). The Gaussian pdf models the color components of a face area and shape constraints are applied on the resulting color-compatible regions, using a binary template matching technique. Then human body is localized using a probabilistic model, the parameters of which are estimated according to the center, height and width of the face region. Based on the new training set, network weight adaptation is very quickly performed, since the retraining method results in the minimization of a convex function subject to linear constraints. After retraining the network is applied to the new environment and extracts the VOs. Experimental results on real videoconference/videophone sequences indicate the promising performance of the proposed adaptive neural network classifier.

2 Problem Formulation and Retraining Algorithm

2.1 Problem Formulation

Let us consider video object segmentation as a classification problem. Under this consideration each video object would correspond only to one class ω_i out of p available. Then, for each image region the neural network will produce a p-dimensional output vector $\mathbf{z}(\mathbf{x}_i)$:

$$\mathbf{z}(\mathbf{x}_i) = \left[p^i_{\omega_1} \, p^i_{\omega_2} \cdots p^i_{\omega_p} \right]^T \tag{1}$$

where $\mathbf{x}_i$ is a feature vector, describing the content of the i-th block of the image and $p^i_{\omega_j}$ denotes the likelihood of the i-th block to belong to the j-th class ω_j (j-th VO).

Let us consider that the neural network has been initially trained using the training set, $S_b = \left\{ (\mathbf{x}'_1, \mathbf{d}'_1), \cdots, (\mathbf{x}'_{m_b}, \mathbf{d}'_{m_b}) \right\}$, where vector $\mathbf{x}'_i$, $i = 1, 2, \cdots, m_b$ denotes the i-th input training vector, while $\mathbf{d}'_i$ is the respective desired output vector.

Supposing now that a retraining phase is activated, new weights $\mathbf{w}_a$ are estimated by taking into consideration both the former knowledge (of set S_b) and the current knowledge of set $S_c = \{ (\mathbf{x}_1, \mathbf{d}_1), \cdots (\mathbf{x}_{m_c}, \mathbf{d}_{m_c}) \}$ composed of m_c elements. In particular, the new network weights $\mathbf{w}_a$ are estimated by minimizing the following error criteria,

$$E_a = E_{c,a} + \eta E_{f,a} \tag{2}$$

with

$$E_{c,a} = \frac{1}{2} \sum_{i=1}^{m_c} \left\| \mathbf{z}_a(\mathbf{x}_i) - \mathbf{d}_i \right\|_2 \tag{2a}$$

and
$$E_{f,a} = \frac{1}{2}\sum_{i=1}^{m_b}\left\|\mathbf{z}_a(\mathbf{x}_i') - \mathbf{d}_i'\right\|_2 \tag{2b}$$

In the previous equations, $\mathbf{z}_a(\mathbf{x}_i)$ and $\mathbf{z}_a(\mathbf{x}_i')$ are the network outputs when the new weights $\mathbf{w}_a$ are used and feature vectors $\mathbf{x}_i$ (of set S_c) and $\mathbf{x}_i'$ (of set S_b) are fed as network inputs. Therefore, $E_{c,a}$ is the error between the network output and the desired output over all data of set S_c („current" knowledge), while $E_{f,a}$ the corresponding error over training set S_b („former" knowledge); Parameter η regulates the significance between the current training set and the former one.

2.2 The Retraining Algorithm

During retraining new weights $\mathbf{w}_a$ are estimated assuming that a small perturbation of the network weights $\mathbf{w}_b$ is enough to achieve good classification performance. Then
$$\mathbf{W}_a^0 = \mathbf{W}_b^0 + \Delta\mathbf{W}^0 \text{ and } \mathbf{w}_a^1 = \mathbf{w}_b^1 + \Delta\mathbf{w}^1 \tag{3}$$
where $\Delta\mathbf{W}^0$ and $\Delta\mathbf{w}^1$ are small weight increments, $\mathbf{W}_a^0$ and $\mathbf{W}_b^0$ are matrices of the form $\mathbf{W}_{\{a,b\}}^0 = \left[\mathbf{w}_{1,\{a,b\}}^0 \ldots \mathbf{w}_{q,\{a,b\}}^0\right]$ and $\mathbf{w}_a^1$, $\mathbf{w}_b^1$ are vectors containing the weights between the output and the hidden layer neurons after and before retraining respectively. This assumption leads to an analytical and tractable solution for estimating $\mathbf{w}_a$ [9]. Furthermore, in order to stress the importance of current training data, equation (2a) can be replaced by the constraint that the actual network outputs are equal to the desired ones:
$$z_a(\mathbf{x}_i) = d_i \quad i=1,\ldots,m_c, \text{ for all data in } S_c \tag{4}$$

Solution of (4) with respect to the weight increments is equivalent to a set of linear equations [9]:
$$\mathbf{c} = \mathbf{A} \cdot \Delta\mathbf{w} \tag{5}$$
where $\Delta\mathbf{w}=[(\Delta\mathbf{w}^0)^T(\Delta\mathbf{w}^1)^T]^T$, $\Delta\mathbf{w}^0 = \text{vec}\{\Delta\mathbf{W}^0\}$, with $\text{vec}\{\Delta\mathbf{W}^0\}$ denoting a vector formed by stacking up all columns of $\Delta\mathbf{W}^0$.

Uniqueness is imposed by an additional requirement due to $E_{f,a}$ in (2b). In particular, equation (2b) is solved with the requirement of minimum degradation of the previous network behavior, i.e., minimizing the following error criterion:
$$E_S = \left\|E_{f,a} - E_{f,b}\right\|_2 \tag{6}$$
It can be shown [9] that (6) takes the form:
$$E_S = \frac{1}{2}(\Delta\underline{w})^T \cdot \mathbf{K}^T \cdot \mathbf{K} \cdot \Delta\mathbf{w} \tag{7}$$

where the elements of matrix $\mathbf{K}$ are expressed in terms of the previous network weights $\mathbf{w}_b$ and the training data in S_b. The problem results in minimization of (7), which is convex, subject to the linear equalities of (5). Thus, only one global minimum exists and in this paper it is obtained by the gradient projection method.

3 Unsupervised Extraction of the Retraining Set

In the considered videoconference/videophone sequences, retraining set S_c consists of foreground (human) and background information. In particular, human faces are first detected using the chrominance components of the image and shape information of face regions, while human bodies are localized based on a probabilistic model. In subsections 3.1 and 3.2 human face and body detection are presented.

 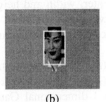 

(a)   (b)   (c)   (d)

Fig. 1. The "Akiyo" Sequence. (a) Original image. (b) Detected face area (c) Retraining sets for the foreground and background VOs (region of uncertainty in gray) (d) Foreground VO.

3.1 Human Face Detection

In the proposed approach, the two chrominance components of a color image are used for face detection, as the distribution of human-face chrominance values occupies a very small region of the color space [10].

Towards this direction, the histogram of chrominance values corresponding to the face class Ω_f, is initially modeled by a Gausssian probability density function (pdf):

$$P(\mathbf{x}|\Omega_f) = \frac{\exp(-\frac{1}{2}(\mathbf{x}-\mathbf{\mu}_f)^T \cdot \Sigma_f^{-1} \cdot (\mathbf{x}-\mathbf{\mu}_f))}{2\pi \cdot |\Sigma|^{1/2}} \qquad (8)$$

where $\mathbf{x}=[u\ v]^T$ is a 2x1 vector containing the mean chrominance components u and v of an examined block, $\mathbf{\mu}_f$ is the 2x1 mean vector of a face area and Σ is the 2x2 variance matrix of the pdf. Parameters $\mathbf{\mu}_f$ and Σ are estimated based on a set of several face images and using the maximum likelihood algorithm [11].

Equation (8) indicates that an image block B_i belongs to the face area, if the respective probability of its chrominance values, $P(\mathbf{x}(B_i)|\Omega_f)$ is high. In the proposed scheme and in order to take more reliable results, a confidence interval of 80% is selected from the Gausssian model, so that only blocks inside this region are considered as face blocks. Therefore, for each test image of size N_1xN_2 a binary mask M is formed, with size $N_1/8 \times N_2/8$ pixels (as block resolution is assumed). However, if only color information is considered, the final binary mask M may also contain non-face blocks, with chrominance properties similar to face regions, e.g. human hands. To confront this problem shape information of human faces is also utilized and the method described in [10] is adopted, where rectangles with certain aspect ratios approximate the shape of a face and are used to filter non-face blocks in mask M.

In particular, the aspect ratio for face areas is defined as:

$$R= H_f / W_f \qquad (9)$$

where H_f is the height of the head, while W_f corresponds to the face width. After several experiments R was found to lie within the interval [1.4 1.6] [10].

3.2 Human Body Detection

Human body is detected by exploiting information derived from the human face detection task. In particular, initially the center, width and height of the face region, denoted as $\mathbf{c}_f=[c_x\ c_y]^T$, w_f and h_f respectively are calculated. Human body is then localized by incorporating a probabilistic model, the parameters of which are estimated according to $\mathbf{c}_f$, w_f and h_f.

In particular let us denote by $\mathbf{r}(B_i)=[r_x(B_i)\ r_y(B_i)]^T$ the distance between the i-th block, B_i, and the origin, with $r_x(B_i)$ and $r_y(B_i)$ the respective x and y coordinates. In this paper the product of two independent 1-dimensional Gaussian pdfs is used to model the human body. Then for each block B_i of an image, a probability $P(\mathbf{r}(B_i)\,|\,\Omega_b)$ is assigned, expressing the degree of block B_i belonging to the human body class Ω_b:

$$P(\mathbf{r}(B_i)\,|\,\Omega_b) = \frac{\exp(-\dfrac{1}{2\sigma_x^2}(r_x(B_i)-\mu_x)^2)\exp(-\dfrac{1}{2\sigma_y}(r_y(B_i)-\mu_y)^2)}{(2\pi)\sigma_x\sigma_y} \tag{10}$$

where μ_x, μ_y, σ_x and σ_y are the parameters of the human body location model; these parameters are calculated based on the information derived from the face detection task, taking into account the relationship between human face and body.

$$\mu_x=c_x,\ \mu_y=c_y+h_f \tag{11a}$$
$$\sigma_x=w_f,\ \sigma_y=h_f/2 \tag{11b}$$

Similarly to human face detection, a block B_i belongs to class Ω_b (body class), if the respective probability, $P(\mathbf{r}(B_i)\,|\,\Omega_b)$ is high. Again a confidence interval of 80% is selected so that blocks belonging to a body region are more reliably detected.

(a)   (b)   (c)   (d)

Fig. 2. The „Silent" Sequence. (a) Original image. (b) Detected face area. (c) Retraining sets for the foreground and background VOs (region of uncertainty in gray) (d) Foreground VO.

4 Experimental Results

Initially a new sequence was formed by joining together the 'Akiyo', 'Silent' and 'Trevor' standard videoconference sequences. As in videoconference environments little changes occur, retraining was performed only when a new shot was detected. For this reason the automatic shot cut detection algorithm described in [12] was incorporated due to its efficiency and low complexity. Figures 1, 2 and 3 illustrate the performance of the proposed unsupervised video object segmentation scheme. In particular in Figures 1(a), 2(a) and 3(a) one frame from each of the standard sequences is depicted while in Figures 1(b), 2(b) and 3(b) the performance of the human face detection module is illustrated. It should be mentioned that in 'Trevor' and 'Silent' sequences, additional blocks are also selected as candidate face blocks, since their chrominance characteristics are close to those of face regions, e.g., the hands. For clarity of presentation, background blocks are depicted with gray color. Finally the adopted binary template-matching algorithm for face detection is incorporated. In our implementation, valid values for the aspect ratio R of the bounding rectangles were in the interval [1.2 1.7], also addressing cases of covered foreheads and exposed necks.

As shown in Figures 1(b), 2(b) and 3(b) a part of the human face is extracted in all cases regardless of its size, located inside the area, depicted by a white bounding rectangle. As observed, non-face regions are discarded, as they do not satisfy shape constraints. Figures 1(c), 2(c) and 3(c) illustrate the estimated retraining sets, for the foreground and background video objects. In these figures, gray color is used to represent the region of uncertainty. As it can be observed, the region of uncertainty contains several blocks, located at the boundaries of the face and body areas and thus it protects the network from ambiguous regions. Afterwards foreground/background blocks are selected using a PCA method, to retrain the neural network classifier. During retraining the DC coefficient and the first 8 AC coefficients of the zig-zag scanned DCT for each color component of a block, (i.e., 27 elements in total) are used as feature vector $\mathbf{x}_i$ of the respective block.

      (a)   (b)   (c)   (d)

Fig. 3. The „Trevor" Sequence. (a) Original image. (b) Detected face area. (c) Retraining sets for the foreground and background VOs (region of uncertainty in gray) (d) Foreground VO.

Finally, the retrained network performs the final video object segmentation task. Figures 1(d), 2(d) and 3(d) depict the final segmentation provided by the neural network classifier. As it can be observed in all cases the foreground object is extracted with accuracy, despite the complexity of the background.

References

1. K. N. Ngan, S. Panchanathan, T. Sikora and M.-T. Sun, „Guest Editorial: Special Issue on Representation and Coding of Images and Video," *IEEE Trans. CSVT*, Vol. 8, No. 7, pp. 797–801, November 1998.
2. ISO/IEC JTC1/SC29/WG11 N3156, "MPEG-4 Overview," Doc. N3156, December 1999.
3. Meyer and S. Beucher, „Morphological Segmentation," *Journal of Visual Communication on Image Representation*, Vol. 1, No. 1, pp.21–46, Sept. 1990.
4. O. J. Morris, M. J. Lee and A. G. Constantinides, „Graph Theory for Image Analysis: an Approach based on the Shortest Spanning Tree," *IEE Proceedings*, Vol. 133, pp.146–152, April 1986.

5. W. B. Thompson and T. G. Pong, „Detecting Moving Objects," *Int. Journal Computer Vision*, Vol. 4, pp. 39–57, 1990.
6. J. Wang and E. Adelson, „Representing Moving Images with Layers," *IEEE Trans. Image Processing*, Vol. 3, pp. 625–638, Sept. 1994.
7. D. Wang, „Unsupervised Video Segmentation Based on Watersheds and Temporal Tracking," *IEEE Trans. CSVT*, Vol. 8, No. 5, pp. 539–546, 1998.
8. T. Meier, and K. Ngan, „Video Segmentation for Content-Based Coding," *IEEE Trans. CSVT*, Vol. 9, No. 8, pp. 1190-1203, 1999.
9. A. Doulamis, N. Doulamis, S. Kollias, „On Line Retrainable Neural Networks: Improving the Performance of Neural Networks in Image Analysis Problems," *IEEE Trans. on Neural Networks*, Vol. 11, No. 1, January 2000.
10. H. Wang and Shih-Fu Chang, „A Highly Efficient System for Automatic Face Region Detection in MPEG Video Sequences," *IEEE Trans. CSVT*, vol. 7, pp. 615–628, Aug. 1997.
11. A. Papoulis, Probability, Random Variables, and Stochastic Processes. New York: McGraw Hill, 1984.
12. B. L. Yeo and B. Liu, „Rapid Scene Analysis on Compressed Videos," *IEEE Trans. CSVT*, Vol. 5, pp. 533–544, Dec. 1995.

Learning Face Localization Using Hierarchical Recurrent Networks

Sven Behnke

Freie Universität Berlin, Institute of Computer Science
Takustr. 9, 14195 Berlin, Germany
behnke@inf.fu-berlin.de, www.inf.fu-berlin.de/~behnke

Abstract. One of the major parts in human-computer interface applications, such as face recognition and video-telephony, consists in the exact localization of a face in an image.

Here, we propose to use hierarchical neural networks with local recurrent connectivity to solve this task, even in presence of complex backgrounds, difficult lighting, and noise. Our network is trained using a database of gray-scale still images and manually determined eye coordinates. It is able to produce reliable and accurate eye coordinates for unknown images by iteratively refining an initial solution.

The performance of the proposed approach is evaluated against a large test set. The fast network update allows for real-time operation.

1 Introduction

To make the interface between humans and computers more pleasant, computers must adapt to the users. One important step for many adaptive applications, like face recognition, lip reading, reading of the users emotional state, and video-telephony is the localization of the user's face in a captured image.

A recent survey on face detection can be found in [4]. Many localization techniques rely on image motion or skin color which are not always available. In [9] multiresolution window scanning in combination with a neural network is used to detect faces in gray-scale static images. Such sequential search techniques are computationally expensive. Many methods preprocess the data intensively to extract facial features and match them with predefined models [5,6].

In this paper, we present a method that uses a hierarchical neural network with recurrent local connectivity to localize a face in gray-scale still images. The network operates by iteratively refining an initial solution. We present images directly to the network and train it to do the job.

2 Face Database and Preprocessing

To validate the performance of the proposed approach for learning face localization, we use the BioID data base [5]. This database consists of 1521 images that

J.R. Dorronsoro (Ed.): ICANN 2002, LNCS 2415, pp. 1319–1324, 2002.

show 23 individuals in front of various complex office backgrounds with uncon-trolled lighting. The persons differ in gender, age, and skin color. Some of them wear glasses or a beard. Since the face size, position, and view, as well as the facial expression vary considerably, the dataset can be considered challenging.

Fig. 1. Some face images from the BioID data set.

Such real world conditions are the ones that show the limits of current local-ization techniques. For instance, while the hybrid localization system described in [5] correctly localizes 98.4% of the XM2VTS database [7] that has been pro-duced under controlled conditions, the same system localizes only 91.8% of the BioID faces. Figure 1 shows some images from the dataset.

The gray-scale BioID images have a size of 384×288 pixels. To reduce border effects, we lowered the contrast towards the sides of the image. To limit the amount of data, the image is subsampled to 48×36, 24×18, and 12×9 pixels as shown in Fig. 2(b). In addition to the images, manually labeled eye positions are available. Fig. 2(a) shows the marked eye positions for a sample image. We produce a multi-resolutional Gaussian blob for each eye (see Fig. 2(b)).

3   Network Architecture

The preprocessed images are presented to a hierarchical neural network that is structured as Neural Abstraction Pyramid [1]. As can be seen in Figure 3, the network consists of four layers. Each layer contains excitatory and inhibitory quantities. Each quantity is computed at a 2D-grid of locations by $\sum$-units that share a common weight template. The resolution of the layers decreases from L_0 (48×36) to L_2 (12×9) by a factor of 2 in both dimensions. L_3 has only one unit

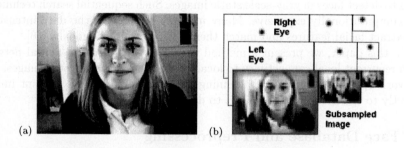

Fig. 2. Preprocessing: (a) original image with marked eye positions; (b) eye positions and subsampled framed image in three resolutions.

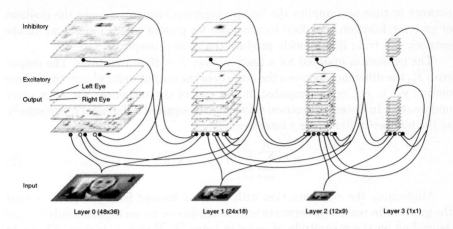

Fig. 3. Sketch of the hierarchical network architecture with local recurrent connectivity.

per quantity. The number of quantities per layer increases when going from L_0 (4+2) to L_2 (16+8). L_3 contains 10 excitatory and 5 inhibitory quantities.

The network's connectivity is recurrent and local. Each unit receives input from only a small window of units that correspond to similar locations in the layer below (forward weighs), in the same layer (lateral weights) and from the layer above (backward weights). Weights from excitatory units are non-negative. Weights from inhibitory units are non-positive. Input weights can have any sign.

The excitatory units of L_0-L_2 receive input from 4×4 windows of the quantities one layer below. They look at 5×5 input units and at a 3×3 neighborhood at the same layer. The backward weights have a window size of 2×2. Connections between L_2 and the topmost L_3 are different. Both, forward- and backward weights have a 12×9 window size. Inhibitory quantities look only at 5×5 windows of all excitatory quantities at the same layer. Of course, in L_3 this reduces to 1×1. The update step $(t + 1)$ of a unit for quantity q at position (x, y) in L_z is done as follows:

$$v^{t+1}_{x,y,z,q} = \sigma \left[\sum_{j \in \mathcal{L}(i)} \mathcal{W}(j) \, v^t_{\mathcal{X}(j,x),\mathcal{Y}(j,y),\mathcal{Z}(j,z),\mathcal{Q}(j)} + \mathcal{B}(i) \right]. \tag{1}$$

$\mathcal{L}(i)$ is the set of links of the associated template $i = \mathcal{T}(z,q)$, and $\mathcal{B}(i)$ is the template bias. $(\mathcal{X}(j,x), \mathcal{Y}(j,y), \mathcal{Z}(j,z), \mathcal{Q}(j))$ describe location and quantity of the input for link j, and $\mathcal{W}(j)$ is its weight. The output function $\sigma(x) = \ln(1 + e^{\beta x})/\beta$ is here a smooth approximation to the rectifying function $\max(0, x)$. In addition, a start value $\mathcal{V}^0(i)$ for initialization at $t = 0$ is needed for each template.

4   Supervised Training

Training recurrent networks is difficult due to the non-linear dynamics of the system. The backpropagation through time algorithm (BPTT) [10] unfolds the network in time and applies the backpropagation idea to compute the gradient of an error function. For face localization, we present a static input $\mathbf{x}_k$ to the network and train it to quickly produce the desired output $\mathbf{y}_k$.

The network is updated for a fixed number $T = 10$ of iterations. The output error δ_k^t, the difference between the activity of the output units $\mathbf{v}_k^t$ and the desired output $\mathbf{y}_k$, is not only computed at the end of the sequence, but after every update step. In the error function we weight the squared differences progressively, as the number of iterations t increases:

$$E = \sum_{k=1}^{K} \sum_{t=1}^{T} t \, \|\mathbf{y}_k - \mathbf{v}_k^t\|^2. \tag{2}$$

Minimizing the error function with gradient descent faces the problem that the gradient in recurrent networks either vanishes or grows exponentially in time depending on the magnitude of gains in loops [3]. Hence, it is very difficult to determine a learning constant that allows for both stability and fast convergence.

For that reason, we decided to employ the RPROP algorithm [8], that maintains a learning rate for each weight and uses only the sign of the gradient to determine the weight change. We modify not only the weights in this way, but adapt the biases and start values as well. To accelerate the training, we initially worked with randomly chosen subsets of the training set, as described in [2].

5   Experimental Results

We divided the BioID data set randomly into 1000 training images and 521 test examples. Figure 4 shows the development of the trained network's output over time when the test image from Fig. 2 is presented as input. One can see that the blobs signaling the locations of the eyes develop in a top-down fashion. After the first iteration they appear only in the lowest resolution. This coarse localization is used to bias the development of blobs in lower layers. After five iterations, the network's output is close to the desired one. It does not change significantly during the next iterations that take 22ms each on a P4 1.7GHz PC.

The generation of stable blobs is the typical behavior of the network. To evaluate its performance, one has to estimate eye coordinates from the blobs and to compute a quality measure by comparison with the given coordinates.

We estimate the position of each eye separately by finding the output unit with the highest activity in the corresponding high resolution output. For all units in a 7×7 window around it, we segment the units belonging to the blob

Fig. 4. Recall. Shown are the activities of the network's output over time.

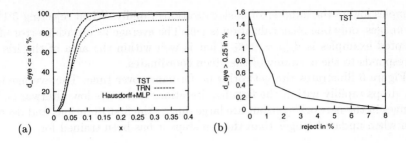

(a)   (b)

Fig. 5. Localization performance: (a) percentage of examples having small d_{eye} for the proposed method (TRN, TST) and for the hybrid system (Hausdorff+MLP)[5]; (b) rejecting the least confident examples lowers the number of mislocalizations.

by comparing their activity with a threshold that increases with greater distance from the center. The weighted mean location of the segmented units is the estimated eye position. After transforming these eye positions into the original coordinate system, we compute a scale independent relative error measure as suggested in [5]: $d_{eye} = \max(d_l, d_r)/\|C_l - C_r\|$, where d_l and d_r are the distances of the estimated eye positions to the given coordinates C_l and C_r. A relative distance $d_{eye} < 0.25$ is considered a successful localization, since $d_{eye} = 0.25$ corresponds approximately to the half width of an eye.

Figure 5(a) shows the network's localization performance for the training set (TRN) and the test set (TST) in comparison to the data taken from [5] (Hausdorff+MLP). All training examples have been localized successfully. The performance on the test set is similar. Only 1.5% of the test examples have not been localized accurately enough. Compare this to the 8.2% mislocalizations of the reference system.

A detailed analysis of the network's output for the mislocalizations showed that in these cases the output deviates from the one-blob-per-eye pattern. It can happen that no blob or that several blobs are present for an eye. By comparing the activity of a segmented blob to a threshold and to the total activity a confidence measure is computed for each eye. Both are combined to a single confidence. In Figure 5(b) one can see that rejecting the least confident test

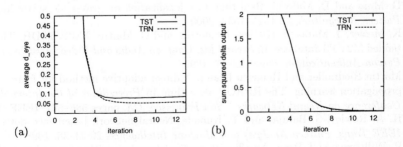

(a)   (b)

Fig. 6. Performance over time: (a) average distance d_{eye}; (b) sum of squared changes in the network's output.

examples lowers the number of mislocalizations rapidly. When rejecting 3.1% of the images, only one mislocalization is left. The average localization error of the accepted examples is $d_{eye} = 0.06$. That is well within the area of the iris and corresponds to the accuracy of the given coordinates.

Figure 6 illustrates the network's performance over time. The average error d_{eye} drops rapidly within the first five iterations and stays low afterwards. The changes in the network's output are large during the first iterations and decrease even when updated longer than the ten steps it has been trained for.

6 Conclusions

In this paper we presented an approach to face localization that is based on a hierarchical neural network with local recurrent connectivity. The network is trained to solve this task even in the presence of complex backgrounds, difficult lighting, and noise through iterative refinement.

We evaluate the network's performance on the BioID data set. It compares favorably to a hybrid reference system that uses a Hausdorff shape matching approach in combination to a multi layer perceptron.

The proposed method is not limited to gray-scale images. The extension to color is straight forward. Since the network works iteratively, and one iteration takes only a few milliseconds, it would also be possible to use it for real-time face tracking by presenting image sequences instead of static images.

References

1. S. Behnke and R. Rojas. Neural Abstraction Pyramid: A hierarchical image understanding architecture. In *Proc. IJCNN'98-Anchorage*, pages 820–825, 1998.
2. Sven Behnke. Learning iterative image reconstruction in the Neural Abstraction Pyramid. *International Journal on Computational Intelligence and Applications, Special Issue on Neural Networks at IJCAI-2001*, 1(4):427–438, 2001.
3. Y. Bengio, P. Simard, and P. Frasconi. Learning long-term dependencies with gradient descent is difficult. *IEEE Trans. on Neural Network*, 5(2):157–166, 1994.
4. Erik Hjelmas and Boon Kee Low. Face detection: A survey. *Computer Vision and Image Understanding*, 83:236–274, 2001.
5. O. Jesorsky, K. J. Kirchberg, and R. W. Frischholz. Robust face detection using the Hausdorff distance. In *Third Int. Conf. on Audio- and Video-based Biometric Person Authentication, Halmstad, Sweden*, pages 90–95. Springer, 2001.
6. D. Maio and D. Maltoni. Real-time face localization on gray-scale static images. *Pattern Recognition*, 33:1525–1539, 2000.
7. K. Messer, J. Matas, J. Kittler, J. Luettin, and G. Maitre. XM2VTSDB: The extended M2VTS database. In *Second Int. Conf. on Audio and Video-based Biometric Person Authentication*, pages 72–77, 1999.
8. Martin Riedmiller and Heinrich Braun. A direct adaptive method for faster backpropagation learning: The RPROP algorithm. In *Proceedings of the International Conference on Neural Networks – San Francisco, CA*, pages 586–591. IEEE, 1993.
9. H. A. Rowley, S. Baluja, and T. Kanade. Neural network based face detection. *IEEE Trans. Pattern Analysis and Machine Intelligence*, 20:23–38, 1998.
10. R. Williams and J. Peng. An efficient gradient-based algorithm for on-line training of recurrent network trajectories. *Neural Computation*, 2(4):491–501, 1990.

A Comparison of Face Detection Algorithms

Ian R. Fasel[1] and Javier R. Movellan[2]

[1] Department of Cognitive Science, University of California, San Diego
La Jolla, CA, 92093-0515
[2] Institute for Neural Computation
University of California, San Diego
La Jolla, CA, 92093-0515 {ian,javier}@inc.ucsd.edu

Abstract. We present a systematic comparison of the techniques used in some of the most successful neurally inspired face detectors. We report three main findings: First, we present a new analysis of how the SNoW algorithm of Roth, Yang, and Ahuja (200) achieves its high performance. Second, we find that representations based on local receptive fields such as those in Rowley, Baluja, and Kanade consistently provide better performance than full connectivity approaches. Third, we find that ensemble techniques, especially those using active sampling such as AdaBoost and Bootstrap, consistently improve performance.

1   Introduction

Face detection is a crucial technology for applications such as face recognition, automatic lip-reading, and facial expression recognition (Pentland, Moghaddam, & Starner, 1994; Donato, Bartlett, Hager, Ekman, & Sejnowski, 1999). One aspect that has slowed down progress in this area is the lack of baselining studies whose goal is not just the development of complete systems but the analysis of how the different pieces of a system contribute to its success. The goal of this study then is to perform a systematic comparison of techniques used in three of the most successful neurally inspired face detection systems reported in the literature (Rowley et al., 1998; Roth et al., 2000; Viola & Jones, 2001). In particular, we focus on the different uses in these papers of high dimensional input representations, local versus global connectivity, and active sampling and ensemble techniques such as AdaBoost and Bagging.

2   Face Detection Framework and Image Database

The face detector used throughout this paper is based on the system described in Rowley et al. (1998). A small window is scanned across each image and a classifier is applied to each window, returning *face* or *nonface* at each location. This is repeated at multiple scales. Finally, nearby detections are suppressed using the clustering and overlap removal techniques described in Rowley et al. (1998). For training, we randomly selected 443 frontal faces from the FERET database

J.R. Dorronsoro (Ed.): ICANN 2002, LNCS 2415, pp. 1325–1330, 2002.

containing a variety of different individuals and facial expressions, and carefully aligned them within 20x20 pixel image patches. Following Rowley et al, small amounts of translation, scale, and rotation were randomly added to these images, resulting in a total set of 8232 training images. For negative examples we used 20,000 windows taken from scenery images known to contain no faces. Finally, to compensate for differences in lighting and camera gains, logistic normalization (Movellan, 1995)[1] is performed on each image subwindow before classification, except for an oval mask which blocks out background pixels. This normalization step was also performed for each window in the detection phase.

3   Factors for Comparison

We constructed sixteen experimental classifiers, each using a combination of the factors used in the Rowley et al. (1998), Roth et al. (2000) and Viola & Jones (2001) face detectors. The goal of these experiments was to clarify which particular techniques were responsible for the success of these algorithms.

Component Classifiers

There were two component classifiers. (1) **Ridge Regression** (Hoerl & Kennard, 1970) This method was used for training classifiers directly on real valued pixel inputs. Ridge regression is equivalent to linear backpropagation networks with weight decay. (2) **SNoW** This classifier first transforms the pixel inputs into a sparse binary representation and then uses the Winnow update rule of Littlestone (1998) for training. In effect, the resulting network performs an arbitrary function on each input pixel, then combines the function outputs linearly and applies a threshold. While this high-dimensional representation is counterintuitive to traditional neural network researchers (Alvira & Rifkin, 2001), Roth et al. have nevertheless reported the most accurate face detector in the literature. It is thus important to replicate Roth et al.'s results in order to form a better understanding of how SNoW produces such impressive results.

Each of these is optionally enhanced with **Bootstrap** . Rowley et al. (1998), Roth et al. (2000) and Viola and Jones (2001) all used a "Bootstrap" technique based on Sung and Poggio (1994). The Bootstrap technique is an active sampling technique for expanding the training set of a classifier during training. Bootstrap begins by training a classifier on the full set of face examples and a random set of 8000 nonface examples. This classifier is then used as a face detector on a set of unseen scenery images, and 2000 of the false alarms are randomly selected and added back into the training set. The existing classifier is then discarded, and a new classifier is trained on this expanded training set. The process repeats until the classifier has satisfactory performance.

Full vs. Local Connectivity

Rowley et al. (1998) used a standard multilayer perceptron, with receptive fields localized over 26 rectangular subregions inspired by Le Cun et al. (1989). The regions were 4 10x10 pixel patches, 16 5x5 pixel patches, and 6 overlapping

---

[1] $X = 1/(1 + e^{-\pi(.8)\mu/\sqrt[3]{\sigma}})$, where μ and σ are the mean and variance of the window.

20x5 horizontal stripes. In our experiments, component classifiers (trained with ridge regression or SNoW) received input from either the entire image or one of these subregions, which were then combined using an ensemble technique.

Ensemble Classifiers

We used several ensemble methods. (1) **Bagging** In this ensemble method, multiple instances of a classifier are trained on random samples from the training set. The final hypothesis of each classifier is then combined with a unity vote. This procedure has been shown to improve performance in many types of classifiers (e.g., Breiman, 1996; Opitz & Maclin, (1999). (2) **AdaBoost** A modification of Bagging, AdaBoost (Freund & Schapire 1996) trains an ensemble of classifiers sequentially. For each round of boosting, a distribution over the training set is modified so that examples misclassified in previous rounds of boosting receive more emphasis in later rounds. This procedure guarantees an exponentially decreasing upper bound on training error, and in practice AdaBoost is reported to be resistant to overfitting (Opitz & Maclin, 1999; Schapire & Singer, 1998. (3) **AdaBoost and Bagging for Feature Selection.** Tiu and Viola (2000) and Viola and Jones (2001) used AdaBoost as a method for selecting a few key features from a large set of possible features by constraining the weak learners to make their decision using only one feature at a time. Using this technique, Viola and Jones (2001) were able to select about 200 features from their initial set of 45,396 to build a high performance face detector. We tested the flexibility of this technique by using the Rowley rectangular regions trained with SNoW or AdaBoost as the basic features. We also tried replacing AdaBoost with Bagging in this algorithm.

Ensemble Bootstrap [Bootstrap + Bagging]. We experimented with a novel condition in which the classifier at each round of Bootstrap is saved, and the resulting classifiers are combined with a unit vote. This is similar to Bagging, but with active sampling instead of random sampling, and makes for a fairer comparison with other ensemble techniques.

We trained different classifiers from different combinations of these methods in order to tease apart the role each method plays in the success of a face detector. Not all possible combinations of these methods could be practically tested and thus we focused on experiments that tested the following questions: (1) Is the patch-based representation proposed by Rowley helpful? (2) Does SNoW really work? How? (3) How helpful are ensemble methods in the face detection task? (4) How crucial is the Bootstrap method?

4   Results and Discussion

For each experimental classifier, our performance measure was the total error rate on a cross-validation set of 4434 unseen face and nonface examples withheld from the training set from FERET; this measure seems appropriate since the classifiers were trained to minimize overall error. Table 1 shows these results for all the conditions, sorted in order of decreasing error rate. The three main findings were (1) SNoW consistently performed among the best classifiers, confirming the results of Roth et al. (2000). In addition, we found an intuitive explanation for how SNoW works, which we describe below. (2) The local receptive fields used by Rowley consistently improved performance over equivalent classifiers that used the full 20×20 input. (3) Active sampling consistently improved performance as well; AdaBoost was always superior to the equivalent network using Bagging, and Bootstrap was usually superior to the equivalent networks that didn't use Bootstrap.

Table 1. Performance on generalization set of withheld faces from FERET.

Condition	Total Error	Hit Rate	False Alarm Rate
13) Patches + Ridge + Bagging + Bootstrap	47.85 %	52.75%	48.21%
1) Global + Ridge	21.95 %	96.00%	32.46%
2) Global + Ridge + Bagging	21.86 %	96.00%	32.00%
11) Global + Ridge + Bootstrap	11.53 %	92.75%	14.03%
8) Patches + Ridge + Bagging	6.47 %	99.75%	10.09%
12) Global + Ridge + Bagging + Bootstrap	2.28 %	97.75%	2.30%
4) Global + SNoW	0.64 %	98.50%	0.15%
5) Global + SNoW + Bagging	0.46 %	99.00%	0.15%
14) Global + SNoW + Bootstrap	0.35 %	99.75%	0.40%
15) Global + SNoW + Bagging + Bootstrap	0.21 %	99.50%	0.04%
3) Global + Ridge + AdaBoost	0.18 %	99.50%	0.00%
6) Global + SNoW + AdaBoost	0.16 %	99.75%	0.15%
10) Patches + SNoW + Bagging	0.16 %	99.75%	0.11%
16) Patches + SNoW + Bagging + Bootstrap	0.12 %	99.75%	0.04%
9) Patches + SNoW + AdaBoost	0.12 %	99.75%	0.04%
7) Patches + Ridge + AdaBoost	0.09 %	99.75%	0.00%

SNOW: This architecture appeared in four of the best five experimental classifiers, demonstrating its strength in the face detection task (the differences between the classifiers in experiments 13-16 are not significant, $Z_{test} = -2.83$, $p = 0.93$). Figure 1 shows different visualizations of the representation learned by SNoW. The left image shows the intensity corresponding to the peak weight in each pixel of the SNoW network. This image represents the SNoW's "favorite" face, i.e., the pattern of pixel values that maximizes the output of the SNoW model. At the surface level, SNoW has learned a favorite face that looks very much like a template, and is very similar to the favorite face of the linear network. The center image shows the sum of the weights for each pixel, which are shown in greater detail in the callouts to the right. This image represents the importance, or attentional strength assigned by SNoW to each pixel region. Clearly, the areas where SNoW has developed large weights correspond closely to recognizable facial features while surrounding weights have been lowered close to zero. The callouts display the tuning curves learned by SNoW for several different pixel positions. We found that all the important pixels have unimodal tuning functions with a range of preferred intensities. The fact that the tuning curves developed by SNoW are unimodal is interesting, because SNoW could have developed arbitrary functions, such as linearly increasing or decreasing weights (which would be identical to the ridge regression solution). This also suggests a possible architecture for an improved face detector: since the SNoW weights resemble bandpass tuning functions, a classifier that explicitly uses such tuning functions in training may be able to perform even better.

Local Connectivity is Better than Full Connectivity: Ensemble classifiers that split the input into the Rowley et al. patches typically performed better than classifiers that used full connectivity. The average performance of the four best classifiers using local patches (experiments 13, 14, 15, and 16) was significantly better than the performance of equivalent classifiers using global connectivity (experiments 8, 10, 11, and 12, $Z_{test} = -2.83$, $p < .005$). The conclusion we can draw from this is that the face detection task truly does benefit from the use of local receptive fields like the ones used in Rowley. Adding local connectivity to the ridge regression based ensemble classifiers even improved performance enough to produce the best overall classifier in the study.

Active Sampling: The active sampling done by AdaBoost and the Bootstrap method improved performance over their random sampling counterparts in all but one condition. Classifiers using AdaBoost (11, 12, 15, 16) performed significantly better than equivalent classifiers using Bagging (3, 5, 8, 13, $Z = -16.57$, $p < .005$). Interestingly, while AdaBoost helped in all cases, it was only slightly better than Bagging when used on SNoW. It seems that SNoW is able to account for most of the variation in the training set on the first round of Boosting, so that the impact of the active sampling done by AdaBoost is minimal. In contrast, AdaBoost provided huge benefits to the ridge regression classifiers.

This study provides clear evidence of the usefulness of some of the techniques used in face detection systems and suggests several areas for future improvements. First, we found that SNoW is indeed a promising classifier for face detection. The analysis of the way SNoW solved the problem suggests that a powerful face detectors may be built using explicit intensity tuning functions. Second, the superiority of sparse local representations, especially when used with the AdaBoost feature selection method, supports the exploration of other localist representations. Finally, the improvements provided by active sampling methods

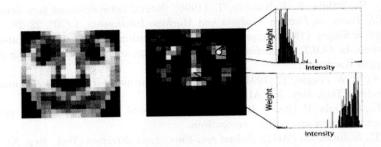

Fig. 1. SNoW generated weights from experiment (4). While at the pixel level the weights have learned the most likely intensities (left), most of the weights are close to zero (center). However for pixels in which the weights are not close to zero, the weights form a bandpass tuning curve. The callouts show the weights for two individual pixels which have this property. While eye pixels favor dark intensities, bridge of nose pixels favor high intensity.

like Bootstrap have exciting implications for the role of active sampling in other machine perception tasks.

References

Alvira, M., & Rifkin, R. (2001). *An empirical comparison of SNoW and SVMs for face detection* (Tech. Rep. No. CBCL Paper #193/AI Memo #2001-004). Massachussetts Institute of technology, Cambridge, MA.

Breiman, L. (1996). Bagging predictors. *Machine Learning*, 24 (2), 123–140.

Donato, G., Bartlett, M., Hager, J., Ekman, P., & Sejnowski, T. (1999). Classifying facial actions. *IEEE Trans. on Pattern Analysis and Machine Intelligence*, 21 (10), 974–989.

Freund, Y., & Schapire, R. E. (1996). Experiments with a new boosting algorithm. In *Proc. 13th international conference on machine learning* (p. 148–146). Morgan Kaufmann.

Hoerl, A. E., & Kennard, R. W. (1970). Ridge regression: biased estimation for nonorthogonal problems. *Technometrics*, 12, 55–67.

Le Cun, Y., Boser, B., Denker, J., Henderson, D., Howard, R., Hubbard, W., & Jackel, L. (1989). Backpropagation applied to handwritten zip code recognition. *Neural Computation*, 1, 541–551.

Littlestone, N. (1988). Learning quickly when irrelevant attributes abound: A new linearthreshold algorithm. *Machine Learning*, 2, 285–318.

Movellan, J. R. (1995). Visual speech recognition with stochastic networks. In . T. G. Tesauro, D. Toruetzky (Ed.), *Advances in neural information processing systems* (Vol. 7). MIT Press.

Opitz, D., & Maclin, R. (1999). Popular ensemble methods: An empirical study. *Journal of Artificial Intelligence Research*, 11, 169–198.

Pentland, A., Moghaddam, B., & Starner, T. (1994). View-based and modular eigenspaces for face recognition. In *IEEE conference on computer vision and pattern recognition*.

Roth, D., Yang, M., & Ahuja, N. (2000). A snow-based face detector. In *NIPS-12*. To Appear.

Rowley, H., Baluja, S., & Kanade, T. (1998). Neural network-based face detection. *IEEE Trans. on Pattern Analysis and Machine Intelligence*, 1 (20), 23–28.

Schapire, & Singer. (1998). Improved boosting algorithms using confidence-rated predictions. In *COLT: Proceedings of the workshop on computational learning theory*. Morgan Kaufmann.

Sung, K. K., & Poggio, T. (1994). *Example based learning for view-based human face detection* (Tech. Rep. No. AIM-1521).

Tieu, K., & Viola, P. (2000). Boosting image retrieval. In *Proceedings ieee conf. on computer vision and pattern recognition*.

Viola, P., & Jones, M. (2001). *Robust real-time object detection* (Tech. Rep. No. CRL 20001/01). Cambridge Research Laboratory.

Part IX

Special Session: Adaptivity in Neural Computation

Adaptive Model Selection for Digital Linear Classifiers

Andrea Boni

University of Trento
Department of Information and Communication Technologies
Via Sommarive 14,
38100 Povo (Trento), Italy
andrea.boni@ing.unitn.it

Abstract. Adaptive model selection can be defined as the process thanks to which an optimal classifiers h^* is automatically selected from a function class H by using only a given set of examples z. Such a process is particularly critic when the number of examples in z is low, because it is impossible the classical splitting of z in $training + test + validation$. In this work we show that the joined investigation of two bounds of the prediction error of the classifier can be useful to select h^* by using z for both model selection and training. Our learning algorithm is a simple kernel–based Perceptron that can be easily implemented in a counter–based digital hardware. Experiments on two real world data sets show the validity of the proposed method.

1 Introduction

In the framework of classification problems, the machine learning community refers to model selection as the task through which an optimal function class H_k is selected from a given set H with the goal of optimizing the prediction error of the classifier h^* extracted from H_k. Usually such a task involves a tradeoff between the capability of the model to learn the given set of labeled examples $z = \{(x_i, y_i)\}_{i=1}^{m}$, and a measure of the complexity of the model itself [11]. Each (x_i, y_i) is supposed to be designed independently and identically distributed over an unknown distribution P; we use the following notation: $x_i \in X$, $y_i \in Y$, $z_i = (x_i, y_i) \in X \times Y = Z$; $z \in Z^m$; $X \subset \Re^r, Y = \{-1, 1\}$. In this paper we consider digital learning algorithms $\mathcal{A}$ as operators that map z to a set of parameters A, $\mathcal{A}:Z^m \to A$, $A \subset (N^+)^m$; in particular we consider the Digital Kernel Perceptron with Pocket (DKPP), which leads to the following function class: $H = \{h : \mathcal{K} \times A \times X \to Y : h(\alpha, x) = \text{sgn}(\sum_{j=1}^{m} \alpha_j y_j K(x_j, x)), K \in \mathcal{K}, \alpha \in A, x \in X\}$. $K(\cdot, \cdot) : X \times X \to \Re$ is a kernel function that realizes a dot product in the feature space [1]. The choice of the kernel perceptron (KP) is justified by recent studies that have revalued its role in classification tasks, showing the relation between its generalization ability and the sparsity of the final solution [9]. Furthermore its *simplicity* suggests that a very simple digital hardware can be designed. In this paper we propose an adaptive automatic procedure to select the

J.R. Dorronsoro (Ed.): ICANN 2002, LNCS 2415, pp. 1333–1338, 2002.

best function in H. Our approach is based on recent studies that have proposed uniform (over the distribution P) and data–dependent bounds for the error rate of the KP [9] and for a given learning algorithm [3], respectively. In the next section we briefly introduce the DKPP learning algorithm, while in section 3 we describe uniform and data–depended bounds; based on such bounds we propose a procedure for adaptive model selection. In the end, in section 4 we comment some experiments on two real–world data sets.

2   The DKPP Algorithm

It is well known that the KP algorithm belongs to the class of kernel-based approaches, in which a non linear classifier is built after a mapping of the input space to a higher, possibly infinite, feature space via a function Φ. This is a well-known theory that exploits the Reproducing Kernel Hilbert Space (RKHS) framework [1], and recently has been applied with success by the machine learning community. Support Vector Machines (SVMs), designed by Vapnik [11], represent one of the most successful examples of a learning machine based on a kernel method. In practice, the kernel function $K(\cdot, \cdot)$ acts as an operator that hides the non linear map in the feature space, thus avoiding to work directly on ϕ. For classification purposes, the most used kernel functions are: $K(x_i, x_j) = x_i \cdot x_j$ (linear), $K(x_i, x_j) = (1 + x_i \cdot x_j/r)^p$ (polynomial) and $K(x_i, x_j) = \exp\left(-\|x_i - x_j\|^2/2\sigma^2\right)$ (Gaussian); p and σ^2 are usually indicated as the *hyperparameters* of the problem. The KP has been applied with success for classification and function approximation tasks [6], [7], and its generalization capabilities has been reported for example in [9]. It can be considered as the dual formulation of the Rosenblatt's learning algorithm which, as known, can be applied to any choice of the updating step η. The advantage of the dual formulation consists in the fact that one can exploits the theory of kernels, thus allowing an implicit non linear transformation; furthermore the choice $\eta = 1$ leads to the fact that $\alpha \in A$. Table 1 sketches the flow of our algorithm, called DKPP. We implemented the pocket algorithm [8] in order to accept errors on the training set. This is a crucial point for optimal model selection, as will be clear in the next sections. It is important to point out that the on-going research on kernel-based methods suggests that many classifiers perform comparably, provided that a good kernel has been chosen [9]; therefore classifiers like the DKPP, are very appealing and show their superiority when targeting VLSI implementations, as simple counter–based architectures can be designed. Figure 1 shows such an architecture, where $q_{ij} = y_i y_j K(x_i, x_j)$ and MSB_i indicates the Most Significant Bit of the accumulator, in a 2's complement coding. In this paper we explore the way of automatic tuning the model of the DKPP.

3   Uniform and Data-Depended Bounds for the DKP

In general, the error rate (or prediction error), of a classifier can be defined as the probability for (x, y) drawn randomly over P that h is wrong, that is $h(\alpha, x) \neq y$:

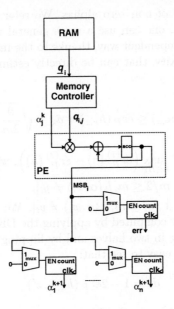

Fig. 1. A counter–based architecture for the DKPP.

$er_P\left(h\left(\alpha_{k,z}, x\right), y\right) = \Pr\left\{(x,y) \in Z : h\left(\alpha_{k,z}, x\right) \neq y\right\}$ [2], where $\alpha_{k,z} = \mathcal{A}_k(z)$ is the vector obtained by $\mathcal{A}_k$ after the observation of z and over the function class H_k. It is well known from the Probably Approximately Correct (PAC) learning framework that the behavior of er_P can be represented in terms of a bound which holds with a given probability; in practice, usually one asserts that the following inequality:

$$er_P\left(h\left(\alpha_{k,z}, x\right), y\right) \leq \hat{er}_P\left(h_{k,z}, z\right) + \varepsilon\left(m, d_k, \delta\right) \tag{1}$$

holds with a given probability $1 - \delta$; $\hat{er}_P(h_{k,z}, z)$ is the error rate of $h_{k,z}$ on z (also known as empirical error rate), that is:
$\hat{er}_P\left(h_{k,z}, z\right) = \frac{1}{m}\left|i : 1 \leq i \leq m, h\left(\alpha_{k,z}, x_i\right) \neq y_i\right|$ and d_k is a measure of the complexity of the function class H_k (for example the Vapnik–Chervonenkis dimension [11])[1]. Exploiting the well-known compression scheme by Warmuth et. al. [5], a uniform bound for the DKPP has been recently proposed [9]; according to such a theorem, with probability at least $1 - \delta$, the prediction error of the DKPP, as long as $\hat{er}_P = 0$, is:

$$er_z\left(h_{k,z}\right) \leq \frac{1}{m - d_k}\left(\ln\left[\binom{m}{d_k}\right] + \ln m + \ln\frac{1}{\delta}\right) \tag{2}$$

---

[1] For simplicity, during the text we will use the notation $er_P(h_{k,z}) = er_P\left(h\left(\alpha_{k,z}, x\right), y\right)$ and $\hat{er}_P(h_{k,z}) = \hat{er}_P\left(h_{k,z}, z\right)$, to indicate the prediction error and the empirical error rate of $h_{k,z}$; furthermore, given a set S with $|S|$ we intend its cardinality.

where d_k is the number of non zero alphas. We refer to eq. (2) as B_1. In the agnostic case ($\hat{er}_P \neq 0$), one can use a more general result which permits the calculus of ε in a data–dependent way, thanks to the introduction of a measure, called penalized complexity, that can be directly estimated on the basis of the given training set z [3]:

$$er_P(h_{k,z}) \leq \hat{er}_P(h_{k,z}) + d_k + \sqrt{\frac{9}{2m}\ln\frac{1}{\delta}} \tag{3}$$

where, in this case, $d_k = \max_{h \in H_k}\left(\hat{er}_P^{(1)}(h) - \hat{er}_P^{(2)}(h)\right)$, with:

$\hat{er}_P^{(1)}(h) = \frac{2}{m}|i : 1 \leq i \leq m/2 \leq m, h(\alpha, x_i) \neq y_i|$,

$\hat{er}_P^{(2)}(h) = \frac{2}{m}|i : m/2 + 1 \leq i \leq m, h(\alpha, x_i) \neq y_i|$. We refer to (3) as B_2. The measure d_k, can be easily computed by applying the DKPP on a new trining set z' obtained by splitting z in two halves and by flipping the labels of the second half [3]. In practice, it is easy to see that:

$$d_k = 1 - 2\hat{er}_P(h_{k,z'}, z') \tag{4}$$

The main goal to design an *optimal* learning systems, is to bound the right side of equation (1), in order to bound the prediction error. This leads to the Structural Risk Minimization (SRM) principle; according to SRM one looks for a function h having lower empirical error over a fixed function class H_k, characterized by the complexity d. From a formal point of view, one should build several function classes in increasing size, that is increasing complexity ($H_1 \subset \cdots \subset H_k \subset \cdots$), and then pick a function h which has small training error and comes from a classes H_k having lower complexity (that is lower ε according to (1)). Our idea for adaptive model selection consists in building several H_k and then in measuring their *richness* on the basis of $B_{1,2}$. In practice we choose the function $h^* = \arg\min_k [\hat{er}_P(h_{k,z}) + \varepsilon(m, d_k, \delta)]$. The general procedure is sketched in table 2.

Note that, theoretically, one could only use B_2. Actually, it is known that several function classes, such as the Gaussian one, have typically an high complexity, and the corresponding measure d_k, found according to the maximal discrepancy estimate, does not give any useful information on the problem ($d_k = 1$); on the other hand, in these cases we have $\hat{er}_P = 0$, thus allowing the use of B_1. Our experiments show that the joined investigation of $B_{1,2}$ can assure an optimal model selection for different real–world problems.

4   Experiments

We test our approach on two well–known datasets from [4,10], that is sonar (SNR) and pima indian diabetes (PIMA), respectively. The results are summarized in table 3. Both data sets are two–class problems with $r = 60$ and $r = 8$ input features respectively, whilst the number of training samples are $m = 104$

Table 1. The DKPP algorithm.

Step	Description
1.	set $\alpha = 0$, $\alpha_{opt} = 0$, $nerr = m$
2.	Repeat until no mistakes occur or a max number of steps has been reached
3.	for $i = 1$ to m do
4.	compute $\text{MSB}_i = y_i \left(\text{sgn} \left(\sum_j \alpha_j y_j K(x_i, x_j) \right) \right)$
5.	if $\text{MSB}_i < 0$ then
6.	$\alpha_i = \alpha_i + 1$
7.	compute $err = \lvert i : y_i \left(\text{sgn} \left(\sum_j \alpha_j y_j K(x_i, x_j) \right) \right) < 0 \rvert$
8.	if $err < nerr$ then $\alpha_{opt} = \alpha$, $nerr = err$
9.	end for
10.	$\alpha_{k,z} = \alpha_{opt}$

Table 2. A procedure for adaptive model selection.

Step	Description
1.	set $k = 1$
2.	Select a kernel function (es: linear, polynomial or gaussian)
3.	Select a value for the hyperparameter (build H_k)
4.	Apply DKKP and find $\alpha_{k,z} = \mathcal{A}_k(z)$, d_k
5.	Compute $B_{k,\{1,2\}} = \hat{er}_P (h_{k,z}) + \varepsilon (m, d_k, \delta)$
6.	$k = k + 1$
7.	Choose $h^* = \arg \min_k B_{k,\{1,2\}}$

Table 3. Experiments for the sonar and pima indian diabetes datasets.

SNR	$\sigma^2 = 0.1$	$\sigma^2 = 0.5$	$\sigma^2 = 1.0$	$p = 2$	$p = 3$	$p = 4$
$B_1(TS)$	1.86(11)	1.82(7)	1.96(10)	6.58(19)	4.8(15)	4.8(13)
$B_2(TS)$	–	–	–	–	–	–

PIMA	$\sigma^2 = 0.05$	$\sigma^2 = 0.1$	$\sigma^2 = 0.5$	$p = 2$	$p = 3$	$p = 4$
$B_1(TS)$	1.41(53)	1.6(50)	2.16(60)	–	–	–
$B_2(TS)$	–	–	–	0.5(41)	0.51(40)	0.6(43)

and $m = 576$; TS indicates the number number of errors on 104 and 192 validation patterns respectively. We applied the model selection method to select one among two kernel functions (Gaussian and polynomial) and the associated hyperparameter (σ^2 or p). As expected, there is a substantial agreement between the bounds $B_{1,2}$ and TS; as reported in table 3, our criteria chooses the best model, that is a Gaussian kernel with $\sigma^2 = 0.5$ for SNR and a polynomial with $p = 2$ or $p = 3$ for PIMA. As a final remark note that with '–' we mean that corresponding bound does not give any useful information for model selection. For example, in the case of B_2, it means that $d_k = 1$ for every value of the hyperparameter, while in the case of B_1 it simply means that we cannot apply equation (2) as we are in presence of an agnostic case.

5 Conclusion

In this paper we have propose an efficient method for adaptive model selection. We used a digital kernel based perceptron with pocket, but our approach could be applied to other learning systems as well, as long as good bounds of the prediction error are provided; in particular, in this work we have investigated the use of two different kinds of bounds: a uniform bound, that can be applied to systems characterized by a high level of complexity, and a data–dependent bound. Experiments on two real world problems have demonstrated the validity of the method.

References

1. Aizerman, M. A., Braverman, E. M. and Rozonoer, L. I.: Theoretical Foundations of the Potential Function Method in Pattern Recognition Learning. Automation and Remote Control, **25** (1964) 821–837.
2. Bartlett P.: Neural Networks Learning: Theoretical Foundations, Cambridge University Press, 1999.
3. Bartlett, P.L., Boucheron, S., and Lugosi, G.: Model Selection and Error Estimation. Machine Learning, **48** (2002), 85–113.
4. Blake, C., Keogh, E., and Merz, C.J.: UCI Repository of Machine Learning Databases, *http://www.ics.uci.edu/ mlearn/MLRepository.html.*
5. Floyd, S. and Warmuth, M.: Sample compression, learnability and the Vapnik–Chervonenkis dimension. Machine Learning, **21** (1995) 269–304.
6. Freund Y. and Shapire, R.E.: Large Margin Classification Using the Perceptron Algorithm. Machine Learning, **37** (1999) 277–296.
7. Friess, T.T. and Harrison, R.F.: A Kernel-Based Adaline for Function Approximation. Intelligent Data Analysis, **3** (1999) 307–313.
8. Gallant, S.I: Perceptron–Based Learning Algorithms. IEEE Transaction on Neural Networks, **1** (1990) 179–191.
9. Herbrich, R., Graepel, T., and Williamson, R.: From Margin to Sparsity. NIPS 13, 2001.
10. J.M. Torres Moreno, M.B. Gordon. Characterization of the Sonar Signals Benchmark. Neural Processing Letters, **7** (1998) 1–4.
11. Vapnik, V.N.: The Nature of Statistical Learning Theory. John Wiley & Sons, NY, USA, 2nd edition, 1999.

Sequential Learning in Feedforward Networks: Proactive and Retroactive Interference Minimization

Vicente Ruiz de Angulo and Carme Torras

Institut de Robòtica i Informàtica Industrial (CSIC-UPC)
Llorens i Artigas 4-6, 08028-Barcelona, Spain.
{ruiz, torras}@iri.upc.es,
http://www-iri.upc.es

Abstract. We tackle the catastrophic interference problem with a formal approach. The problem is divided into two subproblems. The first arises when one tries to introduce some new information in a previously trained network, without distorting the stored information. The second is how to encode a set of patterns so as to preserve them when new information has to be stored. We suggest solutions to both subproblems without using local representations or retraining.

1 Introduction

The degradation in performance suffered by a neural network when a new pattern is introduced isolatedly is usually called *interference* or *catastrophic forgetting*. Its extent depends basically on two factors: the type of representation and the training scheme applied.

- Distributed representations are known to be prone to catastrophic forgetting. For the study of interference, we consider that an input is encoded distributedly if most of the network derivatives at the input with respect to the parameters have a significant magnitude [4]. Thus, each item is represented by several parameters and each parameter of the neural net supports partially several items. Local representations can avoid more easily the problem, but unfortunately they exploit memory resources much worse and also are reputed to interpolate less adequately (See [4] for more on this).
- The two basic training schemes are the introduction of the new item isolatedly and the join training of the new item with the old ones. The first one quickly learns the new item at the cost, in many cases, of catastrophic forgetting. The second requires a complex optimization process, especially with distributed representations.

The different previous works on interference can be placed in an imaginary plane whose axes refer to the locality of the representation and the extent of retraining with old items. For example, French's approach [7] is situated at the extreme

J.R. Dorronsoro (Ed.): ICANN 2002, LNCS 2415, pp. 1339–1344, 2002.

of localized solutions. He uses sigmoidal units with a $[0,1]$ range, saturating them and favoring the zero states. Hetherington and Seidenberg [8] suggest to retrain new and old items together after the introduction of the new pattern. This approach falls in the join training extreme in the retraining axis, where information about old items is used in a delayed mode. In [9], the new item is trained with a (fixed or, preferably, randomly changing) *subset* of the previously trained items. Thus, it can be situated in the middle of the retraining axis.

We have contributed to the field using neither local representations nor retraining. Theoretical reasons forbid a perfect solution to the interference problem under these conditions [3]. As a consequence, our goal has been limited to trying to palliate interference as much as possible. In this paper we will place in a common context the results we have obtained with different algorithms.

We consider that the interference problem must be handled in two stages separated by the arrival of the new item:

- A priori stage: encoding the old, known, items in the best way to prepare the network to minimize the impact of learning new items. We call this stage **proactive interference minimization**.
- A posteriori stage: learning the new items while minimizing the damages inflicted on the old items already stored in a given weight configuration. We call this stage **retroactive interference minimization**.

We will concretize the scenario in a set of items $1 \ldots N-1$ which are stored prior to the arrival of the new item N. Let $E_{1\ldots p}$ denote the error in items $1, 2 \ldots p$. We will reflect the calculations of the a priori stage in the weight variable W and those of the a posteriori stage in the variable vector ΔW, so that the value of the weight vector after the introduction of the new item will be $W + \Delta W$. The minimization of $E_{1\ldots N}(W)$ is approximately equivalent to:

$$\min_{W, \Delta W} [E_{1\ldots N-1}(W) + \Delta E_{1\ldots N-1}(\Delta W; W)]$$

$$\text{subject to } E_N(W + \Delta W) = 0.$$

For clarity of explanation, we make all the derivations assuming a perfect encoding of the new item ($E_N(W + \Delta W) = 0$). However, the N-th item can also be introduced partially within this approach. We need to express $\Delta E(\Delta W + W)$ in a manageable way through a truncated Taylor series expansion. A linear model would be too rude (and unfeasible [3]). Since the computation of the Hessian requires very costly computations, we include only the diagonal terms. Thus, the most faithful problem formulation we can reasonably aspire to deal with is:

$$\min_{W, \Delta W} \left[E(W) + \sum_i \frac{\partial E}{\partial W_i} \Delta W_i + \frac{1}{2} \sum_i \frac{\partial^2 E}{\partial W_i^2} \Delta W_i^2 \right]$$

$$\text{subject to } E_N(W + \Delta W) = 0.$$

The problem is how to carry out this constrained minimization on W (interference prevention) and on ΔW (retroactive interference minimization).

2 Minimizing Retroactive Interference

2.1 Problem Formulation

Here we consider W as a constant. Note that, by introducing the first $N - 1$ items, we should have minimized E and, therefore, also the absolute value of its first derivatives. Thus we assume that the linear terms are zero (but see below), and our definitive formulation for the retroactive interference subproblem is:

$$\min_{\Delta W} \sum_i c_i \Delta W_i^2 \tag{1}$$

$$subject\ to\ E_N(\Delta W) = 0.$$

This formulation, with different c_i assignments, was used in previous works by the authors [3] and others [2]. Among them, $c_i = 1$ is interesting for its cheap computational cost, and because it is intuitively appealing, since (1) calculates in this case the nearest solution for the new item in parameter space. One can wonder what the best c_i assignment is, outside the minimum of the old items. The answer is shown to be $c_i = \frac{\partial^2 E}{\partial W_i^2}$ [4], even in this general case.

2.2 Minimization Method

The question remains of how to solve (1) efficiently. A usual way to tackle a constrained optimization problem is to linearly combine the cost function with the deviation of the constraint from zero, and then minimize this new error function. An appropriate scheduling biasing progressively the tradeoff between the cost function and the constraint satisfaction towards the latter is needed. More sophisticated algorithms from the theory of constrained optimization can also be applied, but they can be complex and computationally expensive.

In [3], an algorithm to solve (1) that exploits the structure of multilayer neural networks is developed. The drawbacks of solving a constrained minimization problem are here avoided through the transformation of retroactive interference into an unconstrained minimization problem. This transformation has also other advantages, like a number of variables much lower than in the original problem, and an always perfect encoding of the new item.

2.3 Experimentation

The aforementioned algorithm, called LMD, was applied in an experimental comparison of options for assigning c_i. One result is that indeed $c_i = \frac{\partial^2 E}{\partial W_i^2}$ is the best option in general, even outside the minimum. Another conclusion emerged: the difference between using $c_i = 1$ and $c_i = \frac{\partial^2 E}{\partial W_i^2}$ decreases with the number of items stored in the network. As the number of stored items increases, all weights tend to have similar, high-magnitude second derivatives and, thus, using either of the assignments tends to yield the same results. Experiments with backpropagation showed that, when the learning rate tends to zero, it approximates LMD with $c_i = 1$ and, thus being much more inefficient than LMD to obtain the same results.

2.4   Relation between Retroactive Interference Minimization and Pruning

To palliate retrograde interference, one of the more important issues is the determination of the less significant parameters for the encoding of a number of stored memories. These are the parameters that should support most of the necessary changes to introduce new information. Instead, pruning detects the less profited parameters to eliminate them. This antithetic relation suggests that advances in pruning techniques can be incorporated into retroactive interference minimization algorithms.

3   Interference Prevention

3.1   Problem Formulation

Now, we have to minimize (1) in W, taking ΔW as an unknown constant. Before knowing the new item, we cannot assume any particular value for ΔW and, therefore, the ideal solution is that which, in average, best solves the problem, taking into consideration the distribution of ΔW:

$$\min_{W} \left[E(W) + \left\langle \frac{1}{2} \sum_i \Delta W_i^2 \frac{\partial^2 E}{\partial W_i^2} + \sum_i \Delta W_i \frac{\partial E}{\partial W_i} \right\rangle \right].$$

where $< \cdot >$ denotes expectation. Using results from [6], we get:

$$\min_{W} \left[E(W) + \frac{\sigma^2}{2} \sum_i \frac{\partial^2 E}{\partial W_i^2} \right], \tag{2}$$

where σ^2 is the variance of the ΔW distribution. Reassuringly, this expression can also be derived without supposing a particular shape for $\Delta E_{1...N-1}$ $(\Delta W; W)$, by formulating interference prevention as the search for a W such that, after being modified by the introduction of the new item, it would still be able to reproduce old ones. This can be expressed as minimizing $E(W)$ stably with respect to the random perturbations produced by the new unknown items:

$$\min_{W} \int E(W + \Delta W) \, \mathcal{P}(\Delta W) \, d\Delta W. \tag{3}$$

Using results in [6] again, this expression can be made exactly equivalent to (2).

σ^2 could be approximately deduced from the expected error of future items. However, there is another important issue about the selection of σ^2 to be dealt with: the way in which the items $1 \ldots N-1$ are encoded determines the answer of the network to all possible inputs, i.e., generalization. This could be priorized over interference prevention in tuning σ^2. In conclusion, either by error or intentionally, the parameter σ^2 used in (2) could be significantly different from real weight variances. A detailed mathematical analysis [1] concludes that, if σ^2

is smaller than the real variance, the minimization (2) is always beneficial. The opposite case is also safe if the remaining error $E(W)$ in the minimum of (2) is not much higher than in the minimum of $E(W)$. $E(W)$ in the minimum of (2) has a sigmoidal shape when considered as a function of σ^2 and, therefore, in practice σ^2 can be increased until the error begins to grow quickly.

3.2   Minimization Method

It is possible to minimize (2), or equivalently (3), by adding noise to the weights while minimizing $E(W)$, so that a sample of the gradient distribution of $\frac{\partial E}{\partial W_i}(W + \Delta W)$ is calculated in each iteration. However, it is extremely inefficient to sample the high-dimensional parameter space of a neural network, in order to obtain the averages in the optimization steps. In [6], a method based on the gradient of (2), especially adapted for feedforward networks, is developed. It has the advantage of being deterministic and much more stable. In addition, it is easily computable with an algorithm of the same order as the backpropagation of the gradient of $E(W)$. An important property of this algorithm is that it produces results that tend to be independent of the number of hidden units used [5].

3.3   Relation between Interference Prevention and Generalization

We have assumed that the variance of ΔW, which is directly related to the expected error for the new item, does not change while performing the minimization (3), i.e., it is independent of W. But the error in the new items is also controlled by the selection of W. Thus, there exists an alternative way to prevent interference, namely improving generalization. The minimization (3) serves also this purpose, since the term $\frac{\sigma^2}{2} \sum_i \Delta W_i^2 \frac{\partial^2 E}{\partial W_i^2}$ is a regularizer that constrains the network function to be simple [5,6].

3.4   Experimentation

The experimental results in [4] show that our approach to interference prevention alleviates interference while at the same time improving generalization. This is in contrast with other strategies for interference avoidance based on local representations (e.g., saturating the hidden units). Nevertheless, we must point out the limitations in the application of this approach: the algorithm minimizing (2) makes the network output insensitive to changes in the weights for the stored items, but this insensitiveness is transmitted or generalized to the rest of the input space. Because of this, it is also necessary to modify more the weights to introduce the new items, and the potential benefits of the strategy get limited.

4   Conclusions

The amount of information available clearly determines the nature of the interference problem. When the new pattern is known, it can be used as a constraint in the minimization of the damage inflicted by its introduction. When it is still not available, the problem is the minimization of the expectation of the error increment that the possible new patterns can produce. For both problems we developed algorithms very well suited to feedforward networks. The experimental results were as good as could be expected, given the limitations on the solution derived theoretically.

References

1. Ruiz de Angulo V., 1996: Interferencia catastrófica en redes neuronales: soluciones y relación con otros problemas del conexionismo, Ph. D. Thesis, Universidad del Pais Vasco.
2. D.C. Park, M.A. El-Sharkawi, and R.J. Marks II, 1991: An adaptively trained neural network, *IEEE Transactions on Neural Networks* **2(3)**: 334–345.
3. Ruiz de Angulo V. and Torras C., 1995: On-line learning with minimal degradation in feedforward networks, *IEEE Transactions on Neural Networks* **6(3)**: 657-668.
4. Ruiz de Angulo V. and Torras C., 2000: A framework to deal with interference in connectionist systems. *AI Communications* **13(5)**: 259–274.
5. Ruiz de Angulo V. and Torras C., 2001: Architecture-independent approximation of functions, *Neural Computation* **13(4)**: 1119–1133.
6. Ruiz de Angulo V. and Torras C., 2002: A deterministic algorithm that emulates learning with random weights. *Neurocomputing*, In Press.
7. French R.M., 1994: Dynamically constraining connectionist networks to produce distributed, orthogonal representations to reduce catastrophic interference, in: *Proc. of the 16th Annual Conf. of the Cognitive Science Society*, Erlbaum, Hillsdale: 335–340.
8. Hetherington P.A. and Seidenberg M.S., 1989: Is there catastrophic interference in connectionist networks?, in: *Proc. of the Eleventh Annual Conf. of the Cognitive Science Society*, Erlbaum, Hillsdale, NJ: 26–33.
9. Robins A., Catastrophic forgetting, 1995: rehearsal and pseudorehersal, *Connection Science* **7(2)**: 123–146.

Automatic Hyperparameter Tuning for Support Vector Machines

Davide Anguita, Sandro Ridella, Fabio Rivieccio, and Rodolfo Zunino

DIBE - Dept. of Biophysical and Electronic Engineering,
University of Genova, Genova, I-16145, Italy,
{anguita,ridella,rivieccio,zunino}@dibe.unige.it

Abstract. This work describes the application of the Maximal Discrepancy (MD) criterion to the process of hyperparameter setting in SVMs and points out the advantages of such an approach over existing theoretical and practical frameworks.

The resulting theoretical predictions are compared with a k–fold cross–validation empirical method on some benchmark datasets showing that the MD technique can be used for automatic SVM model selection.

1 Introduction

Support Vector Machines (SVMs) are a new class of algorithms for classification, regression and novelty detection tasks [12,9] and one of the most successful algorithms recently appeared in the machine learning literature.

SVM *training* is performed by solving a linearly constrained quadratic programming problem, whereas the *design* of an SVM requires one to tune a set of quantities ("hyperparameters") that ultimately affect its generalization error.

In order to effectively adapt, in an automatic way, the SVM hyperparameters to optimally solve a particular problem, is therefore necessary to find reliable methods for predicting the generalization error. We propose here the use of the Maximal Discrepancy (MD) method [2] for performing this estimate.

In the following section we revise briefly the SVM algorithm; in Section 3 a simple procedure for applying the MD method to SVM hyperparameter tuning is detailed and, finally, in Section 4 some experimental results are reported.

2 The Support Vector Machine

We review here, very briefly, the SVM in order to introduce both the notation and the general framework of subsequent sections.

Given a training set $\{(x_i, y_i)\}$, the SVM for classification tasks is defined as:

$$f(x) = \text{sign}\left(\sum_{i=1}^{n_p} y_i \alpha_i K\left(x_i, x\right) + b\right) \tag{1}$$

J.R. Dorronsoro (Ed.): ICANN 2002, LNCS 2415, pp. 1345–1350, 2002.

where n_p is the number of training patterns, α_i are the parameters of the SVM, $K(\cdot, \cdot)$ is a suitable kernel function, and b is the bias term [5].

The parameters of the SVM are found by solving the following linearly constrained quadratic programming problem for which a global solution exists and can be found in a finite number of steps [7]:

$$\min_{\alpha} \alpha^T Q \alpha - e^T \alpha \tag{2}$$

$$0 \leq \alpha \leq C \tag{3}$$

$$y^T \alpha = 0 \tag{4}$$

where $e_i = 1 \; \forall i$ and $q_{i,j} = y_i y_j K(x_i, x_j)$. Note that the bias term b does not appear explicitly in the above problem: it can be found by applying one of the Karush–Kuhn–Tucker conditions at optimality [4] or is directly provided by the optimization routine, if based on double–dual methods [9].

We denote as hyperparameters the set of variables that affect the behavior of the SVM, but do not derive directly from the solution of the above problem. One of them is the hyperparameter that weights the misclassification errors (C), others are all the quantities that shape the kernel function. In this paper we will focus on RBF kernels $K(x_1, x_2) = \exp(-\gamma \|x_1 - x_2\|)^2$, therefore γ will be treated as an additional hyperparameter.

3   Automatic Hyperparameter Tuning for SVM

The problem of selecting effective hyperparameters is of paramount importance in building a SVM with good generalization properties. Several methods have been proposed so far, mainly relying on two different approaches. The first family of methods are of theoretical nature and try to derive the expected error of the SVM by providing generalization bounds based on statistical and combinatorial arguments, taking into account any possible distribution of test data that could be encountered [8,12]. The second, somewhat empirical, approach relies on the assumption that the training data is a good representation of the underlying distribution and performs an estimate of the generalization error by splitting all the available data in training and test sets: the first ones are used to train the classifier and the second ones to estimate the generalization performance [1,6].

Unfortunately, it is well known that the bounds obtained with the first approach are often too loose to be useful in practice, while the second ones can be misleading if the actual data distribution is not very well represented in the training set.

We present here a new approach to the hyperparameter tuning of SVM that inherits the advantages of both methods. This technique allows estimating the complexity of a learning machine but considers, at the same time, the actual distribution of the training data.

The method requires to randomly split the available data in two halves. Let us denote the empirical error performed on each half of the data set as:

$$\nu_1 = \frac{2}{n_p} \sum_{i=1}^{n_p/2} L\left(f(\boldsymbol{x}_i), y_i\right) \qquad \nu_2 = \frac{2}{n_p} \sum_{i=n_p/2+1}^{n_p} L\left(f(\boldsymbol{x}_i), y_i\right) \qquad (5)$$

where $L(f(\boldsymbol{x}_i), y_i) = 0$ if $f(\boldsymbol{x}_i) = y_i$ and $L(f(\boldsymbol{x}_i), y_i) = 1$ otherwise. Then, for a given set of hyperparameters values, the classifier complexity can be estimated by computing the MD value:

$$\max_{\alpha} (\nu_2 - \nu_1) \qquad (6)$$

This value can be used for estimating the generalization ability of a learning machine by means of the bound:

$$P\left\{\pi_{MD} - \nu > \max_{\alpha} (\nu_2 - \nu_1) + \epsilon\right\} \le e^{-2\epsilon^2 n_p/9} \qquad (7)$$

that can be easily rewritten as a usual generalization bound

$$\pi_{MD} \le \min_{\alpha} \nu + \max_{\alpha} (\nu_2 - \nu_1) + 3\sqrt{\frac{-\ln \delta}{2n_p}} \qquad (8)$$

that holds with probability $1 - \delta$ and where ν is the empirical error performed by the classifier on the original dataset [2].

Note that this approach is similar to the method used for estimating the effective VC–dimension of a classifier [11]. However, it is worthwhile noting that the MD technique estimates directly the generalization error without passing through the computation of the VC–dimension or the Growth Function [12], that could lead to unnecessary overestimations of its value.

The actual procedure for applying the MD method to SVM hyperparameter tuning is the following:

1. Split the data into two halves (assuming n_p even)
2. Flip the target values of the second half, obtaining a modified data set $(X, \tilde{Y})$
3. Train the SVM on $(X, \tilde{Y})$
4. Compute the empirical loss

$$\tilde{\nu} = \frac{2}{n_p} \sum_{i=1}^{n_p} L\left(f(\boldsymbol{x}_i), \tilde{y}_i\right) = \qquad (9)$$

$$= \frac{1}{2} + \frac{1}{n_p} \sum_{i=1}^{n_p/2} L\left(f(\boldsymbol{x}_i), y_i\right) - \frac{1}{n_p} \sum_{i=n_p/2+1}^{n_p} L\left(f(\boldsymbol{x}_i), y_i\right) = \qquad (10)$$

$$= \frac{1 + \nu_1 - \nu_2}{2} \qquad (11)$$

5. Compute the maximal discrepancy value

$$\max_{\alpha} (\nu_2 - \nu_1) = 1 - 2\tilde{\nu} \tag{12}$$

The above procedure can be iterated for several (C,γ) pairs, for example on a discrete lattice with a given step. The optimal hyperparameters correspond to the pair for which $\nu + (1 - 2\tilde{\nu})$ is minimum.

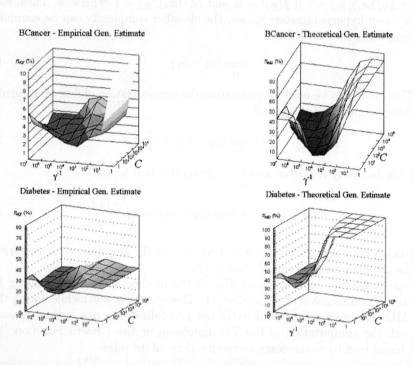

Fig. 1. Empirical error for WBC (up) and PID (low).

Fig. 2. MD error estimate for WBC (up) and PID (low).

The MD method described above is very attractive for its simplicity, however, it is important to note that its application is, in effect, an approximation; in fact, the SVM algorithm does not provide the minimum classification error. This is the price to pay for the smart formulation of the SVM learning as a quadratic programming problem. The alternative approach would be to implement an expensive (NP) integer–programming technique to look for the minimum classification error. Despite this approximation, our approach is supported by similar target–flip methods used by V.Vapnik [11] and V.Cherkassky [10] for measuring the effective VC–dimension and by the outcome of the experimental results.

4   Experimental Results

We compared the MD method, as described above, with a k–fold cross–validation method on several datasets. Due to space constraints we report here the results obtained on two of them.

For each dataset, we performed a 5–fold cross–validation, as described in [6], for several values of the hyperparameters in the ranges $\gamma^{-1} \in [1, 10^7]$ and $C \in [1, 10^5]$ and recorded the generalization prediction π_{KF}. Analogously, we computed π_{MD} for the same values of the hyperparameters, using Eq. (8).

The Wisconsin Breast Cancer (WBC) dataset [3] is composed by 683 samples (after removing 16 cases with missing values), each one represented by a 9-dimensional feature vector.

The 5-fold procedure yields the generalization estimates displayed graphically in Fig. 1, while the outcome of the MD method is showed in Fig. 2 (upper row). The comparative analysis of the obtained results shows a substantial fit between the empirical and theoretical approaches (Fig. 3).

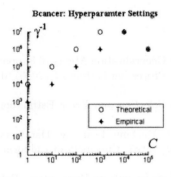

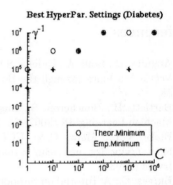

Fig. 3. Optimal hyperparameter values for WBC.

Fig. 4. Optimal hyperparameter values for PID.

The Pima Indians Diabetes (PID) database [3] provides another clinical testbed. The data set includes 768 patient descriptions, each characterized by 8 features; the two classes are quite intermixed and make this sample rather difficult to discriminate.

Fig. 1 (lower row) provides the experimental results obtained by processing the PID dataset by the 5–fold empirical procedure for generalization assessment. Note the large values of π_{KF} that witness the difficulty of the benchmark.

Fig. 2 shows the result of the MD method. The difficult testbed distribution is confirmed by the fact that a significant portion of the hyperparameter configurations does not allow a meaningful prediction of the generalization error (in several cases, $\pi_{MD} = 100\%$). In spite of the fact that theoretical predictions for large values of the γ hyperparameter appear useless, the values in the "valid" region match empirical evidence quite well. More importantly, the two methods agree satisfactorily in the best-performing configurations that yield the smallest generalization estimates for possible values of C. From this viewpoint, Fig. 4 shows that the results of the hyperparameter-design criteria appear quite correlated.

5   Conclusion

A crucial issue in setting up an effective classifier that must adapt to a new dataset or a new environment often concerns the method for the eventual hyperparameter setting.

The methods considered here split the available sample into two subsets; the theoretical one exploits both subsets to derive a *bound* to the generalization error, whereas the empirical one uses a subset for training and the remaining data for *estimating* it. The estimates obtained by the two methods span quite a narrow interval; thus the overall framework turns out to be of high practical interest.

References

1. Anguita, D., Boni, A., Ridella, S.: Evaluating the Generalization Ability of Support Vector Machines through the Bootstrap. Neural Processing Letters **11** (2000) 51–58.
2. Bartlett, B., Boucheron, S., Lugosi, G.: Model Selection and Error Estimation. Machine Learning **48** (2002) 85–113.
3. Blake, C.L., Merz, C.J.: UCI Repository of Machine Learning Databases. http://www.ics.uci.edu/ mlearn/MLRepository.html, University of California, Irvine (1998).
4. Burges, C.: A Tutorial on Support Vector Machines for Pattern Recognition. Data Mining and Knowledge Discovery **2** (1998) 121–167.
5. Cortes, C., Vapnik, V.: Support Vector Networks. Machine Learning **20** (1995) 273–297.
6. Duan, K., Keerthi, S.S., Poo, A.: Evaluation of a Simple Performance Measure for Tuning SVM Hyperparameters Tech. Rep. CD–01–11, Dept. of Mech. Eng., University of Singapore, 2001.
7. Keerthi, S.S., Gilbert, E.G.: Convergence of a Generalized SMO Algorithm for SVM Classifier Design. Machine Learning **46** (2002) 351–360.
8. Ridella, S., Rovetta, S., Zunino, R.: K–Winner Machines for Pattern Classification. IEEE Trans. on Neural Netwoks **12** (2001) 371–385.
9. Schölkopf, B., Smola, A.: Learning with Kernels. The MIT Press, 2002.
10. Shao, X., Cherkassky, V., Li, W.: Measuring the VC–dimension Using Optimized Experimental Design. Neural Computation **12** (2000) 1969–1986.
11. Vapnik, V., Levin, E., Le Cunn, Y.: Measuring the VC–dimension of a Learning Machine. Neural Computation **6** (1994) 851–876.
12. Vapnik, V.: Statistical Learning Theory. Wiley, 1998.

Conjugate Directions for Stochastic Gradient Descent

Nicol N. Schraudolph and Thore Graepel

Institute of Computational Science
ETH Zürich, Switzerland
{schraudo,graepel}@inf.ethz.ch

Abstract. The method of conjugate gradients provides a very effective way to optimize large, deterministic systems by gradient descent. In its standard form, however, it is not amenable to stochastic approximation of the gradient. Here we explore ideas from conjugate gradient in the stochastic (online) setting, using fast Hessian-gradient products to set up low-dimensional Krylov subspaces within individual mini-batches. In our benchmark experiments the resulting online learning algorithms converge orders of magnitude faster than ordinary stochastic gradient descent.

1 Introduction

Conjugate gradient. For the optimization of large, differentiable systems, algorithms that require the inversion of a curvature matrix (*e.g.*, Levenberg-Marquardt [1,2]), or the storage of an iterative approximation of that inverse (quasi-Newton methods such as BFGS [9, p. 425ff]), are prohibitively expensive. Conjugate gradient methods [3], which exactly minimize a d-dimensional unconstrained quadratic problem in d iterations without requiring explicit knowledge of the curvature matrix, have become the method of choice for such problems.

Stochastic gradient. Empirical loss functions are often minimized using noisy measurements of gradient (and, if applicable, curvature) taken on small, random subsamples ("mini-batches") of data, or even individual data points. This is done for reasons of computational efficiency on large, redundant data sets, and out of necessity when adapting online to a continual stream of noisy, potentially non-stationary data. Unfortunately the fast convergence of conjugate gradient breaks down when the function to be optimized is noisy, since this makes it impossible to maintain the conjugacy of search directions over multiple iterations. The state of the art for such *stochastic* problems is therefore simple gradient descent, coupled with adaptation of local step size and/or momentum parameters.

Curvature matrix-vector products. The most advanced parameter adaptation methods [4,5,6,7] for stochastic gradient descent rely on fast curvature matrix-vector products that can be obtained efficiently and automatically [8,7]. Their calculation does *not* require explicit storage of the Hessian, which would

J.R. Dorronsoro (Ed.): ICANN 2002, LNCS 2415, pp. 1351–1356, 2002.

be $O(d^2)$; the same goes for other measures of curvature, such as the Gauss-Newton approximation of the Hessian, and the Fisher information matrix [7]. *Algorithmic differentiation* software[1] provides generic implementations of the building blocks from which these algorithms are constructed. Here we employ these techniques to efficiently compute Hessian-gradient products which we use to implement a stochastic conjugate direction method.

2 Stochastic Quadratic Optimization

Deterministic bowl. The d-dimensional quadratic bowl provides us with a simplified test setting in which every aspect of the optimization can be controlled. It is defined by the unconstrained problem of minimizing with respect to d parameters $\boldsymbol{w}$ the function

$$f(\boldsymbol{w}) \;=\; \frac{1}{2}\,(\boldsymbol{w} - \boldsymbol{w}^*)^T \boldsymbol{J}\boldsymbol{J}^T\,(\boldsymbol{w} - \boldsymbol{w}^*)\,, \tag{1}$$

where the Jacobian $\boldsymbol{J}$ is a $d \times d$ matrix, and $\boldsymbol{w}^*$ the location of the minimum, both of our choosing. By definition the Hessian $\bar{\boldsymbol{H}} = \boldsymbol{J}\boldsymbol{J}^T$ is positive semidefinite and constant with respect to the parameters $\boldsymbol{w}$; these are the two crucial simplifications compared to more realistic, nonlinear problems. The gradient here is $\bar{\boldsymbol{g}} = \nabla f(\boldsymbol{w}) = \bar{\boldsymbol{H}}(\boldsymbol{w} - \boldsymbol{w}^*)$.

Stochastic bowl. The stochastic optimization problem analogous to the deterministic one above is the minimization (again with respect to $\boldsymbol{w}$) of the function

$$f(\boldsymbol{w}, \boldsymbol{X}) \;=\; \frac{1}{2b}\,(\boldsymbol{w} - \boldsymbol{w}^*)^T \boldsymbol{J}\,\boldsymbol{X}\boldsymbol{X}^T \boldsymbol{J}^T(\boldsymbol{w} - \boldsymbol{w}^*)\,, \tag{2}$$

where $\boldsymbol{X} = [\boldsymbol{x}_1, \boldsymbol{x}_2, \dots \boldsymbol{x}_b]$ is a $d \times b$ matrix collecting a *batch* of b random input vectors to the system, each drawn i.i.d. from a normal distribution: $\boldsymbol{x}_i \sim N(\boldsymbol{0}, \boldsymbol{I})$. This means that $E[\boldsymbol{X}\boldsymbol{X}^T] = b\,\boldsymbol{I}$, so that in expectation this is identical to the deterministic formulation:

$$E_{\boldsymbol{X}}[f(\boldsymbol{w}, \boldsymbol{X})] \;=\; \frac{1}{2b}\,(\boldsymbol{w} - \boldsymbol{w}^*)^T \boldsymbol{J}\,E[\boldsymbol{X}\boldsymbol{X}^T]\boldsymbol{J}^T(\boldsymbol{w} - \boldsymbol{w}^*) \;=\; f(\boldsymbol{w})\,. \tag{3}$$

The optimization problem is harder here since the objective can only be probed by supplying stochastic inputs to the system, giving rise to the noisy estimates $\boldsymbol{H} = b^{-1}\boldsymbol{J}\boldsymbol{X}\boldsymbol{X}^T\boldsymbol{J}^T$ and $\boldsymbol{g} = \nabla_{\boldsymbol{w}} f(\boldsymbol{w}, \boldsymbol{X}) = \boldsymbol{H}(\boldsymbol{w} - \boldsymbol{w}^*)$ of the true Hessian $\bar{\boldsymbol{H}}$ and gradient $\bar{\boldsymbol{g}}$, respectively. The degree of stochasticity is determined by the batch size b; the system becomes deterministic in the limit as $b \to \infty$.

Line search. A common optimization technique is to first determine a search direction, then look for the optimum in that direction. In a quadratic bowl, the step from $\boldsymbol{w}$ to the minimum along direction $\boldsymbol{v}$ is given by

$$\Delta \boldsymbol{w} \;=\; -\,\frac{\boldsymbol{g}^T \boldsymbol{v}}{\boldsymbol{v}^T \boldsymbol{H} \boldsymbol{v}}\,\boldsymbol{v}\,. \tag{4}$$

---

[1] See http://www-unix.mcs.anl.gov/autodiff/

Hv can be calculated very efficiently [8], and we can use (4) in stochastic settings as well. Line search in the gradient direction, $v = g$, is called *steepest descent.* When fully stochastic ($b = 1$), steepest descent degenerates into the *normalized LMS* method known in signal processing.

Choice of Jacobian. For our experiments we choose J such that the Hessian has a) eigenvalues of widely differing magnitude, and b) eigenvectors of intermediate sparsity. These conditions model the mixture of axis-aligned and oblique "narrow valleys" that is characteristic of multi-layer perceptrons, and a primary cause of the difficulty in optimizing such systems. We achieve them by imposing some sparsity on the notoriously ill-conditioned *Hilbert matrix*, defining

$$(J)_{ij} = \begin{cases} \frac{1}{i+j-1} & \text{if } i \bmod j = 0 \ \vee \ j \bmod i = 0, \\ 0 & \text{otherwise}. \end{cases} \tag{5}$$

We call the optimization problem resulting from setting J to this matrix the *modified Hilbert bowl.* In the experiments reported here we used the modified Hilbert bowl of dimension $d = 5$, which has a condition number of $4.9 \cdot 10^3$.

Stochastic Ill-Conditioning. Such ill-conditioned systems are particularly challenging for stochastic gradient descent. While directions associated with large eigenvalues are rapidly optimized, progress along the floor of the valley spanned by the small eigenvalues is extremely slow. Line search can ameliorate this problem by amplifying small gradients, but for this to happen the search direction must lie along the valley floor in the first place. In a stochastic setting, gradients in that direction are not just small but extremely *unlikely*: in contrast to deterministic gradients, stochastic gradients contain large components in directions associated with large eigenvalues even for points right at the bottom of the valley. Fig. 1 illustrates (for $b = 1$) the consequence: although a line search can stretch the narrow ellipses of possible stochastic gradient steps into circles through the minimum, it cannot shift any probability mass in that direction.

3   Stochastic Conjugate Directions

Looking for ways to improve the convergence of stochastic gradient methods in narrow valleys, we note that relevant directions associated with large eigenvalues can be identified by multiplying the (stochastic estimates of) Hessian H and gradient g of the system. Subtracting the *projection* of g onto Hg from g (Fig. 2, left) then yields a *conjugate* descent direction c that emphasizes directions associated with small eigenvalues, by virtue of being orthogonal to Hg:

$$c = g - \frac{g^T H g}{g^T H H g} H g$$

Fig. 2 (right) shows that stochastic descent in direction of c (dashed) indeed sports much better late convergence than steepest descent (dotted). Since directions with large eigenvalues are subtracted out, however, it takes far longer to reach the valley floor in the first place.

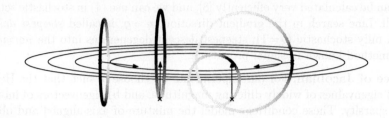

Fig. 1. Distribution of stochastic gradient steps from equivalent points with (circles, right) *vs.* without (ellipses, left) line search in ill-conditioned quadratic bowl. Black is high, white low probability density. Compare to deterministic steepest descent (arrows).

Two-dimensional method. We can combine the respective strengths of gradient and conjugate direction by performing, at each stochastic iteration, a two-dimensional minimization in the plane spanned by g and Hg. That is, we seek the α_1, α_2 that produce the optimal step

$$\Delta w = \alpha_1 g + \alpha_2 Hg. \tag{6}$$

Using $g \stackrel{\text{def}}{=} H(w - w^*)$ gives $\Delta g = \alpha_1 Hg + \alpha_2 HHg$. We can now express the optimality condition as a system of linear equations in the quadratic forms $q_i \stackrel{\text{def}}{=} g^T H^i g$:

$$g^T(g + \Delta g) = q_0 + \alpha_1 q_1 + \alpha_2 q_2 \stackrel{!}{=} 0 \tag{7}$$

$$g^T H (g + \Delta g) = q_1 + \alpha_1 q_2 + \alpha_2 q_3 \stackrel{!}{=} 0 \tag{8}$$

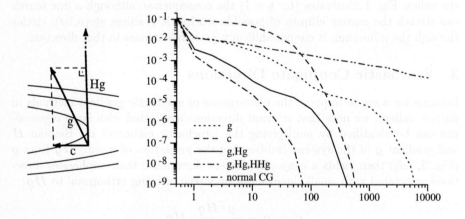

Fig. 2. Left: construction of conjugate direction c via projection of gradient g onto Hg. Right: log-log plot of average loss (over 100 runs) *vs.* number of stochastic iterations ($b = 3$) in modified Hilbert bowl when minimizing in: direction of g (dotted), direction of c (dashed), plane spanned by g and Hg (solid), and subspace spanned by g, Hg, and HHg (dash-dotted). Compare to normal conjugate gradient (dot-dash-dotted).

Solving this yields

$$\alpha_1 = \frac{q_0 q_3 - q_1 q_2}{q_2^2 - q_1 q_3}, \qquad \alpha_2 = \frac{q_1^2 - q_0 q_2}{q_2^2 - q_1 q_3} \tag{9}$$

Fig. 2 (right) shows that this approach (solid line) indeed combines the advantages of the gradient and conjugate directions.

Stochastic Krylov subspace. This approach can be extended to minimization in the m-dimensional, stochastic *Krylov subspace* $K_m \stackrel{\text{def}}{=} [\boldsymbol{g}, \boldsymbol{H}\boldsymbol{g}, \dots \boldsymbol{H}^{m-1}\boldsymbol{g}]$ with $m \leq \min(d, b)$. The expansion of $\Delta \boldsymbol{g}$ in K_m is given by

$$\Delta \boldsymbol{g} = \sum_{i=1}^{m} \alpha_i \boldsymbol{H}^i \boldsymbol{g}; \tag{10}$$

for optimality we require

$$\boldsymbol{q} + \boldsymbol{Q}\boldsymbol{\alpha} \stackrel{!}{=} 0 \quad \Rightarrow \quad \boldsymbol{\alpha} = -\boldsymbol{Q}^{-1}\boldsymbol{q} \tag{11}$$

with

$$\boldsymbol{q} \stackrel{\text{def}}{=} \begin{bmatrix} q_0 \\ q_1 \\ \vdots \\ q_{m-1} \end{bmatrix}, \quad \boldsymbol{\alpha} \stackrel{\text{def}}{=} \begin{bmatrix} \alpha_1 \\ \alpha_2 \\ \vdots \\ \alpha_m \end{bmatrix}, \quad \boldsymbol{Q} \stackrel{\text{def}}{=} \begin{bmatrix} q_1 & q_2 & \cdots & q_m \\ q_2 & q_3 & \cdots & q_{m+1} \\ \vdots & \vdots & \ddots & \vdots \\ q_m & q_{m+1} & \cdots & q_{2m-1} \end{bmatrix} \tag{12}$$

$\boldsymbol{Q}$ and $\boldsymbol{\alpha}$ can be flipped to bring (11) into the form of a standard Toeplitz system, which can be solved in as little as $O(m \log^2 m)$ operations [9, p. 92ff]. The quadratic forms q_0 through q_{2m} can be calculated efficiently as inner products of the m fast Hessian-vector products $\boldsymbol{H}^i \boldsymbol{g}, 0 \leq i \leq m$. Fig. 2 (right) illustrates the rapid convergence of this approach for $m = 3$ (dash-dotted).

Relation to conjugate gradient methods. It has not escaped our notice that on quadratic optimization problems, instead of solving this linear system of equations explicitly, we can equivalently perform m steps of ordinary conjugate gradient within each mini-batch to find the optimum in the Krylov subspace K_m. Instead of a single m-dimensional optimization, we then have m successive line searches according to (4). The initial search direction is set to the gradient, $\boldsymbol{v}_0 := \boldsymbol{g}$; subsequent ones are calculated via the formula

$$\boldsymbol{v}_{t+1} = \frac{\boldsymbol{g}_{t+1}^T \boldsymbol{H} \boldsymbol{v}_t}{\boldsymbol{v}_t^T \boldsymbol{H} \boldsymbol{v}_t} \boldsymbol{v}_t - \boldsymbol{g}_{t+1} \tag{13}$$

or one of its well-known variants (Fletcher-Reeves, Polak-Ribiere [9, p. 420ff]). The crucial difference to standard conjugate gradient techniques is that here we propose to perform just a few steps of conjugate gradient *within* each small, stochastic mini-batch. A reset to the gradient direction when moving on to another mini-batch is not only recommended but indeed mandatory for our approach to work, since otherwise the stochasticity collapses the Krylov subspace. To illustrate, we show in Fig. 2 (dot-dash-dotted) the inadequate performance of standard conjugate gradient when misapplied in this fashion.

4   Summary and Outlook

We considered the problem of stochastic ill-conditioning of optimization problems that lead to inefficiency in standard stochastic gradient methods. By geometric arguments we arrived at conjugate search directions that can be found efficiently by fast Hessian-vector products. The resulting algorithm can be interpreted as a stochastic conjugate gradient technique and as such introduces Krylov subspace methods into stochastic optimization. Numerical results show that our approach outperforms standard gradient descent for unconstrained quadratic optimization by orders of magnitude in a noisy scenario where standard conjugate gradient fails.

At present we only consider sampling noise due to small batch size. Future work will address the question of noise in the target vector w^*, corresponding to unrealizable approximation problems, which may require the incorporation of some form of step size annealing or adaptation scheme. We are also investigating the extension of our techniques to nonlinear optimization problems, such as online learning in multi-layer perceptrons. In this case, conjugate gradient methods are *not* equivalent to the explicit solution of (11), and it is an open question which approach is preferable in the stochastic gradient setting.

References

[1] K. Levenberg. A method for the solution of certain non-linear problems in least squares. *Quarterly Journal of Applied Mathematics*, 11(2):164–168, 1944.

[2] D. Marquardt. An algorithm for least-squares estimation of non-linear parameters. *Journal of the Society of Industrial and Applied Mathematics*, 11(2):431–441, 1963.

[3] M. R. Hestenes and E. Stiefel. Methods of conjugate gradients for solving linear systems. *Journal of Research of the National Bureau of Standards*, 49:409–436, 1952.

[4] G.B. Orr. *Dynamics and Algorithms for Stochastic Learning*. PhD thesis, Department of Computer Science and Engineering, Oregon Graduate Institute, Beaverton, OR 97006, 1995. ftp://neural.cse.ogi.edu/pub/neural/papers/orrPhDchi-5.ps.Z,orrPhDch6-9.ps.Z.

[5] T. Graepel and N. N. Schraudolph. Stable adaptive momentum for rapid online learning in nonlinear systems. In *Proceedings of the International Conference on Artificial Neural Networks* (to appear), Lecture Notes in Computer Science. Springer Verlag, Berlin, 2002.

[6] N. N. Schraudolph. Local gain adaptation in stochastic gradient descent. In *Proceedings of the 9th International Conference on Artificial Neural Networks*, pages 569–574, Edinburgh, Scotland, 1999. IEE, London. http://www.inf.ethz.ch/~schraudo/pubs/smd.ps.gz.

[7] N. N. Schraudolph. Fast curvature matrix-vector products for second-order gradient descent. *Neural Computation*, 14(7), 2002. http://www.inf.ethz.ch/~schraudo/pubs/mvp.ps.gz.

[8] B. A. Pearlmutter. Fast exact multiplication by the Hessian. *Neural Computation*, 6(1):147–160, 1994.

[9] W. H. Press, S. A. Teukolsky, W. T. Vetterling, and B. P. Flannery. *Numerical Recipes in C: The Art of Scientific Computing*. Cambridge University Press, second edition, 1992.

Part X

Special Session: Recurrent Neural Systems

Architectural Bias in Recurrent Neural Networks – Fractal Analysis

Peter Tiňo[1] and Barbara Hammer[2]

[1] Neural Computing Research Group, Aston University, Birmingham B4 7ET, UK
[2] University of Osnabrück, D-49069 Osnabrück, Germany
tinop@aston.ac.uk, hammer@informatik.uni-osnabrueck.de

Abstract. We have recently shown that when initiated with "small" weights, recurrent neural networks (RNNs) with standard sigmoid-type activation functions are inherently biased towards Markov models, i.e. even prior to any training, RNN dynamics can be readily used to extract finite memory machines [6,8]. Following [2], we refer to this phenomenon as the *architectural bias of RNNs*. In this paper we further extend our work on the architectural bias in RNNs by performing a rigorous fractal analysis of recurrent activation patterns. We obtain both lower and upper bounds on various types of fractal dimensions, such as box-counting and Hausdorff dimensions.

1 Introduction

There is a considerable amount of literature devoted to connectionist processing of sequential symbolic structures. For example, researchers have been interested in formulating models of human performance in processing linguistic patterns of various complexity (e.g. [2]).

It has been known for some time that when training RNNs to process symbolic sequences, activations of recurrent units display a considerable amount of structural differentiation even *prior to learning* [2,4,5,7]. Following [2], we refer to this phenomenon as the *architectural bias of RNNs*.

We have recently shown, both empirically and theoretically, the meaning of the architectural bias: when initiated with "small" weights, RNNs with standard sigmoid-type activation functions are inherently biased towards Markov models, i.e. even prior to any training, RNN dynamics can be readily used to extract finite memory machines [6,8]. In this study we further extend our work by rigorously analyzing the "size" of recurrent activation patterns in such RNNs. Since the activations are of fractal nature, the size is expressed through fractal dimensions.

J.R. Dorronsoro (Ed.): ICANN 2002, LNCS 2415, pp. 1359–1364, 2002.

2   Recurrent Networks as IFS

In this paper we concentrate on first-order (discrete-time) RNNs of N recurrent neurons driven with dynamics

$$x_n(t) = g\left(\sum_{j=1}^{D} v_{nj} i_j(t) + \sum_{j=1}^{N} w_{nj} x_j(t-1) + d_n\right), \tag{1}$$

where $x_j(t)$ and $i_j(t)$ are elements, at time t, of the state and input vectors, $\mathbf{x}(t) = (x_1(t), ..., x_N(t))^T$ and $\mathbf{i}(t) = (i_1(t), ..., i_D(t))^T$, respectively, w_{nj} and v_{nj} are recurrent and input connection weights, respectively, d_n's constitute the bias vector $\mathbf{d} = (d_1, ..., d_N)^T$, and $g(\cdot)$ is an injective non-constant differentiable activation function of bounded derivative from $\Re$ to a bounded interval Ω of length $|\Omega|$. Denoting by $\mathbf{V}$ and $\mathbf{W}$ the $N \times D$ and $N \times N$ weight matrices (v_{nj}) and (w_{nj}), respectively, we rewrite (1) in matrix form

$$\mathbf{x}(t) = G(\mathbf{V}\mathbf{i}(t) + \mathbf{W}\mathbf{x}(t-1) + \mathbf{d}), \tag{2}$$

where $G : \Re^N \to \Omega^N$ is the element-wise application of g.

Assume RNN is processing strings over a finite alphabet $\mathcal{A} = \{1, 2, ..., A\}$ of A symbols. Symbols $a \in \mathcal{A}$ are presented at the network input as unique D-dimensional codes $\mathbf{c}_a \in \Re^D$, one for each symbol $a \in \mathcal{A}$. RNN can be viewed as a non-linear iterative function system (IFS) [1], i.e. as a collection of A maps

$$\mathbf{x}(t) = F_a(\mathbf{x}(t-1)) = (G \circ T_a)(\mathbf{x}(t-1)) = G(\mathbf{W}\mathbf{x}(t-1) + \mathbf{d}_a), \quad a \in \mathcal{A}, \tag{3}$$

acting on Ω^N, where $\mathbf{d}_a = \mathbf{V}\mathbf{c}_a + \mathbf{d}$.

For a set $B \subseteq \Re^N$, we denote by $[B]_i$ the slice of B defined as

$$[B]_i = \{x_i | \ \mathbf{x} = (x_1, ..., x_N)^T \in B\}, \quad i = 1, 2, ..., N. \tag{4}$$

When symbol $a \in \mathcal{A}$ is at the network input, the range of possible net-in activations on recurrent units is the set

$$C_a = \bigcup_{1 \le i \le N} [T_a(\Omega^N)]_i. \tag{5}$$

Recall that singular values $\alpha_1, ..., \alpha_N$ of the matrix $\mathbf{W}$ are positive square roots of the eigenvalues of $\mathbf{W}\mathbf{W}^T$. The singular values are lengths of the (mutually perpendicular) principal semiaxes of the image of the unit ball under the linear map defined by the matrix $\mathbf{W}$. We assume $\mathbf{W}$ is non-singular and adopt the convention that $\alpha_1 \ge \alpha_2 \ge ... \ge \alpha_N > 0$. Denote by $\alpha_+(\mathbf{W})$ and $\alpha_-(\mathbf{W})$ the largest and the smallest singular values of $\mathbf{W}$, respectively, i.e. $\alpha_+(\mathbf{W}) = \alpha_1$ and $\alpha_+(\mathbf{W}) = \alpha_N$. An easy lemma gives us conditions under which the maps F_a are contractive.

Lemma 1: If

$$k_a^{max} = \alpha_+(\mathbf{W}) \cdot \sup_{v \in C_a} |g'(v)| < 1, \tag{6}$$

then the map F_a, $a \in \mathcal{A}$, is contractive.

Proof: The result follows from two facts:

(1) a map f from a metric space[1] $(U, \|.\|_U)$ to a metric space $(V, \|.\|_V)$ is contractive, if it is Lipschitz continuous, i.e. for all $\mathbf{x}, \mathbf{y} \in U$, $\|f(\mathbf{x}) - f(\mathbf{y})\|_V \leq k_f \|\mathbf{x} - \mathbf{y}\|_U$, and the Lipschitz constant k_f is smaller than 1.

(2) The Lipschitz constant k of a composition $(f_1 \circ f_2)$ of two Lipschitz continuous maps f_1 and f_2 with Lipschitz constants k_1 and k_2 is equal to $k = k_1 \cdot k_2$.

Note that $\alpha_+(\mathbf{W})$ is a Lipschitz constant of the affine maps T_a and that $\sup_{v \in C_a} |g'(v)|$ is a Lipschitz constant of the map G with domain $T_a(\Omega^N)$. $\quad\square$

By arguments similar to those in the proof of lemma 1, we get

Lemma 2: *For*

$$k_a^{min} = \alpha_-(\mathbf{W}) \cdot \inf_{v \in C_a} |g'(v)|, \tag{7}$$

it holds: $\|F_a(\mathbf{x}) - F_a(\mathbf{y})\| \geq k_a^{min} \|\mathbf{x} - \mathbf{y}\|$, *for all* $\mathbf{x}, \mathbf{y} \in \Omega^N$.

3   Fractal Dimensions

Let $K \subseteq \Omega^N$. For each $\delta > 0$, define $N_\delta(K)$ to be the smallest number of sets of diameter $\leq \delta$ that can cover K (δ-*fine cover of* K). Since the RNNs state space Ω^N is bounded, $N_\delta(K)$ is finite for each $\delta > 0$. The rate of increase of $N_\delta(K)$ as $\delta \to 0$ tells us something about the "size" of K.

Definition 1: The *upper* and *lower box-counting dimensions* of K are defined as

$$\dim_B^+ K = \limsup_{\delta \to 0} \frac{\log N_\delta(K)}{-\log \delta} \quad \text{and} \quad \dim_B^- K = \liminf_{\delta \to 0} \frac{\log N_\delta(K)}{-\log \delta}, \tag{8}$$

respectively.

Definition 2: Let $s > 0$. For $\delta > 0$, define

$$\mathcal{H}_\delta^s(K) = \inf_{\Gamma_\delta(K)} \sum_{B \in \Gamma_\delta(K)} (diam\ B)^s, \tag{9}$$

where the infimum is taken over the set $\Gamma_\delta(K)$ of all countable δ-fine covers of K. Define $\mathcal{H}^s(K) = \lim_{\delta \to 0} \mathcal{H}_\delta^s(K)$. The *Hausdorff dimension of the set* K is

$$\dim_H K = \inf\{s|\ \mathcal{H}^s(K) = 0\}. \tag{10}$$

To find out more about fractal dimensions, we refer the interested reader to [3]. It is well known that

$$\dim_H K \leq \dim_B^- K \leq \dim_B^+ K. \tag{11}$$

---

[1] the metric is expressed through a norm

4   Fractal Dimension Estimates for RNNs Driven by Bernoulli Source

Consider a Bernoulli source $\mathcal{S}$ over the alphabet $\mathcal{A}$. Without loss of generality assume that all symbol probabilities are nonzero. When we drive a RNN with symbolic sequences generated by $\mathcal{S}$, by action of the IFS $\{F_a\}_{a\in\mathcal{A}}$, eq. (3), we get a set of recurrent activations $\mathbf{x}(t)$ that tend to group in well-separated clusters [4,5,7]. In case of contractive RNNs (that are capable of emulating finite-memory machines, see [6,8]), the activations $\mathbf{x}(t)$ approximate the unique attractor $K \subset \Omega^N$ of the IFS $\{F_a\}_{a\in\mathcal{A}}$ via the so-called chaos game algorithm [1]. In this situation, we can get size estimates for the activation clusters.

Define

$$\ell = \sup_{v\in\bigcup_a C_a} |g'(v)| = \max_{a\in\mathcal{A}} \sup_{v\in C_a} |g'(v)|, \qquad q = \min_{a,b\in\mathcal{A}, a\neq b} \|\mathbf{V}(\mathbf{c}_a - \mathbf{c}_b)\|. \qquad (12)$$

Theorem 1:
(i) *Suppose* $\alpha_+(\mathbf{W}) < \ell^{-1}$. *Let* s_{max} *be the (unique) solution of (see (6))*

$$\sum_{a\in\mathcal{A}} (k_a^{max})^s = 1. \qquad (13)$$

Then, $\dim_B^+ K \leq s_{max}$.
(ii) *Suppose*

$$\alpha_+(\mathbf{W}) < \min\left\{\frac{1}{\ell}, \frac{q}{\sqrt{N}\,|\Omega|}\right\}$$

and let s_{min} *be the (unique) solution of (see (7))*

$$\sum_{a\in\mathcal{A}} (k_a^{min})^s = 1. \qquad (14)$$

Then, $\dim_H K \geq s_{min}$.
Proof: Since $\alpha_+(\mathbf{W}) < \ell^{-1}$, by (12) and lemma 1, the IFS $\{F_a\}_{a\in\mathcal{A}}$ is contractive. Furthermore, the contractions F_a have Lipschitz constants k_a^{max}. Statement **(i)** follows from [3], proposition 9.6.

For the lower bound **(ii)**, we need the open set condition (OSC)[2]:
(j) for all $a \in \mathcal{A}$, $F_a(\Omega^N) \subseteq \Omega^N$ and
(jj) $F_a(\Omega^N) \cap F_b(\Omega^N) = \emptyset$, for all $a, b \in \mathcal{A}$, $a \neq b$.

(j) is automatically satisfied, since all F_a's are contractions. To see that **(jj)** holds, note that $\alpha_+(\mathbf{W}) < \frac{q}{\sqrt{N}\,|\Omega|}$, which implies[3]

$$q = \min_{a\neq b} \|\mathbf{V}(\mathbf{c}_a - \mathbf{c}_b)\| = \min_{a\neq b} \|\mathbf{d}_a - \mathbf{d}_b\| > 2\,\alpha_+(\mathbf{W})\,\frac{diam\ \Omega^N}{2},$$

---

[2] It is not necessary to apply the OSC to the whole state space, as done here. An open subset of Ω^N, such that **(j)** and **(jj)** hold, would suffice.
[3] radius of the sphere encircling Ω^N is $(diam\ \Omega^N)/2 = \sqrt{N}\,|\Omega|/2$

and so $T_a(\Omega^N) \cap T_b(\Omega^N) = \emptyset$, for all $a, b \in \mathcal{A}$, $a \neq b$. But since G is injective, $(G \circ T_a)(\Omega^N) \cap (G \circ T_b)(\Omega^N) = \emptyset$, which, by (3), is equivalent to **(jj)**. The statement **(ii)** then follows from [3], proposition 9.7.   □

Using (11), we get

Corollary 1: *Under the assumptions of Theorem 1* **(ii)**, $s_{min} \leq \dim K \leq s_{max}$, *where s_{max} and s_{min} are determined by (13) and (14), respectively, and 'dim' is any of the fractal dimensions[4] $\dim_H$, $\dim_B^-$ and $\dim_B^+$.*

4.1 Closed-Form, Less Tight Bounds

Theorem 2: *Let*

$$k^{max} = \max_{a \in \mathcal{A}} k_a^{max}, \quad k^{min} = \min_{a \in \mathcal{A}} k_a^{min}, \quad g'_{max} = \sup_{v \in \Re} |g'(v)|, \quad w_{max} = \max_{1 \leq i,j \leq N} |w_{ij}|.$$

(i) *Suppose $\alpha_+(\mathbf{W}) < \ell^{-1}$. Then*

$$\dim_B^+ K \leq s_{max} \leq -\frac{\log A}{\log k^{max}} \leq -\frac{\log A}{\log N + \log w_{max} + \log g'_{max}}.$$

(ii) *Suppose*

$$\alpha_+(\mathbf{W}) < \min\left\{\frac{1}{\ell}, \frac{q}{\sqrt{N}\,|\Omega|}\right\}$$

Then

$$\dim_H K \geq s_{min} \geq -\frac{\log A}{\log k^{min}}.$$

Proof: Fractal dimension s_+ corresponding to the IFS with A *similarities* of contraction ratio k^{max} is the unique solution of

$$\sum_{a \in \mathcal{A}} (k^{max})^s = 1, \tag{15}$$

and so is given by $s_+ = -\frac{\log A}{\log k^{max}}$. Subtracting (13) from (15) and denoting $\Delta = s_{max} - s_+$, we obtain

$$\sum_{a \in \mathcal{A}} (k^{max})^{s_+} - (k_a^{max})^{s_+} (k_a^{max})^{\Delta} = 0.$$

Since for all $a \in \mathcal{A}$, $k^{max} \geq k_a^{max}$, it must be that $(k_a^{max})^{\Delta} \geq 1$. Because the IFS $\{F_a\}_{a \in \mathcal{A}}$ is contractive, Δ must be ≤ 0, and so $s_{max} \leq s_+$. **(i)** follows from Theorem 1(i), and from realizing that $N \cdot w_{max} \cdot g'_{max} \geq k^{max}$.

To prove **(ii)**, note that the fractal dimension of IFS with A *similarities* of contractive ratio k^{min} is $s_- = -\frac{\log A}{\log k^{min}}$. Similarly to the case **(i)**, we write

$$\sum_{a \in \mathcal{A}} (k^{min})^{s_-} - (k_a^{min})^{s_-} (k_a^{min})^{\Delta} = 0, \quad \Delta = s_{min} - s_-.$$

Since $k^{min} \leq k_a^{min}$, and $k_a^{min} < 1$, for all $a \in \mathcal{A}$, it must be that $(k_a^{min})^{\Delta} \leq 1$, and so $\Delta \geq 0$, which means $s_{min} \geq s_-$.   □

---

[4] and in fact many more, e.g. packing dimension, see [3]

5   Discussion and Conclusion

Recently, we have extended the work of Kolen and others (e.g. [2,4,5,7]) by pointing out that in dynamical tasks of symbolic nature, when initiated with "small" weights, recurrent neural networks (RNN), with e.g. sigmoid activation functions, form contractive IFS. Such networks are inherently biased towards Markov models, i.e. even prior to any training, RNN dynamics can be readily used to extract finite memory machines [6,8]. In other words, even without any training, the recurrent activation clusters are perfectly reasonable and are biased towards finite-memory computations. This paper further extends our work by showing that in such cases, a rigorous analysis of fractal encodings in the RNN state space can be performed. We have derived (lower and upper) bounds on several types of fractal dimensions, such as Hausdorff and (upper and lower) box-counting dimensions.

The theory presented in this paper is general, but can be readily applied to any RNN with dynamics (1) and injective non-constant differentiable activation functions of bounded derivative, mapping $\Re$ into a bounded interval. For example, when the standard logistic sigmoid $g(u) = 1/(1+\exp(-u))$ is used, initiating recurrent weights of N recurrent neurons from the interval $(-r/N, r/N)$, $0 < r < 4$, will insure contractive symbol maps F_a. By theorem 2 (i), the upper box-counting dimension of recurrent activations is upper-bounded by $\log A/\log(4/r)$, where A is the number of symbols in the input alphabet $\mathcal{A}$.

Many extensions and/or refinements are possible. For example, the theory can be easily extended to second-order RNNs and can be made more specific for the case of unary (one-of-A) encodings $\mathbf{c}_a$ of input symbols $a \in \mathcal{A}$.

References

1. M.F. Barnsley: Fractals everywhere. Academic Press, New York (1988)
2. M.H. Christiansen, N. Chater: Toward a connectionist model of recursion in human linguistic performance. Cognitive Science, **23** (1999) 157–205
3. K. Falconer: Fractal Geometry- Mathematical Foundations and Applications. John Wiley & Sons, Chichester, UK (1990)
4. J.F. Kolen: Recurrent networks: state machines or iterated function systems? In: Proceedings of the 1993 Connectionist Models Summer School. Lawrence Erlbaum Associates, Hillsdale, NJ (1994) 203–210
5. J.F. Kolen: The origin of clusters in recurrent neural network state space. In: Proceedings from the Sixteenth Annual Conference of the Cognitive Science Society. Lawrence Erlbaum Associates, Hillsdale, NJ (1994) 508–513
6. B. Hammer, P. Tiňo: Neural networks with small weights implement finite memory machines. Technical Report P-241, Dept. of Math./Comp.Science, University of Osnabrück, Germany (2002)
7. P. Manolios and R. Fanelli: First order recurrent neural networks and deterministic finite state automata. Neural Computation **6** (1994) 1155–1173
8. P. Tiňo, M. Čerňanský, L. Beňušková: Markovian Architectural Bias of Recurrent Neural Networks. In: 2nd Euro-International Symposium on Computational Intelligence. Springer-Verlag Studies in CI (2002) to appear

Continuous-State Hopfield Dynamics Based on Implicit Numerical Methods*

Miguel A. Atencia[1], Gonzalo Joya[2], and Francisco Sandoval[2]

[1] Departamento de Matemática Aplicada. E.T.S.I.Informática
[2] Departamento de Tecnología Electrónica. E.T.S.I.Telecomunicación
Universidad de Málaga, Campus de Teatinos, 29071 Málaga, Spain
matencia@ctima.uma.es

Abstract. A novel technique is presented that implements continuous-state Hopfield neural networks on a digital computer. Instead of the usual forward Euler rule, the backward method is used. The stability and Lyapunov function of the proposed discrete model are indirectly guaranteed, even for reasonably large step size. This is possible because discretization by implicit numerical methods inherits the stability of the continuous-time model. On the contrary, the forward Euler method requires a very small step size to guarantee convergence to solutions. The presented technique takes advantage of the extensive research on continuous-time stability, as well as recent results in the field of dynamical analysis of numerical methods. Also, standard numerical methods allow for synchronous activation of neurons, thus leading to performance enhancement. Numerical results are presented that illustrate the validity of this approach when applied to optimization problems.

1 Introduction

The Hopfield network [1,2] has been proved to be a valuable resource for the solution of a variety of computational tasks, among which optimization problems [3] are specially interesting. Several models with significant differences have been proposed. Networks with discrete states ($s \in \{-1, 1\}$ or $s \in \{0, 1\}$) have been shown to be very sensitive to local minima [4,5], so they will not be further addressed in this contribution. Therefore, in the sequel, we restrict ourselves to continuous-state networks $s \in [-1, 1]$ or $s \in [0, 1]$) and the words "continuous" and "discrete" are referred to the manner time is advanced. Continuous networks are represented by -vectorial- ordinary differential equations (ODE) while discrete networks are formulated as difference equations. Finally, in discrete networks, neuron states may change simultaneously for all neurons – synchronous networks – or a neuron at a time – asynchronous networks –. The study of continuous networks has received important theoretical contributions ([6,7] and references therein) and its application to optimization has been extensively studied [8,9]. On the contrary, the results obtained for discrete networks [10,11] are

---

* This work has been partially supported by the Spanish Ministerio de Ciencia y Tecnología (MCYT), Project No. TIC2001-1758.

J.R. Dorronsoro (Ed.): ICANN 2002, LNCS 2415, pp. 1365–1370, 2002.
© Springer-Verlag Berlin Heidelberg 2002

comparatively scarce, because they do not benefit from the powerful mathematical tools that are applied to continuous systems. Notwithstanding, the analysis of discrete networks is worthwhile because, in practice, continuous networks are programmed on digital -discrete- computers, and the continuous analysis is not valid for the implemented discrete system [4]. Moreover, discrete networks have been proved to be stable only in the asynchronous case, but synchronous networks are preferable due to better performance and easy parallelization.

In this work, we discretize the ODE that represents the continuous model with an implicit numerical method: the backward Euler rule. The rationale behind this decision is found in recent results in the field of dynamics of numerical analysis: in [12] both the backward Euler rule and the forward Euler rule are proved to inherit the Lyapunov functional of the continuous ODE. Furthermore, a bound on the step size is given, but it depends on the used method: in the backward Euler rule, the allowable step size is significantly larger than in the forward Euler rule. The main drawback of the backward rule is being an implicit method, so a nonlinear system of algebraic equations must be solved at every time step. We show that the step size enlargement compensates for this disadvantage and the backward Euler rule can outperform the simpler forward rule, while preserving the qualitative behaviour of the original continuous network.

The Hopfield model is briefly recalled in Sect. 2, together with an example that illustrates the flaws of the forward Euler rule. Section 3 presents the proposed technique, based upon results on numerical methods for gradient systems, which are reproduced for completeness. In Sect. 4 the implementation of the novel method and numerical results that show the validity of our approach are presented. Finally, Sect. 5 provides some conclusions and future directions.

2   Implementation of Continuous Hopfield Networks

The continuous Hopfield model is a fully connected neural network where the input to each neuron is given by the usual summation:

$$net_i = \sum_j w_{ij}s_j - I_i \tag{1}$$

where s_i is the state of neuron i. For simplicity, we adopt the first order formulation, but high order generalization is well known (e.g. [4]). The network evolution is defined by the ODE:

$$\frac{du_i}{dt} = net_i \; ; \qquad s_i(t) = g\left(\frac{u_i(t)}{\beta}\right) \; ; \qquad g(x) = \tanh(x) \tag{2}$$

in the Abe formulation [13], which has the advantage that the Lyapunov function has the simple form $V(s) = -1/2 \sum_i \sum_j w_{ij} s_i s_j + \sum_i I_i s_i$, where w_{ij} and I_i are the weights and biases, respectively. The application of the Hopfield model to optimization is a consequence of its stability: since the network seeks a minimum of its Lyapunov function, it can be regarded as a minimization method, as long as the target function can be identified with the Lyapunov function.

The analog hardware implementation of continuous systems faces obvious drawbacks, such as lack of flexibility and cost. Therefore, the results on continuous networks are usually illustrated by simulations on digital computers, where the differential equation 2 is simply replaced by the difference equation $\Delta u_i / \Delta t = net_i$. Frequently, Δt is chosen by empirical intuition or $\Delta t = 1$ is assumed for simplicity. Some authors have pointed out that the resulting discrete system does not possess the same Lyapunov function [14] and it may even present cycles, thus destroying stability. The discrete network has been proved to work in asynchronous activation mode [10], but this usage discourages any implementation by parallel algorithms on multiprocessor computers. Also, advances in numerical methods for ODE are not applicable, because in these methods all states are simultaneously changed. These restrictions have reduced the ability of Hopfield networks to compete with classical optimization methods.

In order to show the significance of these results, we have built a very simple example with two neurons and the following parameters:

$$W = \begin{bmatrix} 0 & -0.35 \\ -0.35 & 0 \end{bmatrix} \quad ; \quad I = \begin{bmatrix} -0.2 \\ -0.1 \end{bmatrix} \quad ; \quad \beta = 1 \quad ; \quad \Delta t = 1 \quad (3)$$

When neuron activation is performed synchronously, simulation shows that the states $s_1 \approx [0.3042 - 0.5317]^T$, $s_2 \approx [-0.9565 - 0.9947]^T$ form a period two cycle, so the network is no longer an asymptotically stable system, it can not possess a Lyapunov function, and it is useless as an optimization method. Moreover, for some initial states, the system trajectory escapes to infinity so stability is destroyed and all the desirable properties of the continuous system are lost.

3   Discretization by the Backward Euler Rule

We proceed to implement a discrete Hopfield network with favourable properties in two stages: first, a numerical method for generic ODE is chosen, according to its ability to preserve the stability of the ODE; second, equation 2 that defines the continuous Hopfield network is expressed in the general form $dx/dt = f(x)$. Among the simplest numerical methods for ODE we find the forward Euler rule, defined by the difference equation $x_{n+1} = x_n + \Delta t \, f(x_n)$, and the backward Euler rule, given by $x_{n+1} = x_n + \Delta t \, f(x_{n+1})$. In the latter the value of x_{n+1} is only found after the solution of an equation, thus it is called an implicit method. Although this fact increases the computational cost of implicit methods, they are known to possess favourable stability properties, and the benefits from using implicit methods for Hopfield networks have been shown [15].

Between the two described methods, we select the backward Euler rule, based upon results in [12] that we summarize in a simplified form, useful for our aim:

Definition 1. *The ODE $dx/dt = f(x)$ is said to define a gradient system if there exists a function $F(x)$ satisfying: i) $F(x) \geq 0$. ii) $F(x) \to \infty$ as $\|x\| \to \infty$. iii) $F(x(t))$ is non-increasing in t for a solution $x(t)$ of the ODE. iv) if $F(x(t)) = F(x(t_0))$ for all $t \geq t_0$ then $f(x(t_0)) = 0$. Also, for a gradient system f is assumed*

to satisfy a one-sided Lipschitz condition, i.e. $\langle f(u) - f(v), u - v \rangle \leq c\|u - v\|^2$
for all u, v and $c > 0$. F is called a Lyapunov function.

Theorem 1. *Let $dx/dt = f(x)$ be an ODE that defines a gradient system. If the ODE is approximated by the backward Euler rule with $\Delta t < 1/c$ then the resulting discrete system is stable and it possess the same Lyapunov function as the original ODE.*

Theorem 2. *Let $dx/dt = f(x)$ be an ODE that defines a gradient system, with f globally Lipschitz with Lipschitz constant L. If the ODE is approximated by the forward Euler rule with $\Delta t < 1/L$ then the resulting discrete system is stable and it possess the same Lyapunov function as the original ODE.*

When searching an appropriate numerical method for a gradient system, it must be noted that the global Lipschitz condition in Theorem 2 is far more restrictive that the one-sided Lipschitz in Theorem 1. From the computational perspective, this means that the backward Euler rule is able to mimic the gradient nature of the continuous network with a large step size. In contrast, the application of the forward Euler method requires either significantly decreasing the step size or losing the Lyapunov function preservation. The former option leads to performance degradation whereas the latter impairs the optimization ability of the model. Consequently, we propose the backward Euler rule as the most appropriate method for discretizing the continuous Hopfield network.

The second step, formulating the Hopfield dynamical equation 2 into a single ODE, results in $f(s) = 1/\beta (1 - s^2) net(s)$ where the functional relation $g' = (1 - g^2)$ has been used. Obviously, the usefulness of the above results for Hopfield networks stems from the gradient nature of continuous Hopfield systems. Although this fact was implicit in the definition of a Lyapunov function by Hopfield, the first author to explicitly regard Hopfield models as a subset of the wider class of gradient systems was Vidyasagar [6], to the best of our knowledge.

Finally, assembling the previous ODE and the backward Euler rule results in the proposed discrete dynamical equation:

$$s_i(n+1) = s_i(n) + \Delta t \frac{1}{\beta} \left(1 - s_i(n+1)^2\right) net_i(s(n+1)) \tag{4}$$

4   Implementation Results

In this section we build a modest scale optimization problem by choosing random matrix W and vector I of dimension 9. The ODE of the continuous network, with $\beta = 1$ is discretized both with the forward and the backward Euler rules. The nonlinear equation that appears in the implicit method is solved by functional iteration. By means of numerical simulations, the value of the Lipschitz constants is approximately calculated: $c \approx 500$, $L \approx 10^6$. Thus, according to the results of the previous section, the step size would be required to be $\Delta t \leq 10^{-6}$ for the forward Euler rule to preserve stability. This causes excessive computational cost and the extremely slow convergence is impractical. Subsequently, we define $\Delta t = 10^{-3}$ and proceed to compare both methods, by performing 100 runs on each model. The networks are started with random initial values and they are left to evolve until they attain a stable state. The value of the target function at the final state is registered and it is compared with the global optimum, which has been calculated by an exhaustive procedure. These results are shown in Table 1.

Table 1. Results after 100 runs. The global minimum is $x_{opt} = -9924$.

	Average solution	Percentage of runs that reach the global minimum
Forward Euler	-8575.78	28 %
Backward Euler	-8974.48	41 %

A glance at Table 1 reveals that the backward Euler rule not only achieves the global optimum more often than the forward method but also obtains a better average solution. Although the proposed method gets sometimes stuck into local minima the results must be considered promising, in view of the fact that local minima are a common problem in the continuous Hopfield network and, indeed, in any gradient-like optimizer, such as the Newton method. A question that arises from results is how the forward Euler rule is able to achieve a minimum, sometimes even the global one, if the step size condition of Theorem 2 is not satisfied. This issue is illuminated by noting that the condition $\Delta t < 1/L$ is a sufficient one, but it is not necessary for stability. Numerical results suggest that the forward Euler rule leads to a stable system, but the Lyapunov function is different from the original, thus causing poor optimization performance. Clearly more work needs to be done in this direction.

5   Conclusions and Future Directions

The efficiency of the Hopfield network as an optimization method is enhanced by discretizing the continuous network with a numerical method that preserves its stability. Concretely, we have chosen the backward Euler rule because the expression of the Lyapunov function is then exactly matched. Also, the usual discretization -the forward Euler rule- preserves the Lyapunov function but in doing so it requires a very small step size, which results in unaffordable computational cost. Besides, the usage of a standard numerical method allows for synchronous neuron updating, which is a more efficient procedure that is easily implemented on a parallel computer. Numerical results have shown the effectiveness of the proposed technique, as well as several open ways for refinement.

Current research is directed towards extending our knowledge of discrete Hopfield networks with regard to their optimization ability. The tuning of the program that implements the network has not been addressed in this work, but several enhancements are possible. A lower bound on the step size or a necessary condition for gradient preservation would be desirable. If the forward Euler rule does not possess the same Lyapunov function but a different one, the explicit expression of this perturbed function should be obtained in order to determine the importance of the deviation from the original function. All these advances could be combined so that discrete Hopfield networks could compete with classical optimization algorithms.

References

1. Hopfield, J.: Neural networks and physical systems with emergent collective computational abilities. Proc. Natl. Acad. Sci. USA **79** (1982) 2554–2558
2. Hopfield, J.: Neurons with graded response have collective computational properties like those of two-state neurons. Proc. Natl. Acad. Sci. USA **81** (1984) 3088–3092
3. Tank, D., Hopfield, J.: 'Neural' computation of decisions in optimization problems. Biol. Cybern. **52** (1985) 141–152
4. Joya, G., Atencia, M.A., Sandoval, F.: Hopfield neural networks for optimization: Study of the different dynamics. Neurocomputing **43** (2002) 219–237
5. Vidyasagar, M.: Are analog neural networks better than binary neural networks? Circuits, Systems and Signal Processing **17** (1998) 243–270
6. Vidyasagar, M.: Minimum-seeking properties of analog neural networks with multilinear objective functions. IEEE Trans. On Automatic Control **40** (1995) 1359–1375
7. Chen, T., Amari, S.I.: New theorems on global convergence of some dynamical systems. Neural Networks **14** (2001) 251–255
8. Smith, K., Palaniswami, M., Krishnamoorthy, M.: Neural techniques for combinatorial optimization with applications. IEEE Trans. On Neural Networks **9** (1998) 1301–1318
9. Bharitkar, S., Tsuchiya, K., Takefuji, Y.: Microcode optimization with neural networks. IEEE Transactions on Neural Networks **10** (1999) 698–703
10. Wang, L.: On the dynamics of discrete-time, continuous-state Hopfield neural networks. IEEE Trans. On Circuits and Systems-II **45** (1998) 747–749
11. Tino, P., Horne, B., Giles, C.: Attractive periodic sets in discrete-time recurrent networks (with emphasis on fixed-point stability and bifurcations in two-neuron networks). Neural Computation **13** (2001) 1379–1414
12. Stuart, A., Humphries, A.: Dynamical systems and numerical analysis. Cambridge University Press (1996)
13. Abe, S.: Theories on the Hopfield neural networks. In: Proc. IEE International Joint Conference on Neural Networks. Volume I. (1989) 557–564
14. Galan-Marin, G., Muñoz Perez, J.: Design and analysis of maximum Hopfield networks. IEEE Trans. On Neural Networks **12** (2001) 329–339
15. Atencia, M.A., Joya, G., Sandoval, F.: Numerical implementation of continuous Hopfield networks for optimization. In: Proc. European Symposium on Artificial Neural Networks. (2001) 359–364

Time-Scaling in Recurrent Neural Learning

Ricardo Riaza and Pedro J. Zufiria

Departamento de Matemática Aplicada a las Tecnologías de la Información
ETSI Telecomunicación, Universidad Politécnica de Madrid
Ciudad Universitaria s/n - 28040 Madrid, Spain
{rrr,pzz}@mat.upm.es

Abstract. Recurrent Backpropagation schemes for fixed point learning in continuous-time dynamic neural networks can be formalized through a differential-algebraic model, which in turn leads to singularly perturbed training techniques. Such models clarify the relative time-scaling between the network evolution and the adaptation dynamics, and allow for rigorous local convergence proofs. The present contribution addresses some related issues in a discrete-time context: fixed point problems can be analyzed in terms of iterations with different evolution rates, whereas periodic trajectory learning can be reduced to a multiple fixed point learning problem via Poincaré maps.

1 Introduction

Recurrent Backpropagation (RBP) has become an important paradigm for invariant learning problems in neural networks and parameterized dynamical systems [3,6,7]. In continuous-time systems, the RBP scheme for fixed point problems can be naturally formalized as a differential-algebraic equation or DAE [4,8], which in turn makes it natural to introduce singularly perturbed regularizations in the learning model. These regularizations clarify the relative time-scaling between the network dynamics and the adaptation process, and allow for rigorous local convergence proofs based on the results discussed in [10].

A natural analog in the discrete-time context is defined by the use of different rates in the iterations describing both the network evolution and the weight adjustment. This is similar to the approach supporting Predictor-Corrector (PC) schemes in continuation problems [1], which underly usual implementations of RBP for fixed point learning. Local convergence of such discrete-time schemes can be obtained in a manner similar to the one employed in the continuous-time setting. Also, this approach leads to a interpretation of certain RBP implementations for periodic trajectory learning in terms of (multiple) fixed point problems, obtained via Poincaré maps [2]. In this direction, the purpose of the present contribution is to discuss the presence of multiple time scales in both continuous and discrete-time schemes, their role in the local convergence properties of the learning process, and also the above-mentioned reduction of periodic trajectory learning to multiple fixed point problems, together with the convergence of the resulting scheme. Accordingly, the paper is structured as follows:

J.R. Dorronsoro (Ed.): ICANN 2002, LNCS 2415, pp. 1371–1376, 2002.

Section 2 presents some background on RBP schemes for fixed point and trajectory learning problems. Differential-algebraic and singularly perturbed models for RBP in continuous-time fixed point problems are discussed in Section 3, whereas discrete-time analogs are considered in Section 4. Finally, the reduction of trajectory learning problems via Poincaré maps is addressed in Section 5.

2   RBP for Fixed Point and Trajectory Learning

The learning problems tackled in this paper will consider generic recurrent networks defined by a system of the form

$$\dot{u} = F(u, w),$$

where $u \in \mathbb{R}^n$ represents the network state, $w \in \mathbb{R}^m$ encompasses all network parameters, and $F \in C^2(\mathbb{R}^{n+m}, \mathbb{R}^n)$. This comprises neural systems such as, for instance, Hopfield networks or Cohen-Grossberg models [5].

2.1   Fixed Point Learning

A fixed point learning problem is defined by the goal of determining a parameter vector w^* such that the trained network $\dot{u} = F(u, w^*)$ has an asymptotically stable equilibrium point at a prefixed point u^* or set $U^* = \{u^{[1]*}, \ldots, u^{[p]*}\}$.

If the existence of (at least) p asymptotically stable equilibrium points $u^{[1]f}(w), \ldots, u^{[p]f}(w)$ for any fixed value of w is assumed, the learning problem may be formulated as the minimization of the following function $E : \mathbb{R}^m \to \mathbb{R}$:

$$E(w) = \sum_{i=1}^{p} \frac{1}{2} \|u^{[i]f}(w) - u^{[i]*}\|^2,$$

where $\| \cdot \|$ denotes the Euclidean norm in $\mathbb{R}^n$.

In the sequel, we focus on single pattern problems, although the extension to multiple pattern problems ($p > 1$) is straightforward. The gradient-descent rule

$$\dot{w} = -E_w(u^f(w)),$$

leads to the Recurrent Backpropagation scheme [3,7]. The computation of the gradient can be performed using the chain rule, which yields the adaptation law

$$\dot{w} = (u^f(w) - u^*)(F_u)^{-1}(u^f(w), w)F_w(u^f(w), w). \tag{1}$$

Adjoint-based approaches can also be found in [3].

2.2   Trajectory Learning

In trajectory learning problems, w^* should force the neural network to approach a prescribed trajectory $u^*(t)$ over an interval $[t_0, t_1]$. This problem can be formalized as the minimization of the following integral error function

$$E = \frac{1}{2} \int_{t_0}^{t_1} \|u(t) - u^*(t)\|^2 dt. \tag{2}$$

Now, the gradient-descent rule reads

$$\dot{w} = -E_w = -\int_{t_0}^{t_1} (u(t) - u^*(t))v(t)dt, \tag{3}$$

where $v(t)$ is given by the sensitivity system [2,3]

$$\dot{v} = F_u v + F_w, \ v(0) = 0. \tag{4}$$

Let us assume that the learning goal $u^*(t)$ is a T-periodic trajectory. Among others possibilities [3], the error function E in (2) can be approximated through a numerical estimation of the integral using I points $u(ih), i = 0, \ldots, I - 1$, where h is chosen so that $I = T/h$ is a positive integer. This approximation defines a discrete error

$$\tilde{E} = \frac{1}{2} \sum_{i=0}^{I-1} \|u(ih) - u^*(ih)\|^2 h, \tag{5}$$

which transforms the problem into another one where the learning goal is the approximation of a finite number of points in the trajectory. This leads to the following version of the adaptation law (3)

$$\dot{w} = -\sum_{i=0}^{I-1} [u(ih) - u^*(ih)]v(ih)h. \tag{6}$$

3 Singularly Perturbed Schemes for Continuous-Time Fixed Point Learning

The adaptation law given by (1) can be rewritten as the semi-explicit differential-algebraic equation (DAE)

$$\dot{w} = E_u(u)F_u^{-1}(u, w)F_w(u, w) \tag{7a}$$
$$0 = F(u, w). \tag{7b}$$

If we relax the assumption that the network is in equilibrium all along the training process, we get the following *singularly perturbed* regularization

$$\dot{w} = E_u(u)F_u^{-1}(u, w)F_w(u, w) \tag{8a}$$
$$\varepsilon \dot{u} = F(u, w), \tag{8b}$$

where ε models the relative time-scaling between the network dynamics and the adaptation process [9]. Local convergence properties of this scheme can be obtained from the results in [10]: namely, if $u^f(w)$ is uniformly exponentially stable for the dynamics of F near a solution (u^*, w^*), then the singularly perturbed scheme (8) is locally convergent to a solution for $\varepsilon > 0$ small enough. More precisely, there exists a value ε^* below which the system is guaranteed to converge: this shows that, if the network (8b) approaches a steady state with a sufficiently large rate (when compared with the rate of the adaptation process (8a)), the process is locally convergent.

In multiple pattern problems, $p > 1$ fixed points $u^{[1]*}, \ldots, u^{[p]*}$ must be learned by the network, and the existence of p asymptotically stable equilibria $u^{[1]f}(w), \ldots, u^{[p]f}(w)$ must be assumed for every w. The regularized model presented above for the single pattern case may then be extended to p patterns assuming that p systems evolve in parallel, that is, the state vector u now leads to p vectors $u^{[1]}, \ldots, u^{[p]}$. The resulting singularly perturbed system is then

$$\dot{w} = \sum_{i=1}^{p} (u^{[i]} - u^{[i]*}) F_u^{-1}(u^{[i]}, w) F_w(u^{[i]}, w)$$

$$\varepsilon \dot{u}^{[1]} = F(u^{[1]}, w)$$

$$\vdots$$

$$\varepsilon \dot{u}^{[p]} = F(u^{[p]}, w).$$

4 Discrete-Time Fixed Point Learning

A discrete-time analog of system (8) can be easily obtained if we consider the network and the adaptation process to be defined by two iterations which do not necessarily evolve at the same rate. Consider the neural network to be defined by a discrete-time system of the form

$$u_{r+1} = G(u_r, w). \tag{9}$$

In particular, G might arise as the numerical discretization of a continuous-time network. The fixed point learning goal would be now the location of a w^* such that a prescribed u^* satisfies $u^* = G(u^*, w^*)$.

A discrete-time gradient-descent scheme reads

$$w_{k+1} = w_k - \eta E_w(u^f(w_k), w_k),$$

where $u^f(w)$ stands for a fixed point of (9) at a given w, and the error gradient can again be computed as $(u^f(w_k) - u^*) u_w^f(w_k)$. Now, the derivative $u_w^f(w_k)$ is defined by the fixed point condition $u^f(w_k) = G(u^f(w_k), w_k)$. Implicit derivation yields the adaptation law

$$w_{k+1} = w_k - \eta(u^f(w_k) - u^*)(I - G_u)^{-1}(u^f(w_k), w_k) G_w(u^f(w_k), w_k)$$
$$= w_k - \eta H(u^f(w_k), w_k).$$

Here, the adjustment function H amounts to the error gradient E_w, but in other cases it might describe more general adaptation laws. When G arises as the Euler discretization of a continuous time network, this law can be easily proved equivalent to the Euler discretization of its continuous-time counterpart (1).

Again, there is no need to assume that the network has reached the fixed point $u^f(w_k)$ before applying the adaptation rule. Although details are beyond the scope of the work, it seems reasonable to conjecture that a large number K of iterations in (9) with fixed w_k, prior to each adjustment step, should be enough to guarantee local convergence of the scheme. With abuse of notation, this can be formalized as

$$u_{k+1} = G^{(K)}(u_k, w_k) \tag{10a}$$

$$w_{k+1} = w_k - \eta H(u_{k+1}, w_k), \tag{10b}$$

where $G^{(K)}(u_k, w_k)$ stands for the K-fold iteration of G from u_k with fixed w_k. In this discrete-time context, K plays the role of $1/\varepsilon$ in continuous-time problems, and cases with $K \neq 1$ can be understood as describing two time-scales in the adaptation process.

5   Periodic Trajectory Learning and Poincaré Maps

5.1   A Discrete-Time, Multiple Fixed Point Problem Reformulation

Consider a T-periodic trajectory learning problem, and let us assume that the neural design guarantees the existence of an asymptotically stable T-periodic trajectory for every w. Any point u of such a trajectory would then be a stable fixed point of the so-called *Poincaré map* [2]

$$\mathcal{P}(u, w) = \phi_w(T, u),$$

where $\phi_w(T, \cdot)$ stands for the T-time flow defined by the neural dynamics with fixed w. The discrete error formula (5) allows one to reformulate the problem as a multiple fixed point one in a discrete-time context; namely, we may consider the learning goal as the approximation of I stable fixed points $u^{[i]} = u(ih)$ towards prescribed values $u^{[i]*} = u^*(ih)$, for the dynamics defined by

$$u_{r+1} = \mathcal{P}(u_r, w).$$

5.2   Learning Strategy

The discussion in 2.2 naturally leads, in T-periodic trajectory learning problems, to an adaptation scheme defined by the evolution of the network

$$\dot{u} = F(u, w_k),$$

and the sensitivity system

$$\dot{v} = F_u(u, w_k)v + F_w(u, w_k), \tag{11}$$

over a period T, keeping track of the values $u(ih)$, $v(ih)$ for $i = 0, \dots, I - 1$. Subsequently, parameters are updated through a law of the form

$$w_{k+1} = w_k - \tilde{\eta} \sum_{i=0}^{I-1} [u(ih) - u^*(ih)]v(ih),$$

where $\tilde{\eta}$ denotes the product ηh.

Nevertheless, from the discussion in previous sections, it is reasonable to assume that the network should have approached a stable invariant before applying the weight adjustment step, in order to guarantee convergence of the scheme. If, accordingly, we let u and v evolve for KT units of time, keeping track of the values $u(ih), v(ih)$ only in the last period, and consider the I Poincaré maps

$$u_{r+1}^{[i]} = \mathcal{P}(u_r^{[i]}, w), \quad i = 0, \ldots, I - 1,$$

the learning process can be formalized as

$$u_{k+1}^{[i]} = \mathcal{P}^{(K)}(u_k^{[i]}, w_k), \quad i = 0, \ldots, I - 1,$$

$$w_{k+1} = w_k - \tilde{\eta} \sum_{i=0}^{I-1} [u_{k+1}^{[i]} - u^{[i]*}] v_{k+1}^{[i]},$$

together with the auxiliary system (11). This process has essentially the same form as (10). Again, if K is large enough, convergence of the scheme (in the norm defined by the discretized error (5)) may be locally guaranteed.

Acknowledgements. This work has been partially supported by Project BFM2000-1475, Programa Nacional de Promoción General del Conocimiento, Ministerio de Ciencia y Tecnología, Spain.

References

1. Allgower, E. L., Georg, K.: Numerical Continuation Methods. An Introduction. Springer Verlag, 1990
2. Amann, H.: Ordinary Differential Equations. Walter de Gruyter, 1990
3. Baldi, T.: Gradient descent learning algorithm overview: A general dynamical systems perspective. IEEE Trans. Neural Networks **6** (1995) 182–195
4. Brenan, K. E., Campbell, S. L., Petzold, L. R.: Numerical Solution of Initial-Value Problems in Differential-Algebraic Equations. SIAM, 1996
5. Cohen, M. A., Grossberg, S.: Absolute stability of global pattern formation and parallel memory storage by competitive neural networks. IEEE Trans. Systems, Man and Cybernetics **13** (1983) 815–826
6. Jin, L., Gupta, M. M.: Stable dynamic backpropagation learning in recurrent neural networks. IEEE Trans. Neural Networks **10** (1999) 1321–1334
7. Pineda, F. J.: Generalization of back-propagation to recurrent neural networks. Phys. Rev. Let. **59** (1987) 2229–2232
8. Riaza, R., Campbell, S. L., Marszalek, W.: On singular equilibria of index-1 DAEs. Circuits, Systems and Signal Processing **19** (2000) 131–157
9. Riaza, R., Zufiria, P. J.: Rates of learning in gradient and genetic training of recurrent neural networks, in A. Dobnikar et al, eds., Artificial Neural Nets and Genetic Algorithms (ICANNGA'99), pp. 95-99. Springer Computer Science, 1999
10. Saberi, A., Khalil, H.: Quadratic-type Lyapunov functions for singularly perturbed systems. IEEE Trans. Aut. Cont. **29** (1984) 542–550

Author Index

Lecture Notes in Computer Science

For information about Vols. 1–2341
please contact your bookseller or Springer-Verlag

Vol. 2383: M.S. Lew, N. Sebe, J.P. Eakins (Eds.), Image and Video Retrieval. Proceedings, 2002. XII, 388 pages. 2002.

Vol. 2384: L. Batten, J. Seberry (Eds.), Information Security and Privacy. Proceedings, 2002. XII, 514 pages. 2002.

Vol. 2385: J. Calmet, B. Benhamou, O. Caprotti, L. Henocque, V. Sorge (Eds.), Artificial Intelligence, Automated Reasoning, and Symbolic Computation. Proceedings, 2002. XI, 343 pages. 2002. (Subseries LNAI).

Vol. 2386: E.A. Boiten, B. Möller (Eds.), Mathematics of Program Construction. Proceedings, 2002. X, 263 pages. 2002.

Vol. 2387: O.H. Ibarra, L. Zhang (Eds.), Computing and Combinatorics. Proceedings, 2002. XIII, 606 pages. 2002.

Vol. 2388: S.-W. Lee, A. Verri (Eds.), Pattern Recognition with Support Vector Machines. Proceedings, 2002. XI, 420 pages. 2002.

Vol. 2389: E. Ranchhod, N.J. Mamede (Eds.), Advances in Natural Language Processing. Proceedings, 2002. XII, 275 pages. 2002. (Subseries LNAI).

Vol. 2391: L.-H. Eriksson, P.A. Lindsay (Eds.), FME 2002: Formal Methods – Getting IT Right. Proceedings, 2002. XI, 625 pages. 2002.

Vol. 2392: A. Voronkov (Ed.), Automated Deduction – CADE-18. Proceedings, 2002. XII, 534 pages. 2002. (Subseries LNAI).

Vol. 2393: U. Priss, D. Corbett, G. Angelova (Eds.), Conceptual Structures: Integration and Interfaces. Proceedings, 2002. XI, 397 pages. 2002. (Subseries LNAI).

Vol. 2395: G. Barthe, P. Dybjer, L. Pinto, J. Saraiva (Eds.), Applied Semantics. IX, 537 pages. 2002.

Vol. 2396: T. Caelli, A. Amin, R.P.W. Duin, M. Kamel, D. de Ridder (Eds.), Structural, Syntactic, and Statistical Pattern Recognition. Proceedings, 2002. XVI, 863 pages. 2002.

Vol. 2398: K. Miesenberger, J. Klaus, W. Zagler (Eds.), Computers Helping People with Special Needs. Proceedings, 2002. XXII, 794 pages. 2002.

Vol. 2399: H. Hermanns, R. Segala (Eds.), Process Algebra and Probabilistic Methods. Proceedings, 2002. X, 215 pages. 2002.

Vol. 2401: P.J. Stuckey (Ed.), Logic Programming. Proceedings, 2002. XI, 486 pages. 2002.

Vol. 2402: W. Chang (Ed.), Advanced Internet Services and Applications. Proceedings, 2002. XI, 307 pages. 2002.

Vol. 2403: Mark d'Inverno, M. Luck, M. Fisher, C. Preist (Eds.), Foundations and Applications of Multi-Agent Systems. Proceedings, 1996-2000. X, 261 pages. 2002. (Subseries LNAI).

Vol. 2404: E. Brinksma, K.G. Larsen (Eds.), Computer Aided Verification. Proceedings, 2002. XIII, 626 pages. 2002.

Vol. 2405: B. Eaglestone, S. North, A. Poulovassilis (Eds.), Advances in Databases. Proceedings, 2002. XII, 199 pages. 2002.

Vol. 2406: C. Peters, M. Braschler, J. Gonzalo, M. Kluck (Eds.), Evaluation of Cross-Language Information Retrieval Systems. Proceedings, 2002. pages. 2002.

Vol. 2407: A.C. Kakas, F. Sadri (Eds.), Computational Logic: Logic Programming and Beyond. Part I. XII, 678 pages. 2002. (Subseries LNAI).

Vol. 2408: A.C. Kakas, F. Sadri (Eds.), Computational Logic: Logic Programming and Beyond. Part II. XII, 628 pages. 2002. (Subseries LNAI).

Vol. 2409: D.M. Mount, C. Stein (Eds.), Algorithm Engineering and Experiments. Proceedings, 2002. VIII, 207 pages. 2002.

Vol. 2410: V.A. Carreño, C.A. Muñoz, S. Tahar (Eds.), Theorem Proving in Higher Order Logics. Proceedings, 2002. X, 349 pages. 2002.

Vol. 2412: H. Yin, N. Allinson, R. Freeman, J. Keane, S. Hubbard (Eds.), Intelligent Data Engineering and Automated Learning – IDEAL 2002. Proceedings, 2002. XV, 597 pages. 2002.

Vol. 2413: K. Kuwabara, J. Lee (Eds.), Intelligent Agents and Multi-Agent Systems. Proceedings, 2002. X, 221 pages. 2002. (Subseries LNAI).

Vol. 2414: F. Mattern, M. Naghshineh (Eds.), Pervasive Computing. Proceedings, 2002. XI, 298 pages. 2002.

Vol. 2415: J. Dorronsoro (Ed.), Artificial Neural Networks – ICANN 2002. Proceedings, 2002. XXVIII, 1382 pages. 2002.

Vol. 2417: M. Ishizuka, A. Sattar (Eds.), PRICAI 2002: Trends in Artificial Intelligence. Proceedings, 2002. XX, 623 pages. 2002. (Subseries LNAI).

Vol. 2418: D. Wells, L. Williams (Eds.), Extreme Programming and Agile Methods – XP/Agile Universe 2002. Proceedings, 2002. XII, 292 pages. 2002.

Vol. 2419: X. Meng, J. Su, Y. Wang (Eds.), Advances in Web-Age Information Management. Proceedings, 2002. XV, 446 pages. 2002.

Vol. 2420: K. Diks, W. Rytter (Eds.), Mathematical Foundations of Computer Science 2002. Proceedings, 2002. XII, 652 pages. 2002.

Vol. 2421: L. Brim, P. Jančar, M. Křetínský, A. Kučera (Eds.), CONCUR 2002 – Concurrency Theory. Proceedings, 2002. XII, 611 pages. 2002.

Vol. 2423: D. Lopresti, J. Hu, R. Kashi (Eds.), Document Analysis Systems V. Proceedings, 2002. XIII, 570 pages. 2002.

Vol. 2430: T. Elomaa, H. Mannila, H. Toivonen (Eds.), Machine Learning: ECML 2002. Proceedings, 2002. XIII, 532 pages. 2002. (Subseries LNAI).

Vol. 2431: T. Elomaa, H. Mannila, H. Toivonen (Eds.), Principles of Data Mining and Knowledge Discovery. Proceedings, 2002. XIV, 514 pages. 2002. (Subseries LNAI).

Vol. 2436: J. Fong, R.C.T. Cheung, H.V. Leong, Q. Li (Eds.), Advances in Web-Based Learning. Proceedings, 2002. XIII, 434 pages. 2002.

Vol. 2440: J.M. Haake, J.A. Pino (Eds.), Groupware – CRIWG 2002. Proceedings, 2002. XII, 285 pages. 2002.

Vol. 2442: M. Yung (Ed.), Advances in Cryptology – CRYPTO 2002. Proceedings, 2002. XIV, 627 pages. 2002.

Vol. 2444: A. Buchmann, F. Casati, L. Fiege, M.-C. Hsu, M.-C. Shan (Eds.), Technologies for E-Services. Proceedings, 2002. X, 171 pages. 2002.